What is Tourette syndrome?

A common, hereditary, behavioral disorder. Its n
prominent symptoms are motor and vocal tic
p11

Motor tics include eyeblinking, facial grimacing, .
jerking, shoulder shrugging and other tics. p 13

Vocal tics include throat clearing, spitting, barking, squeaking, humming and other noises. p17

Swearing in a compulsive manner is present less than a third of the time and is not necessary for the diagnosis. p19

The tics can be *suppressed*, but only for short periods of time. p23

The symptoms start before age 21 but may persist throughout adulthood. p24

Tourette syndrome is common

Up to 1 in 100 school boys have the disorder. p615

Boys are affected more often than girls. p25

The diagnosis is often missed. p655

What other symptoms are associated with Tourette syndrome?

TS is a complex behavioral disorder. Other symptoms include:

- Short attention span, poor concentration and inability to sit still, i.e. ***attention deficit hyperactivity disorder (ADHD)***. p73, 99
- ***Learning disorders, dyslexia,*** poor handwriting, trouble with math or reading. p105
- ***Obsessive-compulsive behaviors,*** touching things with both hands, counting or checking things, perfectionism, hand washing. p113
- ***Conduct disorder,*** oppositonal and confrontive behavior, talking back, can't take "no" for an answer, discipline problems, won't listen, aggressive, temper tantrums over nothing. p131
- ***Depression, mood swings, irritability, short temper***. p183,191
- ***Anxiety with panic attacks, phobias,*** school phobia, and emotionally hypersensitive. p173,183
- ***Speech problems,*** stuttering, hesitancy, delayed speech, talking too loud or too fast. p293
- ***Sleep problems,*** unable to get to sleep, bedwetting, night terrors, sleep apnea. p249
- ***Inappropriate sexual behaviors*** including compulsive swearing, sexual touching, crotch touching, and exhibitionism. p114,163
- ***Addictive behaviors*** including ***alcoholism, compulsive eating,*** compulsive shopping, and pathological gambling. p225,235,239

(over)

What causes Tourette syndrome?

Several brain chemicals have been implicated including dopamine, serotonin and norepinephrine. A dysregulation of *serotonin* can account for all of the symptoms. p429,445,457

How is it inherited?

- Many observations suggest TS subjects have inherited a Tourette gene from both parents. p513
- About half of the parents, who carry the gene, have one or more of the above symptoms themselves. p641
- Between 5 and 20% of the general population may carry this or similar genes. p615

Is Tourette syndrome treatable?

Yes. Many or all of the symptoms can be eliminated with medications. p533

Why is knowledge about Tourette syndrome important?

Because many children and adults suffer from Tourette syndrome and don't know it.

Because it is treatable.

Because the symptoms are often misdiagnosed as due to psychological causes, poor parenting, parental abuse or divorce, and thus incorrectly treated or not treated.

Because teachers or therapists often assume the child is severely emotionally disturbed, trying to get attention or just a behavior problem.

Because child protective services may assume the inappropriate sexual behaviors are due to sexual abuse in the home when this may not be the case.

Because the learning disabilties, behavioral problems, compulsive and addictive behaviors cause low self esteem, school drop outs, drug and alcohol abuse and result in dysfunctional adults. Early recognition and treatment is critical.

Because many of the these behaviors are rampant in our society and it is important to understand when they are genetically and not environmentally caused.

Because Tourette syndrome can have positive as well as negative effects on behavior.

The book also provides the reader with a basic knowledge of:

- ***Human genetics*** — what genes are, what mutations are, how they are inherited, how they affect behavior
- ***Brain structure and function*** — what the limbic system is, why the frontal lobes are important, and how they control emotion and behavior.
- ***Brain chemistry*** — what dopamine, serotonin, norepinephrine, GABA and endorphins are and do.
- ***Chemical frontal lobe syndromes*** — how a genetic disorder can affect frontal lobe function and lead to impulsive and addictive behaviors.

Tourette Syndrome and Human Behavior

David E. Comings, M.D.

City of Hope National Medical Center

Hope Press
Duarte, California 91009-188

Tourette Syndrome and Human Behavior

by

David E. Comings, M.D.
Department of Medical Genetics
City of Hope National Medical Center
Duarte, CA 91010

Published by: **Hope Press**
P.O.Box 188
Duarte, CA 91009-188 U.S.A.

Other books on Tourette syndrome by Hope Press:
see order forms on backleaf

First Printing February, 1990
Second Printing November, 1992
Third Printing July 1994
Fourth Printing August 1997

Printed in the United States of America

Library of Congress Cataloging-in-Publication Data

Comings, David E.
Tourette syndrome and human behavior / David E. Comings
p. cm.
Includes bibliographical references
ISBN 1-878267-27-2 : $49.95 ISBN 1-878267-28-0 (pbk.) : $39.95
1. Gilles de la Tourette's syndrome. 2. Human behavior.
I. Title
RC375.C66 1990
616.8'3 --dc20 89-83294
CIP

Tourette syndrome is a common, hereditary disorder that provides insight into how we and our children behave and misbehave and why some of us can't read, learn, or pay attention; compulsively do things including eating and abusing drugs or alcohol, spouses or children and are angry, short-tempered, anxious, afraid, depressed, or feel different and all alone.

This book is written for both the lay reader and the health or education professional as a guide to the exciting story of how a common gene may control a wide range of human behaviors.

Disclaimer

This book is designed to provide information in regard to the subject matter covered. It is sold with the understanding that the publisher and author are not engaged in rendering medical, psychiatric, psychological, legal or other professional services to the reader. If such are required, the services of a competent professional in the appropriate field should be sought.

Every effort has been made to make this book as complete and accurate as possible. However, there may be mistakes both typographical and in content. Therefore, this text should be used only as a source of general information and specifics relating to a given individual should be obtained from the services of a professional. Furthermore, this book contains information only up to the printing date.

The purpose of this book is education. Neither the author nor the publisher shall have or accept liability or responsibility to any person or entity with respect to loss or damage or any other problem caused or alleged to be caused directly or indirectly by information contained in this book.

About the Author

David E. Comings, M.D., is a human geneticist. He obtained his M.D. degree at Northwestern University Medical School in 1958 and took a rotating internship and residency in internal medicine at Cook County General Hospital in Chicago. After serving in the Army Medical Corps for two years, he took post-doctoral speciality training in medical genetics at the University of Washington in Seattle from 1964 to 1966. Since then he has been the director of the Department of Medical Genetics at the City of Hope National Medical Center in Duarte, California.

For many years Dr. Comings was primarily involved in basic research on human chromosome structure and was author of over 200 research articles in this and similar areas. During this time his clinical work was related to individuals with hereditary neurological disorders such as Huntington's disease. He served on many national committees including National Institutes of Health grant review study sections, the Hereditary Disease Foundation, the Neurofibromatosis Foundation, The National Foundation of the March of Dimes, and others. From 1978 to 1986 he was editor of the *American Journal of Human Genetics*, the official publication of the American Society of Human Genetics. In 1988 he served as president of the American Society of Human Genetics.

Beginning in 1980 Dr. Comings and his psychologist wife, Brenda, began seeing a few patients in the Genetics Clinic with what was then thought to be a very rare disorder — Tourette syndrome. Through the combined efforts of a geneticist taking detailed family histories and a psychologist, interested in the emotional and behavioral problems, they quickly recognized that Tourette syndrome was much more than simply a tic disorder. Because of the effect on learning, they felt it was important to identify affected children early and discussed it on a local television show. This resulted in the identification of many new cases and in the past ten years they have treated over 1,400 Tourette syndrome patients and their families and over 600 patients with related problems such as attention deficit hyperactivity disorder. This led to the publication of over 30 articles relating to different aspects of Tourette syndrome.

This book summarizes the clinical, genetic and biochemical aspects of Tourette syndrome and related disorders based on extensive clinical, genetic and research experience with these two thousand patients and their relatives.

About the Fourth Printing

When *Tourette Syndrome and Human Behavior* was first published in 1990 I never dreamed that it would be selling as briskly in 1997 as it did in 1990. Many have asked me if it is out of date. My response is that 95 percent of the book is just as valid today as it was in 1990. This is because the basic facts about TS have not changed. It is still a spectrum disorder, consisting not only of motor and vocal tics but of the same wide range of associated comorbid conditions listed inside the front cover. The only things that have changed are that I now believe that TS is a polygenic disorder, due to the coming together of a number of variant genes from both parents, instead of a single gene. The second thing that has changed is the availability of many new medications for the treatment of TS. Instead of producing a new edition, I have chosen instead to write a second, shorter book summarizing the new thinking about the genetics of TS, including some of the genes involved, and an appendix to bring the newer medications up to date. This book is entitled *Search for the Tourette Syndrome and Human Behavior Genes*. A companion book entitled *The Gene Bomb: Does Higher Education and Advanced Technology Accelerate the Selection of Genes for Learning Disorders, ADHD, Addictive, and Disruptive Behaviors?* was published at the same time. These and other books printed by Hope Press (see order form in the back) will bring the reader the latest knowledge about the clinical and basic research aspects of TS and ADHD.

To ***Brenda****, my beautiful wife and partner,*

who —

"dragged me kicking and screaming" out of my laboratory
and into the clinic,

there to find a fascinating, complex, intriguing disorder —
Tourette syndrome,

providing a personal, emotional and intellectual treasure
beyond imagination.

Table of Contents

Part I
INTRODUCTION

Chapter 1

The Spectrum of Tourette Syndrome

What do all the following people have in common?

A bright 9 year old who is failing in school because he can't concentrate on his work.

A boy who is reprimanded by the teacher because he is "acting out" by constantly making noises in class.

A teenager who is driving his parents to divorce because they don't know how to handle his confrontive personality and compulsive lying and stealing.

A young adult who has been addicted to alcohol and drugs since he was 14.

A kindergarten child who has just gotten allergy medicine from the eye doctor because of rapid eyeblinking.

A 13 year old boy who used to be a star pupil but has begun to hate school and is now failing all his courses.

A 23 year old woman who began having panic attacks and is now afraid to go out of the house alone.

A brilliant 45 year old engineer with top secret clearance who can't stop exposing himself in public.

A housewife who compulsively vacuums the house eight times a day.

A young bride whose husband is threatening to leave her because of her violent temper.

A young man who has been in individual psychotherapy for a severe conduct disorder. Despite treatment for two years his behavior has not changed.

All of these individuals are carrying a gene for Tourette syndrome. This disorder, which was once thought to be extremely rare, is in fact one of the most common genetic diseases affecting man and many people carry the trait. Oftentimes health care professionals, if they have heard of Tourette syndrome, believe it to be a rare disorder in which the individual must sit in their office swearing and having violent muscle tics and vocal noises to make a diagnosis

The purpose of this book is to pass on the experience that my wife, Brenda, and I have had over the past ten years with over 1,400 Tourette syndrome families and thousands of their relatives. This experience includes over sixty hours a week of direct one to one contact with patients and with running a support group every two months for 50 to 100 patients and family members. Individuals who carry the gene can range all the way from being totally normal to totally incapacitated. Tourette syndrome can be visualized as a treatable "disinhibition" disorder affecting a primitive part of the brain known as the limbic system.

The book is also devoted to *the power of studying a single genetic disorder for understanding complex biological phenomena.* Translated, this means that there are often two ways to study the complex subject of the influence of genes on our behavior. One is to take a group of people behaving in a certain way, such as being depressed or abusing alcohol, and ask if other members of the family have similar problems and try to determine if the trait is clearly inherited. Unfortunately, this often does not provide clear answers. A second way is to start with a clearly defined genetic disorder and ask if such individuals tend to behave differently than other people. Tourette syndrome is idealy suited to the latter approach. It is as if with a blink of the eye, a jerk of the head or the utterance of a noise; these individuals are saying "I have this gene — come study me," and we have.

I hope to provide the reader with a thorough understanding of all aspects of Tourette syndrome and other impulse disorders — symptoms, diagnosis, behavior, genetics, cause, and treatment.

The book is intended for the general reader who is interested in learning how a common genetic disorder can affect a wide range of behaviors. It is also written for those with Tourette syndrome, their parents and other family members; for teachers; and for psychologists, physicians and other professionals who are, or should be, diagnosing this disorder. This encompasses a very broad range of sophistication, from the hyperactive, learning disabled Tourette syndrome patient, to whom school and all its associated features has been a lifelong problem, with reading being a least favorite pastime, to the academic physician who likes to see references to journals documenting every sentence. To attempt to bridge this gap, I have chosen the following strategy. For the more academically oriented, references are included to the medical literature listed in the bibliography at the end of the book. These are listed as a small superscript number[4]. A reference to something covered in more detail in another part of the book will be given as a superscript page number[p20]. For those who find the terminology new, I will attempt to make your journey as painless as possible by clearly explaining new words and concepts, providing many figures often intimately intermixed with the text, placing important information in boxes, and providing a glossary.

The book is divided into six parts.

Part I — **Introduction**, is a general introduction to the most obvious aspects of Tourette syndrome — the motor tics, vocal noises and swearing.

Part II — **Genetics**, provides enough information about genetics and pedigrees so the reader can understand the family histories and related information presented in Part III.

Part III — **Behavior**, constitutes the bulk of the book and reviews the many behavioral aspects of Tourette syndrome and related disorders. No one should have any problem up to this point.

Part IV — **The Cause**, delves into what is in many ways the most exciting aspect of the book, namely the insights that Tourette syndrome provides into the chemistry of behavior. Although, for some, there will be more new terminology and more new concepts in this part of the book, every new word, structure or chemical is first introduced and explained. I have made every possible attempt to avoid jargon and abbreviations.

Part V — **Treatment**, discusses medical and pyschological treatments. Although readers can skip the part on Cause and jump to the Treatment section if they wish, the treatments are much easier to understand if Part IV has been read.

Part VI — **Conclusions**, presents additional material that became available as the book was being written, and presents some implications the study of TS has for individuals and society.

Each chapter has a summary at the end that allows you to skim any part you wish, and the last chapter is a summary of the book. For confidentiality all the names in the case reports and pedigrees have been changed. The extensive index allows the reader to look up any specific behavior of particular interest.

A Final Point — Evolution of the Book

I started this book in August 1986. At that time I didn't have the slightest idea about what caused Tourette syndrome. I decided to write the book because our experience indicated there were many more behavioral problems, in both the patients and their relatives, than was generally appreciated. There was so much to say, a book seemed the best vehicle. However, the act of putting things down on paper helps to clarify one's thinking about a subject. Writing the chapters about the behavioral problems associated with TS forced me to appreciate that each one of them had been proposed by others as being due to defects in the brain chemical serotonin. Because of this I started examining the blood levels of serotonin and its precursor, tryptophan, in all old and new patients and their immediate relatives. Also, when Part III on Behavior was finished and it was time to start Part IV, The Cause, I chose to first write the section on serotonin.

By that time it became apparent that my subject was rapidly evolving under me, as the book was being written. So the reader could participate in the same shifts in thinking about TS that I was going through, the new information was incorporated into the book *as it was being written.* I did not go back and change the old, but just added on the new. Thus, by mid-1987, as the section on serotonin was being written, the preliminary data on the blood serotonin and tryptophan in TS patients were added. Many subsequent chapters contained other new information as it became available.

Come then and join in the excitement of watching behavior, genetics, psychology, neurology and chemistry meld together in this complex, frustrating, often maddening, yet infinitely intriguing and important disorder called Tourette syndrome.

Chapter 2
Gilles de la Tourette

Gilles de la Tourette (1857-1904)
Courtesy of Dr. Alexander R. Lucas, Mayo Clinic, Rochester, Minn.

Over a hundred years ago, in 1885, Gilles de la Tourette, a French neurologist at the Salpêtrière Hospital in Paris, described a condition characterized by multiple muscle tics, vocal noises and compulsive swearing[715,716,744]. He described nine patients and reviewed earlier references to others with this[955,1961] and related problems[135,714,845,1421]. The primary case he described was a woman who had first been reported 60 years earlier by Dr. Itard in 1825[955]. She died in 1884.

"Madame de D. . . at the age of 7 was afflicted by convulsive movements of the hand and arms. . . After each spasm, the movements of the hand became more regular and better controlled until a convulsive movement would again interrupt her work. She was felt to be suffering from over excitement and mischief, and because the movements became more and more frequent, she was subjected to reprimand and punishment. It soon became clear that these movements were indeed involuntary... in nature... involved the shoulders, the neck, and the face, and resulted in contortions and extraordinary grimaces.

"As the disease progressed, and the spasms spread to involve her voice and speech, the young lady made strange screams and said words that made no sense. However, during all this, she was clearly alert, and showed no signs of delirium or other mental problems. Months and years passed with no real change in her symptoms. It was hoped that with puberty these might naturally abate, but this did not occur.

"In the midst of an interesting conversation, all of a sudden, without being able to prevent it, she interrupts what she is saying or what she is listening to with horrible screams and with words that are even more extraordinary than her screams. All of this contrasts deplorably with her distinguished manners and background. These words are, for the most part, offensive curse words and obscene sayings. These are no less embarrassing for her than for those who have to listen, the expressions being so crude that an unfavorable opinion of the woman is almost inevitable."

In addition to the involuntary motor and vocal tics and swearing, Gilles de la Tourette also noted that:

a) the disorder had an onset in childhood, usually between 7 and 10 years of age,

b) affected males more than females, and

c) was hereditary;

d) the tics usually started in the face or upper extremity;

e) the symptoms waxed and waned spontaneously;

f) the tics were made worse by stress and diminished in sleep and occasionally during fevers; and

g) it was not a progressive degenerative disorder.

He correctly distinguished tics from other similar disorders in that the movements were

"short lived, and extremely brisk . . . They are intermittent, never continuous, so that they neither prevent normal eating nor limit independent ambulation... Their mental state is normal and most of them are highly intelligent."

Echolalia. Gilles de la Tourette also described the tendency of these patients to mirror the speech of others — a characteristic called echolalia.

"A patient may be speaking to a person in a group and find himself repeating, along with a body jerk, the last word or words of a sentence that he has just heard. He is perfectly aware of the situation but when asked to control himself, he cannot."

One patient recalled:

"In listening to a discussion I was seized by the almost irresistible need to repeat a word or the end of a sentence. I needed all my strength and sense of propriety to hold back from repeating this word out loud; as I could only restrain myself part way, I saw on different occasions how people around me clearly heard my noises.

"Echolalia can occur not only when the patient hears an extraneous sound, but also when he reads a word or even thinks of a word or of the object which the word represents."

Palilalia, or the tendency to repeat one's own words was also described.

Echopraxia, is the mimicking of other people's actions. Gilles de la Tourette described this as a subtype of echolalia as follows:

"Echolalia should not be considered in its most restricted sense, since these people also will imitate a gesture or an act, even when it is rather complex, as a manifestation of this same disorder. The echolalia of activity can be developed to an extreme degree and can appear in the form of quite bizarre behavior. This latter echolalia may be so bizarre, in fact, that if we had not witnessed it ourselves we would be inclined to doubt the description of the former authors. But as our own case corroborated:

"S. is in the courtyard of the Salpétrière in his usual way. He is moving about, making a few contortions and a few strange sounds, all of which is his ordinary practice. Another patient approaches him and decides to imitate one of his stranger movements, which consists of lifting the right arm and leg and stomping with the left foot, a peculiar position that one can clearly see is apt to cause a loss of balance. At the same time, he mimics one of our patient's more bizarre and characteristic utterances. Soon S., who has been rather calm, starts to imitate the screams and strange mannerisms of his fellow patient,

> and does it so vigorously that he falls, fortunately not hurting himself and hospital guards have to intervene to stop this game, which could become dangerous. This incident, sadly, has become the source of many cruel episodes where other patients take sadistic advantage of S.'s irresistible imitative compulsion."

Coprolalia is the involuntary utterance of swear words. Gilles de la Tourette expressed the same amazement that we all have when first confronted with this symptom.

> "One could understand how a 19-year old lad could have obscene ideas and translate them into words. But that women, young girls, and boys of superior background and upbringing should change inarticulate screams into obscene expressions is particularly unusual and entirely unexplained. Furthermore, nothing changes this vocabulary, neither scolding nor attempts by the patient to substitute these expressions. Perhaps the willpower of the patient can force the obscenities to be momentarily suppressed, but this control is short-lived and usually followed by an exaggerated explosion of obscenities."

He recognized that coprolalia was not present in every patient, usually appears well after the beginning of the illness, and could not be prevented by disciplinary measures.

Summary: Gilles de la Tourette described the syndrome that bears his name over 100 years ago. He showed it was a hereditary disorder and clearly identified many of its major features.

Chapter 3

The Diagnosis

For many years diagnosis in the field of psychiatry was a very subjective art which was practiced differently in different places. In an effort to standardize psychiatric diagnoses, the American Psychiatric Association sponsored the development of the Diagnostic and Statistical Manual, termed the DSM. The first was published in 1952 and quickly became the psychiatrist's diagnostic bible, especially after insurance companies insisted that a DSM standardized diagnosis had to be made before they would pay the bills. The second edition was printed in 1968 and is known as the DSM-II. The DSM-III[490] was published in 1980. The latter was very important for TS since it was the first time that the diagnostic criteria of Tourette syndrome were listed in any of the DSM manuals. It was listed as Tourette's Disorder 307.23 and had the following 6 diagnostic criteria:

Each of these aspects of the diagnosis is discussed in future chapters. The criteria of requiring that the symptoms be present for more than one year distinguishes TS from **Transient Tic Disorder,** which refers to individuals in whom muscle tics are present in childhood and last for at least one month but less than one year. It is clear from family histories that some cases of Transient Tic Disorder are caused by the Tourette syndrome gene.

If all the criteria of Tourette syndrome are present except for the absence of vocal tics, the individual is said to have **Chronic Motor Tic Disorder**; and by contrast, if there are vocal tics but no motor tics it can be called **Chronic Vocal Tic Disorder**. It is clear from family studies that both of these are genetically related to Tourette syndrome. This is discussed in more detail in the chapter entitled The Genetics of Tourette Syndrome[P45].

The DSM-III Criteria for the Diagnosis of Tourette Syndrome

1. Age at onset between 2 and 15 years.

2. Presence of recurrent, involuntary, repetitive, rapid, purposeless motor movements affecting multiple muscle groups.

3. Multiple vocal tics.

4. Ability to suppress movement voluntarily for minutes to hours.

5. Variation in the intensity of the symptoms over weeks or months.

6. Duration of more than one year.

In 1987, the DSM-III was revised and appropriately called the DSM-IIIR[491]. There were only minor changes in the diagnostic criteria for Tourette syndrome. The time of onset was changed to read before age 21. The two previously separate criteria of motor and vocal tics were collapsed into a single criteria and only one vocal tic was required. The description of the year of symptoms, and the waxing and waning, was made more elaborate. The criteria of suppressibility was moved into the definition of a tic. A **tic** was defined as:

> "An involuntary, sudden, rapid, recurrent, nonrhythmic, stereotyped, motor movement or vocalization. It is experienced as irresistible, but can be suppressed for varying lengths of time."

Finally, some exclusions, such as being on major tranquilizers or having other neurological disorders, were added.

The DSM-IIIR criteria are shown in the box below.

Summary: The essence of the diagnosis of Tourette syndrome is the presence of both motor and vocal tics, occurring almost every day for a period of at least one year. Since these tics wax and wane in severity and are suppressible they may be absent during a visit to the doctor. Swearing is not necessary for a diagnosis.

The DSM-IIIR Criteria for the Diagnosis of Tourette Syndrome

A. Both multiple motor and one or more vocal tics have been present at some time during the illness, although not necessarily concurrently.

B. The tics occur many times a day (usually in bouts), nearly every day or intermittently throughout a period of more than one year.

C. The anatomic location, number, frequency, complexity, and severity of the tics change over time.

D. Onset before age 21.

E. Occurrence not exclusively during Psychoactive Substance Intoxication or known central nervous system disease, such as Huntington's chorea and postviral encephalitis.

Chapter 4
The Motor Tics

Tics are the major symptom of individuals with Tourette syndrome. They may involve virtually any muscle in the body and consist of a sudden involuntary contraction of that muscle. The frequency of the most common tics [358,176] are shown in the following box.

other member of the family had already been diagnosed as having TS. Parents are usually told that their child has an allergy or needs glasses. Despite the fact that the child now wears glasses or has drops in his eyes, the tics persist.

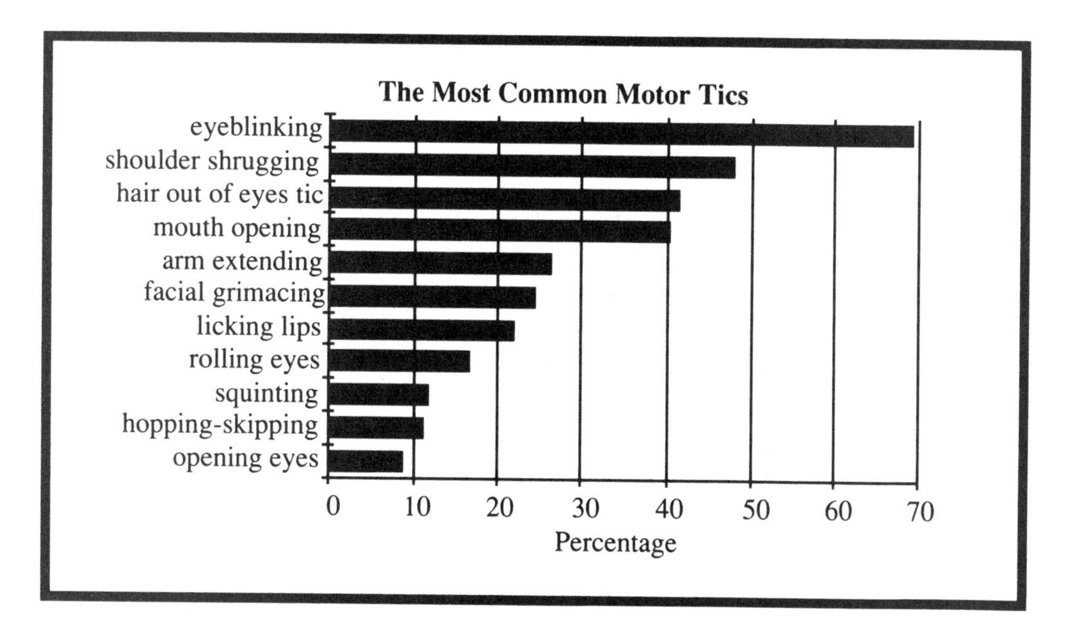

A more complete list of the different tics are shown in the table. Some deserve special comment.

Eyeblinking The most common first symptom heralding the onset of TS is rapid eyeblinking. This usually causes the parents to take their child to the eye doctor or the pediatrician. I have never yet heard of the correct diagnosis being made at that time, unless an-

Eye tics Other common tics of the eyes include squinting, turning or rolling the eyes upwards, and opening the eyes wide. A more uncommon tic that we have seen now in three people is that of closing their eyes while driving. Whether this is a tic or a compulsion is not clear, but one young man said it only came on when he was driving the most dangerous roads such as mountain curves. This so frightened his wife that she refused to let him drive when she was in the car.

Facial grimacing There are so many tics of the face that it is easiest to just lump them together as facial grimacing. These include pulling the mouth upward at the corners, bearing the teeth, wrinkling of the nose (bunny rabbit tic), squinting, pulling the mouth to one side and similar variations. Some children have been dubbed the class clown because of their grimacing.

Lips Repetitive sticking out of the tongue is often mislabeled by teachers or parents as acting out behavior. In some children a tic of licking their lips is so severe that they develop a discolored rash and get nicknamed "dirt lips"[p291]. One child with an infected rash around the mouth had been to six different dermatologists without success. After being treated with small doses of medication for TS the rash disappeared.

Mouth Repeatedly biting the tongue or the inside of the cheek or chewing on the inside of the lip are common tics. In some mild TS cases this may be one of the only tics present. In one youngster tongue biting was so severe that he bit off a third of his tongue and finally required a dental prosthesis to stop the biting.

Neck What we call a "hair out of the eyes" tic is one of the most frequent tics. This often causes the mother to take them to the barber shop for a haircut, a maneuver that is rarely successful. When the child finally comes home with a crew cut and is still ticing, the parents realize the hair is not the problem. There also may be a slower turning of the head upward and backward in either direction.

Legs Various types of leg tics can result in a bizarre gait. When this happens to children they may be labeled as "queer" by their peers. This can be a particularly cruel and difficult stigma to live down.

Feet Parents have often complained that their TS children wear out shoes at a rapid rate because of a tic that results in turning of the foot and scraping off part of the shoe.

In summary, the following are the **most common motor tics**:

- eyeblinking
- eyes rolling up
- squinting
- facial grimacing
- mouth opening
- head tic — hair out of eyes tic
- shoulder shrugging
- pulling at clothes
- arm or leg jerking
- touching self in the crotch

Complex tics Complex tics involve more than one muscle and include what are often called **stereotyped movements.** These are complex movements that are always done in the same manner. Some complex tics are passing the hand through the hair in a grooming or combing fashion, skipping, jumping, dropping down on one knee, hitting or biting one's self, picking at sores, foot stamping, smelling objects, complex facial gestures, and combinations of tics such as a simultaneous facial grimace and flailing an arm. Sometimes the combinations can be quite complex and border on compulsive behaviors. These include things like turning around three times and squatting down, taking two steps forward and one step backward, crossing one's self and kneeling, and bending over to smell the grass. Oftentimes vocal and motor tics occur together in a stereotyped fashion such as barking and shrugging a shoulder, or throat clearing and hopping together.

Pulling at clothes TS patients often state that their clothes feel too tight and to relieve this tension they pull at their clothes. A combination of excessive touching or pulling at their crotch can be particularly embarrassing and their clothes may become discolored or worn in this area.

The following table shows a more complete list of the motor tics seen in TS.

The Motor Tics in TS

Face
- eyeblinking
- eyes rolling upward
- opening eyes wide
- squinting
- closing eyes while driving
- facial grimacing
- sticking tongue out
- licking lips
- smacking lips
- licking shoulder
- biting tongue or cheek
- looking at the sun
- grinding teeth

Head and Neck
- hair out of eyes tic
- vertical neck jerking
- touching shoulder with chin
- throwing head back

Shoulders
- shoulder shrugging

Arms
- extension of arms at the elbow
- flexion of arms at the elbow
- flailing arm out or up

Hands
- biting nails
- finger sign (copropraxia)
- flexing fingers
- piano fingers
- picking at skin
- poking
- popping knuckles
- waving

Diaphragm
- inhaling
- exhaling
- gasping for breath

Legs
- kicking
- hopping
- skipping
- jumping
- bending at one or both knees
- stooping down
- stepping backward

Feet
- flexing ankles
- extending ankles
- turning foot in
- dragging foot
- shaking foot
- stamping feet
- tapping feet
- tripping
- toe curling
- walking on toes

Others
- banging table
- blowing on hand
- chewing on clothing
- flapping arms
- hitting self
- kissing hand
- kissing others
- picking at lint
- pressing rectal sphincter
- pulling at clothes as if too tight
- scratching self
- shivering
- smelling fingers or objects
- sticking finger in throat
- twiddling thumb on nose
- twirling hair
- whole body jerking
- hunching over while walking

Summary: Motor tics are sudden, jerk-like contractions of any muscle. The most common ones occur in the head and include eyeblinking, facial grimacing, and head jerking. The most common first tic is rapid eyeblinking.

Chapter 5

The Vocal Tics

Vocal tics are involuntary, repetitive noises or utterances. Virtually any non-word vocal noise can be a vocal tic. The frequency of the most common vocal tics[358,1767] is shown in the box.

Coughing, burping, hiccuping Vocal noises such as these which can be due to other disorders, can be easily overlooked as being true vocal tics. If a patient has typical motor tics plus chronic coughing without any lung disease, they may only be convinced that the coughing was a vocal tic when it goes away with treatment by haldol or a related medication.

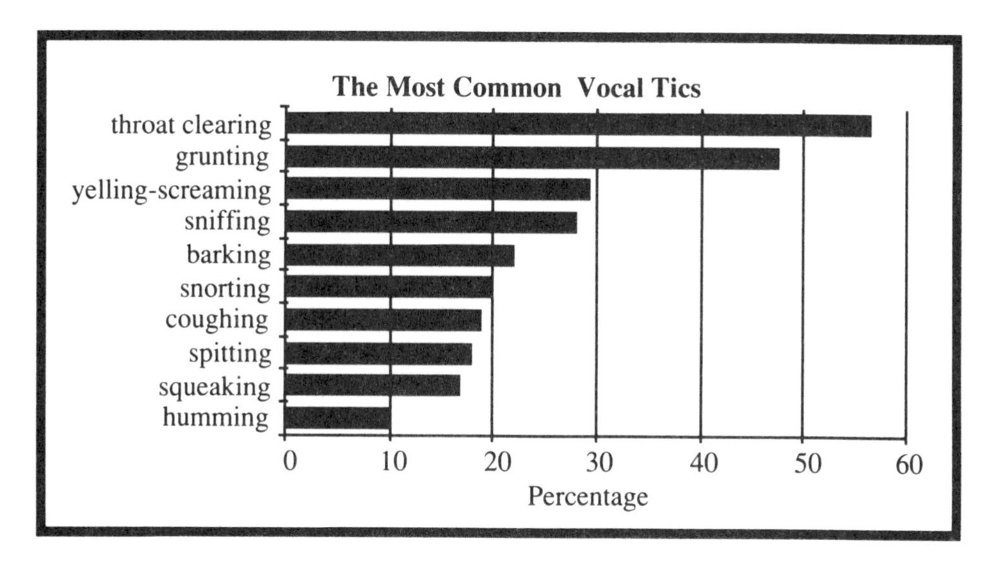

A more complete list of the different vocal tics is given in the table. Some deserve special comment.

Throat clearing This is the most common vocal tic. It is often passed off as the result of an allergy or post-nasal drip. If either of these conditions are present the critical factor is whether the throat clearing continues when the allergies are better or when there is no longer any throat infection.

Sniffing is a common vocal tic that is also often mistaken as being due to an allergy.

Barking This can sometimes be so loud that it will virtually jolt people nearby.

Animal and motor noises, humming and whistling Some patients emit noises that mimic the real thing so precisely that people think the animal is just around the corner. Some vocal noises may be difficult to distinguish from normal childhood behavior.

The feature that distinguishes them from the usual childhood noises is their constant, repetitive, inappropriate, almost compulsive nature.

Parents and spouses often tell us it is not the noises *per se* that drive them crazy but their constant repetitive nature — like the Chinese water torture.

Spitting This is one of the most difficult tics for parents, teachers or spouses to deal with since it is so socially objectionable. One mother complained that she had to place plastic sheets on and around the furniture her son sat in because he would cover everything in sight with spit. It was even dripping down the curtains in the kitchen.

Uneven modulation of the voice Some TS children will be speaking in a normal voice when suddenly the volume is turned up like someone fiddling with the controls on a radio.

The following is a **more complete list of the vocal tics**:

animal noises
barking
belching
blowing breath out
burping
coughing
deep breathing
grunting
hiccups
hissing
honking
"huh"
humming
motor or jet noises
screaming
smacking lips
sniffing
snorting
spitting
squeaking
stuttering or stammering
sucking breath in
throat clearing
uneven modulation of voice
whistling
yelling

Summary: Vocal tics consist of virtually any non-word noise made in a repetitious fashion. The most common vocal tic is repeated throat clearing.

Chapter 6

The Swearing

The compulsive swearing in Tourette syndrome is one of its most dramatic symptoms. The medical term used to describe it is **coprolalia**, coming from the roots *copro* meaning feces and *lalia* meaning lips. When Tourette syndrome was first becoming more publicized in magazines and newspapers, this was often used as the grabber to catch the reader's attention with titles like, *The Cursing Disease* or *The Foulmouth Syndrome*. Although this publicity helped many patients to first diagnose their problem, it unfortunately further established in the minds of many professionals the idea that coprolalia had to be present to make a diagnosis of TS. However, even from the first descriptions of the disorder it was clear that not all patients had coprolalia. In early reports, where many of the cases were severe, it was present in about 60 percent of patients [1767]. In later series, where many milder cases were included, its frequency dropped to between 8 percent and 33 percent[358,550,1761,1771]. The characteristic of coprolalia that sets it apart from simple swearing is that the words come out in a compulsive, repetitive, almost ritualistic fashion even when the individual is not angry. The words are sometimes said over and over just under the breath so they are heard only when one is close to the individual.

The Words

By far the most common word used is "fuck," followed by "shit" and "piss." The words are often not completed so that one hears only "fu" or "sh," almost like the person is trying to suppress saying it, but is only partially successful. More complex phrases such as "cocksucker," "motherfucker," or "fuck a duck" may be used. One patient, whenever he was introduced to a girl, such as Julie, used her name to the yell out, "Fuck you Julie"—hardly leaving an endearing first impression.

Rather than listing all these words, just imagine any swear word or obscene saying you have ever heard and it would be on the list[1767]. We had a dear friend with severe coprolalia who used to leave messages on the answerphone. She always started with "fuck, shit, piss, fuck, shit, piss," followed by "This is Amy." We would pick up the phone and say "Yes, Amy, we know who it is."

It is not difficult to understand that this symptom is the one parents and patients dread the most. It can be particularly troublesome in a home where such words are forbidden and never used. Every time we hear obscene words coming from the mouth of a two or three year old we never cease to be amazed at how quickly and at what an early age children hear and pick up on these terms. Parents often react in astonishment, "Where did he hear that, we never say those words!"

Before a diagnosis has been made and the parents understand it is a symptom of a genetic disorder, coprolalia has often elicited severe punishment and even child abuse. After the diagnosis, parents are often extremely guilty about the punishment they have meted out to their child. When undiagnosed, teachers also have a great deal of difficulty with this symptom and may label the child as "acting out" or a "severe behavioral problem."

In adults, coprolalia is also the most difficult of the TS symptoms to live with. Many times they will either take jobs that isolate them from the public, or jobs where a lot of swearing is accepted. An alternative is to simply be open with peers about the diagnosis. One college girl had such severe coprolalia that at the beginning of each new class she gave a brief description of TS to her classmates. After the first day they learned to ignore her outbursts and accept her as any other student. She never got on the bus unless she was armed with pamphlets about TS, which she handed out to everyone around. Her boyfriend told us a story that was indicative of the increased public awareness of TS. After a movie, he was waiting for her outside the ladies room when a girl came out and told her companion, "There is a girl in there swearing like crazy." The companion said, "Oh, she probably has Tourette syndrome."

Like tics, the coprolalia can sometimes be suppressed. This has allowed some patients to hold it off until they are in a private place where they can let it out. One housewife said she was able to keep her husband from knowing she had TS by waiting until she was in the shower to release her tics and obscenities. All her husband knew was that "My wife is a very clean person, she is always in the shower."

Many times the words are racial or ethnic slurs such as "nigger" or "wop." When a child prances around the playground using these words he can get himself beaten up very easily.

It is clear that TS patients often simply swear more than others when angry or trying to emphasize a point. Although not true coprolalia, this can become a distinctive part of a person's personality. When taking histories we often hear the term "he swears a lot," or "she swears like a sailor" even though more careful questioning indicates it is not real coprolalia.

Mental coprolalia Some patients who do not have audible coprolalia have mental coprolalia, that is, obscene words swirl around in their head as obsessive thoughts. Sometimes this can be the only symptom of TS. We have known several mothers of TS children, who we know carry the gene since they also have a parent or sibling with TS, whose only symptom is mental coprolalia. Until this symptom was explained they thought they were "crazy" or "evil" because of their apparent obsession with dirty words.

Age of Onset

As originally noted by Gilles de la Tourette, the coprolalia usually comes on sometime after the onset of the motor and vocal tics. We found that the average onset was at ten years of age, and the average duration from the onset of tics to the onset of coprolalia was four years[358]. Shapiro[1767] found the average age of onset to be 12 years with an onset five years after the beginning of the tics. This onset with the beginning of puberty does not mean there is any correlation between coprolalia and increased levels of the sex hormones. We often see coprolalia in grade school children.

Why Coprolalia?

There have been many efforts to explain the coprolalia of TS. Early psychoanalysts had a field day with this symptom, claiming it was obviously the result of repressed rage buried in the unconscious mind and boiling up like volcanic ash in an explosive release. A much more likely explanation is that Tourette syndrome is a disorder of disinhibition, that is, the inhibition of actions and thoughts is an active process in the brain, and this activity is impaired in this genetic disorder, i.e., there is poor inhibition or *dis*—inhibition. This concept is elaborated on in more detail in later chapters. In simple terms there seems to be a "box" in the brain where words and thoughts that we have been told are bad are placed and the lid is shut. In TS the lid is poorly shut. Freud talked of the *id, ego* and *superego.* A friend of mine once characterized TS as "a lid off the id"[530].

Copropraxia This term refers to "dirty movement." This is primarily the excessive use of the finger sign. It can be distinguished from regular adolescent behavior by the inordinate frequency and inappropriateness of its use.

Coprographia This refers to dirty writing or drawing. TS patients may sometimes write dirty words on their school work or other inappropriate places. One seven year old boy used a pad of yellow stickers to write the word "fuck" on. He would then pass these out to his mother, brother and other people.

Other forms of inappropriate sexual behavior are described in the chapter on Sex and Exhibitionism.

Summary: Compulsive swearing or coprolalia is one of the most highly publicized symptoms of Tourette syndrome and many professionals have the mistaken idea that it must be present to make a diagnosis. However, it is not one of the diagnostic criteria and is present in less than one-third of TS patients. It is one of many examples of the fact that TS is a "disinhibition" disorder.

Chapter 7
Characteristics of the Tics

The most frequent statement that TS patients make in trying to explain the tics is that they feel a sudden tension building up that must be relieved by the tic. Brain wave tests done at the time of muscle ticing show no unusual activity[1218,1426]. This indicates that the tics are elicited by nerve activity that does not involve the conscious part of the brain. This interpretation is consistent with the generation of the tics from a more primitive part of the brain[p401]. Some TS patients have described various sensations preceding the tics[180,1771].

Suppressibility

A major characteristic of the tics is that they are suppressible for varying periods of time. Some patients can only suppress their tics for a few minutes while others can suppress them for most of the day. Thus, school children may suppress tics throughout the school day and adults can suppress them while at work. However, the longer the tics are suppressed the more the tension builds up and the more there is a need to release them at a later time in the day. Because of this there is often a period of one or two hours after a child returns home from school when the tics are severe. This was well expressed by a young girl who said:

> "It seems that I am constantly covering up or controlling symptoms. As I became older I seemed to be better able to control my tics in public and I am able to hold down a job or go to school without drawing attention to myself. However, it seems that I make up for this when I am alone. When I come home it feels as if all my tics assault me and I am too weak to do anything but surrender to them."

While the ability to suppress the tics is very helpful in allowing patients to cope in society there are several problems with this aspect of the disorder.

Suppressibility inhibits the diagnosis The ability of patients to suppress their tics accounts for what we call "the doctor's office syndrome." Parents often complain that their child has had severe tics and vocal noises in the car on the way to see the doctor, but that all the symptoms disappear the minute they walk in the door. If the doctor or psychologist is not familiar with this aspect of TS they may conclude that the parents are just a little bit crazy. Many professionals, if they have ever heard of Tourette syndrome, have the distorted view that patients must tic in their presence to truly have TS. In fact, suppressibility is one of the diagnostic criteria of TS — not a reason to reject the diagnosis. Some have referred to TS as "a corridor diagnosis" since it is only after the doctor has left the room to stand in the corridor that the patient stops suppressing and the vocal noises or the swearing can be heard.

Suppressibility can be misinterpreted by school psychologists We have seen many reports by school psychologists who have assumed that since the tics are virtually nonexistent at school, but reported to be severe at home, "there must be terrible tensions in the home causing these symptoms." Some have even gone so far as to recommend that the child be taken out of the home. In fact, just the opposite is the case. Since the child feels comfortable at home, they can finally relax and let the tics out.

Waxing and Waning of Symptoms

In addition to suppressibility, a second major characteristic of TS is a waxing and waning of the symptoms. This includes both a change from one tic to another, to the complete cessation of symptoms for varying periods of time. Since the tics are significantly *exacerbated by stress*, they frequently get worse when school first starts in the fall, during Christmas vacation (which can be a very stressful time), during tests, and with emotional upheavals in the home. The flip side of this is that the symptoms often significantly improve during the summer when the children are out of school. This may be so marked that the medication can sometimes be decreased or stopped during the summer.

This waxing and waning makes the evaluation of any type of treatment difficult. The improvement of symptoms may be accidental rather than due to the new medicine. Many patients must be followed over a prolonged period of time, or placebos used, to avoid mistakenly thinking some new medication is effective.

Grades of Severity of TS

A number of important questions can be answered about TS by dividing the cases into groups according to severity. However, to do this effectively requires that the division be as unbiased as possible. Toward this end we have divided our cases into only three groups. Grade 1 cases (mild) are defined as meeting the criteria for diagnosis but having symptoms too mild to justify medical treatment. Grade 2 cases (moderate) differ from grade 1 simply by being severe enough to require treatment. Grade 3 cases (severe) are those in which there is a significant interference in the individual's life, either by having persistent problems in school, severe difficulties with interpersonal relationships, or trouble holding a job.

The percentage of these three groups in 500 cases[358,360] was as follows:

Grade	Percentage
Grade 1	15%
Grade 2	60%
Grade 3	25%

There has been no significant change in these percentages over a period of eight years. It should be kept in mind that these represent individuals who have been sufficiently concerned about their symptoms to seek medical help. Family and epidemiological studies indicate that when everyone in the population is counted, including those who have not sought medical help, the percentage of mild grade 1 cases is much higher.

Age of Onset

The tics and vocal noises of TS usually begin when children are starting in kindergarten or the early grade school years. The distribution of age at onset is shown on the next page.

Twenty percent start before kindergarten, and 15 percent start between 10 and 16 years of age. Sometimes the tics may begin in the first year of life[p271]. Although the revised DSM-III criteria indicate symptoms can start as late as 20 years of age, in our experience onset past age 16 is rare. This onset of symptoms in childhood is consistent with the continued maturation of some aspect of the nervous system up to and through puberty.

Is Tourette syndrome more severe when the symptoms start early? We can answer this by asking if the symptoms start later in the mild grade 1 cases compared to the more severe grade 3 cases. The following are the

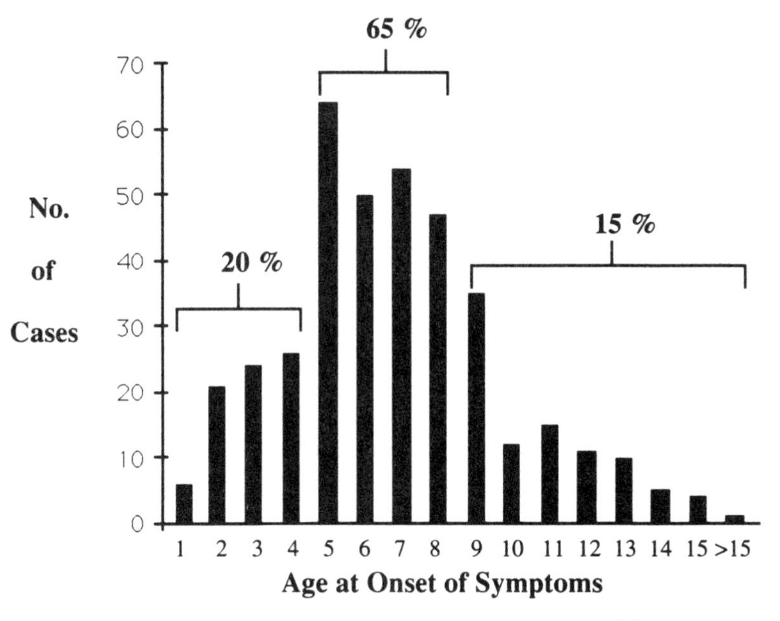

results based on the age of onset of 178 cases[360].

Grade 1	7.36 years
Grade 2	6.77 years
Grade 3	5.81 years

This indicates that indeed the onset of symptoms is 1.5 years later in mild compared to severe cases.

Who Gets Tourette Syndrome?

Anyone can. For a number of years it was thought that TS might be more common in Jews, especially Ashkenazi (European) Jews[530,531,1767,2029]. This, however, was due to what we call ascertainment bias in that some studies were done in areas with a large Jewish population. In California, only about 10 percent of the population is Jewish and only 10 percent of our cases are Jewish[358]. In Washington, D.C., 10 percent of the population is Jewish and 10 percent of TS cases were Jewish[1397]. In Texas, only 1percent of the population is Jewish, and there, about 1 percent of TS individuals were Jewish[754]. We have seen TS in blacks, Hispanics, Filipinos, Vietnamese, Chinese and Japanese. The frequency in different national and racial groups reflects the frequency of these individuals in the population. The intriguing question is whether these all are due to the same gene, or whether different genes are involved.

The Sex Ratio

There are 3 to 4 males with TS for every female. The possible reasons for this are discussed later.

Precipitating Factors

When mothers have a child with a birth defect they often look back on their pregnancy for some cause like, "It must have happened when I tripped and fell at Judy's party." They don't realize that life is full of little accidents that actually have nothing to do with causing anything. TS mothers do the same thing. "The tics started after my husband and I had a big argument, so we are to blame." If a child is destined to have TS the tics have to start sometime and it will always be possible to pick out some life stress that occurred around that time to blame. Having said this, I should point out that if TS is simmering and almost ready to start, some stresses can be the final straw that initiate the symptoms. A frequent stress is the beginning of school and many children have their first tics in August or September. Minor irritants can sometimes focalize the site where the tics will start. One boy was playing football with a loose helmet and kept flipping his head up. When football season was over his first symptom was a vertical head tic, as if he was still adjusting his helmet. Several months later he joined the track team and had a loose shoe. When track season was over his head tic was replaced with a foot tic. Another 13 year old boy never had tics until he got on a plane by himself to visit his grandmother. He had multiple tics coming down the ramp at his destination. His mother received a frantic call from grandma, "What is wrong with your son, he is jerking all over the place?"

The important point is *these events do not cause the TS, genes cause TS*. No event and no person should be blamed.

TV and Tics

One of the most commonly mentioned events that seem to bring out motor and vocal tics is watching TV. When a patient is concentrating on something the tics are minimal, but as soon as they relax and watch the tube, out come the noises and tics. Sometimes these get so bad that other members of the family find they cannot watch TV in the same room.

Summary: The motor and vocal tics of Tourette syndrome wax and wane, and one tic may be replaced by another. Tics can be suppressed for varying periods of time. Stress makes them worse. The average onset is at seven years of age and three to four times as many males are affected as females. Individuals of all races can have Tourette syndrome.

Part II
GENETICS

Chapter 8

Pedigrees

A pedigree is a diagrammatic presentation of a family history. Since I will make liberal use of pedigrees in this book to illustrate the many relationships between Tourette syndrome and other behavioral disorders, an explanation of the symbols used in pedigrees is necessary.

A male is represented by a square:

A female is represented by a circle:

Ages are placed in the upper right corner:

If an individual has died they have a diagonal slash:

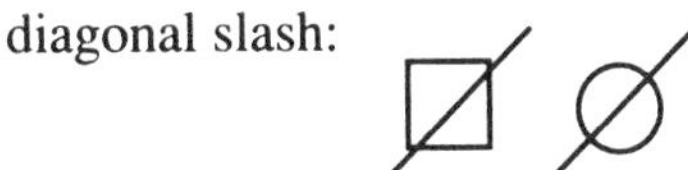

The initial patient, also called the **proband** or **propositus**, is indicated by an arrow:

For a marriage a male and a female are joined by a line:

A divorce is:

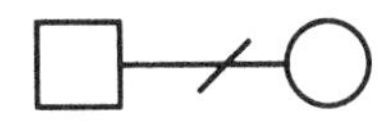

A union without marriage is a broken line:

Children, such as a boy and a girl, are:

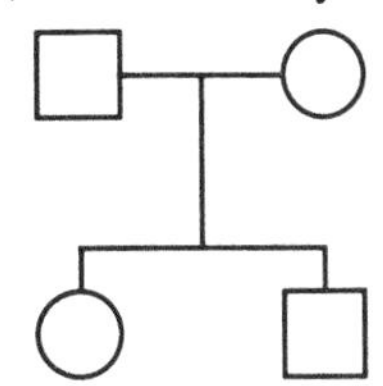

A divorce and remarriage to a younger woman, with the older children living with the first wife is:

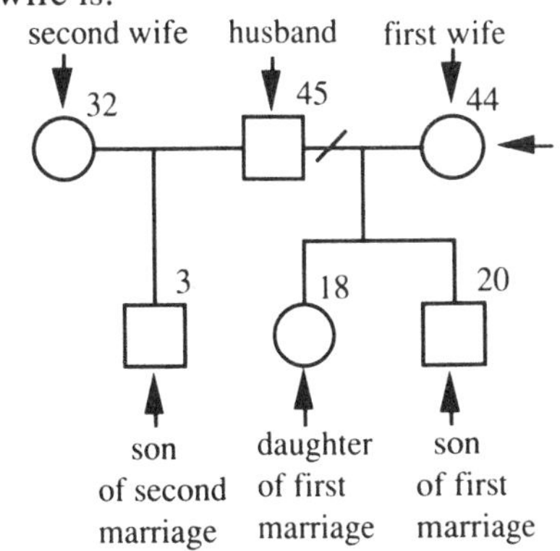

Males or females with motor tics are:

Males or females with vocal tics are:

Males or females with motor and vocal tics, i.e., Tourette syndrome, are:

A presumptive Tourette syndrome gene carrier is:

Additional information can be entered with letters and explained with a letter:

A **A = alcoholism**

If the man shown on the last page had Tourette syndrome and had a son with Tourette syndrome and a daughter with motor tics by the first marriage and a son with vocal tics by the second marriage, it would be shown as:

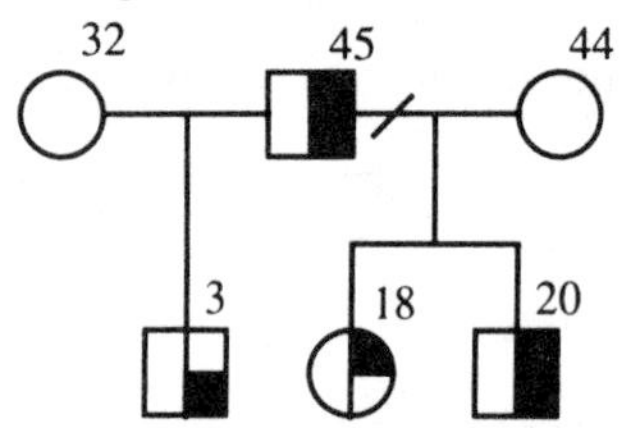

Thus, the pedigree allows us to see at a glance a great deal of information about families and how genetic diseases affect different individuals. Taking a detailed pedigree and obtaining information not only from relatives but from the individuals themselves can be critical in making a diagnosis. For example, if a teenager comes in with a history of a severe conduct disorder and a history of throat clearing for eight months when he was in the first grade, the diagnosis would simply be conduct disorder.

C **C = conduct disorder**

However, if a careful family history shows that his father and brother have Tourette syndrome, and we know that conduct disorder is common in individuals with Tourette syndrome [p131], then our diagnosis is changed to *conduct disorder probably due to a Tourette syndrome gene.*

46 42

15 18

C

This change in diagnosis can be very important since if we believed the conduct disorder had an organic rather than a purely psychological basis then we might try medication before committing him to long-term psychotherapy.

How to Take a Pedigree

Taking a good pedigree is a critical part of the evaluation of a patient with Tourette syndrome or any other behavioral disorder. Physicians and psychologists who are not interested in genetics have a tendency to simply ask a general question such as, "Does anyone else in your family have a similar problem?" While this may sometimes work for the closest relatives, and the specific problem at hand, it is usually totally inadequate for examining a wide range of behavioral problems. There is often a tendency to ask if the relatives are affected only with the problem that is present in the patient. As shown throughout this book, the *Gts* gene may present as one disorder in one member of the family and as a completely different disorder in another member of the family. Standardized naming of genes requires that gene names be italicized. ***Gts*** is the symbol for the presumed **G**illes de la **t**ourette **s**yndrome gene.

I have found it is most effective to take a pedigree in five steps. First, determine the name, sex and age, living or dead, of all the known relatives on the mother's side of the family and diagram them into the pedigree. Second, ask detailed questions about the personality and behavior of each of these individuals. The third and fourth steps are to do the same for the father's side of the family. The fifth step is to go back over all members, asking additional specific questions. This last step will often bring out things that were initially forgotten. It is necessary to ask about each type of behavior. For example, we are unlikely to find out about Uncle Ben's alcoholism and wife abuse, or the fact that Aunt Jane weighs 310 pounds, if specific questions about alcoholism, wife abuse and obesity are not asked. Obviously, obtaining a good pedigree can take a lot of time. It is a very important but often badly ignored part of a good psychiatric history.

Summary: Obtaining a detailed pedigree and diagraming it, so the entire family can be seen at a glance, is a critical part of evaluating a family with Tourette syndrome or any other behavioral disorder.

[Those who have a deep seated phobia of "genetics" now have enough information to jump to the section on behavor, if they wish. Those who really want to understand the subsequent chapters should finish this section.]

Chapter 9

A Very Brief Course in Genetics

The substance that carries the genetic information that is unique to each individual is deoxyribonucleic acid or DNA. Its structure was first described by Watson and Crick[2032].

DNA

DNA is a long twisted ladder-like compound which looks like this:

The secret of the ability of DNA to carry information is the fact that there are two types of rungs in the ladder. These are A-T rungs and G-C rungs. Just as in a computer, where all the information is ultimately broken down into a series of 0's and 1's, all the information in DNA is ultimately broken down into a series of A-T's or G-C's. Thus, in a computer a specific piece of information may read:

0 1 0 0 1 0 1 1 0,

while in DNA a specific piece of information may read:

A A G T C T C T T
T T C A G A G A A

In the computer the code is electrical where the 0's or 1's represent the presence or absence of an electrical charge. In DNA the code is chemical where the A, T, G and C are chemical compounds. These compounds are as follows:

A = adenine,
T = thymine,
G = guanine,
C = cytosine.

The genetic code is, in essence, a four letter alphabet. The critical feature of these letter compounds is that A always pairs with T to form an A-T pair:

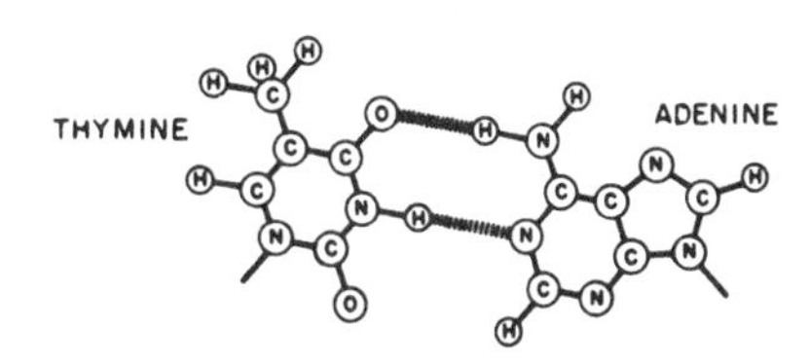

and G always pairs with C to form a G-C pair:

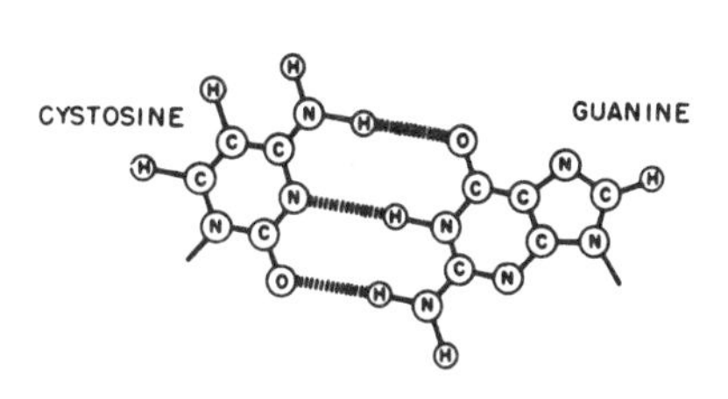

The A, T, G and C are called bases and the pairs are called base pairs. There are three billion (3,000,000,000) base pairs in human DNA. This is capable of carrying an enormous amount of information. In order for DNA to serve its critical function of making each of us a unique individual, it must be able to do three things:

1. Store information. We have just seen how it does that by using the letters A, C, G and T.

2. Pass that information from one generation to the next.

3. Pass that information from the nucleus of the cell to the cytoplasm where it tells the cell how to make the different proteins and

enzymes that regulate the chemistry of the body.

All three of these tasks are made possible by the fact that A always pairs with T, and G always pairs with C. Lets examine its second task.

Passing information from one generation to the next In order to pass the coded information from one generation to the next, or from an old cell to two new ones, one DNA molecule must be able to divide into two identical DNA molecules. To do this the rungs split down the middle:

AAGTCTCTT
—— split ——
TTCAGAGAA

and new pairs are formed from a pool of A's, T's, G's and C's, as follows:

AAGTCTCTT
TTC - - - - - -

and

AAG- - - - - -
TTCAGAGAA.

As the process continues it forms two new identical pieces:

AAGTCTCTT
TTCAGAGAA

and

AAGTCTCTT
TTCAGAGAA.

The site where this takes place is called the replication fork, and looks like this:

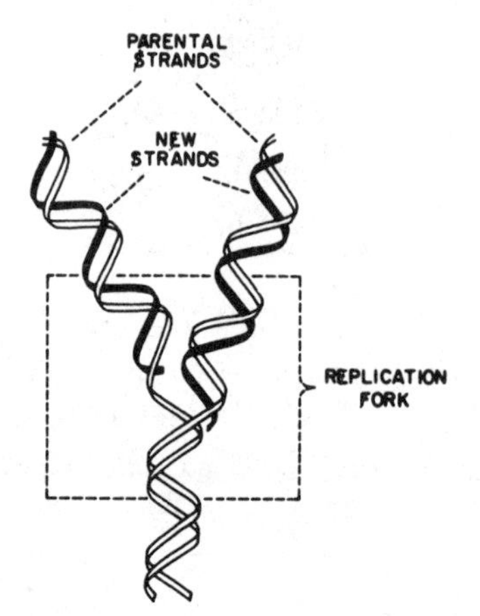

In greater detail it would look like this:

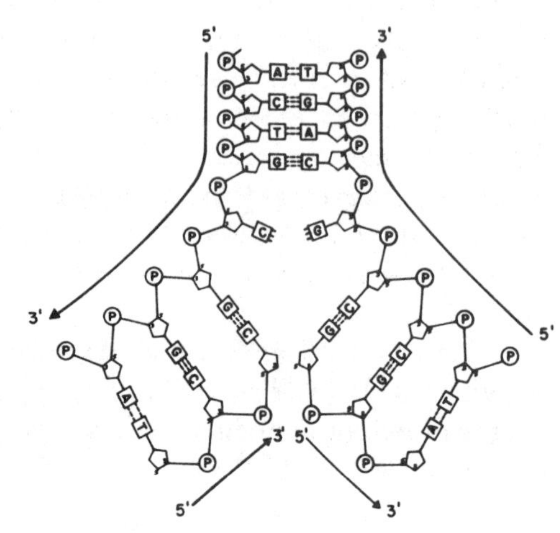

As mentioned before, the rungs of the ladder are made up of A-T and G-C pairs. The sides of the ladder are composed of sugar molecules, called deoxyribose, connected by phosphorous molecules, P.

For centuries man has puzzled over how the unique identity of individuals and species is passed from generation to generation. When Watson and Crick discovered the structure of DNA in 1952, this problem was solved.

Passing information from the nucleus to the cytoplasm The third task is to use the information coded in the DNA. To do this that information must be sent to a different part of the cell. Cells have two major parts: a nucleus where the DNA is stored, and a cytoplasm where the information from the DNA is used to make proteins.

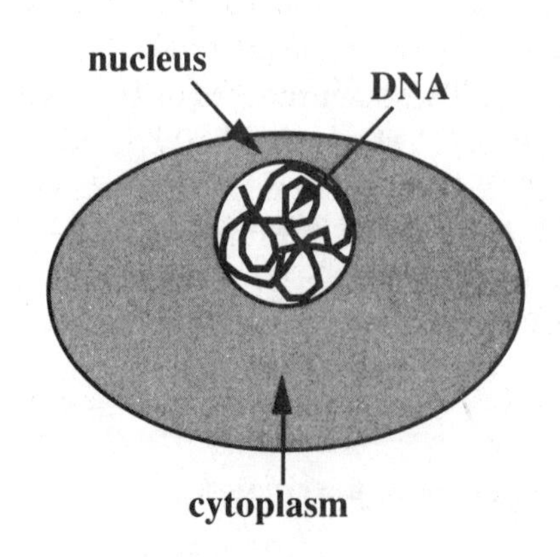

The genetic information from the DNA is carried into the cytoplasm by a messenger appropriately called messenger RNA, or **mRNA**. RNA, or ribonucleic acid, is identical to DNA except that the sugar molecules in the sides of the ladder are slightly different (ribose instead of dexoyribose) and U or uridine is used instead of T or thymine. Thus, the base pairs in RNA are A-U and G-C instead of A-T and G-C.

The way that mRNA is made is similar to the way new DNA is made. Again the rungs split apart and one side of the ladder is read into mRNA:

AAGTCTCTT DNA
UUCA- - <-mRNA being formed

The completed mRNA molecule then passes into the cytoplasm where it is used as a code for making proteins and enzymes.

Proteins and Enzymes

The human body and brain basically function through a complex series of chemical reactions. For example, when the amino acid, tryptophan, enters the brain the first thing that is done is to add a hydroxyl group (oxygen+hydrogen, —OH). This reaction is:

tryptophan + —OH -> tryptophan—OH

However, this reaction doesn't just happen it must be made to happen by an enzyme which serves as a catalyst. Every one of the thousands of chemical reactions that take place in the body and brain are driven by a separate enzyme. Thus, thousands of different enzymes are needed. The enzyme that adds the hydroxyl group to tryptophan is called tryptophan hydroxylase. Thus, the reaction actually looks like this:

tryptophan hydroxylase
tryptophan + —OH ————> tryptophan—OH

Enzymes are proteins. Like DNA, proteins are also made up of a long string of compounds called amino acids. There are 20 primary amino acids. Some examples are alanine, tryptophan, tyrosine, and phenylalanine. This long string folds in a specific way to form a three-dimensional protein structure which contains an active site where the action of the enzyme takes place.

enzyme

Each enzyme works on a specific chemical compound called a substrate.

The substrate fits into the active site.

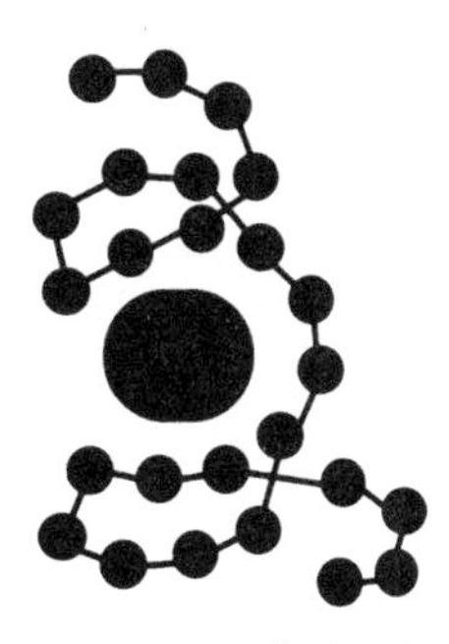

enzyme + substrate

Holding the substrate in the active site is like holding a piece of wood in a vice. This allows it to be worked on or, in this case, chemically modified. After the work is finished, it is released and the enzyme is now ready to work on the next molecule. In this manner, enzymes drive the chemical reactions of the body.

Each set of three bases in the mRNA codes for a single amino acid. For example, the code for the amino acid phenylalanine is UUU, and the code for tryptophan is UAU. Thus, if a section of mRNA contained the sequence UUUUAU it would code for a section of protein containing the amino acids phenylalanine and tryptophan. In this manner the mRNA dictates the sequence of amino acids in the enzyme. The sequence of amino acids in the enzyme in turn determines which specific chemicals the enzyme works on. This can be summarized in the following way:

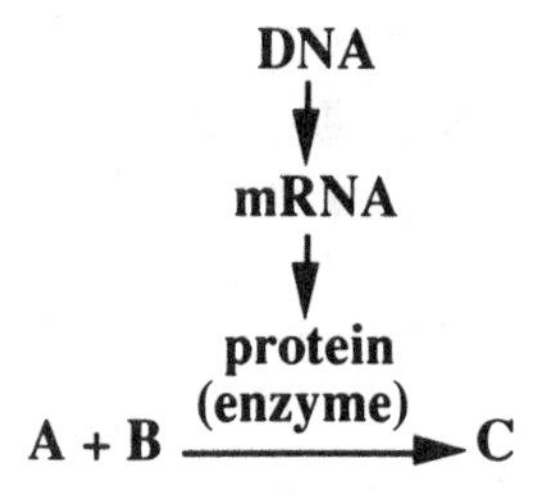

Mutations and Mutant Genes

Since the major purpose of this book is to show how mutations affect human behavior, it is critical to understand what a mutation is. A mutation is any type of alteration in the sequence of base pairs in DNA. The simplest type of mutation is a change in the sequence of one of the base pairs. This, logically enough, is called a **single base pair mutation**. For example, in the sequence

AAG TCT CTT
TTC AGA GAA

a single base pair mutation would produce the following sequence:

AAG TTT CTT
TTC AAA GAA

Here the base pair T-A has replaced the base pair C-G. If this mutation occurred in a part of the DNA that was coding for an enzyme, the mRNA would have a different sequence of bases and thus the enzyme it coded for would have a different sequence of amino acids. If the sequence of the active site in the enzyme was changed the enzyme might not work. This is shown as follows:

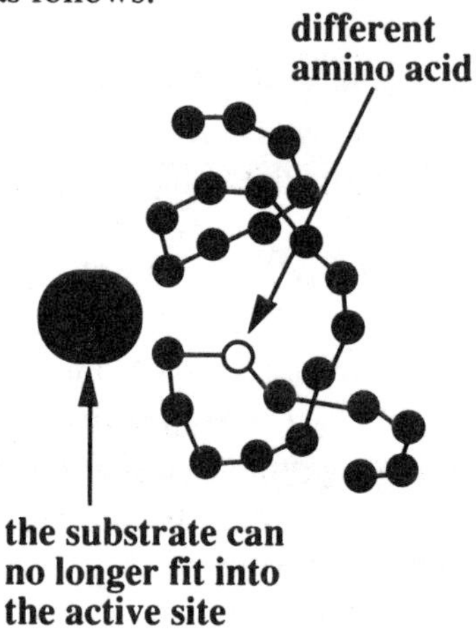

The change in one amino acid in the active site caused the protein to fold differently, and now the substrate cannot fit into the active site and the enzyme doesn't work anymore. Thus, *mutations in the DNA can change the chemistry of the body by preventing or altering the way certain enzymes and chemical reactions work.*

Genes

Genes are those segments of DNA that carry the code for proteins and enzymes. There are somewhere between 30,000 and 100,000 genes in humans. Despite this large number they still represent only a small fraction of the total DNA. Thus there are long stretches of DNA between the genes.

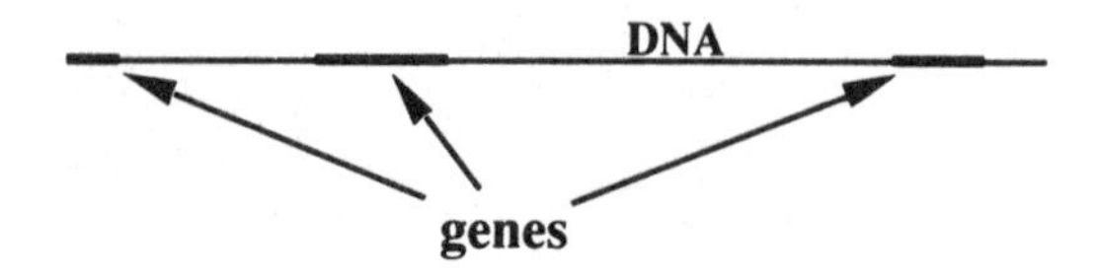

We once thought that genes were continuous and that mRNA was simply read from one end to the other. However, we now know that the genes themselves are broken up into segments called **exons** (**ex**pressed regi**ons**) that actually code for proteins and other segments called **introns** (**int**er regi**ons**) that act as spacers.

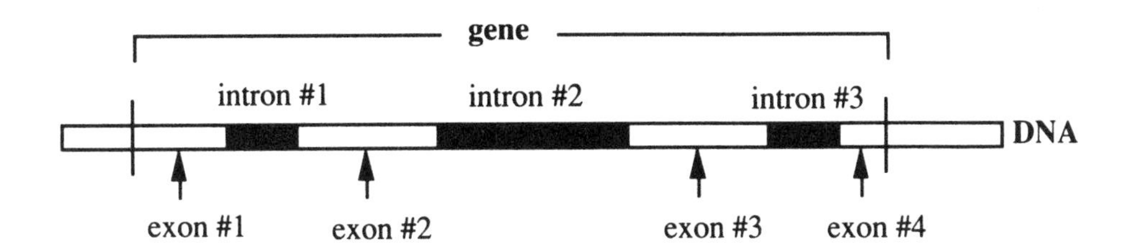

The entire gene is read into a large piece of RNA called hnRNA. The RNA corresponding to the introns is then spliced out to result in the final mRNA.

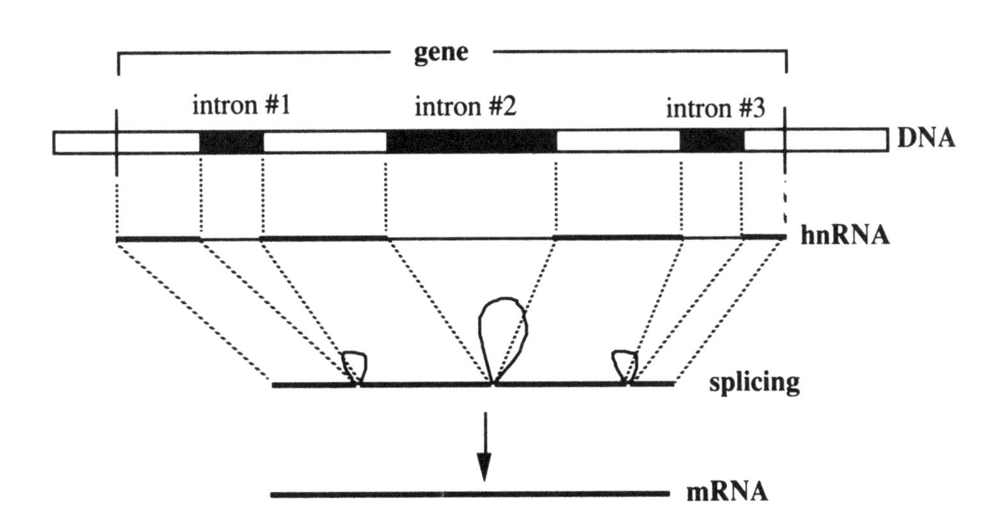

This may seem like a bizarre way to do things. However, from an evolutionary point of view it is probably an efficient way of making new genes. Once exons are perfected to make particular sections of a protein, nature can rearrange them to form new genes in the same manner that a new house can be made up of prefabricated kitchens, bathrooms and bedrooms.

Some genes have no introns, others have many. As a result some genes are very small, consisting of only a thousand base pairs, while others are huge consisting of several hundred thousand base pairs.

Chromosomes

Chromosomes represent a large collection of genes on one long piece of DNA. They have short and long arms that come together at a centromere and look like this->.

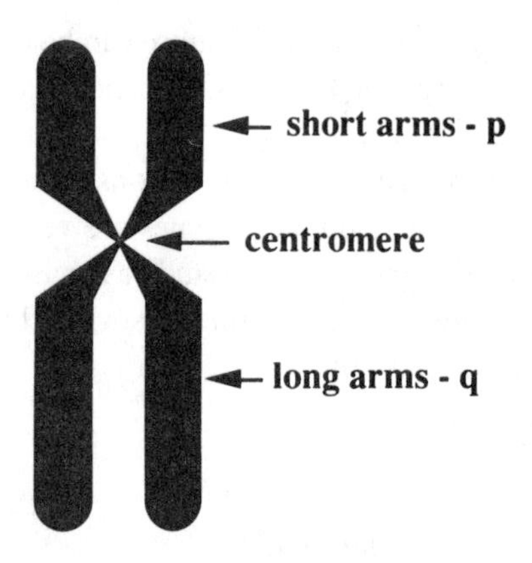

By convention the short arms are denoted p and the long arms q. Thus 4p would refer to the short arm of chromosome #4. The chromosome arms are composed of highly folded, compacted DNA associated with proteins. These DNA-protein fibers are shown in the following electron microscope picture of a human chromosome.

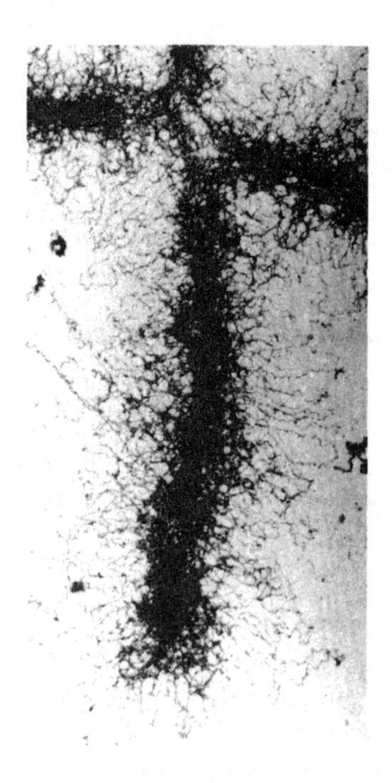

Electron microscope picture of a human chromosome at a magnification of 17,000 times. From D.E. Comings and T.A. Okada, Chromosoma[372].

While there are up to 100,000 genes, in man there are only 46 chromosomes. These are divided into one pair of sex chromosomes, XX in females and XY in males, and 22 pairs of non-sex chromosomes called autosomes. Since each individual has only 23 different chromosomes there are thousands of genes on each chromosome.

If we wish to examine the individual chromosomes, blood is drawn and the white cells (lymphocytes) are stimulated to grow in a test tube for several days. They are then treated in special ways to bring out a pattern of bands that is specific for each chromosome.

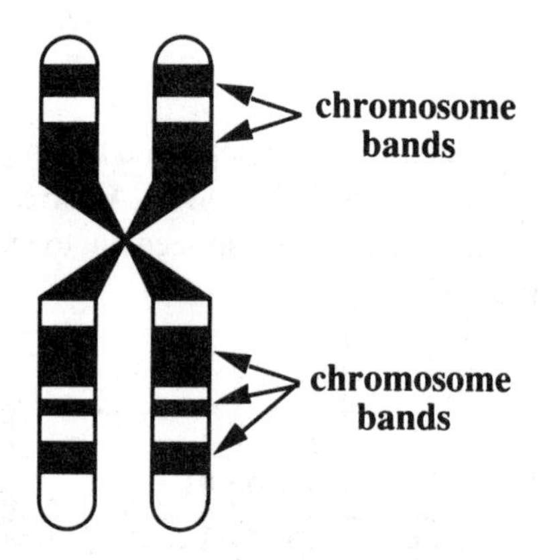

The chromosomes are cut out and arranged according to size and banding pattern. The **following page** shows a set of normal banded human chromosomes and a diagram-showing how the bands are labeled. As shown in the following chapters, genes causing some behavioral disorders have been localized to specific chromosome bands.

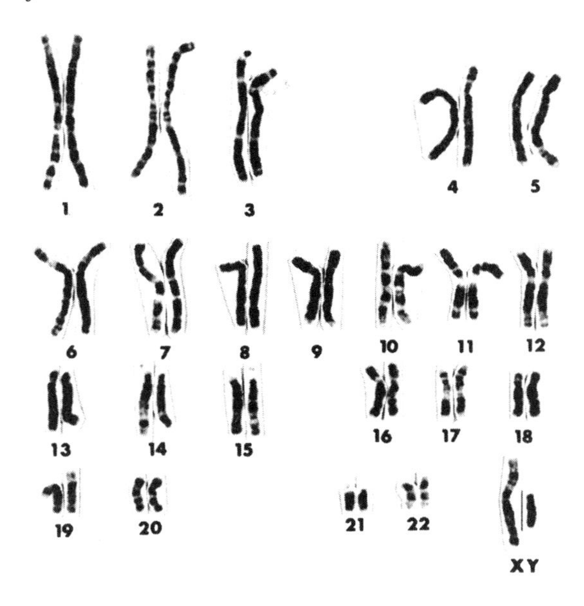

Normal human banded chromosomes. Courtesy of Dr. Dean Stock, Department of Cytogenetics, City of Hope Medical Center.

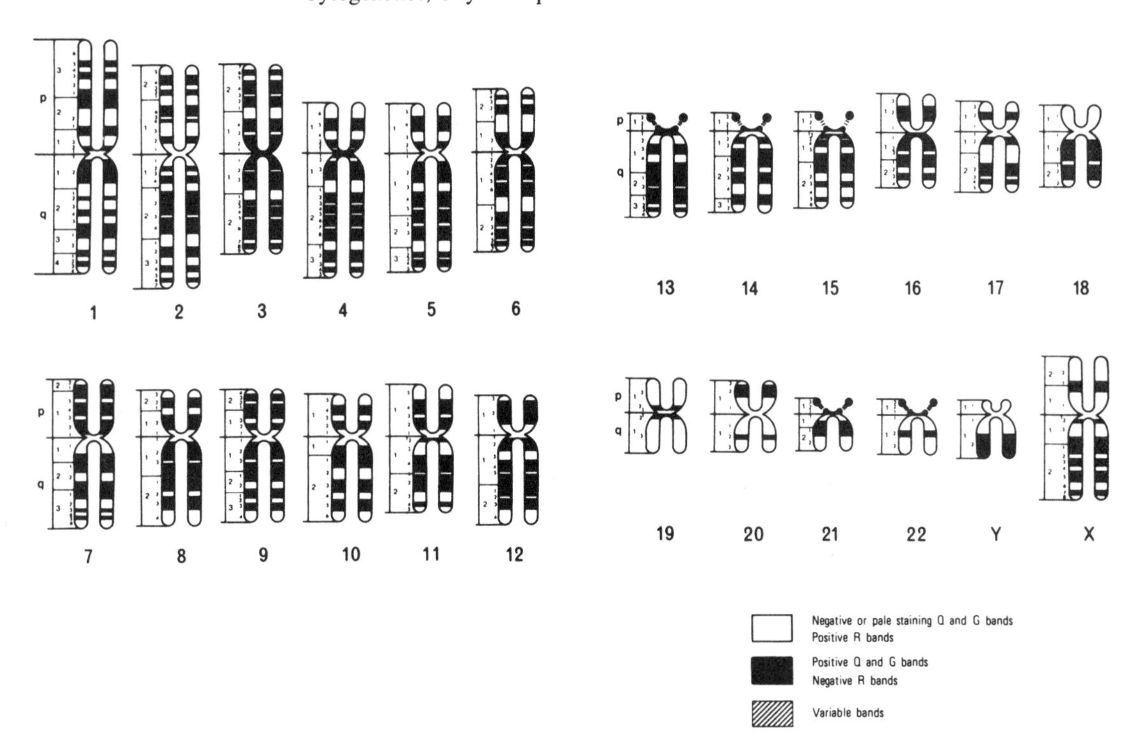

A diagram of the banding pattern of human chromosomes. From the Paris Conference (1971) The National Foundation[1463].

Summary: DNA is the basic genetic material of the cell. Its secret for carrying information is that of its four bases, adenine is always paired with thymine and cytosine is always paired with guanine. The sequence of these bases provides the code for the unique information in each gene. The genes in turn determine the type of enzymes that are made and the enzymes control the types of chemical reactions that occur in the body and the brain. A mutation causes a change in the base sequence of the DNA of the gene. This may cause the enzyme the gene produces to be ineffective. As a result, the specific chemical reaction controlled by that enzyme no longer works.

Chapter 10
Types of Inheritance

While pedigrees are useful for visualizing how a disease is distributed among different family members, they also help to visualize how the genes themselves are distributed throughout the family. As discussed in the previous chapter, there are 46 human chromosomes divided into 23 different pairs. Chromosomes come in pairs because one is inherited from the father and one from the mother. When there is a genetic disease in the family, such as Tourette syndrome, one of the chromosomes will carry the abnormal or *Gts* gene, shown here as a dark band.

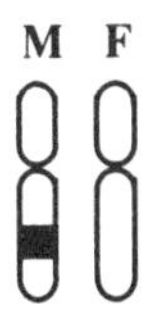

The name of a gene is always italicized; thus — *Gts* gene. When the mutant gene is dominant, or semidominant[p513], it is capitallized and the normal, non-mutant gene is in lower case — *gts*.

Autosomal Dominant Inheritance

This form of inheritance is called autosomal because the gene is on one of the autosomal chromosomes, as opposed to a sex chromosome. It is dominant because only one abnormal gene is required to produce the disease, that is, the abnormal gene is dominant over the normal gene. In this type of inheritance there is a 50 percent chance that any child will inherit the abnormal gene, and males and females are equally affected. Multiple generations are often affected in genetic disorders that are inherited in an autosomal dominant fashion.

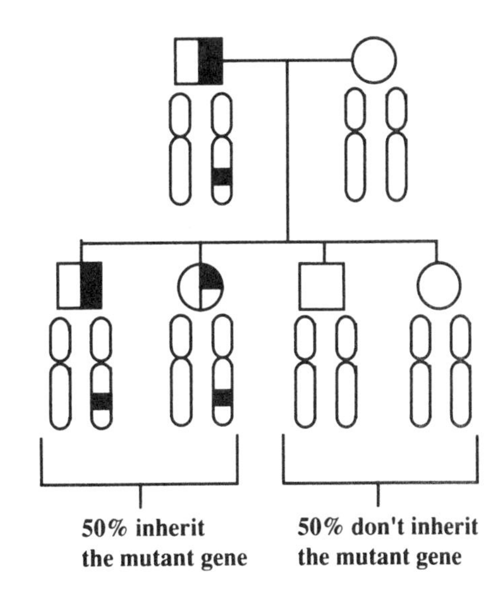

The above pedigree also illustrates the concept of **variable expression**. The father and his son with the *Gts* gene have fully expressed TS with both motor and vocal tics, while the daughter has an incompletely expressed *Gts* gene with only motor tics.

The ability of the *Gts* gene to show variable expression is a central message of this book. This expression may be as full-blown TS with motor and vocal tics, or motor or vocal tics alone, or attention deficit disorder, dyslexia, phobias, panic attacks, obsessions, compulsions, conduct disorder, depression alone, or in many other ways.

A second concept related to expression is that of **penetrance.** Non-penetrance means that a person can carry a gene but show no signs of it. But if a person shows no signs of TS, how do we know they carry the gene? We know it in the following way. If a woman has a child with TS and a parent or sibling with TS, and there is no evidence of TS in the spouse, then that woman is carrying a TS gene that is non-penetrant. This is illustrated as follows:

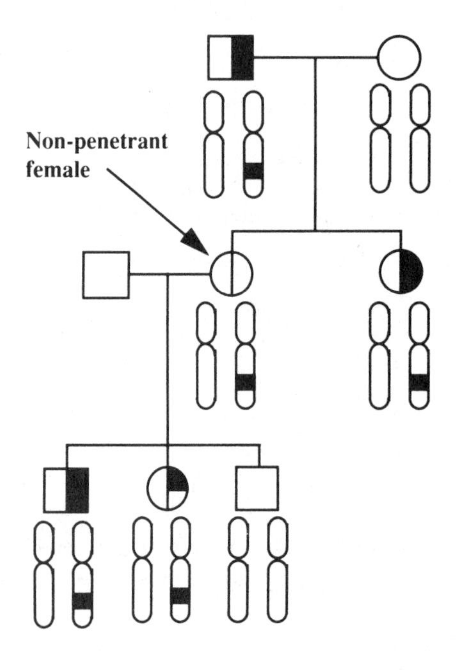

Males or females who are carriers of a *Gts* gene (but don't express it) are shown as:

Autosomal Recessive Inheritance

A second type of autosomal inheritance is recessive. Here the abnormal gene must be on both the father's (paternal) and mother's (maternal) chromosome for the individual to have the disease. This is because the gene is recessive to the normal gene. In order to have the gene on both chromosomes both parents have to carry the gene. They do not have the disorder since only one of their genes is affected. This is illustrated in the following pedigree for cystic fibrosis.

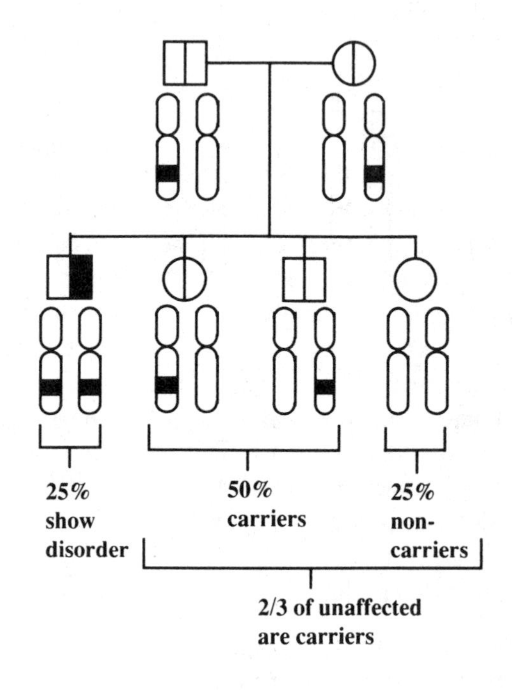

In recessive inheritance, on average, 25 percent of the children, regardless of sex, are affected with the disorder, 50 percent are unaffected gene carriers, 25 percent are neither affected nor gene carriers, and two-thirds of the *unaffected* individuals are gene carriers. Since it requires two genes to produce recessive diseases, usually only siblings are affected. Because both parents have to carry the same gene, recessive diseases are more likely to occur when the parents are related to each other (first, second or third cousins). This is termed **consanguinity**. The rarer the recessive disease the more often the parents are consanguineous.

X-linked Recessive Inheritance

In X-linked inheritance the mutant genes are on the X-chromosome. Since females are

XX, a recessive gene on one X does not cause the disorder and they can be carriers. By contrast, since males are XY, and have only one X chromosome, all males carrying the mutant gene will have the disorder. This is illustrated in the following pedigree for hemophilia.

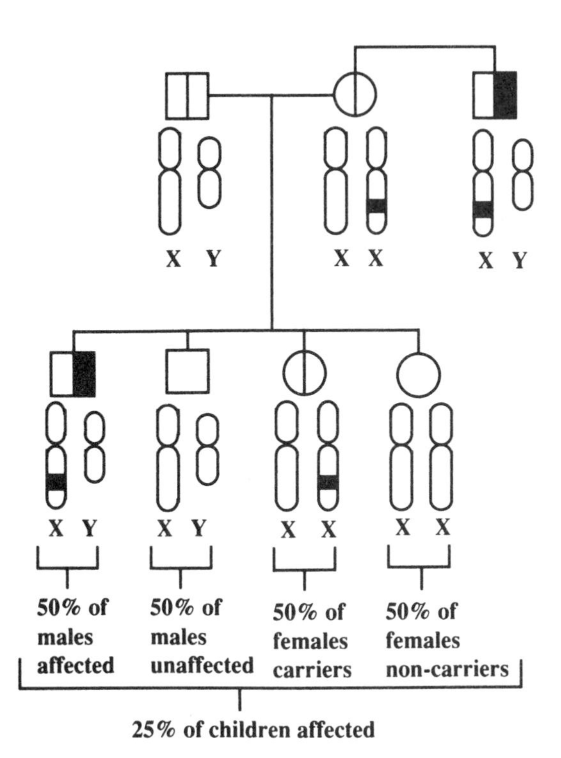

One of the characteristics of X-linked inheritance is that since males only pass their Y chromosome to their sons, there can be no male-to-male transmission of X-linked genes.

Multifactorial Inheritance

Since it will be discussed later, a final mode of inheritance needs to be mentioned. Some disorders are caused by multifactorial inheritance, in which two or more genes are inherited from one or both parents. To illustrate, let us assume that three different pairs of genes, or six genes in all, contribute to a certain disease and in order to get the disease one has to get at least three of these genes. In the following pedigree the father has only two and is thus unaffected, the mother has only one and is thus unaffected, but one child inherited three and thus has the disease. In this part of the family the disease appears recessive. However, the affected daughter marries a man with two of the genes and has a son who also has the disease because he inherited four of the genes. Now this part of the pedigree looks like dominant inheritance.

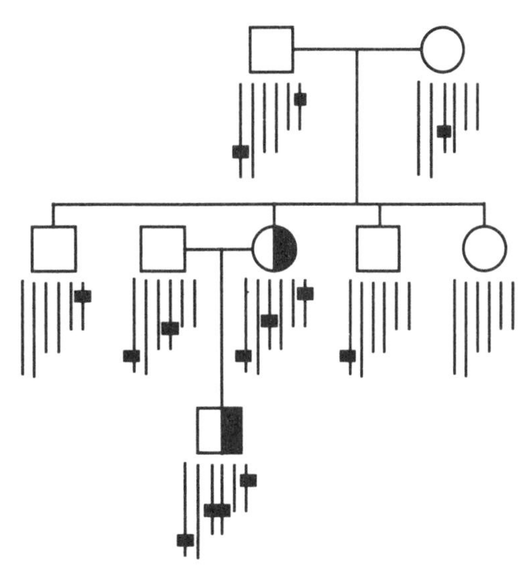

If these genes are common it is easy to see how this mode of inheritance can complicate the task of finding out the precise mechanism of inheritance of some diseases.

Summary: There are several different ways that disorders can be inherited depending on the type of chromosome the gene is on, and whether one or two mutant genes are required to produce the disorder. A disorder is dominant if it can be caused by only a single mutant gene, and recessive if it only appears when the genes from both parents are mutant.

Chapter 11
The Genetics of Tourette Syndrome

Gilles de la Tourette, in his original description of Tourette syndrome, concluded it was a hereditary disorder. However, in the early part of this century, following the enormous influence of Sigmund Freud and the development of psychoanalysis, many disorders of unknown cause were considered to be psychological in origin. Tourette syndrome, with its bizarre muscle movements, coprolalia, and compulsive behaviors, all increased by stress, provided an irresistible setting for the psychoanalysts to claim that these symptoms were all the result of repressed primeval rage stemming from some dire event in early childhood. Numerous single case reports were published claiming success with psychoanalysis, not recognizing that the disorder had a naturally waxing and waning course. Dr. Arthur Shapiro, a psychiatrist at Mount Sinai Medical School in New York, was a pioneer in recognizing that TS had an organic basis[1767,1770]. However, he met tremendous resistance to this concept among his psychoanalytic colleagues.

Pedigree Studies

The single most powerful piece of evidence that Tourette syndrome was an organic disorder would come from the clear demonstration that it was a genetic disease. Perhaps contributing to the bias toward a psychological cause of TS was the fact that, other than a few references to its familial nature in the 19th century[715,1072,1961], the next reference to a positive family history was by Eisenberg and colleagues in 1959[528]. Following this, scattered reports of multiple affected members in one family began to appear[189,512,553,580,589,646,649,746,1219,1346,1524,1688]. Summaries of larger numbers of patients with tics or TS reported a positive family history in about one- third of cases with a range of 10 to 67 percent [399,1347,1767,1956,2147].

In 1977 Roswell Eldridge from the National Institutes of Health and his colleagues were the first to present a series of pedigrees on multiple families [530]. They specifically selected 21 families believed to have a positive family history from a roster of 120 cases. The number of affected relatives in a family ranged from zero to twelve. Similar studies of fourteen families from the midwest were reported in 1979 [2029] in which eight had a positive family history of TS or multiple tics, and in 1980 Nee and colleagues[1397] reported 50 cases where 32 had a positive family history. These reports clearly established that motor tics alone and vocal tics alone were genetically related to TS. Despite these demonstrations of many affected individuals in one family, the highly selected nature of the families made it difficult to determine the precise mode of inheritance. Both recessive[531] and autosomal dominant inheritance with reduced penetrance [371a] and non-genetic[2080] mechanisms have been suggested.

Twin Studies

Identical twins are formed by the separation of the very early embryo into two parts. Since this occurs after conception, these twins carry identical sets of genes. By contrast, fraternal twins don't share genes any more than regular brothers and sisters. Because of this, the comparison of identical with fraternal twins has long been a favorite tool of geneticists for sorting out the relative role of genetics versus environment, or "nature versus nurture." If a disorder is purely genetic, every time it occurs in one identical twin it will occur in the other one as well. This is called **concordance.** The concordance rate of purely genetic disorders is 100 percent for identical twins, and depending upon the mode of inheritance, closer to 25 to 50 percent for fraternal twins. If a disorder is caused purely by the environment, then the concordance rate for identical and fraternal twins is approximately the same, although it may be somewhat higher for identical twins since they tend to more intimately share environments than fraternal twins. If a disorder is believed to be predominantly genetic, but is influenced by environmental factors, then the concordance rate will be less than 100 percent for identical twins, but still much greater than for fraternal twins. Studies of identical twins reared apart have demonstrated the remarkable role of genes in determining behavior and personality traits[911]. An example of the similarity between two identical twins raised apart, then interviewed many years later, is as follows:

Similarities in Identical Twins Reared Apart

	Jim Springer	**Jim Lewis**
Best subject	math	math
Worst subject	spelling	spelling
Training	law enforcement	law enforcement
Car	Chevrolet	Chevrolet
Dog	"Toy"	"Toy"
First wife	Linda	Linda
Second wife	Betty	Betty
Son's name	James Allen	James Allen
Hobbies	drawing carpentry	drawing carpentry
Habits	chews nails to nub	chews nails to nub
Medical	hemorroids	hemorroids
	same pulse and blood pressure	
	same sleep pattern	
	both gained ten pounds at the same age	
	both had mixed headache syndrome (tension + migraine) at age 18	
Differences	hair over forehead	hair slicked back
	expresses best verbally	expresses best by writing

What has the study of twins told us about the genetics of Tourette syndrome? In 1980 the Shapiro's summarized the twins known to them[1760] and observed that seven of nine pairs of identical twins were concordant for TS (78 percent), while one of four fraternal twins were concordant (25 percent). This is consistent with a major genetic influence but some role of the environment in the expression of TS, since not all the identical twins were concordant. In the most extensive study yet done on twins and TS, Price and coworkers[1548] mailed out 8,000 questionnaires to registered TS patients and found 43 pairs of same-sexed twins. These are summarized as follows:

	Identical	**Fraternal**
Number	30	13
Concordant for TS	53%	8%
Concordant for tics	77%	23%

In a subsequent study they reported being unable to find any identical twins where one had TS and the other had no evidence of any tics[1474]. Thus, their revised figures would be:

Concordant for TS	53%	8%
Concordant for tics	100%	23%

When obsessive compulsive behavior was examined, 89 percent of the identical twins were concordant.

This is similar to our own experience with twins which is summarized as follows:

	Identical	Fraternal
Number	6	6
Concordant for TS	83%	0%
Concordant for tics (and ADHD)	100%	50%

In one of our sets of identical twins, one twin had TS and the other twin had only attention deficit hyperactivity disorder (ADHD), indicating the *Gts* gene can be expressed as ADHD only[p259].

In a follow-up of the study by Price and co-workers, in the six cases where one identical twin had TS and the other had only motor or vocal tics, the non-TS twin always weighed more than the TS twin[1146]. This suggests that environmental factors can play a role in the expression the *Gts* gene. While this is consistent with one study showing that mothers of children with tics had more complications of pregnancy than control mothers[1469], in another study no differences were found when TS mothers were compared to mothers of children with a different genetic disorder[1075].

In the Price study the penetrance for TS or tics for male identical twins was 76 percent and for females was 80 percent. They also asked whether the tics tended to come on at the same age in the identical twins. The answer was a resounding yes. For the identical twins concordant for TS, the similarity score [correlation coefficient, see glossary] for age of onset was 0.93 [1.00 being maximum]. For those concordant for TS or tics it was 0.80. There was also a similarity in the severity of symptoms. Here the similarity score was 0.70. Finally they examined the concordance for obsessive-compulsive behaviors. Here the concordance rate was 53 percent for identical twins compared to 15 percent for fraternal twins.

Narrative accounts of the similarities in symptoms in identical TS twins is also interesting, even though it is subject to a great deal of selective bias in that cases showing similarity tend to be reported while the cases showing dissimilarity are not. Jenkins and Ashby[979] reported a set of 16 year old male twins concordant for TS.

> "Both twins showed destructiveness early. They occupied two cribs in the same room and were so destructive of curtains and window shades that these were abandoned. . .
>
> "Twin A used foul language, vocalized grunting sounds, was under achieving at school and was mildly paranoid. Also, ritualistic behavior and phobic avoidance were observed in the boy.
>
> "Twin B constantly expressed anger verbally and reported that tics helped him relax. Phobias developed as well as obsessive thoughts, avoidance behaviors and a feeling that 'I will not be clean.' He was noted to have self-induced vomiting resulting in a loss of weight that required hospitalization and was resolved in the hospital.
>
> "Both boys believed other people's actions could weaken them; they became obsessively unkempt, psychologically close to each other, yet highly competitive. Both boys were hospitalized without resolution of behavioral problems or tic behaviors."

Segregation Analysis

The difficulty of determining the precise mode of inheritance in TS from single or a few selected families can be appreciated by looking at a few pedigrees.

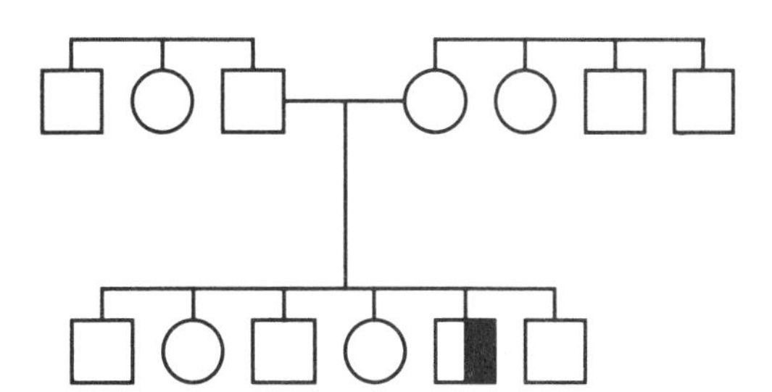

This type of pedigree is typical of 10 to 30 percent of TS families. They are sporadic cases with a negative family history of tics. These pedigrees are most consistent with either a recessive mode of inheritance, polygenic inheritance or a non-genetic disorder.

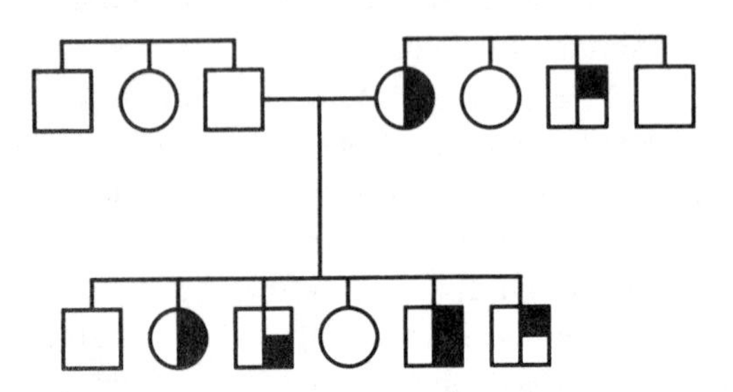

This type of pedigree, with many members of the family affected with either TS or chronic motor or vocal tics, and a positive history on only one parent's side, is consistent with autosomal dominant inheritance.

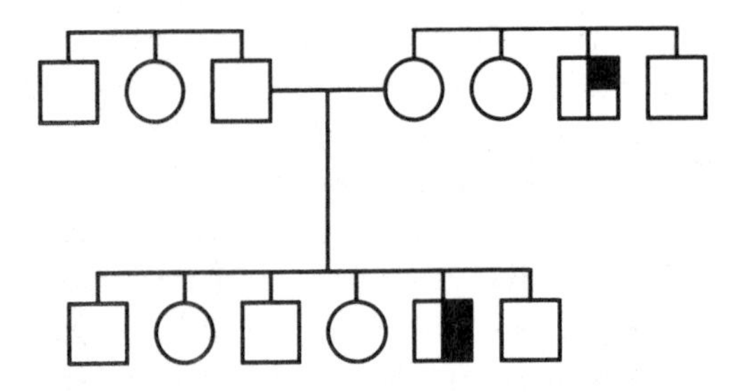

Finally, this type of pedigree, which is also commonly seen, is consistent with recessive inheritance of a common gene, with polygenic inheritance, with autosomal dominant inheritance with reduced penetrance, with purely environmental factors, and even with X-linked inheritance. In short, it is consistent with everything.

How do we sort all these pedigrees out to determine the real mode of inheritance? This is done by a procedure called **segregation analysis** in which data from a large number of non-selected or minimally selected cases are examined by various computer programs[535,536,1367,1112-1114]

The first report of such an analysis was by Kidd, Prusoff and Cohen from Yale, in 1980[1043]. They studied 67 families based on questionnaires mailed to 200 patients. While they were unable to state the mode of inheritance they did conclude that TS was genetically inherited and confirmed the earlier conclusions of Eldridge and co-workers[530] that chronic motor tics were genetically related to TS. They also felt that females with TS had more *Gts* genes (more genetic loading) than males. Such a concept has validity primarily under the hypothesis that TS is due to multifactorial inheritance and affected females have inherited more *Gts* genes and have a higher threshold before they express the disorder. Similar results were obtained when a group of 52 TS clinic patients were studied[1477].

Baron and colleagues[128] examined 136 families where 70 percent of the probands were classified as mild or very mild. Although they were able to exclude a polygenic mode of inheritance and demonstrate a single major gene, 79 percent of the males and 53 percent of the females were considered to be sporadic, non-genetic cases, hardly a ringing endorsement of a genetic cause for Tourette syndrome. The frequency of the *Gts* gene for the genetic cases was estimated to be .003 or 0.6 percent of the population being gene carriers. In the genetic cases, the penetrance for males was estimated at 81 percent and for females at 64 percent. They also concluded there was increased genetic loading in females, a concept that is difficult to reconcile with a single major gene model.

In 1984 my colleagues and I[376] examined 246 consecutive, unselected TS families of all grades of severity. Using a technique of segregation analysis called POINTER[1112] we were able to rule out an environmental and a polygenic model. The results were consistent with TS being inherited as an **autosomal dominant trait with reduced penetrance.**

The penetrance for TS or chronic motor or vocal tics was 70 percent in males and 30 percent in females, and the frequency of the *Gts* gene was estimated to be 0.006, or 1 in 83 people being carriers. This analysis suggested that only 35 percent of cases were not genetic. This was the first demonstration that the majority of TS cases were due to a single major gene. Using POINTER to examine the cases reported in the literature, Devor[485] also concluded TS was inherited as an autosomal dominant with reduced penetrance. Because of the highly selected nature of the pedigrees, only 0.6 percent of the cases were estimated to be sporadic.

POINTER was called into use once again in a study by Pauls and Leckman[1473] of 30 families.The unique aspect of this study was that many of the relatives were individually interviewed[1479], rather than relying on the information from one or two members of the family, and obsessive-compulsive behaviors were included, in addition to motor and vocal tics. The following penetrances were obtained:

	males	females
Tourette syndrome	45 %	17 %
TS or CMT	99 %	56 %
TS, CMT or OCB	100 %	71 %

The marked increase in penetrance for females, when obsessive-compulsive behaviors were included, suggested that 15 percent of females carrying the *Gts* gene manifest it only as obsessive-compulsive behaviors. This concurs with our observations[376] that some relatives of TS patients can express the gene as obsessive-compulsive behavior without tics[p113]. In agreement with our earlier studies, Pauls and Leckman[1473] also concluded that TS is inherited as an autosomal dominant trait with reduced penetrance and also estimated the gene frequency as 0.006, or 1 in 83 people being carriers.

The importance of interviewing all family members When pedigree information is obtained from a single or only a few members of the family, it is often incomplete[1473,1479]. This is well illustrated in the pedigree of a young man with compulsive exhibitionism and very mild tics who was certain no one, with the possible exception of this mother, had any TS symptoms. When the individual family members were questioned, all of his four siblings and both his parents had significant TS symptoms[p165].

Tourette Syndrome Is a Common Disorder

Two different epidemiological studies have suggested that between 1 in 1,000 to 1 in 1,400 male children have TS[255a,277,p615]. In our experience, finding individuals with TS is a constant ongoing process. In a study in which a school psychologist familiar with TS monitored 3,000 school children over a period of two years, definite TS was diagnosed in 1 out of 100 male students[p615]. In this sense, the story that Oliver Sacks tells in his book, *The Man Who Mistook His Wife for a Hat,* is relevant[1677].

> "The day after seeing Ray [a TS patient], it seemed to me that I noticed three Touretters in the street in downtown New York. I was confounded, for Tourette syndrome was said to be excessively rare. It has an incidence, I had read, of one in a million, yet I had apparently seen three examples in an hour. I was thrown into a turmoil of bewilderment and wonder: was it possible that I have been overlooking this all the time, either not seeing such patients or vaguely dismissing them as 'nervous', 'cracked', or 'twitchy'? Was it possible that Tourette's was not a rarity, but rather common — a thousand times more common, say, than previously supposed? The next day, without looking, I saw another two in the street. At this point I con-

ceived a whimsical fantasy or private joke: suppose (I said to myself) that Tourette's is very common but fails to be recognized, but once recognized is easily and constantly seen."

Summary: Tourette syndrome is a common genetic disorder. Studies of many families are consistent with it being inherited as an autosomal dominant trait in which some who carry the gene have no symptoms. When case finding is maximized by constant monitoring of grade schools, approximately 1 in 100 male students have TS.

[Recall from the Introduction that if new information became available while this book was being written, it would be added at that time, rather than going back to change chapters already written. Significant new information concerning the way TS is inherited did become available and was added in Chapters 69 and 90.]

Chapter 12

Are Girls Smarter Than Boys?

We all have the perception that boys are more poorly behaved and have more problems with school than girls. Is this simply our social prejudice or is it real? There are three to five times as many males with TS as females and this is not unique to TS. The following is a list of other disorders that are much more common in boys:

	M:F ratio	**Approx. % of the Population**
Autism	3.8:1	0.01
Tourette syndrome	3.5:1	0.5
Delayed speech	4.0:1	0.6
Stuttering	3.8:1	1
ASP*	4.0:1	3
Dyslexia	3.5:1	5
ADHD	4.0:1	5
Conduct disorder	6.0:1	5

*ASP = antisocial personality disorder

All of these disorders have some effect on speech, learning or behavior. Since some children have more than one disorder they do not completely add up to 20 percent, but they do account for approximately 10 percent of the population. Since on the average four males are affected to one female, this explains why boys seem to have more problems with conduct and school than girls.

Why are boys the frailer sex when these problems are concerned? Norman Geschwind, a neurologist at the Massachusetts General Hospital in Boston, proposed that learning disorders, left handedness and certain immune disorders were related to each other[695]. He proposed that males had more learning disorders because the male hormone (testosterone) altered the rate of growth of the left half of the brain in boys and increased their susceptibility to left handedness, dyslexia and learning problems. We might call this a developmental hypothesis.

We cannot explain the higher frequency of TS in males by proposing that the *Gts* gene is on the X-chromosome because X-linked traits are not passed from father to son, and TS is often passed from father to son. A polygenic hypothesis with a greater threshold in females[1043,1477] also doesn't work since this has been excluded in later genetic studies[371,1473]. One possibility is that there may be a gene on the X-chromosome that controls or modifies the expression of the *Gts* gene[359] or that hormone levels affect the expression of the *Gts* gene.

How do we explain the predominance of males in the other disorders listed above? All of these disorders have a strong genetic basis. This is true for autism[619], stuttering[1041,1042], antisocial personality[191,432,945,331,334], dyslexia[603-605,1816], attention deficit disorder[69,268,285-288,1363-1365,1482,1881,2054] and conduct disorder[69,1881,1969]. As will be discussed in other parts of this book, all these disorders can also be caused by a *Gts* gene. What percentage are due to a *Gts* gene, even if they have no tics, remains to be determined. Whatever causes TS to be more common in males could also explain why these other disorders are also more common in males.

An additional explanation is that females may also be having trouble with the same genes but expressing them as different disorders[p235,p644].

Summary: Many of the disorders that cause problems with learning, speech or behavior, are both genetic and occur in males about four times more often than they occur in females. The presence in the population of a common Gts gene, causing these disorders in the absence of tics, could be the common thread between them.

Chapter 13

Linkage and DNA Markers

The concept of linkage is important to understand because it represents the major tool for answering a number of critical questions about TS, such as: What chromosome is the *Gts* gene on? What chromosome band is it on? Is there more than one *Gts* gene? What is the normal function of the *Gts* gene?

The principle of linkage works as follows. Let us suppose the *Gts* gene and the *ABO* blood group genes are located close to each other on the same chromosome. In the following individual the *A* gene, for the type A blood group, is next to a *Gts* gene for Tourette syndrome,

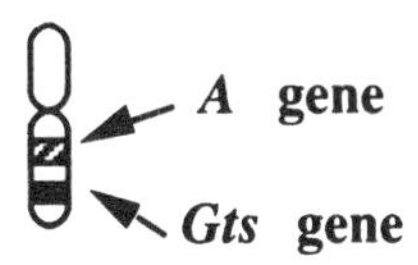

while on the other chromosome the *B* blood group gene is next to the normal *gts* gene.

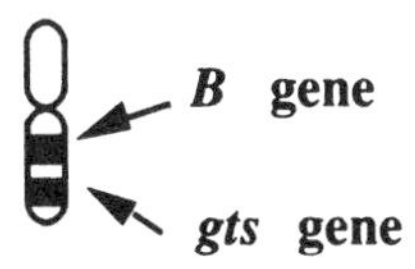

To make things easy we will give the *O* blood group gene to other members of the family. If the *A* blood group gene is very close to the *Gts* gene, the two will always travel together in the family.

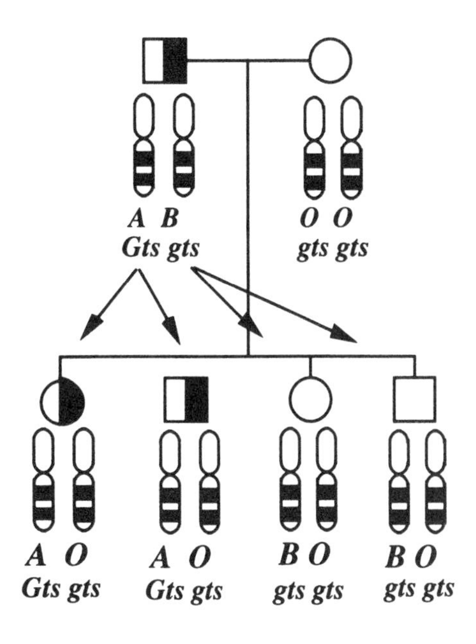

Thus, every individual in this family with Tourette syndrome, and carrying the *Gts* gene, also carries the *A* blood group gene. If enough individuals in this family or several similar families gave the same results we could confidently say that the *ABO* gene and the *Gts* gene were closely **linked**. However, if these two genes are not very close together then during the formation of the egg or sperm, a **crossover** would occur separating the two genes.

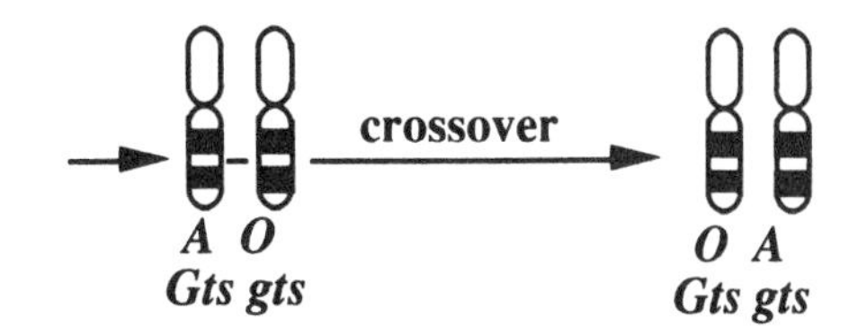

The frequency with which such crossovers occur is a measure of the distance between the two genes. The units of this distance are called **map units.** If two genes are separated by 5 map units then they are separated by a crossover about 5 percent of the time. One map unit is approximately 1 million base pairs long. If the two genes are on different chromosomes then the two will occur randomly in different family members and the apparent "crossover" frequency will be around 50 percent. Such genes are **unlinked**.

In the above example, if we didn't know where the *Gts* gene was located, but did know the *ABO* gene was on chromosome band 9q34, and found that the *ABO* gene and the *Gts* gene were separated by only three map units, then we would have also identified the location of the *Gts* gene as being on band q34 of chromosome 9. Thus, linkage studies have the potential of answering the first two of the above questions: What chromosome is the *Gts* gene on? and, What chromosome band is it on?

In addition, the third important question, Is there more than one *Gts* gene? could also be answered. For example, if we studied linkage in twenty TS families and found that TS was linked to the *ABO* gene in only ten of them, this would indicate that there were at least two *Gts* genes, one on 9q34 and another one, or several genes, at other places. If further linkage studies found another *Gts* gene on chromosome 4, and every family studied showed linkage to either 9q34 or chromosome 4, we would know that there were two distinct *Gts* genes. If all families studied showed linkage to 9q34, this would indicate there was only one *Gts* gene.

DNA Markers

As powerful as linkage studies are, for many years they were of limited value because there were only a few blood groups or other markers that could be used in the linkage studies. This stagnated state was dramatically reversed by the development of DNA markers. These work as follows.

First, DNA is divided into short but non-random pieces by enzymes that cut DNA only at a specific sequence of base pairs. These enzymes are called **restriction endonucleases.** The term endonuclease is used because the nucleic acid cutting [nuclease] is at the inside [endo] rather than at the ends of the DNA. There are hundreds of these enzymes, each cutting at a different sequence. For example, one commonly used enzyme called TaqI cuts DNA only at the following site:

In the following diagram, **x** represents the places on a set of short pieces of DNA where the TaqI cutting sites are located, and the heavy line ▬ represents a specific DNA sequence that I will call the **A** gene.

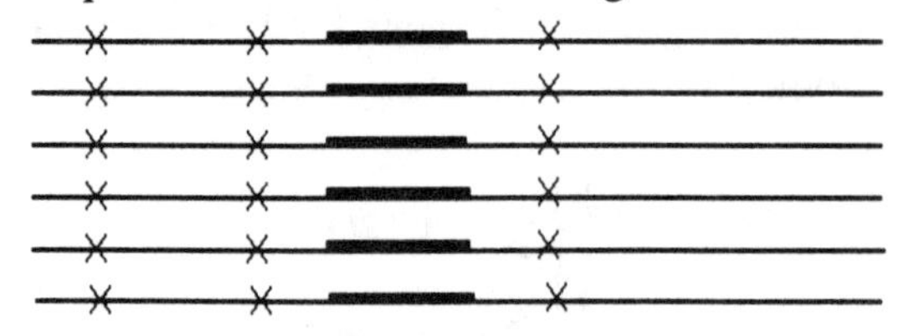

Treatment with TaqI would produce four sets of very short, short, intermediate and long pieces of DNA with the **A** sequence on the intermediate set.

Next, these pieces of DNA are separated by placing them in a small well in a jelly-like substance and then placing this in an electrical apparatus. This technique is called **electrophoresis.** Since the long pieces of DNA travel slowly, while the short pieces travel more rapidly, the different size pieces are separated as follows:

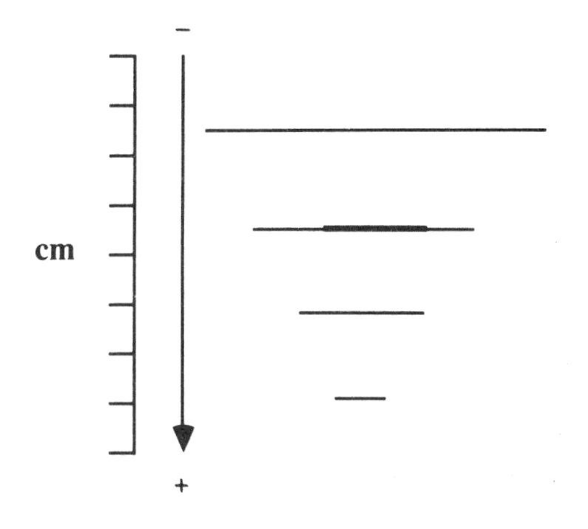

Obviously, the DNA does not remain as nice long pieces. Instead it coils up but is still separated according to size:

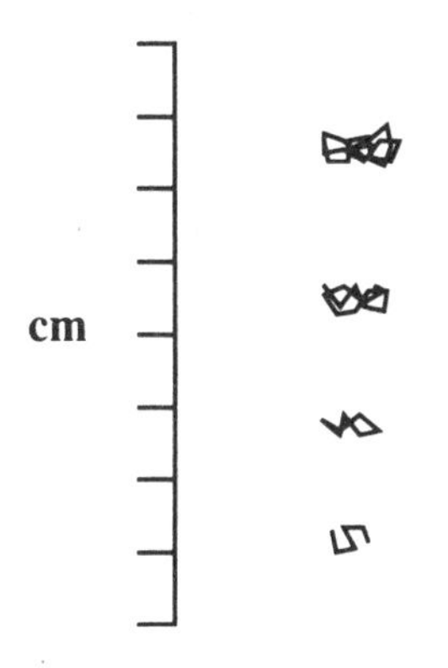

In reality, when we look at the gel, the only thing that can be seen is the well where the DNA sample was placed.

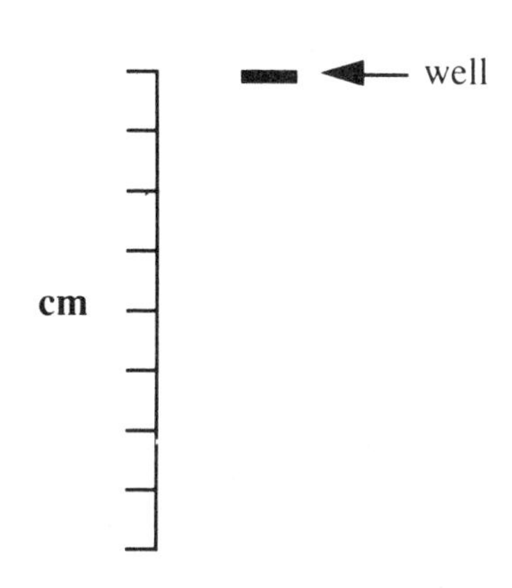

If we have isolated or cloned the **A** gene, it can be labeled with radioactivity to produce a radioactive probe.

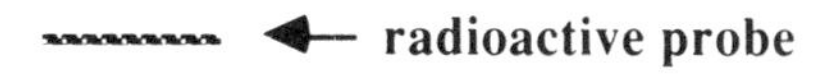

Under the right conditions this probe will specifically bind to the **A** gene by a process called **hybridization** in which base pairs of the probe align with the identical base pairs of the **A** gene. This then makes the piece of DNA carrying the **A** gene radioactive,

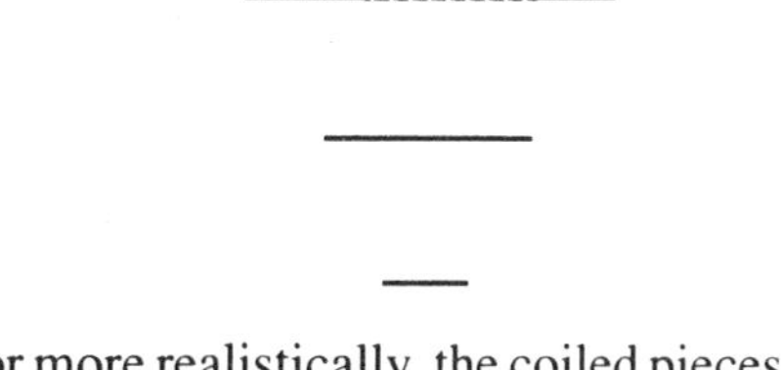

or more realistically, the coiled pieces of DNA radioactive.

Finally, in a process called **autoradiography**, a piece of photographic paper is placed on top of the gel and left in the dark. After the film is developed a heavy band occurs where the radioactive DNA was located in the gel.

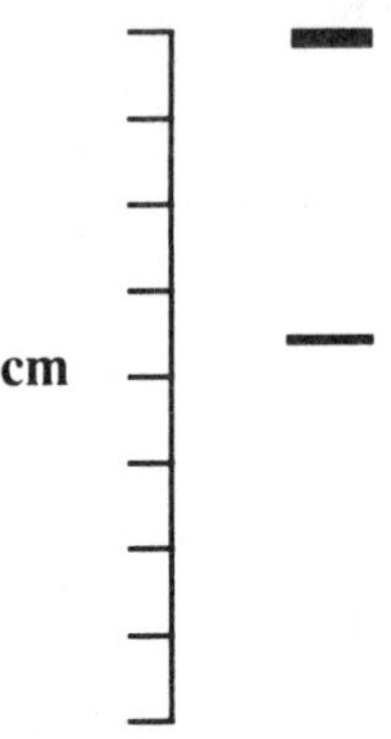

If a mutation happened to produce a TaqI site in the middle of the **A** sequence,

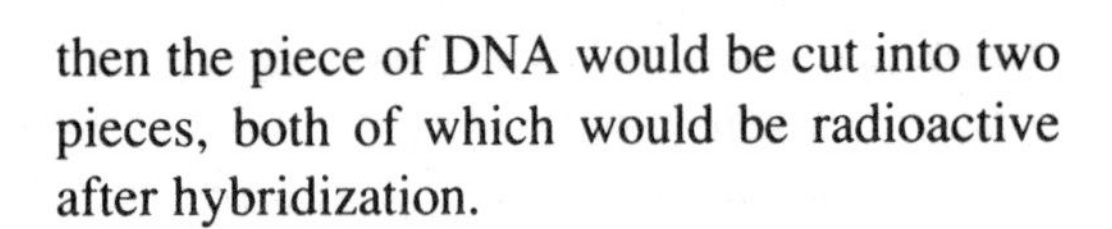

then the piece of DNA would be cut into two pieces, both of which would be radioactive after hybridization.

When this is electrophoresed and autoradiographed, two shorter pieces will be seen instead of one large piece.

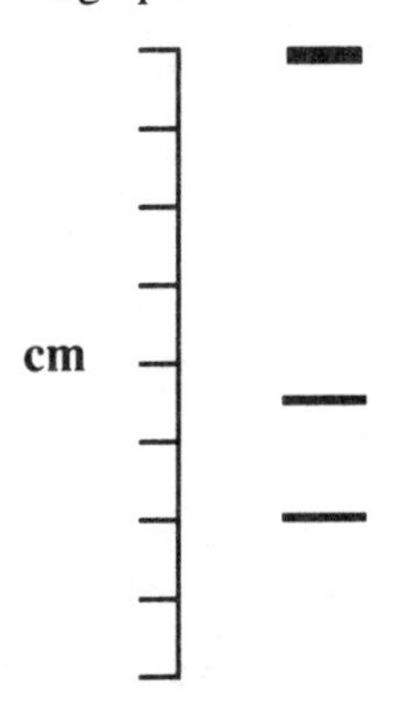

Everyone has two genes, one from their father and one from their mother. These are called **alleles**. We can call the **A** gene without the TaqI site the **A** allele,

A

and the **A** gene with the TaqI site the **B** allele.

B

If DNA from **A/A**, **B/B,** and **A/B** individuals are placed in adjacent wells the following pattern would be produced:

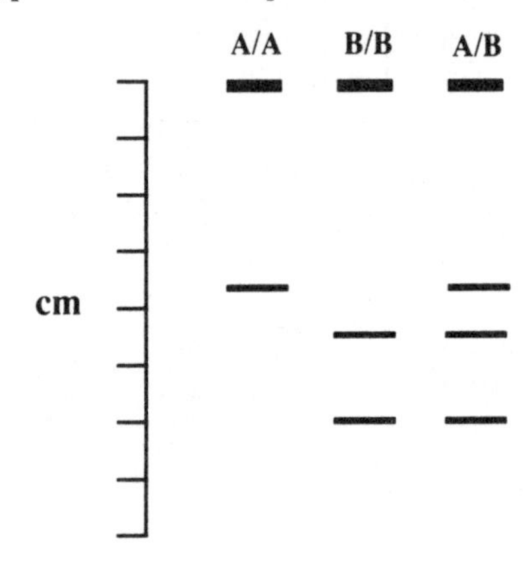

Since this polymorphic marker was produced by the TaqI enzyme, it is called a **TaqI polymorphism**. A polymorphism is any genetic variation that is present in more than 1 percent of a given group of people. Since these polymorphisms are due to different fragment lengths of DNA produced by restriction endonucleases they are called **restriction fragment length polymorphisms**, or **RFLPs** for short[198a]. The analogy to the A/A, B/B and A/B blood groups is apparent. Since the DNA sequence of every person is different, there are thousands of these DNA markers, many of which have now been localized to specific chromosome bands. These markers have revitalized the use of linkage to determine the location of genes causing various disorders.

If we didn't have a probe to the **A** gene but knew that a specific RFLP was closely linked to it, then following an RFLP around in a family can tell us who has different types of mutations of the **A** gene.

A —————————————— **RFLP**

The same would be true if we had an RFLP linked to the *Gts* gene.

This illustrates the enormous power of these **molecular biology** techniques. If a probe of the *Gts* gene was available, by simply drawing a small amount of blood, isolating the DNA from the white blood cells, and following the above procedure, we could easily determine if an abnormal *Gts* gene was present. If we had no information about the *Gts* gene, but had a closely linked RFLP, we would be able to determine who in a family carried the abnormal *Gts* gene.

Summary: Linkage studies allow us to determine on which chromosome, or part of a chromosome, a given gene is located. The development of thousands of DNA markers has made linkage a powerful technique. If a probe to the Gts gene itself was available, simply by drawing a small sample of blood and testing the DNA, it would be possible to determine who might develop Tourette syndrome.

Part III. BEHAVIOR

Chapter 14

What's Wrong with My Son?

There is no better way to introduce some of the behavioral problems seen in Tourette syndrome than to let a mother tell you in her own words of her personal nightmare. I have italicized some of the key phrases.

"I'm at a loss to describe my son. What I have to say seems unfair. It took me a long time to believe my own perceptions of him. Two more children and exposure to life reassured me I was right—*something is wrong with my son.* But it's a frustrating something, an often undefinable something, a fluctuating something. My son looks normal—he's bright and good looking—but he's full of gaps and inconsistencies. Nobody has been able to tell me what's wrong, but I've come to the conclusion that he has Tourette's syndrome, perhaps among other things.

"Jim was conceived within weeks of our wedding. I was ecstatic. I wanted a child more than anything in the world. I did all the "right things" during my pregnancy and was disappointed when his birth proved difficult. In a twenty-hour labor, I finally accepted some Demerol and agreed to a saddleblock. The doctor used forceps to usher Jim into the world.

"Still, I was in heaven. I had my baby. Then when Jim was three weeks old, I went into the hospital for kidney stone surgery. Although I pumped my breasts while in the hospital, I was so full of medication when I returned home ten days later that it seemed best to keep Jim on the bottle. I watched my breasts dry up with great sadness.

"Jim was my sunshine. I played house with him day and night. Our marriage was in trouble. My husband and I argued violently. But I watched my little son for signs of distress and was relieved to find none. He ignored our behavior. I might have realized that was a sign in itself.

"Jim was self-possessed even as an infant, intent on doing his own thing. He was a serious child, observing life but not interacting with it as other babies might. I had no one to compare him with. I thought he was perfect. Then when he was eighteen months old, he began to throw *violent temper tantrums.* I believed he was finally reacting to the stress in our marriage.

"'It's one thing to hurt yourself or even another adult,' I'd tell myself, 'but it's unacceptable to hurt a child.' I was pregnant with my second son by now and I knew something had to change. I told my husband I would leave him if we weren't in therapy before the birth of our next child.

"After trying a psychiatrist in Pasadena, we left to join a Christian community in the midwest.The community had a reputation for helping troubled marriages. It was a financial sharing community where members gave up control of their individual lives for the good of the group. We stayed in the community for six years. Some of the decisions that were made for us proved to be disastrous for Jim.

"Upon our arrival at the community, the elders counseling with us agreed that Jim's problems stemmed from the conflict in our marriage. Bill was born in December of 1978. Jim, busy with his inventions, mostly ignored his new brother. Jim would sit at a small table in front of public television kiddie shows and make things with blocks, paper, and masking tape. At two and a half, he was intent on building a robot. This goal consumed him until he discovered computers in the seventh grade.

"When Jim was three and a half, our counseling elders decided I was a negative influence and sent me to work outside of my home. Jim went to the community nursery school. I was certain this was wrong, but it took me two years to convince the decision-makers. Jim had always been intent on a world of his own making—he reminded me of an old bachelor—suspicious of new people and experiences. The nursery school was full of live wires that year, and it blew Jim's circuits.

"He had been *rigid* and *demanding* before, but now he became *unreasonable*. I had always tried to humor him—after all, there was no real reason we couldn't drag all his stuffed animals along when we went shopping—but now he had to 'fit.' Jim didn't fit. And he resisted the teacher's attempts to help him get along. He began to *blink his eyes continually*. His *fears of crowds, loud noises, and being touched* increased. If an adult touched him, he would *blow on them and scratch where he had been touched*. In church, he would plug his ears against the singing. And when he walked, he would push at the air as if to stretch his space.

"Jim's tension rapidly increased to the point where he refused to smile or admit he was ever happy. If he was tempted to smile, he would suck in his cheeks until the urge passed. Finally, he refused to let even his father and mother touch him. When he was five, the nursery school teachers finally recognized he needed a calmer atmosphere. He was allowed to stay home most of the day, and his worst symptoms disappeared.

"Also at this time, I convinced our counseling elders that something else might be wrong with Jim. I was allowed to take him to a pediatrician. He told me Jim would grow out of his *tics* and *phobias*. I then took him to a specialist—I don't remember his exact title—who believed in whole foods. It was a relief to have something concrete to do for Jim. I learned to cook everything from scratch.

"At this point, our marriage had improved greatly. We moved into a large household—four couples, their children, and two singles living together in a large Victorian house. One of our counseling elders was the head of the household. After observing Jim up close for some time, he began to agree with me that something besides our marriage was wrong with Jim. Jim began play therapy with a trained children's counselor in the community and saw her once a week for a year. He stonewalled every session.

"Jim's kindergarten teacher was gentle and sympathetic. He did well that year in school. But *first grade was a disaster*. His *eye tic* returned within a couple of weeks, and when I observed his classroom I could tell why. *The teacher was on the verge of a nervous collapse*. It took me until Christmas to get him transferred out of the room. His new teacher was a little high strung, but sympathetic to Jim and he was able to stay with her in the second grade. She told me once, however, that having Jim in her class was 'like watching someone be tortured every day.'

"It was a rowdy school, and Jim *seemed to attract the school bullies*. It wasn't unusual to find he had been roughed up on the playground or on the way home. The school's attitude was that kids need to learn to take care of themselves. I wondered if Jim ever would and wished I could send him to private school. The community wouldn't OK this because I couldn't prove Jim had a special need.

"By now, Jim had begun to *roll his eyes* as well as blink them. He had also begun to *sniff, cough, and yell out inappropriately.* He was *obsessed* with building a robot, protecting himself from being touched although he was now accepting hugs from his parents—and compulsively worried about his rights. He was *negative, fearful, and angry.* Yet, abruptly, it would all go away. I could almost pinpoint the hour. *Suddenly, he would be 'normal'* and we would praise the Lord, figuring that we had finally hit on the right combinations of love, discipline, etc. *The good times never lasted for long.*

"When Jim was seven, our last boy was born. Jim adored him from the moment he set eyes on him, and for the first time, we had the fun of watching Jim care for someone other than himself. Until then, the only compassion and tenderness Jim had been able to express was to a Dr. Dan rag doll and his kitten.

"Also during this time, Jim had a spiritual awakening of some kind. Although he had been born into a God-directed family, he had been precociously cynical—saying at five that God was just a big foot and we were ants in his backyard so God could just step on us. He was unable to show the openness to God most young children have. While Jim was sleeping, he dreamed that angels came to him, lifted up his arms, and told him he would be OK. From then on, Jim called himself a Christian.

"We moved back to California when Jim was eight. A *pattern of missing school* had already been established, and even though he now went to small private schools, he still couldn't make it through a school year. His tics grew so bad and the *teasing so unbearable* that we would pull him out the end of November for a rest. Then again, he would leave school early in May. In spite of this, Jim has stayed at grade level plus.

"After our move, Jim's phobias seemed to gravitate more and more toward his closest brother, Bill. He became *convinced that Bill was using his elbows to destroy him.* Finally, he wore a hat pulled down so low on his forehead that he couldn't see Bill, or anything else. The damage to Bill and to our family unit, because of this obsession, has been hard to describe. It's so bizarre no one can imagine it.

"Jim's progression of tics and other difficulties continued, as did our efforts to help him. He went from school to school hoping to find one where he could last the school year out. At one point, I took him for private tutoring. We tried disciplines, health foods, vitamins, behavior modification, counseling, exercise, prayer, exorcism, medications and testing. Sometimes a new tactic would coincide with one of Jim's 'remissions' and we would believe we'd 'finally found it,' only to be crushed when his negative behavior and strange symptoms returned.

"When Jim was eleven, we took him to a children's guidance center. The doctor there was the strangest man I have met in the counseling profession—and I have met some doozies—but he suggested Mellaril. It was a miracle drug for Jim. Combined with a good diet, vitamins, lots of sleep, and a low input life—things I had already learned helped Jim—it seemed to blunt the world enough for him to handle it.

"Seventh grade was his best year ever. The elbow phobia was almost completely hidden, although I knew it was still lurking in the background because it came out in times of stress. Also in the seventh grade, Jim discovered computers. Finally, he had something to channel his robot passion into which wouldn't end in failure the way all the masking tape, cardboard, tin cans, and used electrical parts had.

"The Mellaril began to lose its effectiveness in eighth grade. This coincided with a tremendous growth spurt and I realized that his dosage wasn't adequate for his size. Still, I was reluctant to up the medication because no one had given me an intelligent answer about Mellaril and why Jim was taking it. I didn't want him on medication if there was another way. I didn't know if time would prove the medicine had lasting side effects. And I was still hoping that Jim would grow out of his chemical imbalance, *minimal brain dysfunction*, *attention deficit disorder*, neurotic syndrome, demons, or whatever the heck he had. I had been told often enough that he would probably outgrow his childhood problems as he entered his teenage years. For the first time, I had experienced a relatively normal firstborn son—seventh grade—and I hoped his difficulties in early eighth were only temporary.

"I am bringing Jim to City of Hope now because I believe his disability is long lasting. He won't grow out of it. I was reluctant to label Jim with Tourette's syndrome. I shared with him that the disease existed and he was relieved to learn he wasn't the only one with these problems, but we both agreed he must have a mild version which would take care of itself with time.

"This year has been hell. We moved to Pomona in January. *Changes have always been difficult* for Jim, but this one was heightened because the only school we could locate was the public middle school. Jim *went from a B+ average to C's and D's*. In addition, he began to *challenge our parental authority* in daily matters. I had learned to control his life very closely so he could function, and now he wanted to make these decisions himself.

"When he turned 14 in March, we made a deal with him. He could make basic decisions—bedtime, etc.—and we would let him learn by his mistakes—rather than stepping in to dictate as we had in the past when he messed up. In turn, he would respect our rules and lifestyle, trying to match them with his decision. Also, he would not blame us when he suffered for a bad decision he had made. Jim turned into the slob of the year, and we *found ourselves hating him*. Finally, he was such a mess that we took him out of school for fear he would have a nervous breakdown.

"Who is Jim today? He is creative, stubborn, determined, and single-minded. He would make a marvelous eccentric genius. Right now, he is living in a trailer in the yard because we *can't handle his behavior*. He still *has times when the clouds part and a wonderful human being shines through*, but we haven't seen this in a long time.

"He is *obsessive, narcissistic, negative, argumentative, angry, and unrealistic*. I can't leave him alone with either of his brothers or his father. He will hurt his brothers and *drive his father into a yelling match*. I have even seen him *hurt his animals*. *His world revolves exclusively around himself*. Although he can keep up his grades in school, he *often can't add 6 and 8*. He can remember complicated computer codes, but he *can't follow two consecutive directions in a daily living situation*. He *has almost no memory*. He *doesn't understand time, distances, and relationships*. He appears *lazy, irresponsible, ungrateful, self-conscious, and self-centered*. He is the kind of kid who grows up to shoot presidents.

"I stopped long ago allowing my world to disintegrate because Jim was having a bad day. But it has been only recently that I have wondered if he is hopeless. He is my son, but I have to admit I don't like him. He *blames others constantly*. *Nothing is ever his fault*. I want to help Jim be a successful human being, but if I have to choose between him and my other children, I will choose them. They have suffered enough already because of their older brother's behaviors and have understood and forgiven more than could be expected of them.

"Even Jim's faith and occasional acts of giving seem self-serving. His God is a God of rules, not of grace. And when he gives, it is with a 'look at me' attitude. Yet when we try to explain these things to him, it's *as if we're talking to a deaf person*. Jim would make a good Nazi. I feel desperate right now. *All my golden hopes for my son are gone*. Right now, all I feel is that I want us to survive until he is on his own. I have a brother who is brain damaged. Jim is so emotionally handicapped that his ability to interact with people rivals my brother's.

"We have decided that Jim *can't hope to cope in a public high school*. We have located a private teacher who will tailor make a high school program for him. But we can't let him live in a social desert. We have told him he needs to join the church in town that has a large, active youth group. We have also encouraged him to develop some hobbies beyond his computer, biking or tennis, and join a club where he can meet people. He is finishing eighth grade this summer with the private teacher.

"If things are resolved enough by the fall so that he can live in a house with us, we will be grateful. If not, Jim will have to continue in the trailer or live somewhere else. It is very difficult juggling our family this way, but although it's never been this tense before, I have done it for many years. I hate it.

"At times, Jim has *delusions of grandeur*, of extrasensory perception and powers. This worries me. He is also *unrealistic about himself*. Since his time concept and memory are confused, if he does something once, it becomes always. Last night he walked in the brush barefoot believing he had power to withstand pain. He must have done it for ten minutes, but he believed it was two hours. Today, he believes it was all day and that I probably resent him because he has this power and I don't.

"If he can't be realistic about his disabilities, he will never be able to overcome them. Filling out the questionnaire was very difficult for him because he didn't want to admit anything was wrong. When he goofs up, he *won't take responsibility*. When he has a bee in his bonnet about something—most of the time—he pursues it with such tenacity, you want to find a bridge and jump to escape him. Although we have praised him and reassured him that he was a wonderful person, he hasn't heard it. This year—for the first time—we are giving him consistent messages that we can't stand him. This worries me. I have slapped him twice in the past two weeks. Maybe Jim shouldn't live with us, but I think he would drain the patience out of Mother Theresa."

The essence of this boy can be summarized in the italicized words and phrases:

something is wrong with my son
violent temper tantrums
rigid
demanding
unreasonable
blink his eyes continually
fears of crowds, loud noises, and being touched
blow on them and scratch where he had been touched
tics
phobias
first grade was a disaster
eye tic
the teacher was on the verge of collapse
seemed to attract the school bullies
roll his eyes
sniff, cough and yell out inappropriately
obsessed
negative, fearful, and angry
suddenly, he would be "normal"
the good times never lasted for long
pattern of missing school
teasing so unbearable
convinced Bill was using his elbows to destroy him
minimal brain dysfunction

attention deficit disorder
changes have always been difficult
went from a B+ to C's and D's
challenge our parental authority
found ourselves hating him
can't handle his behavior
has times when the clouds part and a wonderful human being shines through
narcissistic, negative, argumentative, angry, and unrealistic
drive his father into a yelling match
hurt his animals
his world revolves exclusively around himself
often can't add 6 and 8
can't follow two consecutive directions in a daily living situation
has almost no memory
doesn't understand time, distances, and relationships
lazy, irresponsible, ungrateful, self-con scious, and self-centered
blames others constantly
nothing is ever his fault
as if we're talking to a deaf person
can't cope in a public high school
delusions of grandeur
won't take responsibility
all my golden hopes for my son are gone

Jim's mother has eloquently and poignantly described the types of problems that we hear over and over again in dealing with individuals with Tourette syndrome.

The phrases:

blink his eyes continually
tics
eye tic
roll his eyes
sniff, cough, and yell out inappropriately

define the common ground for these children and the features that allow a diagnosis of Tourette syndrome—multiple motor and vocal tics.

The phrases:

violent temper tantrums
fears of crowds, loud noises and being touched
phobias
obsessed
convinced that Bill was using his elbows to destroy him
minimal brain dysfunction
attention deficit disorder
delusions of grandeur

illustrate some of the associated behaviors and psychiatric syndromes linked with Tourette syndrome — attention deficit disorder, phobias, panic attacks, obsessions and compulsions, conduct disorder, paranoid thoughts, depression and mania.

The phrases:

pattern of missing school
teasing so unbearable
attention deficit disorder
went from a B+ average to C's and D's
can't follow two consecutive directions in a daily living situation
has almost no memory
doesn't understand time, distances, and relationships
can't hope to cope in high school

illustrate the learning disorders and school problems.

All the other statements show the many behavior problems and parental frustrations of living with these children. I will constantly point out that *not all Tourette syndrome patients have all these problems.* They are not unlike the rest of us — some of us have emotional problems and nerdy behavior and some of us don't. The problem is that these behaviors are just *much* more common in TS. The whole of Part III. BEHAVIOR deals with the *associated behaviors* in TS — defining what they are and showing by comparison to controls and by pedigree studies that they are an integral part of the total spectrum of Tourette syndrome.

Summary: This mother's story of her son with Tourtte syndrome illustrates many of the clinical aspects of this disorder that we have heard over and over in 1,400+ cases.

Chapter 15
A Controlled Study of Tourette Syndrome

It is impossible to see even a few individuals with Tourette syndrome without realizing some of them have a lot of behavioral problems. However, if I say, for example, "It is my impression that many Tourette syndrome patients have a short temper," a natural rejoinder for the skeptic is to say, "So what, a lot of people have a short temper." Put in a different way, any serious attempt to determine whether certain behaviors are more common in TS individuals requires a comparison with a normal population. These are called **controls.**

The study by Nee and coworkers[1397] illustrates the problem. They reported 50 TS cases studied at the National Institutes of Health and found a high percentage of behavioral disorders including sleep disturbances, obsessive-compulsive, antisocial and self-destructive behaviors, and inappropriate sexual activity. However, these patients were highly selected in that a) they had to be referred by a physician (which selects for severity), b) had to be able to pay their own way to Washington, D.C. (which selects for middle and upper class) and c) came to the National Institutes of Health, a place often considered the last resort (which also selects for severity). Also, as the authors pointed out, "Our data give some support to an association between Tourette syndrome and other diagnosable behavior disturbances; but without a highly controlled comparison population, it is impossible to state definitely that these behaviors occur with unusual frequency."

Another issue concerns when a diagnosis was made. Many studies of TS are done after the diagnosis has been made. This is called a retrospective study and is subject to a number of biases. However, if all the information is collected before any diagnosis is made it is called a **prospective** study. Prospective studies avoid many subtle biases.

Thus, to provide an adequate determination of whether certain behaviors are more common in TS patients than the general population, four requirements needed to be satisfied:

a) there should be as little artificial selection of cases for severity as possible,

b) there should be a control population that is as unselected and normal as possible yet derived from the same patient base as those with TS,

c) the study should be prospective, and

d) a significant number of patients should be studied.

To specifically examine behavior in Tourette syndrome we met these four requirements. After describing our first 250 cases of TS[358], we studied the next 246 patients in the following manner. Any patient calling for an appointment to be evaluated for TS, or possible TS, was sent a detailed questionnaire covering over 425 different aspects of TS and behavior. Many of these questions were modeled after the National Institutes of Health Diagnostic Interview Schedule (DIS) designed to make 32 different DSM-III psychiatric diagnoses for epidemiological studies[1630-1632]. These questionnaires were stored away and not used in the evaluation of the patient. After the clinic visit, if the patient satisfied the criteria for TS [p11], they were entered into the series. Thus, all the information had already been gathered before any diagnoses were made — i.e., it was a prospective study.

The majority of the cases were self-referred or referred by school teachers or nurses, and had not been previously diagnosed as TS or treated for TS. All patients were seen regardless of ability to pay, sex, age, race or socioeconomic status. Thus, the series consisted of 246 consecutive, unselected Tourette syndrome cases of all ages and grades of severity.

All individuals were interviewed and examined by myself using a structured outline for history taking, and a detailed pedigree was obtained on each family using information from one or more family members. This eliminated variability due to different interviewers.

Finally, a great deal of thought went into selecting a control group. We wanted patients that would:

a) be motivated to fill out the extensive questionnaire,

b) be normally coming to the City of Hope Medical Center for care, and thus match the TS referral base,

c) be randomly selected,

d) be of a similar age to the TS patients, and

e) have no illnesses.

All of these conditions were satisfied by utilizing our prenatal diagnosis clinic. Pregnant women 35 years of age or older were referred to this clinic for amniocentesis to make sure their unborn child did not have Down's syndrome. If they had a child in the age range of 7 to 18, the mother and the child were asked to work together to fill out the questionnaire. If there was more than one child, the one nearest age 15 was chosen. In some cases the siblings of the mother were asked to participate to obtain some older subjects. The mothers were requested to return the questionnaire when they came in for their amniocentesis. This resulted in 47 randomly selected, normal controls motivated to fill out the questionnaire. Ninety-eight percent of the questionnaires were returned. The average age of the controls and TS patients was similiar[355,360-364].

To determine whether a given behavior was more common in TS than controls, the frequency in the two groups was compared by a statistical procedure known as the Chi square test. This allows us to determine the probability that the differences are real or by chance. These **probabilities** are called **p values.** In the following chapters many p values will be quoted. If the p = N.S. this means the difference between the TS patients and controls was **Not Significant.** For any p value of .05 or less the TS and control groups are considered to be significantly different from each other. The smaller the p value the more significant the differences. In the sometimes contorted world of statistics things seem to be stated backwards. Thus, a p value of .05 means there is only a 1 in 20 chance that the two groups are actually <u>not</u> different from each other. The following table illustrates this:

p value	chance that the TS and control group are <u>not</u> significantly different
.05	1 / 20
.01	1 / 100
.005	1 / 200
.001	1 / 1,000
.0005	1 / 2,000

For most of us it is easier to think of these figures in a more natural order. Thus,

p value	chance that the TS and control group <u>are</u> significantly different
.05	19 / 20
.01	99 / 100
.005	199 / 200
.001	999 / 1,000
.0005	1,999 / 2,000

Viewed in this way, the normal cutoff for significance of p = .05 means there are 19 chances out of 20 that the differences seen are real and not by chance. P values of .005 to .0005 are considered to indicate very significant differences, especially if a lot of comparisons have been done. Many of the p values in the following chapters are < (less than) 0.0005 which means the chance that the differences observed between the controls and the TS patients are real is greater than 1,999 chances out of 2,000.

In addition to the TS patients and controls, reference will be made to 17 patients with attention deficit disorder (ADD), and 15 patients with ADD and a family history of TS, who were also in the controlled study.

There are three major ways these results will be presented:

a) **summary boxes**,

b) **tables**, and

c) **figures**.

In the following sections on behavior in Tourette syndrome, four sources of information are used:

a) the existing literature on TS,

b) the controlled study,

c) our clinical experience, and

d) detailed pedigrees of over 2,000 families with TS and related problems.

On the **next page** show are two sample diagrams and an explanation of the way results will be presented.

Summary: To evaluate the frequency of behavior problems in Tourette syndrome 246 consecutive TS patients of all grades of severity were questioned about a wide range of symptoms and compared to 47 random controls. In future chapters this will be referred to as the controlled study and it will be used, along with other data, to define the spectrum of behavior in TS.

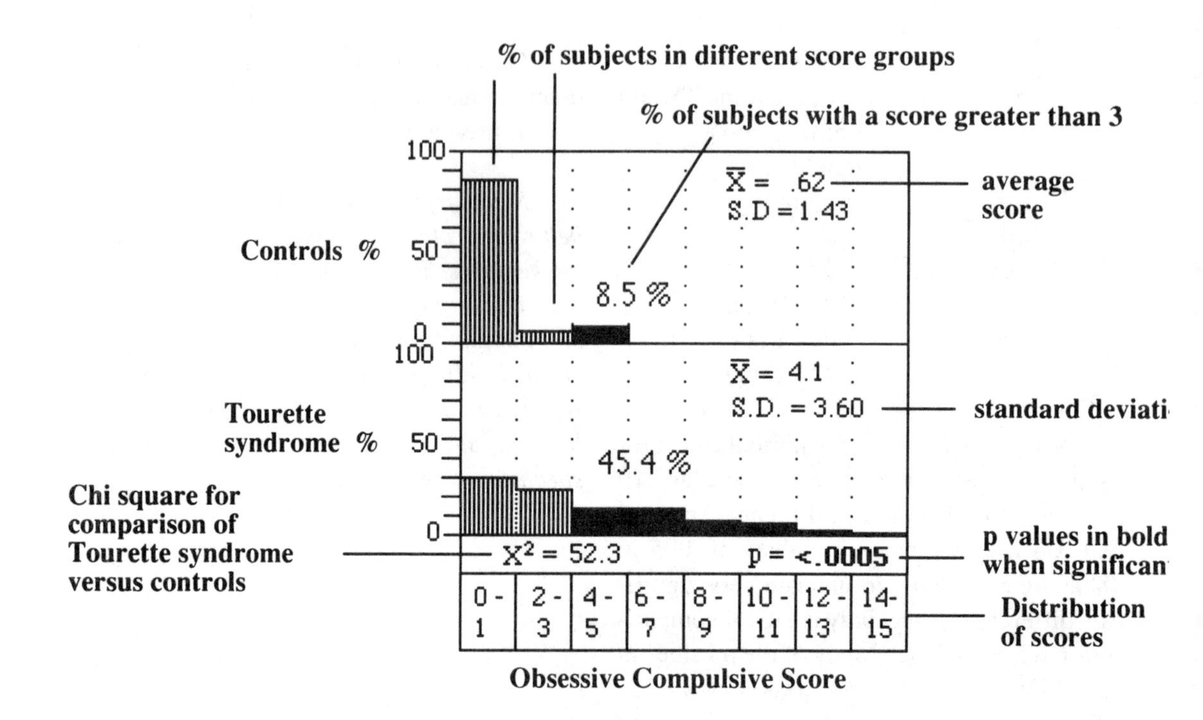

% of subjects in different score groups
% of subjects with a score greater than 3
100
Controls %
50
0
X̄ = .62
S.D = 1.43
average score
8.5 %
100
Tourette syndrome %
50
0
X̄ = 4.1
S.D. = 3.60
standard deviati
45.4 %
Chi square for comparison of Tourette syndrome versus controls
X² = 52.3
p = <.0005
p values in bold when significan
0 - 1
2 - 3
4 - 5
6 - 7
8 - 9
10 - 11
12 - 13
14- 15
Distribution of scores
Obsessive Compulsive Score

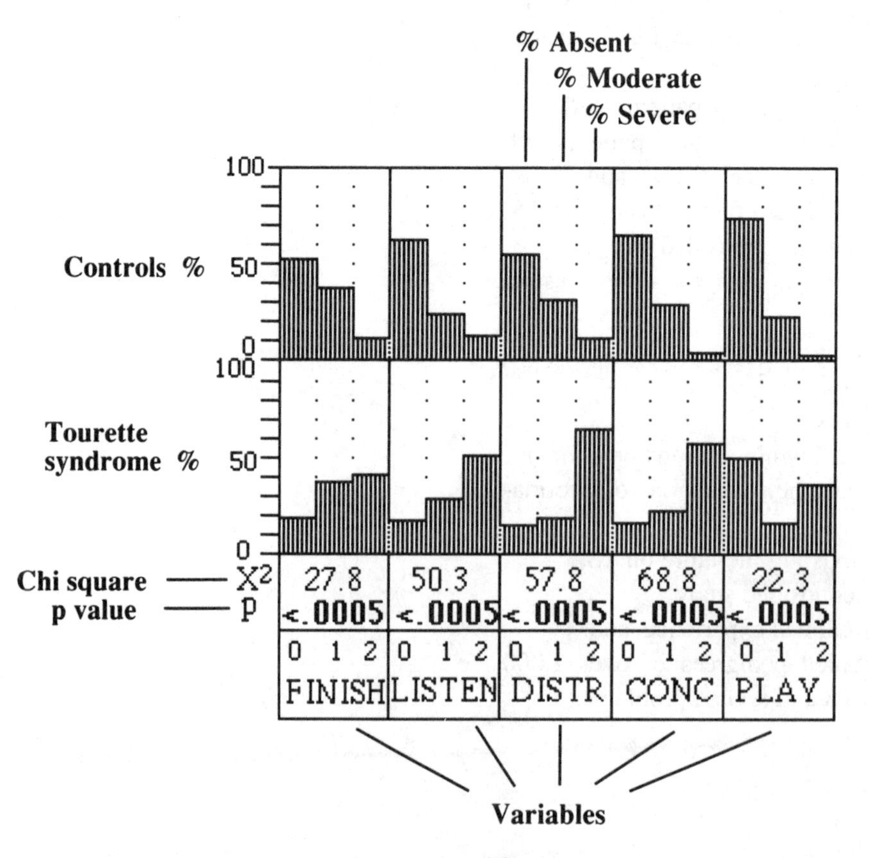

% Absent
% Moderate
% Severe
100
Controls %
50
0
100
Tourette syndrome %
50
0
Chi square
p value
X2
p
27.8
50.3
57.8
68.8
22.3
<.0005
<.0005
<.0005
<.0005
<.0005
0 1 2
FINISH
LISTEN
DISTR
CONC
PLAY
Variables

Chapter 16

Attention Deficit Hyperactivity Disorder

Attention — *"is the taking possession by the mind, in clear and vivid form, of one out of what seem several simultaneous possible objects or trains of thought. Focalization, concentration of consciousness are its essence. It implies withdrawal from some things in order to deal effectively with others."* William James[973]

Attention deficit hyperactivity disorder (ADHD) refers to a syndrome that begins early in life, is more common in boys, and is characterized by **inattention** and **impulsivity** and in most cases **hyperactivity**. Other than motor and vocal tics, it is the most common symptom of Tourette syndrome. Since ADHD is such an integral part of TS and is itself associated with many behavior problems, an important question is— Are the behavior problems in TS all due to the associated ADHD? Since a disorder can best be understood by knowing the characteristics required to make a diagnosis I will first examine these. In our controlled study of TS, started in 1983, we used the criteria based on the 1980 DSM-III[490] for a diagnosis of ADHD. These were as follows:

The presence of problems with **inattention** required at least three of the following:

1. Often fails to finish things he or she starts.
2. Often doesn't seem to listen.
3. Easily distracted.
4. Has difficulty concentrating on schoolwork or other tasks requiring sustained attention.
5. Has difficulty sticking to play activity.

The presence of problems with **impulsivity** required at least three of the following:

1. Often acts before thinking.
2. Shifts excessively from one activity to another.
3. Has difficulty organizing work.
4. Needs a lot of supervision.
5. Frequently calls out in class.
6. Has difficulty awaiting turn in games or group situations.

The presence of problems with **hyperactivity** required at least two of the following:

1. Runs about or climbs on things excessively.
2. Has difficulty sitting still or fidgets excessively.
3. Has difficulty staying seated.
4. Moves about excessively during sleep.
5. Is always "on the go" or acts as if "driven by a motor."

These characteristics must come on before seven years of age and last at least 6 months. Retarded children with an IQ of less than 80 are usually excluded from most studies. If the criteria are met only for inattention and impulsivity then a diagnosis of attention deficit disorder or **ADD** is made. If the criteria for hyperactivity are also met then it is termed ADD with hyperactivity or **ADDH**.

These criteria were revised for the 1987 revision of the Diagnostic and Statistical Manual, DSM-IIIR[491], and at least eight of the following were required to make a diagnosis:

1. Often fidgets with hands or feet or squirms in seat (in adolescents, may be limited to subjective feelings of restlessness).
2. Has difficulty remaining seated when required to do so.
3. Is easily distracted by extraneous stimuli.
4. Has difficulty waiting turn in games or group situations.
5. Often blurts out answers to questions before they have been completed.
6. Has difficulty following through on instructions from others, e.g., fails to finish chores.
7. Has difficulty sustaining attention in tasks or play activities.
8. Often shifts from one uncompleted activity to another.
9. Has difficulty playing quitely.
10. Often talks excessively.
11. Often interrupts or intrudes on others, e.g., butts into other children's games.
12. Often does not seem to listen to what is being said to him or her.
13. Often loses things necessary for activities at school or at home (e.g, toys, pencils, books, assignments).
14. Often engages in physically dangerous activities without considering possible consequences (not for the purpose of thrill seeking), e.g., runs into street without looking.

These have been listed in decreasing order of relative importance.

History

Over a century ago a German physician, Heinrich Hoffman, wrote a verse for children:

Fidgety Phil,
He won't sit still;
He wriggles,
And giggles...

The naughty restless child
Growing still more rude and wild.

This was a prelude to a burgeoning literature on hyperactive children. In 1902 Still[1886] reported a series of psychological symptoms in brain damaged children. This was followed by a report by Ebaugh in 1923[518] of similar behaviors in children affected by an epidemic of encephalitis. This included motor and vocal tics and the total picture was so similar to that seen in Tourette syndrome that it is described in more detail later[P400]. This seemed to establish a connection between these behaviors and brain damage[323,524,946,1065,1130,1447,2064]. Although the brain damage was not always obvious it was assumed by the term **minimal brain damage** or **minimal brain dysfunction (MBD)** [323,1452]. The suspicion that the term MBD was an inappropriately severe term began to surface when children with similar behaviors were observed with no evidence of brain damage. For example, in 1960 Chess[312] described 82 hyperactive children in which there was evidence for brain damage in only 16 percent. Others agreed and observed an even lower percentage with evidence for brain damage[1334,1884]. Thus, the term **hyperactive**, hyperactivity syndrome or **hyperkinetic** child was substituted for MBD [312,1130,1884,1878]. With the publication of the DSM-III in 1980, the term **attention deficit disorder (ADD)** was used. It was more precise and focused on the major difficulties of inattention and distractibility and their attendant learning disabilities. In the DSM-IIIR its name was changed for a fourth time to **attention deficit hyperactivity disorder (ADHD).** The major problem with this version is that the distinction between ADD without hyperactivity and with hyperactivity was eradicated. This was unfortunate since many children, especially girls, can have significant problems with ADD but not have motor hyperactivity. Throughout the rest of this book I will conform to the DSM-IIIR guidelines and use the terms ADHD except in those places where I want to specifically refer to ADD.

Diagnosis

A number of checklists have been designed to aid in the diagnosis of ADHD [4,5,382,459,1563,2069]. A straight forward approach is to ask the same questions as given in the criteria above. Since all observers do not always agree, it is important to get input from as many sources as possible, especially both parents and teachers[2157].

More Males Have ADHD Than Girls

The increased preponderance of males is just as striking as that for TS with four to ten times as many males affected as females [312,1884,1750]. However, when ADD without hyperactivity is examined the sex ratio is closer to 1:1[p86]. This is important to remember since it indicates that the *male preponderance is for motor hyperactivity not ADD itself.* Since the motor hyperactivity is often the most striking symptom, boys are more likely to be labeled than girls. This, plus cultural considerations of more concern for an adequate education in boys than girls, contribute to the preponderance of males. However, the converse of this is that many girls with learning problems due to ADD are being overlooked. Girls with ADD are no more likely than boys to have problems as adults, suggesting they do not have more ADD genes than boys[1254].

ADHD is the most common behavior problem in childhood. The exact prevalence of ADHD in children has been difficult to determine but most estimates suggest about 5 percent of boys and 1 percent of girls are affected [1884]. This agrees with the opinion of many kindergarten and first grade teachers that on the average there is one hyperactive child in every classroom[1884]. ADHD is the most common diagnosis for children referred to child guidance clinics, often constituting half to three-quarters of the patients[807,2058].

The Symptoms Often Start in Infancy

Although the DSM-IIIR criteria require that the symptoms start prior to seven years of age, many mothers relate that their child was different in infancy and some say the child was hyperactive while still in the uterus. Mothers often state their ADHD child did not like to be cuddled, cried all the time, was diagnosed as having colic, didn't take naps, slept poorly, rocked the crib and banged their head, had temper tantrums and "was awake when I went to bed and awake when I woke up."

Many mothers with a single child have thought "this was the way all children are" and did not become aware of the problem until their child entered kindergarten and they began receiving frantic phone calls from the teacher complaining about their child's behavior.

Behavior and Conduct in ADHD

As in Tourette syndrome the most effective way of determining the type of behavior problems associated with children with ADHD is to compare them to a group of normal control children. In such a study Stewart and colleagues[1884] examined 37 children between five and eleven years of age who manifested overactivity and short attention span, were attending school and were free of chronic medical or neurological disease. The 36 controls were selected at random. Table 16.1 shows the results of interviews of the mothers of the children. Such a study is important because it directly addresses one of the criticisms of the whole concept of hyperactive children—namely that it is a mythical disorder since all children tend to be very active and fidgety.

Table 16.1 Behaviors (%) in ADHD Children and Controls
(from Stewart et al., 1966[1884])

	ADHD	Cont
Number of subjects	37	36
Hyperactivity		
Overactive	100	33
Fidgets	84	30
Talks too much	68	20
Gets into things	54	11
Can't sit still	81	8
Wears out toys, furniture	68	8
Restless in doctor's office	38	3
Inattention		
Doesn't complete projects	84	0
Doesn't stay with games	78	3
Doesn't listen to whole story	49	0
Doesn't stay on task	46	6
Doesn't follow directions	62	3
Impulsivity		
Can't tolerate delay	46	8
Unpredictable	59	3
Leaves class	35	0
Unpredictable affection	38	3
Reckless	49	3
Accident prone	43	11
Conduct		
Teases	59	22
Fights	59	3
Constant demand candy, etc	41	6
Irritable	49	3
Can't accept correction	35	0
Temper tantrums	51	0
Destructive	41	0
Unresponsive to discipline	57	0
Defiant	49	0
Lies	43	3
Steals	43	3
Unpopular with peers	46	0
Vandalism	22	3
Fire setting	11	3
Cruelty to animals	11	3
Other		
Enuresis (bedwetting)	43	28
Poor speech	54	25
Delayed speech	35	6
Poor coordination	62	8
Strabismus	19	0
Colic	24	8
Infant feeding problems	27	8
Infant sleep problems	22	3

As can be seen, the characteristics that least distinguished the two groups were features associated with hyperactivity such as "overactive,""fidgets" and "talks too much". By contrast, symptoms relating to inability to pay attention, impulsivity and conduct were strikingly different in the two groups. The important point is that the crucial aspect of attention deficit disorder is the inability to concentrate, distractibility and impulsivity — not hyper-activity *per se*.

Other associated problems include conduct disorder, antisocial behavior, learning disabilities[607,1031,1649,1797] and depression[475,808,1307]. These are discussed in subsequent chapters.

In Table 16.2 (pages 79-81), I have included a complete listing of various symptoms in children with ADHD. This is the result of a task force convened in 1966 by the National Institutes of Health to clarify and define what was then called MBD and is based on 100 published reports on MBD to that time[325]. Since a high percentage of children with TS have ADHD these two tables accurately portray many of the problems seen in TS patients.

Inattention and impulsivity are the core problems Many of the criticisms of the concept of ADHD come from its earlier names[503,607,1665]. Minimal brain damage was a terrible term since it frightened parents. In fact there is no evidence for brain damage. Hyperactivity was almost as bad because it carried the implication that there is something wrong with a normally rambunctious child and was not very specific. Some have also criticized the concept because it is often hard to distingush from conduct disorder[1665,1791]. I feel this actually strengthens the concept when ADHD is viewed as a genetic disorder of disinhibition that results in inattention, impulsivity and conduct disorder. The emphasis on the core problems of inattention and impulsivity have resulted in a more clearly defined syndrome and eliminated many of the objections to the previous terms.

Inattention and impulsivity — mirror images of each other. The often heard statement that a child can't have ADHD if he is not hyperactive or restless during a brief visit to the doctor ignores the core problem of inattention and impulsivity. A diagnosis of ADHD should be based on a long-term history of a wide range of symptoms, not on what the doctor sees in a brief visit to the clinic. In attempting to find psychological tests that would distinguish children with ADHD from normals, Douglas[503,505,506] found that tests which required sustained attention and impulse control were the most sensitive, and concluded that inattention and impulsivity were reciprocal aspects of the same process.

One such test was the Matching Familiar Figures Test. Here the child was asked to pick which of several figures most closely matched a test figure. Since the answer was not obvious, it required careful scanning and comparison to pick the right one. The ADHD child tended to respond impulsively with little critical evaluation. The designers of the test suggested it distinguishes between a reflective (thoughtful) versus an impulsive style of solving problems[999]. A second similar test, the Children's Embedded Figures Test[2095], required the child to find a figure in a confusing background or field and also differentiated ADHD from control children.

ADHD children may also do poorly on tests which require eye-hand coordination. A frequently used one is the Bender Visual-Motor Gestalt Test, or Bender-Gestalt, which requires the child to copy several figures. A related test is the Goodenough-Harris Draw-a-Person Test.

The common feature running through all of these tests is the need to stop-look-listen and move[505]. *These tests require careful attention to detail, suppression of impulsive choices, and fine motor coordination,* all problems for the ADHD child.

ADHD children may do well in one-on-one testing. As mentioned above, it is not unusual for a child's family physician to say to the mother, "Your child isn't hyperactive, look how calm he is in my office." It is also common for a child tested one-on-one in a quiet room to do very well on academic tests despite the fact he is flunking out of school. These two common errors overlook the fact that the arena where the ADHD child does poorly is in a classroom, over an extended period of time. In situations where the child is highly motivated and receiving individualized attention they can do very well.

ADHD and moral judgement. Although it will be discussed in more detail in later chapters, conduct problems including lying, stealing and cheating are not uncommon in ADHD and TS children. Ease of frustration is also a common theme in ADHD and TS. Douglas[504,506] found that story completion tests could provide insight into the maturity of a child's response to frustration. When ADHD children were confronted with stories depicting inescapable frustrations they were unable to change their expectations to accept a compromise solution. They also gave more aggressive responses than normal children. Tests that include cheating behavior indicate that cheating involves an inability to resist a quick effortless solution. It is thus not surprising that there was a positive relationship between measures of attention and ability to resist temptation and moral behavior[805].

The Frontal Lobe and ADHD

The importance of impaired functioning of the frontal lobe in ADHD and TS is described in detail in the chapter on the frontal lobe[p341]. I bring it up here since while we are describing psychological tests in ADHD, the Wisconsin Card Sorting Test[789] should be added, since it is sensitive to a rigid style and inability to change a course of action[1331,1332]. The Porteus Maze[1527] test is another test of frontal lobe function[p356].

Brain Wave Tests

Brain wave testing (electroencephalograms, EEGs) has been performed extensively in ADHD children, in part because of its historical background of being linked to "brain damage." There is much controversy as to whether EEGs are significantly more abnormal in ADHD compared to normal children. Since some studies of normal children have shown a remarkably high frequency of abnormal or borderline EEG, the answer to this question depends as much on the choice of controls as on the choice of the patients. One of the most impressive series was reported by Gross and Wilson[807]. They reported on 1,056 consecutive admissions of children, 18 years old or younger, to a mental health clinic in suburban Chicago. Of these, 77 percent were diagnosed as ADHD. Of all the children, both with and without ADHD, 50 percent had an abnormal EEG. To obtain normal controls they examined 160 children of all ages especially selected by teachers as being "normal." Thus, this was a "supernormal" group of children, since instead of being a random cross section of youngsters, they were non-randomly chosen as "normal." Only 3.8 percent of these children had an abnormal EEG. The most common abnormal pattern was a positive spike or positive spike and wave (6 Hz or 14/6 Hz) pattern.

There are two important implications of these observations. First, they tend to verify that some subtle structural or biochemical problem is going on in the brains not only of ADHD children but children with a wide range of psychiatric problems. The second intriguing implication has to do with the nature of the 14/6 Hz and 6 Hz phenomena. These waves were discovered in 1951 by Gibbs and Gibbs and they considered them to originate in the thalamus[708,709]. The description and possible role of this part of the brain in ADHD and TS will be described later[p389].

Several authors have commented on the behavioral aspects of individuals showing these waves[41,704,937,941]. In one case, a woman was operated on who had intractable headaches, an irritable bowel and emotional instability. A wire was introduced into the anterior thalamus[p368] and showed a 14 Hz positive spike which spread to the frontal and temporal lobes. Destruction of this part of the thalamus resulted in the complete alleviation of her complaints. A biochemical abnormality in the thalamic-frontal lobe structures may be involved in ADHD[p341].

[In relationship to this woman's irritable bowel, in a psychiatric study of 22 adults with irritable bowel syndrome, 27 percent had adult ADHD[2055].]

In my experience, the results of EEGs in children with ADHD are usually normal or borderline and do not contribute as much to the diagnosis as a good history.

Hypoactive ADD

Some children with ADD, instead of being hyperactive are hypo- or underactive[2058]. They have all the other symptoms of ADD except they are listless and may appear depressed. The diagnosis and appropriate treatment of a hypoactive ADD child may be even more dramatic than for a hyperactive child. The response to medication can be as dramatic as a switch, overnight turning a veritable zombie into a normally functioning child.

Additional aspects of ADHD are presented in subsequent chapters.

Summary: ADHD is a very common childhood disorder characterized by inattention, impulsivity and hyperactivity, and is associated with learning disabilities and conduct problems.

Table 16.2 Symptoms in Children with ADHD (MBD) from Public Health Service Monograph 1966[325]

(Based on review of 100 articles to that date)

A. Test Performance Indicators.

1. Spotty or patchy intellectual deficits. Achievement low in some areas; high in others.
2. Below mental age level on drawing tests (man, house, etc.).
3. Geometric figure drawings poor for age and measured intelligence.
4. Poor performance on block design and marble board tests.
5. Poor showing on group tests (intelligence and achievement) and daily classroom examinations which require reading.
6. Characteristic subtest patterns on the Wechsler Intelligence Score for Children, including "scatter" within both Verbal and Performance Scales; high Verbal - low Performance; low Verbal - high Performance.

B. Impairment of Perception and Concept-Formation.

1. Impaired discrimination of size.
2. Impaired discrimination of right-left and up-down.
3. Impaired tactile discrimination.
4. Poor spatial orientation.
5. Impaired orientation in time.
6. Distorted concept of body image.
7. Impaired judgement of distance.
8. Impaired discrimination of figure-ground.
9. Impaired discrimination of part-whole.
10. Frequent perceptual reversals in reading and in writing letters and numbers.
11. Poor perceptual integration. Child cannot fuse sensory impressions into meaningful entities.

C. Specific Neurologic Indicators.

1. Few, if any, apparent gross abnormalities.
2. Many "soft," equivocal, or borderline findings.
3. Reflex asymmetry frequent.
4. Frequency of mild visual or hearing impairments.
5. Strabismus (eye deviated to one side).
6. Nystagmus (fine rapid jerking of eyes on lateral gaze).
7. High incidence of left, and mixed laterality and confused perception of laterality (confusion of right and left).
8. Hyperkinesis (hyperactive).
9. Hypokinesis (hypoactive).
10. General awkwardness.
11. Poor fine visual-motor coordination.

D. Disorders of Speech and Communication

1. Impaired discrimination of auditory stimuli.
2. Various categories of aphasia.
3. Slow language development.
4. Frequent mild hearing loss.
5. Frequent mild speech irregularities.

E. Disorders of Motor Function.

1. Frequent athetoid, choreiform, tremulous, or rigid movements of hands.
2. Frequent delayed motor milestones.
3. General clumsiness or awkwardness.
4. *Frequent tics and grimaces.*
5. Poor fine or gross visual-motor coordination.
6. Hyperactivity.

F. Academic Achievement and Adjustment. (Chief complaints about the child by his or her parents and teachers).

1. Reading disabilities.
2. Arithmetic disabilities.
3. Spelling disabilities.
4. Poor printing, writing, or drawing ability.
5. Variability in performance from day to day or even hour to hour.
6. Poor ability to organize work.
7. Slowness in finishing work.
8. Frequent confusion about instructions, yet success with verbal tasks.

G. Disorders of Thinking Processes.

1. Poor ability for abstract reasoning.
2. Thinking generally concrete (takes things at their literal meaning).
3. Difficulties in concept-formation.
4. Thinking frequently disorganized.
5. Poor short-term and long-term memory.
6. Thinking sometimes autistic.
7. Frequent thought perseveration (stuck on one thought).

H. Physical Characteristics.

1. Excessive drooling in the young child.
2. Thumb-sucking, nail biting, head-banging, and teeth-grinding in the young child.
3. Food habits often peculiar.
4. Slow to toilet train.
5. Easy fatigability.
6. High frequency of enuresis (bed wetting).
7. Encopresis (soiling clothes with stool).

I. Emotional Characteristics.

1. Impulsive.
2. Explosive.
3. Poor emotional and impulse control.
4. Low tolerance for frustration.
5. Reckless and uninhibited; impulsive then remorseful.

J. Sleep Characteristics.

1. Body or head rocking before falling into sleep.
2. Irregular sleep patterns in the young

child.

3. Excessive movement during sleep.

4. Sleep abnormally light or deep.

5. Resistance to naps and early bedtime, e.g., seems to require less sleep than average child.

K. Relationship Capacities.

1. Peer group relationships generally poor.

2. Overexcitable in normal play with other children.

3. Better adjustments when playmates are limited to one or two.

4. Frequently poor judgement in social and interpersonal situations.

5. Socially bold and aggressive.

6. Inappropriate, unselective, and often excessive displays of shyness.

7. Easy acceptance of others alternating with withdrawal and shyness.

8. Excessive need to touch, cling, and hold onto others.

L. Variations of Physical Development.

1. Frequent lags in developmental milestones, e.g., motor, language, etc.

2. Generalized maturational lag during early school years.

3. Physically immature; or

4. Physical development normal or advanced for age.

M. Characteristics of Social Behavior.

1. Social competence frequently average for age and measured intelligence.

2. Behavior often inappropriate for situation, and consequences apparently not foreseen.

3. Possibly negative and aggressive to authority.

4. Possibly antisocial behavior.

N. Variations of Personality.

1. Overly gullible and easily led by peers and older youngsters.

2. Frequent rage reactions and tantrums when crossed.

3. Very sensitive to others.

4. Excessive variation in mood and responsiveness from day to day and evey hour to hour.

5. Poor adjustment to environmental changes.

O. Disorders of Attention and Concentration.

1. Short attention span for age.

2. Overly distractible for age.

3. Impaired concentration ability.

4. Motor or verbal perseveration.

5. Impaired ability to make decisions, particularly for many choices.

Chapter 17
The Genetics of ADHD

There are many theories concerning the cause of attention deficit disorder with hyperactivity or ADHD. These include the following:

a) ADHD, or the hyperactive child, is a mythical beast conjured up by distraught teachers, parents and physicians to provide an excuse to medicate rambunctious children into oblivion [1724].

b) A hyperactive child is the result of marital discord or a psychiatric illness in a parent[1328.1665.1667].

c) Hyperactive children with behavior problems are simply the result of poor parenting skills and it is all the parent's fault.

d) ADHD is due to some type of brain damage[P74].

e) ADHD is a genetic disorder.

In a later chapter I discuss ADHD in Tourette syndrome. The very high frequency of ADHD in this clear-cut genetic disorder is one of the strongest pieces of evidence that ADHD can be genetically caused. In this chapter I discuss the evidence, independent of TS, that ADHD is a genetic disorder. Before we get into this, however, it is necessary to define some psychiatric disorders that will be mentioned because they are important to the concept that ADHD is a genetic disease.

Antisocial Personality Disorder

The essential feature of individuals with this disorder is a history of continuous and chronic antisocial behavior in which the rights of others are violated. These individuals were previously termed psychopaths or sociopaths. The symptoms begin before age 15 and there is a persistence of antisocial behavior into adult life, including failure to sustain a good job performance over a period of several years.

> "Lying, stealing, fighting, truancy, and resisting authority are typical early childhood signs. In adolescence, unusually early or aggressive sexual behavior, excessive drinking, and use of illicit drugs are frequent. In adulthood, these kinds of behavior continue, with the addition of an inability to sustain consistent work performance or to function as a responsible parent and failure to accept social norms with respect to lawful behavior. After age 30 the more flagrant aspects may diminish, particularly sexual promiscuity, fighting, criminality, and vagrancy"[490].

Associated features include:

> "inability to tolerate boredom, depression, and the conviction (often correct) that others are hostile toward them. Almost invariably there is markedly impaired capacity to sustain lasting, close, warm, and responsible relationships with family, friends, or sexual partners"[490].

Individuals must be 18 years of age or older. The prevalence of antisocial personality disorder has been estimated at 3 percent for males and 1 percent for females.

Hysteria

One of the difficulties of comparing psychiatric studies through the years is that the terminology keeps changing. As used in this chapter it refers to women with a dramatic or complicated medical history beginning before age 35 and consisting of multiple physical symptoms in the absence of demonstrable disease. Particularly common are chronic headache, blindness, paralysis, weakness, a lump in the throat, shortness of breath, stomach and intestinal problems, dysmenorrhea, frigidity and thoughts of wanting to die[1497].

Hysteria is almost exclusively seen in women. Its prevalence in the general population is between 1 and 2 percent. Family studies suggest it is a genetic disorder since 14 percent of the mothers and sisters of hysteria patients also have hysteria[53]. This is about ten times the frequency in the general population.

Histrionic Personality Disorder

Previously termed hysterical personality, the essential feature of histrionic personality disorder is an overly dramatic, reactive, and intensely expressed behavior.

> "Individuals with this disorder are lively and dramatic and are always drawing attention to themselves. They are prone to exaggeration and often act out a role, such as the 'victim' or 'princess' without being aware of it.
>
> "Behavior is overly reactive and intensely expressed. Minor stimuli give rise to emotional excitability, such as irrational, angry outbursts or tantrums. Individuals with this disorder crave novelty, stimulation, and excitement and quickly become bored with normal routines.
>
> "Interpersonal relationships show characteristic disturbances. Initially people with this disorder are frequently perceived as shallow and lacking genuineness, though superficially charming and appealing. They are often quick to form friendships; but once a relationship is established they can become demanding, egocentric, and inconsiderate; manipulative suicidal threats, gestures or attempts may be made; there may be a constant demand for reassurance because of feelings of helplessness and dependency"[490].

They often experience periods of intense dissatisfaction and a variety of dysphoric moods, tend to be impressionable and easily influenced by others or fads. They are apt to be overly trusting of others, suggestible, and show a positive response to any strong authority figure who they think can provide a magical solution for their problems. Interpersonal relationships are usually stormy and ungratifying[490]. Many hysteric women also have a histrionic personality.

Genetics of ADHD

The following summarizes some of the evidence suggesting that ADHD is a genetic disorder.

The fathers of ADHD children tend to have antisocial personality, the mothers histrionic personality If ADHD is a genetic disorder we would expect to find that many of the parents also had ADHD when they were children. In addition, if ADHD doesn't always go away there should also be a higher frequency of such problems among parents of ADHD children. Several studies have examined these points.

The first was in 1971 by James Morrison and Mark Stewart[1363]. They examined the *parents of 59 hyperactive children* and 41 control children. The percent of mothers and fathers with alcoholism, hysteria or antisocial personality were as follows:

Parents of:

	ADHD children		Control children	
	Mo	Fa	Mo	Fa
% with ADHD	5	**15**	2	2
% with alcoholism	5	**20**	0	10
% with ASP	0	**5**	0	0
% with hysteria	**10**	0	0	0

Thus, ADHD, alcoholism and antisocial personality were more common in the fathers and hysteria and alcoholism more common in the mothers of ADHD children than in the controls. In 21 of the 59 families which had a hyperactive child, at least one parent was alcoholic, hysteric, or sociopathic. By contrast only 4 of 41 control families were so affected ($p < .025$). Morrison and Stewart concluded:

> "These familial associations of adult and childhood psychiatric disorders indicate that childhood hyperactivity may be etiologically related to alcoholism, hysteria and sociopathy, and that the hyperactive child syndrome is transmitted genetically or socially from parent to child."

Dennis Cantwell[285] examined the *parents of 50 hyperactive children* and 50 matched control children. He also found an increased frequency of antisocial personality, alcoholism and ADHD among the fathers and other male relatives, hysteric personality, alcoholism and ADHD among the mothers. In all, 10 percent of the parents of ADHD children were considered to have ADHD themselves and 10 percent were psychiatrically ill adults with either alcoholism, antisocial personality or hysteric personality.

This interrelationship between antisocial personality and alcoholism in men and hysteric personality in women has been noted independent of ADHD. In a family study of the *parents and siblings of felons*, there was an increased frequency of antisocial personality, alcoholism and drug addiction in the male relatives and histrionic personality in the female relatives[828]. In a study of 78 convicted women felons, sociopathy or hysteria-histrionic personality was present in 80 percent, 20 times that of the general population[327].

On the other side of the coin, in a study of *relatives of histrionic women,* there was an increased frequency of hysteric personality in the mothers and sisters, and an increased frequency of alcoholism and antisocial personality in the fathers and brothers[53].

Evidence that like tend to marry like comes from the observation of an increased frequency of hysteric personality in the *wives of men with antisocial personality*[827] and of antisocial personality and alcoholism in the husbands of hysteric women[2096]. In genetic terminology this is called **assortative mating.** The children of such parents are in double jeopardy to inherit an ADHD gene, and one-fourth of such children could be homozygotes, that is, have received the ADHD gene from both parents. Such a child would probably be more severely affected than either parent.

The brothers and sisters of children with ADHD also tend to have ADHD If attention deficit hyperactivity disorder (ADHD) is a genetic disorder we would expect brothers and sisters of affected children to have more ADHD than children in the general population. Numerous studies[69,258,288,1482,2054] indicate that 20 to 30 percent of siblings of children with ADHD also have ADHD. This is two to seven times the frequency in control children.

In one study where all first-degree relatives [siblings and parents] of ADHD children were examined, 31.5 percent had ADHD compared to 5.7 percent in control families. These relatives of ADHD children were also seven times more likely to have oppositional disorder and five times more likely to have major depression than in the control families[165,166].

Two things are apparent from these studies. First, the frequency of ADHD is much greater in siblings of ADHD children than in the general population, and second, the ratio of affected brothers to sisters varies with the definition of ADHD. For example, in two studies using a definition of ADHD that emphasized hyperactivity, Cantwell[285] reported that 22 percent of brothers and eight percent of sisters of hyperactive children had hyperactivity, and Welner and co-workers[2054] found that 26 percent of brothers and nine percent of sisters had hyperactivity. However, when the DSM-III criteria were used, emphasizing inattention and impulsivity, Cantwell[288] found that now 24 percent of brothers and 19 percent of sisters had ADD, and August and Stewart[69] found equal rates in brothers and sisters. These results suggest that *boys are more likely to have ADD with hyperactivity while girls tend to have ADD without hyperactivity.*

In a study of 72 ADHD children, it was found that if *neither parent had ADHD* then 11 percent of the siblings had ADHD, while if *one parent had ADHD* then 34 percent of the siblings had ADHD[1482]. If some cases of ADHD are due to environmental causes these would tend to be in the group where neither parent was affected, while the group where one parent was affected would contain the genetic cases. In this group the frequency of ADHD among the siblings was eight times that of the general population.

Siblings of ADHD children have more ADHD than half-siblings The demonstration that the parents of an ADHD child also had ADHD as children doesn't absolutely rule out the possibility that ADHD is a learned rather than a genetically caused behavior. The increased frequency of ADHD in uncles and cousins favors a genetic factor since such individuals are usually not in close life-long contact with the ADHD child or his family. One method of sorting out a genetic versus a learned condition is to compare siblings with half-siblings. Both are raised in a similar environment so if ADHD is learned its frequency should be the same in both. However, half-siblings have only half the genetic similarity that siblings do. A genetic disorder should be less frequent in half-siblings of an ADHD child than in siblings. The results of such a study, done by Daniel Safer[1678b] are as follows:

	Sibs	Half-Sibs	Sibs/ Half-Sib
Hyperactivity	47	23	2.0
Short attention span	47	14	3.4
Behavior problems	53	18	2.9
Diagnosis of ADHD	47	9	5.2

The significantly decreased frequency of ADHD symptoms in half-siblings compared to siblings is strong evidence for a genetic cause.

Twin studies There have been few twin studies of ADHD. In a very limited one Lopez[1205] reported 100 percent concordance for three identical twins and 17 percent concordance for six fraternal twins. In a study which simply looked at similarities in behaviors in 93 pairs of twins of the same sex[2078], the similarity in "activity level" was much higher (0.92) for identical twins than for fraternal twins (0.60). The similarity between eight identical twins for simply being in the top 20 percent of the activity score was very high (0.7) while it was very low (0.0) for 16 fraternal twins.

Adoption studies One of the best techniques for studying the effect of genes versus environment is to study children adopted at birth. If ADHD is a genetic disorder then the biological parents should show a higher frequency of ADHD or antisocial personality than controls. If it is a learned behavior, then there should be a high frequency of ADHD or antisocial personality in adoptive as well as biological parents who do not give their children up for adoption.

To determine whether the transmission of ADHD from parent to child was by genetic or social-environmental factors, both Morrison and Stewart[1364] and Cantwell[286] also examined a set of ADHD children raised by adoptive parents. They both found that the frequency of ADHD, sociopathy, alcoholism and hysteria was greater in the biologic as opposed to the adoptive parents. The fact that the adopted children still developed ADHD despite being raised by normal parents suggests the transmission was by genetic rather than social-environmental factors. Since the biologic parents of adopted children are hard to find, in both of these studies the true biological parents of the adopted children were not examined, i.e., the biological, adoptive and control parents were separate groups. Both studies had the disadvantage that the individuals interviewing the parents knew whether the child was in the ADHD or control group, and the diagnosis of ADHD in a parent required a rather subjective look backward[1293].

Cadoret and co-workers[270,272,441] examined 59 adopted away children whose parents had a known psychiatric problem, the majority having *antisocial personality disorder*. The predominant problems were with the 32 adopted males of the antisocial parents. These children had significantly more problems with ADHD, reacting violently to things, having frequent temper tantrums, and other antisocial behaviors than the control children. Among the adopted children of antisocial parents, 37 percent required professional help for behavioral disorders compared to only 14 percent in the controls.

If ADHD is a lifelong disability we would expect that some ADHD children would still have problems after they had grown up. The following chapter examines what happens to ADHD children as they grow up.

Summary: Attention deficit hyperactivity disorder is not simply a random group of rambunctious children whose behavior happens to irritate parents and teachers. It is a distinct genetic, behavioral syndrome which is expressed as ADHD in children, and may be a lifelong problem resulting in antisocial personality and alcoholism in adult men, and in a hysteric-histrionic personality in adult women.

Chapter 18
ADHD Children Grown Up

It Doesn't Always Go Away

A common misconception about ADHD children is that "they grow out of it." Unfortunately, this is not always the case. At the opposite extreme there is also the misconception that all ADHD children grow up to be criminals, alcoholics or drug addicts. This is equally incorrect. Where does the truth lie? There are two ways to determine what happens to ADHD children when they grow up — by looking backward and looking forward. Looking backward involves examining a group of subjects who someone else, years earlier, had labeled as ADHD. These are called **retrospective studies.** Such studies are easier because they are faster, but this approach is fraught with many potential biases. Looking forward involves actively following a group of ADHD children into adulthood. This is the most accurate, but it obviously takes a long time. These are called **prospective studies.** Let us first look at the retrospective studies.

Looking Backward

One of the earliest long- range retrospective studies was by Lee Robins[1627] of Washington University School of Medicine. She examined 524 individuals who had attended a child guidance clinic 30 years previously. As adults 34 percent had disabling symptoms including antisocial behavior and overt psychosis. Although many of these had ADHD as children, the exact percentage is not known. Of 76 girls referred to the clinic between the ages of 12 and 16 because of antisocial behavior, 20 were diagnosed as having hysteria as adults.

The most pessimistic and grim view of what happens to ADHD children came from a 25 year follow-up study of 14 ADHD children by Menkes and co-workers[1313]. They found that a third were psychotic, a fifth still had ADHD, a fifth were retarded, and a fifth had been institutionalized for a criminal offense. Only 29 percent were normal and self-supporting. This is an alarmingly high frequency of problems and the difficulty with this study is that the numbers are small and only individuals who were hyperactive and had some evidence of brain damage and from a low socioeconomic class were included.

Stewart and co-workers[1307,1883] studied a group of teenagers who had been diagnosed as ADHD when they were two to five years of age. Approximately *half were much improved.* A fourth remained unchanged and the other fourth were in between. However, *despite improvement there were many problems with learning disabilities, inattention, impulsivity, poor conduct, and delinquent behavior such as temper tantrums, irritability, lying, fighting, and stealing*. These are shown in the following table:

ADHD Children as Teenagers
(Mendelson et al.[1307])

General	%
Impulsive, impatient	85
Poor concentration	77

Defiant, rebellious	74
Hyperactive, restless	71
Irritable	67
Temper tantrums	56
Low self-confidence	54
Loner, no friends	46
Antisocial Behavior	
Frequent lying	83
Incorrigible	66
Cursing excessively	59
Destructive	52
Fighting	51
Stealing	51
Bad associates	34
Threatened to kill parents	34
Reckless and irresponsible	22
Excessive drinking	15
Setting fires	15
Poor employment record	13
School Record	
Hard to sit and study	78
Flits between projects	71
Poor concentration	70
Discipline problem	59
Failed one or more grades	58
Reading difficulty	57
Gets into fights	45
Math difficulty	44
Poor relations with teachers	39
Skips classes	18

A more optimistic report of the outcome of hyperactive children grown to adulthood was presented by Borland and Heckman[197]. They interviewed a group of 20 men who had been seen 25 years previously in a guidance clinic for symptoms that would have led to a diagnosis of ADHD had that entity been recognized at that time. The control group was their brothers. They found that the men who had ADHD as children were not experiencing serious social or psychiatric problems as adults. A large majority had completed high school, a few had gone to college, each was steadily employed and self-supporting, and most had achieved a middle-class stature. However, *half of them continued to show a number of major symptoms of hyperactivity such as restlessness, nervousness, and difficulty with temper.* A substantial minority continued to be *impulsive, easily upset, or often sad, blue or depressed.* As shown below they never achieved a socioeconomic status equal to their brothers.

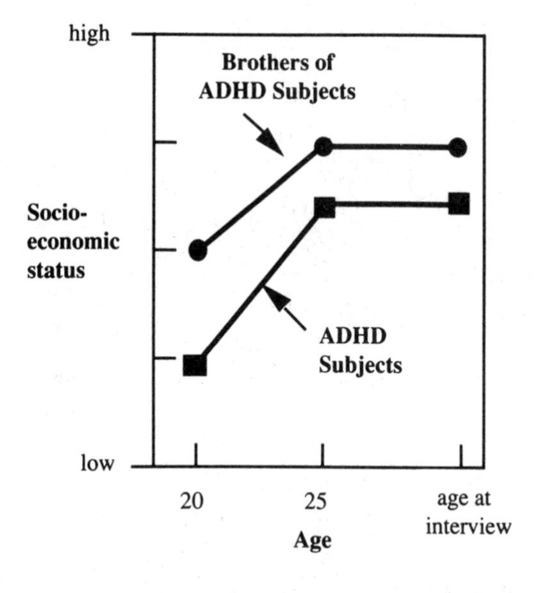

This was because they entered the work force at a lower level and never caught up to their brothers. They also worked more hours each week and changed jobs more often than their brothers. Half of them had part-time or extra jobs that incorporated musical or mechanical interests or work they had done as a hobby. They regarded this work as a means of avoiding feelings of restlessness and nervousness in periods of inactivity. There was a strong tendency for them to regard their work as unexciting and repetitious. They seemed to *tire of their jobs easily* and became impatient with their prospects for advancement. There were no problems with alcoholism. Although there were no stated problems with antisocial behavior, a later report[2046] indicated that 23 to 25 percent of the ADHD subjects did have such problems.

Studies looking backward from antisocial personality and alcoholism show that a high percentage of such individuals were hyperactive, impulsive and aggressive as children[988,1277,1565].

Looking Forward

To examine the question of whether ADHD children subsequently had greater problems with *alcohol abuse*, Blouin and co-workers[187] compared 23 ADHD subjects to 22 matched control children *also* with school difficulties but without ADHD. After a five year follow-up, at an average age of 14 years, *55 percent of the ADHD children as opposed to only 20 percent of the school problem children, were were found to use alcohol more than once a month.* As teenagers these children still showed signs of being hyperactive and were rated as having more conduct problems by their parents than the controls.

Huessy and co-workers[931,935,936] followed 84 ADHD children to ages 9 to 24. Their *rate of institutionalization for delinquency was 20 times that of the general population.* In another study[933,935] 500 children were rated by their teachers for ADHD symptoms in the second, fourth and/or fifth grade. By the time they were in the ninth grade, of those children who had the most ADHD symptoms, *62 percent had flunked one or more grades and 42 percent had poor or very poor social adjustment.* By contrast, of those children who had the least ADHD symptoms, *0 percent had flunked a grade and 0 percent had a poor or very poor social adjustment.*

Ackerman and co-workers[7,515] examined 23 hyperactive *and* learning disabled boys in grade school, then again at age 14. About *57 percent of the hyperactive-LD boys had experienced major conflicts with authority compared to only three percent of the controls.* Although overt symptoms of motor hyperactivity had disappeared in 43 percent, only nine percent were so much improved that they presented no problem for the home, school or community. These 21 who continued to have problems could be divided into ten who had only behavior problems, and eight who had only learning problems.

An intriguing *study of delinquency in ADHD children as teenagers* was reported by Satterfield and co-workers[1703]. They examined 110 ADHD children and 88 normal controls from childhood through adolescence by studying official arrest records on both groups. To control for environmental factors, the groups were divided into lower, middle and upper socioeconomic class. The results were:

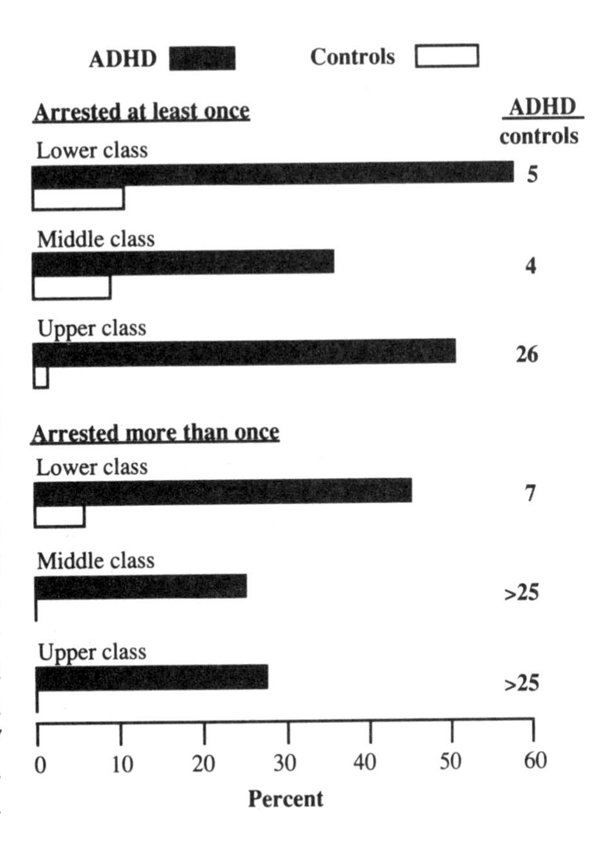

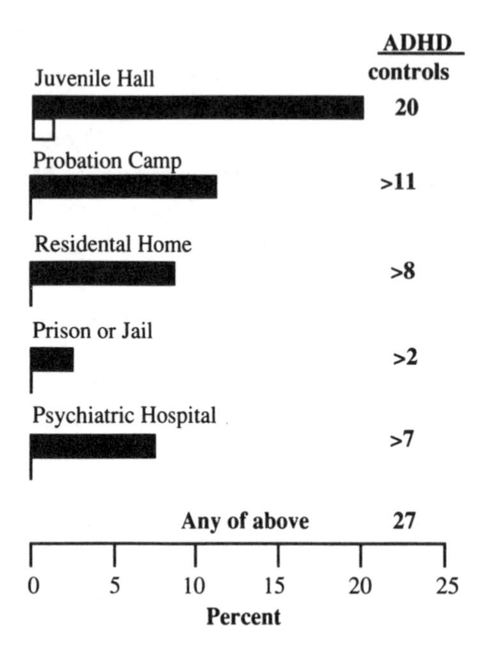

The *effects of childhood ADHD on middle- and upper-class delinquency are particularly striking. Being arrested more than once occurred 25 times more often in the middle and upper-class children with ADHD than controls.* The frequency of problems for all three classes are higher than other studies. The authors point out that they were able to follow-up 100 percent of their children, while many studies lost up to 50 percent of their cases to follow-up. *Individuals lost to follow-up are twice as likely to have an arrest record as those who are easy to find.* They concluded that the presence of *childhood ADHD identified one group of children who are at increased risk for serious teenage delinquency.* It was emphasized that to prevent such an outcome, a multimodal treatment of childhood ADHD is necessary, including not only medication, but treatment aimed at poor peer relationships, poor self-image, antisocial behavior and learning disabilities[1073a].

Weiss and co-workers[872-875,1333-1335,2045-2049] have presented 5, 10 and 15 year follow-up studies on a group of children diagnosed as ADHD at an average age of eight years. After five years (average age 13 years), although the hyperactivity and aggressiveness had diminished with age, *many continued to have problems with emotional immaturity, inability to maintain goals and low self-esteem and 25 percent had demonstrated antisocial behavior.* Poor academic functioning was the feature that most clearly characterized the ADHD group as a whole and was present in 80 percent. Seventy percent had repeated at least one grade and 35 percent had repeated two or more grades.

At ten years, after the turbulent adolescent years were over, things seemed to look up. Although the ADHD subjects were more impulsive and had higher scores on anxiety, tension, and hostility, only a small minority showed severe psychological problems or antisocial behavior. There was no difference from the controls for problems with drug abuse[872,873,1985,2047]. The ADHD subjects were rated as markedly inferior to controls by teachers, but not by employers[2043]. This is probably a reflection of the fact that one can make many choices to find satisfactory employment while school is school wherever you go.

However, this apparent *healing trend did not continue* and at 15 years, as adults averaging 25 years of age, the *ADHD subjects had significantly more antisocial personality disorders* than their controls (23 percent versus 2 percent), and 66 percent still had at least one disabling symptom of ADHD compared to only 7 percent of the controls. However, the ADHD group did not have a significant excess of drug or alcohol abuse[2046].

Jan Loney and co-workers, from the University of Iowa, have been following a group of 200 boys in which the diagnosis of ADHD had been made when the boys were between 4 and 12 years of age. By 1984, 65 were restudied when they were 21 to 23 years old[1201,1203]. Of these, there were 22 in whom an unaffected brother was also evaluated. *At 21+ years of age there was a high frequency of antisocial personality and alcoholism.* The interplay of the two are shown as follows:

29% Alcoholic
of which 79% also have Antisocial Personality

37% Antisocial Personality
of which 62% are also Alcoholic

Using the unaffected brothers as controls, the significant differences were as follows:

Behavior	**ADHD**	**Brothers**
	(%)	(%)
Frequent truancy	41	14
Expulsion from school	54	18
Persistent lying	50	14
Thefts	36	14
Unemployment	50	18
Time in jail or prison	41	5
ASP*	45	18

* meet criteria for antisocial personality

These figures are particularly significant since the "unaffected" brothers, who also share some of the genetic and environmental factors of their ADHD siblings, show rates for these behaviors that are significantly higher than normal controls.

This can be appreciated in the following table showing the significant differences in *substance abuse* for the ADHD (**A**) subjects, their brothers (**B**) and a group of 935 18 to 25 year old controls (**C**):

Substance abused	**A**	**B**	**C**
	%	%	%
Inhalants (glue,etc)	36	27	16
Cocaine	50	36	28
Sedatives	45	32	22
Stimulants	45	41	22

Of interest was the finding that those ADHD children who responded best to stimulant medication were less likely to show subsequent irritable behavior and illegal drug abuse. *The use of stimulant medication in treatment actually decreases the risk of subsequent use of illegal drugs.*

Another major prospective study passing from ADHD children into young adulthood was by Gittelman and co-workers[723]. ADHD with hyperactivity was still present in 31 percent versus 3 percent of the controls. The only other two conditions that distinguished the two groups were *conduct disorder and substance abuse*. These were primarily present in those still troubled with ADHD. Substance abuse followed the onset of conduct disorder in the overwhelming majority of the cases. They conclude that *the greatest risk factor for the development of antisocial behavior and drug abuse is the persistence of ADHD symptoms.*

It is clear that while different studies come up with somewhat different conclusions the trends in all are similar. *Not all children with ADHD outgrow their symptoms. Somewhere between a fourth and a half continue to have problems with impulsivity, distractibility, antisocial behavior, short temper and alcohol or drug abuse in later life.* This can be summarized as follows:

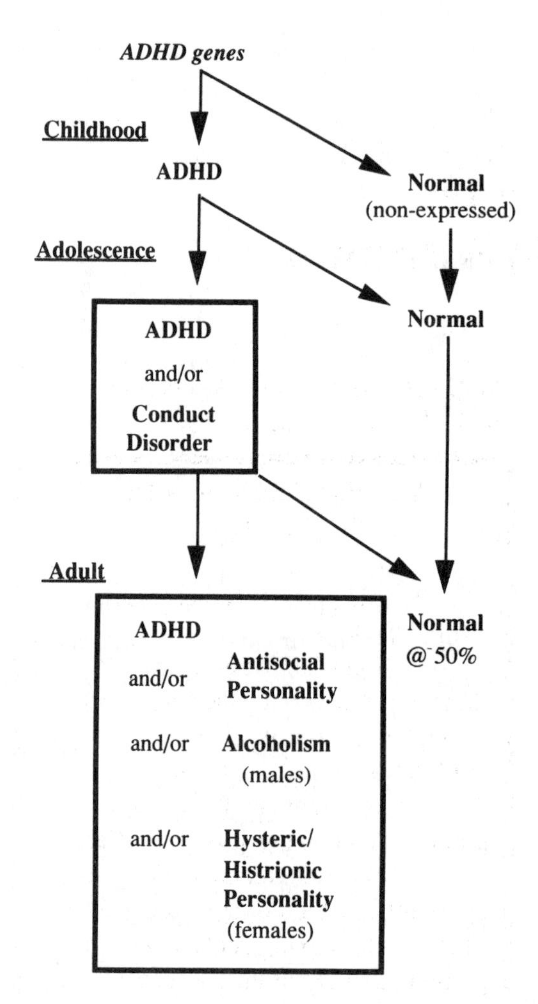

Such a relatively pessimistic outlook on the long-term prognosis for ADHD raises the next most natural question — which of the individual symptoms of childhood ADHD are most likely to predict a poor outcome in adolescence and adulthood? Studies[874,1201,1203] have shown that *low social class, amount of aggressive behavior in early childhood, and a family history of similar problems* were the most important predictors of a poor outcome. By contrast, the severity of childhood motor hyperactivity *per se* did not predict a poor outcome, and a family history free of such problems, positive childhood achievements and a good response to medical treatment were correlated with a good outcome.

Summary: Children with severe aggressive behavior in early childhood, driven by genetic factors and compounded by poor socioeconomic environment, are at a high risk for delinquent behavior as young adults.

Although about half of the children with ADHD outgrow it as they get older, half do not and they are at significant risk for problems with alcohol, drug abuse and antisocial behavior.

Middle- and upper-class children with ADHD are 4 to 28 times more likely to be arrested or institutionalized than their peers without ADHD.

Long-term studies of ADHD emphasize a major theme of this book, that genes which lead to disinhibited behavior in childhood can result in a lifetime of impaired functioning.

Chapter 19
ADHD in Adults

By now it should be obvious that adults can also suffer from ADHD. This concept began to surface as the studies on ADHD children growing up were being published. The idea is well summarized in a brief note by Huessy in 1974[934]:

> "Over ten years ago, in desperation, I treated one adult diagnosed as schizophrenic, who had a rather typical childhood history of hyperkinesis, with amphetamine — with surprisingly good alleviation of his most bothersome symptoms of impulsivity and emotional overreaction. I [have since found] many young adults who had been through prominent psychiatric institutions... with the diagnosis of schizophrenia, but who showed few evidences of primary symptoms of schizophrenia and responded poorly to [Thorazine]. These patients did show typical childhood histories of hyperkinesis. I have begun treating these patients with imipramine hydrochloride (Tofranil)... and have seen definite reduction of impulsivity and emotional overreaction.
>
> "Improved self-control, control of temper outbursts, greater thoughtfulness, increased perseverance, and decreased impulsivity can be achieved in some of these patients."

This concept has now reached the level of respectability and is included in the DSM-III[490] as attention deficit disorder, residual type, (ADD-RT). Its diagnosis depends upon meeting the following criteria:

Attention Deficit Disorder Residual Type

A. Once met the criteria for ADD with hyperactivity.

B. Signs of hyperactivity are no longer present, but other signs of the illness have persisted to the present without periods of remission, as evidenced by signs of both attentional deficits and impulsivity (e.g., difficulty organizing work and completing tasks, difficulty concentrating, being easily distracted, making sudden decisions without thought of the consequences).

C. The symptoms of inattention and impulsivity result in some impairment in social or occupational functioning.

D. Not due to other disorders.

As early as 1944 Hill[895] described the successful use of amphetamines in the treatment of adults characterized as aggressive, ill-tempered, hostile, impulsive, irritable, fickle, irresponsible, easily frustrated, antisocial and alcoholic. They also had problems with enuresis and deep sleep. Similar problems were noted in other family members. Between then and the time of the above note, several others have described the usefulness of stimulants or antidepressant medications in the treatment of adult impulse character disorders[858,1234,1565,1836]. Gross and Wilson[807] reported a 38 year old mother of three hyperactive children who

> "talked in short, disconnected phrases, frequently flitting from one thought to another with no connecting bridge. She gave the appearance of being scatterbrained because of staccato, disorganized speech, yet it was clear she was an intelligent, thoughtful woman."

Treatment with imipramine led to considerable improvement.

Paul Wender and co-workers[2107] described a group of 15 adults who had histories of long-standing impulsiveness, inattentiveness, restlessness, short temper and emotional lability. As children they were unable to sit still, disobedient, had difficulty concentrating, failed to finish things, were slow learning to read, were unpopular with other children and general trouble makers. As adults they showed the following characteristics:

Restless, always on the go	100%
Anxious, overreactive	87%
Moody, short-tempered	80%
Low self-esteem	77%
Trouble concentrating	73%
Irritable	73%
Impulsive	67%
Dissatisfied with life	67%
Very demanding	60%
Trouble seeing the other person's point of view	47%

To objectively evaluate the response to medication they were treated in a double-blind fashion, where neither the patient nor the doctor knew whether they were receiving the active drug or a placebo. Overall, 53 percent responded well to Ritalin Those who responded were less anxious, irritable, impulsive, and angry and their mood stabilized. Two women who had been child abusers stopped when their irritability and anger diminished. Five of the patients with unstable, unhappy marriages reported improvement there.They had previously received individual psychotherapy or marital counseling with no appreciable benefit.

Of particular interest, *none of the patients showed any tendency to abuse the use of Ritalin.* Their response to the medication differed from normal people in that it did not make them euphoric and they did not show tolerance, or a need to progressively increase the dose. Those who responded to the antidepressants also responded differently in that *the medication was effective immediately* rather than the three weeks it takes to improve depression, and the dosages required were much smaller than those required in the treatment of depression.

There were no patients 40 years of age or older, suggesting that like antisocial personality, adult ADHD tends to burn itself out in middle age.

Case Report

A case report[2107] can illustrate many of the above points. Items of particular note are in italics.

> "A 32 year old mother of four, had problems with *anxiety, irritability, emotional instability, intolerance of minor frustrations, and chronic marital difficulties.* She wept continuously throughout the first interview, claiming that her life had brought her little happiness. She came to the clinic on the recommendations of a child psychiatrist who, while evaluating one of her sons for 'hyperactivity' had noticed similar difficulties in her.
>
> "During grade school she did well in arithmetic, but *reading and spelling*

were problems. She remembers being so "fidgity" in the first and second grade that the teacher had to turn the pages of the books since she could not do this herself. Her academic problems worsened, and in the third grade she was placed in a *special school* for children with learning difficulties, despite an IQ in adulthood of 115. Throughout grade school she had difficulties not only academically, but with her peers because of a *quick temper*. There was improvement in high school where she began to have friends. Reading continued to be difficult and she avoided it as much as possible. Although she participated in many social activities, she had the nagging feeling that she derived *less pleasure* from them than did others. She married when she was 22, only to have problems with her husband from the very start. These were compounded by the birth of four children in rapid succession. Since marital problems persisted, both she and her husband began psychotherapy. Despite intermittent individual and couple therapy over a ten-year period, there was little improvement in the marital relationship. Problems diminished somewhat when her husband learned the *efficacy of leaving her alone when she became angry, since arguments only increased their difficulties*.

"Her *parents had separated* while she was an infant. Her father, whom she did not know, was *quick-tempered*. Her younger brother, although intelligent, was a *poor student, a behavioral problem in school, and later had a variety of difficulties with the police*.

"After treatment *with Ritalin she felt much calmer*, no longer overreacted to minor problems, felt happier and organized her household. Her husband was gratefully aware of a marked change almost immediately. Even though his wife still becomes angry at disagreements with him, she seem *better able to control her temper*. Buoyed by her own improvements from the drug, she has suggested that her hyperactive son receive the same treatment, a remarkable turnabout, as she had previously been adamantly opposed to its use for her boy."

Since these early descriptions, many reports of adult ADHD have appeared[818,1104,1251,1366]. This description, including the short temper, irritability and strong family history, is identical to what is often seen in one or both of the parents of TS children. This is understandable because, as described later, ADHD is very common in TS[p99] and many cases of ADHD may be due to a *Gts* gene[p259].

Men and Women Are Equally Affected

In contrast to childhood ADHD, there is no significant predilection for males in adult ADHD[817,2107]. As mentioned in a previous chapter, this is probably a reflection of the fact that in childhood motor hyperactivity appears to be more common in males, thus increasing the probability that they would come to the attention of mental health workers. When this burns itself out, and inattention, impulsivity and irritability become the focus of attention, females are affected as often as males.

Adult ADHD Responds to Stimulants and Antidepressants

As with children, ADHD in adults responds to stimulants (Ritalin, Dexedrine, cylert), antidepressants (imipramine, nortriptyline, desipramine) and in some cases anticonvulsants (Dilantin, Tegretal, Valium). Not all individuals who fit the criteria of ADD-RT respond to medication or feel that their problems are sufficiently severe to justify daily medication[817].

ADHD in Adults — A Complex of Symptoms

The core of ADD-RT is the persistence of childhood symptoms of inattention and impulsivity into adulthood. Like Tourette syndrome, there are a plethora of associated behaviors and problems including irritability, explosive personality, antisocial behavior, depression, anxiety, mania, alcoholism and drug abuse.

There are two interpretations of these findings.

1. Many different disorders have been lumped together under the umbrella of ADHD, conduct disorder and antisocial personality, or

2. A relatively small number of genes produce disinhibited behaviors and these genes manifest themselves in different ways in different people depending upon their genetic background and environmental factors. The model presented by Tourette syndrome suggests the latter.

Summary: ADHD can persist into adulthood where all the symptoms of childhood ADHD are present including poor concentration, distractibility, short temper, irritability, depression, mood swings, and chaotic relationships. The same treatment used for children with ADHD is effective in adults. Adult ADHD is common in one or both parents of TS children.

Chapter 20

ADHD in Tourette Syndrome

Now that we have a clear picture of what ADHD is we can ask — What percent of TS patients have ADHD or ADD? In a total of 1,500 TS cases from nine different reports, an average of half had ADHD[368].

In our controlled study we asked about the occurrence and severity of each of the criteria needed to make a diagnosis of ADHD[360]. Records were also obtained from the teachers and school psychologists.

Inattention

The frequency in percent and severity (0 - none, 1 - occasional, 2 - often) of the five symptoms of inattention for the controls and all TS patients was as follows:

Controls

Tourette syndrome

	FINISH	LISTEN	DISTR	CONC	PLAY
X^2	27.8	50.3	57.8	68.8	22.3
p	<.0005	<.0005	<.0005	<.0005	<.0005
	0 1 2	0 1 2	0 1 2	0 1 2	0 1 2

[Refer to Chapter 15 for an explanation of the figures.]

In TS patients, all five characteristics of inattention, fails to FINISH things, seems not to LISTEN, easily DISTRacted, has difficulty CONCentrating, and has difficulty sticking to PLAY activity were significantly worse than in the controls. The Chi square values (X^2) can tell us the relative severity of these traits, and indicate that *seeming not to listen, being easily distracted, and having difficulty concentrating* were the most severe of the five symptoms.

It is also informative to examine the symptoms of inattention in the three grades of TS.

These symptoms were significantly worse in all three grades of TS compared to the controls. It is of particular interest that even in grade 1 TS, where the tics are too mild to treat, the symptoms — doesn't seem to listen, is easily distracted and has difficulty concentrating were all much worse than in the controls. This is important because it indicates that ADHD in TS is not simply an artifact of selecting the most severe cases, as some have suggested[1472]. The presence of these symptoms, even in very mild cases, indicates it is an integral part of the disorder.

Impulsivity

The frequency and severity of five symptoms of impulsivity in controls compared to TS were as follows:

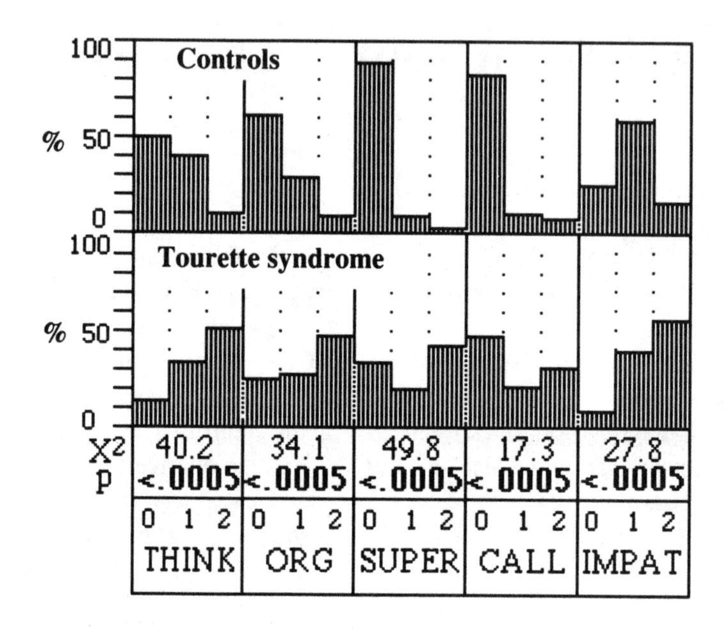

Again all five variables—often acts before THINKing, has difficulty ORGanizing work, needs a lot of SUPERvision, frequently CALLs out in class, and is IMPATient—were very significantly different from controls. The Chi square values indicate that — acts before thinking, has difficulty organizing work and needs a lot of supervision — were the most severe of the five problems.

The distribution of symptoms of impulsivity in the three grades of TS was similar to that of symptoms of inattention. For grade 1 TS, all symptoms, except calling out in class, were significantly more common than in the controls. The significant presence of these symptoms, even in grade 1 TS, again indicates that impulsivity is an integral part of TS.

In addition to the presence of individual symptoms, it is important to know how many symptoms a given individual has. This is the basis for making a diagnosis of ADD. To evaluate this, the severity scale 0, 1 or 2 for each symptom of inattention and impulsivity were added up for each individual to give an ADD score, with a minimum of 0 and a maximum of 20. A score of 9 or greater was considered to represent the presence of ADD. The ADD scores for the controls versus TS patients are shown below. The areas in black = ADD scores indicative of the presence of ADD.

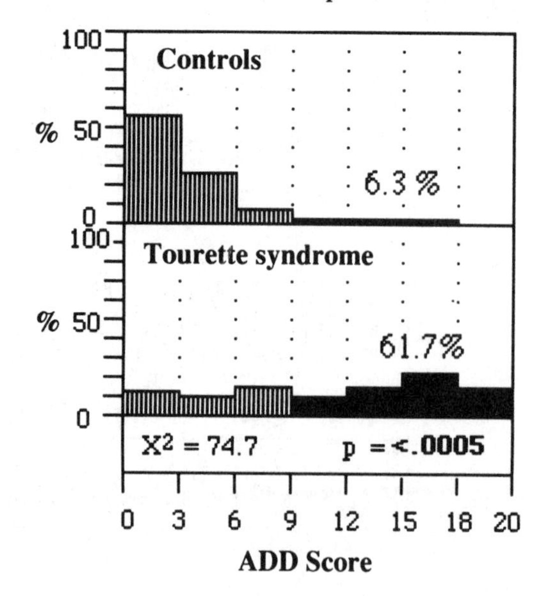

Here 61.7 percent of the TS patients compared to only 6.3 percent of the controls had a score indicating the presence of ADD.

The distribution of the **ADD scores in the three grades of TS** was as follows:

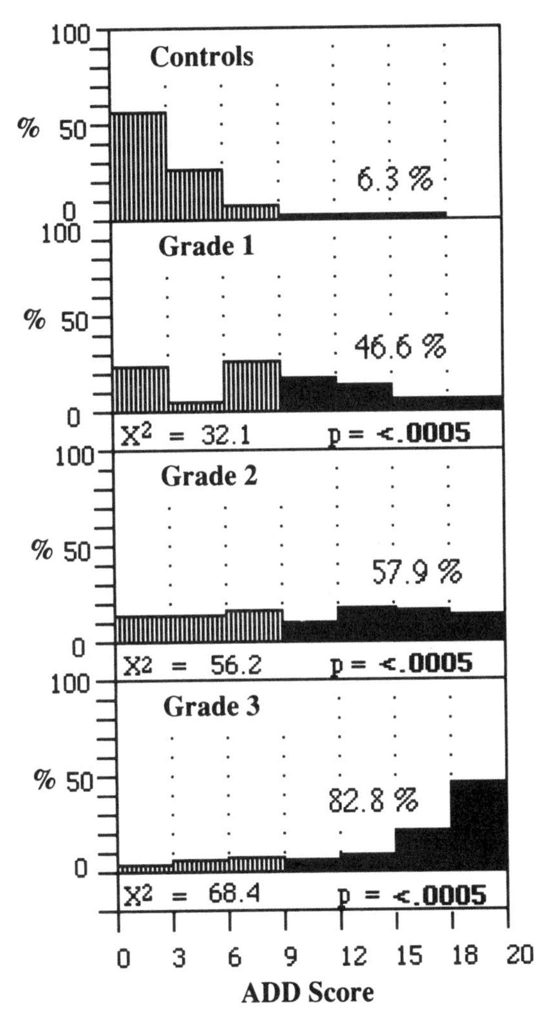

The vast majority of the grade 3 TS patients (83 percent) had ADD. Almost 50 percent had the maximum attainable ADD score. Here 58 percent of grade 2 and 47 percent of grade 1 TS patients had ADD. The finding that almost 50 percent of the grade 1 TS patients (whose tics are too mild to treat), had ADD, further indicates the intimate intertwining of ADD and TS.

Hyperactivity

To make a diagnosis of ADHD, motor hyperactivity also has to be present. The frequency and severity of symptoms of hyperactivity in controls was compared to TS patients.

All five symptoms of hyperactivity — runs about or climbs on things excessively, has difficulty sitting still, has difficulty staying seated, moves about excessively during sleep, and is always on the go as if "driven by a motor" —were significantly more frequent in the TS patients than in the controls. Of the five, has difficulty sitting still, and has difficulty staying seated, were the most troublesome.

For grade 2 and 3 TS all five symptoms of hyperactivity were significantly more frequent than in the controls, while in grade 1 individuals only three symptoms; constantly running about, unable to sit still and can't stay in their seat, were significantly more common than in the controls.

As with the ADD score, it is also possible to add all the scores together to form an **ADHD score**. This showed that ADHD was present in 48.7 percent of the TS patients compared to only 4.2 percent of the controls. These values are lower than those for ADD because not all individuals with ADD also had hyperactivity.

ADHD was present in 71 percent of grade 3, 46 percent of grade 2 and 33 percent of grade 1 TS patients. Again this illustrates that while the frequency of ADD or ADHD depends somewhat on the severity of the TS, it may be present in even the mildest of cases.

How constant are these variations in frequency of ADHD with severity of TS? The following table compares the frequency of ADHD in our first 250 patients[357,358] compared to the second 246 in the controlled study[360]:

	Gr. 1	Gr. 2	Gr. 3
lst Study	33	51	69
2nd Study	33	46	71

The frequency of ADHD in the different grades of TS was stable over time.

Previous Diagnoses

An additional way of determining if minimal brain damage (MBD), hyperactivity, ADHD or behavioral problems are more common in TS is to ask which of these diagnoses, if any, had been made before coming to the clinic. This was done for TS patients and the controls.

A prior diagnosis of MBD, hyperactivity, ADHD, or severely emotionally disturbed (SED) had been made much more often for the TS patients than the controls. The category ALL included those where any of the above diagnoses have been made. Here 50 percent of the male TS patients had previously had one of these diagnoses compared to only 4 percent of the controls.

Twenty percent of the grade 1 TS patients had received a prior diagnosis of one of the four disorders, 40 percent of the grade 2, and 80 percent of the grade 3 patients. Remarkably, 35 percent of the grade 3 TS patients had been given a prior diagnosis of "severely emotionally disturbed." While some of this is the result of the old incorrect assumption that if an individual is ticing it must be due to an emotional disturbance, in the majority of cases there actually were severe behavioral problems [360].

Previously Prescribed Drugs

To further determine how many new TS patients had problems with ADHD we can ask what percentage of the different groups had previously been prescribed various drugs used to treat ADHD. The **following box** shows these results:

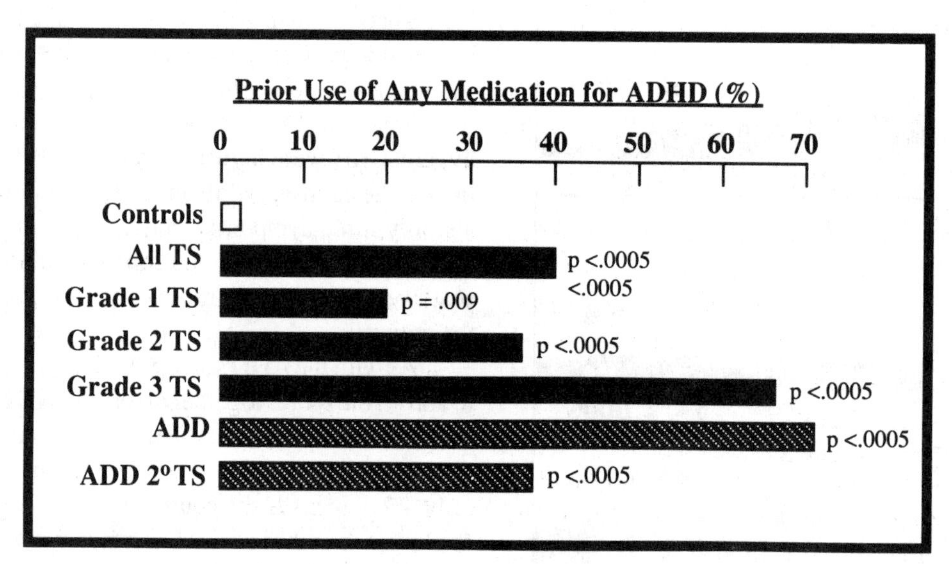

It is not too surprising that 70.6 percent of the ADD group had previously received medications for ADD. However, the grade 3 Tourette syndrome patients were a close second with 66 percent. On average 39 percent of all TS patients had previously been treated compared to only 2 percent of the controls[360].

The Natural History of TS: ADHD-> Tics

When relating the histories of their TS children, most mothers describe an initial period of irritability, impulsivity, short attention span, temper tantrums, and hyperactivity — otherwise known as ADHD — before the onset of motor and vocal tics. The following are some examples:

> "As soon as he was able to crawl I had to put locks on all the doors. He would crawl out of his crib before we got up, open all kitchen cupboards and pull all the pots and pans onto the floor."
>
> "Even as a toddler he never slept. He was awake when I went to bed at midnight and awake when I got up in the morning."
>
> "I knew she was going to be different. When I was pregnant she was much more active than any of my other children."
>
> "At six months of age he suddenly began to have screaming temper tantrums at the slightest frustration."

This raises the question: On average what is the duration between the onset of ADHD and the onset of tics? The average age of onset of ADHD was 4.2 years and of motor or vocal tics, 6.6 years. If we subtract one from the other we get an average of 2.4 years from the onset of ADHD to the onset of TS. However, in some patients they come on at the same time and in others ADHD may be present for five to ten years before the onset of tics.

Motor tics tended to come on about one year before vocal tics. The onset of TS was later in grade 1 cases (7.36 years), intermediate in grade 2 (6.77 years) and earliest in grade 3 cases (5.81 years). Since the onset of ADHD was not that different in the three grades, this meant that the interval between the onset of ADHD and the onset of tics was longest for grade 1 cases (3.45 years), intermediate for grade 2 (2.37 years) and shortest for grade 3 cases (1.56 years). Taken together these figures indicate that the more severe cases of TS with ADHD tend to have an earlier age of onset of both ADHD and tics, and the duration between the beginning of ADHD and tics is shorter.

This data can be better appreciated in visual form:

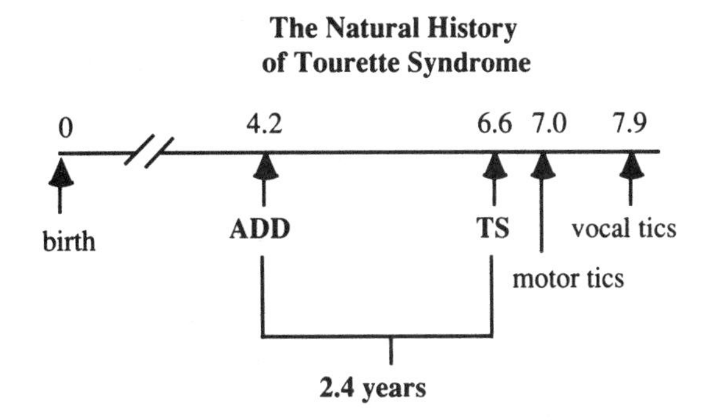

It should be remembered that these are just averages and every child is different.

ADHD Secondary to a *Gts* Gene

If many children carry a *Gts* gene and many present as ADHD, there should be children with ADHD due to a *Gts* gene. These would be of two types: those who are in the ADHD phase before the onset of tics, and those in the ADHD phase who never develop tics. The former are illustrated by the following pedigrees:

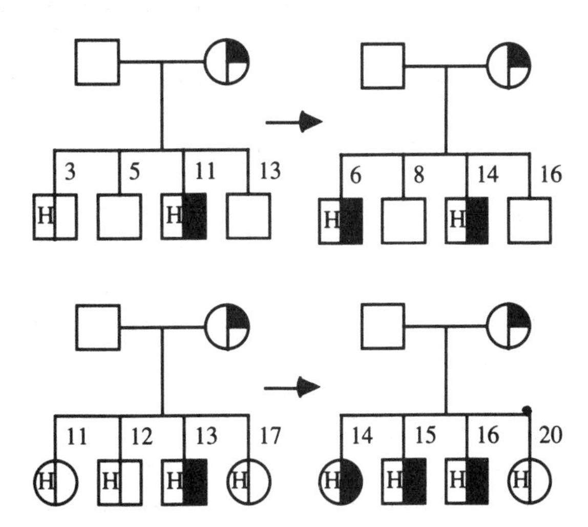

The pedigrees on the right represent the status of the family three and four years later than the pedigrees on the left. At the time they were initially seen they were classified as ADHD probably due to a *Gts* gene, and this was proven when they subsequently developed TS. In the bottom pedigree the 17 year old girl with ADHD never did develop TS, and she probably represents the partial penetrance of the *Gts* gene as ADHD only. Just as *Gts* gene carriers may present as motor tics only or vocal tics only, they may also present as ADHD only, without tics. We see many families like the one above, where one or more individuals have full blown TS with ADHD while another sibling, past age 16, has only had ADHD. An even more convincing example that this can occur is a pair of twins we proved were identical, by blood and tissue typing, where one had TS + ADHD while the other had only ADHD.

When an individual has ADHD and a family history of TS I call this ADHD probably due to a *Gts* gene or *ADHD 2°TS* for short. In practice this refers to children who *satisfy the DSM-IIIR diagnostic criteria for ADHD and have either a family history of TS, or some mild motor or vocal tics, but not both.* This concept is important since such children are very likely to develop tics later on. If they are given stimulant medication it should be carefully explained to the parents that they may subsequently develop tics and may need additional treatment with other medications.

Since there are many children with ADHD 2°TS, it is not surprising that if they are treated with stimulant medication they may subsequently develop motor or vocal tics and when the stimulant medication is stopped the tics persist. Some physicians have concluded from this that the medications caused the TS, while in fact it is simply part of the natural history of the disorder. The subject of the use of stimulant medications, such as Ritalin and Dexedrine, in the treatment of ADHD in TS is discussed in Chapter 81.

Do ADHD and TS Go Together in Families?

Given all the above evidence that ADHD is an integral part of TS, we would expect that in families with TS and ADHD, the ADHD would tend to occur predominately in those individuals with TS or carrying the *Gts* gene. Despite this, Pauls and co-workers[1472], on the basis of a small study, concluded they did not run together. From this, they went on to conclude that when ADHD and TS occurred together it was simply due to chance and bias — that is, a child with TS and independent ADHD is more likely to come to the attention of a physician than a child with TS alone. There is no question that the latter is true — the more severe the symptoms the more likely it is that a person with any disorder will come seek help. But is the statement that they don't run together valid? To examine this we looked at a total of 199 individuals in many families. Here 34 percent of those who were not the proband but had TS also had ADHD. By contrast, only 4.6% of those who did not have TS had ADHD. These differences were highly significant ($p < .0005$). This study clearly indicated that the *Gts* gene and ADHD did travel together in families.

Summary: Other than the motor and vocal tics, attention deficit disorder is the most common symptom in individuals with Tourette syndrome. ADHD often has a far more disruptive impact on the life of the TS patient than the tics. In many cases ADHD can be the only manifestation of a Gts gene. The evidence presented in Chapter 17, and the intimate association between ADHD and Tourette syndrome, indicates that in most cases ADHD is a genetic disorder.

[Other evidence on the intimate relationship between TS and ADHD is given in Chapters 42, 56, 57, and 90.]

Chapter 21

Learning Disorders and Dyslexia

How come I'm so dumb?
Why can't I get this when everyone else does?

This is a frequent lament of children with ADHD and Tourette syndrome. Learning disabilities are common in TS[253a,340,358,360,755,834,972,1221,1465,1858,1903] and once the tics are brought under control with medication the school problems often become the major focus of attention. One mother epitomized the problem by saying, "Jack has *a Teflon brain, nothing sticks.*" The majority of children in special education classes in the United States have ADHD or ADHD and TS.

A learning disorder is assumed to be present when a child with a normal or high IQ is two or more years behind his classmates, and there are no other obvious causes such as a prolonged physical or mental illness. Another definition of learning disabilities has been relegated to the public law books:

Learning Disabilities

Those who have a disorder in one or more of the basic psychological processes involved in understanding or in using language, spoken or written, which disorder may manifest itself in imperfect ability to listen, think, speak, read, write, spell, or do mathemetical calculations.

U.S. Public Law 94-142
Federal Register, Section 121a
Volume 5, p9, 1977

The vast majority of TS patients have a normal IQ[198,834,1767,1771,1903]. Although learning and school problems are frequently mentioned in TS, there have not been many studies of this aspect of the disorder. Stefl and Rubin[1858] mailed a questionnaire to 431 members of the Ohio Tourette Syndrome Association and found that 25.8 percent of TS patients had a diagnosis of ADHD, MBD or hyperactivity, 30.5 percent a diagnosis of learning disability, and 25.1 percent a diagnosis of severe behavioral problem. A total of 47.7 percent had been given at least one of these diagnoses. The following table lists the school-related problems they found.

Problem	Often	Some	Never
Paying attention	48.1	32.3	19.7
Trouble concentrating	49.5	33.3	17.2
Poor at Math	31.1	26.9	42.0
Poor at Reading	26.2	30.3	43.5
Poor at P.E.	12.5	23.6	63.9
Trouble with teachers	14.6	33.8	51.6
Trouble with students	25.2	40.8	34.0
Trouble with timed tests	45.6	28.0	34.0
Behavior problems	26.7	33.3	40.1

Problems paying attention and concentrating were common, and problems with academic subjects such as math and reading were much more common than problems with physical education. Although there were no questions sent to non-TS individuals as a control, almost half of the TS students had trouble with timed tests, and a fourth often had problems with behavior. In a review of 200 TS patients at the Cleveland Clinic, Erenberg and colleagues[550] found that 58 percent had learning problems, behavior problems, or both. Problems with math are particularly common[592a].

There is only one prior study where specific tests of learning in TS patients have been quantitatively compared to normal controls. Sutherland and co-workers[1903] showed the presence of significant problems with short term memory and delayed recall in TS patients compared to controls. The following paragraphs present some of the results of the controlled study relevant to learning disorders.

Special Classes

One simple method of determining the frequency of learning problems in TS patients is to simply ask if they ever had to be placed in any special classes. Different schools have different names for these classes such as EH for educationally handicapped, LH for learning handicapped, and SED for severely emotionally disturbed. A high frequency of TS children of all grades of severity had been placed in severely emotionally disturbed (SED) classes. Even among the grade 1 TS children, with tics too mild to treat, 18 percent had been in an SED class, compared to 24 percent of grade 3 TS children and only 2.1 percent of the controls. The frequency with which the different groups of patients were ever in any special class[360] is **summarized on the right->:**

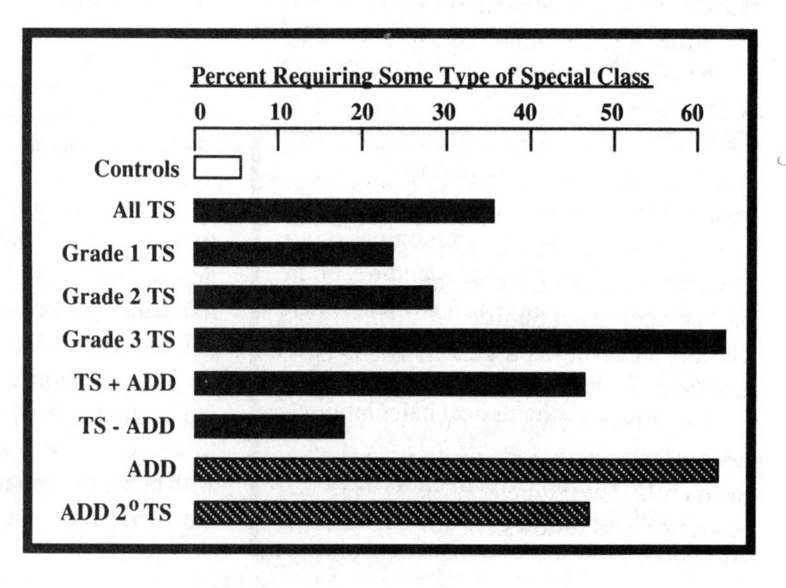

Thus, based on the criteria of whether the learning disorders are disabling enough to require special classes, 35 percent of TS patients compared to 6.4 percent of controls had learning problems this severe. This is actually an underestimate since many had learning problems but were not placed in special classes. The following table lists some other problems in TS patients.

Frequency of Other School Problems

	Cont.	TS	Gr 1	Gr 2	Gr 3
Need a home teacher					
%	0.0	**13.0**	4.6	**11.0**	**24.1**
p	—	.009	—	.02	<.0005
Held back one grade					
%	8.5	**26.4**	23.3	22.7	**37.9**
p	—	.02	.10	.07	.001
Skipped one grade					
%	4.2	4.88	7.0	4.8	3.5
p	—	.77	.56	.76	.77
Test anxiety					
severe	0.0	**17.1**	**16.3**	**15.2**	**22.4**
p	—	.0015	.002	.002	.0005

A severe type of school problem is *requiring a home teacher*. This usually indicates that there were so many problems that the school would rather send a teacher to the child than the child to the school. Here none of the controls required a home teacher while 13 percent of all TS patients and 24 percent of grade 3 TS patients required a home teacher.

Flunk a Grade, Skip a Grade

If school was a problem, how often did TS children flunk a grade? Among all TS patients, 26 percent had flunked a grade, and among grade 3 TS, 38 percent flunked a grade, compared to 8.5 percent of controls. The converse of this is skipped a grade. Here 4.2 percent of controls skipped a grade while 4.9 percent of all TS patients and 7 percent of grade 1 TS patients skipped a grade. This is testimony to the fact that TS children are very bright and when they do not have ADHD or a learning disorder they may do very well.

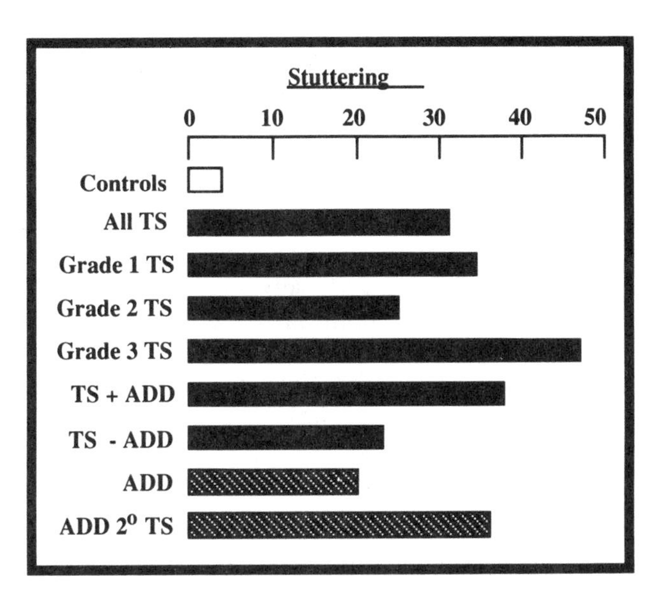

Severe Test Anxiety

One of the complaints I hear often is that a TS child may study at home and know the material well, then panic or freeze and flunk the test. This led us to ask about problems of test anxiety? Seventeen percent of all TS patients complained of severe test anxiety compared to none of the controls. This is consistent with the high frequency of panic disorder in TS[p175].

Stuttering — An Inherited Disorder

Stuttering has often been assumed to be a psychological disorder, caused by such things as childhood stress or attempting to change handedness. It is, in fact, a genetic disorder. Like TS, three times as many males stutter as females and 22 percent of the first-degree relatives of male stutters also stutter, compared to 5 percent of the general population[1041,1042]. Segregation analysis of families of stutterers indicate that stuttering is a genetic disorder, due either to a single gene or multiple genes[1041]. In our controlled study, 32 percent of all TS patients stuttered compared to 6.4 percent of the controls. The frequency of stuttering for all the groups is as follows:

Stuttering was significantly more common than in the controls in all except the ADD group. It was present in almost half of grade 3 TS patients. Obviously, *the Gts gene is one of the genes that causes people to stutter.* In family pedigrees, I often see first-degree relatives of TS patients who have had problems with stuttering but never had motor or vocal tics. This suggests that stuttering can sometimes be the only way a *Gts* gene is expressed.

DYSLEXIA

The ability to read is a critical skill in our complex civilization. Some children with no other apparent problems, are consistently poor readers. This has been termed *dyslexia*, *specific reading disability*, *congenital word blindness*, or *developmental reading disorder*[490]. To make a diagnosis the following must apply[491]:

> **Developmental Reading Disorder**
>
> Performance on standardized, individually administered tests of reading skill is significantly below the expected level, given the individual's schooling, chronological age, and mental age (as determined by an individually administered IQ test). In addition, in school the child's performance on tasks requiring reading skills is significantly below his or her intellectual level.

The difficult part is defining significant. Significant usually means two or more years behind grade level. However, since children in first and early second grade obviously can't be two years behind grade level, it is necessary to waffle a little for younger children. Specific problems include the reversal of letters and words, omission or insertion of words, slow reading, poor comprehension, many spelling errors that are not explained by attempts at phonetic spelling, and difficulty putting words into sentences. Many of these problems are common in children who are first beginning to read. It is the frequency, degree and persistence of these problems that is apparent in dyslexia. There is sometimes an overlap with symptoms of ADHD since many children with ADHD have dyslexia and many children with dyslexia have ADHD and behavioral problems. However, "pure" dyslexia without ADHD or behavioral problems is common. The above definition specifically excludes mental retardation and cultural deprivation as a cause of poor reading skills.

Dyslexia is a common problem, estimated to occur in 5 to 10 percent of children[156,158]. Like TS, ADHD, and stuttering, it is three to four times more common in males than females.

An Example

To illustrate the problem the following is a sample of a writing lesson the mother of a nine year old TS student gave to me. The sentences he was asked to write are written below his writing.

I wot to plea

I want to play

you orh the dest aunt

You are the best aunt

I lie to et salab

I like to eat salad

This shows the spelling errors and letter reversals common in dyslexic individuals.

The Genetics of Dyslexia

Although there had been sporadic reports of dyslexia occurring in families[425,1263,1870] the first comprehensive examination of genetic factors was published in 1950 by Hallgren from Sweden[840]. He examined 116 reading disabled children from Stockholm schools and found 160 additional cases in the families of these children. In many cases, three generations were found to be affected. He suggested that dyslexia was inherited as an autosomal dominant trait with reduced penetrance. To explain thesmaller number of affected females, he suggested that additional factors were influencing their decreased susceptibility. One problem with this study is that the diagnoses were not made blindly or by objective testing. As a result the known occurrence of dyslexia in one member of the family might influence making the diagnosis in another member.

The few twin studies that have been done are summarized as follows giving the **percent concordance for dyslexia** in identical versus fraternal twins:

Twin Study	Identical		Fraternal	
	No.	%	No.	%
Hermann & Norrie[883a] and Zerbin-Rudin[2150a]	17	100	34	35
Bakwin[90a]	31	84	31	29
Total No. and Ave %	48	90	65	32

The fact that the concordance rate for identical twins (90 percent) was 3 times that of fraternal twins (32 percent) strongly supports a genetic cause of dyslexia. A twin study of reading scores has shown a much greater correlation for identical than for fraternal twins[464]. Other studies have also emphasized the genetic role of dyslexia[464,465,604,885,1168,1439].

Linkage Studies in Dyslexia

Some preliminary studies have suggested that the gene for one form of dyslexia, occurring in 20 percent of patients, may be linked to the centromere region of chromosome 15[1816].

Another form may be linked to chromosome 6, and still another form is not linked to either of these chromosomes. This suggests there are at *least three different genetic types of dyslexia*. The report of five dyslexia families from Denmark that were not linked to chromosome 15[170] further suggests the presence of several genetic types.

Dyslexia as a Defect in Word Memory

There have been many theories of the cause of dyslexia including defects in vision, visual processing, or the inner ear. However, recent studies strongly suggest the sensory input is normal and the primary defect is in the proper storage of images of words in memory[2007]. In dyslexic individuals, words are stored without a complete set of phonetic clues and when dyslexic children are asked to recall a word, they cannot because they have not retained enough clues to do so. This suggests that treatment programs should emphasize one on one tutoring, remedial instruction or compensatory training rather than various exercises or procedures designed to improve the eyes or ears.

Reading Problems in Tourette Syndrome

Many parents of our TS patients complain that their children have problems learning to read. In some families dyslexia is present in multiple members. Some may also have motor or vocal tics while others only have dyslexia. An example of such a family is shown on the **next page**.

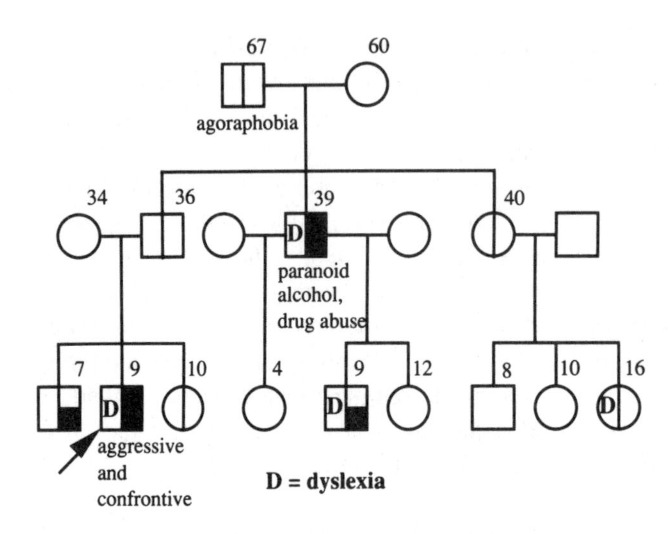

Here in addition to the propositus (arrow) an uncle and two cousins had dyslexia. The two males with dyslexia also had motor or vocal tics, while the female had only dyslexia.

In our controlled study, although we did not have the resources to individually examine each TS child and each control for dyslexia, we were able to get a good approximation of the frequency of this problem by asking, "Did you ever have frequent problems with any of the following?"

1) Letter reversal (p for q, b for d, etc.),
2) Number reversal,
3) Word reversal (saw for was, etc.),
4) Drop or insert words while reading aloud,
5) Read very slowly (word by word) when your peers were reading at a normal speed, and
6) Unable to retain the meaning of what you just read.

All of these problems were significantly more common in TS than controls, except for dropping or inserting words.The most severe problem in all groups was *poor retention,* with 42 percent of TS patients having poor retention compared to only 8.3 percent of the controls (p <.0005). This problem increased to 55 percent in grade 3 TS. The cause and significance of this poor retention will be discussed in Chapter 51 on frontal lobe disorders.

To examine how often patients had more than one of these problems we made a reading problem score, where each "no" answer to the above six questions was given a 0, and each "yes" answer a 1. Adding these together the minimum reading problem score was 0 and the maximum was 6. The distribution of these scores (**see below**) showed that the controls had a score of 3 or more, compared to 27 percent of all TS patients:

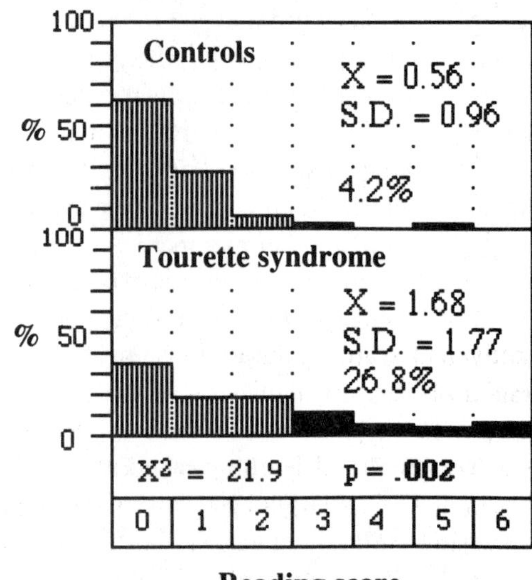

Reading score

A reading problem score of three or more was present in 21 percent of grade 1, 26 percent of grade 2 and 35 percent of grade 3 TS patients. For TS patients with ADHD, 38 percent had a reading problem score of three or more compared to only 9.6 percent of those without ADHD.

The final arbiter of whether children are having learning or school problems is to simply ask if their performance in grade school and high school was below average, average, or above average for math, reading and writing. The results showed that TS patients had significantly more problems in all three areas compared to controls[360].

An Individual Family

The following pedigree provides an example of how learning disabilities (**LD**) and ADHD (**H**) frequently occur in the relatives of patients with TS.

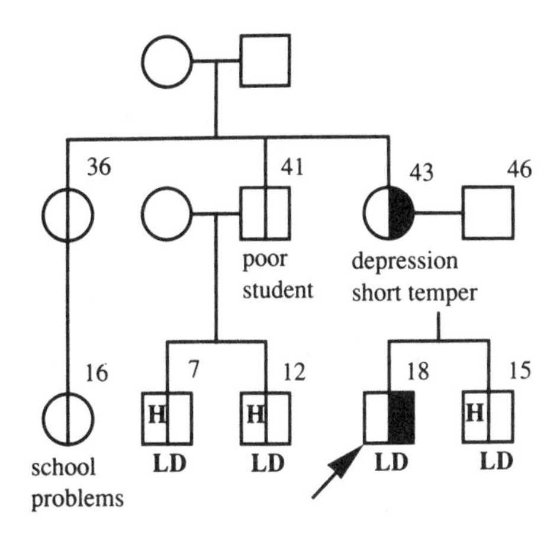

The teenager with TS (arrow) in this family had learning disabilities and was in special classes all of his school career. His younger brother had ADHD, presumably due to a *Gts* gene, and learning disabilities. His mother had TS and lifelong mood swings with depression. Her brother was a poor student and both his children had ADHD and learning disabilities. Her sister's daughter had multiple problems in school. Each of these relatives with learning problems probably carries the *Gts* gene.

Dyslexia, Aphasia, and Learning Disabilities Preceding TS

When Jerry was 5 years old he was saying only a few words. His teachers insisted that he repeat nursery school and be involved in speech therapy before starting kindergarten. Extensive testing at age seven showed *severe dyslexia, visual-perceptual and auditory-perceptual disability, poor articulation and ADHD*. When he finally entered the first grade at age eight, he had trouble keeping up with the other children and was placed in a special class. Because Jerry's father was physically abusive, his mother divorced him and she and Jerry moved to the west coast. At age 10 he again had an extensive battery of tests. These reported that he had *"receptive and expressive aphasia with disnomia and articulation problems, visual-spatial disorientation, poor motor coordination, poor memory, severe adjustment reaction of childhood and poor auditory discrimination."* Translated this means he had difficulty hearing, seeing, writing and expressing words. At age 11 Jerry and his mother were planning to visit his father in their hometown. One week before they left he began to have severe jerking of his head, facial grimacing, jerking of his arms and legs, sniffing, and high pitched noises in his throat. These persisted throughout the trip and after he returned. A family history showed that his father had eyeblinking and picking at his skin for years, and the short temper and the violent disposition described above.

The message of this case is that *many of the sensory perception and motor coordination problems common in children with TS may precede the onset of motor and vocal tics by many years*. The neurological basis of these problems is discussed in Part IV.

Summary: Individuals with Tourette syndrome may have significant learning disabilities throughout their school years including difficulties with reading, writing and mathematics, and visual and auditory perception problems. They often have dyslexia and problems with retaining information. A Gts gene can express as dyslexia alone, and a percentage of individuals with dyslexia may carry a Gts gene.

Chapter 22

Obsessions, Compulsions, Repetitive and Ritualistic Actions

Other than the motor and vocal tics and ADHD, obsessions and compulsions are the most frequent behavioral manifestations of a Tourette syndrome gene. Such behaviors have been reported in 30 to 90 percent of individuals with TS[358,363,365,440,589,590,627,784a,1149,1353,1360,1397,1398,1473,1483,2022a] and may be present without tics particularly in women carrying the *Gts* gene[365,1473]. Studies in five cities indicate that 1.2 to 3.3% of the population have an obsessive-compulsive disorder[1015a].

Compulsions

Compulsions are repetitive, seemingly purposeful behaviors that are performed in a stereotyped fashion. The acts are done with a sense of subjective compulsion coupled with a desire to resist them (at least initially). Although the individual usually recognizes the senselessness of the behavior it may provide a release of tension[490].

Some of the most common compulsive symptoms in TS are shown in the following list:

Compulsive behaviors:

touching things just right
eating
constant fiddling with objects or clothes
dressing perfectly
erasing mistakes
if something is touched with one hand having to touch it with the other (**even up**)
placing objects in just the right place
pulling hair out
pulling up socks
rechecking things many times
smelling things
spending
talking
touching objects
touching things a exact number of times
washing food before eating
washing hands until raw

Many of these deserve special comment.

Touching objects Many of the compulsions in TS center around not only touching objects and people but the manner in which things are touched. Parents often complain their TS children cannot keep their hands off things. They enter a store and must finger everything. While this can be a normal childhood behavior, in TS it may be carried to the extreme.

A mother reported,

> "I couldn't imagine what made him want to touch everything. As he walked he touched the ground, and at school, he constantly reached out and touched the floor by his desk. When the students were lined up at school he would keep touching them. They would hit him in return, since they thought he was just

being pesky."

"At age 5 I became obsessed with beach balls and wanted to sit on them naked. They had to be 22 to 24 inches in diameter and blown up tight to make them firm. They reminded me of the female form, especially breasts."

Samuel Johnson's compulsive habits included never walking in the cracks of paving stones, and touching every post along the street as he walked. If he missed a post he would keep his friends waiting until he went back to touch it. He would also go in or out of a passage by a certain number of steps from a certain point, or at least so that his left or right foot constantly made the first movement when he came close to a door[1288,1382].

The touching is not always with the hands.

"I remember touching my nose to the desk at school and to all of my books. I tried to suppress it, but the urge was there. Even now, if I think about it, the urge arises and I catch myself touching things to my nose."

Counted touching Touching is often done a certain number of times, often by two's until "it feels just right."

"I have this compulsion to touch the ground two times. If it wasn't right I would touch it two more times, then two more, sometimes up to 32 times before it felt right."

Even-up A common feature of the touching is the necessity to even-up, that is, if something is touched with one hand it must also be touched with the other. While this usually simply involves touching things, it may sometimes become more elaborate.

"When I am playing football if I spin around to the left three times as I fall down I have the get up and spin myself around three times to the right."

"He goes twice around a street light and has to turn around and come back the same way."

The ultimate in even-up compulsions was a 16 year old boy who began to take an interest in helping his mother load the dishwasher. Since this was refreshingly out of character his mother was delighted. Soon, however, he insisted on doing it all himself and each plate and spoon on the left had to be balanced by a plate and a spoon on the right. This got to be so bad that he would take clean dishes out of the cupboard to make things balance.

Touching oneself Since the object most consistently close at hand is one's own body self-touching is understandably common. This may be of any area, especially the face, arms and crotch.

Clothes A frequent feeling of TS patients is that their clothes are too tight or don't fit right. Thus, a common tic or compulsion is to pull at their clothes.

"His shoe laces have to be perfect so the shoe is just the right amount of tightness. He can't have wrinkles in his clothes or bed. He once went berserk because he couldn't get a wrinkle out of his bed."

Sexual touching of oneself The most frequent example is touching oneself in the crotch. This is particularly common in boys and may occur so often that their pants become discolored in the crotch area. This is one of the many TS compulsions that may lead to negative comments by teachers and peers. In extreme examples, the touching becomes a more violent obvious hitting.

"He hits his penis because he gets a strange feeling there and it makes him feel better."

One particularly severe 20 year old male walked into the office screaming "fuck, fuck, fuck" while pounding himself in the crotch. He had such difficulty controlling this that he was afraid to leave the house.

Touching other people The urge to touch other people can present social difficulties which patients may go to elaborate lengths to cover up.

> "I would have an urge to touch a stranger at a very specific place on their back. To cover it up I would pretend I tripped-then touch them. Sometimes I would follow a person for blocks until I found the right opportunity to cover up touching them"

A 25 year old girl recounts her very first TS symptom,

> "I was 3 years old and in the bank with my mother. I pulled away from her and ran to a man standing at the counter and pushed at his feet. I said to him, 'You *have* to stand with your feet together.'"

Sexually touching others TS boys may touch their mothers or sisters on the breast, buttocks or pubic area. The urge to touch others and sexual touching may manifest itself as an exceptionally "touchy-feely child."

Self-destructive impulses These may take many forms from constantly picking at one's skin or sores to self-hitting or overtly dangerous activities. The picking at one's skin is so common that we have a rule that *any skin lesion in a TS patient is self-induced until proven otherwise.*

The idea that something has to hurt before the tension is relieved is common. It is most often expressed as pinching or hitting oneself but may take more perverse forms.

> "If the stove was hot I had to put my finger close to it and then touch it to get a relief of tension. Touching something that didn't burn didn't help."

> "I would touch the tip of a knife then press until I drew blood."

A potentially suicidal compulsion was closing one's eyes while driving.

> "I would have an urge to close my eyes while driving and see how long I could keep them closed. This scared my girlfriend so much she refused to let me drive."

One variation of this was:

> "My eyes would dart back and forth off the road. This would only occur in the most dangerous situations such as going around a curve on a mountain road. Sometimes, if I was driving next to a large truck in my sports car, I would have an urge to just drive right under it."

Or,

> "I kept thinking of pulling the steering wheel to one side while driving."

Nail biting Nail biting is a common symptom and is sometimes so severe that the fingers bleed. This may be the first symptom to disappear upon treatment with Haldol. On a follow-up visit one mother excitedly exclaimed,

> "He stopped biting his nails for the first time in years."

Thumb sucking Thumb sucking is such a common behavior in young children that it might seem strange to be on this list. However, when it continues far past the time when it is no longer age appropriate, it takes on the characteristics of a compulsive behavior. Some TS children have problems with compulsive thumb sucking into grade school and must wrestle with the social ostracism this causes.

Cracking knuckles This is also common and often goes away after treatment.

Vomiting Periods of repeated vomiting without any physical cause may occur in some TS children. It is usually associated with periods of stress.

Mimicking This can take several forms. **Echolalia** is the parrot-like repetition of words or phrases that other people have said. A variant is **palilalia** or the repetition of one's own words. **Echopraxia** is the compulsive repetition or mimicking of other peoples' physical actions. These can sometimes be all used together in a pernicious compulsion to mimic everything another person does.

"Between age nine and ten he would mimic everything his sister said. He would repeat her sentences several times, then go to his room and continue to repeat them, sometimes for 45 minutes, until he felt he could exactly mimic every inflection in her voice. From age 11 to 12, he then began to mimic his father in the same way, ignoring his sister. Finally, from age 13 to 18 he mimicked his mother, ignoring both his sister and father. He says he began doing this because he didn't like the tone of her voice."

"He compulsively traces over numbers and makes crosses and erases them, and moves his chair ten times before sitting down. He follows his brother around and imitates his movements."

Stealing The subject of stealing will be discussed more in the chapter on conduct. However, it is clear that much of the stealing is compulsive in nature. This is well illustrated by the following case:

A 15 year old male with TS began stealing at the age of 12. He would steal his sister's earrings, but only stole one of each set. He would use the pin on the earring to attempt to pierce his own ears. When that didn't work he would place the single earring on top of the bookcase in the living room, then steal a single earring from another sister and try that one. Eventually he had a set of single earrings from all his sisters and his mother. He would then start stealing a new set.

Obsessions

Obsessions are involuntary, recurrent, persistent *ideas, thoughts, images or impulses* that invade the consciousness and are usually experienced as senseless or repugnant[490]. The following is a list of some of the most common ones seen in TS patients.

Obsessive thoughts in TS

- about excretions
- about sex
- about violence
- counting by two's
- counting objects
- mental coprolalia
- spelling things forward or backward

These can best be explained by a series of examples.

Narcissistic obsessions The original definition of narcissism came from the mythological story of Narcissus who couldn't stop staring at his reflection in the pond. The following compulsion is strikingly similar:

"One day I suddenly became concerned about my looks and three seconds later it became an obsession. I began to stare at mirrors for hours each day. When I went to the library I would spend five hours there and get no studying done because all I could think about was my looks."

As the above example shows, many of the obsessions in TS center around one's self-image. In the following example one boy was concerned about being too small:

"He was on the football team and got the idea he would get bigger by lifting weights. He attacked this compulsively, doing it every spare moment. We told him they might make his muscles larger but would not make him grow bigger. There was no reasoning with him and he became frustrated that he wasn't betting bigger."

The mother of a 15 year old girl relates,

"She was a compulsive exerciser and did aerobics, sit ups, swimming, and weight lifting. She would do each a certain number of times and the next day tried to do more, always looking at the clock. If something stopped her before she did it a certain number of times she would throw a temper tantrum. She later became anorexic and had to be hospitalized twice to save her life."

Cleanliness and order

"She is very fastidious. Everything has to be in just the right place."

"He erases his school work so compulsively there are holes in his paper."

A 6 year old boy was described by his mother,

"He is a perfectionist and if things don't go just right he has a tantrum."

Obsessions about violence are frequent.

"In one of my frequent obsessions I would have vivid ideas of slitting the bank teller's throat. The entire scene was played out in my mind including all the hysteria afterward."

"I am always thinking something bad is going to happen, like my sister is going to die."

"I keep thinking about holding my baby under the water."

"He constantly thinks about death."

We have followed a 26 year old girl with severe TS for a long time. She kept telling me that she had vivid obsessions about violence and only after many years did I finally convince her to write some of them down for me.

"I am standing in a large cell and about seven or nine policeman come in and form a circle around me. They each are holding a rubber hose or truncheon in one hand. The one facing me comes forward while I start backing up. When he pauses to aim a backhand smack at me, I back off (my hands are still cuffed behind my back). Whereupon one behind me smacks me and knocks me down. Then they all begin beating me and kicking me and I edge over to the corner of the room or cell, facing the corner in a futile attempt to protect myself from the blows. They sometimes smack me so as to cause my back to arch and my front becomes exposed.

"Soon I am partly walking, mostly dragged, into a large room that is full of torture devices. I am lifted onto one and tied, or chained down to it, spread eagle, and lying on my stomach. Then I am branded with an iron all over my back.

"Afterwards I am hooked up, this time lying on my back and tied, like before, spread eagle, to a mechanism containing all kinds of wires and alligator clamps. They are hooked up to all parts in sensitive areas of my body, except for my head (my head is to be protected so they can successfully get whatever information from me, without my knocking out on them)."

An analyst would have a field day with that.

Knives Some TS children have an obsession with knives.

"He loves knives. He wanted to make one 20 inches long with razor blades in the side."

"He can't walk past a knife without picking it up. He has threatened his sister with a knife and collects and stockpiles them in his room."

"She has had a preoccupation with knives since she was 13 [now 16]. I have found pocket knives in her purse, her dresser drawer and the nightstand. We came home once to find a large butcher knife on the table near where she was sitting. It was, 'to protect herself.'"

This fascination with knives has been noted by others[512,1523,1625].

Obsessions about sex Although everyone has sexual thoughts, they become obsessions when they are constant and intrusive and as psychiatrists put it "ego-dystonic," that is, different from what that person feels is normal for them.

"I have this obsession my father is going to rape me. I know it is silly but I can't get it out of my mind. I also keep thinking my mother is going to tickle me. It is so bad I can't stand to look at her hands and insist that she keeps her hands below the dinner table."

"He talks about masturbation over and over like a stuck record."

A 21 year old male with multiple motor and vocal tics recounts his development of obsessions and compulsions,

"A age seven I started having rapid eyeblinking and a bouncing walk. At the same time I developed a desire to hurt little girls by hitting them during recess, but I never acted on it. At 14 I began to compulsively wash my hands until they began to crack and bleed. At 16 I became a 'religious freak' and tried to lead a perfect life. In a vain attempt to keep evil thoughts away I would rake my fingernails over my skin until it bleed. I became obsessed with hurting myself and would press on my eyes or throat until I felt pain. I also became obsessed with evil bizarre thoughts which came over me for no apparent reason, 'thoughts so strange no normal person would conjure them up.' Some attacks would last for three hours. One I particularly remember is thinking how long it would take to commit a particular act and how many people I could hurt. For example, I thought it would take 15 minutes to disguise myself, 15 minutes to drive to the location, 15 minutes to find the person, 15 minutes to rape them, and 15 minutes to call the police and escape. I sometimes wanted to commit suicide just to control these thoughts."

Scatology (noun. Preoccupation with excrement or obscenity.)

"He is obsessed with farts. After farting he will laugh for hours. He even records them and plays them back, laughing."

Mind racing A frequent complaint of TS patients is that their mind is always racing. Sometimes this interferes with sleep, speech and other activities. There is a fine line between racing thoughts and intrusive thoughts, i.e. obsessions.

"My thoughts race and I say the first thing that comes into my mind. As a result my conversation seems to jump all around. People think I am sort of flaky."

"My thoughts come so fast I can't seem to grab onto any one of them."

"My brain works so fast that my mouth can't keep up. As a result I seem to mumble and people are always saying to me, 'What did you say?'"

Numbers TS patients are often obsessed with words and numbers. Counting things is a frequent obsession. They count everything including cracks in the floor, tiles in the ceiling, words in sentences, letters in words, steps, etc.

Words "I am constantly spelling things backwards."

"When I hear certain words I have to put them into the form of a star in my mind."

"Everytime I write an O, A or I, I have to put a dot over it."

"I have to say rest in peace everytime I hear someone died, even if I just read it in the paper."

Sometimes the obsessions with manipulating words are so severe that reading becomes impossible.

"He has to spell out every second word. As a result he hates reading."

Writing One father said of his obsessive compulsive son:

"He has many sheets of paper in his room that he fills with rows of dots. I have seen over 100,000 dots. If we are in a hurry to leave the house and ask him to stop he gets furious that he was interrupted. He says, 'Wait, I have to finish.'"

Food and eating "If I have any compulsion it is eating."

This can be such a problem that an entire chapter has been devoted to it. [See Chapter 36.]

"Every time I take a drink of water the last swallow has to hit just right."

Textures "If the food doesn't have just the right texture he refuses to eat it."

"Because he doesn't like the texture of the bags, he refuses to help bring in the groceries."

"I am very sensitive to the texture of things."

Perseveration refers to the constant repetition of a thought, phrase, question or action. It is like a palilalia gone berserk. In its extreme form it can be one of the most annoying symptoms of TS, driving parents, teachers and peers to distraction. The following are some examples:

"He began to ask questions over and over. They had to be answered in a certain way or he would keep asking. For example, if he asked how did I do at the baseball game and I answered 'OK,' he would say, 'What do you mean.' He kept asking until the questions were answered in excruciating detail."

"He would ask the same question many times without listening to the answer."

Pedigrees

In addition to the above anecdotal information, we can also learn a great deal about obsessive-compulsive behavior in TS by examining family pedigrees. It is clear that TS patients themselves have a lot of obsessive-compulsive behaviors, but what about relatives of TS patients? The existence of first-degree relatives who are obligatory carriers of a *Gts* gene (have a child and a parent or sibling with TS), and have severe obsessive-compulsive behaviors, but no motor or vocal tics, provides additional evidence that the *Gts* gene can manifest as obsessive-compulsive behavior only (i.e., incomplete expression of the *Gts* gene). The following pedigrees are illustrative examples of this phenomena.

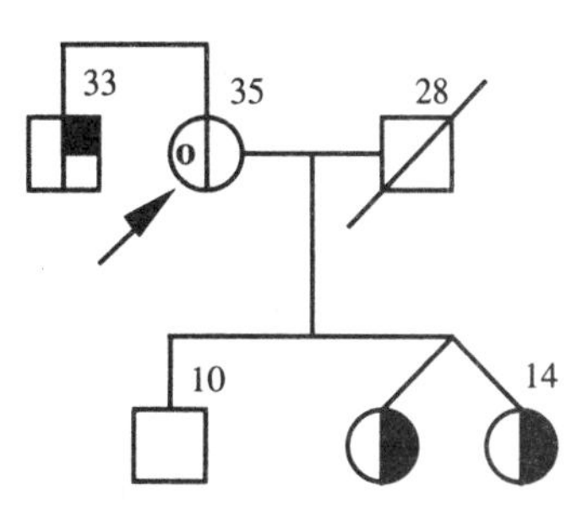

o = obsessive compulsive

The mother of these identical twin girls with TS is an obligatory carrier since she also has an affected brother. She has an interesting combination of obsessions and compulsions. She has a compulsion to fold and unfold the pages of books and can sit doing this for hours. If she goes somewhere without a book she feels very uncomfortable until she gets one in her hands again. In addition, she has an obsession with dates and will fix on the day's date and say it over and over in her mind hundreds of times. She combines the compulsion and obsession by carrying around a date book with her at all times. She has also had problems with severe depression, mood swings, premenstrual tension, and panic attacks[364].

The 36 year old mother (arrow, **next page**) of a daughter with TS had eyeblinking tics for a few years as a child. If she touches one finger she has to touch them all, and if she touches one tooth with her tongue she has to touch them all. She also has frustrating obsessive thoughts in

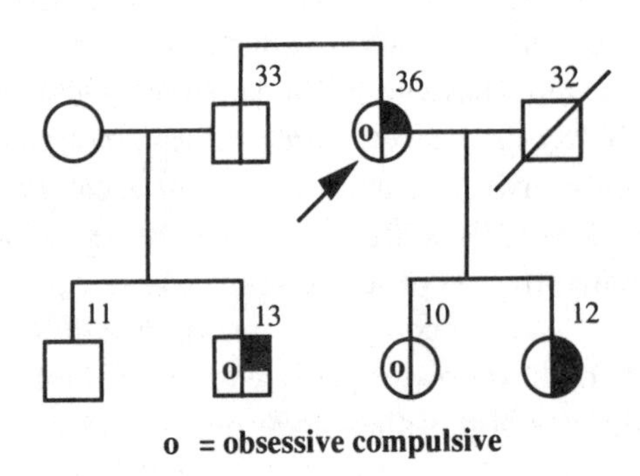

the form of drawing in her mind a square with a cross inside; however, she also has a rule that she cannot lift the imaginary pencil from the imaginary paper and can go over a given line only once. Since this is impossible she dwells on it constantly, 50 times a day. Her ten year old daughter has to count to ten many times a day. Her brother's 13 year old son has finger tics and takes many small steps and counts them[364].

Controlled Studies

Despite the large number of reports of obsessive-compulsive behaviors in TS patients, there have been few studies that included control groups. Interestingly, the first such study concluded that there was no increased prevalence of obsessive-compulsive behavior in Tourette syndrome[1767,1769,1771]. However, while the control group consisted of non-TS patients, they were all from a psychiatric clinic and also had a high frequency of obsessive-compulsive behaviors, and the test instruments were not specifically designed for the elicitation of obsessive-compulsive symptoms. By contrast, Frankel and co-workers[627] using an extensive obsessive-compulsive disorder inventory examined 63 Tourette syndrome patients and 41 *normal* controls. They found obsessive-compulsive behaviors in 52 percent of TS patients compared to 12.2 percent of the controls.

The results of our controlled study of compulsions and repetitive behaviors in TS are summarized in the following box[363]:

Compulsive and Repetitive Behaviors in Tourette Syndrome

Behavior	TS (%)	Cont (%)
Shouting inappropriately	40	4
Excessive touching	39	2
Palilalia	33	0
Counting things	32	9
Echolalia	30	2
Excessive touching others	29	0
Mimicking others	25	0
Biting or hurting self	23	0
Rocking	21	2
Touching a # of times	19	2
Head banging	12	2

Further details about the frequency of these different symptoms in the three grades of TS are given in the following table:

Compulsive and Repetitive Behaviors in TS and Controls

	Controls	Gr 1	Gr 2	Gr 3
N	47	43	145	58
Echolalia				
Yes (%)	2.1	9.3	**29.0**	**48.3**
p	—	.10	<.0005	<.0005
Palilalia				
Yes (%)	0.0	**9.3**	**35.2**	**46.5**
p	—	.03	<.0005	<.0005
Shouting inappropriately				
Yes (%)	4.2	**27.9**	**36.5**	**56.9**
p	—	.0015	<.0005	<.0005
Touching things excessively				
Yes (%)	2.1*	**18.6**	**37.2**	**56.9**
p	—	.0015	<.0005	<.0005
Touching things a certain number of times				
Yes (%)	2.1	7.0	**19.3**	**27.6**
p	—	.26	.002	<.0005
Touching others excessively				
Yes (%)	0.0	**11.6**	**29.0**	**39.7**
p	—	.02	<.0005	<.0005

	Controls	**Gr 1**	**Gr 2**	**Gr 3**
Touching sexually				
Yes (%)	4.3*	**20.9**	**24.8**	**37.9**
p	—	.002	.0006	<.0005
Biting or hurting self				
Yes (%)	2.1*	**14.0**	**22.1**	**32.8**
p	—	.009	<.0005	<.0005
Head banging				
Yes (%)	4.3*	9.3	**18.6**	**24.1**
p	—	.15	.006	.001
Rocking				
Yes (%)	4.2*	7.0	**17.9**	**37.9**
p	—	.27	.008	<.0005
Mimicking others				
Yes (%)	0.0	**9.3**	**25.5**	**34.8**
p	—	.03	<.0005	<.0005
Counting things				
Yes (%)	8.5	16.3	**33.8**	**41.4**
p	—	.26	.001	<.0005

* includes the TS patient among the controls

Several generalities can be made.

First, all the symptoms listed were very significantly more common in the TS patients than controls.

Second, all symptoms increased in frequency with increasing grades of TS. For some symptoms, such as echolalia, the frequency varied greatly between grade 1 (9.3 percent) and grade 3 (48.3 percent). This indicates why selective bias in obtaining patients can make such a difference in the published frequencies of any symptoms in TS patients. If only severe patients are present in a series, a given symptom is likely to be more common than if many mild cases were present. The complete tables are published elsewhere[363] and also show that these symptoms were more common in TS patients with ADHD and were also increased in patients with pure ADHD.

In making a diagnosis of **obsessive behaviors** by standardized criteria[490,1630-1632] we ask about *unpleasant thoughts* and *silly thoughts*. The following two questions were asked[1630]:

1. "Have you ever been bothered by having certain unpleasant thoughts all the time? An example would be the persistent idea that you might harm or kill someone you loved, even though you really didn't want to. Have you ever been bothered by that or by any other unpleasant and persistent thought?"

2. "Other thoughts that keep bothering some people, even though they know they are silly, are that their hands are dirty or have germs on them no matter how much they wash them, or that relatives who are away have been hurt or killed. Have you ever had any kind of unreasonable thought like that?"

To gauge the severity of these thoughts we also asked if they had been consistently present for three weeks or more, and did these thoughts keep coming into your mind no matter how hard you tried to get rid of them? The results are summarized in the following box:

Obsessions in Tourette Syndrome

Behavior	**TS** (%)	**Cont** (%)
Obsessive unpleasant thoughts	31	6
Last three weeks or more	15	6
Cannot get rid of	22	6
Silly or unreasonable thoughts	20	6
Last three weeks or more	9	4
Cannot get rid of	11	4

As the box shows, 31 percent of TS patients have obsessive unpleasant thoughts and 22 percent felt they could not get rid of them. Both of these are significantly different from controls.

The occurrance of these symptoms in the three grades of TS patients is shown in the following table:

Obsessions in Tourette Syndrome and Controls

	Con.	Gr 1	Gr 2	Gr 3
N	47	43	145	58
Obsessive unpleasant thoughts				
Yes (%)	6.25	**20.9**	**27.6**	**46.5**
p	—	.04	.001	<.0005
Last three weeks or longer				
Yes (%)	6.25	7.0	13.1	**25.9**
p	—	.75	.20	.008
Cannot get rid of thoughts				
Yes (%)	6.25	11.6	**19.3**	**37.9**
p	—	.37	.04	<.0005
Silly or unreasonable obsessive thoughts				
Yes (%)	6.25	**23.6**	15.7	**25.7**
p	—	.02	.10	.009
Last three weeks or longer				
Yes (%)	4.2	4.7	6.1	**20.7**
p	—	.91	.62	.014
Cannot get rid of thoughts				
Yes (%)	4.2	9.3	7.6	**22.4**
p	—	.33	.42	.008

In addition to the frequency of individual symptoms, another dimension is whether an individual has one or several obsessive-compulsive behaviors. To examine this we made an **obsessive-compulsive score** consisting of the following variables: touching things excessively, touching things a certain number of times, touching others, touching sexually, biting or hurting self, mimicking others, counting things, and all six of the obsession variables from the above table. The results of the obsessive-compulsive scores for all TS patients versus controls were as **shown on the right->**:

Of all TS patients, 45.4 percent had a score of four more greater and the scores ranged to the maximum of 15. By comparison only 8.5 percent of the controls had score of four or greater and none had a score of more than five. Statistically these were enormously different.

These results showing obsessive compulsive symptoms in 45 percent of TS patients compared to 8.5 percent of controls are very similar to those of Frankel and coworkers[627] of 52 percent for TS patients compared to 12.5 percent

The distribution of the obsessive-compulsive score in the three grades of TS is **shown on the next page**. Here 28 percent of grade 1 TS patients, and 69 percent of grade 3 TS patients, had a score of greater than three.

Multiple Obsessions and Compulsions

The difference between an occasional obsessive-compulsive behavior and the presence of multiple symptoms in some TS patients is illustrated in the following report by a very bright 27 year old TS patient:

> "At age 21 I began to develop obsessive thoughts about many things. These included a) constant dwelling on math especially long division, b) constant thoughts of how traffic lights work, c) constant thoughts of how fractions increase and decrease with changes in the numerator and denominator, d) a desire to keep my teeth in perfect order, e) buying things like a bicycle just to take it

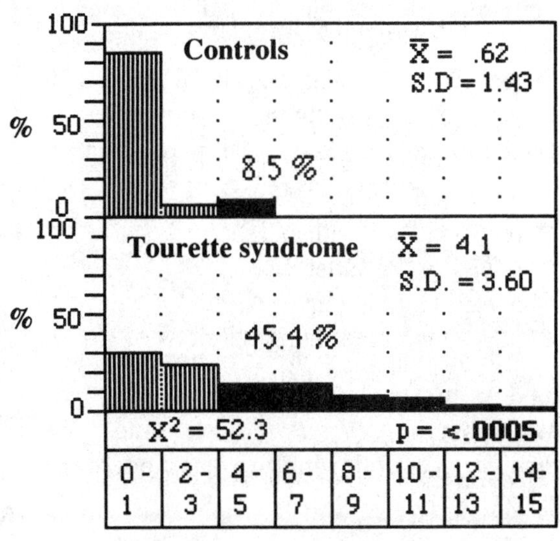

Obsessive - Compulsive Score

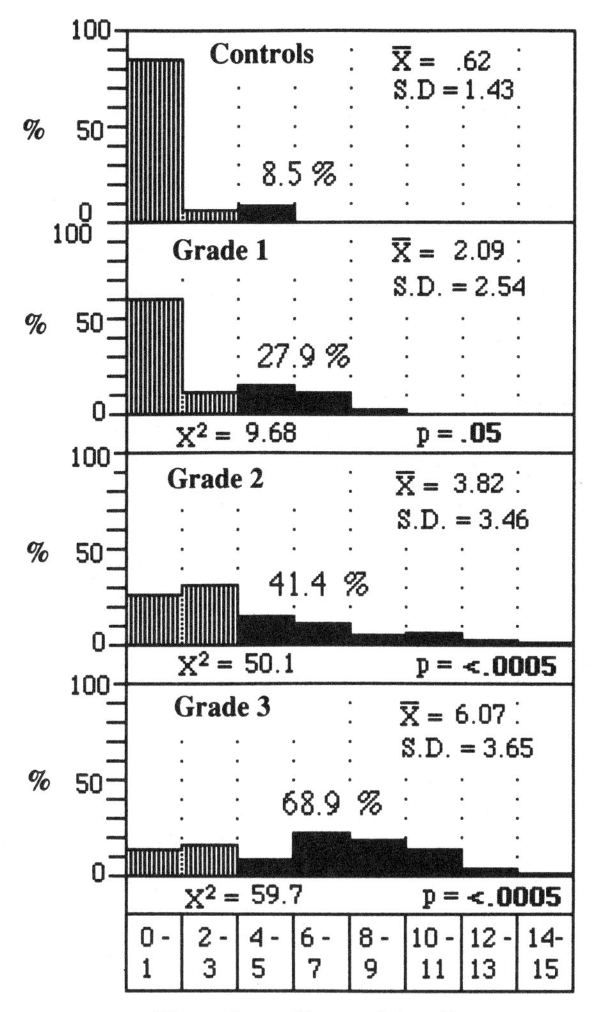

Obsessive - Compulsive Score

apart and see how it works, f) worrying about whether traffic lights really went green and backing up to check, and g) a preoccupation with measurements of all kinds. In the latter case, I would find myself climbing up to the top of buildings to measure the dimensions of the structures, taking exact measurements for hours. Someone once asked me why I was doing this and I replied, 'In case my future employer or wife wants the information.'"

The fine line between complex tics and ritualistic compulsions The list of complex tics[p14] indicates the similarity between complex tics and compulsive ritualistic actions. When does a multiple string of tics, done in a definite reproducible manner, become a ritualistic compulsion? The similarity of the two is illustrated in this lucid account by a patient[363].

"During the last ten years my tics have become quite sophisticated and complicated and are almost impossible for me to articulate. Part of the reason for this is that so many different tics occur together. They come in groups. Oftentimes three or four different tics accompany each other and occur in a matter of seconds.

"As an example, and a very elementary one: I am brushing my teeth. The need to use my right hand to hold the toothbrush causes me to begin jerking my right hand. This in turn triggers my whole right arm to begin flinging outward. While this is happening I am blinking my right eye and opening my eye wide. Also I am bouncing my head up and down while I am repeating, often times aloud and in groupings of four, the words, 'Amen, Amen, Amen, Amen' and then another series, over and over. I also am tapping my right temple with my second and third fingers, again in some series of four, and pounding my hands and repeating 'Amen.' They are somehow all coordinated.

"Another reason my tics have become so difficult to articulate is that in the last ten years I have developed what I call 'mental tics.' These are quite different from the motor tics and verbal tics. These are the tics that drive me to seek help. They are immeasurably more complicated and intricate. I do not really know how to describe them. They are like rituals, but rituals that have somehow become actual motor tics. I have watched this development of my tics with fascination. What was once merely a ritual that was used to either comfort me or ward off danger, has become a tic which I feel I must do as much as eyeblinking. That is, the ritualistic reason that was originally associated with it is no longer there. The tic has now become one which I *physically* feel I must do, not mentally."

The similarities between pure obsessive-compulsive disorder (without motor or vocal tics) and Tourette syndrome have been emphasized by Cummings and Frankel[440] and are listed as follows:

Similarities between Tourette Syndrome and Obsessive-Compulsive Disorder

1. Onset in childhood, adolescence, or early adulthood.
2. Lifelong course.
3. Waxing and waning course.
4. Involuntary, intrusive, ego-alien behaviors.
5. Bizarre sexual, aggressive, and scatalogical (feces) themes.
6. Occurrence in the same families.
7. Worsened by depression and anxiety.
8. Respond similarly to some types of therapies such as clonidine.
9. May occur in the course of other neurological conditions such as postencephalic states, amphetamine intoxication, and levodopa administration.
10. Have an increased incidence of neuropsychological and EEG abnormalities.

We know from the genetic studies that the *Gts* gene may be expressed as obsessive-compulsive behavior without tics. The tics seem to predominately involve the neurotransmitter dopamine, while the neurotransmitter serotonin is more involved in obsessive-compulsive behaviors (see Part IV).

Tics in Obsessive-Compulsive Disorder

The emphasis above has been on the high frequency of obsessive-compulsive behavior in patients with tics. The opposite has also been reported, a high frequency of tics in patients with a primary diagnosis of obsessive-compulsive disorder. Of 16 adults with obsessive-compulsive disorder, 69 percent either had a history of tics or a family history of tics[797,1520a]. In Pierre Janet's classic book *Les Obsessions et al Psychasthénie (Obsessions and Psychasthenia)* written in 1903, he described the frequent presence of tics in his patients[1520].

Psychoanalytic Theory versus a Genetic Cause

Classical psychoanalytic theory has provided a rich lode of varying explanations for obsessive-compulsive behaviors. Much of this has derived from Freud who believed that most psychological symptoms were produced by blocked or repressed drives primarily of a sexual or aggressive nature. The following is a typical psychoanalytic explanation of obsessive-compulsive behavior:

> ". . . an obsessive-compulsive neurotic might unconsciously want to do away with his burdensome wife and children. This wish might be manifested, every time he left his home, by the need to return to make sure that the gas stove had been turned off and the door locked. He would return immediately to check, turning the gas on again and making sure he had turned it off, and likewise unlock and lock the door. Ostensibly, these compulsive acts gratified his wish to obliterate the family (he was going through the motions of administering lethal gas and leaving the family vulner-

able to thieves) and at the same time kept the wish from consciousness by symbolizing extra care"[2056,p28].

Other explanations are,

"... the origin of obsessive-compulsive behavior is believed to rest in the early struggle of the growing child between his drives for omnipotent self-assertion and the necessity to conform to the demands of his parents in order to maintain their love and respect. The early period of toilet training initiates the conflict. Often the latter obsessive-compulsive individual is one brought up by a rigid compulsive mother, who insistently demands compliance and threatens both loss of love and various forms of punishment for failure to behave properly. This train of behavioral responses, derived from the long-continued relationship with such parental figures, becomes internalized and unconscious so that each thought reflecting rage or hostility sets in action the psychodynamic train of ambivalently balanced action and counteraction symbolizing rage, gratification, compliance and expiation"[1074].

An alternative, less contorted explanation, is that most obsessive-compulsive behaviors are due to a defective gene resulting in an imbalance in a neurotransmitter[p441]. The high frequency of obsessive-compulsive behaviors in TS, a clear genetic disorder, provides very strong evidence for the genetic, neurochemical cause of these behaviors. In a summary of studies of obsessive-compulsive behavior in 93 twins[303,1968], 70 percent of identical twins were concordant compared to only 15 percent for fraternal twins. Both of the concordant identical twin pairs reported by McGuffin and Mawson[1286] were separated prior to the onset of symptoms and neither were aware of the others problems. Despite this, the obsessive-compulsive symptoms started at similar ages and followed a similar course in both pairs. The fathers in both sets of twins were compulsively neat. Interestingly, one of the twins had childhood tics and two of the four sets of identical twins with obsessive-compulsive disorder described by Inouye[947] had motor and vocal tics and coprolalia — indicating they had Tourette syndrome. An analysis of obsessional traits in twins[419] showed the strong interaction between genetic and environmental factors[322a].

In a study of 70 children or adolescents with obsessive-compulsive disorder, 25 percent had a parent or sibling with obsessive-compulsive disorder[1909b].

These observations support the significant role of genetic factors in obsessive-compulsive behaviors. An interesting question is — How many genes are involved and what percentage of them are the *Gts* gene? The answer to this will have to await the development of a genetic marker for the *Gts* gene.

Summary: Many different types of obsessive - compulsive behaviors and ritualistic movements occur in about half of the individuals with Tourette syndrome. They may be present in the absence of any tics. The contrary is also true: many adults with a primary diagnosis of obsessive-compulsive disorder have had tics or a family history of tics. This association between the Tourette syndrome and obsessive-compulsive symptoms provides strong evidence that these behaviors are genetic and biochemical in nature, rather than due to psychological conflicts.

Chapter 23

Conduct — A Mother's Story

Before discussing conduct in Tourette syndrome I would like to have you again listen to a mother — often a great source of wisdom. This is the story of her struggle with how to handle the behavior associated with her son's Tourette syndrome.

As before, some phrases that we hear over and over, from other mothers, are italicized.

"I had a child for whom praise was a trigger for him to do something horrible. He *did not cuddle, did not like to be touched,* was, as my grandmother puts it, 'So contrary that if he fell in the river and drowned we'd have to look for him upstream.' I am, by nature, a loving, hugging, kissy mother. I was thwarted by this child at every turn. He would permit hugs and caresses on rare occasions, if we caught him between sleep and waking, early in the morning. We could not show him empathy or sympathy or he went off the deep end. He seemed to have to prove that he was a bad child. At the end of every single day of his life, for twelve years, he left a wake of exhausted, trampled adults, at home and at school. He would turn around, look at the prostrate bodies, and ask 'What's the matter with you?' He had no realization of what he had done, or *any feeling of responsibility for any of the results of his own actions.*

"We tried everything. We took two semesters of behavior modification. We took P.E.T. classes. We went to a psychiatrist for $85.00 an hour over a five year period. We read every book on child rearing we found. In fact, I am sitting here looking at my library ranging from *How to Parent* to *Tough Love* [2132].

"We read all of these books, all of the magazine articles, and library books, took classes, workshops, and advice from family and friends. It all made me a wonderful mother — for my other son. He was the exact opposite of David, a lap junkie with winning ways, hugs, snuggles, and kisses. He used to bribe me with 'If I can have some apple, I'll give you ten sousand kisses!' Who could resist?

"As far as our studying to find 'the answer' to help with David, we could have saved our time and money. We came to develop our own way of dealing with the day to day confrontations, tantrums, and the impulsive, obnoxious behavior. We found that *steady, firm, unbending, unrelenting rules were essential.* We had a child who refused to accept rewards, seeming to understand only negative consequences. In fact, we didn't dare praise him for anything well done. We had to bite our collective tongues and that was difficult and painful for us. Do you have any idea what it is like to be *afraid to even speak to your child because you know he is going to explode* and that it will go on for hours or maybe even days? I would lie in bed in the morning, dreading having to see that David ate breakfast, chose a shirt, dressed, and went out the door. I couldn't say 'Good morning' without eliciting *a screaming rage.* We could not discuss anything with David unless it was in the realm of his interest. We had to give him direct instructions without allowing him to protest. *Small protests always escalated into uncontrolled rages.* If we caught the tantrum early, we could stop it. We had Haim Ginott spinning in his grave.

"There were periods of time when I could see that there was a neat and wonderful child underneath. At first, he showed us an hour maybe every three months. Then it was half a day, then maybe a day, and the times between became shorter. I was blessed with an incredible strength and with the ability to separate the child from the behavior. I kept saying to him that underneath he was wonderful but that I could not understand the behavior. We relished the good times and during the bad times, we gritted our teeth, cried, laughed, screamed, talked to each other, planned in bed at night the next day's strategy, and we made it.

"I can honestly say that I didn't even like my child until he was about seven years old and we began to see a bare hint that we were making headway. As a baby, he would sit in his high chair, ask for milk, and I would give it to him and he would say 'I don't want a yellow glass, I want a green one.' I would give him choices, he would make a choice, then refuse to take what he had chosen. That happened, not once in a while, but every hour of every day. He *threw public tantrums, screaming at the top of his lungs.* I have left every restaurant in our city in humiliation. I couldn't stay home during the day after he went to school because the school would call me about some brutal thing he had done to another child. I couldn't let him play outside with the other children because my doorbell would ring constantly. Nothing we ever did satisfied him. We went to Disneyland once and did everything except the Pirates of the Caribbean ride. That was quite an accomplishment, actually. We had to listen to how we hadn't done that one thing for the ninety miles it takes to get home. We would swear to ourselves that this was the last time we would try to do anything for him or take him for fun. Then, like fools, we would try it again.

"I cannot tell you how many times we had to stay home or go home, or how it feels to go to bed every night and cry yourself to sleep because you try so hard and want so much to show affection and love, and you know you are a complete failure. You see other mothers whose happy children see the positive side, who mind, who care about other family members, who succeed in school, whose houses are not in constant turmoil. You *know that you are a lousy mother and you don't know why.* This is your first child so you have no way of knowing that you have not caused some obscure psychological damage to your own child.

"My husband and I waited seven years to have children and this is what we were handed. I couldn't possibly relate to you the sense of failure and frustration or the daily exhaustion of living with a Tourette syndrome child with a severe conduct disorder. I felt guilty, cruel, frustrated, deprived of being able to wrap my arms around David and cuddle and show my love. We figured out on our own that we had to do it this way but we didn't like it. We wondered constantly about what sort of parents we were.

"I would like to tell you that the happiest day of my whole life came when, at the age of eleven, David had done something and I had his face in my hands as I always did, going through the usual routine of 'Whose responsibility was it?' When he answered 'mine.' I burst into tears, and hugged him. It was the first sign that he was going to be a social being who could be allowed to enter the outside world someday. It took ten years of constant, endless going over of events and his part in these events — what had happened? How had it started? How could we have handled it better? Of course, we had to assume he was a liar, because he always did lie. Maybe he just didn't perceive the situation the way others did. We had to always get the other side. He had to prove that he could tell the truth. Now what kind of life is that?

"Our son has 'mild' tic symptoms. Most of his tic activity is on the inside. His outward signs are limited, generally to *head shaking, eye blinking, squinting, winking, and periodic involvement of his hands. His verbal tic is a low volume 'Humph' or a rare powerful expiration sound.*

"There has been, in the past few years, much written about the child whose tics are difficult to ignore; those whose tics throw them around, who screech, swear, stamp their feet, all of the symptoms that are hard to miss. These people obviously have a problem. The cause is usually misdiagnosed, or misunderstood, but everyone can see that something is amiss.

"In talking to other parents of TS children, I have found that there is another type of child, who in many ways is working under more of a handicap because of the 'mild' symptoms. This type of child is the one with the behavior problems who has impulsive reactions to the world. Reference to these people are made, through all the literature, but one has to read everything to find them. Teachers, neighbors, and friends are not interested in reading stacks of written material to filter out which things may refer to our child.

"Our son's physical symptoms are not obvious to the casual observer. He draws no more attention to himself, in a classroom full of boys, than the average boy. However, he is *in constant motion.* Even when he sits quietly, his muscles are taut, and he *cannot sit still for more than seconds.* He constantly *fidgets, jumping out of his seat as if spring loaded. He touches others, touches other people's things, is inattentive in class, has great difficulty grasping verbal instructions, has difficulty staying on task.* He is the picture of a 'typical' hyperactive (much over-used term) child.

"There are differences which are so subtle and maddening, that parents are driven to distraction. We have been fortunate, although we certainly didn't think so at the time. The sleep disorder, which so many of these children have, came with his birth and disappeared when he was 3-1/2. He has the *low frustration level, the preseveration, the peer problems,* and hardest of all, no day is ever the same as the last. *One day he is crystal clear in his thinking and the next day, he can barely function.*

"He is now entering puberty, with all of the normal, all encompassing problems of puberty, plus TS. I have just survived another parent-teacher conference. His teachers inform me that they 'Don't see any of the tics you describe, and if he has Tourette syndrome, then so do a lot of other seventh grade boys.' How does a mother or father handle this type of closed-minded ignorance?

"I cried. I cried for my son who must feel so totally alone and abandoned that he doesn't feel safe to tic in this teacher's room. He is concentrating so hard on controlling the tics that he cannot read, listen or produce work. He misses hearing instructions for assignments, or misunderstands them, loses his papers, or does his homework and then forgets to turn it in.

"My husband and I went to this conference with two team advocates. There were five teachers and two counselors. We were told that the principal and the school nurse had previous appointments. I know how a piranha victim must feel. I sat there and listened quietly, as one teacher after another, aired his or her grievances against my son. All I heard was negative, of course. 'He is always out of his seat, he doesn't turn in his homework, he bugs the other kids, *he is belligerent and won't leave the room when I tell him to, he speaks constantly, he always has one-liner jokes* (it is funny, but I can't allow him to disrupt my teaching), he refuses to ignore the kids when they 'aggravate him,' ad nausem.

"I think I will make a recording and just play it at each meeting and they can save their breath. Maybe they needn't come at all. In one hour and 15 minutes I heard not one word of understanding, not one word of empathy for what this child is feeling — a child whose school life is hell.

"The school looks to us, as parents, for some sort of switch we can throw to tame this child to sit still, be quiet, and cooperate. Don't they understand that if we had that power, we would surely use it?

"We have discipline at home. It is discipline tempered with love and empathy, however. It is essential that these children have help in controlling the impulses which they find so difficult to control alone.

"Children such as ours are wonderful victims. The *other children learn quickly, that a look, a quick punch when no one is looking, taking a book when the book report is due in a few days, snatching their lunches, all bring about violent reactions*. Our children are soon invited to leave the class. The teachers deny the connection between TS and the behavior problems. Why not? Even physicians pass out false information, based on ignorance. These people are self-appointed experts.

"The problems these children encounter, for whatever reason, are related to Tourette syndrome. Why else are they so changeable, so awful one day, and gone the next? Why doesn't psychotherapy effect miraculous cures?

"As parents, we live day to day, coping the best way we can. We need to talk to each other, gather data, educate the physician, and the public. The parents we have met who have TS kids are just like other parents who love and care and survive. We are not responsible for the behavior problems. How many parents feel as alone and abandoned as my son feels? How many times have some of us asked ourselves, 'Why me? What am I doing wrong? What did I do to deserve this? How can I fight off the whole world to protect this child?'

"Our son is an intelligent, funny, creative, loving, caring, sensitive, problem solver. He had better be. He is fighting, not only his own impulses, but the prejudices, ignorance, and lack of empathy in the world he encounters.

"At fifteen years old, David is now a different kid. There are times when he is sullen and negative, and has moments of stubborn refusal to do whatever has to be done. However, now his logic works and we can talk to him. He has a sense of humor. He has learned to do socially acceptable things in public. He is working on his eagle in scouts. He has friends, jobs mowing lawns, and can put a lawn mower engine together from scratch. He is a computer nut and a math wiz.

"David has one of the most fascinating brains I have ever known. On a subject in which he is interested, he can take in information, process it, and come up with several solutions to a problem. He watches me in the grocery store to make sure that I get the best bargains. He gives me a Christmas list on which is listed, part numbers or order numbers, how much it should cost, and where to find it the cheapest. I have fought long and hard for this kid, and with this kid, and I am determined he go on with a productive life. School is still a problem since he is so arbitrary and visually distractible, and his organizational skills are poor. I suspect that his room is an outer manifestation of the mess in his head.

"TS or no TS, these children must learn to function within the framework of the society in which they live. *Feeling sorry for them or excusing them because of TS ruins their chances of internalizing social behavior which will help them to lead normal, happy lives.* I consider myself a damn good mother. I have a head full of gray hairs, and I wear them proudly. After all, I earned every one of them, one day at a time."

Anne Poffenberger, San Diego

Chapter 24
Conduct

By the time we had only examined our twentieth Tourette syndrome patient, we had begun to see and hear certain patterns of behavior occurring over and over again. The previous mother's report is just one example. Now, more than a thousand patients later, this pattern has not changed. As intriguing as all the other symptoms of Tourette syndrome are, it is the conduct that has held our greatest interest and was one of the driving forces behind writing this book. The reason for this is twofold:

First, other than the tics and the school problems, it is the conduct that most often drives the parents to seek help.

Second, the clear association between conduct problems and a common genetic disorder, Tourette syndrome, sheds a new light on behavior problems that stands in marked contrast to the standard psychological orientation that all poor conduct is learned behavior and mostly the result of poor parenting or some type of emotional stress in the family.

In this chapter I examine the question — What are the types and frequencies of conduct disorder in Tourette syndrome? In other chapters I discuss the possible genetic and neurogenetic causes of these behaviors.

I have outlined in the following box, some of the phrases that parents have most often used to describe the behavior of their TS child. Again I emphasize that *not all TS individuals have these problems*. This chapter is devoted to the question, "What percentage do?"

Behavior Problems in Some Tourette Syndrome Patients

angry
argumentative
confrontive
everything has to be their way
everything is someone else's fault
Jekell and Hyde personality
lies
one minute fine, next minute in a rage
relentless teasing of siblings
short temper
smart mouth
steals
talks back
temper tantrums over nothing
won't take 'no' for an answer

As an example, the following is a fairly typical teacher's report brought in by a parent concerning their child with ADHD and conduct problems:

> "Cary has a difficult time starting, attending to and completing a task. He requires many prompts to hold his attention for even two minutes. He does not accept corrections when he gets an answer wrong, gets easily upset, and will even deny he made a mistake. He complains and threatens the other children if

he loses a game, does not accept, 'being wrong' for any situation, and feels that everyone is after him."

Conduct in TS—The Observations of Others

Although there have been many anecdotal statements about behavior in TS there have been few quantitative studies.

In a study of 15 TS patients Moldofsky[1347] reported that 10 or 66 percent had aggressive conduct disorders consisting of frequent fighting, vandalism, theft, fire setting or threatening with a dangerous weapon.

Wilson and coworkers[2081] examined behavior in TS children using a Behavior Problem Checklist and other tests. The 21 TS patients showed considerably more disturbance than 17 unselected public school children. The TS scores were comparable to those of aggressive, hyperactive and withdrawn children.

In a study of our first 250 TS patients[358] we noted that varying degrees of discipline problems were present in 44 percent and problems with anger or violence in 42 percent. Combined, 61 percent had problems in one or both of these areas. We suggested that the features shown in the box at the beginning of this chapter were characteristic personality traits of some TS patients.

In a study where questionnaires were returned from 431 TS patients in Ohio, Mary Stefl[1859] obtained the following results:

Problem Area	Frequency in %		
	Often	Some	Never
Extreme mood swings	32.7	42.8	25.5
Obsessive-compulsive	32.3	41.7	26.0
Extreme anxiety	31.6	48.3	20.0
Hyperactive behavior	29.2	43.8	27.1
Extreme temper	28.9	46.4	24.6
Aggressive behavior	24.8	39.2	36.0
Lying and stealing	14.1	32.3	53.6
Self-abusive behavior	7.2	27.1	65.7

Although there was no control group, the frequency of many of these behaviors seemed significantly outside the realm of normal.

In a summary of 200 TS patients Erenberg and co-workers[550] found "a strikingly large number (58 percent) had learning problems, behavior problems, or both."

The Controlled Study

Since all children may show some of these behaviors it was important to determine if they were really more common in individuals with Tourette syndrome or if this was just our imagination? To do this, in our controlled study we asked a series of questions that both related to the DSM-III criteria for conduct disorder in children[361] and that seemed to be occurring with undue frequency in our experience with TS patients.

Each question was to be answered with a no (1), occasionally (2), or often (3), and where appropriate input was obtained from both the parent and patient. The first set of questions was, Did you ever run away from home? Did you ever lie about things in school or at home? Did you ever steal things from school, home or stores? Did you start fires? Did you ever vandalize or destroy property? The results comparing the controls to TS patients are show on the next page:

There was no difference between the controls and TS individuals in regards to running away from home. This is in part due to two things. First many TS children have a lot of separation anxiety and do not like being separated from their parents, and second, because of problems with peers they often have no one to run away to. By contrast, there were significant problems with lying, stealing, starting fires and vandalism.

How severe were these problems for the three grades of TS? None of the grade 1 TS patients had significant problems, the grade 2 TS patients had problems in all areas except running away, and the grade 3 TS patients had significant problems in all areas.

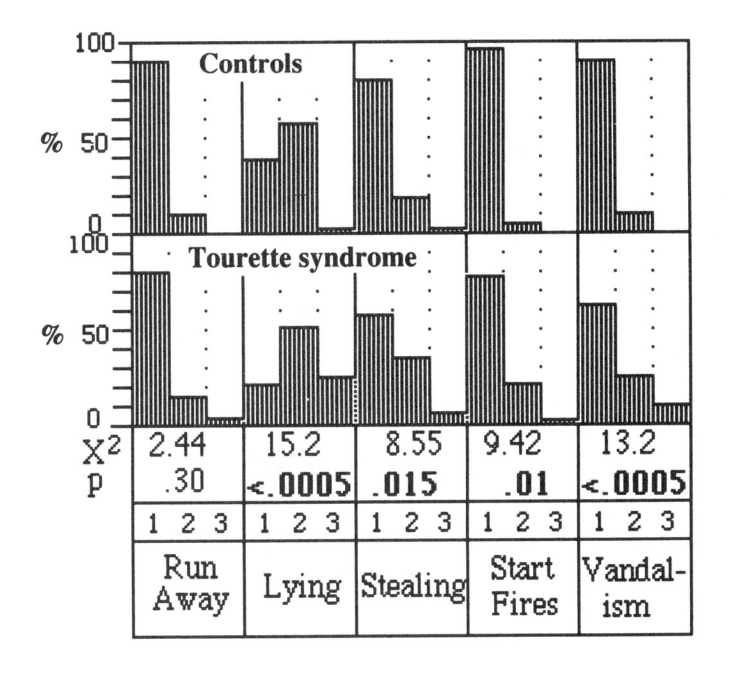

Here there were no problems with getting in trouble with the law, but very significant problems with often getting into fights with peers, shouting at or attacking parents.

Looking at the three grades of TS we found that even in grade 3 there were no significant problems with getting in trouble with the law. The behavior problems in TS tend to occur within the families and usually do not to translate into violence against public property. Problems with the law can occur if fighting in the home becomes too violent. In grade 3 TS getting into fights with peers, shouting at parents and attacking parents were all much worse than in the controls.

The second set of questions was, Were you ever in trouble with the law? Did you get into fights with your peers? Did you shout at your parents very much? Did you ever physically attack your parents? The results were as follows:

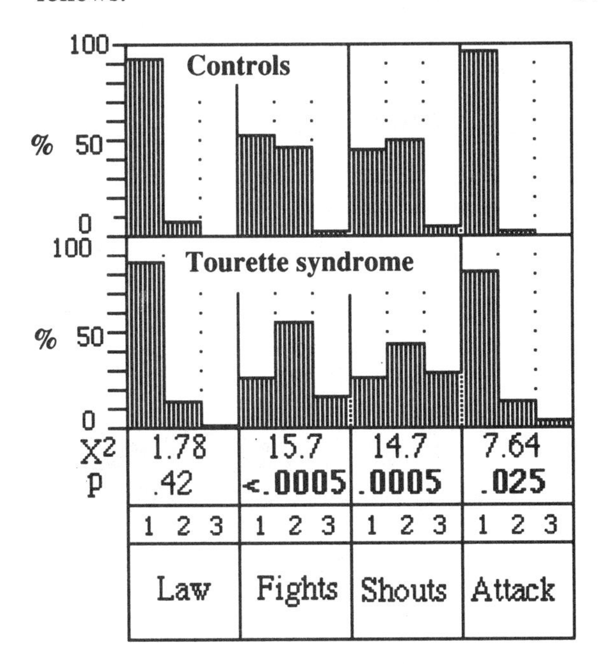

The final set of questions was, Did you have a lot (1), some (2), or no (3) respect for adults? Do you have a short temper? Did you ever hurt animals such as your pets? Were there times when you felt full of hate for others? When involved in fighting with others, did you ever get to a point where you couldn't seem to stop? Have you ever had problems with drug or alcohol abuse? The latter was restricted to individuals 16 years of age or older. The results are shown on the following page.

All of these categories were significant problems for TS patients. The most striking was short temper, followed by often feeling full of hate and by getting so involved in fighting they couldn't stop. Hurting animals was next followed by drug or alcohol abuse.

When the different grades of TS were examined, it was not surprising that short

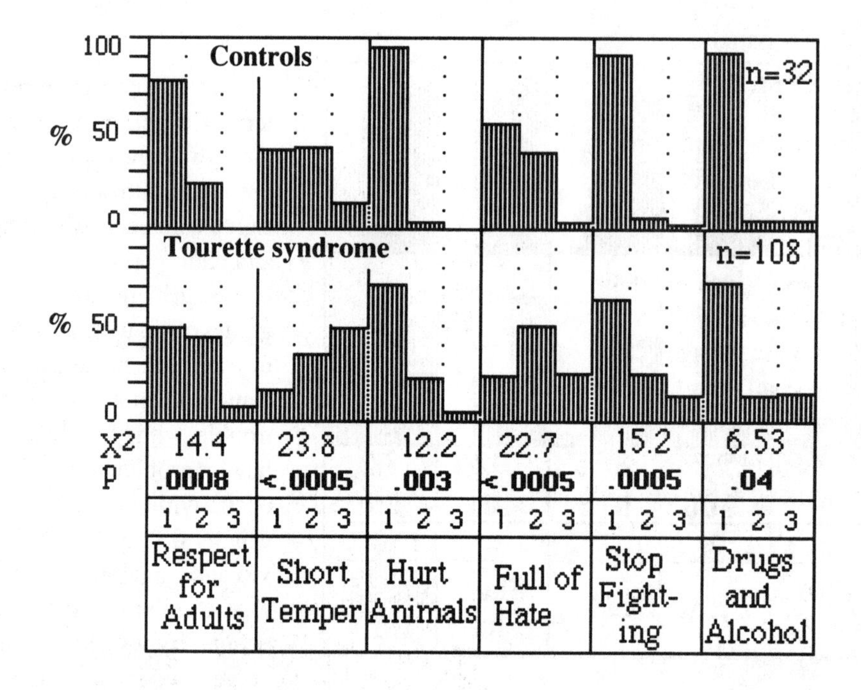

temper was a problem even in grade 1 patients. It was surprising that drug or alcohol abuse was a significant problem in grade 1 patients. In grade 3 patients all categories were very different from controls. *Here 40 percent the patients, 16 years of age or older, had problems with drug or alcohol abuse.*

[Additional information on drug and alcohol abuse in TS is given in Chapters 35, 90, and 98.]

As with the other behaviors, it is important to ask not only how frequent are the individual problems, but how often do many of them occur together. The results were added together to produce a *conduct score* (where 1 = 0, 2 = 1 and 3 = 2 for a maximum score of 30). For controls compared to all TS patients, the results were as **shown on the right->:**

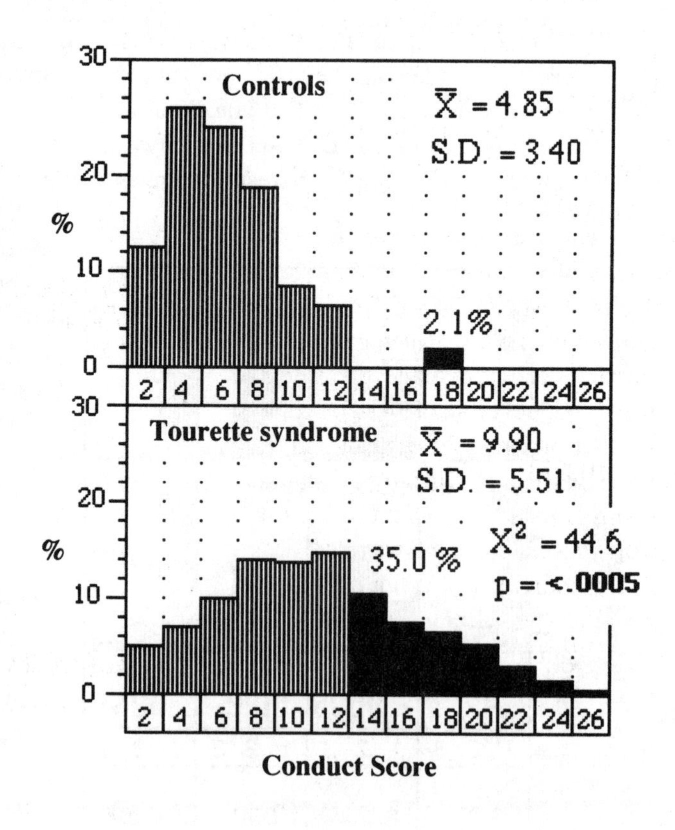

Among the TS patients 35 percent had a total conduct score of 14 or greater with a gradual decrease to a maximum score of 26. By comparison, only one of the controls had a score of 18. It was of considerable interest that this individual had Tourette syndrome. He was a 15 year old male with an onset of ADHD at age two and was treated with Ritalin from ages 2 to 7. He was in educationally handicapped classes throughout his schooling. He had problems with impatience and short temper and was suspended from school twice for fighting. He touched things excessively, including his crotch, and occasionally exhibited himself. He had some minor motor and vocal tics starting at age 14. The fact that the only individual in the control group with significant conduct problems also had TS testifies to the contribution of this disorder to conduct disorder in the general population, and to the frequency of the disorder, since 1 in 47 randomly chosen controls had TS.

The distribution of the total conduct scores in the different grades of TS is shown on the right. Here 16 percent of grade 1, 30 percent of grade 2 and 64 percent of grade 3 TS patients had a conduct score of 14 or more. For grade 1 patients, the distribution was not significantly different from the controls if the TS patient in the controls was included, but was significant if this individual was not included (p = .04). The differences were highly significant for the other two groups. In the grade 3 patients the distribution was highly skewed to the right where the most frequent conduct score was 20 and the average was 14.

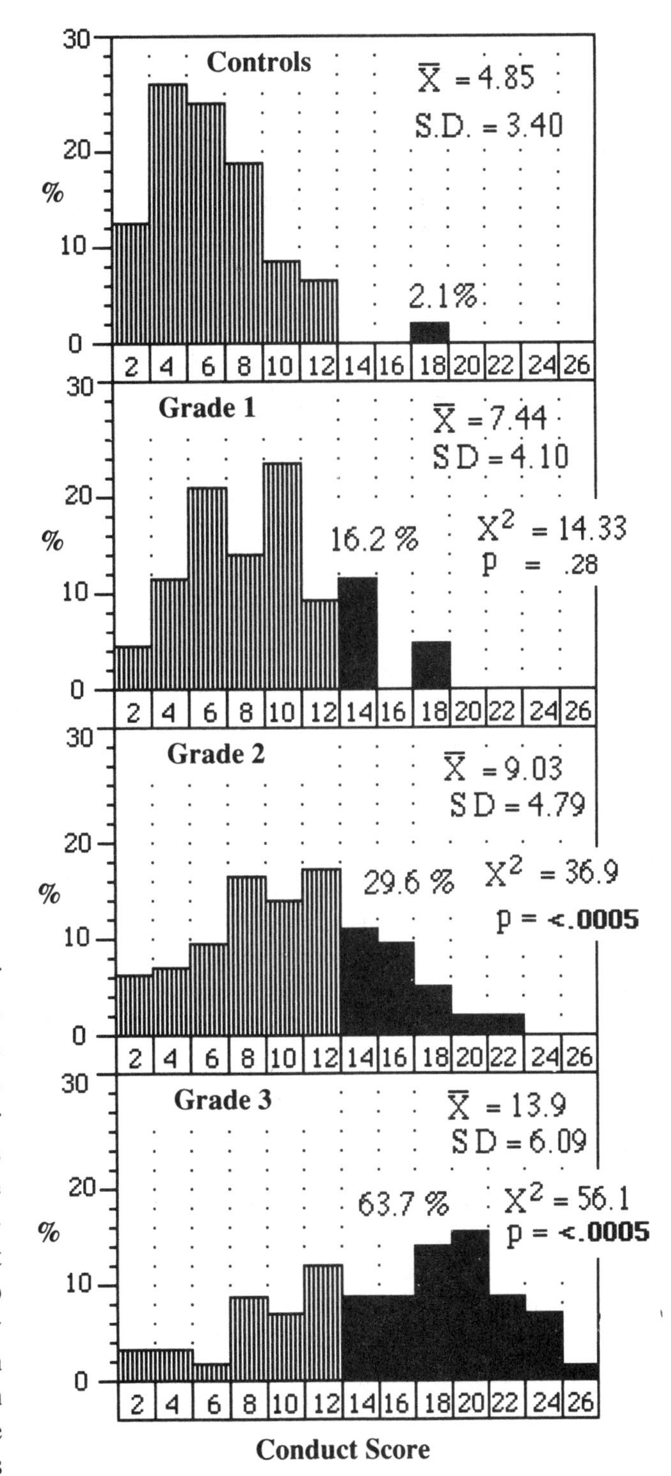

As discussed before[p51], males often have more problems with conduct than do females. In this study almost as many females had conduct problems (30 percent) as males (36 percent), indicating conduct problems can be a significant problem in both males and females with TS.

The Contribution of ADHD to the Conduct Score

Since many of the conduct problems discussed are also seen in children with ADHD alone, it is reasonable to ask, Are these behaviors only seen in those TS patients who also have ADHD? It was clear that TS patients with ADHD had more frequent and more severe conduct problems than those without ADHD. Specifically, 48 percent of TS patients with ADHD had a total conduct score of 14 or more, compared to 15 percent of TS patients without ADHD.

The Role of TS in Conduct Disorder in General

From our experience with both TS and conduct disorder, we conservatively suggested that in non-economically disadvantaged children, 10 to 30 percent of conduct disorder may be due to the presence of a *Gts* gene[361]. Since that time, Sverd and colleagues[1908] examined this from the viewpoint of children referred for conduct disorder without a prior diagnosis of TS. They found that *11 percent of children presenting with conduct disorder actually had undiagnosed TS* [and this does not include those who may have a *Gts* gene with few or no tics]. They agreed with our earlier statement that this "underscores the high prevalence of this condition and the necessity of inquiring about tics in children who present with behavioral and school related problems"[358]. They found that regardless of the reason for referral, TS boys as a group presented with attentional and behavioral problems. In their series, 94 percent of TS patients had attentional problems and 77 percent were diagnosed as oppositional.

Are the Conduct Problems Due to the Tics?

Most of the literature on TS has emphasized the tics and corpolalia. Behavior problems, if recognized, have usually been assumed to be a direct result of trying to cope with having the tics. Is this valid or are the behavior problems independent of the severity of the tics? Our clinical experience tells us several things:

First, the severity of the conduct problems bears little relationship to the severity of the tics, and,

Second, sometimes the short temper, anger and aggressiveness get worse when the tics get worse, and sometimes they get worse during periods when the tics are getting better. On the surface these may seem contradictory, but not really since the tics and the conduct may progress as independent entities.

We can ask whether significant conduct disorder can exist in the face of mild tics. Put statistically, what is the correlation between the number of tics and the conduct score? Remember if $r = 1$ there is perfect correlation, if $r = 0$ there is no correlation. The results showed that for tics versus the conduct score the correlation was only 0.15, indicating *there is little correlation between conduct and the number of tics.* Erenberg and co-workers[550] came to the same conclusion —

> "... the presence or absence of behavior and learning problems did not correlate significantly with the severity of the tics."

In another study of 29 TS children, parent ratings of aggression, somatic complaints, hyperactivity, obsessive-compulsive behavior, and depression were all significantly above these for non-TS children[1890a]. The magnitude of the scores were unrelated to the severity of the tics.

This lack of correlation between tics and behavior problems is a very important point since it means that *some children with significant learning and behavioral problems that are due to a Gts gene may have subtle tics, or in some cases the tics may have been present for several years but are now gone, even though the behavioral aspect of the disorder persists.*

We have had many cases where a child with a history of significant motor and vocal tics, and a strong family history of TS, enters a phase where the tics for all practical purposes disappear, but the behavior is still atrocious. Convincing a mental health worker that the core problem in such a child is Tourette syndrome, rather than a purely psychological problem, is often difficult or impossible.

These concepts can be illustrated as follows. The usual assumption is that the behavior problems are a result of trying to cope with the tics.

***Gts* gene → tics → 2^{0} behavior problems**

By contrast we believe the two can occur independently with only a small proportion of the behavior being a direct result of the tics.

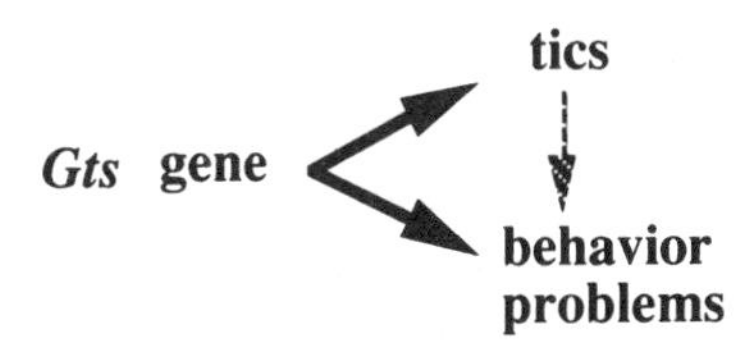

Other Correlations

In other portions of Part III on behavior, additional scores are developed for other behaviors. The degree of correlation between these scores and the total conduct score give some estimate of the degree to which these behaviors contribute to the conduct. These results were as follows:

Score	Correlation
ADDH	.490
ADD	.476
Mania	.467
Compulsions	.450
Depression	.395
Schizoid behaviors	.387
Obsessions	.264
Panic attacks	.208
Phobias	.163
Number of tics	.147
Age of onset of TS	-.131

The age of onset of TS was negatively correlated with the conduct score, indicating that conduct was slightly worse in those individuals with an earlier onset of TS. Other than age of onset, *conduct showed the poorest correlation with the number of tics.* As was discussed above, the presence or absence of ADHD strongly influences conduct and thus it is not surprising that ADDH showed the highest correlation with conduct. ADD without hyperactivity was next, followed by the mania, compulsion, depression and schizoid scores. Each of these features contributed to the severity of the conduct disorder.

Denial

TS patients often deny both their tics and behavior problems. The following is an example:

A physician mother and her 16 year old daughter, Julie, sat down in the clinic for an initial interview. Julie was pleasant, smiling and seemed to enjoy the visit. However, as the mother began to describe some of her daughter's behaviors Julie's entire continence changed. She became very angry and hostile and began screaming at her mother, "I did not do that, I never did any of those things." As her mother continued Julie began to sob hysterically. Variations on this theme are common for TS patients.

The angry denials from TS children occur in three settings.

First, the child is embarrassed when parents tell others about their tics or compulsive behaviors. Here the responses are mostly from acute embarrassment over tics and compulsive actions they know they cannot control and do not understand.

This type of reaction was also described by Parker[1465] who related the story of an 8 year old boy, Ryan, whose parents were telling the doctor about his symptoms.

As they began to discuss the motor tics and grunting noises, Ryan, reserved up to this point, suddenly yelled "You're lying." "Why are you doing this?" "You just want to make me look bad." "It's not true." "You're lying." His face was angry and flushed. His mother apologetically reached out, but he turned quickly and ran away, crying with his head buried in his arm.

Gentle reassurance that we have heard all these symptoms before and they are nothing to be embarrassed about is usually sufficient to calm such a disturbed child.

A second type of denial is a much more severe problem. This is illustrated by a mother who saw her son steal some change from her purse and put it in his pocket. Seconds later, while he was still in the room, she confronted him saying, "Billy, it is not right for you to steal money from my purse." At this point, virtually with the smoking gun in his hand, he vehemently denied that he had stolen anything. These children can be so totally convincing in their denial that the mother almost began to disbelieve her eyes, until she reached into his pocket and produced the change. Despite being caught red-handed the denials continued, "I didn't do it, that was my money." With this type of denial it is easy to understand how difficult it can be to attempt to blame such a child for stealing something when no one saw him take it, even though the circumstantial evidence is overwhelming to rational beings.

Once a pattern of compulsive stealing has become apparent, in order to avoid constant confrontations and arguments, the family's only out may be to set up the rule that "Billy is guilty until proven innocent," rather than vice versa. This might sound unfair but it can be time limited. If the child can demonstrate that he can tell the truth, the rule can be turned back to its more normal form.

The third type of denial occurs in association with rage attacks. In this setting, Billy was asked to wash the dishes and immediately went into a severe temper tantrum which lasted for two hours. The next day, after calming down, he claimed to have no memory of the event and denied that it ever occurred. This denial can be so convincing that we have often wondered if the TS children were having some type of seizure with subsequent amnesia. However, when discussed matter of factly they are usually able to recall the event.

Some Examples

Statistics on a large group of patients can give the overall picture but only the stories of individuals give a real feel for the human side of these problems. The following are some clinical vignettes to provide this flavor.

1. An adoption study in miniature A classic method of dissecting out the role of genes versus environment is to examine adopted children and ask — Does their behavior more closely match that of their biological or adoptive parents? One family presented us with a classic version of this approach. Dan was 15 when we first saw him. He was hyperactive from birth and had an onset of motor and vocal tics at age seven. His mother, Ann, states that his primary problem now is uncontrolled anger and aggressive behavior. Many times when playing with other children he would lose his temper and physically attack them. He once threw a knife at his sister and on another occasion almost choked a playmate to death. We saw him for several months believing this was the pedigree:

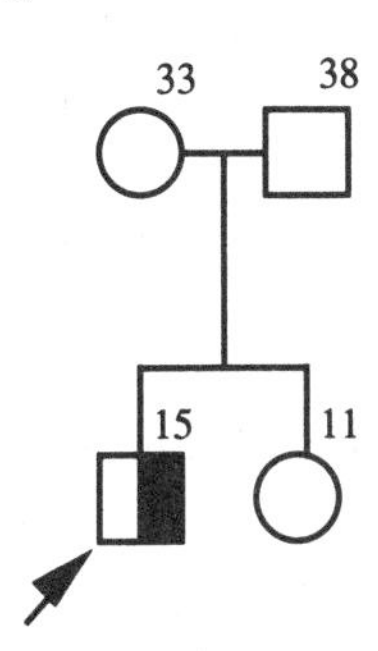

The family history was more extensive than this, but appeared to be completely negative for anyone else with TS or ADHD or behavioral problems. One day the mother took us aside and told us that she had been briefly married before and that Dan was not her present husband's son. The real pedigree was as follows:

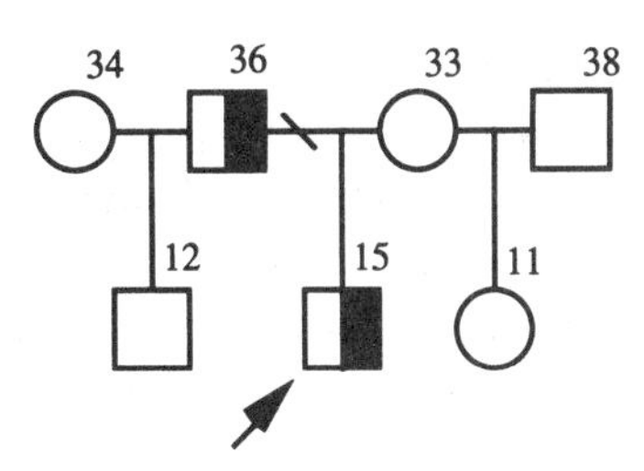

When Ann married her first husband he had already spent time in Juvenile Hall. He was "very nervous" and had several motor and vocal tics. He physically abused Ann, especially when he was drunk, and they were divorced only weeks after Dan was born. Within three months she had remarried and Dan never knew her present husband was not his father. In the meantime, Ann's first husband was in and out of jail for burglary, assault and dealing in drugs, and was finally permanently incarcerated for murder.

In the next several years Dan was constantly in trouble with the law and at 18 drove a stolen car across state lines. After spending several months in jail he was placed on probation. His parents asked him to leave home because of his aggressive behavior. At age 20 he surfaced again asking us for help. He now had a girlfriend he loved very much but was afraid he was going to lose her because of his temper. The precipitating event was that after an argument with her he went to get a gun and came back in a rage planing to kill her. Fortunately, he didn't carry it out but was concerned that unless he had help he would eventually murder someone.

This miniature adoption study is informative not only for its illustration of the genetics of TS, but for the genetics of sociopathic behavior, in this case secondary to the TS. Dan's life is becoming almost a mirror repeat of his fathers—*disinhibited behavior driven by a biological ghost he doesn't even know exists.*

2. Out of Control Tim is 12 year old boy with an onset of ADHD at age 5 and tics at age 10.

> "He has an extremely short temper, and argues with everything his mother says. During temper tantrums he has broken many objects including throwing things at the TV set and smashing it, and poking holes in the walls of his room. He has made statements such as 'I would love to kill someone.' He was so mean to the dog they had to remove it from the home. He claims such a fear of some textures, such as paper, that he refuses to carry in the groceries. He talks while others are talking, interrupts everyone, lies and steals constantly, and wants instant gratification of his wishes. Sometimes after being extremely angry he claims to have no memory of the event.

He is very demanding and refuses to take responsibility for himself such as to take showers or clean up his room. He says to his working mother, 'Why should I clean up my room, that's what you are for.' He has often said 'I cannot stand to be told what to do.' He talks disrespectfully to people of all ages. Every problem is someone else's fault. He teases his siblings incessantly, despite being told to stop. He has no friends. His mother says, 'After he spends five minutes with some of the calmest people I know, they want to choke him. He brings out the worst in people.' He once so threatened his mother with a knife she had to knock it out of his hand with a broom. However, in some situations he can be an absolute ange[1361].

3. Rage attacks A 15 year old male had an onset of ADHD and motor and vocal tics at 2 years of age.

"He had a life long history of severe behavioral problems. Multiple mental health workers had characterized him as argumentative, prone to violent temper tantrums, aggressive outbursts, physically and verbally abusive, having severe learning disabilities, having disorganized thought processes, refusing to accept responsibility for his own behavior, and frequently threatening teachers and parents. In spells of rage he has poked holes in walls and knocked the door off its hinges. He has been in and out of special schools and residential treatment facilities all his life."

Diagnoses prior to TS were childhood schizophrenia and undersocialized, aggressive conduct disorder.

4. Wonderful at school, a monster at home This is the story of a 14 year old girl whose brother has TS. She has only minor motor and vocal tics. Although she is anxious before tests, she is an excellent student and has received many scholastic and performance prizes. Despite these many positive strokes in the school setting, at home she was a completely different person. The following is her mother's description of her:

"She's been angry since the fourth grade. She verbally attacks people, is very manipulative and everything leads to a confrontation where the parents end up arguing. She constantly smells things, touches and pushes on her teeth, pops her braces and is a "motor mouth." She puts things on her desk in a certain order, makes sarcastic faces, is hostile, manipulative, and threatens others. She doesn't go out unless someone else initiates it, picks her up and brings her back. She has an obsession with her hair, wets it, dries it, and wets it again over and over. The curling iron is on three to four hours a day. She constantly thinks people are talking about her and interrupts saying 'What did you say about me?' She has come back to the house after walking miles to get something she already had with her. She sticks pens in her ears, nose or braces, and bruises her legs while shaving. She drops and spills things more than others. She wears some clothes over and over despite a complete wardrobe. Sometimes she is afraid to look nice because she doesn't accept compliments easily. At one time she sucked on her lips so much she had 'bongo red lips.' Sometimes when I ask for a kiss she burps or yawns in my face. She has a short temper, slams doors, throws dirty clothes in my face and takes pictures of me off the wall, lying them face down when she is angry. She is demanding and expects immediate gratification. She never says I'm sorry, and never apologizes."

Treatment of this case was compounded by an autocratic father who refused to bring her in for medical treatment, saying either, "It's her mother's problem," or "There's nothing wrong with her."

She is characteristic of the female TS patient with mild tics presenting predominately with behavior and emotional problems, including obsessive-compulsive behavior. *This is the type of case where the correct diagnosis of TS is almost never made because the tics are mild and specific questions about them and the family history have to be made to identify the problem.* The family pedigree was as follows. Her brother (arrow) was the initial case.

The presence of phobias, alcoholism, depression, and schizophrenia-like symptoms in TS patients and their relatives will be covered in later chapters.

5. Trivial things set him off One mother related the following about her 10 year old son with TS:

> "Trivial things set him off. The other day he was reading a book on the table and a page flipped back. He immediately began screaming and in a rage said, 'I can't do this, nothing is working right for me.' He constantly teases other children by waving his hand in their face. When they finally get upset with him he can't understand why they are angry.

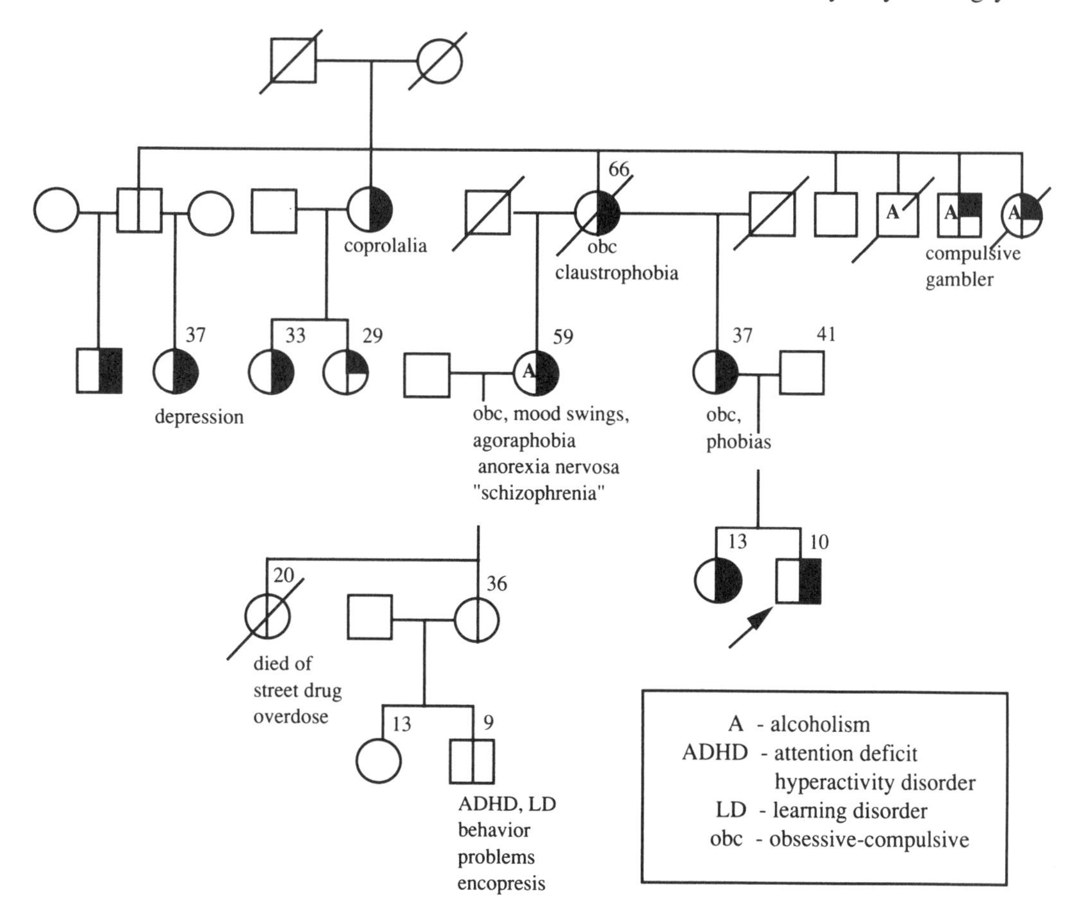

"'What did I do?' He will be very confrontive with me, but as soon as his father gets home he is an angel, because he knows his father won't take his guff. He pushes and pushes to get a certain toy and as soon and he gets it he immediately loses interest and goes onto the next toy he wants. He wants to be the leader, but is so rigid about wanting his way that he can only play with children younger than he is, and they tend to just watch him rather than play with him. He doesn't understand why children his age don't want to play with him. He has no friends."

6. I'm going to get mad and I can't help it Ray was a wide-eyed, smiling blond boy of eight who sat playing with a coloring book as his mother related the following story:

"Other than being late in walking and talking, everything seemed to be going well until he began to develop terrible temper tantrums over nothing. During these episodes he would scream at the top of his voice and start ripping buttons off his clothes. At 2 1/2 he began to repeat things so much my mother called him 'Her little parrot.' He also began to complain that his clothes were too tight, especially at the waist and neck. He refused to wear long pants and we had trouble getting him to use underwear. I finally realized there was something seriously wrong when he was five and I went to get him a half an hour early at kindergarten one day. I was planning to take him out to lunch and thought he would be pleased. When I walked in the door the children were watching a movie. When Ray saw me he started screaming that he didn't want to leave, he wanted to see the rest of the movie. I said, fine I'll wait for you. By this time he was too upset and had caused such a commotion he wanted to leave. I put him in the back of our station wagon, with a seat belt on. While I was driving he took the seat belt off, climbed into the back and began throwing shoes at my head. He then picked up a lounge chair and threw that at me. I stopped the car, put him back in his seat and fastened the seat belt again. After I started driving he once again undid the seat belt, got behind me, put his hands around my neck and tried to strangle me. The car was weaving all over the road. Fortunately we were only a few blocks from home. When we got there he ran inside, picked up his lunch box and slammed it thru the front window, breaking it and the wooden frame."

His mother said he would sometimes say,

"I'm going to get mad. I don't want to get mad, but I can't help it."

Although Ray showed clear signs of ADHD from age one, other than the echolalia and clothes feeling tight at three, he did not develop overt signs of TS until age six when he started some throat clearing, and age eight when the whole spectrum of multiple motor and vocal tics began. In addition to showing how violent these episodes can become, this case also illustrates something we see often, that the *severe temper tantrums can precede the onset of motor and vocal tics, clearly indicating that they are not the result of having tics, but rather an integral part of the entire spectrum of behavioral disinhibition.*

7. Other examples Another mother said of her son,

"He says things with a sharp edge on them, always has a chip on his shoulder and is ready to explode at the slightest provocation."

And another,

"I was standing outside with another mother and her son when out of the blue Jeff picked up a bottle, broke it on the step and went after the other boy with the broken end. After that I could never leave him to play by himself. I can't even leave a knife by his plate since he once attacked his three year old brother with it. He has destroyed furniture in our apartment several times. He sometimes seems to have superhuman strength. When he was only six he picked up our color TV and carried it across the room, then smashed it on the floor. Time outs for inappropriate behavior helped a great deal. It extinguished his swearing in two weeks. Positive rewards for a period of good behavior worked for awhile but soon he was upping the ante, trying to get more and more rewards for less and less good behavior."

And another,

" At home it is a constant battle to get him to do anything. He refuses to do his home work even though he gets good grades when he does. He would walk completely across the school yard just to hit someone he didn't even know. He is constantly fighting with his sister. If there is no fight, he will start one. At age four he began stealing things and now steals money anywhere he can find it."

A father reports,

"He left his toy airplane on the floor of the bus and someone stepped on it. He went crazy with rage and started screaming and pounding on the side of the bus. I slapped him and told him that was totally inappropriate behavior. He looked at me like I was from Mars and couldn't fathom that there was anything wrong with how he was acting and how dare I reprimand him."

And another father,

"His sister accidentally picked up his fork at the dinner table. Even though we had guests he leaped out of his chair at her and literally tried to scratch her eyes out."

And another,

"He is very irritable and argumentative. He constantly picks on his brother for no reason. Recently he tossed a two by four through the kitchen window, pounded on our pickup truck with a baseball bat, and destroyed several of his brother's toys. He keeps doing things he knows he shouldn't do. He is constantly correcting us, even when we are not talking to him. If we say it is a clear day, he says it is raining. On the other hand there are some days when he is well behaved and couldn't do enough for me."

And other parents,

"He never simply accepted a request to do something. He would ask over and over 'why' he had to do it."

"I told him he couldn't get his drivers license until he cleaned up the back yard, a 20 minute task. He waited eight months to do it simply because he didn't like being told what to do."

"He always wants a little more than he can get, that 'extra step.' If I give him 100 pieces of candy, he wants 101."

"He loses his temper when he is pushed in any way."

"His flame is easily lit."

"He has radical mood changes, that come on as fast as snapping your fingers."

"He delights in tormenting me. He

flips himself on the bed and laughs at me, and teases the dog so much that the dog bit him."

"His anger is like a white heat."

During an episode of severe tics a 17 year old girl said:

"I was overreacting to everything. I felt like I was inside a shell watching it do things I couldn't believe were happening. I felt hateful to people for no reason. Someone would just say 'Hi' and I would get angry and scream at them."

A proper 46 year old woman with severe motor and vocal tics related the following:

"I have always been guilty and afraid of something I did as a child. I had a pet chick and one day I went out and brought it up to my room. I took pins out of my mother's pin cushion and stuck dozen's of them into the chick until it was dead. Does that have anything to do with this disorder or am I a horrible person?"

I reassured her she was not an evil person and that children with TS sometimes do these cruel, impulsive acts.

8. Short temper with marital problems A 22 year old recent bride came in with her husband. At age seven she began stuttering a lot coincident with the onset of eyeblinking and a sudden change in behavior. She became bossy, short tempered and began masturbating excessively. By age 11 the motor tics had increased and included grunting, mouth opening and neck stretching. In junior high school she had trouble with friendships because she didn't trust her peers. She was always wondering what they were thinking of her. This continued to cause trouble in college because she constantly thought her roommates had something against her. She has panic attacks when she has to do something new. Because of financial problems she quit college after two years and began working. She began dating her present husband at age 19 and in six weeks was pregnant. They were married four months later. Presently her tics are very minor and consist of biting her lip and inside of her tongue and stopping in the middle of sentences to whistle.

Her husband was on the verge of divorcing her. He says she is very hard to get along with. She makes mountains out of molehills. When they have a fight he gets over it quickly while she mopes about it for hours. He says they go after each other in a fashion that he never does with other people. She makes him very angry, has a very short temper, compulsively spends money and compulsively talks on the phone. Without thinking of its effect on the other person, she frequently blurts out embarrassing thoughts and often makes scenes in public over minor frustrations.

Both were surprised that there might be an organic basis and medical treatment for her problems. If something wasn't done to change things he was filing for divorce. Treatment with imipramine 25 mg three times a day resulted in a dramatic improvement.

9. More marital problems in a presumptive *Gts* gene carrier A 24 year old sister of a TS patient was hyperactive as a child, "my mind was always racing." At age 13 she began to be easily upset and was always fighting with her father.

"When I came home in the evening I felt compelled to say mean things to my father. I would get onto something and couldn't let go. I thought I was going crazy. I was never satisfied with what I had and if my father didn't give me a present then he didn't love me. I constantly wanted new things but once I got them they never satisfied me. In high school I felt like an outcast and people

> were always staring at me [despite no motor or vocal tics]. I got very possessive about my friends and was very upset if they had another friend. I would panhandle at lunch to get money from others even though I had plenty, and used boyfriends to get them to buy clothes for me. Even though they were nice to me I was mean to them. I have a very short temper and once stuck my arm through a window when I was mad."

She had daily mood swings and by age 18 these problems were so severe she was placed in a psychiatric hospital for six months. At age 19 she moved in with a new boyfriend and they were married when she was 22.

> "As I got more secure I felt I could be mean to him. I began losing my temper over minor things and we were constantly at each other's throat. At work I would behave myself and hold my negative feelings in until I got home and take them out on him."

Her husband's complaint was,

> "She would get onto something and not let go of it until I wanted to scream. I would try to get out of the room and she would complain that I wasn't staying to hear her out, but if I stayed we would just end up shouting at each other."

Her response was,

> "He is right. Sometimes I say something like 'pick up that sock' and I would say it over and over until he just got mad. Then I would cool down and say to myself, 'Why did I do that?'"

As one can imagine, this marriage was in trouble. The husband left once and has often thought of divorce. Low doses of haldol significantly improved her irritability.

10. A brother with conduct disorder, a sister with compulsive masturbation We first became aware of Bill when his mother attended a support group with a friend. By the end of the meeting she approached us for help. She told us that Bill was incarcerated in juvenile hall for stealing a car. He was coming up for sentencing in a few days and could we help in any way? We pointed out that even if he had TS it could not be used as a legal defense, but might provide some mitigating circumstances. Since we were not allowed to see him in person before the sentencing we agreed to at least sit down with the mother to get a detailed history. We were willing to submit our provisional findings to the court.

By one year of age Bill was into everything and never sat still. His mother couldn't take him to the store because he would get into everything. In nursery school he was disruptive and became the class clown. At age five he developed an eye tic, shoulder shrugging, and repeated throat clearing. These began to subside a year later. In second grade he was given an IQ test because of poor school performance and inability to concentrate. His IQ was 130. By third grade he had severe behavioral problems and was constantly suspended from school for disruptive behavior. The IQ test was repeated and this time gave a result of 80. At eight years of age he was evaluated at a neuropsychiatric institute. Behavioral therapy was recommended but was ineffective. Problems continued and at age 11 he was diagnosed as conduct disorder, unsocialized, aggressive. In order to get him into a residential treatment facility, it was recommended that his mother relinquish custody, which she did. He was sent to a program out of state. He hated it and realized that the only way he could get out was to commit a crime. He burglarized a nearby house and, sure enough, he was expelled and returned to his mother. For the next several years he was in public school and was constantly being suspended and shuffled in and out of juvenile hall for disruptive behavior. These included shooting his English teacher in the face with a water gun, and riding his bike up and down the hall of the school. He was also compulsively stealing and lying. As a prank he drove a teacher's car several blocks away and left it. This resulted in his arrest on grand theft auto. He is now 17 years old, and his only significant Tourette symptom was moderate snorting noises. He was put in jail for two weeks then released on his mother's recognizance. She was to bring him back for sentencing in two weeks. Since she knew he would be sent to the Youth Authority, and felt he needed psychiatric help rather than imprisonment, she took him to a child guidance clinic instead. Here she was promptly turned in for child neglect, for not taking him in for sentencing. Bill felt it would be easiest on his mother if he was in jail, so he stole a purse and turned himself in.

The parents had divorced seven years previously. Bill's father had an eye tic and a very short attention span. His brother had exposed himself in school and was presently institutionalized for "impulsive-compulsive behaviors," a euphemism for exhibitionism.

The pedigree after the first visit was as follows:

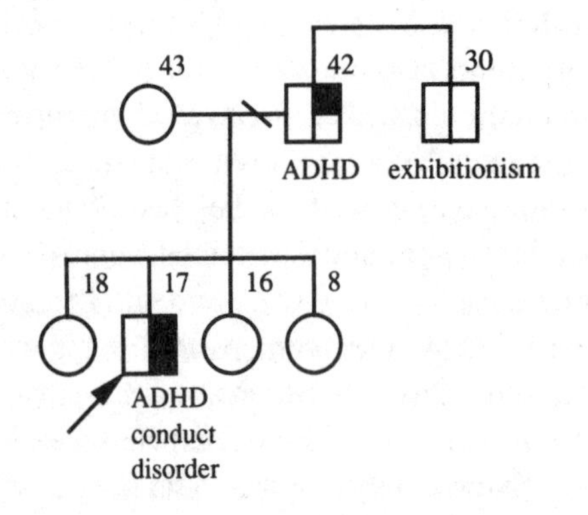

Initially this was the extent of the family history. We wrote a letter to the judge who was going to sentence Bill, suggesting that his conduct disorder was probably due to a *Tourette syndrome* gene and that medication and psychiatric care would be more beneficial than incarceration. His TS symptoms were subtle. The past history of motor tics, the presence of vocal tics and a suggestive family history were sufficient for us to be convinced that this was a case where the primary expression of the *Gts* gene was as conduct disorder. The response of the judge was,

> "I have never heard of Tourette syndrome, how can these people make a diagnosis without even seeing him? He is an incorrigible adult and I am sentencing him to three years in prison."

Three months later the mother called, very upset because her eight year old daughter, Kim, had been snorting for the past two months and it finally hit her that she had Tourette syndrome.

Kim was a very bright girl with an IQ of 141, and had been hyperactive and impulsive since age two. At age three she began to be argumentative and confrontive and constantly masturbating. By six her open masturbation in class had become such a problem that she was taken to a child guidance clinic. Here they felt this must have been the result of early molestation. Kim is very verbal and told us, "I told them it was true just to get them off my back." She didn't understand why she was masturbating so much, but inherently recognized it was a compulsive behavior she could not stop. It bothered her a great deal and she desperately wanted to stop. She had no motor tics, but had been snorting, sniffing and throat clearing for the past two months.

Treatment with small doses of clonidine completely extinguished the compulsive masturbation. As we got to know this family the pedigree continued to expand and now looks like this (**see next page**):

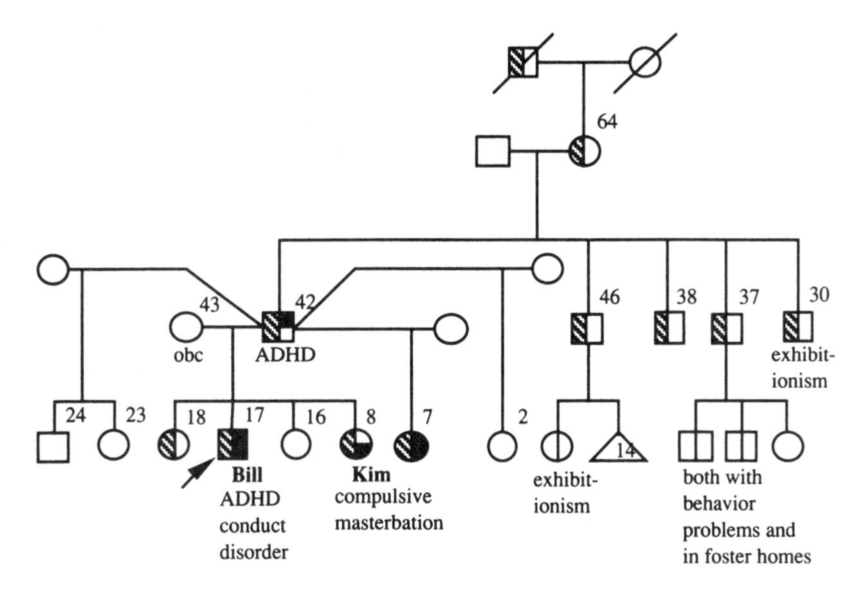

There were multiple members with exhibitionism and other behavioral problems. The symbol

indicates the excessive intake of water or **polydipsia**. This presence of polydipsia in some TS patients will be discussed later[p255].

This family illustrates several points:

a) Severe conduct disorder can be present in Tourette syndrome despite minimal motor or vocal tics. This is verified by the poor correlation between the total number of tics and the total conduct score.

b) Inappropriate sexual activity such as exhibitionism and excessive masturbation is a compulsive behavior that can be associated with Tourette syndrome[p163].

c) There is a need to foster a greater awareness of Tourette syndrome in the juvenile judicial system and among counselors involved in child behavior.

11. Onset of conduct disorder prior to tics One of the most convincing pieces of evidence that conduct disorder is both an integral part of TS, and is not secondary to the tics, comes from the many cases where conduct problems preceded the tics. The following case is an example:

Up to age 15, Dan had no problems. He was a straight A student and well behaved. However, in September, shortly after his 15th birthday, he began to be very confrontive with his parents, was irritable, flew off the handle at minor things, picked on other family members, didn't want to got to school, and procrastinated in doing both school work and tasks around the house. His grades dropped drastically and by the end of the quarter he got 1 F, 3 Ds, and 1 C instead of his usual string of As. In December, three months after the personality change he began to show a neck tic with stretching and jerking. He said he couldn't help it.

Dan was quickly diagnosed because two other members of the family were already being followed in the clinic for TS. The pedigree was as follows:

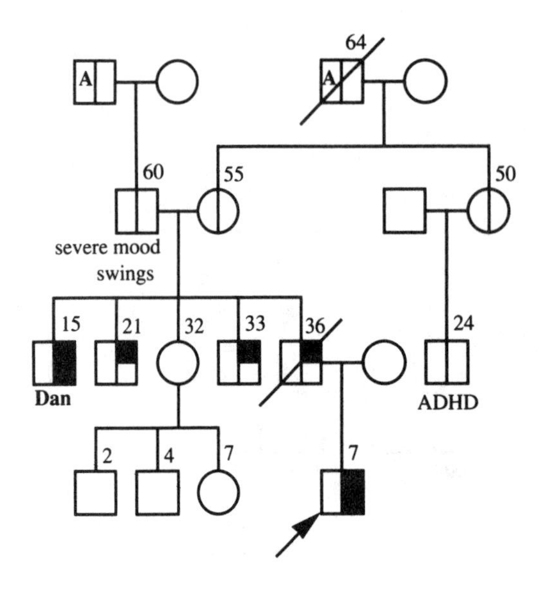

The seven year old boy was the propositus, and his 33 year old uncle, with chronic motor tics, was the second patient. Dan was the 15 year old in the pedigree.

12. "He sees red and goes crazy"— familial conduct disorder with no evidence of TS This 10 year old boy was brought in by his mother who had written the following note to me:

> "This boy is impossible to live with. I just can't take it anymore. No matter how much time I spend with him he will inevitably aggravate his parents, friends and sister to the point of exasperation. Even if you let him know you are upset it seems to egg him on, instead of stopping. He always has to be number one and constantly talks about how smart he is. Nobody wants to be around him. He is cruel to his baby sister and insanely jealous. He cannot be in any situation with other children without physically or verbally provoking them to the point of having to separate them. He gets into numerous fistfights or will provoke others by calling them extremely foul names. At home it takes him three hours to do a 30 minute chore. He cannot function at school, camp or home and is extremely disruptive 16 hours a day. The sad part is that he is very bright. We have four children and this child has made our life a living hell. We're even considering giving him up for adoption. We can't live this way anymore. Please help us."

She related an incident at school during P.E. when the children were throwing the ball to each other. He thought a little girl threw the ball at him too hard and promptly went over and beat her up. *He sees red and goes crazy.*

His 32 year old father was,

> "just like him. He had a short temper and frequently beat me. He constantly moves from place to place, sometimes six times a year, and from job to job, if he has one. I left him because he was very controlling and jealous. If I went to the store, when I got back he would ask where I have been and check the gas gauge to see if I was telling the truth."

There was no family history of motor or vocal tics or other obsessive-compulsive behaviors. This could reasonably be classified as a hereditary disinhibition disorder, or antisocial personality disorder, non-TS in type. I present it to illustrate that these familial behavior disorders can be heterogeneous in type. From a genetic point of view it will be interesting to determine if the problems in this family are due to a *Gts* gene that happens not to express with motor or vocal tics, if it is an allele of the *Gts* gene, or if is is an entirely different gene.

13. Joel—child abuse or parent abuse? Joel was a seven year old boy brought into the clinic by his stepfather. His real father, the presumptive gene carrier, was characterized as an alcoholic and chronic liar and abandoned his mother when she was six months pregnant with Joel. He has never been seen since. At two years of age Joel began to have temper tantrums during which he would get so excited he would either hyperventilate and pass out or vomit. If he didn't get his way he became very defiant and would scream at his mother, "You hate me," or "You don't love me." A short time later a diagnosis of ADHD was made and he was treated with Ritalin. Even this resulted in fewer problems in preschool and a significant decrease in temper tantrums, it was discontinued after one year because his new pediatrician didn't believe in Ritalin. Aggressive behavior with kicking other children continued throughout his preschool years. Just before entering kindergarten he began to develop motor tics consisting of eyeblinking and facial grimacing and throat clearing. In kindergarten, he was the class clown, didn't listen, and didn't mind. One day a teacher found him on the play-ground cutting himself with a piece of broken glass. When asked why he replied, "I was mean to my kitty this morning. Since my kitty didn't scratch me and I deserve to be punished, I am scratching myself." The stepfather came into the picture when he was six, showed Joel a lot of love and caring and Joel accepted him as his father. Both parents found that his behavior improved considerably if he was occasionally paddled for inappropriate, aggressive acts. One day when his stepfather was paddling him he pulled away and was hit on his leg. Two days later, while in a swimming class at school, a teacher noticed the bruise. When Joel said his father was spanking him, the authorities were notified and Joel was taken out of the home. When questioned in private Joel stated that his parents loved him, he loved them and only wanted to go home. In Joel's presence, the judge said Joel could only go home if the parents "promised never to spank him again." Ever since then, Joel taunted his parents. One time he asked his stepfather to open his mouth and show him his teeth, at which point Joel spit in his mouth and laughed, knowing he wouldn't be spanked. Attempts to control his behavior by time outs and restrictions were ineffective and his behavior continued to deteriorate. At present, Joel still doesn't pay attention in class, can't sit still, doesn't listen, and is very disruptive. However, all of his behavior problems are blamed on his parents and the stepfather is stereotyped into the uncaring, abusive "stepfather syndrome."

This is not an isolated case. These children are very bright and can be very manipulative. They quickly learn to use the system to their own advantage. Many parents have told us that their efforts to discipline their out of control children have been thwarted by the child's threat, "You do that and I'll call the child abuse hotline."

In this modern day of almost obsessive concern about child abuse, there should at least be some recognition of the fact that there is also a lot of "parent abuse."

Ways of handling the conduct problems are discussed in Part V on treatment.

Summary: Conduct disorder can be a manifestation of a Tourette syndrome gene and can be severe despite mild tics or only a past history of tics.

Chapter 25
The Genetics of Bad Conduct

This phrase may seem incredulous to some people — bad conduct a genetic disorder? Obviously there are many reasons and causes of bad conduct — social, environmental, parental, the children themselves and genetic factors. Dissecting these out can be difficult. However, the mental health professionals of this generation have grown up in a era not far removed from the pervasiveness of the Freudian principle that disturbed behavior is a result of repressed conflicts, or the Skinner behaviorism philosophy that how an individual acts is purely the result of what they have learned and experienced. However, we have just seen in the previous chapter that Tourette syndrome, a single gene disorder, can be associated with a high frequency of inappropriate, impulsive and aggressive conduct. Independent of TS, what other evidence is there for genetic factors in bad conduct? The longitudinal studies of ADHD described in a previous chapter have already shown that ADHD and childhood conduct disorder are intertwined entities and that aggressive conduct disorder in childhood is the strongest predictor of conduct disorder in adults — otherwise known as antisocial personality disorder. Some other studies directed primarily toward the genetics of antisocial behavior are described here. First, however, I will examine some aspects of conduct disorder itself.

Conduct Disorder

Conduct disorder refers to a constellation of behaviors, in individuals less than 18 years old, including lying, stealing, destructiveness, vandalism, setting fires, truancy, running away from home, disobedience, defiance of authority, fighting, and poor interactions with others including failure to make lasting friends, failure to be helpful to others, lack of guilt or remorse, cruelty, blaming others for their problems, and lack of concern for the welfare of others[139,140, 893,1822]. In the DSM-III[490] these are divided into aggressive versus non-aggressive and socialized versus undersocialized types of behaviors to give four sub-types of conduct disorder. These tend to reflect variations in severity, with socialized, non-aggressive the least severe and undersocialized, aggressive the most severe[893,1882]. A more complete list of the typical behaviors in conduct disorder is given as follows:

Behaviors in Aggressive Conduct Disorder
(Based on Stewart et al.[1882])

Aggression
- Fights with peers
- Attacks adults
- Shouts at parents
- Extremely competitive
- Cruelty to peers or pets

Antisocial behavior
- Lies
- Steals at home
- Steals outside the home
- Fire setting

Vandalism

Reactivity

Impatient
Impulsive
Reckless
Easily upset
Excitable

Depression

Low mood
Cries often
Sleep problems
Low opinion of self
Few friends

Noncompliance

Ignores directions
Resents discipline
Oppositional
Stays out late

Egocentricity

Excessive need for attention
Projects blame on others
Insensitive to other's feelings
Lack of remorse

Anxiety

Worries
Fearful
Nervous
Stomachaches
Scared of new experiences

To investigate whether conduct disorder was a distinct entity or simply a random collection of "bad boys," Behar and Stewart[139] compared 58 children with aggressive conduct disorder to 33 patients also admitted to a psychiatric ward but with other disorders. The results are shown as follows:

Comparison of Children with Conduct Disorder (CD) to Others (Cont) in a Psychiatric Hospital

(Behar and Stewart[139])

	CD	Cont	p
General features			
age at onset (yrs)	3.36	7.21	.001
no. of mother's marriages	1.50	1.15	.03
age of mother at birth of first child (yrs)	18.7	23.0	.001
child was abused (%)	37.9	12.1	.02
mother abused by father (%)	22.4	12.1	N.S.
child removed from home (%)	17.2	0.0	.03
child illegitimate (%)	19.0	6.1	N.S.
parents separated before child was 10 (%)	67.2	33.3	.004
stepfather in the home (%)	36.2	12.1	.03
Scores			
aggression score	6.52	0.82	.0001
egocentricity score	4.10	1.03	.0001
reactivity score	6.53	1.36	.0001
antisocial behavior	3.10	0.48	.0001
Symptoms (%)			
firesetting	34.5	3.0	.002
accident proneness	25.9	3.0	.02
precocious sexual activity	24.1	6.1	N.S.
hyperactivity	69.0	0.0	.0002
attention deficit disorder	79.3	9.1	.0001
Developmental problems (%)			
fine-motor coordination	53.4	12.1	.0002
articulation problems	39.7	15.2	.03
expressive language prob.	25.9	3.0	.02
receptive language prob.	12.1	3.0	N.S.
reading disability	25.9	12.1	N.S.
arithmetric disability	27.6	15.2	N.S.
bedwetting	37.9	25.0	N.S.
encopresis (fecal soiling)	22.4	4.0	.05
I.Q.			
full scale	90.7	104.0	.001
verbal	92.3	99.9	.025
peformance	91.6	106.5	.0001

Most of the scores were all significantly higher for the children with conduct disorder. As shown under general features these children had an earlier age of onset of their problems and came from more chaotic homes, with higher frequencies of child abuse, wife abuse, illegitimacy, parental separation and need to remove the child from the home. Although it has been assumed by some[1278,1663] that these things cause the conduct disorder, genetic studies (see below) suggest that the family chaos is due to the fact that at least one of the parents suffers from the same hereditary disorder.

Conduct disorder occurs in 2-3 percent of males and about 0.6 percent of females[1282,1707]. The male:female ratio is about 4:1. The symptoms usually start in early childhood before the fifth birthday[1879].

Is conduct disorder in childhood simply a minor inconvienence that the child will grow out of, or does it forbode a poor outcome later in life? The best answer comes from asking what happens to such children when they grow up?

Deviant Children Grown Up

An early and classic study of conduct disorder and later antisocial personality was by Dr. Lee Robins[1627] of the University of Washington School of Medicine in St. Louis. In 1922 the St. Louis Psychiatric Clinic was opened in the Municipal Courts Building with the purpose of demonstrating the presumed effectiveness of psychiatric treatment of juvenile delinquents. The clinic continued to operate for 22 years, finally closing its doors in 1944. For many years the records were stored until the demands for space resulted in the proposal to burn them. Foresighted intervention got them transferred to the University of Washington. A sampling showed that they included a large number of severely antisocial children, and thus provided "a treasure trove of research materials representing the first step in the study of the natural history of the development of adult antisocial behavior." The plan was to take a population of children seen in the clinic 30 years previously, match them with a sample of normal school children, and find both groups as adults. Matching for sex, socioeconomic status, and other factors would allow the effects of these variables to be sorted out. Children with IQ's under 80 and blacks were eliminated. This left a group of 524 white children seen in the clinic for behavior problems, of which 406 were referred for antisocial behavior.

In this study, children were considered to have been referred for antisocial behavior if there was a history of theft, burglary, robbery, forgery, truancy, chronic tardiness at school, running away, sexual perversion, public masturbation, excess heterosexual interest or activity, vandalism, false fire alarms, carrying deadly weapons, incorrigibility, refusal to work, lying, fighting, or physical cruelty.

Robins states,

> "We had expected that the deviant children referred for antisocial behavior would provide a high rate of antisocial adults, but we had not anticipated finding differences invading so many areas of their lives. Not only were antisocial children more often arrested and imprisoned as adults, as expected, but they were more mobile geographically, had more marital difficulties, poorer occupational and economic histories, impoverished social and organizational relationships, poor Armed Service records, excessive use of alcohol, and to some extent, even poorer physical health. The control subjects consistently had the most favorable outcomes. Those referred to the clinic for reasons other than antisocial behavior were intermediate.
>
> ". . . That the tendency toward deviant behavior pervades every area in which society sets norms, strongly suggests that the occurrence of deviance is a unitary phenomenon. No clear connections were found between the type of deviance in childhood and the type of

deviance in adults."

Antisocial behavior in childhood predicts no specific kind of deviance but rather a generalized inability to conform and perform in many areas. The best predictor of adult antisocial behavior was *the total number of antisocial behaviors as a child, not the type of antisocial behaviors.* Although the average age of referral was 13 years, most of the children had a history of behavior problems dating back many years. *The median age of onset for boys later diagnosed sociopathic personality was seven years.* It was also striking that *no child without serious antisocial behavior became a sociopathic adult.* In summary —*the more severely disturbed and aggressive the child's conduct the more likely he or she will become an antisocial adult.*

These findings indicate that *adult antisocial personality disorder has its origins in early childhood,* and if a boy or girl passes through childhood without symptoms of antisocial behavior it is unlikely they will develop subsequent problems. One explanation is that antisocial behavior is a genetic disorder which begins to manifest itself at an early age relatively independent of environmental factors. If this were the case we would anticipate that family studies would be informative and that individuals with antisocial personality disorder would be more likely to have similarly affected parents than those without this disorder. This, in fact, proved to be the case. Sociopathic and alcoholic fathers produced a significantly higher rate of sociopathic patients (32 percent) than fathers without this diagnosis (16 percent). The *highest risk factor for producing an adult with antisocial personality was the presence of ten or more antisocial behaviors as a child plus an alcoholic father.* By striking contrast, *when the parents had no problems, or problems that were not of an antisocial type, there were was no increased risk of the child becoming an antisocial adult.*

Discipline An interesting finding was that when fathers were described as cold toward their children without physically abusing them, the frequency with which their children became antisocial adults was very low. This could be best explained if "coldness" meant strict disciplinarians. In this regard the following table relating the outcome to parental discipline is interesting:

Parental Discipline	%*
both parents adequate or strict	9
one or both parents lenient	29
one or both parents exerted no discipline	32

* % of time the child became an antisocial adult

Genetic explanations One obvious way to interpret this is that disciplining children keeps them from becoming antisocial adults. This may be the major truth of this finding. However, another variable may also play a role — perhaps the antisocial parent is less likely to care about disciplining the child and children of such an antisocial parent are more likely to inherit that parent's gene for antisocial personality. In this regard it was found that if the father was antisocial himself he provided adequate discipline or supervision only 21 percent of the time. By contrast if the father was not antisocial himself, he provided adequate discipline and supervision 62 percent of the time. Both the discipline itself and the genetic constitution of the parent are important.

This is the first of several examples of a situation where what appears on the surface to be purely an environmental or learned influence, in this case discipline, may also be explained by genetics — a non-disciplining antisocial parent passing a gene for antisocial behavior on to their child. This study itself provided some additional examples. While adequate discipline appears to help prevent antisocial personality in predisposed adults, Glueck and Glueck[740] included over strict discipline as one predictor of delinquency. From a sociological point of view one might say that over strict discipline produces a rebellious child who is more likely to become a juvenile delinquent. However, again there is an alternative, genetic point of view. The Gluecks included in their definition of strictness physical abuse, a common behavior by sociopathic fathers. It may be that the sociopathic father passing on his gene for antisocial behavior is more important for his child becoming antisocial than his excessive strictness.

Another example was provided by Robins[1627,p179]:

> "Having an antisocial father has, as one of its consequences, a high rate of quarreling, separation, and divorce between the parents. Such evidences of discord do not appear themselves to increase the child's probability of adult antisocial behavior. The relation between broken or discordant homes and delinquency, or adult criminality so often interpreted in the literature as showing that broken homes 'cause' delinquency or criminality may well be a spurious relationship occurring only because having an antisocial father simultaneously produces adult antisocial behavior in the children and marital discord between the parents."

Thus, we have three examples — too little discipline, too strict discipline, and broken homes — where an apparent environmental factor seems to control subsequent behavior in adulthood. However, on closer examination all three are better explained on a genetic basis of antisocial behavior being passed from father to child.

One could still object that antisocial children are simply learning their behaviors from antisocial parents and to become an antisocial adult you have to learn the behavior at a very early age. In this regard, it was interesting that of all the children whose fathers were diagnosed antisocial, 10 percent lived with the father either not at all or only during infancy. Of these 50 percent were diagnosed as antisocial adults compared to 30 percent where the child remained with his antisocial father. These percentages are not significantly different and suggest that being an antisocial adult is not something learned from an antisocial father at an early age, but rather something inherited from an antisocial father at conception.

Socioeconomic status Many sociological studies have blamed juvenile delinquency and antisocial behavior on poverty, poor housing and poor jobs. However, when Robins dissected out this variable she found that *having a sociopathic or alcoholic father was a far better predictor of a child developing an antisocial personality than was the earning power or socioeconomic status of the father*. She points out that: "Families can live in slums and be on relief rolls *without* their childrens responding to the frustrations of poverty or to the examples of delinquency in their neighborhoods with sociopathic behavior *if* the families' poverty stems from [factors other than a parent's] antisocial behavior. It is unreasonable then to attribute a crucial role in the production of adult antisocial behavior to the frustrations consequent to low prestige and poverty in childhood. The practical implication of this conclusion is that there is little hope of preventing sociopathy through slum clearance, more adequate relief payments, or the breaking up of juvenile gangs. While an improved level of living for the poor is unquestionably a humane goal in its own right, it should not be expected to materially reduce the numbers of criminal, alcoholic, and maritally unstable persons in the population."

These observations agree with those of

Scott[1738a] who found that the personality of the delinquent was the same whether he came from a high or low socioeconomic status. They were all unpopular with their peers and maladjusted in the school room. Thus, delinquency was not "normal" behavior in "bad" neighborhoods and "sick" behavior in "good" neighborhoods. He pointed out that the *higher frequency of antisocial personality in poor neighborhoods is not because the neighborhood produced the personality, but rather because those with antisocial behavior gravitate to the poorer sections of the city.*

Robins concluded that

> "Our data certainly do not permit a complete rejection of the hypothesis that social class plays a role in sociopathic personality. But its contribution was certainly trivial as compared with the effect of antisocial behavior in the child and in his father. It appears to be the father's *psychiatric* status rather than his *socioeconomic* status that predominately determines the child's outcome."

A remarkably similar story has been related by Morris and co-workers[1361]. They checked up on 90 patients who had been *treated 20 years earlier* in a Pennsylvania psychiatric hospital between the ages of 4 and 15 *for antisocial behavior*. This consisted of at least four of the following: repeated truancy, stealing, lying, cruelty including marked teasing or bullying, disobedience, marked restlessness or distractibility, wanton destruction, or severe temper tantrums when crossed. The authors characterized these children as "acter-outers," a commonly used term even now that carries the implication that something is being done to them to which they are responding. The authors stated that "in the majority of cases there was open rejection of the child on the part of one or both parents." However, when the reasons for this rejection are listed— desertion of the child, placement in a foster home or institution, the court removed the child because the parents were "unfit," open and constant expression of dislike for the child, and marked preference for another sibling— four out of five suggest sociopathic behavior on the part of one or both parents. In fact when the parents were examined, 85 out of 180 "showed rather marked evidence of some type of personality disturbance." Although it was stated that "immediately after placement in the hospital, 66 percent of the children showed considerable improvement in their behavior, 20 years later they could no longer be so sanguine about the improvement. The results were as follows:

Outcome for children with conduct disorder as adults at age 19-25[1361]	**No.**	**%**	
Doing well	13	22	
Never adjusted	23	40	78 (Never adjusted + Psychotic + Borderline)
Psychotic	13	22	
Borderline	9	16	
Total	58	100	

Of 58 children who could be found 20 years later, 78 percent had a poor outcome. Socioeconomic factors did not play a role since the children represented all social strata. Sexual acting out — open sex play, open masturbation and exhibitionism — were indicative of a poor prognosis.

Psychological explanations of this data would include the following:

1. Children deprived of their parents' love, support and food experience this deprivation as a direct attack with destructive tendencies to which they react with aggressiveness. They try to get satisfaction from other sources and start with a destructive search, or,

2. Parents vicariously achieve gratification of their own poorly integrated forbidden impulses through a child's acting out.

However, these studies also provide all the elements expected of a *genetic explanation* of the data. These are:

1. Socioeconomic factors are not involved.

2.There was a very high frequency of psychiatric disorder or antisocial personality in one of the parents.

3. The onset of symptoms started at an early age.

4. The symptoms persisted for life despite attempts at treatment.

Other studies have been remarkably consistent in showing that early childhood aggressive behavior predicts abuse of alcohol in adolescent boys[1020,1557,1558] and adult men[1020,1557,1558,1629,2070], and criminal or antisocial behavior in adult men[1420]. *Aggressiveness in childhood tends to be a stable trait that persists into adolescence and adulthood* [1197,1438].

A study comparing full to half-siblings of 151 boys with aggressive conduct disorder showed that 20 percent of full brothers as opposed to 10 percent of half-brothers and 0 percent of controls had aggressive conduct disorder[1969]. This is best explained by aggressive conduct disorder being a genetic trait.

Where psychiatry and social problems meet Because of the importance of the conclusions reported by Robins in her book *Deviant Children Grown Up,* she undertook some additional studies to see if these findings could be replicated[1628]. As she states, this is an "area where psychiatric diagnosis meets social problems—the area of conduct disorders, antisocial personality, alcoholism and drug dependence. . . . May we be just using pseudomedical jargon to refer to behavior we disapprove of? Are we victims of a secular religion that substitutes 'sickness' for 'sin'? Or are we 'blaming the victims' of a sick society by designating as evidence for an illness . . . behaviors that are caused by the experience of poverty and discrimination? If the latter is the case, are these behaviors 'normal' for the poor but pathological when they occur in the middle class?"

Additional longitudinal studies were carried out. One involved 223 blacks chosen from public elementary school records. The other was a follow-up of a random sample of Army enlisted men who left Vietnam in September 1971 and were interviewed one and three years later. There was a control group of non-veterans. The conclusions from all of these studies were remarkably similar[1628].

1. Antisocial behavior in adults virtually *requires* childhood antisocial behavior. Adults who were diagnosed as having antisocial personality had almost all been antisocial children. Antisocial behavior in adults almost never arose *de novo* in adulthood.

2. *Although most antisocial children do not become antisocial adults, a significant percentage do.* The following shows the results of the three studies:

	No. Studied	**% of children with conduct disorder that became antisocial adults**
Study I	157	36
Study II	86	41
Study III		
A. Veterans	208	41
B. Non-veterans	48	23
Total	499	38

Thus, on average, 38 percent of children with conduct disorder become antisocial adults while essentially 0 percent of non-antisocial children become antisocial adults.

3. *The variety of childhood antisocial behaviors was a better predictor of severe adult antisocial behavior than was any particular childhood behavior.*

4. A *child's own behavior was a better predictor of his adult behavior than were his family characteristics or social status.* Social class was unimportant in predicting outcomes as compared with childhood behavior.

5. Family variables matter more for moderately than for severely antisocial children. These studies examined severely antiso-

cial adults. *Socioeconomic status and environmental variables do play a role in children and adults with moderate antisocial behavior.*

The Genetics of Criminal Behavior

Although some individuals with antisocial personality commit crimes, the majority do not and many crimes are perpetrated by individuals who do not have antisocial personality. Antisocial personality and conduct disorder are set in terms that apply to behaviors both within and outside the home, while criminal behavior is predominately set in terms of behavior outside the home. Thus, criminal behavior represents a subset of antisocial personality. Although the possible inheritance of criminal behavior has been a subject of great interest, it is likely to be more affected by environmental variables than antisocial personality. Keeping that distinction in mind, it is of interest to examine some of the studies of the genetics of criminal behavior.

Twin studies A summary of 11 twin studies of criminal behavior, examining a total of 590 twin pairs, showed that 51 percent of identical twins were concordant, compared to 20 percent of fraternal twins[1990]. In some of the studies there was only a minimal difference between identical and fraternal twins suggesting environmental factors.

Adoption studies Crowe[432] examined 46 children, 18 years of age or older, whose mothers had been imprisoned for criminal offenses and had given them up for adoption in infancy. There were 46 adopted controls. Here, 13 percent of the children of imprisoned mothers were diagnosed as having antisocial personality compared to 0 percent of the controls. When 37 of each group were examined for criminal offenses, seven of the children of imprisoned mothers had been convicted of a crime compared to one of the controls[433]. In a similar study, Cadoret [272] found that 4 of 18 or 22 percent of adopted-away children of an antisocial parent had antisocial personality compared to 0 percent of adopted controls. A similar trend but less impressive results were obtained by Schulsinger[1718]. He compared the biological parents of 57 adoptees with antisocial personality to biological parents of non-sociopathic control adoptees. Among the antisocial adoptees 4.3 percent of the biological parents had antisocial personality compared to 0.9 percent of biological parents of the controls.

There are two studies which allow four-way cross comparisons of the effect of a criminal versus non-criminal biological parent and criminal versus non-criminal adoptive parent. One was by Mednick et al.[1296-1298] examining 1,145 males adopted in Copenhagen. The results were as follows:

Biological parent with criminal record	Adoptive parent with criminal record	Frequency of petty crime in adoptees
No	No	13.5%
No	Yes	14.7%
Yes	No	20.0%
Yes	Yes	24.5%

Here, if the biological parent was non-criminal, having a criminal as opposed to a non-criminal adoptive parent made little difference in convictions of the adopted child (14.7 vs. 13.5 percent). However if the biological parent was criminal the conviction rate increased to 20 to 24 percent.

Even more impressive results were obtained by Cloninger and co-workers[326,334] using 862 male and 913 female Swedish adoptees.

The following are the results in terms of the percent of male adoptees having problems with petty crimes.

Biological parent with criminal record	Adoptive parent with criminal record	Frequency of petty crime in male adoptees
No	No	2.9%
No	Yes	6.7%
Yes	No	12.1%
Yes	Yes	40.0 %

Here there was a significant influence of both the biological and adoptive parent. If neither the biological nor the adopting parent had a criminal record, only 3 percent of the adopted children had a criminal record. If the biological parent was criminal but the adoptive parent was not, 12 percent of the adoptees had problems. However, this shot up to 40 percent if the adoptive parent was also criminal. This is a classical example of how the predisposing nature of a genetic background (criminal biological parent) can be compounded by an adverse environment (criminal adoptive parent).

In female adoptees, the results were less striking but the environmental effects were even more important.

Biological parent with criminal record	Adoptive parent with criminal record	Frequency of petty crime in female adoptees
No	No	0.5%
No	Yes	2.9%
Yes	No	2.2%
Yes	Yes	11.1%

Here, if the biological parent was criminal but the adoptive parent was not, only 2.2 percent of the female adoptees have a criminal record. If the adoptive parent had a criminal record, the frequency of a criminal record in the adoptees increased to 11.1 percent.

There have been three explanations of the mechanisms involved. Zuckerman[650,2158] proposed that the amount of sensation sought by an individual is genetically determined and this determines the level of thrill seeking behavior the person engages in and the degree of disinhibition or not being restrained by social conventions. Eysenck[569] suggested that extroversion (impulsivity), neuroticism and psychoticism are the core problems of criminals and juvenile delinquents. Extreme extroverts were thought to be more difficult to socialize than others. Mednick and co-workers[1296-1299] suggested there is an abnormal functioning of the autonomic nervous system in sociopaths with a slower recovery from stress.

When crime was associated with alcohol abuse, the crimes were usually against people. When crime was not associated with alcohol abuse, the crimes were almost always against property only[191]. The former is what we see in some patients with TS where the aggressive conduct is against other family members with less of a tendency to commit criminal acts for profit.

A Miniature Adoption Study of Antisocial Behavior

Sam was a 14 year old teenager. As he sat

watching his mother, Adrienne, tell his life history he had an unpleasant snarling look that suggested that at any minute he would either bite her or me. By six years of age he had demonstrated a short temper, aggressive behavior, and began stealing and lying. These problems continued throughout his life. He had been in and out of many schools, suspended innumerable times, and spent four years in various residential treatment programs including Boy's Town. At home he was constantly confrontive. Neither he nor any other members of the family had motor or vocal tics.

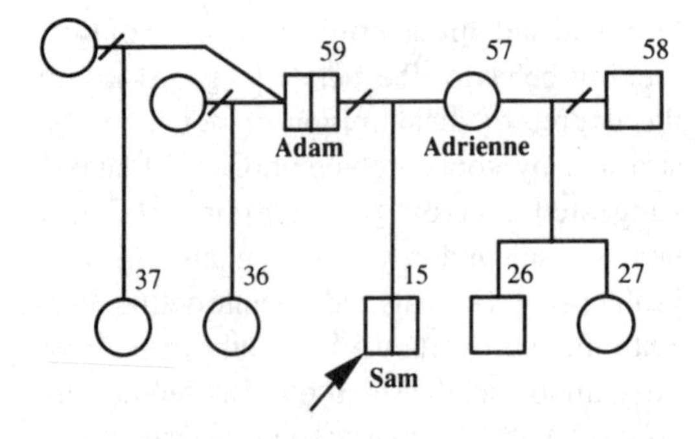

His father, Adam, had an identical personality. Although personally likable he was a compulsive liar, stole things, and had been in and out of prison many times for forgery. His chaotic lifestyle had resulted in the breakup of three marriages. Adrienne's other two children by a previous marriage had no problems.

One possible explanation of Sam's behavior is that he simply learned it from his father. However, Adam left Adrienne before Sam was born and never returned. Shortly after this breakup Adrienne's first husband returned and raised the child. Sam has only recently learned he was not his real father. This family can be best explained by assumingAdam carried a gene for a disinhibition disorder resulting in short temper, confrontive personality, lying, stealing and other antisocial behavior, and passed this gene to Sam.

ADHD, Conduct Disorder and Antisocial Behavior

Previous chapters have shown the interrelationship between these disorders. However, we will take one more look at them, this time from the prospective of conduct disorder and ask, What percentage of children with conduct disorder have ADHD? In a study of 175 consecutive admissions to a child psychiatry clinic, excluding mental retardation and neurological diseases, Stewart and co-workers[1880] at the University of Iowa found that three out of four children with aggressive conduct disorder had ADHD, and two out of three children with ADHD had a conduct disorder. This can be illustrated as follows:

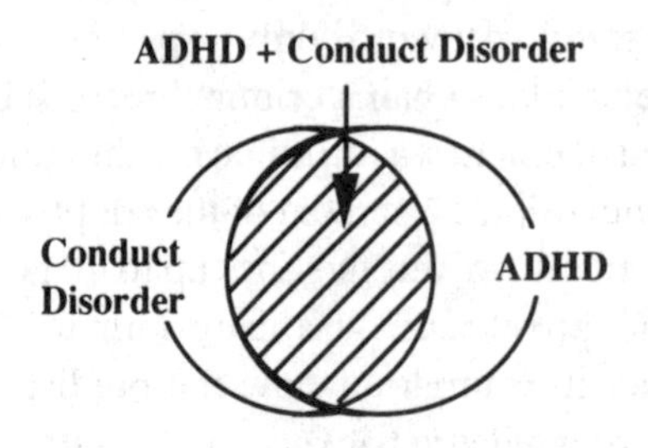

This is similar to our observations of 32 percent of ADHD patients, with or without hyperactivity, having conduct disorder. Depending upon the severity (grade 1-3), 16 to 64 percent of our TS patients had conduct disorder. Others have also found a high frequency of conduct disorder in ADHD and a high correlation between ADHD and conduct scores [1109,1282,1551,1564,1687]. Some believe there is a separate group of ADHD children without conduct disorder[69,1282,1880,1882] while others do not[1180,1564]. In a comparison of three groups of boys, aggressive, hyperactive, and hyperactive-aggressive, McGee and co-workers[1282] found there were more learning problems, especially in reading, in the latter two groups. In our TS patients we found a correlation coefficient of 0.49 between ADHD and conduct scores. In ADHD patients others have found correlations between hyperactivity and conduct scores of 0.7[1180], 0.61[1280] and 0.47[1551].

Attempting to further dissect out the relationship between ADHD and conduct disorder, August and Stewart[69] interviewed the parents of 67 ADHD children and divided them into two groups. One group consisted of 36 children where at least one parent had antisocial behavior (either antisocial personality, alcohol or drug abuse, or hysteria). The other group consisted of 31 children where neither parent had antisocial behavior. While the children in both groups were similar in regard to inattention, impulsivity, anxiety and depression, there were greater problems with aggression, failure to take responsibility for their actions, oppositional and antisocial behavior in the group with a positive family history of antisocial behavior. If this difference is inherited it should show up in the siblings. The frequency of ADHD in siblings was comparable in both groups (8-17 percent). By contrast 23 percent of the siblings of the group with a family history of antisocial behavior had conduct disorder while 0 percent of the group without such a family history had conduct disorder. These results suggest that *the strongest predictor for conduct disorder associated with ADHD was the presence of antisocial behavior in one or more of the parents.* This is consistent with two types of ADHD gene, one type producing ADHD without significant conduct disorder and one producing ADHD associated with conduct disorder. Earlier studies by the same group[1882] also indicated that the findings of alcoholism and antisocial personality in fathers and hysteria in mothers[285,1363] was predominately true of children with ADHD + conduct disorder rather than ADHD alone. These findings agree with those described above and in previous chapters that *ADHD + childhood aggressive conduct disorder is a better predictor of adult antisocial behavior than ADHD alone.*

Conduct Disorder and Associated Behavior Problems

Tourette syndrome is not the only disorder with a wide range of "associated" behaviors. In a study of 55 girls with conduct disorder[1636a] the following additional problems were present: drug or alcohol abuse or dependence (24 percent), major depression (31 percent), suicide attempts (62 percent), phobic disorder (45 percent), obsessive-compulsive disorder (16 percent), anorexia nervosa (7 percent), and bulimia (5 percent). In all, 84 percent of the girls had five or more complaints about their health (somatization[p293]), or a mood, anxiety, or eating disorder, or had attempted suicide. As noted before, childhood ADHD is very common. Conduct disorder is a spectrum disorder with many similarities to Tourette syndrome.

TS, Conduct Disorder, and Antisocial Behavior

Tourette syndrome provides a model for understanding the above. TS is a genetic disorder passed from parent to child. Two-thirds of the cases have ADD, half have ADHD, and one-third have conduct disorders identical to those discussed above. In addition, half of the TS patients with ADHD have conduct disorder while one in seven without ADHD have conduct disorder. With all of this variation in a single genetic disorder no wonder it is difficult to dissect out the genetic relationship between ADHD and conduct disorder.

Lessons for society in general are also relevant. A father with antisocial behaviors may pass this *disinhibition* gene on to his son who then may also develop antisocial behaviors. The son doesn't necessarily have these behaviors because he learned them from his father, or because his father has a poor paying job and lives in a poor neighborhood; he has these behaviors because he inherited a *disinhibition* gene from his father. Furthermore, the father doesn't have these behaviors because he lives in a poor neighborhood; he lives there because he has these behaviors.

Summary: Conduct disorder refers to a repetitive and persistent pattern of behavior in children in which either the basic rights of others or major age-appropriate norms or rules are violated. It has an early age of onset and may persist, resulting in antisocial behavior in adulthood. The prognosis depends upon the number of symptoms presenting in childhood, especially in regard to aggressive behaviors. Conduct disorder has no single characteristic but instead represents disinhibited behaviors covering a broad range of social interactions. It has a strong genetic basis. As will be discussed in later chapters, TS, ADHD, conduct disorder and antisocial personality can be thought of as a group of closely related genetic disorders resulting in disinhibited behaviors.

Chapter 26

Sex and Exhibitionism

One of the features of Tourette syndrome that clearly puts it into the categoryof a generalized disorder of disinhibition (loss of normal inhibitions) is its effect on sexual behavior. Here the record is paradoxical. On the one hand coprolalia, or the involuntary yelling of obscenities, has been so overemphasized that many professionals have the mistaken idea that this must be present to make a diagnosis. In fact it is present in less than a third of the cases. On the other hand there are other aspects of sexual behavior in TS which can be even more troublesome, that many are unaware of.

The first clear indication of these came from a study reported by Roswell Eldridge and colleagues in 1977[530]. He convened a single Saturday clinic to examine 21 TS families. Since he was a geneticist at heart, a major emphasis was on the family histories which showed that in most families there were other members with TS, or motor or vocal tics. The intriguing observation was that twelve of the 21 index cases "had troublesome sexual and aggressive impulses, differing only quantitatively from normal." Comments on some of these individuals were as follows:

> "He began exposing himself in the family setting. This was worse at about age 12 and stopped at about age 14. This activity was replaced by frequent touching of his mother's breasts, which lasted less than a year."
>
> "Episodes of exposure began at age 21."
>
> "At age eight he would have erections and play with himself in class. Exposes himself in the home setting and will 'pop in' on mother when he is naked while she is fixing lunch."
>
> "At age 5, he had the urge to kiss girlfriends but altered activity, instead kissing the air, bus, or other substitute. He now has the urge to touch the genital area of both men and women but substitutes touching the arm."
>
> "Beginning at age 16, had a 2-year period during which he would expose himself in public."

The aggressive urges especially involve hitting ones' self or hitting others.

Having been aware of these behaviors, when these workers examined a different set of 50 patients three years later[1397] they asked the necessary questions and found the following:

Behaviors in TS Patients	%
Obsessive-compulsive behaviors	68
Coprolalia	58
Family history of alcoholism	52
Self-destructive behaviors	48
Sleep disturbance	44
Learning disability	40
Inappropriate sexual activity	32
Speech problems	32
Antisocial behavior	26

These figures may be somewhat high because they tended to be grade 2 and grade 3 cases. Nevertheless, the frequency of aggressive, antisocial and inappropriate sexual activities seemed to be much higher than in the general population. These stories are typical of the ones we also hear — a precocious interest in sexual things, compulsive sexual touching of their own crotch or the genitalia or breasts of others, exhibitionism in the home and sometimes in public, excessive "accidental" walking in on mother or sister when she is dressing, compulsive masturbation, and a general preoccupation with sex. The following are some examples:

> "At age 12 I developed the urge to expose myself but I never did it in public. I sometimes drop my pants at home when no one is around. I was afraid to go to the beach because of the urge to take down my swimming suit."
>
> "Although I am very heterosexual I have the urge to kiss men on the shoulder."
>
> "Since he was five years old he has been excited about sexual things. He touches his brother on the rectum and penis and masturbates. He runs around the house without clothes on and is constantly touching his mother's breasts every chance he gets."
>
> "When my girlfriends come over he [5 years old] likes to feel their legs and sometimes humps them like a dog."
>
> "He pulls his pants down at home and pretends he is playing a guitar but strums his penis instead."
>
> "He [her 15 year old son] is always trying to crawl in bed with me and feel me up."
>
> "I'm a dental technician and when I work at the chair I find myself staring at the man's crotch, trying to outline his penis in my mind. They see me doing it. It usually bothers them but I can't help it."

In a review of our first 250 cases of TS we found that 6 percent had problems with public exhibitionism[358]. Brunn[241] observed public exhibitionism in 5 percent of 140 patients over 16 years of age.

Since some of sexual problems in TS are only extremes of normal behavior it is important to compare them to a control population. This will be covered later in the chapter.

First, I will present some typical cases.

Exhibitionism

In 1981 a 32 year old male was referred to our clinic by a prison psychiatrist because he had motor and vocal tics. He was in prison because of repeated public exhibitionism and sexual abuse of his son. His story was as follows[356]:

At age seven Jack began to have eyeblinking, lifting of his eyebrows, jerking of his head to one side, facial grimacing, grunting and fears of being alone. These and similar tics have continued all his life. At age 15 he began to expose himself to children in his neighborhood. He was married at 19 and began to expose himself at home when his wife was away. When she was home he would go to the park and expose himself to children. The exhibitionistic urges came on about once a week and were so strong he could think of nothing else until he had performed some overt act. Some other person usually had to see him in order for him to feel relieved. He acted on the urges hundreds of times and had been arrested and placed on probation twice.

One night he was home alone with his children and saw his nine year old son expose himself to his 11 year old son. He took the boy to the bedroom to punish him, then developed an urge to feel him, which he did and they engaged in mutual masturbation. This continued for four months until his guilt about it caused him to seek help. He was arrested, imprisoned and his wife obtained both a divorce and a court order for him not to see his children.

Treatment with haloperidol extinguished his tics and exhibitionistic urges. This treatment has continued to be successful to the present time.

The pedigree was as follows:

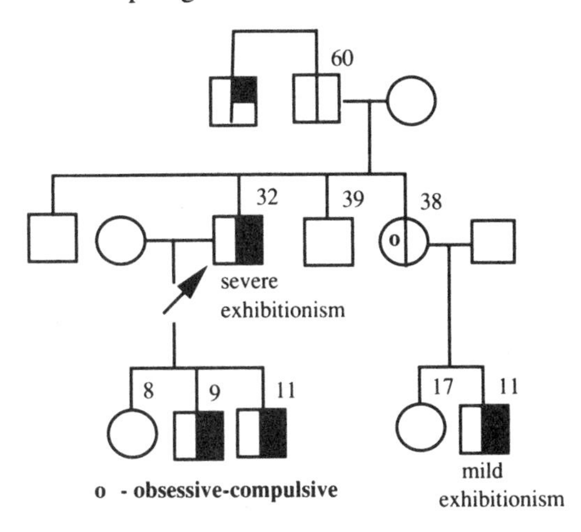

Since we reported this family[356] and showed that some cases of exhibitionism respond to haloperidol, twelve individuals with "pure" exhibitionism have been referred for evaluation. These were individuals whose primary problems were legal difficulties with public exhibitionism. It came as no surprise to us that nine of the twelve in fact had Tourette syndrome. The following are some of their stories:

James came to the clinic to see if we could help him control his exhibitionism.

> In grade school he and his brother would pull down their pants in front of his sister, and he would unzipper and expose himself in class. He then began to walk past his window at home with no clothes on. This behavior continued into college by which time the urge to do it was much greater. After he graduated he began to expose himself in the car and did this hundreds of times. He both wanted people to see and was afraid of what might happen if they did. Someone finally called the police and he was arrested and placed on probation. He has not repeated the exhibitionism since being on probation but came in for help because the urge to do it was still strong.
>
> He had no vocal tics but some mild motor tics in the form of arm and leg jerking, and some minor eyelid tics. These were not prominent enough to be totally convincing that his exhibitionism was due to a *Gts* gene. The day before coming to the clinic he talked to his mother who told him she had some throat clearing. He insisted that there was no one else in the family with any symptoms of TS.

Interviewing the rest of the family resulted in the following pedigree:

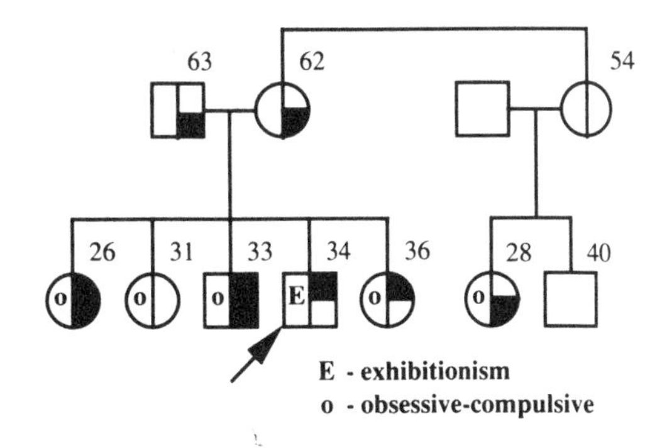

The family was riddled with TS. His brother had muscle tics in his legs throughout high school and college, throat clearing every day, and was always counting things — the number of people in a room or in a line and the number of objects on a shelf. One sister had eyeblinking and twitching, and compulsive counting. Another sister had no tics but in grade school had to touch everything an equal number of times with both hands, and take as many steps in a block of the sidewalk with one foot as with the other. Now she ritualistically checks the door each night over and over and checks that the oven is off. His youngest sister had occasional eyeblinking and throat clearing and touched things an even number of times as a child. His cousin had motor and vocal tics from age 5 to 10, and described herself as an extreme perfectionist with mood swings.

His exhibitionistic urges disappeared on 0.5 mg of Haldol per day.

Sam, was a middle-aged electronics engineer.

He developed rapid eyeblinking and foot tapping in grade school. The eyeblinking continued through high school at which time he also began to have repeated spitting, throat clearing and shouting which he had to do "to release tension." In college he developed head tics, and grunting and barking noises. For work he commuted long distances and began masturbating in the car. Two years later he was caught in the parking lot of a shopping center. He admits there were definite exhibitionistic aspects of this behavior and he wanted people to see him. Probation and the fear of being caught inhibited this activity for two years, but when he began commuting again he became obsessed with looking at women in the cars beside him and began masturbating again. He was caught once more and even though he was placed on probation, in six months he was doing it again. He was treated with 2 mg of pimozide which extinguished all these behaviors.

Lou was a 30 year old computer genius in a sensitive security job.

Between the ages of 6 and 14 he had the compulsion to take two, four or eight steps before reaching a crack in the sidewalk. At 10 to 11 he developed a hand washing compulsion. From age 9 to 25 when anyone touched him it would sometimes set him off in spells of laughing that could last 40 minutes. He also had a habit of blurting out things without thinking and was frequently beaten by his father for this behavior. At 22 he began compulsively touching his hair and drumming his fingers on the table, always an even number of times. At 14 he began thinking about being in public places without his clothes on and at 16 he would take off his clothes and run out of the house a short way and come back. He did this about 12 times then stopped. After finishing college there was a period of one month when almost every night he would drive around in the car without clothes on. Shortly afterward he was arrested for exposing himself on a beach. There have been intermittent episodes of public exhibitionism ever since. Although he had no motor or vocal tics there were multiple members of the family with compulsive behaviors and an aunt with chronic motor tics. Treatment with only 0.5 mg of haloperidol extinguished this behavior.

ADDH and Familial Exhibitionism

Gordon was "very active as a child" and was behind the others academically. In his mid 20's when alone in his sister's apartment he once took off his clothes and walked around for a few hours. In his early 40's he walked around nude in his house with the side door open. After this he developed more of an urgency to expose himself and almost everyday would drive around in his car at night naked or during the day without pants on. It "never occurred to me that I might get caught." One day he got out of his car in a parking lot with no pants on and was reported. He denied it and was not arrested. One month later he was arrested after exposing himself to a girl in the park. His exhibitionism also responded to haloperidol.

The pedigree was as follows:

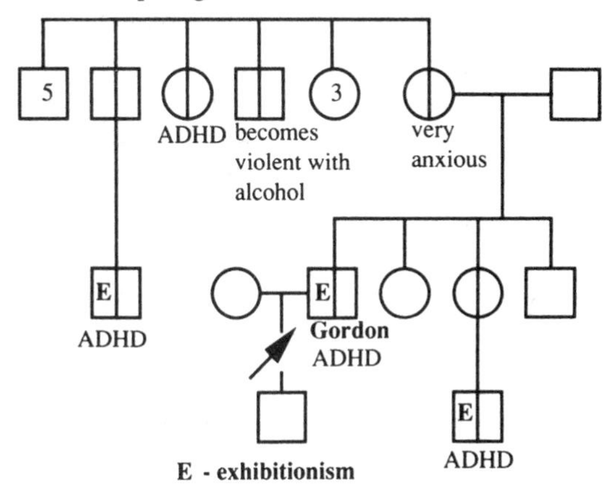

His sister had a teenage son who had ADHD and had been arrested twice for indecent exposure. He had run away from home several times, physically attacked his parents, and been in trouble in school for fighting. He had no history of motor or vocal tics. Gordon's uncle had a son with ADHD who had also been arrested for public nudity. No one in this family clearly had TS. This family represents familial ADHD with exhibitionism. As such, it represents a TS-like variant of a hereditary disinhibition disorder.

Hypersexuality

In addition to exhibitionism, some TS patients simply have trouble with hypersexuality, although they themselves may not see it as a problem. The following pedigree is an example:

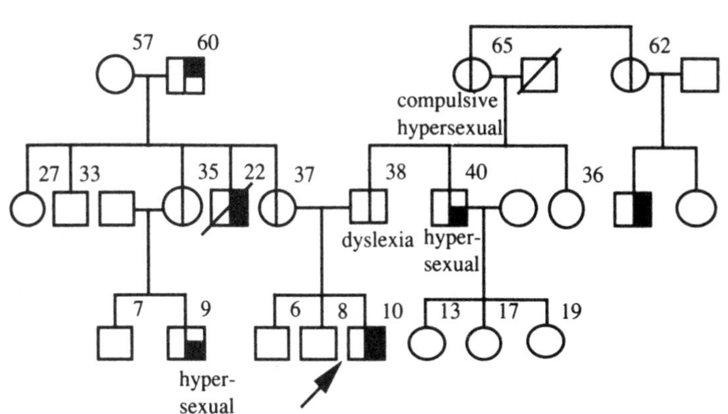

The index case was a 10 year old boy with TS. His 40 year old paternal uncle was characterized as follows:

> "He likes to feel other people sexually and touches the breasts and legs of women even when he doesn't know them. He has an increased sex drive and is obsessed with pornography." His mother "has a strong sexual drive, has many compulsive behaviors, and has problems with drug abuse and has wrecked six cars. To her everything is someone else's fault and she has sued people eight times. Tourette syndrome was also present on the mother's side. His nine year old cousin "is very sexual. At five years of age he was caught fellating his brother. He is obsessed with his mother's underwear and is constantly touching people sexually."

Masturbation

While taking pedigrees of TS patients the subject of compulsive masturbation may come up. The following are two examples.

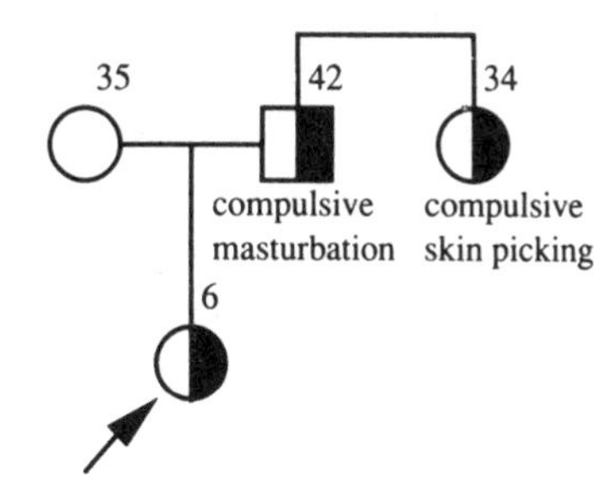

The father of this six year old girl stated that for two years, from age 10 to 12, he compulsively masturbated up to 10 times a day, even though he had no ejaculation. "I didn't know why I was doing it, I just had to." His sister had problems with compulsively picking at her skin.

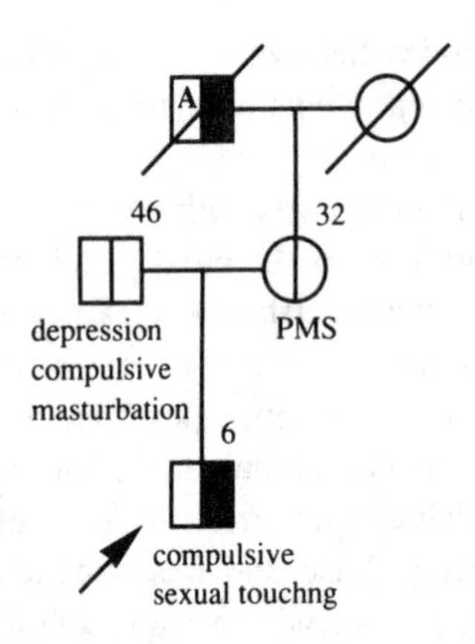

This six year old boy had such severe problems with wanting to touch the genitalia of other children that he could not be left alone in the classroom. His father had a long history of compulsive masturbation of three to four times a day. This disappeared with appropriate medication. The mother was also a presumptive *Gts* gene carrier since she had severe PMS[p243] and a father with TS and alcoholism[p225].

Gerald, another 6 year old boy with TS, was being cared for by his grandmother. He was compulsively masturbating several times a day and everytime he approached a female would attempt to feel her buttox or breasts, often reaching down the top of their blouse, rather than simply touching through their clothes. Another 3 year old grandchild was often present in the home and at every opportunity Gerald would lure him into a back room of the house and attempt to have sex with him. It quickly became the rule that whenever the younger boy was around no one was to let Gerald out of their sight. One day after a temporary distraction everyone noticed both boys were gone. A quick search found that Gerald had taken him into a back closet, removed his clothes and was attempting anal intercourse. Appropriate medication eliminated this compulsive sexual behavior.

Molestation

The hypersexuality, impulsiveness, disinhibition and poor ability to connect actions with consequences in some TS patients may combine to produce problems with molestation. Several examples have been given. The following is one more:

> Dan is now 32 years old. He developed motor and vocal tics at age 15. At age 21, for reasons he did not understand, he would go out into the desert and take his clothes off where no one could see him. Two years later he began to stand naked on the river bank so that people in boats could see him. He felt this was safe since he had time to drive away if they came toward him. He did this about ten times. He was in the Army and moved to Germany with his wife. While there he would go into the forest and walk around without clothes on. There were also periods when he felt compelled to watch pornographic movies. During these times his demands on his wife for sex increased greatly and he was more violent in his lovemaking. These periods would last for about three weeks and occurred three times a year.
>
> Upon returning to the States, he began to molest his sister's daughter who was five at the time. This consisted entirely of oral sex but continued for a period of five years. At age 28 he also began molesting his son. His exhibitionism continued during this time and he also began to search trash cans for women's underwear which he would wear. He finally turned himself in and is presently in prison.
>
> The pedigree is as follows:

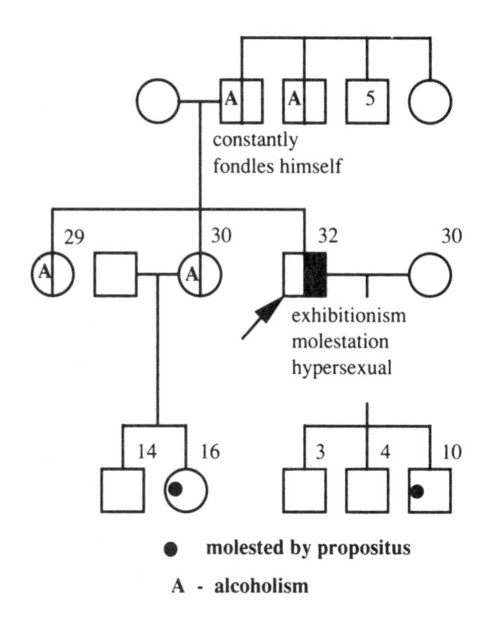

Many individuals in the family have problems with alcoholism. His father is alcoholic "and is always fondling himself at home. He always had his hand in his pants."

Misunderstood by Social Service Agencies

The disinhibited sexual behaviors seen in children with TS are often misinterpreted by social service agencies as indicating that there must be sexual abuse in the household. The following is an example:

Carlos This nine year old boy was very active even at ten months of age. In kindergarten he was constantly in and out of his chair, stubborn and very aggressive toward other children. By the end of first grade he began to make cat and dog noises in class associated with grunting and throat clearing. He had to repeat first grade. By the middle of second grade he was switched to half-days because he was too restless and aggressive. He began touching the genitalia of other children in the class. Because of this the school district insisted that he obtain psychological counseling. The psychologist immediately assumed that the behavior must have been the result of sexual abuse at home. The mother vehemently denied this. They threatened to not only take him out of the home but his three siblings as well. This caused the mother a tremendous amount of stress and she lost 20 pounds and was crying constantly. The mother kept insisting that there was "something wrong with my boy." After she made an appointment to be seen in the Tourette syndrome clinic, she was encouraged by the psychologist not to keep the appointment, claiming "That was not the problem." After the diagnosis of chronic motor tics with attention deficit disorder he was treated with 0.5 mg of haloperidol per day. This resulted in a cessation of all his vocal noises, a cessation of his sexual touching and a significant improvement in his school performance. A letter and phone call to the social worker and psychologist explaining that the sexual touching and aggressive behavior was due to the Tourette syndrome resulted in a discontinuation of all threats to remove the children from the home.

Kim Kim's story has already been told in the chapter on conduct[p145]. Recall that her initial manifestation of TS was as ADHD and compulsive masturbation. The latter was interpreted by a prominent child guidance clinic as indicative of sexual abuse by someone in the family and they were on the verge of taking her from the parents. Treatment with only .025 mg of clonidine a day extinguished this behavior. Two other distant members of the family had problems with exhibitionism.

Results of the Controlled Study

The above vignettes indicate that inappropriate sexual activity, including exhibitionism, can occur in Tourette syndrome. The next question becomes — Are these behaviors more common in children with TS, and if so, how much more common? To answer this we need to go to the controlled study. Although it was covered previously in the chapter on obsessions and compulsions, let us look again at sexual touching. This refers to excessive touching of one's own genitalia. This sometimes can be so severe that a child's pants are discolored in the crotch. The results were shown as follows[363]:

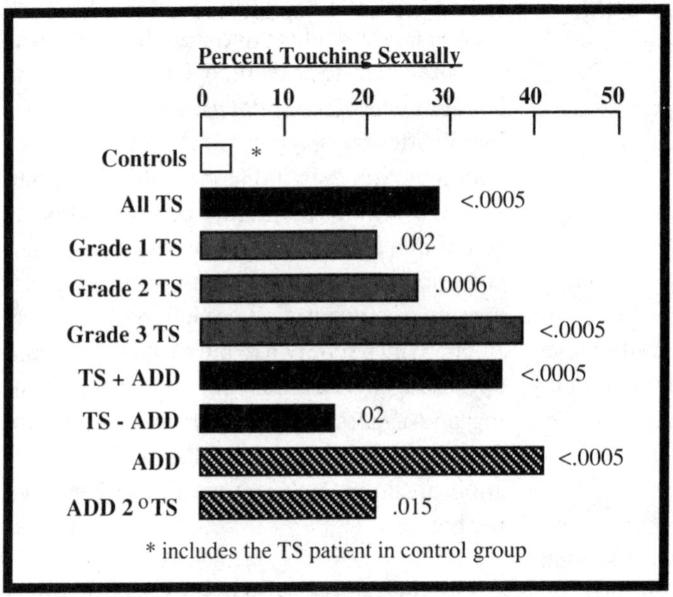

Percent Touching Sexually

0 10 20 30 40 50

Controls *
All TS <.0005
Grade 1 TS .002
Grade 2 TS .0006
Grade 3 TS <.0005
TS + ADD <.0005
TS - ADD .02
ADD <.0005
ADD 2° TS .015

* includes the TS patient in control group

All groups showed significantly more sexual touching than the controls. The important role of ADHD is indicated by the fact that children with ADHD showed a lot of sexual touching and the TS patients with ADHD did it twice as often as those without ADHD.

A related question is — How often do TS patients touch other people? These results are **shown on the right->.**

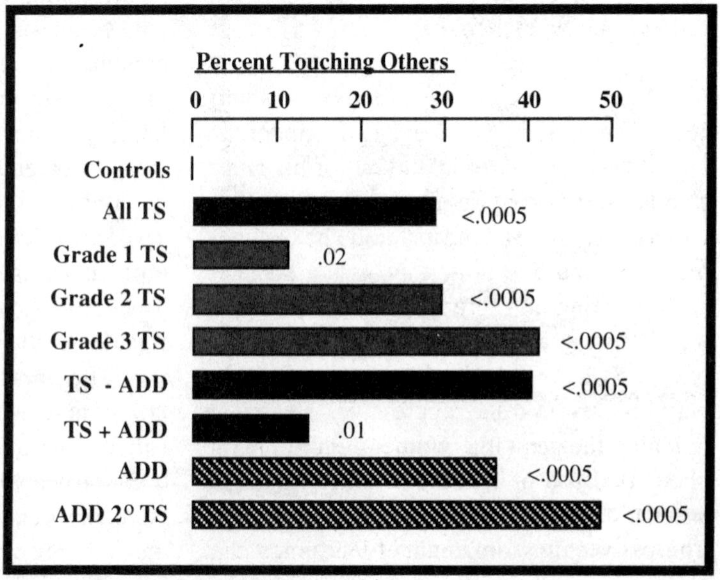

Percent Touching Others

0 10 20 30 40 50

Controls
All TS <.0005
Grade 1 TS .02
Grade 2 TS <.0005
Grade 3 TS <.0005
TS - ADD <.0005
TS + ADD .01
ADD <.0005
ADD 2° TS <.0005

Now none of the controls compared to 29 percent of the TS patients were compulsively touching others including sexual touching. Again this was related to ADHD in that the frequency was 3 times higher in the TS patients with ADHD (38 percent) than those without ADHD (13 percent). In addition it was present in 35 percent of these with only ADHD and showed the highest frequency in those with ADHD associated with TS (47 percent).

We can conclude from this that teachers and school psychologists should remember that *whenever a child in a classroom exhibits compulsive sexual touching of themselves or other children, that cannot be extinguished with simple reprimands, it might be prudent to have an appropriately trained professional rule out Tourette syndrome before involving social service agencies.*

Since interest in sex is a natural and common phenomena, the controlled study was critical in determining if a precocious interest in sex was more common in patients with TS. In response to the question "Did your child show a precocious or unusually early interest in sex?" the results were significant only for the grade 3 TS patients (yes for 30 percent of TS versus 14 percent of controls, p = 0.05). When asked if there were excessive thoughts about sex the results were again significant only for the grade 3 TS patients (yes for 36 percent of TS vs. 12 percent of controls, p = 0.007).

Exhibitionism

Since an interest in sex and thoughts about sex are normal human activities it is not surprising that they would not be a particularly sensitive indicator of sexual disinhibition in TS. A more sensitive variable would be an activity that is less commonly accepted as normal behavior. The results of the question about whether exhibitionism occasionally or often occurred were as follows:

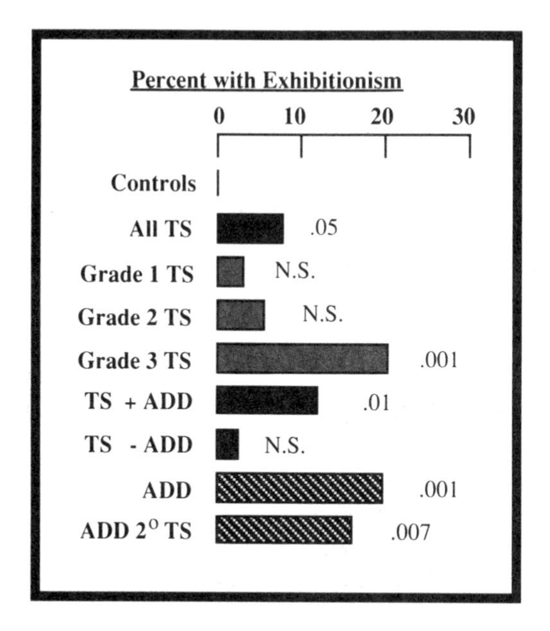

Here, none of the controls occasionally or often exhibited themselves compared to 7.7 percent of the TS patients. This increased progressively to a maximum of 19.2 percent in grade 3 TS. ADHD was an associated factor in that 19 percent of patients with only ADHD, and 15 percent of patients with TS and ADHD, exhibited themselves. Among the TS patients who did not have ADHD, only 2.2 had exhibitionism, while if they had ADHD this increased to 12 percent.

It is important to point out that *the majority of TS patients, 92 percent, do not have problems with exhibitionism.*

Coprolalia and Inappropriate Sexual Behaviors

It has been suggested that the coprolalia in Tourette syndrome can be explained in terms of a computer model of speech which randomly spews out a disproportionately high frequency of obscene words[1415a]. While this is an interesting observation, I do not believe it is a realistic model for TS for two reasons. First, with coprolalia the words that come out are pure obscenities, not a random set of phrases that happen to contain some swear words. Second, given the other types of disinhibited sexual and aggressive behaviors the coprolalia is best understood as being just one part of a generalized disinhibition disorder.

Summary: In addition to coprolalia, some individuals with Tourette syndrome have disinhibited or compulsive sexual behaviors such as sexual touching, an excessive preoccupation with sex, compulsive masturbation or exhibitionism.

Chapter 27
Phobias

While all of us have some reasonable fears or uneasiness about certain things, such as speaking in public, heights, spiders and snakes, we recognize that these are rational feelings and the fears are in proportion to the danger. By contrast a phobia is a "persistent and irrational fear of a specific object, activity or situation that results in a compelling desire to avoid the dreaded object or situation. . . The fear is recognized by the individual as excessive or unreasonable in proportion to the actual dangerousness of the situation"[490]. When the phobia is a source of such distress that it begins to interfere with one's life it is considered a phobic disorder. Such disorders come in three types: agoraphobia (which will be covered in the next chapter), social phobias, and simple phobias. Social phobias refer to situations where we will be scrutinized by others and we are afraid of being humiliated. This involves speaking or eating in public, or involvement in other public events. Simple phobias cover irrational fears of objects such as spiders, snakes, insects, or of situations such as heights or closed spaces.

Phobias in Tourette Syndrome

In talking with Tourette patients we noted that many seemed to have multiple phobias. Phobias are fairly common, as indicated by a recent extensive survey showing that 7 percent of the general population have a phobic disorder[1632]. Thus it was necessary to use our controlled study to determine if they were really more common in TS[362]. The questions were the same as those used in the above study of the general population[1630-1632].

Some phobias, such as fear of spiders and speaking to strangers, were so common in the controls that they were not significantly more common in TS patients.

Those phobias that were much more common in TS patients were:

	Cont %	**All TS** %	**p**
of being in a crowd	2.1	15.8	**0.01**
of being on public transportation	2.1	19.1	**0.002**
of speaking to a small group	10.4	22.8	**0.05**
of being alone	8.3	16.0	**0.009**
of being in water	2.1	15.8	**0.025**

The first three of these might be explained as simply being afraid of being seen in public because of the tics. However, if this was true we would expect that there would be a correlation between the number of tics and the phobia score. This was not the case since the correlation between tics and phobias was low. In addition, these phobias persist even when the tics are well controlled with medication. This explanation would also not account for the significant fear of being alone. Here only 8.3 percent of the controls were afraid of being alone compared to 26 percent of all TS patients and 33 percent of grade 3 TS patients.

In addition to looking at individual phobias a second dimension is — Are TS patients more likely to have multiple phobias than controls? To examine this we simply counted up the number of phobias for each person to obtain a phobia score. The distribution of the phobia scores for controls versus all TS patients was a follows:

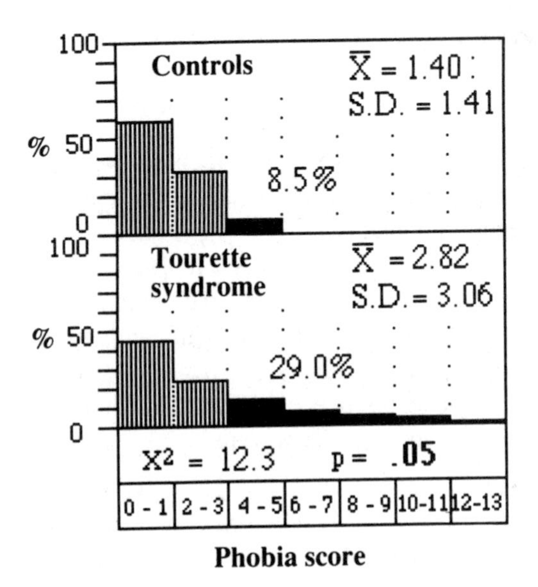

Phobia score

Some TS patients had up to 13 different phobias.

Among the three grades of TS, three or more phobias were present in 28 percent of grade 1, 29 percent of grade 2, and 38 percent of grade 3. Since "the weaker sex" is often thought of as having more phobias, we also examined this. Here 55% of the female TS patients had more than 3 phobias compared to 28% of the males.

Finally, as stated at the beginning of the chapter, the real factor which determines whether these phobias are a serious problem is the degree to which they interfere with life. None of the controls had both a phobia score of more than three and phobias that significantly interferred with their life compared to 19 percent of all TS patients and 31 percent of grade 3 TS patients ($p < 0.0005$). There was no correlation between the ADHD score and the phobia score. Thus, while the ADHD is a contributing factor to many behaviors in TS it does not contribute to the phobias.

The Inheritance of Phobias

Independent of agoraphobia, which is discussed in the next chapter, there has been relatively little evidence for a significant genetic component to simple and social phobias. Most of the information has come from studies of twins. In a study by Torgerson[1952] of 50 identical and 49 fraternal twins, there was a high estimate of heritability for phobias involving animals and bodily mutilations such as castration, but low estimates for social phobias. Rose and Ditto[1643] in a study of 354 like-sex twins not only obtained high estimates of heritability for fears of spiders, snakes and other small animals, but also for fear of death, dangerous places such as heights and social situations. The fact that dizygotic twins often show a high concordance for some fears[303,1643] indicates that phobias may be learned. This has led some to suggest there is only a minimal genetic component to simple and social phobias[1990]. The evidence for a significant genetic component to these phobias has been so limited that there has been little speculation as to the type of inheritance, if any. These results with Tourette syndrome suggest for the first time that a tendency to multiple phobias can be inherited.

Summary: Multiple simple and social phobias that can interfere with life are common in Tourette syndrome. This indicates that a tendency to multiple phobias can be inherited.

Chapter 28
Panic Attacks

A young lady down on her luck with love once disparagingly said to me, "Men are all the same, only their names change." I think of that statement when the subject of anxiety neuroses comes up — the symptoms people have are the same only the names change. When I took abnormal psychology in college, half of the book was filled with a discussion of the neuroses. Today that diagnostic entity does not exist in the DSM-III[490]. Did the neuroses go away? Of course not, just the name went away or more accurately — it was subclassified out of existence. What used to be anxiety neuroses are now:

a) obsessive-compulsive disorder (obsessive-compulsive neuroses),

b) phobic disorders (phobic neuroses),

c) generalized anxiety disorder,

d) panic disorder (anxiety neuroses), and

e) agoraphobia with or without panic attacks.

I have already discussed the presence of obsessive-compulsive behaviors[p113] and phobias [p173] in Tourette syndrome. In this chapter I discuss the remaining three, especially panic attacks and agoraphobia.

Panic Attacks and Agoraphobia

Anxiety is one symptom that everyone can relate to — we have all had it. Before a big test or an important date our heart beats fast, we perspire, have clammy hands, a dry mouth, feel short of breath, are fidgety, restless, and may go the bathroom every chance we get. Once the identified threat goes away — we pass the test or the date turns out to be just as we dreamed — the symptoms go away. The anxiety neuroses are characterized by the presence of these feelings of anxiety without an identifiable threat. In generalized anxiety disorder these symptoms persist for at least a month. In panic attacks they come out of the blue, last a few minutes or rarely a few hours, and go away. They are characterized by a sudden onset of intense apprehension, fear, or terror often manifested with a feeling of doom[490]. They may develop into agoraphobia — a fear of being alone or in public places. Often this develops over a period of multiple panic attacks. For example, if the first panic attack comes on when the person is on an escalator, they may associate the distress with escalators and avoid them like the plague. If the next one occurs when they were in a department store they then avoid department stores. Pretty soon their world has contracted down to the point they are afraid of leaving the house or even their room. It is often associated with a fear of being left alone. Sometimes this fear of being alone or leaving the house can develop in the absence of panic attacks and is referred to as agoraphobia without panic attacks.

We became intrigued with the relationship between panic attacks and Tourette syndrome when we noticed that an unusually large number of relatives of TS patients had agoraphobia. For example, of 90 females over the age of 18, presenting with TS or motor or vocal tics, 10 percent had agoraphobia with panic attacks[362].

The following are some examples:

Mary Ann was a 37 year old, attractive, vivacious woman who had an onset of facial tics, grunting and squeaking at 4 years of age. Repeated streams of obscene words started whirling through her mind (mental coprolalia) at age 10. At the time she was seen she had a wide range of problems including motor and vocal tics, obsessive thoughts of sex and violence, such as killing herself and her son by driving into a concrete wall at 100 miles an hour, kleptomania, and self-destructive acts such as cutting off her hair. She had a responsible, well-paid job which involved a lot of travel. She began to have panic attacks, especially on the freeway. This progressed into such severe agoraphobia that she had to quit her job.

Her pedigree was as follows:

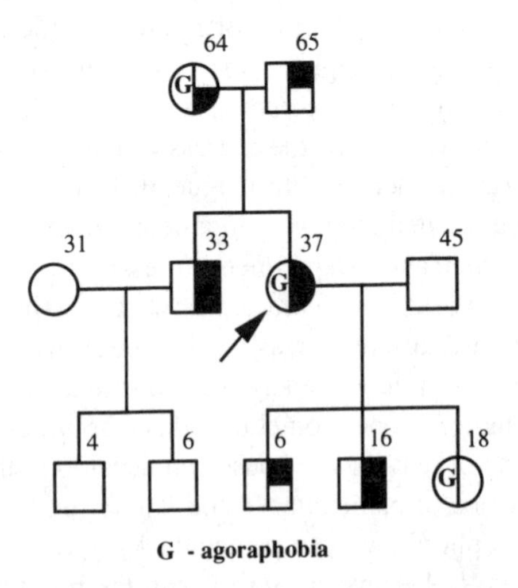

Her mother had repeated throat clearing, panic attacks and agoraphobia for many years. Her thoughts raced and she was forgetful. Her 18 year old daughter began to have panic attacks on escalators and while driving, and has become increasingly reluctant to leave the house.

This family had three generations of women with panic attacks and agoraphobia. The usual psychologial interpretation would have been that each successive generation learned the behavior from their mother. However, these are such random, spontaneous, and uncomfortable symptoms that this explanation is inherently difficult to accept. Here the presence of the *Gts* gene suggests these are inherited predispositions to panic attacks. In the daughter they were the only manifestation of the *Gts* gene.

Other families have also suggested that panic attacks and agoraphobia may be the only manifestation of a *Gts* gene. The following family is one example[365]:

A 16 year old boy was brought in by his parents. He began to have eyeblinking, eyes rolling backward, a neck tic and throat clearing at age four. By the time he entered kindergarten he had ADHD and the tics were quite severe. These problems have continued to the present.

The updated pedigree is as follows:

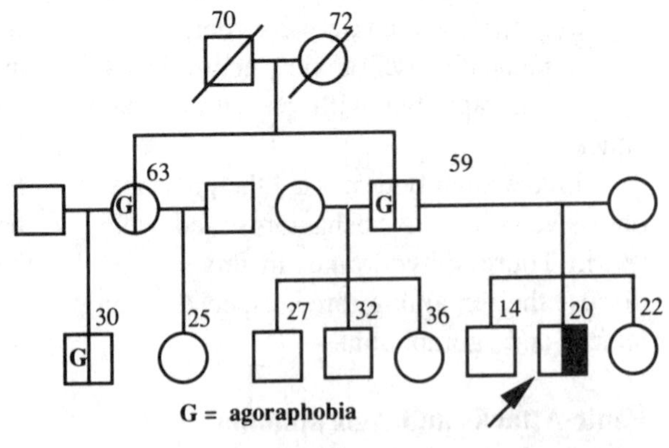

His father had no motor or vocal tics but had a brief panic attack at age 24 and again at 32. By 35 years of age he had a very successful business.

One day he simply walked into his office, developed shortness of breath and passed out. After excluding a heart attack he returned home and subsequently didn't leave his room for a year and a half because of intense fear of leaving the house. A psychiatrist visited him at home and made a diagnosis of paranoid schizophrenia even though he had no paranoid thoughts and was able to successfully run his business from his home. At age 37 he read about panic attacks and agoraphobia in a magazine and realized this was his problem. Slowly over the years he was able to occasionally get out of the house and be in a car as long as others drove. He is still agoraphobic but less incapacitated than before.

His sister also has severe agoraphobia. Her symptoms started at age 32 and persisted to her present age of 63. She has a son who had to have home tutors because of severe school phobia and at age 30 is still agoraphobic. This family suggests there were three individuals who were manifesting a *Gts* gene only as agoraphobia with panic attacks.

This is also suggested in the following pedigree where the proband was referred because of familial agoraphobia[365].

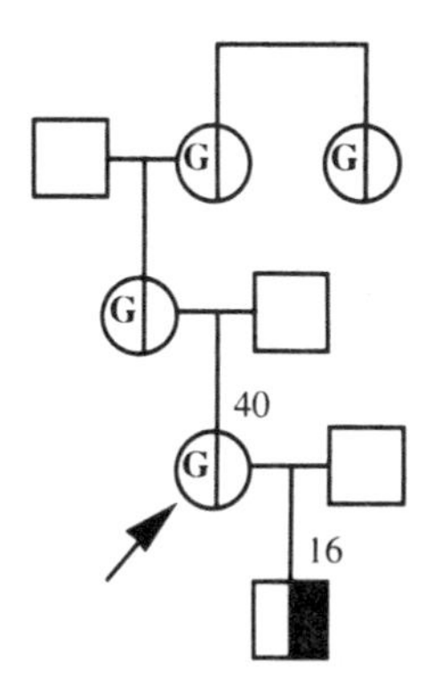

This 40 year old mother (arrow) had an onset of panic attacks with agoraphobia at age 22. These have continued almost daily ever since. Her mother had an onset of agoraphobia with panic attacks at age 24, and the maternal grandmother and grandaunt also had agoraphobia. The proband had a 14 year old son who showed an onset of eyeblinking, facial grimacing, head jerking, and hyperactivity at two years of age. He subsequently developed ritualistic licking of his shoulders and feet, palilalia, numerous phobias and obsessive-compulsive behaviors.

Here is a family referred because of familial agoraphobia in which the fourth generation presented as a typical, previously undiagnosed, case of Tourette syndrome. This suggests that in three generations of women the *Gts* gene manifested itself as panic attacks with agoraphobia, then finally produced a typical case of TS.

In another example, the phobias and panic attacks evolved into general anxiety.

This 32 year old *Gts* gene carrier began pulling her mouth to one side, eyeblinking and stretching her neck at 9 years of age. These symptoms were severe for two years then began to subside. By the time she entered high school all her tics had completely disappeared. She married at age 19. Her first child was born when she was 25 and the second two years later. About six months after her second child was born she began to develop phobias about many things she hadn't been afraid of before, including heights, bridges, small places, being alone, storms, water, spiders and freeways. One day she suddenly began feeling overwhelmed. "I felt like I was in a tunnel and my children's voices kept getting further and further away. I started shaking, having chills and my heart began pounding." These episodes began occurring daily and lasted 5 to 30 minutes. Within a year she was feeling constant panic and anxiety virtually every day. At times her pulse was as high as 160. She visited many doctors, including several cardiologists, who simply told her she was fine. During some of her panic attacks her hands and feet began to tingle and she was afraid she was going to die. She also had severe premenstrual tension for the week prior to her period. During these times she became paranoid and very irritable. She would sometimes become so angry at trivial things her husband did that she would begin to beat on him physically.

The pedigree was as follows:

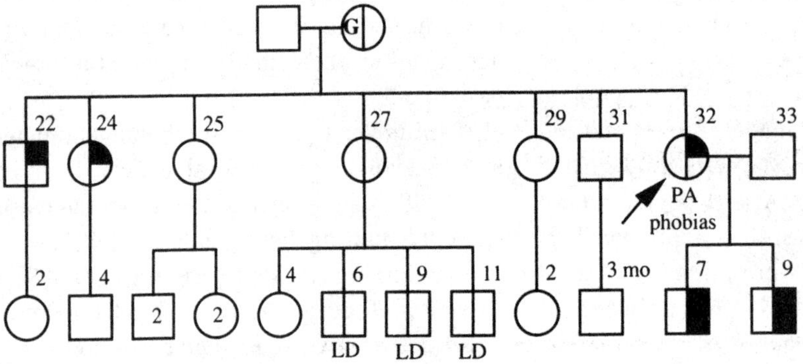

It was not until Tourette syndrome was diagnosed in both of her boys that her panic attacks were diagnosed. Treatment with a minute dose of imipramine (10 mg/day) resulted in a significant leveling out of her mood and a decrease in her panic attacks.

Here again is a woman who has had no motor tics since she was a young girl. Her *Gts* gene didn't begin to re-express itself until she was in her late 20's. Her natural history was as follows:

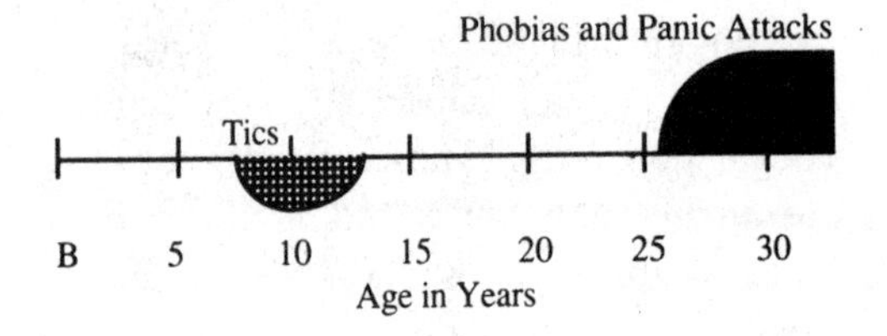

She was doubly misdiagnosed. She suffered from severe panic attacks for years before they were diagnosed, and their relationship to a *Gts* gene was recognized only because of her two affected children. Only then did questioning about her early childhood indicate she also had TS.

Other Studies

There have been only occasional references to TS and panic attacks in the literature. In 1962 Baker[88] described a 22 year old male with TS characterized by motor and vocal tics since age nine, longstanding complaints of genital and rectal autostimulation, sadistic "torture" fantasies referable to adult females, "hatred" of his father and severe panic attacks beginning at age 20. He was treated with a bilateral frontal lobectomy (leucotomy) which resulted in significant decrease in the tics and panic attacks. Much later Montgomery and co-workers[1353] described the first-degree relatives of 15 TS patients. Although none of the probands with TS were listed as having panic attacks, four of the 30 first-degree relatives had panic attacks. After our initial study on panic attacks and TS appeared[362,365], Sverd[1907] described a preteenage boy with TS and ten panic attacks occurring over a three month period. These and our own observations[365] led us to ask in the controlled study — What is the frequency of panic attacks in patients with Tourette syndrome?

The Controlled Study

In the controlled study we used the following questions from the Diagnostic

Interview Schedule[1630]:

> "Have you ever had a spell or attack (not due to a physical illness) when all of a sudden you felt frightened, anxious, or very uneasy in situations where most people would not be afraid?"

The results were as follows:

One-third of the TS patients stated they had panic attacks compared to only 8.3 percent of the controls. This increased to 55 percent in grade 3 TS.

If they answered yes to having panic attacks the next question was, "How many such attacks have you had in your life?" While none of the controls had three or more panic attacks, 16 percent of the TS patient did ($p = 0.002$). This increased from 7 percent in grade 1, to 12 percent in grade 2 and 31 percent in grade 3 ($p < 0.0005$).

To define the types of symptoms the next question was, "During one of your worst spells, which of the following problems were present?" These were: shortness of breath, rapid heart beat, light headedness, tingling of the arms or legs, tightness in the chest, choking sensation, feeling faint, excessive sweating, feeling shaky, hot flashes, unreal feeling, and being afraid of dying. Virtually every one of these symptoms of panic attacks were significantly more common in the TS patients than in the controls[362].

Finally, to determine what percent had both panic attacks and phobias we asked, "If you also had some of the phobias listed above, have the phobias and the anxiety attacks sometimes occurred together?" This occurred in 4.2 percent of the controls and 14 percent of all TS patients ($p = 0.06$) and 27 percent of grade 3 patients ($p = 0.001$).

When the number of panic attacks that each person had was placed in one of six categories of frequency the comparison of the controls to all TS patients showed the following:

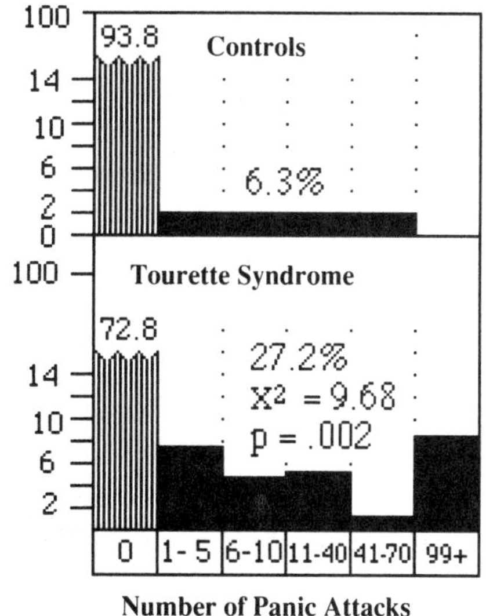

Number of Panic Attacks

Eight percent of the TS patients had more than 99 panic attacks compared to none of the controls. Even in grade 1 TS, 7 percent had more than 99 panic attacks. This is consistent with the pedigree data indicating that *panic attacks can be the only manifestation of a Gts gene.* In these patients the panic attacks were far more troublesome than the tics. If they had sought help for the panic attacks and were not specifically questioned about tics, the true diagnosis would have been missed.

These results indicate that a susceptibility to panic attacks can be inherited. What is the previous evidence that panic attacks are genetic in origin?

The Genetics of Panic Attacks

One of the greatest advantages of separating the anxiety neuroses into generalized anxiety disorder versus panic attacks is that this has significantly clarified the role of genetic factors. This is because there is a strong genetic influence on panic attacks but a relatively weak one on generalized anxiety disorder.

Prior to the time that this distinction was made there were a number of studies on the inheritance of the "anxiety neuroses"[227,332,344,1289,1412,1990]. These all agreed that about 15 percent of the first-degree relatives of patients with anxiety neuroses had a similar diagnosis compared to 1 to 3 percent of the general population. In the report by Cohen and co-workers[344] 55 percent of the mothers had an anxiety neurosis, reflecting the higher rate of anxiety neurosis in females. Slater and Shields[1810] studied 142 twins and found a concordance rate for marked anxiety of 65 percent for monozygotic twins compared to 13 percent for dizygotic twins.

The separation into panic attacks and generalized anxiety disorder allows these two subgroups to be examined separately. There was initially some confusion about the role of a cardiac condition called *mitral valve prolapse* and panic attacks since many patients with panic attacks were found to have mitral valve prolapse[1014,2112]. Because mitral valve prolapse causes heart palpitations, it was thought this disorder might be causing the panic attacks. However, subsequent studies showed the two disorders traveled separately in the families (segregated independently)[434,438,1476,1478,1481] indicating that the mitral valve prolapse was not causing the panic attacks. When individual family members were interviewed 31 percent of the first-degree relatives of probands with panic attacks also had panic attacks, compared to 4 percent of relatives of the controls[438]. The frequency of the panic attack gene was estimated at 0.005, indicating 1 percent of the general population carried the gene. Although they initially proposed that the gene acted as an autosomal dominant trait, in a subsequent study of 41 panic attack families, these investigators[436] were unable to distinguish between a single major gene versus polygenic inheritance. In this study 24.7 percent of the first degree relatives had panic attacks compared to 2.2 percent for the controls, and 61 percent of the families had another member with panic attacks. Among the probands who also had agoraphobia, 31 percent of the first-degree relatives had panic attacks.

In a **twin study**, Torgerson[1953] found that when generalized anxiety disorder was excluded there was a 45 percent concordance for anxiety disorders in monozygotic twins compared to 15 percent for dizygotic twins. When grouped as panic disorder, possible panic disorder or agoraphobia with panic attacks, 31 percent of identical twins were concordant compared to 0 percent of fraternal twins. He found no evidence for a significant genetic factor in generalized anxiety disorder.

Panic Attacks Progressing to Generalized Anxiety

Although family studies indicate that panic attacks and generalized anxiety disorder are separate entities, some TS patients have both. A young woman eloquently describes her progression from panic attacks to generalized anxiety.

> "One of the most severe and pervasive of my symptoms is a feeling of extreme anxiety, morbidness and feelings of doom and disorientation. They appear to surface for no apparent reason at times and yet, although often I can trace them back to a source, the feelings which I experience are totally disproportionate to their cause.
>
> "At first I experienced them rarely, but within the last eight years they seem to have become almost my natural state and I no longer think of them as 'attacks.' This is because they often stay for days at a time or weeks, sometimes months, and nothing I can do seems to be able to alleviate or appease them.
>
> "I feel so frightened that I find myself physically working my body into itself as if I am quite literally holding myself together. Also, I feel extremely disoriented and foggy as if I am in slow motion. I have also noticed that on these days it becomes very difficult for me to shift my eyes from a fixed state. I tend to fix my eyes and not want to move, as if moving will disrupt some delicate balance which I maintain. This feeling is often accompanied by what I might describe as a rumbling sensation inside. Also I feel as if I am 'boxed off' from my surroundings, like I am running parallel alongside myself."

School Phobia in Tourette Syndrome

School phobia is very common in Tourette syndrome. Given the number of problems TS children have with ADHD, learning disorders, motor and vocal tics, and compulsive and related behaviors it is no wonder that getting up in the morning and going to school is often looked upon with less than enormous enthusiasm. It is included in the discussion of panic attacks rather than the chapter on phobias because it is more like an anxiety or panic attack than a simple phobia. It has been reported to especially occur in the children of agoraphobic women and of women who were school phobic themselves when they were children[157]. Severe separation anxiety can contribute to the problem. In some cases every morning is a major war between the parents and the child. Headaches, stomachaches, vomiting and other somatic complaints rapidly proliferate and mysteriously disappear if the child is allowed to stay home. Oftentimes the battle of wills continues after they have gotten to school with visits to the nurses office and calls home with a multitude of ills. *The best approach is firm resolve*. Every time the parent or the school gives in the subsequent encounters will be even more difficult. The school needs to know that the parent will support them in not tolerating the constant complaints and phone calls home. I recall particularly one mother telling me that she became fed up with her TS son constantly calling her with various ills, trying to get her to take him home. She finally screamed into the phone, "Get back in the class — now." He docilely replied, "Oh, OK Mom," and it never happened again. One set of rules that often helps is to make it clear to the child that only severe illness will get them out of going to school. While school phobia can sometimes be precipitated or exacerbated by treatment with Haldol it is also common in TS children receiving no treatment.

Panic Attacks and Inhibited Children

When young children are examined for different personality traits, some are shy, inhibited or introverted. They react to new people or new situations with restraint and withdrawal. When the parents of these children were examined for the presence of panic attacks, agoraphobia or depression a very high percentage of parents with panic attacks or agoraphobia had inhibited children, while children of parents without panic attacks or depression were not inhibited[1643].

Precipitation of Panic Attacks by Lactate

When patients with a history of panic attacks are given sodium lactate intravenously, all the symptoms of the panic attack can be precipitated[602,1021,1182,1521]. This does not occur in normal controls. The two drugs most used to treat panic attacks, monamine oxidase inhibitors and tricyclic antidepressants[p906], block the precipitation of panic attacks by lactate[1021,1182].

A Brain Abnormality in Panic Attacks

Further evidence that panic attacks are an organic and not a psychological disorder comes from the observations of Reiman and co-workers[1596,1597]. These investigators examined blood flow in the brains of ten patients with panic attacks and six normal controls using the technique of positron emission tomography, otherwise known as PET scans. As I will discuss later, Tourette syndrome may be due to an abnormality of the limbic system of the brain. Reiman et al. looked at areas of the limbic system thought to be involved in anxiety, panic and vigilance[804,1458,1699]. Even when they had no symptoms, the seven panic attack patients whose symptoms could be precipitated by lactate infusion showed a significant abnormality of blood flow in the parahippocampal region with the blood flow on the right being greater than on the left. The three panic attack patients whose symptoms were not precipitated by lactate, and the six controls, did not show this asymmetry. This part of the brain represents the major input and output to the hippocampus, a region felt to be involved in the expression of emotion[804].

Panic Attacks, Depression and Alcoholism

Many studies have reported an increased frequency of depression or alcoholism, or both, in the patients themselves or the relatives of patients with panic attacks or agoraphobia [216,438,854,1376]. In one study 68 percent of 60 patients with panic attacks or agoraphobia had a past or current episode of major depression, most of whom were diagnosed as endogenous-type (organic) major depression. Leckman and co-workers[1147] found that when patients with major depression also had panic attacks their relatives were twice as likely to have major depression, panic attacks, phobia, and/or alcoholism than those with depression but no panic attacks. Patients with borderline personality disorder[p221] also have an increased frequency of depression, panic attacks and agoraphobia[815]. These studies are consistent with the presence of genes in the population that can be expressed in a variety of ways including depression, panic attacks, agoraphobia, alcoholism or borderline personality disorder. The theme of this book is that the *Gts* gene is a clear example of one of these genes.

Summary: Panic attacks, with or without agoraphobia, are common in Tourette syndrome, occurring in about one-fourth of patients. Pedigree studies suggest that panic attacks can occur as the only manifestation of the Gts gene. This high frequency in TS complements other evidence that panic attacks and agoraphobia are often the result of a genetic disorder.

Chapter 29
Depression

Other than love and hate, depression is one of the most universally experienced human emotions. We all have periods of feeling blue, down in the dumps and wanting to go to bed and let the world pass us by for awhile. Fortunately, for most of us, these are short-lived episodes that occur in response to life's minor reverses, such as failing an important test, not getting a particular job or date, or more serious ones such as losing someone close through death or divorce. These are often referred to as *exogenous or reactive depressions*. Unfortunately, there are many individuals for whom depression is more severe and longer lasting and comes out of the blue without any obvious precipitating event. These are *endogenous depressions*. In the DSM-III severe depression, whether endogenous or exogenous, is termed *major depression*. The flip side of such severe depression is *mania*. Together these are the *affective disorders*. The combination of depression and mania is *manic- depressive disorder* or *bipolar disorder*. This is covered in the following chapter.

Although there are many causes of endogenous depression, there is increasing evidence that a genetic predisposition plays a major role.

The Diagnosis of Depression

The diagnosis of major depression depends upon the presence of a pervasive sadness, feeling blue, or hopeless that persists for at least two weeks or more and is associated with at least four of the following symptoms: significant weight loss or gain, too little or too much sleep, restlessness or being slowed down, loss of interest in pleasure including sex, fatigue, feelings of worthlessness or inappropriate guilt, or recurrent thoughts of death and suicide attempts[490,491].

A somewhat less severe form of depression is called *dysthymia* [491]. Here fewer of the above symptoms are required but they must be present more days than not for at least two years. Dysthymia thus refers to chronic low-grade depression.

Depression in Tourette Syndrome

We have observed that many individuals with Tourette syndrome have significant problems with depression. This raises three questions:

1. Is depression more common in patients with TS than controls?

2. Is the depression related to the severity of the tics?

3. Can a *Gts* gene be expressed as depression only, that is without any tics?

Basically, the major question is whether the depression is simply due to having motor and vocal tics (exogenous depression) or is depression one more of the many effects that the *Gts* gene has on the brain (endogenous depression)?

The Controlled Study

The results of questions asked in the

controlled study to access the presence of depression in TS are shown in the following table.

Symptoms of Depression in Tourette Syndrome Patients Compared to Controls (%)

	Cont.	TS	Gr 1	Gr 2	Gr 3
N	47	246	43	145	58
Two weeks of depression?					
Yes	21.3	41.5	27.9	37.9	60.3
p	—	**.002**		**.02**	**<.0005**
Two years of depression?					
Yes	6.4	22.0	11.6	17.2	41.4
p	—	**.015**			**<.0005**
Two weeks or more of lost appetite?					
Yes	2.1	17.9	11.6	17.9	22.4
p	—	**.006**		**.007**	**.0016**
Loss of weight over several weeks?					
Yes	2.1	11.4	4.6	11.7	15.5
p	—	**.05**		**.05**	**.02**
Gained weight?					
Yes	2.1	21.9	11.6	17.9	39.7
p	—	**.001**		**.007**	**<.0005**
Two weeks of sleep problems?					
Yes	17.0	48.8	46.5	41.4	69.0
p	—	**<.0005**	**.001**	**.001**	**<.0005**
Two weeks of sleeping too much?					
Yes	14.9	24.0	18.6	21.4	34.5
p	—				**.01**
Two weeks of feeling tired out?					
Yes	27.7	41.5	32.6	40.7	50.0
p	—	**.04**			**.01**
Two weeks of talking or moving slowly?					
Yes	6.4	19.9	6.9	20.7	27.6
p	—	**.03**		**.02**	**.005**
Two weeks of unable to sit still, pacing up and down?					
Yes	2.1	37.4	23.6	31.7	62.1
p	—	**<.0005**	**.0015**	**<.0005**	**<.0005**
Several weeks of decreased interest in sex?					
Yes	8.5	11.0	6.8	11.7	12.1
p	—				
Two weeks of feeling worthless, sinful or guilty?					
Yes	12.8	27.2	14.0	24.8	43.1
p	—	**.02**		**.04**	**<.0005**
Two weeks of trouble concentrating?					
Yes	10.6	41.1	30.2	35.8	62.1
p	—	**<.0005**	**.001**	**<.0005**	**<.0005**
Thoughts slow or mixed up?					
Yes	10.6	32.1	25.6	26.2	51.7
p	—	**.001**	**.04**	**.01**	**<.0005**
Two weeks of thinking a lot about death?					
Yes	19.1	37.8	23.3	35.7	53.4
p	—	**.015**		**.04**	**<.0005**
Two weeks of wanting to die?					
Yes	12.8	22.8	14.0	17.2	43.1
p	—				**<.0005**
Felt so low thought of committing suicide?					
Yes	8.5	26.8	16.3	24.1	41.4
p	—	**.002**		**.009**	**<.0005**
Attempted suicide?					
Yes	0.0	8.94	7.0	6.9	15.2
p	—	**.035**			**.005**

Although 21 percent of the controls reported having 2 weeks or more of depression, this increased to 42 percent in all TS patients and 60 percent in grade 3 TS. An indicator of more severe and prolonged depression was to ask if it was present for two years or more. Now only 6 percent of the controls responded yes compared to 22 percent of all TS patients and 41 percent of grade 3 TS. Symptoms of loss of appetite, weight loss, weight gain, sleep problems, tiredness, being slowed down, agitation, feeling worthless, trouble concentrating, mixed-up thoughts, thinking of death, thought of committing suicide, and attempting suicide were all significantly more common in TS patients than in the controls. None of the controls compared to 9 percent of all TS patients and 15 percent of grade 3 TS patients had attempted suicide.

These results indicated that as far as individual symptoms of depression are concerned, TS patients have them much more often than controls. However, to make a diagnosis of major depression requires that an individual have multiple symptoms for two weeks or more. To examine this, a depression score was made with a 0 for the absence and 1 for the presence of a symptom. Since there were 18 symptoms listed the maximum score is 18. Since this was a self-report questionnaire we made the criteria for major depression even more strict and required that subjects have more than eight of the above listed symptoms.

Thus, a score of nine or more (in black) was considered to represent major problems with depression. The results of the distribution of the scores for each individual in the controls versus all TS patients was as follows:

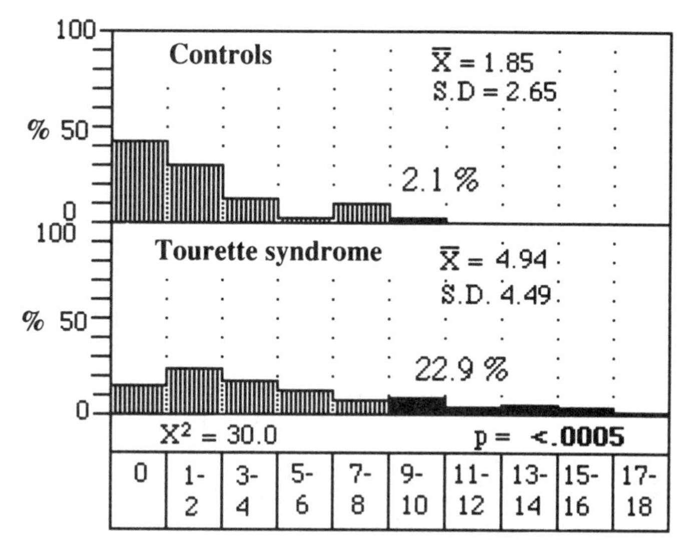

Depression Score

Only 2.1 percent of the controls had a score of nine or more and none had a score of greater than 10. By comparison 23 percent of all TS patients had a score of nine or more and they ranged all the way up to a maximum score of 18.

The distribution of scores for the three grades of TS showed that 14 percent of grade 1, 16 percent of grade 2, and 47 percent of grade 3 TS patients had problems with major depression.

Since these questions were framed in terms of "Have you ever. . ." the results referred to a history of depression anytime in one's life. To determine if the individuals were depressed at the time of filling out the questionnaire, a modification of what is called the Beck depression inventory[136] was also done. These questions are framed in terms of, "Are you now feeling. . ." Among the controls 6.3 percent had a Beck score of nine or more and none had a score of greater than 16. By comparison, 21 percent of all TS patients had a score of nine or more and some had a score as high as 31.

When TS patients with or without ADHD were examined, it was clear that the presence or absence of ADHD made no difference in the frequency of depression. For those without ADHD 24 percent had a depression score of 9 or more compared to 22 percent for those with ADHD. Further evidence that ADHD did not contribute to the depression in TS is indicated by the fact that the correlation between the depression score and the ADD score was a negligible 0.066. By contrast, as expected, the correlation between the depression score and the Beck score was high at 0.63.

Depression in *Gts* gene carriers One of the observations that led to our interest in the association of depression and TS was the frequent occurrence of depression in a parent who carried the *Gts* gene. These observations are important for the second and third questions —Is the depression related to the severity of the tics, and can a *Gts* gene be expressed as depression only, that is without any tics?

The following are some examples:

Elaine

A ten year old boy had typical

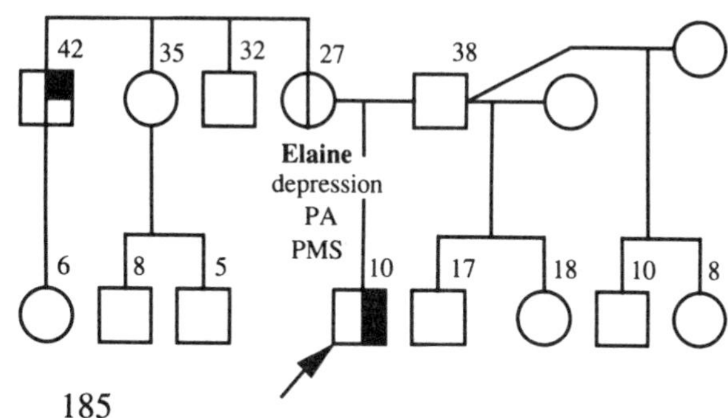

Tourette syndrome with motor and vocal tics and ADHD. His mother, Elaine, had no tics, but the fact that her brother had chronic motor tics indicated she was carrying the *Gts* gene. She had a five year history of severe **premenstrual tension** and would cry continuously for three days before each period. At age 25 Elaine developed a severe major depression with constant continual crying for months at a time and a weight loss of 20 pounds. She also had numerous panic attacks.

Dianne

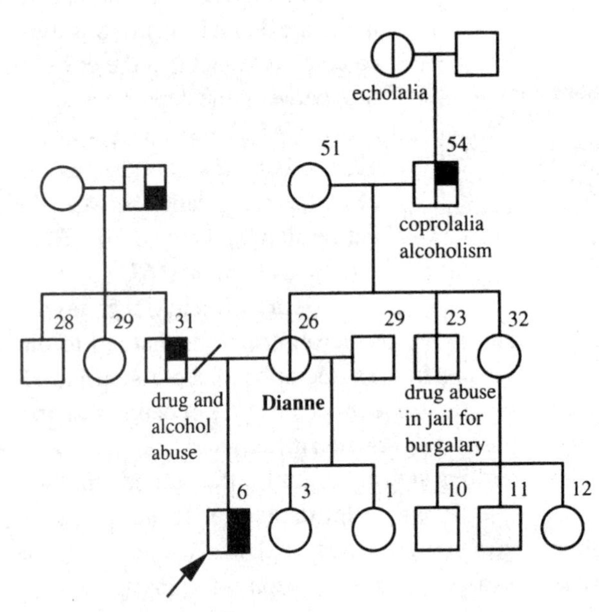

This six year old boy, Tim, was initially brought to the clinic at age four by his stepfather for evaluation because of severe behavioral problems.

His mother, Dianne, was in the hospital because she had just attempted suicide. At 6 months of age he began biting, tearing things apart and was very aggressive. By 18 months he began smearing feces on himself, the bed and walls. At this time, Dianne divorced her first husband because he was physically abusing both her and her son. This was so bad that he would sometimes beat Dianne in public. He had eyeblinking tics and abused drugs and alcohol. His father had repeated throat clearing.

She remarried a year later and despite a more stable environment Tim continued to be very aggressive and continued smearing feces. At age four he had only subtle tics consisting of biting the inside of his cheek, hand flapping, flipping his fingers, and a vocal "whoop" sound.

Our initial diagnosis for Tim was ADHD probably due to a *Gts* gene. He subsequently developed more prominent tics and the diagnosis was changed to TS with ADHD.

Dianne denied having tics as a child but had occasional throat clearing when nervous. She had an onset of major depression at age 16 and was hospitalized then for an attempted suicide. She blamed herself for her son's problems and had just attempted suicide for the second time when her present husband first brought Tim to the clinic. In the past two years she has had a very stormy course. Soon after having her third child she separated from her second husband and has been in and out of mental hospitals six times since with attempted or threatened suicide. Although she has occasional manic periods her primary problem has been agitated depression and panic attacks. In the past six months she has developed severe motor and vocal tics requiring Haldol to control them. Her father had facial grimacing and coprolalia and was an alcoholic.

This family presents several lessons. Virtually every one of dozens of mental health professionals who have seen Dianne had diagnosed her as major depression or manic depressive and all had missed the connection with Tourette syndrome. Even when we talk to them they have still denied this connection until her last hospitalization when she developed severe motor and vocal tics. Now, finally, her discharge diagnoses include TS. After she developed the tics her memory was jogged and she now recalls having similar tics as a child. Thus, her TS started in a typical fashion with motor and vocal tics in childhood which then went away and the picture evolved into one of manic-depression with predominant depression.

Her natural history was as follows:

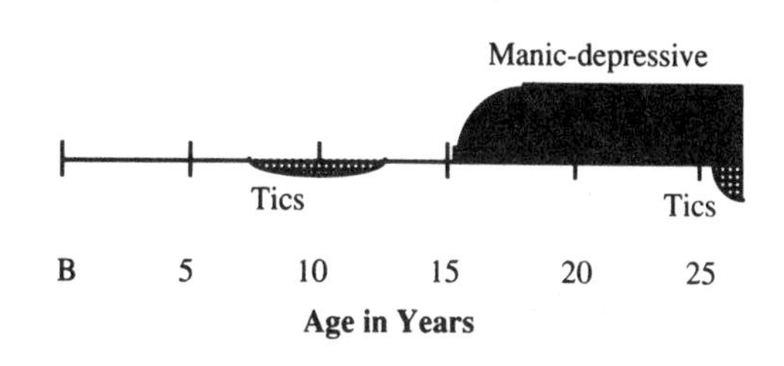

Although the very early onset of severe aggressive behavior and encopresis in Tim may be partly due to his abusive father, it is more likely that Tim is a homozygote for the *Gts* gene. His behavior continued to be so aggressive that he had to be placed out of the home. Finally, we see again, the not infrequent association between TS and physical violence and drug and alcohol abuse. The *Gts* gene truly brought much suffering and chaos to this family, as it does to many.

Suicide in a *Gts* Gene Carrier

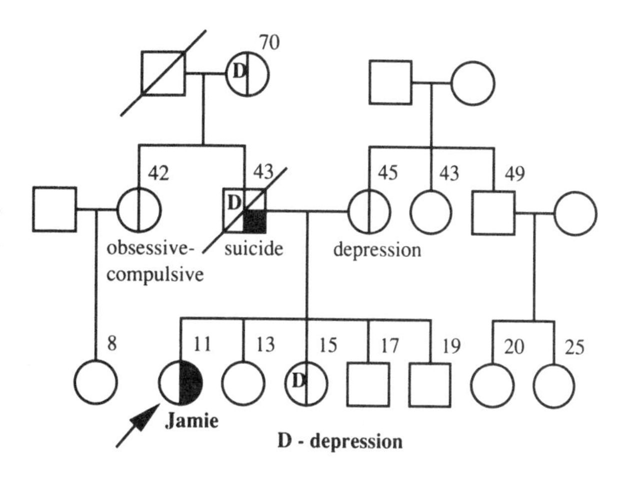

When Jamie was three years old she still was not speaking, but was a great screamer, constantly having temper tantrums and pulling at her clothes. At nine she developed rapid eyeblinking and began pulling at her hair. At 10 she began scratching her hands until they were raw. By the time she was seen at age 11 she had a wide range of motor and vocal tics, compulsive behaviors, overreaction to minor things, and in a rage once tried to choke her sister.

Her father had chronic grunting vocal tics and a lifelong history of depression requiring hospitalization and many years of therapy. He had a short temper and many panic attacks. At 43 he hung himself. His mother had also been hospitalized several times for depression.

David

David was 47 when he came to the clinic. He had just learned of TS from a friend and wanted to know if it was contributing to his problems. He recalls having many motor and vocal tics as a child, especially facial grimacing and throat clearing. These spontaneously improved when he was in high school. Since then he has had only occasional eyeblinking. At age 19 he began having panic attacks and these have recurred ever since. A year later he began having intermittent periods of depression during which he would hide out in his room for long periods of time. He got married for the first time at age 40. At age 44 he became so depressed that he drove his car to a deserted woods and intermittently put a gun to his head over a period of four hours before finally deciding not to pull the trigger. A year later he repeated this vigil in the woods. Antidepressants have improved his depression.

David's life can be diagramed as follows:

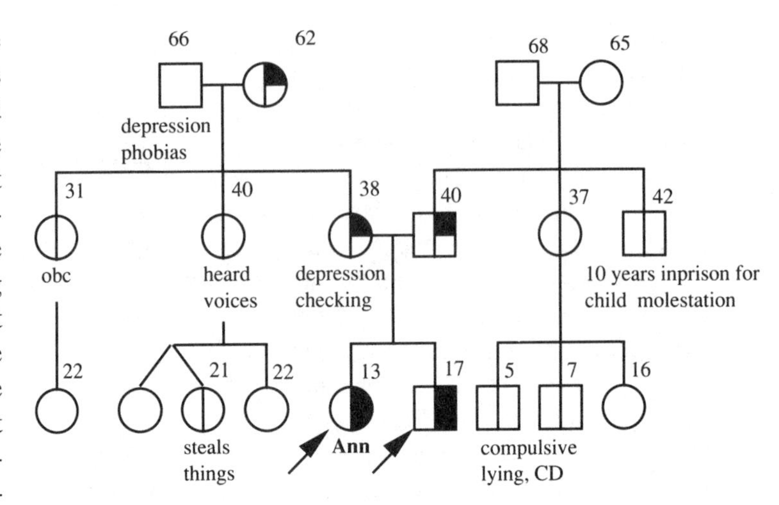

The 17 year old male (right arrow) was the initial proband in this family and had been followed for six years in the clinic, with severe grade 3 TS. At one of our support groups his mother took me aside saying, "I think my daughter, Ann, is developing TS." When seen

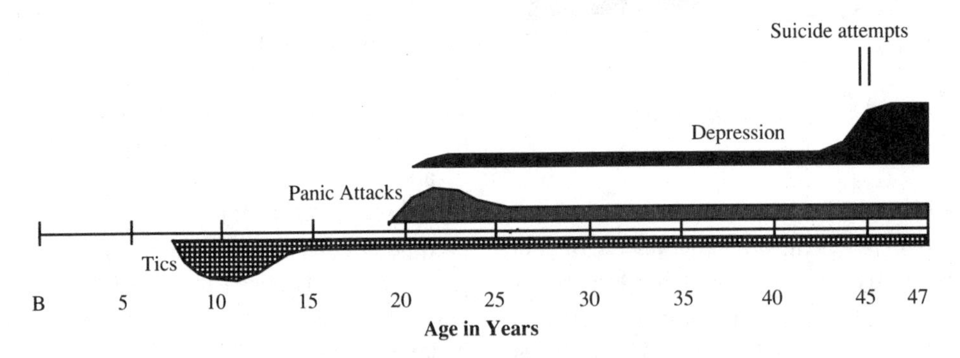

The tic phase of his TS was largely over by the time he was a teenager. The manifestations of his *Gts* gene as panic attacks and depression were far more devastating.

Depression in a pre-teenager One of the notable features of many TS individuals who speak of their depression is how early it often starts. *Many cannot remember a time when they did not having feelings of depression and of not enjoying life*. The following case illustrates this early onset:

in the clinic a short time later the following history was obtained:

> Ann recalls that at age eight she had periods of throat clearing that were noticeable enough that her fellow students often told her to keep quiet. Her mother never knew this as they soon subsided and were virtually forgotten. At age 11 her mother noted that Ann began to be very depressed. She always seemed sad, rarely smiled, was never happy, would cry at trivial things, and her teachers complained that she didn't play with

other children. A year later she developed rapid eyeblinking and squinting tics only when playing video games. These lasted until school was out in June. Three months later she developed severe stretching tics of her neck and torso, breath holding tics, throat clearing, checking compulsions and began hearing voices which told her to do good things such as go to bed on time. They also whispered lottery numbers to her, which unfortunately were never winners. Her appearance was that of a very depressed young girl who seemed on the verge of crying and who talked only when spoken to.

This case is particularly relevant in that except for a brief period of almost forgotten throat clearing, the severe depression started a year before minor tics and two years before major tics and persisted unabated for three years. This depression cannot be attributed to having the tics but rather to being an integral part of the expression of the *Gts* gene.

This is only a small sampling of many similar observations.

Depression in Mothers of Children with Tics

After we had reported these studies of the relationship between depression and Tourette syndrome[355], I became aware of the study of Corbett and colleagues[399] published almost 20 years previously. They examined 180 children with tics. Most of these would now be classified as chronic motor or vocal tics since only 11 percent had both, i.e., Tourette syndrome. Thus, the majority would be milder than most of the TS cases we studied. Despite this, they found that of a total of 184 parents of these children, 31 percent had a history of psychiatric illness, compared to 6.2 percent of 145 parents of children attending a dental clinic (p <.0001). They stated that over half of these were mothers suffering from mood disorders. This is summarized as follows:

History of Psychiatric Illness:

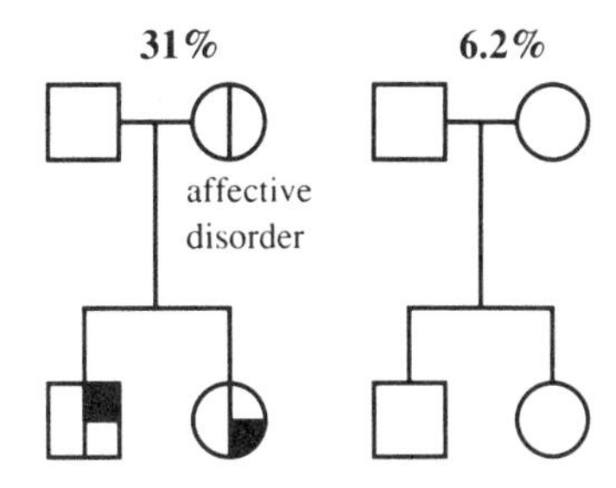

In psychiatric interviews on 30 of the children, 1 to 18 years later, 33 percent had depressive symptoms despite the fact that in many the tics had disappeared. Other problems in the follow-up group included aggressive and antisocial behavior, anxiety, phobias and schizophrenia. As children, 20 to 40 percent of the 180 patients had problems with sleep and speech disorders, aggression, short temper, lying, stealing, disobedience and anxiety.

The skeptic could propose that the depression in the mothers of these children was simply due to having a child with tics. Our family studies make this explanation unlikely since they show that the depression is often present before the children were born.

Depression in Attention Deficit Disorder

Like TS, relatives of children with ADHD are also more likely to have problems with depression than relatives of control children[164a]. In a study of the families of 22 children with ADHD, 31 percent of their relatives had a depressive illness compared to 6 percent of the relatives of 20 control children[166]. Among the ADHD children themselves 32 percent also had a diagnosis of a major mood disorder compared to 0 percent of the controls.

These studies suggest that in some cases the same gene that is causing ADHD is also causing the depression. This agrees with other studies that depression is common in children with ADHD[290]. Among the relatives of ADHD children who were depressed there was a significant increase in antisocial personality disorder (21 percent), separation anxiety (16 percent), obsessive-compulsive disorder (18 percent), oppositional disorder (21 percent), and drug dependence (25 percent). In a study of adults with ADHD[1782a] 45 percent had generalized anxiety disorder, 28 percent alcohol abuse or dependence, 26 percent drug use or dependence, and 40 percent mood swings. Only a small percent had ADHD only.

Depression and Obsessive-Compulsive Disorder

The prevalence of obsessive-compulsive symptoms in individuals with acute depression ranges from 25 to 40 percent[727,1026]. Usually the obsessive-compulsive symptoms appear or increase during the height of the depression, then decrease or disappear after the depression improves[1025]. The converse, episodes of depression complicating the lives of obsessive-compulsive individuals, can also occur[1869,1180]. This interweaving of the two and the exacerbation of both by external events suggests they are biologically related disorders. Viewed in this manner, the frequency of depression in Tourette syndrome, a common cause of obsessive-compulsive behavior, is not surprising.

Other Studies

Around the time we reported our observations on depression in TS[355], a study was reported of 90 TS patients by Robertson and colleagues in England[1625]. This study showed an increased prevalence of psychiatric problems in patients with TS, "in particular depressive illness, aggression and obsessive-compulsive disorder." Seventeen percent had major depressive illness and 28 percent had severe anxiety. A report on 50 TS patients in Holland also showed they were significantly more anxious, depressive, aggressive and obsessive-compulsive than control children[1991].

Depression in TS — Exogenous or Endogenous?

The questions I started with can now be answered. Depression *is* significantly more common in patients with TS than controls. Depression *is not* related to the severity of the tics. This is indicated by the fact that the correlation between the depression score and the number if tics was low, 0.27[355], and by clinical observations indicating that depression may develop long after the tics have disappeared. These same pedigree studies indicate that a *Gts* gene may be expressed as depression only, that is, without any tics. Finally, as shown in the next chapter, the depression is often a manic depressive disorder, rather than simple [unipolar] depression. Since we would not expect to see euphoric highs as a result of having tics, the most reasonable conclusion is that the *Gts* gene, in addition to many other behavioral problems, can cause mood swings associated with depression. Thus, the depression in TS is predominantly an endogenous depression.

Summary: Depression is common in individuals who carry a Tourette syndrome gene whether they have tics or not.

[See Chapters 90 and 98 for further evidence on the role of the Gts gene in depression.]

Chapter 30

Mania and Manic-Depressive Disorder

Mania is one of the most intriguing of the psychiatric disorders. Moderate degrees of mania, or hypomania, can be an intoxicating state. Such individuals tend to be active, energetic, happy, full of ideas, optimistic, creative, charismatic, enthusiastic about life and generally in excellent spirits. If this state can be maintained, these people can lead very productive and influential lives. However, hypomania can spill in both directions. If the high gets too extreme it can turn into a full manic episode. Afflicted individuals can't stop talking and show a pressure of speech, racing thoughts, hyperactivity, irritability, a decreased need for sleep, hypersexuality, overly expansive ideas and grandiosity. People in such a state may show poor judgment, go on buying sprees they can't afford, become involved in business ventures that are poorly planned and ultimately disastrous, and engage in promiscuous sexual activity.

In the other direction the manic highs may give way to depressive lows creating the roller-coaster course of the manic-depressive disorder. Many famous and very creative people have been manic-depressives including Balzac, Van Gogh, Handel, Rossini, Schumann, Joshua Logan, Ernest Hemingway, Theodore Roosevelt and others[598].

The Controlled Study

Since many of the symptoms of mania are similar to those that some TS patients report, we included in our controlled study of Tourette syndrome the questions about mania[355] from the Diagnostic Interview Schedule[1630].

For the TS group as a whole, questions relating to two weeks or more of being very high or excited, and a week or more of being very active, of talking so fast people could not understand you, of racing thoughts, and of being very distractible were all significantly greater than the controls. Only 4.3 percent of the controls had more than five manic periods compared to 19.6 percent of all TS patients and 37.9 percent of the grade 3 TS patients.

To examine how many of these symptoms were present in a given individual, the results of each question were added to give a mania score. The distribution of these scores for the controls and for all TS patients was:

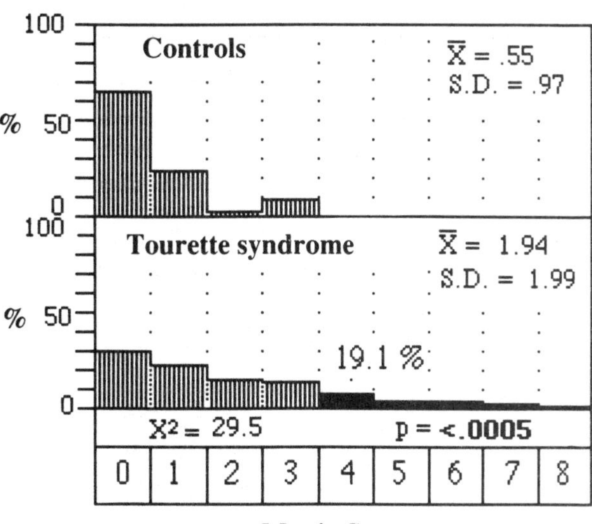

A score of four or more indicated problems with manic symptoms. None of the controls and 19.1 percent of the TS patients had manic episodes. A score of 4 or more was present in 0 percent of grade 1, 14 percent of grade 2 and 41 percent of grade 3 TS patients. Eleven percent of those without ADHD had a mania score of 4 or more. This increased to 24 percent for TS patients with ADHD. There was a high correlation, $r = 0.63$, between the mania and the depression score. This indicates that most of the individuals with a mania score of four or more also had a depression score of nine or more—that is, they had severe mood swings.

Manic Depression in TS Pedigrees

The association of manic-depressive disorder with TS was also apparent from our studies of individual families. The following are some examples:

Sue

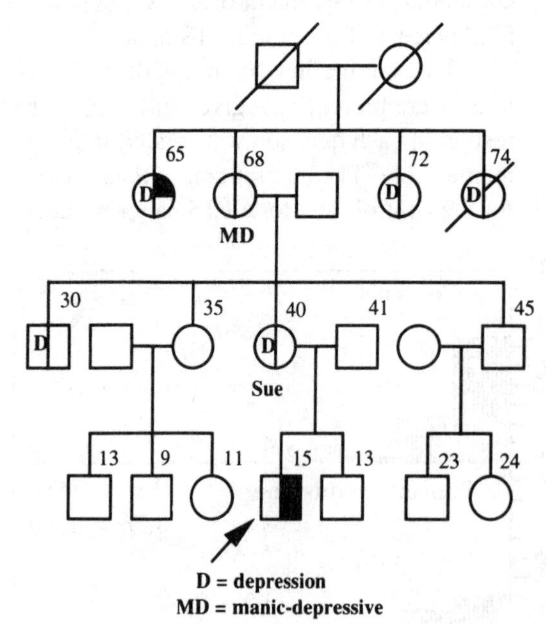

The proband was a 15 year old boy with motor and vocal tics and ADHD who responded well to 1.5 mg of haloperidol and imipramine. His mother, Sue, had severe depression following the birth of her second son. This **postpartum depression** lasted for six months. Four years later, while on a new medication for migraine headaches, she developed a severe depression which lasted for one and one-half years. Her depression was unresponsive to many medications and finally abated with Nardil, a monamine oxidase inhibitor[p192]. She had no motor or vocal tics but "was very organized" and got upset if things were not done on an exact time schedule. She is the presumptive carrier for the *Gts* gene since she had a maternal aunt with depression so severe that she was treated twice with electroshock, not realizing she actually had TS. She also had motor tics and depression. Sue's mother, also a presumptive gene carrier, had a life-long history of manic-depressive disorder. She had about 15 manic episodes in her life alternating with severe depression and periods of being normal. During her manic periods she was so severe (raving) that she needed to be physically taken to the mental hospital for treatment with lithium. Her two sisters also had episodes of depression. The uncle had several years of depression as a teenager and six months of depression following a divorce.

Thus, in this family there were three individuals who carried the *Gts* gene and in each one the major symptom has been depression or manic depression. In addition there are three other individuals with depression who probably carry the gene.

Father with a manic-depressive disorder

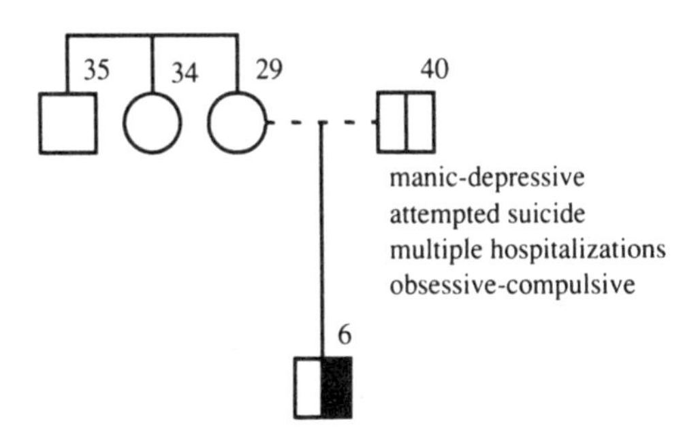

Here, the father of a boy with TS had been treated for years with lithium for manic-depressive disorder. He had been hospitalized several times for attempted suicide and had obsessive-compulsive behaviors. He left his wife shortly after his son was born.

Manic-depressive disorder and sexual abuse

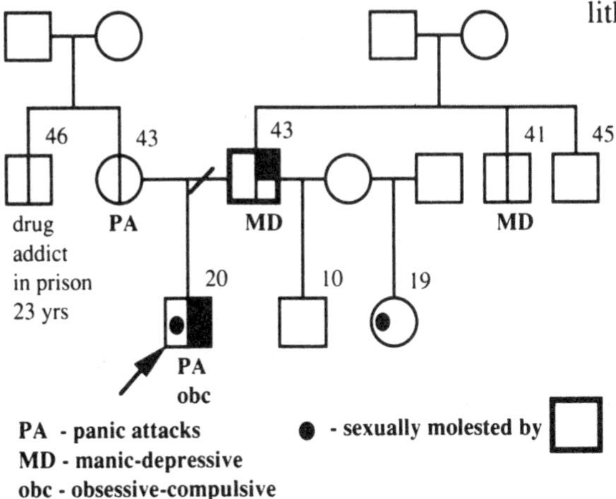

The proband, George, was a 20 year old male with an onset of motor and vocal tics at age 8, and panic attacks and depression at age 18. His mother had panic attacks and her brother had spent 23 years in prison for crimes related to drug abuse. The father and his brother both had been diagnosed elsewhere as manic depressive and both had been treated with lithium. The parents were divorced when George was in grade school. His father is now in prison for sexually molesting his stepdaughter over a ten year period, between the times she was 6 and 16 years of age. George had also been sexually molested by his father before his parents were divorced.

The relationship between sexual abuse and TS is discussed later[p263].

Other studies A number of individuals with TS who have also had manic-depressive disorder have been reported[176a,250a,1033a,b]. Burd and Kerbeshian[1033a] described three boys with TS and ADHD. At ages 8, 12 and 13 they also met the criteria for a manic episode or bipolar disorder. Relatives of these boys had problems with tics, depression and alcoholism. In another report of ten TS boys who also had bipolar disorder, they found that half responded to lithium[1033b].

Genetics of Depression and Manic-Depression

Depression and manic-depressive disorder, collectively known as affective disorders, have long been recognized to run in families. Whether this is a learned psychological behavior or a genetic disorder has been hotly debated. Let us look at the three major methods of dissecting out the relative role of environment versus genetics — family studies, twin studies and adoption studies.

Family studies When the parents, brothers and sisters of patients with major depression or manic-depressive disorder are examined, 5 to 32 percent also suffer from depression compared to 0.4 to 3.3 percent of normal controls[44,45,694,983,1310,2152]. This suggests genetic factors are involved but does not clearly distinguish between the environmental versus the genetic component.

Twin studies For manic depressive disorder the concordance rates average 60 to 70 percent for identical twins compared to 15 percent for fraternal twins[124,162,694,1847,1954]. Somewhat lower rates for both types of twins are noted for pure depression[1954]. The fact that the concordance rate is not 100 percent in identical twins indicates that some environmental factors are involved, and the fact that the concordance rate is 2 to 3 times greater for identical compared to fraternal twins clearly indicates that genetic factors play an important role.

Adoption studies The most diehard environmentalist could argue that the identical twins are more concordant because they share some mysterious bonding not present in fraternal twins. To counter this argument, adoption studies have also been done. Here manic depressive individuals who were adopted at birth were the subject of study, and the question asked was, which set of parents has more manic depression — the biological parents or the adoptive parents? If the biological parents have the more depression, then genetic factors are the only adequate explanation. If the adoptive parents have the more manic-depressive disorder then the children learned their manic-depressive behavior from them or from a shared environmental factor. A study by Mendlewicz and Rainier[1310] showed that the biological parents of adopted individuals with manic-depressive illness had a significantly higher frequency of manic-depressive disorder than did the parents who adopted them. In a second adoptive study, Cadoret[269] found that the incidence of depression was significantly higher, 37.5 percent, in children of depressed biological parents, than in children whose biological parents who were psychiatrically well, 6.8 percent.

Are manic-depression and depression different genetic disorders? Manic-depression has been referred to as a **bipolar** disorder because of the ups and downs, while pure depression has been called **unipolar** or **monopolar** disorder because there is only depression. Bipolar disorder appears to be a distinct genetic entity since individuals with bipolar disorder have many relatives with both bipolar and unipolar disorder while individuals with unipolar disorder have many relatives with unipolar but not bipolar disorder[44,45,693,1786,2052,2087]. The alternative explanation is that bipolar and unipolar disorder are on a continuum, with bipolar being the more severe form of the illness[1813].

Autosomal dominant inheritance of depression in the Amish Up until 1987, despite the extensive genetic studies of depression and manic-depression, there had been no clear-cut evidence of the type of inheritance—i.e., autosomal dominant with reduced penetrance, autosomal recessive or multiple genes[124,242,435,762,1423,1813].

In 1987 Dr. Janice Egland and her colleagues[520] reported on the result of years of study of manic-depressive disorder in the Amish. The Amish represent a **genetic isolate** in that the 12,000 Old Order Amish, living predominantly in Pennsylvania and surrounding states, were all descended from 30 individuals who emigrated from Europe in the early eighteenth century. Because of their religious beliefs there was very little intermarriage with anyone other than the descendants of these original people. This results in what is termed an **inbred population.** Because of this, certain recessive genes are quite common and dominant genes tend to be of a single type. Thus, if there were five different manic depressive genes in the general population it is quite likely that only one of these would have been present in those original 30 settlers and thus all the descendents would have only one genetic type of manic-depressive disorder. This simplifies linkage studies since only one gene locus is involved. When Dr. Egland and colleagues examined manic-depressive disorder in this isolate they found that it was inherited as an autosomal dominant trait with reduced penetrance. By the time people reached 30 years of age, 65 percent of those carrying the gene had symptoms of manic depressive or a depressive disorder and it was estimated that four percent of the Amish carried this gene.

They also looked at many markers to see if they could determine by linkage studies[p53] which chromosome and what chromosome segments this manic-depressive gene was on. The initial results showed no crossovers[p53] between this manic-depressive gene and a gene called *Hras*, a so-called oncogene involved in the production of some types of cancer (*onco* = cancer). Other genes close-by were the insulin and the globin (of hemoglobin) gene. These genes were on the short arm of chromosome 11.

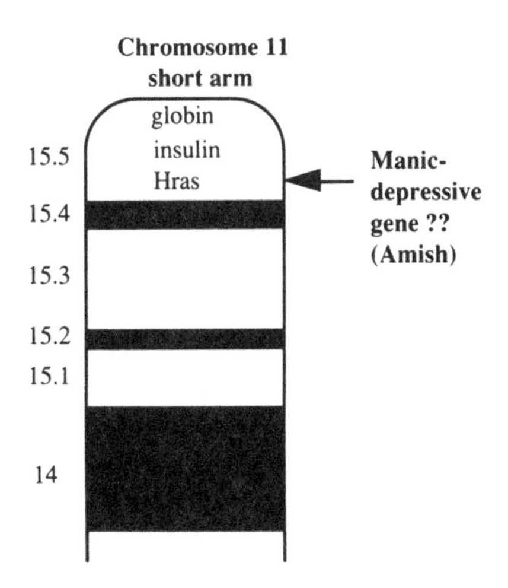

Less than a month later, in same journal, Baron and co-workers[127] reported finding another manic-depressive gene on the X-chromosome[1311] linked to colorblindness at the tip of the long arm.

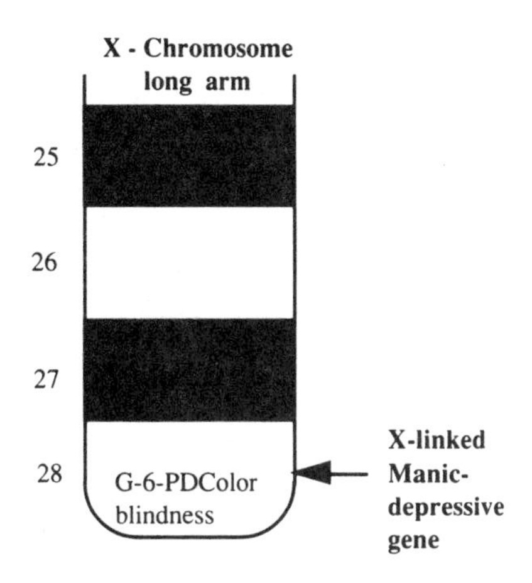

Others found that some families did not show linkage to the X-chromosome or chromosome11[482,903] and as a demonstration of how difficult these studies are, when additional family members were added to the Amish pedigree, and the results reanalyzed, the linkage to the short arm of chromosome 11 was no longer significant[1023a].

In three families a suggestion of linkage between manic-depressive disorder and stuttering was found[864a]. This is consistent with the involvement of a *Gts* gene since it is associated with both.

Depression Spectrum Disorder

In studies of the genetics of human behavior, the term **spectrum disorder** has often been used. For example, in the chapters on longitudinal studies of ADHD and ADHD children grown up, we saw that conduct disorder, antisocial personality, alcoholism and hysteria often clustered in families. If we wished to do genetic studies on ADHD, and limited our diagnoses to ADHD, fewer relatives would seem to be affected than if we expanded our net to include those with conduct disorder, antisocial personality, alcoholism and hysteria. Thus, we could speak of pure ADHD and ADHD spectrum disorder.

One of the best- known examples of using

the concept of a spectrum disorder came from studies by Winokur and colleagues[2085]. They looked at the family histories of 100 patients with depression and found evidence for at least two types of depressive illness. The first they called **depressive spectrum disease.** The prototype of this would be a female with an onset of depression prior to age 40 in whose family more depression is seen in female relatives than male relatives and the deficit in males is made up by alcoholism and antisocial personality. The second type is that of **pure depressive disease** in which the proband is a male whose illness starts after age 40, and in whom there are equal amounts of depression in both male and female relatives, and no increase in alcoholism or antisocial personality in the males. This study generated a lot of interest because it added the dimension of depression to the apparent link between alcoholism and antisocial personality[p83].

Although some studies have not supported this concept, others have. For example, in a study of relatives of patients with unipolar depression, Mendlewicz and Baron[1309] found that families of *individuals with an early onset of depression had more relatives with depression, alcoholism and antisocial personality than families of individuals with a late onset of depression.* Others have also found that the risk of having other family members with depression or alcoholism was greater for patients with an early onset of depression[176,2052], or with recurrent episodes of depression[176]. In the latter study, if an individual had a single episode of depression later in life only 3.4 percent of the relatives had problems with major depression. On the other hand, if an individual had recurrent episodes of depression starting early in life, 17.4 percent of the relatives had depression[176]. A similar trend has been noted in manic depressive disorder with increased risk of affective disease[973,1927] and alcoholism in relatives with early age of onset in the proband[1927]. The major message is that there is a subgroup of patients with familial depression where the gene produces depression in some members of the family and alcoholism and antisocial personality in others. Some of these latter individuals may be carrying a *Gts* gene.

Atypical Depression

A form of depression resembling Winokur's depression spectrum disorder is atypical depression or bipolar II disorder[38,478,513,514,539a,1829]. These individuals are both anxious and depressed. The mood swings are less severe than in manic-depression (bipolar I), and the manic episodes are milder and do not require hospitalization. The similarities between TS and atypical depression are shown in the **box on the next page.**

Although they respond to the usual antidepressants (tricyclics [p415]), they may respond even better to monamine oxidase inhibitors [p415,1585].

These similarities could be due to the fact that many or most bipolar II individuals may actually carry a *Gts* gene. If they are different genes, bipolar II disorder would be another example, in addition to TS, of a condition manifesting as a wide range of behavioral problems. The role of the neurotransmitter serotonin, in all of these behaviors, is discussed later[p429].

Secondary Depression

So far we have been concerned with individuals in whom depression or manic-depression is the primary diagnosis. There is another large category called secondary depression where the patient has some other primary psychiatric diagnosis associated with depression. These individuals "tend to have more of almost everything" — anger, hostility, phobias, anxiety, sleep disorders, multiple bodily complaints, verbal expansiveness and social obstreperousness. In a sense, the depression

Similarities Between Tourette Syndrome and Bipolar II Disorder (Atypical Depression)

	Tourette Syndrome	Bipolar II
Hereditary	+	+
Labile mood	+	+
Onset in childhood	+	+
Associated features		
Anxiety	+	+
Alcohol or drug abuse	+	+
Eating disorders	+	+
Enuresis	+	+
Irritable	+	+
Fatigue	sometimes	+
Learning disorders	+	+
Migraine	+	+
Nail bitting	+	+
Panic attacks	+	+
Phobias	+	+
Poor concentration	+	+
Premenstral tension	+	+
Self injurious behavior	+	+
Sexual disorders	+	+
Sleep problems	+	+
Waxing and waning	+	+
Worse with stress	+	+
Basic mechanism proposed		
Defect in serotonin	+	+

associated with Tourette syndrome is also a secondary depression in that the primary diagnosis is TS, although sometimes the tics are such a minimal symptom that the depression or manic-depression becomes the primary diagnosis. These distinctions can become blurred in other cases as well. For example, is an alcoholic with depression an alcoholic because he is depressed or depressed because he is an alcoholic? This is discussed further in the chapter on alcoholism. In children there is a strong link between educational failure and depression[2038], but it is not clear which came first or possibly, as in TS, the same gene causes both. These confusing inter-relationships will be sorted out as genetic markers become available for the different types of affective disease.

Endogenous versus Exogenous Depression

Another frequently used set of words in relation to depression is endogenous and exogenous. Endogenous depression refers to depression due to genetic factors while exogenous depression refers to depression clearly due to environmental factors such as the death of a loved one or feeling trapped in a relationship or job with no way out. The depression in carriers of a *Gts* gene is an endogenous depression[p183].

Winter Depression — SAD

An interesting form of depression comes on in the fall and winter and disappears in the spring and summer. This has been termed seasonal affective disease or SAD[1648]. It may be related to decreased light due to a shortening of the days and, in fact, patients often respond to exposure to bright artificial light in the home. Carbohydrate craving and abnormalities in serotonin have been implicated in SAD[478a,2113d]. Both of these features also occur in TS patients[p466].

Depression in Children

Many of our TS patients with depression have told us that their symptoms began in childhood. While depression is often thought of as an adult disorder, there is wide recognition that it is common in adolescents and in children and that the symptoms are similar to depression in adults. In a study comparing the symptoms of depression in children compared to adolescents, there were no significant differences in most of the symptoms. However, the children tended to show a greater depressed appearance, have more bodily complaints, hyperactivity, separation anxiety, phobias and hallucinations, while the adolescents showed a greater tendency to take no pleasure in life, sleep a lot, have feelings of hopelessness, and abuse alcohol and drugs[1669]. It is important to be alert to these symptoms in children and adolescents since depression is a treatable disorder.

Rapid Mood Swings in Tourette Syndrome

Many of our TS patients or their parents complain of mood swings. These are often very rapid with ups and downs occurring within hours of each other and many times in one day. Like more classic bipolar disorder they can also occur over a period of days or even months.

Summary: Wide mood swings are often present in individuals with Tourette syndrome. The fact that both depression and mania occur, often despite few or no tics, indicates that these mood disorders are an integral part of Tourette syndrome and not just secondary to having tics. The association of depression and manic-depressive disorder with Tourette syndrome indicates that these disorders can be inherited as a genetic trait.

Chapter 31
Schizoid Behaviors

Schizophrenia is a complex mental disorder. In the acute phase symptoms include the presence of bizarre delusions, illogical thinking, paranoid ideas, feelings of persecution and auditory hallucinations or hearing voices. While Tourette syndrome and schizophrenia are clearly distinct entities, there are scattered reports of individuals with TS and schizophrenia[399,1033d,1767,1918], occasional comparisons of the similarities between the two disorders[1918,2110], and sometimes TS patients are misdiagnosed as having childhood schizophrenia[358,1767]. We have observed that some TS patients have schizophrenia-like symptoms that are milder versions of those seen in schizophrenia itself. Specifically these are paranoid ideas, feelings of persecution and hearing voices. The following are some examples.

A 26 year old girl with TS and many obsessive-compulsive symptoms recalled that:

> "I was experiencing feelings of disorientation and confusion. I felt as if I was 'fading away' and became more and more dissociated mentally from my surroundings. Everything began to take on a dream-like state and my thinking became quite irrational and absurd. For example, I would think that the devil was interfering with my life and my thoughts. I must caution, however, that even at the most severe point in my illness, I always realized that this was absurd. Yet on another level I believed it — and believed it so strongly that I thought I would have to kill myself for all the things I thought I had done, yet I never totally lost contact with reality. I was, throughout, always painfully aware of what was happening to me, and I watched myself degenerate, completely helpless and unable to save myself."

The theme of **hearing voices** telling them to do bad things is not uncommon. A 16 year old girl related that:

> "I get scary paranoid thoughts that someone is going to kill me, or that I am going to kill somebody. These are worse when I am upset or angry. I sometimes hear two voices— a good one and a bad one. One tells me to do good things, the other bad things and I often hear them fighting with each other. They are often critical of me, critical of how I look and act."

Another teenage girl wrote:

> "I hate the voice. I hate listening to him and I hate what he makes me do. He's mad that I'm writing this but I've got to get what I feel out of my system. There is no way to get out of myself. I'm just so tired of obeying the voice. He's so mean. Sometimes he's funny and makes jokes about others. But he's always scary, and often makes me cry. He says he is going to grind me up in a blender or kill me in other ways. He puts me down by telling me that people don't

like me. That's why I want to get out of my body for awhile, so I can be free of myself, and of my thoughts. I feel like I'm always miles away from everyone else. I'm never 'with it' because I'm always somewhere with the voice."

After she had been taking clorimipramine[p583] (40 mg/d) for her compulsive behaviors the voices also began to diminish and her feelings toward them changed.

"His voice is now there one-third of the time when I want it to be, one-third when I don't, and it is gone the rest of the time."

"When do you want him to be there?" I asked.

"When I am feeling insecure and lonely and we can talk back and forth, and we are on good terms."

"What happens when you are not on good terms?"

"He says I'm wrong and a bad person."

TS patients may have thoughts that seem logical to them but which often result in inappropriate or antisocial behaviors. For example, a ten year old boy justified an act of stealing money in the following way:

"We were all supposed to have a dollar to go to a movie at school. This girl had $5 and I stole it because she shouldn't have more money than I did."

The illogical thinking and hearing voices is further illustrated by a seven year old boy who threw himself in front of a truck because a voice told him to and he wanted to see what it felt like. He didn't think he would get hurt and felt that after the experience would be able to get up and walk away. Fortunately the truck was able to stop seconds before it hit him.

The Controlled Study

To evaluate the extent of these symptoms in TS we asked a series of questions from the Diagnostic Interview Schedule[1630] concerning schizophrenic symptoms. Those that were significant are shown in the following box[363].

Summary of Schizophrenia-like Symptoms in Tourette Syndrome

Symptom	Controls %	T.S. %	p
Believe people are watching or spying on you	8.5	32.5	**.0009**
Believe someone is plotting against you	0.0	11.4	**.016**
Hearing voices	2.1	14.6	**.02**

The most significant feature was the thought that people were watching them or spying on them. While this is in part a result of feeling self-conscious about the motor and vocal tics, the degree of this perception is often out of proportion to the severity of the tics. For example, one teenager with relatively moderate motor tics and no vocal noises went to a restaurant with his parents. He felt that everyone in the restaurant was watching him and when he got home he was so overwhelmed by this feeling that he attempted suicide by overdosing on his medication. This feeling of being watched often persists despite the complete cessation of the tics with medication. The presence of this feeling in 28 percent of the grade 1 TS patients further indicates how this occurs despite very mild symptoms. It was present in 48 percent of grade 3 TS patients.

If the feeling that people were watching them was simply a result of having tics we would expect it to be limited to that. However, there was also a significant increase in patients who felt that someone was plotting against them or trying to hurt or poison them. None of the controls versus 11.4 percent of the TS patients had this feeling.

The final significant feature was that of hearing voices. This was present in 2 percent of the controls and 14.6 percent of the TS patients. It was present in 16 percent of the grade 1 TS patients and increased to 31 percent in the grade 3 patients. These voices are often identified as having come from the devil. Often the devil is telling them to do bad things.

Although the variables of believing people were following them, thinking someone was reading their mind, that others were controlling their thoughts or stealing thoughts out of their mind, thinking they were receiving special messages through the TV or radio, having visual hallucinations, and thoughts of violence or someone hurting them were all increased in the TS patients. These were not significant for the total TS group.

To determine if several of these features were present more often in TS patients than controls, they were added together to form a "schizoid score." This had a maximum value of 11. The results comparing controls to all TS patients were as **shown on the right->**.

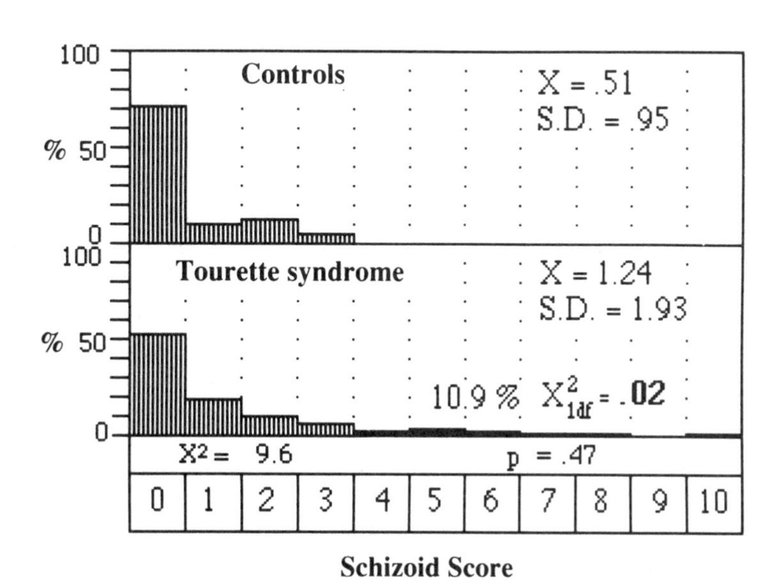

Schizoid Score

None of the controls had a score of four or more compared to 10.9 percent of the TS patients. In examining the three grades of TS, a schizoid score of four or more was present in 10.6 percent of grade 1, 7 percent of grade 2, and 21 percent of grade 3 TS. The presence or absence of ADHD had little effect on the frequency of these symptoms.

The other behavior score that showed the highest correlation with the schizoid score was the obsessive-compulsive score where $r=0.547$. The interplay between obsessive-compulsive symptoms and schizoid symptoms is shown in the following case[363]:

> The patient was a 15 year old girl whose brother and mother had Tourette syndrome. She never had motor or vocal tics, but she touched things a certain number of times and if she touched something with one hand she had to touch it with the other. Her mother says she seems to search for things to be upset about and is then relentless and doesn't drop it until she finds something else to be upset about. This drives the family to distraction. "She is inflexible." If her mother asks her to do something she repeatedly asks detailed questions about how to do it to the point her mother finds it easier to do things herself. She is confrontive and argumentative, and talks very rapidly. She has paranoid feelings

that people are looking at her "critically." In the clinic her mother pointed out that not everyone looks at her critically because she doesn't, at which point her daughter said, "Yes you do mother, you just don't realize it." She has no best friends. They somehow start out being friendly, but within 6 weeks can't stand to be with her anymore. This, of course, is a fault with the friends who are out to get her.

This case demonstrates various aspects of some patients with Tourette syndrome. It illustrates how females often present only as obsessive-compulsive behaviors and how these can paralyze not only family interactions, but personal relationships. It illustrates that the frequent paranoid feelings they have, such as that people are looking at them, are not simply because they have tics. Precisely the same feelings are present in individuals without tics and only with obsessive-compulsive behavior, or sometimes without any of the other manifestations of TS. Finally it illustrates the "no win" situation people often feel when interacting with some TS patients, i.e., there is always an explanation of why nothing is their fault, yet chaos reins around them[363].

The following are some of the responses elicited by the other questions:

Do you feel you have a special gift that others don't possess?

> "I think I have ESP where I can read minds and move things and make people do things they wouldn't ordinarily do."

Do you think that someone is reading your mind?

> "Some power from somewhere else has tapped into my mind and taken my ideas."

Have you been sent special messages?

> "People I dislike would get hurt and I would get the message on the news that they were hurt." "I couldn't help it, a voice made me do it."

Does anyone put thoughts into your mind?

> "The devil is placing thoughts in my head that I don't want." "I think people put bad things in my mind and then I do them. I think it is the devil."

Did you ever see things that others couldn't see?

> "I often see people, giant animals, cars, and trees that others don't see. Once I thought I saw a terrible car crash and reported it to the police — just to be told there was nothing there."

The Misdiagnosis of TS as Schizophrenia

Children with TS are sometimes misdiagnosed as having childhood schizophrenia. This usually occurs for two reasons. Most commonly, if the examining physician is not familiar with Tourette syndrome he may label the child as schizophrenic simply for lack of any other explanation for bizarre behaviors. Other times, there are a sufficient number of schizophrenic symptoms such as paranoid ideas, talking to themselves, hearing voices, or an intense involvement with fantasy playmates, to justify such a diagnosis. In the absence of awareness about the full range of symptomatology of TS, such a misdiagnosis is understandable.

In a series of 95 TS patients from North Dakota, 12 percent were found to have schizophrenic-like symptoms[1032,1033]. This was virtually identical to the 11 percent we observed. These studies also emphasized that many TS children may be mislabeled as childhood schizophrenia.

In a study of the Minnisota Multiphasic Personality Inventory (MMPI) in 29 TS patients and 29 controls, the TS subjects scored significantly higher on the schizophrenia score[811]. This indicated that TS patients "entertain occasional bizarre perceptions." They also scored higher on the scales for depression, hypocondriasis, psychopathic deviate and psychasthenia. The psychopathic deviate scores suggested a strong sense of social alienation. The psychasthenia scores reflected a preoccupation with anger and guilt.

Schizophrenia and Obsessive-Compulsive Behaviors

The high correlation between the schizoid and the obsessive-compulsive score, r = 0.55, is of interest in regard to reports of schizophrenic symptoms in individuals with obsessive-compulsive disorder[978]. Such patients had an extremely high rate of treatment failure. For example, if schizophrenic symptoms were not present, 90 percent improved. If some schizophrenic symptoms were present, only seven percent improved. The converse has also been reported[586] where persistent obsessive-compulsive symptoms were a powerful predictor of a poor prognosis in patients with schizophrenia.

The increased frequency of schizophrenia-like behaviors in TS patients is not entirely coincidental. As will be discussed in more detail in the next chapter, TS and schizophrenia have a number of intriguing similarities.

Summary: Schizophrenia-like symptoms of paranoid ideas, hearing voices, or having bizzare thoughts are present in some TS patients. The voices may often tell them to do bad things, or they may hear both a good and a bad voice. Lack of awareness of this aspect of TS can lead to a misdiagnosis of childhood or adult schizophrenia.

Chapter 32

The Symptoms and Genetics of Schizophrenia

Schizophrenia and Tourette syndrome have several things in common. They are both:

- genetic,
- involve abnormalities in dopamine[P373],
- as children, show deficits in attention with hyperactivity and conduct disorder,
- share some symptoms such as paranoid ideas and hearing voices,
- have both positive and negative symptoms,
- are common, and
- show a wide range of expression of the involved gene.

Because of these many similarities, knowledge about one may help us to understand the other. For this reason I will examine certain aspects of schizophrenia in some detail.

History of Schizophrenia

Emil Kraepelin was a German psychiatrist who in the latter part of the 19th century wrote an influential series of editions of his textbook *Psychiatrie*. In the fourth edition, published in 1893, to delineate a syndrome distinct from depression and mania he coined the term **dementia praecox**. This described a chronic, irreversible, slowly degenerating disorder of young adults characterized by impoverishment of emotions (affect), disturbance in interpersonal relationships, lack of motivation, loss of personal identity, and stereotyped behaviors[1089,1090]. In 1911 the Swiss psychiatrist, Eugen Bleuler, wrote a book entitled *Dementia Praecox oder Gruppe der Schizophrenien* (Dementia Praecox or The Group of Schizophrenias)[177,178]. In it he stated:

> "I call dementia praecox schizophrenia because I hope to show that the split of the several psychic functions is one of its most important characteristics."

This has led to the public perception of schizophrenia as a "*split personality.*" This is a poor concept since it mistakenly suggests there are two personalities, one intact and one separate or split off. It is much more accurate to think of schizophrenia as a *generalized disintegration of psychic functions*. Some of the characteristics that Bleuler ascribed to schizophrenia were fairly nonspecific:

> ". . . in schizophrenia. . . there are a number of symptoms which fall within the wide frame of what one calls, if not exactly 'health,' nevertheless not 'mentally ill.' Personality anomalies, indifference, anergia [lack of energy], querulousness, obstinacy, moodiness, . . . hypochondriasis etc, that need not be symptoms of a mental disease; but only too often they are the only visible signs of schizophrenia"[906].

This type of imprecision has plagued the diagnosis of schizophrenia for years. Schneider [1725-1727] attempted to make the definition more specific by listing a set of first and second rank symptoms. There were 11 first rank symptoms similar to those listed on the next page.

The Diagnosis of Schizophrenia

If you are still a little confused about what schizophrenia is, I will summarize the modern diagnostic criteria based on the DSM-III[490]. It consists of three parts:

A. There must be at least one of the following during some phase of the illness: bizarre delusions such as feelings of being controlled by outside forces, objects broadcasting thoughts into their head or withdrawing thoughts from them; various delusions such as delusions of grandeur or of persecution; hearing voices; incoherent speech; illogical thinking; or a marked decrease in spontaneous talking.

B. Deterioration from a previous level of functioning.

C. The presence of these problems for at least six months.

Following an acute phase, there are residual symptoms of social isolation, marked inability to function in society or hold a job, peculiar behavior such as talking to themselves in public or collecting garbage, poor personal hygiene, blunted inappropriate affect (mood), vague or strange speech, bizarre ideas and illusions. While further modifications of the definition were made in the DSM-IIIR[491] the essence is the same.

Positive and Negative Symptoms

While the most striking symptoms of schizophrenia are the delusions, hallucinations and bizarre speech, an equally important part is the apathy, lack of motivation and social withdrawal. To emphasize these differences Crow[429] speaks of positive and negative symptoms of schizophrenia, forming type I and type II syndromes. These syndromes have the following characteristics:

	Type I	**Type II**
Positive vs. negative	positive	negative
Symptoms	hallucinations delusions thought disorder	flat mood little speech loss of drive
Course	acute	chronic
Response to medication	good	poor
Intellectual impairment	no	sometimes
Cause	dopamine abnormality?	cell death?
Prognosis	good	poor

Both of these syndromes may be present in the same individual either simultaneously or at different times, usually starting with positive symptoms and evolving into negative symptoms[34,38]. While the subject of dopamine abnormalities will be discussed in more detail in Part IV it is relevant here to point out that while the positive symptoms are probably due to increased numbers of dopamine receptors in some nerve cells, the negative symptoms may be due to too little dopamine in other nerve cells[25,319,1148,1233]. This is particularly relevant to Tourette syndrome since while some of the TS symptoms appear to be due to dopamine hypersensitivity or increased number of receptors, other symptoms may be due to a functional deficiency of dopamine (see Part IV).

The Frontal Lobe and Negative Symptoms

Individuals who have experienced lesions in the prefrontal portion of the frontal lobe display symptoms that are remarkably similar to the negative symptoms of schizophrenia[36]. These are diminution in spontaneous movement and speech, loss of creativity, impaired attention and concentration, excessively concrete thinking, blunting of emotional response, and profound apathy[653]. Because of its importance in understanding the symptoms of TS, the frontal lobes are discussed in detail in Part IV.

Schizotypal Personality Disorder

A related entity is schizotypal personality disorder[490,1027]. Here the essential feature is the presence of various oddities of thought, perception, speech and behavior, similar to those described above, but not severe enough to meet the criteria for schizophrenia.

> "Not infrequently. . . among the brothers and sisters of the patients [with schizophrenia] there are found striking personalities, criminals, queer individuals, prostitutes, suicides, vagrants, [and] wrecked and ruined human beings. . ."[1090].

This poignant description of some of the relatives of schizophrenic patients was given by Kraepelin in 1909. He termed them "latent schizophrenics." This concept is similar to what Kety and coworkers were later to call "schizophrenia spectrum disorder"[1037-1039] and what was formalized in the DSM-III as **schizotypal personality disorder.** These might be termed "almost but not quite schizophrenics." Bleuler[177,178] spoke of them in the following terms:

> "If one observes the relatives of our patients, one often finds in them peculiarities which are qualitatively identical with those of the patients themselves, so that the disease appears to be only a quantitative increase of the anomalies seen in the parents and siblings. . ."
>
> ". . .schizoid characters, people who are shut-in, suspicious, incapable of discussion, people who in a narrow manner pursue vague purposes."

The term **schizotypal** was first coined in 1953 by Rado[1566] as a shorthand expression for a real psychiatric mouthful, "the psychodynamic expression of the schizophrenic genotype."

In summary, the term schizotypal personality disorder is meant to describe the "aberrant but non-psychotic relatives of schizophrenics and as a disorder that presents, in subtle form, the symptoms considered fundamental to classic schizophrenia"[1027].

I have examined the subject of schizotypal personality disorder in some detail since short of out and out psychosis it, more than classic schizophrenia, shows what effects a gene causing disordered functioning of dopamine nerve cells can do to behavior. If we are looking to compare the effects of the schizophrenia gene to that of the *Gts* gene, this is the best arena. The following are a collection of some of the statements, given by various writers to describe the group of schizotypal personality disorder syndromes that are most reminiscent of behaviors that we frequently see in grade 3 TS:

> Some schizoid persons seemed truly devoid of sensitivity and empathy and therefore capable of extreme cruelty to animals or man[1094].
>
> "Sudden surges of temperament and inappropriate motor responses to emotional stimuli. . ."[1009]

The paranoid traits which were one of their most marked characteristics are described in the following terms:

> "suspicious, sensitive, sullen, touchy, grouchy, morose, resentful, unforgiving, difficult, quarrelsome, self-conscious, jealous, litigious, critical, takes things the wrong way, has rows with all the family, doesn't get on with people, makes heartless accusations."[1811].
>
> "an inner life suffused with hatred, a profound anger. . .[1566,2153,2154]

Finally, Hoch[901,902] described a syndrome he termed **pseudoneurotic schizophrenia** that also contains many of the elements of some grade 3 TS patients. These individuals experienced persistent diffuse anxiety pervading all aspects of their life and multiple neurotic symptoms including obsessions, phobias, depression, feelings of unrealness, acting out aggressive behaviors, poorly controlled anger, episodes of rage, paranoid thoughts, and sexual preoccupations. Hoch[902] described them as "polymorphous perverse" manifestations, a term that has also been used by Shapiro to describe a subset of TS patients[1757]. None of the five cases he initially described[902] had a family history of schizophrenia, furthur raising the suspicion that some actually had TS.

Tic-like Movements in Schizophrenia

I have pointed out that the disturbed thoughts in some patients with Tourette syndrome are similar to those seen in schizophrenia. The comparison between these two disorders also works in reverse. Some of the abnormalities of motor or muscle movements seen in schizophrenia bear a resemblance to those seen in TS. Since the introduction of thorazine and related drugs in the treatment of schizophrenia, there is a tendency to assume that any motor abnormalities in schizophrenia are a side effect of the medication. However, long before such drugs were available Kraepelin [1091,1450] noted that:

> ". . .the spasmodic phenomena in the musculature of the face and of speech, which often appear as being extremely peculiar disorders, reminiscent of the corresponding disorders of choreic patients and which in no way bear the stamp of voluntary movement."

Bleuler[178,1450] also wrote of patients:

> "...performing all kinds of manipulations with their teeth..with *grimaces of all kinds and extraordinary movements of the tongue and lips.*"

Owens and colleagues[1450] in Middlesex, England, had the rare opportunity of comparing the muscle movements of 47 severe chronic schizophrenics who had never been treated with Thorazine-like medications and comparing them to 364 similar patients who had been treated. They found that there were a significant number of involuntary motor movements including movement of the tongue, puckering and smacking the lips, *facial grimacing, eye blinking,* eyelid tremors, finger movements, and other movements of the upper and lower extremities. The following figure shows the frequency of these movements according to severity and location:

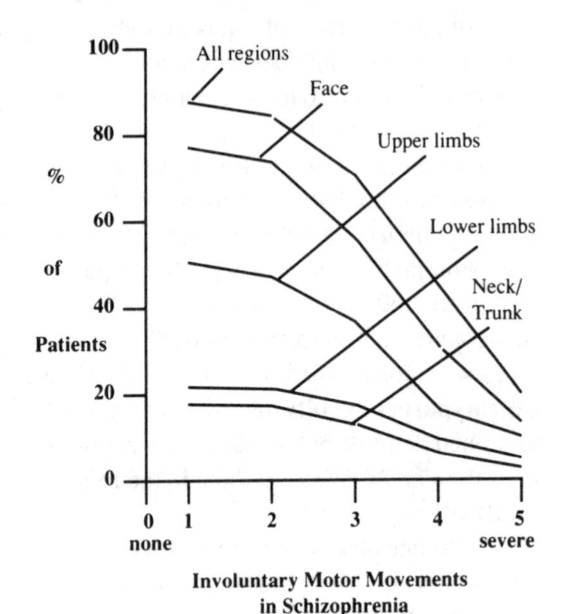

Involuntary Motor Movements in Schizophrenia

The striking feature about these motor movements is that they are comparable in frequency and distribution to those in Tourette syndrome, being most prominent in the face and decreasing in prominence in the extremities[1256].

A second type of motor problem in schizophrenia is the presence of *poor coordination* and other neurological soft signs in some of the young children of a schizophrenic parent[608,1258,1611]. *Specific reading disabilities* and *perceptual disorders* were also present[606]. This resembles what is seen in children with ADHD.

A third motor problem is a disturbance in smooth *eye tracking* in schizophrenia[918-922]. This was also present in relatives without schizophrenia and might be a marker of the schizophrenia gene.

Attentional Deficits in Schizophrenia

Attention deficits appear to be a core problem in schizophrenia. They are present in active schizophrenia[882,1082,1137,1285,1446], schizophrenia in remission[55,2097], unaffected first-degree relatives of schizophrenics[918-922], and children of schizophrenics[56,400,1414]. Schizophrenics with predominantly positive symptoms have more attentional deficits than those with predominantly negative symptoms[400,796]. Those with negative symptoms tend to have more problems with processing information[400].

The Genetics of Schizophrenia

Schizophrenia has been the granddaddy for studies of the genetics of behavior. The first twin studies of schizophrenia were carried out by Luxenburger in 1928, and the famous studies of Semour Kety, David Rosenthal and Paul Wender were the prototype for the use of adoptees to dissect out the effects of genetics versus environment.

Schizophrenia occurs in 0.85 percent of the population. On average 6 percent of the parents, 14 percent of the children and 10 percent of the siblings of schizophrenics have schizophrenia[781,1645,2151]. It is thus approximately 12 times more common in first-degree relatives than in the general population.

Twin studies. Many twin studies have been done[781,1847,1645,1646]. In the six studies done between the years of 1928 and 1961, which included only "severe" cases, the concordance rate in identical twins ranged from 58 to 77 percent compared to 2 to 18 percent for fraternal twins. In more recent times, where less severe cases were studied, and both schizophrenia and schizophrenia-like cases were included, the concordance rates in identical twins ranged from 6 perent to 48 percent while that of fraternal twins was 4 percent to 20 percent[1646]. In a subset of these studies, there were 17 cases where the identical twins were reared apart. Here the concordance rate was 65 percent. This counters the objection that the concordance rate in identical twins is higher simply because they more intimately share their environment.

Adoption studies Two major adoption studies of schizophrenia have been done, both using the extensive twin registries of Denmark. In one of these[1295,2061], 32 percent of children of a schizophrenic parent had a schizophrenic spectrum disorder compared to 18 percent of adopted children without a known schizophrenic parent. This latter value was greater than that expected of the general population. A study by Horn et al.[926] sheds some light on this. Using the MMPI test they studied 363 women who had given up their children for adoption. When compared to mothers who had not given up their children for adoption, they scored significantly higher on the schizophrenia or *sc* scale. This suggests that women with a predisposition to schizophrenia are more likely than average to give their children up for adoption.

The second major adoption study was by Kety and co-workers[11038,1039]. This study began with adopted individuals who had been diagnosed as schizophrenic and examined the biological and adoptive parents. Of the 173 biological relatives 21 percent had schizophrenia or schizophrenia-like disorder compared to 11 percent of the controls. This was significant at $p < 0.006$. When these data were independently re-examined by the modern DSM-III criteria of schizophrenia and schizotypal personality disorder, the conclusions remained unchanged and showed a clear genetic relationship between schizotypal personality disorder and chronic schizophrenia[1028]. Among the biological parents 10.5 percent had schizotypal personality disorder compared to 0 percent among the relatives of the adoptive parents (p = 0.002).

A clue that a schizophrenia gene may be on chromosome 5 came from a family in which two members with a partial duplication of this chromosome both had symptoms of schizophrenia[132]. A study using linkage analysis of families from Britan and Iceland showed that one gene for schizophrenia was on the long arm of chromosome 5 near the centromere[1784a].

One of the interesting aspects of this study was the observation that the evidence for linkage was strongest when individuals with other psychiatric disorders such as alcoholism, phobias, and depression were included. This schizophrenia gene appears capable of causing a spectrum of disorders wider than schizophrenia alone, similar to what appears to occur with the *Gts* gene. Linkage studies on families from Sweden were negative for these chromosome 5[1030a] markers, indicating there is more than one gene for schizophrenia.

Depression in Schizophrenia (Schizoaffective Disorder)

When depression is a prominent part of schizophrenia, the diagnosis is usually changed to schizoaffective disorder, meaning there are elements of both depression and schizophrenia. Is schizoaffective disease a disorder separate from schizophrenia? The easiest way to investigate this is to ask if there is a difference in the incidence of depression or schizophrenia in the relatives of patients with schizoaffective disorder compared to schizophrenia alone or depression alone. What most studies showed was that there was an increase in both depression and schizophrenia. For example, Mendlewicz and co-workers[1312] found the following frequencies of affective disease and schizophrenia in the families of the following types of probands:

Diagnosis in the proband	**N**	**% of relatives with affective disorder**	**% of relatives with schizophrenia**
schizophrenia	55	8.6	**16.9**
unipolar disorder	55	**28.5**	3.2
bipolar disorder	55	**39.4**	1.8
schizoaffective	55	**34.6**	**10.8**

Schizophrenia and the unipolar and bipolar affective disorders tended to breed true in that the relatives of schizophrenics had a much higher frequency of schizophrenia than affective disorder, and the relatives of patients with affective disorder had a much higher frequency of affective disorder than schizophrenia. By contrast, the patients with schizoaffective disorder had a high frequency of relatives with either schizophrenia or affective disorder. However, the frequency of schizoaffective disease in the relatives was too small to tabulate. Others have reported similar results[126,402,692,693,1964].

These studies suggest that schizoaffective disorder is not a distinct genetic entity in that it does not breed true. It actually has characteristics of both schizophrenia and affective disease in that the frequency of relatives with schizophrenia or affective disorder is comparable to that of patients with schizophrenia only or affective disease only. One possible explanation is that individuals with schizoaffective disorder are **mixed heterozygotes,** that is, they represent the coming together of two genes in one person, one for schizophrenia and one for affective disease.

I have discussed schizoaffective disorder in some detail because it is one of the better examples of what can happen when the genes for two different behavioral disorders come together. This can result in what appears to be a new disease, while in fact it is simply the result of having one gene for two different disorders.

This has relevance to Tourette syndrome in that some of those patients with TS plus other severe problems, such as overt schizophrenic symptoms, may be due to the fact they have inherited genes for two disorders. The following case illustrates the probable interaction of *schizophrenia* and *Gts* genes:

> Sam was a two year old mulatto boy brought in by his foster mother. She could not understand the boy's behavior and had been told by a neurologist that all he needed was a good spanking. She inherently understood that this was not his problem. She took over the care of him when he was 1 year old. At that time he was a severe head banger and had been doing this since he was 6 months old. He also pulled his hair and hit himself. She described him as "always unhappy, withdrawn, 'spacy' and very hyperactive." She initially felt that he would get better with love and affection but instead he continued to get worse. By 14 months of age he began self-destructive acts such as biting his arms and legs and putting his hands on the oven door, sometimes burning them. He also began making strange gurgling, grunting and growling noises. He often went into uncontrollable rages out of the blue or for minor things such as dropping something. When angry he would bite anything in sight including himself, other people or door knobs. By 2 years of age he was only saying "ma ma and da da." He usually did not like to be held, but occasionally would cuddle.

The family history provided important clues as to what was going on.

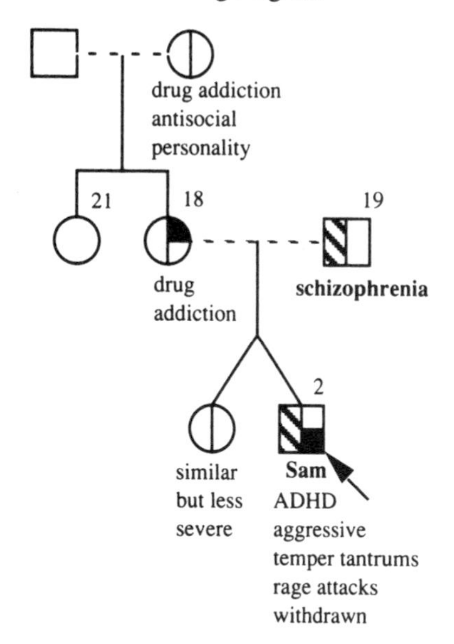

The Caucasian mother was described as "real mixed up." She was taking cocaine daily for the first five months of her pregnancy but claimed to have stopped. She "drank some." She had migraines, rapid eyeblinking and was "very fidgity and twitchy" and did very poorly in school. She was very irresponsible and would leave her children with friends for days at a time. Her mother was also a drug addict and so irresponsible that the grandmother had to raise the children. Sam has a twin sister, also cared for by the foster mother. The sister had problems similar to Sam's, but less severe. The father was black and schizophrenic. He once took a gun to a social worker and once tried to kill Sam by holding him under the water in the bathtub. He "acted like a zombie."

The presence of eyeblinking and other "twitches," and learning disorders in the mother suggests she carried a *Gts* gene. The father had schizophrenia. The severe ADHD, temper tantrums and rage attacks, self-destructive behavior, strange vocal noises, delayed speech, social withdrawal and early onset are consistent with the probability that this child is a mixed heterozygote, carrying both a *schizophrenia* and a *Gts* gene. The prognosis is guarded to say the least.

Behavior in Children with a Schizophrenia Gene

Four methods have been used to examine the effects of a schizophrenia gene (or schizophrenic parents) on young children — to look backwards, asking what adult schizophrenics were like as children, to look forward to see which children studied in childhood became schizophrenic as adults, to look at children of schizophrenic parents, realizing that about 50 percent of them should have a schizophrenia gene, and to look at children diagnosed as having childhood schizophrenia.

Several studies have looked at the childhood of individuals subsequently diagnosed as having schizophrenia[202,1392]. These showed that on average 20 percent had conduct disorder, 50 percent were quite withdrawn and 30 percent had no problems. Looking backward to elementary and high school records showed that 52 percent of male schizophrenics had "unsocialized-aggressive" conduct disorder as children[2034]. Comparing male to female schizophrenics, as children the boys tended to have conduct disorder while the girls were socially withdrawn.

In addition, we can turn again to the study of Robins'[p153,1422,1627]. She found that about 20 percent of the antisocial boys seen at the child guidance clinic had a diagnosis of schizophrenia as adults. In a ten year follow-up study[1295,1300] on children of schizophrenic mothers, those who subsequently developed schizophrenia had:

> "disturbed the class with inappropriate behavior and were characterized as being violent and aggressive and a disciplinary problem for the teacher."

In a study of 29 *children of a schizophrenic parent*[1611] 25 percent showed ADHD-like behavior with increased activity, impulsivity, distractibility and emotional lability. Some of them also had soft neurological signs. Finally, using the data from the Danish adoptive study Kendler and co-workers[1029] describe the presence of a *childhood social withdrawal syndrome* (shy, sensitive, insecure, few friends) and *antisocial behavior* (disobedient, short temper, stealing, and truancy) among the biological relatives of adopted schizophrenics compared to controls.

In looking at *schizophrenic children* themselves, Bender[149] described over 100 preadolescent children with childhood schizophrenia. They showed *motor awkwardness, impulsive behavior, facial grimacing, vocal noises, biting, spitting, excessive swearing, 'occasional explosive expletives, aggressive or obscene in nature,' echolalia, perseveration, obsessive-compulsive behavior, marked anxiety, concrete thinking, and sometimes excessive and open masturbation and preoccupation with the functions of elimination.* Since this study was done in 1947 and these descriptions are so similar to those of some TS children I strongly suspect many unrecognized Tourette patients were mixed in with this group especially since her definition of childhood schizophrenia was simply "abnormalities in behavior at every level and in every area of functioning."

The above observations indicate that like the *Gts* gene, the *schizophrenia* gene or genes can also predispose children to hyperactivity, attentional problems, conduct disorder, antisocial behaviors, muscle movements and incoordination. The added dimension in schizophrenia is a much greater degree of social withdrawal.

Superphrenics

In most studies of schizophrenia the emphasis has been on its negative psychotic aspects. However, some investigators have brought up the possibility that a person carrying a schizophrenic gene may in fact benefit from it by being a more creative super achiever. Karlsson[1015] called these individuals *superphrenics.* In the process of examining some extensive pedigrees of schizophrenic families in Iceland he noted that individuals with outstanding achievements and individuals with schizophrenia tended to cluster in the same families. In one extended pedigree he identified seven individuals who were considered gifted in the sense of being scholars, political leaders or financially successful community officials. If we assume that the schizophrenia in this family, affecting at least 12 individuals, was due to a partially dominant gene, then several of these seven individuals had to carry that gene, since both ancestors and children had schizophrenia.

This theme was also emphasized by Heston[890-892] who studied 47 children of schizophrenic mothers and compared them to 50 controls. Schizophrenia was present in five of the children of schizophrenic mothers and none of the controls. There was also a significantly increased frequency of antisocial personality, mental deficiency, and felons among the children of schizophrenics. However, among the 21 children who had no psychological problems, there were more individuals holding creative jobs such as painters, musicians, and designers, and more individuals with creative hobbies, than among the controls.

Others have made similar observations about children of schizophrenics. They have been characterized as "invulnerable"[51,675,676], of "outstanding ability"[1664], and "superkids"[1016]. Simply having to cope with adversity at an early age does not explain the superachievers among children of a mentally ill parent since the most competent children were more likely to have a schizophrenic mother, while the least competent children were more likely to have a severely depressed mother[1016]. This suggests this characteristic may be unique to the schizophrenic gene and, like TS, to changes in brain dopamine.

These types of studies have often led to the question — "Is creativity the flip side of madness?" The possibility that the genes for TS may be advantageous in some individuals is discussed in the chapter on *The Positive Features of Tourette syndrome*[p295].

Comparison of the Hereditary Dopamine Syndromes That Affect the Frontal Lobe and Limbic System

The purpose of presenting the clinical and genetic aspects of schizophrenia in some detail is to lay the groundwork for the concept that there is a group of genetic syndromes that appear to involve abnormalities in dopamine nerve cells distributed in the frontal lobe and limbic system. The chemistry, structure and function of these areas will be described later. The table on the following page summarizes the three major syndromes, ADHD, TS and schizophrenia, which together affect approximately 7 percent of the population.

Summary: Schizophrenia is a hereditary disorder characterized by bizarre thoughts, paranoid ideas and hallucinations. These are often termed positive symptoms. Negative symptoms consist of social withdrawal, lack of motivation and underachievement in life. A less severe form of schizophrenia is called schizotypal personality disorder. There are intriguing similarities between some patients with schizophrenia or schizotypal personality and some patients with TS. Both appear to involve abnormalities in dopamine in the frontal lobe and limbic system, and both show problems with attention, learning, speech, conduct, motivation, sleep and motor movements.

Comparison of ADHD, Tourette Syndrome and Schizophrenia

Feature	**ADHD**	**Tourette syndrome**	**Schizophrenia**
Cause	genetic	genetic	genetic
Neurotransmitter involved	dopamine	dopamine	dopamine
Site of involvement	frontal lobe limbic system	frontal lobe limbic system striatum	frontal lobe limbic system striatum
Motor coordination	poor	poor	poor
Other motor problems	hyperactive	tics especially facial	movements especially facial
Positive and negative symptoms	+-	+	+
Reading disability	+	+	+
Learning disorders	+	+	+
Attentional deficits	all	50-90%	common
Speech problems	+	+	+
Sleep disorders	+	+	+
Poor motivation	+	often	+
Hearing voices	-	occasional	+

Chapter 33
Autism

"She looked through human beings as if they were glass.
She created solitude in the midst of company,
silence in the midst of chatter." [1464]

The core symptom of autism is a severe withdrawing into one's own world. There are many causes of autism and the frequent presence of stereotyped motor movements such as arm-flapping and grimacing suggest that a Tourette syndrome gene may be one such cause. About 2 percent of our TS patients have such severe autistic symptoms that they carry a dual diagnosis of autism and Tourette syndrome. Autism provides many important insights into understanding Tourette syndrome.

Kanner Blames the Parents

Autism was first described by Kanner in 1943[1012]. He described his first patient as follows:

> "I was struck by the uniqueness of the peculiarities which Donald exhibited. He could, since two and one-half years, tell the names of all presidents and vice-presidents, recite the letters of the alphabet forwards and backwards, and flawlessly, with good enunciation, rattle off the Twenty-Third Psalm. Yet he was unable to carry on an ordinary conversation. He was out of contact with people, while he could handle objects skillfully. His memory was phenomenal. The few times when he addressed someone—largely to satisfy his wants — he referred to himself as 'You.'"

Kanner was struck by the

> "extreme aloneness from the beginning of life and an anxiously obsessive desire for the preservation of sameness."

Unfortunately, Kanner[1013] also believed that autism was a psychological disorder caused by having cold, rejecting parents, so called "emotional refrigeration" and the autistic's behavior was a rage reaction against the mother's unconscious wish that the child not exist. In a *Time* interview (July 25, 1960) he stated that:

> "children with early autism were the offspring of highly organized, professional parents, cold and retinal, who just happened to defrost long enough to produce a child."

He thought many of the parents were obsessive perfectionists. For many years this was the prevailing view[1663] and a parent has written eloquently of the pain such a theory can cause[1464]. However, by the mid 1970s it became apparent that this was a biased view, and that the environment of autistic children was the same as non-autistic children[419,471,1663].

The lessons learned with autism also apply to Tourette syndrome when the conduct disorder they sometimes have is blamed on poor parenting skills. The fact that the parents may have raised several well-behaved children before being presented with a TS child is often overlooked. An unfortunate maxim is — "the child is misbehaving—blame the parents."

Autism as a Neurological Disorder

As experience with autistic children grew the psychodynamic explanation of their behavior gave way to the realization that this was a neurological disorder, paralleling the experience with Tourette syndrome. The same areas of the brain and the same neurotransmitters involved in TS are also involved in autism. This will be discussed later[p499].

The Diagnosis of Autism

Autism is a triad of three sets of symptoms:

1. a profound failure to develop social relationships,
2. defective speech and language, and
3. ritualistic or compulsive behaviors.

More detailed aspects of these symptoms are shown in the box on the right[491,1445].

The following are some of the features that autistic children have in common with Tourette syndrome:

anxieties and fears
attention deficits
echolalia
echopraxia
facial grimacing
hand flicking
hyperactivity
inappropriate anger or giggling
labile mood
obsessive-compulsive behaviors
onset in childhood
panic at trivial things
perseverations
poor control of speech volume
sniffing and smelling things
stereotyped movements
whole body movements

There are, of course, significant differences in that social withdrawal is severe in autism and unusual in TS. Like many disorders there is a continuum of severity and autistic children may be severely "socially retarded" or a milder "socially impaired"[2082,2083]. In the latter, imaginative activity was replaced by a narrow range of stereotyped, repetitive pursuits. Autism and TS obviously have overlapping symptoms.

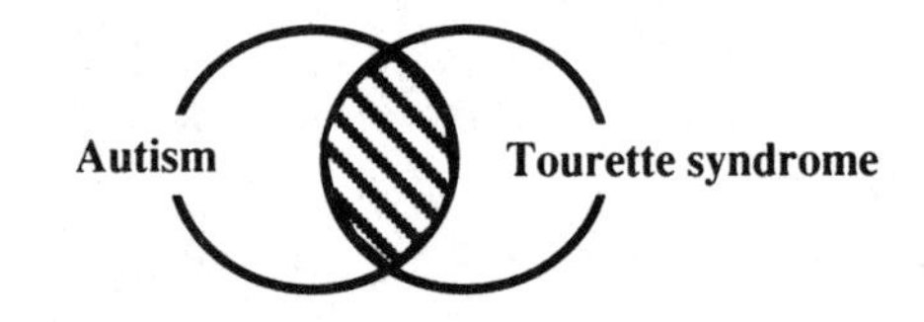

Characteristics of Autism

Poor social interactions
- **-unaware of other's feelings**
- **-does not seek comfort when upset**
- **-mechanical imitation of other's actions**
- **-no social play**
- **-no friends**

Poor communication
- **-no speech or body language**
- **-no eye contact**
- **-absence of imaginative play**
- **-abnormal speech (monotonous, high pitched)**
- **-echolalic and repetitive speech**
- **-unable to carry on a conversation**

Restricted activities and interests
- **-stereotyped body movements (hand flapping, spinnning, twisting, head-banging), sniffing, smelling, repetitive feeling, spinning of objects**
- **-upset over trivial changes**
- **-compulsive routines**
- **-narrow range of interests**

Onset in infancy or childhood

Autistic Child with a TS Parent

The following is an example of one of over a dozen autistic children we have seen who have a parent with Tourette syndrome or chronic motor tics:

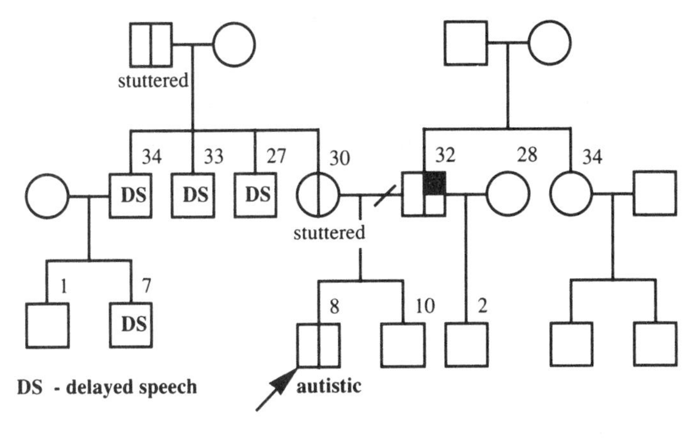

The proband was an eight year old boy with all the features of autism including severe speech delay, stereotyped twirling and handwaving movements, and social withdrawal. His father had chronic motor tic disorder with eyeblinking, eyerolling, arm tics, hyperactivity, learning disabilities, short temper, and alcohol abuse. Multiple individuals on the mother's side had a history of stuttering or delayed onset of speech.

Asperger's syndrome About the same time that Kanner was describing autism in America, Hans Asperger[64] in Europe was describing a group of children with developmental delay and bizarre speech with a flat, emotionless, monotonous, dull quality. Nonverbal communication was poorly done and often inappropriate to what was being said. Like autism, social interactions are poor, withdrawn and clumsy, with no empathy for others. Monotonous fixation on the same subject and resistance to change was often present along with stereotyped body movements. Onset was at about four years of age.

Despite these problems rote memory was often focused with obsessive fixation on a single, specialized subject in which the person may show great mastery. The diagnostic criteria of Asperger's syndrome[2083,2084] are shown in the **box below**:

Many of these features are also present in Tourette syndrome, especially repetitive activity, resistance to change, and stereotyped movements. While some TS patients have great trouble with memorization skills, other do not. Thus it is not surprising that in a recent study of six cases of Asperger's syndrome, three developed into Tourette syndrome and a fourth developed tics but didn't meet the full criteria of TS. The symptoms in one patient, called the pinball wizard, were as follows[1032a]:

> **The Diagnosis of Asperger's Syndrome**
>
> **1. Bizzare, pedantic, monotonous speech**
> **2. Clumsy non-verbal communication**
> **3. Clumsy social interaction**
> **4. No empathy for others**
> **5. Repetitive activity**
> **6. Resistance to change**
> **7. Good to excellent rote memory**
> **8. Narrow interests**

When seen at 13 years of age he was a very visually-oriented person, highly interested in such things as pinball machines. He drew detailed pictures of pinball machines and had, in fact, made his own. He spoke very little, did not relate well to adults, and did not have close friends. He admitted to ritualistic and obsessive concerns such as compulsively drawing schemata for pinball machines. The impression at that time was of a schizoid personality or borderline personality in early adolescence.

When seen next he was 18 years of age. His mother indicated that since he was last seen he had become obsessed with certain things such as computers, and behaved in a "goofy" fashion when something struck him as funny. He had inappropriate body movements and when nervous would pick up pieces of fluff and examine them carefully. Throughout the interview he avoided direct eye contact and when eye contact was made he would avert his gaze. He was restless, shifting his position often, and was distractible. He engaged in multiple motor tics which had started several years previously.

Some autistic children have remarkable talents. These are often in the area of art, music, math, or memory skills[1612]. These individuals have sometimes been called *idiot savants*, because the high level of these narrow skills is so disproportionate to their apparent general intelligence. Rather than being the result of some unexplainable neurological phenomenona, many of these skills may be largely a reflection of the combination of an obsessive compulsively narrow focus in combination with a withdrawal from usual types of social activity.

Wing[2083,2084] subdivided her autistic children into three groups: *aloof* — reacting only to satisfy their needs, *passive* — interacting only upon the approach of others, and *active but odd* — initiating their own interactions but in peculiar ways. Her active group is similar to children with Asperger's syndrome. From the above it is clear that there is a great deal of overlap between higher functioning, active, Asperger's type autism and Tourette syndrome.

Tom the Twiddler The following represents an example of our experience with Tourette syndrome in individuals with an Asperger's-like syndrome.

Tom is 16 years old. In preschool he would carry around an object such as a shoestring and play with it using complex stereotyped movements that got faster and faster as he continued. This was associated with grunting, squeaking and loud exhaling noises. He was withdrawn and hard to hold. The parents were concerned that he was autistic. These stereotyped hand movements have continued to the present. If he could not do these movements in school he would come home and "make up his twiddling." He would go out in the backyard with a long string, and twiddle it going faster and faster and getting red in the face with concentration and exertion and grunting. He often spent up to eight hours a day with moderate twiddling and one hour of intense twiddling.

Other compulsive behaviors included such things as "if someone touches my ear I have to lick my wrists, wet my fingers, and touch my ear." He also smells all his food before eating it. He eats constantly but is not overweight. If he wishes to discuss something he does so in great detail and will return to the subject again and again even though the topic of the conversation has changed. It is like he has to exhaust all thought on the matter before he can disengage and move onto another.

Tom is quite bright. In first grade he

showed word recognition skills of a fifth grader. He skipped the first grade but this later caused trouble because he was so socially immature. Psychological testing at age 6 showed that "Tom tends to focus very narrowly on any problem or piece of information with which he is dealing."

In the third grade his teacher noticed eye, head and arm tics.

The pedigree was most informative.

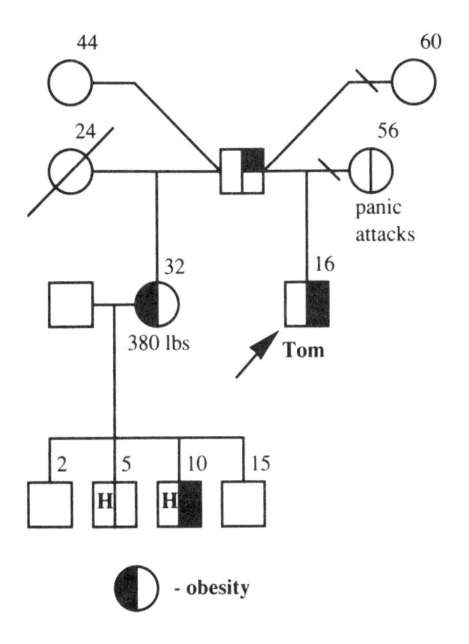

Although the stereotyped hand movements overshadowed the other motor tics the presence of a *Gts* gene was verified by the presence of motor tics in the father. Tom's mother had 20 episodes of severe panic attacks starting at age ten and ending at age 17. This suggests that Tom could have been a homozygote for the *Gts* gene. Tom's half-sister had no motor or vocal tics but was a compulsive eater and developed massive obesity — a staggering 380 pounds. This association of compulsive eating with obesity in *Gts* gene carriers is discussed later[p235]. Further evidence that she carried the *Gts* gene was indicated by the fact that one of her four sons had TS with ADHD and another had ADHD. Both required treatment with Ritalin.

The presence of autism or autistic-like symptoms in conjunction with Tourette syndrome has been noted by others[109,1587]. Burd and colleagues[251,253,610] reported that of 59 patients whose early history was consistent with autism or other pervasive developmental disorder, 20 percent developed symptoms of Tourette syndrome. Those who developed TS were functioning at a higher level than those who did not. They had higher IQ's and better development of speech. Sverd[1907a] described three children with TS and pervasive developmental disorder and suggested there was a causal relationship.

Summary: While the impairment of social interactions in autism is far more severe than that usually seen in Tourette syndrome, about 2 percent of Tourette syndrome children start life with a autistic-like behavior, milder forms of autism show remarkable similarities to TS and some may be due to a Gts gene.

[For a discussion of the possibility that severe autism and TS may represent different mutations of the same gene, see Chapter 67.]

Chapter 34
Borderline Personality
Life on the Edge of Sanity

If schizotypal personality disorder qualifies as an intriguing entity because of its similarities to grade 3 Tourette syndrome, then borderline personality disorder is equally or even more so. This term refers to a psychiatric condition that is more severe than a "neurosis" but not as severe as a "psychoses" — thus the term *borderline*.

History of Borderline Personality Disorder

In 1938 Stern[1871] was the first person to give a formal status to the term borderline[1892]. He described a group of office patients who were "too ill for classical psychoanalysis." Some of their characteristics were:

- alternating idealization and contemptuous criticism of the analyst and other individuals in their life,
- narcissistic (engrossed in themselves),
- poor response to stress,
- overreaction to mild criticism or rejection that sometimes bordered on paranoia,
- depression or outbursts of rage,
- some *hereditary factor* deeply embedded in the personality of the patient causing poor self-esteem,
- projection as a defense mechanism (everything that goes wrong is someone or something else's fault), and
- poor ability to have real empathy for other people.

Features added by latter psychiatrists included:

- an "as if" quality of adopting or acting out the personalities of other persons to gain love rather than being truly integrated into their own genuine personality[483].
- engage in petty criminal acts,
- fears of aggression, closeness, responsibility and change[806],
- heightened use of primitive defenses of denial (it never happened) and projection,
- late and unreliable about appointments,
- lead chaotic lives with something dreadful always happening,
- obsessions and phobias[1064],
- poor impulse control[648,1892],
- poorly motivated to obtain treatment,
- show poor insight into their own behavior [1721],
- tendency to break many rules of social convention, and
- unable to tolerate routine and regularity.

Grinker and colleagues of the Chicago Psychoanalytic Institute described four subtypes of borderline patients but the characteristics they all had in common were anger as a major emotion in their lives, poor interpersonal relationships, absence of self-identity and chronic depression[806].

Otto Kernberg's approach to borderline personality is even more heavily engrossed in psychoanalytic terms than his predecessors[1035,1036]. He particularly talked about *"faulty ego integration and splitting."* These refer to the absence of a firm sense of self and the ability to maintain sharply contradictory attitudes about themselves, side by side, without producing a sense of uneasiness. An example would be a patient who one minute is telling the therapist how wonderful, saintly and unselfish her mother is and shortly later what a cold, selfish bitch she was, without recognizing the inconsistency of the two. The patient may at one time project this "good self" onto the therapist or spouse, idealizing them, then at some later date project the "bad self," and be supercritical and denigrating. This black and white, all good or all bad approach to relationships can obviously lead to chaos in one's life.

To attempt to place the diagnosis of borderline personality disorder (BPD) on a more objective basis Gunderson and colleagues[820-822] developed a Diagnostic Interview for Borderlines containing 125 items in five different areas of function.

The Diagnosis of Borderline Personality

By 1980 the diagnosis of borderline personality disorders appeared for the first time in the DSM-III[490]. The criteria were derived by consulting a large number of practicing psychiatrists who were treating such patients, from the work of Gunderson[820-822] and Spitzer[1835] and by an analysis of the relatives of individuals in the Danish Schizophrenia Adoption study[1038,1039]. From these data emerged two distinct syndromes: schizotypal personality disorder, described in the previous chapter and genetically related to schizophrenia, and borderline personality disorder. The latter was described as follows:

> "The essential feature is a personality disorder in which there is instability in a variety of areas, including interpersonal behavior, mood, and self-image. No single feature is invariably present. Interpersonal relations are often intense and unstable, with marked shifts of attitude over time. Frequently there is impulsive and unpredictable behavior that is potentially physically self-damaging. Mood is often unstable, with marked shifts from a normal mood to a dysphoric (depressed) mood or with inappropriate, intense anger or lack of control of anger. A profound identity disturbance may be manifested by uncertainty about several issues relating to identity, such as self-image, gender identity, or long-term goals or values. There may be problems with being alone, and chronic feelings of emptiness or boredom."

Features of Borderline Personality Disorder
(DSM-III)

Impulsivity or unpredictability
Unstable personal relationships
Inappropriate, intense anger
Poor self identity
Marked mood shifts
Intolerance of being alone
Suicidal gestures, self-mutilation
Chronic feeling of emptiness or boredom

The Genetics of Borderline Personality Disorder

Studies of the relatives of borderline patients can answer the following questions: Is borderline personality a genetic disorder or primarily environmental? If genetic, is it related to schizophrenia, or depression, or manic- depressive disorder, or is it unique into itself?

Because of the strong psychoanalytic origins of the borderline concept it is not surprising that many considered it to be primarily psychological in origin. It has, for example, been proposed to be the result an arrest or fixation of the development of an infant when they should be separating from their parents and becoming independent personalities[1247,1614]. This, in psychoanalytic terms, is called *separation-individuation*. This has been taken even further — in the usual direction of blaming the mother — by stating that the problem was that the mother had withdrawn her love (libidinal availability) at this critical separation-individuation stage[1270].

When the relatives of 83 patients diagnosed as borderline by the DSM-III criteria were examined, none had schizophrenia, 0.5 percent had manic-depressive disorder, 6.4 percent unipolar depression and 11.6 percent bipolar personality disorder[1207]. There was a tenfold increase in the frequency of relatives with borderline personality among the relatives of borderline patients compared to the relatives of schizophrenic and manic-depressive patients.

Other studies have shown that if the borderline patients also have manic or hypomanic symptoms then significant numbers of their relatives may have manic-depressive disorder[18,1892]. However, these appear to be independent entities since there was no increase in the frequency of borderline individuals among the relatives of 100 manic-depressive patients[1207]. Others have found significant mood disorders among relatives of borderline patients and agreed that the relatives do not show an increased prevalence of schizophrenia[1823]. This suggests that the categories of schizotypal personality and borderline personality have been successfully teased apart into those personality disorders related to a schizophrenia gene (schizotypal) and those due to other genes and other factors (borderline).

The term *"emotionally unstable personality"* provides a reasonable alternative name to borderline[1207]. Just like the only thing that is constant in the world is change, the only thing that is stable in a borderline is unstability. They are *"stably unstable"* people.

Borderline Syndrome in Children

While it is somewhat of a nebulous concept, "borderline" has also been used to describe some children[146,314,529,540,576,647,682,2036,2037]. These children are characterized by seeming to be psychologically intact most of the time, but under periods of stress they are capable of severe temper tantrums with uncontrolled aggression, a paranoid reaction to others and a psychotic-like world of fantasy[529]. The latter can be terminated quickly by reassurance and support. Phobias, obsessions and severe anxiety are common. The anxiety tends to be rooted in fears of self-annihilation, body mutilation, or world catastrophe[540]. They may show very concrete thinking, have learning disabilities, poor attention span, reading disabilities, motor hyperactivity, poor speech and even tics[146].

> "They may bully or torment smaller children while being fearful of children their own age. . .have difficulties controlling anger, delaying gratification, containing frightening fantasies and appropriately channelling motoric impulses. Experiences of frustration quickly mount into temper tantrums or frantic, aimless hyperactivity. They easily work themselves into a state of rage or panic with little provocation. They have difficulty modulating stimulation from without and excitement from within. Some children are especially sensitive to tension"[146].

In a series of 24 children diagnosed as borderline, Bemporad and co-workers[146] found that 37 percent had ADHD and 63 percent had motor incoordination or movements *including tics*.

Grade 3 TS patients and "borderline"

children are similar in almost every type of behavior problem. Both clearly qualify as disorders of "disinhibition." A significant number of "borderline" children may carry a *Gts* gene.

Borderline Personality and ADHD

As described in Chapter 18, children with ADHD may grow into adults with explosive and aggressive personalities, and borderline-like symptoms are often reported in the adult type of ADHD[858,1251,1979,2107]. It is thus not unexpected that some ADHD/LD (learning disorder) children might present as borderline personality disorder not only as children but also when they grow to adulthood. This, in fact, has been described. Adult borderline patients were divided into three subtypes:

1. No childhood history of any brain damage,
2. A childhood history of brain damage (trauma, encephalitis or epilepsy) and,
3. A childhood history of ADHD/LD[39].

Forty percent of the adult males and 14 percent of the females with borderline personality disorder fell into the third group of having a childhood history of ADHD or LD. The ADHD/LD and brain damaged group frequently displayed *episodic dyscontrol syndrome*, consisting of intermittant rage attacks[1351], adult ADHD, or a limbic system[p321] disorder[40].

Dopamine and the Treatment of the Borderline Personality

Despite the rich involvement of psychoanalysis in the history and theory of borderline patients, they are relatively refractory to change despite intensive psychoanalysis[2020]. In a rare double blind study (where neither the patient nor the doctor know which medication was being used), Soloff and colleagues[1824] found that Haldol, a drug that inhibits dopamine function[p397], produced significant improvement of borderline symptoms, while antidepressants were no better than placebo. This suggests that like TS[399] some imbalance in dopamine neurons may be involved.

Borderline Personality and Tourette Syndrome

I have included a chapter on borderline personality disorder because of some of the similarities between it and TS. The features many of these individuals have in common are: impulsivity, aggressive outbursts, self-abusive behaviors, inappropriate intense anger, frequent display of temper, mood swings, low self-esteem, feelings of rage, emptiness and loneliness, and defects in serotonin metabolism[p457]. A least a dozen of the adult women in our Tourette syndrome clinic have previously been disgnosed as borderline personality disorder and do in fact satisfy the criteria for this disorder. A potential relationship between a *Gts* gene or genes, ADHD and borderline personality disorder is shown as follows:

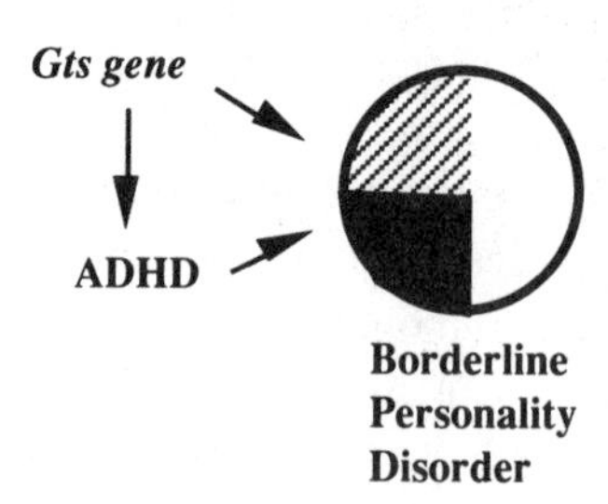

Summary: Borderline personality disorder is characterized by a long-term unstable personality, labile mood, and intermittent episodes of intense anger. It is one of a group of genetic disinhibition syndromes and some cases may be due to a Gts gene.

Chapter 35
Alcoholism — Born to Drink?

Alcoholism is a common disorder affecting at least 5 percent of the population. Since it is so common, and exerts such an enormous economic burden on society, it has been the object of many studies. A significant number of these have been devoted to the question of what causes alcoholism. In the chapters ADHD Children Grown Up, and Conduct, it was clear that children with ADHD and conduct disorder have a high risk of becoming alcoholics as adults. Since two-thirds of TS individuals have ADHD and one-third have a conduct disorder it would not be too surprising if the *Gts* gene was associated with a higher than normal frequency of alcohol abuse. Is this in fact true?

From the controlled study we found that 15 percent of all adult TS patients stated they often had problems with drug or alcohol abuse compared to five percent of the controls. This was barely significant. However, when the grade 3 TS patients were examined this increased to 40 percent and this was highly significant (p = 0.003).

In taking TS pedigrees I have observed that a significant number of the relatives have problems with alcohol and/or drug abuse. Many of these definitely carry a *Gts* gene because they have chronic tics.

To determine if this clinical impression was real I examined 20 TS pedigrees and compared then to 13 control pedigrees. The latter were taken from women coming to the clinic for prenatal diagosis and were asked the same range of questions I asked of the TS families. The following shows the frequency of alcoholism or drug abuse severe enough to significantly interfere with life, in these two sets of pedigrees[374].

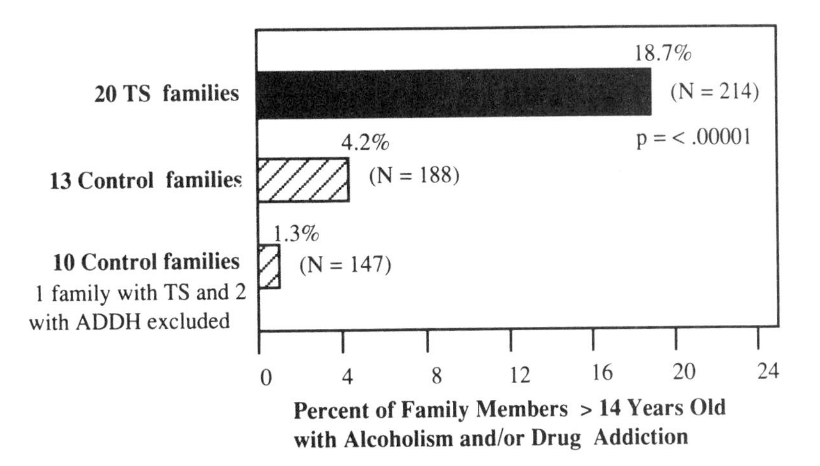

Of the 20 TS families 18.7 percent of the relatives (sisters, brothers, parents, grandparents, uncles, aunts, maternal and paternal nieces and nephews) over 14 years of age had significant problems with alcoholism and/or drug addiction. This compared to only 4.2 percent of the control families. The difference was highly significant ($p < .00001$). If one family containing a member with TS, and two families containing a member with ADHD were removed from the control group the percentage dropped to 1.3 percent.

The following shows the number of individuals with just alcoholism in these families[374].

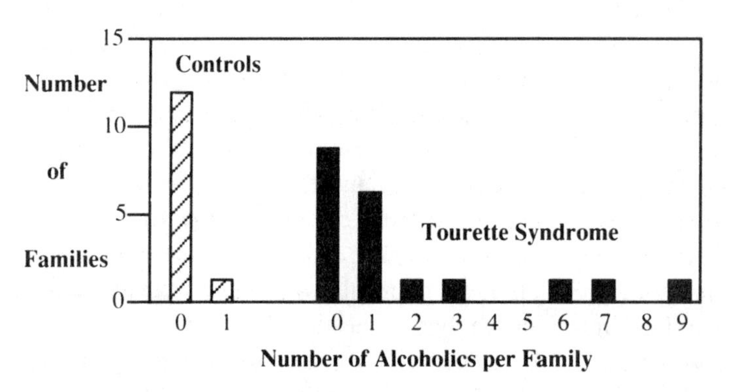

This shows that many of the alcoholics in the TS families came from five families each containing two to nine members with alcoholism, i e., familial alcoholism. The following are some typical TS pedigrees[374]:

Barry

67 60
A
LD
42 37 40 45 43 37
A A
obc
19 16
A
Barry
ADHD

Barry is 16 years old and has already spent a year away from home in a treatment facility for alcoholism. At four years of age, in nursery school, he was hyperactive and aggressive and didn't get along with other children. Motor and vocal tics started at age six. By 14 years of age he was getting into trouble because of intermittent episodes of violent behavior. He began drinking excessively, using pot and "crystal" and was suspended from school multiple times. Because of his rapid downhill course he was placed in a residential treatment facility and remained there for a year. Only a highly structured setting allowed him to remain off alcohol and drugs for a full year.

His maternal grandfather had motor tics and was a lifelong alcoholic. His mother had no tics but was an obligate *Gts* gene carrier. She had a history of alcohol abuse. The mother's sister was also an alcoholic. Thus, of the four members of this family with significant alcohol abuse or alcoholism, three clearly carried a *Gts* gene and the fourth was a probable carrier[374].

Larry

This 37 year old male with TS became heavily involved with drug use at age 18, especially marijuana and cocaine, and by age 30 he was addicted to heroin. He soon began robbing banks to support his habit. He stated, "I wanted to get caught because I needed help." His wish came true and he was caught and put in prison. Since he had no prior record, and was willing to enter a drug rehabilitation program, he was released on probation in five years. "Now I drink a lot of alcohol because it helps to replace the heroin." He sadly recalled, "I used to be a wealthy man, prominent in the community and owned a lot of real estate. The heroin took all that away and put me in prison."

Keith

This 35 year old male had an onset of multiple motor and vocal tics at age eight. They were never severe enough to require medication. At age 14 he began to use speed, marijuana and LSD. Within a few years he was using cocaine and heroin. By age 28 he was physically addicted to heroin. When first seen he was 30 years old and addicted. He reappeared four years later telling me that he had entered a drug program, was now clean and had been off street drugs for two years. He had a lifelong history of having trouble with his short temper and was very angry that when he had to go to an emergency room recently, for a severe infection, he was treated like an addict. When seen again, just a month later, he confessed that in the past several weeks he had gone back to using cocaine and heroin.

Chris

This 9 year old boy was brought to the clinic because he was having motor and vocal tics and behavioral problems just like his 12 year old cousin who had been diagnosed as having TS in the clinic two months before.

The pedigree is **on the right**. The fathers of both boys, and their grandfather, had severe problems with alcohol abuse since they were 18 years old. The mother's brother also had an onset of alcohol abuse at 18, and her sister had never married but lived with four different men and had two children by one of them. One of these children had severe stuttering.

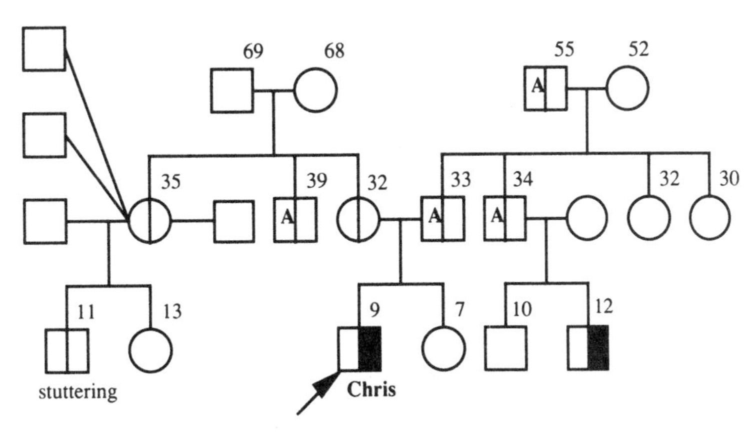

A - alcoholism - onset age 18

Gary

In the chapter on autism some pedigrees were shown illustrating the close relationship between autism and TS. In the following pedigree, the propositus was a 26 year old male with autism, TS and self-abusive behavior. The parents gave relatively few details about the family history and it was not until Gary's sister made an appointment for her TS son to be seen in the clinic that the extensive family history of alcoholism, on both sides of the family, came to light. Altogether there were 19 relatives with alcoholism, and/or drug abuse. [**Gary's pedigree is on the following page.**]

In addition to the results from the pedigrees shown above, the following pages present additional observations that suggest the *Gts* gene is one of the major genes predisposing to alcohol and drug abuse.

Attention Deficit Disorder as a Prelude to Alcoholism

In previous chapters we saw that a significant proportion of children with ADHD had problems with alcohol abuse as adults. Starting from alcoholism we can look at this

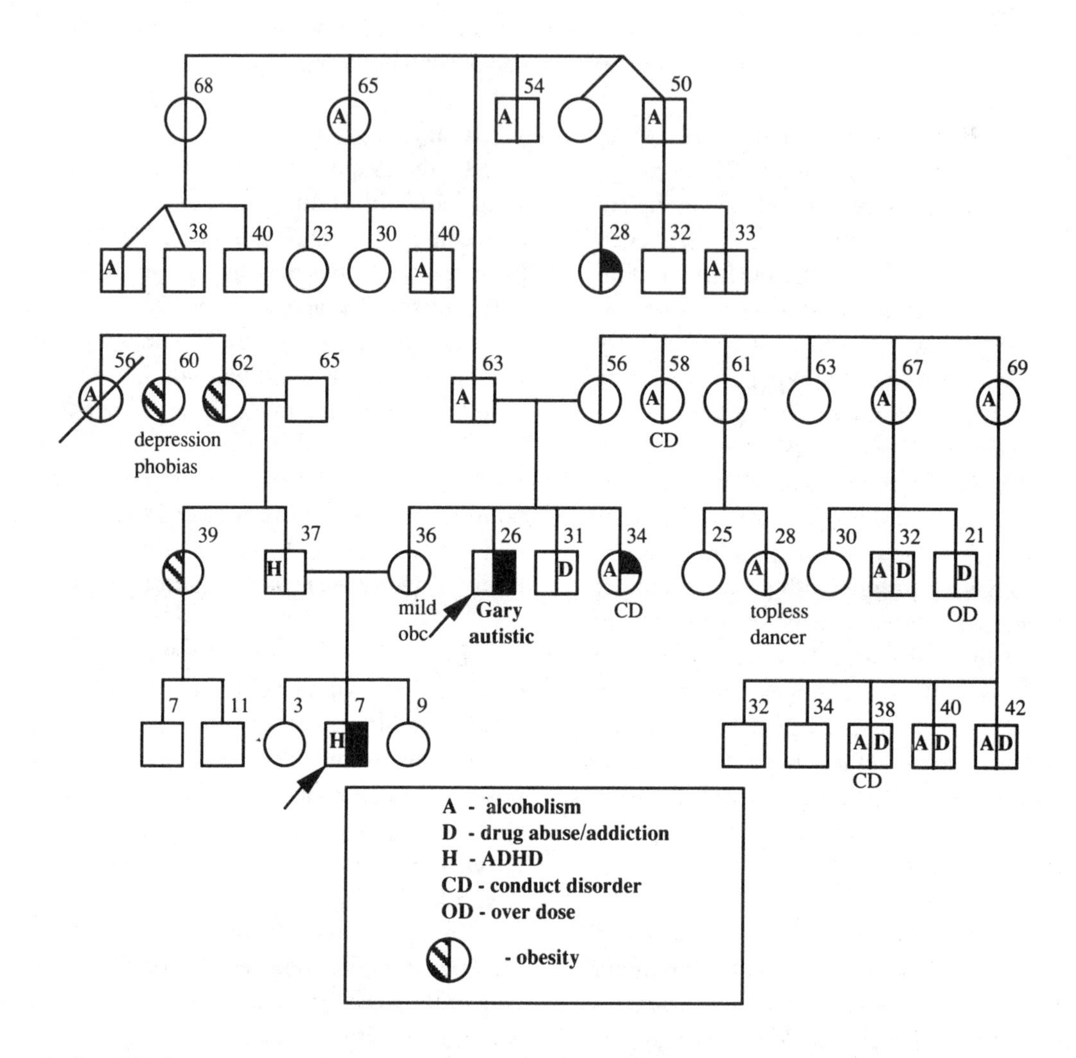

relationship from the opposite direction — Do adult alcoholics often have a childhood history of ADHD?

In their study, *The Origins of Alcoholism,* McCord and McCord[1277] noted that alcoholics tended to be impulsive, restless and aggressive in their youth. To study this in a manner that allowed the separation of genetic and environmental factors, Goodwin and colleagues[778] examined 133 *adopted males* in Denmark. Of these 14 were diagnosed as alcoholic while the remaining 119 were non-alcoholic. The results of an examination of childhood and adult problems and biological parents in these two groups is shown in the **table on the following page**.

Although these were retrospective studies it was clear that *the triad of hyperactivity, learning disabilities and antisocial behavior were significantly more common in the alcoholic than non-alcoholic adoptees.* There were many other characteristics, including birth complications, that were not different for the two groups. As adults the alcoholics differed from the non-alcoholics only in the fact that they also abused drugs more, and often felt angry and expressed their anger. The advantage of studying adoptees was that the

Adoption Studies of Childhood and Adult Behavior of Alcoholics and Non-alcoholics (Goodwin[778])

Behavior	**Alcoholics** %	**Non-alcoholics** %	**p**
As Children			
Hyperactive	50	13	.01
Below average in school	43	15	.05
Antisocial behavior	21	2	.01
Any combination of the above	57	15	.01
Shy, sensitive, insecure	64	20	.01
Aggressive, hot tempered	50	18	.05
Disobedient	29	4	.01
As Adults			
Used sleeping pills	50	17	.05
Used marijuana	50	14	.01
Frequently felt angry	39	3	.01
Frequently expressed anger	23	2	.01
Biological Parents			
At least one alcoholic	71	0	<.0005

biological parents could also be examined. Any characteristics that were present in the adoptees and their biological parents, but absent from the adoptive parents, would be due to genetic factors. This showed a marked difference in that *71 percent of the alcoholics had at least one alcoholic parent while none of the non-alcoholics had an alcoholic parent.* By contrast there were no differences in alcohol use, socioeconomic status, education, or emotional problems between the adoptive parents of the two groups.

In a different study of *the childhood of alcoholic*s, Tarter and colleagues[1924] studied four groups of individuals:

1. A group of primary alcoholics defined as those with no known precipitating cause for their excessive drinking, plus the presence of other characteristics such as development of tolerance (needing increasing amounts of alcohol), withdrawal symptoms (such as delirium tremens), and loss of control.

2. Secondary alcoholics who did not fulfill the above criteria.

3. A group of psychiatric patients without alcoholism.

4. Normal controls.

They were all asked to fill out a detailed questionnaire concerning 50 different behavioral characteristics of their childhood before they were 12 years old. As shown in the **table on the following page**, the primary alcoholics were significantly different from the other three groups in 12 of these characteristics.

The primary alcoholics were significantly different from the secondary alcoholics in the seven additional characteristics shown in the above table.

Many of these behaviors are characteristic of ADHD. These studies indicated there is a subgroup of alcoholism characterized by having ADHD in childhood. Such individuals become addicted to alcohol earlier in life and manifest a more severe form of alcoholism, but do not have other psychiatric problems.

There were several other interesting

Study of the Childhood of Alcoholics and Controls (Tarter et al[1924])

Childhood Characteristic (%)	Primary Alcoholics (n = 38)	Secondary Alcoholics (n = 28)	Psychiatric (n = 49)	Normal Controls (n = 27)
1° Alcoholics significantly different from all other groups in:				
Daydreams	78	46	51	22
Feels left out	73	25	44	14
Impulsive	71	21	30	25
Not working up to ability	68	14	42	14
Easily frustrated	63	14	40	0
Can't sit still	52	10	22	7
Can't accept correction	52	21	24	11
Poor handwriting	50	14	22	18
Short attention span	44	7	24	0
Fidgits	44	7	18	11
Doesn't complete projects	44	7	22	0
Overactive	36	10	16	11
1° Alcoholics significantly different from 2° alcoholics in:				
Withdrawn	52	14	32	14
Demands attention or affection	42	10	36	14
Gets into things	36	10	20	7
Unpopular with peers	31	3	14	3
Unresponsive to discipline	28	3	18	0
Vandalism	26	3	14	3
Accident prone	21	0	14	3

aspects of this study. It has often been suggested that alcoholism is a form of self-treatment for individuals whose primary problem is depression. The Tarter study indicated that the secondary alcoholics were normal as children and then became psychiatrically disturbed in adulthood and turned to alcohol relatively late in life to obtain symptomatic relief from distress. In addition, among the psychiatric group, those reporting a family history of alcohol abuse had twice as many ADHD symptoms as those without a family history of alcohol abuse.

In a forward looking study, Cloninger and co-workers gave 431 eleven year old Swedish children a detailed behavioral assessment and re-examined them when they were 27 years old. This showed that those children who ranked high on novelty seeking and low on avoiding harm were *20 times* more likely to develop alcoholism as adults than those with the opposite personality[333]. The personality characteristics of high novelty seeking included *"extreme overactivity, cannot sit still, unable to concentrate, disorganized, easily provoked, quick to loose temper, and disruptive."* Features of low harm avoidance included *"extremely overactive, highly uninhibited, distractible and little reluctance to avoid personal injury."* These characteristics describe the ADHD child.

Adult ADHD in Alcoholics

Examining the interaction of adult ADHD and alcoholism does not require looking backward. Approximately one-third of adult alcoholics meet the criteria for ADHD residual type[p95,929,2108]. Given the relationship between

ADHD and borderline personality, discussed in the previous chapter, it is interesting that one-third of the fathers of borderline patients were found to be either alcoholics or heavy drinkers[1208].

Learning Disorders as a Prelude to Alcoholism

Since ADHD can be a prelude to alcoholism and learning disorders are common in children with ADHD, it is not surprising that alcoholics and children of alcoholics often have a history of poor academic performance in school [414,477,558,658,877,1066,1067,1923]. As discussed previously[p77] children with ADHD often show an impulsive as opposed to a reflective learning style as tested by the Matching Familiar Figures Test. The same defects are seen in children of alcoholics[984].

Antisocial Personality as a Prelude to Alcoholism

This has been discussed in some detail previously[p151]. It is clear from many studies that antisocial personality is frequently associated with alcoholism (and drug abuse)[778,788,881,888,889,971,1167,1713,1848,1922,2011]. However, antisocial behavior and alcoholism can also be inherited independently of each other[190-192,272-274,330,1973]. The additive effects of gender, antisocial personality and family history of alcoholism on the lifetime risk of becoming an alcoholic were shown by Stabenau[1849] as **shown on the right->**.

This shows that males and females with no antisocial behavior and no family history of alcoholism have the lowest lifelong risk of becoming alcoholic. As antisocial personality and/or a family history of alcoholism are added the risk progressively rises and is always higher for men than women.

Two Types of Alcoholism

I have discussed early onset, primary and late onset secondary alcoholism and have repeatedly referred to one type of alcoholism associated with ADHD and antisocial behavior. Robert Cloninger in St. Louis[329,330] has built on these differences to distinguish two types of alcoholism, type I and type II.

Type I was characterized by an onset after age 25 of increasing difficulty in controlling alcohol intake. While these individuals were able to tolerate more alcohol than average persons (behavioral tolerance), they tended to drink in binges, and had guilt feelings about their alcohol abuse. Personality traits showed a minimal tendency for sensation seeking behavior, avoidance of risk, and a dependence on rewards.

Type II was characterized by an onset of alcohol abuse before age 25, occurrence of fighting, reckless driving, auto accidents and arrests while drinking (behavioral intolerance), and hospitalizations for alcoholism. Their personality characteristics were high sensation seeking behavior, high risk taking, and minimal dependence upon rewards. These are the

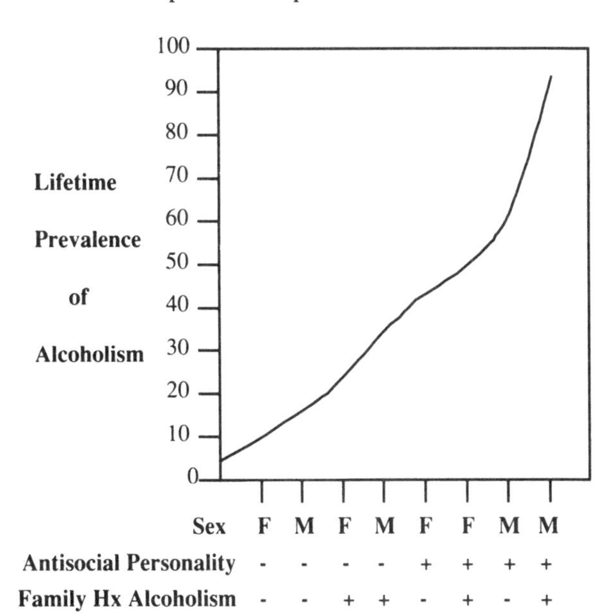

alcoholics with impulsive-compulsive personalities.

These types are summarized in the table at the bottom of the page. I have added the characteristics we have noted for many TS patients to illustrate the marked similarity between TS and type II alcoholism[374].

Although the daughters of type II alcoholics did not have alcoholism, they did have many complaints about their health — so-called somatizers or hypochondriacs. In Tourette syndrome, the females show a tendency to compulsive eating. This is discussed in more detail in the next chapter[p235].

The Genetics of Alcoholism

Many studies over the past 80 years have shown that alcoholism is present in approximately 25 percent of the parents, siblings and children of alcoholics[414]. This is three to five times the frequency of alcoholism in the general population[1632]. Although alcoholism tends to run in families, it is such a common disorder that a more formal approach is necessary to dissect out the genetic from the environmental factors. The best approach to this is through adoption studies. Taking advantage of this powerful tool, Goodwin and colleagues[779] utilized a Danish adoption registry to examine children of an alcoholic parent. They compared 78 control men where neither parent was alcoholic with 55 men who had an alcoholic parent but were adopted early in life and raised by a non-alcoholic parent. Of the adopted away sons of alcoholics, 27 percent became alcoholic themselves. By contrast, only 6.4 percent of the sons of non-alcoholic parents became alcoholic.

In a second study, also by Goodwin[780], 30 sons of an alcoholic parent, raised by their alcoholic parent, were compared to 20 of their brothers adopted out of the family before six

Comparison of Two Types of Alcoholism With Tourette Syndrome

Features	Type I	Type II	Tourette syndrome
General personality	passive-dependent	impulsive-compulsive	impulsive-compulsive
Alcohol-Related Problems			
Age of onset	after 25	before 25	before 25
Ability to abstain	good	poor	poor
Aggressive when drunk	rare	common	common
Guilt about alcohol use	common	rare	rare
Personality Traits			
Sensation seeking	low	high	high
Risk taking	low	high	often high
Reward dependence	high	low	low
When Abstinent			
Personality	worriers anxious apprehensive	distractible impulsive easily bored	distractible impulsive easily bored
Genetics			
Sons	alcoholic	alcoholic	alcoholic
Daughters	alcoholic	non-alcoholic	obesity

years of age. If alcoholism was a learned behavior the brothers adopted into non-alcoholic families should have a much lower incidence of alcoholism than those who remained with their alcoholic parent. However, the incidence of alcoholism was similar in both groups: 17% in the sons raised by their alcoholic parent and 25% in the sons adopted out. These studies indicate that the *frequency of alcoholism among sons of alcoholics was the same whether the child was raised with alcoholic or non-alcoholic parents, i.e., the primary determinant is a genetic factor*. Other adoption studies have come to a similar conclusion[190,268,271,330]. Again, conduct disorder in childhood was the only other variable predicting alcoholism in the adoptees[271].

In a study of half-siblings, the results were equally striking[1714]. If a biological parent was alcoholic and the child was raised by an alcoholic parent, 46 percent of the children became alcoholic, while if the child of an alcoholic was raised by non-alcoholic parents, 50 percent still became alcoholic. If neither biological parent was alcoholic and the child was raised by an alcoholic parent, 14 percent became alcoholic, while 8 percent of those raised by a non-alcoholic parent became alcoholic. This study also showed that conduct disorder in childhood predicted alcoholism as adults.

Adoption Studies in Type I and Type II Alcoholism

An important question raised by the above separation of alcoholism into two groups is whether both types are genetic. This can be answered since **cross-fostering**, adoption studies have been done on both types. Cross-fostering refers to the examination of the outcome in children of both alcoholics and non-alcoholics raised in families with both high and low alcohol use. This allows both genetic and environmental variables to be investigated. The results with severe type I alcoholics were as follows[192,326,330,335]:

Has a parent with type I alcoholism	Alcohol use by adoptive parents	Frequency of alcohol abuse in male adoptees
No	No	4.3%
No	Yes	4.2%
Yes	No	6.7%
Yes	Yes	11.6% p = <.02

This indicated that if neither parent had type I alcoholism then the risk of becoming alcoholic was no greater than for the general population, even if the adopting parent used alcohol. However, if either parent had type I alcoholism then the risk of becoming an alcoholic increased only moderately in the absence of environmental exposure to alcohol, but increased three fold if there was environmental exposure to alcohol. *Thus, type I alcoholism required both genetic and environmental factors to be expressed.*

The results with type II alcoholics[192,330] are shown as follows.

Has a parent with type II alcoholism	Alcohol use by adoptive parents	Frequency of alcohol abuse in male adoptees
No	No	1.9%
No	Yes	4.1%
Yes	No	16.9% p = <.01
Yes	Yes	17.6% p = <.01

Here again, if neither parent had type II alcoholism then the risk of becoming alcoholic was no greater than for the general population, even if the adopting parent used alcohol.

However, if either parent had type II alcoholism then the risk of becoming alcoholic was many times greater even in the absence of significant alcohol use by the adopting parents. *Type II alcoholism is strongly genetic.* Here, as with the Goodwin study, environmental factors played a minor role. Thus, the impulsive-compulsive, ADHD, anti-social personality individual has a high inherent risk of becoming alcoholic regardless of environmental factors.

Depression in Alcoholism

It has often been suggested that alcoholism is a form of depression in which the patient is attempting self-medication, through alcohol, to relieve the pain of depression. Does alcoholism cause depression or does depression cause alcoholism? One of the best studies attempting to dissect out this interrelationship[2111] examined three groups of patients, a group with depression without alcoholism, a group of alcoholics without depression, and a group of alcoholics with depression. The important observation was that the aspects which most distinguished these three groups were:

- an alcoholic father, (or other alcoholic relatives)
- outbursts or rage, and
- into many fights in school.

In each case the depression alone group had the lowest frequency of these features, the alcoholism plus depression the highest frequency, and the alcoholism alone intermediate frequency. The critical feature was that these distinguishing features were present *before* the onset of alcoholism. This suggests that depression and alcoholism can be independent entities and that the antecedents of alcoholism and alcoholism with depression occur in childhood and adolescence *before* the onset of alcoholism itself. Put in a different manner, there is a biological- genetic predilection to antisocial behavior which evolves into antisocial behavior plus alcoholism with or without depression.

A different study[1315] showed that the relatives of depressed patients did not have an increase frequency of alcoholism while the relatives of individuals with depression *and* alcoholism did have many alcoholic relatives.

As with antisocial behavior and alcoholism, the simplest explanation of these findings is that some genes lead to depression alone, some genes lead to alcoholism alone, and some genes lead to depression and alcoholism.

OtherDisorders Associated with Alcoholism.

The frequency of other mental disorders in alcoholics is very high. In one study of 501 patients with alcoholism and/or drug abuse, 78 percent had a lifetime diagnosis of at least one other psychiatric disorder[1650a]. The most common were antisocial personality disorder, phobias, sexual dysfunctions, and depression.

Summary: Type II alcoholism occurs predominantly in males, has an early onset of compulsive drinking, and is strongly genetic. There is a high frequency of type II alcoholism in TS patients and their relatives.

[Further evidence for the relationship between TS and alcoholism are given in Chapters 61-64, 89, 90 and 98.]

Chapter 36

Obesity — Born to Eat?

Many of our TS patients have eating disorders. While the most common is compulsive eating with obesity, compulsive not-eating (anorexia nervosa) or compulsive eating followed by vomiting (bulimia) also occur. The following pedigree illustrates a family in which three different members carrying a *Gts* gene all had a different eating disorder.

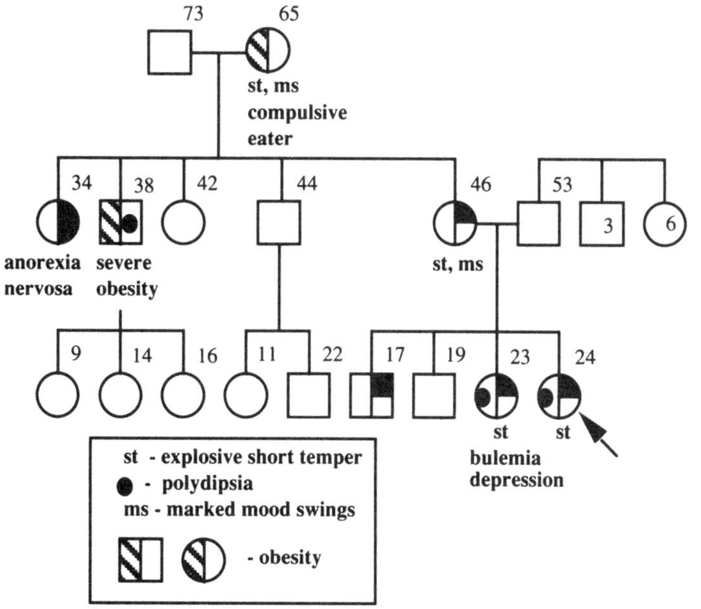

st, ms
compulsive
eater
anorexia nervosa
severe obesity
st, ms
st
st
bulemia
depression
st - explosive short temper
● - polydipsia
ms - marked mood swings
- obesity

The proband had rapid eyeblinking and a severe hair out of eyes tic since she was two years old. At age 12 she began to deliberately lie about little things to make herself seem to be better than she was. On the rare occasions when she drinks alcohol she starts swearing and has a loud barking and screaming vocal tic that lasts for about one hour.

Her sister has motor tics, low self-esteem, intermittent severe depression, and explosive temper. Since age 14 she had bulimia that frequently required medical help. Her 17 year old brother has motor tics, constantly lies to his parents and devalues everything they try to do for him. At age 14 he began to constantly play hooky and although he is very handsome he feels he is ugly.

The mother's 34 year old sister had motor and vocal tics and many discipline problems when she was a teenager including lying, stealing money from her parents and drug abuse. She had anorexia nervosa from age 25 to 30.

By contrast to the compulsive attempts to remain thin by the females in the family, the mother's 38 year old brother was a compulsive eater and was massively overweight at 350 pounds. All of his children were also overweight. His mother was also a compulsive eater.

Altogether, this family presents with a museum of TS-related behaviors of motor and vocal tics, explosive temper, polydipsia, compulsive behaviors, eating disorders, drug abuse, school problems, poor self-esteem, denigration of parents, and conduct disorder.

One of the most famous TS patients, Dr. Samuel Johnson[p114], also had bulimia[1982]. Families such as these, and others, suggest that the obsessive-compulsive behaviors of TS patients often spill into the area of eating disorders — both too much and too little. Others have also reported TS patients with anorexia nervosa[47,2126] and suggested that the obsessive-compulsive aspects of TS played an important role[47]. It is interesting that in addition to Tourette syndrome, one of Gilles de la Tourette's major interests was anorexia nervosa[47]. Studies of twins have provided evidence for the role of genetic factors in anorexia nervosa since 56 percent of 16 identical twins were concordant for the disorder compared to only 7 percent of 14 fraternal twins[912].

Obesity

Although we have noticed for years that mothers of TS patients were often overweight, the significance of this did not begin to sink in until they kept giving the same answer to the question — Do you have any compulsions? "Yes, eating," was the frequent reply. We then realized that the pedigrees were telling us that the *Gts* gene tended to express as compulsive over eating with obesity in females and compulsive over drinking with alcoholism in males. The following pedigrees illustrate the point:

Three generations of 300+ lbs The eight year old boy with typical TS was the proband. His pedigree was impressively "large."

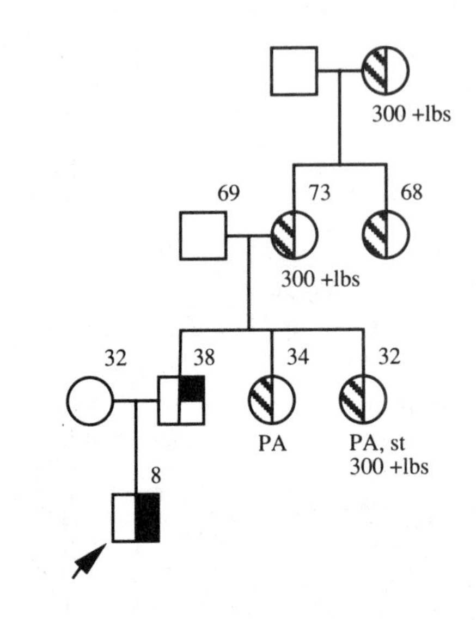

Both of his aunts had panic attacks and both were quite obese, one weighing over 300 pounds. The grandmother and great-grandmother also weighed over 300 pounds.

Obesity and bulimia

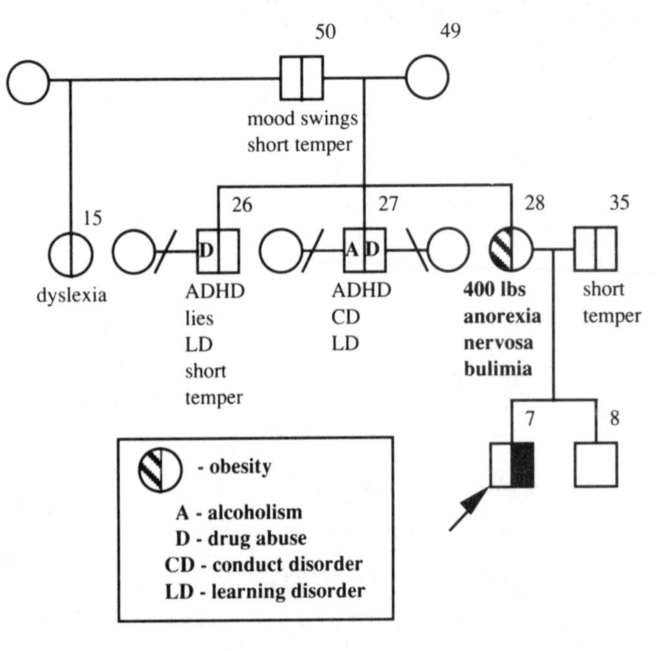

In the above pedigree the mother of the propositus was a compulsive eater all her life and weighed 240 pounds when she was a teenager. During this period, in an effort to loose weight, she would sometimes starve herself for a week at a time. As she grew older she began to binge and vomit and continues to do so. After the birth of her son she compulsively put on another 100 pounds.

Both of her brothers had ADHD as children, learning disorders, and other behavioral problems.

Obesity and spastic colon The propositus in the following pedigree was the 51 year old man with severe TS who had been followed for years in the clinic. As he began to learn more about TS he remarked one day, "I think my daughter also has symptoms of TS." When seen, her most striking feature was her weight. Although she was only 5 feet 1 inches tall she weighed 225 lbs. Her weight problem began when she was 13. She had chronic depression as long as she could remember and at age 23 developed a spastic colon. She had such a poor memory that at work she would write things on her arm to avoid forgetting them. The pedigree showed that her father's two sisters and mother

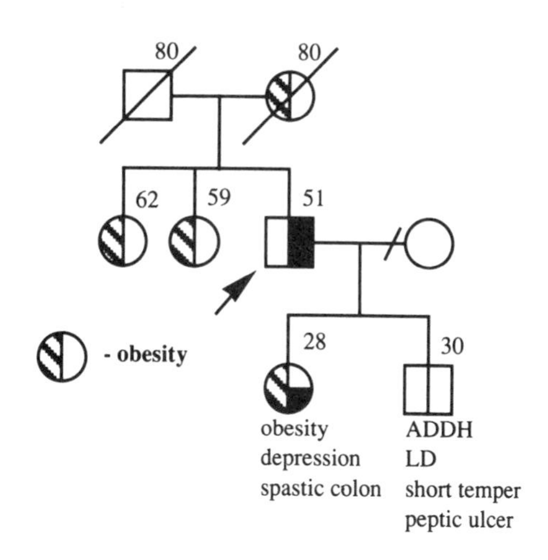

were also quite overweight.

Her brother had ADHD, learning disorders, short temper and chronic peptic ulcer. The presence of "psychosomatic" disorders such as spastic colon and peptic ulcer may also be more common in TS. However, a controlled study of this has not yet been done.

Other examples of many presumptive *Gts* gene carriers weighing over 300 pounds are shown elsewhere[p629].

Genetics of Obesity

Obesity has been shown to have a strong genetic influence[1546,1547]. In a study of weight (corrected for height) of children separated from their mothers at birth, the *correlation between weight of the adult daughters to their biological mothers was high* while there was no correlation with the weight of the mother who adopted them[1546]. There was much less correlation between the weight of daughters and their biological fathers, or sons with their biological mother, and no correlation between the weight of adopted sons and their biological fathers. The **correlations for weight** for the eight different possibilities are shown on the **next page**[1546].

	Biologic Parents		**Adopting Parents**	
	Mo	**Fa**	**Mo**	**Fa**
Adopted daughters	**.40**	**.18**	.06	.09
sons	**.15**	.08	.04	-.09

In another adoption study there was a strong correlation between the weight of adoptees and their biological mothers ($p < .0001$), less so with their biological fathers ($p < .02$) and no correlation with the weight of either adopting parent[1899]. This means that a girl without a genetic predisposition to obesity adopted by food guzzling parents would be unlikely to become obese, and in reverse, a girl with a genetic predisposition to obesity would probably become obese even if the adopting parents did not overeat. The similarities to Goodwin's[P232] adoption studies of alcoholism are striking. The TS pedigrees, showing that the *Gts* gene often expresses as alcoholism in males but as obesity in females, are consistent with these adoption studies showing a high correlation between the weight of daughters and their biologic mothers, and a high correlation between the drinking habits of sons and their biologic fathers.

Serotonin and eating It will be shown later that serotonin plays a critical role in the

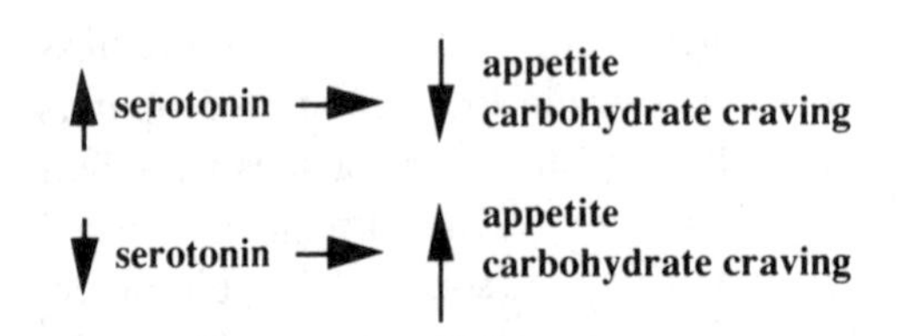

control of the appetite and carbohydrate intake[P495]. This can be summarized as follows:

Serotonin also plays a critical role in alcoholism[P436]. The observation that the *Gts* gene appears to cause low serotonin levels[P457] is consistent with the presence of both alcoholism and obesity in *Gts* gene carriers.

Summary: Eating disorders are common in individuals carrying the Gts gene. While this is most often expresed as compulsive eating with obesity, it can also take the opposite form of bulimia and anorexia nervosa. The Gts gene appears to cause an appetitive compulsion which is expressed as over eating with obesity in women and over-drinking with alcoholism (or drug abuse) in men. The common thread between these three factors, alcoholism, compulsive eating disorders and the Gts gene, may be low brain serotonin.

[For more information on the role of serotonin in alcoholism, eating disorders and TS see Chapters 61-64. For additonal evidence on the occurrence of obesity in TS see Chapter 90.]

Chapter 37

Pathological Gambling

Pathological gambling is an intriguing impulse disorder that has psychological and biochemical ties to depression, addiction, obsessive-compulsive and sensation-seeking behavior. While it is the most common of the impulse disorders, it has received relatively little attention since it is more socially acceptable than alcohol or drug abuse or antisocial behavior. Gambling is considered pathological when the individual is chronically and progressively unable to resist the impulse to gamble, and it disrupts or damages personal and family life[490]. It is estimated to occur in well over three million individuals in the United States[800]. As with other impulse disorders there is an increased sense of tension before, and a re-experience of pleasure or release upon committing the act[2101]. Unlike impulsive exhibitionism, the act may be consciously desired and may or may not be followed by guilt.

Like alcoholism there are "social gamblers" who gamble for relaxation, diversion, and profit[800,1355]. By contrast the pathological gambler is driven and unable to stop when winning, and experiences a more intense cycle of pleasure and painful tension[159]. Gamblers have personality features in common with other habitual risk-takers[977]. Personality characteristics include high intelligence, energy, workaholic, insomnia and a low threshold for boredom[449].

Gambling in Tourette Syndrome

I have included a chapter on gambling because of the presence of compulsive gambling in some of our TS patients and their relatives. This is consistent with the obsessive-compulsive and risk- taking behavior in many TS subjects.

> J.J., a 26 year old sporadically employed male, has had severe motor and vocal tics since age nine. They had not responded to a wide range of medications and he had been off drugs for many years. At age 24 he began to gamble and this soon became an irresistible compulsion. By the time he sought help he was $50,000 in debt and had his life threatened if he didn't pay this off.

The following is another representative TS pedigree:

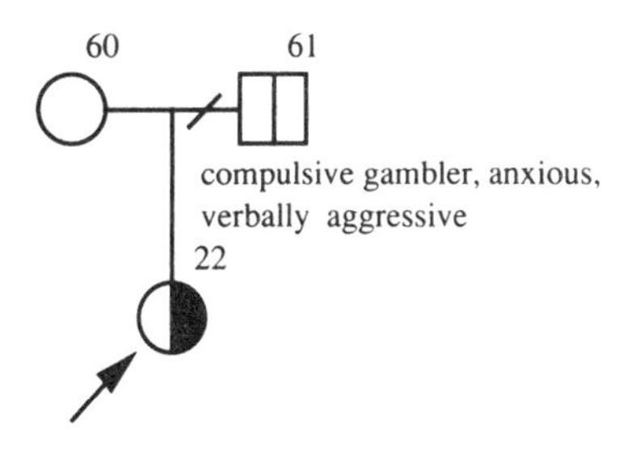

Here the parents of a TS girl were divorced because the father was a compulsive gambler and verbally aggressive.

In the following pedigree (**next page**) the father was an alcoholic and compulsive gambler and the uncle was an alcoholic.

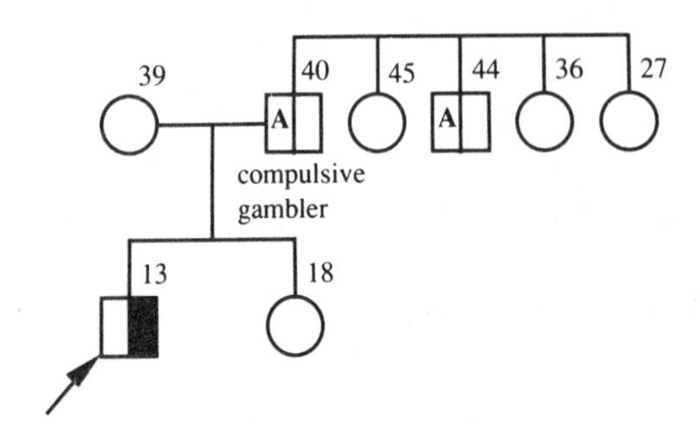

Depression and Gambling

Although it was a selected group, three-fourths of gamblers seeking medical help met the criteria for having a major depression and a third were hypomanic[1278]. Like the proposal that alcohol may be used to self-medicate depression, for many gamblers, gambling was the only activity that elevated their depressed mood[1278]. Drugs used for depression, such as lithium and antidepressants, have been effective in treating pathological gambling[1278,1368,1355].

Gambling as an Addiction

Gamblers repeatedly describe an "action high"[1683]. It has been suggested that any often repeated behavior that is associated with a high or feeling of arousal, whether drug-induced or not, may be addictive and associated with withdrawal-like symptoms when the behavior is stopped[2113]. The ending of a love affair is a classic example. Gamblers who have stopped gambling report feelings of depression, irritability, restlessness, inability to concentrate, insomnia, headaches, sweats, diarrhea, tremors, nightmares and obsessive thoughts of gambling[450,2113]. The similarity to drug withdrawal is striking.

Gambling and the sensation-seeking personality The psychology and biochemistry of the sensation-seeking, risk-taking personality will be discussed later after the reader has been introduced to the chemistry of the brain[p303]. Gamblers fit into this category of individuals[2158]. Such risk-taking is well-described by Dostoevsky in the *Gambler* [1355]:

> "I wanted to astonish the spectators by taking senseless chances and — a strange sensation — I clearly remember that even without any prompting of vanity I really was suddenly overcome by a terrible craving for risk. Perhaps the soul passing through such a wide range of sensations is not satisfied but only exacerbated by them. . ."

Gambling and neurotransmitters While the subject of neurotransmitters will be discussed in more detail in Part IV, a study of the spinal fluid of pathological gamblers showed increased levels of norepinephrine[1658]. This is consistent with sensation-seeking behavior.

ADHD and Gambling

Since pathological gambling can be viewed as an impulse disorder, the possible correlation between childhood ADHD and pathological gambling is of interest. Pathological gamblers had significantly more self-reported symptoms of ADHD as children than controls[296] and had some subtle brain wave findings similar to those in children with ADHD[294-296,769]. A role of low brain serotonin has also been suggested[295]. This is of particular interest given the presence of compulsive gambling in some TS patients, the high incidence of ADHD in TS, and the important role of low serotonin in TS[p457].

Summary: Pathological gambling is in part an impulse disorder and part a compulsive behavior. Like alcohol and drug abuse it can be addicting and shows many features of the more classic addictions. Some TS patients and Gts gene carriers have problems with compulsive gambling.

Chapter 38
Periodic Behaviors in Tourette Syndrome

The world in which all earthly organisms evolved is filled with periodic events. The most striking are daily and monthly rhythms.The monthly rhythms are most classically represented by the menstrual cycle. So many chemical events are keyed into daily rhythms that most plants and animals have their own internal clocks. The mechanism by which organisms can keep track of time, independent of environmental signals, remained a puzzle until recently when "clock genes" were identified and isolated[133].

Some TS patients have shown a striking periodicity to their behavior. The following are some of the more intriguing examples.

Periodic Rage Attacks

A science writer called me one day to discuss his wife, and told me the following story, illustrating an incredible periodicity of rage attacks in TS.

> Sally is now 48 years of age. At age 5 she began to have severe temper tantrums associated with the shouting of obscenities. These attacks continued and at age 16 her parents took her to a psychiatrist who said it was the parent's fault for picking on her. She admits it was the other way around, they only picked on her after violent temper tantrums. She married at age 35. Out of nowhere she would start to viciously pick on her husband and say terrible derogatory things about him. He would become angry and hit her. The marriage broke up in two years. At age 41 she met her present husband. With his science background he began to take detailed notes on her behavior. Like clock-work, every sixth day, she began to verbally attack him, berate him and shout obscenities at him. This would last about an hour then progress into physical violence where she began to throw things at him, smash up furniture, and poke holes in the walls. This would last about four hours. She would then wander off into the bedroom, sleep for six hours and wake up depressed and remorseful about what she had done. The rest of the week she was the most kind, loving and gentle person one could hope to know.
>
> Extensive studies including EEGs were negative. Dozens of drugs were ineffective. One day her husband read about TS and asked the psychiatrist if this might be her problem. She was treated with 0.5 mg Haldol and the rage attacks disappeared. Many years later a prescription was misfilled with Halcion instead of Haldol and in two weeks the rage attacks and severe depression reappeared. They immediately disappeared when she was placed back on Haldol.
>
> At age 47 she developed severe panic attacks with agoraphobia and began to abuse alcohol. She has always been a

compulsive eater and weighed 325 pounds when I first saw her. Two years later, on appropriate medical therapy, she was down to 180 pounds, had stopped drinking and no longer had panic attacks or agoraphobia.

Monthly Periods of Bizarre Behavior

Jamie was a 12 year old quiet girl who always had a wide-eyed look of innocence to her. She was very irritable as an infant and was "active" in nursery school. By first grade she had to be placed in special education classes because of learning problems. For the next four years she did satisfactory work and was well behaved. At ten her parents began to notice facial grimacing, but only when she was under stress. She touched others a great deal and had a tic-like movement of flopping her arms up and down. On April 5th of her 11th year, she suddenly began to talk about killing her bird and herself, swore excessively, scribbled on her clothes, screamed out in anger, talked constantly to others and to herself, and was very hyperactive. After five days she suddenly returned to her previous well-behaved self. On May 5th she again became cranky, had screaming temper tantrums, talked constantly, and made weird squeaking noises. This again suddenly stopped after five days. These intermittent episodes of bizarre, almost psychotic behavior, continued for the next six months. Her pediatrician then treated her with Ritalin and this resulted in considerable improvement of the symptoms. The periodic monthly episodes continued but consisted predominantly of irritability and short temper. There were some associated behaviors that were not periodic, such as enuresis, echolalia and compulsive behaviors consisting of a need to do her homework on a rigid schedule, and becoming very upset if she couldn't eat her food in a specific order or take a shower at exactly five o'clock. When I saw her the episodes were getting more severe again and I switched her to imipramine which once again controlled the episodes to the point she could be tolerated. A year later her parents took her off the medication for one month and during that time she was much more hostile, beat on the dog with a spoon until he was bleeding, and strangled a pet duck. She showed no remorse for these events and the parents were especially concerned about the "cold blooded manner" in which she did these things.

At one clinic visit her mother brought in a calender that Jamie kept. She had a stamp of a smiling kitten and each evening she compulsively stamped each date. On the days she was feeling angry or down she stamped the date only once. On the days she felt up or hypomanic she stamped the date many times. The result for one month, the month of August, is shown as follows:

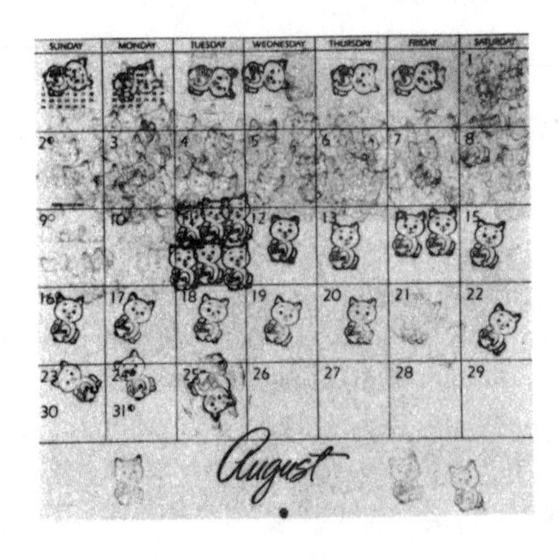

This dramatically shows the periodic, monthly nature of her mood swings. From the 1st to the 11th she was so up the date is a blur of stamping. On the 11th her mother gave her a new ink pad. From the 12th on she was feeling down again.

The family history was negative except for a maternal grandfather who was described as "moody, periodically depressed, short tempered and had problems in school."

The striking monthly periodicity to these bizarre episodes, beginning in a young girl at age 11, suggested that even though she did not have menstrual periods, the irritable behaviors might be triggered by hormonal cycles. Since the motor tics were not prominent she had been carried with a diagnosis of probable Tourette syndrome. The associated features of obsessive-compulsive behaviors, learning disabilities, echolalia, palilalia, periodic coprolalia, high pitched squeaking vocal noises and cruelty to animals, were consistent with Tourette syndrome.

A less striking but more common example of periodicity in TS occurs on a yearly basis. Some cases seem to start up slowly with tics for several weeks or months, at a certain time of the year, followed by no symptoms for the rest of the year. For example, one child had an onset of motor tics in April, lasting for two weeks. They spontaneously disappeared, but started up again the following April, this time lasting for one month. In the third year they again started in April, but lasted for two months. By the following year the symptoms were constant throughout the year.

Premenstrual Syndrome (PMS)

By far the most common behavioral periodicity seen in TS is the significant exacerbation of irritability, hostility, suspiciousness, moodiness, depression, rage and paranoia, in women during the days or even weeks prior to a menstrual period. The individual can be like two different people, one normal during most of the month, the other almost impossible to live with. This syndrome, called PMS, seems to be particularly common and severe in many TS patients or their mothers. Among women who seek help for psychiatric disorders, premenstrual exaggeration of symptoms is the rule rather than the exception. The possible role of serotonin in premenstrual tension[p241] is discussed later.

Summary: Part of the definition of Tourette syndrome is that the tics wax and wane in severity. However, in addition there may also be a marked waxing and waning of mood, confrontive and oppositional behavior, and general irritability. These cycles can vary from minutes, to hours, days, months, or years. These cycles may be a reflection of the susceptibility that Gts gene carriers seem to have to manic-depressive disorder[p191]. Premenstrual tension appears to be common in women who carry the Gts gene.

Chapter 39
Bedwetting and Soiling Enuresis and Encopresis

Enuresis refers to repeated, involuntary voiding of urine in the day or night, occurring past a specified age, usually five years, and in the absence of any physical disorder[490].

Since parents often complain of this problem in their TS children we investigated this in the controlled study[364].

cantly greater than the controls in all grades of TS and categories except TS patients without ADHD.

Other studies have reported enuresis present in 27 percent of TS patients[972]. Among children with ADHD, up to 50 percent have enuresis and 25 percent encopresis[120]. This is consistent with our finding that TS children with ADHD had a much higher frequency of enuresis than those without ADHD.

Enuresis and Encopresis in Tourette Syndrome

	Controls	TS	Gr 1	Gr 2	Gr 3	+ADHD	-ADHD
Enuresis (% with frequent bedwetting past age 5)							
%	10.6	22.2	25.5	14.4	37.9	27.6	12.7
p	—	**0.05**			**0.001**	**0.009**	
Age toilet trained (years)							
Average	1.38	1.97	1.95	1.93	2.07	2.15	1.68
p	—	**0.002**	**0.02**	**.004**	**0.005**	**<.001**	
Last age of poor bowel control (years)							
Average	0.91	2.17	2.25	2.11	2.24	2.59	1.48
p	—	**<.001**		**0.002**	**<.001**	**<.001**	**0.04**

Among the controls, 10.6 percent had problems with enuresis. This doubled to 22.2 percent in all TS patients, and tripled to 38 percent in grade 3 TS. It was very common even in grade 1 TS. ADHD was a major contributing factor in that 28 percent of TS patients with ADHD had enuresis while it was not very different from the controls for TS patients without ADHD.

The average age of becoming toilet trained gave the same results. This was signifi-

The Genetics of Enuresis

A significant role of genetic factors in enuresis is indicated by the much higher concordance rate[P46] in identical compared to fraternal twins. This is shown in the following two studies giving the percent of time when both twins had enuresis.

	N	Identical	Fraternal
Hallgren[842]	40	70%	0%
Bakwin[90]	104	68%	36%

This would predict that enuresis runs in families, and this is true. When both parents had a history of childhood enuresis, 77 percent of the children had enuresis. If only one parent had enuresis, 44 percent of the children had enuresis[90].

Behavior Problems and Enuresis

The relationship between emotional disorders and enuresis have been controversial. While many studies have found such a relationship, some have not[1678c]. In one study, where children were evaluated by a behavioral questionnaire that included antisocial conduct and poor relationships with other children, enuresis was found to be three to four times more common in "deviant" children[1668]. Anxiety-producing events in a young child's life, such as the parents divorce, can increase the frequency of enuresis. Several early studies have suggested that enuresis was especially common in children with tics and temper tantrums[841,1324,2016] and conduct disorder[1441,1865,1866]. Many of these children would now be diagnosed as having Tourette syndrome.

While enuretic children are more likely to have various behavioral disorders than those who do not have enuresis, it is important to point out that the majority of children with enuresis do not have significant behavior problems.

One of the most effective treatments of enuresis is the antidepressant drug imipramine (Tofranil)[1678a]. This medication is also quite effective in the treatment of behavioral disorders of childhood including ADHD, temper tantrums, and borderline symptoms. Since the enuresis often disappears overnight after the first dose of imipramine, it is unlikely that the medication eliminates the enuresis by improving the behavioral disorders. It is most likely that the same neurotransmitter imbalance that produced the behavioral problems also produces enuresis. Imipramine appears to work by affecting the brain's influence over bladder control or sleep[1579].

Encopresis — No Bowel Control

In children, the bowel equivalent of enuresis is encopresis—meaning lack of bowel control. Its definition is similar to that of enuresis — repeated, involuntary passage of feces in inappropriate places. Since bowel control usually occurs before control of urine, the cutoff age is four years[490]. A common symptom is apparent lack of normal sensitivity of the rectum, with the children often reporting that they do not feel the stools coming. There are two forms, retentive and nonretentive. In the retentive form the colon may become distended and lead to fecal leakage. Some psychoanalysists, feeling it is purely psychological in origin, have claimed that the retentive encopresis is due to fixation at the anal-retentive phase of Freudian development[50,1294].

Although encopresis is less common in TS than enuresis it is sometimes quite striking. When the average age of last poor bowel control was examined, all categories of TS and ADHD (except grade 1) were significantly different from the controls. This even included the TS patients without ADHD. As mentioned above, ADHD alone is often associated with encopresis. Child psychiatrists feel that most children with encopresis have some type of psychological problem[145,1294,2028]. In our experience, imipramine can eliminate the encopresis of some TS children.

The following families give a flavor, or more appropriately, a smell, of the problem.

The Black family

The 11 year old boy in the Black family was very active from birth and slept so poorly his mother had to walk him to get him to sleep. At eight he developed throat clearing, snorting, and grunting, and a year later a horizontal head tic, rapid eyeblinking and stereotyped flapping of his hands. He could not accept "no" for an answer and didn't accept responsibility for his own actions. His tics responded well to Haldol. It was his two brothers that were of interest in regard to encopresis and Tourette syndrome.

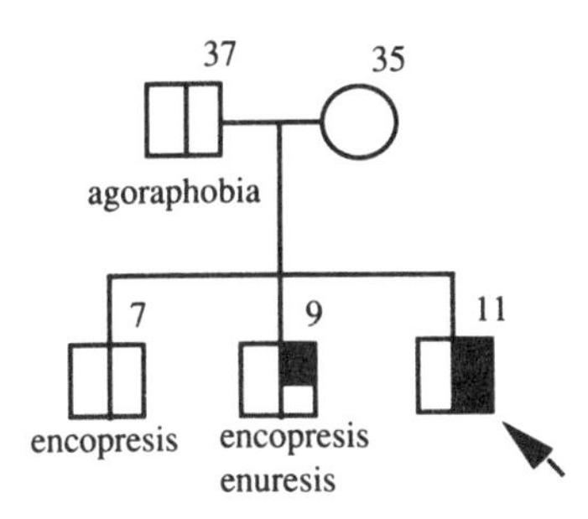

His nine year old brother developed a horizontal head tic at age seven and at the same time began to have nighttime and daytime wetting and bowel movements in his pants. He claimed he did not remember having the bowel movements. Initially they happened twice a week, but after several months this was happening up to twice a day. The encopresis, enuresis and daytime wetting continued until he was first seen at nine years of age. In the six months prior to being seen he became more and more aggressive.

The other seven year old brother had no problems until he started kindergarten at five years of age. The teachers noted that he had difficulty following instructions, seemed not the hear, and could not sequence more than two instructions. He began to say "you fucker" whenever he was angry. At six years of age, in the first grade, his mother noticed that he seemed to be sloppy in wiping himself, since his pants were often soiled. After several months he began to have full bowel movements in his pants almost daily, and would swear it was not him. He would hide his soiled pants under the bed.

Randy

Randy is five years old. He has had ADHD since he was six months old. Encopresis started at age three and one-half. He would do fine in preschool but on the way home in the car he would soil his pants. This has continued to the present. When told that he smelled he would deny that this was him. He would often play with the feces sticking it on the ceiling tiles, furniture, or smearing it on the bathtub. Motor tics and vocal tics started at age four and one-half. Thus, in this case *the encopresis preceded the onset of motor and vocal tics*. Associated problems were a precocious interest in sex, exhibitionism, oppositional behavior and fire starting. His father had motor and vocal tics, compulsive behavior, short temper and was manic-depressive, and sexually abused both his daughters and granddaughters.

Is Encopresis a Tic?

One mother of a TS boy with encopresis reported,

> "It came on suddenly, lasted two weeks and went away just as suddenly, almost like a tic, but is not related to his motor tics."

Unfortunately, the encopresis does not

always go away that easily. In some it is a severe, chronic problem lasting for years. One of the most remarkable aspects of the soiling is the tendency for the affected children to totally deny that they are the one responsible for the smell. They can walk around with their pants full of stool and show angry incredulity when anyone suggests they are the source of the smell.

There is a marked increase in the frequency of encopresis in children who also have enuresis[841]. The significant increase in enuresis and encopresis in Tourette syndrome, a genetic behavioral disorder, is consistent with the previous findings that these symptoms have a genetic basis. The most logical explanation of the association with behavioral problems is that the chemical imbalance in the brain caused by the *Gts* or other *disinhibition* genes causes both the behavior problems and enuresis and encopresis, rather then the enuresis and encopresis being caused by the behavior problems.

Summary: Bedwetting, or enuresis, is very common in both Tourette syndrome and ADHD. Poor control of stool, or encopresis, may also occur. The most logical explanation of the frequent association of both enuresis and encopresis with behavior problems is that the chemical imbalance in the brain, caused by Gts, ADHD, or related genes, causes both the bad behavior and the enuresis and encopresis.

Chapter 40

Sleep Problems

One of my favorite questions when interviewing parents about their problem children is, "What is the earliest thing you can remember that suggested something was different about your child?" Some of the most frequent answers were:

"He never went to sleep."

"She never took naps."

"He was awake when I went to bed at night and awake when I got up."

"For the first two years I never saw him in a sleeping state."

Various sleep disturbances are common in children with ADHD and TS. Here I address the questions — How common and what types of sleep problems occur in TS patients?

Sleep problems are frequently mentioned in TS[111,112,358,860,974,975,1308,1347,1397]. In a very early report[1347] 12 of 15 TS patients were found to have difficulty getting to sleep or waking up early, and in a series of 50 patients, 22 had undefined sleep problems[1397].

Some Characteristics of Sleep

In the mid 1950s sleeping infants were noted to have quiet periods of no body movements and no eye movements alternating with periods of active body and eye movements[63]. The latter was called a stage of rapid eye movement or REM sleep. When the brainwaves and eye movements of adults were recorded during sleep it was found that they too had stages of REM sleep alternating with non-REM or NREM sleep[470]. The brain wave studies showed that the NREM sleep could be divided into four stages of depth from stage I (light) to stage IV (deep), each with a characteristic brain-wave pattern. The stages of deep sleep are characterized by slow, high voltage brain-waves called *alpha waves*. This is referred to as slow wave sleep. By contrast the early stages of sleep are characterized by rapid, low voltage waves.

In infants about half of the sleep is REM sleep. By adulthood this has dropped to 20 percent[581]. About four to six cycles of REM/NREM occur during a typical night's sleep. Dreaming occurs in REM sleep.

Sleep Disturbance Versus Disorders of Arousal

After an initial period of deep sleep, there is typically a "lightening" or arousal into the lighter stages and REM sleep. It is during this transition period that bed wetting, sleep walking, sleep talking and night terrors occur.

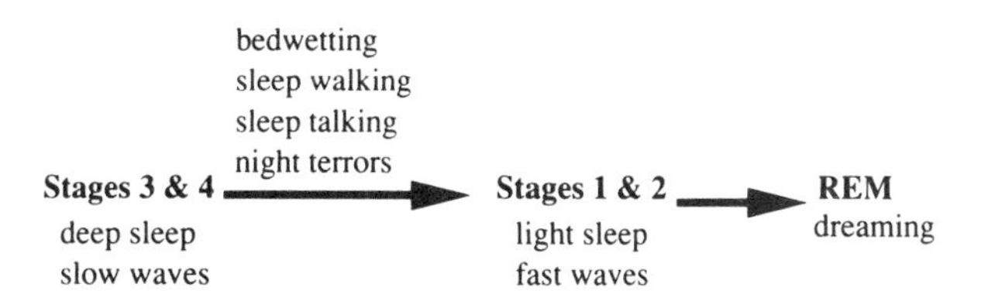

These are thus termed disorders of *arousal*. Contrary to a frequent misconception, they do not occur during dreaming or REM sleep. Other problems such as nightmares, trouble falling asleep, early awakening and narcolepsy are true sleep disturbances.

The distinction between disorders of arousal and sleep disturbances can be seen by examining the difference between night terrors and nightmares. Children with night terrors sit upright in bed and scream hysterically with an expression of intense fear. They are difficult to arouse at the time and usually do not recall the event. Night terrors are not associated with REM sleep. By contrast, nightmares are associated with REM sleep and are often vividly recalled[31].

Individuals with disorders of arousal often have a family history of other relatives with the same problems[89,1264,1512,1513]. In naval recruits with enuresis, 25 percent had a positive family history of sleep walking[1512] and identical twins show a six-fold increase in concordance for sleep walking compared to fraternal twins[89].

Disorders of Arousal in TS

In one of the only controlled studies, Barabas and co-workers[111,112] compared the frequency of these three disorders of arousal in 57 TS patients, 58 epileptics, and 53 children with learning disorders. The results were as follows:

	Sleep-Walking %	Night Terrors %	Enuresis %
Tourette syndrome	**17.5**	**15.8**	18.9
Epilepsy	1.7	1.7	7.4
Learning disorders	3.8	3.8	15.4

Sleepwalking and night terrors were significantly more common in TS than the other two groups.

Tics occur during sleep There are two ways to find out if an individual has sleep problems — ask them or watch them. Sleep laboratories do the latter. Subjects are hooked up to brain-wave leads and studied throughout the night as they sleep. Glaze and colleagues[732] studied TS patients that were 23 years of age or less. They found an increased number of awakenings, decreased total sleep, decreased REM sleep, and observed *motor tics in all stages of sleep*. This is important since some of the earlier criteria for Tourette syndrome included the statement that "symptoms always disappear during sleep"[1767]. We have had some parents tell us that their pediatrician told them their child could not have TS since he had tics in his sleep. This is no longer valid.

Some TS patients had *unusual behavioral episodes during sleep*. These were characterized by sudden and intense arousal, apparent disorientation, confusion and combativeness[732]. In another sleep lab study of TS patients 53 percent showed reduced REM sleep, 29 percent abnormal arousals, and 23 percent sleep apnea[975]. Fewer sleep problems were noted in older TS patients[1308,2146].

The Controlled Study

Since sleep problems of all types are fairly common, a controlled study was critical for determining if in fact they are more common in Tourette syndrome.

Trouble getting to sleep At times this is a common problem for everyone and is generally not considered to be hereditary. However, mothers of TS and ADHD children often complain of this in their children. This was verified in the controlled study showing that this was a significant problem in all the TS and ADHD groups[364] (**see figure next page**).

Trouble getting to sleep was just as common in the mild grade 1 cases as in the grade 2 cases. Surprisingly it was almost as common in TS patients without ADHD as those with ADHD. By comparison, only 15 percent of the controls had problems getting to sleep.

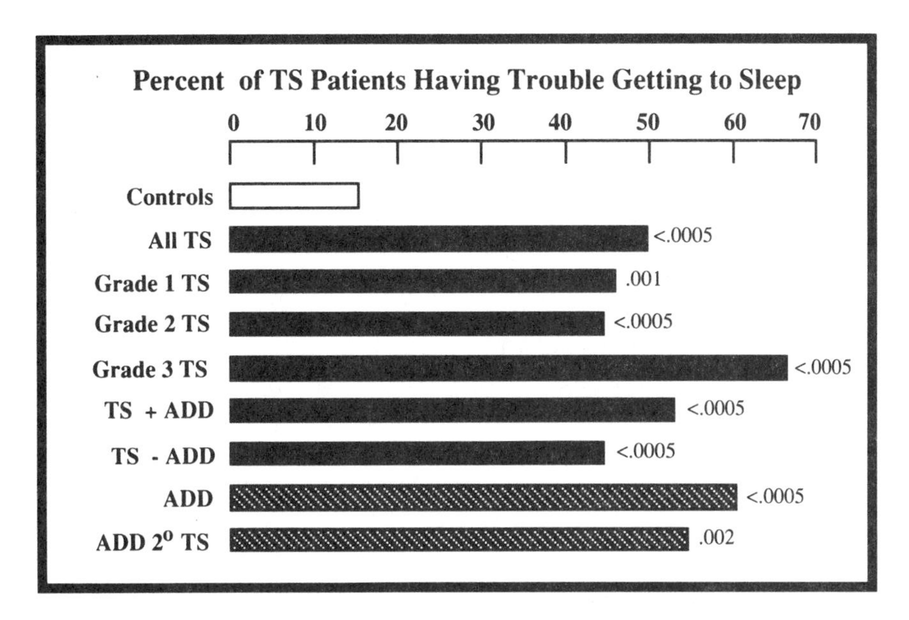

Sleepwalking (somnambulism) During sleepwalking the individual often abruptly sits up with open glassy, unseeing eyes. Although asleep, doors may be opened and furniture avoided[31]. Efforts to talk to the person get only mumbles and single-syllable answers. By the next morning there is no memory for the event. Like enuresis, sleep-walking usually occurs in the first one to three hours of sleep. In the controlled study the queries about sleep walking were divided into no, occasionally and often. Occasional or frequent sleep-walking was present in 46 percent of all TS patients compared to 21.2 percent of the controls, and was significantly more common in all TS and ADHD groups. Again there was only a moderate difference between TS patients with and without ADHD.

Sleep apnea Sleep apnea refers to a tendency to stop breathing for short periods of time during sleep. Patients often wake up or raise to much lighter levels of sleep during these periods. When this occurs often, sleep is no longer restful even though patients don't realize why. This problem is best diagnosed in a sleep laboratory. *Sleep apnea was found to be present in 23 percent of TS patients* based on sleep laboratory studies[974]. In one of our families four different individuals had sleep apnea. Thus, it can often be hereditary and TS may be one of the major genes responsible.

Night terrors Night terrors are not simply severe nightmares. As discussed above, they are disorders of arousal and are not dreams. They do not occur during REM sleep. Children with night terrors wake up screaming in fear, often breathing heavily and sweating. Although they are difficult to comfort they usually spontaneously relax after about ten minutes and fall back to sleep[31].

In the controlled study, night terrors were significantly more common in all groups of TS patients except mild grade 1. They were often present in 18 percent of all TS patients compared to two percent of the controls. This increased to 32 percent in grade 3 TS.

Early awakening Like trouble getting to sleep, early awakening was also common in TS, occurring in 32.5 percent of all TS patients compared to 4.3 percent of the controls. This increased to 50 percent in the grade 3 patients.

No naps This, of course, is a symptom relating to the early years when mothers are supposed to get some relief from child care in the middle of the afternoon. Many mothers of TS children don't get this luxury. An inability to take afternoon naps was often seen in 17.5 per ent of all TS children compared to 4.3 percent of the controls. The only groups where it was not a problem were the mild grade 1 and the TS patients without ADHD.

Sleep problem score To determine how often subjects had multiple problems with sleep, the sleep problem score was determined by adding up the scores for the individual symptoms. The results of comparing all the TS patients to the controls were as follows:

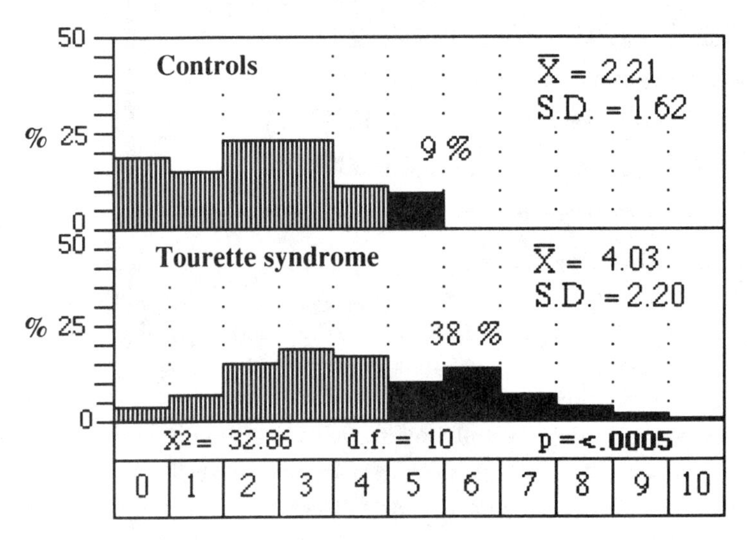

Sleep Problem Score

Nine percent of the controls had a sleep problem score of five while 38 percent of the TS patients had a score of five to ten.

There was a progressive increase in the number of individuals with a sleep problem score of five or more with increasing grades of severity of TS, with 21 percent of grade 1 TS, 37 percent of grade 2 TS and 45 percent of grade 3 TS patients having problems. The contribution of ADD was shown by the fac that sleep problems increased from 20 percent in TS patients without ADD to 48 percent in those with ADD.

A TS Mother's Story

The following is a report by a mother with TS who had two boys with TS. It graphically illustrates many of the sleep problems discussed.

"Sleepwalking and sleeptalking runs rampant on both sides of my family. My father's brother used to awaken, lock the bedroom door, hide the key, and go back to sleep. The next morning there was no memory of the hiding place so the boys had to crawl back in through the kitchen window.

"My mother's mother was once stopped at the front door, stark naked, yelling, 'there he goes, there he goes!' Who 'he' was, is still a mystery to all. My mother's father used to awaken, get into his car, and drive off. He would come back, park the car, go back to bed, and remember nothing.

"I had grown up with these stories and had accepted them as being amusing but mostly harmless. I first became aware of talking in my sleep while in college. I was living in the dorm and it was the first time that I had shared a room with anyone. I was majoring in nursing, which was a very strenuous program, requiring much study and producing incredible stress. When I sat up straight in bed and announced to my roommate that 'my Angel Sister told me to take the cadaver home and study it,' I assumed it was due to fatigue and strain. My roommate, who had a very well developed sense of humor, said, 'Would you mind repeating that?' I did, and then laid back down. When I mentioned the episode to my mother she said, 'Oh that. You've done it for years.'

"I became aware of sleepwalking about four years later when I was again under stress as an assitant head nurse on a medical isolation unit with dying children. I would run down the hall from my bedroom into the kitchen, thinking that one of the children's IVs had run out and I had to replace the bottle. I would wake up, feel like a fool and go back to bed. My new roommate laughed, describing the incidents as sounding like 'patter, patter, patter, Oh, Damn! , patter, patter, patter.'

"As I grew older, the incidences appeared to increase and for 24 years my husband lived with nighttime rousings to "look for the spiders crawling all over", or the 'bugging wires behind the bed', or to hear the 'very important information' that he 'must know about' at three o'clock in the morning.

"My wonderful husband is a very sound sleeper and has the capacity to drop back to sleep as soon as I awaken enough to know that I have done it again. At first, he tried to tell me that I was asleep which infuriated me, leading me to yell at him that I was not asleep and that I knew what I was doing. This outburst would wake me up. The awakening was always accompanied by a feeling of embarrassment for having acted like a fool. I could never quite capture what I thought had been the dream that triggered the behavior. I was frustrated constantly by my inability to recall and especially since so many of the incidents were recurrent. The room bugging idea lasted about eighteen months but who would bug the room and why always eluded me.

"I then began to start my sentences with a biligerent, 'Now, don't tell me I am asleep, because I know that I am awake and I have to tell you this!' By the time I said all of this, I was awake, feeling foolish, once again.

"There was a slow but definite progression of these behaviors as I aged. I began to wake up panicky, with my heart pounding and a definite feeling of flight or fight. I would sit up, feeling disoriented and frightened, realize that I was safely in bed, and would go back to sleep. The sleep-walking continued and became more frequent. I woke up at various times, standing in front of the refrigerator downstairs or at the open front door. I did not necessarily remember doing these things but if something was said during the next day, I would recall being in another room and I would remember what I had said. I never remembered walking down the stairs or walking back up the stairs to bed. I never tripped or fell even if there were things lying on the floor.

"I also began to combine the sleepwalking and the talking. If my husband was still up after I had gone to sleep, I would walk downstairs and talk to him. He had no way of knowing if I was awake or asleep. I remember one incident when I stood on the stairs, announced that I didn't know if I was awake or asleep but that I was looking for my canvas bag with the blue handles. My husband told me that he had not seen it. I don't remember walking down the stairs and I don't remember walking back up to bed. I don't own a canvas bag with blue handles.

"During a nap one afternoon, I

awoke, beating on the top of my clock radio. I awoke one night, flaying around on the floor next to my bed where I had knocked over the plant stand, broken a manicured fingernail, and awoke, feeling like an idiot and realizing I then had a mess to clean up in the morning.

"I eventually made my way to an endocrinologist, ostensibly for estrogen therapy. When I related my nighttime behaviors to him, he treated me with 75 mg each night of a tricyclic. The terrors decreased in frequency but seemed to intensify in severity. The dose was increased to 100 mg which seemed to calm them. I also began to dream dreams which I can remember which is a new experience for me. I am also less fatigued, take fewer afternoon naps and generally feel more energetic. I always said that I was born tired and never got rested. My husband always retorted that if I would stay in bed and sleep like other people, I wouldn't BE so tired. He was right."

Emotions versus Genes

Prior to the sleep laboratories, most of the above disorders were considered to be emotional in origin. While this can still be valid for nightmares, especially those with a bizarre content, the sleep studies have shown that the disorders of arousal occur at specific phases of the sleep cycle and are often associated with various disorders of the nervous system[31]. Since even these individuals may show an increased frequency of various psychological problems[1512] the implication that emotional problems cause the sleep disorders persists. However, the TS model sheds considerable light on these associations. Here a genetic disorder causes both the behavior and sleep disorders, rather than the behavior problems causing the sleep problems. This is not meant to deny the fact that being acutely emotionally upset can't disturb sleep. However, when the sleep problems persist for years they may have a biological basis.

As described in Part IV *sleep is regulated by serotonin-containing nerve cells in the brain.* Disorders of sleep in TS may be due to abnormalities in serotonin levels[P448].

Summary: Sleep problems, especially trouble getting to sleep, early awakening and disorders of arousal such as sleep-walking, night terrors and enuresis are common in Tourette syndrome and ADHD. While it has often been assumed that these sleep disorders are caused by psychological problems, they can be better explained as the result of specific genes causing both behavioral problems and sleep disorders.

Chapter 41

Excessive Drinking of Fluids — Polydipsia

The excessive drinking of fluids is known as polydipsia. *Poly* for many and *dipsia* for drinking. We first became aware of polydipsia in Tourette syndrome from the family described in the chapter on *Conduct*[p147].

The proband was a 17 year old male with motor and vocal tics and conduct disorder. His three full siblings, half-sister and both parents had some TS symptoms. Although not having motor or vocal tics, the father's siblings had many and varied types of behavioral and sexual problems. The interesting feature of this family was that many members had polydipsia — consuming unusual quantities of fluids in the form of ice water, cokes, or ice chips.

As a result of this family, we subsequently began asking all new TS patients about fluid intake. About 15 percent stated that they had almost a compulsive need to drink fluids. The following family is an additional example.

Julie

Julie is 11 years old and has been very active since she was two. She never took naps, slept poorly, and was always on the go. In nursery school they called her "the terror." In kindergarten she tried to run the class and was so involved in what other children were doing she didn't do her own work. At age seven a psychologist thought she might have ADHD and she was placed in a resource room. At age nine she began lying a lot. She had no motor or vocal tics. Other behaviors included rocking while holding her crotch, enuresis, constantly talking to herself, having an imaginary friend, and many confrontations at home including refusing to do what she was told. She drinks water constantly and likes to put salt in it.

The family history was as follows:

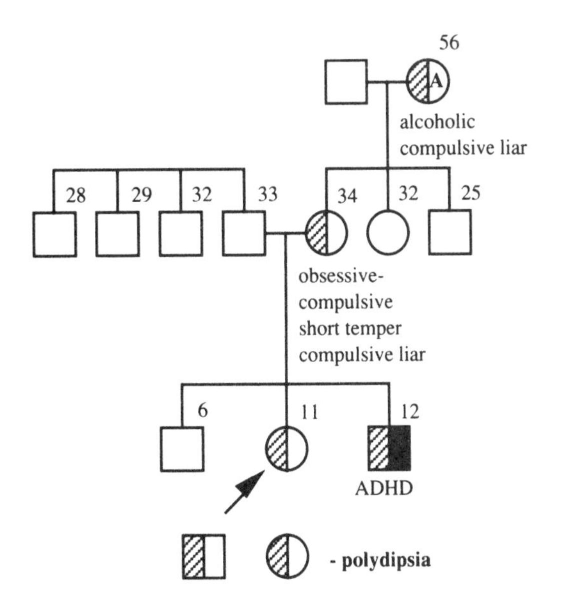

Her 12 year old brother "fills a quart glass with ice water and lays on the living room floor drinking it. In 10-30

minutes he will refill it and drink a second glass. On the weekends this happens six to seven times. He always feels like he is hot. He turns the air conditioner to cold when everyone else in the house is freezing." His parents estimate he drinks a gallon of water a day. He has motor and vocal tics.

Her mother also drinks water excessively. In her early 20s she had problems with depression, and now has numerous obsessive-compulsive behaviors. She has a short temper and problems with compulsive lying. Her mother also drank water excessively, was an alcoholic, a compulsive liar and "a very manipulative person."

The Hormonal Control of Water Intake

One of the major mechanisms of controlling the balance of fluids is through a hormone secreted by the intermediate lobe of the pituitary. This hormone is called antidiuretic hormone or **vasopressin**. Diuretic means to excrete fluid so antidiuretic hormone suppresses the excretion of fluid by the kidneys. When the pituitary is damaged so it fails to excrete vasopressin the kidneys excrete large amounts of water — a condition called **diabetes insipidis** (not related to sugar diabetes or diabetes mellitus). To determine if the production of vasopressin was abnormally low in TS patients I examined the blood level of vasopressin in some who complained of polydipsia. In about a third it was abnormally low relative to the amount of water excreted by the kidneys[1633].

Vasopressin and Memory

One of the most intriguing features of this observation was the fact that vasopressin has frequently been reported to have an effect on memory and memory is poor in many TS patients. Humans with diabetes insipidis often have problems with memory[1107], and rats with a hereditary defect in the vasopressin gene[1719] have specific memory defects[486,487]. Both normal and diabetes insipidus rats, and humans, when injected with vasopressin show improvement in certain learning and memory tasks[138,986,1088,1107,1999,2040]. Some have suggested that the improvement in learning is secondary to the ability of vasopressin to enhance arousal and attention[138].

Dopamine and Polydipsia in TS

Moldofsky and co-workers[1347] also described polydipsia in some TS patients. They stated:

> "The auxiliary features in multiple tic syndrome, of childhood hyperkinesis and persistence into later years of restlessness, hyperactivity, pathological aggressiveness, and frequency of drinking and urinating might be related to CNS [central nervous system] dopaminergic disturbance. A comparable animal model of increased aggressiveness, restlessness, chattering, irritability, hypersensitivity, and increased water intake had been induced with agents that stimulate dopamine receptors in monkeys[770]."

This is consistent with the observation that dopamine can regulate the secretion of vasopressin[245,1184], and neurons which appear to use vasopressin as a neurotransmitter project into the dopamine neurons of the limbic system[245]. The subject of dopamine, the limbic system and TS is discussed later[p363].

Is Tourette syndrome Due to a Mutation in the Vasopressin Gene?

These observations raise the obvious question — Are some cases of Tourette syndrome due to a genetic defect in the vasopressin genes, resulting in increased fluid intake, poor memory with learning disorders, and defective regulation of dopamine neurons in the brain? The vasopressin genes are located on the short arm of chromosome 20[1609]. To date we have not found any evidence for linkage of TS to the vasopressin gene. This suggests that the polydipsia in TS is a secondary effect due to dopamine or other abnromalities. A relative deficiency of vasopressin could be a contributing factor to the learning disorders in TS.

Serotonin and Vasopressin

The other neurotransmitter that plays an important role in the control of vasopressin is serotonin[P455]. Defects in serotonin can explain both the behavior and the polydipsia in Tourette syndrome. Polydipsia might also be simply another compulsive behavior in TS similar to alcoholism[P217] and compulsive eating[P227].

Summary: About 15 percent of patients with Tourette syndrome drink excessive amounts of fluids. This may be related to the basic brain chemical imbalance in TS which then causes changes in the hormone vasopressin, which regulates water input and output, or it may be another compulsive behavior.

Chapter 42

ADHD Secondary to a *Gts* Gene

While this has already been discussed in the chapter on ADHD in Tourette syndrome, it comes up so often that some additional emphasis is useful. This diagnosis refers to those children whose attention deficit hyperactivity disorder is probably due to a *Gts* gene. The criteria for the diagnosis of ADHD 2^0 TS or 2^0 to a *Gts* gene are[357,358,360,366,368]:

a. Meets the DSM-IIIR criteria for ADHD.

b. One or both of the following are true:

1. Has a positive family history of TS or chronic motor tics.
2. Has one or two mild or subtle motor or vocal tics, but not both.

This designation is useful for several reasons. It allows us to inform parents that we believe their child's ADHD is due to the partial or early expression of a *Gts* gene and that he or she might develop motor or vocal tics. If stimulant medications are prescribed they are told of the possible exacerbation or precipitation of tics by the medication. Depending upon the severity of the particular case the medication can either be discontinued or another medication added to control the tics[p575].

Since this group of patients has been mentioned in relationship to all the different behaviors seen in TS, it is now time to summarize and see if there are any clinical features that distinguish "pure" ADHD from ADHD 2^0 to a *Gts* gene. First, let us look at a typical case.

John

At four years of age John's mother noticed that he was very aggressive in groups, cried easily, and had a very short attention span. This continued until at age seven the parents began therapy in a behavior modification program with positive rewards for good behavior. This worked for awhile but he soon began to manipulate them to get bigger and better rewards for less and less good behavior.

At age nine he was treated with Ritalin and this resulted in real improvement for the next two years. The summer before coming to the clinic the parents stopped the Ritalin and decided to try a Tough Love[2132] approach. He was also placed in a summer camp but was thrown out after one month because he was lying, stealing, and pulling girls pants down. When his mother locked him out of the house for several hours in the afternoon, because of his aggressive behavior, he broke the windows to get back in. When school started he didn't want to go and frequently smashed things in fits of anger. When seen in the clinic he fit all the criteria for ADHD and was treated with imipramine (Tofranil), which resulted in a striking improvement in his attention span and decrease in his short temper and aggressive behavior.

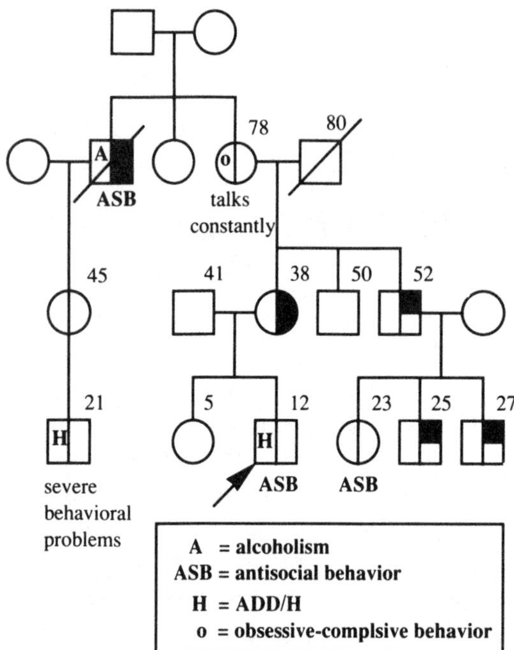

He never had motor and vocal tics and would normally have been diagnosed simply as ADHD and conduct disorder. However, the family history was as shown above. The family was riddled with other members with Tourette syndrome (mother, uncle, two cousins, and grand uncle). His grandmother had severe obsessive-compulsive behaviors, another cousin showed antisocial behavior (juvenile delinquency and drug abuse) and a second cousin had ADHD and severe behavioral problems. Thus, the diagnosis became *ADHD 2⁰ to a Gts gene.*

Are conduct problems and other behaviors more severe in children with ADHD 2⁰ TS compared to ADHD? The **figure on the right** summarizes the frequency of various behaviors in Tourette syndrome, ADHD, ADHD 2⁰ TS and controls.

Surprisingly, there were no significant differences in the frequency of obsessive-compulsive behaviors, conduct disorder, or exhibitionism in the ADHD compared to the ADHD 2⁰TS groups. While stuttering, panic attacks, and depression seemed to be more common in ADHD 2⁰TS, the two groups were small and further studies are necessary.

Late Onset ADHD 2⁰ *Gts* gene

Part of the definition of ADHD in the DSM-III is that the symptoms must begin by seven years of age[490]. However, we have seen many cases of ADHD 2⁰TS where the symptoms did not come on until after seven, sometimes being delayed until 11 to 16 years of

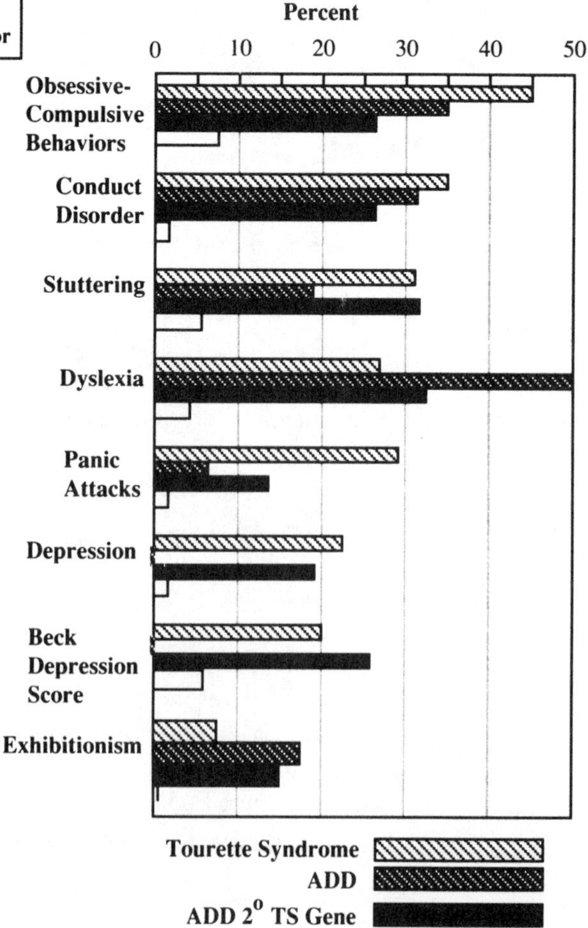

age. The following is an example.

> D.P. was a 14 year old boy. He had a brother with TS who we had been treating for years. D.P. got straight A's in grade school and was well behaved with a good sense of humor. However, after six months into the seventh grade his grades suddenly dropped to D's and F's and at home he was always angry about everything. For the first time his teacher began to complain that he was not doing his homework, was not paying attention, had a poor attitude, was resistant to the teacher's authority and was failing his classes. The usual explanation for such a sudden change at this age is that the student is taking drugs. However, D.P. vigorously denied this, he was not "hanging out" with the wrong crowd, and drug testing was negative. The diagnosis of **late onset ADHD 2^0 *Gts* gene** was made. He was treated with imipramine with excellent results.

This diagnosis requires a high index of suspicion and a careful family history is often critical. A proper diagnosis backed up by a drug test can avoid the destructive effects of parents, teachers and others insisting the problem is due to street drugs.

Conduct in ADHD

I have repeatedly emphasized the role of ADHD in contributing to conduct disorders in childhood and the above results showing such problems in about 30 percent of the TS, ADHD and ADHD 2^0 TS groups is further evidence of this association. Many years ago, when the concept of minimal brain dysfunction and hyperkinesis was still being formed Eisenberg[524] wrote of the psychiatric implications of this disorder as follows:

> "Hyperkinetic children are constantly on the move, unable to sit still, fingering, touching, mouthing objects. They are frequently destructive, at times by design, at others inadvertently, because of impulsive and poorly controlled movements. Their overactivity thrusts them into the center of the group so that they are typically described as attention-seeking. This aspect of their behavior has been vividly described by Kahn and Cohen[1000] as *organic drivenness.* These authors ascribe the constant activity to a *surplus of inner impulsiveness.*"
>
> There is "marked lability of mood. Frustration threshold is reduced. When this threshold is exceeded, outbursts of angry behavior result. Mercurial changes of mood from tears to laughter are seen. Unprovoked frenzies of rage, in which for no apparent reason the child strikes out blindly at all about him, often inflicting harm on others, can be noted. When these attacks terminate, the child may be bewildered by what he has done and genuinely apologetic for it — only to undergo another uncontrollable crisis not long afterward."
>
> "Antisocial behavior, in the form of lying, stealing, truancy, cruelty and sexual offenses, may be a prominent feature. Tasks requiring abstract thought are particularly apt to be difficult for these children, whereas concrete problems may be successfully completed[768]. They may exhibit perseveration in responses and often display meticulous and pedantic behavior as if they are desperately trying to keep a chaotic inner world in order by limiting outer stimuli[1896,1897]. The distinction between figure and ground appears to be blurred.. Anxiety may mount to. . .panic. . .[and there is an] inability to inhibit, or even delay, upsurging inner impulses. Such children have great difficulty in postponing

present gratification for future gain."

This lucid description of conduct problems in the ADHD child is as valid today as it was over a quarter of a century ago. In many respects, whether due to a *Gts* gene or other genes, ADHD is a syndrome of widespread behavioral disinhibition.

How Much ADHD Is Due to a *Gts* Gene?

The definitive answer to this question will have to await the development of specific genetic tests for the *Gts* gene. However, until then some estimates are possible. In our experience, when a careful clinical and family history is taken about half of ADHD is ADHD 2^0 TS. Studies of blood serotonin and tryptophan in the parents of ADHD children[P649] suggest this figure may be even higher.

Summary: The concept of ADHD secondary to a Gts gene is re-emphasized. The behavior of children with ADHD or ADHD 2^0 TS is very similar. A significant percent of children with ADHD may actually be carrying a Gts gene.

[Further evidence on the relationship between TS and ADHD is presented in Chapter 90.]

Chapter 43
Physical and Sexual Abuse

It should not come as any great surprise that a genetic disorder that can predispose a person to short temper, periodic outbursts of rage, antisocial acts and disinhibited sexual behavior could also predispose some individuals to physical abuse of spouses, or sexual abuse of children or siblings.While we have observed some of these problems in individuals with TS and ADHD, they are not rampant and can also occur in individuals without these genetic disorders. I distinctly do not want to stigmatize everyone with TS or ADHD as a potential wife or or child abuser — that would be blatantly overstating the case. However, these problems are sufficiently common in the general population that any insight we can obtain into factors that cause them is a worthy subject of discussion. There is much emphasis in the psychological literature on the philosophy that "an abused child may grow up to be an abusing parent." I think it is important to present an alternative view that is rarely considered: "an abusing parent may have a gene for a disinhibition disorder that is passed on to the child who then also becomes an abusing parent." The following are some examples.

Some Pedigrees

The Taylor Family

The following is a family illustrating the complex question of whether sex abuse is a learned or a genetic behavior or some of both.

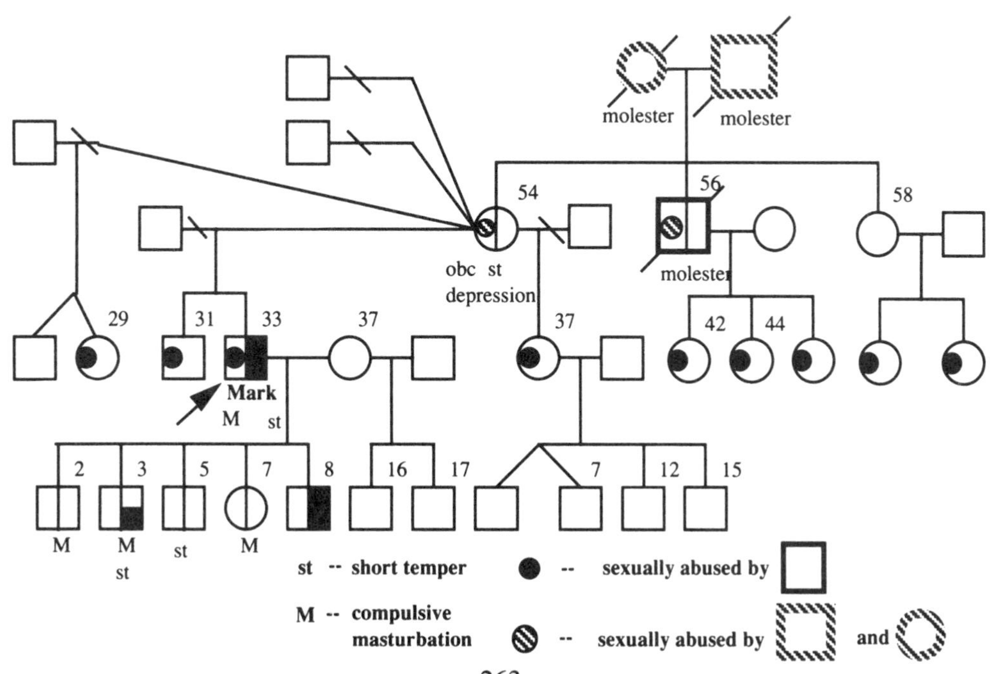

Mark was a 33 year old male who was referred by his therapist because he had a very short temper, constantly complained that his clothes were too tight and had repeated throat clearing and coughing. He was in therapy because he had been abused as a child and had repressed all memory of his life prior to 12 years of age. When he was 16 he noticed some faint scars on his back and asked his mother what they were. His mother stated that she had physically abused him when he was a child. Other symptoms he manifested were a tendency to blow up at minor things, temper tantrums even as an adult, compulsive masturbation beginning at age 14, and obsessive thoughts of a violent and sexual nature. These included thoughts of raping women and sexually abusing his children. He never acted on any of these thoughts. He said his early family life was one of constant conflict and everyone was always yelling at each other. It was not unusual for different members of the family to go after each other with knives.

His mother, age 54, led a chaotic life. She was married five times and had many more boyfriends. She had an explosive temper, chronic lifelong depression and was a compulsive spender and eater. Her brother, now 56 (heavy square), at one time or other sexually molested virtually every female and several males in the extended family [indicated by ● in the pedigree], including his three daughters, all of his nieces, two out of three of his nephews and Mark. This individual and Mark's mother had in turn been sexually abused by both of their parents [stippled square and circle].

This family shows the classic pattern of the sexually abused becoming abusers themselves. But was this a learned or a genetically transmitted impulsive-compulsive behavior? This family shows the presence of Tourette syndrome in Mark, who himself had obsessive thoughts of sexual abuse, and passed the *Gts* gene onto at least two and probably all of his children. The gene was presumably present in his mother, who had an explosive temper, chronic lifelong depression, chaotic interpersonal relationships, and compulsive behaviors. While there was not enough history available from the uncle's side to be certain, because of the driven compulsive nature of his behavior it is probable that he and at least one of his parents also carried the *Gts* gene.

The Jamison Family

Mike was a 13 year old boy brought in by his distraught mother with the following story:

> At five years of age when Mike began to attend kindergarten he was noted to be very disruptive and couldn't sit still. At seven a diagnosis of ADHD was made but since Ritalin caused him to act "weird" he was treated with Cylert with considerable improvement in his behavior and attention span. At 10 years of age, he began talking out loud for no apparent reason, and developed sniffing vocal tics. At 12 he also began to have barking noises. By the time I saw Mike he had many behavior problems consisting of being very disruptive in the class-room, talking out loud, standing up and walking around without permission, truancy, stealing and lying. At home he was staying out late at night, talking back and smarting off and wanted to be in control of everything. The vocal tics were still present. A diagnosis of chronic vocal tics and ADHD due to a *Gts* gene was made and he was treated with slowly increasing doses of Haldol. At a dose of

2 mg every morning his vocal noises had stopped, he was attending school and getting As and Bs and his behavior had markedly improved.

The family history was interesting:

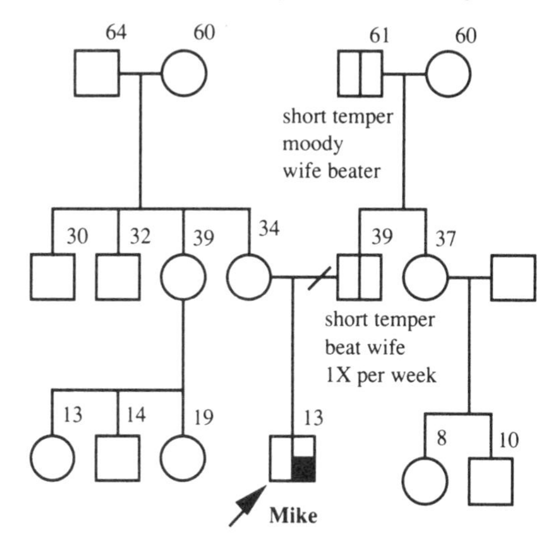

His mother had divorced his father because, "He was violent, had a very short temper and beat up on me about once a week." His father also physically abused his wife. In time, without treatment, Mike probably would have married and also become a wife beater. To behavioral psychologists, this two generation family of wife beaters would have provided strong evidence for the statement that children of abusers become abusers themselves. However, in this case, the best explanation is that each of these individuals inherited a gene for a genetic disinhibition disorder, and this, rather than learned behavior, was the cause of two generations of spouse abusers.

The Samson Family

Jan was brought in by her single mother for evaluation because her teacher said she was too aggressive to keep in the class.

During her first year she was a head banger and cried constantly. In nursery school she fought a lot and couldn't get along with the other children. Throughout kindergarten and the first three grades she was noted to be "very distractible" and by the fourth grade the school district had placed her in an SED class (severely emotionally disturbed). Just before coming in she had been expelled from SED class for being too aggressive toward other children, and was given a home tutor. The mother stated that "the kids know how to push her buttons and she would explode and beat them up. If she hears other children laughing she will walk over and say 'why are you laughing at me,' and start a fight." There was a history of facial grimacing in third and fourth grade.

The family history was as follows:

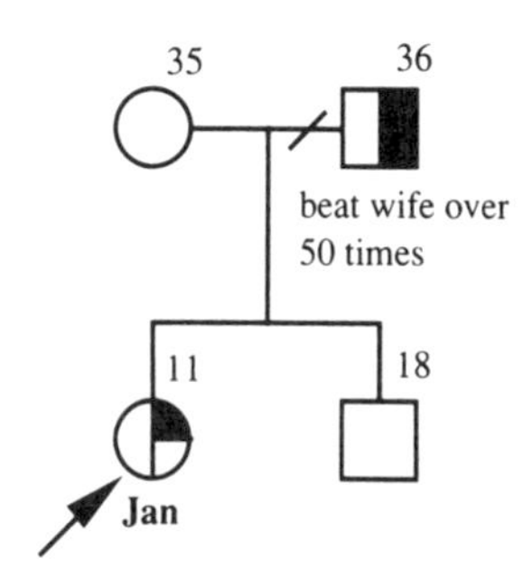

The mother had divorced her husband because he had beaten her up over 50 times. He had migraine headaches and constant facial tics and throat clearing that were worse when he was nervous. "He would be calm one minute then suddenly be in a rage and tear up the house. He once cut up all my clothes." The father clearly had Tourette syndrome with associated explosive behavior.

The Saunders Family

John is now six years old. He began being very aggressive at two and for a full year cried hysterically every time his father left him at nursery school. He has had repeated throat clearing for the past

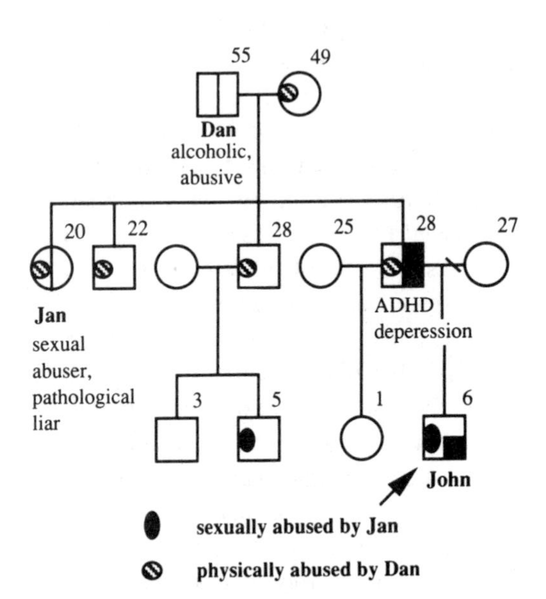

year. His father had ADHD as a child, and motor and vocal tics and problems with depression since he was a teenager. His grandfather, Dan, was an alcoholic, had a short temper, couldn't get along with other people and regularly beat up on his wife and all four children.

One day after the father's sister Jan had baby sat John, he began to imitate movements of sexual intercourse. His father asked him where he learned that and he related a story of Jan lying on him and making these movements. They subsequently banned Jan from ever coming to their house again, but said nothing to other members of the family. A year later they found that Jan had also sexually abused the five year old cousin and a five year old neighbor girl. Jan had no history of motor or vocal tics but was a pathological liar.

In this family, although neither Dan nor Jan had motor or vocal tics, they both probably carried the *Gts* gene.

The Hallman Family

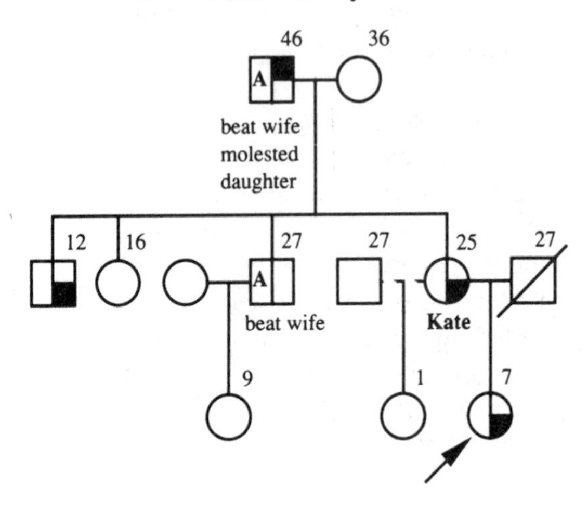

Mrs. Kate Hallman's seven year old daughter developed ADHD at age three and was having severe temper tantrums by age four. For the previous year she had constant grunting vocal noises and was constantly lying about not doing things even when caught in the act. She had echolalia, palilalia, constant talking, excessive touching things, night terrors, sleep-walking, and insomnia. She would often lift her dress when boys were around. Kate brought her to the clinic because of discipline problems. She laughs and talks back at her mother when she tries to discipline her. Everything is always someone else's fault. She has been stealing money from her mother.

Kate is 25 and has had vocal grunting tics for years, she often ditched school because she couldn't keep her mind on her work. She states that from ages seven to 13 her father molested and raped her, and that he beat her mother almost daily. He was so jealous that every time her mother went to the store he would put his hand up her dress, when she returned, to see if she had sex when she was gone. Her 27 year old brother also beat his wife constantly. Kate often had to pull him off his wife when she was visiting. His wife finally got a court order to keep him out of the house.

The Wiley Family

Jake was a nine year old boy with TS who came to the clinic with his mother and stepfather. After specific questioning the mother reported that she had divorced her first husband he had a short temper and had beaten her many times. He never abused their son. He had facial tics, symptoms of adult ADHD and had a brother with ADHD.

19 26 29 32 29 29 26 13 9
ADHD
ADHD
short temper
beat up his
wife many
times
9 4
Jake

ADHD, Learning Disorders and Abusive Behavior

The previous families have all shown the potential for physical of sexual abuse in some individual with Tourette syndrome or presumably carrying the *Gts* gene. However, in some cases other disinhibition disorders are involved.

Alex

Alex was diagnosed as having only ADHD. An extensive pedigree showed a remarkable incidence of childhood and adult ADHD in other members of the family .

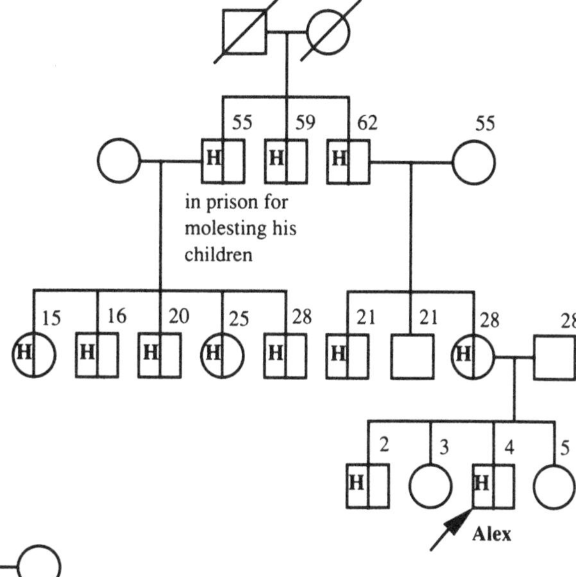

The mother's 55 year old uncle and his five children, all had histories of ADHD. The uncle was presently in prison for sexually abusing his daughters.

Charles

Charles was six years old. His mother said he was kicked out of preschool at two years of age because they couldn't control him. If he was not the center of attraction he had a temper tantrum. From three to five years of age he was violent, hurting people and breaking things. In the past year, in first grade, he has done poorly, interrupted the teacher and bothered other children. His behavior is getting worse at home. He satisfied all the DSM-IIIR criteria for ADHD. After treatment with Ritalin 10 mg twice a day his academic performance and behavior both at home and at school improved remarkably.

The pedigree was as follows:

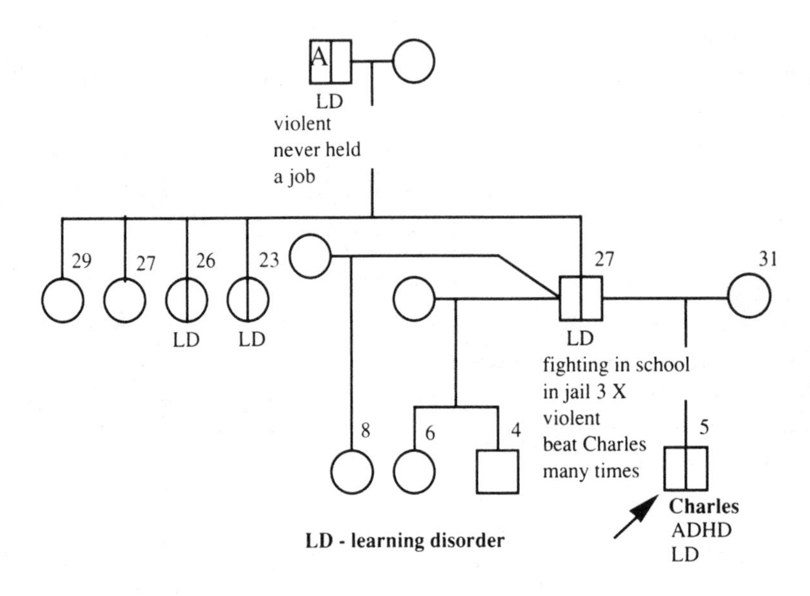

The biological father had children by three different women, but married none of them. He had a learning disorder and was constantly fighting in school. He had a violent temper and repeatedly beat Charles during the three years his mother was with him. His father was also a violent man, an alcoholic, could never hold a job and had learning disorders as a child.

This provides an example of a hereditary disinhibition disorder probably secondary to an ADHD gene. The severe behavioral problems in Charles, in the third generation, only improved following treatment with Ritalin.

A Multigeneration Problem

These families indicate that TS and ADHD genes which result in disinhibited behavior can sometimes explain physical and sexual abuse. Because these are autosomal dominant genes these behaviors can be passed from generation to generation. This concept is in striking contrast to the usual *generational transfer theory* that abused children grow up to become abusing parents. This theory has been challenged on other more practical grounds of simply examining the facts. Stacy and Shupe[1850] studied a large sample of wife batterers. Of these 40 percent had never witnessed physical violence between their parents, and 60 percent had not been physically abused as children or neglected by their parents. Most of the women who abused their children had not been abused by their parents. Although environmental factors can definitely play a role in multiple generations of physical and sexual abuse, genetic factors should also be carefully considered and in some cases may often be much more important contributing factor.

Child Abuse or Parent Abuse?

So far I have been emphasizing the role of genetic factors as a cause of physical or sexual abuse. However, it is equally important to look at the opposite side of the coin — some times parents of TS children whose behavior is totally out of control, are accused of child abuse after administering discipline to their child. These accusations often arise from complaints by the child, frequently highly distorted, that their parents have been abusive to them. These children quickly learn to use and manipulate the system to threaten their parents.

Adam

When Adam was two years old he was "thrown out" of nursery school because of violent aggressive behavior. At three he was placed in a special class. After two weeks he threw a chair at a teacher and was considered too dangerous to have in a class. This type of behavior was also going on in the home. At age four a diagnosis of ADHD was made, however, neither Cylert nor Ritalin helped. Because of some arm tics and stereotyped hand waving movements he was then treated with Haldol. Even though "this helped him more than any other medication" it was soon stopped by another doctor. At age five he was placed in a residential treatment facility because his mother couldn't handle his aggressive behavior. A year later there was an onset of hooting, whistling and blowing vocal tics. At age seven he returned home.

I first saw him because he was still having problems with violent behavior. *He often attacked his mother when asked to do something.* Other behaviors included lying, stealing, running away from home, setting fires, drastic mood changes, confrontive behavior and constantly getting into fights. He had coprolalia and compulsive behaviors such as getting up in the middle of the night and counting money over and over. The family history was as follows:

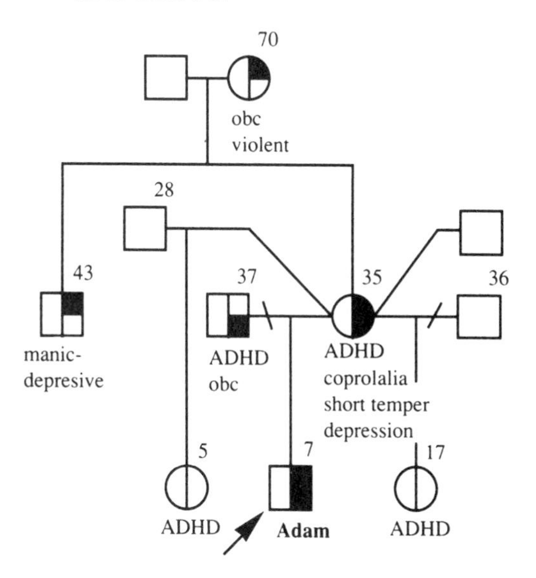

Both parents were carrying a *Gts* gene. The mother had motor and vocal tics, ADHD in school, coprolalia, multiple episodes of depression and a short temper. The father had sniffing vocal tics and obsessive-compulsive behaviors. Adam seemed to clearly be a homozygote, that is he received a *Gts* gene from both parents[p630].

A second aspect of this case, one that has occurred in many similar cases, is the interaction between the mother and Child Protective Services. While Adam was in the residential treatment home the social workers there assumed that the degree of violent behavior Adam was showing *must have been due to the fact that his mother physically abused him.* They repeatedly threatened to take him out of the house and prevent her from seeing him. There was no objective evidence of this and it was vehemently denied by the mother and her present husband.

Treatment with Haldol helped somewhat but he still had explosive aggressive outbursts. Despite a negative EEG I treated him with Tegretal 100 mg 3 times a day[p585]. This resulted in a significant decrease in the aggressive violent behaviors.

This case has four important lessons:

1. TS with severe aggressive behavior starting early in life may be due to a homozygous dose of the *Gts* gene.

2. There are explanations other than child abuse for aggressive behavior in young children.

3. Tegretal can sometimes extinguish such behaviors.

4. Parent abuse may sometimes be a more severe problem than child abuse.

A final lesson from this family is that the question of "who did what to whom" is often complex. An aggressive, out of control TS child by definition has at least one parent who is also carrying the *Gts* gene. *If* that gene also expressed itself as short temper and low tolerance to frustration, a truly explosive situation can develop. Oftentimes both medication and intensive family therapy[p603] is required to lower the level of chaos in these families.

Summary: In some individuals genes for disinhibited behavior are the cause of wife or child abuse and the transmission of these genes to their offspring is the reason why these behaviors may occur in succeeding generations.

Although the most common problem is abuse of the child by an adult, sometimes, what on the surface may appear to be child abuse by the parent can more accurately be portrayed as parent abuse by the child.

Chapter 44

Homozygotes, Assortative Mating and Gene Selection

Families Where Both Parents Have Tics

As the number of TS pedigrees continued to mount, we began to notice a significant number in which both parents had tics. The following pedigree shows two sisters with chronic tics who have each married men with tics.

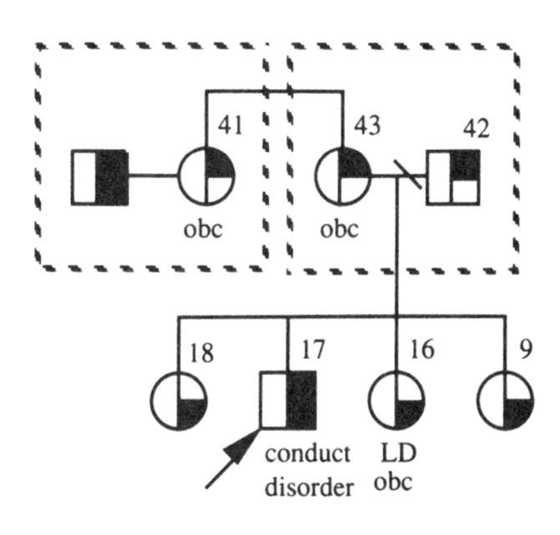

This is part of the pedigree shown previously[p146]. Each time the family returned, more was learned about the pedigree. Both parents carried the *Gts* gene and the mother's sister realized she had motor tics and had married a man with TS, only after the diagnosis became known in her sister's family.

A series of additional pedigrees illustrating the occurrence of a *Gts* gene in both parents are as follows:

A baby with TS — A probable homozygote

The following report[377] illustrates that the symptoms of TS can be striking, even before one year of age. The pedigree is as follows:

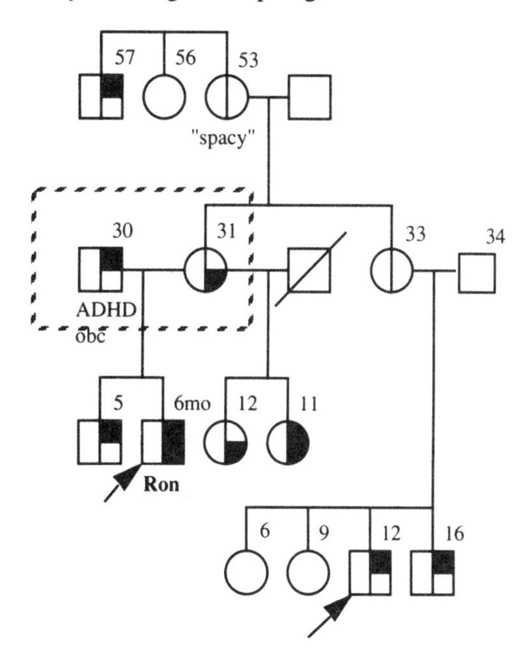

The mother wrote that:

"Ron had repetitious movements in the womb that I didn't experience with any of my three other children. After he was born he screamed for seven hours straight and disrupted the whole nursery. The nurses taped a pacifier to his face, but to no avail. He continued to scream for four months as if he were in horrid pain. He only gained one and one-half pounds for the first four months. I then started feeding him whenever he would eat. His crying decreased and he began to gain weight. He almost never slept. The only thing that seemed to calm him was if I rocked him. His emotions seemed uncontrollable, one minute he was laughing, the next screaming. I noticed muscle tics of his face at four months and frequent grunting sounds. Ron growls at people and makes weird noises. A neighbor who had raised five boys noticed the strain on me and offered to baby sit. After three times she refused to do it again.

Various doctors have told me that there is nothing wrong, that he just has colic, that he is a brat and needs a lot of spankings, that I'm a poor mother, and that he needs a hernia operation."

The mother's 33 year old sister had come to the clinic four months previously because her 12 year old son had motor tics, ADHD, and conduct disorder. She was the one that suggested that Ron may have TS. It is clear that Ron's mother carried the gene since she had two children with TS or chronic vocal tics, by her first husband, and had vocal tics and some compulsive behaviors herself. Ron's father also had a *Gts* gene as indicated by the fact that he had multiple motor tics and some learning problems in school. This suggests that the best explanation for Ron's very early onset of severe TS is that he inherited a *Gts* gene from both parents and is a homozygote.

Four TS x TS marriages.

The initial proband in this family [below] was the eight year old boy with TS (right arrow). His mother had TS and

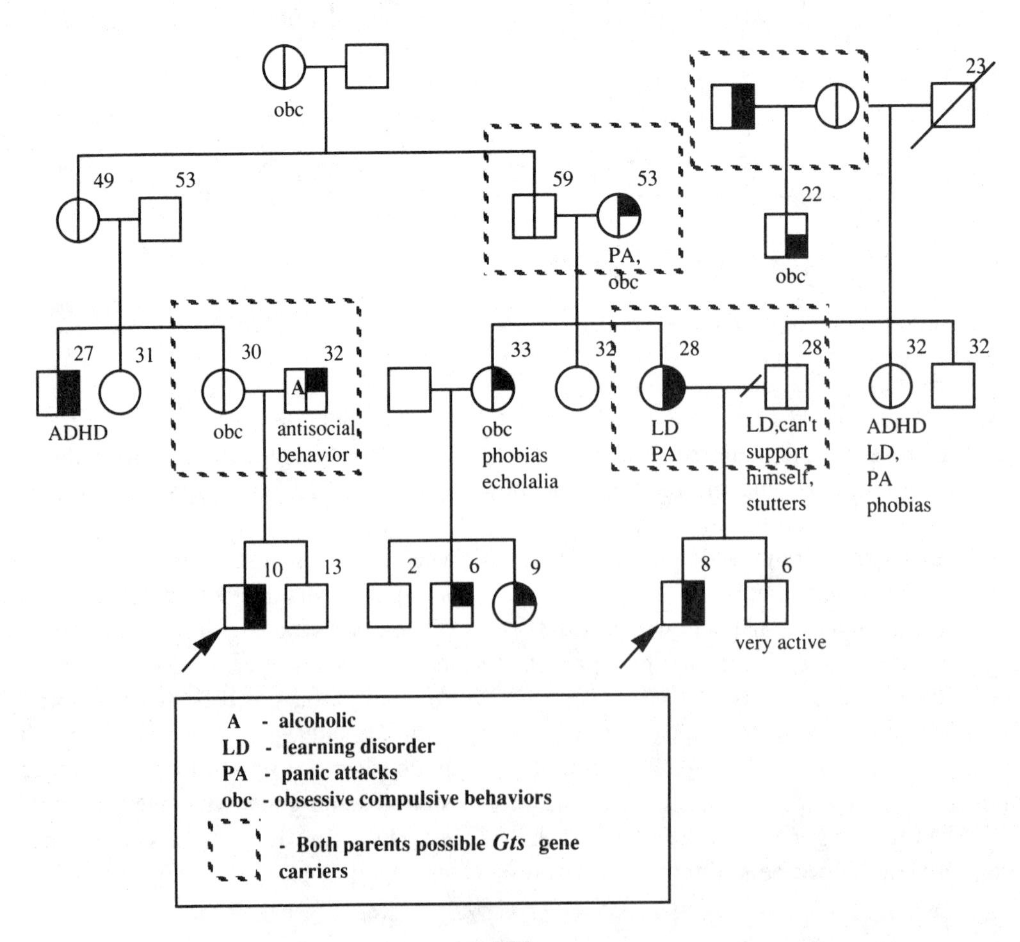

his father had learning disabilities and stuttering and was never able to hold down a job. The father's sister had ADHD, learning disabilities, panic attacks and phobias. After the father's father died, the man his mother married had TS and they had a son with vocal tics and obsessive-compulsive behaviors. There were four other individuals with motor tics on the mother's side including the mother's mother. The mother's father's sister had a son with TS and ADHD and another daughter married a man with motor tics, alcoholism and antisocial behaviors. They had a son with TS, the second proband in this family. All together there were four marriages where both parents appeared to carry a *Gts* gene.

There are so many marriages in this pedigree where TS is on both sides that one wonders about the possibility that the *Gts* gene is even more common than previously thought. [For further discussion of this see Chapter 69.]

TS x chronic motor tics

In the following pedigree the presence of a *Gts* gene in both parents was clear-cut since the mother had both motor and vocal tics and the father had motor tics and obsessive-compulsive behaviors.

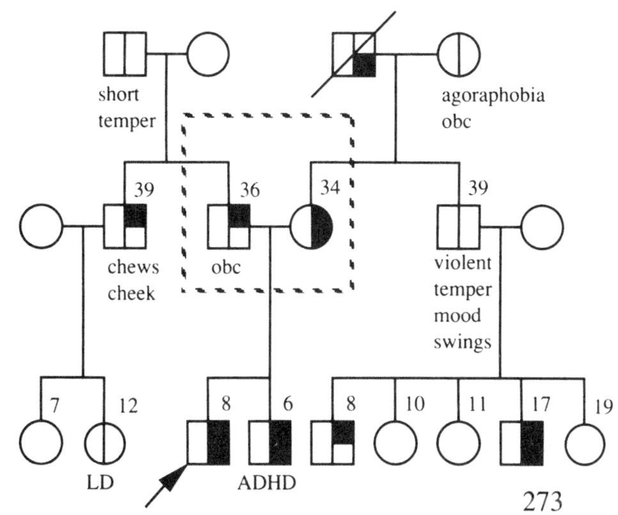

At three years of age the proband began having violent temper tantrums, crying, beating up on his baby brother, and destroying toys. When first seen at age eight he had multiple motor and vocal tics, was constantly pulling at his groin, squeezing his penis, exposing himself to other children and was showing rapid mood swings in which trivial things would set him off into rages. He had been previously diagnosed as having a childhood form of manic-depressive disorder. His brother also had TS.

TS x manic-depressive

The proband is a 15 year old male. At age two and one-half he was very active, and had a rich lexicon of dirty words. At five years of age, in kindergarten, he had an onset of eyeblinking and head jerking and would run around the room making animal sounds like a coyote. First grade was a disaster. He was constantly touching and pushing other children, running off at the mouth, making farting noises. He did poorly academically and had a poor memory. Successive years only got worse. By the sixth grade he was asked to leave school because of willful defiance and extremely disruptive behavior in the class room.

The **pedigree is on the next page**. His mother had eyeblinking tics, pulled all her eyelashes and eyebrows out between ages seven and eight, and is always counting things. His maternal grandfather's sister had two sons with TS and alcoholism. His father had extreme mood swings and had been diagnosed as manic-depressive. His father's sister had four children. One son had a learning and conduct disorder, the other son is presently in jail for murder and was a member of the

Manson cult.

The proband probably represents homozygous *Gts*/*Gts*.

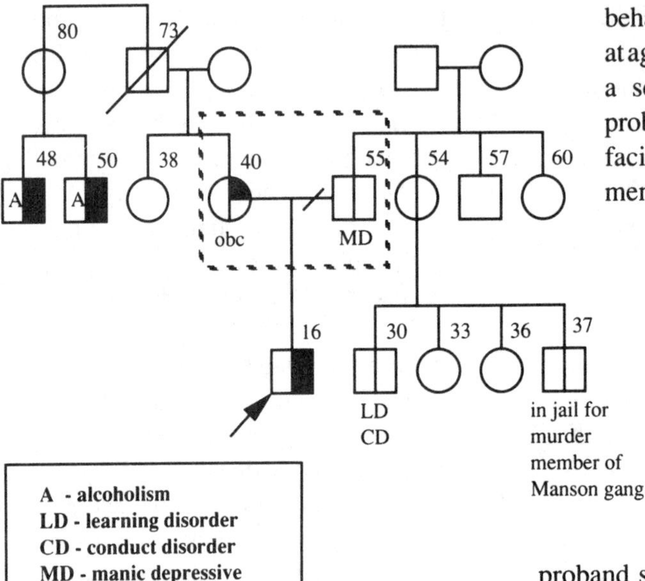

Chronic motor tics x obsessive-compulsive

At the age of four, the proband began to ask questions over and over and the answers had to be given a certain way, very specific and detailed. At age eight he developed mouth opening tics, and at nine began to mimic his sister, and then his mother[p116]. By age 15 he became obsessed with weight lifting and looking at himself in the mirror in a repetitious, ritualistic manner that was repeated over and over sometimes for hours at time. Severe obsessive-compulsive behaviors have continued ever since, at times requiring hospitalization. Treatment with MAO inhibitors, pimozide, clomipramine and other medications have only partially controlled the symptoms. His father had some moderate compulsive behaviors such as keeping files on all sorts of things. His father's brother had more severe obsessive-compulsive behaviors and a "nervous breakdown" at age 18. He was diagnosed as having a schizophrenia-like disorder. The proband's mother had eyeblinking and facial grimacing and many other members of her family had TS or motor or vocal tics.

Since neither the father nor his brother had motor or vocal tics it is not clear whether their obsessive-compulsive behaviors were due to a *Gts* gene or a "pure" obsessive-compulsive behavior gene. The extreme severity of the obsessive compulsive behaviors and tics in the proband strongly suggests he inherited a *Gts* gene from both parents, or a *Gts* gene from his mother and a related *Obc* gene from his father.

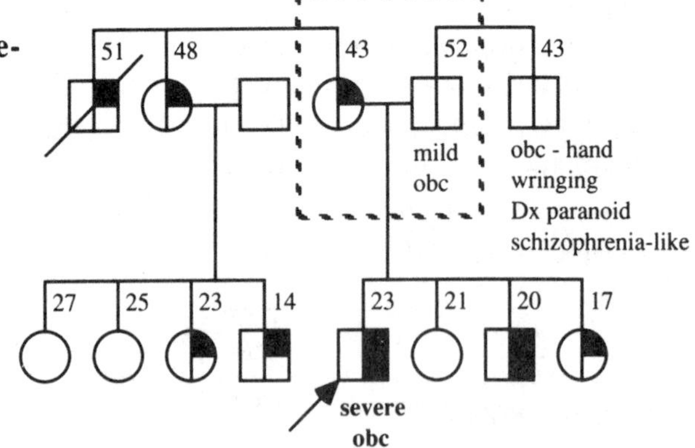

Obsessive compulsive x obsessive compulsive

In the above pedigree it was not clear whether the father's obsessive-compulsive gene was a *Gts* gene or an obsessive-compulsive gene. The following pedigree is an example where both parents of a TS child had obsessive compulsive behaviors that were more clearly due to a *Gts* gene since both had brothers with TS.

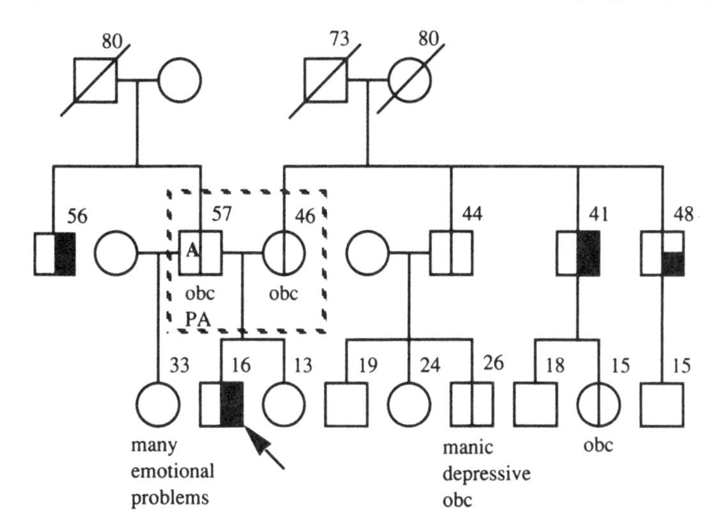

The father of this 16 year old boy with TS had obsessive-compulsive behaviors, panic attacks and problems with alcohol abuse. His mother had obsessive-compulsive behaviors. Although neither had motor or vocal tics, they both had brothers with TS, strongly suggesting that their behaviors were due to a *Gts* gene. The 26 year old nephew of the mother had been hospitalized for attempted suicide, had obsessive-compulsive behaviors and was diagnosed a manic-depressive. He is a presumptive *Gts* gene carrier.

Why Are There So Many Cases with Both Parents Affected?

This is just a small sample of the total number of pedigrees where both parents appear to be carrying a *Gts* gene. Why is this occurring? There are three possible explanations:

1. Assortative mating Our initial assumption was that is was due to **assortative mating** or a tendency for like to marry like. For example, tall men tend to marry tall women, Catholics marry Catholics, Methodists marry Methodists, and so on through a wide variety of physical, cultural, religious, and other characteristics. Assortative mating has often been noted for several personality traits or behavioral disorders. Some examples are that individuals with schizophrenia, alcoholism, antisocial personality or other mental disorders tend to mary each other[625,870,1051, 839,1051,1595,1846].

2. Selective bias A second possible explanation is **selection or ascertainment bias**. For example, 25 percent of TS patients seen in the clinic have severe grade 3 TS[p24] while a much smaller percent of their relatives are grade 3 cases. Because of ascertainment bias, TS patients seen in the clinic are more severe than those in the community as a whole. Since severe cases are more likely to be homozygotes there would also be a selection for cases where both parents carried a *Gts* gene.

3. All TS cases are homozygotes. A third and particularly intriguing explanation is that all, or almost all, TS patients carry a double dose of the *Gts* gene. This would presume that *both parents always carry a Gts* gene but since the penetrance of the gene is low it would only be apparent in some pedigrees.

[As this book was being written evidence accumulated that this was the correct explanation. For further details see Chapters 69, 89 and 90.]

The Spider People

In the process of obtaining pedigrees on TS families I not infrequently find individuals with a history of many alliances, marriages and divorces.The appearance of these pedigrees with mating lines going off in all directions makes me think of them as spider people because of the resemblance of the pedigree to a spider web. The following are two examples.

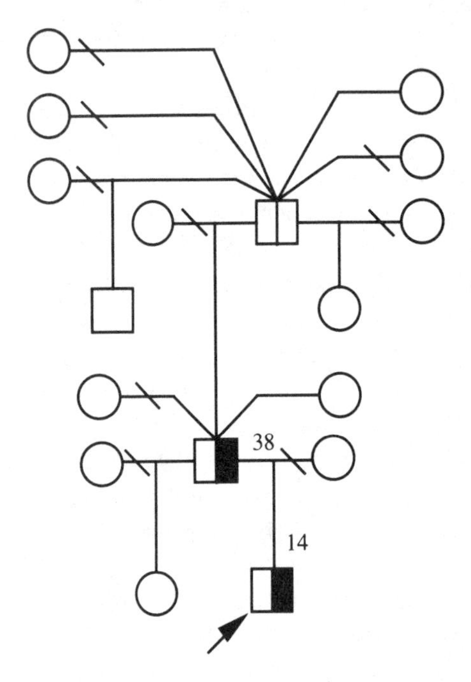

A 14 year old boy with TS was accompanied by his mother. His parents had been divorced for five years. His father had married for the fourth time and his paternal grandfather had a history of seven marriages, six divorces and a unknown number of other brief relationships.

In the above family there was no direct history available to determine if the grandfather had motor and vocal tics and the assumption was made that he was the *Gts* gene carrier. In the family **on the right**, direct evidence was available.

Here the maternal grandmother had six children by six different men and was not married to any of them. It was clear she carried the *Gts* gene since her mother and grandson both had TS.

A Miniature Adoption Study of Hypersexuality

At seven years of age Jerry stood on a picnic table at school and began rubbing himself all over and moaning in a sexually explicit manner. At eight his schoolwork began to deteriorate and he constantly needed help to keep up. While his mother was visiting friends he began to molest their two year old son. Despite reprimanding this occurred again the next day and the mother was asked to never bring him back. Within a month he began to molest his own two year old brother, pulling his pants down and fondling him. At nine years of age his mother was afraid to let him out of the house. However, at the encouragement of a therapist, who felt it was unfair to keep

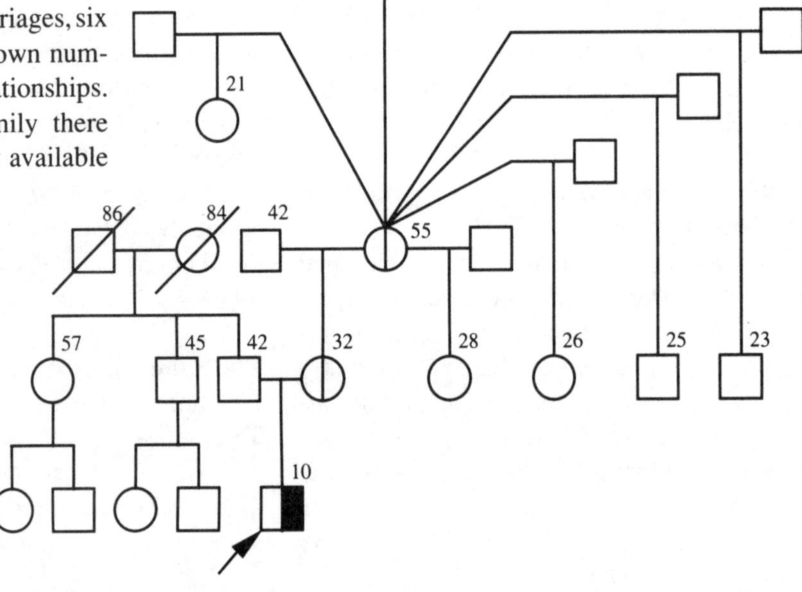

him inside, his mother let him out one Saturday. Twenty minutes later he returned with 15 other children chasing after him. They reported that he had grabbed a six year old girl, pushed her up against the wall and put his hand into her pants, in front of all the others.

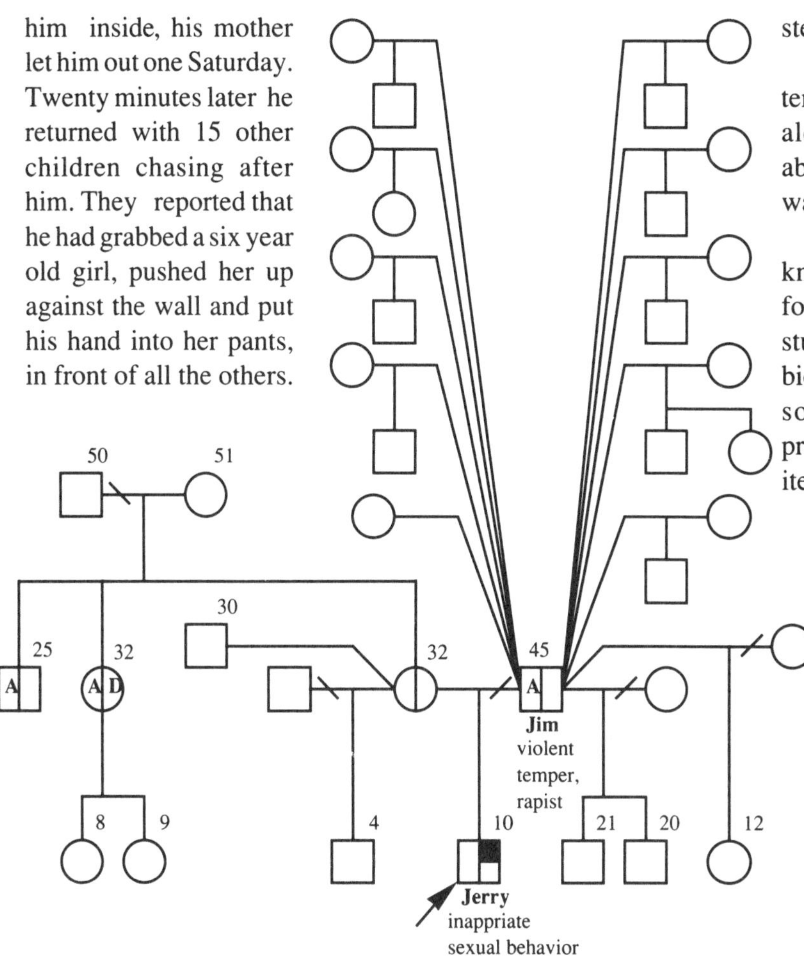

In addition to eye tics he was compulsively standing on his head up to 50 times a day, lying, and stealing constantly.

The **pedigree is shown above**. Because he often beat her, Jerry's mother left his father, Jim, before he was born. Jim had a violent temper, was into constant trouble by the time he was 13 and has been in jail numerous times. Problems with alcohol abuse started when he was a teenager. He was married four times and divorced three times. In addition, he had ten children by a number of different women he never married. He is now separated from his present wife because he assaulted and raped his stepdaughter.

The mother's sister had problems with alcoholism and drug abuse and her brother was an alcoholic.

Since Jerry never knew his father, this forms a minor adoption study in which both the biological father and his son had significant problems with disinhibited sexual behavior.

These pedigrees are a graphic illustration of the chaotic lives of serial broken relationships that can occur in some individuals carrying genes for disinhibition disorders. The webs spinning off in all directions each indicate a union formed among great expectations of love and passion that soon dissolved in a disillusion of short temper, sometimes physical abuse, drug or alcohol abuse, and narcissistic self-involvement.

Does the *Gts* Gene Produce a Selective Reproductive Advantage?

When Darwin wrote his classic book on the evolution of species by natural selection, he proposed that the most fit were more likely to survive and reproduce. The emphasis was on the selection of the whole organism. After Mendel's laws of genetic inheritance were rediscovered at the turn of the 20th century, it became apparent that single mutant genes, if they produced a reproductive advantage in the organism, would be preferentially selected. As a result, such genes would gradually increase in frequency in the population. The key was reproductive fitness. A gene which was responsible for allowing the organism to produce more progeny would increase in frequency.

These observations of multiple children from multiple matings in some *Gts* gene carriers led me to suspect that one of the reasons the *Gts* gene is so common is that disinhibited behaviors, especially sexual ones, could result in a selective reproductive advantage. The **figure on the right ->** shows how potent even a moderate degree of reproductive advantage can be in increasing the frequency of a gene. For example, if 1 in 83 people carry a *Gts* gene and they have 5 percent more children than those without the *Gts* gene (i.e., 2.1 instead of 2.0 children) over a period of 50 generations the frequency of the *Gts* gene would increase to 1 in 8 persons[571].

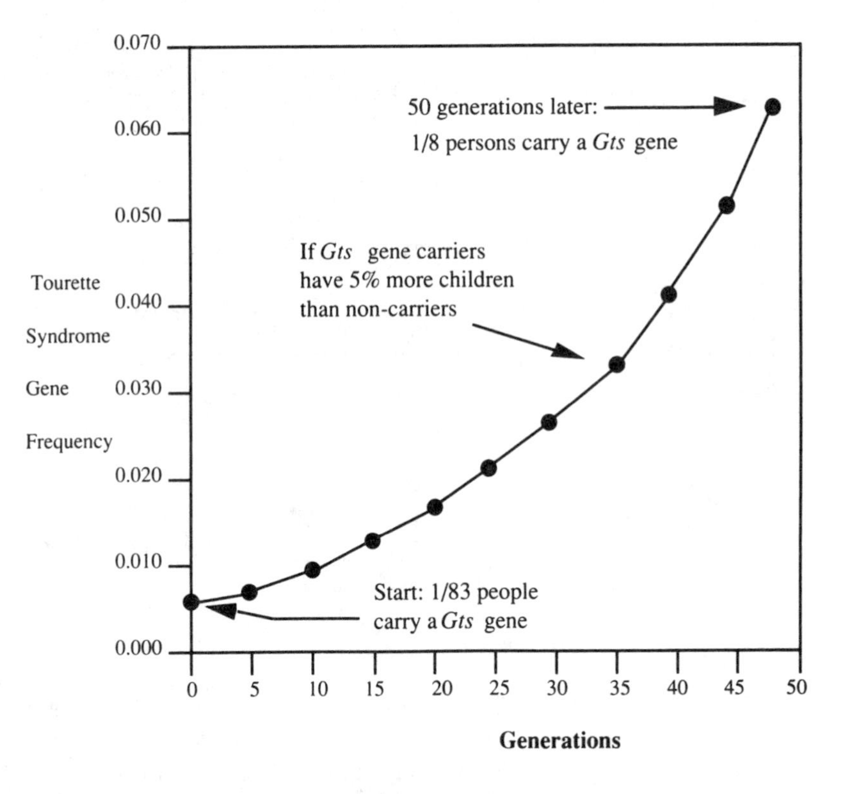

In addition to a possible selective advantage on the basis of *Gts* gene carriers having more children, there is a second more subtle form of selective advantage. In the study of Behar and Stewart, comparing children with conduct disorder to those with other psychiatric disorders, the average age of the mother at the birth of her first child was 18.7 years for those with conduct disorder versus 23.0 years for the controls [p152]. This was a difference of 4.3 years and was very significant ($p = .001$). If this difference continued throughout multiple generations, and both groups of mothers's had only one child, this would represent an 18.7 percent increase in the number of offspring by the conduct disorder group. Even if only part of the conduct group carried a *Gts* gene, this would still result in a rapid increase in the frequency of the Gts gene.

The important point is that the disinhibition disorders are associated with a wide variety of associated behavioral problems such as ADHD, learning disorders, conduct disorder, panic attacks, phobias, depression and mania. The additional feature of disinhibited sexuality in some individuals is the one behavior that may be dragging the others into increased frequency in the general population. This concept is diagramed as follows:

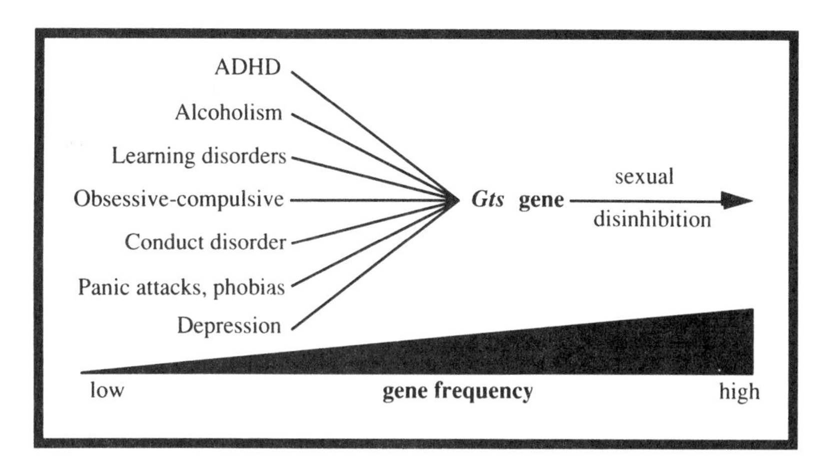

This proposal does not "stigmatize" all TS patients as being sexually disinhibited[1480]. The equation shows that the actions of a small minority of gene carriers over a number of generations can result in a significant increase in the frequency of the gene[367].

A Bit of Everything

The **family on the following page** illustrates a number of the concepts discussed in Part III.

Jim (4), was a 23 year old male. As an infant he was diagnosed as having ADHD. During the grade school years he developed motor and vocal tics. Although he was getting B's and C's in school his behavior "was atrocious." He was fighting and hitting other students, and talking in class. His behavior in the home was good until age 17 when it changed drastically. He would pound brick walls in fits of angry rage and threaten others in the family. At 18 he joined the Army but was dishonorably discharged six months later for hitting his drill sergeant. For the next several years he wandered around the country abusing drugs and alcohol. He recently returned home and was so abusive he was asked to move out of the house. He lost four jobs in a period of several months because of his aggressive behavior and short temper. His sister (5), half-sister (6), mother, and aunt had problems similar to Jim's. They all had motor and vocal tics, short temper, polydipsia, violent outbursts of temper, paranoid ideas, compulsive behaviors including eating, alcohol and drug abuse. All had been diagnosed as having borderline personality disorder or paranoid schizophrenia. In addition to her three living children, Jim's mother had six other pregnancies that she had aborted. She killed one of them after it was born alive and was sent to prison for manslaughter. His maternal grandmother had 27 children by at least four different men, including a remarkable series of nine identical twins. She is reported to have tried to make money by pushing some of her daughters into prostitution. At one point a social services department made unsuccessful attempts to have her sterilized.

This family illustrates the ravages of a disinhibition gene at its worst — sexual disinhibition with multiple children by multiple marriages, chaotic lifestyles, antisocial behavior, borderline personality disorder, short temper, drug and alcohol abuse, and even polydipsia.

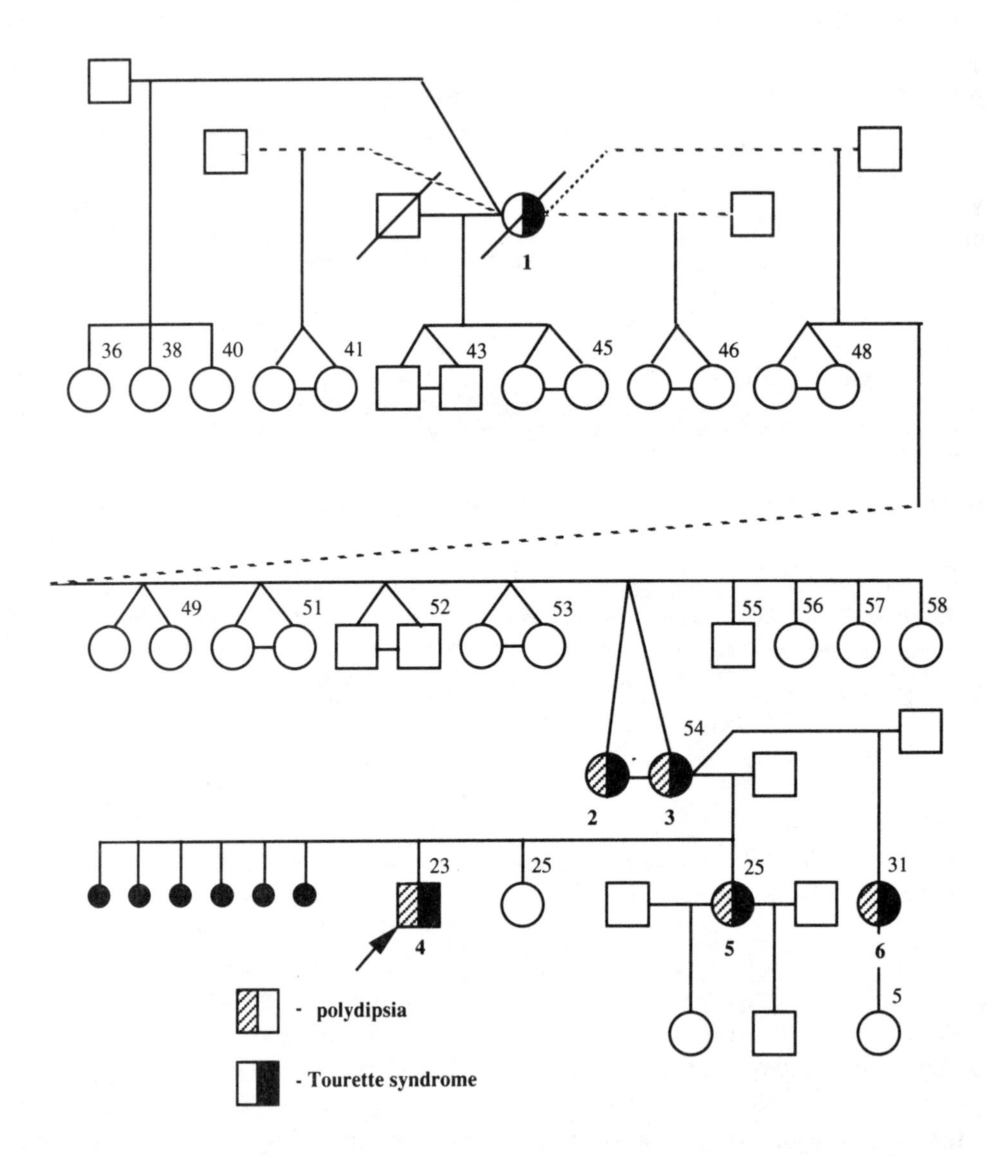

Is Needing to Place a Baby for Adoption the Result of an Impulse Disorder?

An unusual number of TS patients I see are adopted. Of 950 TS patients from completely different families, 41 or 4.3 percent were adopted. The generally quoted figure for the frequency of all adopted children is 1%. There is a significant difference in these two figures. A disproportionate frequency of adopted children among children attending psychiatric facilities has been noted before with frequencies ranging from 2.4 to 13.5 percent. The average was 5.2 percent of a total of 5,192 patients[1592]. Aggressive and antisocial behavior was especially common. One psychological explanation has been that this is the result of "an unconscious and unresolved aversion toward parenthood in one or both of the adoptive parents, particularly the mother"[1957]. This demonstrates the frequent propensity to blame the mother for the misbehavior of the child.

To me, the most likely explanation is that teenagers who themselves have a hereditary impulse disorder, such as TS or ADHD, are more likely than others to get pregnant, or get others pregnant, under circumstances where they cannot keep the baby. Since there is a significant chance that the gene will be passed onto the child, it is not surprising that a disproportionate number of children placed for adoption carry a *Gts* or *ADHD* gene.

The **following pedigree** provides a look at this question from the viewpoint of the family placing the child up for adoption.

In summary: Family studies suggest that in many cases TS patients have received a TS gene from both parents. Selective bias and like personalities marrying each other provide two explanations for this.

[A third explanation, homozygosity of all TS patients and high frequency of the TS gene, is discussed later[p513]*.]*

Tourette syndrome is a disinhibition disorder and the spectrum can include sexual disinhibition. This is likely to result in selection for the Gts gene and may account for the high frequency of disinhibition disorders in the general population.

An unusually high number of TS children are adopted (4 percent). This is best explained by assuming one or both parents carried a Gts gene and the product of an impulsive pregnancy had to be adopted out. That adopted child was then at risk for inheriting the Gts gene.

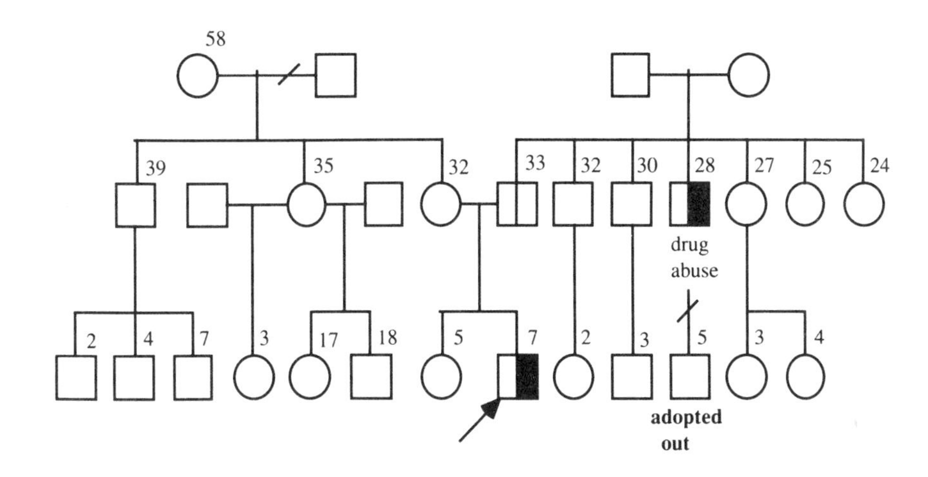

The proband with TS had one sister and 11 nephews or nieces. Of these, the only one that was placed for adoption was the son of the only other member of the family with TS, a 28 year old male who was into drug and alcohol abuse as a teenager and got a girl pregnant. That child was the only one of 12 children in this family placed for adoption.

In a study relevant to this, psychological testing was done on unwed mothers and compared to married pregnant women and 18 year old nonpregnant women. The unwed mothers had significantly higher scores on five of nine scales. The scales testing psychopathic behavior and schizophrenia were especially high[926].

Chapter 45
Miscellaneous Features

This chapter covers a potpourri of miscellaneous clinical observations about patients with Tourette syndrome, listed in alphabetical order. These features are not present in all TS patients, but occur often enough to call special attention to them. Some have been mentioned before but are included here for re-emphasis. Sometimes it is the little observations that hold the key to understanding the fundamental cause of a disorder. Miscellaneous doesn't mean unimportant.

They are:

Allergies and handedness
Body odor
Can't take no for an answer
Colic
Coprolexia
EEGs — brain wave tests
Explosive personality
False paralysis
Fear of intimacy
Heat intolerance
Intolerance to stress
Jealousy
Jekyll and Hyde personality
Lying and confabulation
Migraine headaches
Myoclonic jerks and narcolepsy
Picking on mother
Picking on siblings
Plucking eyebrows
Poor handwriting
Poor memory
Self-induced lesions
Sequencing problems
Slow-down spells
Somatization
Speech problems
Sweating

Allergies and Handedness

A common theory states that hyperactivity is the result of allergies to food or other agents[521,601,1582,1782,2005], and left-handedness has often been reported to be more common in children with learning and reading disabilities[46,1554]. In 1982, a Boston neurologist, Norman Geschwind, suggested that there might be a link between left-handedness, allergies (auto-immunity) migraine headaches, stuttering, learning and reading disabilities[695]. The left half of the brain normally controls speech while the right half controls spatial abilities. Geschwind suggested that if, during development, the right brain gains ascendency over language, this anomalous control carried an increased risk of problems with reading and speaking. Comparing left-handed to right-handed people, learning disabilities were 12 times more common in left-handers. Autoimmune disorders such as thyroiditis and ulcerative colitis were also more common in left-handers. The ascendancy of the right hemisphere was felt to be due to testosterone slowing the growth of the left hemisphere. This was proposed to also explain the greater frequency of learning disorders in males (but see[p51]) and the effect of testosterone on immune lymphocytes caused the greater frequency of autoimmune disorders.

Since learning disorders are such an integral part of TS, we were curious to know if there was a higher frequency of left-handedness and allergies in TS patients. The controlled studies indicated that neither was true — left-handedness and allergies were not more common in TS patients[364].

Body Odor

Parents occasionally complain about a peculiar odor to their TS children. This is usually expressed in terms of a locker room, sweaty smell. One mother reported that "When he sweat his sweater smelled so bad that his teacher wrapped it up in a plastic bag and sent it home to me. I tried to wash it out but gave up and threw it out. He has to shower every day to control the body odor."

Can't Take No for an Answer

While this has been emphasized many times, I hear it so often it deserves repeating. The TS and ADHD child hates the word "no" and it is the most frequently cited precipitating event that sends them into a temper tantrum. One mother had a sign on her refrigerator that read, "What part of the word **NO** don't you understand?"

Colic

Colic is a catchall diagnosis that is used to explain why some newborn babies are irritable, cry constantly and sleep poorly. The term suggests some sort of intestinal spasms that cause the infant pain. The combination of sleeping poorly and being labeled a "colicky" baby are the most common early symptoms we hear describing TS children with early onset ADHD. Although our viewpoint may be biased from looking backward from a child diagnosed as ADHD or TS, I think it is reasonable to suggest that colicky babies are probably at a high risk of developing ADHD, and that some colic is the result of irritability due to a brain dysfunction, rather than a dysfunction of the intestines.

Coprographia

Some TS children express their coprolexia in the form of writing dirty words — coprographia. The following is an example:

The mother was tutoring her TS son by saying some words and having him practice writing them. She had given him the words "lays, pays, and ways" when the phone range. When she came back he had written in his dyslexic way, "Ryan is a fart shit ass damn. Butt. Dammit. Bitch."

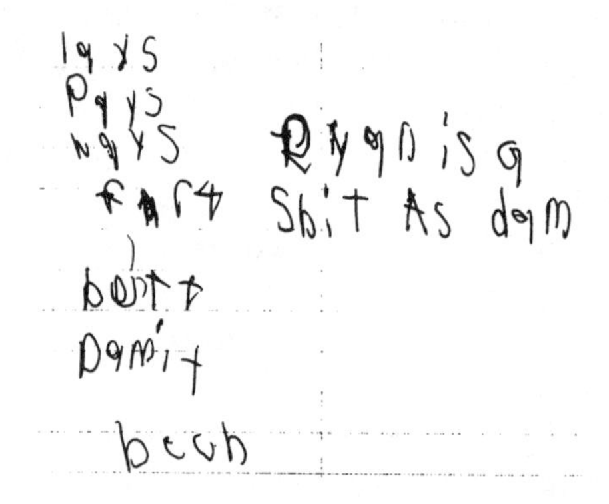

One might call him a *dyslexic coprolexic*.

EEGs — Brain Wave Tests

Brain wave tests or EEGs are predominantly used in the diagnosis of seizure disorders. Since the motor tics in TS are sometimes mistaken as a form of a seizure disorder, it is not uncommon for EEGs be ordered during the evaluation of the child with TS. How useful are they? A summary of many studies indicates that the EEG shows some type of abnormality in 25 percent to 65 percent of TS subjects[1767]. More specifically, about 45 percent of TS children show moderate to marked EEG abnormalities compared to 45 percent of children with emotional disturbances without TS and 15 percent of normal children[1767,1877]. EEG abnormalities are neither diagnostic nor specific for TS. In our experience, the primary indication for an EEG in a child with TS is the presence of explosive temper tantrums that have not responded to standard medications for TS and ADHD or of seizure-like episodes.

Explosive Personality

A number of names have been used to describe a group of individuals who show occasional explosive outbursts of rage or aggressive behavior. In the DSM-II, the term explosive personality was used to describe a behavior pattern characterized by "gross outbursts of rage or of verbal or physical aggressiveness. These outbursts are strikingly different from the patient's usual behavior, and he may be regretful and repentant for them. These patients are generally considered excitable, aggressive and overresponsive to environmental pressures. It is the intensity of the outbursts and the individual's inability to control them which distinguishes this group. A related term is *episodic dyscontrol syndrome*. In the DSM-III[490] the term *intermittent explosive disorder* has been used with the condition that impulsivity or aggressiveness is absent between episodes.

There can be many causes of an explosive personality, including various forms of brain damage and/or epilepsy. Our studies indicate that some first-degree relatives of individuals with ADHD or TS have episodic dyscontrol syndrome. They are under control most of the time but occasionally stresses cause them to blow.

False Paralysis

Children with TS will sometimes drag a foot or droop a shoulder to the extent that they may mistakenly be thought to have some type of paralysis. The presence of typical TS symptoms should alert the physician to the correct diagnosis[255].

Fear of Intimacy

Some Tourette syndrome patients seem to have a problem with intimate relationships. The combination of narcissistic self-involvement, short temper, needing to control others, and to have things their own way — frequently interferes with another person's desire for a warm, caring, and considerate partner. Part of this is fear of intimacy or a fear of committment. An example was a very good looking, vivacious, blonde 16 year old girl with Tourette syndrome who related the following:

> "I start out thinking a guy is really gorgeous, a real hunk, but after a few weeks I become bored with him. I don't even want to hold hands with them and they keep wanting to kiss me. I start out wanting them to treat me nice but when they do I think they are just a wimp and I don't like them. This keeps happening with all the boys except Jack."
>
> "What is different about Jack?"
>
> "Jack is a real challange. He says he will pick me up at eight and then never shows. He calls me up and yells at me telling me I'm a horrible person. He is not a wimp."

She could only feel good about someone she knew would never place demands on her. Although part of her fear of getting close was related to some obsessive-compulsive behaviors such as washing her hands constantly and not wanting anyone to touch her for fear of "getting her dirty," similar problems occur in some TS patients without the obsessive-compulsive element. An opposite fear in other TS individuals is fear of abandonment, once a tie is made.

Heat Intolerance

About 20% of our TS patients, both children and adults, report feeling warm or hot when others in the house are not.

> "We are constantly battling over the air conditioner. She [the Tourette patient] always wants it on while I'm freezing to death."
>
> "He doesn't wear a jacket when others are cold."
>
> "He always feels hot."

This intolerance to heat may be accompanied by excessive sweating.

Intolerance to Stress

Poor tolerance of stress is a common problem for TS patients. This, in combination with adult ADHD, has often led adults with TS to pass up promotions that would have placed them in more stressful positions. One patient described it as follows:

> "I have tense feelings caused by events, people, and pressures of daily life. These make me feel like I'm being funneled thru the eye of a needle into a black hole. It causes confusion, lack of control. I try to talk myself out of the problem but can't."

Jealousy

The subject of intense jealousy has come up enough to deserve mention. The right combination of short temper, fear of abandonment, a need to control and obsessive thoughts can produce intense feelings of jealousy in some TS patients. The following is one of many similar comments:

> "I was insanely jealous of my first wife. I was so hot headed about it I came within inches of killing two people I thought were interested in her."

Other examples are given elsewhere [p148,266]. This jealousy may play a role in triggering wife abuse.

Jekyll and Hyde Personality

The emotional instability that may occur in TS patients has often been referred to by parents as a Jekyll and Hyde personality[358]. They can be wonderful and charming one minute and explode in anger the next. ADHD children have also been characterized in this fashion[1884].

Lying and Confabulation

This has been mentioned before but deserves emphasis. The lying is often about things that are relatively unimportant. In addition to vigorously denying doing something that they obviously did, they may also make up stories that seem designed to make them larger or more impressive than life. While young children may normally do this, what distinguishes the TS child is doing it at an age when they should know better, doing it repeatedly and compulsively, and the vociferous denials when caught. Parents are often at a total loss to understand this aspect of their child's behavior. The following are some reports of parents:

> "I found three empty beer cans in my 17 year old son's room. When I asked him about them he vigorously denied that they were his. 'They are not mine! They have been under my bed for years.' I threw them away and two days later found two fresh cans in his room with beer still in them. I got the same crazy story and he became so enraged he grabbed my hair in his fists and began banging my head against the wall. It reminded me of movies of angry lions grabbing their victims in their jaws and shaking them to death. If my husband hadn't been there I don't know what would have happened."
>
> "I heard him talking to a neighbor relating how we used to live in the snow and how much he enjoyed making ice houses just like the Eskimos. Since we

have never lived outside of Southern California I asked why he said that. This only brought on vigorous denials that he had said this and even denied talking with the neighbor, even though he had just walked in from doing so. 'I didn't say that. I didn't. I didn't.'"

The following is from the report of a school psychologist:

"By far the most notable aspect of his behavior was his almost ceaseless fabrication. Some of these tales were rooted in reality making it hard to tell if they were true or not when heard: his mother was a nurse; he had bought a stop watch like the examiner's; his father's knee was shot off in the Navy, but there was no scar. Others were outlandish: he was a black belt in karate; he leaped 15 feet into a tree, dropped a rope ladder to some friends and they built a tree house; he was president of the student body in the second grade; he broke the leg of a second grader he didn't like; he was in a near miss crash at an air show; he shot a grizzly bear and moose within fifteen minutes, pumping 17 shots into each with a 22 rifle; he caught a fifty pound bass and a five-foot sand-tiger shark. When confronted with the truth about one of his stories and asked why he made it up, he finally admitted its falsehood, but then he said, 'But my dad did catch a 500 pound shark with a fishing pole.'"

Migraine Headaches

A study by Barabas and colleagues[109,110] showed that 27 percent of 60 children with TS had migraine headaches compared to 11 percent of children with epilepsy and 6 percent of children with learning disabilities. This compares to estimates of about 4 percent of grade school children[167,1978] and 7 percent of adolescents[1978] with migraine. When the TS patients were divided into those with migraine or a family history of migraine, and those without migraine, 77 percent of those with migraine compared to only 22 percent of the controls had sleep disorders of arousal (sleepwalking and night terrors). Since sleep is regulated by serotonin nerves cells[p448] and migraine headaches have been proposed to be due to changes in serotonin chemistry the authors felt this supported the proposal of an abnormality of serotonin in TS[p457]. We have also observed a high frequency of migraine headache in TS and related disorders. The following pedigree is an example:

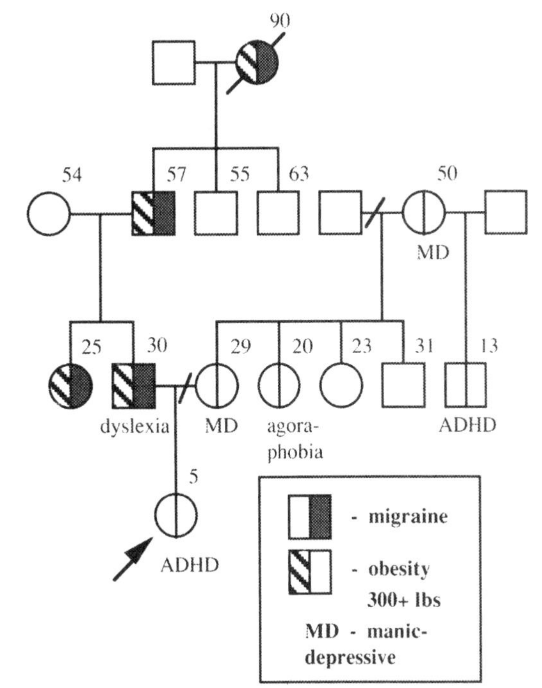

The proband had severe ADHD. Non-stop speech started as soon as he began to talk at age two. By four years of age he was extremely hyperactive, always interrupting, constantly needed his parents or teachers attention. There was striking improvement after he was treated with Cylert. His mother and mother's mother were both manic depressive. His father had dyslexia, at times had weighed over 300 pounds, and had episodes of migraine associated with nausea, vomiting, blurred vision and numbness on one side of the body. These headaches started at age 13. His sister had migraine headaches starting at age 18 and weighed 330 pounds. His father and his father's mother had similar headaches and weight problems.

In this family, the gene on the father's side, probably causing low brain serotonin, was associated with compulsive eating, severe migraine, and dyslexia. Since there was evidence for a *Gts* gene in the families of both parents, the proband probably had a double dose of the gene

Myoclonic Jerking and Narcolepsy

Narcolepsy is a condition where the subject suddenly falls asleep in inappropriate places. If standing they will lose muscle tone and slump to the ground. The question of the possible relationship between TS and narcolepsy came from a case of a young girl who had three conditions. Typical TS symptoms had started when she was eight years old. Within a year she also had myoclonic jerking and a year later this was complicated by narcolepsy. She had a remarkable cycle of being totally asymptomatic in the summertime, needing no medication. By November of each year she began to develop tics, then myoclonic jerking and then narcolepsy, each starting about a week apart and remaining intermixed such that all three symptoms could occur almost simultaneously. One day when I saw her she showed ticing interspersed with vigorous myoclonic jerking. As she sat in her mother's arms she suddenly fell asleep and soon her hands began to clench. When aroused by pinching hard on her shoulder, she woke up and immediately developed facial grimacing tics.

I have never seen this triple combination in anyone else but I have one other TS patient with the combination of severe myoclonic jerking and TS. In both patients the myoclonic jerks were controlled with Tegretal[p585]. In addition, the combination of narcolepsy and TS has occurred in eight other TS patients or family members. This has varied from one person who could fall asleep anywhere and anytime she wished, one who frequently fell asleep at the wheel with the result that over the years she totally destroyed six different cars, to several patients with more typical narcolepsy including falling down.

The combination of TS and myoclonic jerking was described once before[583]. I am unaware of any previous descriptions of narcolepsy in TS. As discussed later[p429] all three disorders are consistent with a defect in serotonin.

Picking on Mother — Mother Abuse

This urge to bug someone is oftentimes taken out on mother. This can sometimes escalate to the point of physically hitting and abusing her and constitutes the reciprocal of child abuse, i.e., parent abuse[p268]. In single parent homes this has occasionally resulted in needing to place the child in a foster home for the mother's safety. For awhile we attempted some deep psychoanalytic explanations of this until we had a family where the father was in a wheel chair. In this case it was the father who was abused. This led us to the conclusion that it is simply the parent who is perceived as the weakest who receives most of the abuse.

Picking on Siblings

Anyone who has had a brother or sister knows that one of the great joys in life can be to bug them in a multitude of little ways. This is a given and is virtually written into your contract at birth about how you are supposed to treat siblings. However, the way some TS children compulsively pick on their siblings is far above this background of normal behavior. Parents often report that the TS child seems incapable of being in the same room with their siblings without hitting them or trying to irritate them in some manner. Some typical comments are:

> "Billy can never walk by his brother without hitting him."
>
> "I have this need to hit my sister every time I see her. It's almost a compulsion."
>
> "When he gets in fights with his younger brother he sometimes goes into a rage. In the past he has tried to drown or kill his brother and he really means it. He is out to kill when he fights."
>
> "When playing with children he gets angry too fast and finds himself wanting to physically hurt others in a blind rage and with no remorse. On one occasion he choked a friend until he turned purple. After a good friend taunted him about his tics he wanted to kill him, not just figuratively speaking. His father had a similar personality and is presently in prison for violent behavior."

One 13 year old TS boy says to his mother:

> "I'll agitate you until you commit suicide," and, "I can get you to spaz out."

The following psychiatrist's report accompanied a patient with multiple motor and vocal tics:

> "The patient had a several month history of throwing temper tantrums at home of a fairly serious nature with apparently minimal provocation. He also had a tendency to tease his siblings to the point where he was being extremely destructive to the entire family. The parents found it very difficult to control the patient's behavior and because of this hospitalization is recommended. Diagnosis — Conduct disorder, socialized, aggressive."

One 23 year old young man had no siblings and took his need to bug someone out on his dog.

> "I love my dog but sometimes I come in the room and hit or squeeze him until he squeals. I feel terrible but I just can't help wanting to hurt him."

This problem is not unique to children and siblings. Adults with TS have told of similar urges to hurt others. For example, a 25 year old male with TS whom we have followed for years moved to another city, but would call frequently for support and general ventilation. He began seeing a new girl friend with several young children. He called somewhat desperately one day, and in his typically colorful language, left the following message on the answerphone:

> "I have this problem and I want to know if I am in regression or what. This gal I'm seeing has got three young kids right, and they are all crying every time I see them. The 1-1/2 year old, I just want to fucking grab him by the neck and punch him and shit you know. And its getting worse and worse. It's getting to the point where I am afraid to be alone in the room with him. And the three year old, every time I see him I just want to walk up and smash him upside in the back of the head you know. I used to feel this way when I was about seven years old walking home from school. I used to hit two and three year olds just to watch them cry. And this is starting to come back to me and this is fucking bugging me. I haven't done it, but I want to do it. I don't know if it is his looks or me or what, just every time I see him I feel like I could walk up and kick him in the stomach or something and just watch him cry you know. I don't know if I want to give him pain or if I want to see fear in their faces or what you know, but the feelings are getting stronger. I even stayed away for three to four days and she's wondering what the fuck is my

> problem. Am I going back into regression or maybe I have some stress or something and I am trying to release the stress by putting it on these kids or what?"

This was a transient problem and the next time we talked to him it was over. These feelings are obviously relevant to the discussion in the chapter on physical and sexual abuse[p263].

Plucking Eyebrows

Plucking out eyebrows is a common practice in women as part of their beauty care. However, when a child or a male adolescent plucks out all their eyebrows, it is an abnormal behavior that should raise suspicions about TS. When confronted there is often vigorous denial about doing it. One mother stated she discovering one morning that her nine year old TS son had symmetrically plucked out the lateral one-third of the hairs of both eye brows. When confronted he denied it vigorously, saying:

> "I was standing at the sink and saw them just floating down the drain."

This compulsive hair plucking has been given the mouthful name of *trichotillomania*. It may go away after treatment with Prozac[p584] or Anafronail[p583] or Haldol[p547].

Poor Handwriting and Poor Fine Motor Coordination

Children with ADHD or TS often have problems with fine motor coordination. They often seem to trip over their own feet, constantly break things, "can't be trusted with good toys," have trouble learning to ride a bicycle, and are often the last to be chosen for teams where they may become the bartered booby prize — "If I take Sam you *have* to take Johnny." In school, this incoordination often shows up as poor handwriting and inability to copy letters or draw figures. Since ADHD usually starts in early childhood, this has often been assumed to be some type of delayed maturation of the nervous system. Our experience with TS suggests it is more likely to be due to the same imbalance in brain chemistry that causes the rest of the symptoms. The best way to illustrate this is by the following figure. Adam had no symptoms of either ADHD or TS for the first 9 years of his life. A sample of his handwriting in January 1986, is shown on the left below and is better than mine. His TS symptoms started abruptly in September of 1986 with motor and vocal tics. A few months after the onset of TS his handwriting had severely deteriorated (right side of figure).

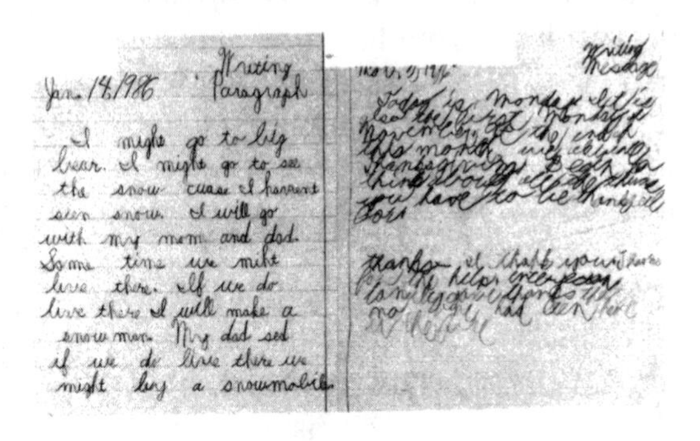

Handwriting before and after the onset of Tourette syndrome. Left — handwriting on Jan. 14, 1986 before TS. Right—handwriting on Nov. 7, 1986, two months after the onset of TS.

The tics themselves were not causing the poor handwriting since they could be suppressed for significant periods of time. This clearly shows that his nervous system had matured at the same rate as his peers and that his handwriting was excellent. It deteriorated only after the onset of TS symptoms. This is consistent with the important role of dopamine in the control of motor movements (see Part V).

Poor Memory

A feature related to the attention deficit disorder is poor memory. Some TS patients complain vigorously about this while others have no problem. The following are some characteristic statements:

"Is poor memory common in TS patients? Mine is terrible."

"I got lost twice just driving home yesterday on roads I know very well."

"I often can't remember whether I took my pills."

"You would think that because I repair Fax machines, my memory would have to be excellent. But unlike my buddies who keep it all in their heads, I have to carry the repair manuals everywhere I go and constantly refer to them."

"When I tell people how bad my memory is they wonder how I got through college. I did it by studying hard the night before and taking the test while it was still fresh."

"I teach Bible class each Sunday and I can't remember what I did last week."

"I look at the clock and after I look away I can't remember the time."

While the tests for immediate memory are usually normal in TS patients, our experience suggests some have significant problems with storing recent or short-term memory[p333].

Self-induced Lesions

Various type of self-abusive behaviors are not uncommon in TS. These include head banging especially as young children, hitting one's self, licking lips sometimes until they are bleeding and infected, washing hands until they are raw, constant picking at sores, grinding and pulling teeth, and biting their hands, lips, cheeks or tongue. These occur in 12 to 50% of patients[530,972,1347,1859,2001]. I have a rule that *any skin lesion in a TS patient is self-induced until proven otherwise*. Other than the tics and vocal noises, these often produce the only physical signs of the disorder.

The most common self-induced facial lesions are due to compulsive licking of the lips. This can produce problems varying from discoloration around the lips, which sometimes results in nicknames like "dirt lips," to open infected sores.

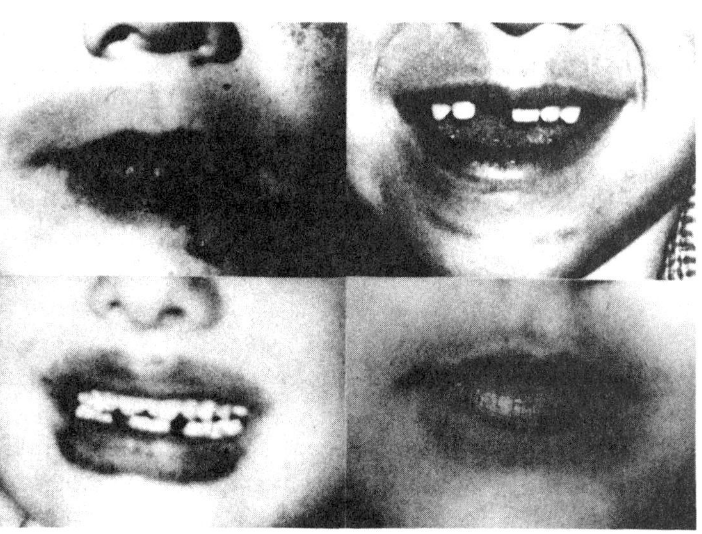

Four children with Tourette syndrome and lesions around the mouth due to constant, stereotyped, licking of the lips.

One of our TS patients with chronic excoriation around his mouth for many years had been to dozens of dermatologists for a multitude of treatments — none of which were effective. The lesions disappeared two weeks after treatment with 0.5 mg of Haldol.

TS children often bite their nails and other parts of their hands. The following figure shows the results of nail biting and biting the second knuckle of both hands.

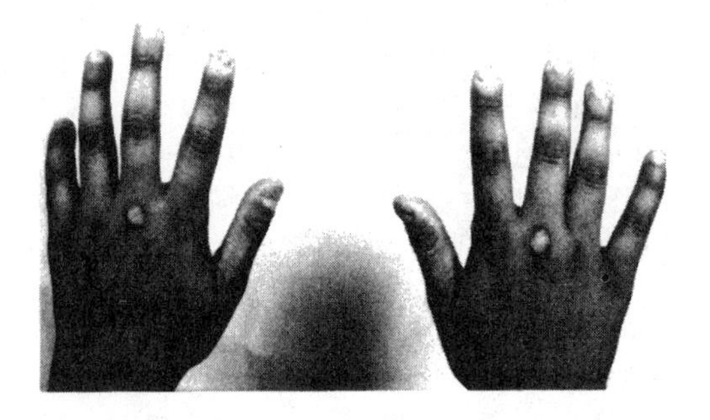

Sometimes the the self-abuse is taken out on the skin. The following figures show how a girl with TS carved initials into her skin resulting in permanent scars.

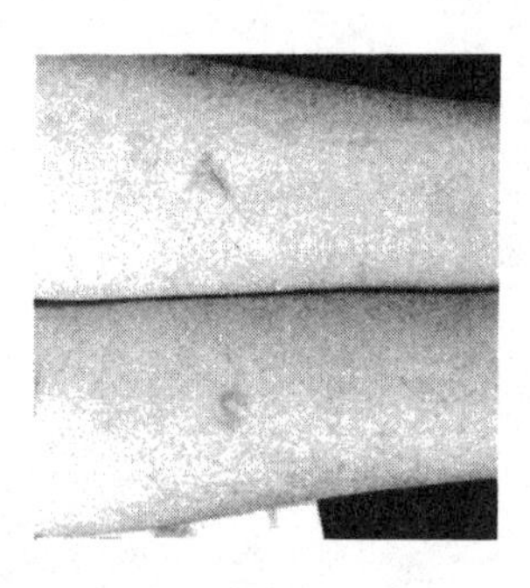

Another girl repeatedly scratched herself.

The "lesions" can also be on the clothes. The following shoes are only two weeks old and belong to a young TS boy whose tics including scraping the tip of his right shoe along the sidewalk.

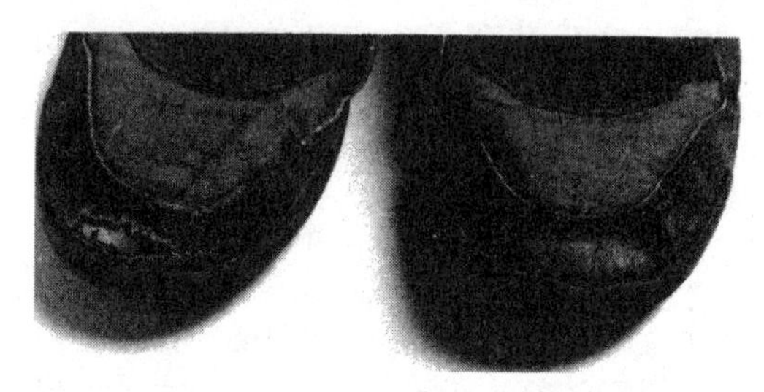

Sequencing Problems

Sequencing refers to the ability to carry out a series of more than one instruction. Children with TS and ADHD often have difficulty carrying out sequential instructions and sometimes even have difficulty with even a single request. This difficulty is illustrated in **the cartoon on the following page**.

Slow-down Spells

At a group support meeting a TS patient stated he had occasional episodes lasting only a few minutes when "everything seemed to be happening in slow motion." His mother asked if this was part of TS. Since I had never heard of it before I asked the group of some 70 individuals if anyone else had ever experienced this and surprisingly five people put up their hands. They all described the same experience of occasional spells, lasting only a few minutes, of everything seeming to be slowed down. One noticed it while reading. There was no loss of consciousness. Others then began to describe other spells — some felt that everything was suddenly smaller, or suddenly larger. One described hers as "Alice in Wonderland" spells when she felt she was looking at the world through a looking glass.

Somatization

This fancy word refers to people who seem to take their anger, frustration and anxiety out on their own bodies (soma), resulting in peptic and duodenal ulcers, spastic colon, ulcerative colitis, asthma and other problems with body organs. While such symptoms seem to occur in our TS patients more often than in the general population, studies specifically devoted to this question are needed.

Speech Problems

TS children have many speech problems[364,1033e,1223a]. These range from delayed onset of speech, active choice not to speak (mutism), stuttering, stammering, talking too fast or too loud, poor modulation of the voice such as starting to speak too loudly and ending words too loudly. Stuttering is the most common and occurs sometime during life in a third of TS patients[364].

Sweating

Some TS patients or their parents report problems with excessive sweating. This may contribute in part to the occasional complaints of an unusual body odor[p284]. I have found only a single prior reference to this problem[627a].

Summary: Additional aspects of the behaviors present in some TS patients are listed in this chapter. Many of these, such as can't take no for an answer, EEG abnormalities, explosive personality, heat intolerance, Jekyll and Hyde personality, migraine headaches, myoclonic jerking, narcolepsy, picking on siblings, plucking eyebrows, poor memory, and certain compulsive behaviors, are consistent with the serotonin theory of TS [p457].

Cartoon illustrating problems with sequencing common in TS children. King Features, by Permission.

Chapter 46
The Positive Features of Tourette Syndrome

In the previous chapters I have discussed the negative aspects of having Tourette syndrome or other types of impulse disorder. However, many people who carry a *Gts* gene are in fact well served by it. It raises them, or drives them, above the level of mediocrity. In the previous chapters the right side of the graphs have been emphasized. For example, in the chapter on conduct disorders[p134] the blackened portion illustrated that 35 percent of TS patients seeking medical help had a conduct disorder.

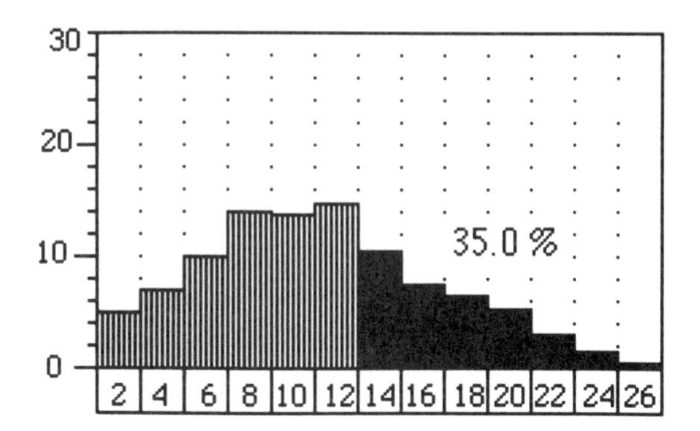

However, it could also have been presented as follows:

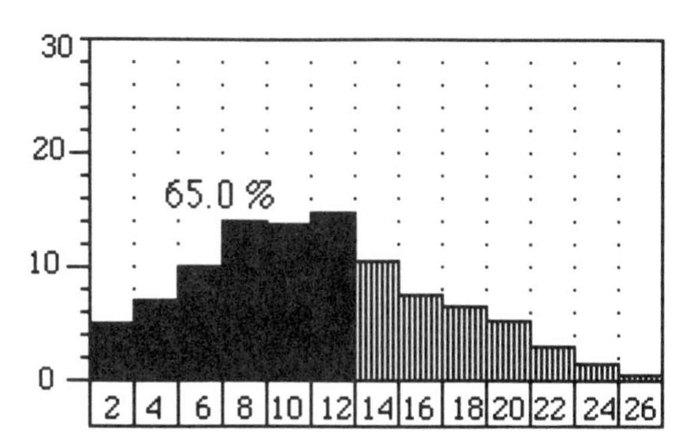

emphasizes that 65 percent do not have such problems. The following case is an example:

> **Jamie** is the most verbal, animated, funny, delightful life-of-the-party type of person one could ever want to know. In her younger years she was an actress and a marvelous mimic. I have heard her hold forth at parties telling joke after joke with the most complex and perfectly rendered series of different ethnic accents imaginable. One day she came to us concerned about her 30 year old son who had never gotten far in life because of a learning disorder and inability to stick with anything for long. She knew of our interest in Tourette syndrome and ADHD, so half in jest we asked, "Jamie have you ever had any motor or vocal tics." It turned out that in her childhood she had had such severe problems with eyeblinking and vocal noises that her mother had dragged her to half a dozen doctors trying to find out what was wrong. These eventually went away except for a some head jerking that she suppresses in company. The primary legacy her Tourette syndrome left was an undying almost hypomanic enthusiasm for life, a verbal delightful "on" personality, and incredible talent as a mimic and actress.

Many of our TS patients have a similar uninhibited, delightful demeanor, a tendency to say what they mean, language more colorful than most of us would dare utter, and often just enough compulsiveness to get things done well. The situation was well described by Oliver Sacks' account of Witty Ticcy Ray[1677].

> **Witty Ticcy Ray** "When I first saw Ray, he was 24 years old, and almost incapacitated by multiple tics of extreme violence coming in volleys every few seconds. He had been subject to these since the age of four and severely stigmatized by the attention they aroused, though his high intelligence, his wit, his strength of character and sense of reality enabled him to pass successfully through school and college, and to be valued and loved by a few friends and his wife. Since leaving college, however, he had been fired from a dozen jobs — always because of tics, never for incompetence — was continually in crises of one sort and another, usually caused by his impatience, his pugnacity, and his coarse, brilliant 'chutzpah,' and had found his marriage threatened by involuntary cries of 'Fuck!' 'Shit,' and so on, which would burst from him at times of sexual excitement. He was (like many Touretters) remarkably musical, and could scarcely have survived — emotionally or economically — had he not been a weekend jazz drummer of real virtuosity, famous for his sudden and wild extemporizations, which would arise from a tic or a compulsive hitting of a drum and would instantly be made the nucleus of a wild and wonderful improvisation, so that the 'sudden intruder' would be turned to brilliant advantage. His Tourette's was also of advantage in various games, especially ping pong, at which he excelled, partly in consequence of his abnormal quickness of reflex and reaction, but especially, again, because of 'improvisations,' 'very sudden, nervous, *frivolous* shots' (in his own words), which were so unexpected and startling as to be virtually unanswerable. The only time he was free from tics was in post-coital quiescence or in sleep; or when he swam or sang or worked, evenly and rhythmically, and found 'a kinetic melody,' a play, which was tension-free, tic-free and free."

When Sacks treated Ray with Haldol, a small dose made him tic free and without side effects, except for the elimination of his witty flamboyance which he sorely missed. He solved that problem in the following manner.

> "During his working hours, and working week, Ray remains 'sober, solid, square' on Haldol — this is how he describes his 'Haldol self.' He is slow and deliberate in his movements and judgments, with none of the impatience, the impetuosity, he showed before Haldol, but equally, none of the wild improvisations and inspirations. Even his dreams are different in quality: 'straight wish-fulfillment' he says, 'with none of the elaborations, the extravaganzas, of Tourette's.' He is less sharp, less quick in repartee, no longer bubbling with witty tics or ticcy wit. He no longer enjoys or excels at ping-pong or other games; he no longer feels 'that urgent killer instinct, the instinct to win, to beat the other man'; he is less competitive, then, and also less playful; and he has lost the impulse, or the knack, of sudden 'frivolous' moves which take everyone by surprise. He has lost his obscenities, his coarse chutzpah, his spunk. He has come to feel increasingly,

that something is missing.

"Most importantly, and disabling, because this was vital for him — as a means both of support and self-expression — he found that on Haldol he was musically 'dull,' average, competent, but lacking energy, enthusiasm, extravagance and joy. He no longer had tics or compulsive hitting of the drums — but he no longer had wild and creative surges.

"As this pattern became clear to him, and after discussing it with me, Ray made a momentous decision: he would take Haldol 'dutifully' throughout the working week, but would take himself off it, and 'let fly,' at week ends. This he has done for the past three years. So now, there are two Rays — on and off Haldol. There is the sober citizen, the calm deliberate, from Monday to Friday; there is 'witty ticcy Ray,' frivolous, frenetic, inspired, at weekends."

In my experience, such a remarkable disparity between the "on Haldol" and "off Haldol" personality is unusual, but not rare. It seems to require a high native intelligence that is allowed "to soar" by the disinhibitions of the Tourette state. Our Jerry is an example.

Jerry was one of those people you tend to like instantly. As he sat telling his history of involvement with Tourette syndrome he had an infectious dynamic energy that I sensed had been put to good use during his 35 years. At eight years of age he developed rapid eyeblinking and grunting vocal noises. His parents were divorced when he was 10. By age 12, since the symptoms had gotten worse, his mother took him to a psychiatrist who said they were a result of the fact that Jerry hated his mother for taking him away from his father. Jerry related this with twinkle in his eye, indicating that even though, at that age, he didn't know what caused these symptoms, he did know that hating his mother, a famous movie star, was not one of them. It was through his mother that he got into acting, where he found that the symptoms disappeared while he was performing. He could then go into his trailer and let all his tics out. He was able to hide the tics from his girlfriend for hours, but this left him exhausted.

He had a number of compulsive, self-abusive behaviors such as needing to touch hot things until it burned his fingers, pressing the tip of a knife against his skin until he drew blood, and closing his eyes as long as he dared while driving on mountain roads. Other times while driving he had an urge to drive under another car. He related having vivid dreams, often about violent acts such as walking up to a teller and slashing his throat, or about frustrating situations such as having to shoot someone to save his life and having the bullet dribble out of the end of the gun and fall to the ground. He often became obsessed with imaginary but vivid scenes. It came as no surprise to me that in addition to acting he was also a successful screen writer specializing in mystery and horror stories. His ability to call up vivid imagery served him well in his writing. He worked as a private investigator specializing in finding missing persons. He felt that the key to his success in the latter was a bulldog compulsive stick-to-it-ness that allowed him to succeed when others failed.

In 1980 he saw a segment on the David Frost show describing TS and knew instantly what his diagnosis was.

He went to a clinic elsewhere and was treated with Haldol but stopped it, complaining that the medication *stopped all of his creative energies.*

Two PhD's

When looking for the positive aspects of a *Gts* gene we often have to find it in the relatives of TS subjects rather than in the patients who come for help. The following is such an example.

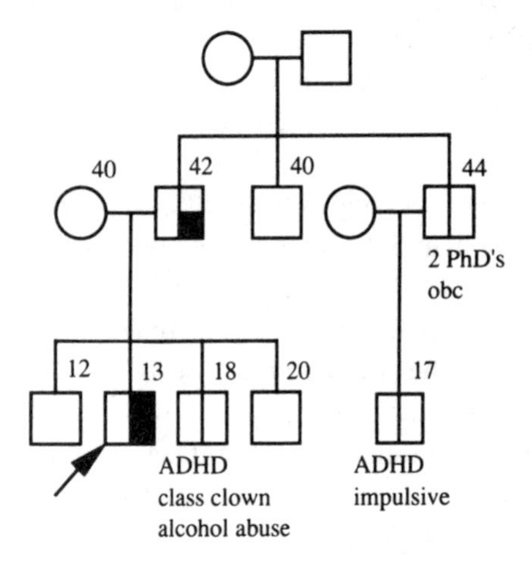

The proband (arrow) had TS with severe conduct disorder. His brother had ADHD, problems with alcohol abuse and was the class clown.

The father, like his son, had many problems with conduct as a teenager. His problems rapidly improved when he discovered in college that he had *eidetic imagery,* that is, he only needed to glance at a sheet of paper full of numbers and then, at any time in the future, he could read them off from memory. Eidetic imagery is a special type of automatic total memory recall, and those that have it can obviously be well served by it. He also had compulsive work habits and had risen to high position in his company.

The father's brother was a compulsive academic. "He seemed to love to study 24 hours a day." He obtained a PhD in anthropology and then a second PhD in public health. Various compulsive behaviors and the fact that he had a son with ADHD and impulsive behaviors strongly suggests he was a *Gts* gene carrier. In this case the gene was utilized to considerable advantage.

Ten degrees The proband, a 25 year old college student, was taken to a neurologist at age five because of poor handwriting. No cause for this was found. At age 14 he developed motor tics and repeated throat clearing began at age 17. At age 25 he was hospitalized for a peptic ulcer, spastic colon and panic attacks. These responded well to diet and medication. He was referred to a psychiatrist to help him deal with stress. He states he "thrives on hard work." Looking at his pedigree it is little wonder he is imbued with the work ethic.

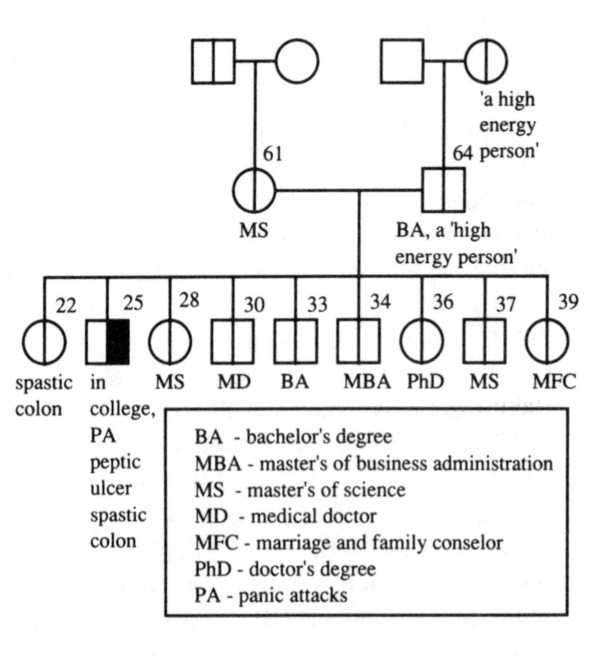

His father had a college degree and was described as a "high energy person." His mother had a master's degree. There were eight siblings. His younger sister was in college and also had a spastic colon. Each of his older siblings had a degree, and most a postgraduate degree. In the pedigree I have speculated that both parents and all the children may be *Gts* gene carriers, manifesting it enough so that the compulsive hard work is predominantly a positive factor in their lives. In view of the proposal that many TS patients carry two *Gts* genes[p271,513], the TS patient may be homozygous while all the siblings carry only a single gene.

Some psychologists might suggest the two siblings had a spastic colon because they were trying to keep up with their older siblings. Spastic colon, peptic ulcer, asthma, headaches, and other disorders have often been called "psychosomatic disorders." The modern term is *"somatization disorder"* meaning "affecting the body." However, I have often seen spastic colon and other psychosomatic disorders in *Gts* gene carriers and somatization disorders may be one more of the many ways a *Gts* gene can manifest itself[p293].

Musical Talent

A number of TS patients have unusual musical talent. This seems to be a combination of an appreciation for repetitous rhythms of music, the ability of music to placate the tics, a keen sense of mimicry and a touch of compulsiveness. The following story illustrates the point.

> A nine year old boy with TS began taking piano lessions in September and by November he was practicing eight hours a day. He became so engrossed in this endeavour that his parents had to bribe him to come to dinner. His teacher characterized him as the most remarkable student she ever had. All she had to do was play a piece for him once and he could immediately sit down and repeat it without missing a note. Although he had difficutly concentrating in school he could sit for hours at the piano in deep concentration during which all his facial tics ceased.

Hyperlexia

Many of the above-average accomplishments of the human brain have been done by individuals with various types of neurological disorders. Some individuals with these above-normal talents have been called *idiot savants* because in other areas they perform below average. A wide range of examples have been reviewed in the book *The Exceptional Brain* [1427]. Significantly above-average performance in reading is called *hyperlexia.* Many children with hyperlexia may also have delayed onset of speech, echolalia, attention deficit disorder, hyperactivity, repetitive stereotyped movements, head banging, compulsive behaviors and a family history of dyslexia and learning disorders. It occurs predominantly in males. Some have eidetic word imagery. "A compulsive preoccupation with reading sharply differentiates hyperlexic children from precocious readers who otherwise are developing normally. The total absorption in reading pervades the case descriptions. For hyperlexics, reading seems to have replaced other play activities" [866,1427]. The presence of autistic-like behavior, TS and hyperlexia has been described in four patients[252].

All of the above features, including the relationship between some high functioning autistics and TS[p217,499], suggests that many hyperlexics may in fact have Tourette syndrome.

The above examples of unusual creativity in some TS patients raise the often debated question of the relationship between "disordered" brain chemistry and creativity.

"Supernormal" Children with Obsessive-Compulsive Disorder

In her study of obsessive compulsive children Judy Rapoport[1576a] noted some that she called "super normal." "These were phenomenally ambitious, energetic youngsters who were carrying out more academic and extracurricular activities and having more responsibilities than most classmates; they led heavily programmed lives, leaving themselves almost no flexibility, sometimes to a remarkable degree. They described a life of teams, jobs, exercise, extra classes, community volunteer work, etc., with extraordinarily high performance in all areas. They were concerned about expectations and achievement and worried that they would not meet their heroic list of commitments. What proportion of super achieving adults are obsessive-compulsives in disguise, and carry a *Gts* gene, remains to be determined.

The Political Radical

In the book *Roots of Radicalism,* Rothman and Lichter[1654] described two somewhat diametrically opposed cultural and personality profiles of individuals drawn to radical causes. Especially focusing on the period of the late 1960's they noted that the activists were drawn mainly from a white, affluent, upper-middle-class Jewish and liberal Protestant constituency. While these activities were often thought of as issue oriented (anti-war, anti-nuke), Mike Goldfield described in *New Left Notes:*

> "You have to realize that the issue didn't matter. The issues were never the issues. You could have been involved with the Panthers, the Weathermen, SLAT, SNCC, SDS. It didn't really matter what. It was the revolution that was everything."

This sounds remarkably like the description compulsive gamblers give:

> "its not the winning or the loosing that is important, its the action."

The Jewish radicals were distinct from the Protestant radicals. Those who were Jewish tended to simply be more radical extensions of a historical, cultural and intellectual lean to the left. They were children of highly educated parents who were also more involved in leftist issues and came from the professional sector — teachers, professors, and educational administrators, more than business executives. Thus, the Jewish activitists were simply a more radical chip off the old parental block.

The Protestant radicals were dramatically different. Their fathers tended to be businessman rather than professionals. The students themselves tended to "drive fast and recklessly, to drink liquor often and heavily, and to be involved in physical fights." They were dangerously aggressive behind the wheel of a car, intemperate in their alcoholic consumption, and physically pugnacious. Strong narcissistic tendencies were present in both groups.

The similarity of the non-Jewish radical to the psychological profile of type II alcoholism[p232] and the impulsive, narcissistic, antisocial behaviors of some TS patients is intriguing. Many parents have stated that their TS child seemed to have an excessive, almost compulsive sense of justice for the underdog. This seemed to be most closely related to a distaste or disrespect for authority figures. While it is not clear whether radicalism should be a positive feature, it is a form of activity not generally considered pathological, but at the fringes of normal behavior, and is some cases may be driven by neurochemical and genetic factors.

Is Creativity the Flip Side of Madness?

A popular and intriguing idea is that creativity or artistry is somehow linked to madness. While one might tend to discard this out of hand as an old wives' tale reminiscent of sour grapes by those not so endowed, there is in fact some evidence that while the raging psychotic mind is dysfunctional, the same "chemical fire storm" at a lower level of intensity can drive considerable degrees of creativity that can raise one above the mainstream of normal mundane thought.

Superphrenics Children of schizophrenics with unusual ability are termed "superphrenics" and have been discussed elsewhere[P213]. There is much to indicate that genes for manic-depressive disorder also play an important role in creativity.

Manic-depression, hypomania and creativity In his popular book, *Moodswing*, Ronald Fieve[598] describes many famous and creative people who were manic-depressive. The list included such notable artists and writers as Handel, Rossini, Schumann, Balzac, Van Gogh, Hemingway, and Logan. During a spell of hypomania, Rossini wrote the *Barber of Seville* in an incredible thirteen days, a rate that is faster than others take simply to copy the score. After that episode he didn't write again for fourteen years.

Joshua Logan, the playwright, describes the creative fire well[598]:

> "Finally, as time passed, the depression gradually wore off and turned into something else, which I didn't understand either. But it was a much pleasanter thing to go through, at least at first. Instead of hating everything, I started liking things — liking them too much perhaps. I swung into a different mood altogether, which I didn't understand, nor did anyone else. At first people thought I was drinking, even though I was seldom around any bar, and I wasn't seen to take a drink of alcohol in front of anyone, so they couldn't quite explain it that way. And yet I was fairly flamboyant in my thoughts, imagination, and speech without really being dangerous. I was certainly very active mentally and physically. I lost weight, dropped down almost overnight to my best weight, like a fighter in good trim. I put out a thousand ideas a minute, things to do, plays to write, plots to write stories about."

This puts in words the typical push of thoughts and driven nature of the creative process of individuals in a hypomanic to manic state. This state serves the bearer well — unless it gets out of hand and transforms into disturbed psychotic thinking.

In a study of 30 writers and poets, Nancy Andreasen of the University of Iowa College of Medicine found that 80 percent had been treated for mood disorders compared to 30 percent of a control group, and 43 percent had some degree of manic-depressive illness compared to ten percent of the controls[37]. Alcoholism was also very common, being present in 30 percent of the writers compared to seven percent of the controls. A study by Kay Jamison of the University of California at Los Angeles confirmed these results. Here a survey of 47 of the top English artists and writers showed that 38 percent had sought treatment for mood disorders. Poets topped the list with 50 percent having been treated with drugs or hospitalization for depression or mania.

In one study, the relatives of patients with just depression were compared to the relatives of patients with manic-depressive disorder. The relatives of manic-depressives showed significantly greater levels of achievement even if they did not have manic symptoms themselves[402a].

As described in the chapters on serotonin, one of the effects of disinhibition of the limbic system is to increase the reactivity to sensory input from the environment[P321]. In many ways this describes the artistic, poetic mind — a hightened sensitivity to people, thoughts and emotions.

The Highs and Lows of Social Achievement

Just as the *Gts* gene can produce mood

swing highs and lows it can also produce highs and lows in the larger arena of social achievement. Individuals without the *Gts* gene follow a normal bell-shaped curve with some individuals chronically unemployed, the majority functioning in an average fashion and a small number involved in outstanding scientific, artistic or business achievements. This is illustrated as follows:

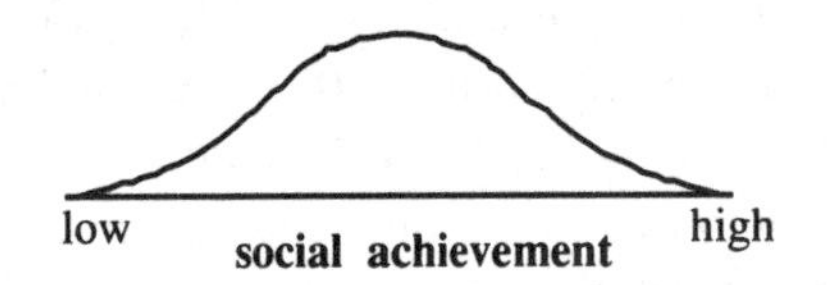

While it cannot be proven until a marker for the *Gts* gene is available, it is my impression that individuals with TS, or carrying the *Gts* gene, show a greater tendency to migrate to the extremes than those who do not carry the gene. Whether the extreme is to greater or lesser achievement depends upon which of the spectrum of symptoms occurs. This is illustrated as **shown below:**

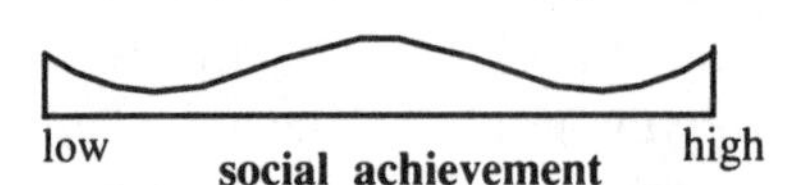

low	high
severely obsessive-compulsive	moderately obsessive-compulsive
mania and/or depression	hypomanic
ADHD	hyperactive
learning disorder	restless
dyslexia	moderately aggressive
conduct disorder	high IQ
antisocial personality	want things their way
angry and abusive	controlling

Summary: When the chemical fires responsible for many of the important psychiatric disorders of man— manic depressive disorder, schizophrenia, Tourette syndrome and obsessive-compulsive behavior — burn at a moderate flame there may be a positive effect of increased drive, a need to excel, an independence of thought, a nonstop restlessness and a creative energy that can lift individuals above the level of the commonplace and mundane.

Part IV
CAUSE

Part III was devoted to understanding the broad spectrum of behavioral problems that may be present in subjects with Tourette syndrome and related disorders.

In Part IV I ask, What causes these aberrant behaviors? Since TS is so clearly a genetic disorder, a major question is — Which one of the thousands of human genes is abnormal? There are two ways to find out. One is the linkage approach[P53], determining which chromosome the *Gts* gene is on, then attempting to find it. This can be a very difficult process. The linkage of the gene for Huntington disease was discovered in 1983 and despite intense effort in many laboratories, years to passed before the gene was identified. A much more common approach is to identify the enzyme or protein defect that causes a given genetic disease. This is called the *candidate gene* approach. For some diseases this is relatively easy. However, for disorders like Tourette syndrome, this has proven to be quite difficult because specific chemical abnormalities are difficult to find.

In Part IV, I review the different physical and chemical parts of the brain that could be involved and suggest a specific candidate gene that might be the cause of Tourette syndrome.

Chapter 47
The Anatomy of the Brain

Just as it was necessary to introduce some basic concepts of genetics before discussing behavior in TS, it is now necessary to introduce some basic concepts of nerve structure and function before investigating the possible causes of TS and related impulse disorders.

The Central Nervous System

The term central nervous system, or CNS, refers to the brain and spinal cord. By contrast, the peripheral nervous system consists of the nerves that are outside these two structures. Figure 1 shows the human brain and spinal cord and a cross section of both. The spinal cord regulates the nervous activity of the body's muscles and organs. The cross section of the brain (Figure 1B) illustrates the general position of some of the major sub-divisions of the brain. These are given the tongue-twisting names of telencephalon, diencephalon, mesencephalon, metencephalon, and myelencephalon. In each of these terms — *encephalon* means brain. These parts reflect the divisions of the brain early in embryonic development.

When the brain is cut so the right and left hemispheres are separated, the relative position of these divisions is seen from a side view (Figure 2).

The Brain Stem

The brain stem contains groups of nerve cells that are important for understanding Tourette syndrome and other impulse disorders. These nerves are rich in different neurotransmitters, such as dopamine, norepinephrine and serotonin, which are discussed in subsequent chapters. The brain stem is made up of the myelencephalon, metencephalon and mesencephalon (Figure 2).

The Diencephalon

The diencephalon contains the thalamus, an important set of nerve cells with many connections between the cortex and the lower parts of the brain. It also contains the hypothalamus which regulates the pituitary — the master endocrine gland of the body.

The Cerebral Cortex

In an evolutionary sense, the cerebral cortex is a late appearing part of the brain and is especially well developed in mammals and man. It forms the major part of the telencephalon or "late-brain."

The surface of the cortex shows many ridges called **gyri**, and valleys between the gyri called **sulci**. The major parts of the cortex, separated by sulci, are shown in Figure 3. These are the occipital, parietal, precentral, temporal and frontal cortex. As will be seen in subsequent chapters, the temporal and frontal cortex play an important role in the impulse disorders.

Areas of the cortex which perform specialized functions are identified in Figure 4, an illustration of the left side of the brain. Note the different regions for specific areas of the body.

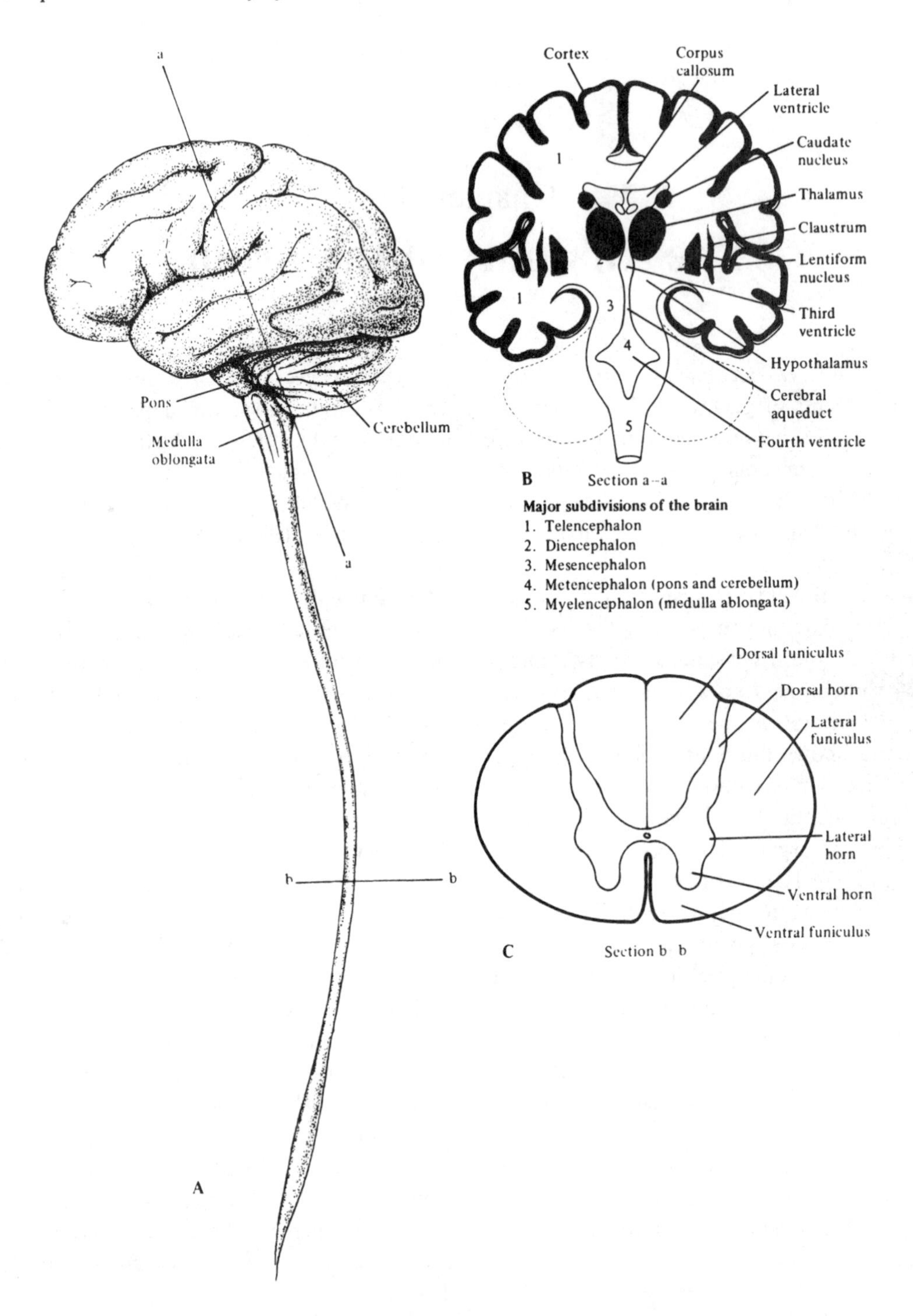

Figure 1. The brain and spinal cord. From L. Heimer *The Human Brain and Spinal Cord.* Springer-Verlag, New York, 1983.

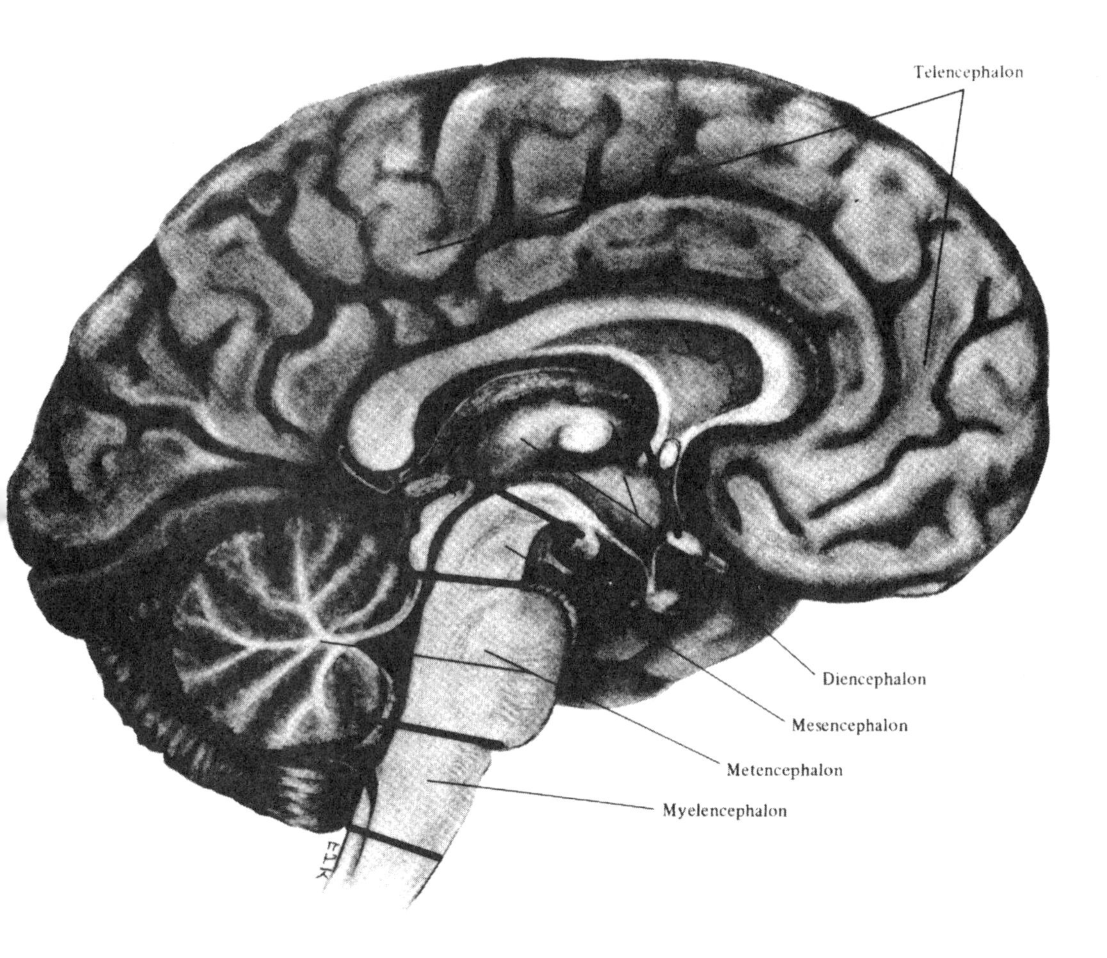

Figure 2. The major divisions of the brain cut down the middle. From L. Heimer, *The Human Brain and Spinal Cord.* Springer-Verlag, New York, 1983.

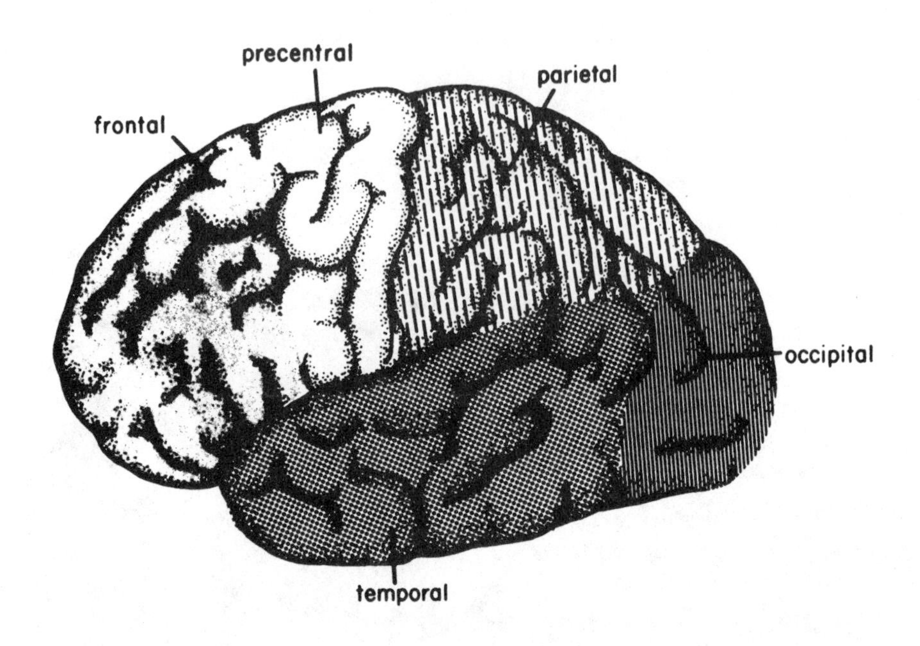

Figure 3. The major lobes of the human cortex. (From Thompson (1975) *Introduction to Physiological Psychology*, New York: Harper Row.)

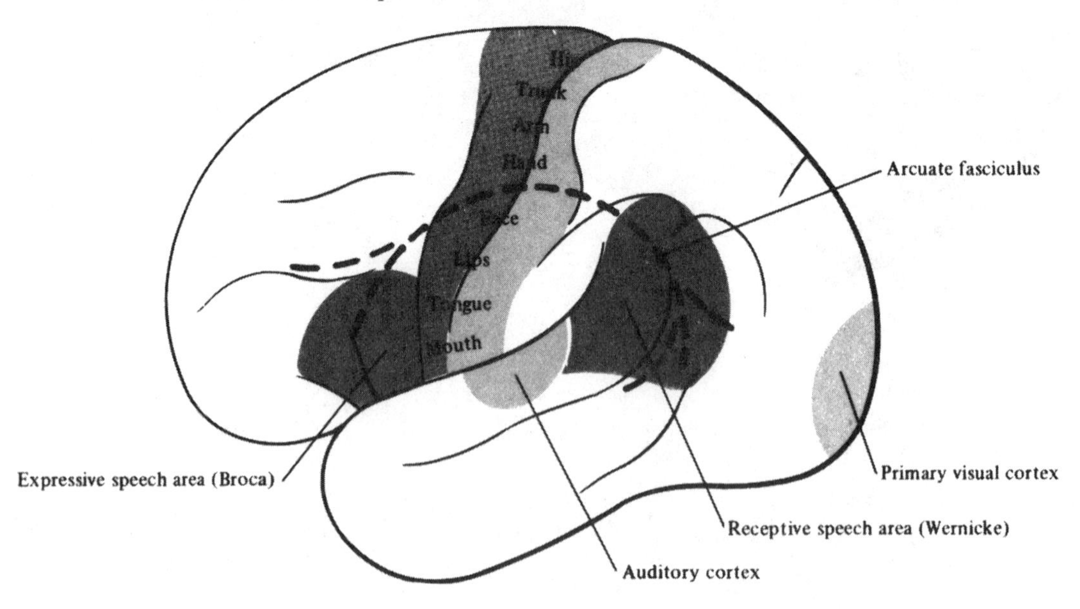

Figure 4. Areas of the cortex involved in movement, expressive speech, receptive speech, hearing and vision. From L. Heimer, *The Human Brain and Spinal Cord*. Springer-Verlag, New York, 1983.

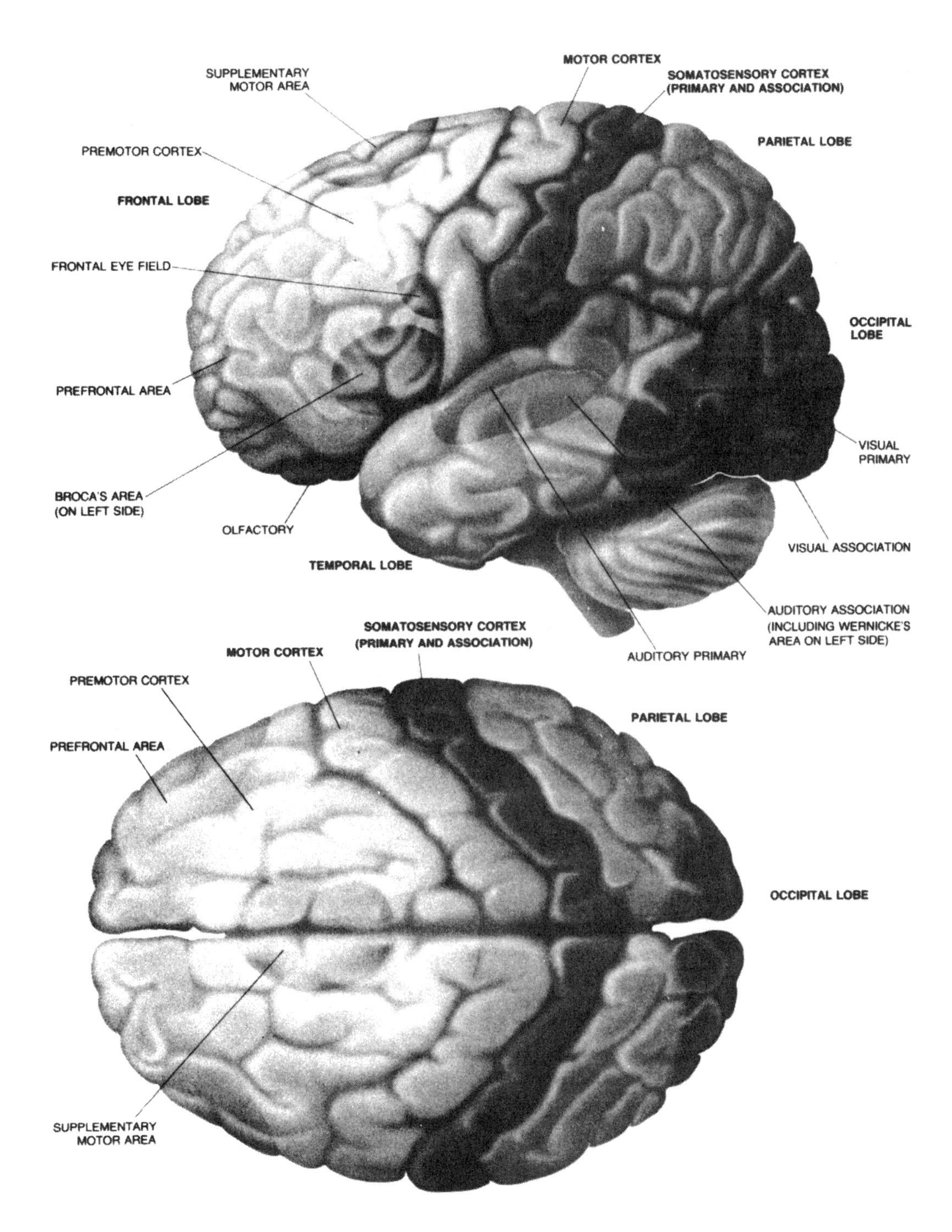

Figure 5. Further details of the organization of the human cortex. From Niels A. Lassen, David H. Ingvar and Erik Skinhøj, "Brain Function and Blood Flow,"[1129] Copyright (1978) by Scientific American, Inc. All rights reserved.

The representation of areas controlling the face and hands is disporporationate to that controlling the arms and legs. This is because the wide variety of facial expressions and extensive use of the hands in humans required many nerves to control them. By comparison, fewer nerve cells were required by the arms and legs. Also important are Broca's area[219] (left brain) which controls the utterance of speech (expressive speech) and Wernicke's area (left brain) which controls the hearing of speech (receptive speech). The occipital lobe at the back of the brain processes visiual stimuli (primary visual cortex).

Some additional aspects of the subdivision of the different parts of the cortex and their functions are shown in Figure 5. Many of these areas will be referred to later and the reader may refer back to this figure for orientation.

Right Brain — Left Brain

There is a great deal in the popular press about the different functions of the right versus the left halves of the brain. As speech began to develop in early man, there was no area of the brain assigned to control this function. Since no new brain parts could be made, the solution was to take over part of one half of the brain and leave the other half to perform its previous function. Since speech required control of the muscles of the larynx and face, the part of the brain close to the motor cortex controlling the tongue and mouth, on the left side of the brain, was used (Broca's area). A related area, close to the part of the cortex for hearing was used for listening to the speech of others (Wernicke's area). Thus, the left half of the brain became specialized in the processing of verbal information and the right half of the brain, *de facto*, became specialized in the processing of nonverbal information (music, visual-spatial relationships, nonverbal ideas). To extrapolate hemispheric functions any further than this is probably stretching the data too far[684].

This is just a brief introduction to some of the major parts of the brain. Some portions not shown or only briefly mentioned, such as the limbic system and the frontal lobes, are so important that whole chapters are devoted to them.

The Nerve Cell or Neuron

Nerve cells are called **neurons**. While they come in an astounding variety of types[246] the basic structure of a neuron is shown below and in Figure 6 (page 312).

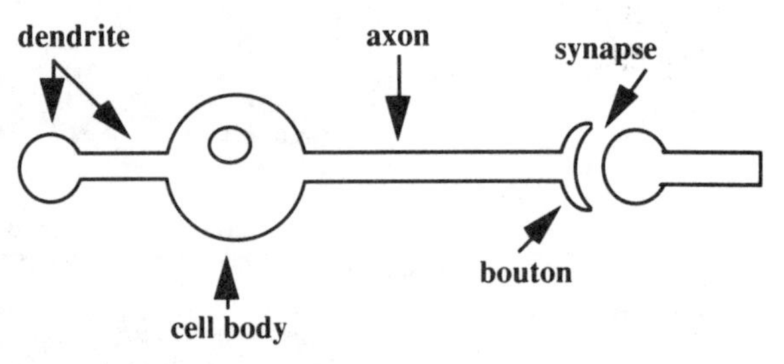

Neurons consist of a **cell body** containing the cell nucleus, **dendrites** and **axons**. Dendrites receive nerve impulses from other cells and axons pass nerve impulses to other cells. The interaction between two neurons is from axon to dendrite. In some neurons the axon is insulated by lipid layers forming **myelin**. These are called myelinated fibers. They form the "white matter" that is seen when the brain is cut in slices. The well-known "gray matter" is made up of cell bodies. The end of the axon contains a swelling called a **bouton**. The site at which axons interact with dendrites is called the **synapse**.

The synapse between two resting nerves can be diagramed in shorthand as follows:

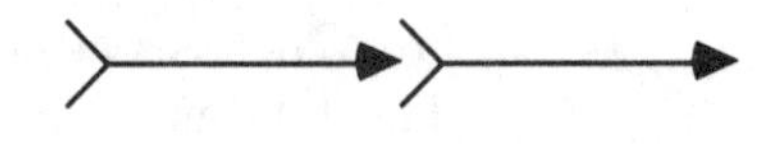

If the effect of the neuron is to **stimulate** the next neuron, the shorthand is:

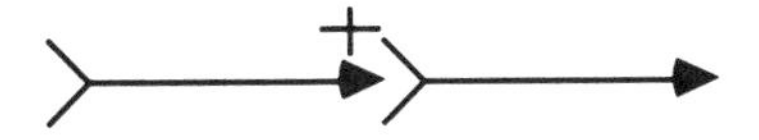

If the effect of the neuron is to **inhibit** the next neuron, the shorthand is:

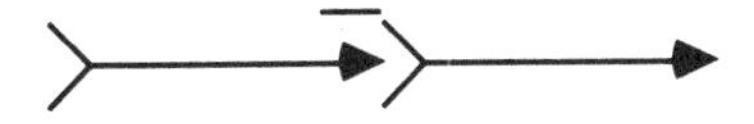

The axon acts like the copper wiring of an electrical apparatus. It is a passive structure. The action, which makes one neuron different from another, takes place at the synapse. The details of this action are discussed in the following chapter.

Summary: This chapter provides a short road map to many parts of the brain that will be important for understanding some of the defects in Tourette syndrome and related disorders. The basic unit of the nervous system is the nerve cell or neuron. Messages coming into the neuron travel on the dendrites. Messages leaving the neuron travel on the axons. The meeting ground between two neurons is called the synapse. It is the site of the chemical imbalances that affect behavior.

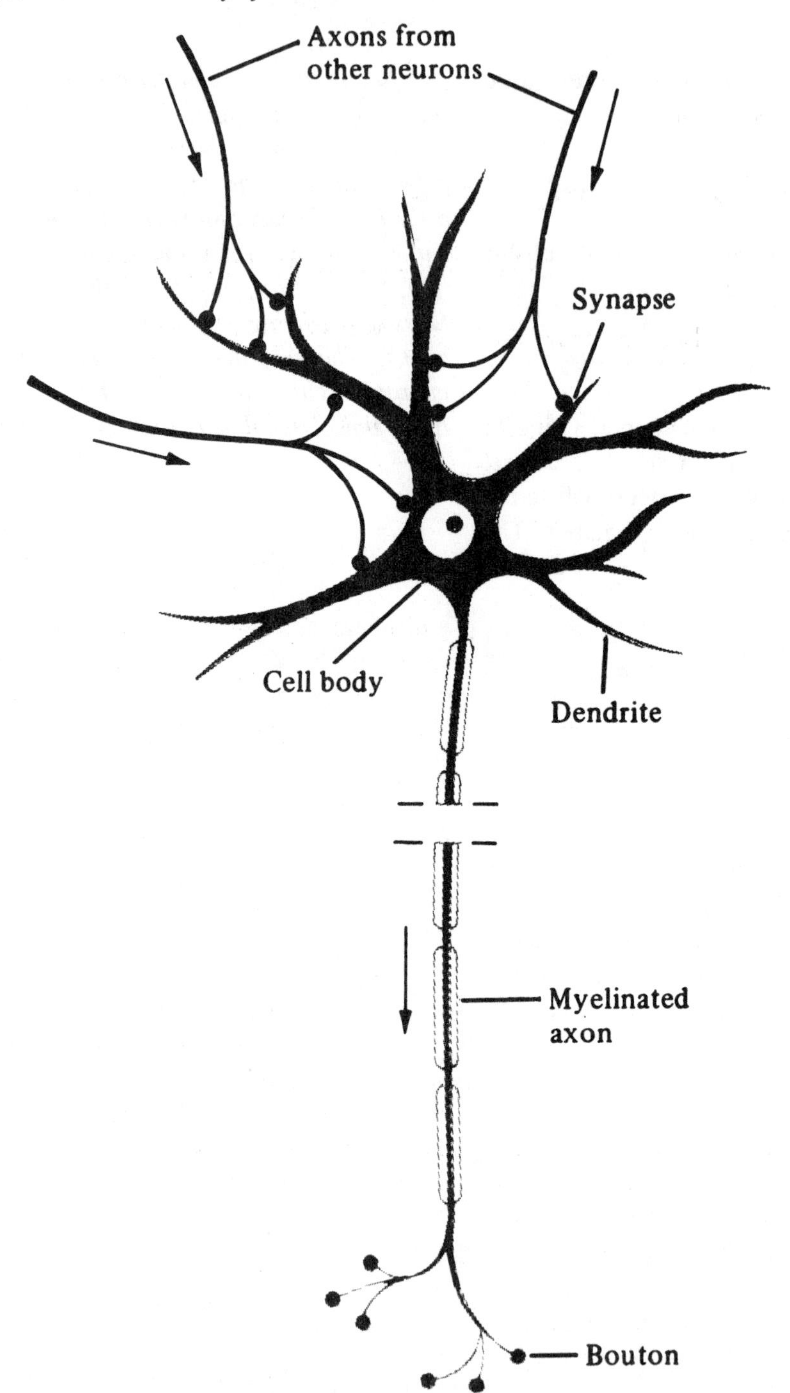

Figure 6. The anatomy of a neuron. From L. Heimer, *The Human Brain and Spinal Cord.* Springer-Verlag, New York, 1983.

Chapter 48

Communication between Nerves: The Neurotransmitters

In the 1870's Thudichum wrote of the chemistry of the brain:

> "I believe that the great diseases of the brain and spine will all be shown to be connected to specific chemical changes in the neuroplasms. It is probable that by the aid of chemistry many derangements of the brain and mind which are presently obscure will become accurately definable and amenable to precise treatment"[533].

This prediction is in the process of coming true. To understand brain chemistry it is necessary to understand a few basic concepts about the chemistry of nerves — at their surface (the membranes), and at their ends (the synapses).

The Chemistry of the Nerve Surface — The Membrane

Membranes are thin layers of fatty material that separate the inside from the outside of cells. Although they are only a few molecules thick, membranes are very important. They are like doors to a very exclusive club — some people are let in while others are not, and some people who get in are thrown out.

The unique aspect of nerve cells is their electrical excitability. The chemicals involved in this excitability are simple — sodium (Na^+), potassium (K^+), chloride (Cl^-) and calcium (Ca^{++}). These charged compounds are called **ions.** Each ion has a different charge, namely +, ++, -, --. The total number of ions determines the electrical chage produced. Consequently, the electrical charge inside a cell will be different than the charge outside, since the number of ions and the charge of each is different. For example, there is much more sodium on the outside of the nerve cell than the inside, and the rate at which sodium is allowed into the cell is very low (small arrow).

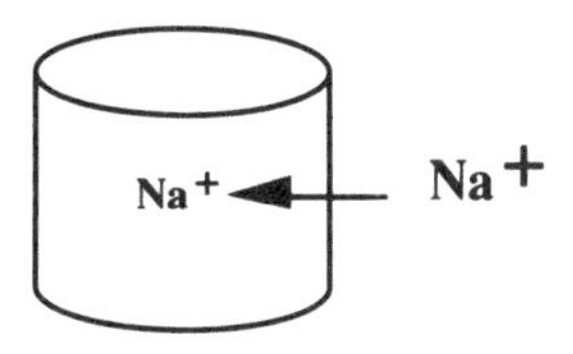

By contrast, there is much more potassium on the inside of the nerve cell than on the outside, and the rate at which potassium is allowed out of the cell is 50 times greater than the rate at which sodium is allowed into the cell (large arrow).

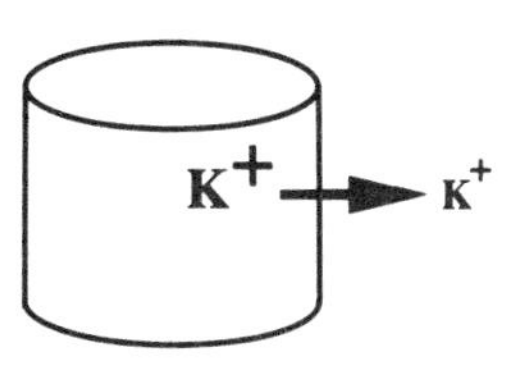

Finally, there is more chloride and calcium on the outside of the nerve cell than on the inside, and negatively charged proteins (anions, A^-) are much more common inside the cell. This can all be summarized as follows:

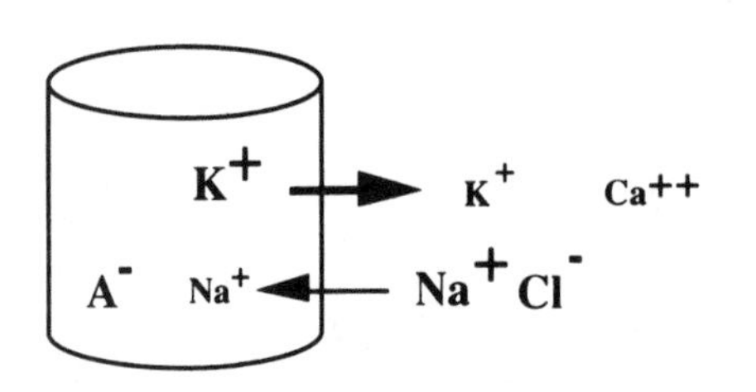

The net result is that there is a negative charge on the inside of the cell equivalent to -60 millivolts.

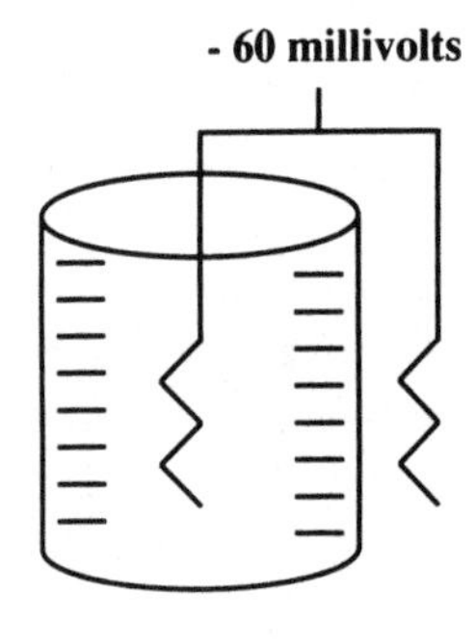

This potential difference is called the **resting potential.**

If the membrane in one spot is altered or stimulated the relative rate at which potassium and sodium cross the membrane is reversed.

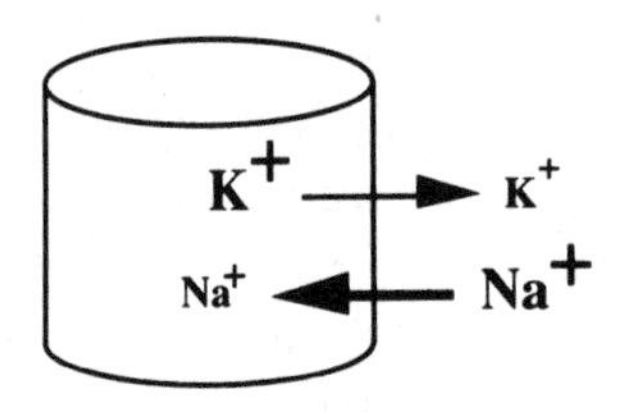

This results in a local reversal of the membrane potential to +50 millivolts.

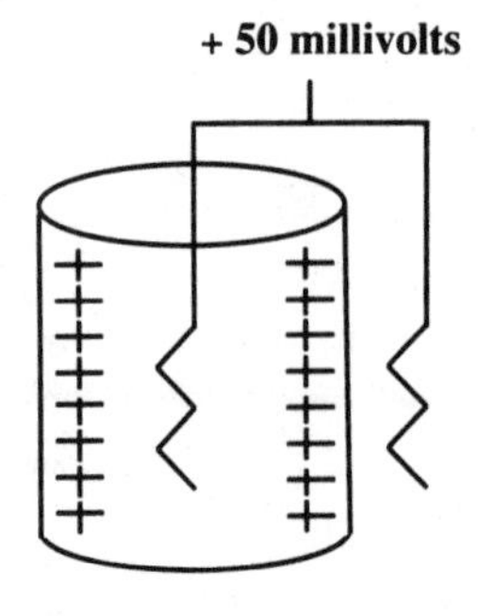

When this event is recorded the sudden change in voltage passing down the nerve cell produces an **action potential.**

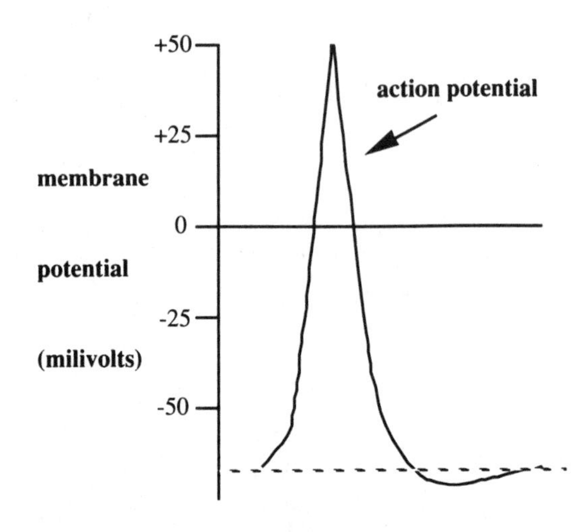

This wave of electrical charge is what passes down the axon of the neuron when it is excited.

Ion channels The ions don't just randomly pass in and out of the cell membrane. They must go through specific round doors or channels — appropriately called ion channels. Since each ion has a different channel there are sodium, potassium, chloride, and calcium channels. These channels are complex holes in the membrane made up of different subunits of protein. An example of the structure of one of these channels, the chloride channel, is shown in **Figure 1 on the following page**.

The holes in the channels can be opened both electrically and chemically. When the action potential reaches the end of the neuron, it doesn't just pass onto the next neuron. The communication between neurons is done chemically rather than electrically. These chemical messengers are called **neurotransmitters.** Neurohormones, neuropeptides, and neuromodulators are other groups of chemicals that play a role in controlling the communication between neurons[909,1215,1723]. The difference between them has grown increasingly

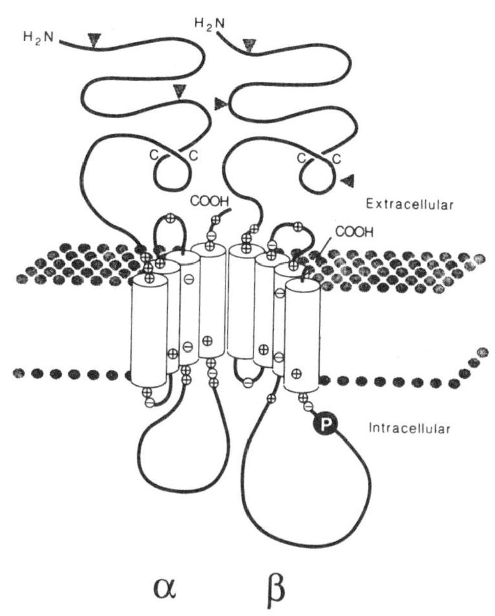

Figure 1. Diagram of the chloride ion channel regulated by the neurotransmitter GABA. From Schofield et al.[1730].

less distinct. Table 1 lists some of the most important ones. Throughout this book, those I will discuss most are dopamine, norepinephrine and serotonin. Others will be mentioned where appropriate.

Table 1 Neurotransmitters, Neurohormones and Neuromodulators

Chemicals
- acetylcholine
- aspartate
- dopamine
- gamma aminobutyric acid (GABA)
- glutamate
- glycine
- histamine
- norepinephrine
- serotonin

Small Proteins
- α-melanocyte-stimulating hormone (α-MSH)
- adrenocorticotropin (ACTH)
- angiotensin (I, II and III)
- bombesin (gastrin-releasing peptide)
- ß-endorphin
- bradykinin
- calcitonin
- carnosine
- cholecystokinin octapeptide (CCK-8)
- corticotropin-releasing factor (CRF)
- enkephalin (methionine and leucine)
- gastrin
- glucagon
- insulin
- luteinizing hormone-releasing hormone (LHRH)
- melantonin
- neuropeptide Y
- neurotensin
- oxytocin
- prolactin
- secretin
- somatomedin
- somatostatin
- substance P
- thyrotropin-releasing hormone (TRH)
- vasoactive intestinal peptide (VIP)
- vasopressin

The Synapse

Understanding the synapse is central to understanding the chemistry of the brain. Neurotransmitters remain in small bag-like storage areas or **vesicles** until an electrical impulse reaches the synapse at the end of the axon. This impulse causes the vesicle to open, releasing the neurotransmitter into a gap between the neurons known as the **synaptic cleft**. The released neurotransmitters excite the next neuron. These features are shown in Figure 2.

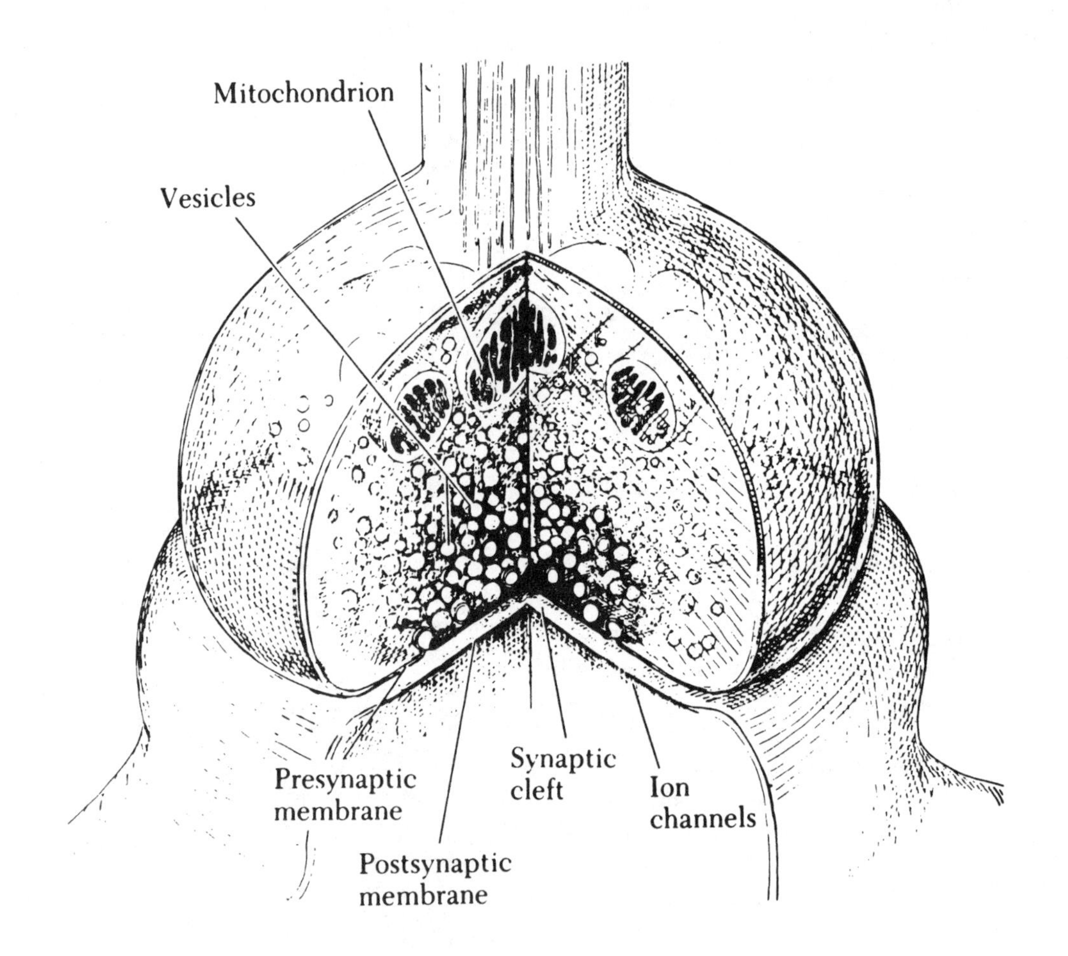

Figure 2. The synapse. From C. Stevens, "The Neuron." Copyright (1979) by Scientific American, Inc. All rights reserved.

The mitochondria are the source of the energy needed for the chemical reactions in the synapse. The **presynaptic membrane** is on the axon side of the **cleft**. The **postsynaptic membrane** is on the dendrite side The receptors that receive the released chemicals are in the postsynaptic membrane and are called **postsynaptic receptors**. These relationships are shown diagrammatically as follows:

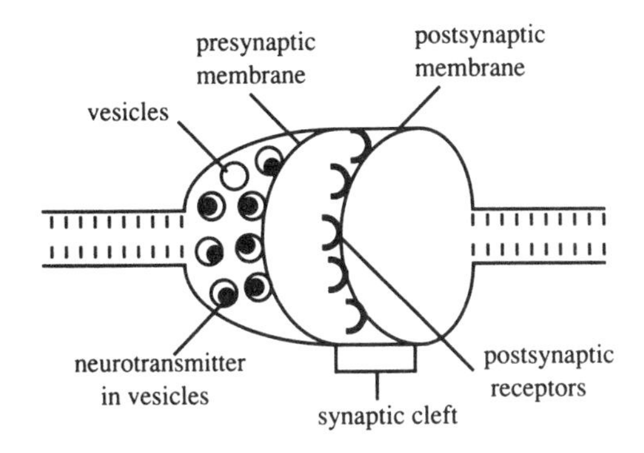

If the nerve is stimulated an action potential makes its way down the axon.

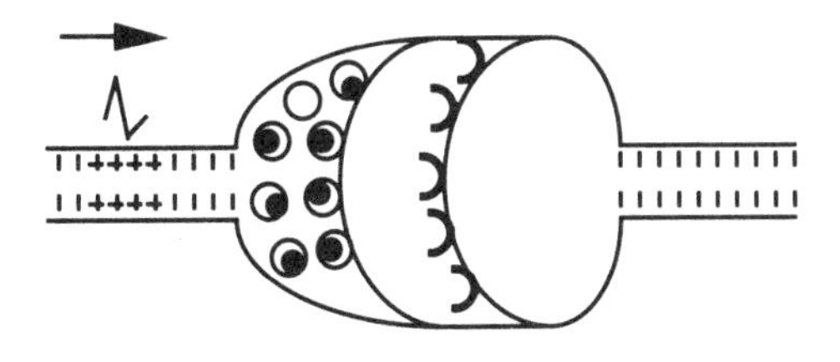

When the excitation wave reaches the synapse it sets in motion a complex series of events which result in the vesicles opening up and releasing the neurotransmitter they contain into the synaptic cleft.

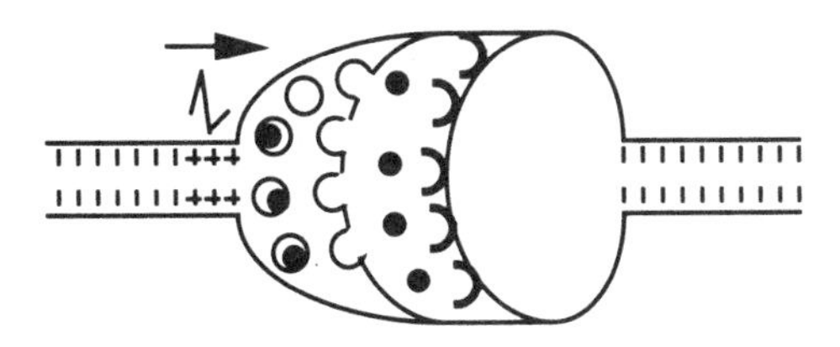

The neurotransmitter binds to the postsynaptic receptor that is specific for that neurotransmitter and this sets off the action potential in the next neuron.

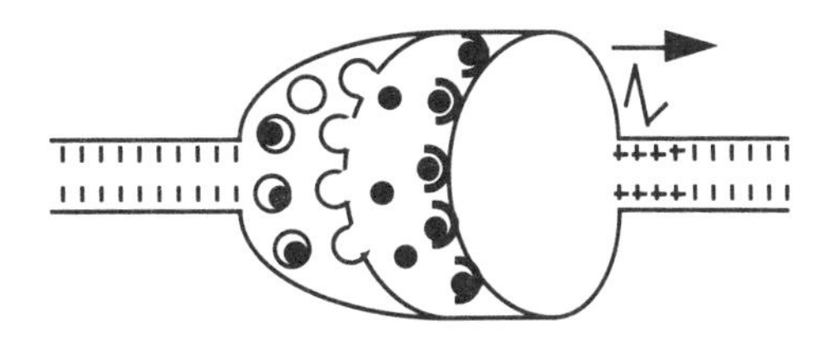

This is the manner in which nerve impulses are passed from one neuron to the next.

Neurons are named according to their neurotransmitter While there are some exceptions, each neuron has one type of neurotransmitter stored in its synaptic vesicles and the postsynaptic receptors respond to that neurotransmitter. This allows us to name neurons according to their neurotransmitter. Thus, neurons containing dopamine are called **dopamine neurons** (dopaminergic neurons); those containing serotonin are called **serotonin neurons** (serotonergic neurons), and so on. In the same fashion, the receptors are named after the neurotransmitter they bind. Thus, there are **dopamine receptors**, **serotonin receptors**, and so on.

Haldol and postsynaptic dopamine receptors To illustrate the importance of understanding how synapses work, let us examine the effect of Haldol (haloperidol), a drug often used to treat TS. In the next few chapters we will see how many of the symptoms of TS can be understood on the basis of overactive dopamine neurons. Both dopamine and Haldol bind to the dopamine postsynaptic receptor. However, unlike dopamine, Haldol does not stimulate the receptor to set off an action potential in the next neuron. As a result the total activity of the dopamine neurons is watered down or inhibited. Thus, *Haldol is a specific chemical treatment for a specific chemical imbalance in the brain.*

What happens to the neurotransmitter in the synapse? If the neurotransmitter that is released remained in the cleft, the nerve would continue firing like a motion picture stuck on one frame. This is prevented by removing the neurotransmitter from the cleft. This is done in two ways:

a) by destroying it, or

b) by readsorbing it into the presynaptic membrane.

Neurotransmitters are destroyed by specific enzymes. One of the most important of these is **monoamine oxidase (MAO)**. Three of the major neurotransmitters, dopamine, norepinephrine and serotonin are all broken down by monoamine oxidase.

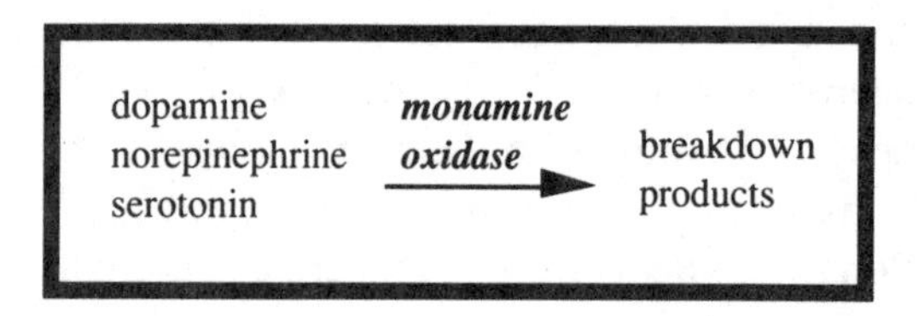

Any drug that inhibits this enzyme would increase nerve activity. The whole group of **MAO inhibitors,** used in the treatment of depression, act by increasing the level of dopamine, norepinephrine and serotonin in the synapse by preventing their breakdown

Neurotransmitters are taken up by binding to **presynaptic receptors.** These receptors are located in the presynaptic membrane.

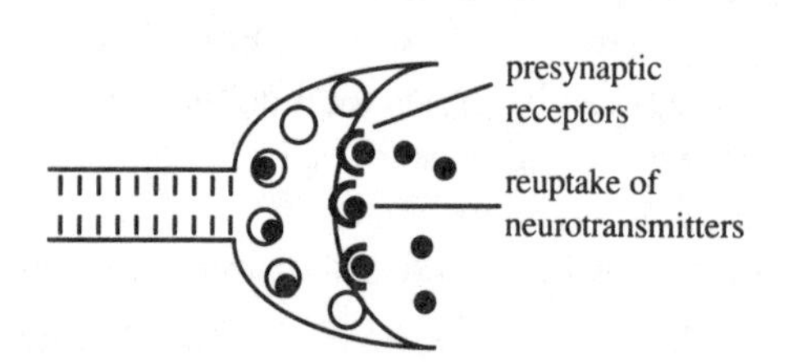

The other major class of antidepressant drugs, the tricyclics, work by inhibiting this reuptake process. Thus, the two major types of antidepressant drugs, the MAO inhibitors and the tricyclics, work by increasing the concentration of various neurotransmitters in the synaptic cleft, thus enhancing the activity of dopamine, norepinephrine and serotonin neurons.

Agonists versus Antagonists

These are two terms that are frequently used in discussing the neurochemistry of the brain. **Agonists stimulate specific receptors. Antagonists inhibit specific receptors.** For example, apomorphine stimulates dopamine receptors and is thus a dopamine agonist. Haldol inhibits dopamine receptors and is thus a dopamine antagonist.

Stimulating and Inhibiting Neurotransmitters

When a receptor is activated it does not always result in the stimulation of the nerve involved. Some neurotransmitters tend to stimulate neurons while others tend to inhibit neurons. For example, aspartate and glutamate stimulate neurons while GABA and serotonin inhibit neurons.

Presynaptic Inhibition

Another important concept is that of presynaptic inhibition. The following diagram illustrates this.

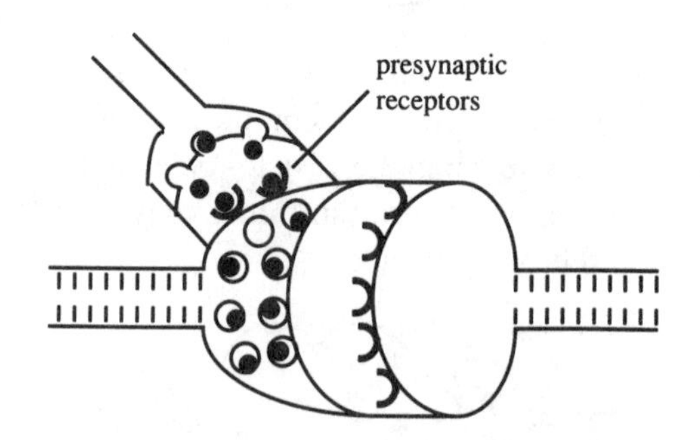

When the axon of one neuron interacts with the axon of another, an impulse coming down the second neuron can be inhibited from releasing the neurotransmitter at its synapse. Thus, depending on the location of the interactions and the type of neurotransmitter, one neuron can either stimulate or inhibit another neuron.

Summary: When neurons are excited an electrical impulse travels down the cell membrane. When these impulses get to the end of the nerve they cause the release of chemicals, called neurotransmitters, from small storage bags. These neurotransmitters pass across the gap between the first and the next neuron. They then bind to specific receptors in the dendrites of the next neuron and the electrical impulse continues. Thus, communication within nerves is electrical, while communication between nerves is chemical. The large number of different neurotransmitters, some of which stimulate and some inhibit, provides an enormous number of ways that neurons can interact.

Chapter 49

The Limbic System

One of the major themes of this book is that many of the behavioral disorders are caused by a genetic disinhibition of the limbic system. Thus, it is critical to understand what the limbic system is and what it does.

Papez — A Proposed Mechanism of Emotion

In 1937 Papez[1458] wrote a paper entitled *A Proposed Mechanism of Emotion.* "Emotion can be considered as *a way of acting* and *a way of feeling*. The former is designated as *emotional expression* and the latter as *emotional experience,* a subjective feeling." Papez assumed that the cerebral cortex had to participate in these activities in order for the human to experience emotional phenomena[953]. He considered that the cingulate cortex was the "emotional cortex" that could pass information onto the other parts of the cerebral cortex for "emotional coloring." The synthesis of the two parts of emotion required structures that could receive and integrate sensations and pass these onto higher cortical functions — the psyche. He proposed that the following structures:

- hypothalamus,
- thalamus (anterior thalamic nuclei),
- cingulate cortex,
- hippocampus,
- and their interconnections,

"constitute a harmonious mechanism which may elaborate the functions of central emotion, as well as participate in emotional expression." He asked,

> "Is emotion a magic product, or is it a physiologic process which depends on an anatomic mechanism?"

His idea that there was a neurological structure for the emotions was so radical for his time that the heresy caused an uproar that almost cost him his academic post[533].

Papez proposed a circuit of nerve impulses connecting the cortex (psyche) to the senses. This circuit was as follows:

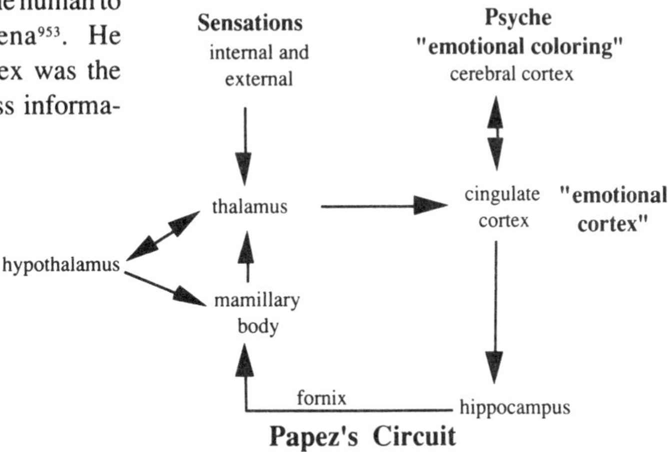

Papez's Circuit

In subsequent years this basic concept has been retained in a modified and expanded form. One of the major expanders was Paul MacLean[1236-1241], director of the Laboratory of

Evolution and Behavior at the National Institute of Mental Health. He[1237] named these parts of the brain the "limbic system" after the work of the French neuroanatomist Paul Broca. Broca first described the "limbic lobe," so named because the tissues involved surrounded the middle portions of the brain, like an encircling limb. He pointed out that the microscopic structure of these areas was primitive compared to the rest of the cortex, that there were strong connections to the smell or olfactory apparatus, and that these structures were all preserved in mammals. Because of this, the limbic lobe has also been called the rhinencephalon or "smell brain." The following figure shows the location of these structures.

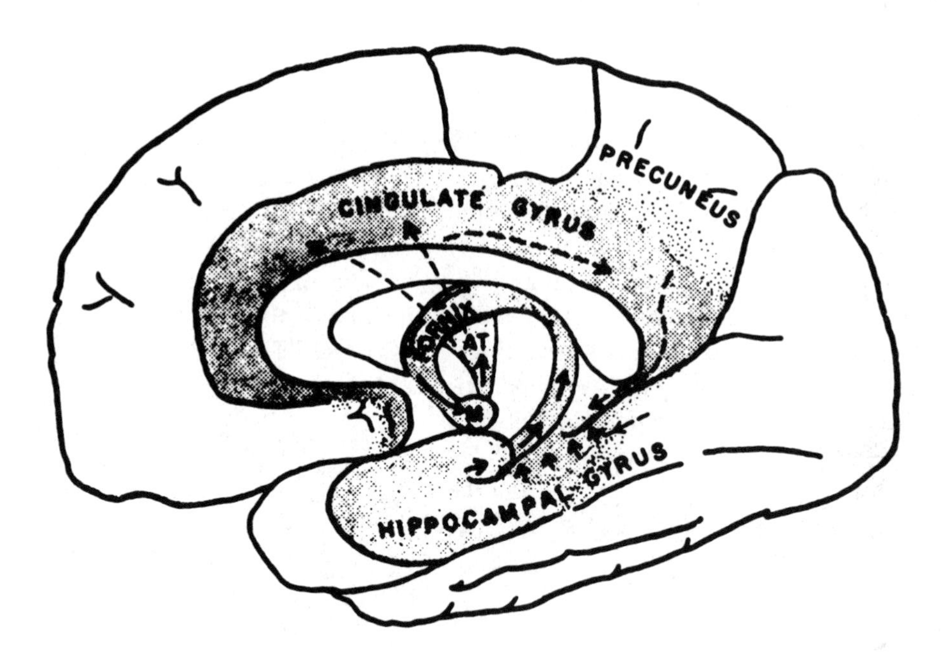

Figure 1. The shaded area of the cortex, seen from the center of the brain, represents major portions of the limbic system. From MacLean[1236]. AT = anterior thalamus.

This figure shows, in a more anatomically correct fashion, the Papez circuit, with the information coming from the cingulate gyrus -> hippocampus -> making a great loop in the fornix -> mammary bodies -> anterior thalamus -> back to the cingulate cortex. Additional players in this scenario are the amygdala, the septum, the olfactory bulb, nucleus accumbens, and the frontal lobe. Some of these are shown in Figure 2.

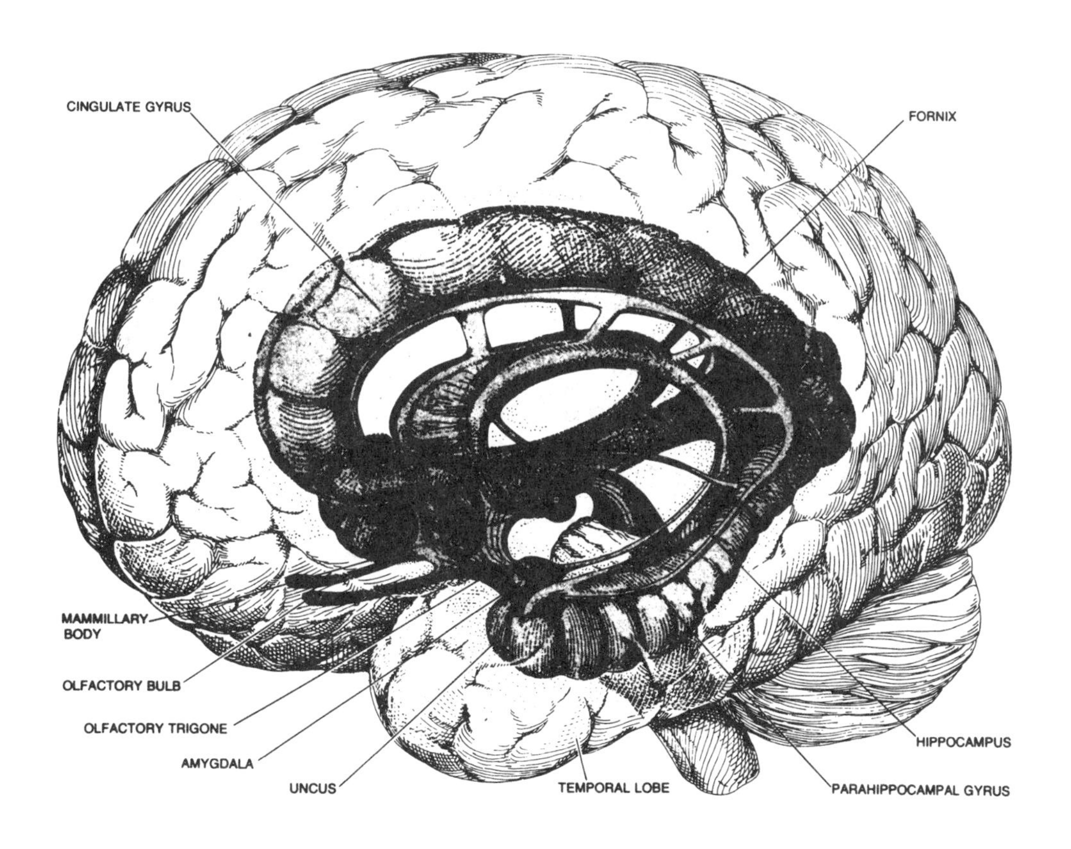

Figure 2. The human limbic system. From S. Snyder, "Opiate Receptors and Internal Opiates"[1819] Copyright (1977) by Scientific American, Inc. All rights reserved.

The complete cast for the limbic system is shown in the following box:

Structure of the Limbic System
amygdala
cingulate cortex
fornix
hippocampus
hypothalamus
nucleus accumbens
olfactory tubercule
prefrontal lobe
septum
thalamus

Don't be put off by the strange names. Think of them as the cast of players in the drama of emotion. Later, I will discuss in more detail how they interact, but first a brief introduction to their roles. If we had to design a system for the emotions, there are several jobs that would have to be done. The following are those jobs and the limbic structures that handle them.

Major job 1: Information input

Subcontract - external senses: sight, smell, hearing, touching, taste:

-> **amygdala, olfactory tubercle**

Subcontract - internal senses: pain, temperature, sexual feelings:

-> **thalamus, hypothalamus, septum**

Major job 2: Data storage

Subcontract - memory (Is this new or old data? What happened before?):

-> **hippocampus**

Major job 3: Formulation of possible responses

Subcontract - stop and think. (What should I do?):

-> **prefrontal lobe**

Major job 4: Action

Subcontract - contact the motor system (caudate and putamen):

-> **nucleus accumbens, frontal lobe**

Subcontract - contact the endocrine system and bodily functions:

-> **hypothalamus**

Thus, the limbic system can be thought of as the central processing system for emotional experience and expression. Let us look at it from an evolutionary point of view.

The Triune Concept

Paul MacLean[1240] has emphasized that there are three major portions of the brain: the reptilian or R-complex, the limbic system and the neocortex. During evolution these were layered on top of each other.

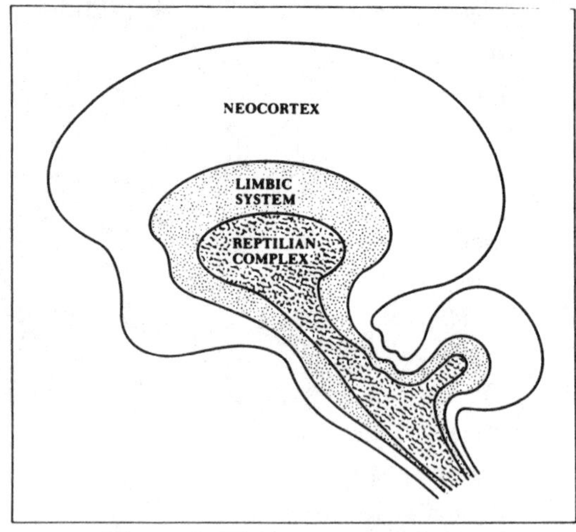

Figure 3. The concept of the evolution of the R-complex, limbic system and neocortex. After MacLean[1679].

The **reptilian complex** derives its name from the fact that in reptiles this constitutes the major portion of the brain. It is composed of the brain stem, midbrain, reticular system, and the basal ganglia (caudate and putamen). Reptilian behavior follows rigid, unthinking schedules, triggered off by environmental cues. In higher organisms such as birds and mammals, this portion of the brain controls mating and breeding rituals, imprinting, hunting, fighting, territorial defense, nesting, and the formation of social hierarchies. In modern computer terms these behaviors can be considered as containing hardwired programs for stereotyped, instinctive behaviors. They require no thought. Once triggered they are played out to completion like a mindless computer program. For example, a Tom turkey can be triggered to perform a copulatory act, performed without a partner, by a mere phantom image of a female turkey[1240]. The obsessive-compulsive and stereotyped behaviors in Tourette syndrome are reminiscent of such mindless behaviors. The reptilian mind "is not characterized by powerful passions and wrenching contradictions but rather by a dutiful and stolid acquiescence to whatever behavior its genes and brain dictate"[1679]. To add "the powerful passions and wrenching contradictions" required the addition of the limbic system.

The limbic system removed animals from robotic, reflexive action to an emotional involvement in life. Love, pleasure, grief, anger, fear, jealousy, and commitment were now possible. The third part of the triune brain concept is the *neocortex. This added an inhibiting, reflective, rational control over the storming limbic passions.* The relative size of the limbic system compared to the neocortex in different mammals illustrates the increasing importance of the neocortex. In the rabbit the limbic system is comparatively much larger than the neocortex, while this is reversed in monkeys and man, **see figure at right**.

Lesions of the Limbic System

Once the concept of the limbic system was developed many people attempted to understand the function of its different parts by surgically removing them. The resultant behaviors gave some indication of what the regions were doing. The following summarizes some of these results for different limbic structures.

The Amygdala

Because this region of the temporal lobe resembles an almond, it was named after the Greek word for almond — amygdala. One of the most intriguing studies of the limbic system involved the removal of the amygdala from monkeys. This resulted in a very unusual syndrome.

The Klüver-Bucy syndrome Two years after Papez proposed his mechanism of emotion, Klüver and Bucy[1060] described the behavior of monkeys following the removal of the temporal lobes containing primarily the amygdala. They developed what was termed "psychic blindness"or "visual agnosia" (not knowing what was seen). As a result they would approach all objects without hesitation even when they had previously been afraid of them. They would also indiscriminately examine every object in sight by putting it in their mouth, rather than examining it by hand. They would also reexamine the same object many times as if it had never been seen before. The oral examination consisted of putting the

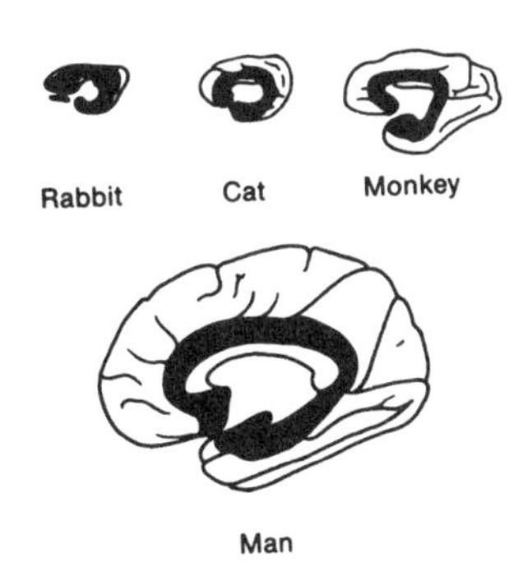

Figure 4. The relative size of the limbic system and neocortex in rabbits, cats, monkeys and man. Upper - lateral view, lower - medial view. (MacLean, 1977[1239].)

object in their mouth, biting gently, chewing, licking, and smelling. Their behavior seemed to be controlled by irresistible impulses, hyperactivity, and a compulsion to examine all objects in sight. These monkeys were also docile, showing none of the motor or vocal reactions associated with fear or anger in normal monkeys. Finally, they were also hypersexed. There was excessive penile erection and manipulation. There was such prolonged licking and sucking of the penis that the animals would sometimes fall asleep with their penis in their mouth. There was both homosexual and increased heterosexual behavior. A lesioned monkey might copulate continuously for 30 minutes, then leave the female only to immediately mount her again.

In summary:

The Klüver-Bucy Syndrome

- lack of visual recognition of objects
- examine objects by putting them in the mouth with licking, biting, chewing and smelling
- react to all objects seen
- hypoemotional
- no fear or anger
- hypersexual

The syndrome varies somewhat in different species. Removal of the amygdala in cats often caused them to become aggressive and show rage reactions[1753,1834,2105].

The amygdala compares the senses Neurons from all the major sensory areas of the cortex (sight, hearing, taste, smell and touch) pass to the amygdala[9,1339].

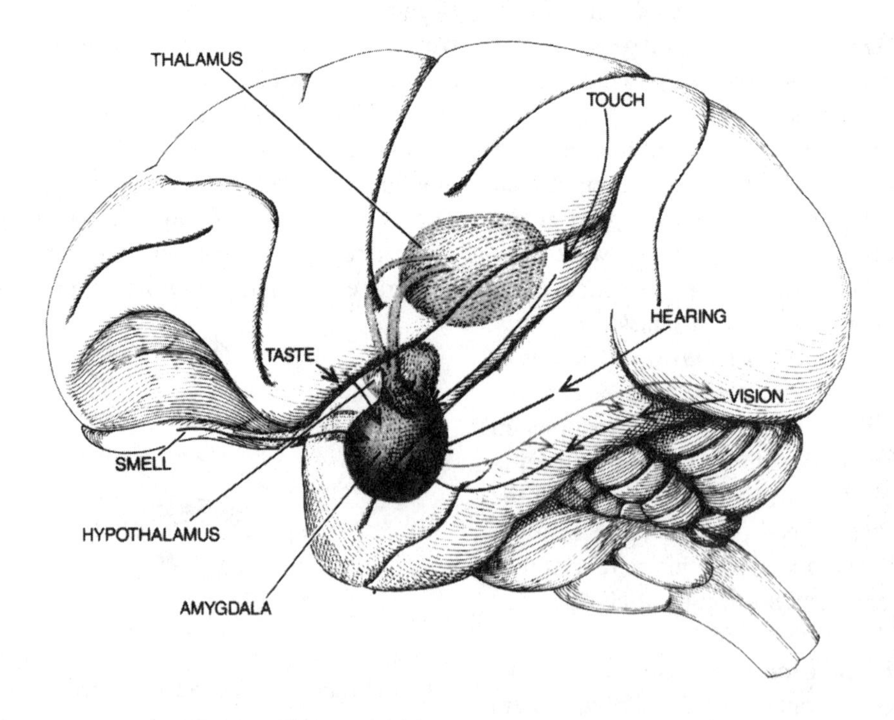

Figure 5. The amygdala receives inputs from the major areas of the sensory cortex. From Mishkin and Appenzeller, "The Anatomy of Memory"[1339]. Copyright (1987) by Scientific American, Inc. All rights reserved.

Neurons from the amygdala then pass to other limbic structures including the thalamus, septal area and prefrontal cortex.

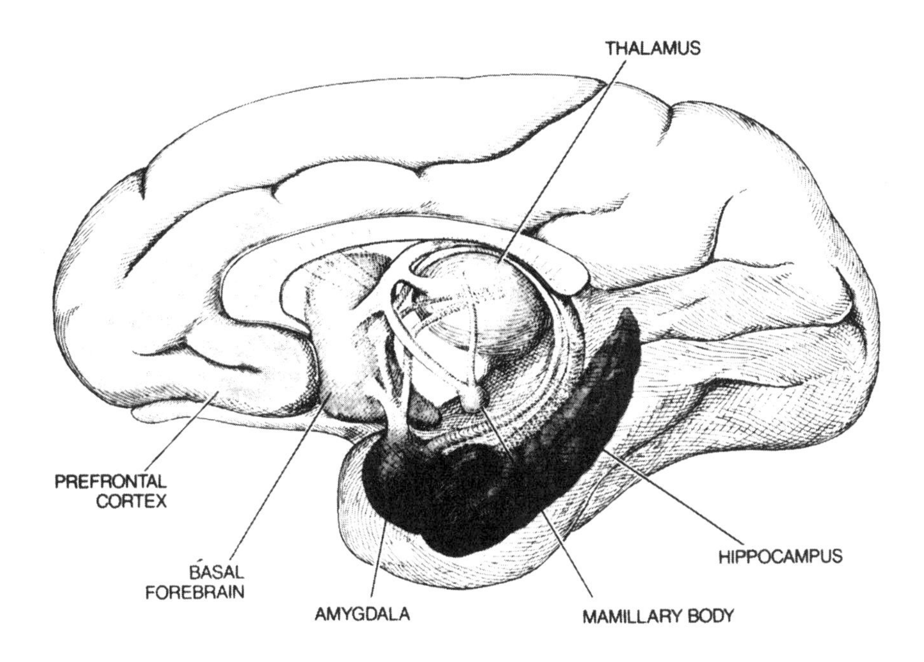

Figure 6. The amygdala sends connections to the thalamus, septum (basal forebrain) and prefrontal cortex. From Mishkin and Appenzeller "The Anatomy of Memory" [1339]. Copyright (1987) by Scientific American, Inc. All rights reserved.

Trained monkeys who have had their amygdala removed were able to recognize objects by sight or touch but could not recognize by sight an object they had just examined by touch[1380]. This indicates that the amygdala is critical for cross-modal association, that is, *sharing and comparing information received from the different senses*. This function is consistent with the fact that the cortical areas of all the major senses feed into the amygdala.

This provides an explanation for the symptoms of the Klüver-Bucy syndrome. When the *amygdala is removed the ability to share information from the different senses is lost.* The monkey sees an object but is unable to recall how the object feels or smells, and even after feeling and smelling is unable to recall how it tastes. As a result he picks up every object he sees and places it in his mouth and smells it. In addition, this combined information cannot be fed into the limbic system for an emotional reaction. Thus, the animal is then not afraid of anything and the emotions are flat since nothing is coming in to color or excite them. No information in — no emotion out. *Removing the amygdala provides a perfect example of what it is like to live without the emotional content of life, provided by the limbic system.* It is not surprising that the amygdala is often the target of psychosurgery done to control emotionally unstable and violent behavior[1260].

The Cingulate Cortex

Problems with sequencing In animals, lesions of the cingulate cortex lead to the disruption of orderly sequencing of behaviors. Motherly tasks such as retrieving pups, gathering nesting materials and nest building may be started but not completed[1854]. Disruption of the cingulate cortex has occasionally been used for the relief of pain in humans. Following surgery these patients also showed problems with temporal ordering or sequencing[570]. ADHD and Tourette syndrome patients often have problems with sequencing[p293].

The Septum

The septal area lies at the center of the limbic system and receives massive inputs from other limbic structures.

Septal rage When the septum of rats or mice is destroyed the animals immediately exhibit rage reactions, hyperactivity and hyperemotionality[209,210]. They show a picture of striking alertness and sudden noises produce an explosive startle reaction. Rapidly approaching objects are attacked immediately with vigorous biting. Attempts to capture or handle the animals are responded to by fierce attacks. Castration of the animals early in life prevents the enhanced emotionality after septal lesions[1181]. Castration of adult animals has no effect, suggesting the action of testosterone on brain development rather than testosterone level per se was the important factor. This could have some bearing on the greater frequency of TS, ADHD and conduct disorder in males compared to females.

Water consumption Animals with septal lesions drink more water than do controls[176,500,859]. This may result from a primary effect on the sensation of thirst[176].

Perseveration An additional effect of septal lesions is the presence of perseveration or inability to "unlearn" a task. These animals have trouble suppressing responses when no response should be made[1,452,453].

Ignoring subtle cues Rats with septal lesions tend to focus rapidly on the most obvious aspects of the environment and ignore subtle cues[501,1093]. In normal animals an intact septal area helps the individual pay attention to the less significant stimuli in the environment, adding subtlety to the emotional climate.

The Hippocampus

The hippocampus is so intimately involved with memory that discussion of the effects of removing of this structure will be covered in the next chapter on memory and learning.

The Hypothalamus

The hypothalamus is a complex area, sitting on top of the pituitary gland, containing many different groups of neurons (nuclei) that control a wide range of functions. Many of these nuclei are *thermostats or regulatory centers concerned with monitoring basic bodily functions* such as eating, drinking, urinating, temperature control, hormonal balance, lactating, control of labor, menstruation and sexual behavior. Lesions of these different nuclei would obviously disrupt these housekeeping functions. Some of the nuclei of the hypothalamus are shown on the next page.

Ventromedial nucleus — hunger, obesity and rage Lesions of this area induce increased hunger and obesity[868] and savage aggression[952]. Sometimes the food intake is voracious and ravenous and may begin even before the anesthesia has completely worn off. Despite this desire for food the animals are

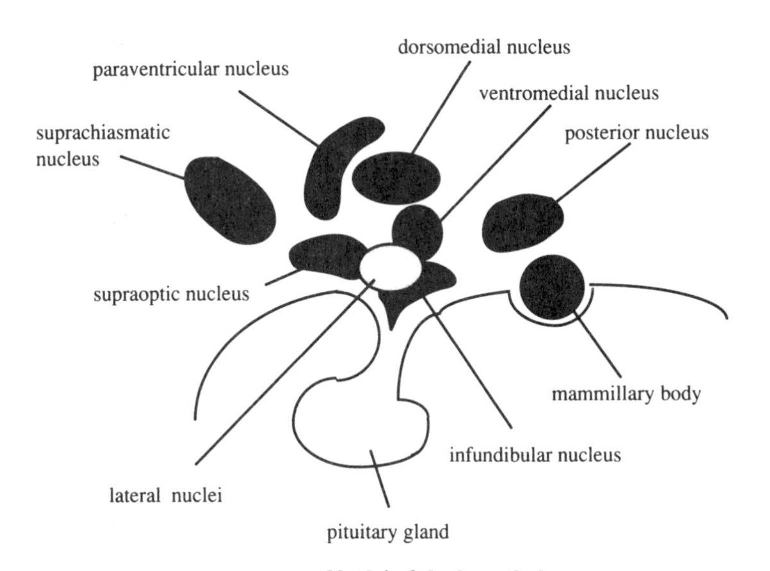

Nuclei of the hypothalamus

sometimes finicky about the taste of the food. They may reject food whose taste has been altered while control animals continue to eat it[1929].

Lateral nuclei — no eating, no drinking and sensory neglect Lesion of the lateral hypothalamus produces the opposite reaction to those of the ventromedial nucleus in that there is a cessation of eating (aphagia) and of drinking (adipsia) to the extent that they will die if they are not force-fed[812,1927]. When the lesion is made on only one side of the hypothalamus, the animals ignore stimuli coming from the opposite side of the body. This is termed *sensory neglect.* In humans it has been described after lesions of the cingulate, frontal, or parietal lobe [p353].

Suprachiasmatic nucleus — the biological clock Many functions of the body change in a rhythmic fashion throughout the day. To time these events the body needs an internal clock. This clock function resides in the suprachiasmatic nucleus and destruction of this area results in abolishing the rhythms of activity, eating, drinking, sleeping and hormone levels[1662].

Anterior versus posterior hypothalamus As a rough generalization the anterior hypothalamus (paraventricular and dorsomedial nuclei) regulate muscle activities associated with the conservation of bodily energies and with digestion and elimination, while the posterior hypothalamus (posterior nuclei) regulate muscle activities directed toward the mobilization and expenditure of the body's resources[952]. These are partly analogous to the parasympathetic and sympathetic parts of what is called the autonomic nervous system. In fact, the hypothalamus has sometimes been called the "head ganglion" of the autonomic nervous system.

This is not a complete listing of the effects of lesions and functions of the hypothalamic region but it is sufficient to illustrate the point that this region is especially important in the regulation of bodily functions involved with emotional reaction (fight or flight). It is a critical member of the limbic system team.

A Summary of Limbic Studies in the Rat

The variation seen in different animals makes it difficult to compare the effects of lesions of different limbic structures across species. Because of this, it is informative to pick a single species and compare the effects of different lesions. This is possible with the rat because of the large number of studies done on this animal. Here the typical responses were hyperactivity to touch, hyperdefensiveness (biting the gloved hand of the experimenter) and increased aggressiveness as measured by killing of mice placed in the cage. A careful review of limbic system lesions in rats gave the following results[20]:

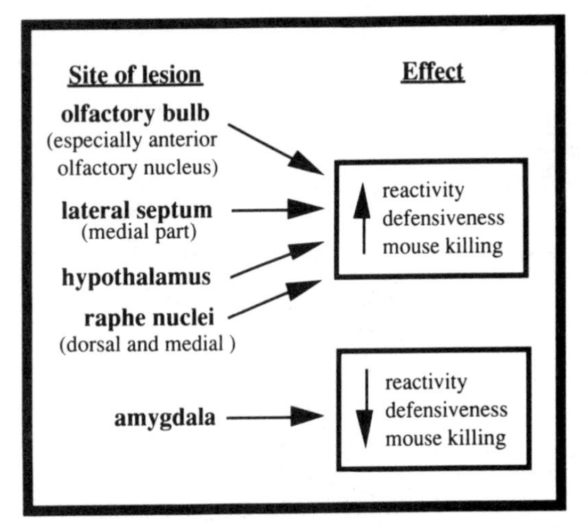

This indicates that the olfactory bulb, lateral septum, hypothalamus and raphe nuclei are all involved in *suppressing the limbic system* behaviors (hyperactivity, anger, rage and aggressiveness), and when these structures are destroyed, this inhibition is lost and these behaviors come out. By contrast, the amygdala *activates the limbic system* and when this structure is destroyed, these behaviors are reduced. Serotonin plays a major role in controlling the function of the raphe nuclei[p445].

Limbic System Lesions in Man

Over the years there have been intermittent reports of the effect on behavior of different limbic lesions in man[155,456]. Since these are rarely specific to one structure their interpretation in terms of learning the role of different structures is more tenuous than in animal studies. However, they are interesting for the insight they provide into the general effects of limbic diseases on human behavior.

One of the first such reports concerned a 19 year old boy who had both temporal lobes removed to control temporal lobe seizures[1955]. These were characterized by terrifying visual hallucinations and paroxysms of aggressive and violent behavior. Several times during these attacks he attempted to strangle his mother or crush his brother under his feet. After the surgery, which included removal of the hippocampus, he could not recognize anyone, including his mother, and had serious problems with both recent and long-term memory. He took possession of all objects he saw, mimicked the manners of people around him, and developed an insatiable appetite. He developed severe problems with reading, was not able to distinguish one word from another, and was unable to write anything spontaneously. He developed sexual exhibitionism and homosexual desires. There was a complete loss of emotional reactivity and he didn't get angry about anything.

A remarkably similar case has been reported where all the major structures of the limbic system were destroyed by a viral encephalitis[644].

> This was a 50 year old man. Following recovery from the acute illness he had severe loss of recent memory. Nothing could be retained from one minute to the next. Learning was abolished. To correct for this he would confabulate, or make up answers and events. He could no longer perform even the simplest calculations but continued to use tools correctly and perform tasks such as driving and filling the gas tank. His handwriting was barely legible. Sustained effective activity was lacking. According to his wife, he lost "whatever it takes to go on and move" in the pursuit of specific goals. Repetitious motor activity was almost constant, such as patting his knee, tapping his foot, humming, or singing bits of songs — some slightly lewd. Remarks were generally sexually flavored and composed of both heterosexual and homo-

sexual content. There were brief episodes of exhibitionism. No aggression, violence, or rage ever occurred. Frequent impulsive behavior was displayed in the form of tactless, inappropriate remarks. Mischievous behavior was frequent, as for example, hiding money underneath a rug.

These two cases had in common severe memory loss, inability to read, write or do math, lack of initiative, motor hyperactivity, hypersexuality, exhibitionism, homosexual tendencies, and a very docile nature.

A TS example One of the most memorable sights in my years of caring for TS patients was the day John, a 26 year old man with a very severe tics, came into the clinic spitting saliva and yelling at the top of his voice — "fuck, fuck, fuck" and at the same time compulsively rubbing his crotch and making wildly obscene forward pelvic thrusts, almost throwing himself across the room. Every time I read that "the limbic system controls the motor aspects of sexual behavior" I think of how John so graphically and violently demonstrated the disinhibition of his limbic system and the explosive display of the hard-wired, stereotyped behaviors of his reptilian complex.

An Analogy

This is only a brief review of some of the major functions of the different structures of the limbic system. Much more extensive coverage can be found elsewhere[952]. Some parts stimulate and others inhibit emotional reactivity. The effect of the different limbic structures on behavior is analogous to a committee deciding how to act. Each member brings their own personality, genetic predisposition and life experiences into the decision and each modulates the behavior of the other. In the end, no one member can be blamed — each shares the blame or the credit for the final action.

Another way to understand what the limbic system does is to pretend that you are trying to make a robot as human as possible. You have completed all the circuits so that input of all the sensors for sight, sound, smell, pain, and movement are properly connected to the other circuits that allow the robot to move. Now if you come after it with a baseball bat it can move out of the way. If it sees and smells smoke it can move out of the burning room. But is it human? Let's try some tests. You yell at it, "Your mother is just a stupid pile of nuts and bolts."

No response.

You kick him in his metal shins every day for three days, and he still doesn't get angry or cringe when you come in the door.

What is wrong. He certainly isn't human! Of course not, you have left out his limbic system. This system

a) remembers things,

b) is hooked into all the sensory inputs so it can process and remember where it has been and what it has experienced,

c) is hooked into the motor system so that under appropriate stimuli it can turn on "hard-wired" ritualistic movements such as these involved in mating, and

d) is hooked into the autonomic system so it can control pulse, blood pressure, gastric juices and related functions.

So you put in a limbic system.

Now, after you kick him in the shins three days running, the next time you come in the door he is hiding behind it and clobbers you with a wrench before you know what hit you. When you impinge the moral character of his mother he turns red in the face, his blood pressure shoots up and he returns the insults in kind and one better. Now he is human.

You must, however, be careful with this system and make sure it is properly wired so its functions are usually *inhibited* and become activated only under socially appropriate situations. To make these decisions, you add a frontal cortex. Without this precaution, the limbic system can become *disinhibited* and its functions performed at inappropriate times. That is what this book is about, *genetically caused disinhibition of the limbic system*: situations where the wiring or chemical control of the limbic system has not been done right.

Summary: Emotion can be considered a way of acting and a way of feeling. The limbic system is the part of the brain that controls our emotions. It receives information from the senses to the outside world (sight, smell, hearing, taste, touch, pain, temperature, and others) and from the senses to the inside world (sex, hunger and others). It compares this recent information to past experiences in memory, thinks about what should be done, and then initiates what is believed to be an appropriate response to both the outside world (action) and the inside world (arousal, anger, fear). Normal behavior requires a normally functioning limbic system. Genetically altered levels of neurotransmitters can lead to inappropriate behavior by causing the limbic system to react too quickly, or not quickly enough.

Chapter 50

Memory and Learning

Memory and learning are intimately related. We acquire new information about our environment through learning. We retain that information through memory. Memory is also a critical aspect of emotion. Since both memory and learning are affected in Tourette syndrome it is important to understand how they normally work.

Types of Memory

There are four types of memory: immediate, recent, long-term and habit.

Immediate memory is very short and constitutes the immediate reservoir of incoming information. It can be tested by determining the ability to immediately recall a sequence of digits. The maximum capacity of the immediate memory is about eight digits.

Recent or short-term memory refers to memory of recent events, that is, things that were done minutes or hours ago.

Long-term memory Items in recent memory must be stored in long-term memory in order for them to be recalled days, months or years later. Cramming for an exam puts things in recent memory. In order to really know the material it should be put in long-term memory.

Habit memory This is a motor type memory. When we practice and re-practice our tennis strokes, the improvement is due to habit memory.

Each of these different types of memory is handled by different nerves and circuits. As a generalization, short-term memory is handled by the limbic system, long-term memory by the cerebral cortex in general, and habit memory by the caudate, putamen, cerebellum[1340,1943] and related structures. The relation among these different types is shown as follows:

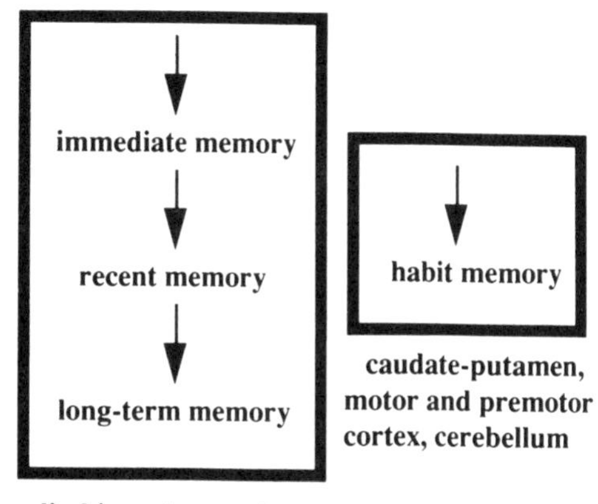

Recent Memory

There are three major regions involved in recent memory — the hippocampus, the diencephalon (mammary bodies, thalamus) and the prefrontal cortex. The hippocampus is particularly important.

The Hippocampus

In 1954 Scoville[1740] described two patients who had brain surgery to control severe seizures or psychotic behavior. Part of the temporal lobe of both sides of the brain had been removed. The left part of Figure 1 shows the part of the temporal lobes, containing primarily the hippocampus, that were removed while the right-hand side shows the normal structures (remember the surgery was done on both sides)[1739]:

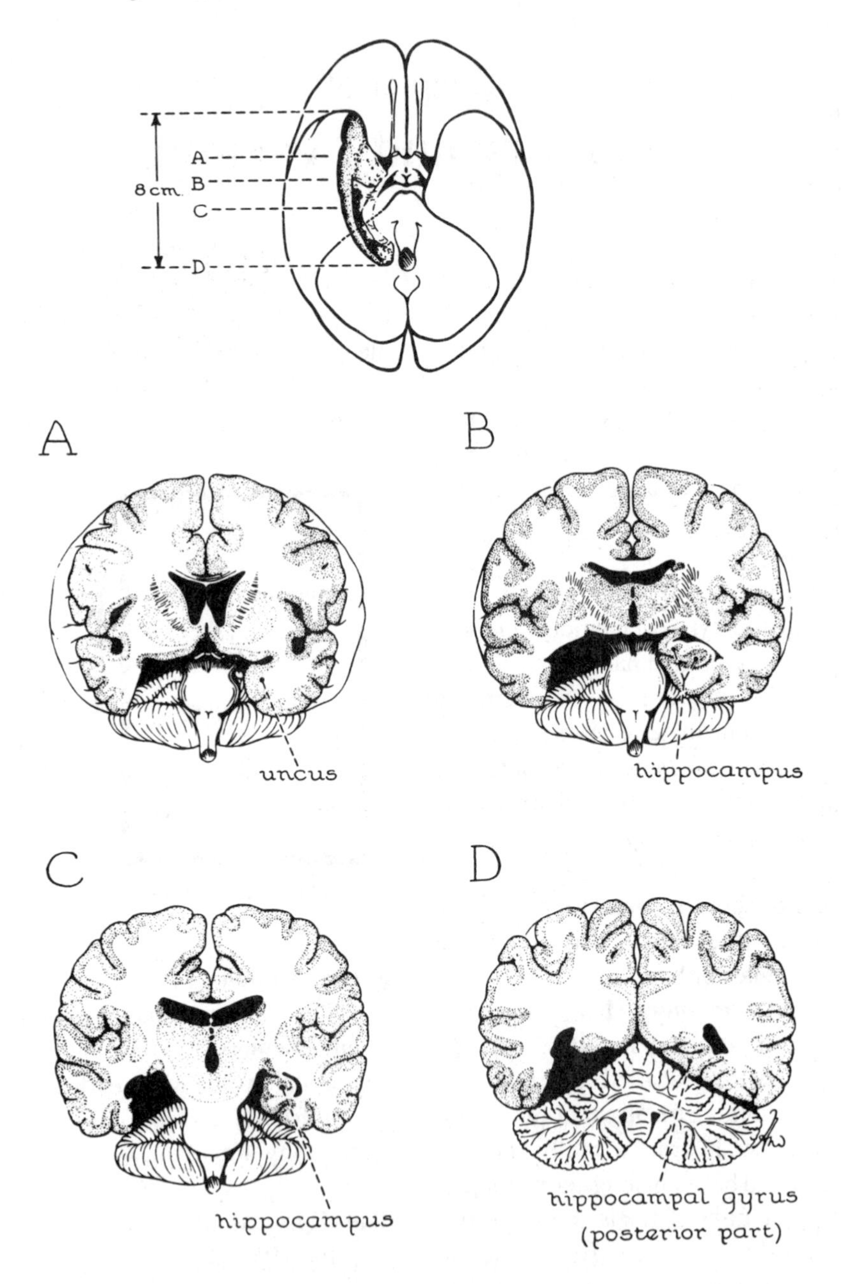

Figure 1. Diagram of the portions of the temporal lobe that were removed (arrows) in the patients described by Scoville and Milner[1739] that resulted in loss of recent memory.

The following is a description of the memory problems in one of these patients, H.M.:

> Ten months ago the family moved from their old house to a new one a few blocks away on the same street; he still had not learned the new address, though remembering the old one perfectly, nor can he be trusted to find his way home alone. Moreover, he does not know where objects in continual use are kept; for example, his mother still has to tell him where to find the lawn mower, even though he may have been using it only the day before. She also states that he will do the same jigsaw puzzles day after day without showing any practice effect and that he will read the same magazines over and over again without finding their contents familiar. This patient has even eaten luncheon in front of one of us without being able to name, a mere half-hour later, a single item of food he had eaten; in fact he could not remember having eaten luncheon at all. Yet to a casual observer this man seems like a relatively normal individual, since his understanding and reasoning are undiminished.

These and similar cases[644,1492] dramatically illustrate the important role of the hippocampus in recent memory.

Alzheimer's disease Other than the surgical removal of the hippocampus the other medical condition which most clearly illustrates the importance of the hippocampus in recent memory is Alzheimer's disease[95]. This is a form of premature senility whose initial major symptom is loss of recent memory. It can be hereditary. The mutant gene is on chromosome 21[761,1885,1921].

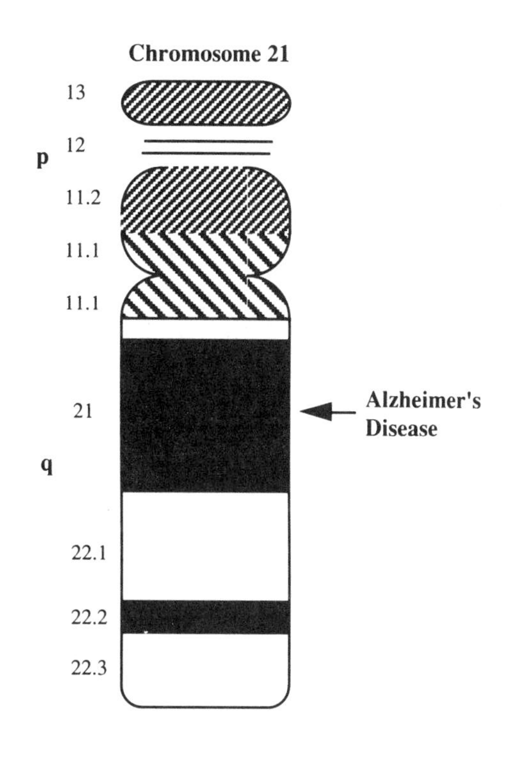

A helical protein is preferentially deposited into parts of the hippocampus, compromising its function and leading to memory loss. The nerves that are destroyed contain acetylcholine[782,2076], i.e., the cholinergic neurons. This emphasizes the important role of this neurotransmitter in memory. The importance of this chemical in memory can also be demonstrated by giving the drug scopolamine. This is an anticholinergic medication that results in temporary amnesia. Because of this it is often used as a preoperative medication before surgery. When patients wake up they have no memory of the operation. Medications that inhibit the breakdown of acetylcholine show promise in the treatment of Alzheimer's disease[1902].

Structure of the hippocampus The hippocampus is a strange looking structure that resembles a snail stuck into the middle of the temporal lobe.

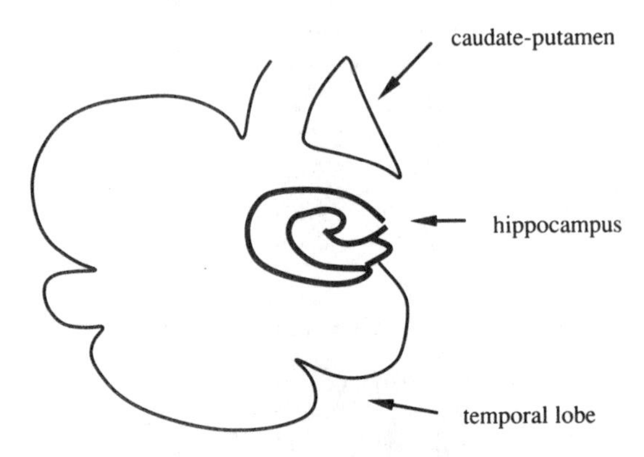

The small, snail-shaped part of the brain has been the focus of an enormous amount of study, with whole books devoted to it. The circuitry is very complex. The multiple overlapping layers are designed not only as a central exchange for recent memory but as a way of comparing new experience to old experience to decide whether an emotional reaction should be allowed or inhibited[790a,2010]. This can be diagramed as follows:

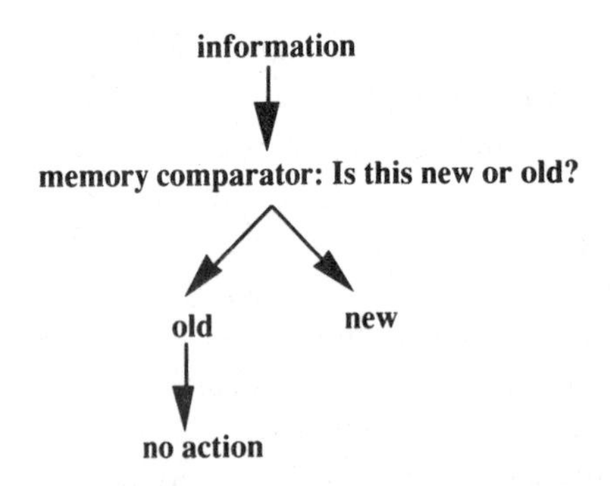

If the information from the sensory inputs is old, no response is required. However, if it is new, a decision must be made as to how to respond.

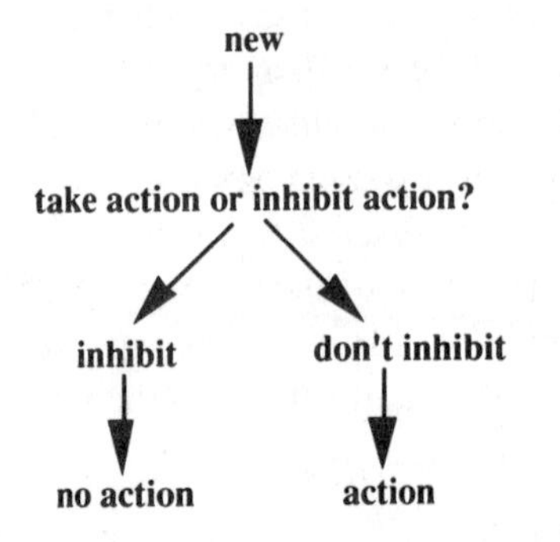

As an example, constant sensory scanning of the environment may tell us that nothing new or unusual is occurring and thus we have no reason to be frightened, anxious, fight or run. However, if the sensory scanners show that a lion has just walked into our tent, that is new. Should we react or not. Again memory stores tell us a reaction is needed — the emotion of fear boils up and we run for our lives. If the limbic system is not functioning properly (disinhibited), this feeling of fear may be present even in the absence of the lion or other dangerous object. If this occurs all the time, it is called general anxiety. If it occurs sporadically and briefly, it is called a panic attack.

Full appreciation of the important role of the hippocampus in recent memory was delayed by fact that removal of the hippocampus in animals did not affect recent memory[1047]. However, when both the amygdala and the hippocampus were removed there was a loss of recent memory[1339].

The Diencephalon

As shown in the chapter on the limbic system, the original Papez circuit shows fibers passing from the hippocampus via the fornix to the mammillary bodies and from there to the thalamus. It is thus not surprising that lesions of the latter two structures can also cause defects in recent memory.

Korsakoff syndrome Chronic alcoholics have much more difficulty remembering recent events than events from their remote past. When such individuals die, examination of their brains shows widespread lesions almost always including the mammillary bodies. This combination of alcoholism, memory defect and lesions of the mammillary bodies is called Korsakoff syndrome, first described 100 years ago[1081]. In addition to memory problems, such individuals tend to exhibit apathy, indifference, loss of initiative, and have difficulty with forming concepts, problem solving, and the meaning of words[264,307]. These features are absent from the memory loss due to destruction of the hippocampus. The dorsomedial nucleus of the thalamus is also involved in Korsakoff syndrome[1842].

The Frontal Lobes

The third major portion of the limbic system involved in memory is the frontal lobe. Lesions in one portion of the frontal lobe (sulcus principalis) result in significant defects in short-term memory[640,1543]. This represents one of the rare instances where similar lesions produce the same effect in both monkeys and humans[1330]. This part of the frontal lobe receives neurons from the portions of the cortex involved in hearing and sight[763-765]. The frontal cortex is clearly one region involved in short-term memory. The frontal lobes play a critical role in the emotions and impulsive behavior and are discussed in the next chapter.

The Cortex and Long-term Memory

The storage area for long-term memory has been more difficult to pin down. Most evidence suggests it utilizes many different regions of the cerebral cortex[954,982]. Studies with PET scanners have shown that so many different neurons are involved in the storage of some events that new concepts of the storage and retrieval of memories may be needed[982]. This widespread location of long-term memory explains why specific memories are not lost following the removal of parts of the cortex. An individual memory may be stored diffusely, like a book with different pages in different rooms, or memory of an individual event may be stored many times, each in a different place, like a whole copy of the same book in different rooms.

Continuing this analogy further, the limbic structures may be like the librarian. She knows where to go in the stacks to get books in the long-term memory (stacks), but can carry only a few books at a time. When the librarian is gone, the visitor may know where to find the volumes that have been there for a long time (long-term memory) but not know where the newest novels (short-term memory) have been put if, in fact, they have been put anywhere.

Habit Memory

Habit memory is closely linked to structures controlling movement. The diagram on the following page shows these structures.

As described in the chapter on the limbic system, this is similar to the primitive R-complex of Paul MacLean. Habit memory, such as how to ride a bicycle or drive a car, remains intact in individuals whose recent memory is completely destroyed by removal of the hippocampus or related limbic structures, indicating that two independent systems are involved [899].

An important aspect of distinguishing these two systems is that the behavioralist school of psychology based much of their conclusions on the use of stimulus-response, and stimulus-response utilizes the habit memory system. Thus, it completely bypassed the limbic system of memory and its intimate links to emotion. No wonder the behavioralist approach was devoid of individuality and feeling and ignored the "mind," "knowledge" and "memory"[1339].

Molecular Mechanism of Memory

Knowing what structures are involved in memory does not tell us how memory takes place. At the level of the nerve cell, memory means that after a given neuron is stimulated, it responds more easily the next time it is stimulated.

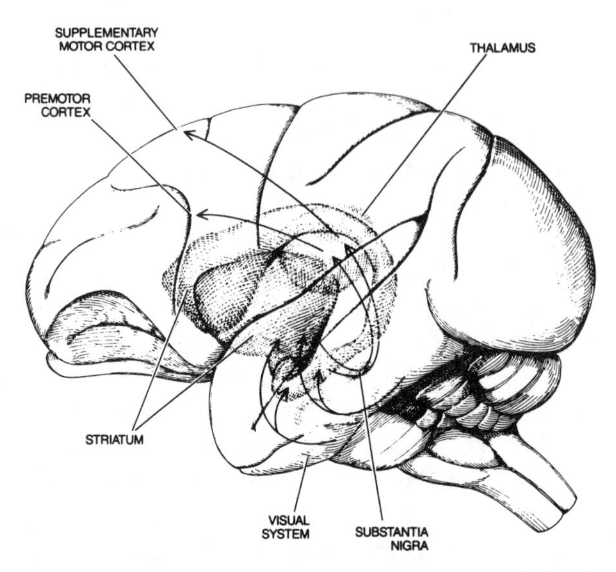

Figure 4. Diagram of the structures involved in habit memory. This includes the dopamine neurons for motor control, i.e., the caudate and putamen, their projection to the motor and premotor cortex and the cerebellum. From M. Mishkin and T. Appenzeller, "The Anatomy of Memory."[1339] Copyright (1987) by Scientific American, Inc. All rights reserved.

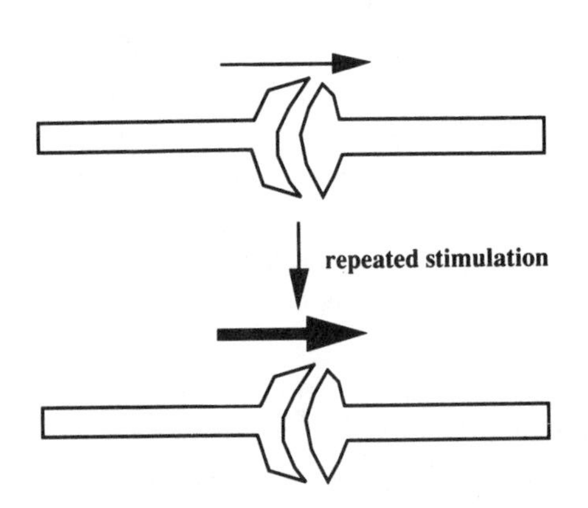

One form of this has been given the term — *long-term potentiation*. Here a brief electrical stimulus produces an increased excitability of the synapse that can last for days or weeks[32,181].

Many theories about the mechanism of short- and long-term memory have been proposed. A recurrent theme is that short-term memory involves temporary changes in the efficiency of events by adding a phosphorous group to one or more molecules in the synapse. This changes the efficiency with which calcium can enter the nerve cell. Long-term memory requires permanent structural changes in the synapse, due to the synthesis of new molecules[173,746,1010] and changes in the shape of the nerve cells. These are called *plastic changes* [1943] and the phenomena is called *neuroplasticity*.

Neuroplasticity

In 1949 Donald O. Hebb[870] proposed that memory and learning are associated with structural changes in the synapses that allow them to function more efficiently. Many studies since then have verified this hypothesis. Lynch and Baudry[1228] have proposed the following mechanism for long-term memory. Multiple nerve stimuli result in the synapse becoming flooded with calcium.

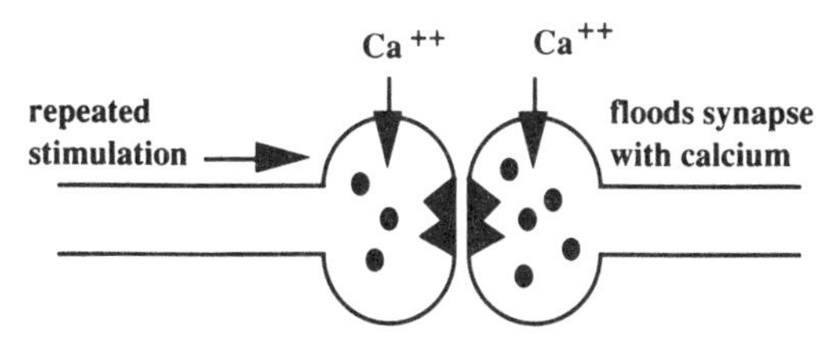

This calcium activates a specific protein degrading enzyme called calpain which then causes an important structural protein of the synapse called fodrin to breakdown.

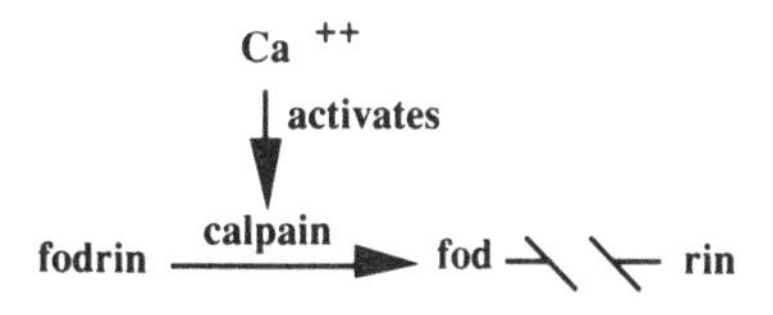

This results in many changes in the structure of the synapse. These changes result in more exposed glutamic acid receptors [N-methyl-D-aspartate or NMDA receptors] and this increases the efficiency at which the dendrite responds to stimulation.

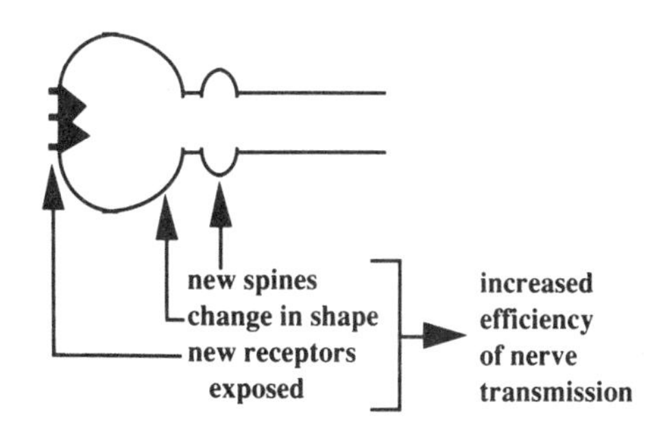

This theory was strengthened by the observation that leupeptin, a compound that specifically inhibits the action of calpain, also inhibits memory[487]. Other structural changes that occur include the formation of new synapses[1507].

Learning is associated with these changes that increase the efficiency of specific nerve pathways. Forgetting, on the other hand, involves the disassembly of these changes. With approximately a hundred trillion (100,000,000,000,000) synapses in the brain, there is plenty of space to store images of past events.

Summary: There are four types of memory: immediate, short-term, long-term and habit. Habit memory is the type used in learning new motor skills and involves the motor parts of the brain. Short-term memory is a function of the limbic system, especially the hippocampus, amygdala, mammary bodies, thalamus and frontal cortex. Long-term memory uses the cortex in general and is brought about by structural changes in the synapses of the nerve pathways involved.

Chapter 51
The Prefrontal Cortex and the Frontal Lobe Syndromes

It is impossible to truly understand Tourette syndrome, attention deficit disorder, conduct disorder and the antisocial personality without understanding the frontal lobes. In many ways they are even more important than the limbic system itself. The prefrontal lobes are the portions of the frontal lobes that lie in the most forward part of the brain in front of the motor areas. In the evolution from mammals to man, the prefrontal cortex has grown more than any other part of the brain. As a result, it is relatively huge in man, constituting a quarter of the entire cerebral cortex. Because of this, it was once thought that the prefrontal lobes must be the site of man's relatively high intelligence. However, since destruction of the prefrontal lobes has little effect on intelligence, they must have some other function. For years this function was shrouded in mystery. It is now clear that the *prefrontal lobes are involved in paying attention, and in the motivation, planning and execution of thoughtful behavior.* As a preview of this chapter, the box on the upper right shows some of these functions.

Prefrontal Lobe Functions

- Attention to tasks
- Analyze consequences of behavior
- Analyze input from the sense organs
- Control the limbic system
- Correct errors
- Goal-directed activity
- Motivation
- Preparation for motor actions
- Programming of complex behaviors
- Recognize mistakes

Because many of these functions are quite subtle it is no wonder it took years to understand the function of this part of the brain. In this chapter I will review many of the studies that led to this understanding.

The Association Areas

Before concentrating on the prefrontal lobes, it is important to introduce two topics — association areas and attention. Different parts of the cortex receive input from the various sense organs, such as vision to the occipital lobes, hearing to the temporal lobes, and touch to different parts of the parietal lobes. These regions are called the *primary sensory cortex. Association areas* refer to regions where neurons converge from one part of the cortex to another (cortical-cortical connections), outside of the primary sensory and motor areas[697]. There are several hierarchies of association areas.

The *primary association areas* allow the processing and integration of information from a single sense organ. The primary visual association areas allow the comparison and integration of visual impulses coming from different fields of vision to build up the whole object seen. These are called *intramodal* associations since a single sensory mode is involved.

The *secondary association areas* allow

the comparison and integration of sensations coming from different primary association areas. Thus, comparing the input from both sight and hearing would help to identify that the animal observed is barking and thus a dog. These are termed *intermodal* or *cross-modal* associations since they compare across different sensory modalities. Language and the ability to name objects depend on cross-modal associations between sight and hearing the word[696].

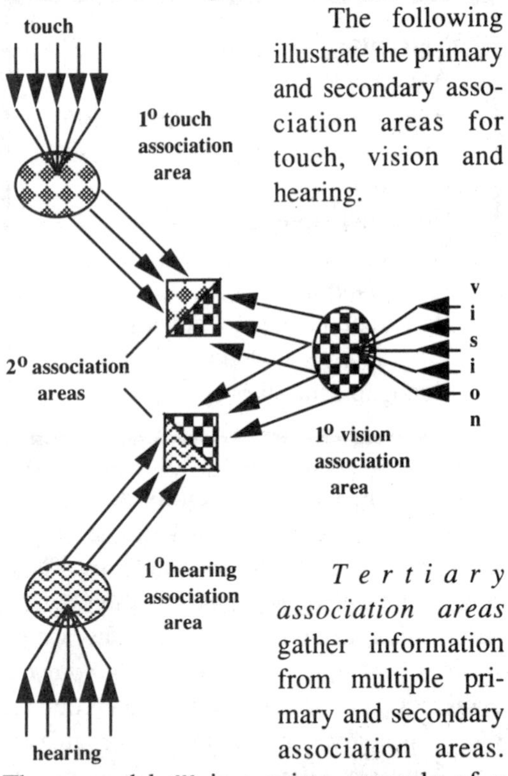

The following illustrate the primary and secondary association areas for touch, vision and hearing.

Tertiary association areas gather information from multiple primary and secondary association areas. The amygdala[p325] is a prime example of a tertiary association area. There is also a major association area in the inferior part of the parietal lobe at the region of the angular gyrus. The association of the primary and secondary regions with the tertiary association regions is **shown at the upper right**.

The inferior parietal area is an association region for sight, hearing and touch. As such it is critical for language and speech and in fact is the *Wernicke receptive speech area*. Lesions in this part of the brain result in *disturbances*

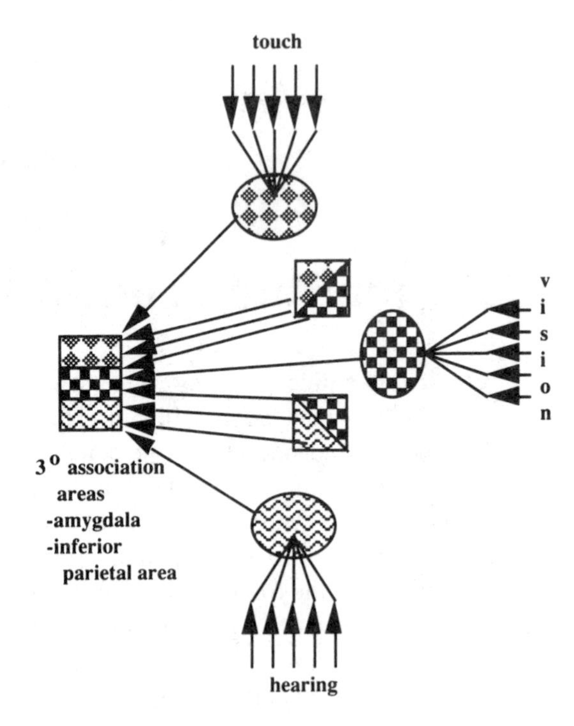

in the comprehension of spoken language, word deafness, difficulties in reading (alexia) and in *handwriting* (agraphia). Spoken speech is normal and fluent except when searching for nouns. The speech problems have been called Wernicke's aphasia[2065].

Finally, the ultimate association area that receives input from all others is the *prefrontal lobe*. The farther an association area is from the primary sensory input, the higher its level of integration. All association regions interact with this grand master association area, the prefrontal cortex (**see next page**).

The layered arrangement of the association areas One of the most striking features of the anatomy of the association areas is their arrangement in layers or modules. This was first noted as a feature of the association areas for touch and sight[932,1369] and is also a characteristic of the limbic and prefrontal lobe association areas[764]. For example, fibers coming from other cortical association areas merge in modular arrays with fibers from the opposite

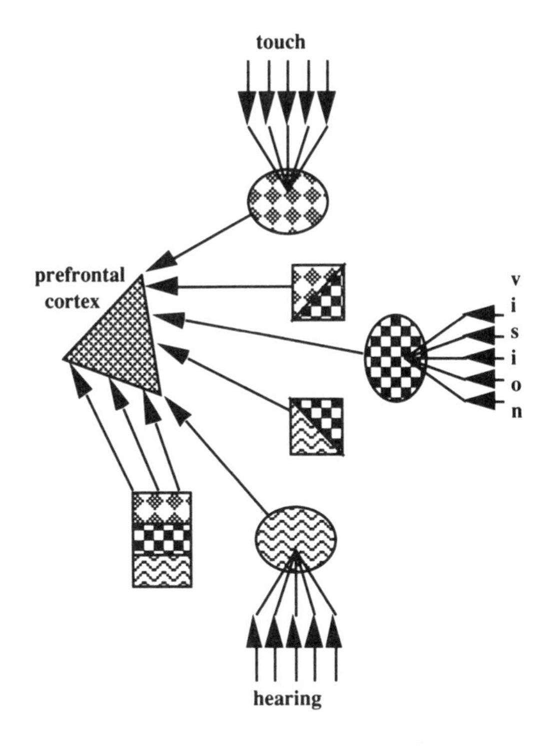

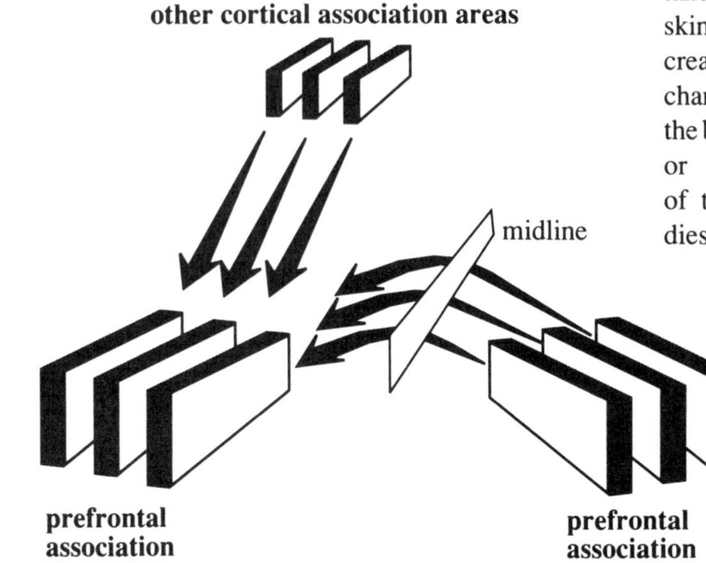

prefrontal lobes **as shown above**.

Like hanging sheets out to dry on rows of clotheslines, such an arrangement provides a large, ordered surface area for the communication and processing of information from two or more areas of the cortex.

> "The increase in number of modular units [in the prefrontal cortex] with increasing cortical surface area may be relevant to the extraordinary expansion of computational and information processing capacity in the human brain"[765].

Attention

It is a giant step between stating what attention means at an intuitive level[p73] and understanding how the brain "pays attention." Significant understanding was provided by the discovery of an arousal reaction[1774] or orienting response[1822].

The orienting response refers to a variety of responses in the body which occur with the appearance of any new or significant stimulus. They include changes in skin resistance to electrical current (galvanic skin response), head turning, decreases in the alpha brain waves, changes in pulse rate, narrowing of the blood vessels of the extremities or widening of the blood vessels of the neck[1049,1287]. This response dies out (habituates) after repeated exposure to the same stimulus, but is reactivated with new stimuli.

Of the various ways to measure the orienting response, galvanic skin response is the easiest[1822]. When parts of the frontal lobes were removed in monkeys they showed hyper responsiveness to new stimuli and the orienting response was absent[814,1048]. Humans with frontal lobe lesions also show a diminished orienting response[1225]. This and other evidence indicate that the orienting response occurs when attention is focused and its absence indicates an inability to focus attention. This response provides an objective measure of the attention deficit disorder present in frontal lobe lesions.

Attention and the association areas The parts of the brain which are most involved in attention are the higher association areas, that is, the temporal, parietal and prefrontal cortex [697,1318,1322]. People with strokes involving these areas have significant problems paying attention and maintaining vigilance. They become distracted by minor stimuli, are unable to maintain a coherent stream of thought or perform specific goal-directed behavior. These patients show behaviors ranging from senseless agitation to apathy and sluggish indifference[1318,1322].

Attention and the right hemispheres The left hemispheres, with Broca's area and Wernicke's area, play a critical role in speech and language. Strokes of these areas affect language but do not cause significant attentional defects. This has lead to the conclusion that the right hemisphere is particularly involved in attentional duties[1318,1322]. This is supported by the observation that when sight or touch is stimulated, brain wave recordings over the touch or visual association areas show greater amplitude over the right side of the brain. The amplitude of these waves are an indication of the degree of attention[835].

In addition to disorders of attention, mood changes are also common in strokes of the right hemisphere associations areas[1318]. This and other evidence[134a,661,1737] have suggested that the right hemisphere is the "emotional brain." There appears to be a parallelism between attention and emotion. Mesulam and Geschwind[1318] have argued that it is no coincidence that the same hemisphere regulates both attention and emotion since the,

> "fundamental role of 'attention' is to direct vigilance toward events which are of emotional or motivational significance to the organism. If it is also assumed that the limbic system is principally responsible for the elaboration of emotion and motivation, then the proper and efficient direction of attention at a cortical level becomes dependent upon the integrity of anatomical connections which convey limbic information to the neocortex"[1318].

Such connections are profuse.

Connections between the Frontal Lobe and the Limbic System

Just as the connections between the amygdala and the different sensory regions of the cortex gave us insight into the function of the amygdala, the connections to and from the prefrontal lobes give us insight into its function. Briefly these connections are as follows:

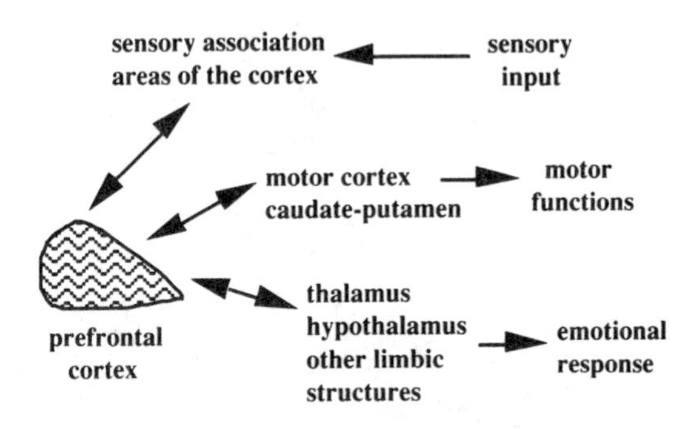

As shown above, the prefrontal lobes receive input from the sensory association areas. They also receive information from many of the limbic structures[1396] and the *pre-frontal lobe is the cortical association area for the limbic system.*. After processing, if a response is required the prefrontal cortex has input to the motor cortex and caudate-putamen for a response requiring movement, to the limbic system for an emotional response and the sensory systems for heightened sensory attention.

Sub-parts of the Frontal Lobe

One of the major components of the cerebral cortex is a triangular cell called the pyramidal cell. The prefrontal lobes are anatomically distinct from the rest of the frontal lobes in that they do not contain these pyramidal cells. Because of this it is often called the *granular frontal cortex,* since under the microscope it has a granular appearance.

The prefrontal lobe is so large that to understand it well we must look at its sub-parts and their different functions. The three major areas are shown in the following figure:

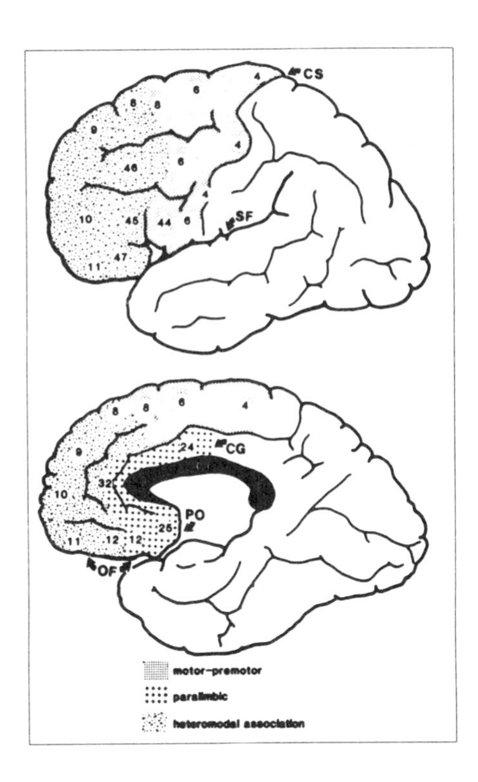

Figure 1. The parts of the frontal lobe. CC = corpus callosum; CG = cingulate region; CS = central sulcus; OF = orbitofrontal region; PO = paraolfactory region; and SF = sylvian fissure. From Mesulam, M.M. Ann. Neurol. 19:320-325,1986[1321].

Motor-premotor region These are areas 4 and 6 and are also called the lateral zones. Since they are closely linked with the part of the brain that controls muscle movement, lesions of these areas result in disturbances in the planning and execution of complex muscle movements such as grasping, lack of movement (akinesia), and uncoordinated movements [237,688,1224,1321]. Since speech is controlled from the left hemisphere, lesions of this area on the left side of the brain cause problems with speech including failure to talk (mutism), stuttering, and poor control of the loudness and rate of speech.

Paralimbic region This is the region that is closely linked to the limbic system and the reticular apparatus (areas 24, 25, and 12). It contains the frontal part of the cingulate cortex (CG) and the smell or olfactory cortex (PO), and is sometimes called the caudal or medial part of the frontal cortex.

Heteromodal or association region This is the high-order association area of the frontal cortex (areas 8-12, 45-47). Lesions of this region show *generalized disinhibition and gross changes in mood.* Luria[1224] described these individuals as follows:

> "The affective disorders, in the form of lack of self-control, violent emotional outbursts and gross changes in character, are among the clearest symptoms. Their intellectual functions remain potentially intact, although actually they are severely disturbed by this increased *disinhibition of their mental processes, which leads to uncontrollable impulsiveness and fragmentation, so that they cannot carry out planned and organized intellectual activity.*"

The prefrontal lobes consist of the last two parts of the frontal lobe — the paralimbic and the association areas.

Prefrontal Lobes in Evolution and Development

The prefrontal lobes were the latest part of the brain to develop in evolution and they are the latest to mature and have their nerves covered with myelin. The other latest developing and latest myelinating region is another major association region, the inferior parietal area[696]. This, in fact, occurs so late that it is not complete until several years after birth. The prefrontal lobes are not fully matured until the child is four to seven years old. This late appearance on the evolutionary and developmental scene is consistent the fact that the prefrontal lobes

> "are in fact a superstructure above all other parts of the cerebral cortex, and perform a far more universal function of *general regulation of behavior*" [1224].
>
> As stated by Grey Walter [2025]:
>
> "with its immensely rich connection to all sources of sensory information, its capacity for economical storage and cross-correlation of information, and its exquisite sensitivity to subtle shades of social and semantic implication, the frontal cortex provides the essential link between the elementary stages of structural evolution and the present phase of psychosocial development of the human organism."

The Frontal Lobe Syndromes

The history of the slowly evolving understanding of the functions of the frontal lobe began in 1868 when Harlow[852] reported a patient, Phineas Gage, who was an efficient and trustworthy railroad foreman.

> He was "a shrewd and smart business man, very energetic and persistent in executing all his plans of operation." One day while blasting a new roadway, an iron tamping bar shot completely through the front part of his head.

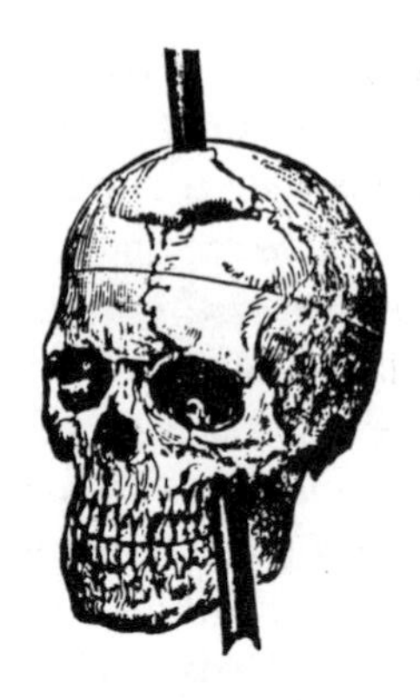

> He "gave a few convulsive motions of his extremities but spoke in a few minutes." His men carried him back to his hotel. "He got out of the cart himself, with little assistance from his men, . . . walked up a long flight of stairs and got in bed..." After a prolonged and stormy course he eventually recovered. Although all of his motor functions were intact his friends stated he was "no longer Gage." "He is fitful, irreverent, indulging at times in the grossest profanity, (which was not previously his custom), manifesting but little deference for his fellows, impatient of restraint or advice when it conflicts with his desires, at times pertinaciously obstinate, yet capricious and vacillating, devising many plans of future operation, which are no sooner arranged than they are abandoned in turn for others appearing more feasible."

World War I injuries The first world war produced many injuries of the frontal lobes. The resulting behavioral changes included childish excitement, impatience, slowing and apathy, both euphoria and depression, short attention span and a deterioration in intellectual capacity[172]. Physical symptoms included urinary and occasionally fecal incontinence[1476].

Pick's disease Like Alzheimer's disease,

Pick's disease is a form of presenile dementia. However, instead of affecting the hippocampus, the primary site of the lesions is the frontal lobe and the tip of the temporal lobe. As such it also serves as a model of frontal lobe pathology. The latest parts of the frontal lobe to develop in evolution (43 and 46) are the earliest to atrophy in Picks disease[1440]. Symptoms include a lack of initiative and spontaneity, disturbance of recent memory, and a disorder of speech. Instinctive grasping and sucking are occasionally seen. Restlessness is prominent, and often there are active erotic inclinations. Stereotyped speech and habit-movements such as swaying rhythmically or smacking the thighs are frequent in the early phases. Perseveration of speech and movements are common[476]. In one report, two of three cases of Klüver-Bucy syndrome reported in humans had Pick's disease[1516].

Removal of the frontal lobes By the late 1930's surgical techniques had improved enough to allow the safe removal of the frontal lobes when indicated because of tumors or other diseases. In 1939 Rylander[1671] reviewed 32 such cases. Of these, 30 showed changes in their behavior. The most common changes were an elevation of mood, a tendency to joke, increased talkativeness, loss of inhibitions, increased concrete thinking with an impaired capacity to make generalizations and do abstract thinking. There may also be a marked loss of motivation and socially inappropriate behavior[559]. One man with bilateral removal of the frontal lobes showed no significant mental changes[869].

Prefrontal lobotomy—Psychosurgery When the frontal lobes are not removed, but the connections with the rest of the brains severed, the operation is called a lobotomy. In 1935, in Lisbon, Portugal, a psychiatrist and a neurosurgeon, Moniz and Lima, performed their first operation on the frontal lobes of a patient in an attempt to cure his psychotic symptoms. When this operation had been completed on 20 patients, seven were recovered from their psychosis, seven were improved, and six were unchanged[1115]. On the basis of this and subsequent work, Moniz was awarded the Nobel prize in medicine in 1949. Similar work was initiated in the United States[638].

The basic procedure was done by drilling holes in the skull and cutting the connection between the frontal lobe and the thalamus and other parts of the brain. This is shown as follows:

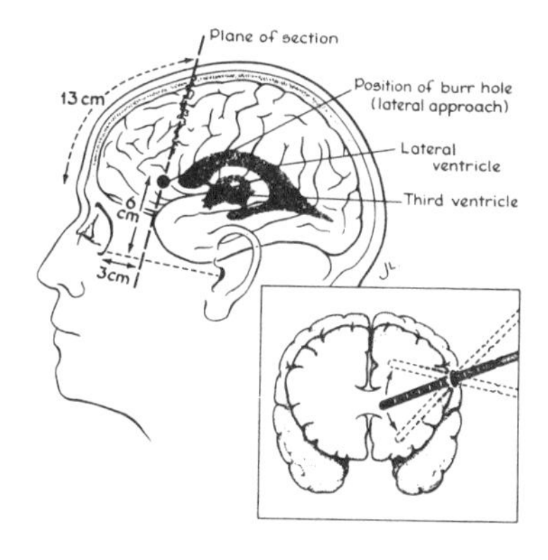

Figure 2. The technique of prefrontal lobotomy. From K. Pribram, *Languages of the Brain.* [1971]

The most uniform result of the surgery was a striking relief of the "psychic pain." Some patients who were totally disabled by their psychotic symptoms and hospitalized for years were able to leave the hospital and return to their former jobs. Patients with chronic anxiety, agitated depression and obsessive-compulsive behavior were particularly improved, and some schizophrenic patients were better. After surgery, some remarked, "I'm well." "The torment is gone." "The voices are quiet"[1115].

In a carefully controlled study of 48 chronically hospitalized psychotic patients, all were anesthetized but only half were actually operated on. Four months later, all were evaluated for release from the hospital by a group of psychiatrists who did not know who had received the surgery. Twenty of the 24 who were operated on were recommended for release compared to only four of those who were not operated on. A wide range of psychological testing of patients after surgery, including IQ tests, showed no defects[1115]. Thus, despite the harsh public verdict about psychosurgery, as epitomized in the movie *One Flew over the Cukoo's Nest,* in the era before antipsychotic drugs were available, the operation benefited many.

Independent of the controversy over ethics, the literature on this procedure, now over 40 years old, provides us with insight into the function of the prefrontal cortex in man. Common themes were that these patients showed little motivation, were often outspoken saying the first thing that came into their minds. They were aware of the fact that they were hasty, undiplomatic and tactless, but couldn't seem to help it. Their families often reported that they were lazy and careless. A leading symptom was a tendency to be governed by the most immediate stimuli. They were "stimulus-bound"[844,1323]. There was an overresponsiveness to immediate impressions and a decrease in the regulation of behavior that normally comes from learned past experience. *The normal bond between immediate stimuli and the maturity of past experience seemed broken.* In this fashion, the loss of anxiety, the bluntly spoken words, and the lack of foresight seem understandable[1115].

Luria

One of the richest sources of information on the function of the frontal lobes comes from the studies of the famous Russian neuropsychologist Aleksandr Romanovich Luria[1224]. The following are some of his observations.

Initiation of programs and attention Because of its rich connections, the prefrontal lobes are in a particularly favorable position to receive the impulses coming in from all the senses and "examine" and synthesize this information preparatory to making complex plans. Luria stated that the *prefrontal lobe* plays "a decisive role in the formation of *intentions* and *programs*, and in the regulation and verification of the most complex forms of human behavior"[1224].

For any mental task to take place, a certain level of cortical tone is necessary and this cortical tone must be modified in accordance both with the task to be accomplished and the stage of activity reached. The first important function of the frontal lobe is to regulate this state of activity[1224]. The frontal lobes participate in the regulation of the activation processes lying at the basis of voluntary attention.

Speech Luria was intrigued with the role of *speech* in human behavior. Of all the behaviors that distinguish man from other animals, speech is one of the most striking. The frontal lobes are critical to the planning, programming and initiation of actions, and one behavior that especially requires thought and planning is speech.

Grey Walter, a Russian physiologist, showed that every act of expectancy evokes a characteristic slow wave (expectancy waves) which increases in size while awaiting an expected stimulus. These waves appear first and primarily in the frontal lobes. Luria concluded from these and other studies that in man *the frontal lobes participate directly in the state of increased activation which accompanies all forms of conscious activity. It is the prefrontal lobes which evoke this activation and enable the complex programming, control, and verification of human conscious activity, which require the optimal tone of cortical processes to take place* [1224].

Speechless apathy The most massive frontal lobe damage is associated with an inert, speechless passivity. Such patients "usually lie completely passively, express no wishes or desires and take no necessary action. This is not a cessation of all behavior since they may join in a conversation between other people. It is the higher order function of the initiation of action that is severely impaired[1224].

Echo movements A variation of this picking up on action initiated by others is the presence of echo movements or echopraxia. For example, if the examiner says to a patient, "When I raise my fist you must raise your finger," the patient will raise his fist, echoing the motion rather than initiating raising his finger.

Perseveration Perseveration refers to the inability to change a course of action with the result that the same movement, or behavior, or words are done over and over. Luria[1224,1225] gives two examples of perseveration in patients with frontal lobe lesions. One, when asked to draw a circle continued drawing circle after circle.

Another was asked to draw first a square then a circle. He could draw squares but could not then change his course of action and instead continued to draw another square.

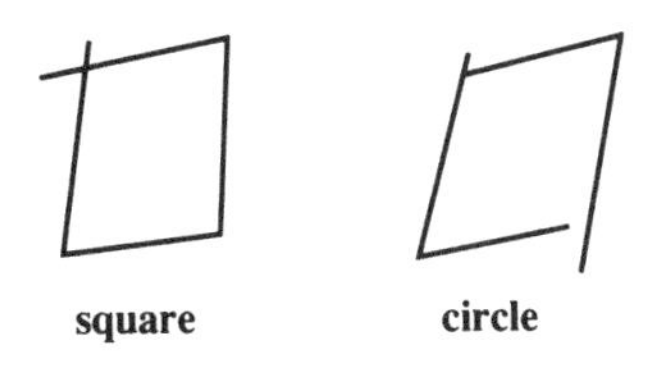

Luria's comment[1224] on such behavior was that these patients "not only lose their assigned program, replacing it by 'basic' or 'echopraxic' action or by pathologically inert stereotypes, but they also fail to notice their mistakes. In other words, they not only lose control over their actions, but also lose the ability to check their results."

Thus, there is a disturbance of the ability to compare the result of one's action with one's original intent. The intriguing feature of this defect is that if the patient is made to think that the behavior was done by someone else, he can recognize the mistake. *He is only incapable of recognizing his own mistakes.* What a distinctly human failing and how reminiscent of the denial commonly seen in some Tourette syndrome patients.

Problems with complex programs A patient was asked to raise his hand. Although he initially did so he was unable to sustain the movement and it soon fell. If his hand was first placed under a blanket, so that a much more complex set of actions was necessary, he could not carry out the movement. Another patient was asked to fetch some cigarettes from the ward. He began carrying out this instruction but when distracted by patients walking in the opposite direction, turned and followed them[1224].

Problems with complex memory While simple memory tasks are not affected by lesions of the frontal lobes, more complex ones are. For example, normal individuals can memorize a list of many nouns after several practice rounds, each of which adds some additional words until they are all memorized. By contrast, in frontal lobe lesions, instead of progressively increasing, the number of retained words remains at around four. This is shown in the following figure[1224].

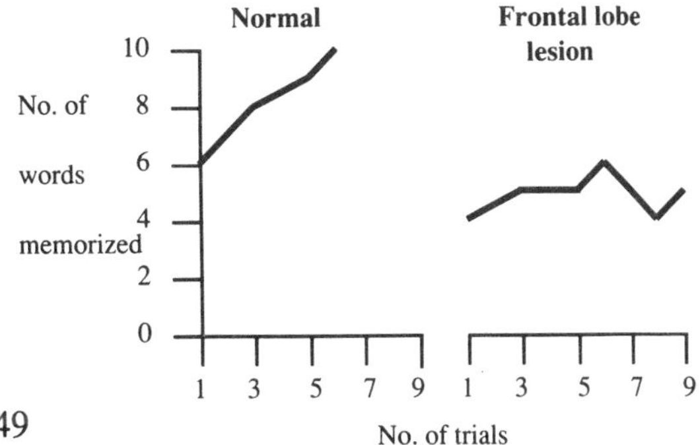

The perseveration causes problems with complex recall tasks. If a patient is asked to remember the phrase, "the cat's name is Frisky," then "her mother went to town," and is then asked to recall the first phrase, the patient with a frontal lobe lesion will perseveratively repeat "her mother went to town."

Problems with complex arithmetic Although simple addition and subtraction cause no problems, frontal lobe lesions result in difficulties with more complex arithmetic which involves the need to hold some intermediate results in their head. For example, the request to successively subtract 7 from 100 may result in perseveration on the first answer 93...83...73.

Teuber and Attention

Hans-Lukas Teuber[1934] suggested that some of the functions of the frontal lobe could be better understood by inverting our thinking from the usual thought that sensory inputs lead to motor movements (sensory -> frontal lobe -> motor) to the reverse idea that the frontal lobe prepares the sensory areas for anticipated action (frontal lobe -> sensory area sensitization). This is closely allied to the idea of selectively focusing our attention on a specific type of sensory input.

An objective example of this is the description by two Russian neurologists[1396] who showed that verbal instructions to wait for a visual signal resulted in increased brain waves over the sensory cortex. In patients with frontal lobe lesions this activation of the sensory cortex was absent. In more recent times these have been described as "expectancy waves"[2025] and "preparatory sets"; the latter is defined as a state of readiness to receive a stimulus or make a response[563]. Behaviors requiring preparatory sets rely on an intact lateral frontal region[765,1900].

Teuber's views help to explain some earlier thoughts about the function of the frontal lobe described in terms of the "will" where will subjectively involves our,

> "anticipation of the future course of events coupled with an awareness of the role of the self in bringing these events about."

Teuber would restate this by saying that the patient with frontal lobe lesions

> "is not altogether devoid of the capacity to anticipate a course of events, but cannot picture himself in relation to those events as a potential agent."

In relation to the problems with lying in some TS patients, Teuber also described a frontal lobe patient with such pathological lying that he called it *mythomania,* telling tall tales without any obvious striving for gain.

Lhermitte's Studies

After a study of 75 patients with frontal lobe lesions, the French neurologist Lhermitte and his colleagues[1171-1173] described two types of related behavior. One they called *utilization behavior,* referring to an exaggerated dependency on the immediate environment. When objects are presented to patients with frontal lobe lesions they magnetically grasp them and use them. The other was called *imitation behavior* or a tendency of patients to imitate the examiner's gestures and behavior, even when asked to stop. Both are variations of the concept of "sensory binding" described under prefrontal lobotomy. A distinction was made between imitation behavior and echopraxia in that echopraxia is an automatic, reflex imitation of movements of others, while imitation behavior had more voluntary control in that the patients did it because they thought they were supposed to. The sight of a movement is perceived in the patient's mind as an order to imitate it. The sight of an object implies an order to use it. There was often a lack of self-criticism that allows patients to perform purposeless gestures and imitate ridiculous or socially unacceptable acts.

These two behaviors combined to produce what Lhermitte[1172] termed an *"environmental dependency syndrome."* Some of the behaviors of individuals with frontal lobe lesions were best observed by using their daily activities as a laboratory. A 52 year old women had her left frontal lobe removed because of a tumor. After surgery she resumed work but showed a *lack of initiative.* The following are some examples of her behavior.

When taken to the examiner's apartment and shown a table with

> "makeup products, she used the powder and eye makeup immediately. She got up to put on the lipstick and to look in the mirror to examine her appearance. She then saw the balls of yarn and knitting needles and began to knit. When she spotted sewing needles, spools of thread, and pieces of fabric, she put on her glasses and began sewing in a precise manner. In the kitchen, she swept the floor, after spotting the broom; when she saw dishes in the sink, she washed them."

Another patient walked into the bedroom.

> "The bedspread had been taken off and the top sheet had been turned back in the usual way. When the patient saw this, he immediately began to get undressed. He got into bed, pulled the sheet up to his neck, and prepared to go to sleep."

Some of the features of the environmental dependency syndrome were: 1) A dependence of the patient's behavior on the social and physical environment. 2) A mental intertia and apathy in the sense that without directions or cues nothing happened, while with directions or cues everything happened. 3) There was a loss of self-criticism as indicated by the patient inappropriately removing his clothing and going to bed.

These symptoms were interpreted as follows: sensations pour into the sensory association centers in the parietal and temporal lobes, and pass to the frontal lobe where they are filtered out and rational decisions are made about which ones to react to.

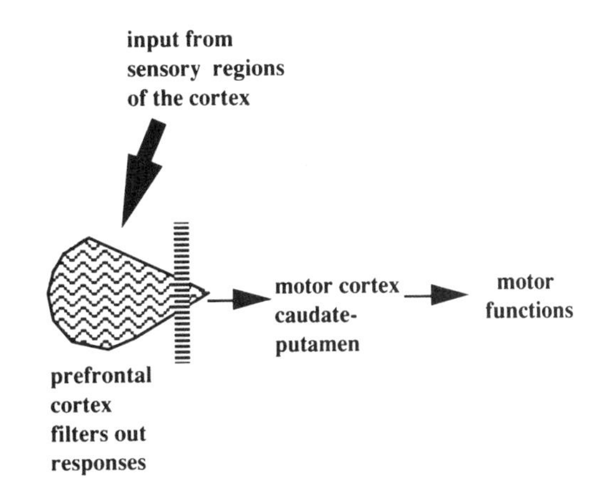

When the frontal lobes are not functioning well, all stimuli are reacted to in an uncontrolled manner.

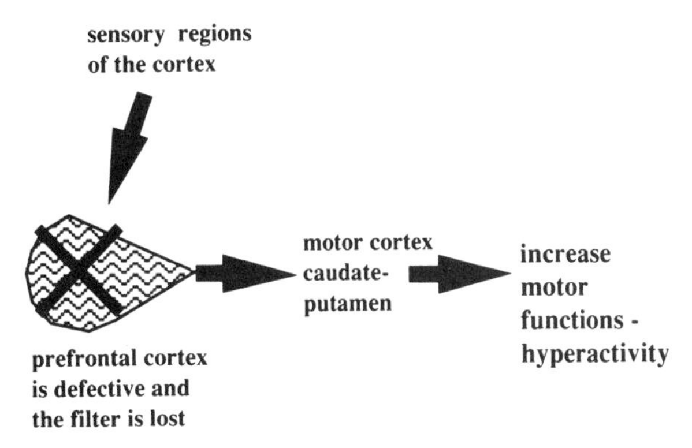

The result is highly imitative, stereotyped, repetitious, sensory-bound, impulsive behavior and hyperactivity. This is similar to children with ADHD or TS. The role of frontal lobe dysfunctions in these syndromes is discussed later.

Memory for time and space While memory per se is usually well preserved in patients with frontal lobe lesions, memory for objects oriented in space or time is often impaired[1708].

Frontal Lobe Syndromes in Animals

Pavlov, the Russian physiologist who brought us the conditioned reflex[1464], also studied the effect of the removal of the frontal lobe in dogs. Such dogs respond to irrelevant stimuli. When they see leaves which have fallen on the ground they seize on them, chew them and spit them out. They do not recognize their master and are distracted by irrelevant stimuli. They respond to any element of the environment by uninhibitable reflexes and these distractions disturb the plans and programs of their behavior, making it fragmentary and uncontrolled. Sometimes the animal's planned, goal-directed behavior is interrupted by the senseless repetition of established inert stereotypes. Based on these observations, Pavlov stated that the frontal lobes play an essential role in the "*synthesis of goal-directed movements*."

Much earlier in 1905, another Russian, Bekhterev[141], had come to a similar conclusion that the frontal lobes play an important role in

> "the correct assessment of external impressions and the goal directed choice of movement in accordance with such an assessment,"

thereby performing a "psychoregulatory activity"[1224,1225].

Delayed responses In an early of study of the effects of frontal lobe removal in primates, Jacobsen[967] concluded that such animals can perform simple behavioral acts following direct instructions but cannot synthesize information arriving from a variety of sensory inputs, that is, behaviors that rely on the synthesis of sensations stored in memory cannot be performed.

The testing of the monkeys was done as follows: The monkey was allowed to watch as food was placed under one of two or more cups on a table. An opaque screen was then lowered in front of the cups for one to two minutes. The screen was then raised and the monkey had to pick up the cup under which the food was placed. This part was easy, indicating memory was intact. However, if the food was alternated between the right and the left cup, the operated monkeys tended to perseverate on a particular choice, picking the cup under which the food was first placed.

Stimulus-bound In a related experiment, if different shapes of cups were used the operated monkeys preferred any novel object over the tried and familiar one, even if the experiment was arranged so that the novel cup never contained the food. This result indicated the monkeys were "stimulus-bound," always preferring novel stimuli[1249,1539,1540,2044] and unable to ignore distracting stimuli. These observations indicate that *destruction of the prefrontal cortex leads to a profound disturbance of complex behavioral programs and a marked increase in immediate responses to irrelevant stimuli* [237,238,1076,1224,1225].

Pribram examined the role of the prefrontal lobes in monkeys. One of his interesting observations was that animals without frontal lobes were unable to maintain a state of "active anticipation" and *could not tolerate long pauses while awaiting* certain experimental stimuli[1539,1540]. He concluded that the frontal lobes are responsible for the orientation of an animal's behavior not only to the present, *but also to the future*. As such, it regulated one of the most complex forms of active behavior. It was also shown that the destruction of the frontal lobes made the animal *unable to assess and correct errors* it had made. Such an animal loses its organized and purposive character[1539-1541].

Hyperactivity When the frontal lobes are removed from monkeys they display a *driven hyperactivity*, pacing like a caged animal[1030]. When movement was measured by a pressure-sensitive pad on the bottom of the cage, normal monkeys showed the following:

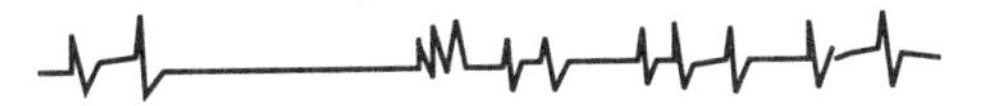

When the frontal association areas (9-12) were removed, the recordings then looked like this:

Similar hyperactivity has been noted following the ablation the frontal association areas in a wide variety of animals including man[218,1030].

The hyperactive monkeys were also highly distractible. While sounds had little effect, there was a remarkable connection between hyperactivity and sight. When the lights were turned out, all the hyperactivity stopped. When the lights were left on but sight prevented, the hyperactivity also stopped. If the eyes were left uncovered but the occipital lobes (the primary sensory area for vision) were removed, the hyperactivity stopped. If the lights were left on for 48 hours, the monkeys were hyperactive for the full time. This suggests that hyperactive children might also do better in more subdued light.

As shown later, the selective destruction of dopamine nerves in the frontal lobes also results in hyperactivity[p389].

Unilateral neglect This is an intriguing variation of the frontal lobe syndrome produced by quite a small lesion on one side of the prefrontal association cortex. The figure **on the right ->** shows how small these lesions can be.

After this lesion, the monkey neglected or seemed not to sense anything that was happening on the opposite side of his body. Objects brought into view on that side were not seen or touched. Sounds were not heard. Food placed in that side of the mouth could not be tasted and enjoyed. Touch was not felt. Dexterity of movement was impaired on the same side as the lesion[2053]. Lesions of the cingulate cortex can produce the same picture[2033].

Unilateral neglect has been described in man with similar small lesions of the frontal and parietal lobes[879]. A case has been described of an alcoholic with severe unilateral neglect. During a withdrawal period he suffered from hallucinations that were seen only on the side not affected by the unilateral neglect[1318]. This syndrome superbly illustrates the importance of the prefrontal lobes in collecting and assimilating different types of sensory information and its role in motor coordination.

Orientation in space A particularly ingenious one is the Morris water maze test[1362]. A large tub is partially filled with cold water and a platform is placed under the surface. When rats are placed in the water, they swim randomly until they find the platform, then climb on it to get away from the water[1058].

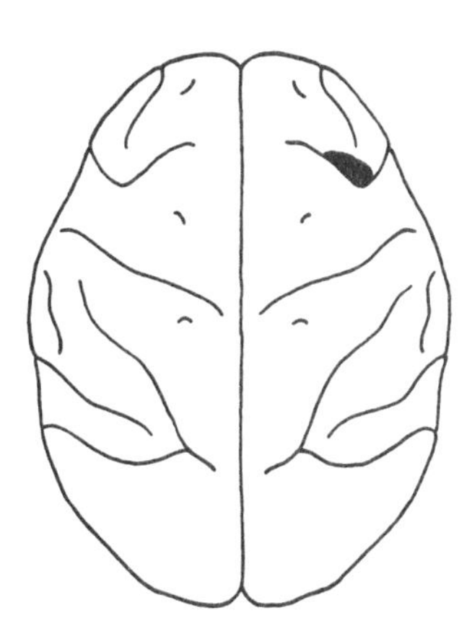

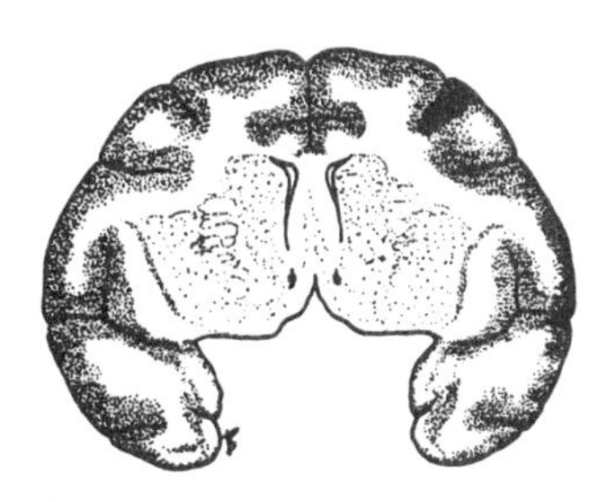

Figure 3. Site of the lesions in monkey prefrontal lobe resulting in unilateral neglect syndrome. From Welch and Stuteville, Brain (Oxford Univ. Press) 81:341, 1958[2053]

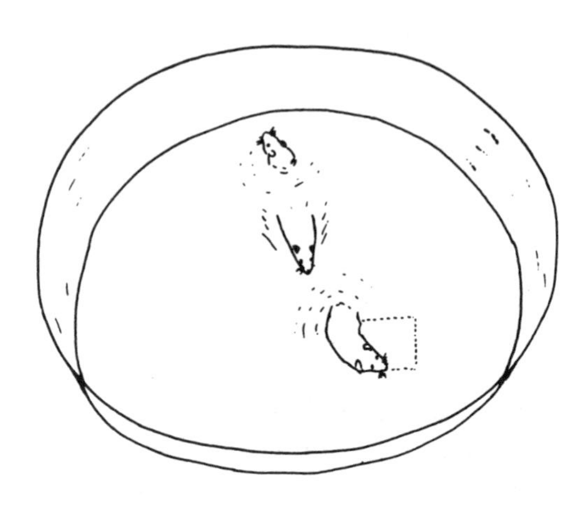

Figure 4. The Morris water maze. (From Klob et al., *Behavioral Neuroscience*, 1983[1058].)

After one trial normal rats quickly find the platform again. However, as shown below, rats with prefrontal lobe lesions require four to five trials to learn the location of the platform[1058].

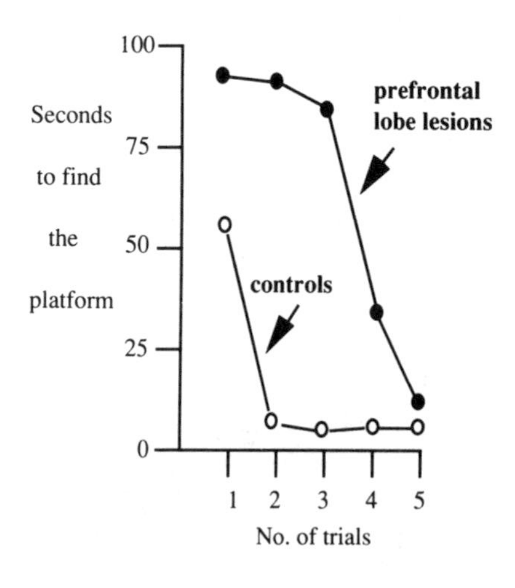

Other tests of spatial orientation were also abnormal[1057a].

The prefrontal lobes and the R-complex In the chapter on the limbic system I discussed the R-complex[p324]. This is synonymous with the caudate and putamen and is often call the *neostriatum*. What the prefrontal lobes do for the integration of sensory information the neostriatum does for the integration of motor functions to produce complex behaviors[493,1133,1652]. The *neostriatum can be considered the association area for the motor cortex.* It is striking that many of the same tests that were abnormal in monkeys after removal of parts of their prefrontal cortex were also abnormal after lesions of the part of the caudate that connects to the prefrontal lobe[493]. In fact, in monkeys, Rosvold[1652] found

> "no exception to the rule that if a frontal lobe lesion results in impairment of a certain test, so does a caudate lesion."

The same is also true for cats and rats[493]. For example, in the following figure, where *either part of the prefrontal lobes or the caudate was removed in cats,* the delayed response task and delayed alternation task was more abnormal with lesions of the caudate than with lesions of the prefrontal cortex[492].

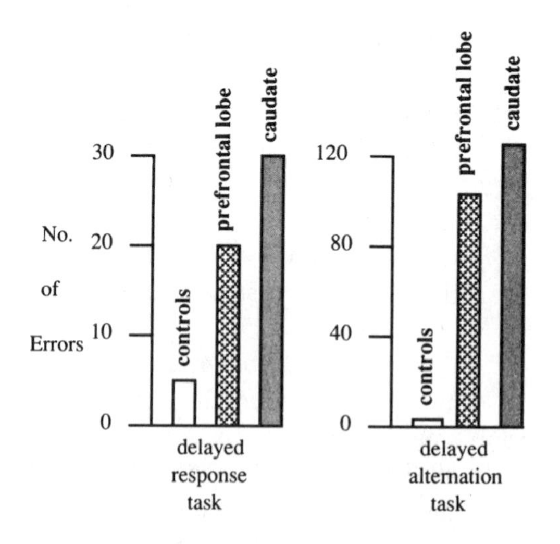

This is a reflection of the important connection of nerve fibers from the prefrontal lobe to the caudate. As will be discussed in the chapters on dopamine, these are dopamine neurons.

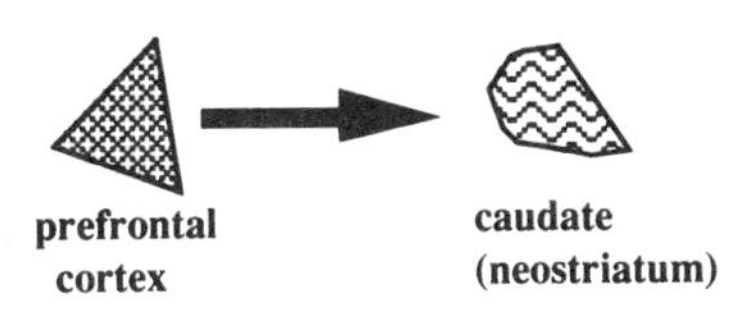

Combining data from many experiments indicates that the prefrontal cortex and the caudate:

- perform different operations, both of which are needed for the same behaviors
- operate in series, that is, the caudate functions requires the input of the prefrontal lobes, and
- some of the functions of the prefrontal lobes and the caudate are interdependent. This especially applies to complex sets of behavior.

Rosvold[1652a] put the relationship between the frontal lobes and the neostriatum eloquently when he said,

> *"both the frontal cortex and the caudate nucleus are parts of a superb guidance system capable of matching against a plan of action the direction, intensity and speed of body movement towards a goal, estimating error, and effecting correction in response mechanisms."*

Tests of the Frontal Lobe

Looking at some of the tests used to detect frontal lobe disorders gives us further insight into the functions of this part of the brain.

The go no-go test One of the remarkable aspects of frontal lobe syndromes is impulsivity or an inability to inhibit an inappropriate behavior. This is well demonstrated in the so-called "go no-go" test. It was first used to test monkeys with various parts of their frontal lobes removed, especially the orbital regions[263]. Before surgery, the monkeys were trained to pick up food after a specific type of sound or tone. This was the "go" task. They were then trained to not pick up the food after a different tone or a visual cue. This was the "no-go" task. When the association areas of the prefrontal cortex were removed the monkeys lost the ability to inhibit responding to the no-go stimulus.

The go no-go test is also sensitive to frontal lobe damage in humans[510,1153,1224]. In one report, a 50 year old woman with a large tumor of the medial side of both frontal lobes was asked to lift her finger in response to a single noise (the go task), but not lift her finger after a double noise (no-go task). She responded normally to the go task but could not inhibit lifting her finger to the double noise. She also had a short attention span and did poorly on the trail making test (see below). After the tumor was removed, both of these tests and her attention span returned to normal[1153,1224].

Luria[1224] noted that lesions of the same medial surface of the frontal lobes produced a disturbance of voluntary attention such that verbal instructions no longer exerted a regulatory influence over motor functions.

Wisconsin Card Sorting Test The subject is given a set of 128 stimulus cards which show objects that differ in three ways: color, shape and number. Four of the cards are as follows:

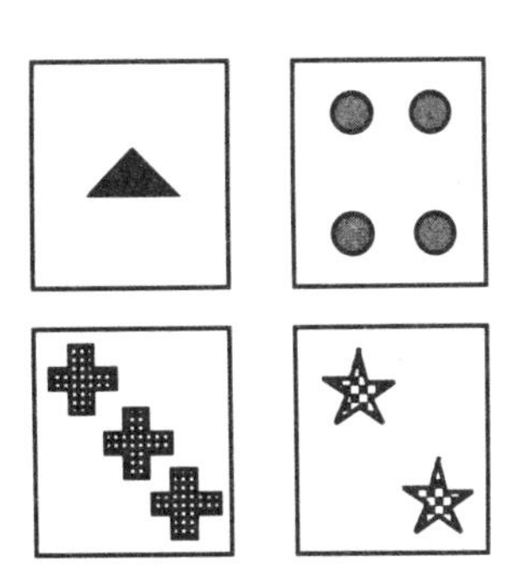

The subject is instructed to place each new card under one of these four cards and the examiner will inform him or her whether the choice is "right" or "wrong." The subject is told they can use this information to try to get as many cards right as possible. The examiner arbitrarily chooses to have the initial "right" response to be what ever criteria the subject first uses, such as color. After there have been ten consecutive correct sorts, the examiner changes to another criteria, such as shape, without telling the subject. After ten more consecutive correct responses, the criteria is switched to number. The process continues until six sorting categories are complete or the cards are used up. Patients with lesions of the lateral portion of the prefrontal lobes tend to perseverate on the first sorting criteria and finish significantly fewer categories and make more total errors than controls or individuals with other types of brain lesions[509,510,1332,1400,1634]. *They are unable to change sorting plans in mid stream.*

The trail making test, part B This is a paper and pencil test in which the subject is required to draw a line connecting numbers 1 to 13 and letters A to G in alternating sequential order (1-A-2-B- etc.). *It requires attention and an ability to switch plans and withhold inappropriate responses* [2043]. It is poorly performed in individuals with frontal lobe lesions and children with ADHD.

The Porteus maze test This test[1527,1528] is similar in principle to the trail making test. The patient is required to follow a correct, visible but circuitous path thru a maze. Individuals with frontal lobe lesions make numerous errors, reflecting an impulsive inability of restraining themselves from rushing head on to the goal. Like the trail making test, it is performed poorly in individuals with frontal lobe lesions and in children with ADHD[505].

Delayed response test This is done in the same manner as with monkeys except that verbal instructions are given and money is used instead of food, an indication of what is more important for people[633]. The subject watches as money is placed under one of two black dishes. A curtain is then lowered, blocking the view for up to 60 seconds. When raised, the subject is requested to lift the dish covering the money. Alternate dishes are used and patients with bilateral frontal lobe lesions perseverate by constantly picking the original dish[633].

The famous French child psychologist Piaget used a similar approach to test infants. This was the AB Object Permanence Test to evaluate the capacity of the subject to *recognize that an object exists in time and space* when it is not in view[633].

Delayed alteration test Here the curtain is lowered and the money is placed under both dishes out of sight of the subject. The curtain is then raised and whichever dish is chosen will contain money. The curtain is lowered again and the money is always placed under the dish opposite to the one last chosen and left there for as many trials as is required for the subject to pick the correct one. The perseveration of patients with bilateral frontal lobe lesions leads to many errors[633].

Word fluency test Here the patient is allowed five minutes to write down as many words starting with S they can think of. They are then given four minutes to write down as many four-letter words starting with C they can think of, and so forth. This test is sensitive to damage of the dominant (usually left) frontal lobe[1332].

Brain Imaging

The above evidence on the functions of the frontal and prefrontal lobes has been based on somewhat artificial situations of the effect of injuries, tumors, and surgical lesions in man and animals. A much more direct way of determining the functions of the frontal lobes, especially in man, is by the technique of brain imaging. This is made possible by the fact that when certain regions of the brain are called upon to function, the rate of metabolism of the cells involved increases. This can be determined by making oxygen, glucose or the blood radioactive, and measuring the rate at which different small areas of the brain accumulate this radioactivity. One of these techniques is called PET scanning, which stands for positron emission tomography.

The beauty of this technique is that the subject can be asked to perform different tasks involving different senses and different responses, and then it can be determined what parts of the brain are called upon to carry out the tasks. When used to examine the function of the frontal lobes, the results were striking and confirmed the conclusions reached by the other techniques[1407,1638-1640]. One of the most remarkable findings was that the frontal lobe can be divided into many sub-divisions and regions, each taking on a slightly different part of the work of the brain. These studies showed at least *17 functionally different areas of the frontal lobe* [1638]. These could be divided into the superior, midfrontal, intermediate, inferior, polar and orbital regions. The functions of these regions is reviewed by a series of diagrams on this and the next page. The granular or prefrontal cortex is shaded more darkly than the remaining dysgranular frontal cortex:

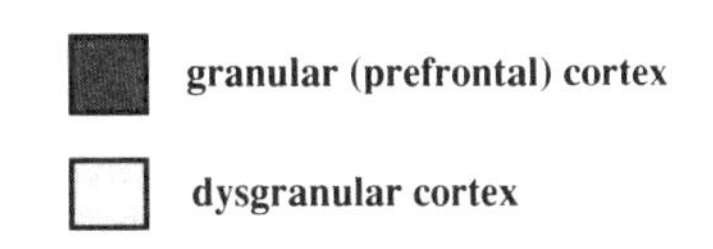

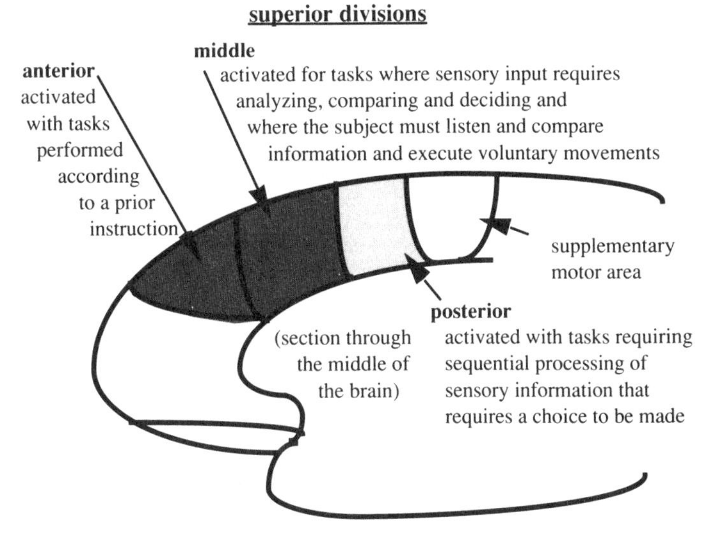

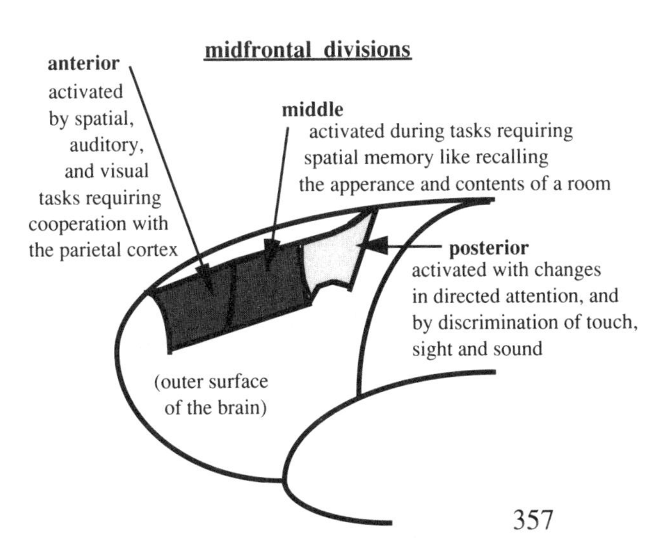

The frontal lobe eye fields[234] (next page) participate in the ability of the eye to quickly fixate on objects such as words on a page. These rapid fixations are called saccades. The partial control of this activity by the frontal lobe is consistent with its role in attention. Abnormalities in these eye movements have been seen in schizophrenia and other psychiatric disorders[p208]. This and the defects of attention in schizophrenia implicate a role of the frontal lobes in this disorder[1160,1161].

Based on the studies summarized in these diagrams, Roland[1638] states:

"What does the human frontal lobe do? The best answer is that it participates in any form of structured brain work a subject can do when awake. Perception, voluntary action, thinking, remembering, calculating,

reading, discriminating, and speaking are all conscious activities that are characterized by an activation of prefrontal areas in conjunction with either motor or sensory areas."

intermediate divisions

anterior activated with tasks requiring discrimination of auditory signals such as listening to a story to be retold later

posterior activated with tasks requiring visual discrimination

(outer surface of the brain)

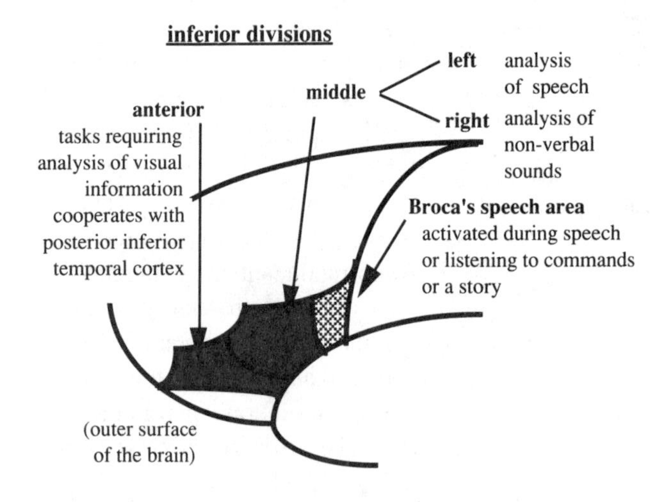

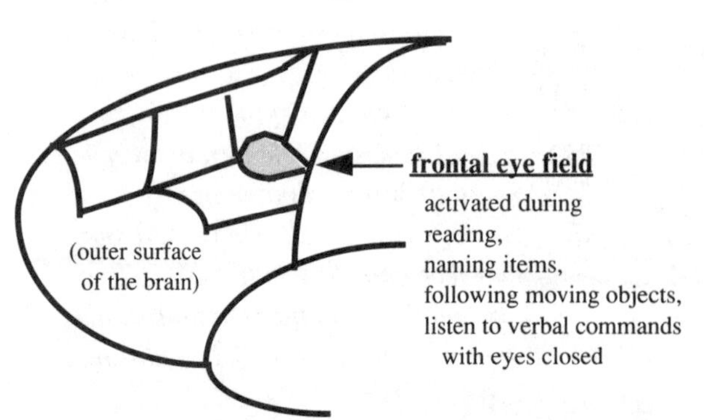

ADHD — Tourette Syndrome and the Frontal Lobe Syndromes

By now you have undoubtedly noticed why I have spent so much time on the frontal lobes and the frontal lobe syndromes. There are many striking similarities between the behavior of children with ADHD or Tourette syndrome and different aspects of the frontal lobe syndrome. This similarity has been commented on many times[345,373,516,561,1271,1526,1853,2143]. Anneliese Pontius[1525,1526] was particularly intrigued by two characteristics of patients with frontal lobe pathology:

1. Their "inability to construct a plan of action ahead of the act, to sketch out a goal of action, to keep it in mind for some time (as an overriding idea) and to follow it through in actions under the constructive guidance of such planning."

2. Their "inability to reprogram an ongoing activity and to shift within principles of action whenever necessary."

She emphasized the observations of Milner[1332], based on the Wisconsin Card Sorting Test, that patients with frontal lobe pathology knew what had to be done, "I knew it had to be the color, the shape or the number," but were unable to do it. *Their perseveration on the first choice overrode all cognitive choice.* In psychiatry this has been called "*ego dystonic*," that is, individuals act even when their ego tells them not to. Impulsive ADHD children may also show a dissociation between "knowing" and "doing." They may say "I know I'm not supposed to do that but I can't help it."

Both the decreased orienting response [345,1702,1841,2141] and decreased brain wave responses to stimuli (evoked potentials)[385,843,1550,2143] indicate that the ability to focus attention is impaired in individuals with ADHD. Both of these responses are modulated by the frontal lobe.

Similarities between the frontal lobe syndrome and ADHD The following is a list of symptoms that are characteristic of both the frontal lobe syndrome and ADHD — Tourette syndrome, as well as conduct disorder and antisocial personality.

Sensory and limbic association areas
(paralimbic and heteromodal regions of the prefrontal lobe)

- cannot forsee the consequences of their behavior
- concrete thinking
- decreased brain wave signs of focused attention (evoked potential)
- decreased orienting responses
- distractible
- do poorly on Porteus maze test
- do poorly on trail making test
- do poorly on Wisconsin Card Sorting Test
- echolalia
- echopraxia
- enuresis and encopresis
- hyperactivity
- impatient and easily bored
- impulsivity
- lethargy (in hypoactive ADD)
- lying
- normal IQ
- perseveration
- PET scan shows decreased activity in frontal lobes (right)
- poor concepts of time and space
- poor judgment
- poor memorization
- poor motivation
- problems with complex math
- sequencing impaired
- short attention span
- stimulus-bound
- unable to change plans
- unable to formulate complex plans

Premotor areas
(motor-premotor areas of the frontal lobes)

- choose not to speak (selective mutism)
- delayed onset of speech
- poor motor coordination
- singsong talking
- stuttering and blocking
- talk too fast
- talk too loud

Some of these features also apply to individuals with schizophrenia[115, 1459,2148].

Some examples in Tourette syndrome While many of these symptoms in TS children have been well covered elsewhere, some additional comments by parents are of interest in that they are so relevant to the frontal lobe.

"He can't tolerate any sudden change in plans."

"If I tell him to do three things, I'm lucky if he even does the first item on the list."

Chemical Lesions of the Prefrontal Lobes

In this chapter I have reviewed the accumulated knowledge about the frontal and prefrontal lobes that came about as a result of studies of various physical lesions of the frontal lobes such as trauma, tumors, surgery, and strokes. These lesions are relatively rare. In the context of this book, *an infinitely more common and important type of lesion is a chemical-genetic lesion of the frontal lobe.* Consistent with the wide range of symptoms of Tourette syndrome, chemical lesions are capable of affecting much larger areas of the brain since many of the chemical neurotransmitters are widespread. Such lesions can also be much more subtle so the tests for frontal lobe dysfunction may also be more mildly affected. It is intriguing that *the two neurotransmitters most often implicated in TS, dopamine and serotonin, are also the two major neurotransmitters of the frontal lobe*. The extremely important role of these neurotransmitters in causing genetic dysfunctions of the prefrontal lobe will be described in the chapters that follow on dopamine and serotonin.

A physical and a chemical frontal lobe syndrome in one family The following family illustrates the similarity between the classical frontal lobe syndrome and the behavior of TS/ADHD individuals.

> Tim was a 20 year old young man who had spitting and snorting tics as long as he or his parents could remember. He didn't speak until he was three years old and then often spoke so fast no one could understand him. In grade school his teachers always said he was very active, didn't pay attention and was not working up to his potential. The problems in school did not seem to be out of hand until one day his mother received a visit from his teacher who handed her of copy of *Dare to Discipline,* and suggested she try it. Unruly behavior and truancy continued until he was expelled from high school as a senior for "irresponsible behavior." His mother's characterization of him was "it is only pure luck that he is not in prison by now."

The pedigree is **on the right:**

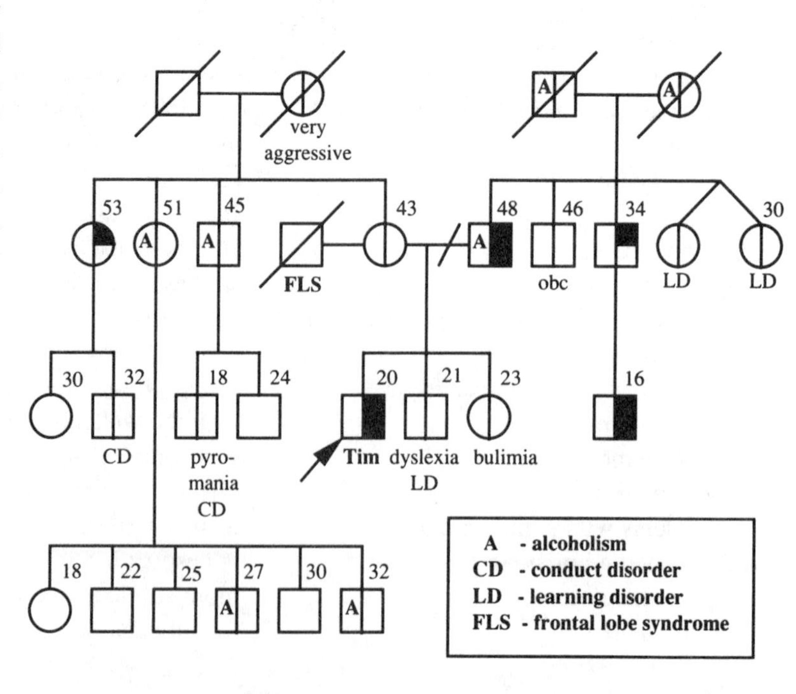

His brother had dyslexia and learning disorders, his sister bulimia, and his father had TS and was a recovered alcoholic. Both of his father's parents were alcoholics. This pedigree further illustrates the interaction of *Gts* genes with alcoholism. On his mother's side there were four individuals with alcoholism and two with conduct disorder, one of whom also had compulsive pyromania and an aunt with chronic motor tics.

After I had discussed the diagnosis and the behavioral spectrum of TS, the mother told me that after her divorce she remarried the local sheriff. Everything went well until he was viciously beaten about the head by an escaping prisoner. This resulted in severe damage to the temporal lobe and the opposite frontal lobe, requiring two brain surgeries to relieve bleeding. After he recovered from the surgery he became very impulsive, irritable and prone to violent temper tantrums. He began to drink excessively, and drive cars very fast. Within a year he was killed in an automobile accident.

His mother said with some wonder that in

the height of her problems she realized that *"my son's and my husband's behavior were identical!"* This is a fortutitous illustration of the similarity of the physical and chemical frontal lobe syndrome.

Summary: The prefrontal lobes perform the highest levels of integration in the brain and act as the central processor for information coming in from the senses. They, along with the limbic system, make plans for action that will be passed out to the motor areas. The prefrontal lobes contain the association cortex that is master over all other lesser association regions.

When the prefrontal lobes don't work properly there is a major impact on the ability to pay attention, to make plans and to change plans. Reactions become thoughtless and impulsive. There may be no motivation to make plans or if made they may be quickly abandoned.

This central role of the prefrontal lobes can be appreciated from the following figure that summarizes the grand design of the human brain.

The Grand Design of the Human Brain

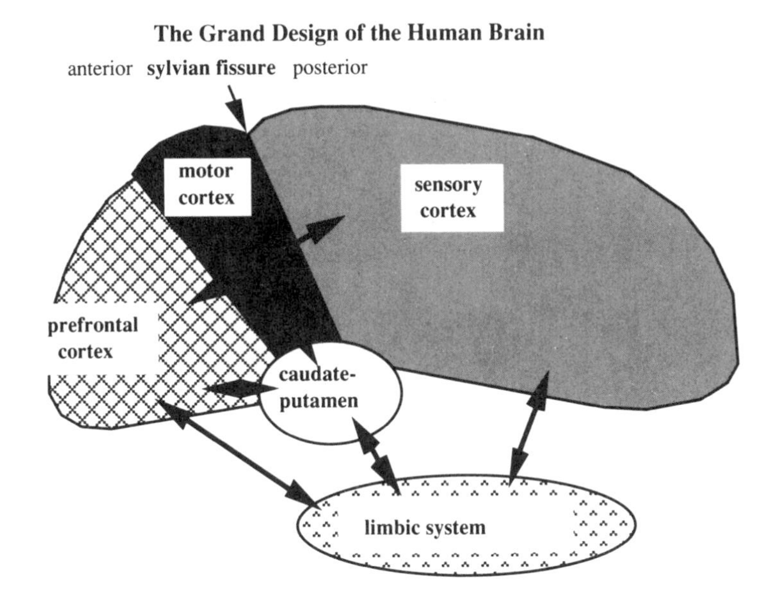

Chapter 52
Dopamine — The Regulator of Motor Activity

Dopamine, norepinephrine and serotonin are the three neurotransmitters most often assumed to play a major role in many different disorders of behavior. As a rough generalization, the caudate and putamen (striatum) and frontal lobes are rich in dopamine, the hypothalamus is rich in norepinephrine and serotonin, and the limbic system and frontal lobes are rich in serotonin. The striatum is particularly involved in the control of complex motor activities.

The Chemistry of Dopamine

Tyrosine is an amino acid that is normally used in making proteins. Amino acids are made up of an acid part, —COOH, and an amino part' —NH_2.With tyrosine the third part is a phenol ring structure. Thus —

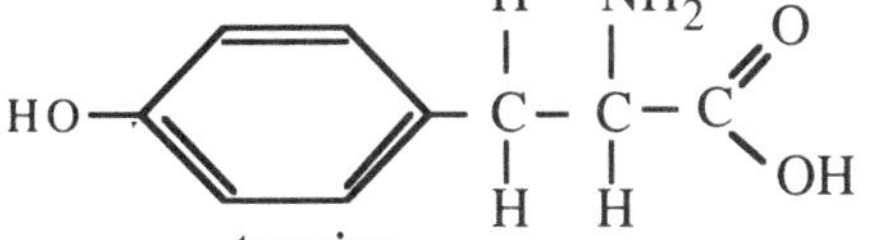

HO—[ring]—C(H)(H)—C(H)(NH_2)—C(=O)OH

tyrosine

Tyrosine also serves as the starting point in the synthesis of dopamine. An enzyme called *tyrosine hydroxylase* adds an —OH group to form DOPA (3,4-di hydroxy phenyl alanine).

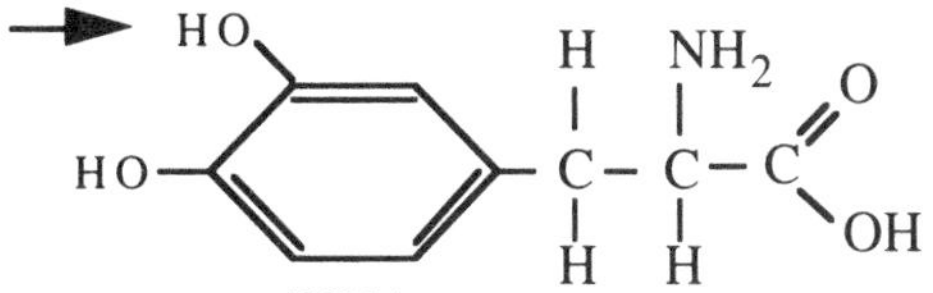

→ HO, HO—[ring]—C(H)(H)—C(H)(NH_2)—C(=O)OH

DOPA

This is what is called the rate-limiting step in the formation of dopamine since once DOPA is formed it'is quickly converted by another enzyme into **dopamine** by the removal of the acid group, —COOH.

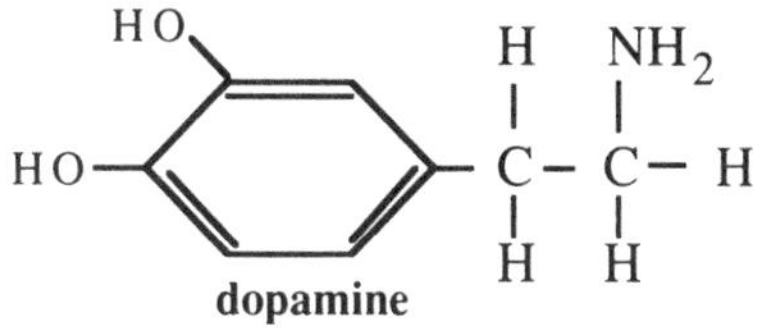

HO, HO—[ring]—C(H)(H)—C(H)(NH_2)—H

dopamine

Since the phenol group with the two —OH groups is called *catechol,* and the catechol is attached to an *amine* group, NH_2, these compounds are often called **catechol amines**.

In order to understand the later chapters, it is necessary to know what happens to dopamine, that is, how is it broken down. This is done by an important enzyme called *monamine oxidase,* which replaces the amino group (-NH_2) with an acid group (-COOH) to form DOPAC (di-hydroxy phenyl acetic acid).

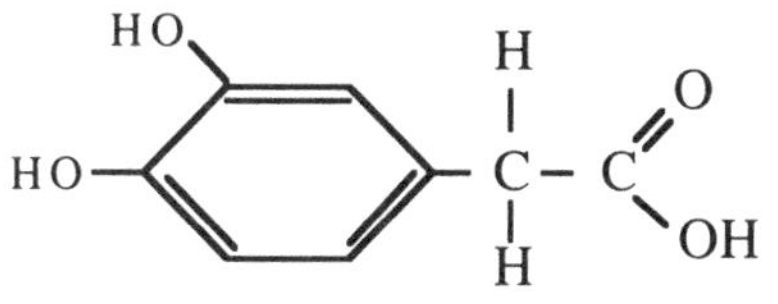

HO, HO—[ring]—C(H)(H)—C(=O)OH

DOPAC

Finally, an enzyme called COMT (catechol-O-methyltransferase) transfers a methyl group, —CH_3, onto the catechol to form HVA or homovanillic acid.

homovalilic acid
(HVA)

The rate at which HVA is formed is a measure of the activity of dopamine neurons. In later chapters you will hear about HVA almost as much as dopamine.

In summary, the formation and breakdown of dopamine is as follows:

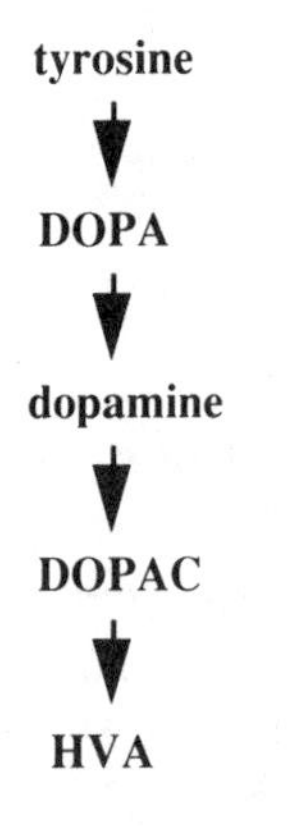

The step between DOPA and dopamine is aided by vitamin B_6.

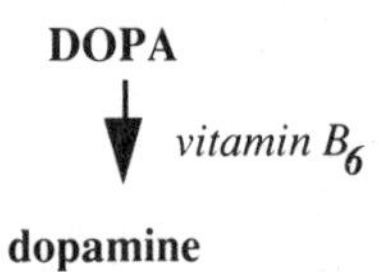

The Anatomy of Dopamine Nerves

There are three major dopamine nerve pathways in the brain. The first leads from an area in the brain stem called the *substantia nigra* to the caudate and putamen or striatum, or R-complex[p324]. This is called the *nigrostriatal pathway.*

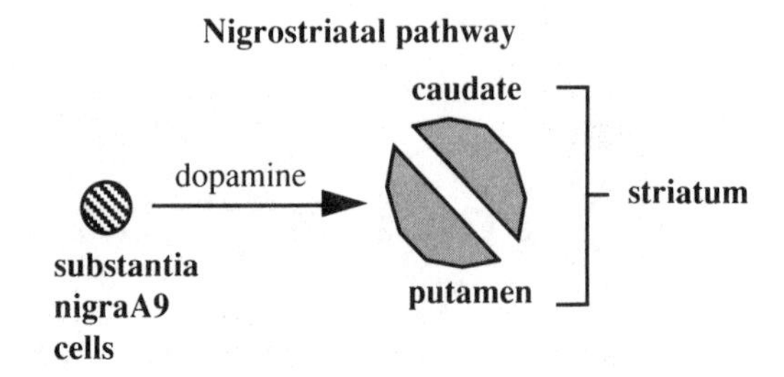

The the other two pathways pass from a region called the *ventral tegmental area* (VTA) in the brain stem to the prefrontal lobes and the limbic system. These are called the *mesocortical* and the *mesolimbic* pathways.

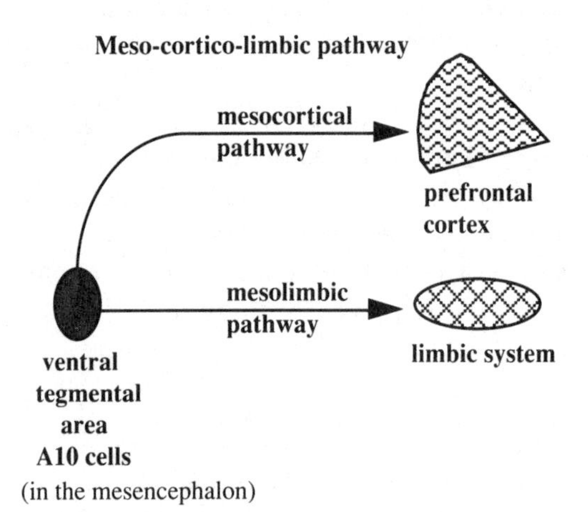

There is a third, minor pathway leading to the hypothalamus and pituitary. This is called the *tuberoinfundibular pathway.* These are illustrated in Figure 1 (**next page**).

The striatum, composed of the caudate and putamen, looks like a germinating seed planted in the middle of the brain as illustrated in Figure 2 (**next page**).

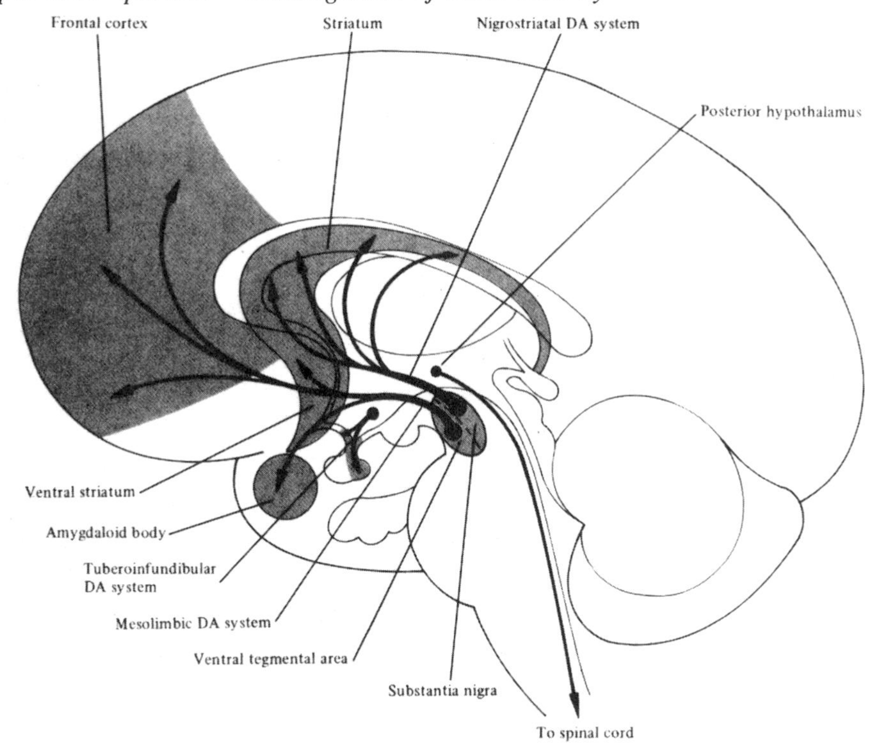

Figure 1. The dopamine pathways in the brain. From L. Heimer, *The Human Brain and Spinal Cord.* Springer-Verlag, New York, 1983.

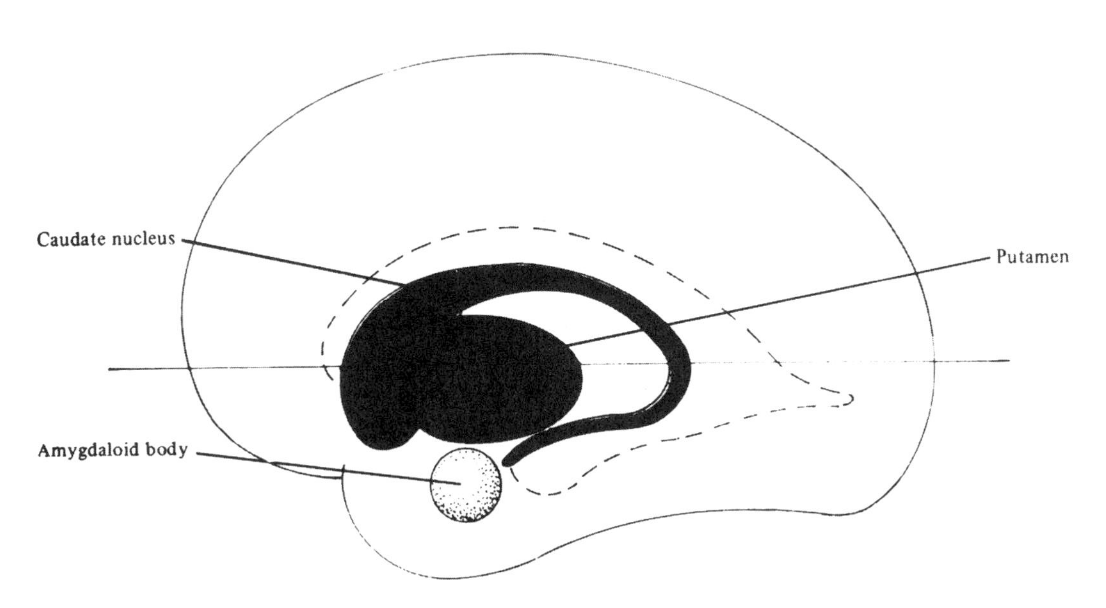

Figure 2. The caudate and putamen viewed from the side of the brain. From L. Heimer, *The Human Brain and Spinal Cord.* Springer-Verlag, New York, 1983.

The following figure of a horizontal section of the brain (see horizontal line in Figure 2) shows the caudate and putamen in black. Abnormalities of dopamine in these pathways are believed to play an important role in schizophrenia, Tourette syndrome, and other behavioral disorders.

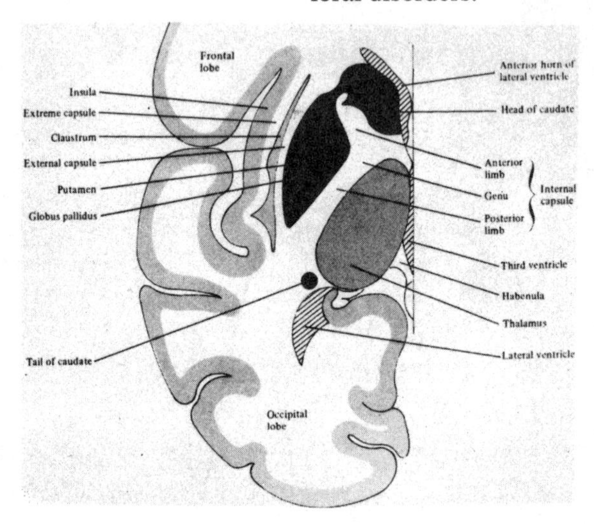

Figure 3. A horizontal section of the brain showing the caudate, putamen and globus pallidus. From L. Heimer, *The Human Brain and Spinal Cord.* Springer-Verlag, New York, 1983.

Dopamine and Movement

The distinction between the nigrostriatal and the meso-cortical-limbic pathways is critical since they have different functions. The striatum exerts an inhibitory action on movement. The substantia nigra, with its rich input of *dopamine neurons to the striatum, stimulates or activates movement.* Thus, when dopamine neurons in this pathway are underactive, it produces muscle rigidity and decreased spontaneous movement and tremors. This is *Parkinson's disease*. When the dopamine neurons of this pathway are overactive, unwanted, jerky movements occur as in Tourette syndrome.

The meso-cortical-limbic system also has an effect on movement, especially hyperactivity. However, because of its major inputs to the frontal lobe and limbic system, it should come as no surprise that these pathways also *play an important role in modulating emotions.*

The Basal Ganglion as a Relay Station

In summary, the basal ganglia, made up of the caudate and putamen, can be thought of as a relay station containing neurons with many different neurotransmitters that regulate and integrate sensory, emotional and voluntary inputs to motor activities.

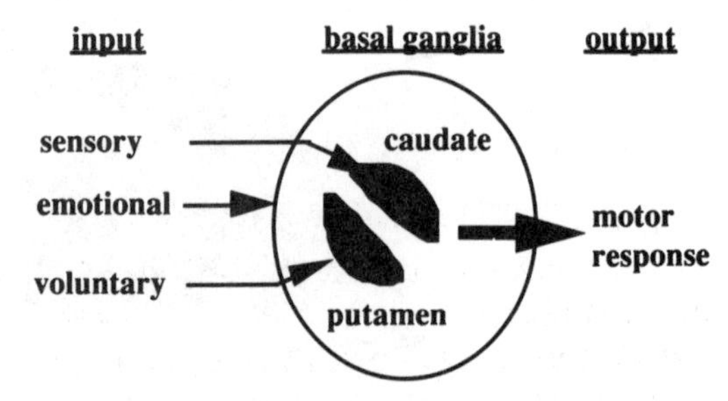

The Pathways in More Detail

The above was a very general description of the major dopamine pathways. The following figure adds a few more details showing that there is some crossover between the contribution of the substantia nigra and the ventral tegmental area to the striatum and limbic system [572,655,1188,1199]. Some of these details will be relevant in future chapters.

complex motor programs. The latter is the system involved in Tourette syndrome.

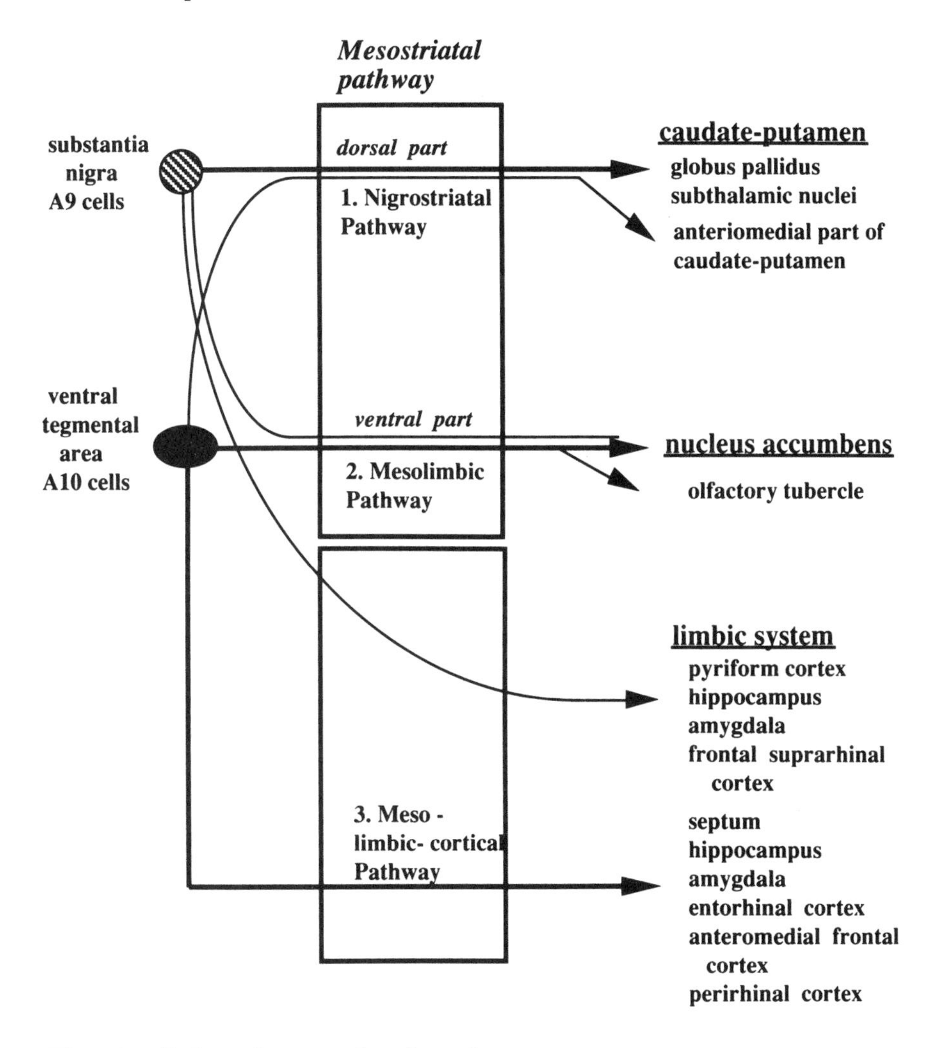

The detailed pathways showing the control of motor movement from the motor and frontal cortex through the striatum and thalamus are synthesized in the diagram on the **next page**[469,562,563].

The details are less important than the conclusion that there are two systems — one through the cerebellum for simple motor movements, and a second through the striatum, substantia nigra and globus pallidus for

Summary: Dopamine is made from the amino acid tyrosine and broken down into HVA. It is important in controlling muscle movements.

There are three major dopamine nerve pathways.

1. The nigrostriatal pathway. This passes from the substantia nigra to the caudate and putamen. When it is defective there is a de-

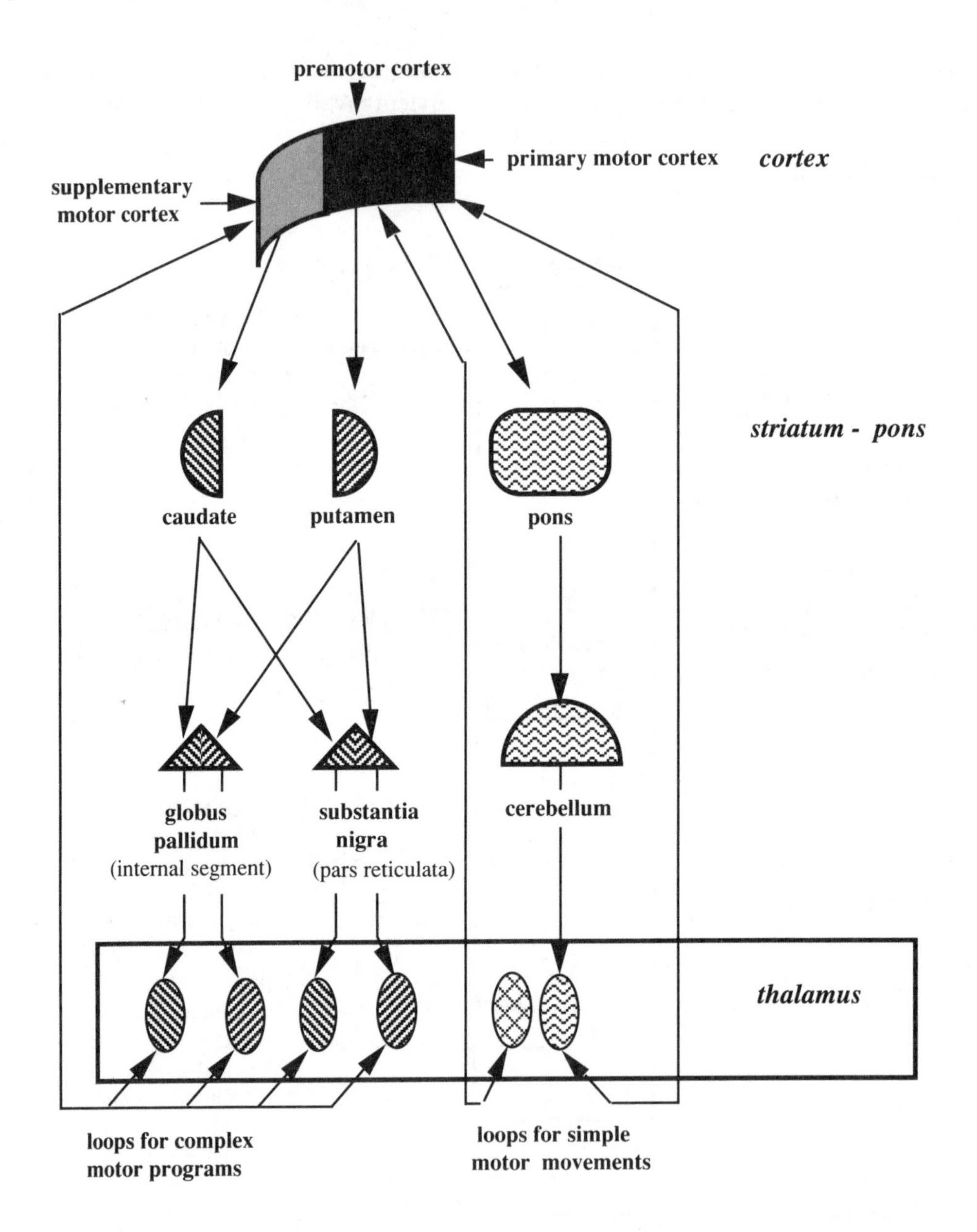

crease in muscle movement and tremors resulting in Parkinson's disease.

2. The mesolimbic pathway. This passes from the ventral tegmental area, in the mesencephalon, to the limbic system.

3. The mesocortical pathway. This passes from the ventral tegmental area to the frontal lobe and limbic cortex.

Since the limbic system and frontal lobes are important in attention and emotion, abnormalities in the function of the dopamine pathways are involved in a number of behavioral disorders including attention deficit hyperactivity disorder, Tourette syndrome, schizophrenia and others.

Chapter 53
Dopamine Receptors

Receptors are proteins that bind tightly to different neurotransmitters. If a receptor can bind a neurotransmitter thousands of times better than other chemicals, then the neurotransmitter can work thousands of times more efficiently. Receptors determine which neurotransmitters a nerve will respond to. For example, if a dopamine receptor is present on a nerve, that nerve can respond to dopamine; if it does not have a dopamine receptor, it will not respond to dopamine. Understanding receptors is critical to understanding how the nervous system works. They are analogous to the locks in our house and car. Without being able to open the locks, nothing works. Without receptors nothing works.

A Brief Course in Receptors

Like a key in a lock each neurotransmitter has a specific shape and the part of the receptor that binds to that transmitter has a similar shape. When the neurotransmitter binds to this site the shape of the receptor changes.

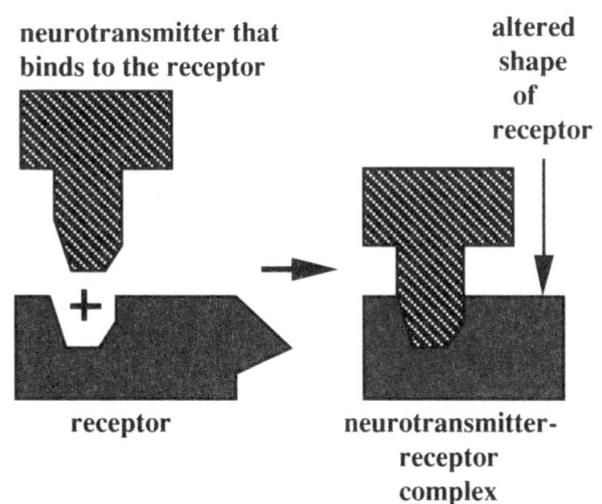

It is this change in shape of the receptor that causes other important things to happen. For example, receptors that are at the edge of a channel or hole in the membrane may keep it closed.

After neurotransmitters bind to the receptors the channel is opened.

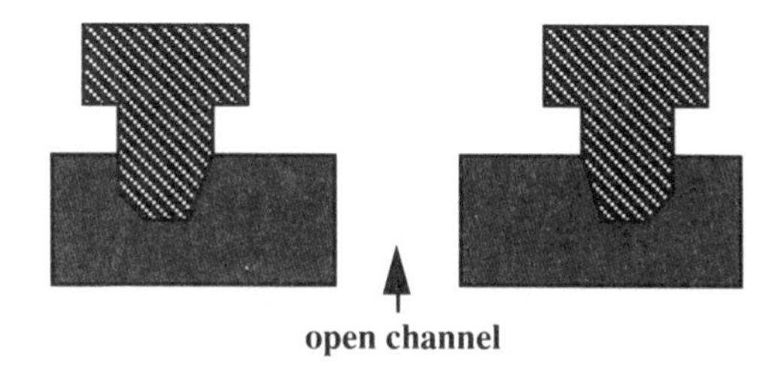

A receptor may be bound to an enzyme.

Changes in the shape of the receptor can cause the enzyme to change shape and become activated so it can now do its work of changing compound A to B.

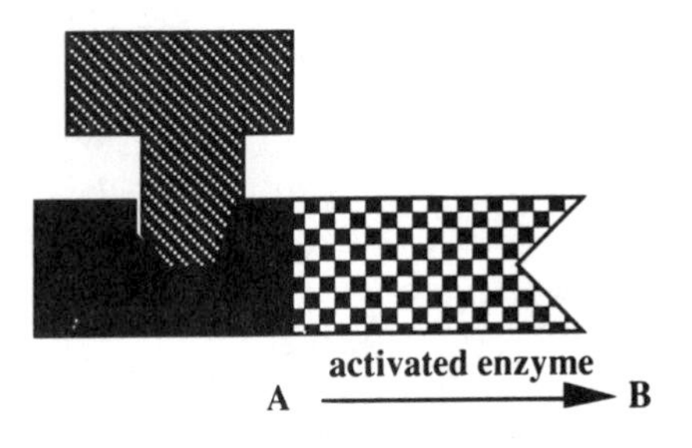

One of the most important of such enzymes is called *adenylate cyclase*. This enzyme changes a compound called *adenosine triphosphate* (ATP) into a circular form called *cyclic AMP* (cyclic adenosine monophosphate). This is often called a *second messenger* (messenger RNA[p34] is the first messenger) because it in turn activates many other enzymes that are not bound to receptors.

Agonists and antagonists Certain compounds, that are similar in shape to the neurotransmitter, can bind to the receptor and change its shape. These are called *agonists*.

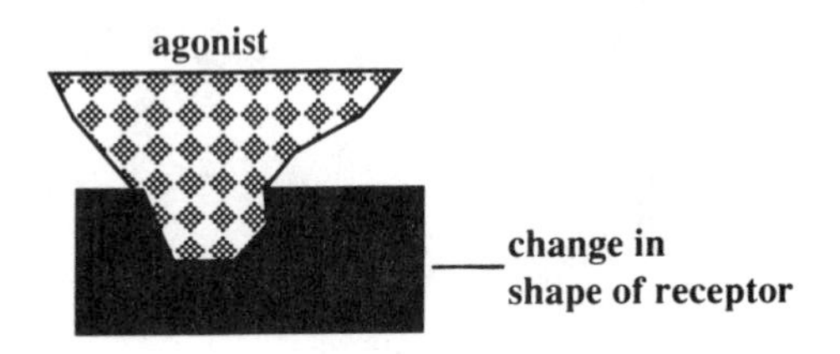

Other compounds may be very similar to the agonists but shaped differently enough that even though they bind to the receptor they don't make it change shape. These are called *antagonists*.

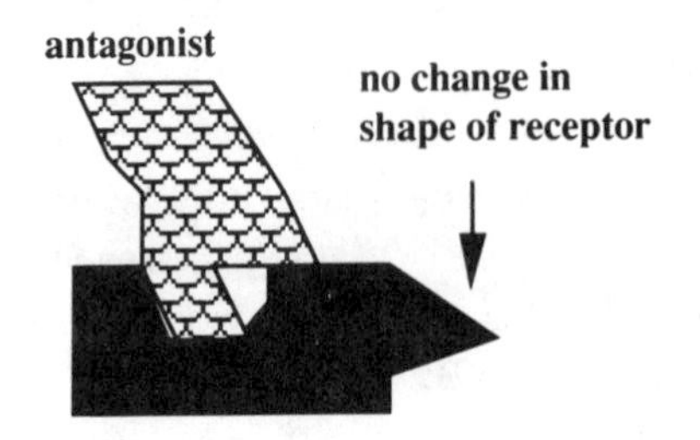

They are important because while they don't cause the receptor to do its work they still bind to the receptor site and *prevent it from binding to the regular neurotransmitter. Haldol is just such an antagonist.* It binds to the dopamine receptor but doesn't activate it and keeps dopamine from binding to the receptor.

Sub-types of receptors There is often more than one receptor for a given neurotransmitter. Since the shape of the site that binds the neurotransmitter and the part of the receptor that changes may be different, the function and the compounds that serve as agonists and antagonists may also be different.

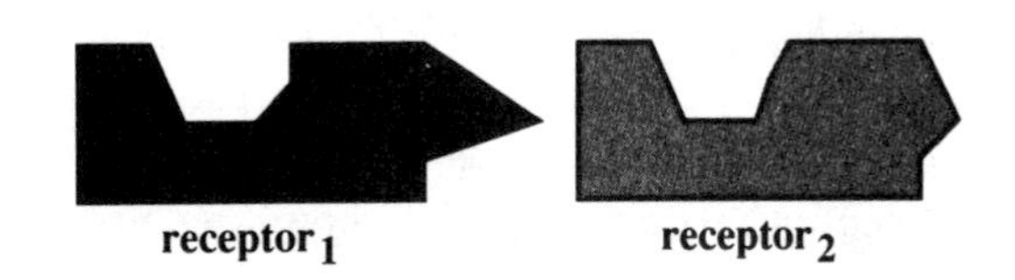

D_1 and D_2 receptors The two major dopamine receptors are called D_1 and D_2. Their differences are shown in Table 1[1019].

Table 1. The Two Major Types of Dopamine Receptors

Type:	D_1	D_2
Use adenylate cyclase:	yes	no
Major location:	striatum	striatum
Stimulation causes:	—	hyperactivity stereotyped moves psychoses, vomiting
Inhibition causes:	—	antipsychotic effects Parkinson's disease
Agonists:	dopamine (weak)	dopamine (strong) apomorphine bromocriptine ergots (LSD) sulphiride

Table 1 (con't)

Antagonists:	ergots (LSD)	Haldol
	Thorazine	Thorazine

This shows that the D_2 receptors are the most important in regards to TS. They are the ones that respond the most strongly to dopamine. When stimulated they cause hyperactivity, stereotyped movements and psychotic thoughts, and they are inhibited by Haldol. While the function of the D_1 receptors in the brain is less clearly defined they are very important outside the brain. For example, stimulation of D_1 receptors increases kidney blood flow and dopamine infusions have saved the lives of many people in shock.

D_3 receptors D_1 and D_2 are postsynaptic receptors. As will be discussed in the chapter *Dopamine and the Frontal Lobes,* some dopamine neurons have autoreceptors. These are presynaptic receptors that inhibit the dopamine neuron. These have been termed D_3 receptors[1947].

D_i and D_e: Inhibitory and excitatory dopamine receptors An alternative way of looking at the sub-types of dopamine receptors is to divide them into those that inhibit and those the stimulate the nerves to which they are attached[387,390] (Table 2).

Table 2. Inhibitory and excitatory dopamine receptors[387,388,1428]

	D_i	D_e
Probable D_1 D_2 terminology:	D_1	D_2
Stimulation causes:	inhibition	excitation
Brain location:	A10 cells ventral tegmental area	A9 cells substantia nigra nucleus accumbens
Dopamine fluorescence:	dotted	diffuse
Dopamine turnover rate:	low	high

Table 2 (con't)

Agonists:	dopamine DPI*	dopamine apomorphine
Antagonists:	ergometrine norepinephrine piribedil	Haldol Thorazine

*(3,4-dihydrophenylamino)-2-imidazoline

This may be particularly relevant for understanding Tourette syndrome and related disorders since it can help to explain why there are two syndromes in one disorder — a frontal lobe syndrome that can best be explained by underactivity of dopamine nerves, and a tic syndrome that can best be explained by overactivity of dopamine neurons. This classification suggests that the nigrostriatal (A9) neurons contain stimulatory dopamine receptors while the mesolimbic and mesocortical (A10) neurons contain inhibitory receptors. The importance of this concept will be discussed later.

Summary: Receptors are critical for nerves to function. They bind neurotransmitters present in small amounts and determine which neurotransmitters a given nerve will respond to, and whether the nerve will be stimulated or inhibited. Many drugs used to treat behavioral disorders work by binding to specific receptors. The two major dopamine receptors are called D_1 and D_2. Stimulation of the D_2 receptors results in hyperactivity, stereotyped movements and psychotic thoughts. Haldol treats these symptoms by specifically blocking the D_2 receptors.

Chapter 54
Dopamine and Human Behavior

There are five human diseases in which abnormalities of dopamine in the brain have been implicated. These are schizophrenia, Parkinson's disease, Huntington's disease, Tourette syndrome and attention deficit disorder. This chapter discusses the first three of these and other aspects of the role of dopamine in human behavior.

Schizophrenia and Dopamine

The essence of the idea that dopamine is involved in schizophrenia[93,292,430,431,1818,1876] comes from the observation that drugs which stimulate dopamine neurons make the symptoms of schizophrenia worse and drugs which inhibit dopamine neurons make the symptoms better. This suggests that some aspect of dopamine is overactive in schizophrenia. Thus, the essence of the dopamine hypothesis of schizophrenia is summarized as follows:

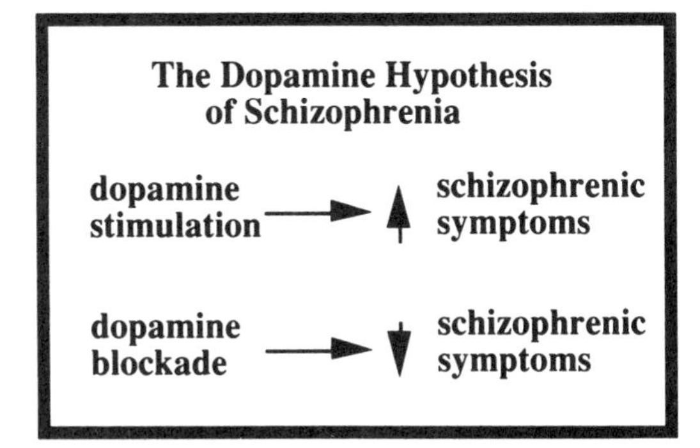

Abnormalities in the mesolimbic and mesocortical dopamine pathways have been especially implicated[1273,1876,1955].

Amphetamine psychosis In 1938 Young and Scoville[2135] reported that when Benzedrine was used to treat people with the sleep disorder called narcolepsy, one side effect was the development of a paranoid psychosis. It was later shown that drug addicts who chronically took high doses of amphetamines (speed) could develop a disorder typical of acute paranoid schizophrenia[43,143,381,532]. This was called *amphetamine psychosis*. This provided a link in the dopamine hypothesis of schizophrenia when amphetamines were shown to release dopamine from nerve endings[1569,1570]. Other dopamine-stimulating drugs such as bromocriptine, cocaine[43,1532,2129], Ritalin[970] and high doses of apomorphine also make the positive[p206] symptoms of schizophrenia worse[849]. Although apomorphine is a dopamine agonist, small doses preferentially stimulate presynaptic dopamine receptors. This inhibits dopamine neurons and improves the symptoms of schizophrenia[849]. In contrast to the worsening of the positive symptoms, all the dopamine-stimulating drugs have been reported to *improve* the negative[p206], frontal lobe syndrome-like[p341] symptoms of schizophrenia[849].

In addition to its effect on dopamine, chronic treatment of cats with amphetamines also produced significant depletion of brain serotonin[1962]. This linkage of dopamine and serotonin is discussed elsewhere[p446].

Thorazine treatment of schizophrenia One of the most remarkable stories in the treatment of behavioral disorders by drugs has been the use of Thorazine and related compounds in the treatment of schizophrenia. Thorazine was first shown to alleviate psychotic symptoms in 1952[466,847]. Because of this effect these drugs were called *neuroleptic,* meaning "to decrease nervousness." The clue to how these drugs were working did not come for another 10 years when Carlsson and Lindqvist[291] found that the turnover of catecholamines was increased. The demonstration that dopamine receptors, especially D_2 receptors, were specifically blocked[1743-1746,1820] by Haldol definitely linked the action of the neuroleptics to dopamine.

The fact that these medications, when used in high doses for prolonged periods of time, can produce significant side effects, such as tardive dyskinesia[p557], has somewhat diminished enthusiasm about their use. Nonetheless, we should not lose sight of the fact that they are still remarkably effective in treating the positive symptoms of schizophrenia and they have allowed hundreds of thousands of individuals to live and work in the community instead of being packed into "insane asylums."

Thorazine (chorpromazine) and related medications are called *phenothiazines.* They owe their effectiveness to the fact that one part of them has the same shape as dopamine and can thus bind to dopamine receptors. However, since they do not activate the receptor they are called *dopamine antagonists* as mentioned previously[p370]. Since drugs such as bromocriptine specifically stimulate the D_2 receptors, and drugs such as Haldol specifically inhibit D_2 receptors, this subtype of dopamine receptor has been implicated in schizophrenia. The more effective a drug is in antagonizing dopamine, the more powerful it is in treating psychotic symptoms[422,423,1498,1743,1744].

The chemistry of the schizophrenic brain. Autopsy studies The natural place to turn to verify the theory that dopamine is abnormal in schizophrenia is in the brains of schizophrenia patients who have died. A problem with many of these studies is that the patients had been on medication and this complicates the findings. Looking at dopamine or HVA in different parts of these brains, compared to controls, has provided little evidence for an increased turnover of dopamine[849]. However, none of these studies looked at dopamine in the ventral tegmental area.

Many studies have been done on dopamine receptors in schizophrenic brains. Most, but not all, showed increased levels of dopamine receptors[849,1744]. However, very few patients had been off medications for a month or more and this led to doubt about the meaning of these results[1233,1234].

Living patients Studying living patients has obvious advantages over studying dead ones. As shown later, if the levels of HVA (the breakdown product of dopamine) are high, then the levels of 5-HIAA (the breakdown product of serotonin[p417]), are also high. The levels of both were found to be increased in the spinal fluid of schizophrenic patients with a family history of schizophrenia[1741,1742] or with poor sexual adjustment[1141]. Even more impressive is the finding of increased levels of these two compounds in the spinal fluid of some well relatives of schizophrenic patients[p440].

Blood flow and PET scans Studies of the blood flow and use of PET (positron emission tomography) scans allow the *function* of different parts of the brain to be examined in living subjects. As such these techniques have considerable advantage over the examination of autopsy material. When blood flow was examined, there was evidence of *decreased flow rate in the frontal lobes* [223,243,574,1105], especially in the dorsolateral prefrontal cortex[2039], and increased blood flow in the left cortex[824,1105] and globus pallidus[517]. When the metabolic rate was examined by determining how fast glucose was used, again *the frontal lobes were* *hypoactive* [2100]. Finally, when the binding of dopamine agonists to the D_2 receptors was examined, there was increased binding in the caudate for schizophrenics both on and off medication[421,1603a,2104]. The results of one study were as follows[2104]:

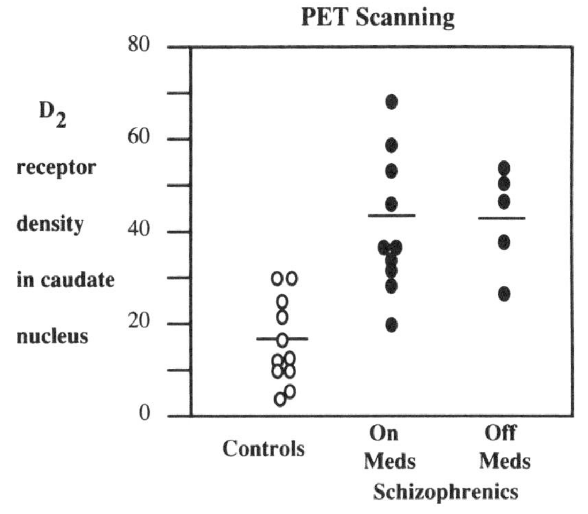

This density tended to be higher on the left than on the right[1603a]. These studies verified the autopsy findings of increased dopamine receptors in schizophrenia.

Frontal lobe stress test In one on the above studies, the blood flow of control and schizophrenic patients was tested before (baseline) and after stressing the function of the frontal lobes by having the subjects perform a Wisconsin Card Sort Test[p355] (WCS)[2039]. In the controls this resulted in increased blood flow in the dorsolateral prefrontal cortex. In the schizophrenic patients this increase did not occur. This is shown as follows:

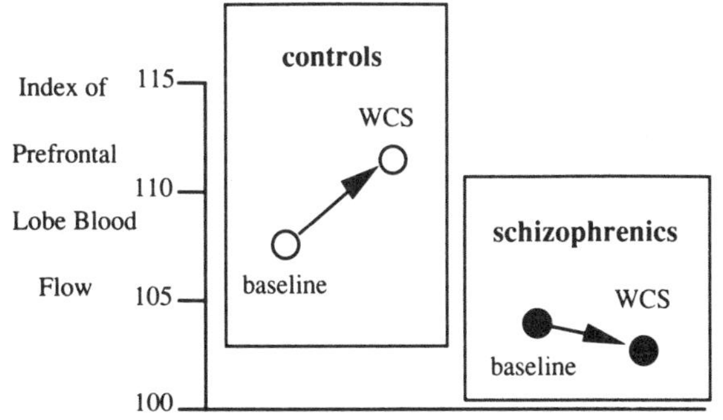

This makes a very important point that some biochemical defects can be subtle and may show up much better when the function of the affected part of the brain is stressed. The same principle holds for the well-known stress test for coronary heart disease. This is an advantage of PET scans and other tests that can be done on living subjects.

Is schizophrenia a dopamine disease? As strong as the dopamine hypothesis of schizophrenia may seem, it does not mean that the genetic defect in schizophrenia directly involves dopamine or dopamine receptors. All of the changes may be secondary to a defect elsewhere, in some other neurotransmitter. The complex interactions between the neurotransmitters is re-emphasized in the chapter called *The Dance of the Neurotransmitters* [p509].

The observations noted above and elsewhere are consistent with the idea that there are actually two syndromes in schizophrenia, one producing the positive symptoms, such as hallucinations and paranoia, due to overactivity of some dopamine neurons, and a second producing the negative symptoms, such as social withdrawal and impaired attention, due to underactivity of some dopamine neurons [319,1187], especially in the frontal lobe. In agreement with the decreased blood flow in the prefrontal lobes of schizophrenics, specific tests of frontal lobe function, such as the Wisconsin Card Sort, are abnormal in schizophrenics[750]. In many ways this duality is similar to the situation in Tourette syndrome.

Parkinson's Disease: Deficiency of Dopamine in the Striatum

In 1817 James Parkinson[1466] wrote *"An Essay on the Shaking Palsy."* In the opening paragraph he stated these patients have a

> "involuntary tremulous motion with lessened muscular power; with a propensity to bend the trunk forward and to pass from a walking to a running

phase."

This disorder especially occurs in older people and is characterized by a combination of decreased muscle movement with frozen facial features and rigid muscles, and a "pill rolling" tremor of the hands. Examination of the brain showed loss of cells especially in the substantia nigra. Other than the recognition that it could follow encephalitis, for the next 140 years there was little new knowledge about the cause of Parkinson's disease. This changed in 1959 with the demonstration that the caudate and putamen were rich in dopamine[163], and a year later dopamine was found to be markedly decreased in the caudate and putamen of patients who had died of Parkinson's disease[522]. This led to the treatment of Parkinson's disease with L-DOPA, which passed the barrier between the blood stream and the brain and increased dopamine levels in the brain[415]. As effective as this treatment was, the disease eventually still progressed.

The lack of spontaneous movement in patients with Parkinson's disease is a dramatic demonstration of the role of dopamine and the nigrostriatal pathway in controlling the initiation of muscle action. While most Parkinson's disease is sporadic, some hereditary cases have been described[1660].

Parkinson's disease in drug addicts A remarkable advance in our understanding of the cause of sporadic Parkinson's disease came as a nice bit of detective work. In 1982 George Carrillo, a 42 year old drug addict, was admitted to Santa Clara Valley Medical Center in San Jose, California, virtually unable to move or talk[1165]. While the neurologists William Langston and Phillip Ballard were initially baffled as to why such a young person would suddenly develop severe Parkinson's disease, a major clue came when a week later his sister was admitted with similar symptoms. Since both were drug addicts, this led to the suspicion that they had been poisoned by some drug they took. After talking to James Tetrud, a fellow neurologist in a nearby town, two additional cases, both brothers, were uncovered. All four patients had developed severe Parkinson's disease after giving themselves IV injections of a synthetic heroin-like drug[1117].

> "First symptoms included almost immediate visual hallucinations, jerking of limbs, and stiffness. Generalized slowing and difficulty in moving occurred within 4 to 14 days. . .Three of the four patients were hospitalized within 14 days to six weeks of first use of the drug. Examination in each revealed near total immobility, marked generalized increase in tone, complete inability to speak intelligibly. . ."

These patients showed all of the symptoms of Parkinson's disease including the "pill rolling" tremor and "cogwheel rigidity." This is a test in which the examiner passively flexes and extends the arm and notes a cogwheel-like jerking.

A toxicologist involved in the case recalled an earlier report, studied at the National Institute of Mental Health (NIMH)[460], of a 23 year old graduate student who was making his own drug, MPPP (1-methyl-4-phenyl-4-propinoxy-piperidine), a compound similar to Demerol. Initially he had no problems but impatience led him to take shortcuts. This led to the presence of a major impurity, MPTP (1-methyl-4-phenyl-1,2,5,6-tetrahydropyridine). Injections of this mixture caused the same Parkinsonian symptoms. He eventually died of an overdose and at autopsy there was death of the cells in the zona compacta region of the substantia nigra[460,1117]. The most frightening aspect of these cases is that the disorder was permanent, since the MPTP was selectively killing cells.

The Langston cases led to a re-kindling of interest at the NIMH in the role of MPTP in Parkinson's disease. "You couldn't had gotten more activity had you walked up there, pulled a pin from a grenade, and rolled it through the door"[1165]. Giving MPTP to monkeys produced an animal model of Parkinson's disease[260].

How does MPTP kill cells? It was soon found that it was not MPTP itself that was killing cells. Instead the enzyme monoamine oxidase was converting it first to an unstable compound, $MPDP^+$, then to MPP^+. The $MPDP^+$ appears to cause dopamine to be broken down into toxic compounds that kill cells containing dopamine. This is shown as follows:

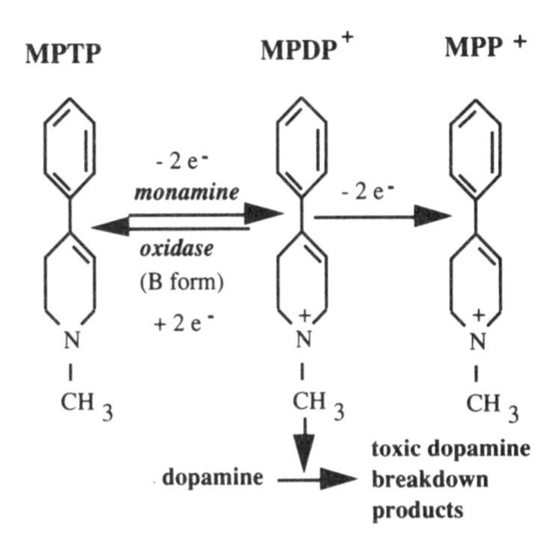

If animals are first given a monoamine oxidase inhibitor, MPTP no longer causes Parkinson's disease[878,1118].

The discovery of the effect of MPTP suggests that most Parkinson's that is not due to encephalitis is the result of a toxin. In this regard it is interesting that Parkinson's disease seems to be a disorder that came in with the industrial revolution. There is no indication it was present prior to James Parkinson's first descriptions in 1817[1165]. There are many MPTP-like compounds in use.

Huntington's Disease

Huntington's disease is the classic example of an autosomal dominant condition. It was first described in 1872 by George Huntington. He, his father and grandfather had studied residents of East Hampton, Long Island, who had a progressive neurological disorder. The symptoms began between 35 and 45 years of age[1266] and were characterized by dementia and grotesque snake-like movements of the extremities, called chorea. Patients usually died 10 to 30 years after onset of the symptoms. Because of its dominant inheritance and the striking nature of the symptoms, it was possible to trace the ancestors back for centuries to immigrants from England[2008].

Early symptoms are often psychiatric in nature and include irritability, depression, outbursts of temper, short attention span and paranoia.

Huntington's disease can be defined as a genetic disorder characterized by programmed premature death of specific nerve cells[1266]. At autopsy the brain shows marked atrophy of the caudate and putamen. Since this part of the brain has a rich collection of different cells, it is not surprising that a wide and confusing array of neurotransmitters are decreased[1266]. In contrast to Parkinson's disease, levels of dopamine are often increased[1169,1302]. While Haldol is effective in partially controlling the movements, nothing has been found to stem the progressive dementia which is just as severe and devastating as that seen in Alzheimer's disease.

Huntington's disease is also the classic example of the use of linkage studies[p53] to identify the chromosomal location of the gene. A genetic marker called G8, and later named D4S10 (which stands for the 10th DNA marker found on chromosome 4), was found be very closely linked to Huntington's disease in a large family from Venezuela[826,1232]. Subsequently DNA markers that were even closer the HD gene, such as D4S43[718], were found (**see next page**).

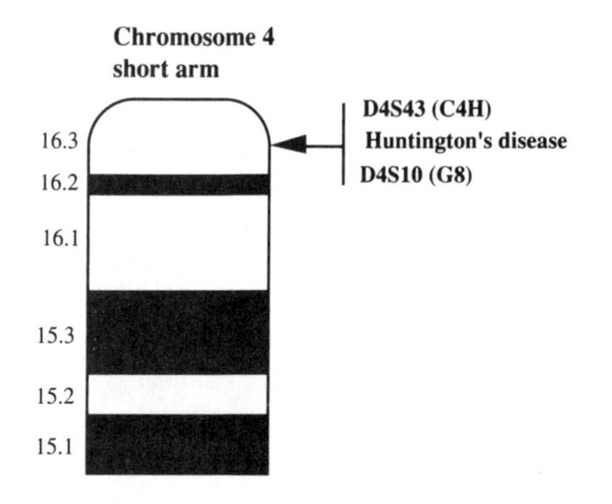

This process allows the HD gene itself to be identified and cloned. While a similar approach will hopefully work for genes causing different psychiatric diseases, the job will be considerably more difficult since the accurate identification of those who carry the genes in question is much more difficult. The mistaken identification of even a small percentage of individuals would make the type of accurate mapping done in Huntington's disease difficult.

Looking at genes that are already identified and suspected to be causing a given disease may be much more efficient. This is called the "candidate gene" approach and is the technique we have used for Tourette syndrome.

Aggression and Sexual Behavior

We can gain much information about the role of dopamine (and serotonin) in aggression and sex by examining the effect of drugs which affect specific neurotransmitters. The **table below** summarizes some of this information[566,698,699].

It is clear that drugs which either *enhance dopamine or deplete serotonin increase aggressiveness and sexual activity, and those which inhibit dopamine or increase serotonin suppress aggressiveness and sexuality.* L-DOPA treatment of patients with Parkinson's disease sometimes has an aphrodisiac effect[698]. Amphetamine given intravenously (I.V.) can cause erections and spontaneous orgasm. Addicts describe the effects of I.V. morphine as a "cerebral orgasm"[565]. Cocaine and amylnitrate are well known among the street drug users for their enhancement of sexual sensations.

Drug	Neurotransmitter affected	Aggression	Sexual activity
L-DOPA	↑ dopamine	↑	↑
amphetamine	↑ dopamine	↑	↑
amantadine	↑ dopamine	?	↑
morphine	↑ dopamine	?	↑
cocaine	↑ dopamine	?	↑
amyl nitrate	↑ dopamine	?	↑
PCPA	↓ serotonin	↑	↑
5-HTP	↑ serotonin	↓	↓
haloperidol	↓ dopamine	↓	↓
phenothiazines	↓ dopamine	↓	↓
reserpine	↓ dopamine, norepinephrine, serotonin	↓	↓

Summary: Dopamine plays an important role in controlling complex muscle movements. In Parkinson's disease there is too little dopamine in the substantia nigra. Patients have a tremor and difficulty starting spontaneous movements. In Huntington's disease cells of the caudate and putamen die. This results in slow snake-like, choreic movements.

While the cause of schizophrenia is unknown, one theory proposes that there is a hypersensitivity to dopamine in the neurons of the striatum and limbic system. Haldol, which blocks D_2 receptors, relieves some of the psychotic symptoms of schizophrenia.

Stimulation of dopamine or inhibition of serotonin nerves causes increased aggression and sexual activity. Inhibition of dopamaine or stimulation of serotonin nerves causes decreased aggression and sexual activity.

Chapter 55
Dopamine and Animal Behavior

Understanding how dopamine affects behavior is critical to understanding Tourette syndrome and many other behavioral disorders. When we study humans we must use the hand that is dealt to us in the form of different diseases. When we study animals we have the luxury of custom making our own diseases. Recalling from Chapter 52, the major dopamine pathways are:

A10 neurons (ventral tegmental area) >
  limbic system and frontal lobes
A9 neurons (substantia nigra) >
  caudate and putamen (striatum)

Most of the induced diseases in animals involve interfering with one of these two pathways (**shown in the box below**).

This interference can take the form of *destroying* or *stimulating* these neurons. First I will examine the effect of destroying dopamine neurons of the A10 cells, then of the A9 cells.

The LeMoal or A10 Syndrome

Since the nerve center of dopamine for the limbic system and frontal lobes is the ventral tegmental area (VTA), the observation of the behavior of animals after this area has been destroyed provides important clues about the function of dopamine in the brain. These studies have been carried out in France by Michel LeMoal and his colleagues over a period of two decades[666-668,1155-1159,1802-1804,1887-1889,1975,1979]. This area has been destroyed either by radio waves or by injecting 6-hydroxydopamine, a compound that selectively destroys dopamine neurons. These animals show a disruption of a wide variety of behaviors (**see box next page**).

You should recognize that these are also symptoms of the frontal lobe syndrome and attention deficit disorder. This is understandable since a significant number of the dopamine nerves of the ventral tegmental area pass to the frontal lobes and many aspects of attention deficit disorder are frontal lobe in nature.

	Mesolimbic, Mesocortical	**Nigrostriatal**
Cell type:	A10	A9
Stimulation causes:	hyperactivity	stereotyped movements
Destruction causes:	hyperactivity	hypoactivity
	frontal lobe syndrome	Parkinson's disease
	LeMoal syndrome	A9 syndrome
	A10 syndrome	
% of brain dopamine:	Less than 5%	More than 90%

The LeMoal or A10 Syndrome
- behavioral disorganization
- decreased response to punishment
- distractability
- emotional hyper-reactivity
- hyperactivity
- impaired learning
- low frustration tolerance
- non-productive activity
- perseveration
- poor sequencing in time
- rigidity
- short attention span
- untidy, messy cage

Some of these behaviors deserve further discussion.

Hyperactivity One of the most striking features of the LeMoal syndrome is motor hyperactivity. This is shown as follows[668]:

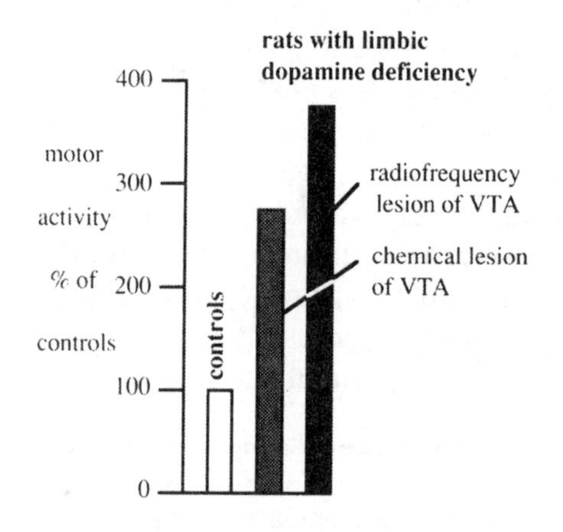

The above results were produced by the *partial* destruction of the A10 cells. If these cells are *completely* destroyed the animals actually become *hypoactive*[1022,1078]. This distinction is important for the following reason. If destruction of the A10 cells produced hyperactivity we would have to conclude that dopamine was acting as an inhibitor and thus destruction of dopamine cells resulted in disinhibition or hyperactivity. However, the complete destruction of A10 cells causes hypoactivity. This implies that *when only some of the cells are destroyed the others become hypersensitive and hyper-reactive and this causes the hyperactivity* [1078,1159].

The **nucleus accumbens** lies at the head of the caudate and putamen and has rich connections with the limbic system. It has been termed the filter[406] or interface[1344] between the limbic and the motor system. Injection of dopamine into the nucleus accumbens in rats causes marked hyperactivity with much running and rearing and sniffing[1515].

The complex effects of lesions of the ventral tegmental area and nucleus accumbens on hyperactivity are more easily understood by a series of diagrams. Some of the dopamine neurons of the ventral tegmental area pass to the nucleus accumbens which itself is dopamine rich.

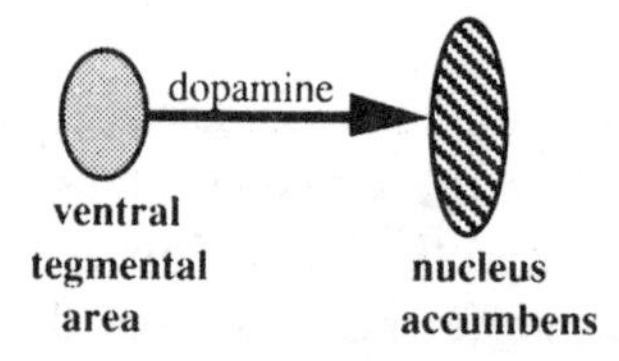

If the nucleus accumbens is stimulated by injecting still more dopamine it produces hyperactivity[409,411,1515].

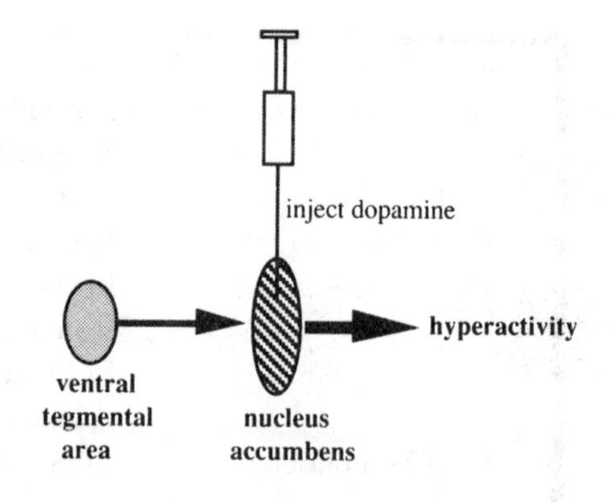

If some of the neurons passing from the ventral tegmental area to the nucleus accumbens are destroyed, the dopamine neurons become hypersensitive, also resulting in hyperactivity.

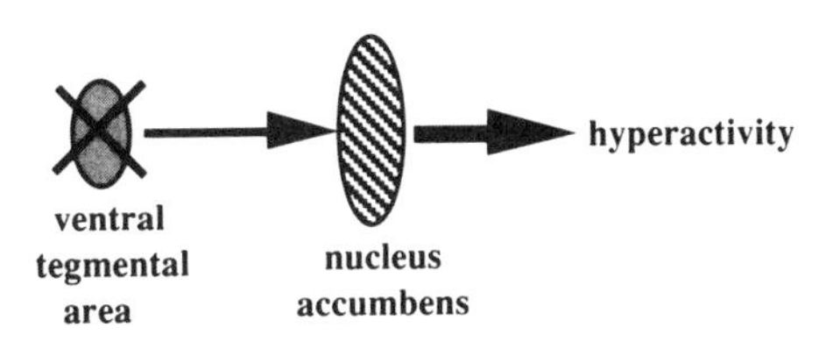

If the hypersensitive neurons of the nucleus accumbens are also destroyed, the animal shows hypoactivity.

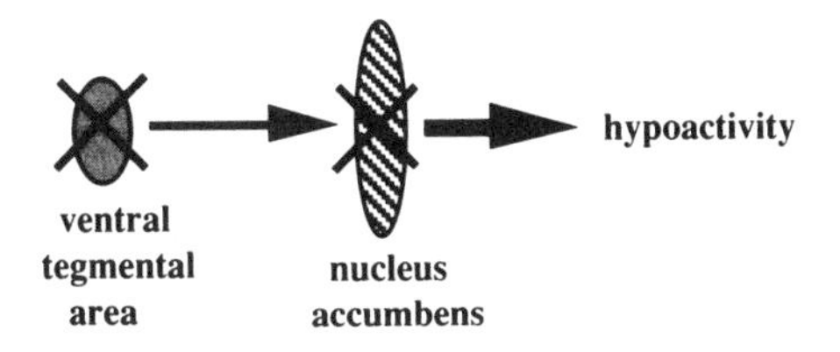

Even if there is no lesion of the ventral tegmental area, destruction of dopamine neurons in the nucleus accumbens will result in hypoactivity.

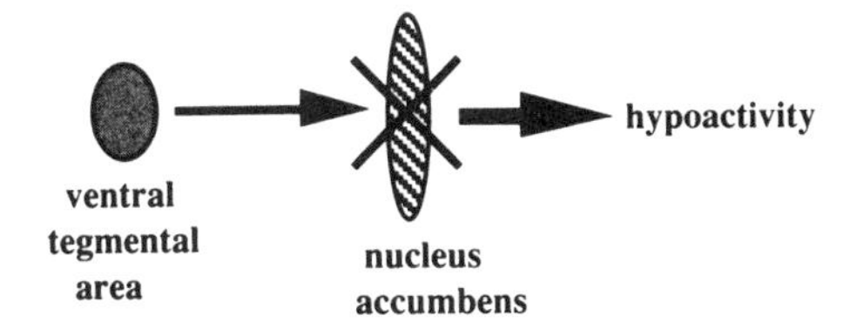

Finally, if all the dopamine neurons of the ventral tegmental area are destroyed, there is no stimulation of the nucleus accumbens and this also results in hypoactivity.

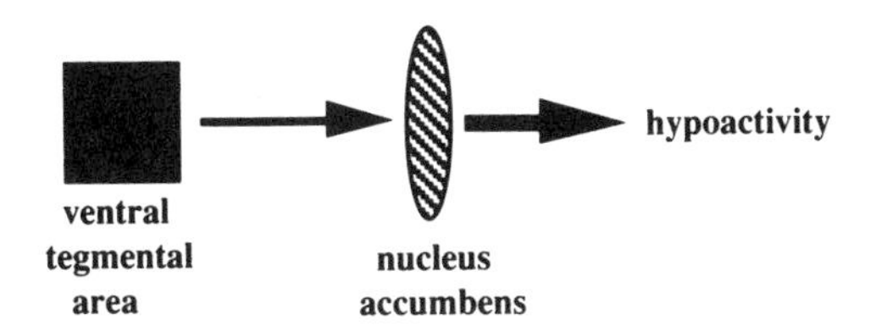

This shows the importance of the nucleus accumbens in motor activity. When it is overstimulated hyperactivity occurs.

The frontal lobes and hyperactivity The amount of hyperactivity produced by lesions of the A10 neurons in the ventral tegmental area is directly related to the amount of decrease of dopamine in the frontal lobes[1926]. A 50 percent decrease in frontal lobe dopamine doubles the activity level, and a 75 percent decrease increases the activity 4-fold. Thus:

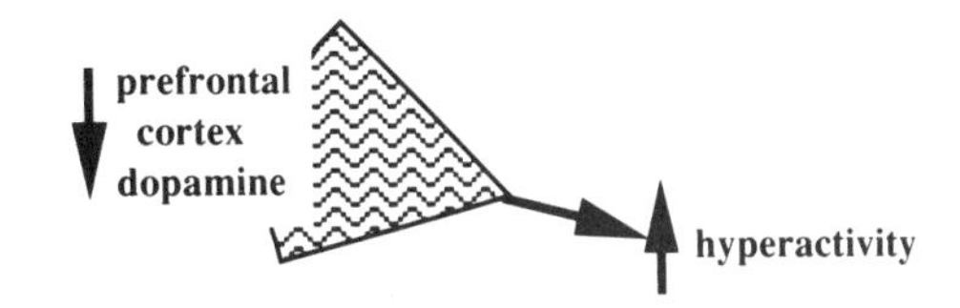

The amount of this decrease is also related to other symptoms of the frontal lobe syndrome such as distractibility and overresponsiveness to stimuli[1803]. Based on this, it is not surprising that destroying the dopamine neurons in the frontal lobe itself also leads to hyperactivity and distractibility[302,1562]. The important role of dopamine in the frontal cortex is described later[p389].

Defective passive avoidance The best way to understand passive avoidance is to give an example. Suppose you hated heights and someone put you on a tall, wobbly stepladder. You would quickly climb down to the floor. But if someone had electrified the floor and gave you a shock every time you stepped on it, you would rapidly learn to stay on the ladder. This is called passive avoidance since as long as you avoided the floor by passively staying on the ladder, nothing would happen to you. When a comparable test was done on rats whose dopamine cells in the ventral tegmental area had been destroyed, they persisted in "climbing down to the floor"[1157]. This is illustrated as follows:

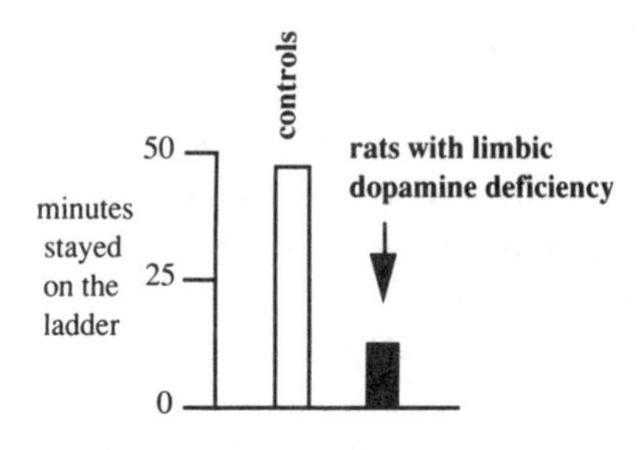

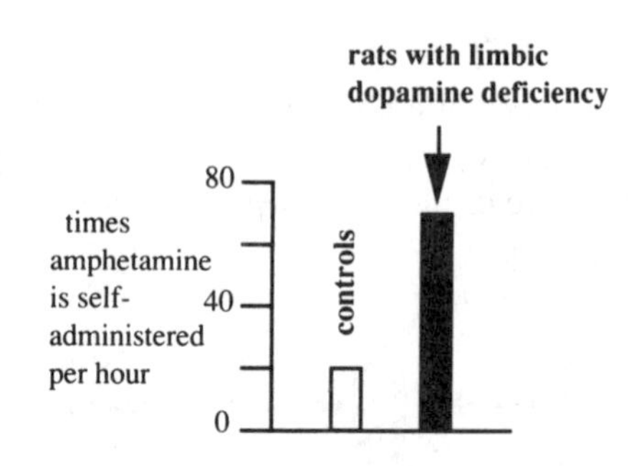

This is reminiscent of the mothers of TS and ADHD children when they say "he doesn't respond to any kind of discipline, nothing seems to phase him."

Sleep It is also important to note some things that don't happen after the destruction of A10 dopamine neurons. For example, there are only minimal alterations in sleep patterns[1158]. Since serotonin controls sleep[p448] and TS patients have significant sleep disturbances[p249] this would suggest that in TS the serotonin defects are primary and the dopamine defects secondary.

These studies indicate that the dopamine neurons coming from the ventral tegmental area modulate the higher functions of the brain such as emotional reactions, learning, sleep and motor activity.

Reward system, drug addiction and dopamine One of the difficulties in the use of animal models in studying susceptibility to drug addiction is that their nervous systems are "in balance" and they do not have any of the genetic predispositions to drug abuse that are present in humans. To circumvent this, LeMoal and co-workers[1159] *partially* destroyed the A10 dopamine cells of the VTA in rats with electric lesions to produce a limbic dopamine deficiency state. When these rats were allowed to self-administer an I.V. fix of amphetamines, they self-injected the drug far more often and used larger amounts than the controls[1159]. This is shown in the following figure:

In addition these rats were *hyper-reactive* to the drug since the effects on motor hyperactivity came on in 12 minutes while this took 45 minutes for the control rats. To make matters even worse, they were *hyposensitive* in that the effect *wore off twice as fast* as in the controls. Thus, they had the lethal combination of more quickly getting a bigger hit and having this wear off more rapidly than in the controls. The presence of some dopamine neurons was critical for this effect since when they were *totally* destroyed the self-stimulation was abolished completely. This is a beautiful model to explain why some individuals with *a partial genetic defect of dopamine may become susceptible to drug addiction.*

This unique model provides another important lesson: *when some dopamine neurons are hypoactive the remaining dopamine neurons become hyperactive*[1159]. This compensatory hyperactivity can be suppressed by chronic low doses of dopamine-like drugs[1887]. An analogy of this can be seen clinically in Tourette syndrome when small doses of Ritalin, a dopamine agonist, sometimes eliminates tics[p575], as well as hyperactivity.

The A9 Syndrome

The destruction of the dopamine neurons of the A9 cells or substantia nigra produces a Parkinson's disease-like syndrome that in many ways is the reverse of the A10 syndrome[1977]. This comparison is shown as follows:

A10 Syndrome	**A9 Syndrome**
hyperactivity	hypoactivity
excessive drinking	too little drinking
over-initiates movement	under-initiates movement
over-reacts	under-reacts
over-eats	under-eats
over-explores	under-explores

When animals with the A9 syndrome were injected with dopamine-like drugs, they showed increased movement, sniffing, licking and biting. They soon developed a furious compulsive gnawing, which was far more violent than that seen in normal animals. Some even chewed up their front paws, bit off their fingers, or ate into their own abdomens[1977].

It is clear that the A9 syndrome tends to be the opposite of the A10 syndrome. This supports the validity of the separation of dopamine receptors into an exciting type and an inhibiting type[P371]. When the neurons of the substantia nigra containing the exciting type are destroyed, the symptoms mimic a dopamine deficiency (A9 syndrome). When neurons of the ventral tegmental area are destroyed, the symptoms mimic dopamine excess (A10 syndrome).

Stimulation of Dopamine Neurons

As a generalization [956,1023,1514,1515]:

dopamine stimulation of the nucleus accumbens -> motor hyperactivity
dopamine stimulation of the caudate nucleus -> stereotyped behavior.

However, since some dopamine neurons of the ventral tegmental areas pass to the caudate nucleus, and some of the neurons of the substantia nigra pass to the nucleus accumbens[P367] these distinctions are not absolute[410].

Stereotyped behaviors have also been called stereotypy. Since in animals they are the closest thing to tics, the ways that stereotypy is produced provides an important understanding of Tourette syndrome.

Stereotypy

Stereotypy refers to the production of repetitive, stereotyped movements in animals following certain types of drugs, usually dopamine agonists. It helps us to understand Tourette syndrome because many of the movements resemble the motor and vocal tics seen in humans. The following box lists some of the typical behaviors:

Stereotypy	
biting	sniffing
gnawing	licking
repetitive head and limb movements	

Many of the early studies on stereotypy were done by Axel Randrup and Munkvad from Denmark[1,1569-1571,1574]. They noticed that when rats were given injections of amphetamines (Dexedrine) in high doses (equivalent to more than 100 mg for a 50 pound child), within 30 minutes they would begin to continuously sniff, lick and bite the cage. This behavior lasted for 2 to 3 hours and went away. Dopamine antagonists such as Haldol and Thorazine inhibited the amphetamine-induced stereotypy [617]. Since specific injection of dopamine and amphetamine into the striatum caused stereotypy[1], and stereotypy was prevented only when Haldol was injected into the striatum[618], this part of the brain seemed particularly involved in stereotyped movements.

In addition to amphetamines, many other drugs have been shown to cause stereotypy. These include dopamine, Ritalin, cocaine, apomorphine, and L-DOPA[347,408,2023]. The feature they all have in common is that they stimulate dopamine neurons.

Many studies have been done on stereotypy with complex and often conflicting results. There is reasonable agreement that biting and licking are separate from sniffing and head and limb movements[411,413,523,617]. Biting and licking were prevented by selectively destroying the dopamine neurons in the caudate and putamen, but sniffing and repetitive head and limb movements were unaffected[411,413]. Stereotypy was readily produced by injection of amphetamine or apomorphine into the olfactory tubercles, which are part of the mesolimbic system[411,413]. Different drugs produce different effects. As discussed below, aggressive behavior can also be produced by amphetamine and other dopamine agonists[862] and prevented by Haldol.

Stereotypy is not unique to animals. It is also seen in humans after taking amphetamines[1572].

Despite these variable results, one important lesson is clear: Tic-like stereotyped movements are due to hyperactivity or hypersensitivity of dopamine neurons in both the nigrostriatal and mesolimbic dopamine systems. These movements can be suppressed by Haldol and pimozide.

Amphetamines, cocaine and stress Like amphetamines, the chronic administration of cocaine can produce a paranoid schizophrenia-like psychosis[1532]. The chronic administration of amphetamines and cocaine increases the sensitivity of animals to amphetamine-produced stereotypy[1054,1242]. This appears to be due to an increased sensitivity of dopamine receptors. Stress may precipitate acute psychotic episodes in individuals with schizophrenia and in individuals who have recovered from amphetamine-induced psychosis[49]. When rats are chronically stressed they subsequently show increased stereotypy when given amphetamines[49] or cocaine[1242]. This has led to the suggestion that amphetamines, cocaine and stress are interchangeable, and all lead to increased sensitivity of dopamine receptors[49,1242].

This helps to explain why speed or cocaine addicts and patients with schizophrenia or Tourette syndrome are so susceptible to stress. This will be discussed again in the next chapter.

Coping and stress Stress tends to be most severe when we feel we cannot control events that influence our lives. Coping involves developing techniques for regaining that sense of control. The above study on cocaine and stress provided some interesting insights on the effect of coping and control in reducing the effects of stress. In the following experiment stress was produced in rats by placing them on a shocking grid. They were divided into three groups —

a) controls receiving no shocks,

b) those who could stop the electrical shocks whenever they wished, and

c) those who had no control over the shocks.

The total amount of shock the latter two groups received was the same. The effect of the stress on dopamine receptors was measured by the amount of stereotypy produced by cocaine. The results are shown on **the following page**[1242].

This indicated that *having some control over the stress prevented it from having any effect on the brain dopamine receptors.*

Electrical stimulation of the VTA Given all the emphasis on the ventral tegmental area as the source of mesolimbic and mesocortical dopamine neurons, what happens when this region is chronically stimulated? This has been done by inserting very thin wires into the VTA of cats and then stimulating the region for two seconds every day for two months[1875]. This resulted in progressive fearfulness, hiding, and loss of

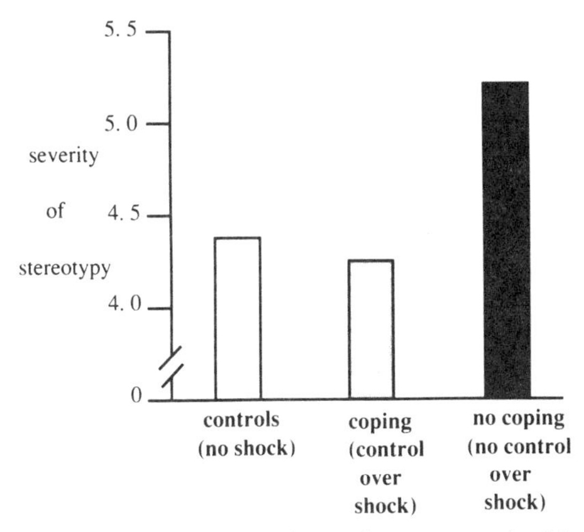

social behavior. The dopamine neurons in this area were also stimulated by blocking GABA [p481], a neurotransmitter that normally inhibits these dopamine neurons. This produced intense arousal, staring, fear, withdrawal, hiding, searching and sniffing[1874].

Other Aspects of Dopamine and Animal Behavior

Dopamine and sex Administration of L-DOPA to animals increases mounting and ejaculation[699]. To determine if this effect was entirely due to the stimulation of dopamine, the drug apomorphine, which is a specific dopamine stimulant, was given to rats. As shown, this resulted in a marked increase in mounting and intercourse from 0 to 76 percent[699].

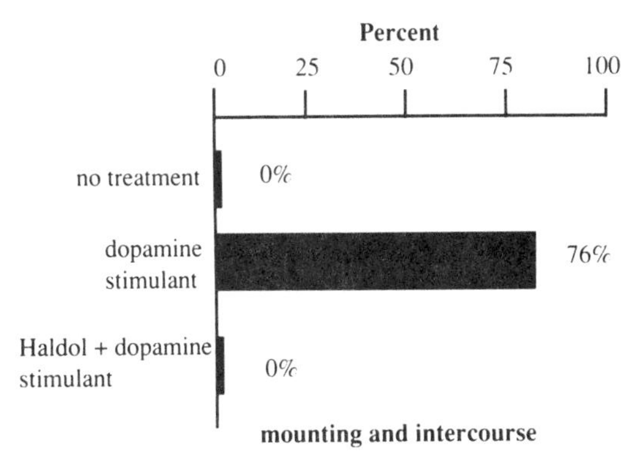

When the apomorphine was given in conjunction with Haldol, which blocks the dopamine receptors, the effect was abolished. Apomorphine increased the percentage ejaculating from 0 to 50 percent. These effects were dependent upon the presence of testosterone, since castration abolished all the effects of apomorphine.

Dopamine and aggression The possible role of the catecholamines in aggression was first realized from the effects of reserpine on taming the aggressiveness of animals. This drug depleted neurons of all three major transmitters — dopamine, norepinephrine and serotonin. Which one was important? This could be answered by replacing some but not others. Only large doses of L-DOPA produced the aggressiveness in reserpine-treated animals[564].

The level of dopamine in the brains of different strains of mice that differ markedly in their degree of aggressiveness has been examined. The more aggressive strains had the highest levels of brain dopamine; the tamer strains had the lowest levels[565].

Dopamine and learning Intact dopamine neurons have been especially implicated in the ability of animals to learn to modify their behavior according to experience[152]. Pimozide and Haldol, which block postsynaptic dopamine receptors, interfere with this type of learning[151,462]. The effect that destruction of dopamine neurons has on learning is covered in more detail in the next chapter.

Summary: Destruction or stimulation of different sets of dopamine pathways in animals has expanded our knowledge of the important role of dopamine in behavior.

When the the dopamine nerves leading to the frontal cortex are destroyed it produces a frontal lobe syndrome with many disinhibited behaviors. When these nerves are stimulated it produces intense arousal, fear and withdrawal.

When the dopamine nerves leading to the basal ganglia are destroyed it produces symptoms of Parkinson's disease. When these nerves are stimulated it produces stereotyped tic-like movements.

Drugs that enhance dopamine activity cause increased aggressiveness and hypersexuality.

Both Tourette syndrome and schizophrenia have symptoms that resemble underactivity of the dopamine nerves leading to the frontal lobe and overactivity of the dopamine nerves leading to the limbic system and basal ganglia.

Chapter 56
Dopamine and the Frontal Lobes

The relationship between dopamine neurons and the frontal lobes is so important to the understanding of Tourette syndrome and ADHD that it deserves a special chapter. It was not until 1973 that dopamine nerve endings and dopamine receptors were found to play a role in the function of the frontal lobes[1938,1939,2017]. While serotonin and norepinephrine are present throughout the cortex[p410,p420], dopamine nerve endings are restricted to the prefrontal cortex[908,910,1188,1189]. This predicts an especially important role of dopamine in the function of the prefrontal lobes[106] and in causing the chemical frontal lobe syndrome referred to in Chapter 51.

The Anatomy of Prefrontal Lobe Dopamine Neurons

As noted before[p364] the dopamine nerves that project to the prefrontal lobes come from the A10 cells in the ventral tegmental area. In the frontal lobes they converge with nerves coming from the mediodorsal nucleus of the thalamus[171,228,494].

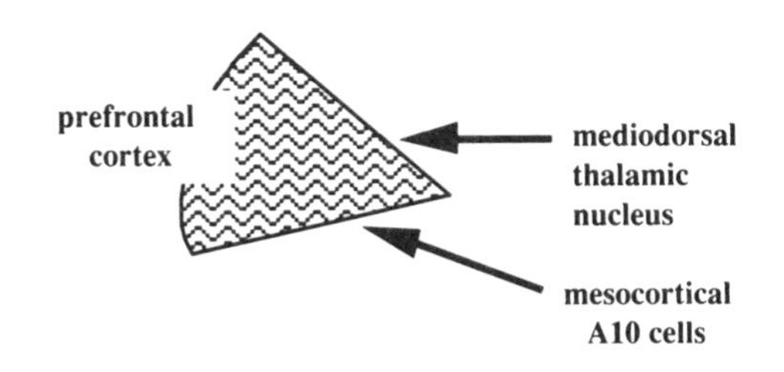

The thalamic to prefrontal lobe pathways may be involved in complex learning tasks[492]. While the nerve endings of norepinephrine nerves are widespread in the cortex, the dopamine nerves end in the lower or deeper layers of the frontal cortex (layer 6)[538,539,739,1938,1939].

High Activity of Prefrontal Lobe Dopamine Neurons

The rate of dopamine turnover, and the firing rate of dopamine nerves, is much higher in the prefrontal lobes than in the caudate, putamen or limbic system[106]. This carries with it the important implication that *anything that affects the rate of dopamine synthesis is likely to have a greater impact on the function of the frontal lobes than on other parts of the brain.* This may explain why the frontal lobe syndrome is so common in many important genetic disorders of the brain, including ADHD, Tourette syndrome and schizophrenia.

Dopamine Is Inhibitory in the Frontal Lobes

In the frontal lobes, dopamine neurons may cause *inhibition* of the regions they supply[247]. This is suggested by the fact that stimulation of the dopamine neurons of the ventral tegmental area causes a *decrease* in the firing rate of neurons in the frontal cortex[738].

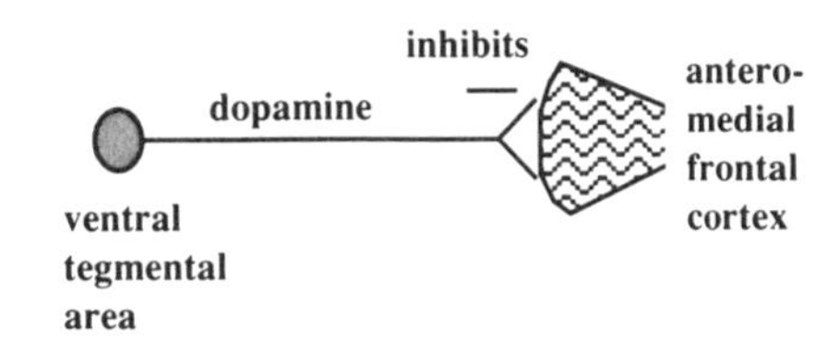

This role of dopamine as an inhibitor of the frontal cortex is consistent with the fact that the predominant receptors are of the D_i inhibitory type[1383]. This contrasts to other regions of the brain, especially the nucleus accumbens and striatum, where dopamine neurons *stimulate* other nerves.

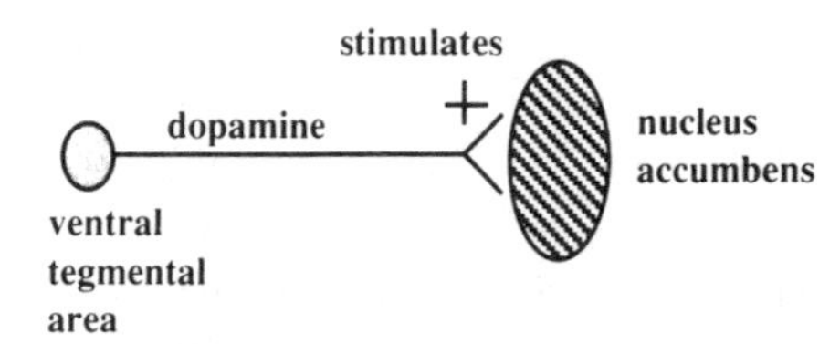

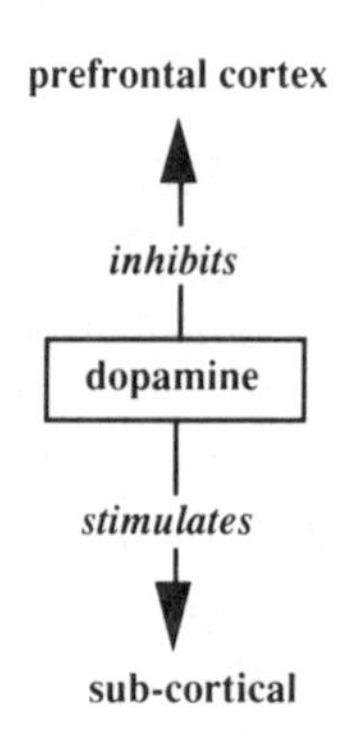

This apparent difference with dopamine stimulating sub-cortical structures but inhibiting cortical structures may provide a key to understanding Tourette syndrome and related disorders.

Prefrontal lobe dopamine is regulated by serotonin and norepinephrine The other two major neurotransmitters yet to be described are serotonin and norepinephrine. For now it suffices to point out that both play a role in regulating prefrontal dopamine neurons. Serotonin nerves interact both directly with the frontal lobes[p426] and indirectly via the dopamine nerves in the ventral tegmental area[1510]. Since dopamine inhibits prefrontal dopamine neurons but stimulates nucleus accumbens neurons, abnormalities in serotonin can cause different effects in the prefrontal lobes compared to the nucleus accumbens. Lesions of norepinephrine nerves leading to the ventral tegmental area also cause a decrease in dopamine activity of the frontal lobe[887].

Decreased Frontal Lobe Dopamine -> Trouble Learning

Difficulties in learning are a prominent part of Tourette syndrome. Given the presence of learning deficits in the frontal lobe syndrome, and the importance of dopamine in the frontal lobe, it is reasonable to suspect there may be a connection between the two. Experiments in animals indicate there is. LeMoal and his co-workers[1803] specifically examined this. Rats were placed in a T maze. This is a box with a right and a left arm.

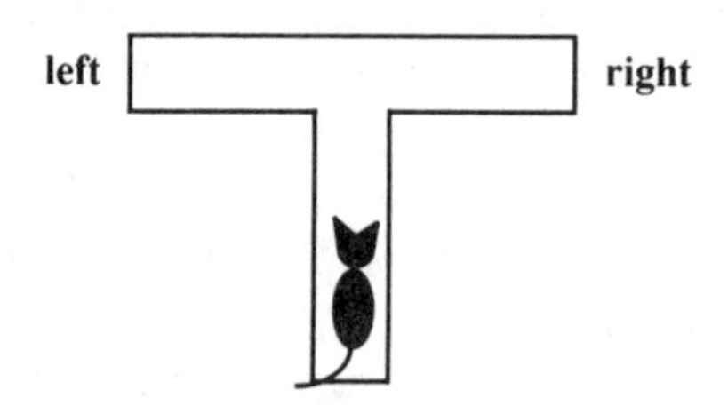

After the animals were fasted for 12 hours they had to learn that food would be presented first in the left and then in the right arm of the maze. Teaching continued until they made five errors or less. Then half of the rats were given an injection in the ventral tegmental area to destroy the dopamine neurons passing to the frontal lobes. This resulted in a 70% decrease in the dopamine content of the frontal cortex. The rats were then re-tested to see how many errors they made. The results were as follows:

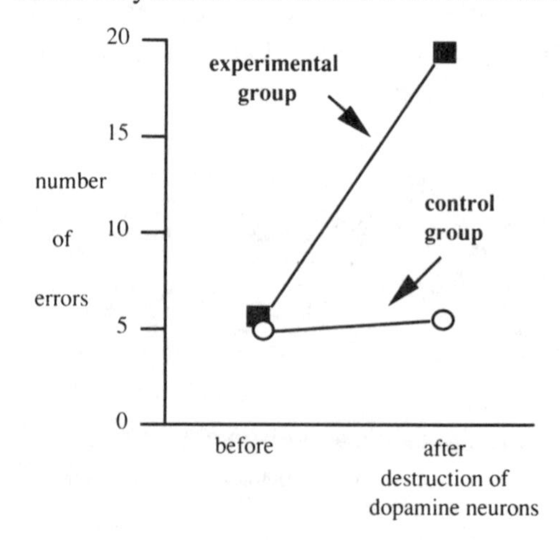

The controls still remembered the task well while the dopamine-depleted rats made an average of 20 errors.

Before the injections it took both groups an average of 147 trials to learn the task. The number of trials it required for both groups to re-learn the task is shown as follows:

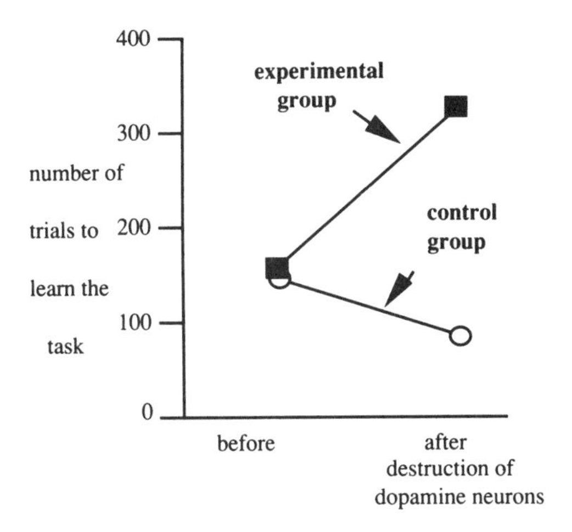

Since the controls had not forgotten the task, it only took 70 trials for them to re-learn it. By contrast, it took the dopamine-depleted rats 313 trials to re-learn a task they previously learned after 147 trials[1803].

Similar results were obtained in monkeys where the dopamine neurons within the frontal lobe itself were destroyed[233]. The learning defects were reversed by giving dopamine agonists such as L-DOPA. Of particular interest, *the defect was also reversed by clonidine,* a norepinephrine agonist[233]. This may explain why this medication is often reported to improve school performance in children with TS[p563].

Thus, defects of dopamine neurons from the ventral tegmental area can cause learning disabilities,

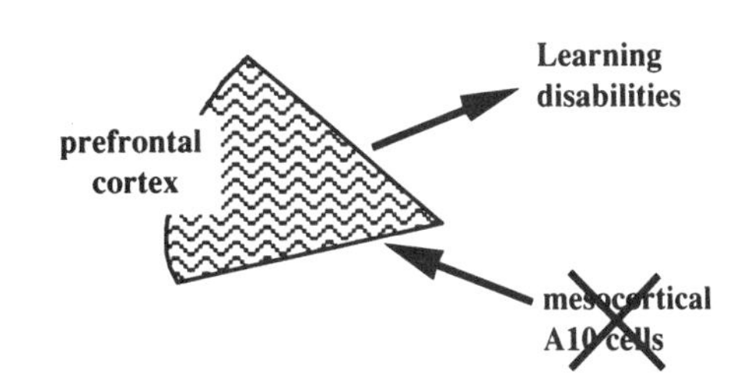

and defects of dopamine neurons in the frontal lobe itself can cause learning disabilities.

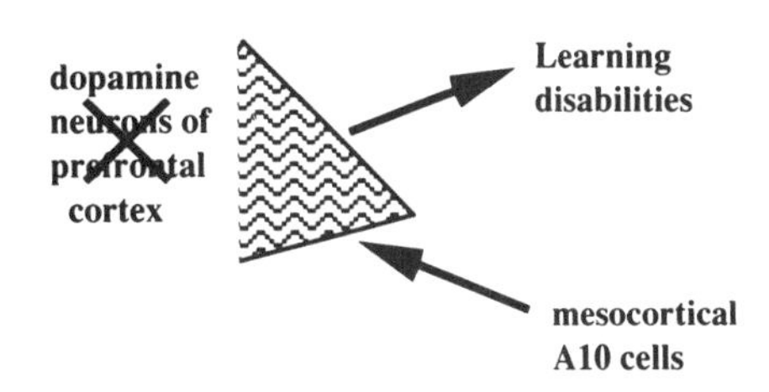

Stress and Frontal Lobe Dopamine

Another striking aspect of Tourette syndrome patients is their susceptibility to stress. All types of stress cause an increase in tics, irritability, short attention span and temper tantrums. Again, studies of dopamine and the frontal lobe give us insights into this problem. When rats are stressed by giving electrical shocks to their feet, there is a marked increase in the activity of the mesocortical dopamine pathways[106,1135,1939a]. This was determined by examining the depletion of dopamine and norepinephrine after the rats were given a drug to prevent the new synthesis of these compounds. The results were as follows[1939a]:

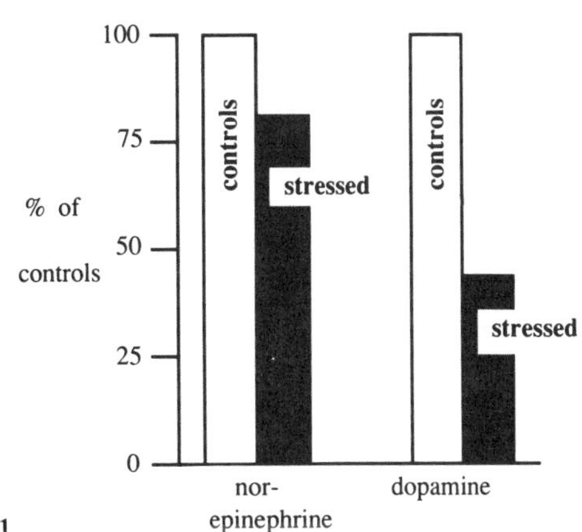

This showed that *stress causes a marked decrease in frontal lobe dopamine*. A moderate decrease in dopamine was also noted in the cingulate cortex and nucleus accumbens, regions that may be particularly involved in producing tics, and the ventral tegmental area [737.739] with its input to the limbic system. This effect was blocked by Valium.

In summary:

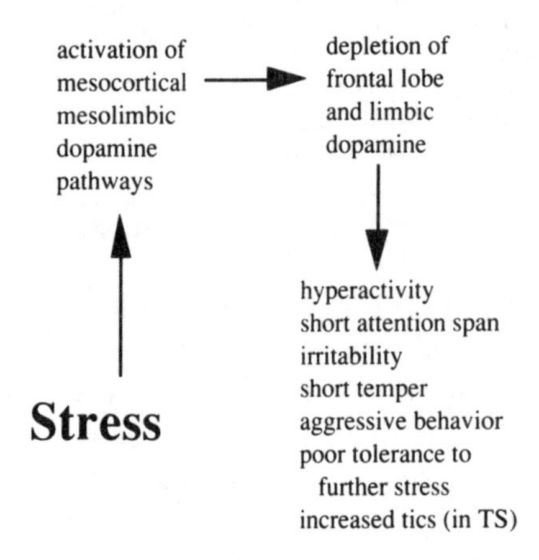

Thus, *when the rate of dopamine production is already low, as in patients with ADHD, Tourette syndrome, and schizophrenia, stress is very poorly tolerated and the above effects are markedly exaggerated.*

The Isolation Syndrome: A Remarkable Model of Tourette Syndrome

Mice and rats that have been isolated from their fellow animals for many weeks provide one of the best animal models of Tourette syndrome. These animals show a behavioral syndrome characterized as follows[213.1981.2042]:

The Isolation Syndrome

- aggressive
- compulsive
- hyperactive
- hyper-responsive to stress
- hypersexual
- increased inner tension
- increased muscle tone
- increased vocalizations
- irritable
- learning defects
- poor memory

The level of dopamine in relation to DOPAC, the breakdown product of dopamine, was examined in various parts of the brains of isolated animals. This showed a significant *decrease of dopamine in the frontal lobe and a significant increase in the nucleus accumbens and striatum* [174]:

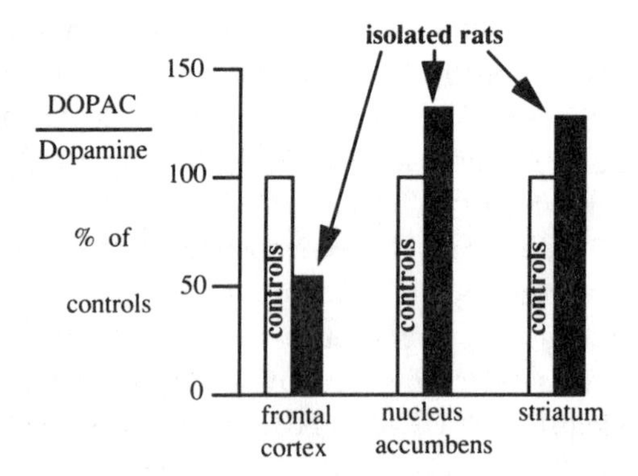

This remarkable observation gives us a lot of insight into the cause of Tourette syndrome. As I have repeatedly pointed out, TS is really at least two syndromes, a frontal lobe syndrome, that is best explained as due to too little dopamine, and a motor and vocal tic syndrome, that is best explained as an overactivity of the dopamine neurons of the nucleus accumbens and striatum. Here we see the identical pattern induced by a manipulation of the environment. This syndrome provides further evidence for a reciprocal relationship between the dopamine neurons of the frontal lobe versus those of the nucleus accumbens and striatum. Thus:

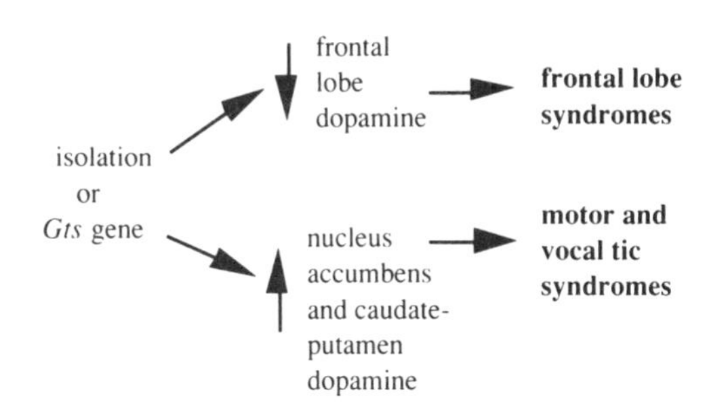

Both an environmental and a genetic factor can cause the same effects. As with Tourette syndrome, these isolated animals are more susceptible to stress. They show a greater increase in dopamine turnover in the frontal lobe after stress than stressed controls[174,738]. As discussed later[p446] the basic defect causing the changes in dopamine may be in brain serotonin.

Increased activity of dopamine in the striatum after frontal lobe lesions In the same year that the studies on the isolation syndrome were being described, Pycock and co-workers[1560] reported complementary studies on the effect that destroying frontal lobe dopamine neurons had on dopamine of the striatum. They chemically destroyed the dopamine neurons of the *medial prefrontal cortex*. A month later there was a 64 percent decrease in dopamine in the prefrontal lobes. The following diagram shows the effect this had on dopamine, HVA and DOPAC (two breakdown products of dopamine), adenylate cyclase (a measure of the number of D_1 receptors), and ^{3}H-spiperone binding (a measure of the number of D_2 receptors), in the nucleus accumbens and striatum[1560].

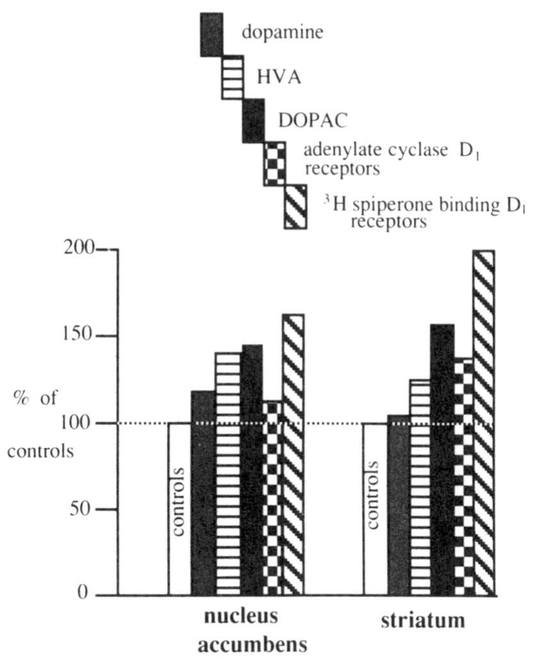

Thus, *when the frontal lobe dopamine neurons are underactive, there is an activation of dopamine neurons in the nucleus accumbens and striatum.* This was both presynaptic (increased dopamine turnover) and postsynaptic (increased density of dopamine receptors) and resulted in hyperactivity and increased sensitivity to amphetamine[735,957,959,1562].

The prefrontal cortex sends nerves to the following sub-cortical structures[1560]:

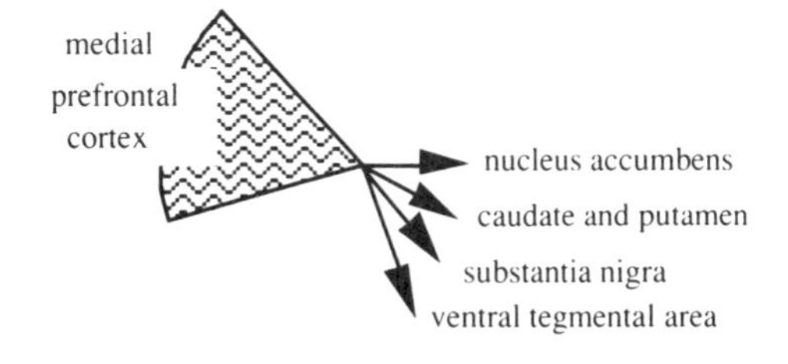

Many of these nerves use glutamic acid as an excitatory neurotransmitter[92,620,1284,1833]. When these nerves from the frontal lobe to the striatum are cut or destroyed it causes an increase in amphetamine-induced stereotypy and hyperactivity[735,959,1229,1706].

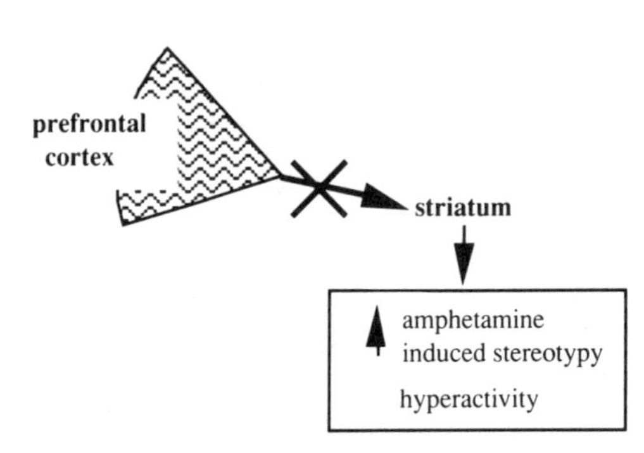

These pathways play a role in regulating dopamine functions in the striatum.

The Shaywitz Syndrome: An Animal Model of ADHD

Another important aspect of Tourette syndrome is ADHD. Just as the isolation syndrome and the LeMoal syndrome provide excellent animal models of Tourette syndrome, the Shaywitz syndrome provides an excellent animal model of ADHD. The LeMoal syndrome was produced in adult animals by injecting a chemical (6-hydroxydopamine or 6-OHDA) into the ventral tegmental area that destroyed its dopamine neurons. Shaywitz and colleagues injected 6-OHDA into the fluid cavities of the brain of rats soon after birth. By the time the rats were 15 days old, they were significantly more hyperactive than the controls. By the time they approached maturity at 26 days, the hyperactivity was no longer apparent[1779]. The ability to learn a task and attention span were also impaired in the hyperactive rats[1778,1937]. When the animals were examined at 35 days, there was 60 percent less dopamine in the brains of the hyperactive compared to the control rats[1779]. Dopamine was especially low in the frontal lobes[615]. If the level of brain dopamine did not drop by at least 50 percent, the rats did not become hyperactive. *If the level of dopamine dropped by more than 65 percent, the hyperactivity did not go away as the rats grew older*[1327].

As in children with ADHD, the hyperactivity decreased and the learning ability increased when treated with stimulants such as Ritalin and Dexedrine[1777,1778]. Similar to the "paradoxical effect" in humans, the stimulants increased activity in the normal control rats.

This experiment was repeated by others with the same results[554,1327,1894]. Injection of the 6-OHDA into the ventricles of adult rats did not cause hyperactivity, indicating that the dopamine neurons had to be destroyed early in life to have this effect[554].

Thus, ADHD in children shows the following similarities to the Shaywitz syndrome in rats:

	ADHD	Shaywitz syndrome
Species:	human	rat
Hyperactivity:	yes	yes
Attention deficits:	yes	yes
Learning deficits:	yes	yes
Improved with amphetamine:	yes	yes
Frontal lobe dopamine:	depleted?	depleted

Dopamine and ADHD in humans In 1971 Paul Wender[2058] proposed that there was a deficiency of one or more of the major neurotransmitters in children with ADHD. He specifically implicated dopamine[2059,2060]. To determine if this was true, Shaywitz and colleagues[1776] examined the level of HVA (the breakdown product of dopamine) in the spinal fluid of children with ADHD. This was done in conjunction with probenecid, a drug which prevents HVA from being removed from the spinal fluid during the test. Since the amount of blockage varies with the amount of probenecid, they looked at the ratio of HVA to probenecid in the spinal fluid of six children with ADHD compared to 26 control children. The results were as follows:

There was significantly less HVA in the

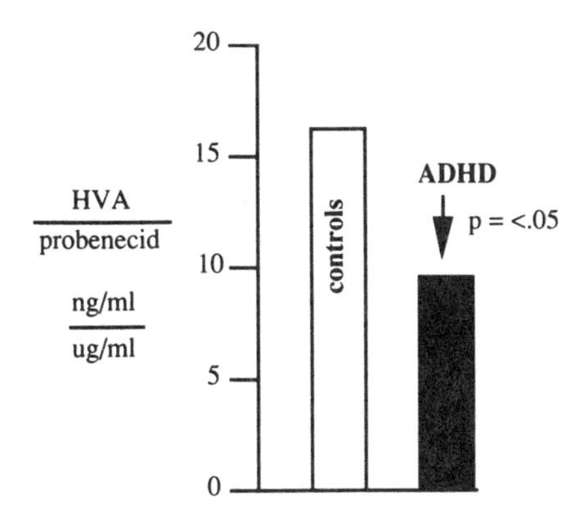

spinal fluid of the ADHD children. There is also a decrease in HVA in the urine of ADHD children[1782]. The role that norepinephrine and serotonin play in ADHD is discussed in later chapters.

Why Children with ADHD Get Better on Stimulants

These observations help us to understand why children with ADHD improve on stimulant medications such as Ritalin, amphetamine, and Cylert. All of these medications act to increase the activity of dopamine neurons. Normal children have normal levels of prefrontal and striatal dopamine and their level of activity and their behavior is normal.

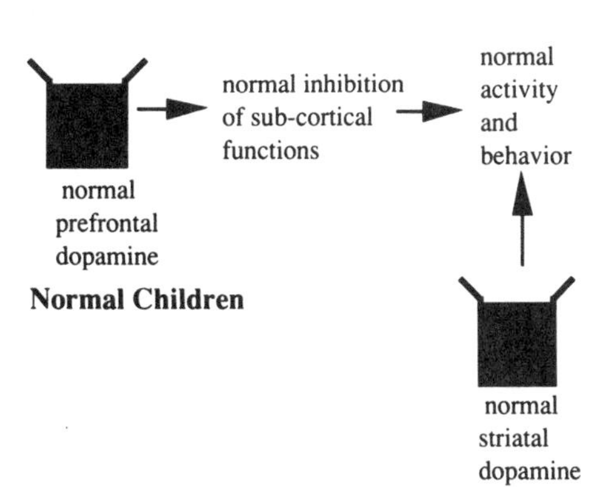

By contrast, children with ADHD appear to have a deficiency of prefrontal lobe dopamine. As a result, the sub-cortical structures such as the striatum and limbic system are disinhibited and these children present with hyperactivity and irritable behavior.

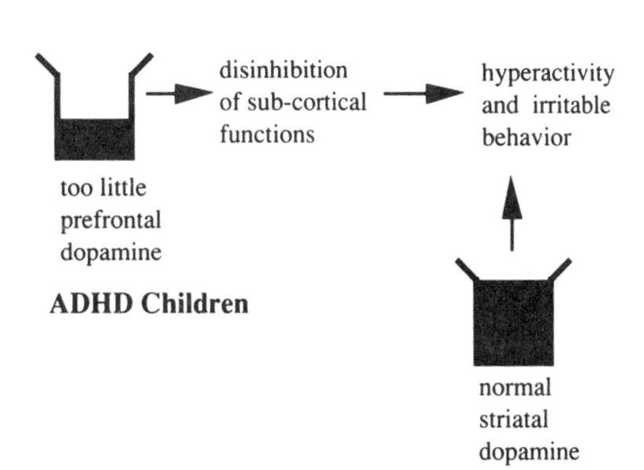

When they are given stimulants, their prefrontal lobe dopamine returns toward normal and the sub-cortical structures are normally inhibited again.

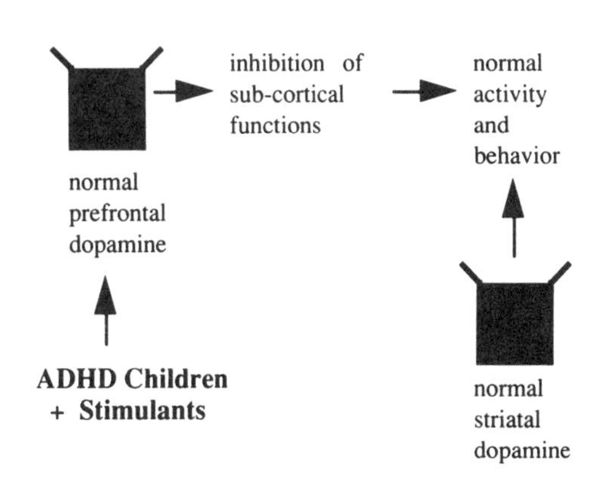

By contrast, when normal children, with normal levels of prefrontal dopamine, are given stimulants, the effect is primarily on the sub-cortical structures and they become hyperactive and irritable.

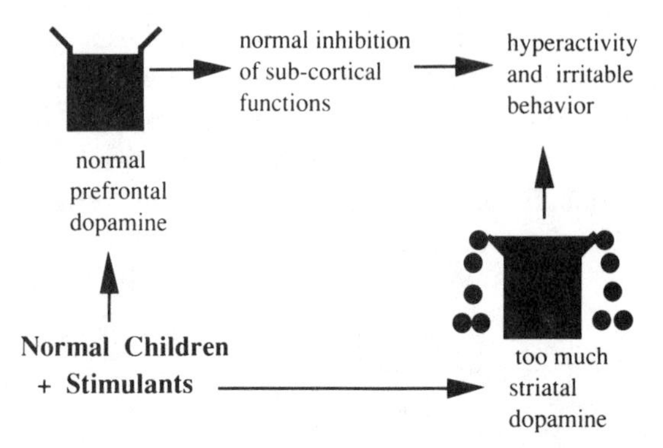

Tourette syndrome has the added problem that in addition to the deficiency of frontal lobe dopamine there is also a hypersensitivity of the sub-cortical dopamine neurons resulting in tics.

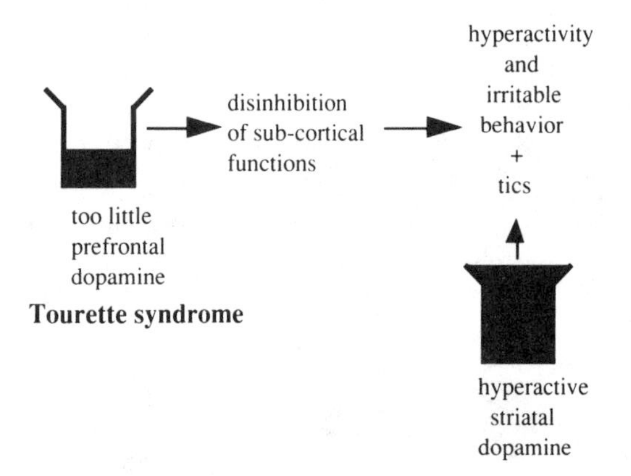

Now, if stimulants are used to try to treat the ADHD, the tics *may* or *may not* get worse.

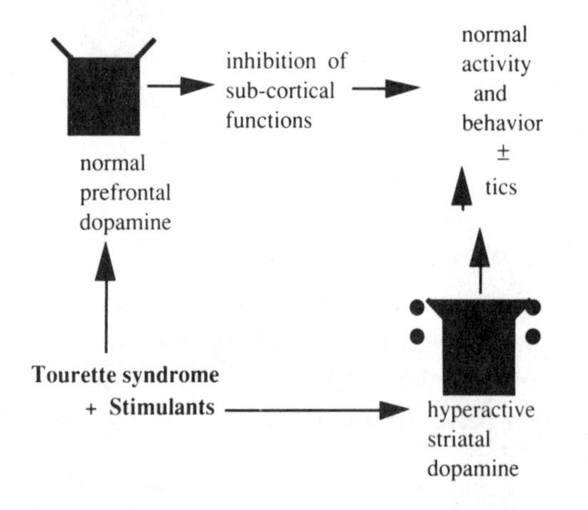

As discussed in the part on treatment[p567], if the tics get significantly worse but the child cannot function without the stimulants to control the ADHD, the tics can usually be handled by adding Haldol to the treatment. While the above emphasizes dopamine, the reason Ritalin and amphetamines are so effective in the treatment of ADHD is probably because they affect both brain dopamine and norepinehprine[2144].

Autoreceptors, the Frontal Lobes, and the Effect of Haldol

Another lesson the frontal lobe dopamine neurons teach us is that they are probably responsible for the effectiveness of Haldol in treating schizophrenia and Tourette syndrome. One might think that Haldol and related drugs used to inhibit dopamine neurons would result in the inhibition of the rate at which dopamine was being made. Wrong! It was noted very early in the use of these drugs that they *stimulated* the synthesis and turnover of dopamine[130,291]. The reason is related to why deaf people turn the TV up louder. The increase in dopamine synthesis is an attempt to override the blockade of the dopamine receptors.

When the postsynaptic dopamine receptor is normally activated, a feed-back loop activates *autoreceptors* on the original nerve body and inhibits further dopamine production as follows:

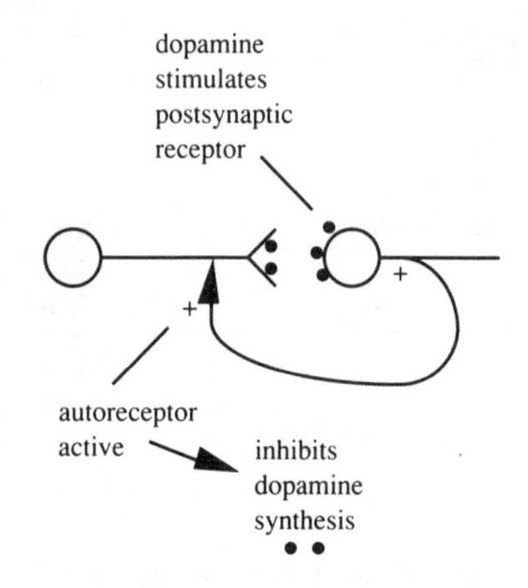

However, when Haldol blocks the postsynaptic dopamine receptors, this feedback loop or governor is not activated and the dopamine synthesis continues out of control.

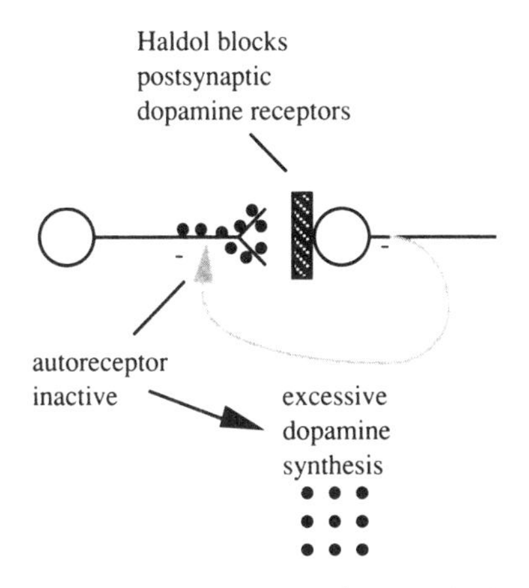

Thus, in the presence of Haldol, there is an increased rate of presynaptic dopamine turnover.

There are no dopamine autoreceptors in the frontal lobe *The frontal lobe dopamine neurons are unique in that they do not contain these autoreceptors*[106,107]. This probably explains the higher activity of the prefrontal dopamine neurons. In addition, when animals are given Haldol there is a marked stimulation of dopamine turnover of the mesolimbic and nigrostriatal pathways, but only a moderate increase in the dopamine turnover in the frontal lobe and limbic cortex. However, if Haldol is given over a long period of time, tolerance develops to this stimulation and the rate of dopamine synthesis drops back to normal[74,106,1705,1108]. This, however, *does not* occur in the frontal lobe. These effects are summarized as follows[106]:

Dopamine turnover

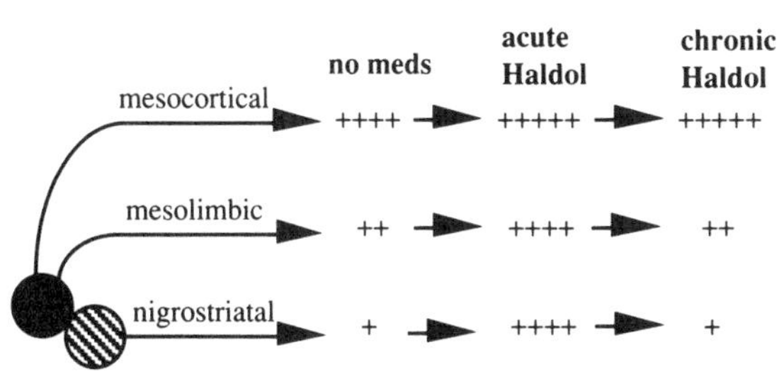

In the frontal lobe, Haldol continues to be effective even when given chronically. This suggests that the anti-psychotic and anti-Tourette *effect of Haldol is due to the effect of the drug on the dopamine neurons of the frontal and limbic cortex.* Postsynaptic receptor hypersensitivity does not occur in the mesocortical pathway, but does occur in the mesolimbic and nigrostriatal pathways. This is apparently what causes *tardive dyskinesia,* one of the side effects of taking Haldol over a long period of time[p557].

The many unique features of the prefrontal dopamine neurons are summarized as follows[106]:

Unique Characteristics of Prefrontal Dopamine Neurons

higher dopamine turnover
higher firing rate
less responsive to dopamine agonists
less responsive to dopamine antagonists
no tolerance to Haldol
marked activation by stress

The Frontal Lobe Matures at Puberty

It is a mistake to think that our brains have finished developing soon after we are born. The prefrontal lobe is the last part of our brain to occur during evolution and the last part to develop as we age. If lesions are made in the dorsolateral prefrontal lobes in newborn monkeys, the impairment in delayed response tests does not show up until after sexual maturity[21]. This may explain why some disorders with prominent frontal lobe symptoms, such as schizophrenia, show the most severe symptoms after puberty.

Summary: Understanding the role of dopamine in the frontal lobes helps us to understand many aspects of Tourette syndrome and ADHD. These include:

- *the learning defects*
- *the susceptibility to stress*
- *the attention deficits*
- *the hyperactivity*
- *the effects of stimulants*
- *the site of action of Haldol.*

The frontal cortex inhibits sub-cortical structures.

The following figure summarizes the effect of lesions of various dopamine neurons:

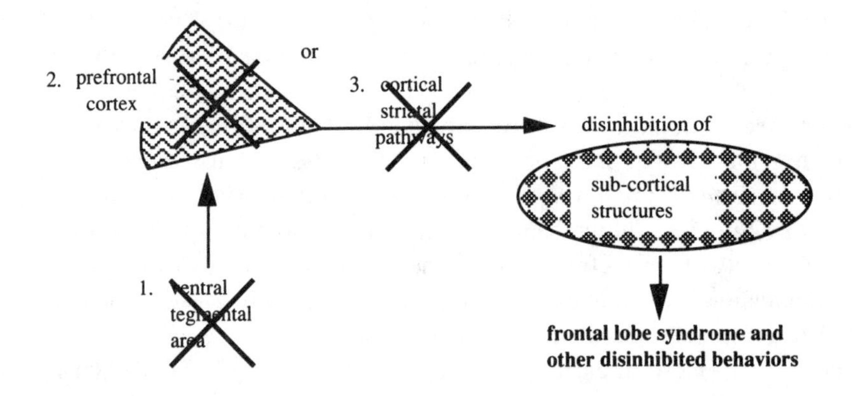

When dopamine neurons to the frontal lobes (1), inside the frontal lobes (2), or passing from the frontal lobes (3) are not functioning properly, the frontal lobe syndrome develops and the sub-cortical structures are disinhibited, resulting in hyperactivity, impulsivity, learning and conduct disorders, and other behavioral problems.

Chapter 57
Dopamine and Tourette Syndrome

The reasons for believing dopamine is involved in Tourette syndrome are similar to those for schizophrenia.

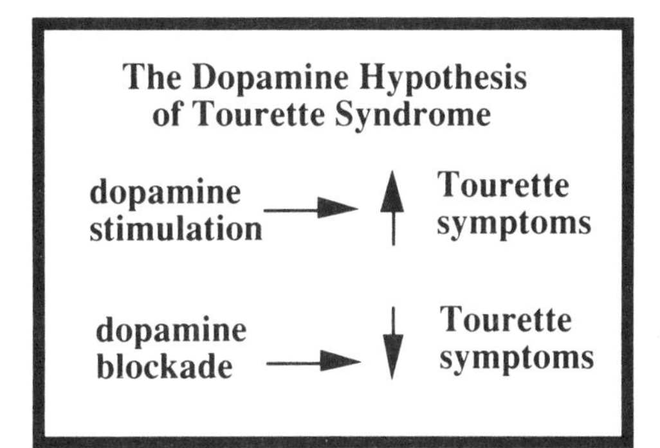

Dopamine Blockade Decreases TS Symptoms

For the first half of the 20th century and well into the 1960's, Tourette syndrome was a gold mine for the psychoanalysts. The swearing and jerking and shouting were seen as clearly an attempt to release primeval rages built up as a result of some deep psychological insult inflicted by mother, or father, or both, or a monstrous attempt to discharge a repressed libido. However, in 1961 Seignot[1749] described the alleviation of TS symptoms with Haldol. Since this is a D_2 dopamine receptor blocker[p370], and since this drug, almost alone among thousands of other drugs, quickly alleviates the symptoms of TS[p547] — it provides strong evidence that TS, is at least in part, caused by hyperactivity of some dopamine neurons or hypersensitivity of some dopamine receptors, especially D_2 receptors[1821].

Dopamine Stimulation Increases TS Symptoms

Drugs which stimulate dopamine neurons or receptors make TS symptoms worse. These include amphetamines, Ritalin, Cylert, L-DOPA, cocaine, and others[p524,1979]. An apparent contradiction to this was the observation that apomorphine, which strongly stimulates postsynaptic dopamine receptors in low doses, improves the symptoms of TS[582]. This actually supports the hypothesis since it has been shown that in small doses apomorphine preferentially binds to presynaptic dopamine receptors and inhibits the activity of the dopamine nerve[10]. In higher doses it binds to the postsynaptic dopamine receptors and then makes the TS symptoms worse.

Dopamine Turnover in Tourette Syndrome

The turnover of dopamine in the brain can be indirectly determined by examining the amount of HVA[p364] in the spinal fluid. This is most accurate when the patient is taking probenecid to block the loss of HVA from the spinal fluid. As with the studies on ADHD, since the level of HVA is related to the amount of probenecid, the ratio of the two provides the most accurate test. The following shows these results in a study where three groups of controls (learning disorders, other neurological disorders and developmental disorders) were compared to six children with Tourette syndrome[341,342].

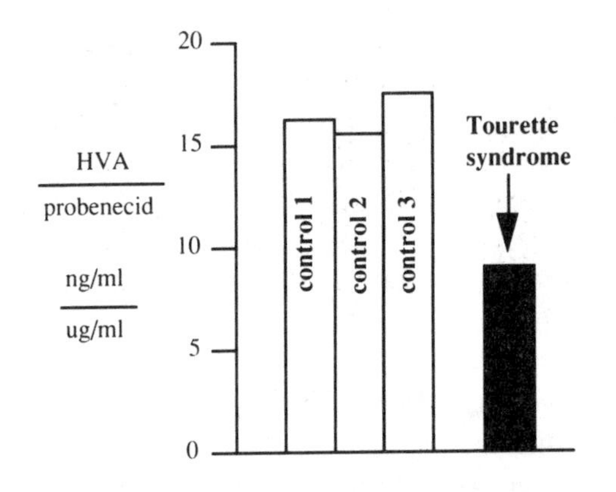

The level of HVA in the TS group was significantly lower than in the controls. Similar results were obtained in a separate study of nine additional TS patients[262].

Hyperactive dopamine or hyper-sensitive receptors? These results help to distinguish between two hypotheses — Are dopamine nerves overactive or are the dopamine receptors oversensitive? The low levels of HVA are most consistent with hypoactivity of dopamine neurons with a secondary hypersensitivity of the postsynaptic receptors.

Sleeping Sickness: Lessons for Tourette Syndrome

Among all the other events of the 1920's there was also a unique medical event — the period of the sleeping sickness. This was called *encephalitis lethargica*—encephalitis meaning infection of the brain, and lethargica meaning sleep. In the latter half of the century we think of encephalitis lethargica as being the cause of Parkinson's disease since many individuals who had this disorder in the 1920's developed Parkinson's disease many years later. However, almost lost to our memory, except in medical journals, are the acute effects of this disorder. Devinsky[484] has reviewed these and pointed out their remarkable similarity to TS. "Along with the symptoms of Parkinson's disease, obsessive-compulsive behaviors, tics, and oculogyric crises[p549] may also develop. Vocalizations similar to those observed in TS (e.g. curses, grunts, barks, and squeaks) and obsessive-compulsive behaviors are especially frequent." At the beginning of the illness they often showed "a mounting excitement of mind and body, violent movements, tics, and compulsions of all kinds."

As mentioned in the chapter on ADHD, in 1923 Ebaugh[512] described a syndrome that followed acute encephalitis in children. They were:

> ". . .unmanageable and unable to progress in school. Frequently they had disturbed the whole class, since they were quarrelsome and impulsive and often left the school building during class exercises. Nor did these children respond to discipline. . . In two cases marked sexual precocity was exhibited. In another case, the patient attempted to kill other members of his family. . . Normal children who were well adjusted in school and home changed abruptly to a state of hyperkinesis, characterized by transient periods of talkativeness, tension states and emotional outbreaks often leading to general incorrigibility and inability to remain in school... They showed marked emotional lability consisting of capricious moodiness, irritability and affability leading to quarrelsomeness and often overaffectionateness. [One girl was] bossy, quick tempered, and impulsive and showed little respect of authority. . . She gave evidence of definite sexual precocity. One boy had streaks of cruelty, and on one occasion stabbed a schoolmate with a knife. Another boy, who formerly had been

> quiet and orderly, became obscene and masturbated in public. He was silly and sang love songs for prolonged periods. Another boy gave evidence of an hysterical makeup and reacted to unpleasant situations by developing many complaints. For example, he became deaf when told to do anything unpleasant . . . He also showed sexual precocity. He frequently kissed girls whom he met on the street and on one occasion attempted intercourse.
>
> "Three children showed depressive tendencies, one boy attempted suicide by drowning, while another stated that he "was no good" and threatened to kill himself with a gun. . . In three of our children marked hysterical phenomena were observed. [One] developed spells of screaming and yelling to avoid the discipline of a correctional school."

The striking feature of these behaviors was that *they are typical of the behaviors common in some TS patients* and they also included motor and vocal *tics*.

> "Irregular movements of the head or one extremity, sniffing of the nose, clicking of the tongue, biting of the finger nails, and obsessive spitting have been observed in these cases. . . all types of incoordinated movements occurred for prolonged periods in seven cases."

Insomnia also developed and was accompanied by motor restlessness. Of these 17 cases, five progressively improved and four others improved slowly. The remaining eight had permanent changes.

This syndrome so closely mimics all the behaviors seen in grade 3 TS that I will call it *Postencephalitic Tourette Syndrome*. If we knew what caused this disorder we would have significant insights into the cause of TS. For this reason, the examination of the brains of children dying of postencephalitic TS should be very informative, and in fact, they were.

Lesions in the ventral tegmental nucleus in encephalitis Since encephalitis lethargica often was a fatal disorder, many studies of the brains of individuals who died in the acute phase were done. These showed *involvement of the region of the ventral tegmental nucleus* and nearby areas[2015]. As mentioned repeatedly, *this is the source of the dopaminergic nerves that pass to the limbic system and the frontal lobes*.

Brain stem injuries in the region of the ventral tegmental area can also cause emotional lability, irritability, short attention span, paranoia, auditory hallucinations, and poor memory[1960].

These observations provide clear evidence that defective mesolimbic and mesocortical dopamine neurons can lead to Tourette syndrome-like behavior and symptoms.

Based on studies in monkeys and other animals, Devinsky[484] has suggested that the cingulate gyrus, part of the limbic cortex, and the area around the ventral tegmental area are involved in the production of vocal tics.

Parkinsonism, L-DOPA and Tourette syndrome In all of medicine there scarcely exists a clearer demonstration of the behavioral effects of a specific chemical, on a specific part of the brain, than that shown by the effects of treating Parkinson's disease with L-DOPA. Years after acute encephalitis individuals may develop Parkinson's disease-like disorder — appropriately called *post encephalitic Parkinsonism*. Oliver Sacks[1676], in his book *Awakenings*, describes how these patients: rigid, immobile, frozen, and speechless for years, almost as if in a time capsule, wake up after being treated with L-DOPA , which makes up for the low levels of dopamine in the substantia nigra. However, like Ritalin in Tourette syndrome, L-DOPA in Parkinsonism can be a double-edged sword. As the rigidity of oculogyric crises improved, not only tics but the whole range of TS symptoms may develop. This further illustrates the maxim — too much dopamine -> TS, too little dopamine -> Parkinsonian symptoms.

The following are some of the descriptions from *Awakenings* that show the range of TS behaviors that can occur *after brain dopamine is increased with L-DOPA* They represent such *an eloquent and explicit demonstration of not only tics but the whole range of TS spectrum behaviors* that they deserve presentation of some detail.

Francis D.

"She seemed somewhat driven and unable to tolerate inactivity."

"She had just washed her hands (she now felt a 'need' to wash them thirty times daily)."

"She exhibited extreme pressure of speech, and now showed, for the first time, an uncontrollable tendency to repeat words and phrases again and again (palilalia)."

"Her gnawing and biting, compulsions, certain violent appetites and passions, and certain obsessive ideas and images — could not be dismissed by her as 'purely physical' or completely 'alien' to her 'real self,' but on the contrary, were felt to be in some sense *releases* or *exposures* or *disclosures* or *confessions* of a very deep and ancient part of herself, monstrous creatures from her unconscious and from unimaginable physiological depths below the unconscious, prehistoric and perhaps prehuman landscapes whose features were at once utterly strange to her, yet mysteriously familiar, in the manner of certain dreams."

Magda B.

". . . developed a curious 'touching tic,' continually touching the rails, the furniture, and above all — various people as they passed in the corridor."

Rose R.

"She became agitated in the middle of the day, asserting that seven dresses had been stolen from her closet, and that her purse had been stolen. She entertained dark suspicions of various fellow patients: no doubt they have been plotting for weeks before. Later in the day, she discovered that her dresses were in fact in her closet in their usual position. Her paranoid recriminations instantly vanished: 'Wow!' she said, 'I must have imagined it all.'"

". . .would experience 'an absolute torrent of thoughts,' rushing through her mind: these thoughts did not seem to be 'her' thoughts, they were not what she wanted to think, they were 'peculiar thoughts' which appeared 'by themselves.'"

"In the past few days, she had recorded innumerable songs of an astonishing lewdness, and reams of 'light' verse all dating from the twenties."

"Her tics of the right hand became almost too fast for the eye to follow, their rate having increased to almost 300 per minute. [Later she] showed very obvious palilalia repeating entire sentences and strings of words again and again: 'I'm going round like a record,' she said, 'which gets stuck in the groove'...her tics became more complex, and were conflicted with defensive maneuvers, counter-tics and elaborate rituals. Thus Miss R. would clutch someone's hand, release her grip, touch something nearby, put her hand in her pocket, withdraw it, slap the pocket three times, put it back in the pocket, wipe her chin five times, put it back in the pocket, wipe her chin five times, clutch someone's hand...and move again and again through this stereotyped sequence."

Hester Y.

"Compulsive tic-like movements made their appearance yesterday. Mrs. Y. slept poorly during the night. Her right hand now shows exceedingly quick darting motions, suddenly touching her nose, her ear, her cheek, her mouth. When I asked her why she had these movements, she said, 'it's nothing, it's nothing. They don't mean a thing, it's just a habit, a habit—like my humming's a habit.' Her movements were extraordinarily quick and forceful, and her speech seemed two or three times quicker than normal speech; if she previously resembled a slow-motion film, or a persistent film-frame stuck in the projector, she now gave the impression of a speeded-up film — so much so that my colleagues, looking at the film of Mrs. Y. which I took at this time, insisted the projector was running too fast. Her threshold of reaction was now almost zero, and all her actions were instantaneous, precipitate and excessively forceful."

"If Mrs. Y. before L-DOPA, was the most *impeded* person I have ever seen, she became, on L-DOPA, the most *accelerated* person I have ever seen."

"...if one tried to prevent her kicking her legs, an unbearable tension developed which sought discharge in pounding of the arms; if these were constrained, she would lunge with her now-immobilized head from side to side; and if this was constrained she would scream."

"Her writings, at this time, ...were long paranoid tirades against various nurses and nursing aids who had 'persecuted' and 'tormented' her since she entered the hospital, and vengeful fantasies as to how she would now 'get back' at them. She reverted, again and again, to a former neighbor in the hospital, a hostile dement who two years before had thrown a glass of water all over her."

"When she was not screaming at this time, she showed gasping and panting, and violent out-thrustings of her tongue and her lips. I requested the nurses to give her 10 mg of Thorazine, intramuscularly, and within fifteen minutes her frenzy subsided, and was replaced by exhaustion, contrition and sobbing, the terror, suspicion and rage went out of her eyes, and were replaced by a look of affection and trust."

"...the torrential successions of tics or bizarre 'behavioral fragments' she may show, are essentially fragmentary and go with a fragmentation of time and space itself."

Rolando P.

"...very flushed, boisterous, insomniac, somewhat manic, and frenzied; his movements, previously so meagre and with so little dynamic background, were now violently forceful, and involved as background the whole of his body; he was intensely vigilant and over-alert."

"His attention was constantly attracted hither and thither, and seemed to be intensified but also short-lived and distractible."

"Sexual and libidinous arousal was still more marked, and the transit of any female personnel across Mr. P.'s field of vision would immediately evoke an indescribably lascivious expression, forced lip-licking and lip-smacking movements, dilation of the nostrils and pupils, and uncontrollable watching; he seemed — visually — to grab and grasp the object of his gaze and to be unable to relinquish it till it passed from his field of vision."

"Powerful sexual urges continued throughout this time, manifest as repeated erotic dreams and nightmares, as frequent and somewhat compulsive masturbation, and (combined aggressiveness and perseveration) as a tendency to curse, to coprolalia, and verbigerative singsong pornoloquies with obscene refrains."

The extraordinarily delicate balance that these patients demonstrated, between too much and too little dopamine, is well described in Mr. P.

"Particularly striking, even on 1 gm daily, was a sudden 'awakening' which occurred every evening — to flushed, bright-eyed, quick-glancing, loud-voiced, forceful, lascivious, expansive, manic-catatonic akathisia — a transformation which often occurred in a minute or less; equally acute (and equally difficult to ascribe to any simple, dose-related effect of the drug) was the reverse change — to compacted, contracted, aphonic akinesia."

Miriam H.

". . .became more demanding and impatient; she is not able to make her needs known in a loud voice, if need be, and if this is not heeded screams shrilly. These occasional screamings are felt as ego-alien, and are followed at once by contrition and apologies."

"I have never seen a human being who could speak like Miss H.: she could easily beat any news-announcer, because she can talk at 500 words a minute without missing a syllable."

". . .a 'nervous' cough and throat-clearing started, associated with a recurrent tic-like feeling of something blocking or scratching her throat; the hiccough disappeared with the onset of throat-clearing and coughing, as if it had been 'replaced' by these symptoms."

Lucy K.

"She was also notorious for her rages, which would come on extremely suddenly, with scarcely a moment's warning; during these she would curse with great violence and fluency in a particularly sarcastic and wounding way. . .The unmistakably murderous quality of these rages, combined with their total unexpectedness, had a peculiarly unnerving effect. These paroxysms of terror or pleasure, of laughter or rage, would rarely last more than a minute or so; they would vanish as abruptly as they had come."

Margaret A.

". . . felt a constant thirst and ravenous hunger — she felt impelled to drink at the water fountain almost incessantly, and her appetite and voracity reminded her of what she had experienced in the early thirties [during the acute stage of her illness]. Her mood became exalted: she felt 'a wonderful flying and floating feeling inside,' became intensely sociable, talked continually, found reason to run up and down the stairways."

"...her dreams were extremely vivid, with a tendency to nightmares; and her mood, though consistently excited, was labile in affect, with sudden veerings from stormy hypomania to fearfulness and agitated depression."

"...she shuffles her feet, crosses and uncrosses her legs, suddenly belches, straightens her dress, pats her hair, belches again, slaps her hands, touches her nose, and belches again."

"Her feeding shows insatiable voracity and hurry: she tears at her food in an animal-like way, grunting with excitement, and stuffing it into her mouth, and when finished gnaws her fingers in an uncontrollable perseveration of greed. I observed too that when she ate, her tongue would shoot out of her mouth as she brought the morsel to her lips. I have the feeling that her tongue was enticed out by the food, and that eating evoked voluptuous pleasure."

"Miss A. has split into a dozen Miss A's — the drinker, the ticcer, the stamper, the yeller, the swinger, the grazer, the sleeper, the fearer, the lover, the hater, etc. — all struggling with each other to 'possess' her behavior. Her real interests and activities, have practically vanished, all have been replaced by absurd stereotypes. She is completely reduced for most of the time, to a 'repertoire' of a few dozen thoughts and impulsions, increasingly fixed in phrase and form, and repeated, compulsively, again and again. The original Miss A.— so engaging and bright — has been dispossessed by a host of crude, degenerate sub-selves — a schizophrenia-form fission of the once-united self."

These eloquent summaries of the effects of too much dopamine in the substantia nigra and ventral tegmental nucleus of patients with postencephalitic parkinsonism provide dramatic verification that TS is a spectrum of behavioral disorders[367], not simply a motor and vocal tic disorder[1480,1771]. When these quiescent Rip Van Winkles are brought to life with L-DOPA, their diseased, dopamine-depleted brains are hypersensitive to the effects of the drug and they are easily pushed into all the features of the entire TS spectrum. These are shown in the following box.

The Spectrum of Tourette Syndrome Behaviors Seen after Treatment of Postencephalitic Parkinsonism Patients with L-DOPA

aggression	impatience
agitation	insomnia
anger	mania
anxiety	<u>motor tics</u>
<u>complex tics</u>	obsessive thoughts
compulsive behaviors	palilalia
compulsive masturbation	paranoia
demanding	perseveration
depression	pressure of speech
driven behaviors	polydypsia
echolalia	rages
hearing voices	stereotyped behavior
hyperactity	touching
hypersexuality	<u>vocal tics</u>

This is a clear demonstration that the associated behaviors of TS are not due to the stress of having the tics. Their rapid onset and offset directly keyed to the dose of L-DOPA, and their appearance, often totally independent of tics, indicate they are but one of the many ways the basic chemical defect of TS can manifest itself.

To me, reading these vignettes was like going backward in time, recognizing in the treated Parkinsonian patients bits and pieces of the past thousand plus TS patients I have seen, struggling with the same symptoms of the chemical monster that had taken over their brains and bodies.

Damage to the Caudate and TS

Another illustration of the importance of the neostriatum in TS can be appreciated by the report[1608] of a 25 year old woman who was an honor student in high school and employed full time when she began to develop severe headaches. She disappeared for three days and when found she had undergone a dramatic personality change.

> "Her abnormal behaviors included vulgarity, impulsiveness, easy frustration, violent outbursts, hypersomnia, enuresis, indifference, wandering, increased appetite, polydipsia, hypersexuality, minor criminal behavior including shoplifting and exposing herself, and poor hygiene."

She had not done any of these things prior to her illness. A CT scan showed destruction of the head of both the right and left caudate nuclei. The cause was not determined.

This remarkable case also dramatically illustrates how a specific lesion of the dopamine-rich caudate nuclei can result in the full behavioral spectrum of TS. It also provides a demonstration of the earlier discussion[p354] that lesions of the caudate can precisely mimic all the symptoms produced by lesions of the frontal lobes.

Both Cocaine and TS Cause a Spectrum of Disorders

Cocaine is a potent dopamine-releasing agent[404,p524]. It is a popular street drug because of the rush of euphoria it produces. However, depending on the amount and duration of use it can cause a wide range of behavioral problems[1532]. These are similar to those produced by the L-DOPA treatment of Parkinson's disease and are shown in the following box.

Cocaine Abuse:
Spectrum of Induced Behaviors

- attention deficits
- compulsions
- depression
- euphoria
- hallucinations
- hyperactivity
- hypersexuality
- insomnia
- irritability
- learning disorders
- mood swings
- obsessions
- paranoia
- short temper
- stereotyped behaviors
- violent behaviors

Each of these symptoms can also be present in patients with Tourette syndrome. In contrast to the popular proposal that a specific neurotransmitter was responsible for a specific behavioral disorder, Post[374] proposed that the lesson from cocaine abuse was that *a single neurotransmitter can be involved in a wide range of different disorders.* The same situation may be occurring in Tourette syndrome[374].

TS patients who take cocaine, often experience a marked worsening of their symptoms[1320].

PET Scanning in TS

As shown with schizophrenia[p374] and ADHD, PET scanning is a sensitive technique for examining the function of the living brain. Only a few studies have been done in TS. When general metabolism was measured based on the uptake of glucose, there was a suggestion of a relative hypermetabolism in portions of the frontal and temporal lobes[310]. In seven TS patients, D_2 receptor activity was normal in the caudate and serotonin receptor activity was normal in the frontal lobes[1807]. One TS brain was examined at autopsy and the level of D_2 receptors was normal in the caudate and globus pallidus and somewhat low in the putamen[1807]. While these results did not support a D_2 receptor hypersensitivity in the striatum, it did not rule out D_2 receptor hypersensitivity in the nucleus accumbens or an abnormality of dopamine autoreceptors[1807].

Frontal Lobe Syndromes

I have emphasized several times that many of the symptoms of ADHD and TS are due to a dysfunction of the frontal lobes. The following diagrams show the important role of the frontal lobe syndrome in human genetic disorders and their animal model counterparts produced by the destruction of dopamine pathways to the prefrontal cortex:

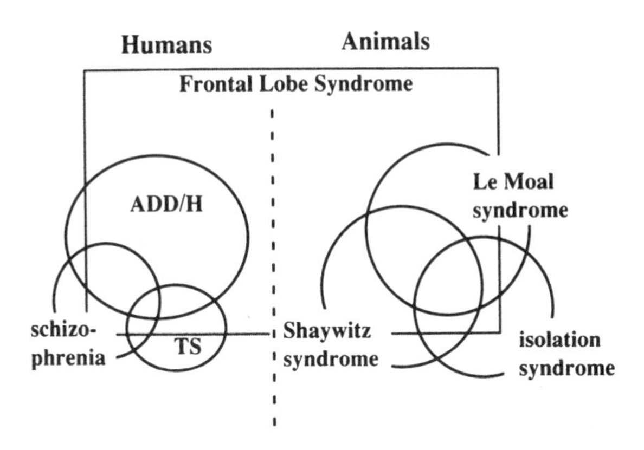

This re-emphasizes the important role that dopamine imbalance plays in ADHD and Tourette syndrome.

Summary: Many of the symptoms of Tourette syndrome can be explained on the basis of abnormalities in dopamine. The presence of symptoms of the frontal lobe syndrome suggests there is a defect in the dopamine pathways to the frontal lobes, while the tics and other disinhibited behaviors suggest a hypersensitivity of the pathways to the limbic system and basal ganglia.

Parkinson's disease is due to a defect of dopamine in the substantia nigra and ventral tegmental nucleus. Patients with Parkinson's disease secondary to encephalitis (sleeping sickness) are very sensitive to treatment with L-DOPA and too much medication produces the full spectrum of TS—not just the motor and vocal tics, but the entire range of behaviors previously described. This provides dramatic verification that TS is a behavioral spectrum disorder, that these other behaviors are not simply an emotional reaction to having tics, and that hypersensitivity to dopamine in the striatum plays an important role in causing TS.

These correlations are summarized in the diagram on the following page.

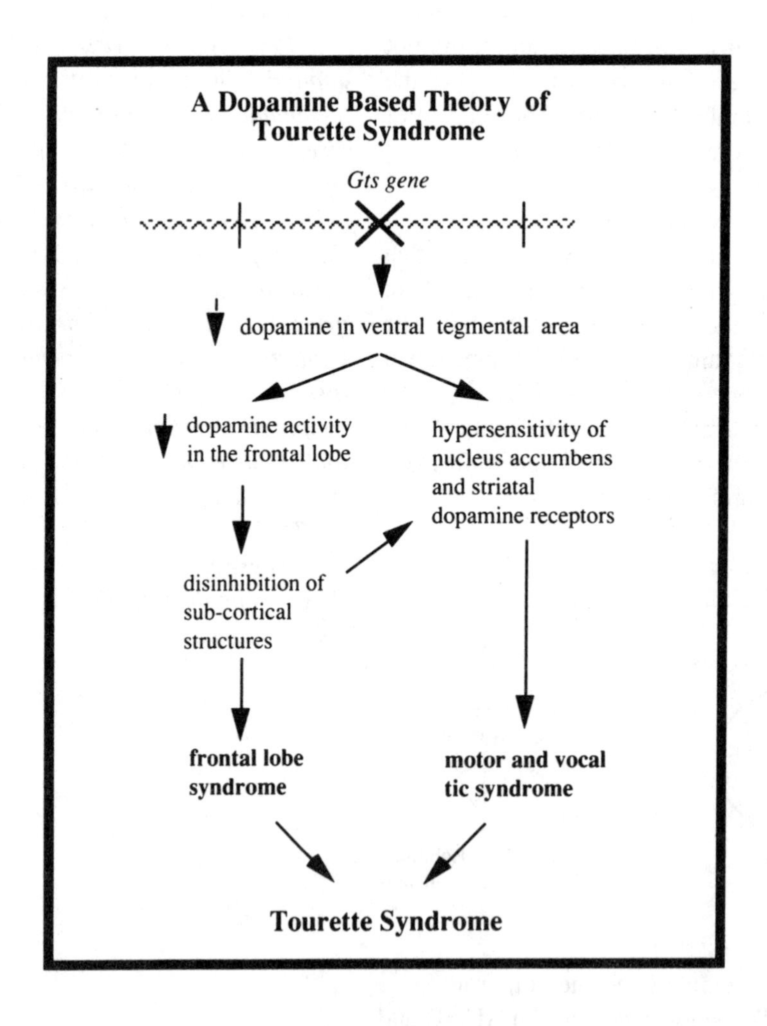
A Dopamine Based Theory of Tourette Syndrome
Gts gene
dopamine in ventral tegmental area
dopamine activity in the frontal lobe
hypersensitivity of nucleus accumbens and striatal dopamine receptors
disinhibition of sub-cortical structures
frontal lobe syndrome
motor and vocal tic syndrome
Tourette Syndrome

Chapter 58

Norepinephrine — The Modulator

We have all heard of adrenalin. "An adrenalin boost" and "running on adrenalin" are common euphemisms in our language. Adrenalin, also known as epinephrine, is made in the adrenal glands. When we are frightened or anxious or simply need extra energy, the adrenalin pours out and speeds everything up. The adrenal glands also make noradrenalin, or norepinephrine, which in many ways functions like adrenalin. The brain also makes norepinephrine which acts as a modulator of the function of other neurotransmitters, especially dopamine. This intimate connection with dopamine is not too surprising since a single minor change in the structure of dopamine turns it into norepinephrine.

Dopamine -> Norepinephrine

The structure of dopamine is:

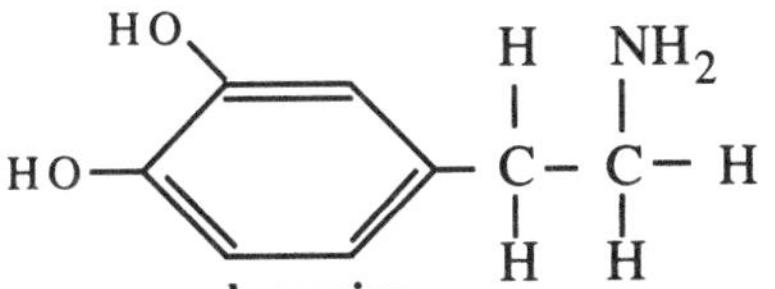

dopamine

The enzyme, *dopamine ß-hydroxylase,* makes a minor change in dopamine by exchanging an —H for an —OH or hydroxyl group as follows:

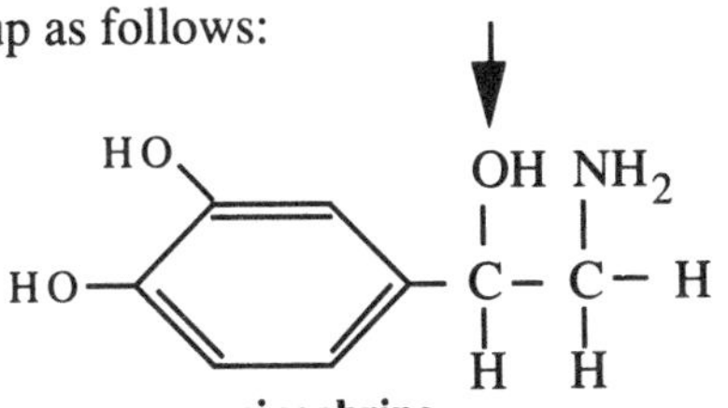

norepinephrine

Adding a methyl group (—CH_3) to norepinephrine converts it to epinephrine or adrenalin:

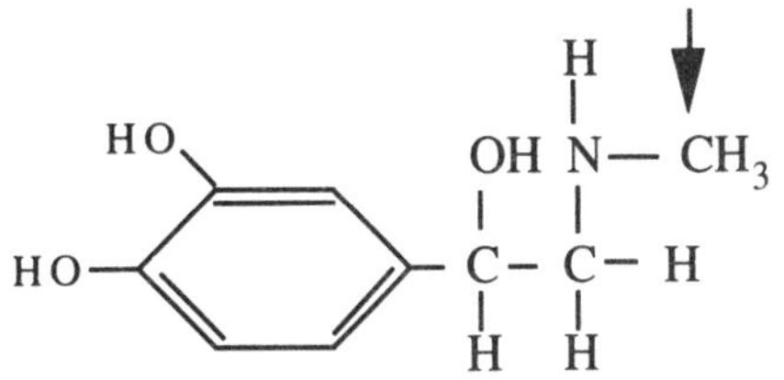

epinephrine (adrenalin)

Like dopamine, two enzymes, *monoamine oxidase* and *COMT* [p364] break down norepinephrine and epinephrine into a variety of other compounds.

The Locus Ceruleus

The brain's adrenal gland is an area called the *locus ceruleus* which produces over 70 percent of the norepinephrine in the brain. This small area is the major source of norepinephrine neurons passing to other areas of the brain. This is shown in Figure 1 on the **following page**.

Norepinephrine nerves pass to the cerebral and cerebellar cortex, the limbic system, brain stem and spinal cord. This massive enervation of norepinephrine fibers to virtually the whole brain is accomplished by less than a few thousand neurons in the locus ceruleus[1354a]. This is made possible by a marked branching of the norepinephrine nerves. In contrast to the extensive network of nerves leading from the locus ceruleus to the rest of

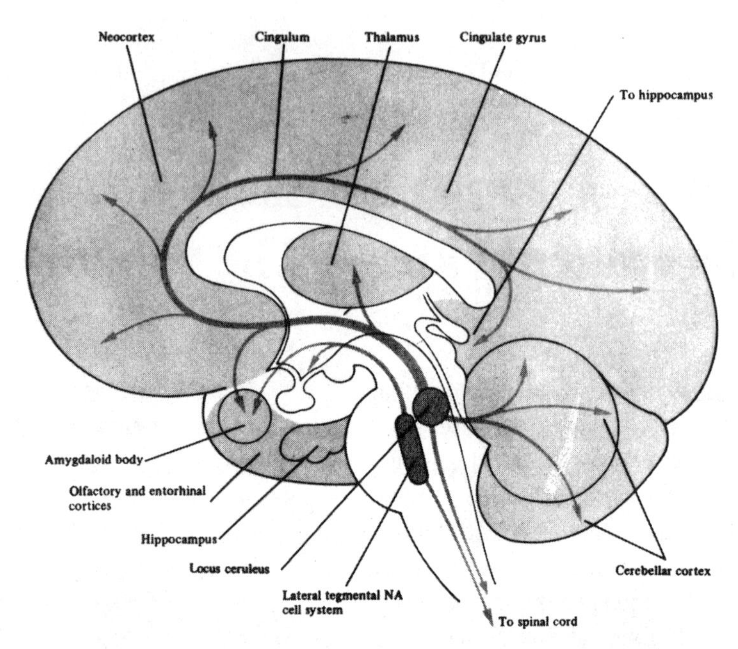

The norepinephrine pathways in the brain. From L. Heimer, *The Human Brain and Spinal Cord.* Springer-Verlag, New York, 1983.

the brain, there are relatively few inputs to the locus ceruleus — the major ones being two nuclei in the medulla closely involved in the sympathetic nervous system[68].

Free nerve endings in the cortex A unique aspect of norepinephrine[481] and serotonin[480] nerve endings in the cortex is that many of the neurotransmitter containing bulges do not end at typical synapses. This raises the possibility that when norepinephrine and serotonin are released in the cortex they act by bathing other nerve cells. In this regard they may work more like localized hormones than neurotransmitters.

Receptors

In order to understand how clonidine works in the treatment of TS, it is necessary to understand the various norepinephrine receptors. Because of the name adrenalin these are called *adrenergic receptors.*

Alpha and beta receptors There are two classes of adrenergic receptors — the α-receptors that *inhibit* and the β-receptors that *stimulate* other nerves. These receptors occur at two locations: the postsynaptic, and the presynaptic or autoreceptors[116,1856,2072]. Autoreceptors respond to the same neurotransmitter that the nerve releases. The α_2-autoreceptors *inhibit* (-) the release of norepinephrine, and β-autoreceptors *stimulate* (+) its release.

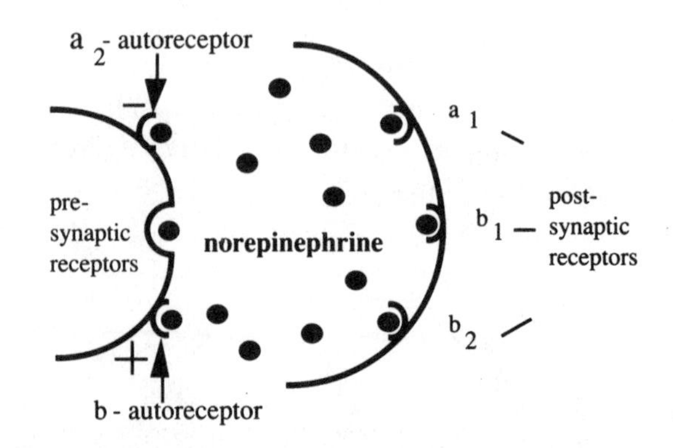

The major α_2-receptor agonists are norepinephrine itself and *clonidine*. When norepinephrine is released it binds to the postsynaptic receptors to activate or inhibit the next nerve, and binds to the α_2-autoreceptors to inhibit further release of norepinephrine. *Clonidine stimulates the α_2-autoreceptors and thus inhibits the release of norepinephrine.*

Norepinephrine and Animal Behavior

In the early years of interest in the role of the catecholamines (dopamine and norepinephrine) in behavior, norepinephrine was considered to play the major role. However, as the years went by, dopamine was implicated first in Parkinson's disease, then in schizophrenia, attention deficit disorder and Tourette syndrome. As the list of disorders attributed to dopamine grew, the list for norepinephrine shrunk. At one point it was suggested that "so much of the behavior previously attributed to norepinephrine now has been found to be mediated by dopamine that questions arise about the role of norepinephrine"[1194]. This emerging view was placed in perspective by the proposal that to understand the role of norepinephrine in behavior it must be examined in terms of a norepinephrine-dopamine balance[48]. Examining norepinephrine in isolation was too confusing; examining it in relation to dopamine finally began to make sense. Norepinephrine neurons are viewed as modulating dopamine neurons and this modulation is different when animals are under various types of stress. The following are some examples.

Stereotypy, hyperactivity, and aggression When stereotypy is produced by giving amphetamines to activate dopamine neurons, some of the dopamine is converted to norepinephrine.

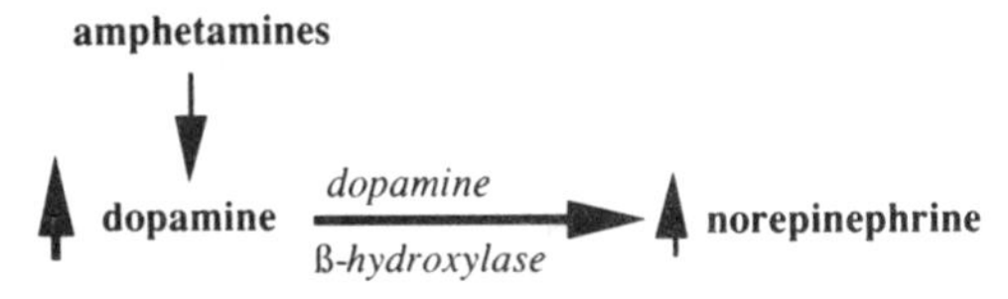

If the production of norepinephrine is blocked by adding disulfram, a drug that blocks dopamine ß-hydroxylase, stereotypy is markedly enhanced[401,532a,1573].

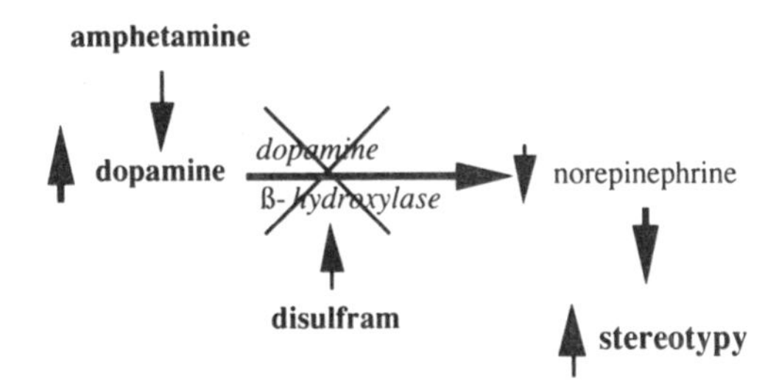

Inhibition of dopamine ß-hydroxylase also increases Ritalin-induced hyperactivity, stress-induced aggression and electrically induced self-stimulation[48]. Amphetamine-induced stereotypy is prevented by chemicals which destroy only dopamine neurons. However, if both dopamine and norepinephrine are depleted, stereotypy persists[913].

These results indicate that *norepinephrine can significantly modulate the effects of dopamine*. The inhibition of dopamine ß-hydroxylase results in the increased synthesis and release of dopamine[28,1573]. This can be explained on the basis of a feedback loop in which norepinephrine inhibits the synthesis of dopamine as follows[71]:

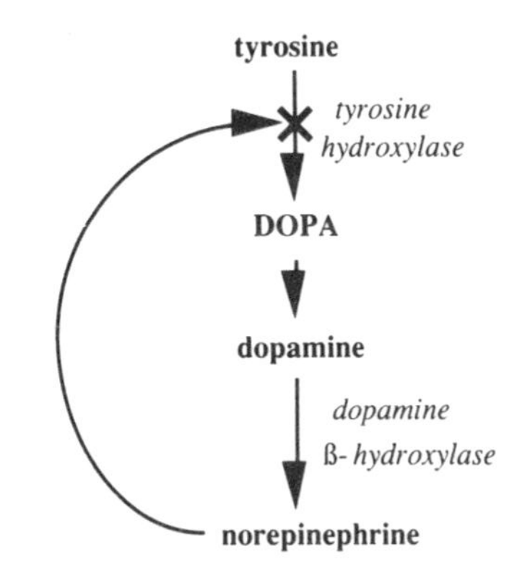

Here norepinephrine inhibits the enzyme tyrosine hydroxylase. This feedback loop regulates the production of both dopamine and norepinephrine and keeps the levels of both in check. Blocking dopamine ß-hydroxylase decreases the production of norepinephrine. This releases the feedback inhibition of tyrosine hydroxylase and stimulates more dopamine activity causing greater stereotypy, hyperactivity, aggression and self-stimulation:

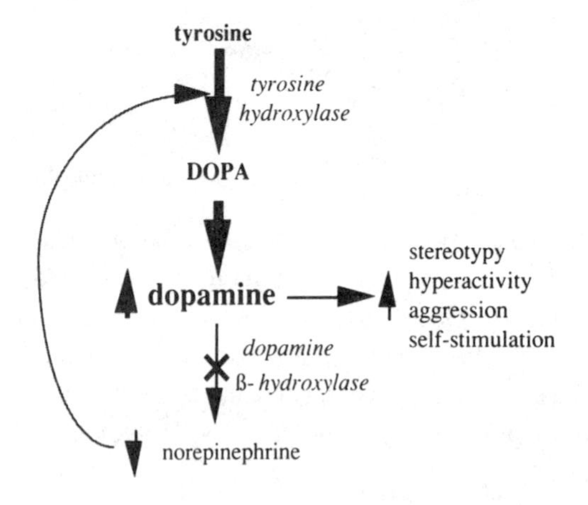

Part of these effects are also due to nerve pathways from the locus ceruleus to the substantia nigra[498,1561].

Stress versus no stress All of the above behaviors are produced by various stressors (amphetamine, Ritalin, electrical shock, self-stimulation). When these stressors are absent, decreasing norepinephrine has the opposite effect and tends to *decrease* different behaviors[148,401].

Norepinephrine and vigilance Vigilance refers to an alertness and sensitivity to the environment. The rate of firing of locus ceruleus increases markedly when animals are presented with *novel* or challenging stimuli[65-68,622]. Such vigilance or *arousal* plays a critical role in learning and survival. An insight into how this vigilance is brought about comes from the observation that a major effect that norepinephrine neurons have on the many brain regions they go to, is to increase the sensitivity of that area to stimuli and decrease the background activity. In electronic parlance this is equivalent to increasing the "signal to noise ratio"[1699]. This is summarized as follows:

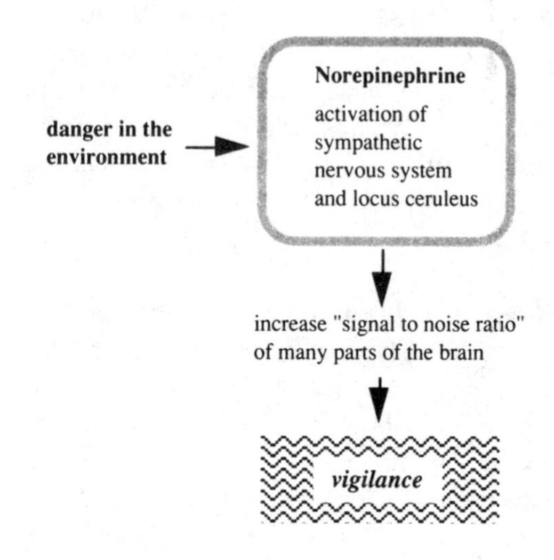

This is consistent with earlier studies that have indicated the norepinephrine system is important in allowing animals to filter out irrelevant stimuli and to habituate, or stop responding to, a stimulus after it has been presented several times[1267-1269].

Norepinephrine and Serotonin

The interactions between norepinephrine and serotonin are also important. These are brought about by the connections between the norepinephrine-rich locus ceruleus and the serotonin-rich raphe nuclei[656]. The norepinephrine nerves usually have a tonic activating effect on the serotonin nerves[1198,1437,1555,1556,1672]:

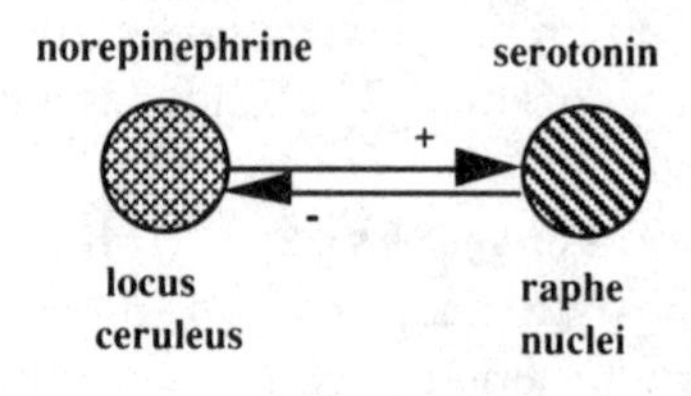

In reciprocal fashion, serotonin neurons from the raphe nuclei have an inhibiting effect on norepinephrine nerves in the locus ceruleus[1086,1151,1511, 1684,1867].

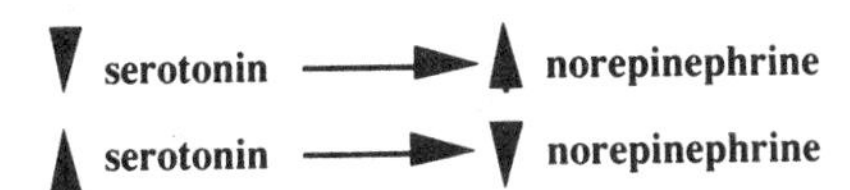

This relationship is important since there is evidence for both a decrease in serotonin activity[p457] and an increase in norepinephrine activity[p561] in Tourette syndrome. These would not be explained if the major effect of increased norepinephrine is to stimulate serotonin activity. If this were the case the levels of both norepinephrine and serotonin should be increased in TS. However, a decrease in serotonin and an increase in norepinephrine would occur *if the primary defect was in serotonin resulting in a disinhibition of norepinephrine neurons*. This *would* explain the results seen in TS.

Clonidine As noted previously, in low doses, clonidine acts as a presynaptic α_2-receptor agonist and inhibits the production of norepinephrine and inhibits serotonin neurons[1598,1636, 1724,1772,1904,1971]. However, at higher doses, it acts as a postsynaptic α_1-receptor agonist[29] and stimulates serotonin neurons. Long before clonidine was ever used to treat Tourette syndrome, the Canadian neurologist Andre Barbeau observed that when high doses of clonidine were givèn to rabbits, in addition to a drop in norepinephrine content of the caudate nucleus, there was a striking increase in serotonin[114]. This is shown in the **figure on the right ->.**

Dopamine remained unchanged. While the effectiveness of clonidine in some cases of Tourette syndrome has led to the proposal that the primary defect is in norepinephrine, this profound effect on serotonin should be kept in mind.

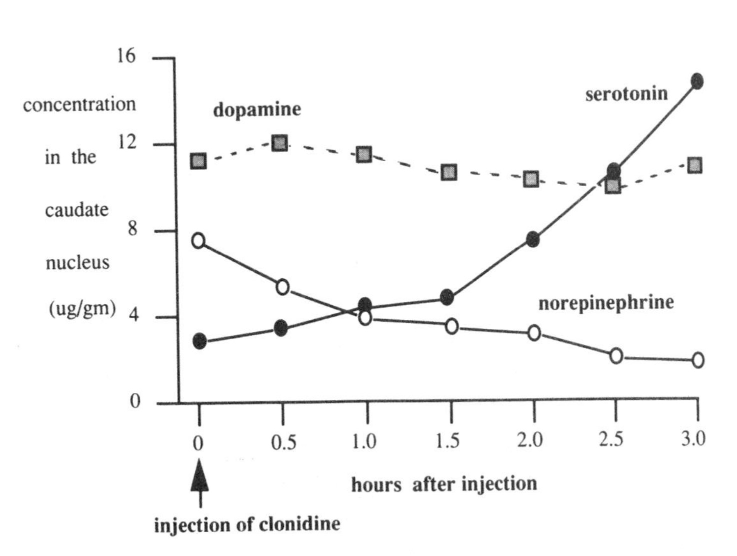

Norepinephrine:dopamine in the prefrontal cortex When the norepinephrine neurons passing from the locus ceruleus to the prefrontal cortex are destroyed there is a marked decrease in norepinephrine in the prefrontal lobe while dopamine increases by up to 40 percent[106,538, 851,1925]:

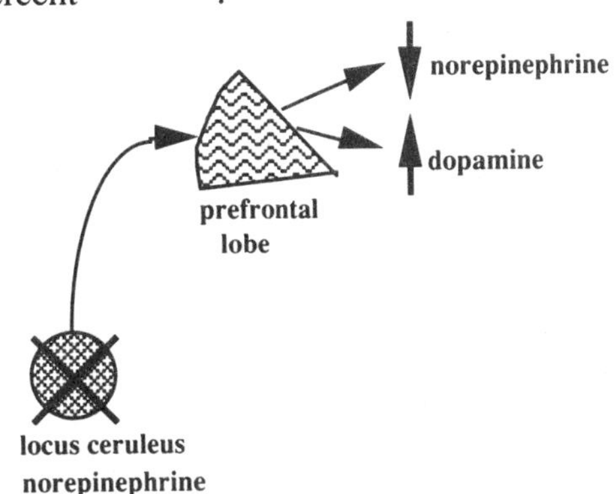

This suggests another way in which lowering norepinephrine with clonidine might be helping children with ADHD, by increasing prefrontal lobe dopamine activity.

Norepinephrine and Human Behavior

The role of norepinephrine in psychiatric disorders, especially depression, is somewhat like the story of two turtles and a hare. The early theories of the role of catecholamines in depression and schizophrenia heavily emphasized norepinephrine[250,1274,1715,1716]. However, with the passing years dopamine and serotonin have slowly overtaken norepinephrine in importance. Still, norepinephrine clearly plays an important role.

Depression, mania and norepinephrine The catecholamine theory of depression and mania suggests that norepinephrine is too low in depression and too high in mania[250,1274,1715,1716].

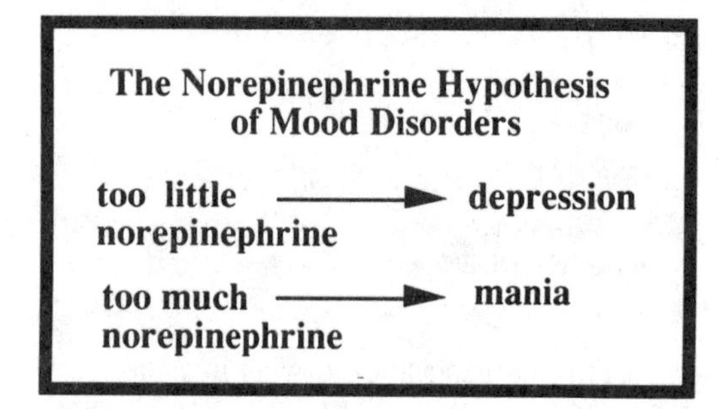

This theory was initially based on the effect of three drugs: imipramine, iproniazid and reserpine.

Imipramine The development of the first effective antidepressant medication was an outgrowth of the successful use of Thorazine (chlorpromazine) in treating schizophrenia[p373]. The Swiss drug company Ciba-Giegy, was making modifications in Thorazine and found a new compound, imipramine (Tofranil) that was well tolerated and had effects on animals that suggested it might have antipsychotic effects. The firm arranged with a German psychiatrist, Roland Kuhn, to try the new drug on some schizophrenic patients. While it had no effect on the delusions and hallucinations, Kuhn noted that severely depressed patients got much better[1101]. The fact that imipramine potentiated the effects of norepinephrine suggested its mode of action was related to this neurotransmitter[1795a]. This was verified when Axelrod and his colleagues[738,886,1717,1795a] showed that imipramine inhibited the re-uptake of norepinephrine in the synapses[1717].

Imipramine has the following structure:

CH_2-CH_2, N, $(CH_2)_3$, N, CH_3 CH_3

imipramine

Because of the three rings, it and other related compounds are often called *tricyclic* antidepressants.

Iproniazid Serendipity plays an important role in medicine and psychiatry and iproniazid is no exception. This was an antibiotic that was being tested on patients with tuberculosis. The major effect it had was to markedly elevate their mood[1204]. It was later shown to be a monoamine oxidase[p363] inhibitor that elevated norepinephrine and serotonin by inhibiting their breakdown in the synapse[1460,1832].

Reserpine Reserpine was the first drug found to be useful in the treatment of hypertension. One of its side effects was the production of severe depression in some patients[1374]. Reserpine causes the release and depletion of all three major neurotransmitters[1790]. The observation that reserpine can improve mania is consistent with the idea of there being too much norepinephrine in mania.

Additional observations consistent with the norepinephrine theory include the fact that monoamine oxidase inhibitors that prevent the breakdown of norepinephrine, and amphetamines that cause a release of norepinephrine, both alleviate depression, and agents that block the action of norepinephrine on its receptors can cause depression[1442].

Studies of the breakdown products of norepinephrine in the urine and spinal fluid of patients with depression and mania have given mixed and conflicting results. There is so much overlap with normal levels that these tests do not aid in the diagnosis.

The problem of the delayed effect of antidepressants Although the observation that tricyclic antidepressants increase the amount of norepinephrine in the synapses makes a beautiful theory of the cause and cure of depression, it is marred by the fact that this effect occurs immediately after the drug is given but it takes 3 to 5 weeks for the depression to go away[1789]. This embarrassing lag has caused problems with a straightforward depression = too-little-norepinephrine hypothesis. Many research studies have attempted to explain this lag. Most concur that *the real effect of tricyclic antidepressants is to slowly decrease the number or sensitivity of β–adrenergic receptors in the brain* [1901a,2022]. This even occurs for those antidepressants whose primary effect is on serotonin[266]. Electroshock treatment has the same effect[1453].

In summary, despite some ripples the norepinephrine theory of depression continues to be quite impressive and explains a wide range of observations[2022]. These neurons are just one part of a complicated circuitry and *changes in norepinephrine may be secondary to changes* in other neurotransmitters.

Schizophrenia and norepinephrine As powerful as the dopamine theory of schizophrenia is, some have suggested that a defect in norepinephrine may be the primary problem [1927,928,1139]. Norepinephrine is also an important neurotransmitter in the limbic system. Depending on the circumstances it can either stimulate or inhibit other neurons[1635,1747]. Because of these characteristics, it has been suggested that norepinephrine may act like the *"gain set"* on a recorder — capable of either amplifying or suppressing responses depending upon how it is set[1354a]. This property allows norepinephrine to exert diverse and often variable effects on the nervous system[927,928] regulating pleasure centers[p522], memory, learning, anger, aggression[701], sleep and many hormonal functions.

Brain and spinal fluid levels of norepinephrine are elevated in some schizophrenics [575,775,1110]. This may cause a hyperarousal state[927,1079] through its effect on the limbic and reticular system[927]. Drugs which block β-adrenergic receptors[2133], and clonidine[636] which decreases synaptic norepinephrine, improve the symptoms of schizophrenia. All of the currently used antipsychotic drugs show more uniform blockage of α_1-noradrenergic receptors than of dopamine D_2-receptors[337]. Drugs which enhance synaptic norepinephrine may[916] or may not[733] make schizophrenic symptoms worse. The abnormalities in norepinephrine in schizophrenia may be secondary to changes in other systems.

Anxiety and norepinephrine Norepinephrine plays a prominent role in the response to stress, both in the body and the brain[904]. Electrical stimulation of the locus ceruleus produces a fear response and destruction of the locus ceruleus almost abolishes fear[1589]. Drugs that activate the locus ceruleus produce anxiety while those that decrease the activity of the locus ceruleus, such as Valium, decrease anxiety[905]. Both clonidine, which decreases synaptic norepinephrine, and tricyclic antidepressants, which increase synaptic norepinephrine, decrease anxiety and panic attacks[905]. This is probably due to the fact that the tricyclics actually suppress the activity of the locus ceruleus[1416], possibly by an effect of norepinephrine on autoreceptors.

The effect of norepinephrine on anxiety appears to be due to the close linkage between norepinephrine neurons and the limbic system[790a].

Tourette Syndrome and Norepinephrine

The primary evidence that changes in norepinephrine may be the major defect in TS comes from the observation that clonidine is effective in treating over half of the cases[p561]. While elevated levels of the breakdown products of norepinephrine in the spinal fluid and urine[341,342,2130] have been reported, low and normal levels have also been described[42,1008,1806,1910]. Elevated levels of brain norepinephrine have been proposed as an alternative to the dopamine theory of ADHD[1008].

Summary: A minor chemical alteration converts dopamine into norepinephrine. This close chemical linkage is reflected in a close functional linkage in which norepinephrine modulates the activity of dopamine neurons. The direction of this modulation can vary according to environmental circumstances, often being different under conditions of stress than in the absence of stress.

Clonidine, one of the most effective medications for treating Tourette syndrome, acts to decrease norepinephrine in the brain. It also increases brain serotonin and may increase dopamine in the prefrontal lobes.

The changes in norepinephrine in Tourette syndrome and related disorders are probably secondary to defects in other neurotransmitters.

Chapter 59
Serotonin — The Great Inhibitor

Serotonin is the third of the triumvirate of major neurotransmitters that play an important role in controlling our behavior. *Serotonin is an especially important neurotransmitter in relation to the impulse disorders* since it normally inhibits the motor actions controlled by the striatum and the primitive emotions regulated by the limbic system[428,534,566,1456,1825]. The concentration of serotonin is particularly high in the limbic system with some regions containing 30 times more serotonin than is present in the motor cortex[1451,1673]. When brain serotonin levels are low, the emotions and impulses come to the fore. Because of this, serotonin has been called "**the civilizing neurohormone**"[566].

The role of serotonin in behavior is central to the theme of this book and the following will show that all of the diverse behaviors in Tourette syndrome could be explained by a genetic defect in brain serotonin.

The Chemistry of Serotonin

Serotonin contains indoles which are composed of two rings of carbon, one of which includes a nitrogen molecule.

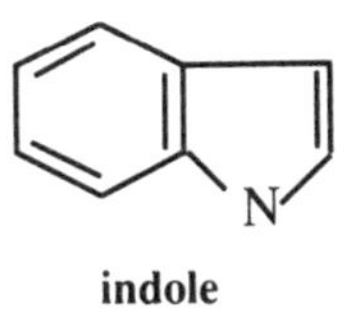

indole

When a hydroxyl,—OH, group and a —$CH_2CH_2NH_2$ group are added it becomes serotonin.

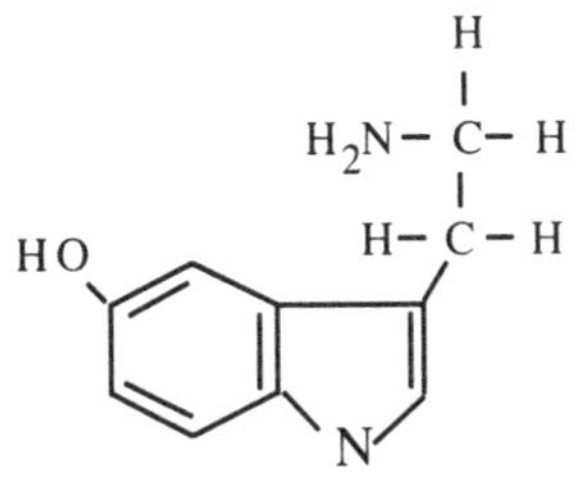

serotonin

In the body, serotonin is formed from a common amino acid, tryptophan.

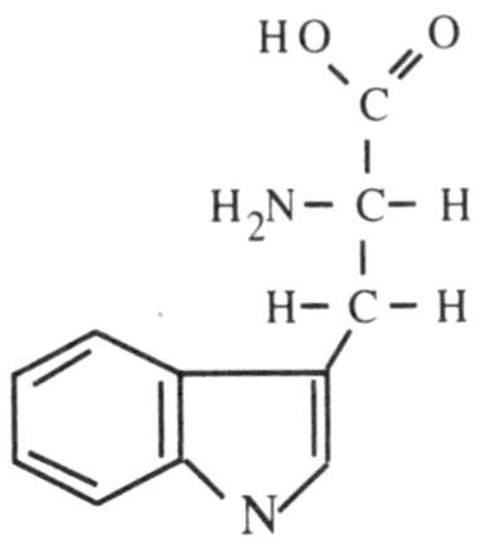

tryptophan

Since the body does not make tryptophan it must all come from the diet. Tryptophan gets into the brain by being actively transported across a barrier between the blood and the brain called the *blood-brain barrier*. Then an enzyme, *tryptophan hydroxylase*, adds a hydroxyl group, —OH, onto the 5 position of the indole to form 5-hydroxytryptophan (5-HTP).

5-hydroxytryptophan (5-HTP)

The conversion of tryptophan to 5-HTP is the step that regulates the rate at which serotonin is formed. Once this step is passed the next step takes place very quickly. Here a second enzyme, *L-aromatic acid decarboxylase,* removes the carboxyl group,

and replaces it with a hydrogen atom, —H, to form serotonin. Vitamin B_6 (pyridoxine) is a cofactor that aids this reaction.

serotonin

Serotonin is broken down by *monamine oxidase* p363 to form 5-hydroxyindoleacetic acid (5-HIAA).

5-hydroxyindoleacetic acid (5-HIAA)

The levels of this compound in the spinal fluid are frequently measured to indirectly determine the amount of serotonin in the brain.

In summary, the formation and breakdown of serotonin is as follows:

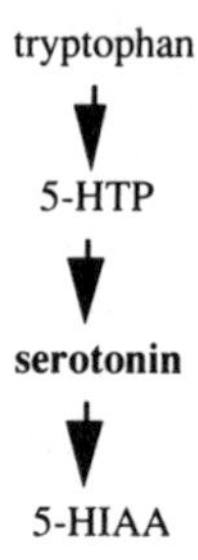

The Anatomy of Serotonin

The dorsal and median raphe nuclei contain the highest concentration of serotonin neurons in the brain[30,451,654] and contain most of the tryptophan hydroxylase of the brain. These nuclei are localized in the midbrain. Some additional raphe nuclei are present in the brain stem and send axons into the spinal cord. The major structures innervated by the serotonin nerves from the dorsal and median raphe nuclei are as follows:

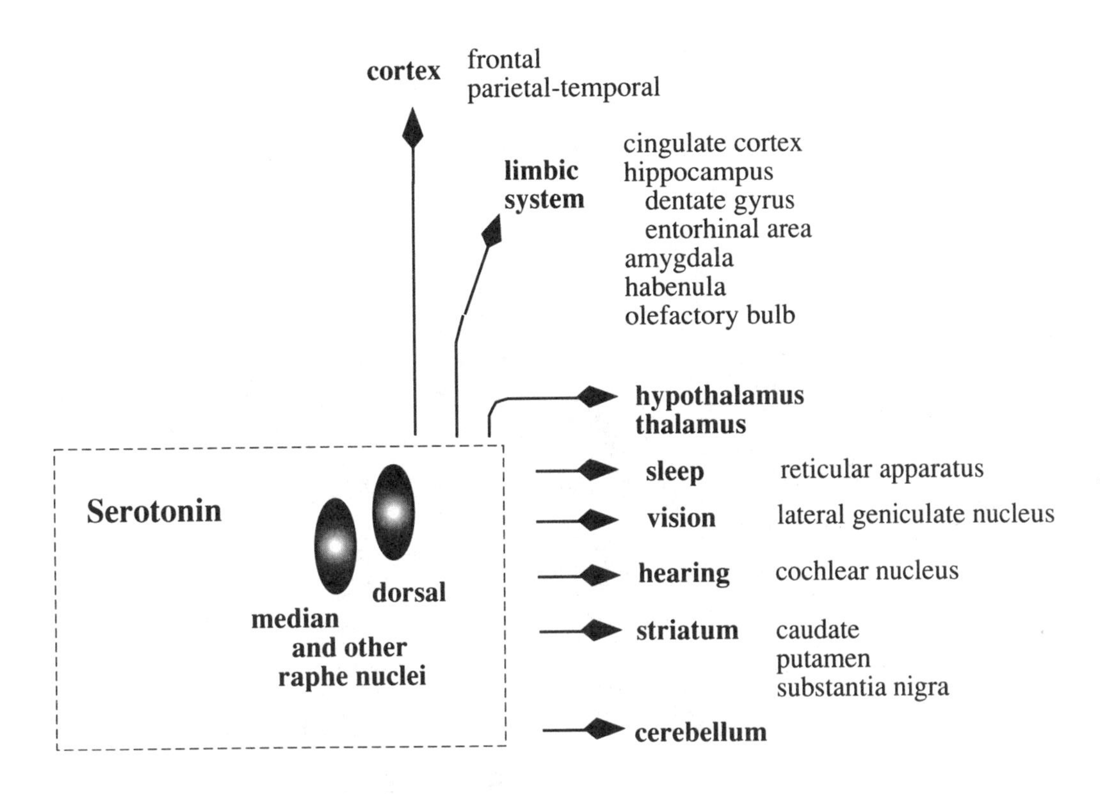

The location of the raphe nuclei and the structures they innervate in the human brain are shown in Figure 1 on the **following page**[880]:

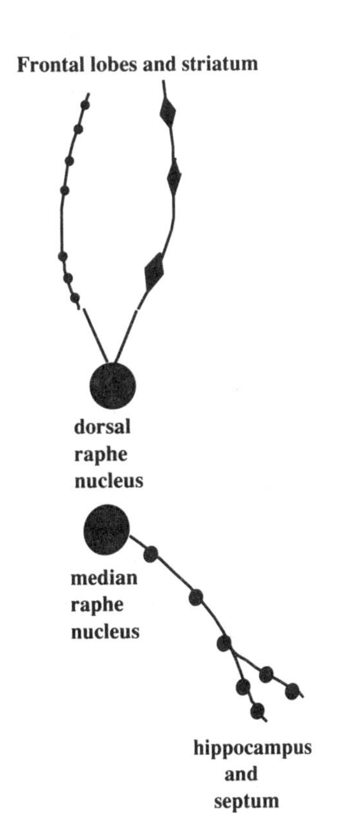

Two Kinds of Serotonin Nerve Fibers

The serotonin nerve fibers that pass from the dorsal raphe nucleus to the frontal lobe and striatum have either small vesicles or elongated enlargements, while those passing from the ventral raphe nucleus to the hippocampus and septum have large vesicles[1083,1350]. This is illustrated **on the right**.

The nerve fibers from the dorsal raphe nucleus are easily destroyed by mood elevating street drugs such as "Ecstasy" (MDMA) while the fibers from the median raphe nucleus are resistant.

In later chapters I will point out that defects in serotonin may play a major role in causing TS. Defects in the above major enervations can account for the ADHD (fron-

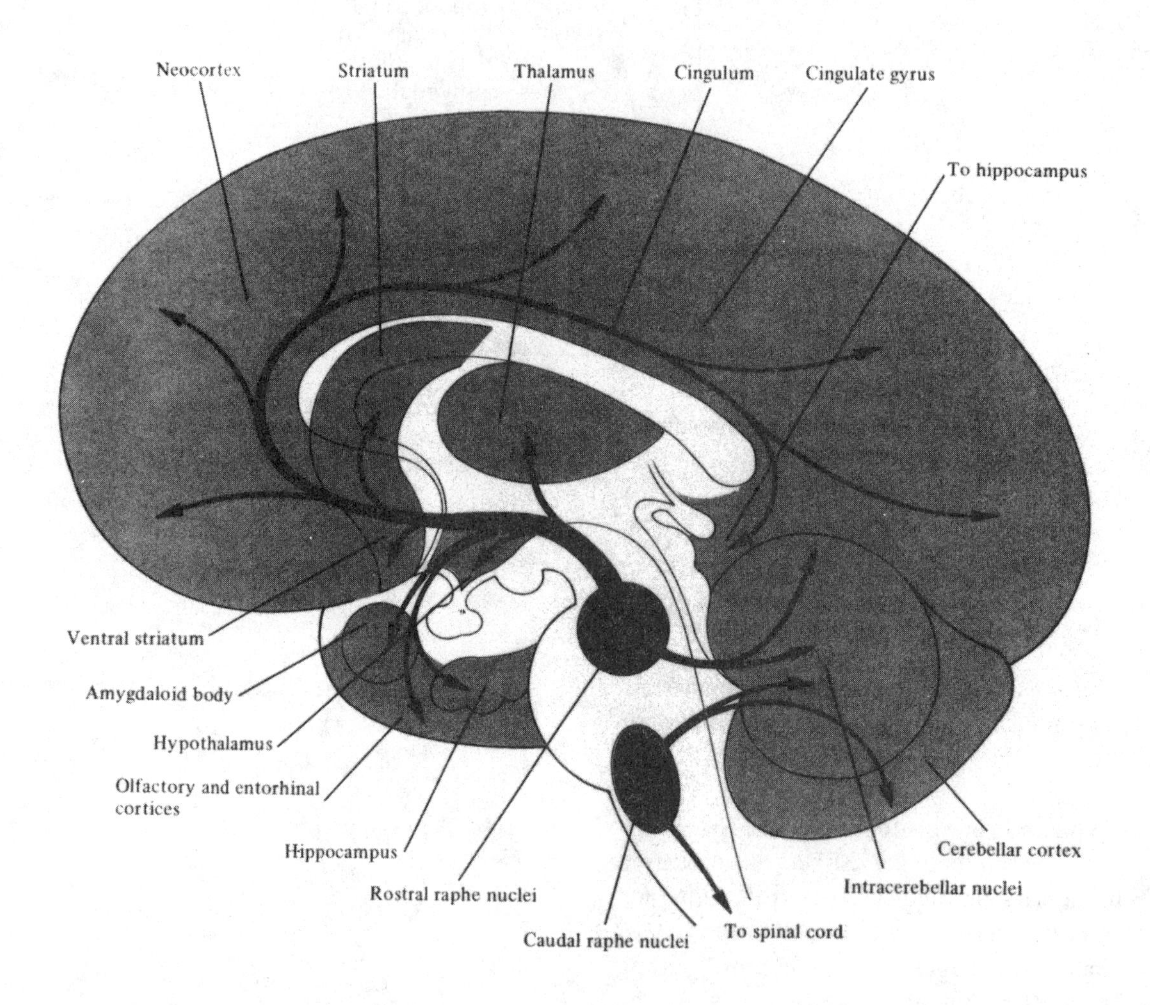

Figure 1. Connections between the serotonin-rich raphe nuclei and other structures of the human brain. (From L. Heimer, *The Human Brain and Spinal Cord*, Springer-Verlag, New York, 1983).

tal lobe), tics (striatum), problems with learning and memory (hippocampus), and emotional lability (septum, limbic system) seen in TS.

Serotonin — The Inhibitor

When serotonin is applied at random to various regions of the brain it can cause either excitation or suppression of activity. However, when applied to regions of the limbic system that receive a dense and uniform input of serotonin neurons, the effect is invariably to produce inhibition of nerve firing[11]. In the following chapters we will see that these serotonin connections, especially to the limbic system, play a critical role in the inhibition of human behavior.

Serotonin and the Prefrontal Lobes

We have already seen the critical role that the prefrontal lobes play in our behavior and especially in the behavior of individuals with various genetic impulse disorders. We can appreciate the importance of serotonin in regulating the functioning of the prefrontal cortex by comparing the *absolute levels* of serotonin to that of norepinephrine and dopamine. Serotonin is present in higher concentrations in the prefrontal cortex than either norepinephrine or dopamine.This is **diagramed below**[1586]:

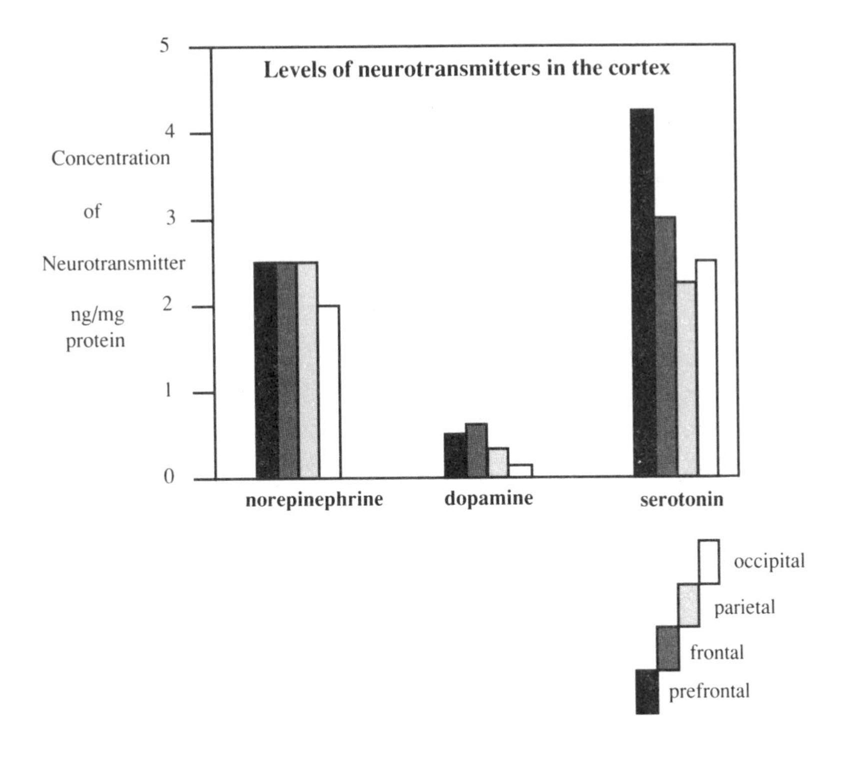

Chapter 59. Serotonin — The Great Inhibitor

Summary: Serotonin is a neurotransmitter that inhibits the functioning of other nerves. It is formed from tryptophan with 5-hydroxytryptophan (5-HTP) as an intermediate step. It is broken down to 5-hydroxyindole acetic acid (5-HIAA). This gives a pathway of tryptophan -> 5-HTP -> serotonin -> 5-HIAA.

The raphe nuclei serve as the major source of serotonin neurons in the brain and pass to many structures of the limbic system and the frontal lobe.

Since serotonin is necessary to inhibit many parts of the brain, the same parts become "disinhibited" when serotonin is too low.

Chapter 60

Serotonin Receptors

In the chapter on the frontal lobe I pointed out the importance of chemical lesions causing a frontal lobe syndrome in TS, ADHD and related disorders. The high density and distribution of serotonin receptors in the frontal lobe emphasizes the importance of serotonin as a major chemical involved in controlling the function of this part of the brain.

When serotonin is applied to the cortex it can cause both excitation and inhibition of nerve activity[991,1624]. This is due to the presence of at least two different serotonin receptors. The major ones in the brain are the S_1 and S_2 receptors[11-16]. The S_1 receptors have been further subdivided into four classes, A-D. **Table 1** summarizes the differences between the S_1 and S_2 receptors[208,541,846,794,1169,1723a,1796]. Don't be put off by the strange names of the different drugs. They are less important than the principles they allow us to examine. These two receptors are usually found on the same cells, except in the cortex where they have distinct locations[1301].

The presence of multiple types of serotonin receptors helps to explain some of the seemingly contradictory effects of different drugs. For example, 5-m-tryptamine and quipazine *stimulate* S_1 receptors and *inhibit aggression* in mice, while mianserin and pirenperone *inhibit* S_2 receptors but also *inhibit aggression* in mice[1186,2006]. This seemingly paradoxical effect can be explained by

Table 1. Serotonin Receptors — S_1 and S_2

Name	S_1 (5-HT_1)	S_2 (5-HT_2)
Sub-types:	1A,1B,1C,1D	—
Brain location:	frontal lobe limbic system (hippocampus) neostriatum	frontal lobe limbic system
Effect of stimulation:	inhibition	excitation
Cell location:	pre-synaptic	post-synaptic
Preferentially binds:	serotonin	ketanserin spiperone
Stimulated by:	5-m-tryptamine 8-OH-DPAT quipazine	serotonin bufotenine
Inhibited by:	(-)propanolol	Haldol siperone cyproheptadine mianserin pirenperone ketanserin methylsergide
Effect of antidepressants:	none	decrease # of receptors
Function:	hyperactivity hyper-reactivity	usual brain serotonin functions contraction of smooth muscle

the fact that different drugs bind to different receptors and some serotonin receptors when activated inhibit nerve activity, while others stimulate nerve activity. The above can be summarized as follows:

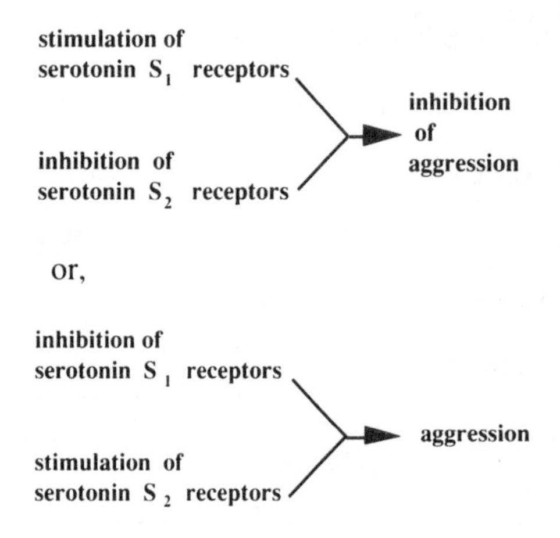

The other important observation is that Haldol and its cousin, spiperone, which are generally considered to bind to dopamine receptors, bind tightly to serotonin S_2 receptors[1170]. Thus, *when Haldol is given to TS patients, part of its effectiveness is due to its interaction with serotonin receptors in the frontal lobe. It inhibits S_2 receptors, which in turn inhibit aggression.*

Serotonin feedback loop Another factor that can complicate things is the existence of *feedback inhibition.* Some drugs which stimulate post-synaptic serotonin receptors actually inhibit serotonin neurons in the raphe nucleus and reduce the turnover of serotonin[12,14]. This occurs through a feebback loop. The full loop appears to act through a part of the brain called the habenula[15]. Thus, *activation* of the serotonin neurons coming out of the raphe nuclei inhibit the neurons in the habenula, which feed back to *inhibit* the serotonin neurons in the raphe nuclei. As will be shown later, this little exercise in circuitry is important since it may have significance in understanding *why both increases and decreases in brain serotonin may cause similar symptoms*[P507]

Distribution of Serotonin Receptors in the Brain

The distribution of various serotonin receptors in the brain can be determined by putting brain slices into solutions containing radioactive compounds that bind to specific receptors, then laying the washed slices onto X-ray films. This process is called *autoradiography.* Results from studies of S_2 serotonin receptors done in this fashion are given on the **following page**[1487].

The strikingly high concentration of S_2 serotonin receptors in layer IV of the frontal lobe and the cingulate cortex can be seen and is further illustrated by the autoradiogram itself (**see following page**). This shows that in addition to the high concentration of S_2 receptors in the frontal lobe there was also a high concentration in the claustrum, part of the cingulate cortex. The claustrum has many connections to the motor cortex and visual and other sensory association areas and may be involved in controlling visually directed movements[424,1607]. The cingulate cortex connects to the amygdala[1454] and may be particularly involved in the motor and vocal tics of Tourette syndrome[193] or in visually precipitated Tourette behaviors such as echopraxia.

A lengthwise cut through the brain shows the striking tendency for these S_2 serotonin receptors to be enriched in the frontal lobe areas of the brain (**see page 426**).

While the highest concentration of serotonin *receptors* was in layers III and IV of the frontal cortex, the highest concentration of serotonin *nerve terminals*, from the raphe nuclei, were in the superficial layer I of the cortex. It is intriguing that *few of these terminals are making true synaptic contacts with other neurons* [479,480]. This may reflect the fact that in the frontal cortex, serotonin is acting

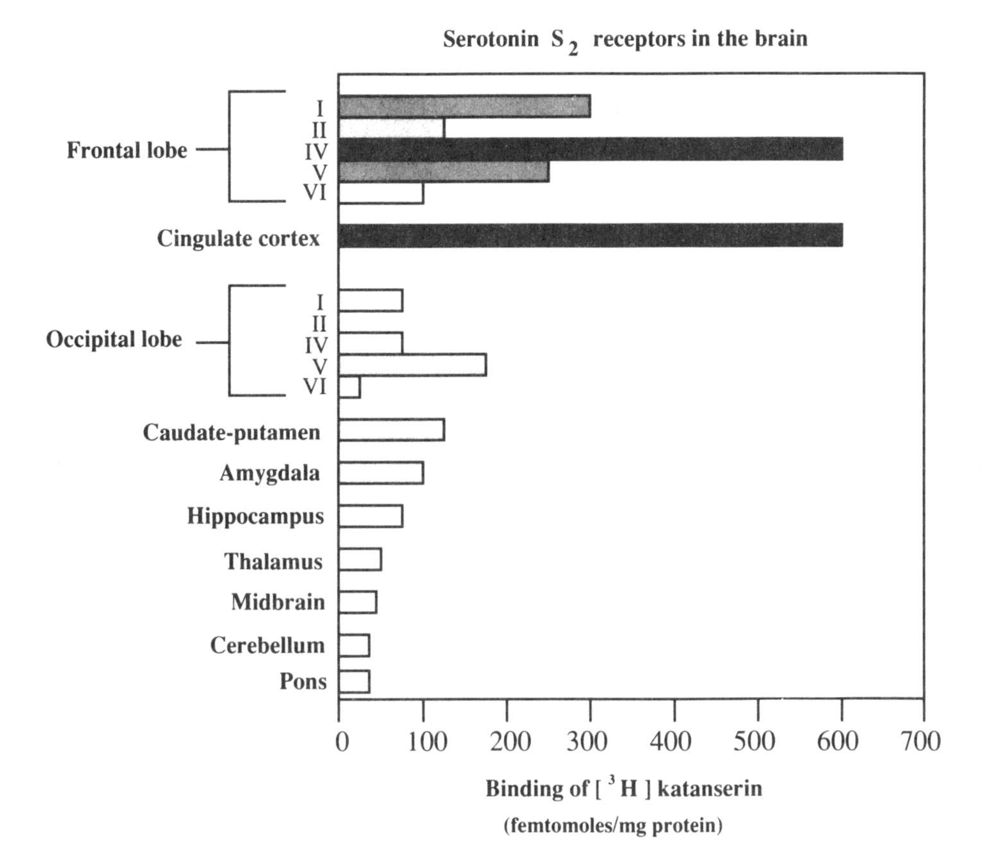

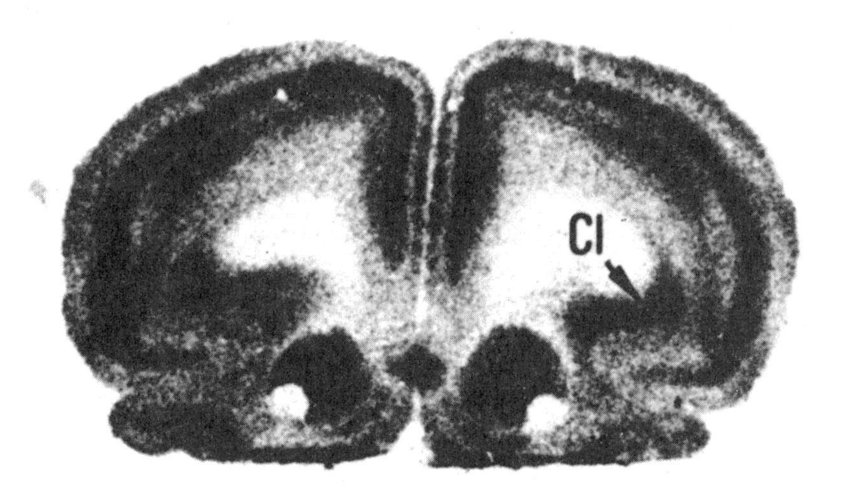

Figure 1. Autoradiogram of the frontal lobes of the rat showing the binding sites of S_2 serotonin receptors in layer III and the claustrum. Cl = claustrum. From Pazos et al.[1487].

more like a hormone than a true neurotransmitter[1301].

The distribution of serotonin S_1 receptors is somewhat different than that of the S_2 receptors. There is a fairly even distribution of the S_1 receptors among the different layers of the frontal cortex. They are densely distributed in the globus pallidus, the dentate gyrus part of the hippocampus, and in the septal nuclei. Lesser concentrations are found in other regions (**see next page**)[1486].

The serotonin nerves passing from the raphe nuclei are critical for normal function of the frontal lobes. These nerves pass to layer I of the frontal cortex and release serotonin which diffuses like a neurohormone to other layers containing serotonin and dopamine receptors.

This is diagramed as follows:

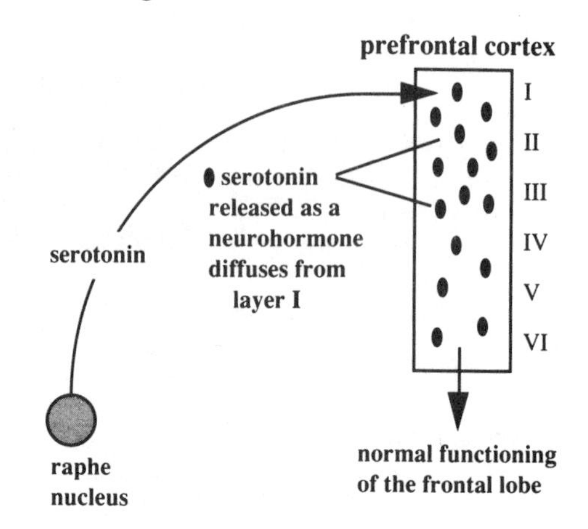

Inadequate function of these nerves can cause a *chemical frontal lobe syndrome*.

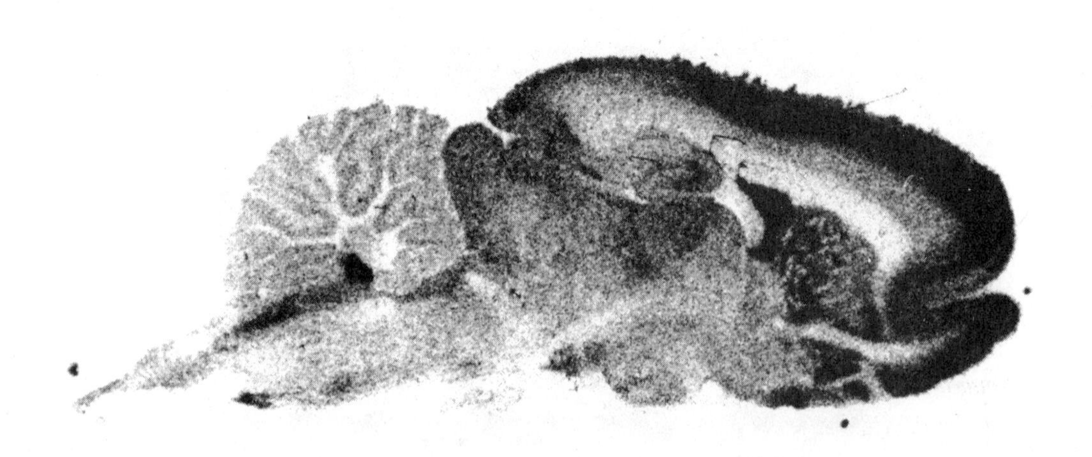

Figure 2. Autoradiogram of the rat brain cut lengthwise shows the striking increase in density of S_2 serotonin receptors in the frontal lobes (on the right part of the picture). (From Pazos et al.[11487].)

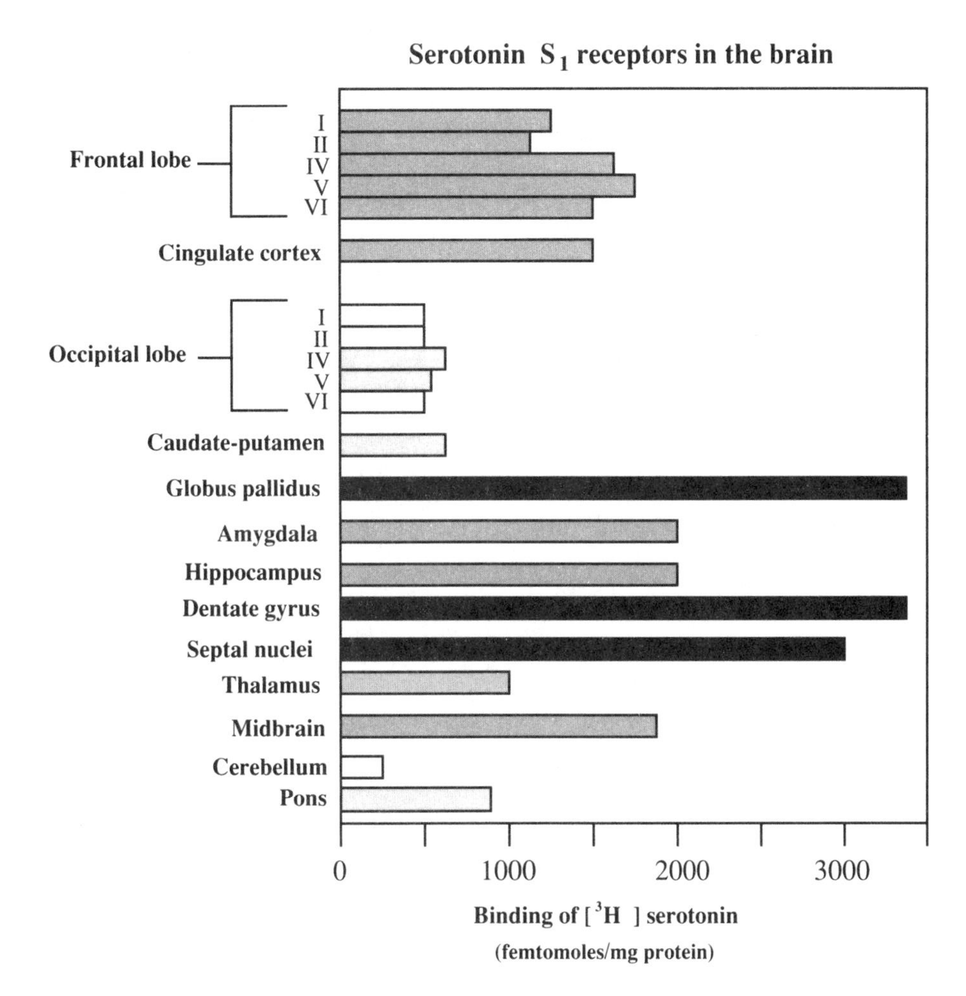

Summary: The importance of serotonin in the function of the frontal lobes is indicated by the fact that they possess a higher concentration of serotonin S_2 receptors than any other part of the brain. Serotonin nerves passing from the raphe nuclei to the frontal lobe are important in maintaining normal frontal lobe function. High concentrations of serotonin S_1 receptors are also present in various regions of the limbic system.

Deficiencies in brain serotonin can explain both the frontal lobe syndrome and the limbic disinhibtion syndrome in Tourette syndrome.

Chapter 61
Serotonin and Human Behavior

When there is not enough serotonin in the brain its ability to inhibit behavior is diminished, leading to a wide variety of disorders. The most important of these are summarized in the following box:

Behavioral Disorders Proposed to be Associated with a Deficiency of Brain Serotonin

- aggression
- alcoholism
- arson
- attention deficit hyperactivity disorder
- borderline personality
- bulimia
- depression
- impulsivity
- migraine headaches
- premenstrual tension
- Tourette syndrome
- violent behavor
- violent suicide

The first behavior disorder to be associated with low serotonin was depression.

Depression and Serotonin

As discussed before[p415], in the late 1950's antidepressant medications came into use and it was soon found that they tend to increase the level of various neurotransmitters, including serotonin, in the synapses of the brain. In addition, treatment of high blood pressure with reserpine, which depletes the synapses of serotonin and other neurotransmitters, often precipitated severe depression. This stimulated the theory that some forms of depression might be due to defective serotonin production[1462,1644].

Spinal fluid 5-HIAA Attempts to prove this by examining the level 5-HIAA[p418] in the spinal fluid suggested there was of a subtype of depression characterized by low serotonin turnover[57,59,60,1994]. Within this group, the lower the 5-HIAA the more severe the depression[60,159]. When the patients were first treated with probenecid, a drug which prevented the loss of 5-HIAA from the spinal fluid, the results were more reproducible. The diagram on the **next page** illustrates one of these studies[1997].

About one-third of the patients with depression had low rates of accumulation of 5-HIAA in their spinal fluid. This has now been observed in many other studies[16,57,62,335a,396,730,1230,1292,1533,1660a,2071].

Among depressed and suicidal patients, *those with low spinal fluid 5-HIAA had significantly more anxiety and hostility and a lower ability to handle conflict* than those with higher levels[1670].

The serotonin in whole blood and plasma has also been reported to be low in depression[1701].

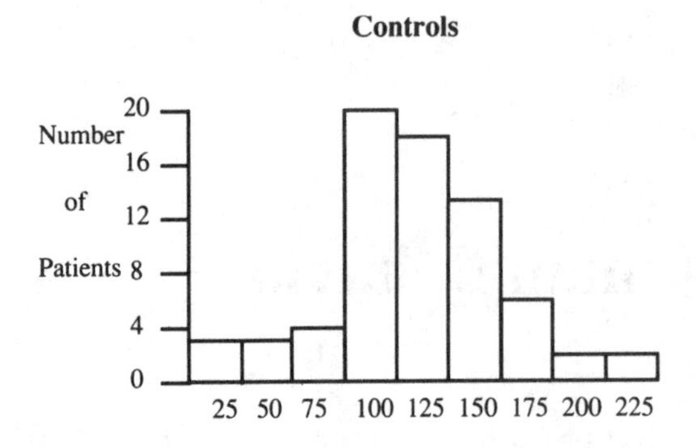

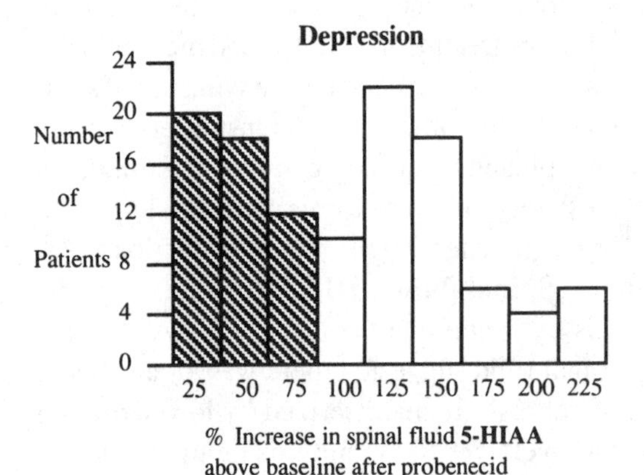

% Increase in spinal fluid **5-HIAA** above baseline after probenecid

Tryptophan and 5-HTP in the treatment of depression If some forms of depression are due to a defective ability to make adequate amounts of serotonin, then increasing the amount of serotonin in the brain should alleviate the depression. Based on the chemistry of serotonin production, we might anticipate that giving depressed patients tryptophan or 5-HTP would improve their depression. Does it?

There have been several studies of the use of tryptophan in depression. Some found it to be effective[393,397,1461,1942,1997] while others did not. One reason why tryptophan may be only partially effective is that tryptophan and tyrosine compete with each other for transport across the blood-brain barrier. As a result, when tryptophan transport is increased, tyrosine is decreased[2115,2116]. Since tyrosine is a precursor to dopamine and norepinephrine[p363] an increase in serotonin may be purchased at the expense of a decrease in these other two neurotransmitters, which may also be important in protection against depression. Thus, any benefit from increasing serotonin may be nullified by decreasing dopamine and norepinephrine.

A carbohydrate-rich meal may increase the effectiveness of tryptophan by increasing blood insulin levels. This lowers the blood level of other amino acids that compete with tryptophan for transport across the blood-brain barrier. This might play a role in the craving some TS patients have for sugar or other carbohydrates. Such food habits might result in increased brain serotonin levels.

5-hydroxytryptophan in the treatment of depression In contrast to tryptophan, 5-HTP increases the metabolism of all three neurotransmitters — serotonin, dopamine and norepinephrine[1995]. Again, some found it to be effective in the treatment of depression, and more effective than tryptophan[1992-1997], while other did not[1993]. It seems to be most effective in those cases of depression with low levels of spinal fluid 5-HIAA[1997]. The advantage of tryptophan is that it can be purchased over-the-counter while 5-HTP is not generally available.

Some antidepressants increase brain serotonin Many antidepressants have been developed. The primary action of some of them is to increase brain serotonin activity by increasing its synthesis or release, or by blocking its removal (reuptake) from the synapse [p318].

Drugs That Affect Brain Serotonin

chlorimipramine clomipramine	(Anafranil, Ciba-Giegy)
fenfluramine	(Pondamin, Robins)
fluoxetine	(Prozac, Lilly)
trazedone	(Desyrel, Mead Johnson)

Fenfluramine increases serotonin release and decreases its uptake[1795]. It is used as an appetite suppressant and in the treatment of autism[p502]. Lithium carbonate increases serotonin activity by a number of different mechanisms[1061,1698,2026].

In the chapter on norepinephrine I pointed out that a major effect of tricyclic antidepressants is to produce a *down-regulation* of β-adrenergic receptors. This same effect is produced by antidepressants that selectively block serotonin uptake[1070].

All treatments of depression increase brain serotonin use A number of treatments of depression, including tricyclics, monamine oxidase inhibitors, lithium, shock therapy, sleep deprivation and others, have in common the production of an increase in serotonin nerve transmission in the brain[179,179a,1997]. Even the change in β-adrenergic receptors is dependent upon intact serotonin neurons[717,1890]. This further supports the idea that low brain serotonin is involved in many forms of depression.

Serotonin and the reversal of the effect of antidepressants One of the more impressive links between serotonin and depression comes from a study in which five depressed patients were treated with imipramine or MAO inhibitor with striking improvement in their depression[1788]. They were then treated with PCPA (parachlorophenylalanine), a drug which inhibits tryptophan hydroxylase[p417] and thus inhibits the production of serotonin in the brain. When these patients were given this drug, even though they were still taking the antidepressant, there was a rapid reccurrence of the depression. As soon as this drug was stopped, the depression again abated. The results for two of these patients are shown in the figure on the **right**[1788]**->**.

Since antidepressants affect both serotonin and dopamine-norepinephrine, it was of interest to try to determine which neurotransmitter was more important. Thus, another study was done using another enzyme inhibitor, α-MPT (alpha-methylparatryosine) which inhibits tyrosine hydroxylase, the rate-limiting enzyme for the production of dopamine-norepinephrine[p363]. In contrast to PCPA, this time there was no return of the depression[1788]. This is a striking demonstration that while antidepressants increase the levels of dopamine, norepinephrine and serotonin in the synapses, *it is their effect on serotonin that is the most important.*

Serotonin, daily rhythms and depression Daily rhythms[p241] have always played a critical role in theories of depression. Depression is often most severe in the morning. Sleep cycles are disturbed and the correction of these alone may relieve depression.

In an attempt to determine the cause of the low blood serotonin observed in many individuals with depression, the rate at which platelets take up serotonin has been studied. It was often shown to be low[398]. While this might indicate that the mechanism by which serotonin is transported into cells is abnormal in depression, an alternative possibility is that the transport mechanism is completely intact but the

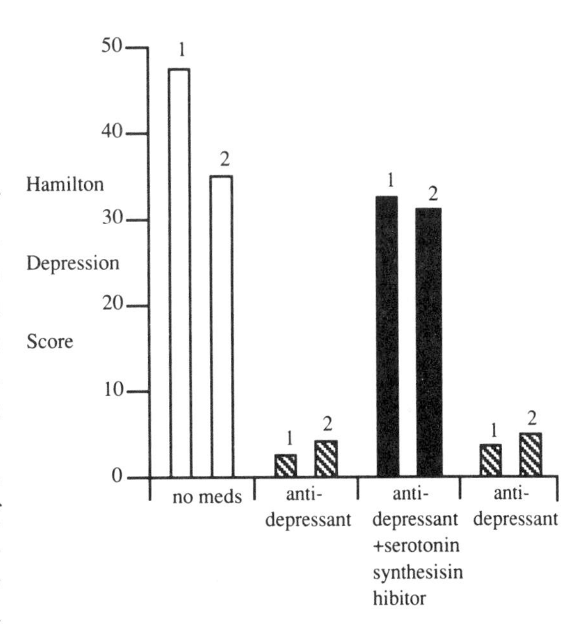

variation in the rate of serotonin transport during the day is out of kilter. The following study of the rate of uptake of serotonin during different times of the day in controls and depressed patients shows that this can be the case[865]:

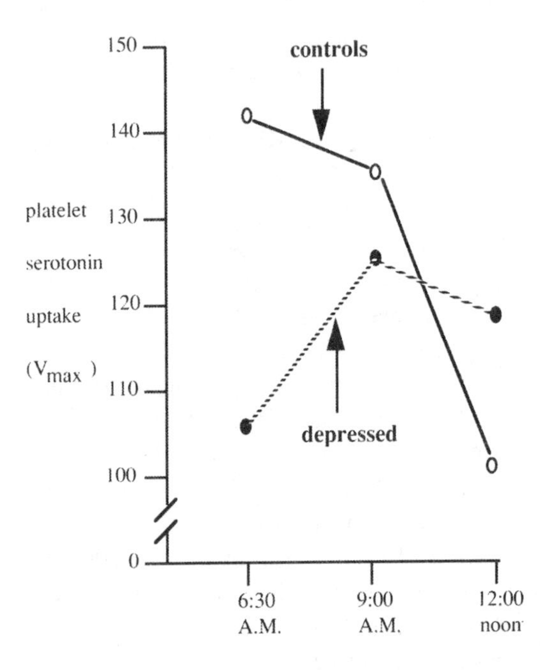

In the controls, serotonin uptake was greatest early in the morning and dropped significantly by noon. By contrast, in depressed patients, the level was the lowest early in the morning and by noon the rate was higher than in the controls, indicating that the timing was off. Those patients that responded to treatment showed a normalization of this pattern and an increase in uptake[865]. Those who did not respond to treatment did not change their daily pattern. It is possible that a defect of brain serotonin alters the daily clock in the hypothalamus[p329] and causes this abnormality in daily rhythms.

Violent Suicide and Serotonin As studies accumulated on the role of serotonin in depression it soon became apparent that the rate of attempted or successful suicide was higher in those individuals with the lowest spinal fluid levels of 5-HIAA[17,59,60,102,103,104,229,1179,1354,1406,1444,1958,1994]. If the suicide attempts were divided into two types: non-violent taking pills and carbon monoxide poisoning, and violent — shooting and hanging, the individuals with the lowest spinal fluid 5-HIAA tended utilize the more violent types of suicide.

A study of 203 patients hospitalized for depression showed that if the patients had a low spinal fluid 5-HIAA they attempted suicide more than twice as often as the other patients and four times as many of these were violent suicides[1994].

Serotonin levels are also lower in non-depressed schizophrenics who attempted suicide after voices told them to[1997]. This suggests that *low serotonin can be a marker for suicide risk* even in the absence of depression[1958]. In one study *of those depressed subjects with abnormally low spinal fluid 5-HIAA levels, 20% had killed themselves within a year of the study*[1958].

A particularly severe form of depression and violence is a suicide attempt combined with the killing of one's own children. The **following figure** shows the results of a study of three such cases[1184]. Their spinal fluid 5-HIAA levels were strikingly low.

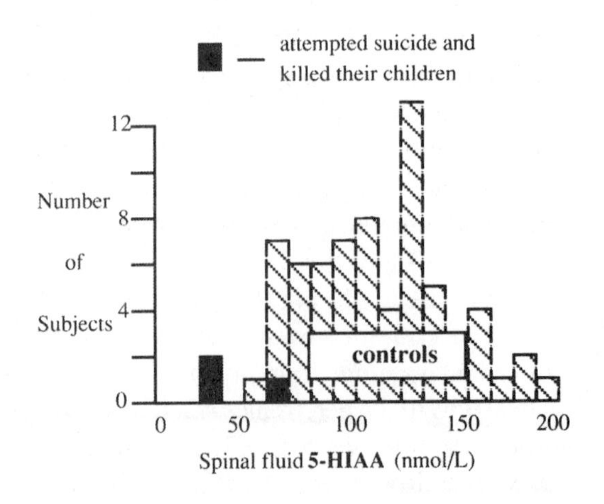

Brain serotonin receptors in suicide The earliest studies of brain tissue in suicide victims used a radioactive antidepressant (imipramine) and found that less of it bound to the tissues of the frontal lobe of individuals who had comitted suicide than in controls[1471,1501]. Imipramine-binding sites appear to be involved in a serotonin transporter function and this would indicate low levels of functioning serotonin in these brains. Stanley and Mann[1855] examined both serotonin S_1 and S_2 receptors in the frontal lobes of 11 individuals who committed suicide and found a significant increase in S_2 receptors. This complemented the imipramine-binding studies since a deficiency of brain serotonin would result in the "up-regulation" and hypersensitivity of the postsynaptic S_2 receptors. Some confirmed these findings[1291] while others did not[1448,1449]. Mann and Stanley[1252,1253] have re-examined a larger number of suicides, carefully matched to controls of the same age and sex, using a sensitive autoradiography technique. Not only did this confirm a significant increase in S_2 receptors in the frontal lobe of suicide victims, but the increase was also where it should be, in layers III and IV of the cortex. They also found an increase in β-adrenergic receptors. This is consistent with a primary defect in serotonin since lesions of the raphe nuclei in rats have been shown to produce an increase in the number of β-adrenergic receptors in the frontal cortex and hippocampus[1890], as **shown on the right ->**.

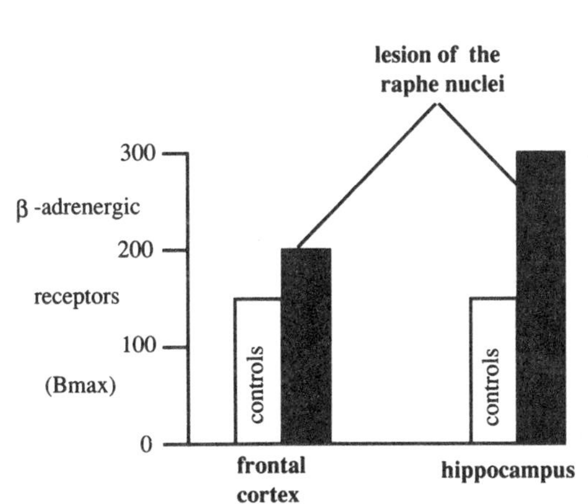

The changes in suicides and patients with depression are summarized as follows:

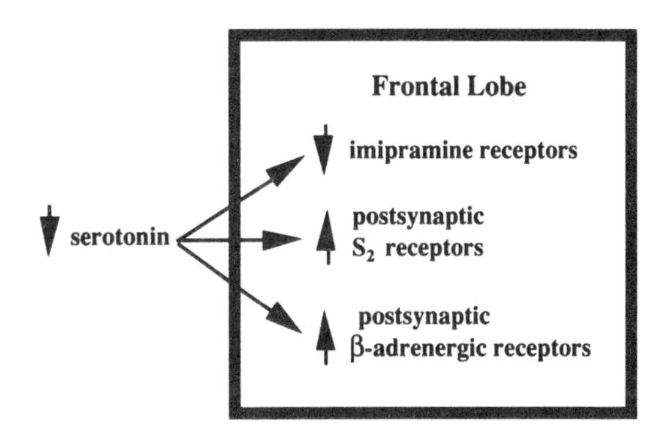

A further indication of the intimate relationship between serotonin and norepinephrine is the observation that the "down-regulation" of β-adrenergic receptors seen after long-term treatment with antidepressants[p414] does not occur if the serotonin pathways are cut[1890].

Aggression, Impulsivity, and Serotonin

Violent suicide can be visualized as severe aggression turned on one's self. Is low serotonin also associated with aggression turned outward?

Several studies have shown that severe aggressive behavior is associated with a low spinal fluid 5-HIAA[168,229,230,1097a,1192]. XYY males with an extra Y chromosome and a lifelong history of aggressive behavior[168] had low spinal fluid 5-HIAA levels. In a study of criminals who had committed or attempted murder, those who did it impulsively had a lower average spinal fluid 5-HIAA than those who premeditated the crime [1192]. *All of those in the impulsive group had a childhood history of ADHD and/or conduct disorder and a history of impulsive suicide attempts.* In a study of military men *those with a life history of aggressive and impulsive behavior had lower spinal fluid levels of 5-HIAA* than those without such a history[230].

A group of 30 retarded individuals in Rosewood State Hospital in Maryland all had hyperactivity, short attention span, low frustration tolerance, and emotional mood swings. *Most also had temper tantrums, engaged in screaming and fighting and were destructive. Of these, 83 percent had low blood serotonin levels* compared to matched normal controls[799]. After treatment with medication that resulted in an increase of the blood serotonin to normal, the hyperactivity and aggressive behavior largely disappeared. When the medications were stopped, the behaviors returned. When the behaviors improved there was a 38 to 200 percent increase in blood serotonin. In the group of patients that did not respond to a specific medication, the blood serotonin increased by only 20 percent.

Attention Deficit Hyperactivity Disorder and Serotonin

In addition to the above study, Wender[2057] also reported low blood serotonin in three siblings with ADHD. In a later study, including some of the same patients reported by Wender, Mary Coleman[349] described blood serotonin levels in 25 hyperactive children. Of these, three had normal values, eight borderline, eight low, and six very low. The one with the lowest blood serotonin had "bizarre posturing and rhythmic movements" and probably had Tourette syndrome. Her report five years later of 11 presumably new patients found a significantly lower blood serotonin in ADHD compared to age-matched controls[164]. A more recent study reported that ADHD children with blood serotonin levels of less than 70 ug/ml were more likely to respond to treatment with Cylert than those with a higher blood serotonin[1704].

Others have not found any significant decrease in blood serotonin in children with ADHD[588,861,1580,1776,1785]. A few have reported increased blood serotonin in some children with ADHD[952] and increased free tryptophan (not bound to albumin)[930,952]. The important factor in these ADHD studies may not be ADHD, but the degree of aggressiveness. In the Rosewood State Hospital study all the patients were institutionalized and very aggressive, and in the study by Mary Coleman the most hyperactive children had the lowest blood serotonin levels. Some additional observations are consistent with a role of serotonin in ADHD[215a]. These include the fact that amphetamines, which are so effective in the treatment of ADHD, increase brain tryptophan, and the fact that children with phenylketonuria (PKU) have low brain tryptophan and serotonin levels and have a behavior syndrome very similar to ADHD children[215a].

Low tryptophan diet and attention Some indirect evidence for the possible role of serotonin in ADHD comes from studies of humans briefly placed on a low tryptophan diet to decrease brain serotonin levels. They reported mild depression and showed the presence of moderate attentional difficulties[2139]. While taking a test requiring sustained attention the subjects listened to tapes that were a) low in distraction (reading statistics), b) high in distraction (playing eyewitness accounts of the bombing of Hiroshima) and c) depressive (themes of hopelessness and helplessness). The results are **shown on the next page**[2139].

With only a moderate decrease in brain serotonin there was a 25 percent decrease in ability to concentrate on a reading task when listening to a depressing tape. Even with minimal distraction there was a 15 percent decrease in ability to concentrate. These studies have the advantage that they were done on humans and suggest that *even mild decreases in brain serotonin can result in a short attention span.*

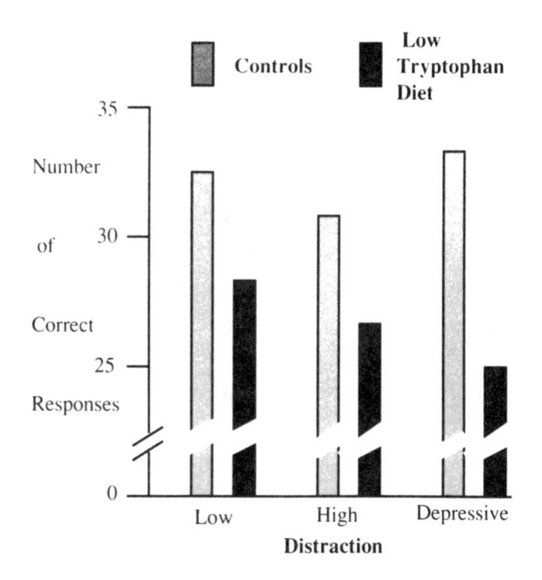

Ecstasy and learning problems A drug variously known as Ecstasy, MDM, MDMA, XTC or Adam has characteristics of both mescaline and amphetamine. It produces a high, increased talking, and a sense of closeness to other individuals. It has been widely used on college campuses. However, in monkeys it has been noted to selectively destroy serotonin nerve endings and deplete brain serotonin[1605]. In addition to euphoria, some of the immediate effects are jaw clenching and teeth grinding. The day after taking the drug about 20 percent of students reported depression and *difficulty concentrating*. "Although students often use the drug at parties, they avoid school nights because the effects of the second day can be so bad"[122]. *This is one of the few direct experiments in humans which shows the effects of acute serotonin depletion as a cause of depression and short attention span.*

Alcoholism and Serotonin

There is a clear connection between alcoholism and suicide in that about 18 percent of alcoholics commit suicide, a rate that is many times greater than the general population[1657]. Alcoholics that developed delirium tremens and other severe problems after they stopped drinking had low levels of spinal fluid 5-HIAA and these levels remained low even after months of not drinking[1917]. These findings were verified by additional studies showing that alcoholics tend to have normal spinal fluid levels of 5-HIAA at the time of admission, but progressively lower levels the longer they remained free of alcohol[98,101]. In addition, patients with depression and a family history of alcoholism had lower levels of spinal fluid 5-HIAA than depressed patients without a family history of alcoholism[1647].

The tryptophan ratio in alcoholics As mentioned above, to gain entry to the brain, tryptophan must compete with six other amino acids including tyrosine.The ratio of the blood level of tryptophan to these six other amino acids is called the tryptophan ratio[215]. The tryptophan ratio was examined in controls and three types of alcoholics: a) those with no depression and no aggression, b) those with depression but no aggression, and c) those with both depression and aggression. The results were **as follows**[215]:

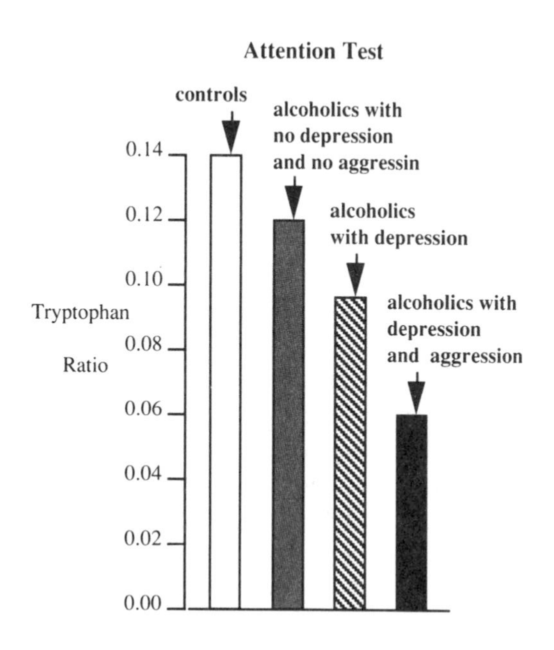

Another study by the same researchers showed that a low tryptophan ratio, aggression and depression were most significant for those whose alcoholism started before age 20[260a,b].

The serotonin theory of alcoholism The above observations indicate that non-drinking alcoholics tend to have low brain serotonin. The fact that alcohol increases the turnover of serotonin in the brain has provided an intriguing hypothesis of the cause of alcoholism. This theory suggests that alcoholics "self-medicate" themselves to raise their low levels of brain serotonin. This, however, leads to a further depletion of serotonin that just exaggerates the problem. This leads to a "vicious cycle" in which the alcoholic attempts to maintain a normal brain serotonin level by the repeated ingestion of alcohol[98].

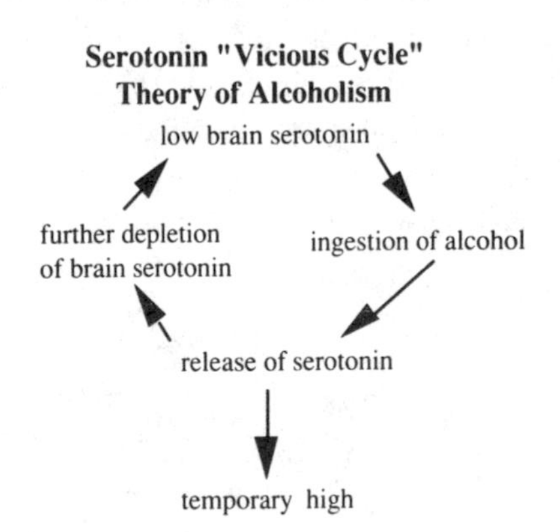

Aggression after alcohol As discussed in the chapter on alcoholism, type II alcoholics are often aggressive or abusive after drinking [p232]. I have previously pointed out the similarities between the personality of type II alcoholics and some TS patients[p232]. If some TS patients have a genetic deficiency of brain serotonin[p457] it could explain the following story related by a 24 year old woman with TS.

> "Whenever I had a drink, I would begin swearing excessively, barking and screaming. This lasted for about one hour and went away. I never did any of these things except after drinking."

Alcohol intake might have transiently exacerbated a preexisting serotonin deficiency and thus brought out symptoms she normally did not have.

Effect of low brain serotonin on alcohol preference The above "vicious cycle" theory of alcoholism would be more plausible if there was direct evidence that low brain serotonin leads to an increased desire for alcohol. Evidence for this comes from several studies.

In one, 5,6-DHT was injected into the ventricles of rats. This compound selectively destroys serotonin neurons, *depletes brain serotonin and results in increased alcohol consumption* [900,1385]. This effect is further verified by the observation that increasing brain serotonin by the administration of 5-hydroxytryptophan and other drugs, either orally or into the ventricles, results in decreased alcohol consumption[685,687,1385-1388]. Consistent with the idea that alcohol raises brain serotonin levels, is the finding that when mice are withdrawing from alcohol the rate of serotonin synthesis is decreased[1914]. This would drive them to return to alcohol use.

In a series of studies by Li and colleagues[1174-1178,1379], rats were bred for alcohol preference and non-preference. The *alcohol preferring rats had 20 to 30 percent lower levels of serotonin in the frontal cortex, nucleus accumbens and striatum and a 25 percent decrease in dopamine in the nucleus accumbens.* The low serotonin verified the role of low brain serotonin in alcohol seeking behavior. Since dopamine neurons play an important role in pleasure centers[p521] in the brain, the low dopamine could also be an important factor in alcohol seeking behavior.

Treatment of alcoholism Alcoholism is often quite resistant to treatment. Drugs which elevate serotonin in the synapse by inhibiting its reuptake have been reported to significantly reduce alcohol intake by alcoholics[1393]. In a similar vein, buspirone, which also affects serotonin, decreases alcohol intake by "alcoholic" monkeys[352] and also inhibits aggressiveness in rats and monkeys[682,1604]. Another serotonin reuptake inhibitor eliminated the memory impairment produced in humans by doses of alcohol[2041], again indicating an important role of serotonin in memory. Alterations of serotonin by other drugs also decrease alcohol preference in rats[1387,1388]. These results provide much hope for the chemical treatment of alcoholism.

All of these observations are consistent with the following relationships:

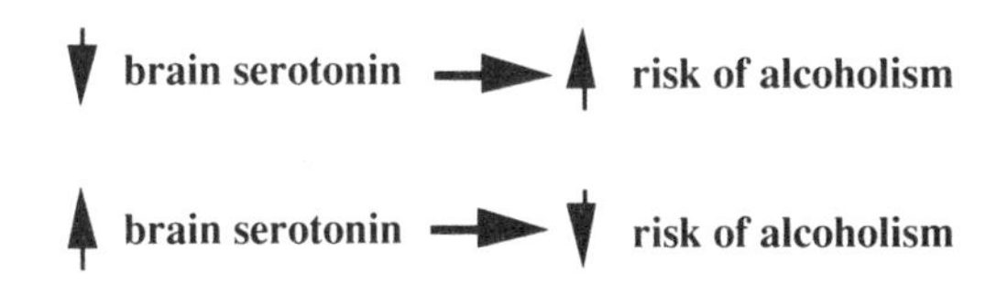

Arson and Serotonin

Arsonists carry out violent, aggressive, impulsive acts against property. A study comparing 20 arsonists to 20 violent criminals and 10 controls[1659,2012] gave the following results:

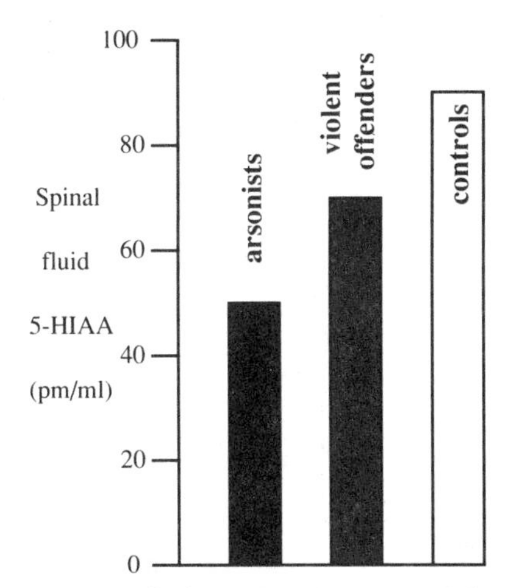

The spinal fluid 5-HIAA was even lower in the arsonists than in the violent offenders and both[2011a,b] were significantly lower than the controls.

Borderline Personality and Serotonin

As already seen in the chapter on borderline personality[p221], these individuals have many of the characteristics we have been discussing —depression, impulsivity, anger, aggression, drug abuse and alcoholism. This raises the question — Do borderline patients have low spinal fluid 5-HIAA levels? While this has not been studied as much as depression, in one set of studies of 24 males with personality disorder and 12 with borderline personality disorder, all without depression, the histories of aggressive behaviors and suicide attempts were associated with each other, and both were significantly associated with lower 5-HIAA levels[229,230]. These results correlating an aggression scale[261] with the spinal fluid 5-HIAA level are diagramed **on the next page**[229]. The effect was specific to serotonin in that the spinal fluid levels of the breakdown products of dopamine and norepinephrine were normal.

In the study of impulsive criminals mentioned above, 20 had intermittent explosive disorder, seven antisocial personality and all met the criteria for borderline personality disorder. Also, in the above study of 20 arsonists, all met the criteria for borderline personality disorder. Thus, 47 *individuals with borderline personality disorder had uniformly low levels of spinal fluid 5-HIAA*. These individuals are the quintessence of the impulsive, aggressive personality.

Bulimia and Serotonin

Anorexia nervosa is an eating disorder characterized by a distorted body image and a fear of getting fat. Such individuals eat very little and have a marked weight loss. Bulimia is a related disorder characterized by binge eating often followed by self-induced vomiting. *Bulimics tend to be more impulsive, outgoing, emotionally labile, and sexually active than*

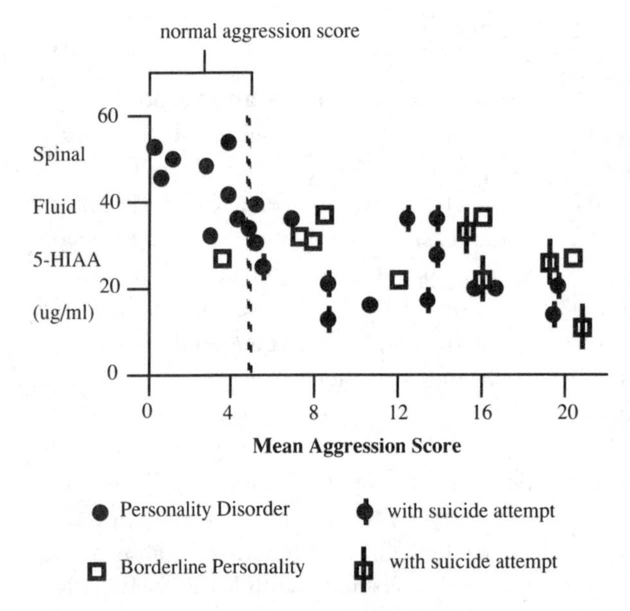

those with anorexia nervosa [306]. *They also have greater problems with compulsive stealing, alcoholism, drug abuse, suicide attempts and self-mutilation* [306,674]. Bulimia is clearly a form of impulse disorder. While individuals with anorexia nervosa have a feeling of mastery over themselves when they are hungry, bulimics tend to be irritable and restless when hungry[306] and this drives them to binge-eating for relief. The possible role of brain serotonin in bulimia was investigated since serotonin has been implicated in appetite suppression[1152], especially for carbohydrates [591,2116]; and bulimia can be considered to be both a mood and a impulse disorder. The **diagram on the right** shows the results of a study of spinal fluid 5-HIAA in controls versus individuals with anorexia nervosa and bulimia[1018]->.

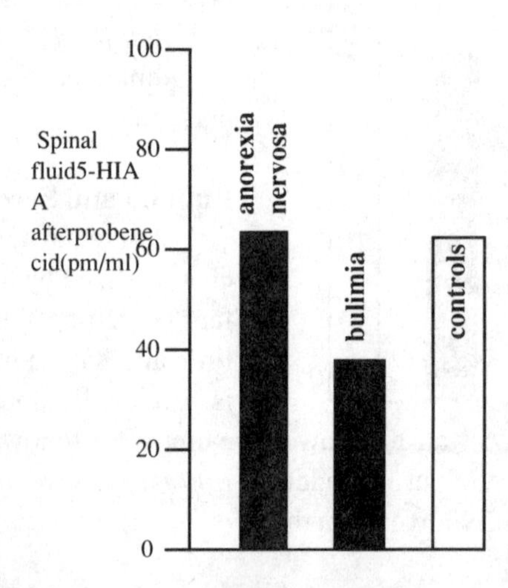

The accumulation of spinal fluid 5-HIAA after probenecid was significantly lower for the bulimic than either the controls or those with anorexia nervosa. When related to spinal fluid probenecid levels, those with anorexia nervosa had higher levels of 5-HIAA than normals.

Treatment of bulimia Just as the urge for alcohol can be attenuated by drugs that modulate serotonin, the urge for eating and binging can also be modulated by the same medications.

Serotonin and Aggressive Behavior in Mental Retardation and Senility

Mentally retarded and senile patients are often very obstreperous and difficult to manage. O'Neal and colleagues[1423] described a man with mental retardation who had disrupted sleep, temper tantrums and intense aggressive behavior directed toward himself and others. His serum serotonin was noted to be very low. Treatment with tryptophan, 2-3 grams per day, and trazodone resulted in a marked decrease in these behaviors. This note stimulated reports by others with similar experiences. A 82 year old demented female was reported who was difficult to control because of mood swings, screaming, echolalia, and banging her chair tray and the sides of her head with both hands[803]. Many different medications, including Haldol and antidepressants, failed to control her behavior. The compulsive, stereotyped and self-abusive nature of the banging movements with seemingly involuntary and echolalic verbal noises are similar to the behaviors seen in autism and Tourette syndrome. When treated with a combination of trazodone and tryptophan, these behaviors were rapidly extinguished. A third report described four senile women with aggressive disruptive behavior who also responded to the trazedone-tryptophan combination[2077].

In **Alzheimer's disease**, a major cause of dementia, the disease process affects the dorsal raphe nucleus[2120], resulting in low brain serotonin. This may cause some of the combativeness seen in some people with this disorder.

In **Down's syndrome**, blood serotonin levels are low due to a defect in the uptake of serotonin by platelets[199].

Lesch-Nyhan syndrome, Myoclonus and Serotonin

Lesch-Nyhan syndrome is a rare neurological disorder characterized by self-mutilation that can be so severe that patients have to have their hands tied and their teeth removed to prevent extensive self-mutilation. These patients may also have a lot of TS-like symptoms such as explosive grunts, cries, head-banging, picking at open wounds and coprolalia[1418]. Treatment of these patients with 5-hydroxytryptophan has been reported to significantly reduce the self-mutilating behavior[1343].

Myoclonus is a neurological disorder characterized by sudden bursts of jerking movements of the extremities that resemble electric shocks. They resemble severe tics. These symptoms also respond well to 5-hydroxytryptophan[2000-2004]. Two of our TS patients had severe myoclonus. A similar occurrence was described in four of 1,377 TS patients[583]. The response to 5-HTP suggests that a brain serotonin deficiency is involved in both Lesch-Nyhan syndrome and myoclonus.

Migraine Headaches and Serotonin

Migraine headaches are the result of painful dilatation of the small arteries in the scalp. This explains the use of ergot-containing medications, since ergots cause the dilated vessels to constrict again. There are many reasons to believe that low serotonin levels play an important role in migraine headaches.

In 1940 a potent blood vessel constricting agent was extracted from the intestine of the rabbit[556]. Nine years later a substance was isolated from the blood that caused constriction of blood vessels and was shown to have the structure of serotonin[1583]. This was shown to be the same substance isolated from the rabbit intestine[555]. Thus, serotonin is a potent constrictor of small arteries. If too little of this substance is present it might cause blood vessels to dilate and cause migraine. Is this true?

Migraine headaches are one of the side effects of the drug reserpine, and reserpine treatment results in the depletion of brain serotonin[1522]. These headaches are relieved by the injection of serotonin or its precursor, 5-hydroxytryptophan[1046]. Blood serotonin drops during migraine headaches[442]. This is followed by an increase in 5-HIAA in the urine. MAO inhibitors, which increase serotonin in the synapses, prevent migraine headaches[643], and medications that stimulate serotonin S_1 receptors relieve the acute pain[495]. Migraine sufferers often report that the headaches stop after they have vomited. Vomiting stimulates intestinal motility and raises blood serotonin[442].

All of these facts suggest that low serotonin levels may predispose to migraine headaches. The observation that migraine headaches are common in TS[p287] supports the proposal that TS may be due to a defective production of serotonin[p457].

Tension headaches are also associated with a blood serotonin level that may be even lower than in individuals with migraine headache[51a].

Normal Individuals with Low Serotonin and Psychiatrically Ill Relatives

Some normal controls also have low spinal fluid 5-HIAA. Are they different in any way from those with normal levels? Sedvall and co-workers[1741,1742] examined a series of controls who had a family history of psychiatric problems and compared them to controls with a negative family history. The results are shown in the following figure.

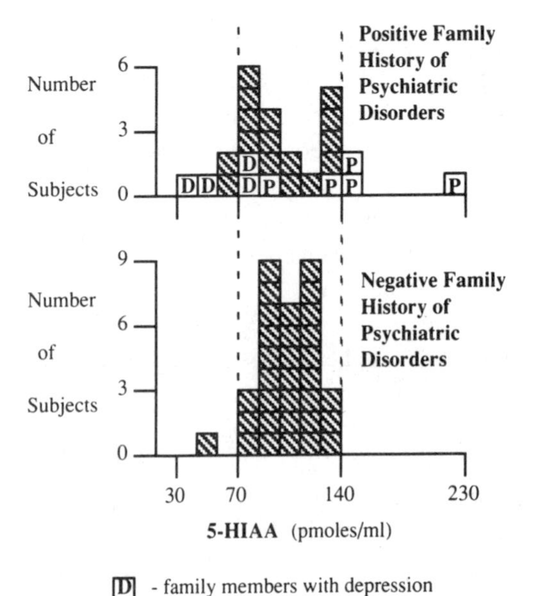

Almost all of the normal controls without a family history of psychiatric disorders had a spinal fluid 5-HIAA level of 70 to 140. However, those with a positive psychiatric family history had a wider spread of values. When the diagnosis in the relative was examined, there was a strong tendency for those with a low spinal fluid 5-HIAA to have relatives with depression (D), while those with a high level had relatives with schizophrenic psychosis (P). This suggests that those controls with low 5-HIAA have inherited a gene for depression which they do not express, but which does cause some central serotonin deficiency. Normal subjects with a below-average spinal fluid 5-HIAA were 2.7 times more likely to have a family history of depression than those with an above-average level[1741]. This further suggests that the low central serotonin level is a *trait marker* for the gene, rather than a *state marker* for the presence of depression. The increased 5-HIAA in those with schizophrenic relatives is consistent with the increased blood levels of serotonin in schizophrenics[467,p443].

A similar relationship was found for spinal fluid homovanlic acid (HVA), the breakdown product of dopamine[p364] as **illustrated below**.

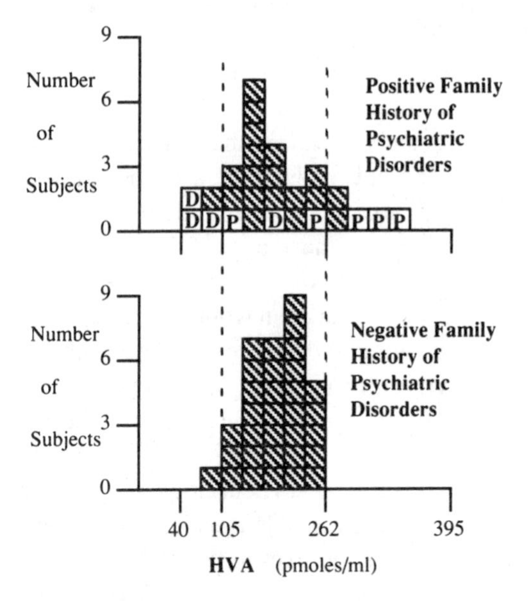

Thus, low spinal fluid levels of 5-HIAA and HVA are related to an increased risk for depressive disorders among family members, and high spinal fluid levels are associated with an increased risk for schizophrenic psychosis among family members[1742]. Increased spinal fluid 5-HIAA correlates well with the number of mannerisms and posturing in schizophrenia[1050].

Premenstrual Tension and Serotonin

I have observed that many women with Tourette syndrome, or mothers of TS children, have problems with premenstrual tension or premenstrual syndrome, usually called PMS. This syndrome is characterized by increased irritability, short attention span, tension, tearfulness, depression, tiredness, and anxiety during the week or two prior to the menstral period. There may also be swelling of the abdomen or feet, breast tenderness and food craving[1591]. The symptoms disappear after menstruation. While there is often question as to whether this syndrome even exists, after great debate it was tentatively placed in the 1987 DMS-IIIR under a special category of "Proposed Diagnostic Categories Needing Further Study." As further camouflage, it was given the jaw-busting name *Late Luteal Phase Dysphoric Disorder*. Since these symptoms are common in all women, the diagnosis is allowed only if the symptoms are severe enough to markedly impair work or social functioning and occur during most menstral cycles.

Studies of blood serotonin levels in women with PMS compared to controls suggest it is another serotonin disinhibition syndrome. The following graph shows that blood serotonin levels tend to increase premenstrally in normal women but decrease slightly in women with PMS[1575].

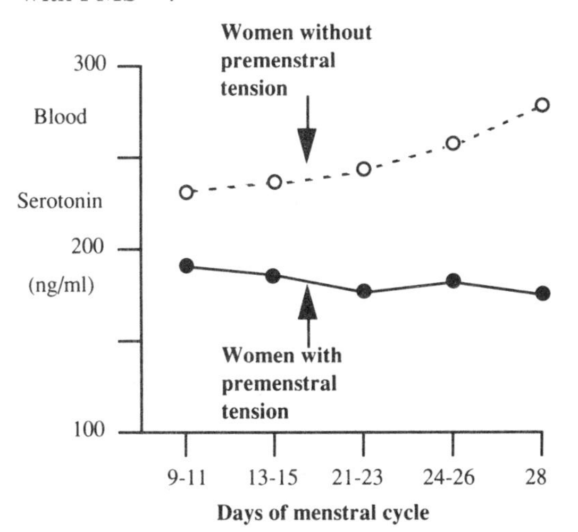

By contrast to serotonin, no significant differences were noted in levels of the female hormones, estrogen or progesterone[1575]. The previous curve is consistent with a defective production of serotonin predisposing to PMS since the normal increase in serotonin during the cycle cannot be accomplished. This is also consistent with the observation that large doses of vitamin B_6, which stimulates serotonin synthesis, may improve symptoms of PMS[3].

Summary: What Does a Serotonin Deficiency Cause?

The way the story has unfolded about the role of serotonin in behavior is not unusual in medicine. It often happens that a biochemical test is found that is initially believed to be relatively specific for a given disease, in this case depression. However, as studies accumulated it became clear that many other related disorders show the same defect. That does not diminish the importance of the discovery. In the case of serotonin it is testimony to the complexity of the human brain. There is no single trait associated with brain serotonin deficiency but rather an increased susceptibility to a complex of related behaviors — *impulsivity, aggression, and depression*[1958].

Disorders Associated with an Increased Sensitivity of Serotonin Receptors

In the chapters on dopamine, we saw that the tics in TS and some aspects of schizophrenia could be explained on the basis of hypersensitivity of dopamine receptors. In a similar fashion there are several disorders that are best explained on the basis of a hypersensivity of serotonin receptors.

Behavioral Disorders Associated with Hypersensitive Serotonin Receptors

obsessive-compulsive disorder
panic attacks

Obsessive-Compulsive Disorder (OCD) and Serotonin

The major evidence that serotonin plays a major role in obsessive-compulsive behaviors comes from the fact that they are best treated by antidepressants, such as clomipramine, that are specific for their effect on serotonin[948,950,1001,1003,1261,1872,1944,1945,2123,2127]. Tryptophan[2122] may also be effective. Most studies of spinal fluid 5-HIAA in obsessive-compulsive individuals have not provided evidence for a decrease in brain serotonin and some suggest an increase[58,950,1944]. Although blood serotonin levels of obsessive-compulsive individuals were no different than controls, those with higher levels responded better to clomipramine than those with low blood serotonin[613,1945].

Abnormalities of the frontal lobes, limbic system and striatum have been proposed in obsessive-compulsive disorder [1040a,1343a,1909c,2037a].

Hypersensitive serotonin receptors in OCD? Indirect evidence that serotonin is involved comes from the finding that when obsessive-compulsive patients are given a drug that stimulates serotonin-1 receptors their symptoms get much worse, and when given a drug that blocks postsynaptic serotonin receptors, their symptoms improve. This is shown as follows[2155,2156]:

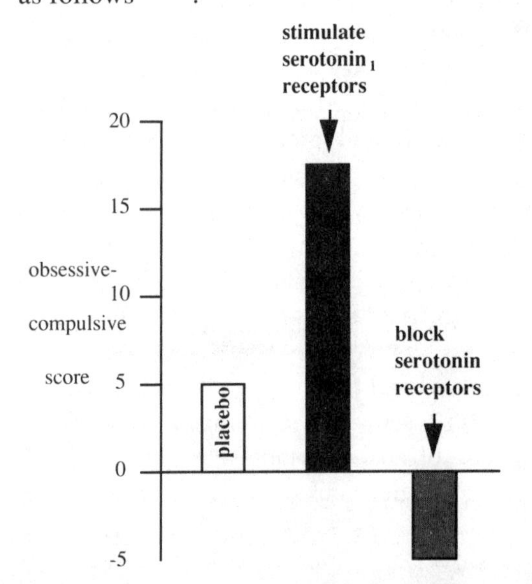

In addition to the increased obsessive compulsive behaviors after stimulation of the S_1 receptors, there was also a marked increase in general anxiety, depression, anger, and irritability, and a decrease in ability to concentrate.

The hypothesis that serotonin metabolism is overactive in obsessive-compulsive behavior conflicts with the observation that tryptophan can improve the symptoms. However, an alternative hypothesis, that serotonin-1 receptors are hypersensitive[1003,2155] is consistent with this. Thus:

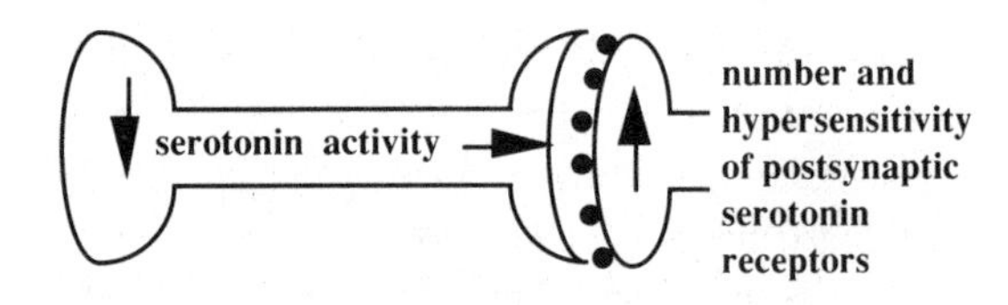

Increasing serotonin activity of the neurons on the left, by giving tryptophan or clomipramine, would down-regulate the number of serotonin receptors of the neurons on the right, thus relieving symptoms. Giving an agent that stimulates the postsynaptic serotonin receptors makes the obsessive-compulsive symptoms much worse[2155,2156].

Panic Attacks and Serotonin

In relation to serotonin, panic attacks bear many similarities to obsessive compulsive behavior. Abnormalities in the frontal lobe, limbic system and striatum have been implicated in both[780a,1343a]. The same antidepressants that have predominately an effect on serotonin uptake and are effective in treating obsessive-compulsive behavior, are also effective in treating panic attacks and phobic anxiety[1560,736,1004,1005]. 5-Hydroxytryptophan is also effective[1005]. While panic attacks have been most often associated with increased norepinephrine from the locus ceruleus[p409], this locus also contains serotonin receptors[1268] and there is an interaction between the locus ceruleus and the raphe nuclei[p413]. Serotonin uptake by platelets is decreased in subjects with panic attacks compared to controls[1490].

There is often a biphasic effect of serotonin agonists on panic attacks wherein the symptoms initially get worse, then after about two weeks, get better[1005]. This initial worsening is consistent with a hypersensitivity of the serotonin receptors. As the treatment continues this hypersensitivity disappears (down-regulation) and the symptoms improve. This lesson should be kept in mind when treating the obsessive-compulsive behaviors and panic attacks of TS patients with clomipramine, fluoxetine and 5-hydroxytryptophan. Here the tics, obsessive-compulsive behaviors and panic attacks may get worse before they get better.

The evidence suggests that *TS patients may have both hypersensitive dopamine and hypersensitive serotonin receptors.*

Generalized anxiety disorder Generalized anxiety disorder differs from panic attacks in that it is present most of the time rather than coming in sudden attacks[p175]. In additon to panic attacks[362,365] we have also observed an increase in generalized anxiety disorder in TS patients. While there is not yet any evidence for hypersensitive serotonin receptors in generalized anxiety disorder, this may be a factor in all of the anxiety disorders[1005].

Clustering of depressive, aggressive and anxiety disorders It has often been noted that depression, aggression and anxiety, in the form of obsessive-compulsive behavior and panic attacks, are often associated with each other[216,728,729,1653,2051]. The common thread may be a defect in brain serotonin.

Disorders Associated with Increased Serotonin

Just as there are a number of important behavioral disorders associated with low brain serotonin, there are also several associated with increased brain or blood serotonin. These are summarized as follows:

Behavioral Disorders Associated with an Increase in Brain Serotonin

- autism
- infantile spasms
- manic-depressive disorder
- schizophrenia
- some types of mental retardation

Schizophrenia and Serotonin

In the studies discussed above, Sedvall and colleagues showed that psychiatrically normal individuals with high spinal fluid levels of 5-HIAA tended to have relatives with schizophrenic psychoses, suggesting that high brain serotonin levels might be a trait marker for schizophrenia. This idea is supported by a number of studies showing that chronic schizophrenics tend to have increased spinal fluid 5-HIAA or elevated blood serotonin levels [467,632,672,1187,1852,1949] and other evidence[1159a]. As with depression some studies have not observed this and others have reported decreased serotonin in the brains of schizophrenics[244]. This suggests there may be chemical subtypes of schizophrenia. As discussed earlier[p206] some schizophrenics show moderate brain atrophy. There is a tendency for those with the greater amount of brain atrophy to have higher levels of blood serotonin[467].

Another link between serotonin and schizophrenia comes from the observation that postsynaptic serotonin receptors were decreased by 40 to 50 percent in the frontal cortex of schizophrenics[154]. No other receptors were so involved. Haldol, an effective treatment for both Tourette syndrome and schizophrenia, results in a significant increase in serotonin level in many different parts of the brain[1584].

Decreased levels of 5-HIAA in the spinal fluid are common in type II schizophrenics with brain atrophy and enlargement of the ventricles[176b]. It has been suggested that serotonin receptor hypersensitivity is present in this group and that the negative symptoms of schizophrenia might respond to drugs that decrease this hypersensitivity[176b].

Autism and Serotonin

The increase in levels of blood serotonin in autism are so striking and important that a whole chapter has been devoted to this subject [p499].

Other Conditions Showing Increased Serotonin

Other neurological or behavioral disorders reported to show elevated brain or blood serotonin include Huntington's disease[27,142], infantile spasms[350], manic-depressive disorder[2089], and some cases of mental retardation [350,1431,1462,1709].

Summary: Low levels of brain serotonin are associated with aggression, depression, violent suicide, alcoholism, arson, borderline personality, bulimia, and other impulsive behaviors. Low brain serotonin may also cause obsessive-compulsive behaviors and panic attacks by producing serotonin receptor hypersensitivity.

Chapter 62
Serotonin and Animal Behavior

Although studies of humans have told us a great deal about the role of serotonin in behavior, animal studies add a unique dimension to this knowledge.

Decreased Serotonin -> Resistance to Punishment

One of the most frustrating features of some TS and ADHD children is their failure to change their behavior despite various types of punishment and discipline, or failure to respond to behavioral modification[p598]. Some experimental work with animals sheds much light on this. Like most humans, if a certain behavior is followed by punishment that behavior stops. However, if animals are given drugs which counteract the action of serotonin[686,786,1864], inhibit the synthesis of serotonin[683,857,1626,2091], or destroy serotonin neurons[1863,1970], punishment is less effective and the behaviors are harder to extinguish. For example, during the control period, one rat stopped trying to obtain food after 11 shocks. However, after depletion of brain serotonin that rat *persisted in attempts to obtain food despite receiving 200 shocks* [1626]. In another study a serotonin-depleted rat continued in his attempts to obtain food after 110 shocks, but had given up after only two shocks during the control period[683]. The administration of 5-hydroxytryptophan "cured" the serotonin-depleted rats of their resistance to punishment[683]. Thus, serotonin maintains the *punishment responding* system of the brain.

Aggression and Serotonin

The previous chapter showed a clear correlation between low brain serotonin and increased aggression in humans. Animal studies confirm this relationship.

Aggression in isolated mice One of the most commonly used models of aggression in animals is the increased aggression seen in mice after they have been isolated from other mice[p392]. Although the level of serotonin is not altered, the rate of serotonin turnover in such aggressive mice is about half that of mice that do not become aggressive after isolation[669,1980-1984,1986]. In the absence of isolation, strains of mice and rats showing the most aggressive behavior had the lowest rate of turnover of brain serotonin[1986]. The lowest turnover of serotonin occurs between 8 P.M. to midnight and this is the time when mice are most susceptible to shock-induced aggression[1986].

Rats that kill mice Another model of aggression in animals is produced by destroying the raphe nuclei[790,1780,2119] or by administering drugs that inhibit serotonin synthesis[315]. This causes rats to kill mice placed in their cages, while normally they do not do so. This behavior is prevented by medications that increase brain serotonin[706,707,1405]. Mouse killing behavior could also be produced by a diet lacking tryptophan and this could be reversed by giving 5-hydroxytryptophan[706]. Clomipramine, a drug which increases the use of brain serotonin[p430], was able to reverse mouse killing behavior only if the serotonin pathways were intact[1262].

Lithium, a medication long known for its ability to decrease aggressive behavior in humans[p585] also inhibits mouse killing behavior. Lithium increases the uptake of tryptophan by the brain[1062,1329] and increases brain tryptophan levels[1495]; this results in an increased production of brain serotonin[1733]. The combination of lithium and tryptophan was more effective in preventing mouse killing than either medication alone[220]. These observations indicate that mouse killing aggression in rats is associated with serotonin depletion and suggest that the anti-aggressive effects of lithium in both animals and man probably work, at least in part, through its ability to stimulate serotonin synthesis.

Aggression and serotonin: A summary of animal and human studies Valzelli[1983-1985] concluded that impairment of inhibitory control by brain serotonin was the major common thread to many different models of aggression in both animals and man. These are summarized as follows[1984]:

The Serotonin — Dopamine Connection

Understanding the intimate connection between serotonin and dopamine in the brain is essential to understanding Tourette syndrome. The chapters on dopamine showed the importance of the substantia nigra and nucleus accumbens to the regulation of muscle movements[p381]. There is a major link between the serotonin containing raphe nuclei (especially the dorsal raphe nuclei), the substantia nigra and the nucleus accumbens:

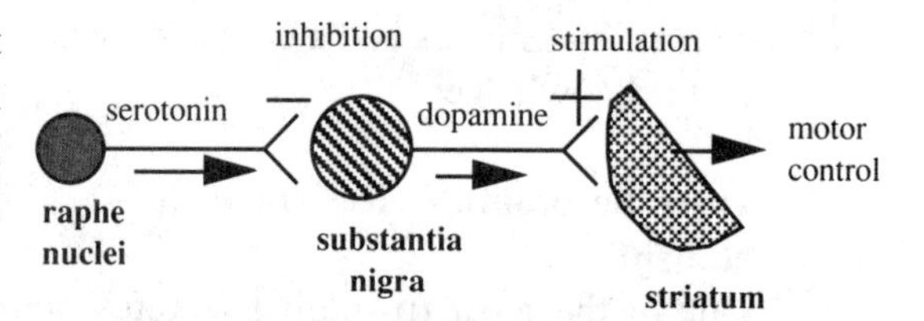

Serotonin neurons exert a tonic inhibition on the dopamine neurons of the striatum [184,389,651,703,1588,1825,1826,1867,1868,2021] and this in turn inhibits muscle movements.

Experimental or clinical situation	Decreased Serotonergic Control	----- Aggressive Behavior ----- Rat or Man	Rat	Man
Raphe lesions	+	+	+	?
PCPA administration	+	+	+	-
Isolation syndrome	+	+	+	?
Deficient tryptophan intake	+	+	+	+
Chronic corn diet (low tryptophan)	+	+	+	+
Chronic amphetamine intake	+	+	?	+
Thiamine deficiency	+	+	+	+
Chronic alcohol intake	+	+	+-	+
“Serotonergic” depression	+	+	+-	+
“Aggressive” schizophrenia	+	+		+
“Aggressive” personality	+	+		+
Dementias	+	+		+
Cornelia de Lange syndrome	+	+		+
De Morsier syndrome	+	+		+
Lesch-Nyhan syndrome	+	+		+
Klinefelter syndrome	+	+		+
Tourette syndrome	+	+		+
ADHD	+	+		+

This affects many aspects of behavior. In relationship to the above experiments on the influence of serotonin on the effectiveness of punishment, if the serotonin nerve endings in the substantia nigra are destroyed, the same resistance to punishment is observed that occurs with destruction of the dorsal raphe nucleus[1936]. *Many of the behavioral effects of destruction of the raphe nuclei are correlated with the resulting increase in dopamine activity in the striatum* [702]. *These relationships play a major role in many impulse disorders.*

Hyperactivity and serotonin As discussed before[p382] dopamine neurons of the nucleus accumbens play an important role in hyperactivity. When dopamine is injected directly into this region animals become very hyperactive. This effect is inhibited if serotonin is also injected into the nucleus accumbens, and markedly enhanced if the serotonin rich raphe nuclei are destroyed[412]. This indicates that *serotonin nerves from the raphe nuclei help to hold motor activity in check* as follows:

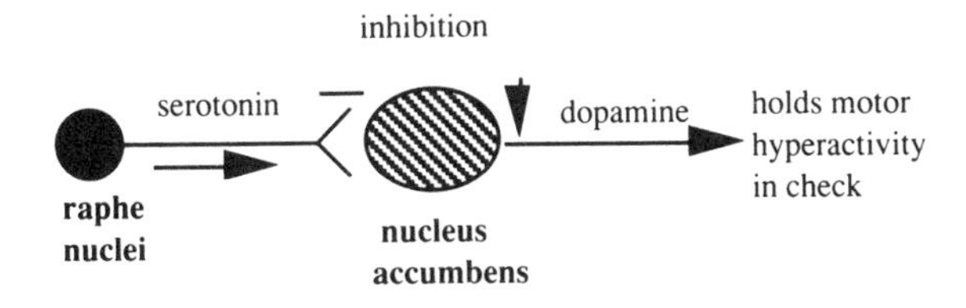

When brain serotonin levels are decreased, or the raphe nuclei are destroyed, it results in disinhibiton of the nucleus accumbens and motor hyperactivity. This is shown as follows:

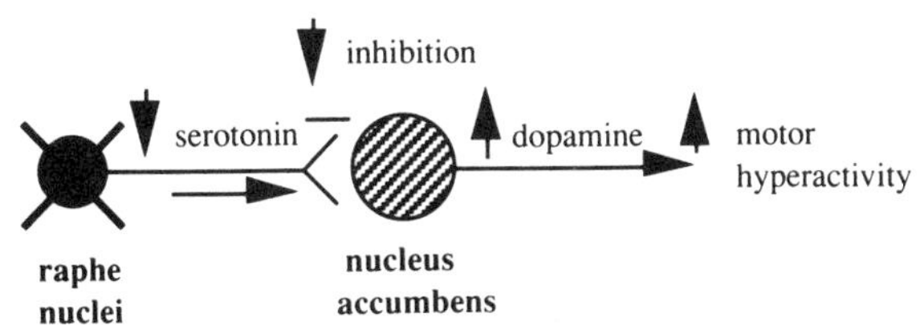

Animals with lesions of the raphe nuclei show increased activity[702,1212,1844] especially in response to new stimuli[702], suggesting that *one of the functions of serotonin is to reduce activity in response to novel stimuli.* This is a very important concept since it defines a mechanism by which *low brain serotonin can be directly involved in motor hyperactivity and distractibility.*

Impulse disorders and serotonin We saw in the chapter on human behavior that many patients with aggressive impulse disorders had low brain serotonin as indicated by low spinal fluid 5-HIAA[p418]. The above relationships between serotonin, the substantia nigra and nucleus accumbens help us to understand this connection. The substantia nigra is the dopamine gateway to the caudate and putamen which regulate movement or action. *Impulse disorders are equivalent to action without thought.* When the raphe nuclei are destroyed or serotonin levels are decreased by other means, there is a *disinhibition* or *disconnection* of thoughtful control over behavior [1946,1825]. *Animals with lowered serotonin levels behave as though they are impelled to act and as though the consequences of that action were unimportant*[1825]. Mothers of ADHD and Tourette syndrome children will recognize these behaviors. Psychologists call this "acting out" behavior. Neurochemically, this is **shown at the top of the next page**.

The decreased serotonin results in increased activity of dopamine neurons in the substantia nigra and nucleus accumbens and decreased control over motor activity or action. This is summarized as:

Put in somewhat more scientific terms:

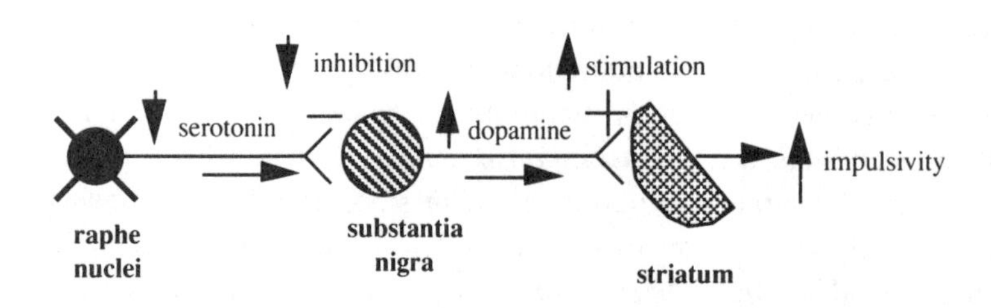

"Serotonergic systems are implicated in a dual process: control of responses and control of information that allows the organism not to enter into an action before the various elements of the situation have been checked"[1825].

TS patients have more anxiety disorders than controls. Normally this might be expected to inhibit them from impulsive actions. However, the depletion of serotonin "produces an organism that is likely to act impulsively *in spite of anxiety*"[2159]. The interrelationship between impulsivity and psychopathic behavior is understandable since in both, individuals "live and behave wholly in the present so as to produce immediate results, with no planning"[1825].

Keeping this relationship in mind helps in understanding many of the behaviors covered in this book.

Behavioral Inhibition and Increased Serotonin

Just as depletion of serotonin leads to behavioral disinhibition, the converse is also true. Stimulation of the serotonin-containing raphe nuclei in rats leads to behavioral inhibition characterized by a marked decrease of muscle movement[785].

Seizures and Serotonin

In line with its role as an inhibitor, raising serotonin levels decrease a tendency to seizures and low serotonin levels increase the susceptibility[309]. Low serotonin in TS could explain the increased frequency of brain wave abnormalities[p284].

Sleep and Serotonin

Sleep was once thought to be a passive process representing the quieting down of active brain functions. This notion began to change when it was found that stimulation of various parts of the brain caused sleep[996]. Sleep can be divided into two types. Initially there is a phase called *slow wave sleep* characterized by slow brain waves. This is followed by *paradoxical sleep*, so-called because the brain waves resemble a waking state, despite being asleep[996]. In young birds, serotonin easily passes the blood-brain barrier. The role of serotonin in sleep was first suspected by the observation that when these birds are given serotonin a natural sleep is induced[1837,1838]. These findings were confirmed in mammals when it was found that compounds that increased brain serotonin led to slow wave sleep while compounds that increased dopamine and norepinephrine led to paradoxical sleep[996,997].

When animals are treated with a compound that inhibits tryptophan hydroxylase, the rate-limiting step in the synthesis of serotonin, brain serotonin levels drop. This is followed by *30 hours of almost total insomnia* and normal sleep patterns only recur after 200 hours[996]. Destruction of the serotonin-containing raphe nuclei also produces insomnia. Thus, serotonin neurons originating in the raphe nuclei play a major role in inducing slow wave sleep. This connection between sleep and serotonin is the reason why 500 to 1,500 mg of tryptophan at bedtime often eliminates insomnia.

As discussed previously, sleep disorders are very common in Tourette syndrome[p249].

Dreams, REM Sleep and Serotonin

Dreams occur during rapid eye movement or REM sleep[p249]. Monamine oxidase inhibitors, which elevate brain serotonin, cause a cessation of both dreaming and REM sleep[996]. During REM sleep the discharges of serotonin neurons from the raphe nuclei completely stop[963,1963]. Thus, increased activity of serotonin neurons blocks dreaming, and *decreased activity of serotonin neurons stimulates dreaming*. Hallucinations can be thought of as vivid dreams occurring when one is awake.

Hallucinations, LSD and Serotonin

One of the strongest pieces of evidence for the involvement of serotonin in alterations of perception and emotion comes from the effect of hallucinogenic drugs on serotonin. The popularity of the street drugs, LSD, psilocin and mescaline, comes from their ability to produce altered states of consciousness and vivid hallucinations. These drugs have a structure very similar to that of serotonin. For example, serotonin and psilocin differ only in the parts shown in the squares.

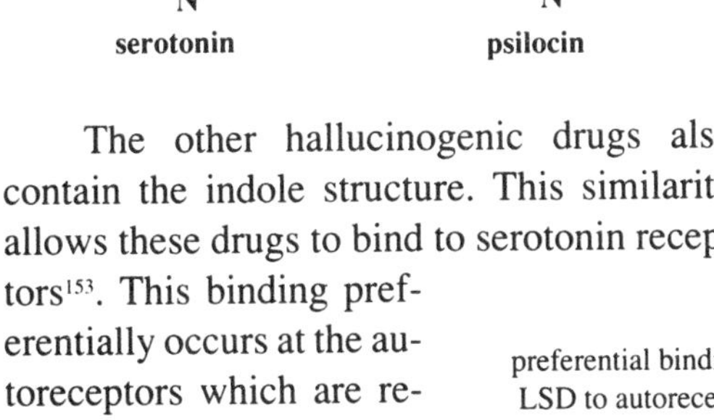

serotonin psilocin

The other hallucinogenic drugs also contain the indole structure. This similarity allows these drugs to bind to serotonin receptors[153]. This binding preferentially occurs at the autoreceptors which are receptors on nerve cells that respond to the cell's own neurotransmitter. This inhibits the activity of the nerve as **shown on the right->**.

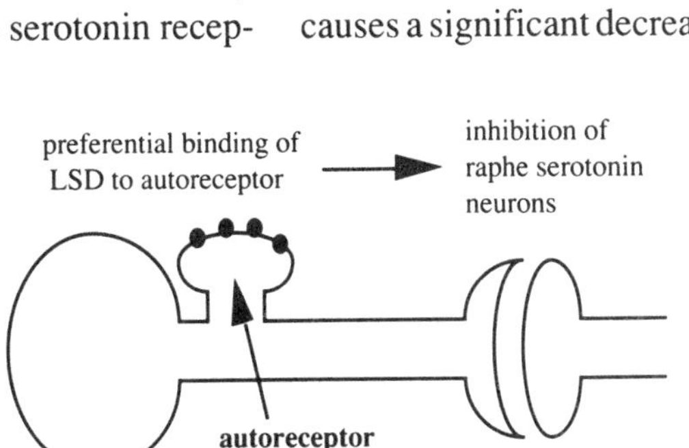

As a result, LSD suppresses the function of the raphe nuclei[14,15,836]. This blocking produces *disinhibition* of the visual and limbic systems controlled by the serotonin neurons and results in the *visual hallucinations and rapid changes in mood typical of these drugs* [963,964]. Part of the effect of these drugs, especially LSD, is due to their ability to stimulate dopamine neurons as well as block serotonin neurons. Since TS is characterized by both a dopaminergic hyperactivity of some neurons and serotonin hypoactivity, it is not surprising that some TS patients report hearing voices, other schizoid symptoms, and rapid mood changes[p191,p200].

The hallucinogenic street drug MDA (3,4-methylenedioxyamphetamine) has been shown to selectively destroy serotonin nerves when given at high doses to rats[1605,p435].

Psychoses and Serotonin

The relationship between serotonin and hallucinations raises the question of whether the hallucinations occurring in psychotic individuals, especially schizophrenics, are related to disturbances in brain serotonin. One of the best animal models of psychoses is that produced by chronic administration of amphetamines. This also produces psychoses in humans — the so-called amphetamine psychoses[p373]. Since amphetamines predominately cause the release of dopamine, this has been widely viewed as strong evidence for the dopamine model of schizophrenia. However, chronic amphetamine administration also causes a significant decrease in brain serotonin and 5-HIAA[963,1963]. As with the effect of LSD, both a decrease in serotonin and increase in dopamine appear to play a role in amphetamine psychosis. The interrelationship among dreams, drug-induced hallucinations, amphetamine psychosis, serotonin and dopamine are summarized in the following table[963].

	Dreams	Drug-induced hallucinations	Amphetamine psychosis
Serotonin activity	decreased	decreased	decreased
Dopamine activity	?	increased	increased

Serotonin + Dopamine Theories of Schizophrenia

These and other findings have led to the theory that changes in both brain serotonin and dopamine cause schizophrenia[1581,1962]. The symptoms of aggression in chronic schizophrenia have been reported to improve following treatment with 5-HTP or tryptophan[1356]. Serotonin receptors measured by LSD binding are decreased by 50% in schizophrenic brains[154], and neuroleptic treatment (Thorazine, Haldol) causes stimulation of tryptophan hydroxylase and increase in serotonin and 5-HIAA levels in the brain[1584]. Thus, part of the effectiveness of Haldol in thetreatment of both schizophrenia and Tourette syndrome may be due to its effect on brain serotonin levels.

Sex and Serotonin

The rich interconnections between the serotonin-rich raphe nuclei and the limbic system suggest that serotonin would also inhibit other limbic-controlled behaviors such as sexual activity. When the level of brain serotonin of rats was depleted to 10 percent of normal, there was a marked increase in **homosexual activity** as indicated by the frequency of males mounting other males[700,1787,1915]. This treatment also resulted in a decrease of brain norepinephrine to 71 percent of normal values. However, when a monamine oxidase inhibitor was added to restore the brain norepinephrine back to normal levels, the amount of sexual activity increased still further. This is shown in the following figure.

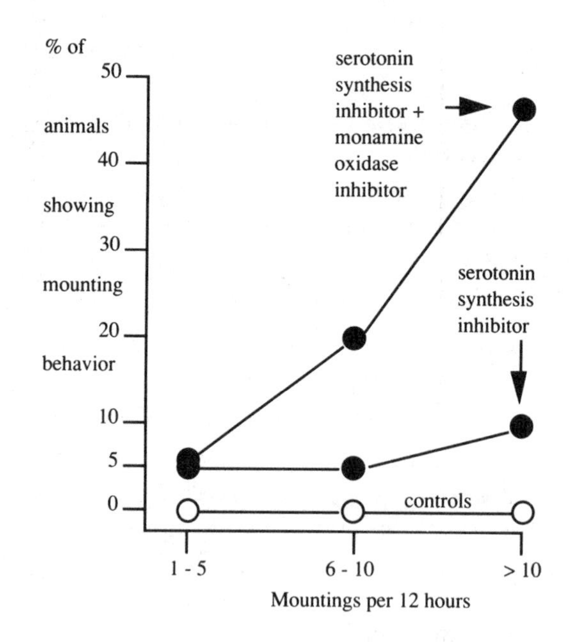

The marked increase in sexual behavior when the brain norepinephrine was increased from 71 percent to 100 percent of normal indicates that both of these neurotransmitters regulate sexual behavior, with serotonin inhibiting it and norepinephrine (catecholamines) stimulating it. When serotonin is markedly decreased and norepinephrine is relatively increased there is a synergistic action with marked disinhibition of sexual activity. A similar effect has been noted with female rats who showed stimulation of sexual behavior following brain serotonin depletion[2149]. The treatment of these rats with 5-hydroxytryptophan completely eliminated the increased sexual behavior.

The sex hormone **testosterone** plays a critical modulating role. If the animals are castrated, depletion of brain serotonin has no effect on sexual behavior. If the missing testosterone is replaced, sexual behavior is markedly increased[700]. This is illustrated as follows:

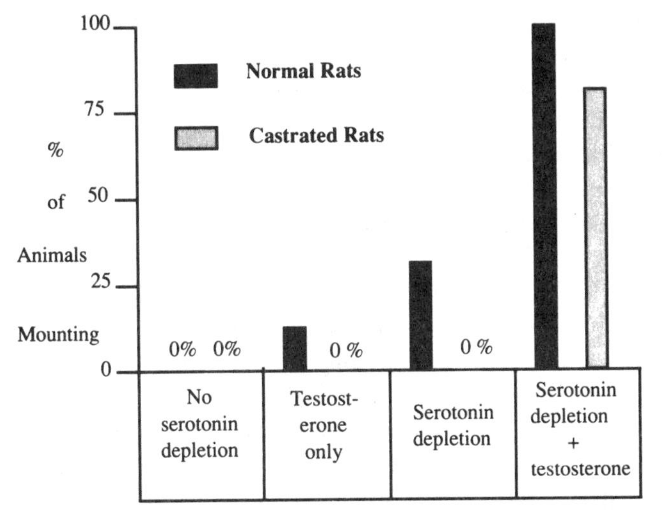

This helps us to understand why the onset of puberty can herald the onset of inappropriate sexual behaviors in human males with genetic disinhibition disorders.

Further evidence for effect of brain serotonin on sexual behavior comes from observing the **heterosexual activity** of male rats whose brain serotonin has been depleted by drugs. Here, the number of times the male inserted or attempted to insert his penis in the vagina of females, and the number of ejaculations was recorded. The results[1827] are **shown on the right->**.

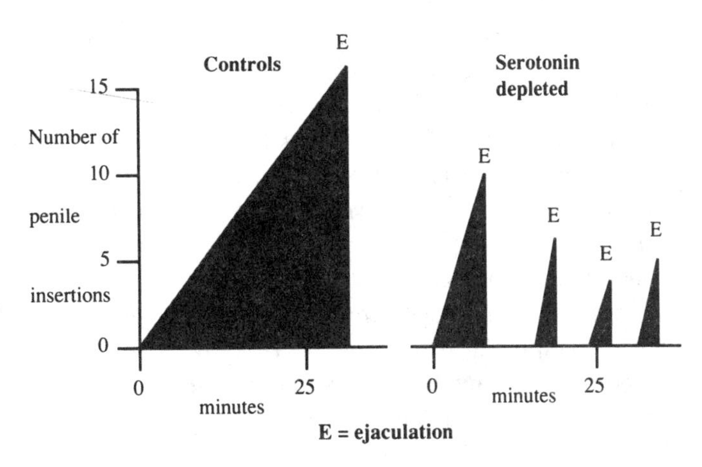

In the serotonin-depleted rats the number of ejaculations was significantly increased, while the number of intromissions required before ejaculation and the recovery time between ejaculations was significantly decreased. *In toto,* there was a marked increase in genital reflexes leading to ejaculation.

Human sexuality and serotonin Several observations indicate that serotonin also plays a role in human sexual behavior. Yohimbine, a serotonin-like compound that interferes with brain serotonin, has long been known to be an aphrodisiac. As perviously discussed, the inhibition of serotonin synthesis reverses the antidepressant effect of imipramine and MAO inhibitors[p431]. In this study a young female was treated with a drug to inhibit serotonin synthesis. During this period she reported a marked sexual arousal and made irresistible, compulsive advances to both males and females[1788].

Reciprocal effect of serotonin and dopamine on sexual behavior These findings and those described earlier for dopamine[p387] indicate that serotonin and dopamine have opposite effects on sexual behavior[1698]. This is shown as follows:

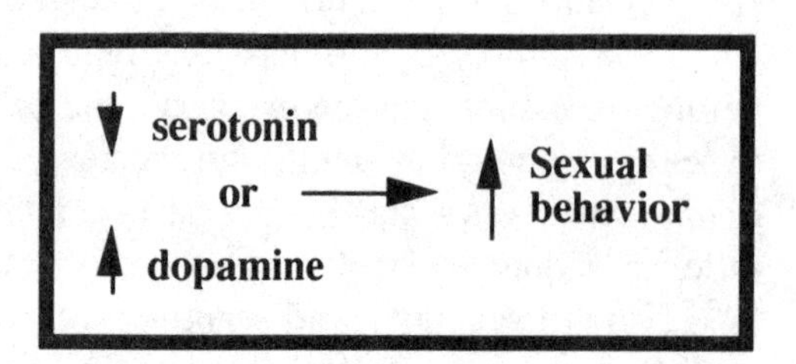

Pleasure and serotonin As discussed elsewhere[p521], the pleasure centers of the brain, as studied by self-stimulation, are dependent on the integrity of the dopamine and norepinephrine neurons. However, depletion of brain serotonin results in an increase in self-stimulation[600,1529,1530]. This indicates that serotonin also controls the pleasure centers and when serotonin is low there is increased pleasure seeking.

Stereotypy, Tics and Serotonin

As an explanation of impulsivity I have emphasized how serotonin inhibits motor activity controlled by the striatum[1825]. When serotonin is depleted by lesions of the raphe nuclei, this disinhibition may explode into the same tic-like movements seen in humans with TS. In one study[1085] destruction of raphe nuclei in rats produced an "increase of spontaneous motor activity including running around the cage, aimless cross-cage movements, occasional *stereotyped sniffing and licking*, slight tremor and rigid posture of the extremities. Hypersensitivity to external stimuli such as clicks and touching was also observed. In another study[599] the rats were described as "*typically jumpy, aggressive and hypersexual.*"

Like the reciprocal effects of serotonin and dopamaine on sexual behavior, there is a similar reciprocal effect on stereotypy. Compounds that inhibit serotonin increase amphetamine-induced stereotypic movements and decrease the ability of Haldol to cause Parkinsonian side effects[100,1082,1098]. This is another aspect of the observation discussed early in this chapter that serotonin nerves from the raphe nuclei pass to the substantia nigra (and nucleus accumbens) where they inhibit dopamine neurons. Thus, when the activity of the serotonin neurons is decreased, this inhibition is released and dopamine neurons become more active and enhance stereotyped and tic-like movements. Lesions of the raphe nuclei also result in increased motor hyperactivity[1825].

These observations provide an important link in understanding how a genetic defect causing low brain serotonin levels could cause the tics and hyperactivity in Tourette syndrome.

The "Serotonin Syndrome"

Animals who are given drugs which cause an increase of brain serotonin, or stimulate serotonin receptors, may show a *serotonin syndrome* consisting of hyperactivity, head weaving, tremor, staggering gait, muscle jerking, and hypersensitivity to noise[787,966,813]. The syndrome is due to the stimulation of both S_1 and S_2 receptors[794]. It has also been reported in humans[951]. It is important to keep this syndrome in mind when individuals are treated with medication that may increase brain serotonin.

Learning, Memory and Serotonin

The role of serotonin neurons in controlling learning and memory is complex[1429]. These skills can be impaired by both increases and decreases of brain serotonin. The following is an example of the effect that decreased brain serotonin can have on learning.

Rats prefer the security of the dark part of their cage. They can, however, be taught to avoid the dark part of the cage by giving them shocks whenever they go there. This learned behavior is called *passive avoidance*. As shown in the following figure, when a rat is treated with a drug that depletes brain serotonin before learning a passive avoidance task, the learned task is intact in the memory after five minutes. However, as time passes much of this learned behavior is forgotten[1430] while it is retained in the controls animals (**see below**).

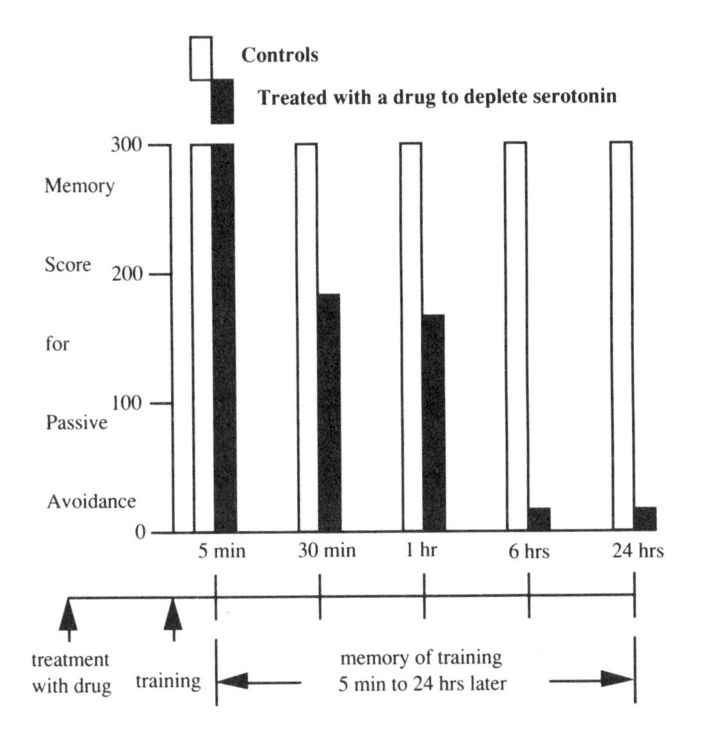

After 24 hours, the memory of this learned task is vitrually gone. This resembles the problems that many TS children have with quickly forgetting things they have just learned.

The Frontal Lobe Syndrome and Serotonin

In the chapter on the frontal lobe syndromes we saw how many of the symptoms of ADHD and Tourette syndrome could be explained by a defect in the dopamine nerves to the frontal lobe. The following shows that the frontal lobe syndrome can be caused by a primary deficiency in serotonin production.

Dopamine, norepinephrine and serotonin neurons all pass to the cerebral cortex, especially the frontal cortex[594]. When rats are treated with drugs which cause *low brain serotonin, the neurons of the frontal cortex show a marked decrease in responsiveness to dopamine and norepinephrine* [594]. There is also a marked decrease in the responsiveness of the frontal lobes to electrical stimulation of raphe nuclei. When the raphe nuclei are stimulated in normal rats, recordings in the frontal cortex show that 4 percent of the cells are inhibited, 15 percent are stimulated and the rest show a biphasic inhibition-stimulation or no response. However, if serotonin is depleted, stimulation of the raphe nuclei results in 56 percent of the cells of the frontal lobe being inhibited and only two percent are excited. This is shown in the diagram on the **bottom of the page**[992].

By contrast, after brain serotonin is depleted, the limbic structures (hippocampus and amygdala) can no longer be inhibited by stimulation of the raphe nuclei

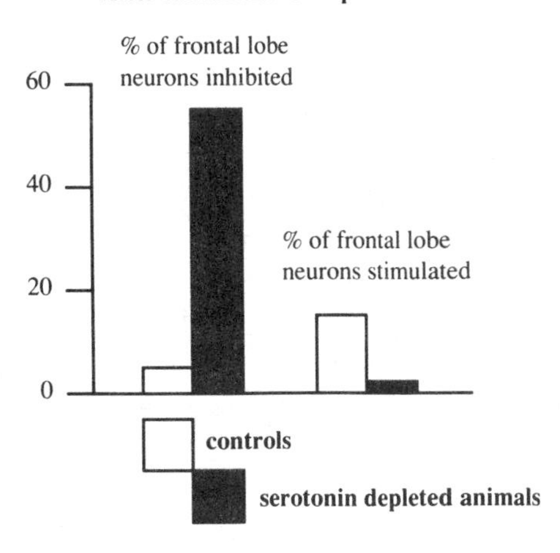

[1746,2027]. *All of these effects could account for a decrease in the function of the frontal lobes (frontal lobe syndrome) and a disinhibition of the limbic system in the face of a genetic deficiency of serotonin production.*

Serotonin inhibits prefrontal lobe glutamic acid neurons A major proportion of the nerves passing from the prefrontal cortex to the striatum use glutamic acid as a neurotransmitter. Since glutamic acid is a stimulating neurotransmitter, this results in activation of the dopamine neurons of the caudate and putamen. When serotonin is applied to these nerves it blocks this dopamine activation[248]. Serotonin diffuses into the cortex from swellings in the nerve endings[p419]. One of its major actions may be to modulate this stimulation of sub-cortical dopamine nerves by glutamic acid-containing neurons as follows:

serotonin
glutamic acid
+ dopamine
prefrontal cortex
caudate-putamen limbic system
normal serotonin → ↓ glutamic acid
↓ dopamine →
normal motor activity and behavior

Thus, there are at least two pathways by which decreased serotonin could cause disinhibited movements, tics and impulsive behavior: a direct effect through the substantia nigra [p446] and an indirect effect through the frontal lobes. This normal state, where serotonin has an inhibiting effect on the motor activites controlled by the striatum, is diagramed as follows:

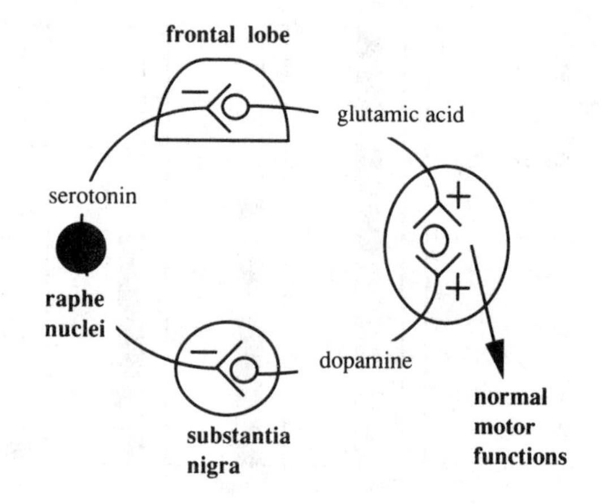

However, when serotonin is depleted, the frontal lobe and the substantia nigra are disinhibited, producing hyperactivity, tics, and impulsivity. This concept is so important to the understanding of Tourette syndrome that it has been summarized in the **box on the following page**.

However, if there is too little serotonin, the glutamic acid neurons may become overactive and overstimulate the sub-cortical dopamine neurons and cause tics, hyperactivity and disinhibited behaviors, as **shown on the right->**.

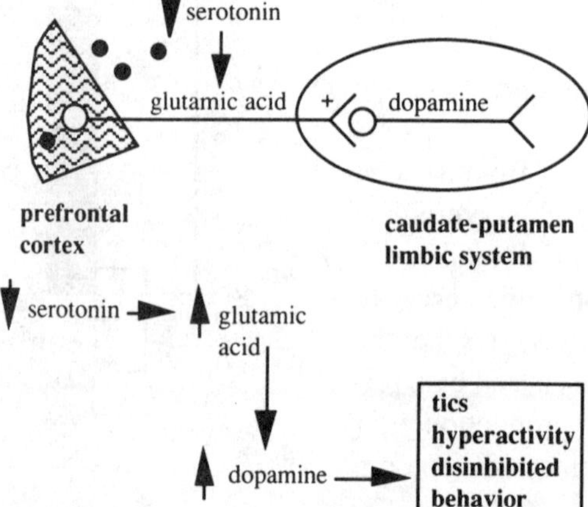

Pain and Serotonin

Neurons in a group of raphe nuclei in the brain stem project into the spinal cord and play a role in regulating pain sensation. Increased serotonin activity decreases sensitivity to pain while decreased serotonin activity increases sensitivity to pain[131,1317]. Chronic pain syndromes in

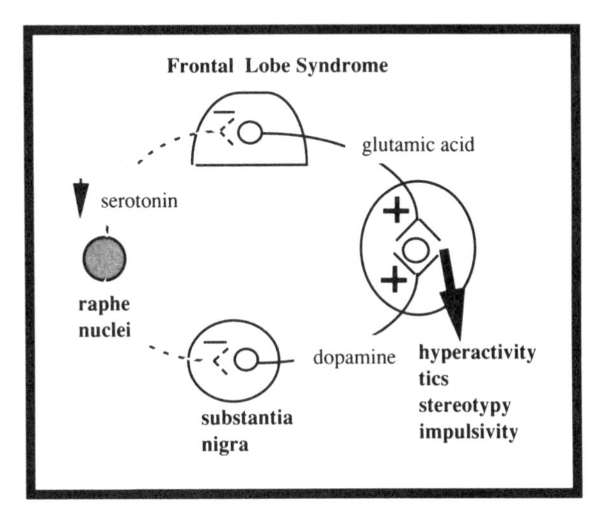

humans, where there is no obvious cause of the pain, are associated with a low spinal fluid 5-HIAA. They are made worse by depleting serotonin, and are improved by serotonin reuptake inhibitors[24].

Temperature and Serotonin

Like a thermostat in the house, the temperature of the body also has a set point. Serotonin plays a role in setting this point which is regulated by a balance among epinephrine, norepinephrine and serotonin in the hypothalamus[585,962]. When rabbits are depleted of serotonin they have problems lowering an induced fever[705]. The difficulty TS patients have with temperature regulation (often feeling hot) could be related to a serotonin deficiency.

Appetite and Serotonin

Decreasing brain serotonin by inhibiting its rate of production results in increased appetite (hyperphagia)[188]. Increasing brain serotonin results in the inhibition of appetite[1152a]. Stimulation of presynaptic serotonin_{1A} receptors inhibits serotonin neurons and compounds that stimulate these receptors result in increased appetite. Stimulation of postsynaptic serotonin_{1B} receptors stimulates serotonin neurons, and compounds that stimulate these receptors cause anorexia[1152a]. The low serotonin in TS[p457] and this important role of serotonin in the regulation of appetite is relevant to the presence of compulsive eating and obesity in some *Gts* gene carriers[p235,p644]. Other aspects of the role of serotonin in regulating appetite are discussed elsewhere[p495].

Drinking Fluids and Serotonin

A decrease in brain serotonin results in increased drinking of fluids[123,403]. This is relevant to the polydipsia seen in TS[p255].

Separation Anxiety and Serotonin

Separation anxiety is common in TS. The observation that separation panic in young animals is decreased when they are given serotonin agonists[1457] suggests this symptom may be related to low brain serotonin.

Serotonin Depletion -> Tourette Symptoms

Although many of the following have been discussed above, I would like to reemphasize that *destruction of the raphe nuclei* results in a marked decrease in serotonin and 5-HIAA in the brain, which in turn can result in the following behaviors [1085,1399,1844] associated with Tourette syndrome:

-> aggression[790],

-> anxiety[958],

-> decrease in the ability of punishment to suppress behavior,

-> hyperactivity[1825],

-> hyperemotionality[463,2019],

-> hyper-reactive to external stimulation — shocks, air puffs, touching and noise [224,461,702],

-> hypersexuality,

-> impulsivity[1825],

-> persistent arousal in the EEG patterns of the frontal lobe,

-> perseveration[702,1212,1844],

-> stereotyped movements,

-> stereotyped licking and sniffing,
-> self-stimulation[1211],
-> increase in water intake[123,1212,1844].

The increase in these behaviors directly correlates with the degree of depletion of serotonin and 5-HIAA[1085].

Some of these behaviors are also seen after lesions of the septal nuclei[p328] and hippocampus[p333]. Since these structures receive serotonin neurons from the raphe nuclei[1965,1085], the hyperactivity[774], hyper-responsiveness to stimuli, and perseveration[452,453] seen especially with lesions of the septum are probably due to the destruction of serotonin neurons[702].

From this list and this chapter, it is apparent that many of the disinhibited behaviors seen in Tourette syndrome could be explained on the basis of a primary defect in brain serotonin. This is summarized as follows:

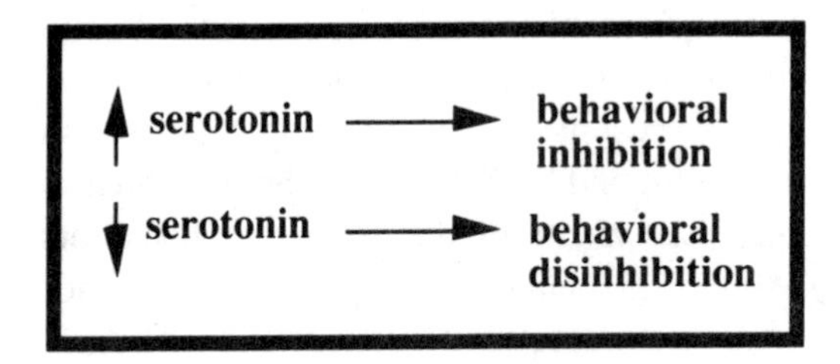

The next chapter examines the question of whether serotonin is in fact deficient in Tourette syndrome.

Summary: Both human and animal studies clearly implicate serotonin as playing a major role in the control of impulsive and aggressive behaviors. Serotonin inhibits the dopamine action system and defects in serotonin result in action without thought, i.e., impulsive behavior. Sleep, sexual activity, and a wide range of other behaviors are also controlled by serotonin. A genetic defect in brain serotonin metabolism may provide the common thread between many behavioral syndromes.

Chapter 63
Serotonin and Tourette Syndrome

By this time if should be clear that all of the behaviors associated with Tourette syndrome have been observed in humans or animals with low brain serotonin level. Is there any direct evidence that brain serotonin is low in TS?

Spinal Fluid 5-HIAA in TS

To examine the possible role of neurotransmitter abnormalities in Tourette syndrome, Cohen and colleagues[341,342] examined the spinal fluid levels of 5-HIAA in six TS children and 27 controls after the administration of probenecid[p394]. Since the level of probenecid in the spinal fluid affects the 5-HIAA level, the results were given both in terms of the level of 5-HIAA itself and the ratio of 5-HIAA to the logarithm of the probenecid concentration.

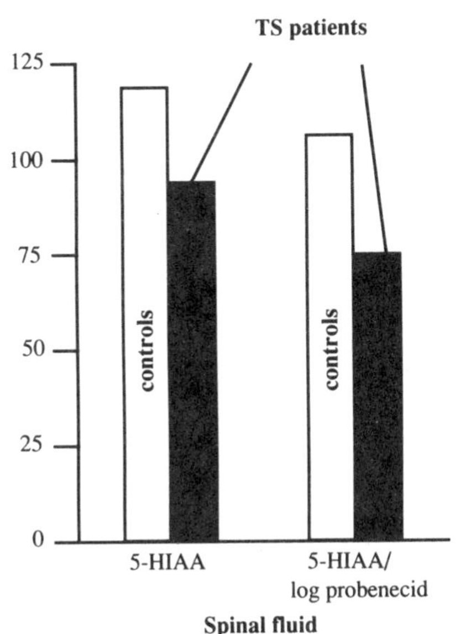

As discussed in the chapter on dopamine, HVA (the breakdown product of dopamine) was also decreased in the spinal fliud[p399]. There was a tendency for the severity of the tics to be worse in the children with the lowest levels of 5-HIAA, especially when related to the level of HVA.

A study one year later by Butler and colleagues[262] gave similar results. They studied nine TS patients and 39 normal controls. The baseline levels of 5-HIAA were low in two patients and the serotonin turnover, based on rates of accumulation of 5-HIAA after probenecid, was impaired in three other patients. Here again, the spinal fluid levels of HVA were significantly lower for the TS patients compared to controls[p399].

There are a number of other reports that suggest a role of serotonin in TS. As was the case with depression, 5-hydroxytryptophan has been reported to be effective in the treatment of TS[2001-2003]. We have also successfully used this serotonin precursor to treat some TS patients[p588]. Conversely, when one TS patient was given cyproheptadine, a drug that blocks serotonin, it caused an "explosion of motor and verbal tics"[426]. Thus, drugs that increase serotonin may improve the symptoms while drugs that block serotonin make the symptoms worse.

Blood Serotonin in TS

The spinal fluid 5-HIAA level is usually considered to be the best way to evaluate the level of brain serotonin. Serotonin in the blood is primarily the result of the uptake into platelets of serotonin made in the gut. However, the reports of low blood serotonin in aggressive, mentally retarded, and senile patients and the improved behavior and increased blood serotonin after treatment[P438] stimulated us to examine the blood serotonin levels in TS patients[375]. The initial results are shown in the following table (bold p values are significant differences):

Blood serotonin in Tourette syndrome, ADHD and controls (ng/ml)

	N	Mean	S.D.	t	p
Controls	79	97.9	36.7		
Tourette syndrome	332	80.7	54.9	2.64	**.0086**
No medications	(171)	82.5	56.0	2.23	**.0250**
ADHD	43	111.7	68.5	1.46	.1442
ADHD 2° TS	22	76.0	51.7	2.25	**.0250**
Parents of TS patients	34	71.7	40.6	3.36	**.0014**
Misc. cases	23	61.6	31.6	4.29	**.0002**
Total	533				

A total of 533 individuals were studied. The average serotonin in 332 TS patients was 80.7 ng/ml, significantly lower than the 97.9 ng/ml for the controls. The level was almost as low in the 171 TS patients who had never been on medication . By contrast, the average serotonin was normal in the individuals with ADHD. However, in those ADHD patients with a positive family history of TS, the average serotonin level was very low, 76 ng/ml. Of particular significance, among 34 parents of TS children who had no tics, the serotonin levels were also very low, 72 ng/ml. This suggests that the low serotonin is a *trait marker* for the presence of the gene, not a *state marker* for the presence of tics. The miscellaneous cases were individuals with a TS-like disorder that did not fulfill the complete diagnostic criteria but probably represented the partial expression of the *Gts* gene. There was no difference in the level of serotonin for patients on Haldol compared to those on no medication. Food was not a factor since the patients had fasted overnight. The effect of medication was excluded by examining 171 patients who had *never* been on medication[375].

The vast majority of the serotonin in the blood is present in the platelets, small elements that assist in the clotting of blood. Since the number of platelets tends to be higher in children than adults, any age differences between the TS patients and controls would be corrected by dividing the serotonin level by the platelet count. The **table on the next page** shows the results of this correction.

The serotonin/platelet ratio was 2.37 for the TS patients, significantly lower than the 3.3 for the controls. It was also much lower for the TS patients on no medication. Again, the ratio was normal for the ADHD subjects, but lower for the ADHD 2° TS. The level for the parents was also low. The distribution of these values for the TS patients on no medications versus the controls is shown in the figure on the **next page**.

These results indicate that blood serotonin levels are significantly decreased in individuals with Tourette syndrome, their parents and miscellaneous cases. The follow-

Blood serotonin/platelets in Tourette syndrome

	N	Mean	S.D.	t test	p[p70]
Controls	36	3.30	1.35		
Tourette syndrome	254	2.37	1.55	3.41	**.0011**
No medications	(137)	2.49	1.61	2.77	**.0063**
ADHD	35	3.39	2.09	0.20	.8205
ADHD 2° TS	16	2.64	1.64	1.51	.1312
Parents of TS patients	28	2.65	1.59	1.71	.0813
Misc. cases	18	1.95	1.17	3.59	**.0011**
Total	387				

ing are some specific examples.

Self-aggressive behavior — tooth pulling A 13 year old boy with TS was admitted on an emergency basis because he was compulsively pulling out his own teeth. He had a history of temper tantrums and ADHD starting at six months of age. Motor and vocal tics began at age seven. He had been treated with low doses of Haldol and Ritalin elsewhere and was doing well until suddenly, the week before admission. At that time he began to be afraid of being left alone, compulsively talk to himself, pace up and down, touch his toes and mouth, pat his teeth and pull at his lips. He next began to grind and then pull at his teeth. Within

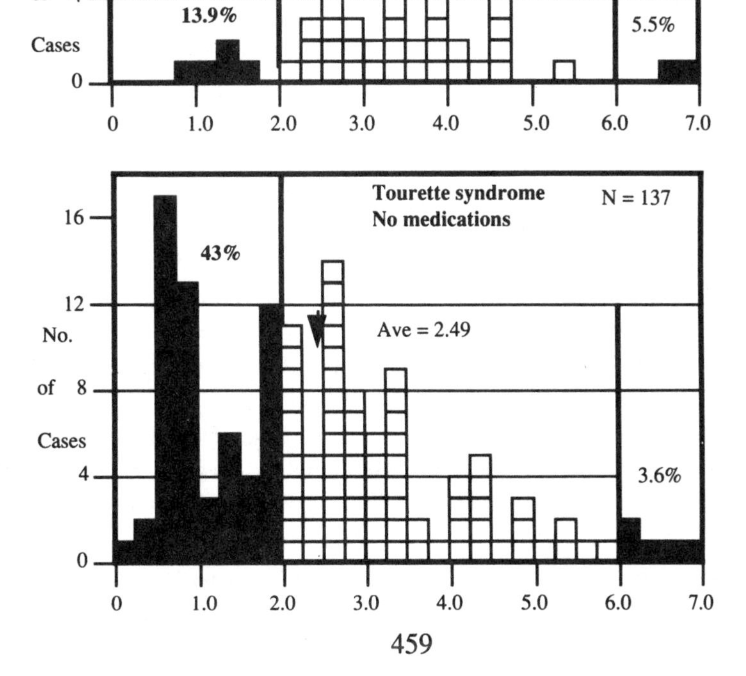

a period of a few days he had pulled out three permanent teeth. His Haldol was quickly increased from 0.5 to 10 mg/d and his compulsive tooth pulling stopped.

His blood serotonin was 42 ng/ml (normal 50 to 200). Despite the fact that increasing his Haldol brought his symptoms under control, his blood serotonin remained unchanged and successive values over a period of six months were 43, 42 and 37 ng/ml.

Aggressive and hypersexual behavior James is nine years old. He was diagnosed as having ADHD at age three. Several weeks after starting kindergarten he was expelled because of "bizarre behavior" consisting of biting other children and threatening them with knives. He also had throat clearing, grunting and snorting. His mother took him to another school and after four months he was dismissed for jumping on desks, biting, and aggressiveness. He was then placed in a kindergarten for severely emotionally disturbed children. He was treated with Ritalin for the third time in his life and again it had no effect. Despite coprolalia and constant vocal noises the diagnosis of TS still had not been made and a regimen of behavior modification was tried without effect on the aggressive behavior. Academic testing showed him to be very bright, in the 98th percentile. At seven years of age a tonsillectomy was recommended to stop the vocal noises — without avail. On the second day of 3rd grade he was removing his clothing, swearing, hitting other children and showed a very short attention span. A contract was tried to control his behavior but within a month he was suspended for stabbing another student with a pencil. A month later a diagnosis of Tourette syndrome was finally made and he was treated with haloperidol 1.5 mg/d. This resulted in the immediate cessation of the vocal tics but the disruptive behaviors continued with lying, stealing and vandalizing property. He often got into fights with his peers and had a very short temper. He has occasionally exhibited himself in public. At five he had a very precocious interest in sex and at six was trying to French kiss girls. He frequently talks about women's breasts and wants to have intercourse as soon as possible. He doesn't listen and does what he pleases. He throws things at his mother, swears at her, and makes sexually explicit comments and gestures to her. Typical panic attacks started at three years of age. If his mother doesn't pick him up exactly on time he is afraid she has been killed. Self-abusive behavior consisting of hitting his genitalia, biting himself, and head banging started at age four. He has often expressed the desire to die, at times talks to an imaginary friend, and sometimes thinks he has special powers. Several times a voice has told him to do mean things. Motor tics are minimal although he had facial grimacing for several years.

The pedigree was negative for motor and vocal tics but the father was an alcoholic and had an alcoholic son by a previous marriage.

The blood serotonin was less than 22 ng/ml on two different occasions. The platelet count was 406,000, giving a serotonin/platelet ratio of 0.54.

There have been only a few other reports of blood serotonin levels in TS. In one[1140], 20 TS patients were compared to 87 normal controls. The blood serotonin levels were lower than the controls but the number of patients was small and this was not significant. They found that treatment with clonidine, a medication that is often effective in the treatment of TS[p561], caused an increase in blood serotonin levels. In one other study of only five TS patients, without controls, the blood serotonin was higher than in subjects with ADHD[1704]. In a single case of TS, the secretion of serotonin in the urinewas half that of normal[2130].

Brain serotonin levels are significantly lower in male rats compared to female rats[293]. If this is also true of humans it could explain why

TS is more easily expressed in males.

These results provide support for the proposal that TS is due to a defect in serotonin metabolism. This serotonin hypothesis of TS can be summarized in the diagram below. The intimate connection between serotonin, dopamine and norepinephrine metabolism[p509] is illustrated. It is not always clear which defect is the primary cause of the symptoms.

Summary: Studies of serotonin in the spinal fluid and blood of Tourette syndrome patients indicate levels are lower than in non-TS subjects. A defect in the metabolism of serotonin could explain all of the symptoms of Tourette syndrome.

[By the time this book was complete, much more extensive studies of blood serotonin had been done. They verified these initial studies and are described in Chapter 90.]

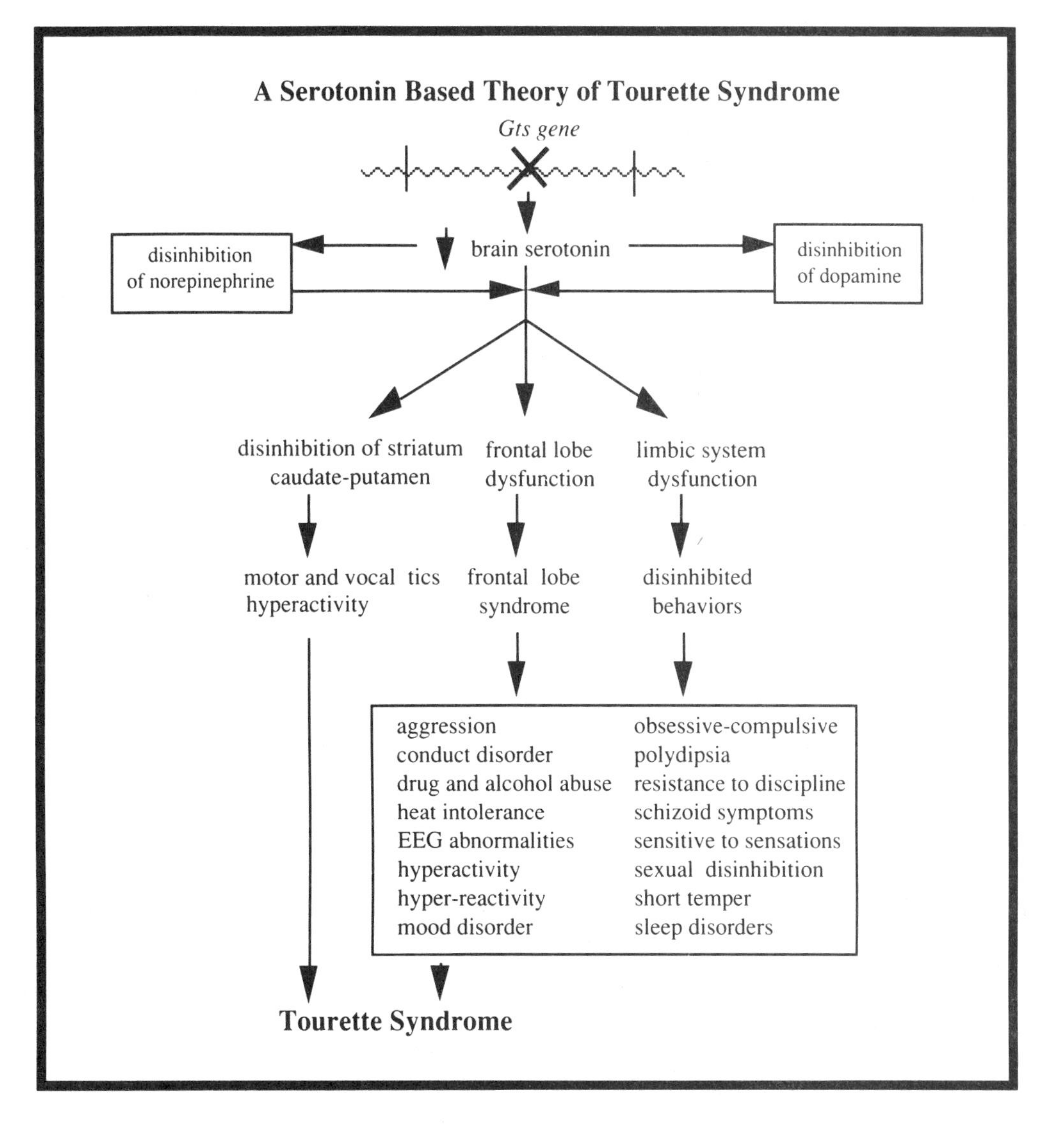

Chapter 64
Is Tryptophan the Key?

In the previous chapters we have seen that low brain serotonin can explain many of the features of TS and other impulsive, addictive behaviors. The next questions is — What causes this serotonin deficiency? Since the body cannot make tryptophan, it must all come from the diet. Once in the body there are two pathways that tryptophan can take — the synthesis of serotonin or the breakdown to kynurenine.

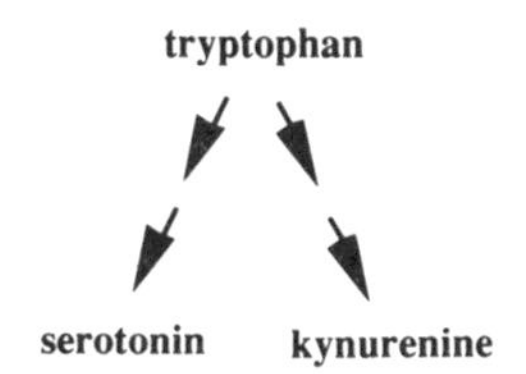

The tryptophan -> serotonin pathway has been discussed. This chapter is devoted to the equally important tryptophan -> kynurenine pathway. While there are many ways in which the brain serotonin may become depleted, two of the major ones are (1) an interruption in the pathway from tryptophan to serotonin;

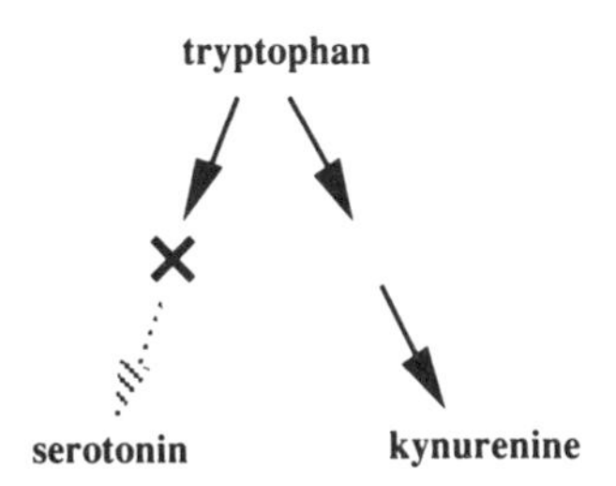

and (2) an increase or acceleration of the breakdown of tryptophan to kynurenine.

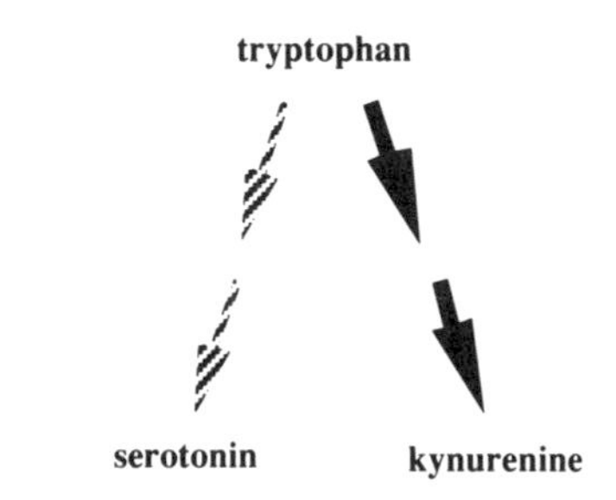

This would shut the tryptophan away from the formation of serotonin. Since more than 90% of tryptophan is broken down by the kynurenine pathway[147a,632] a relatively minor increase in this pathway could have a major effects on serotonin production.

The way to distinguish between these two possibilities lies in examining the level of tryptophan in patients with TS. If the defect in serotonin is due to a block in the tryptophan -> serotonin pathway then tryptophan should be normal or even elevated. However, if the defect is due to an increase in the tryptophan -> kynurenine pathway then tryptophan would also be low. To find out which pathway was involved we examined blood tryptophan in TS patients.

Blood Tryptophan in TS

When the blood tryptophan levels were examined the results were as follows[375]:

Blood tryptophan in Tourette syndrome and controls (mg/dl)

	N	Mean	S.D.	t	p
Controls	108	1.93	.55		
Tourette syndrome	235	1.53	.33	8.36	**.0000**
No medications	(130)	1.51	.31	7.30	**.0000**
ADHD	34	1.72	.65	1.87	.0593
2° TS	14	1.53	.25	2.68	**.0083**
TS parents	27	1.47	.32	4.11	**.0002**
Misc. cases	17	1.60	.60	2.23	**.0259**
Total	435				

The average level of blood tryptophan in untreated TS patients was 1.53 compared to 1.93 for the controls. The level was significantly lower in TS and in parents of TS patients many of whom had no tics. This suggested that the blood tryptophan level was a marker of the *Gts* gene whether it was causing symptoms or not, and suggested that in many cases both parents were carrying the *Gts* gene. [Further studies of tryptophan in TS patients and their relatives are given in Chapter 90.]

The diagram on the **next page** shows the distribution of the tryptophan values for the controls and TS patients. This shows that 33% of the TS patients had a low tryptophan level compared to 4.6% of the controls. The distribution of values in the TS patients was pushed to the left, suggesting that a tendency toward low levels was present in all TS patients.

The only other study of blood tryptophan in TS also showed that tryptophan levels in 20 TS patients were significantly lower than in 87 controls[1142].

These observations suggest that tryptophan -> kynurenine pathway is abnormal in TS. In order to understand these observations it is necessary to discuss what is presently known about blood and brain tryptophan and its relationship to behavioral disorders.

Factors That Affect Brain Tryptophan

The factors involved in the control of tryptophan and serotonin levels in the brain are complex. The three major factors are: free tryptophan, the ratio tryptophan to other large amino acids, and the level of fatty acids in the blood.

Free tryptophan Most of the tryptophan in the blood is attached to albumin. This is called *bound tryptophan.* However, about 10% is not bound and makes up the *free tryptophan.*

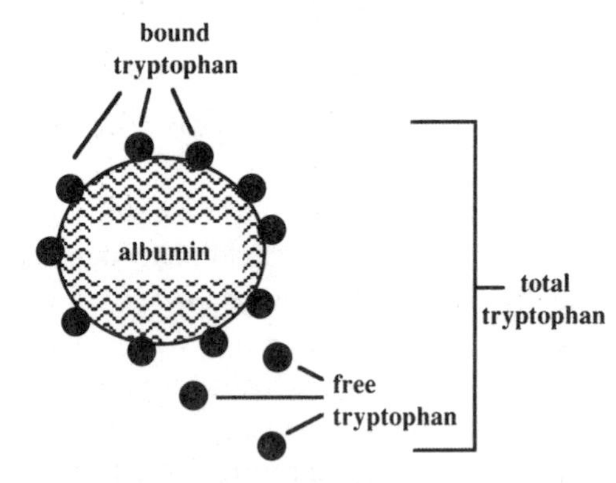

Free tryptophan is determined by passing serum over a filter that allows the small tryptophan molecules to pass but retains the larger albumin molecules. The filtrate is then examined for its tryptophan level. There is some controversy about whether free tryptophan or total tryptophan is more important in controlling brain tryptophan levels[445a,446,448,592,2114]. Tryptophan must be released from the albumin to pass into the brain and there is a better correla-

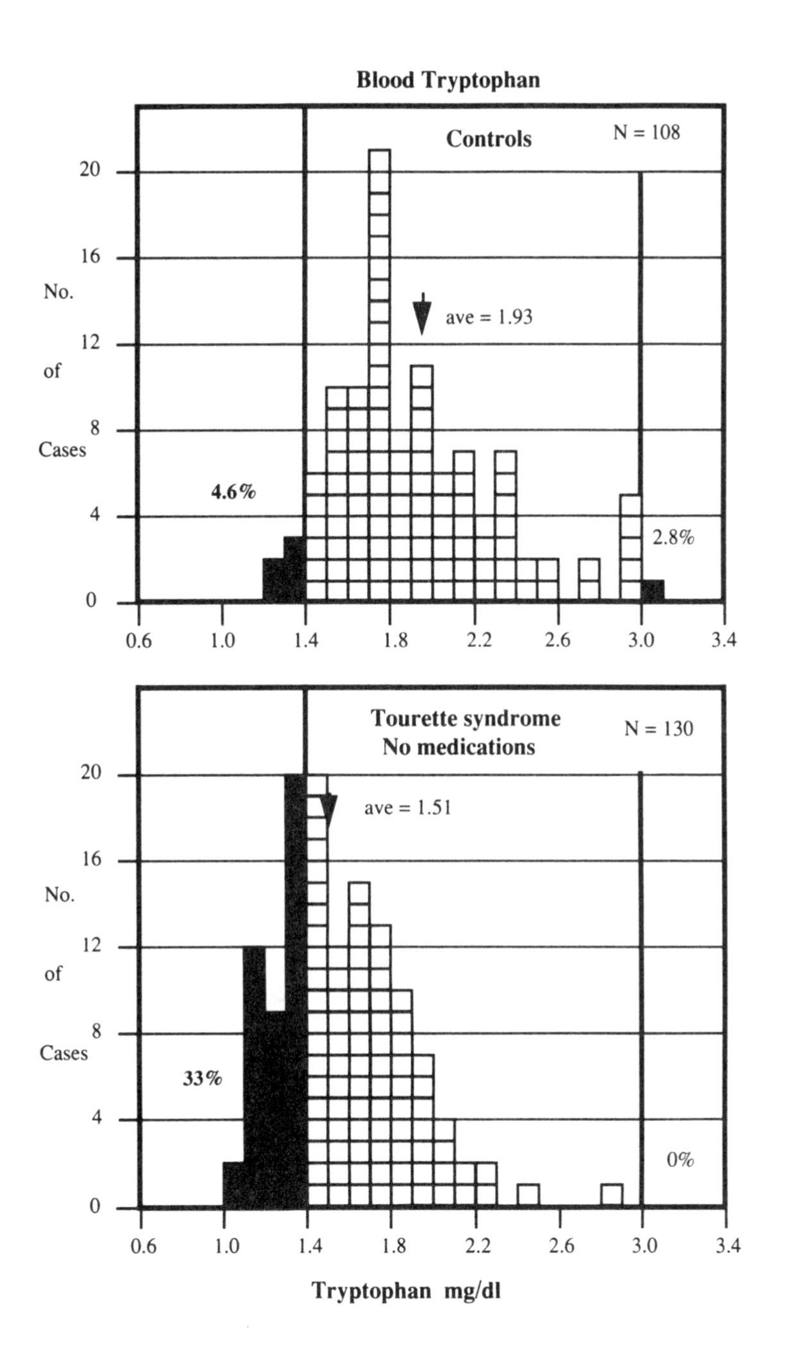

tion between free tryptophan and brain tryptophan levels than between total tryptophan and brain tryptophan[445a]. However, the correlation with total tryptophan is still high and proteins on the blood-brain barrier are able to pull tryptophan off the albumin[2114].

The ratio of tryptophan to other large amino acids Tryptophan does not just diffuse into the brain. It has to enter through a door in the blood brain barrier known as a transport system. The problem with this door is that some of the other amino acids are trying to use it at the same time. These are the large, neutral amino acids (leucine, isoleucine, valine, tyrosine and phenylalanine). The rate that tryptophan gets into the brain is dependent upon the amount of tryptophan in the blood compared to the sum of all the other large amino acids[2114]. Many things, especially the amount of carbohydrate and protein in the diet, control this ratio[2114].

A carbohydrate meal increases brain serotonin A meal rich in carbohydrates results in the release of insulin. Insulin causes the muscles to adsorb the large neutral amino acids except tryptophan. This results in a relative increase in blood tryptophan and brain serotonin[591,592,592a,2113a,b,2114-2116]. Experimentally produced low brain serotonin can cause carbohydrate craving[1152a]. The low serotonin in Tourette syndrome[p457] can explain the *carbohydrate craving, compulsive eating and obesity* seen in some carriers of the *Gts* gene[p236]. This is illustrated as follows:

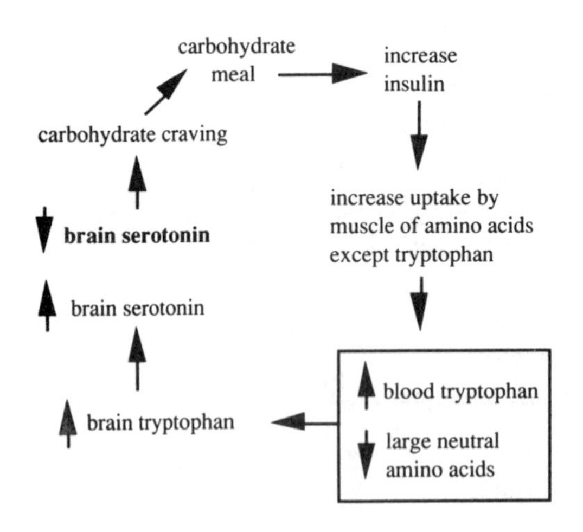

The similarity between this and the serotonin loop proposed for alcoholism[p436] is striking and is consistent with the proposed genetic connection between type II alcoholism in males and obesity in females in TS families[p236].

Since humans cannot make tryptophan it must all come from food. Thus, a low blood tryptophan level is usually the result of an increase in the rate at which tryptophan is broken down. An examinaton of how tryptophan is broken down can give us a clue as to what the defect may be in TS.

The Breakdown of Tryptophan

Recall that the structure of tryptophan is:

NH_2 | $CH_2CHCOOH$ — N

tryptophan

The enzyme *tryptophan oxygenase* (also called *tryptophan pyrrolase, tryptophan 2,3 dioxygenase* or *indoleamine 2,3,dioxygenase*) uses an oxygen molecule to break open the nitrogen ring and form N-formyl kynurenine.

O || $CCH_2CHCOOH$ with NH_2; N— CHO | H

N-formyl kynurenine

The enzyme *kynurenine formylase* removes the —CHO group to form kynurenine.

O || $CCH_2CHCOOH$ with NH_2; N—H | H

kynurenine

The addition of a hydroxyl group (—OH) forms 3-hydroxykynurenine.

O NH2 CCH2CHCOOH NH2 OH

3-hydroxykynurenine

If the genetic defect in TS is in the breakdown of tryptophan, the most likely site would be the enzyme tryptophan oxygenase, since this is the rate-limiting step in the breakdown of tryptophan. Thus, for purposes of subsequent discussion the following simplified path summarizes the critical aspects of tryptophan breakdown.

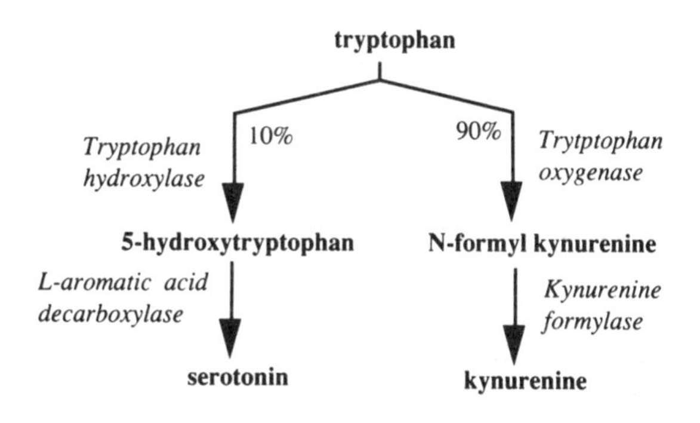

There have been a number of interesting observations about tryptophan, tryptophan oxygenase and kynurenine that have potential bearing on Tourette syndrome and related disorders.

Tryptophan Oxygenase

Tryptophan oxygenase needs more than just the protein part to work. It requires a cofactor called *heme*. It is the combination of heme + globin that makes hemoglobin, the oxygen carrying part of the red blood cells. This need for a cofactor is illustrated as follows:

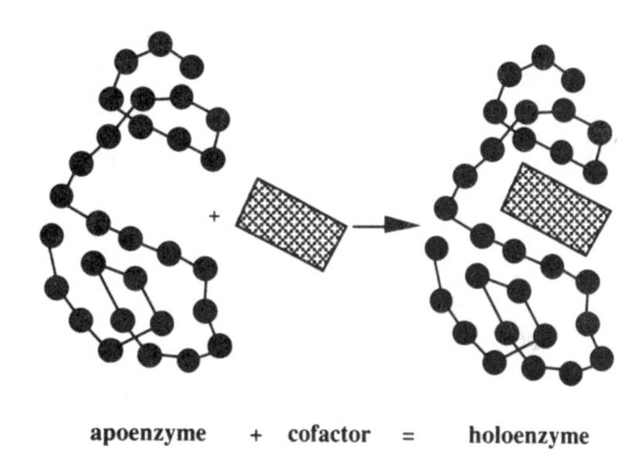

apoenzyme + cofactor = holoenzyme

The complete enzyme is called the *holoenzyme* (like "whole enzyme"). The protein portion alone is the *apoenzyme* (the protein part only).

An inducible enzyme One of the things that makes living organisms so "alive" is that they respond in many different ways to their environment. The amount of messenger RNA produced by some genes can be markedly increased by various stimuli. As a result, the total number of enzyme molecules made by those genes markedly increases. This is called **induction.** Tryptophan oxygenase is one of the best examples of an inducible enzyme. It is strongly induced by steroids[1069]. The activity of this enzyme can be increased or decreased in other ways such as changing the level of certain co-factors necessary for the enzyme to work. The different agents that can increase or decrease the activity of tryptophan oxygenase are as follows:

increase: steroids[1069], acute alcohol, phenobarbital, nicotine, tryptophan, α-methyltryptophan, epidermal growth factor[1044], and γ-interferon[1509].

decrease: nicotinamide, chronic alcohol or morphine, glucose[75], indole-3 pyruvic acid[1134], antidepressants[p457], phenothiazines like prolixin [p457], salicylates, tartar emetic[543a] and others[543a].

Note that tryptophan itself can increase the level of tryptophan oxygenase[1246]. A related compound, α-methyltryptophan causes a marked, long-lasting induction of tryptophan oxygenase. When given to rats this results in a severe depletion of tryptophan in the blood and brain, and of serotonin in the brain[1620,1828]. The concept is illustrated as follows:

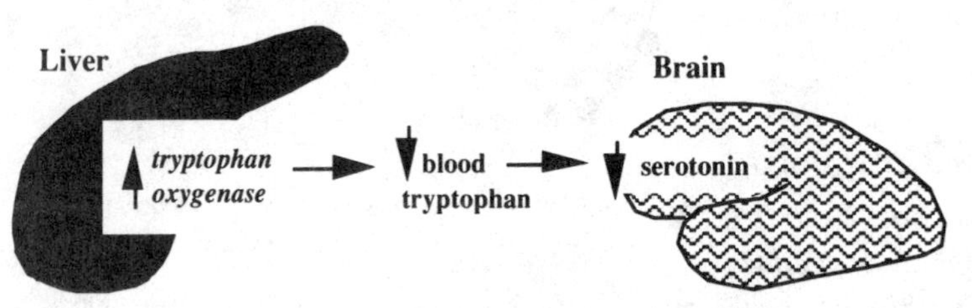

There are many examples of this effect. Some of the following are particularly relevant to human behavior.

Alcohol

In 1972 Badawy and Evans[76] said, *"As far as we know, alcohol is the only drug which is capable of inhibiting tryptophan-pyrrolase activity for prolonged periods."* In later studies it was found that morphine had the same effect. Given the role of serotonin in the impulsive-addictive disorders, *this is a powerful clue to the common ground among alcohol addiction, drug addiction, and serotonin.* If some of these disorders are due to a genetic overactivity of tryptophan oxygenase (pyrrolase), alcohol and the opiates would be the two most readily available "street" drugs for self-medication. Because of the potential importance of this link, I will examine the effect of these drugs on tryptophan oxygenase in more detail.

The effect of chronic administration of alcohol on tryptophan metabolism in rats[77-79,81,83,84,86] is shown **on the right->**:

It can be seen that tryptophan oxygenase is markedly inhibited by a chronic intake of alcohol. This decreases the breakdown of tryptophan and results in an increase in serum and brain tryptophan and brain serotonin. Following withdrawal of alcohol there is a marked rebound increase in the level of tryptophan oxygenase to as much as six times the suppressed level. The enzyme activity then returns to normal levels after one week[79,81,86]. Others have also found that chronic intake of alcohol decreases tryptophan oxygenase levels[645,1357,1358,1567,1568].

This subject is not without controversy. Some have reported that chronic alcohol administration increased the level of tryptophan oxygenase and suggest this could play a role in causing the depression often associated with alcoholism — by decreasing brain serotonin[214]. This was based on tryptophan load studies suggesting tryptophan oxygenase was elevated in five alcoholic patients several days to one month after abstinence[265]. While this was assumed to be due to increased cortisol excretion the *persistent elevation after months of abstnence is consistent with an increase in tryptophan oxygenase as the basic genetic defect in alcoholism.*

The short-term effects of alcohol on brain serotonin have also been confusing with some finding it increased, some finding no effect and others finding it decreased. This

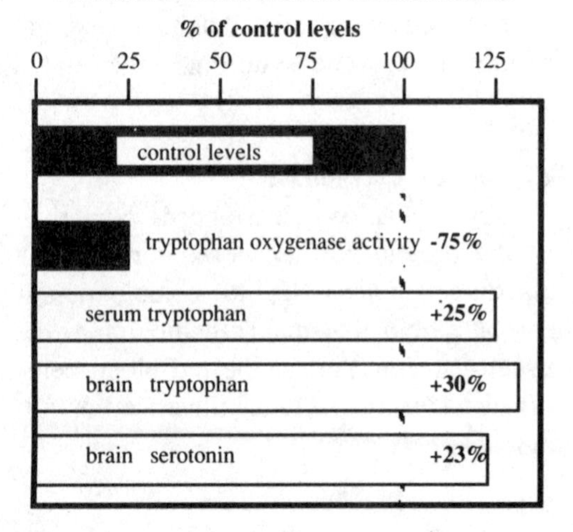

confusion is understandable when the effect is carefully studied over time, as shown in the following study in rats[81b].

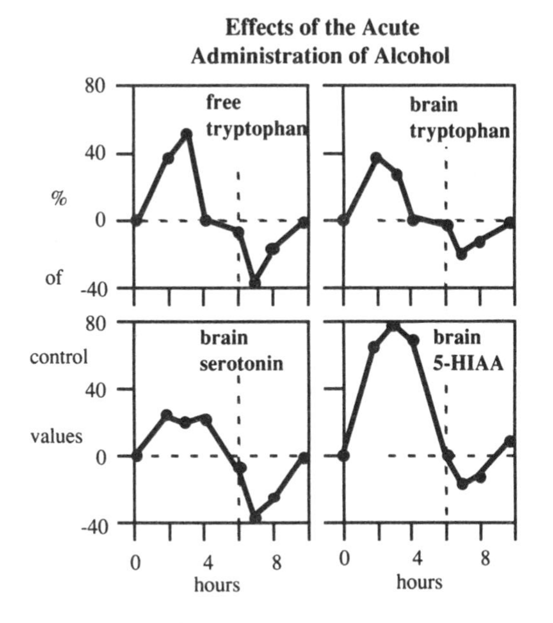

This shows that from two to four hours after drinking alcohol there is an increase in free tryptophan in the blood, and in brain tryptophan, serotonin and 5-HIAA. By six to eight hours there is a compensatory drop in all these values and by ten hours things are back to normal levels. Thus, after a dose of alcohol brain serotonin initially increases, then drops, then returns to normal. The biphasic response is similar to the effect of cocaine on brain dopamine levels[p527].

Drug addiction The effect of chronic administration of morphine on tryptophan metabolism is very similar to that of alcohol[85] (see **figure at the right->**).

When the morphine is stopped, the level of tryptophan oxygenase shoots up, tryptophan levels in the serum and brain drop, and serotonin in the brain also drops[85]. This is similar to that seen with chronic administration and then withdrawal of alcohol, nicotine, and barbiturates[77,78,80,81,84-86].

Implications for addiction These observations are consistent with the serotonin vicious cycle theory of addiction presented previously[p436]. The administration of a wide group of addictive agents, alcohol, morphine, nicotine, and even food, results in the elevation of brain serotonin. Since the sudden cessation of these agents results in an rebound drop in brain serotonin with uncomfortable symptoms (withdrawal) there is a need to return to taking them (addiction). Individuals with various impulse disorders, who already have a low brain serotonin to begin with, are genetically more susceptible to this vicious cycle and such individuals are more likely to seek out these addictive agents in the first place.

Tryptophan Oxygenase Is Also a Brain Enzyme

Tryptophan oxygenase was considered to be primarily a liver enzyme. However, it, or an enzyme very similar to it (indoleamine 2,3-dixoygenase), is also present in the brain[662-665,865,1965]. This enzyme is very important for the following reason. As discussed previously, in order for tryptophan to enter the brain it must be actively carried across the blood-brain barrier. Many compounds compete with this

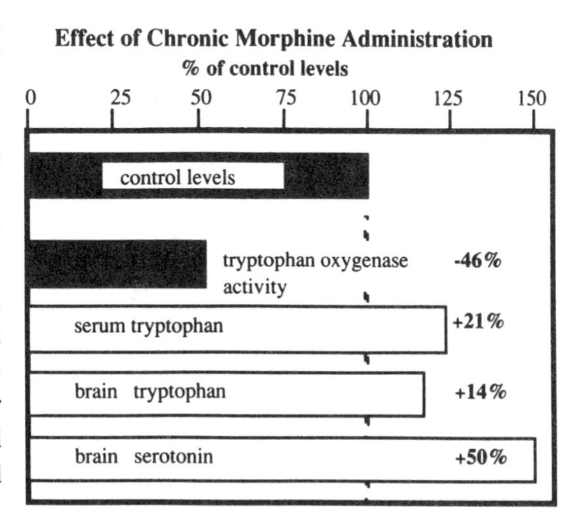

transport, including kynurenine and other amino acids. If tryptophan oxygenase was only present in the liver, then the level of tryptophan in the brain would be almost completely dependent upon the blood level of tryptophan and other compounds that compete for transport into the brain. However, if tryptophan oxygenase activity is also present in the brain, then some of the tryptophan can be siphoned off, after it enters the brain and before it is converted to serotonin. This is illustrated as follows:

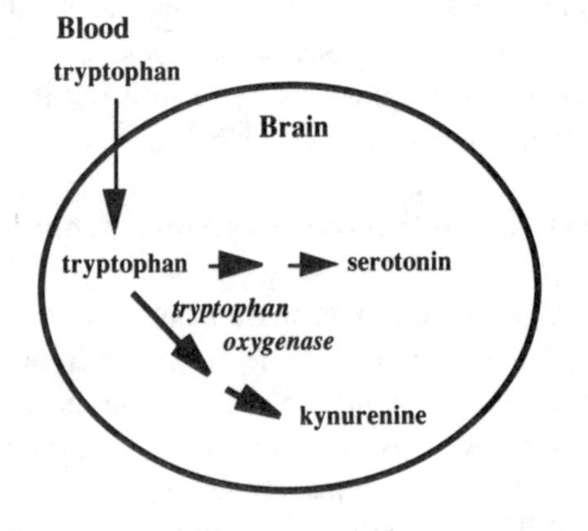

Under normal conditions 70 percent of the brain tryptophan is converted to serotonin and 30 percent to brain kynurenine[662]. *A significant increase in brain tryptophan oxygenase could markedly change this ratio and lower the level of brain serotonin.* Significant changes in the breakdown of tryptophan in the brain could occur with only moderate changes in the blood tryptophan and serotonin.

Differences between brain and liver tryptophan oxygenase There are several similarities and differences between the function of the tryptophan oxygenase in liver and in the brain (and intestine)[662,663,863,894,2121]. These are shown in the following table:

	Tryptophan 2,3-dioxygenase	**Indoleamine 2,3-dioxygenase**
Similarities		
Substrate:	L-tryptophan	L-tryptophan
Product :	N-formyl kynurenine	N-formyl kynurenine
Type of protein:	heme	heme
Inducible by tryptophan:	yes	yes
Differences		
Location:	**liver**	**intestine, brain** and other organs
Type of oxygen:	O_2	$O\text{-}O^-$
Other substrates:	none	D-tryptophan 5-hydroxy-tryptophan serotonin melatonin
Inducible by cortisol:	yes	no

Whether these differences are due to being in different tissues or being produced by different genes is not known[894]. If they are due to different genes, TS would most likely involve the intestine-brain gene since it more easily explains the striking effects of the mutation on behavior.

Distribution of intestine-brain tryptophan oxygenase Although I have termed it the intestine-brain form of tryptophan oxygenase, the diagram on the **next page** shows it is really the intestine form.

The high level of this enzyme in the intestine is due to the fact that enterochromaffin cells, the primary source of serotonin synthesis, are located in the small intestine. While its presence in the brain is of great interest to the genetics of behavior, its concentration in the brain is minute compared to the small intestine[894].

Would Tryptophan Cure TS?

If breaking down tryptophan too rapidly was the basic defect in TS, it would seem that giving tryptophan by mouth would "cure" TS. However, there are several reasons why it is

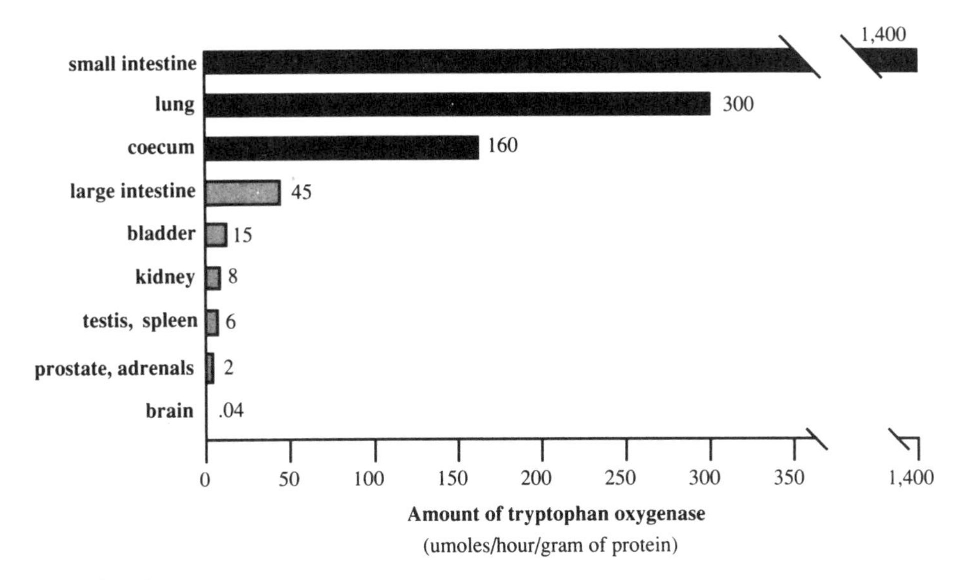

not this simple.

1. Tryptophan induces tryptophan oxygenase. Tryptophan itself can increase the activity of tryptophan oxygenase[662], which increases the breakdown of tryptophan. This can limit the effectiveness of oral tryptophan.

2. Tryptophan oxygenase breaks down serotonin. In addition to breaking down tryptophan, tryptophan oxygenase can also break down serotonin (to 5-hydroxykynurenine)[662,1962], further depleting brain serotonin.

3. Tryptophan inhibits tryptophan hydroxylase. Tryptophan hydroxylase, the enzyme needed for the conversion of tryptophan to serotonin[P417], is inhibited by high levels of tryptophan[639].

4. Tryptophan causes an increased loss of serotonin from the brain. When brain tryptophan increases there is an increased exit of serotonin (and 5-HIAA) from the brain[662].

5. Kynurenine may compete with tryptophan for transport into the brain[662,665,792,1045]. If the basic defect in TS is an overactive tryptophan oxygenase, then added tryptophan will increase the level of kynurenine in the blood and this might inhibit the uptake of tryptophan into the brain. Others suggest this does not occur[665].

The effect of all these factors was shown in a study of rats[665]. A small tryptophan load resulted in a modest increase of brain serotonin while larger amounts actually decreased brain serotonin. This is illustrated in the following [662,665]:

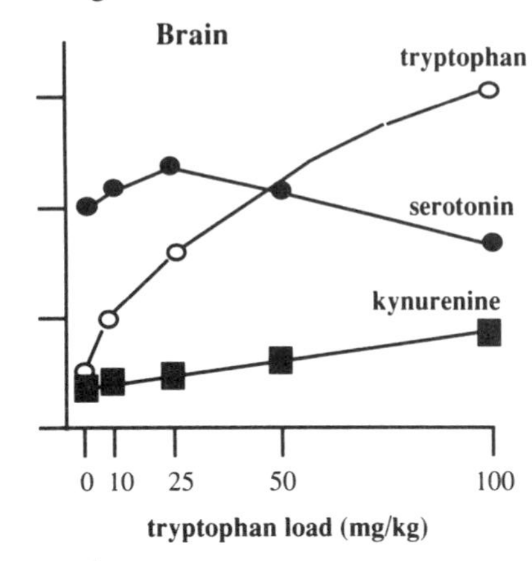

When the tryptophan load exceeded 25 mg per kilogram of body weight, the brain serotonin actually dropped, as the brain tryptophan continued to increase. Others have not seen this leveling off of brain serotonin with increasing doses of L-tryptophan[446].

The basic conclusion from these complex factors is that giving tryptophan by mouth is not the answer to "curing" TS. This is verified by experience with patients who usually report little effect of taking up to 2 grams a day.

Kynurenine as an Anti-serotonin Agent

The above evidence showed how an increase in tryptophan oxygenase could lower the brain serotonin level. A second consideration is the effect of kynurenine per se on behavior. Lapin[1121,1122] found that the kynurenines tended to have an anti-serotonin action, and also enhanced the aggressiveness and stereotyped behaviors produced by amphetamines. Thus, an increase in brain kynurenine could play a role in disinhibiting (over stimulating) the limbic system independent of the serotonin level.

Porphyria, Brain Serotonin and Psychosis

Porphyria is a rare disorder in which there is a defect in the synthesis of heme, the cofactor in tryptophan oxygenase. This results in a decrease in the activity of tryptophan oxygenase, increased blood and brain tryptophan and increased brain serotonin[1195]. This is probably why some individuals with acute porphyria often have schizophrenic behaviors including acute psychotic behavior, delusions of grandeur, hallucinations, distorted senses and paranoia.

In an animal model of porphyria, where tryptophan oxygenase levels were less than 25 percent of normal, the administration of tryptophan resulted in significant increases in brain tryptophan, serotonin and 5-HIAA compared to normal controls[1195]. This further illustrates that if tryptophan oxygenase is defective, it cannot be induced by tryptophan, and tryptophan by mouth can increase the level of brain serotonin. This contrasts to the normal situation where taking tryptophan results in increased (induced) levels of tryptophan oxygenase, and because of this there is much less of an effect on brain serotonin. This indicates that tryptophan oxygenase is ideally suited to protecting the brain from the effects of too much dietary tryptophan.

The Tryptophan Load Test

To examine the hypothesis that increased activity of tryptophan oxgenase is responsible for any disorder it is necessary to examine the activity level of this enzyme in living patients. Since tryptophan oxygenase is not present in the blood its activity has to be determined indirectly. This is done by a tryptophan load test[26,665,994,1545], somewhat similar to a glucose tolerance test. While it can be done by drawing many blood samples during the six hours after giving tryptophan by mouth, it is much easier to simply collect the urine for 12 hours and test it for the excretion of tryptophan breakdown products, especially kynurenine and 3-hydroxykynurenine[23,1545]. The most reliable tryptophan dose is 2 gm for adults[26,418,1545] or 30 to 50 mg per kilogram for children. The production of kynurenine is decreased if subjects are taking the B vitamin nicotinamide[1348].

While the level of excretion of kynurenine after a tryptophan load appears to provide an accurate estimation of the activity of tryptophan oxygenase[25,795], this was not found to be the case in one study[660]. Thus, while the precision of this test as means of determining tryptophan oxygenase activity is not certain, the results in patients with a number of different psychiatric and neurological disorders have been intriguing.

Schizophrenia

The proposal that serotonin metabolism is abnormal in schizophrenia has been a popular one. This theory started because of the hallucinogenic properties of LSD and the knowledge that LSD binds to serotonin receptors[p449,659,2104]. Kynurenine and 3-hydroxykynurenine excretion after a tryptophan load is abnormal in some patients with acute and chronic schizophrenia[1554]. In one study 11 chronic schizophrenics studied showed a marked increase in kynurenine (2.4 times the controls) and in 3-hydroxykynurenine (9 times the controls)[147] while another study was negative[578]. Since different people excrete kynurenine in the urine differently, a tryptophan load followed by testing the blood may be the most accurate test[1348].

If tryptophan oxygenase is overactive in schizophrenia then serum tryptophan should be low. This was found in many studies[496,497,710,1225,2128,1485] but not all[447]. In the negative study there were no differences in total or free plasma tryptophan between controls and schizophrenics, but the variation between tests 1-2 weeks apart was much greater for schizophrenics than controls. This suggests that plasma tryptophan levels may be poorly controlled in schizophrenia.

It would seem that the most direct way to test the proposal that there are serotonin abnormalities in schizophrenia would be to analyze the brains of individuals dying with this disorder. In the two reports where this was done, there was no clear evidence for abnormalities in the level or tryptophan, serotonin or kynurenine in the brain itself[993]. However, the potential problem with these studies is that changes can take place between the time a person dies and when the brain tissue is frozen. These studies do suggest, however, that if there is an abnormality of tryptophan and serotonin in schizophrenia it may be very subtle. As discussed above, eating glucose decreases the level of free tryptophan. The following study of the effect of eating glucose on free tryptophan in acute and sub-acute schizophrenics indicates how subtle the changes can be[2128].

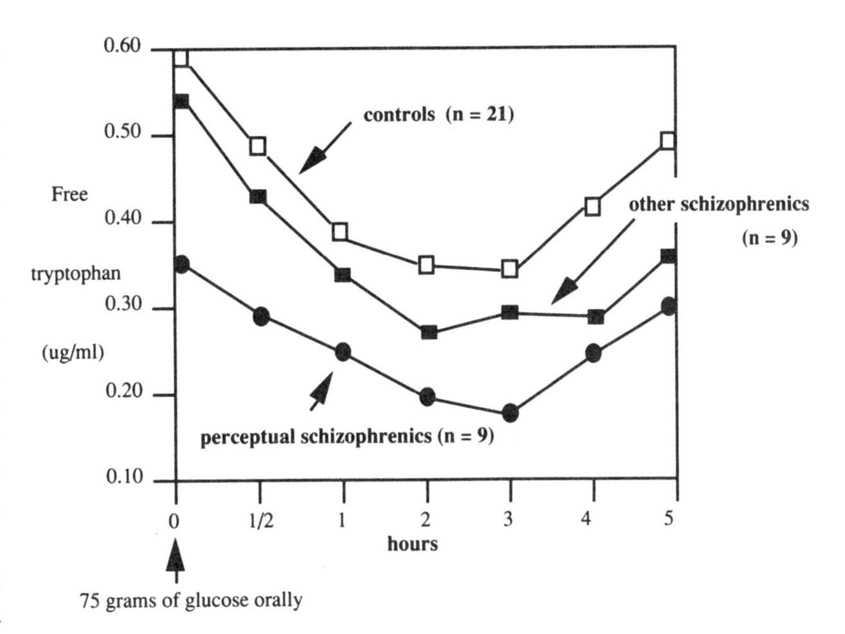

The perceptual schizophrenics with hallucinations had a much lower fasting free tryptophan level, and lower levels after a glucose load, than the controls and other schizophrenics.

Studies of the total plasma tryptophan have shown significant decreases in tryptophan in schizophrenics one week after admission that later returned toward normal but remained significantly lower in females[1255]. Both schizophrenic and control females had lower levels than males. In the females, the increase in plasma tryptophan was associated with an improvement in symptoms[722].

The most viable conclusion from these studies is that while tryptophan oxygenase may be involved in some forms of schizophrenia, examining the gene by genetic techniques may be the only way to prove it.

Depression

As discussed previously, abnormalities in serotonin have also been implicated in depression. The results of the tryptophan load test on women with endogenous depression compared to nine controls are shown below[442a].

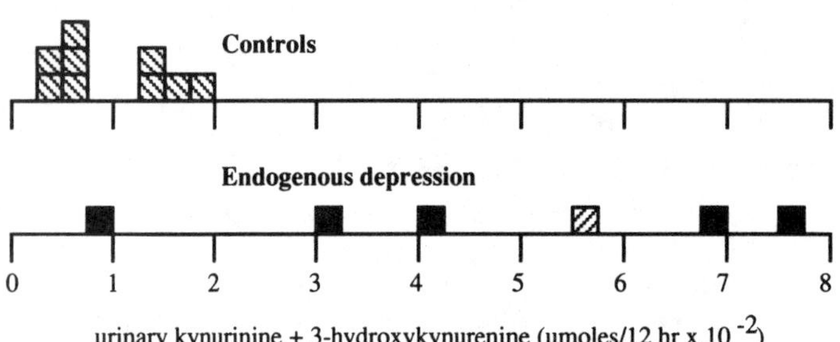

The increase in kynurenine and 3-hydroxykynurenine is striking[442a]. The kynurenine excretion after a tryptophan load remained elevated after recovery in most patients. These observations, although scanty, are consistent with an elevated tryptophan oxygenase as a marker for some types of depression. The increased kynurenine excretion in the patient with hysteria (shaded box) was of interest since that can be an adult symptom of ADHD[p84].

Other studies have not been as striking as this. One examined nine depressed males, five unipolar, four bipolar, and six male controls[630]. All showed normal kynurenine excretion except for one manic-depressive patient. It was concluded that there was no evidence for a role of blood tryptophan in the cause of depression, but "there may be altered metabolism of tryptophan in the brain"[630].

Tryptophan in depression In a study of tryptophan in the spinal fluid, the average level in ten unmedicated depressed patients was 260 ng/ml compared to 488 ng/ml in 14 controls[394]. Depressed patients also have significantly lower levels of free tryptophan in the blood, while the levels of total tryptophan tend to be low or normal[259,392,395,473,1408]. Others have not found a decrease in free tryptophan in depression[259].

The pedigree below illustrates how potentially useful determination of blood serotonin and tryptophan can be in studying depression.

The propositus developed severe depression requiring hospitalization after a divorce at age 37. Treatment with monamine oxidase inhibitors was helpful and she continued this for two years, then stopped it in preparation for surgery. In the following year, the depression returned despite intensive psychotherapy. She came to the clinic to see if this was an endogenous depression.

The family history was replete with others suffering from depression, including both of her parents, and her son, brother and nephew.

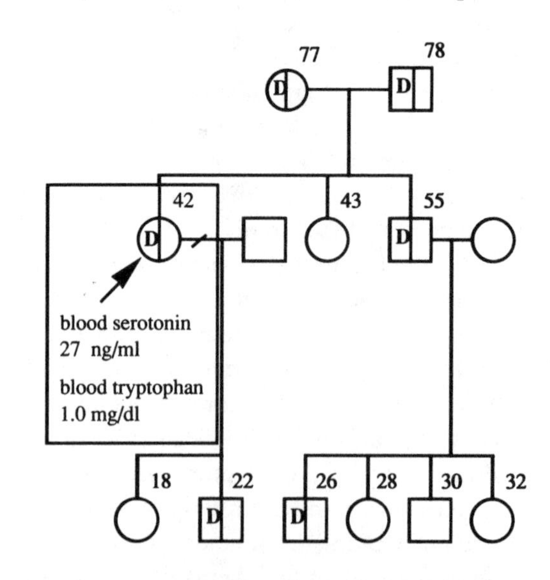

Her blood serotonin and especially tryptophan were very low. Since both her parents had problems with depression she might be a homozygote for a depression gene, possibly for increased levels of tryptophan oxygenase.

(After the fifth visit she related that her son had chronic facial tics, suggesting that a *Gts* gene, expressing mostly as depression, was present in this family.)

Free tryptophan in post-partum depression Depression after childbirth is not uncommon and is called post-partum depression. Eighteen women who had just given birth were asked to fill out a daily depression rating score and blood was tested for free plasma tryptophan on day six. The results were as follows[1976]:

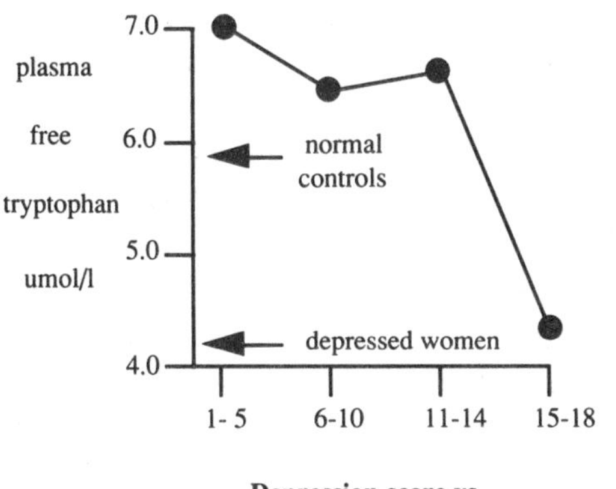

Depression score vs. plasma tryptophan in post-partum women

It was clear that the most severely depressed women had the lowest levels of plasma free tryptophan. The arrows show the average free tryptophan levels for non-pregnant women who were not depressed (normal controls) or depressed (depressed women)[1860]. Those women carrying a gene causing low tryptophan and serotonin levels may be the ones predisposed to post-partum depression.

Antidepressants inhibit tryptophan oxygenase An additional observation linking a tryptophan oxygenase to depression is the finding that virtually all of the commonly used antidepressants (tricyclics[p414]) cause a significant inhibition of tryptophan oxygenase[82,1250]. This is shown as follows:

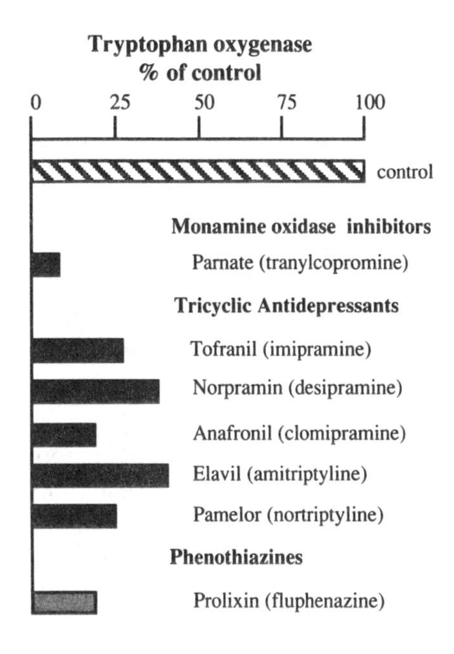

This occurs by inhibiting the enzyme from binding to the cofactor, heme, that is necessary for the enzyme to work. This in turn results in an increase in brain tryptophan and serotonin[82].

There are a number of interesting implications of these observations:

1. Part of the effect of the antidepressants may be through an indirect effect on brain serotonin due to an inhibition of tryptophan oxygenase, rather than the usual assumed mechanism of inhibiting serotonin and norepinephrine uptake at the synapse[p414].

2. These effects occurred only two hours after the drugs were given. While the antidepressant effect of tricyclic medications usually takes three weeks, their effect in calming hyperactive children, increasing attention span, improving behavior and controlling bedwetting often occurs as soon as the drug is adsorbed. This may be the mechanism by which this immediate effect occurs.

3. Prolixin, which is a Haldol-like medication, also inhibited tryptophan oxygenase. Thus, part of the effect of Haldol in TS may

be by an indirect effect on serotonin through its effect on tryptophan oxygenase. Other treatments of depression such as electroconvulsive therapy and lithium may also operate through an effect on tryptophan and serotonin[2138].

Depression—serotonin—cortisol cyle One of the most reproducible chemical abnormalities in depression is the presence of elevated levels of the steroid hormone *cortisol* in the blood stream. These high levels are not suppressed by giving another steroid called dexamethasone. This forms the basis of the often used *dexamethasone suppression test*[298,299]. There is a peptide in the hypothalamus called CRF which stands for corticotropin-releasing factor. It gets its name from the fact that it releases ACTH or **A**dreno-**C**ortico-**T**rophic **H**ormone, or corticotropin, from the pituitary. ACTH passes into the blood stream to the adrenal glands where it increases the production of cortisol. When the blood level of cortisol increases it results in a feedback inhibition of the production of ACTH, thus keeping the cortisol from going too high. These relationships are shown as follows:

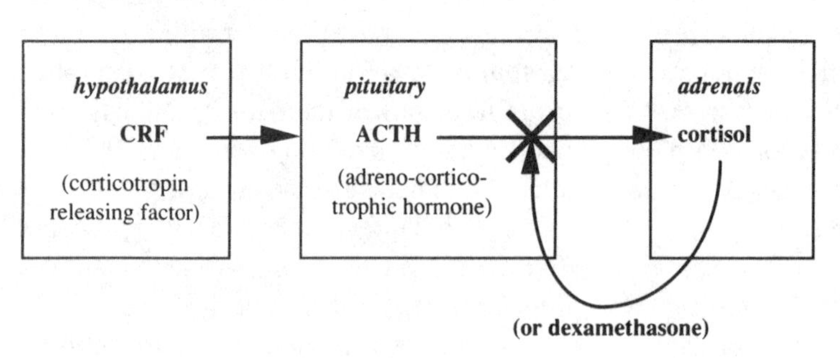

The synthetic steroid *dexamethasone* is like cortisol but stronger. Normally it also turns off ACTH and results in a decrease in cortisol in the blood. However, in depression, and other disorders, it is not able to suppress the cortisol levels. This is because the CRF levels are high in depression[104,1401,1402].

CRF levels are in part regulated by serotonin nerves in the hypothalamus. When serotonin is low the production of CRF is increased.

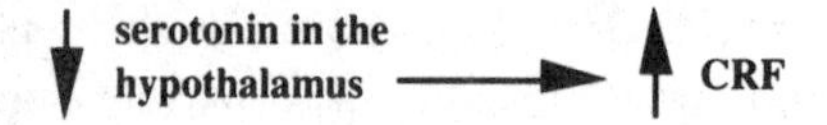

It is been frequently shown that a single administration of steroids (cortisol) produces an increase in tryptophan oxygenase, a decrease in blood tryptophan, and a decrease in brain serotonin in animals[443-445,2137]. Blood tryptophan levels have been noted to be lowest in patients with the most abnormal dexamethasone suppression tests[1243].

Serotonin directly controls cortisol levels in several ways. As noted above it regulates the release of CRF[1111,1304,1305,1415]. However, it also blocks the ability of high levels of blood cortisol to turn off the production of ACTH[1415]. Finally, stimulation of the limbic system, especially the amygdala, can cause marked increases in blood cortisol levels[1661] and decreased brain serotonin can disinhibit or stimulate the amygdala.

These findings raise a chicken or egg type question of what comes first? Does depression cause so much stress that blood steroids are increased, causing increased levels of tryptophan oxygenase, causing low tryptophan[1775] and low brain serotonin? Or, does a genetic increase in tryptophan oxygenase come first, causing cause low blood and brain tryptophan, low blood and brain serotonin, disinhibition of the limbic system with depression, and interference in hormonal control resulting in elevated blood cortisol? These questions can be diagramed as follows[393,1120]:

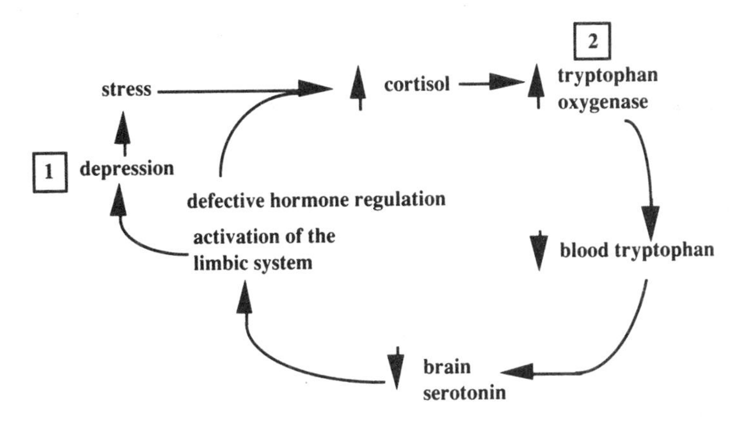

Starting at #1 above represents the depression-first hypothesis, while starting at #2 represents the defective tryptophan oxygenase hypothesis.

In the animal studies, the induction of tryptophan oxygenase was *not* noted after repeated daily administration of steroids[443]. In man, continued high levels of blood steroids occurs in a disorder called Cushing's disease. Depression may be present in this disorder[393], but it is mostly seen in individuals who have problems with depression before they had Cushing's disease[651]. One piece of evidence that #2 is the correct route is the observation that in humans there is a high degree of correlation between the level of serum tryptophan and serum kynurenine (+.81) but no correlation between the level of serum cortisol and kynurenine[75]. These observations suggest that cortisol is not the driving force in causing low blood tryptophan levels in depression, as route #1 would imply. Genetic studies of the tryptophan oxygenase gene will be required to determine with certainty whether #1 or #2 is correct.

Treatment of depression with 5-HTP Given all the above considerations, 5-HTP, the precursor of serotonin, should be effective in the treatment of any behavioral disorder due to a low brain serotonin level, whether secondary to a high tryptophan oxygenase or any other mechanism. There have been numerous reports of the usefulness of 5-HTP in the treatment of depression, either alone[1502] or in combination with antidepressants, especially those affecting serotonin[1057].

Kynurenine in depression On the basis of studies of the effect of kynurenines on rats, Lapin[1122] thought that kynurenines may have *anti-serotonin* properties. His studies suggested that:

1. Kynurenines increased the sensitivity of dopamine receptors in the neostriatum to dopamine released by amphetamine,

2. Kynurenies had effects that were opposite to those of the antidepressant imipramine. In this regard he observed three patients who were resistant to antidepressant treatment and excreted three times more kynurenine than controls.

3. The injection of kynurenine into the brain ventricles caused *motor hyperactivity.*

Because of the effect of stress on steroid production, and the effect of increased steroids causing increased tryptophan oxygenase activity, and thus increased production of kynurenines, Lapin[1122] felt that kynurenines were one of the major bridges between emotional stress and depression.

These studies suggest that an overactive tryptophan oxygenase could cause some of the spectrum of Tourette syndrome symptoms both through an effect on decreasing central serotonin and on increasing the level of blood and brain kynurenines.

Tryptophan, Aggression and Motor Activity

When monkeys were fed a diet low in tryptophan, there was an increase in motor activity in the males. This is shown as follows[308a]:

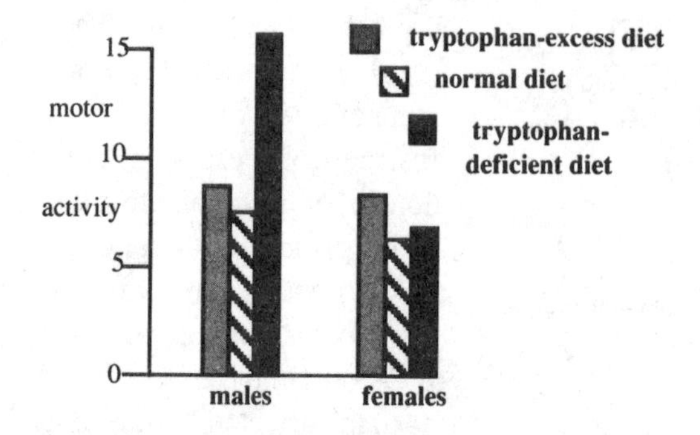

Even more striking results were obtained when the monkeys were scored for spontaneous episodes of aggressive behavior.

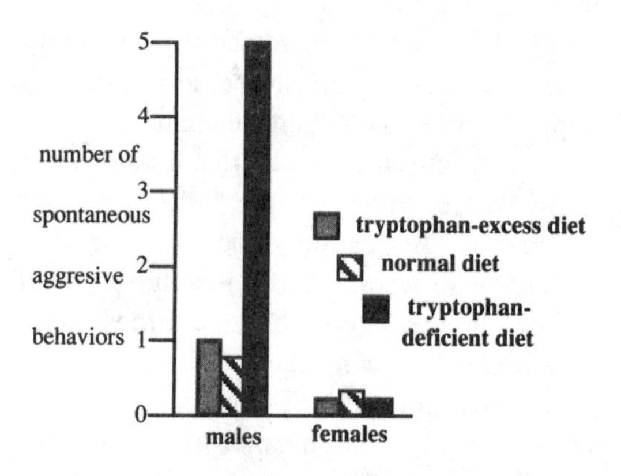

Again the effect was present only in males. Note that a high tryptophan diet did not inhibit the normal levels of aggression. Since this effect tends to mirror the greater expression of TS and ADHD in males it would be of interest to understand the sex difference. The most likely candidate is the added effect of testosterone on aggression[P451]. Another possibility comes from a study of depression in humans where the ratio of tryptophan to other neutral amino acids was determined. The level of leucine was found to be significantly higher in males than females[1243]. Since this amino acid competes with tryptophan for entry to the brain, these higher levels in males would make a low blood tryptophan all the more effective in lowering brain serotonin, resulting in more severe symptoms in males.

The Role of Quinolinic Acid

Quinolinic acid is formed by the further breakdown of kynurenine as follows[147a]:

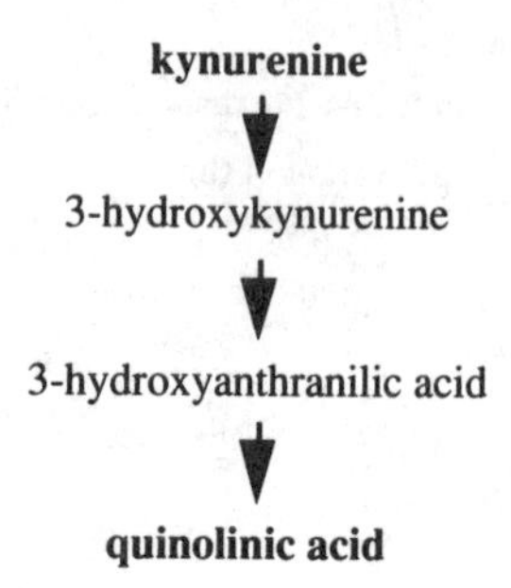

Quinolinic acid may contribute to causing some of the symptoms of TS since it has properties of a neuroexcitant and neurotoxin, that is, it can both stimulate and kill nerve cells[386a,1736,1893]. The above pathway for the formation of quinolinic acid is active in the brain. Quinolinic acid, kynurenine and kynurenic acid can cause convulsions and myoclonic jerking when exposed to nerve cells[1123-1126,1496]. Because of its ability to kill cells, quinolinic acid has been implicated in the death of cells in Huntington's disease[1284]. Kynurenic acid can inhibit the stimulation of cells by NMDA, which plays an important role in memory[P339,1893].

These observations raise the possibility that if a genetic increase in tryptophan oxygenase is involved in TS, it may produce its effects both through disinhibition due to too little serotonin and by direct stimulation of some nerves by kynurenine and quinolinic acid. In this regard, I have found that baclofen occasionally reduces tics and this compound inhibits the action of both kynurenine and quinolinic acid[1126].

The potential of some cell death as a result of increased brain quinolinic acid raises the possibility that some parts of the TS spectrum might not go away even if the basic biochemical defect was corrected.

Regulation of the TO Gene

The rat TO gene is richly endowed with many regions of DNA that control its acitivty. These are summarized as follows[1601]:

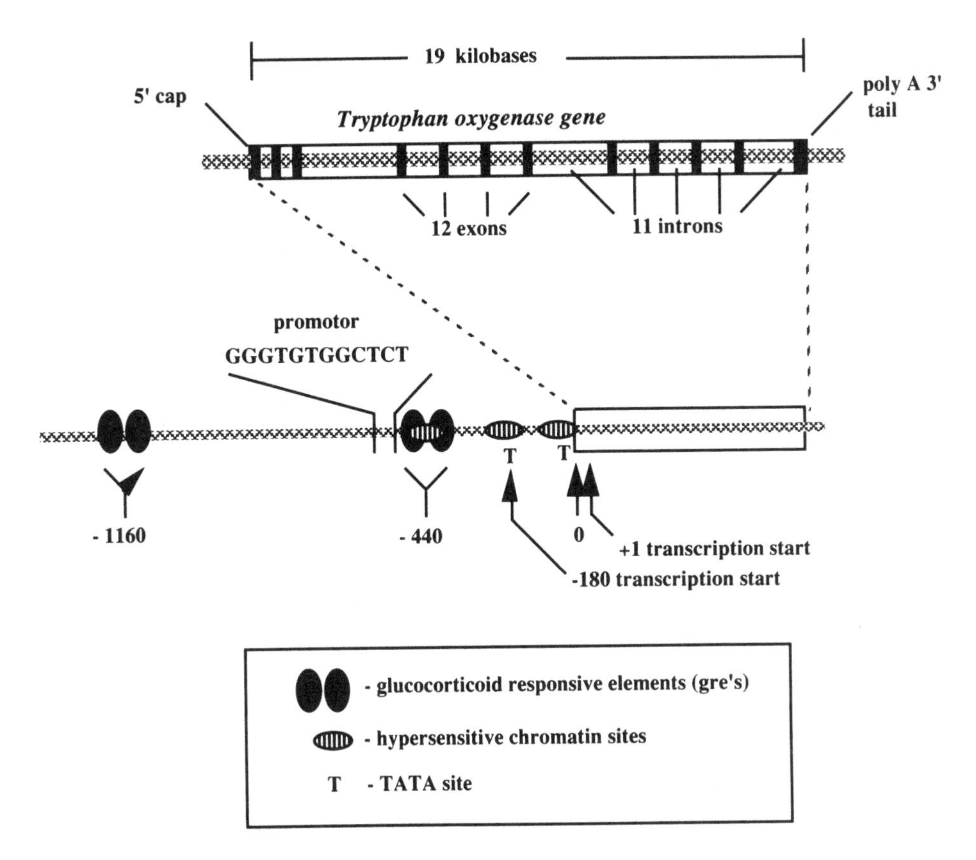

A mutation at any one of these regulatory sites could result in changes in the activity (increase or decrease) of the TO gene.

[The official abbreviation for the human tryptophan dioxygenase gene is TDO2[379a]].

Summary: In the previous chapter, we saw how a genetic defect causing a low brain serotonin could explain all the clinical features of Tourette syndrome. Low levels of tryptophan in Tourette syndrome patients suggest that the basic defect could be due to an overactive or hyper-inducible liver and/or intestine-brain form of the enzyme tryptophan oxygenase. This concept is illustrated as follows:

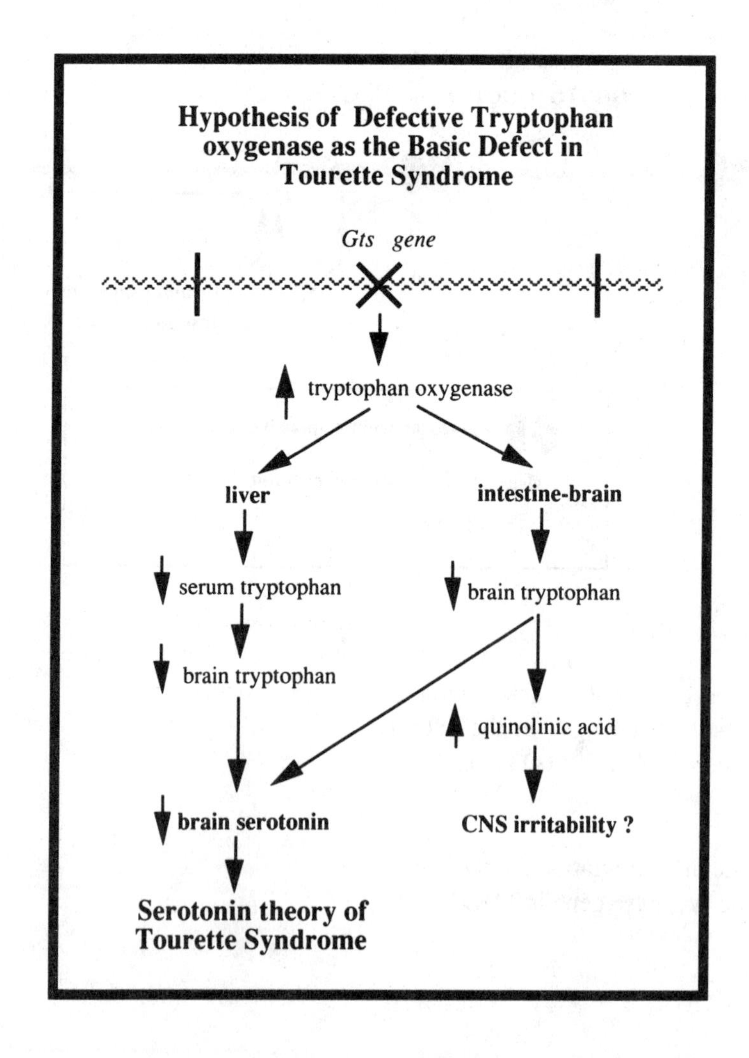

Chapter 65

GABA — Another Great Inhibitor

In 1950, Eugene Roberts was doing cancer research at the City of Hope Medical Center. He was examining the amino acids in different organs using a new technique called paper chromatography. Extracts of tissue were dotted on large sheets of paper and separated by electrical current and different solutions. In the process of examining a brain tumor, normal brain tissue was used as a control. The brain showed a prominent spot that was not present in other tissues. This spot was eventually identified as γ-aminobutyric acid (GABA). Its structure is as follows:

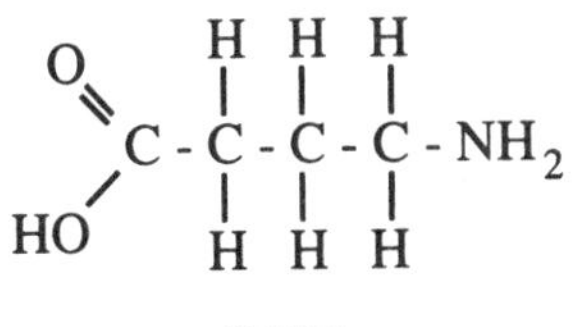

```
O     H   H   H
 \\   |   |   |
   C - C - C - C - NH2
 /    |   |   |
HO    H   H   H
```

GABA

It derives its name from the fact that the amino group, —NH_2, is on the third or γ-carbon of butyric acid, thus γ-aminobutyric acid.

Roberts attended a biochemistry meeting in Atlantic City to present his findings. Roommate assignments were by chance and his was another young scientist, Jorge Awapara from Houston. Ironically, Awapara had also been working on cancer research, had also been using paper chromatography to examine tumors, and had also found the same prominent spot in the brain. More serendipity. Awapara had not yet identified the spot but both had found another amino acid in the brain, taurine. The two independent studies were published in the same journal[70,1670]. The two scientists decided that Roberts would pursue GABA and Awapara would pursue taurine.

Three years later, Florey[614] discovered that extracts from mammalian brains inhibited the activity of certain nerves in the abdomen of the crayfish. This was called factor I, standing for inhibitor. This factor was subsequently found to be GABA[134]. GABA was identified as a neurotransmitter when it was found to duplicate the activity of a natural transmitter in the crayfish[1099]. These combined observations indicated that GABA was an inhibitory neurotransmitter.

The Chemistry of GABA

The precursor of GABA is glutamic acid which has an extra acid or carboxyl group (—COOH).

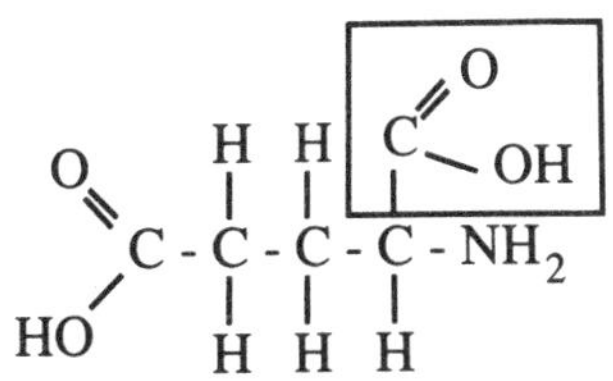

```
                      O
                     //
      H   H       C
O     |   |       |  \ OH
 \\   |   |       |
   C - C - C - C - NH2
 /    |   |   |
HO    H   H   H
```

glutamic acid

This is removed by the enzyme *glutamic acid decarboxylase* to form GABA. This enzyme requires vitamin B_6 to function properly. The importance of GABA as a inhibitor of the nervous system was dramatically illustrated when hundreds of children developed seizures after taking a baby food that was deficient in vitamin B_6, leading to a GABA deficiency. The definitive evidence that GABA was an important inhibitory neurotransmitter in mammalian brains came from the studies on specific inhibitory nerves, called Purkinje cells, in the cerebellum. These cells used GABA as their neurotransmitter[621,1425,1681]. To bring about this inhibition, the GABA-rich nerve endings or boutons from the Purkinje cells literally smother the nerve cells they inhibit.

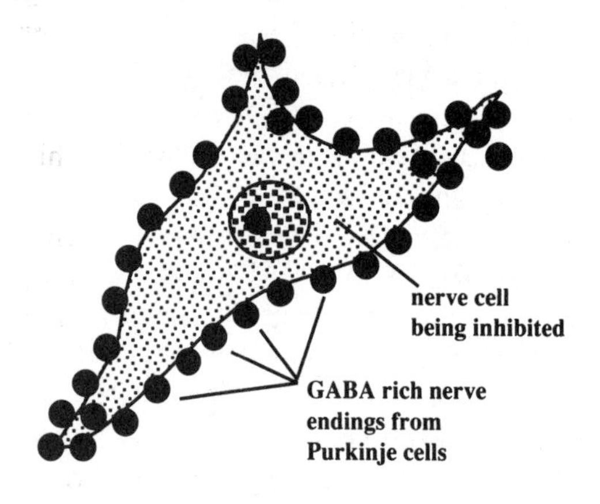

The importance of inhibition in the regulation of the nervous system is indicated by the fact that the brain contains 200 to 1,000 times more GABA than dopamine, norepinephirne or serotonin. Thirty percent of the cells in the brain use GABA compared to 1 percent using dopamine.

GABA Opens Chloride Channels

The clue to how GABA inhibits neurons was provided by the demonstration that it increases the flow of chloride ions (Cl^-) into the nerve[1096]. The receptor for GABA is tightly coupled with the chloride channels in the nerve membrane[p315,212,855].

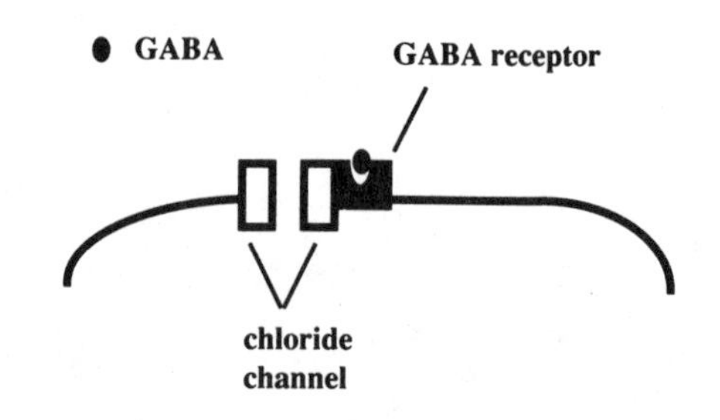

When GABA binds to the receptor, the chloride channels open. As the negatively charged chloride ions move into the nerve, the positively charged bicarbonate ions move out[1006,1940]. This is shown as follows:

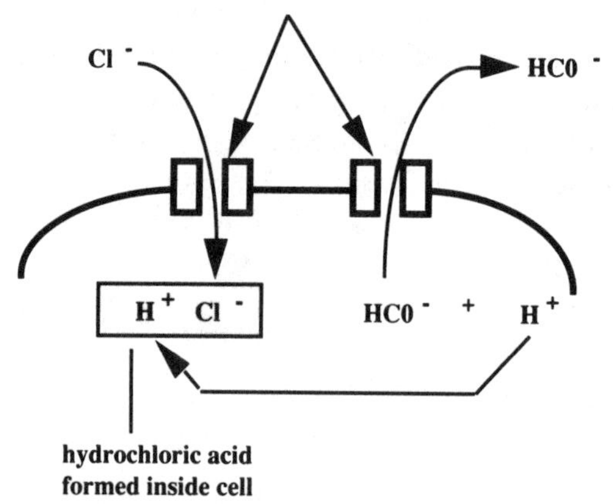

The hydrochloric acid (HCl) formed makes the inside of the cell more acid. The total effect is to make the cell much harder to excite.

GABA Inhibits Dopamine Neurons

I have devoted a chapter to GABA because it plays an important role in regulating dopamine neurons and thus might play a role in Tourette syndrome and related disorders. The ventral tegmental area, substantia nigra, and globus pallidus receive GABA neurons from the striatum and the latter two are particularly rich in GABA[406]. This is shown in the **figure on the next page**.

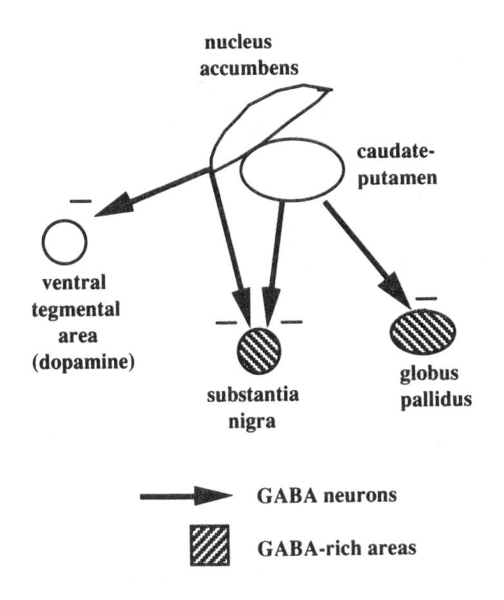

Since dopamine nerves go in the opposite direction, this forms a feedback loop for controlling dopamine activity[1537,2098,2134]. This is shown by the following example:

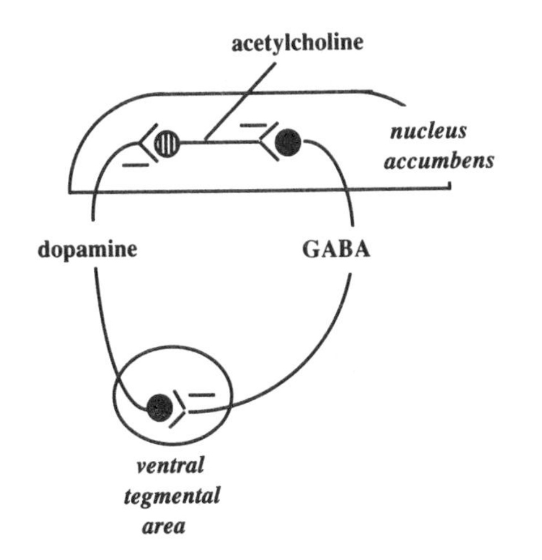

The quickest way to understand this loop is to realize that *two minuses make a plus*. Thus, if the dopamine nerve becomes overactive it inhibits the acetylcholine nerve, which decreases the inhibition of the GABA nerve, which increases the inhibition of the dopamine nerve and brings its activity back to normal. In the chapter on dopamine I often stated that dopamine stimulates nerves in the striatum. Much of this "stimulation" may actually take place by inhibiting the inhibiting acetylcholine nerves.

Evidence for the above inhibitor loop is provided by the fact that electrical stimulation of the nucleus accumbens inhibits the firing of dopamine neurons in the ventral tegmental area[2098] and this effect does not occur when the action of GABA is blocked. When GABA stimulating drugs are placed directly in the substantia nigra[2030] or ventral tegmental area[2098], dopamine nerves are inhibited. When GABA-blocking drugs are injected into these areas they produce a state of "intense arousal, staring, fear, withdrawal, . . .and sniffing. . ."[1874,1875].

Since it plays a major role in regulating the activity of mesolimbic, mesocortical and nigrostriatal dopamine nerves, drugs which alter the activity of GABA could play an important role in treating disorders in which dopamine is not regulated properly. Not surprisingly, there have been theories implicating GABA in schizophrenia.

Schizophrenia and GABA

Since dopaminergic hyperactivity appears to be present in schizophrenia, and since GABA appears to regulate dopamine activity, it is reasonable to propose that the problem in schizophrenia might be too little GABA[1622,1623]. If this were the case then drugs which stimulate GABA activity should improve schizophrenia. Unfortunately, it did not work out this way and in fact schizophrenics so treated often got worse. When the same drugs used in the above experiments were given by way of the blood stream, they often stimulated dopamine nerves[671]. When drugs are in the blood stream they interact with GABA receptors throughout the brain and action in one site may counteract actions in other sites. The lesson from this story is that the brain is a very complex organ and even when you think you have it all figured out, it fools you. Despite its initial promise, manipulation of GABA does not appear to be useful in treating schizophrenia[671].

Depression and GABA

Although less emphasized, a better case can be made for involvement of GABA in depression than in schizophrenia[1196]. Blood levels of GABA were low in six different studies, spinal fluid levels were low in four of six studies and brain levels of glutamic acid decarboxylase, the enzyme responsible for the synthesis of GABA, were low in one study[1196]. In the spinal fluid, GABA levels were lowest in the most severely depressed patients[691].

Drugs which enhance GABA often improve symptoms of both depression and mania, while drugs which inhibit the production of GABA, by inhibiting glutamic acid decarboxylase, cause depression[1196]. As shown below, a wide range of agents and treatments that effectively treat depression all result in a 120 to 187% increase in the binding of GABA to $GABA_B$ receptors.

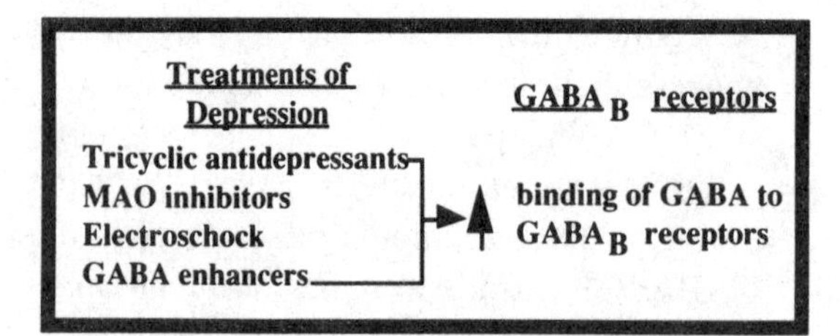

Recall from the chapter on norepinephrine that the only other change that also occurs with such a wide range of treatments is the decrease in β-adrenergic receptors[p414] and increase the effectiveness of serotonin.

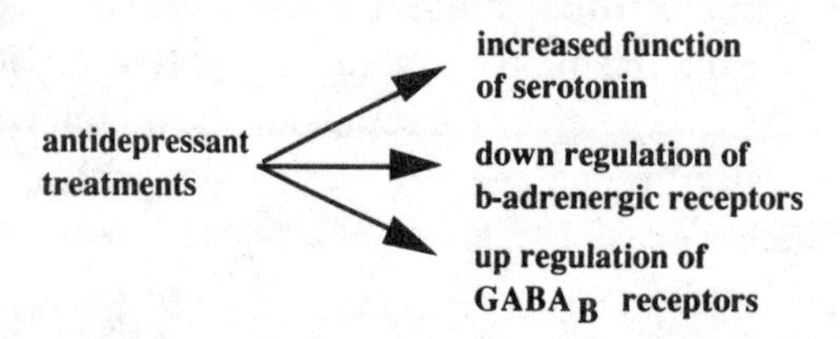

Anxiety, Valium and GABA

Anxiety rates with love and hate as one of the most universally felt emotions. When it warns of impending peril, brief periods of anxiety are helpful. When it is present in the absence of real danger, chronic anxiety can be debilitating. One of the first drugs found to alleviate anxiety without undue sedation was Librium. It belongs to a class of compounds called *benzodiazepines*, popularly called *tranquilizers*. A more potent compound, Valium or diazepam, has largely replaced Librium and is one of the most widely used drugs ever produced. Related compounds such as Dalmane (flurazepam) have sedating properties and are used as sleeping pills. In 1977 over 8,000 tons of benzodiazepines were consumed in the United States[1919].

In view of the massive use of benzodiazepines it is important to know how they work. Extensive studies have shown that there are specific benzodiazepine receptors in the brain. Specific receptors for the benzodiazepines were discovered in 1977[1345,1843]. They are attached to some of the GABA receptors [1919,1929].

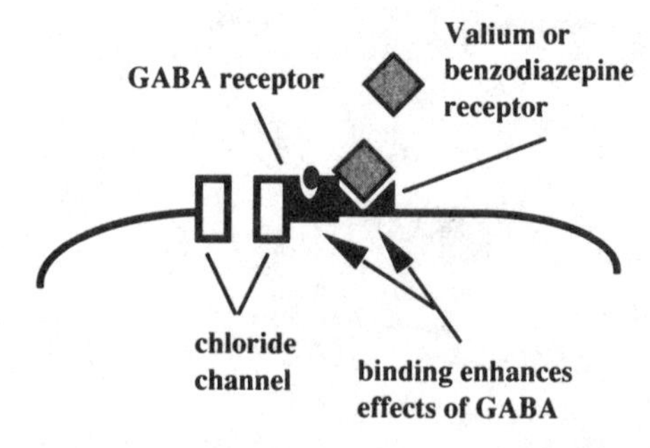

When valium or similar drugs bind to these receptors *the effectiveness of GABA is enhanced*[405,1919]. Thus the effects of benzodiazepines and GABA are similar. The resulting inhibition of nerve activity accounts for their antianxiety and sedative properties. They are also used to control seizures (anticonvulsants)

and as muscle relaxers[802]. One of the most effective anticonvulsants, sodium valproate (Depakane), enhances GABA by inhibiting the enzyme that breaks it down.

Alcohol and the Benzodiazepine Receptor

One of the few "drugs" that is used more than Valium is alcohol. The danger of mixing alcohol and Valium is well known and has contributed to many deaths, including that of Marilyn Monroe. This cooperative effect is due to the fact that alcohol also binds to benzodiazepine receptors and also enhances the effectiveness of GABA[1901].

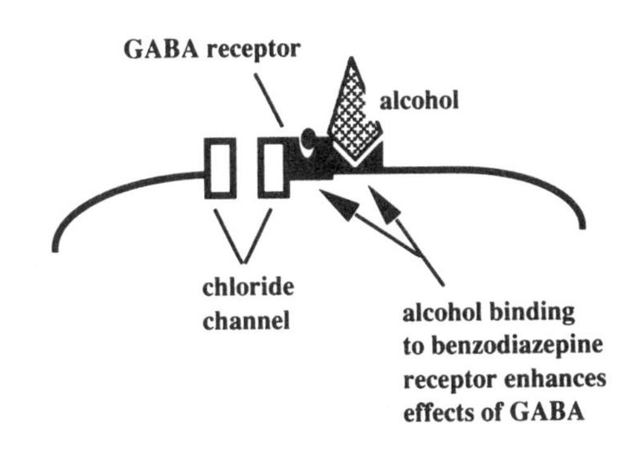

This interaction is so potent that a drug known as RO15-4513, which pushes alcohol off the receptor and inactivates it, is capable of rapidly eliminating the symptoms of drunkenness. This is the so-called "antidrunk pill," which, because of other sides effects, has been used only for research. The interaction of alcohol with the benzodiazepine receptors explains many of the antianxiety and sedative effects of alcohol. Barbiturates, like phenobarbital and Seconal, also interact with this receptor complex and enhance the effect of GABA[1436]. As with Valium, the combination of barbiturates and alcohol is also dangerous.

As with other systems, acute and chronic intake of alcohol produces opposite effects. Acute alcohol intake enhances the effect of GABA, while chronic alcohol intake suppresses GABA functions[1103]. Many of the symptoms of withdrawal from alcohol are due to the decreased functioning of GABA. This is why the benzodiazepines, which stimulate GABA, are effective in treating the symptoms of alcohol withdrawal[802].

The Advantages of Inhibition

We might logically ask, why does the brain put such a massive effort into inhibition? One only needs to have attended a drag race, or used a sling shot, or seen a catapult in action to understand. In a drag race the engine is all warmed, revved and ready to go, but inhibited by the clutch. At the green light, in a fraction of a second, the engine can be "disinhibited" by releasing the clutch. The car leaps ahead. Consider the difference if, instead, at the green light the driver had to climb into the car, start the engine, warm it up and drive off. The awake brain works the same way. It is constantly warmed up and idling, ready to leap into action when disinhibited.

The other aspect of disinhibition that we often loose sight of is the fact that when we take an action it is just as important to be able to stop the action as to initiate it. If we reach to pick up a baby we don't want to punch it in the face because we can't stop reaching. The cerebellum is responsible for this fine balance between action and inhibition of action, and it is little wonder that the inhibiting effects of GABA were first worked out in the cerebellum. The massiveness of the GABA system is again analogous to the race car — the bigger the engine, the bigger the disc brakes need to be.

GABA versus Serotonin

If GABA is such a wonderful inhibitor, and Tourette syndrome is a disinhibition disorder, why isn't a defect in GABA more logical as the cause of TS than a defect in serotonin? The answer lies in the universal presence of GABA. It is "the inhibitor for all seasons."

It inhibits motor nerves, sensory nerves, dopamine nerves, serotonin nerves, and on and on. Serotonin is the inhibitor for one season, the season of the limbic system and frontal lobes. As such, it is more narrowly focused for involvement in behavioral disorders.

Summary: GABA is the major inhibiting neurotransmitter of the brain. The receptors for GABA are intimately associated with the chloride channels. When GABA binds to its receptor, the chloride channels open, chloride enters the nerve, bicarbonate exits, and the nerve is inhibited. The receptor for Valium and related tranquilizers is also intimately associated with the GABA-chloride channel receptor. Valium works by enhancing the effectiveness of GABA. Alcohol works like Valium and binds to the same receptors. Barbiturates also interact with this receptor complex and also enhance the action of GABA. Thus, the three drugs which are most widely used by humans to control stress and anxiety—alcohol, tranquilizers, and barbiturates — all work by enhancing the ability of GABA to inhibit nerves.

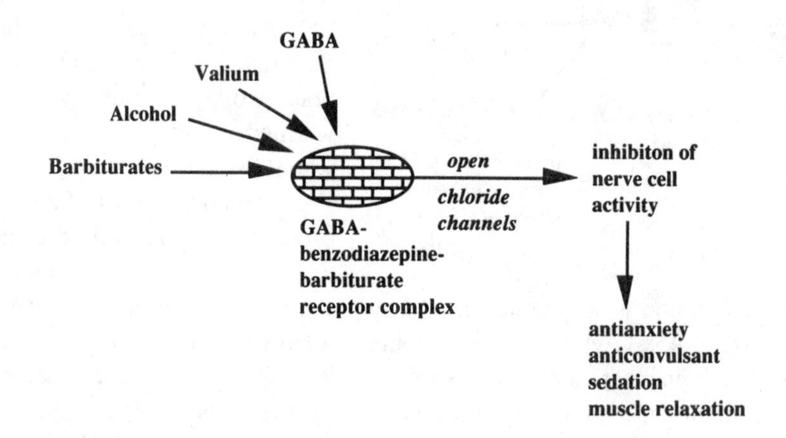

Chapter 66
Opium, Opiates, and Endorphins

For thousands of years opium has been recognized for its dual role of relieving pain and inducing euphoria. It is named after the Greek word *opion*, meaning poppy juice. In 1803 a German pharmacist prepared from poppy juice a compound he called morphine, after Morpheus, the Greek god of dreams[1819]. Because pure compounds are easier to use than crude plant extracts, morphine was rapidly incorporated into the tools of the physician to relieve pain. Use quickly led to abuse and the ease with which one could become addicted to morphine was soon recognized. In an attempt to develop drugs that would relieve pain but not be addicting, related compounds were developed. Two such substances were heroin and Demerol—neither of which solved the problem.

Opiate Receptors

As was the case with Valium in the last chapter, whenever drugs have highly specific effects on the brain it is most likely due to the fact they are binding to specific receptors. Anyone with diarrhea who has used paregoric, which is an alcoholic solution of opium, knows that it slows down the contraction of muscles in the intestine. This effect on the contracture of intestinal muscles is one of several techniques used to identify and study opiate receptors[1503-1506,1801,1931]. The opiate receptors in the central nervous system were found in the highest concentration in three general areas[898,907,923,924,1100,1505,1800]:

1. Along the pathway of *pain fibers* coming from the spinal cord to the brain stem and thalamus,
2. In the *striatum* and
3. In the *limbic system* — especially the amygdala.

The first location relates to the control of pain, and the second to regulation of movement. The latter location is especially relevant to the chemistry of emotions and accounts for the feelings of pleasure produced by these drugs. Opiates may produce euphoria by blocking emotional pain.

The same agonists and antagonists noted for other receptors also occur for the opiate receptors. The agonists, such as morphine, Demerol and heroin, relieve pain. The antagonists, such as naloxone (Narcan) prevent morphine from relieving pain by pushing it off the receptor and binding itself without activating the receptor. This is shown as follows:

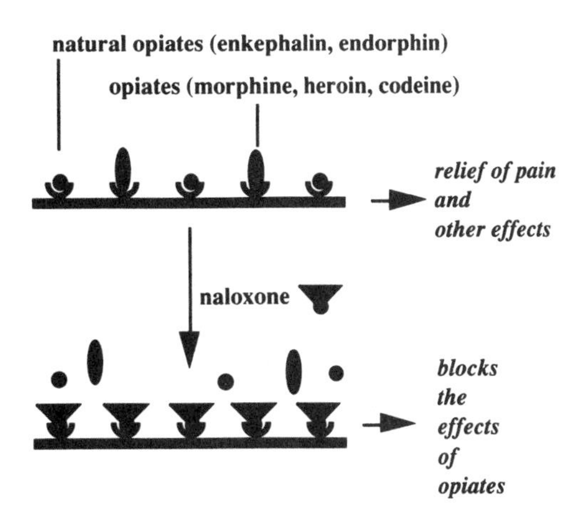

Such antagonists save lives by instantly reversing the effects of an overdose of morphine or other narcotics.

Receptors to opiates, with identical properties to those in mammals, have been found in the most primitive of vertebrates[1504]. Since these animals have been around for hundreds of millions of years, there must be some important natural compound in that brain that binds to the opiate receptors.

Enkephalins — The Brain's Own Morphine

Using the technique of studying the contraction of intestinal muscle, two researchers from Aberdeen, England, John Hughes and Hans Kosterlitz, found that extracts of brain tissue could cause the same contractions as opiates[938,930]. Like opiates, the effect was counteracted by naloxone. They isolated two short peptides, each containing five amino acids. These were called *enkephalins* from the Greek meaning "in the head." Since the sequence of the two was identical except at one position where one had methionine and the other had leucine, they were called *methionine-enkephalin* and *leucine-enkephalin*. The sequences are as follows:

<u>Methionine enkephalin:</u>
H-tyrosine-glycine-glycine-phenylalanine-**methionine**-OH

<u>Leucine enkephalin:</u>
H-tyrosine-glycine-glycine-phenylalanine-**leucine**-OH

Enkephalins are neurotransmitters It is possible to make antibodies to peptides, including enkephalins, and then use these antibodies to study the location in the brain of the enkephalins. These studies showed that enkephalins are localized to nerve endings in the same regions of the brain where the opiate receptors are found. This and other evidence indicated that enkephalins are neurotransmitters and their receptors are the opiate receptors. The enkephalins are present in high concentration in the limbic system — the amygdala and thalamus.

Enkephalins as presynaptic inhibitors In two previous chapters on serotonin and GABA, we saw that their major role is to inhibit nerve activity. Enkephalins also inhibit nerve activity but they do so by an unusual mechanism. They inhibit by stimulating the presynaptic membrane. This is somewhat analogous to having two comedians on stage with the first telling the best jokes, and the second complaining, you stole my best material. In the same fashion, an enkephalin synapse can "steal" the punch of the nerve it is attached to. Illustrated diagrammatically, when the following neuron is stimulated, it releases nine units of neurotransmitter:

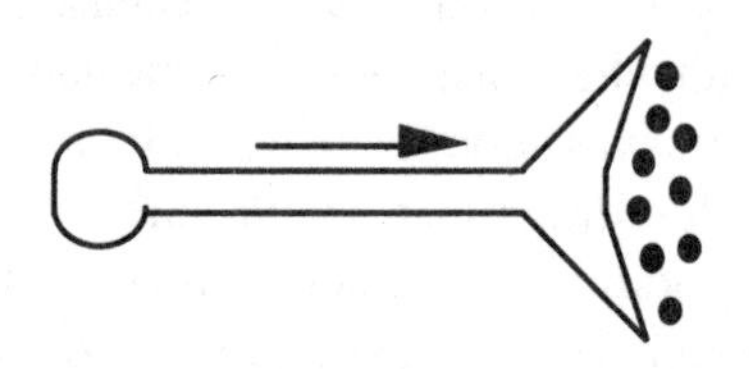

However, when the enkephalin nerve is activated it steals some of the charge on the membrane and the main neuron is inhibited[1819].

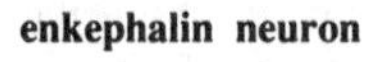

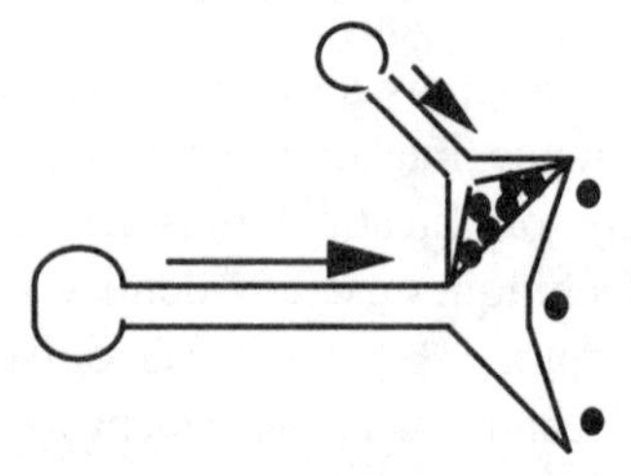

Tolerance, Dependence and Withdrawal

One of the characteristics of addiction is tolerance. This is a situation where a given dose of a drug no longer exerts the same effect and larger and larger doses must be taken. A practical example is the developing addict who initially gets high on a single "dime" of heroin, but with time needs many times that amount to get the same effect. The other components of addiction are dependence and withdrawal symptoms. Dependence exists if severe symptoms develop when the drug is suddenly withdrawn. In opiate withdrawal these symptoms include diarrhea, abdominal cramps, insomnia, irritability, dilated pupils, and gooseflesh. The latter symptom has given rise to the term "cold turkey." Solomon Snyder[1819] from Johns Hopkins proposed the following explanation of tolerance and withdrawal symptoms. During normal function only some of the receptors are occupied by enkephalin and the synapse chugs along at 50% of capacity.

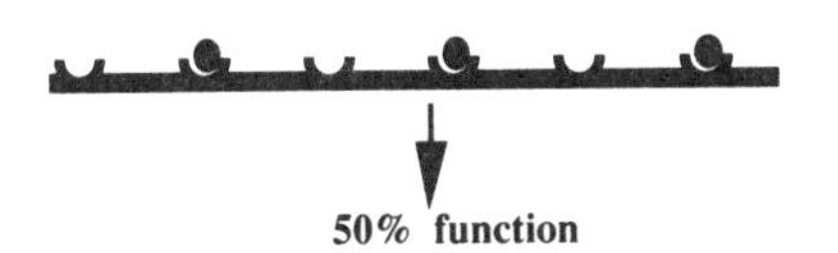

When morphine is added, the function of the system increases to 100% with resulting euphoria.

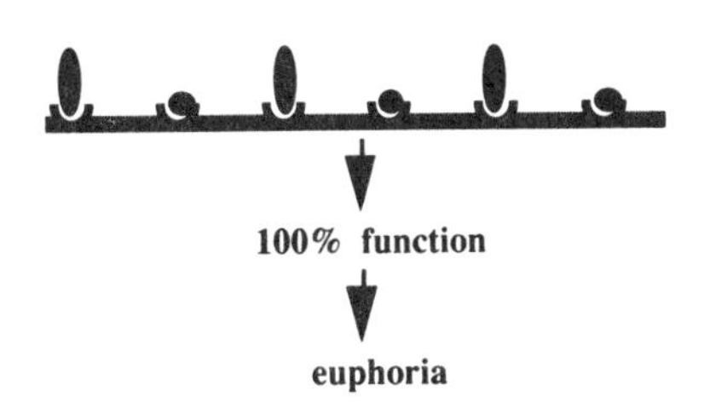

The continued use of morphine sends a feedback signal to stop making enkephalin in an attempt to return the system to normal.

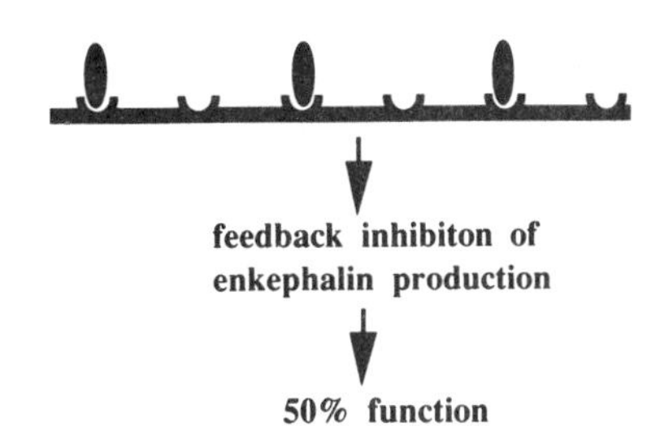

If the drug user prefers the euphoric state he must now use twice as much to get the same level of euphoria.

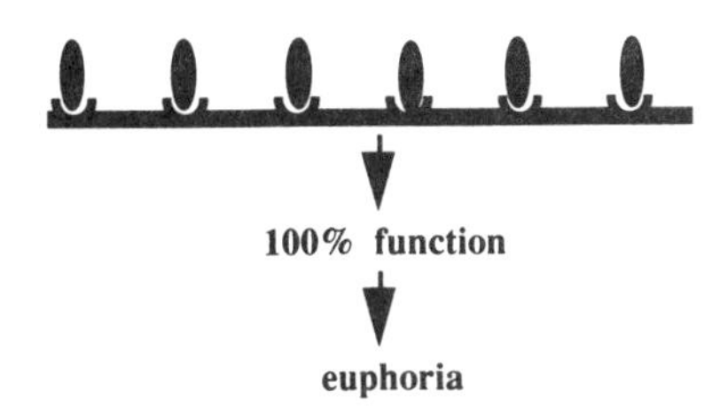

The addict is now **dependent** on the drug. As long as their supply lasts or they don't run out of money they are "fine." Should this happen, however, they now have no morphine and their own natural enkephalin is gone also, and severe symptoms of **withdrawal** occur due to enkephalin-morphine deficiency.

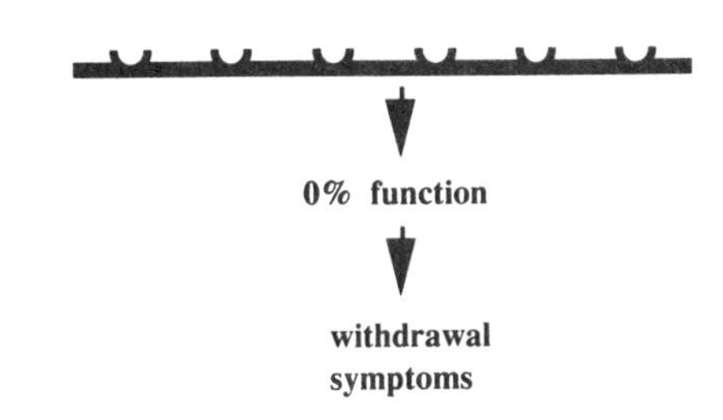

Addiction, dependence and withdrawal at a single cell level In trying to understand complex things, the simplest systems are often the best. When certain nerves become cancerous the tumor cells can be grown outside the body in test tubes. Such a tumor with high levels of opiate receptors was used to study the effect of opiates on cyclic AMP (cAMP), the second messenger[P370]. When no opiates were present in the culture medium, the enzyme adenylate cyclase that produced cAMP was present in normal levels.

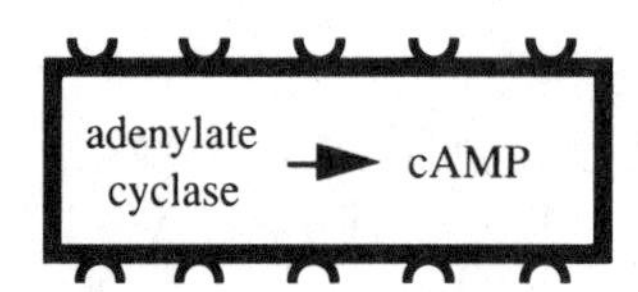

When opiates were added, they bound to the receptors and inhibited adenylate cyclase and decreased the level of cAMP.

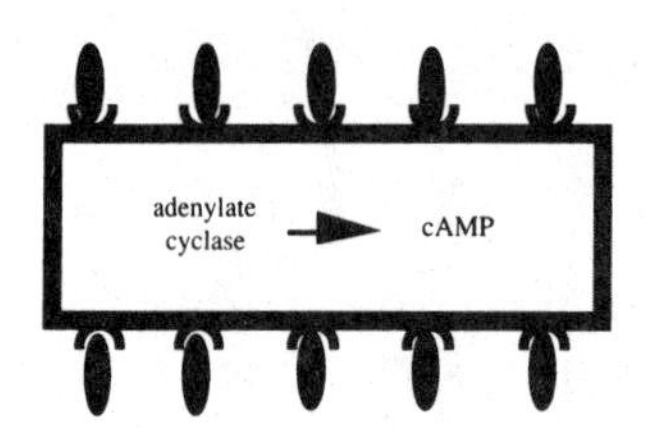

However, if this was continued, adenylate cyclase was produced in excess. Now, even though the opiates still inhibited the production of cAMP, the much larger amounts of the enzyme resulted in normal levels of cAMP.

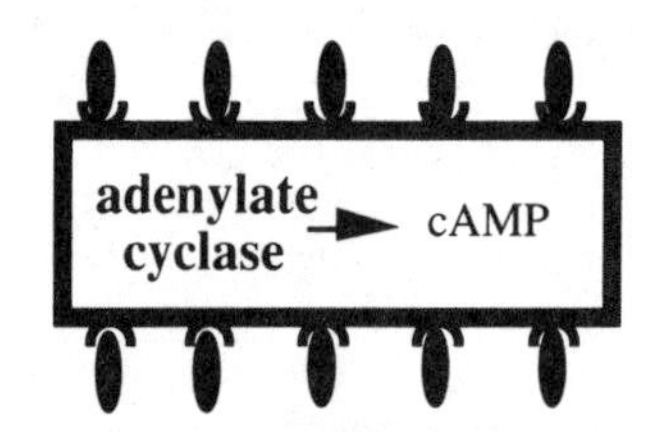

Now, if the opiates are suddenly withdrawn, the adenylate cyclase is no longer inhibited and huge amounts of cAMP are formed.

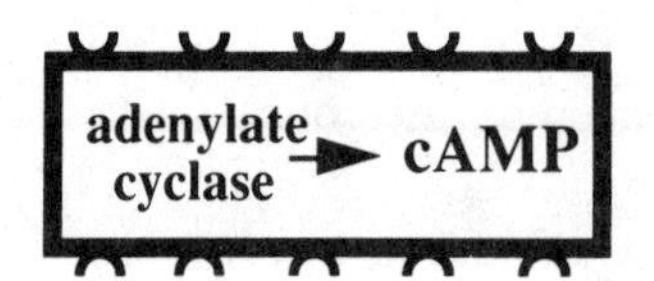

This is equivalent to the withdrawal symptoms of drug addition[1773]. Since increased amounts of an enzyme — *adenylate cyclase,* are formed, this has been called the *enzyme expansion theory* of addiction[766].

This is not meant to imply that all forms of addiction involve cAMP. It is meant to illustrate one of many systems which follow the same principle— *when systems are stressed they slowly change to compensate. When the stress is suddenly removed, it produces withdrawal symptoms due to overcompensation.*

Endorphins

Hughes and Klosterlitz noted that the sequence of their enkephalins was present inside a protein described many years previously by Li[1175], at the University of California in San Francisco, called beta-lipoprotein. This was a peptide containing 91 amino acids isolated from the pituitary gland. Li had also isolated a fragment of beta-lipoprotein with 31 amino acids including the enkephalin sequence[1174]. Since it was 48 times more potent than morphine, it was called *endorphin* [beta-endorphin], meaning "endogenous morphine." Its activity was also blocked by naloxone. Roger Guillemin, who won the Nobel prize for his work on pituitary peptides, isolated two other endorphins from the pituitary: α-endorphin and γ-endorphin[1138]. These relationships are shown on the **next page**.

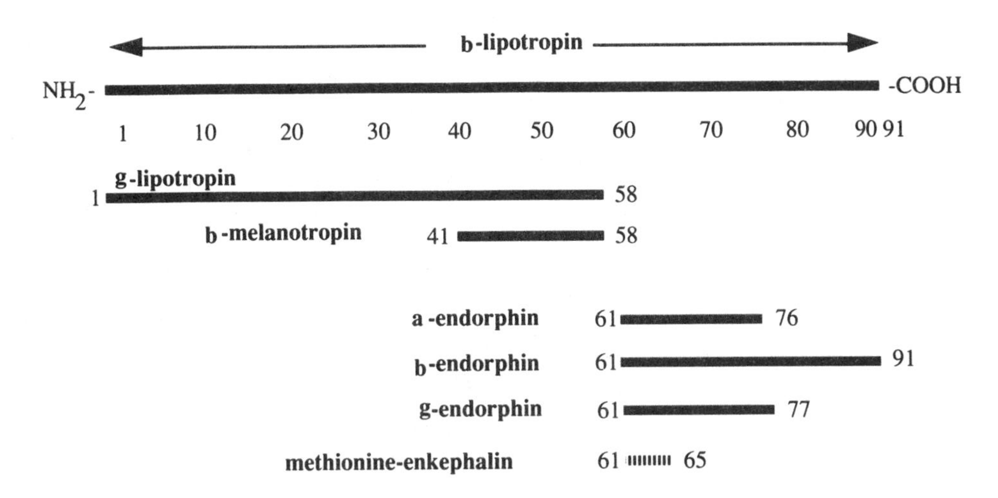

The β-lipotropin in turn was found to be derived from a still larger protein which contained the important peptide ACTH, which stimulates steroid production by the adrenal glands[p476], and two additional proteins (melanotropins) which regulate pigmentation[1389,1390].

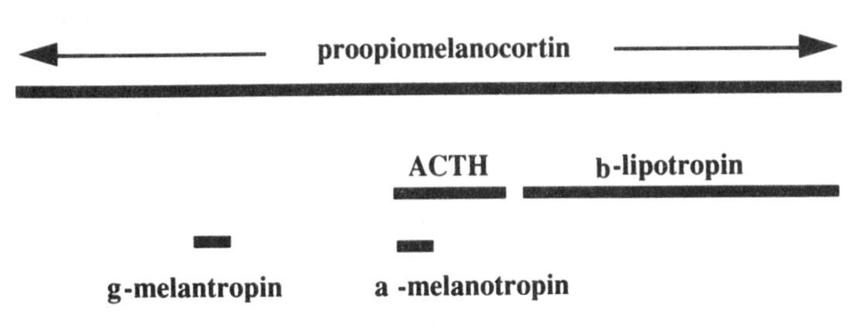

Because this larger protein contains the precursors of opioids, melanotropins, and corticotrophins, it is called proopiomelanocortin or pomc for short. Pomc in turn is derived from a still larger protein that contains a signal sequence that allows it to pass through membranes and a tip on the other end. This protein is called pre-proopiomelanocortin. Fortunately, for the sake of problems with long words, the sequence stops here. The gene for this protein is on the tip of the short arm of chromosome #2.

This cascade of products would seem to present a situation in which pre-pomc could be broken down into different products according to which ones a neuron needed. Wrong! Pre-pomc is produced in the pituitary to produce ACTH and the melanotropins, but it is not produced in other brain cells. Where then do the enkephalins in the limbic system and spinal cord come from? The answer is that these cells use a separate gene to produce their own enkephalins. Not surprisingly this is called proenkephalin. Buried within proenkephalin are four met-enkephalins, two met-enkephalin-like sequences, and one leu-enkephalin[353,1409,1410].

proenkephalin

met-enkephalins

met-enkephalin-like

leu-enkephalin

Dynorphins Finally, there is a third gene that produces additional leu-enkephalins and a different type of peptides called dynorphins[998]. The precursor is called prodynorphin[1651].

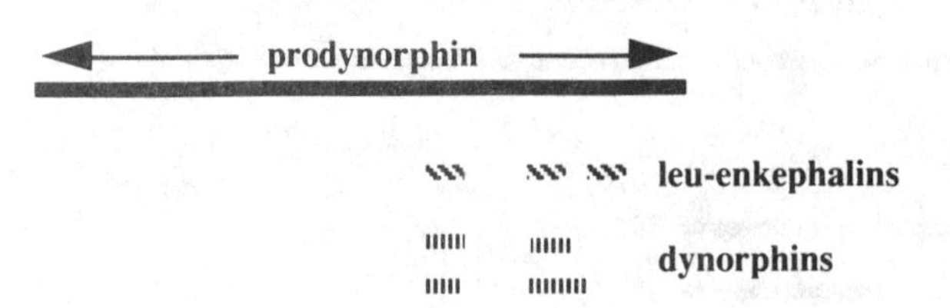

Dynorphin (1-13), containing 13 amino acids, was first isolated from the pituitary gland. Since it was 700 times more potent than leu-enkephalin it was called dynorphin, from the Greek word *dynamis* meaning power[767]. It contained leu-endorphin in its first five amino acids. The eight other amino acids markedly enhanced its potency.

The processing of enkephalins and dynorphins For many years it was thought that the larger precursor molecules were broken into the smaller pieces by the usual enzymes that degrade proteins. However, an enzyme called *enkephalin convertase* was found to be specifically involved in the formation of enkephalins[641,642]. This enzyme was found in the same parts of the brain as the enkephalins[1227]. A similar enzyme has been found for dynorphin[1417]. Specific enzymes were also found for the breakdown of enkephalins and dynorphins. This enzyme already had a function in maintaining blood pressure by converting renin to angiotension, which causes contraction of blood vessels. Thus, this enzyme was known as *angiotensin-converting enzyme*. High levels of this enzyme were unexpectedly found in the substantia nigra where it was found to inactivate enkephalins[1119]. When it acts on enkephalins it has been termed enkephalinase. These enzymes are summarized **on the right->**.

proenkephalin → *enkephalin convertase* → **enkephalins**

prodynorphin → *dynorphin convertase* → **dynorphins enkephalins**

→ *angiotensin converting enzyme* → **inactive fragments**

The reason I have added this detail is that the presence of these enzymes provide the ability to specifically increase or decrease levels of enkephalins or dynorphins in the brain. Inhibiting the enkephalin or dynorphin convertase would decrease levels of endorphins and inhibiting angiotensin-converting enzyme would increase levels. The report of increased levels of enkephalin convertase in morphine addicts[1248] suggests that this enzyme could also be involved in a variant of the enzyme expansion theory of drug addiction[P490].

Types of Opioid Receptors

As with every other neurotransmitter, there are also several types of opioid receptors. They are named for the compounds that specifically bind to them, or for the organ where they were first described[1209,1470]. Thus there are the mu (μ) receptors for morphine binding; kappa (κ) for ketazocine binding; sigma (σ) for SKF-10047 binding; and delta (δ) for vas **d**eferens receptors. Dynorphin binds to the μ receptors and is especially involved in the relief of pain[307a].

Opiates Modulate Norepinephrine Neurons

The norepinephrine-rich neurons of the locus ceruleus[P409] have many opiate receptors on them. When we feel pain, these opiate neurons fire in an attempt to suppress the transmission of pain impulses. Clonidine, which inhibits norepinephrine neurons, blocks symptoms of drug withdrawal[747-749]. This has led to the *hypothesis that opiate withdrawal symptoms are due to disinhibition of the locus ceruleus.* This hypothesis runs as follows[1738].

During normal conditions, with normal amounts of opiates, the norepinephrine neurons are not inhibited and produce a normal response.

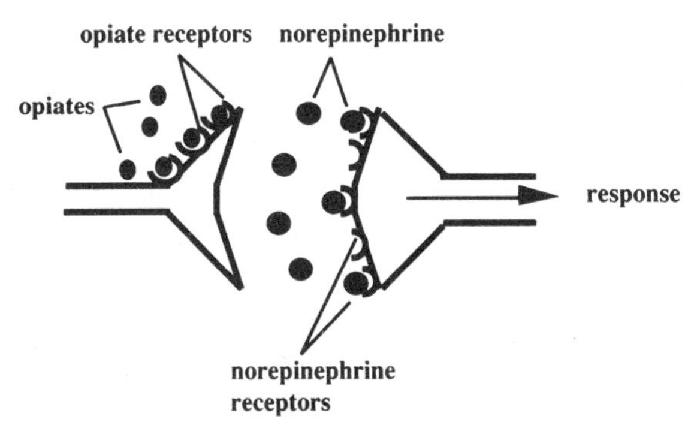

In the presence of opiates, the nerve transmission by norepinephrine is inhibited by the presynaptic opiates giving a much reduced response.

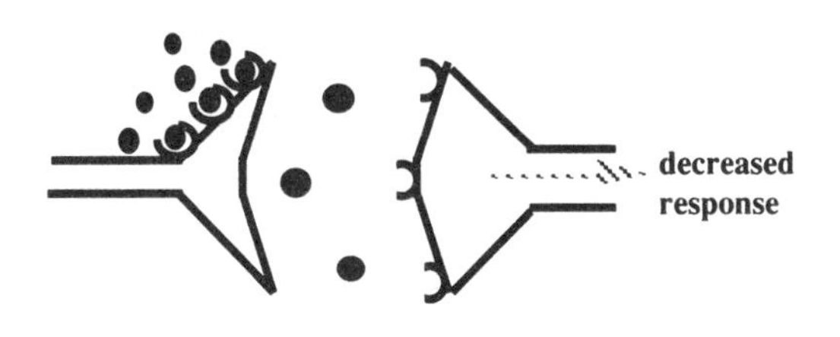

However, after *chronic* administration of opiates, the number of norepinephrine receptors increases to once again produce a normal response.

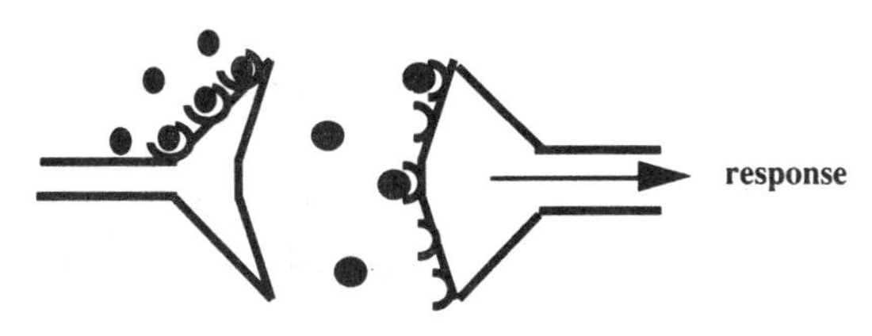

Now, if the opiates are withdrawn, the norepinephrine release is normal again but since there are more receptors there is an exaggerated response.

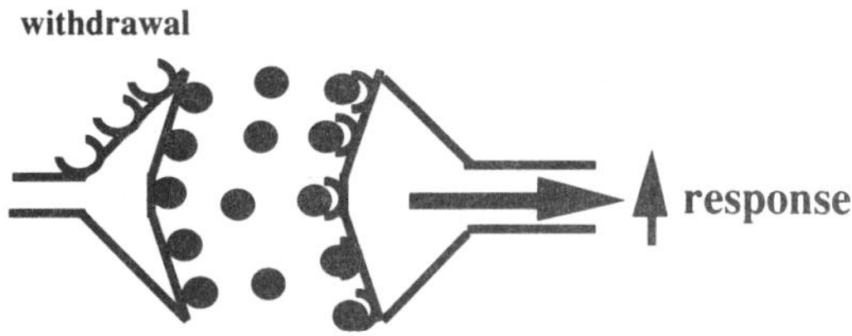

This explains why clonidine, which blocks norepinephrine release in the locus ceruleus, is such a good way to treat withdrawal symptoms. In fact, if clonidine itself is suddenly stopped it can precipitate symptoms like those of opiate withdrawal.

Since clonidine often improves the symptoms of Tourette syndrome by inhibiting norepinephrine neurons, and opiates also inhibit norepinephrine neurons, it might be expected that opiates might also improve the symptoms of TS. This would explain why two of our TS patients have reported that codeine relieves their symptoms, and why some TS patients have problems with drug addiction.

Opiates Modulate Dopamine Neurons

The effects of various opiates on dopamine neurons are complex and vary according to the type of receptors, type of natural opiate, and part of the brain studied[2109]. Opiates have been termed neuromodulators of dopamine neurons[488,1373,1720,2109]. Morphine has been observed to cause a marked increase in spontaneous firing of dopamine (A10) neurons in the ventral tegmental area[829], and when either enkephalins or endorphins are injected into the ventral tegmental area, they stimulate the limbic system and cause hyperactivity[225,1888].

In general, however, the effect of opioids is similar to what would be expected if many dopamine neurons contain presynaptic opioid receptors that inhibit nerve activity as described above. This is consistent with the observation that most opioid agonists cause a decrease in the synthesis, release and turnover of dopamine[221,1830,2109,2131]. Thus, as a generalization:

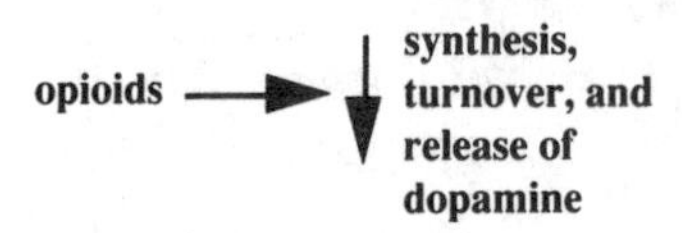

Opioids and Serotonin

In contrast to their effect on dopamine, opioids and endorphins enhance the turnover and release of serotonin[221,1538,1700,2109] and increased levels of serotonin enhance β-endorphin formation. Thus:

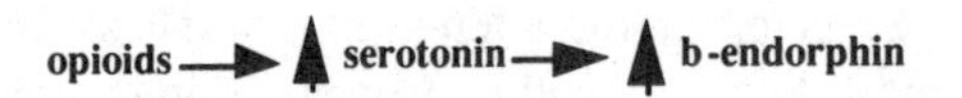

The effects of morphine on both dopamine and serotonin can be summarized as shown in the following study of the rat striatum[p524].

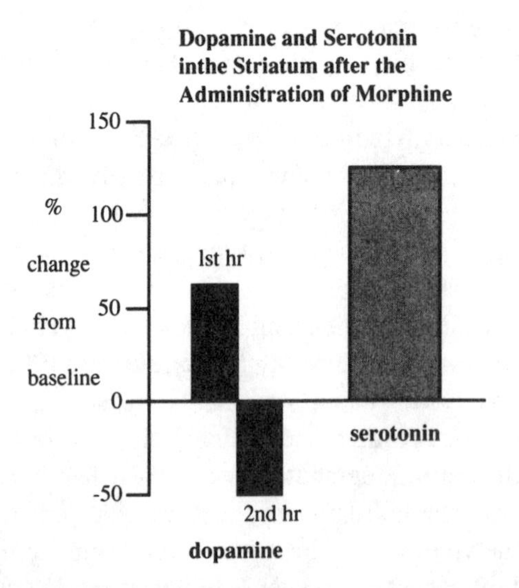

The initial increase in dopamine followed by a rebound decrease is characteristic of the addicting process as discussed later[p521]. The above figure is a biochemical blueprint of a perfect setup for predisposition to addiction. The striking response of serotonin suggests that individuals with low central serotonin, as in TS, may self-medicate themselves with opioids to raise their serotonin levels. The biphasic dopamine response with a "high" followed by a "low" and need for more drug leads to addiction.

Obesity, Bulimia and Endogenous Opiates

The role of endorphins in obesity was first suggested by the finding that mice with a recessive genetic defect that caused them to overeat and become obese had a markedly increased level of β-endorphin in their pituitary glands[1259]. This is shown as follows:

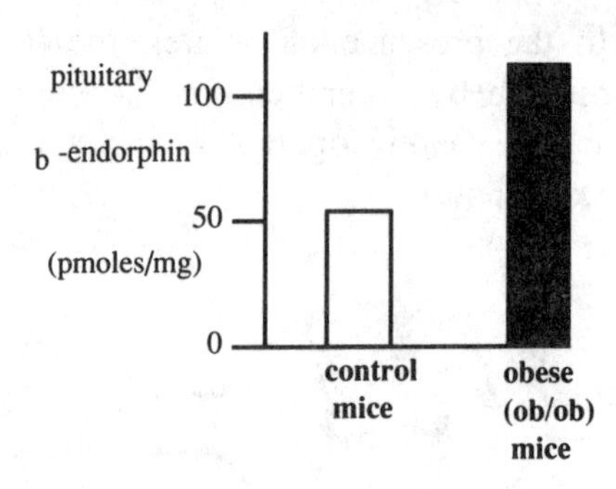

The pituitary was producing twice as much β-endorphin in the obese mice. These mice also had strikingly increased levels of ACTH, which is understandable since pro-popc also contains ACTH. When these mice were *given naloxone to block the effects of the β-endorphin, the overeating stopped*. In fact, the obese mice were ten times more sensitive to the effects of naloxone than the normal mice. The brain levels of the enkephalins were normal, further indicating that the endorphins and the enkephalins are produced by separate genes.

Naloxone treats bulimia This striking relationship between endorphins and overeating suggests that naloxone might be an effective treatment of some types of eating disorders. This turned out to be true. Bulimic patients were markedly improved following treatment with naltrexone, a long-acting form of naloxone[989]. Increased levels of β-endorphin have been found in patients with anorexia nervosa [810,1017].

Opioids and appetite Having a good appetite is critical to the survival of all species. But this, like other urges such as mating, does not just happen—it has to be driven by hormones and neurotransmitters. Because it is so important, hunger is maintained in many different ways to provide a fail-safe system[1359]. Because appetite is involved in many different aspects of TS, from compulsive eating to over- eating as a side effect of Haldol, I will show some of the ways different neurotransmitters, peptides and drugs affect appetite. The core of this diagram is the central effect of β-endorphin on stimulating (+) appetite.

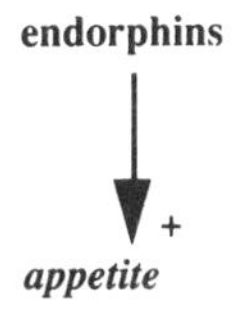

Since too much appetite is just as bad as no appetite, a number of mechanisms, were added to inhibit the system. These are three additional peptides, choleocystokinin (CCK), thyrotrophin-releasing hormone (TRH), and bombesin, which inhibit the effect of endorphins on appetite, and dopamine which inhibits the production of endorphins. All of these inhibiting effects are shown as follows:

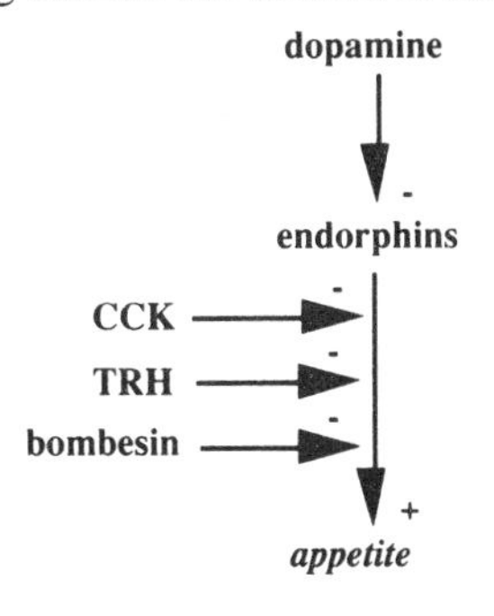

Serotonin is involved through its ability to stimulate endorphin production and to augment the CCK effect:

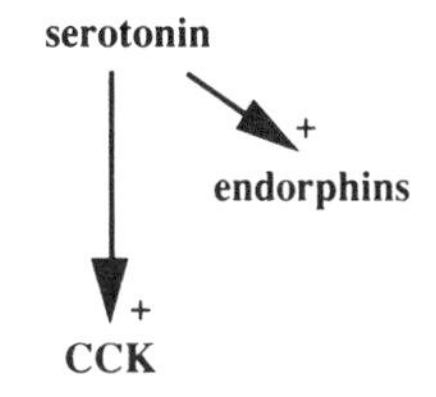

The whole system then looks like this:

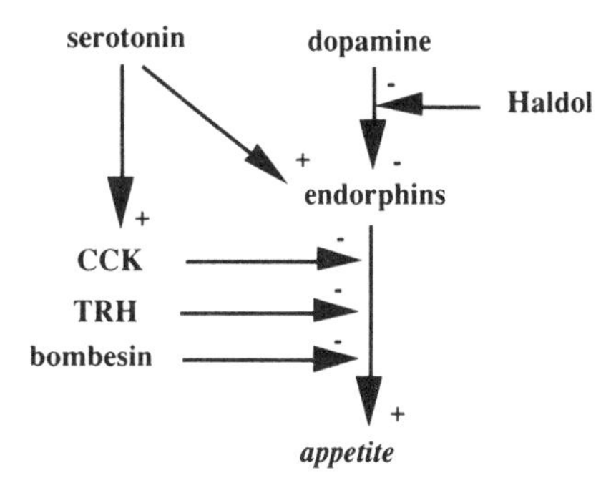

Thus, since two minuses equal a plus, Haldol tends to stimulate the appetite by inibiting dopamine, and serotonin can both stimulate and inhibit appetite depending upon other factors.

Opioids and Sex

Despite the short-term almost orgasmic euphoria produced by opioids[1338], chronic morphine and heroin addicts show a severe inhibition of sexual behavior with elimination of sexual dreams, delayed ejaculation, impotence, loss of morning erections, inability to have orgasms and loss of sexual desire[1508]. The common concept that opioids improve sexual feeling is a myth. In fact, among addicts, the concern about loss of sex drive is second only to the concern about dying of an overdose[1097], and one of the primary reasons for patients abandoning methadone treatment programs for heroin addiction is the loss of sexual performance[670]. Although chronic opioid addiction results in a decrease in the sex hormone, testosterone, this alone does not account for the decrease in sex drive[1508].

By contrast, during withdrawal from opiates, there is a return to normal and sometimes a period of greater than normal sexual activity. Spontaneous ejaculations without sexual stimulation are one of the signs of withdrawal[798,1467]. When naloxone is given to heroin addicts their sexual desire also returns[915].

Many studies in animals have verified that both acute and chronic intake of opioids inhibit sexual behavior and these effects are reversed by naloxone[1508].

Sexual exhaustion When male rats are given unlimited access to a receptive female, they copulate for about 80 minutes until they become sexually exhausted. Subsequent sexual activity is inhibited for about 72 hours. This inhibition may be due to increased levels of endorphins, since when the rats are given naloxone, sexual exhaustion is delayed until 130 minutes[1508]. In male monkeys, naloxone causes increased masturbation[2]. The effect of naloxone on sexual activity in females is less impressive than in males.

As discussed in the chapter on serotonin, decreased levels of serotonin disinhibit sexual activity. Since low serotonin causes low endorphin levels[p506], this may play a facilitatory role in the disinhibition of sexual activity in TS, as follows:

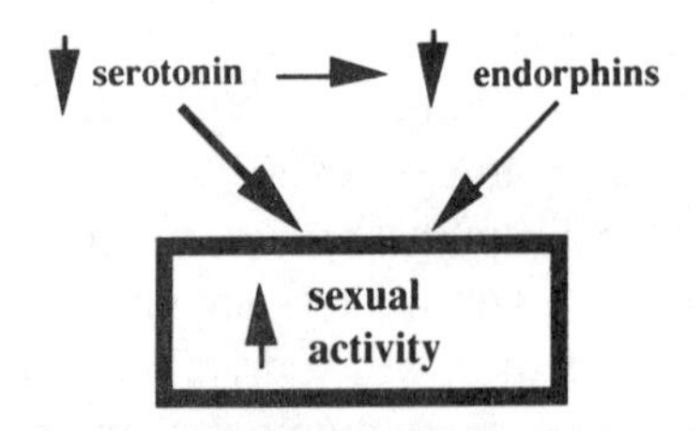

Inhibiting endorphins with naltrexone may be of value in treating some forms of sexual dysfunction in humans.

Endorphins and Enkephalins in Psychiatric Disorders

Given the ability of opiates to inhibit dopamine neurons, it is not surprising that morphine has a long history of being used to treat mental disorders[354,2007a]. Some addicts may have turned to opiates as a form of self-medication for mental disorders. Addicts have reported their use of drugs to decrease anxiety, rages, aggression, paranoia and feeling of inadequacy and depression[2007a]. However, because of their addictive potential, opiates are not an accepted form of treatment of mental disorders.

Opiates and schizophrenia Enkephalins are present in especially high concentration in structures which contain the ends of dopamine neurons[407,925] and in the amygdala and other limbic structures. Because of this, and the fact that enkephalins modulate dopamine neurons[1720], the idea that too much or too little enkephalin may play a role in schizophrenia has been a popular one[185,488,1932]. Despite this, giving enkephalins, endorphins or opiate inhibitors, has not produced clear improvement of schizophrenic symptoms[249,1720].

PCP and schizophrenia PCP is a widely abused street drug. The superhuman strength and acute psychotic state of individuals high on PCP is legend. In many ways the behavior of individuals on PCP resembles schizophrenia[336] and drugs which stimulate δ-receptors have PCP-like effects[211]. PCP also stimulates dopamine neurons through its effect on δ-receptors[1720]. Detection of the natural substances that interact with δ-receptors, and the development of agents that block these receptors, might be beneficial in the understanding and treatment of schizophrenia. PCP receptors have been found[1196a] to be part of the excitatory NMDA[p399] receptors.

Endorphins and Tourette syndrome The most intriguing evidence that opiates may be involved in TS comes from a study of the brain of a TS patient. This showed a virtual absence of dynorphin (1-13) in the globus pallidus[831]. The place of the globus pallidus in the complex pathways for motor control[830,831] was shown previously[p368].

To further examine this, others have examined the levels of dynorphin in the spinal fluid of several TS patients. Although the levels were increased, the assay was for dynorphin (1-8), not dynorphin (1-13). The two fragments may be regulated independently.

Other findings suggest endorphins may be too high in some patients and too low in others.

Evidence for *endorphins being too low in TS* include the observation that nitrous oxide, the dentist's laughing gas, which acts to stimulate opiate receptors, improved the symptoms of TS in one case[720,1691-1697]. This improvement was reversed with naloxone. In agreement with this, I have had two patients who claimed that codeine improved their TS symptoms. The report that naloxone made two patients with obsessive-compulsive disorder worse[949] led to the suggestion that these patients may have a defect in their opiate reward system and thus fail to "reach satisfaction" and so continue to repeat a given behavior. This agrees with the fact that clomipramine both improves obsessive-compulsive behaviors and potentiates the pain-relieving action of opiates. All of these observations are consistent with the following:

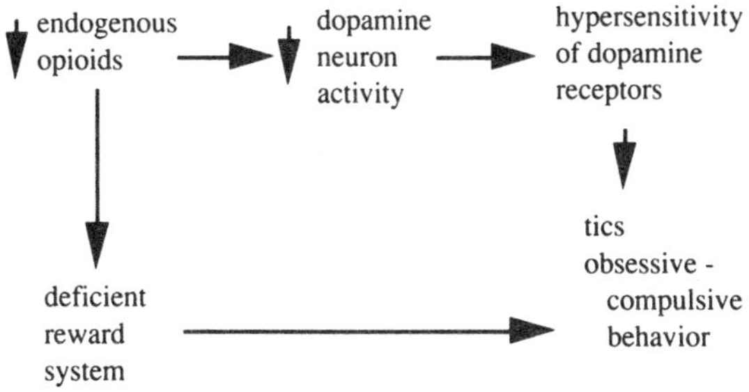

↓ endogenous opioids → ↓ dopamine neuron activity → hypersensitivity of dopamine receptors

↓

tics
obsessive - compulsive behavior

↓ endogenous opioids → deficient reward system → tics, obsessive - compulsive behavior

Evidence that *endorphins may be too high in TS* comes from the observation that naloxone has been reported to improve the obsessive-compulsive behaviors in two cases, improve self-aggressive behavior in two cases, and improve the motor and vocal tics in one case of TS[1691-1697]. Since naloxone is short-acting and has to be given by injection, the effects were short-lived and the symptoms worsen when the drug wears off. Endorphins are often elevated in psychotic or mentally retarded children with self-abusive behavior[346,713]. As in TS, these behaviors also improved with naloxone[1691,1606].

I treated a young woman with TS and severe bulimia with naltrexone, a long-acting naloxone-like medication. This eliminated the bulimia but had no effect on the tics. At the present time too few TS brains have been examined and too few patients have been treated with opiate agonists or antagonists to provide a clear picture of the role of dynorphins and endorphins in TS.

Chapter 66. Opium, Opiates, and Endorphins

Summary: Opiates such as morphine, codeine and heroin control pain and cause a pleasurable sense of euphoria. The brain has its own opiates in the form of a set of short proteins (peptides) called enkephalins, endorphins and dynorphins. Receptors for opiates and these peptides are present in the pain pathways, the striatum, and the limbic system and regulate physical and emotional pain. Through their ability to modulate dopamine and norepinephrine neurons, these peptides regulate motion, emotion, appetite and sexual behavior. They may play a role in Tourette syndrome.

Humans readily become addicted to opiates because of their pleasure-producing potential. Studies of opiates and their receptors have provided insights into the causes of addiction and the phenomena of dependence, tolerance and withdrawal.

Chapter 67
Autism Again — Causes

The symptoms of autism were presented in a previous chapter[p215]. I have put off discussing the possible causes of autism until now, after the subjects of neuroanatomy and serotonin were covered. While Kanner[1012] regarded the autistic child's abnormal social behavior as the primary defect, the neurological model suggests that

> "these abnormalities are of a secondary nature. . .[and] consequent to an organized collection of primary disturbances of. . .attention, and compulsive behaviors, as well as disturbances of communication"[455].

Autism is best explained as a genetic, neurochemical disorder that in some ways is similar to, and in some ways the inverse of, Tourette syndrome.

Autism and the Frontal Lobes

Many of the similarities between ADHD-TS and the frontal lobe syndromes[p359] also apply to autism. This especially applies to the following[455]:

- concrete thinking
- inability to focus attention
- lack of empathy
- lack of initiative
- perseveration
- shallow feelings
- tantrums in the face of change

In a discussion of the neurological basis of autism, Damasio and Maurer[455] state:

> "The tantrums and distress that result when accustomed activities are prevented or familiar settings changed are similar, in nature, frequency, severity, and resistance to modifiability to the 'catastrophic' reactions that have been described in some patients with acute and chronic frontal lobe lesions and that occur when these patients are confronted with situations for which they lack an adequate learned response and in which they are unable to create a new and appropriate response. In this light, 'maintenance of sameness' becomes a form of adaptive behavior aimed at avoidance of 'catastrophe.' Compulsive, stereotyped, ritualistic behaviors may be viewed as ways in which that avoidance is enacted...What autistic children and patients with some frontal lobe syndromes seem unable to do is to teach themselves ways of adapting to modified environmental contingencies. . .Consequently, autistic children, as well as patients with some frontal lobe syndromes, are bound to patterns of behavior that were either previously learned, or explicitly taught to them. In a way, 'perseveration,' another behavioral characteristic that is common to both autistic children and frontal lobe patients, is an aspect of this inability to organize and consolidate new forms of

response. The perseverative tendency permeates tasks of every type and complexity."

As noted before[p354], these stereotypes and stimulus-bound behaviors can also be produced by lesions of the caudate[960,2009].

The supplementary motor area The supplementary motor area is part of the frontal lobe[p368]. In the chapter on autism I pointed out that a unique feature of the lack of speech in autism was that there was also no attempt to compensate for this by using gestures or other forms of nonverbal communication. By contrast, individuals who cannot speak because of a lesion of Broca's area of expressive speech attempt to make up for this deficiency with gestures. Lesions of the supplemental motor area or the cingulate gyrus show autistic-like lack of initiation of speech and also never resort to gestures to compensate for this verbal weakness[455]. Like autistics, they can repeat the speech of others, have trouble controlling speech volume and pitch, and show paralalia or a repetition of their own words[19]. Electrical stimulation of this area causes the same problems[205,1987]. Just as lesions of the caudate could reproduce abnormalities seen with frontal lobe lesions[p354], caudate lesions can also cause these same problems[1988,1989].

Some autistic children also show asymmetry of the lower part of the face when they talk or smile, so-called "emotional facial paralysis"[455]. This can also be produced by lesions of the supplemental motor areas.

Autism and the limbic system Different areas of the limbic system have also been implicated in autism, including the amygdala, hippocampus, thalamus and mesolimbic cortex of the frontal and temporal lobes[348,455,1275,1276]. The amygdala has drawn frequent attention and the following list of behaviors occur in animals with lesions of the amygdala and in children with autism.

- inappropriate response to social cues[1941]
- excessively submissive behavior
- impaired association between the different senses (cross-modal association)[884]
- abnormal reaction to novel or familiar objects[884,1553].

The involvement of these different sites has been unified by the suggestion that they have in common a transitional type of architecture that is old in an evolutionary sense and forms a ring around the middle part of the brain[455].

The Genetics of Autism

Autism in relatives of autistics The frequency of autism in the general population is 1 in 2,500 males. Since the sex ratio is about 4 males for 1 female, the frequency for females is 1 in 10,000. As with other behavioral disorders, one indication that genetic factors might be involved comes from the observation that up to 4.5% of other relatives of an autistic child also were autistic[1616]. This is more than a 100-fold increase over the frequency found in the general population. In an extensive study in Utah, 17 of 187 families had more than one affected child, including one family with four and another with five autistic children[1616].

Twins In a clear demonstration of the role of both heredity and environment, Folstein and Rutter[619] found the following degrees of concordance for autism in identical and fraternal twins.

Concordance for:	Identical Twins	Fraternal Twins	p
autism	36%	0%	.05
autism + learning disorder	82%	10%	.0015

This clearly indicates that genetic factors are very important. However, they also noted that in 12 of 17 pairs where only one twin was autistic, the autistic twin had evidence of brain injury compared to no evidence of such an injury in the nonautistic twin. Other clear-cut environmental causes of autism include measles [313,468,1898] and cytomegalic virus[1898]. Thus, both a genetic predisposition and environmental factors appeared to interact to result in an autistic child. Since about 25% of families with an autistic child had another member with delayed speech[129], the predisposing gene may manifest as delayed speech in the absence of brain injury.

Autosomal recessive inheritance On the basis of a study of 46 families in which there was more than one affected member with autism, Ritvo and co-workers[1617] concluded that these cases were inherited as an autosomal recessive trait. With a frequency of 1 in 2,500 males affected, this would suggest that as many as 1 in 50 people may carry a gene for autism.

Other genetic causes of autism The above cases of autism were "pure" in that they were not associated with other features. There are a number of chromosomal and genetic disorders that have been associated with autism [1600]. Each provides some additional clues to the cause of different autistic syndromes and many produce lesions in the regions of the ventral tegmental area, the raphe nuclei and limbic system[1600]. I will discuss only a few of these. The interested reader may find references to the others elsewhere[970,1419,1600].

Fragile X-syndrome In some people the tip of the long arm of their X-chromosome contains a constriction[1217]. This has been found to be associated with mental retardation in some males. Approximately 10 to 16% of individuals with autism have a fragile X-chromosome[183,231,232,609,712,1312] and most males with the fragile X-syndrome have some symptoms of autism, especially avoiding eye contact, hand flapping, other stereotyped hand movements, language delay and echolalia[833]. Cases of Tourette syndrome with the fragile X-syndrome and autism have been described[1034].

Rett syndrome In 1966 Andreas Rett[1602], an Austrian physician, described an unusual syndrome with autistic-like features and severe mental retardation. It largely languished unappreciated until it was brought to the attention of doctors in the Unites States by a report in 1983 of 35 cases[832]. An unusual feature was that it always occurred in girls who were essentially normal for the first six months to two years of life. Further mental development then stopped. They either lost or never attained speech, and developed stereotyped hand washing or wringing movements. Since their brains stopped growing the head size became small for their age (microcephaly). Walking became very clumsy and they often had seizures. Many of the other features characteristic of autistic children were present[1443].

In genetic terms, Rett syndrome can be best explained as a X-linked disorder in which males die in the uterus and females are too retarded to ever have children. Since the gene is never passed on, all cases represent new mutations[372].

Too Much Serotonin in Autism

One of most remarkably consistent biochemical findings in any of the behavioral disorders is the finding that blood serotonin is abnormally high in about 40% of patients with autism[33,279,771,837,848,1618,1619,1709,1916,2150]. Blood serotonin levels are also significantly elevated in many children with severe mental retardation[1462,1468,1709,1967]. With the possible exception of some schizophrenic and manic-depressive patients, the only other disease with this biochemical finding is carcinoid syndrome, where there is a tumor of the serotonin-producing cells (enterochromaffin cells) of the intestine[848]. Most of the serotonin in platelets is produced by these cells[557,1950], which are derived during early development from nerve cells[1488,1489].

One study of controls, autistic and mentally retarded (M.R.)[848] patients showed the following:

Diagnosis	N	Blood serotonin (ng/ml)	S.D.	% abnormal
Controls	6	56.6	48.8	0
Mild M.R.	23	96.5	38.0	9
Autistic	27	134.5	56.9	30
Severe M.R.	25	156.6	63.3	52

There are a number of important points about this table. First, the increased levels of serotonin in autistic and severely retarded patients (IQ < 20) are not just moderate. They are up to three times that of the normal controls. Second, the more severe the retardation, the higher the serotonin levels.

Many potential causes of this elevation, such as diet, drugs, level of activity, bowel motility, bacteria in the gut, and increased platelet count, have been ruled out[967]. This increased serotonin level is consistently present over months and years[848]. Not surprisingly, autistic patients with increased levels of blood serotonin show increased excretion in the urine of both serotonin[848,2140] and 5-HIAA[848,1336]. The absence of a disproportionate increase in urinary 5-HIAA after a tryptophan load indicates that if there is a defect in the kynurenine pathway it is not complete[848].

The absence of abnormalities in the way serotonin binds to platelets, or in the way platelets accumulate or store serotonin[33,200,848], further points to increased production of serotonin by the gut as the primary cause of the increased serotonin.

Fenfluramine, a drug approved for weight reduction, reduces whole blood serotonin and has been shown to improve sterotypy and hyperactivity in some autistic patients[1615].

Serotonin in the severely retarded The elevation of blood serotonin in many autistic patients has often been minimized by the statement that it also occurs in severely retarded children and is thus not specific to autism. From a geneticist's point of view, this attempt to trivialize a very important biochemical finding looses sight of the fact that outside of carcinoid syndrome, there are only two conditions with such high levels of blood serotonin — autism and severe retardation. A far more likely interpretation is that autism and severe mental retardation tend to be on a continuum of response to the same genetic defect.

Too little serotonin in some autism In contrast to the "pure" autism described above, many of the genetic conditions that can cause autism are associated with low serotonin levels. These especially include autism in phenylketonuria, Down's syndrome, histidinemia[799,2136] and Tourette syndrome[P458]. In several studies of apparently "pure" autism, abnormally low levels of blood serotonin have been present in some patients[771,954]. As discussed below, either too much or too little serotonin may cause similar problems.

The serotonin syndrome and autism In the chapter on serotonin and animal behavior I mentioned the "serotonin syndrome," a behavioral syndrome in animals due to too much serotonin. When a drug that stimulates S_1 serotonin receptors is given to rats, it reproduced the hyperactivity and hyper-reactivity of this syndrome[793]. Since Haldol inhibited this response, dopamine neurons were assumed to be part of the pathway. Drugs that inhibit S_2 receptors enhanced the effect. This and other evidence suggested the following interrelationships between S_1 and S_2 receptors, and dopamine neurons[793]:

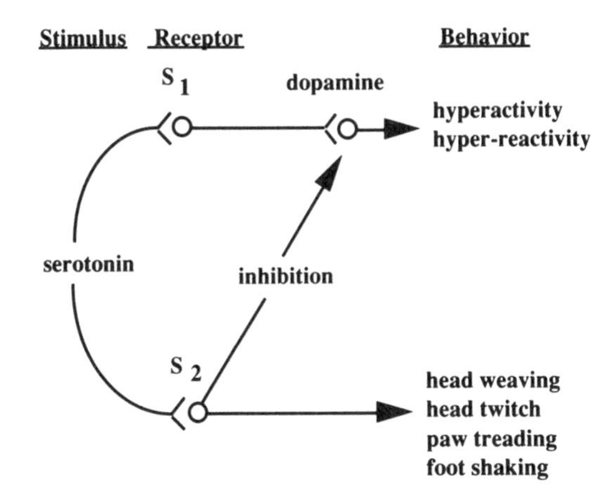

This provides a potential explanation of how *both too much or too little serotonin can cause similar problems.* If serotonin is too high, it overstimulates the S_1 pathway and causes hyperactivity and hyper-reactivity. However, if serotonin is too low, it disinhibits the dopamine neurons and also causes hyperactivity and hyper-reactivity.

Free Tryptophan Is High in Autism

Despite the many studies of serotonin there have been few studies of free tryptophan in autism. We saw how critical this was in providing insight into the potentially important role of tryptophan oxygenase in Tourette syndrome[p467]. One study was done in Japan by Hoshino and colleagues[930]. They studied 37 autistic children and 67 normal controls. While there was no significant increase in the level of whole blood tryptophan, the level of free tryptophan was significantly higher in the autistic patients than in the controls. This is shown as follows:

Subjects	Level of free tryptophan	S.D.	p*
normal children	6.92	1.54	<.05
normal adults	7.47	1.59	<.05
autistic	**9.29**	1.97	

* (versus autism)

The level of free tryptophan was highly correlated (r = .530) with a score measuring hyperactivity in autistic and ADHD children, well correlated with a score measuring different psychiatric symptoms (r = .370), and negatively correlated with a score measuring mental development in infants (-0.285). Thus, *the higher the free tryptophan, the more severe the hyperactivity and psychotic symptoms and the lower the IQ.* Although the blood serotonin was significantly increased, there was no correlation between the level of serotonin and the level of free tryptophan. This suggests there is a close relationship between the free tryptophan and serotonin in the brain but not in the peripheral blood[930]. As with Tourette syndrome, this abnormality in blood tryptophan strongly implicates the role of the tryptophan oxygenase, the enzyme which controls the rate at which tryptophan is removed from the blood or from the cells in the gut where serotonin is made.

This immediately raised the question: Are there any abnormalities in the kynurenine pathway in autism? Very few studies of this have been done. In 1965 Heeley and Roberts[876] reported studies of 16 psychotic children, some of whom were autistic. Ratios of different breakdown products of tryptophan were different than in the controls and treatment with vitamin B_6 resulted in a normalization of these ratios and an improvement in symptoms. They concluded that there was a significant abnormality in the kynurenine pathway.

The Snedden Syndrome

Snedden and co-workers[1817] described two siblings who had markedly elevated levels of tryptophan in their blood, and of tryptophan, 5-HIAA and indole acids in the urine. By contrast, their urinary excretion of kynurenine and related breakdown products was below normal. As shown in the following diagram, this is consistent with a defect in tryptophan oxygenase:

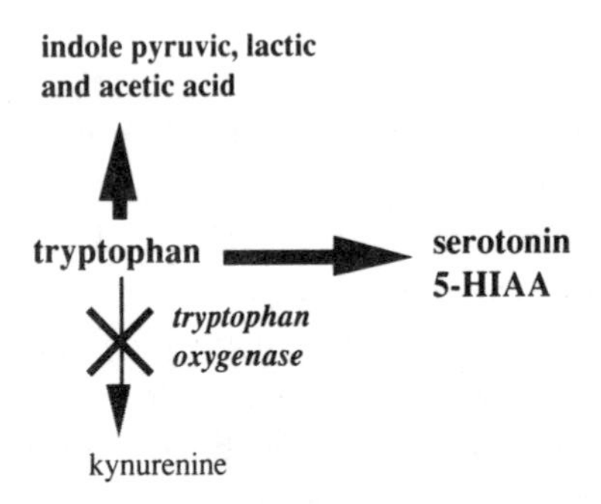

Given this defect, were there any behavioral problems? Yes, very striking ones. In the 23 year old brother there were "rapid and exaggerated changes of affect. *Over the course of a few minutes his mood would change from overt hostility to a cheerful affability and then to suspicion.*" His verbal IQ was 75, and performance IQ 86, with impaired performance of tests of memory. His 22 year old sister was characterized by "*abrupt changes of mood from cheerfulness and affection to marked hostility and depression.*"

The fact that this occurred only in siblings is consistent with an autosomal recessive inheritance[P42], characteristic of enzyme defects. This well-studied case provides many lessons. It verifies that if there is a defect in the breakdown of tryptophan via the kynurenine pathway the blood tryptophan and serotonin is increased. It verifies that a defect in the kynurenine pathway can produce severe mood swings and cognitive defects like those seen in autism.

Schizophrenia and Autism

Autism and childhood schizophrenia have often been linked. In fact there is even a journal called *Autism and Childhood Schizophrenia.* While they are generally believed to be distinct entities they may have some biochemical similarities. Recall that schizophrenia is one of the few other entities that has been reported to show an increased level of blood serotonin[P443] and that the role of tryptophan, serotonin and other indoles has been the subject of much conjecture. One series of studies reported that exacerbations in schizophrenic symptoms were immediately preceded by an increased excretion of tryptamine in the urine[160,161,235]. Tryptamine is a third product that can formed when tryptophan is broken down. A defect in tryptophan oxygenase would be expected result in an increase in tryptamine. This is shown as follows:

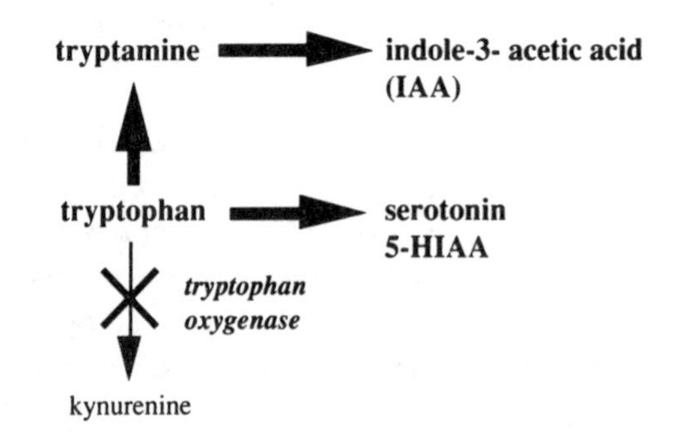

In light of these relationships, a study now over 25 years old is of interest. It showed a significant increase in all three of these compounds during the more actively psychotic stages of individuals with schizophrenia[235]. The blood levels of tryptamine, IAA and 5-HIAA in schizophrenic patients whose psychotic symptoms ranged from inactive to markedly active were as shown on the **next page**. The more psychotic the patients, the higher were the levels of IAA, tryptamine and 5-HIAA.

This, along with the occasional reports of increased serotonin in schizophrenia, are consistent with a defect in tryptophan oxygenase in some patients. Like schizophrenia, abnormalities of dopamine have also been implicated in autism[455,711,1275].

Autism: Too Much Endorphin?

In 1979 Jaak Panksepp[1455,1680] proposed that there were intriguing similarities between

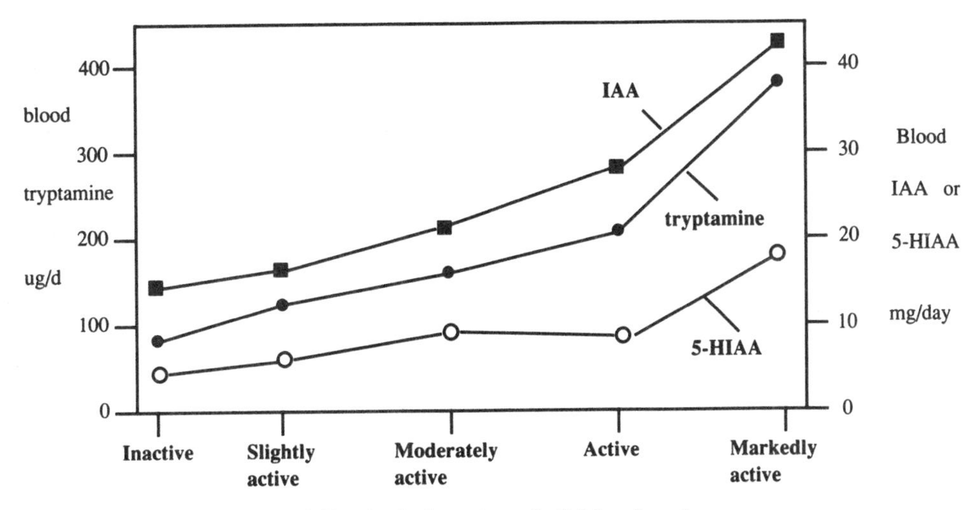

Severity of Psychotic Symptoms in Schizophrenics

symptoms of autism and narcotic addiction. The features in common between animals treated with low doses of narcotics and autistic children were as follows:

- decreased sensitivity to pain
- do not cling to mother
- do not cry spontaneously
- do not desire social companionship
- perseveration
- seizures
- spurts of activity then quiet
- walking on toes

It was suggested that *opiates are the neurochemical equivalent to mother and thus mother is not needed* and the underlying imbalance in autistic children may be excessive activity of their own brain opiates or endorphins. Stated differently, *autistic children are high on themselves*[1257]. Normal children get a surge of euphoria-producing endorphin with every dose of affection, in effect addicting them to their parents. Autistic children, who already have high levels of opioids, do not need this surge. They are already high on themselves[1257].

During pregnancy, the levels of ß-endorphin are very high and this keeps the fetus quiet and happy. During development there is a shift from the long-acting ß-endorphins to the shorter acting enkephalins. The early onset of autism may be a result of the failure to make this switch,

> "leaving the autistic child in the opiate 'bondage' which perhaps all young animals experience, but from which most are gradually liberated. This maturational lag may prevent the brain from becoming appropriately responsive to the sensory and social environment"[1455].

The brain appears capable of rapidly turning on and off the opiate system. The autistic child's dramatic shifts from dreamy, detached states to uncontrollable panic, crying and general emotional turmoil may represent a supersensitive switching process[1455].

The natural conclusion from this is that autistic children should be treated with medications that block the action of endorphins. The best drug for this is naltrexone and this has been effective in decreasing hand flapping and twirling, and increasing eye contact[1257].

The serotonin-endorphin connection If many of the symptoms of autism can be blamed on too much endorphin, but serotonin is also abnormally high, which chemical do we blame? The answer is both. Many different studies have shown a linkage between serotonin and endorphin levels. Increasing serotonin causes an increase in brain endorphin decreased serotonin causes a decrease in brain endorphin[284,1071]. This also explains the observation that antidepressant medications that result in increased serotonin in the synapses also help make morphine more effective in controlling pain[1675]. The ability of clomipramine, a potent serotonin-based antidepressant[p430], to block pain is reversed by blocking endorphins with naloxone, or blocking serotonin with metergoline[1675]. Giving 5-hydroxytryptophan, the immediate precursor of serotonin, also enhances the pain-killing effect of morphine. All of these observations indicate the elevated endorphins in autism may be a direct result of increased brain serotonin.

Serotonin, Endorphins and Addictive Personality

If a high level of endorphins in autism prevented these children from getting a high on motherly love, and doomed them to social withdrawal because they were high on themselves, then the reverse situation, a low level of brain serotonin causing a low level of brain endorphin, would be expected to lead to an increased susceptibility to addictive behaviors — alcoholism, drug and food abuse, i.e., the need to get high on external factors. This is, in fact, what we often see in individuals with Tourette syndrome — an increased frequency

↓ serotonin → ↓ endorphins → addictive personality
- alcoholism
- drug abuse
- overeating

of drug and alcohol abuse. Thus:

While low levels of dopamine and serotonin can play a role in the susceptibility to addiction, independent of their effect on endorphins, the endorphin connection may be a significant contributing factor.

Putting It All Together: Is Severe Autism Due to a Deficiency of Tryptophan Oxygenase?

The most consistent biochemical finding in autism is the presence of significantly increased levels of blood serotonin in almost half of the patients. The demonstration that blood tryptophan levels are also increased suggests that the primary defect may be the opposite of what is postulated in Tourette syndrome, that is, a deficiency of tryptophan oxygenase.

Instead of discarding the increased blood serotonin as unimportant since it is not specific to autism, we should ask if autism and severe mental retardation, the only two brain disorders with such uniquely high blood serotonin levels, are not instead linked in some fundamental way. Since the serotonin levels in severe mental retardation are even higher than those in autism, it is not unlikely that the genetic defect causing the elevated serotonin is also more severe, and this is why they are severely retarded, rather than just autistic. Those with autism may have defective but somewhat higher tryptophan oxygenase levels. Because the defects are enzyme deficiencies, the disorders are autosomal recessive. By contrast, for Tourette syndrome, because the enzyme is presumed to be increased, the trait is autosomal dominant (or semidominant-semirecessive).

This proposal is consistent with the observation that the administration of large amounts of vitamin B_6 (pyridoxine), improves the symptoms in autistic patients[1154,1613]. Vitamin B_6 is a cofactor for many of the enzymes involved in the kynurenine pathway.

Too much or too little serotonin? One question this does not answer is how do we explain the fact that two disorders with similar symptoms, Tourette syndrome and autism, seem to be due on the one hand to too little serotonin and on the other hand to too much serotonin? A reasonable answer is that the basic problem is related more to an abnormal level of serotonin than to whether this level is too high or too low.

Types of autism On the basis of serotonin abnormalities, one could propose the following classification of autism.

1. Too much sertonin
   a. mutations causing too little tryptophan oxygenase
   b. mutations of other genes
2. Too little serotonin
   a. mutations causing too much tryptophan oxygenase — the *Gts* gene
   b. mutations of other genes
3. Normal serotonin — other mutations and environmental factors.

That subgroup of autism due to the *Gts* gene would consist of those cases of TS in which the delay in speech and cognitive defects were so severe that the initial diagnosis was autism and the full picture of TS did not appear until 6 to 9 years of age[p215].

Summary: Autism provides many important lessons for understanding both Tourette syndrome and the role of serotonin in behavior. In many cases of autism serotonin and tryptophan levels are high. One possible explanation is the presence of a defective tryptophan oxygenase gene. This would be consistent with both the autosomal recessive inheritance of autism and the frequent presence of changes in serotonin and tryptophan. Autistic children may also have low serotonin and tryptophan levels and, as discussed previously[p215.], some develop Tourette syndrome as they get older. In both autism and Tourette syndrome the fundamental defect may be a dysregulation of serotonin metabolism.

Chapter 68

The Dance of the Neurotransmitters

We often feel frustrated when a wrong decision has been made and nobody can pin down the blame. The response often is either "A committee made the decision" or "Don't blame me, he made me do it." Trying to pin down the blame for a given psychiatric disorder on a specific neurotransmitter can be equally frustrating. To illustrate this point, I have included this chapter to show the complex manner in which different neurons are interconnected.

Serotonin-Dopamine: Norepinephrine Balance

One of the most important concepts for understanding the brain chemistry of Tourette syndrome and other impulse disorders is that of the reciprocal relationship between serotonin as a behavioral inhibitor and the catecholamines — dopamine and norepinephrine — as behavioral stimulators[222,702,965,1084,1281,1715,1861,2119]. These systems exert this control primarily through their effect on the limbic system.

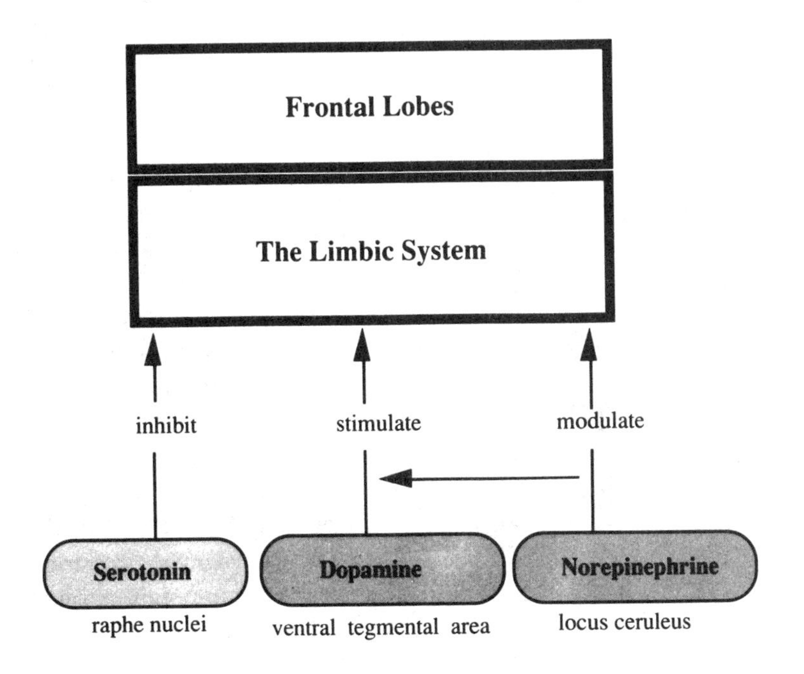

The role of serotonin as an inhibitor is not absolute since in some circumstances it can stimulate neurons[1450,1624].

Interaction of Serotonin (5-HIAA) and Dopamine (HVA)

A common observation of many studies is that whenever spinal fluid levels of 5-HIAA (the breakdown product of serotonin) are decreased, the spinal fluid level of HVA (the breakdown product of dopamine) is also low. This correlation is true in normal individuals as well as those with different psychiatric disorders[57,61,577,690,1302,1411]. The correlation coefficients between these two levels are very high and range from 0.7 to 0.86. This indicates that *any genetic defect that decreases brain serotonin metabolism also tends to decrease dopamine metabolism*, and vise versa.

In one study of this relationship, the activity of brain serotonin was decreased by electrical lesions of both the medial and dorsal raphe nucleus. As shown, this resulted in a significant decrease in dopamine metabolism as indicated by determining HVA levels in the neostriatum[1685].

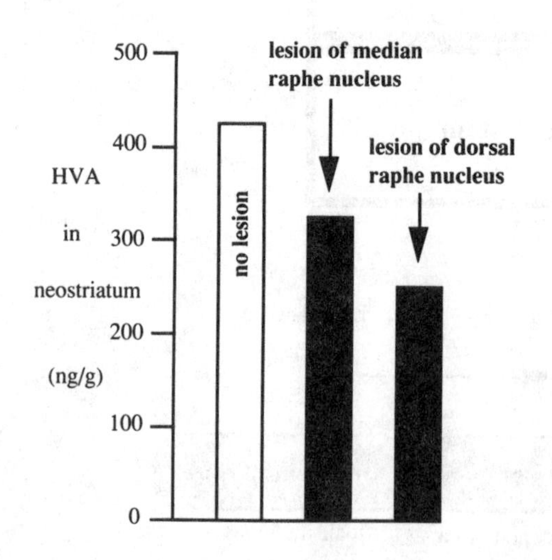

Lesions of the dorsal raphe nucleus caused a greater drop in HVA than lesions of the median raphe nucleus. This relationship was further examined by studying the effect of giving L-tryptophan to increase serotonin synthesis, or a drug which blocked serotonin synthesis[1685].

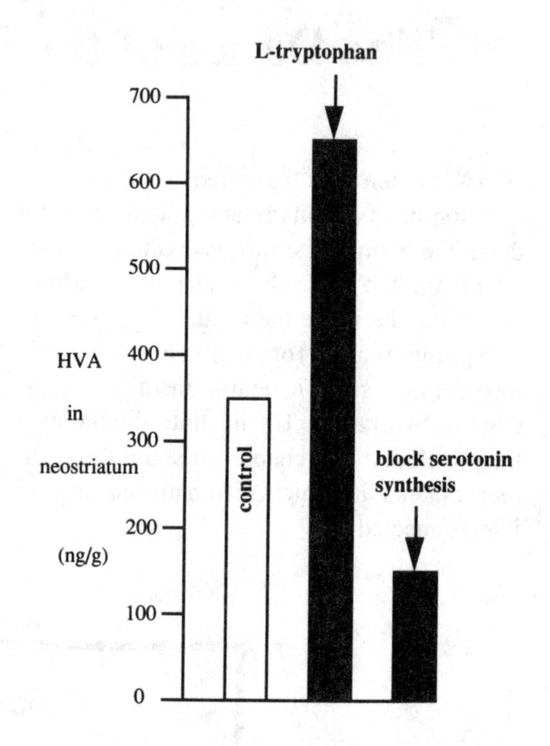

Increasing serotonin with L-tryptophan resulted in a striking parallel increase of dopamine metabolism as measured by HVA in the neostriatum. Blocking serotonin synthesis caused a striking drop in dopamine metabolism and thus HVA production.

This intimate interaction between serotonin and dopamine is an important concept for the following reason. Studies of spinal fluid in Tourette syndrome have shown that both 5-HIAA and HVA are low. The linkage between the two means that a primary defect in serotonin metabolism could explain the low HVA without needing to postulate that there is also a genetic defect in dopamine metabolism. An important question is: What happens first, the decrease in dopamine or the decrease in serotonin? In a careful study of the the high correlation between spinal fluid 5-HIAA and HVA (and DOPAC), in controls and patients with depression and schizophrenia, Faull and co-workers[577] found that *there were no differences in the average values of these compounds between the three groups, but depressed patients showed greater fluctuations in serotonin activity, given a constant change in dopamine activity*. This is important because it indicates that there may be abnormalities in serotonin metabolism even if absolute amounts of serotonin breakdown products are normal. They suggested that *an altered coupling between serotonin and dopamine metabolism may be a problem in manic-depressive individuals.*

The studies shown in the above figures clearly indicate that a defect in brain serotonin can be the primary event resulting in a secondary decrease in dopamine metabolism. In addition to the effect of lesions of the raphe nuclei on dopamine in the neostriatum, there may also be a comparable effect on dopamine in the prefrontal cortex. If the primary defect in TS is defective serotonin production, this would explain the frontal lobe syndromes that are also present. This is shown by the following modification of a previous figure[p426]:

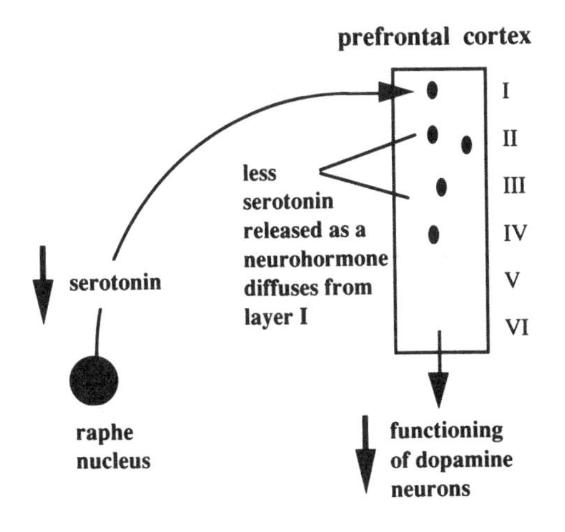

Further complicating things is the fact that a decrease in serotonin activity disinhibits dopamine neurons in the substantia nigra and tentral tegmental nucleus [p446,137,300,301,1971a]:

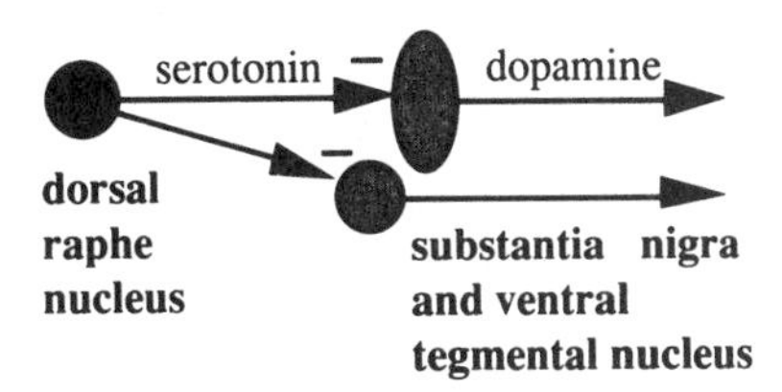

Thus, the hypersensitivity of some dopamine receptors can also be due to decreased serotonin activity. This plays a major role in causing the motor and vocal tics.

All of this is consistent with a defect in serotonin in all TS patients, the abnormal functioning of the frontal lobes, the decrease in spinal levels of both 5-HIAA and HVA, and the low blood serotonin and tryptophan in some but not all patients.

Norepinephrine — Serotonin

An increase in the activity of some norepinephrine neurons stimulates serotonin neurons[108,656,1168,1437,1555,1556,1672]:

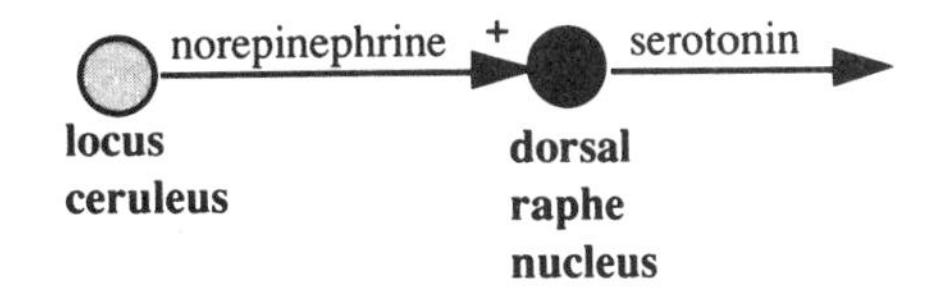

More important for understanding TS is the reverse link. Serotonin inhibits the glutamate-induced excitation of the locus ceruleus and too little serotonin would disinhibit these norepinephrine neurons. This was shown by studies in the rat where depletion of serotonin resulted in increased activity of the norepinephrine neurons of the locus ceruleus. More importantly, this was associated with a decrease in activity of the neurons of the frontal and parietal lobes[594a]. This is shown as follows:

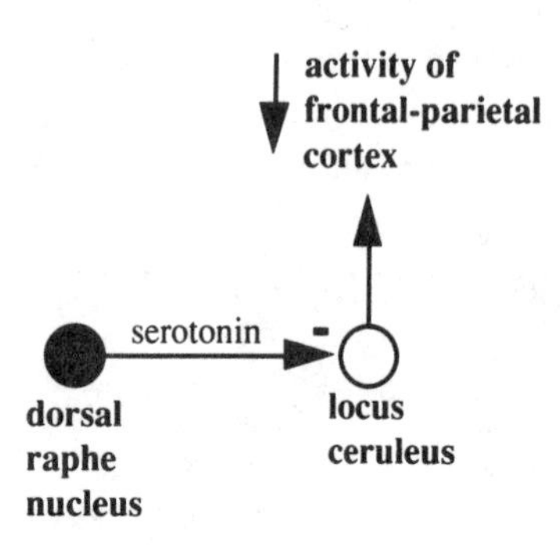

This provides yet another explanation for how low serotonin levels can cause overactivity of norepinephrine in the locus ceruleus and this can cause underfunctioning of the frontal lobes. This helps to understand why clonidine, which suppresses this norepinephirine overactivity is useful in the treatment of both ADHD and TS.

Serotonin and Manic-Depressive Disorder

As described in Chapter 58, there is evidence that norepinephrine plays a role in both depression and mania, with too little norepinephrine in depression and too much in mania. This does not exclude a role of serotonin in mood swings. For example, if the neurons stimulating and inhibiting norepinephrine cells are both inhibited by serotonin neurons, the output is evenly modulated[1273a].

output

NE

serotonin

However, if the serotonin neurons are defective, this moldulation is defective and wide swings in output can occur.

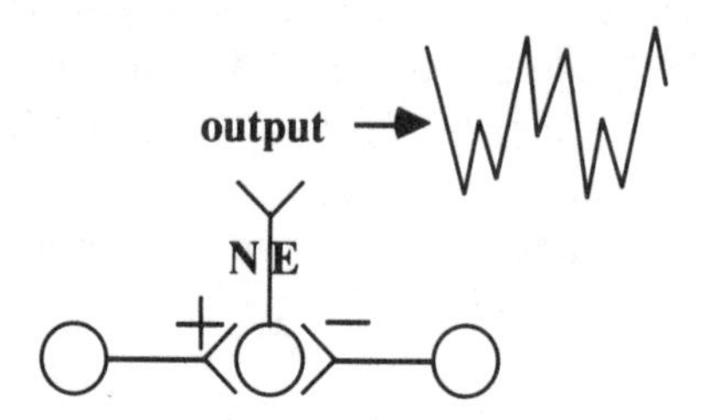

This provides a straight forward explanation for the mood swings that occur in serotonin deficiency disorders such as Tourette syndrome and is consistent with the norepinephrine theory of mood disorders.

These are just a few examples. Since there are billions of neurons in the brain and trillions of synapses per cubic centimeter, the potential for complex interactions is enormous.

The total effect is somewhat like a water bed: if you push down on one part another part pushes up. Patients often feel that if something is missing or too low in the brain, all one has to do is give that substance and everything will be cured. The blood brain barrier interferes with most such plans. Even medications that effectively push the chemical imbalance toward normal run into the problem that the brain constantly tries to maintain sameness, and beneficial changes in one area of the brain can cause detrimental changes in another.

Summary: There is a close linkage between serotonin and dopamine metabolism in the brain. When serotonin turnover decreases the dopamine turnover also decreases, and vice versa. There is also a close interaction between decreased serotonin activity and increased activity of norepinephrine in the locus ceruleus. Thus, a defect in serotonin production in Tourette syndrome could explain why dopamine and norepinephrine metabolism are also affected.

Chapter 69

The Semidominant, Semirecessive Inheritance of Tourette Syndrome

As this book was being written new insights into Tourette syndrome were continuing to occur. As stated in the Introduction, to allow the reader to participate in the interaction between ideas and information, I have attempted to include the more significant new findings so that the sequence in which they occurred was the same as the sequence in which the book was written. This is the first chapter handled in that fashion. As Chapter 68, The Dance of the Neurotransmitters, was being completed, I received a telephone call from Judy Chin, president of the Tourette Syndrome Chapter in San Francisco. Brenda and I had met Judy a few months before when we all participated in a TV show, *People are Talking*, that was devoted to Tourette syndrome. Judy told me of some intriguing work on retinal changes in TS being done by Dr. Jay Enoch, Dean of the School of Optometry in Berkeley. We invited him to spend the day telling us about his studies.

Visual field defects When people have certain types of brain tumors, especially those involving the pituitary, or when they have glaucoma, the function of different parts of the retina may be lost. The patients usually don't notice this since the brain keeps the vision focused on the fovea where the nerves of the retina are the most dense. However, the patient can be asked to keep their vision focused on a point in front of them as an object is slowly moved into their visual field. They are asked to indicate when they first see the object in their peripheral field of vision. When this is done from all different directions, with lights of different intensity, the result is a map of that person's visual fields. In conjuction with other studies[544,545] when Enoch performed this test on TS patients he found their visual fields were not normal [542,543]. Using a small, dimly lit target, the edge of the normal visual field is close to the fovea and forms almost a perfect circle.

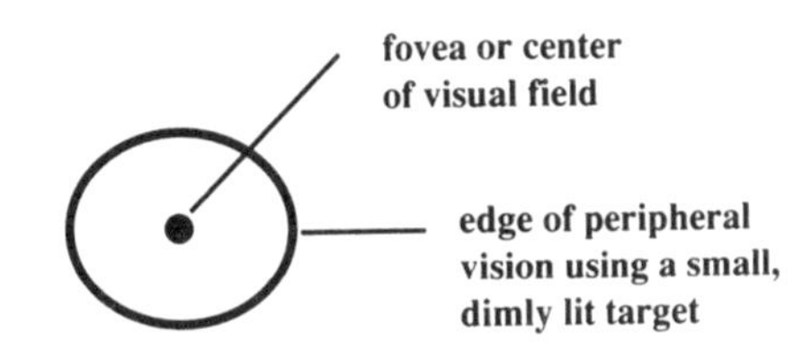

The TS patients had one or more of the following variations. The most frequent was a *step* in the visual fields. This could be facing in one direction,

or the other,

or in both directions,

or facing downward either way.

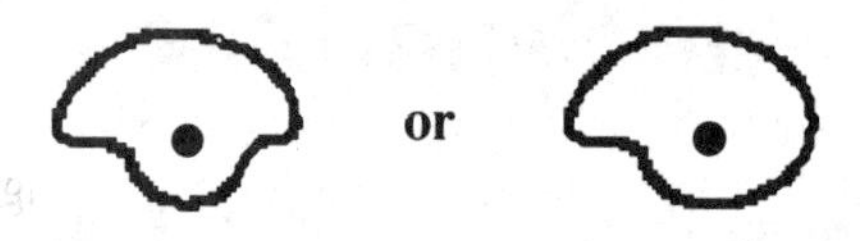

To understand the *steps* it helps to review how the nerves pass from the optic nerve to the fovea (with the visual field superimposed) as follows:

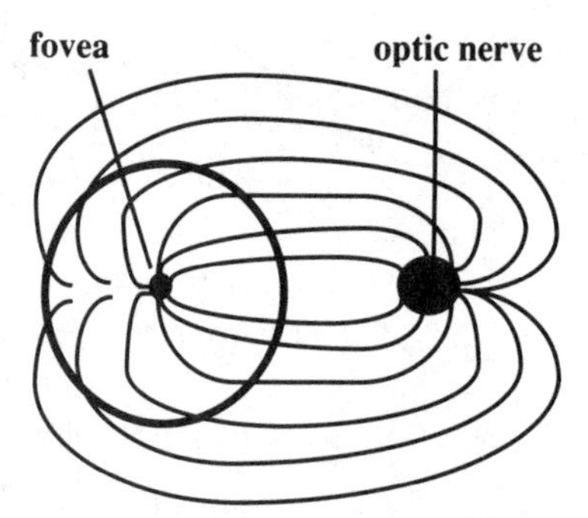

A defect in the nerves occurring at the level of the optic nerve can destroy a selected set of fibers. Depending upon which nerves are affected, different defects in the visual fields are produced. The following shows that a defect (represented by the black wedge) in the outer layer of fibers can produce a step facing upward and away from the optic nerve.

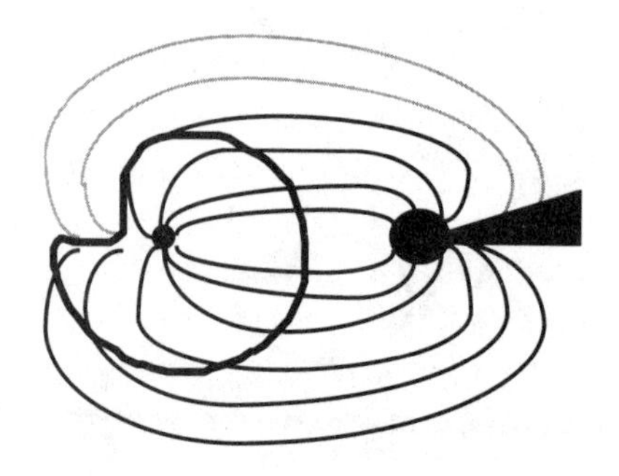

The cause of the defect is not known.

Fatigue Another finding in the TS patients was a phenomena called fatigue. The visual fields are done by progressively going counterclockwise around the fovea. Normally when the examiner gets back to the starting point the edge of the visual field is the same distance from the fovea as where he started. However, in some TS patients a fatigue-like effect is seen "where the island of vision is fading into a sea of darkness." This gives the following result:

To obtain this fine mapping it was critical that the fields be mapped manually and not with the newer computerized machines.

Visual fields as a marker of the *Gts* gene All of the two dozen TS patients Enoch had tested showed one or more of these defects. By contrast, such defects were unusual in normal controls. This suggested these defects were associated with the *Gts* gene. The findings were quite interesting when families were studied. The following pedigree illustrates the point. The propositus was an 18 year old male with Tourette syndrome. His older brother had some motor tics that did not require treatment. Although no one else in the family had tics, the father and the father's mother had some obsessive-compulsive behaviors.

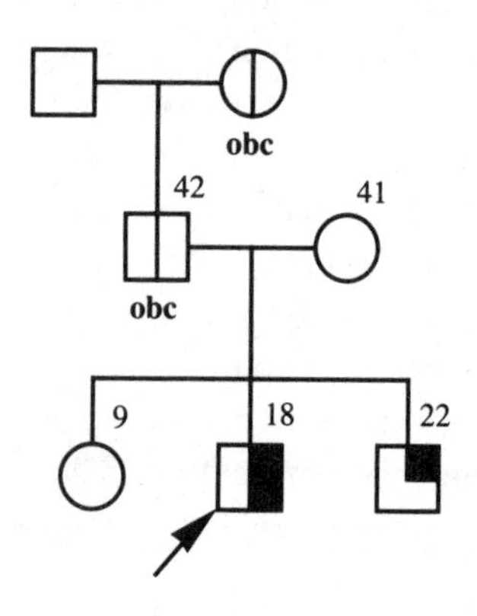

This appeared to be a standard TS family with an autosomal dominant *Gts* gene coming from the father but being expressed only as obsessive-compulsive behavior. However, when the visual fields were examined, a completely different pattern arose with both parents being affected:

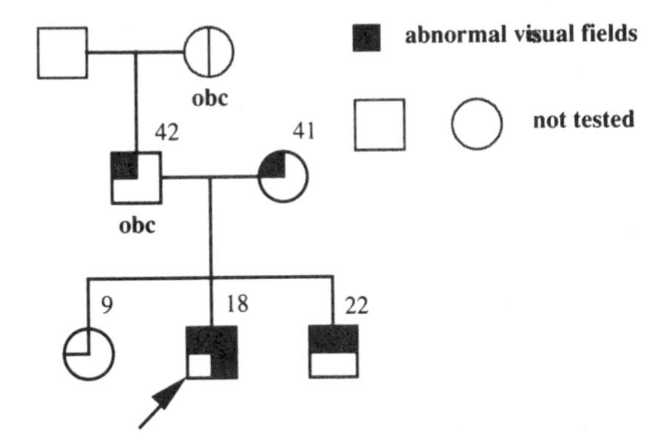

Some families had only a single parent affected.

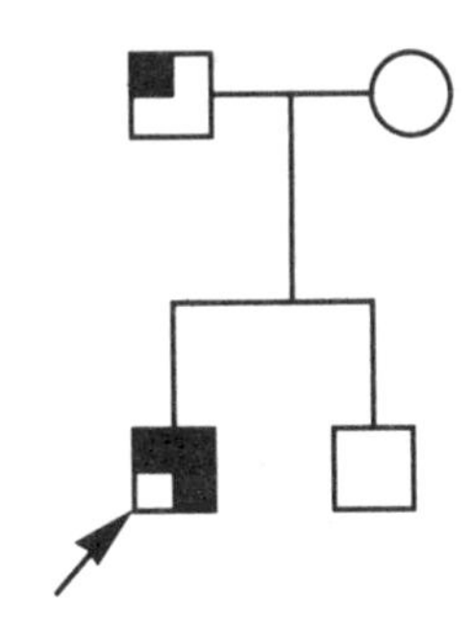

In seven of nine families, both parents had the visual field abnormalities. This is consistent with the earlier chapter on homozygosity in TS[p271], and suggests that even more TS patients are homozygous than we suspected.

Tics and Associated Behaviors in Both Parents and Their Families

When we took pedigrees on the first 250 TS patients[376], questions were primarily asked about the presence of tics in other members of the family. However, as our experience with the broad spectrum of behaviors in TS began to grow questions were asked about these behaviors, as well as tics, in other members of the family. The figure on the **next page** illustrates the results of detailed pedigrees taken more recently on 202 TS families[377].

Of these, the patients were adopted in 14 and no family history was available, and in 18, two probands were present in one family. This left 170 families. The results concerning tics are shown on the left and the results concerning four associated behaviors—obsessive-compulsive[p113], panic attacks[p175], severe alcohol or drug abuse[p225], or ADHD[p99], are shown on the right. In 4.7 percent of the families, *both* parents had motor or vocal tics, and in 8.2 percent, tics were present in *both* the mother's and the father's families. For associated behaviors, in 19.4 percent of the families *both* parents had one or more of the associated behaviors, and in 34.7 percent of the families, the associated behaviors were present in *both* the mother's and the father's families. *In 45.3 percent of the families tics or associated behaviors were present in both the mother's and the father's families.*

To get a feel for whether the frequency of associated behaviors was more than would be expected, 35 pedigrees were taken on families of mothers coming to the clinic for prenatal diagnosis, asking the same breadth of questions as were asked for the TS families. The results for the controls are also shown on the next page.

Here, in only 5.7 percent of the families were the associated behaviors present in both the mother's and the father's families. This was significantly different from the 34.7 percent obtained for the TS families[377].

Like the visual field studies, these observations are consistent with the majority of TS patients having received a *Gts* gene from both parents.

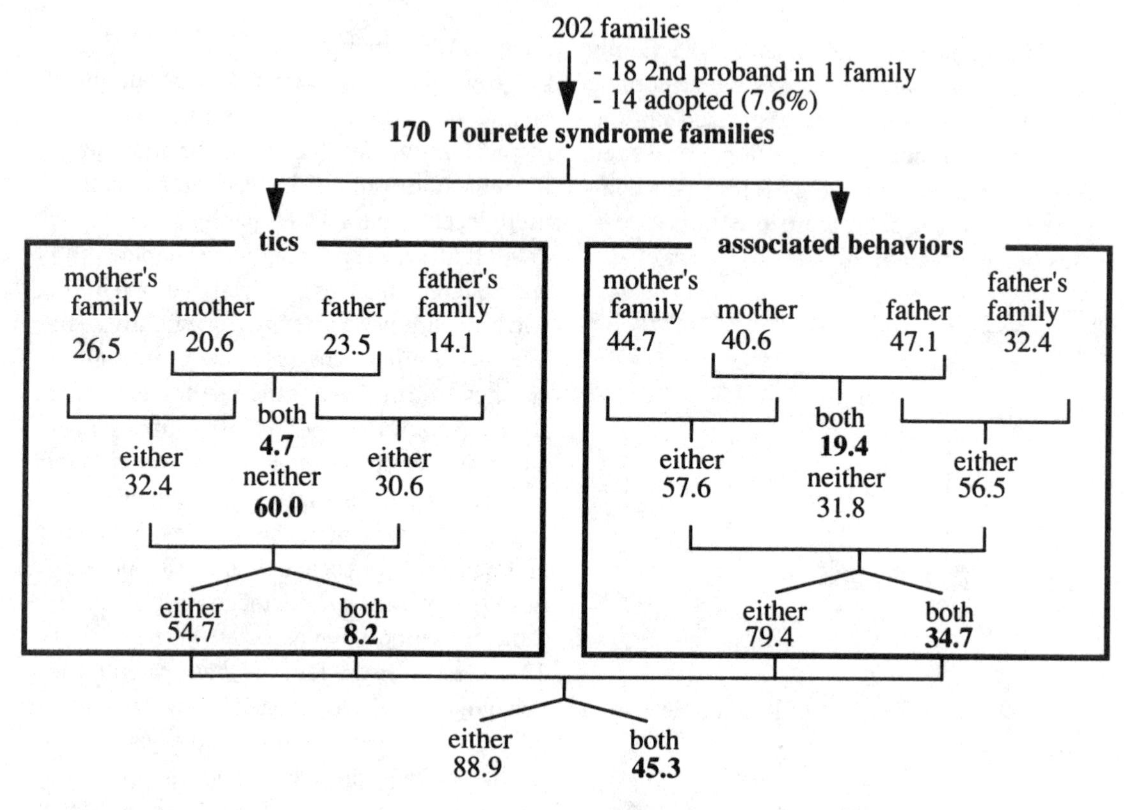

mother with tics or associated behaviors - 48.2
father with tics or associated behaviors - 60.6
both with tics or associated behaviors - **29.4**
neither with tics or associated behaviors - 20.6

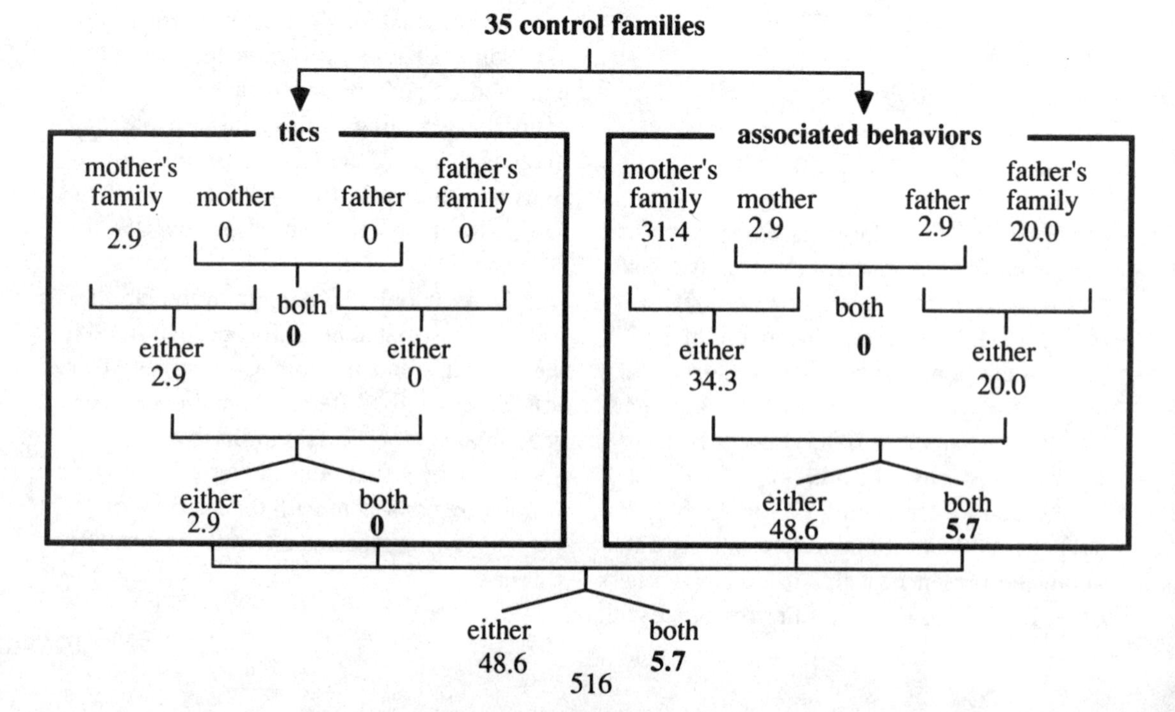

Semidominant—Semirecessive Inheritance in Tourette Syndrome

In pure recessive inheritance[P42], the affected individual has two copies of the abnormal gene, having received one from each parent. The parents who each carry a single copy of the gene have no symptoms. The abnormal gene is recessive to the other normal gene.

In purely dominant inheritance[P41], the affected individual has only one abnormal gene, having received it from one or the other parent, but not both. The symptoms in the parent carrying the abnormal gene are just as severe as in the offspring. Furthermore, for those rare individuals who have receieved two copies of a dominant gene (where both parents have the disease) the symptoms are no worse than for individuals who have received only one abnormal gene.

Semidominant, semirecessive inheritance has characteristics of both. It is like a recessive disorder in that most affected individuals have two abnormal genes. It is like a dominant disorder in that those who carry a single dose of the gene may have some symptoms.

Let us examine the results from the above families to see if they favor TS being inherited as a semidominant, semirecessive trait.

1. If it is not completely recessive, we would expect that the *parents* would have some symptoms That is what was observed in that 20.6 percent of the mothers had tics, 40.6 percent had associated behaviors, and 48.2 percent had tics or associated behaviors. In addition, 23.5 percent of the fathers had tics, 47.1 percent had associated behaviors, and 60.6 percent had tics or associated behaviors. Since the parents have symptoms, it is clear the inheritance is not completely recessive.

2. If it is not completely recessive, in addition to the parents having some symptoms we would expect that the *relatives of the parents* would have some symptoms. This was also observed: in 26.5 percent of the families, someone in the mother's family had tics, and in 14.1 percent, someone in the father's family had tics. Also in 44.7 percent someone in the mother's family had associated behaviors, and in 32.4 percent, someone in the father's family had associated behaviors.

3. If most TS cases are homozygous for the *Gts/Gts* gene, we would expect that despite the fact that the gene is partly recessive many families would show symptoms in *both parents*. That is what was observed. In 4.7 percent of the families, both parents had tics, in 19.4 percent both had associated behaviors, and in 29.4 percent both parents had tics or associated behaviors.

4. If most cases are homozygous we would expect that symptoms would be present either in *both parents or their families*. In 34.7 percent of the families, associated behaviors were present on both the mother's and the father's sides, compared to only 5.7 percent of the control families. Either tics or associated behaviors were present on both sides in 45.3 percent of the TS families compared to 5.7 percent of the control families.

5. If it is not completely dominant, we would expect that *not all the parents would have symptoms*. That is what we observed in that 79.4 percent of the mothers had no tics, 59.4 percent had none of the four associated behaviors, and 51.8 percent had neither tics nor associated behaviors. In addition, 76.5 percent of the fathers had no tics, 52.9 percent had none of the four associated behaviors, and 39.4 percent had neither tics nor associated behaviors. Also, in 60.0 percent of the families neither parent had tics, in 31.8 percent neither had associated behaviors and 20.6 percent neither had tics nor associated behaviors.

Summary of the Evidence

The total evidence in favor of TS being inherited as a semidominant, semirecessive trait can be summarized as follows:

1. Summary of the pedigrees (see above).
2. Individual case reports[p271].
3. The patient is usually more severely affected than either parent suggesting the presence of two *Gts* genes. This was the case in over 96 percent of our families. In those cases where a parent was as severely affected as the proband, there was family history evidence the parent also carried two *Gts* genes.
4. Visual field studies suggesting a defect in both parents.
5. Of the prior studies of TS by segregation analysis[p48], in two[376, 485] out of three[1473] the best solution was semidominant rather than a purely dominant inheritance, that is, the penetrance[p49] was almost 100 percent in individuals who carried a double dose of the gene, but much less in those who carried a single dose of the *Gts* gene.
6. Serotonin and tryptophan studies[p457, p463, p649] show that the average levels of serotonin and tryptophan are low in both parents.

Thus, the inheritance of Tourette syndrome appears to occupy the middle ground between recessive and dominant—i.e., semidominant, semirecessive inheritance. There are several important implications of this[367, 369, 377].

Why Homozygosity in TS Is So Important

***Gts* gene frequency** If both parents of most TS patients carry the *Gts* gene and most patients carry a double dose, then the frequency of this gene is much higher than previously suspected. In studies of the genetics of genes in populations, p = the frequency of the normal gene and q = the frequency of the abnormal gene. In recessive disorders, q is the square root of the frequency of the disease, and the frequency of individuals in the general population who carry the gene is 2pq. The following table shows the implications of this:

Frequency of a recessive disorder	q	% of the general population who carry the gene
1/50	.14	24
1/100	.10	18
1/200	.07	13
1/500	.04	8
1/1000	.03	5

Recall that the best estimate of the percentage of individuals carrying the *Gts* gene, based on the autosomal dominant model, was 1.2 percent[p49]. However, if the majority of TS patients are homozygous and the frequency of moderate to severe TS is 1 in 1,000, then 5 percent of the population would be carrying the *Gts* gene, four times the previous estimate. The table shows the percentage of individuals carrying the gene if the frequency of TS is even higher. If the frequency of TS in male children is closer to 1 in 100, as our studies in Los Angeles suggest[p617], and 1/2 are homozygotes (i.e., 1 in 200), then 13 percent of the general population would carry the *Gts* gene. If all were homozygotes the carrier frequency would be 18 percent.

The penetrance of the *Gts* gene Penetrance refers to the percentage of time that a person carrying a gene shows some signs of it. Penetrances as high as 100% for males and 71% for females have been suggested for TS[1473, 1475]. However, if most cases of TS are homozygous and the frequency of the *Gts* gene is many times higher than we previously thought, then the penetrance of the *Gts* gene for chronic motor tics must be much lower than this. Our experience with over 1,400 families suggests the penetrance for tics or associated behaviors in individuals who carry a single *Gts* gene is around 50 percent[p641]. That is, about half of the individuals carrying the *Gts* gene will have some symptoms.

The expression of TS Expression refers to the different degrees of severity that a genetic disorder may show. Thus, a person with a few motor tics that lasted for only a couple of years would be showing minimal expression of TS, while a person with motor and vocal tics, ADHD, obsessive-compulsive behavior, phobias and panic attacks would have a severe expression of TS. The usual explanation of the variable expression of TS is that other genes are playing a role[p51]. This could still be the case, but a major part of determining whether the TS is severe or mild may lie with the *Gts* gene itself — namely, whether it is present in a single or double dose. Thus, individuals homozygous for a *Gts* gene *(Gts/Gts)* would have moderate to severe symptoms while individuals who are heterozygous *(Gts/gts)* would have no symptoms or mild symptoms. This is summarized as follows:

Genotype	Phenotype	Mode of Inheritance
Gts/gts heterozygous	no symptoms chronic motor tics chronic vocal tics mild TS moderate TS	autosomal dominant
Gts/Gts homozygote	moderate TS severe TS	autosomal recessive

Associated behaviors As striking as these conclusions are, even more remarkable is the implication for associated behaviors. If only 1 in 83 persons carried a *Gts* gene then it would be difficult to assume that it was responsible for very many of the associated behaviors occurring in the general population. To be more specific, if 5 percent of the population had alcoholism, it would be impossible to propose that a *Gts* gene was the cause of it in more than one-fourth of these individuals. However, if 18 percent of the population carried the *Gts* gene, then a significant proportion of alcoholism could be due to a *Gts* gene.

Summary: As information has accumulated it has become increasing apparent that Tourette syndrome is inherited as a semidominant, semirecessive trait. That is, in most cases, both parents carry the Gts gene, the majority of TS patients have a double dose of that gene, and many relatives who carry a single gene have some symptoms of the TS spectrum of behaviors. A significant percent of the population may carry the Gts gene. This raises the intriguing possibility that a percentage of individuals in the general population with ADHD, learning disabilities, obsessive-compulsive behaviors, conduct disorder, phobias, depression, addictive behaviors including alcoholism and compulsive eating, and other related disorders may have these problems because they are carrying a Gts gene.

Chapter 70
Pleasure and Addiction

The concept of pleasure centers in the brain is central to understanding many behaviors. In 1953 Drs. James Olds and Peter Milner[1432] placed wires into the brains of animals so they could stimulate different brain regions to study arousal mechanisms. They noticed that the animals kept returning to the part of the cage where they had been when they received the stimulation. They seemed to "like" this area. A modification of the apparatus allowed the animals to press a level and electrically stimulate themselves (self-stimulation). Comparisons of the frequency of pushing the lever with the site in the brain where the wire was placed allowed the detection of "pleasure centers" in different parts of the brain. This report was a milestone in the history of brain-behavior research.

Prior to this time the behavioral psychologists, or behaviorists, such as John B. Watson[2031] and B.F. Skinner[1809], held sway. John Locke, father of the Age of Enlightenment and Reason in England in the 17th century, proposed the idea that the human mind was nothing but a "tabula rasa" or blank slate that was molded only by the environment in which it was placed. The "behaviorism" school expanded on this idea by using behavior shaping or "operant conditioning" to elicit sometimes very complex behaviors out of otherwise dumb animals. Each thing the animal did that was in the direction of the desired behavior was rewarded, and each thing that was counter to the desired behavior was punished. As a result, birds and other animals could be trained to perform complex human-like tasks such as play a tune on the piano. Waston boasted:

> "Give me a dozen healthy infants, well-formed, and my own specified world to bring them up in, and I'll guarantee to take anyone at random and train him to become any type of specialist I might select — a doctor, lawyer, artist, merchant-chief, and yes, even into a beggar-man and thief, regardless of his talents, penchants, tendencies, abilities, vocations and race of his ancestors."

Behavior was explained in terms of stimuli, responses and rewards. The theories were barren of any reference to the individual's plans, genetic make-up, experiences, or beliefs. Robert Isaacson[953] in his book *The Limbic System,* said the following about the work of Olds and Milner:

> "The discovery of brain regions that, when electrically stimulated, seemed to produce rewarding effects began an era of greater freedom in the scientific description of behavior. It did so because the behavior of the animals seemed most easily explained on the basis that the animals would perform responses of several different kinds to obtain electrical stimulation in certain brain regions. The goal of their efforts was most easily described in terms of obtaining a

pleasurable experience. Thus, the experiences of the organism had to become a factor in describing or explaining behavior. *Pleasure* and *pain* became accepted as legitimate terms, once again, for scientists trying to discover the neural bases of behavior. In short, the results of Olds and Milner helped break down the strong theoretical fortress of artificial concepts developed by learning theorists. Olds and Milner's results justified explanations of experimental data in hedonistic [pleasure seeking] terms."

Correlations between the pleasure centers of the brain and the region rich in specific neurotransmitters helped to place pleasure on a chemical as well as on an anatomical basis[1433,1434].

Electrical Stimulation of the Pleasure Centers

There are a number of different terms used in the studies of the pleasure and pain centers in the brain. These are the following:

Increased self-stimulation	Decreased self-stimulation
pleasure	pain
approach	avoidance or escape
reward	punishment
positive reinforcement	negative reinforcement
addictive	aversive

The variety of these terms indicates the breadth of the possible psychological interpretations. As a generalization, the regions of the brain that elicit increased self-stimulation are the hypothalamus and the median forebrain bundle (carrying the catecholamine neurones), while the regions that resulted in decreased self-stimulation were the thalamic structures (rich in pain fibers)[689,1433,1434]. Predictably, stimulation of these thalamic regions could cause rage reactions. Stimulation of the ascending catecholamine system in man has been reported to be pleasant, even euphoric[867].

Pleasure is more important than food When electrodes were placed in the median forebrain bundle, and the animal given the choice of pressing one lever to receive self-stimulation, or a different level to receive food, they continued to self-stimulate themselves to the point of starving to death[1656]. Analogies to this type of behavior are found in human alcoholics whose preference for alcohol over food leads to cirrhosis and death, and in cocaine addicts who prefer cocaine to anything else in life.

Dopamine and the reward system The initial studies implicated both dopamine and norepinephrine in self-stimulation. However, destruction of the locus ceruleus, the center for norepinephrine neurons[p409], had no effect on self-stimulation, while destruction of the ventral tegmental area and substantia nigra, the centers for dopamine neurons[p364], abolished self- stimulation[597,1656]. Electrodes placed in these areas produced high rates of self-stimulation. Additional evidence for the importance of dopamine was provided by the effect of drugs which either enhanced or blocked dopamine. For example, d-amphetamine, which increases the release of dopamine, enhanced self-stimulation, while Haldol, which blocks dopamine receptors, also blocked self-stimulation [427,596,597,1656,2092].

Serotonin and the reward system Just as dopamine and serotonin have opposing effects on aggression and sexual behavior[p378] they also have opposing effects on self-stimulation. The injection of serotonin in the brain fluids of rats results in a suppression of self-stimulating behavior[1529,2091]. By contrast, a decrease in brain serotonin enhances self-stimulation, a behavior common in TS patients.

The frontal lobes are part of the reward pathway As shown in previous chapters, dopamine neurons pass to the frontal lobes. It is thus not surprising that self-stimulation experiments have shown that parts of the frontal lobes are also involved in the reward system[1655,1656]. These studies further implicate dopamine in the reward system since dopamine fibers pass to only parts of the frontal lobe while norepinephrine fibers pass to all areas. Self-stimulation is produced only when the electrodes are placed in the dopamine-rich areas of the frontal lobes.

Stimulation of the reward system disrupts learning The ectorhinal cortex is one of the regions of the frontal lobe that is rich in dopamine. It also has connections with the hippocampus, the part of the brain involved in memory[p333]. This raises the question of whether stimulation of the reward system would augment or interfere with learning. When various parts of the reward pathways were stimulated it resulted in significant interference in the ability of an animal to remember[1656]. This might seen at odds with the idea of enhancing learning by giving rewards. However, if the animal was allowed to self-stimulate after good performance on a learning task, learning was enhanced.

Chemical Stimulation of the Pleasure Centers

Understanding the pleasure centers of the brain has contributed greatly to an understanding of why people become addicted. The feature that first caught the attention of Olds and Milner was that when the electrodes were in a pleasure center, the rats would return to the part of the cage where they received the stimulus. This is called *place preference*. When animals are given injections of drugs which are addicting to man, they show the same place preference[1370,1734,1831]. When they can give the injections to themselves they often prefer the drugs to food, to the point of starving themselves to death[1731,1831].

Dopamine Release — A Common Denominator of Addicting Drugs

Many lines of evidence have indicated that addicting drugs have in common the ability to cause the sudden release of dopamine. This is illustrated by one experiment in which tubes were placed into rats so that the neurotransmitters released from various parts of the brain could be tested. The primary neurotransmitter released was dopamine and the peak of release was within 30 to 60 minutes. The results with sampling the nucleus accumbens, after injecting several different drugs, are **shown on the next page**[316]:

All of the addicting drugs humans use most — amphetamines, cocaine, nicotine, alcohol, and morphine— caused a 150 to 1,000 percent increase in dopamine in the nucleus accumbens. As described before[p492], morphine binds to the mu opioid receptors. Some other compounds that are not addictive, and actually aversive, bind to kappa opioid receptors. When these drugs were used there was an inhibition of dopamine release.

When animals are given medications that block dopamine receptors, the ability of amphetamine and cocaine to cause self-stimulation is inhibited and animals self-administer more drug to compensate for this[2092]. By contrast, drugs which block norepinephrine do not block the self-administration of amphetamine or cocaine. Both naloxone and Haldol-like medications block the self-stimulation produced by morphine and heroin[204].

Cocaine and the prefrontal lobe. Further evidence that the dopamine neurons of the frontal lobes are part of the reward system comes from the following observation. When rats were allowed to self-administer cocaine into various parts of their brain, there was a 500 percent increase in self-stimulation over baseline, when the needles were in the medial prefrontal cortex[741]. This was significantly inhibited if a Haldol-like drug was injected

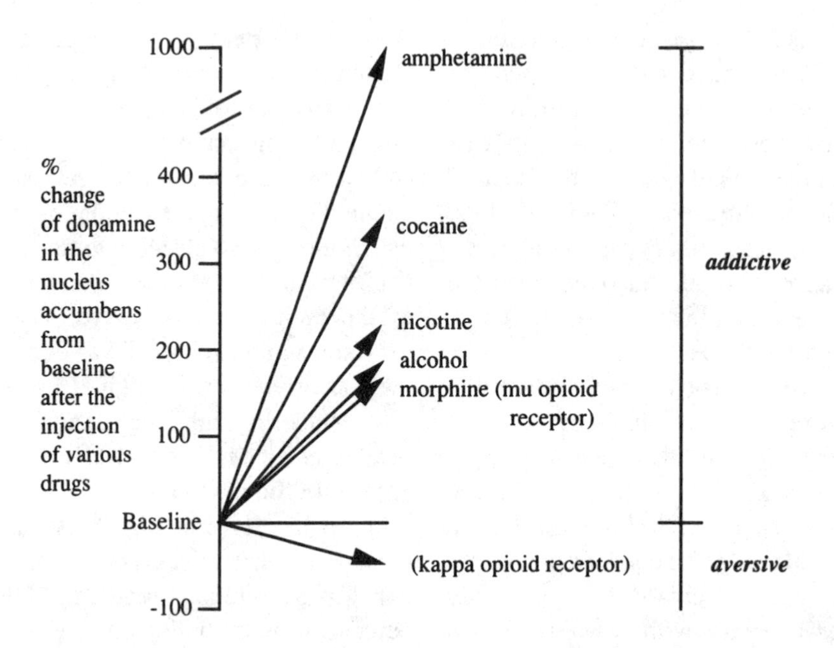

along with the cocaine. This indicated that the mesocortical dopamine neurons are part of the reward pathway.

Food and the reward system It doesn't take a gourmet to realize that food can give pleasure. The rewarding nature of food can also be blocked with dopamine-blocking agents[2093].

Love, sex and others There are many other pleasure-giving activities that can pass beyond normal use and enter into the realm of compulsive addictions. Some people are so preoccupied with sex that the conquest rather than the person becomes the focal point.

Addictive Behaviors in ADHD and Depression

Among cocaine addicts there is a significant increase in frequency of adult attention deficit hyperactivity disorder and mood disorders[678-680,2050]. Approximately half of cocaine addicts have a history of mood disorder, especially manic-depressive disorder[680,2056,2094]. This has led to the suggestion that drug addiction is a form of self-treatment with cocaine used for the treatment of depression, mania and hyperactivity and heroin for the treatment of rage and aggression[1040].

When the effect of cocaine on the brain is compared to the effect of the amphetamines and Ritalin, it is easy to see why cocaine might be used to self-medicate symptoms of hyperactivity, mania and depression. This comparison is shown as follows[678]:

Stimulant	Catecholamine reuptake blockade	Dopamine release	Serotonin release or reuptake blockade
Amphetamines	+++	++	++
Ritalin	+++	++	?
Cocaine	+++	++	++

Addictive Behaviors in TS

A common thread between the addictions, obsessive-compulsive behaviors and Tourette syndrome is that they all may be associated with low serotonin levels and many individuals who carry a *Gts* gene also have obsessive-compulsive disorders and various addictive behaviors. These addictive behaviors, such as alcoholism, drug abuse, overeating and undereating may themselves be viewed as obsessive-compulsive behaviors. Compulsive eating leading to obesity is one form of addiction that is common among some relatives of TS patients. Examples have been given previously[p236]. The following cases further illustrate how the *Gts* gene can be intimately associated with various addictions.

Obesity, alcohol and drug abuse The following family illustrates a remarkable clustering of individuals with TS, chronic motor tics, and addictive behaviors. The propositus (**see pedigree below**) was an 11 year old boy (arrow). Both parents recognized that each side of the family carried the gene. The mother said, "My side tends to be overdrinkers, his side overeats." There were three different sets of parents in this family where a presumptive *Gts* gene was present in both parents.

Sam The story of Sam is another classic tale of "driven," compulsive, substance abuse.

> Sam was adopted at birth and little is known of his parents. At 18 months he began to have frequent night terrors that continued until he was 15 years old. There were no further problems until first grade when he began to have panic attacks. A year later he was using swear words excessively and had a very short temper. When angry he would go to his room and rip up things and punch holes in the wall. He developed a constant eye tic and compulsive spitting.
>
> By 12 years of age he began to use alcohol and pot excessively. He was a chronic truant from school Within two years he was "taking everything but

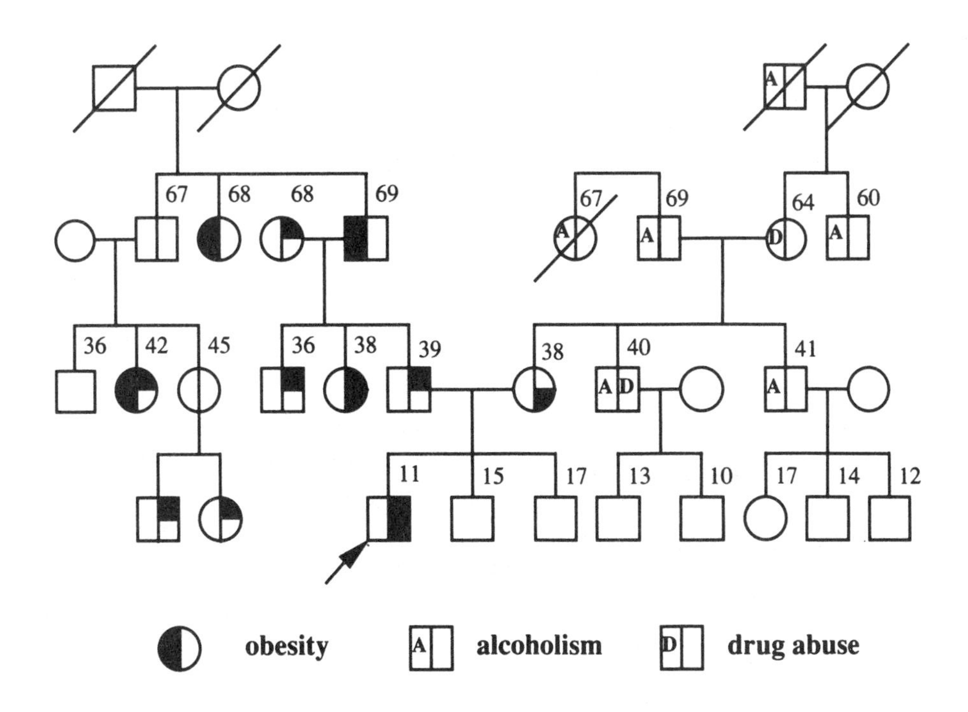

heroin" and drinking alcohol excessively. Sniffing tics also started. By age 15 he was "shooting cocaine." In the next four years he was in and out of five different psychiatric institutions for drug abuse. Each time he was released he would be sober for about two months then back on drugs. After the fourth hospitalization, at age 18, his adoptive parents couldn't tolerate his abusive behavior and he left the house to live on the streets. He next surfaced in the county jail, arrested for grand theft auto, attempting to pay for his drugs. After he was released from jail he returned to an in-house drug treatment program.

At all five psychiatric hospitals Sam was treated as simply an "acting out," severely emotionally disturbed teenager with a drug abuse problem. However, there are clues that he had a genetic impulse disorder — Tourette syndrome. For example, he had many night terrors since he was 18 months of age, an early clue to a brain serotonin deficiency. Panic attacks and severe school phobias were the next symptoms, followed by compulsive swearing, short temper, spitting and sniffing tics, and an eye tic. Only after these started did the drug and alcohol abuse begin.

Like earlier examples, this was also a minature adoption study and socioeconomic factors were not involved. His adoptive parents were financially well off and there were no identifiable environmental factors contributing to Sam's behavior.

The Ups and Downs of Addiction

The one feature that all addicting substances and behaviors have in common is their ability to cause a release of neurotransmitters that stimulate the pleasure centers of the brain. The pleasure is produced by the peak of the release. However, the problem is that there is then a rebound drop to below normal levels of neurotransmitters. This results in a temporary craving to return the levels to normal, if not back to the peak of ectasy. This is illustrated as follows:

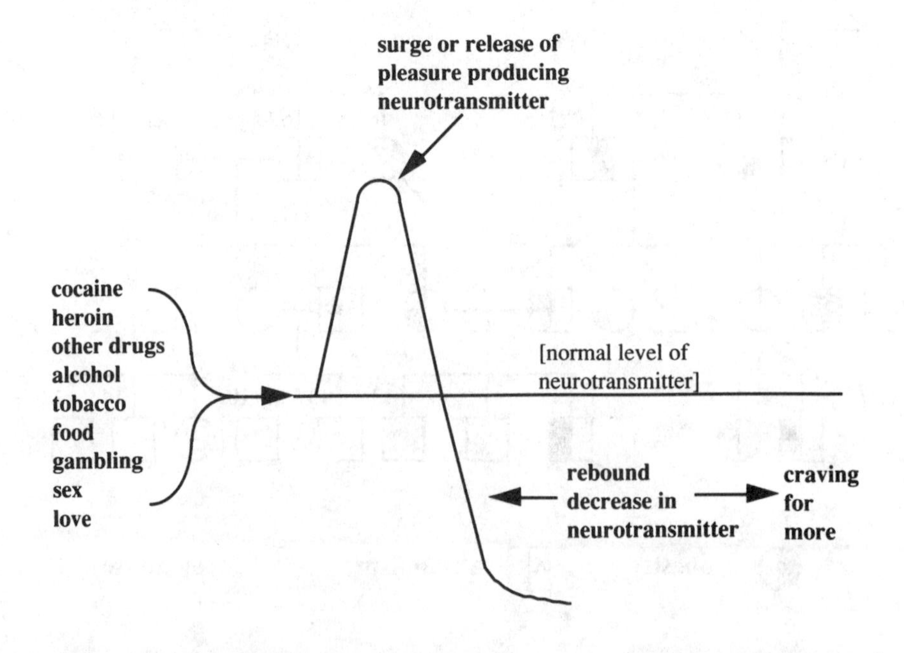

Why Are Some People More Susceptible to Addictive Behaviors?

Most individuals can take these ups and downs in stride and simply enjoy rather than crave the peaks. I have repeatedly pointed out that people who carry a *Gts* gene are often more susceptible to addictive behaviors including alcohol[p225,p644], drug[p225], and food abuse [p236,p644], gambling[p239], and sex[p163]. It is generally felt that there is not one specific "addictive personality." This is consistent with the wide range of expression of the *Gts* gene. Instead of a given type of personality, the common denominator may be given level of a brain neurotransmitter. We have seen that in Tourette syndrome there is a deficiency of brain serotonin, and in some areas, a deficiency of dopamine. This may have a significant effect on the ups and downs of addiction, shown as follows:

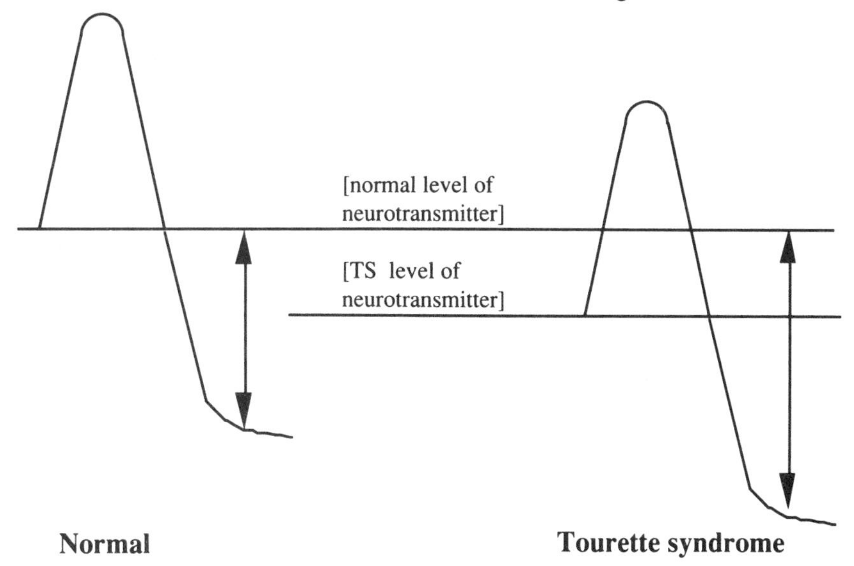

Normal **Tourette syndrome**

If the level of the neurotransmitters are low to begin with, the rebound will sink to a much lower level than normal. This will generate a stronger and more chronic craving since the level is always lower than in normal individuals. This could provide a substrate of an increased need for pleasure-seeking behaviors. *This combined with the preexisting impulsive and compulsive nature of Gts gene carriers provides a combination that leads to a high susceptibility to addictive behaviors.*

While a genetic susceptibility to drug abuse can explain the susceptibility to some addictions, such as to cocaine, certain forms of the drugs, such as crack, are so highly addictive that all users, regardless of their genetic make-up, are susceptible to becoming addicted[678].

Cocaine-Induced Tourette Syndrome

We have all heard stories of individuals who took LSD and then developed a permanent schizophrenia-like disorder. It is possible that these individuals had a gene for schizophrenia to begin with and the LSD just precipitated the symptoms. A similar problem can occur with *Gts* gene carriers. The following is an example.

> Jim never had any motor or vocal tics during his childhood. At age 24 he began to use street drugs heavily, both snorting cocaine and smoking crack. Within months he developed severe motor tics consisting of eyeblinking and

fluttering of his eyes, and violent extension jerks of his elbows and legs. He also developed severe obsessive-compulsive behaviors such a going back and forth and in and out of a door many times. The tics and compulsive behaviors became so severe that he had to be hospitalized and was in and out of mental hospitals five times over the next year. The tics failed to respond to many different phenothiazines. Finally, when his dosage of Thorazine was increased to 800 mg/day the tics stopped. When seen in the clinic the dose was 400 mg/day. He was still tic-free but this level of medication was causing him to have a zombie-like demeanor.

Since the *Gts* gene is common, this is one more of along list of dangers of cocaine abuse.

Summary: In 1953 Olds and Milner discovered pleasure centers in the brain. These areas were found to be rich in dopamine containing neurons. Addicting drugs and pleasant behaviors cause a release dopamine in these pleasure centers. Genetic disorders, such as Tourette syndrome, that result in a decrease in the level of neurotransmitters lead to impulsive-compulsive behaviors and increased susceptibility to a variety of addictions.

Chapter 71
Personality

Psychologists have spent whole careers attempting to define different types of personalities and why people have them. For years the theories of Freud have played a prominent role and still do. However, as our knowledge of the chemistry of the brain increased, there was a burgeoning interest in attempting to dissect out relationships between the levels of certain enzymes and neurotransmitters and different personality traits. First, let us look at the personality traits themselves.

Psychiatric and behavioral problems tend to be described on one of three levels depending on their severity: mental disorder, personality disorder and personality trait. This dependence on severity is illustrated by the definitions from the DSM-IIIR (1987). "*Personality traits* are enduring patterns of perceiving, relating to, and thinking about the environment and oneself, and are exhibited in a wide range of important social and personal contexts. It is only when personality traits are inflexible and maladaptive and cause either significant functional impairment or subjective distress that they constitute *Personality Disorders.*" This is illustrated by some of the classifications given in the DSM-IIIR (**see below**). To further demonstrate that these are considered different diagnoses I have included their code numbers.

Age is also relevant in the classification of personality disorders. A very similar set of symptoms may have one diagnosis for children

Mental Disorder	Personality Disorder	Personality Trait
Obsessive-compulsive disorder (300.30)	**Obsessive-compulsive personality disorder (301.40)**	**Some moderate obsessive-compulsive behaviors**
Delusional (paranoid) disorder (297.10)	**Paranoid personality disorder (301.00)**	**Some moderate paranoia**
Schizophrenia (295.1x-9x) **Schizophreniform disorder (295.40)**	**Schizoid personality disorder (301.20)** **Schizotypal personality disorder (301.22)**	**Some schizoid behaviors**

and adolescents and then magically change to another diagnosis in adults. The following are some examples:

Childhood Disorder		Personality Disorder
Conduct disorder (312.00,20,90)	→	**Antisocial personality (301.70)**
Avoidant disorder of childhood or adolescence (313.21)	→	**Avoidant personality disorder (301.82)**
Separation anxiety disorder (309.21)	→	**Dependent personality disorder (301.60)**
Identity disorder (313.82)	→	**Borderline personality disorder (301.83)**
Oppositonal disorder (313.81)	→	**Passive aggressive personality disorder (301.84)**

The remaining official personality disorders are the following (DSM-IIIR, 1987):

Other Personality Disorders
Histrionic personality disorder(301.50)
Narcissistic personality disorder(301.81)

It is important to point out that these disorders represent man's attempts to put some kind of order into the world of aberrant behavior and are not necessarily distinct disorders. Many individuals may present with symptoms of several personality disorders, and precise diagnoses are often difficult. One of the greatest values of classification is to allow psychiatrists and other mental health workers to know what their colleagues are talking about.

From a genetic point of view we have seen in this book that a specific gene, such as the *Gts* gene, and a specific biochemical defect, such defective brain serotonin, can cause a wide range of symptoms in the same or different individuals. All of the personality disorders listed above are more common in individuals with TS. This suggests that the problem is more one of a "destabilization" or "disinhibition" of the nervous system than a specific chemical change or psychological background causing a specific disorder. Factors such a single or double dose of the *Gt* gene, a mixture of a *Gts* gene and a schizophrenia or other gene, nutrition, the presence or absence of structure in the household, and other environmental effects would determine which set of symptoms would be most prominent and whether a behavior was just a personality trait or had advanced to a personality or mental disorder.

Other Personality Classifications

The above represents the "official" set of personality traits and disorders. Many individuals have used other dimensions to define personality.

Sensation-seeking:extroversion One of the most thoroughly studied dimensions is that of high or low *sensation-seeking* or *extroversion-introversion*. Sensation-seeking is a term popularized by Marvin Zukerman and colleagues[2158,2159] while the Eysencks have used the terms extroversion-introversion[567,568]. The relation between these and other related terms are as follows:

High sensation-seeking	Low sensation-seeking
Extroversion	Introversion
Active	Passive
Type A personality	Type B personality
Impulsive	Controlled

These are not all identical terms since different questionnaires are used to define each. However, for our purposes, the following concept tends to hold for all.

This concept is that the degree to which an individual engages in sensation-seeking behavior depends upon the level of arousal of the nervous system[871], which in turn depends upon a specific level of catecholamines in the brain. The level of the neurotransmitters is dependent upon the level of enzymes such as monoamine oxidase (MAO) and dopamine β-hydroxylase and other factors. The level of neurotransmitters regulate sensation-seeking behaviors such as risky but exciting sports, fast driving, impulsive behavior, social stimulation, gambling[2101], drug abuse[297] or a nonconformist life style.

Many studies have suggested that when *platelet MAO level is low, sensation-seeking is high* [417,624,1377,1710,2018]. The following summarizes the behaviors and personality characteristics found to be associated with low MAO levels.

Behavior in Individuals with low platelet MAO levels

low tolerance for boredom
more alcohol use
more cigarette use
more drug use
more impulsive
more sociable
risky sports
sensation seeking

The fact that women have higher average levels of platelet MAO than males[1375,1635] is consistent with the generally lower level of sensation-seeking in females. High sensation seeking scores also correlate with antisocial personality or psychopathic behavior[2159]. Twin studies indicate that platelet MAO levels are genetically controlled[1404,2159].

In explaining these observations it is assumed that platelet MAO levels are a reflection of brain MAO levels. There are, however, a few of problems with this scheme.

1. There are two types of MAO, type A and B. In the brain, type A is present in the dopamine and norepinephrine neurons while type B is present in serotonin neurons[2073]. Only type B is present in the platelets[1517]. There is little evidence for any correlation between the levels of platelet MAO and brain type A or B MAO[623] or brain dopamine or nonepinephrine levels.

2. The original theories[567,2158] suggested that sensation-seeking or extroverted behaviors were being undertaken to raise the level of arousal in individuals who were chronically underaroused. If that were the case, the expectation should be that high levels of MAO, producing low levels of catecholamines, would be associated with sensation-seeking behavior. Instead it is the other way around.

3. Low levels of β-dopamine hydroxylase in the blood are also associated with high sensation-seeking behavior[1976]. This enzyme converts dopamine to norepinephrine. As shown previously[P411], low levels of β-dopamine hydroxylase result in low levels of norepinephrine and high levels of dopamine. When spinal fluid was examined, norepinephrine was low in those with high sensation-seeking scores, not high as the low platelet MAO theory would propose.

4. Many behavioral disorders other than sensation-seeking have been associated with

low platelet MAO. These include attention deficit disorder[1783,1966], alcoholism[623], manic-depression, depression[8] and some schizophrenia[2117].

These observations suggest that low platelet MAO may have nothing to do with behavior and that some independent factor is causing both the behavior and the low platelet MAO[p664]. Since serotonin is the major neurotransmitter present in the platelets, this other factor may be genes that cause low serotonin in both the platelets and the brain. Since sensation-seeking behavior, ADHD, alcoholism, and mood swings are common in TS, studies of platelet serotonin and platelet MAO (using serotonin as the substrate) in TS would be of interest.

Multidimensional systems Additional scales can be added to studies of personality. In addition to extroversion-introversion, the Eysencks have added neuroticism-stability and psychoticism[567]. Cloninger[328,329] has proposed a three-dimensional model of personality consisting of novelty-seeking, harm avoidance and reward dependence. He equates novelty or sensation-seeking with activation by dopamine pathways. High harm avoidance is equated with active serotonin pathways. As seen in the chapter on serotonin and animal behavior, low serotonin results in the opposite extreme of resistance to punishment. Finally, active norepinephrine pathways are equated with reward dependence.

Summary: A number of personality disorders have been defined by the American Psychiatric Association. In our experience each one of them is more common in patients carrying a Gts gene. This suggests that one of the factors causing these disorders can be a genetic defect which destabilizes the nervous system and results in both a wide range of symptoms and a wide range severity extending from mild personality traits, to personality disorders, and occasionally to more serious mental disorders.

Part V
TREATMENT

Chapter 72
Why Medication?

One-third of this book was devoted to giving the reader an understanding of what is meant when we say that Tourette syndrome is a due to a chemical imbalance in the brain. It was all very academic. Despite the understanding at an intellectual level, at an emotional level I have rarely seen a parent come into the clinic wanting to place their child on medication. A common question is, "Does he or she have to be on medication?" The answer is that the child is only given medication if there is a significant impairment in functioning that cannot be handled by other means such as more structure, tutors or special classes. However, if these approaches do not work, it is *important for the child* that he or she not continue to be out of control but *be allowed to function like a normal child.*

To illustrate this point, I have included the following report that one of our mothers gave as part of introducing us when we were to give talk on Tourette syndrome at her school.

"I would like to give you some insight into the pain, frustration and desperation felt by the parents of a child out of control.

"Jason's problems became apparent at the age of 2 1/2. I enrolled him in a little play school program at the community center for 40 minutes a day, two times a week. After about three weeks in this program the teacher came to me and said, 'You need to use very strict discipline with Jason. He is totally out of control in my class.' She recommended a book on parenting with the philosophy of 'spank him for every offense until the child is under control and submissive to your authority.' I tried this for about four weeks. Jason would break a rule; I would spank him. He would put a whole package of toilet paper in the toilet; I would spank him. He took a permanent ink Magic Markers to every appliance in the house; I spanked him. Our home was a battlefield.

"I decided to look for more effective ways to deal with my son. Jason is an only child so I didn't know if these were normal problems faced by every parent. I began to study child development, parenting, discipline techniques and esteem building so I could develop the skills necessary to work with my son. I read books. I attended classes and seminars. My son's behavior at home improved. But in any social situation he again went out of control.

"One year ago, Jason was attending the Child Development Center. In their evaluation they wrote:

> 'A bright child, totally bored by classroom activities. Extremely sensitive to criticism or rejection. *Makes funny faces and noises to get attention.* Will not comply with rules. Ignores the staff. Tests limits with every adult. Inattentive. Aggressive, disruptive, destructive, a loner, doesn't interact with his peers. Orders other children around. *Uses foul language.* Says things to intentionally hurt other children. Abusive to the

school's animals. Other children have leaned to avoid him. Parents do not want their children to play with Jason.'

"When Jason had broken a rule a teacher corrected him. Jason said, 'You don't make the rules about me.' The teacher said, 'When you are in school, I do make the rules for you.' Jason pulled himself up, put his hands on his hips, and arrogantly replied, 'Then I'll find another school.'

"Because of these problems I took Jason to a pediatrician specializing in behavior problems. He referred me to a psychologist who ran a full battery of nine tests. In her report she wrote,

> 'During the testing Jason refused to follow instructions and said, "I'm going to do it MY way." Then he deliberately went opposite to the test instructions. He asked for a break, then refused to come out of the bathroom for over ten minutes.'

"In spite of these behaviors he obtained a full scale IQ of 124 on the Weschler Preschool and Primary Scale of Intelligence Test. On the Wide Range Achievement Test he rated reading at a 4.4 grade level, spelling at a 2.9 grade level, and math at a 2.3 grade level. This testing was done before Jason was enrolled in an academic program of any kind.

"Based on these findings the psychologist concluded that Jason did *not* have ADHD, or a learning disability, *or a medical problem of any kind.* She referred us for short-term psychotherapy with a psychiatric social worker, with hourly sessions once a week. In therapy we were told that *if we were better parents this wouldn't be happening*. We were too strict — he was rebelling against the teachers because he was afraid of us. We were too lax — he was spoiled and expected his own way with everyone. His school placement was wrong — he was bored and unchallenged. We were expecting too much of him — this was causing stress and depression. 'He's just a little boy.' *If our marital relationship was sound he wouldn't be so insecure.* If she had been familiar with the behaviors associated with TS; if she had been up-to-date on treatments which are effective; if she had been willing to listen to us without making assumptions or jumping to conclusions; if she would have listened to the staff at Jason's school; her therapy could have been so helpful to our family. As it was, her therapy did more damage than good. We were thoroughly confused and didn't know where to go next to get help. In the meantime, Jason was expelled from his school and I had to find a school willing to enroll him.

"He began attending a Montessori Academy. In their report they wrote:

> 'A very disturbed child, devious and destructive. We have to watch him constantly to be sure he doesn't hurt another child or destroy property. Delights in negative behavior and vulgar language. Disrupts the classroom with squealing sounds. Will not accept direction or discipline. Always on guard, observing everything and everyone around him. Always touching everything and bothering other students. Cannot function in a classroom of 18 children and two adults.'

"I would receive up to six calls a day from the school. The school director was being pressured by teachers and parents to expel my son. He was spending most of his ten hour day with her in her 8' by 10' office with her as a favor to me, so I could continue to work. We had no support system whatsoever. Our friends didn't want us around. The neighbors didn't want us around. For a period of ten weeks we couldn't even attend church in case Jason would scream out an obscenity or hurt someone. We had no idea what resources were available to us because the so-called professionals denied that my son had a problem. I again began to educate myself. I was often up until two or three in the morning reading everything I could find on ADHD, seizure disorders, autism, dyslexia, schizophrenia, and many more I can't even remember now. I studied the brain and how the neurotransmitters work. I studied classroom and institutional management of children whose behaviors were out of control.

"At this point, it was suggested that I talk to our district's psychological/health services office. I briefly described Jason's behavior to them and they said, 'That sounds like Tourette syndrome. We see it all the time.' In five minutes they identified what doctor after doctor had missed over a period of nine months. My son started treatment in June, less than five months ago. The first day he was on medication I received about six calls from the director of his school, but this time all good news. She said it was the most dramatic change she had ever seen, and it was all within thirty minutes. He was cooperative, productive, able to follow directions and stay on task. For the first time he had fine motor control and could cut with scissors and color within the lines with crayons. He was helping younger children with their exercises. During recess he actually played with other children, something he had never been able to do before.

"Jason is now enrolled in a K-1 class. In her report the teacher wrote.

> 'Jason has been a student in my kindergarten/first grade combination class for six weeks. He has demonstrated the ability to "get along" or "function" successfully in my classroom atmosphere.
>
> 'Jason is working on level six in reading, the end of the first grade reader. He reads with understanding and has no difficulty in completing his reading assignments. He also works on his problems with very little difficultly and has no problem staying on task.
>
> 'In my classroom, Jason has shown patience, creativity, the ability to compete tasks from start to finish. He has shown me respect, a willingness to assist me as well as other students, and an understanding of my classroom rules. He knows my rules as well as the school rules and several times I have overhead Jason reminding other students of those rules. Jason gets along well with his peers as well as with those a grade older. He shares his ideas and possessions. His social skills are good and he has learned to accept responsibility.'

"[Prior to treatment] because of his behavior, my son was treated as a troublemaker, a nuisance and a brat. He was slapped, pinched and physically restrained by adults who became so angry *they* lost control. He received negative messages all day long from everyone around him: teachers, children, their parents, neighbors, and even his own parents. No one wanted Jason around. I remember once when I asked him, 'Can't you just try to be good? Just for one day, or even a hour if a day is too long?' He answered, 'Mom, there are so many rules, and so many teachers, and I'm so bad I could never be good and I'm too tired to try anymore.' At the age of five he had given up. And looking back now, if I had been subjected to the treatment he was, I don't know if I would have made it as long as he did.

"Now he has more self-control over his behavior. He has the ability to try. We work to accommodate his learning disabilities while building on his strengths and give him work that challenges him. He has friends and is invited to birthday parties. I have to keep a social calender for him so I can chauffeur him to his many activities, and I am more than delighted to do it. People WANT to be with him. I am able to go into a new school, specifically outline his disabilities, and give his teachers specific techniques for classroom management so they don't spend 30% of their time trying to control his behavior and restore classroom order.

"I am excited that we are beginning to recognize and address the needs of learning disabled gifted children like my son."

While not all children respond to medication as well as Jason the point should be clear. *These children have a genetically caused chemical imbalance in the function of their brain and it requires other chemicals to correct that balance.* The objective of medications is to give the smallest dose that provides just enough change in the chemical balance so the child, or adult, can function as near to normal as possible. The following chapters describe the various medications available for treating TS, ADHD, conduct disorder and other behaviors in the TS spectrum—the indications, the way the medications work, and the potential side effects. All medications are a double-edged sword—a balance between benefit and risk, between good effects and side effects. The secret is to find that balance.

Summary: A fairly typical case of Tourette syndrome is presented before and after medication. When an individual has a genetic disorder that severely affects their behavior, proper medication does not "place them in a chemical straightjacket," it removes them from a chemical straightjacket. Such an individual deserves a proper diagnosis and appropriate treatment to allow their most important organ, the brain, to function normally.

Chapter 73
Some General Principles about Medications

Before discussing the different medications used, some general principles are important.

The tics are often the easiest part of TS to treat While motor and vocal tics are the hallmark of the diagnosis of TS, throughout this book we have seen that TS is actually a broad spectrum disorder associated with a wide range of behavioral problems. Many parents have said to us, "The tics are the least of his problems, it is his behavior that is driving us crazy!" There are, of course, some cases where the tics have not responded to medication and they are the major problem. However, as a generalization, the above statement is correct, and the tics are often the easiest part of TS to control.

Brand names and generic names The process that drug companies have to go through to develop and market a drug is a very long one. First, it is necessary to discover a new drug. This is often done by starting with a clue that a certain class of chemicals has a certain effect. Frequently, however, the first drug to be discovered may have side effects or other properties that make it unsuitable for use in humans. Chemists then synthesize many related compounds which are first tested in animals. If the animal work is promising and shows no significant harmful effects, small-scale studies are done in humans. If these look promising, then larger studies are done, all under the guidelines set down by the Federal Drug Administration or FDA. If the company then feels that a drug could be of value, both medically and commercially, they submit an application to the FDA for approval. This entire process usually takes a minimum of ten years and often longer and is very expensive. The drug company protects this investment by taking out a patent on the drug which prevents other companies from selling it. They give the drug a brand name which links it to their company and often spend a lot of money advertising the brand name. The drug also has a chemical or generic name. Thus, Haldol is the brand name that McNeil Laboratories gave to the generic chemical haloperidol. After a number of years, when the patent runs out, other companies can market the same generic medication under their own brand name without having to pay royalties to the original drug company. In this manner, a "generic" form of the medication becomes available. The major advantage of generic drugs is that they are generally much less expensive than the brand name.

Are generic drugs as good as brand name drugs? The usual answer to this is yes. By law, the chemical and the amount has to be the same as in the brand name drug. However, there are no regulations on other aspects of the drug such as the shape, color, coating, or binders. The latter are chemicals that control how a drug is held together and how rapidly it is adsorbed. We have often seen patients begin to do poorly when they switch from brand name to generic drugs. This is only true of some medications, not all. *As a general rule, if the patient has been started on a brand name and gotten a good response, and does not do as well after switching to a generic, going back to the brand name may solve the problem.*

The major types of medications used for treating tics are the dopamine receptor antagonists (blockers) and α_2-norepinephrine receptor agonists. The generic and brand names are as follows:

Medications for Tics

Type	Generic Name	Brand Name
dopamine receptor antagonists (phenothiazines)	haloperidol pimozide fluphenazine	Haldol Orap Prolixin
α_2-norepinephrine receptor agonists	clonidine	Catapres

Medications for ADHD Oftentimes, after the tics have been brought under control, the attention deficit hyperactivity disorder is still such a problem that the patient is having a great deal of trouble concentrating and there are major problems in school. The treatment of ADHD may require a different set of medications. These come in three major categories, the so-called stimulant medications, the tricyclic antidepressants, and clonidine. Clonidine and the antidepressants may be somewhat less effective for ADHD than the stimulant drugs. The generic and brand names of the medications for ADHD are as follows:

Medications for Attention Deficit Hyperactivity Disorder

Type	Generic Name	Brand Name
stimulant medications	methylphenidate amphetamine pemoline	Ritalin Dexedrine Cylert
tricyclic antidepressants	imipramine desipramine	Tofranil Norpramin
α_2-norepinephrine receptor agonists	clonidine	Catapres

Medications for depression Significant problems with depression can occur anytime throughout the lifetime of a patient with TS. The effect of tricyclics and MAO inhibitors on enhancing the effectiveness of neurotransmitters in the synapse has been discussed previously[p413]. Lithium, however, is in a class all by itself.

Medications for Depression

Type	Generic Name	Brand Name
tricyclic and similar antidepressants	imipramine desipramine fluoxetine fluvoxamine clomipramine	Tofranil Norpramin Prozac Faverin Anafronil
monamine oxidase (MAO) inhibitors	phenelzine isocarboxazid tranycypromine	Nardil Marplan Parnate
lithium	lithium carbonate	Eskalith

Medications for obsessive-compulsive behavior As discussed in Chapter 22, many people who carry a *Gts* gene have problems with obsessive-compulsive behaviors. If these are mild they require no treatment. However, for some they seriously interfere with normal functioning and need to be treated. For many years the most effective medications were the tricyclic antidepressants, such as imipramine, and the MAO inhibitors. However, more recently two medications, clomipramine and fluoxetine have been found to be the most effective.

Medications for conduct disorder If significant problems with conduct disorder do not respond to the treatment of the tics and ADHD, in addition to psychotherapy[p595] it is sometimes necessary to add a medication specifically for the conduct disorder. In our experience, the medications that have been most useful are clonidine, Prozac, imipramine, Tegretal, lithium or the phenothiazines.

Medications for Conduct Disorder

Type	Generic Name	Brand Name
α_2-norepinephrine receptor agonists	clonidine	Catapres
tricyclic and similar antidepressants	imipramine desipramine fluoxetine	Tofranil Norpramin Prozac
limbic anti-convulsant	carbamazepine	Tegretal
lithium	lithium carbonate	Eskalith
dopamine receptor antagonists (phenothiazines)	haloperidol pimozide fluphenazine	Haldol Orap Prolixin

Medications in combination You have probably noticed that the same medication often can be used for more than one indication. This can help to keep the number of different medications to a minimum. For example, in some patients clonidine can alleviate tics, improve ADHD, compulsive behavior and conduct. This is the ideal, to find medications that treat several different symptoms. Oftentimes, however, more than one medication is required for the best control of a range of symptoms.

The PDR The PDR, otherwise known as the Physician's Desk Reference, is a valuable book both for physicians and patients. It is updated each year and lists all of the currently available prescription drugs available in the United States. For each drug it gives the brand name, generic name, a description of the chemical nature of the drug, clinical pharmacology or how the drug works, indications for its use, contraindications or situations where it should not be used, warnings if any, adverse reactions, dosage, how it is supplied, that is, in capsules, tablets or liquid, and the amount of medication in the different forms. A particularly useful part of the book is the section that shows pictures of the different drugs and different-size tablets or capsules. Although the number of milligrams of the drug is often written on the tablet, this may be difficult to read. In many cases the different-size tablets are color-coded. This is a particularly useful way for the physician to make sure the patient is taking the proper dose. It is also valuable for the patient who is used to a certain color to make sure the pharmacist has filled the prescription properly. For example, when the brand Haldol is used, the 0.5 mg tablets are white while the 5.0 mg tablets are green. At least once a year we have a parent calling that their child is having side effects, and when checking, we find the pharmacist gave them green pills instead of white ones.

Other sections of the PDR include a list of drugs by drug company (white), in alphabetical order (pink), by type (blue), and by generic name (yellow).

The PDR is a valuable book to have and can be obtained in most book stores. However, *a word of caution. The list of adverse reactions for each drug includes everything that has ever been reported, even though many of the reactions listed may be very rare or unusual.* These long lists on virtually every drug can be very frightening to one who is not used to the PDR. In the following chapters on the different drugs I will be discussing primarily those side effects that in practice cause the most trouble. If you want a comprehensive listing of all conceivable side effects, consult the PDR.

When is medication needed? My general rule is that medication is needed when one or more of the symptoms is significantly bothering someone. The most important person to ask is the patient themselves. Sometimes after discussing with the parents whether or not to use medication and concluding we would not, I have turned to the child to ask their opinion, to be met with the reply, "These tics bother me a lot, I want something to make them better." It can of course work the other way around. A patient with loud vocal noises who is driving everyone around him to distraction may say, "It doesn't bother me at all." Thus, if the symptoms are very disturbing to the parents, siblings, and teachers, and if peers are consistently making fun of the child, medication should be considered. On the other hand, many times after a diagnosis is made and all these people understand why these symptoms are occurring, they sometimes are no longer bothered by moderate tics or vocal noises and no medication is required.

As discussed previously, about 10 percent to 20 percent of the patients whose symptoms are severe enough to come to the clinic end up not being treated, i.e., classified as grade 1[p24]. However, in the community, a much higher percent of *Gts* gene carriers with tics are grade 1 cases. Their symptoms are never severe enough to seek medical help.

Which medication? Which medication to start with depends on many factors. However, as a generalization, unless it is a grade 3 case with severe symptoms and marked ADHD, I like to first try clonidine and usually start with the clonidine patch[p561]. As will be seen in subsequent chapters, clonidine has the advantage that it has few side effects. The tiredness it can cause is often short-lived. It also has the advantage that it can sometimes help with the ADHD and obsessive compulsive behaviors and conduct disorders, as well as the tics. If it works, it is an excellent medication. However, in our experience it is effective in only about 60 percent of the cases. If it doesn't work we try another medication and the only thing that has been lost is some time.

Tourette syndrome patients are sensitive to medication Over the years of treating TS we have repeatedly noticed that they often require lower doses than what is usually recommended for a given medication. The patients themselves often tell us how sensitive they are to different drugs. To give an example, one 38 year old woman with TS, obsessive-compulsive behavior, panic attacks and depression responded well to only 12.5 mg of a tricyclic antidepressant. The recommended adult dose for the medication is 100 to 300 mg per day. She improved further when given 100 mg of lithium carbonate per day, the usual dose of which is 900 to 1,500 mg per day.

We have seen many patients who were adamant about not taking medication because they had previously been overmedicated. A particularly frequent error was to have a psychiatrist, accustomed to treating schizophrenia with large doses of Haldol, start a TS patient on 1 to 10 milligrams the first day. This is 2 to 20 times the starting dose we use. This leads to our next maxim.

Start at low doses and work up slowly This rule will be repeated many times in the next chapters. As a generalization, we start at quite small doses of medication and allow at least a week of observation before increasing the dose. This has several advantages. Foremost, it minimizes side effects by allowing the body to accommodate small amounts of medication without being overwhelmed. Many times side effects disappear after a week or two and this slow increase often eliminates them altogether. Second, since many of the symptoms of TS wax and wane, it gives adequate time to evaluate the results averaged over several days. Third, it helps to keep the doses at the lowest level necessary for a given effect and takes advantage of the fact that many TS patients are quite sensitive to medication.

Don't try to make all the tics go away One of our most fundamental rules is that we never try to make all the tics disappear. Most patients are thrilled with a 70 to 80 percent decrease in their tics. The remaining 20 percent are usually easy to live with or can be readily covered up or inhibited in public. The major advantage of not totally eliminating all tics is that it helps to ensure that the patient is on the lowest dose of medication possible. If all the tics are gone, we don't know whether the dose is just right or whether they are taking much more medication than they need. This leads into our next rule.

If the tics disappear for a number of weeks, decrease the dose The symptoms of TS can spontaneously disappear sometimes for weeks or months at a time. If an individual is in such a remission they should no longer be taking medication for the tics. The easiest way the keep track of this is to keep the medication at a level such that some tics are usually present (as above). Then, if they completely disappear for several weeks, the dose can be slowly decreased on a weekly basis. If, as the dose is being decreased, the tics return, the dose can be adjusted upward slightly and left at that new, lower level. If the dose continues to be dropped until the medication is stopped, and no tics return, then the individual is in a remission and the medication has been appropriately discontinued.

Change only one medication at a time During the course of treatment of TS patients the management of medications can sometimes become rather complex. An important rule in evaluating the effect of starting or stopping medications is to make only one change at a time. If, for example, two medications are started simultaneoulsy, or one is started and the other is stopped, and the symptoms get better, it is difficult to determine which maneuver was the effective one.

The plateau effect Oftentimes when a new patient is just starting on medication they are quite sensitive to it and the tics may disappear at quite a low dose. However, as the body adjusts to this state, in 1 to 4 weeks the tics may slowly come back and the dose will have to be increased again. This is a normal reaction and does not mean that the dose will constantly have to be increased. After one or two of these plateaus, an appropriate maintenance dose is reached and usually no further increases are necessary.

The window effect While eveyone can appreciate the fact that too little medication may be ineffective, it is less well appreciated that too much may be less effective than a smaller dose, even in the absence of side effects. This is called the therapeutic window. It is more true for some medications than others. My general rule is, *If in doubt about whether a patient is getting too much or too little of a medication, decrease the dose.* If the symptoms get worse, we can always turn around and go to a higher dose.

"I'm better, I can stop my medication." Many times a patient will return to the clinic with all of their symptoms back, whether it be tics, depression, panic attacks or any of many others. When asked what happened, the response is often, "My symptoms all disappeared so I thought I was cured and stopped the medication." This frequent bit of wishful thinking illustrates a common misconception about TS and its related behaviors. Although the symptoms of a genetic, neurochemical disorder can be eliminated by chemical treatment, i.e., medication, the basic defect does not go away and long-term continuation of the medication is often necessary. Just as a diabetic is not cured by taking insulin, a TS patient is not cured even though the symptoms can often be eliminated.

Don't suddenly stop medications This is a general rule but applies to some medications more than others. Since haloperidol and pimozide have a long duration of action, if they are suddenly stopped, the major problem is that tics begin to return in 2 to 7 days and may be worse for awhile before settling down to the level present before taking medication. Clonidine, Xanax, Valium, and tricyclic antidepressants are medications that should be stopped gradually to avoid withdrawal effects. The major effect of suddenly stopping moderate doses of Ritalin or Dexedrine is the return of symptoms of ADHD. Sometimes there is a rebound effect of a transient increase in symptoms when medications are significantly decreased or stopped.

"Do I have to take medication the rest of my life?" This is one of the questions most frequently asked when medications are first started. The general response is no. In practice, many TS children begin to experience a lessening of their tics in their mid teens or twenties and during this time the medication is often discontinued. However, there are two major qualifiers of this statement. First, I know of no way to predict who will get better with age and who will not. We have some patients taking medication at all ages including their 70's. Second, the natural history of TS is such that while the tics may get better other problems such as panic attacks, obsessive-compulsive behavior, various addictive disorders and mood swings may take their place. These in turn may require different medications.

Trunk and branches: The use of multiple medications I often think of medications used for TS in terms of a tree. If a drug can correct the basic chemical defect in TS then all of the secondary behaviors should get better. These I call **trunk drugs.** However, different patients have different symptoms and we usually find ourselves treating the different branching effects of the basic defect. These are the **branch drugs**. Usually a branch drug affects only one or a few of the symptoms and patients with multiple symptoms often require multiple drugs. Thus, Haldol and related medications are effective for the tics but usually have little effect on ADHD, and may make depression worse. In a similar fashion, Ritalin is effective for treating the ADHD, but may make the tics worse. Thus, it is not unusual for a patient with a wide range of symptoms to be taking two or even three different branch medications.

Sometimes multiple medications are required for a single branch. Thus, of our 1,400+ patients, there is a number whose tics have not responded to any single medication. When this occurs, a second medication for the tics may be added to the first. A common combination is haloperidol and clonidine. In rare cases even a third medication, such as Clonopin, may be added. Sometimes these very difficult patients who have not responded to any single medication will respond to a combinations of drugs. This is part of the art of TS treatment that comes with experience.

Don't confuse medications with street drugs In an age when "drugs" is a dirty word and every effort is being made to keep children and adults off cocaine, pot, speed and related street drugs, the idea of then treating with "drugs" may be viewed with suspicion, especially by children who do not understand the difference between street drugs and prescription medications. It may be necessary to spend some time with such children and their parents to explain the difference. A useful analogy is with diabetes. The need to use a chemical, insulin, to treat diabetes where there is a genetic-chemical imbalance in the pancreas is easy to understand. TS and related disorders represent a genetic-chemical imbalance of a different organ, the brain, and sometimes it also needs the help of other chemicals to restore that balance.

Summary: A number of basic rules in the use of medications to treat TS patients are given. Some of the most important are to start at low doses and work up slowly, don't try to make all the tics disappear, slowly lower the dose if the tics are virtually gone for several weeks, and do not suddenly stop medications or discontinue them just because the symptoms are better. If the medication needs adjusting, make changes in only one medication at a time.

Chapter 74
Haloperidol

Haloperidol was first reported to be effective in the treatment of Tourette syndome in 1961[289,1749]. In our experience and that of others[1765,1767,1768], it is effective in controlling tics in about 80% of patients. The principal features of haloperidol are shown in the box.

haloperidol (Haldol)

Action:	blocks $dopamine_2$ receptors
Starting dose:	0.25 - 0.5 mg a day
Usual dose:	0.5 - 5 mg a day
Major side effects:	tiredness headache stomachache weight gain Parkinsonian symptoms
Effectiveness:	80% of patients

Initial doses For most children and adults, the starting dose of haloperidol is 0.25 to 0.5 mg each evening. This is continued for one week. If the tics are significantly improved on this dose, no further increases are needed. If after one week, there is little or no improvement, the dose is increased to 0.5 to 1.0 mg each evening for the second week.This process is continued with small weekly increases in dose until the symptoms are improved by 70 to 80% or until side effects prevent any further increases.

Maintenance dose Most individuals require somewhere between 0.5 to 5 mg of haloperidol per day. A typical report is that 0.5 mg produced little effect, that the urge to do the tics diminished at 1.0 mg, and significant improvement of symptoms occurred at 1.5 mg. Then over the next several months, because of the plateau effect (see below), it may be necessary to increase the dose to 2.0 to 3.0 mg.

Haloperidol is long-acting One of the advantages of haloperidol is that it has a fairly long duration of action. When it is discontinued it may take two to seven days for symptoms to return. Because of this it usually needs to be given only once a day. We find that the best time is often just before going to bed. This way, if some tiredness is induced, it tends to occur when the individual is sleeping. If drowsiness in the morning is a problem, the dose can be moved backward to around suppertime to give a few extra hours between the dose and morning. Occasionally splitting the dose into morning and evening doses helps to even the effects throughout the day. Treatment with intramuscular preparations of haloperidol (haloperidol decanoate) may be effective when oral medication is not[322].

Side Effects of Haloperidol

Some patients can take haloperidol without having any side effects. For them it can be the drug of choice. Others have side effects varying from minor to so severe that alternative medications must be sought. The major side effects are the following.

Tiredness The most frequent complaint is feeling tired, groggy or sedated. This may be the most marked in the morning, contributing to a difficulty in getting going, although this can be a problem for TS patients on no medications. As mentioned before, if all the haloperidol is being taken at night, moving the dose back a few hours to suppertime may help the morning tiredness. In older patients, mild stimulants such as coffee may also help.

Cognitive blunting This is similar to tiredness but a little more subtle and involves a decrease in ability to learn. Since learning problems are present with TS itself it can be difficult to detect. It is best handled by trying to keep the dose as low as possible and being alert to any sudden drop in grades or school performance after the medication is started. It is important to point out that sometimes there is a dramatic improvement in school performance after haloperiodol has brought the tics and other behaviors under control. In some of the earlier studies of the use of haloperidol in TS, quite high doses of up to 160 mg per day were used[1765]. It is the higher doses of haloperidol that are most likely to produce significant impairment in learning[773].

Lack of motivation Another symptom related to both tiredness and cognitive blunting is a decrease in motivation. Some patients, especially adults, feel some loss in their "get up and go." This may be due to the effect of haloperidol on blocking dopamine receptors in the frontal lobes[p389].

Depression Depression due to haloperidol is usually fairly clear-cut. While patients may complain of an onset of depression anytime during the course of taking haloperidol, it is most often seen soon after the medication is first started or when the dose is increased. Because of this dependence on the dose it may disappear when the dose is decreased. If the haloperidol is otherwise very effective and the patient is reluctant to change the dose, adding a tricyclic antidepressant can alleviate the depression.

Dysphoria Dysphoria is a symptom similar to depression, but with a wider dimension. It is includes feelings of sadness, heaviness, anger, irritability, "like a shade coming down"[240,275], tearfulness, and lack of motivation. Like depression it responds well to a decrease in dose and may respond to Cogentin. On rare occasions, the aggressive behavior may actually increase after a recent increase in the dose of haloperidol. This is an indication to decrease the dose.

Weight gain Other than tiredness this is the most frequent side effect with haloperidol. Patients gradually gain weight and sometimes the increase can be striking and very disturbing to the individual's self-esteem. Since there is often a constant ravenous hunger, simply asking the patient to stop eating rarely works. It is easier to substitute low calore foods such as celery, jicama, cucumbers and lettuce to fill them up. I often tell parents not to even buy high calorie foods such as potato chips, peanut butter, candy and other delectables. It is much easier to keep from raiding the kitchen if there is nothing there to eat, and it is easier to switch to low calorie foods if that is the only thing in the house.

If this doesn't work and stopping or significantly decreasing the dose is not a viable option, we have found that fenfluramine (Pondimin) 20 mg taken a half hour before meals, and possibly in the evening, often works. This is a non-amphetamine appetite suppressor that works on serotonin pathways to control appetite. I have not observed any increase in tic symptoms with Pondimin, and as mentioned before, it has also been used to successfully treat some of the symptoms of autism[p502].

Anxiety and school phobia Although phobias and anxiety are often an integral part of TS[p175], these symptoms may get worse after haloperidol is started. The triggering of school phobia especially has been reported[1190,1316] and I have also seen this happen. An increase in separation anxiety may also occur.

Nightmares Nightmares may occur. This can sometimes be alleviated by giving the medication in the morning instead of at bedtime.

Hallucinations This has been a rare side effect. It usually occurs soon after starting the medication and responds well to stopping the haloperidol.

Increase in tics *Every medication I have used for treatment of tics has sometimes initially made them worse* and haloperidol is no exception. This is usually, but not always, a reason for switching to another drug. Sometimes this works itself out and I have several patients who are now doing well on haloperidol even though their tics initially worsened.

Decreased libido This can occasionally be a problem for adults taking haloperidol. It disappears when the medication is discontinued.

Parkinsonian Side Effects

The symptoms of Parkinson's disease were discussed previously[p375]. Since this disorder is due to too little dopamine in the neostriatum, and the action of haloperidol is to block dopamine receptors, it is not surprising that some of the symptoms of Parkinson's disease might occur. The symptoms most frequently seen are the following.

Akathisia This refers to a feeling of restless, inner tension, a need to pace, being unable to keep the legs still or an "ants in the pants" feeling[1767]. The name is derived from Greek meaning unable to remain seated[519]. If present it usually comes on soon after haloperidol has been started or soon after the dose has been increased, often the same day. Patients complain they can't sit still. There is often an associated feeling of anxiety and inability to concentrate. Fortunately, this symptom responds very quickly to Congentin (see below) or related medications, and often goes away spontaneously after a few days, weeks or months. It can be distinguished from ADHD or a panic or anxiety attack by the fact that it responds so well to Cogentin. It is important to distinguish akathisia from anxiety or tics since increasing the dose of medication will make the symptoms worse instead of better[2035].

Akinesia This refers to a lack of movement, or a slowing of movement. There is usually a feeling of weakness or fatigue and the muscles or joints may actually ache. It is distinct from drowsiness in that the patient may feel alert but weak. It is also distinct from depression in that it is a symptom in the muscles, not in the mind. Certain movements such as writing or walking may become more difficult. Like akathesia, it also responds to decreasing the dose, to Cogentin or related medications, and usually disappears with time.

Dystonia or muscle spasms In contrast to a tic, dystonia is a slow twisting contraction or spasm of the muscles. While almost any muscle group can be involved, the muscles on the side of the head or abdomen are frequently affected, giving rise to headaches and "stomachaches." If the neck muscles are involved there may be a turning or twisting of the head to one side. Speech and swallowing may become difficult and the tongue may sometimes feel or be swollen. The most striking spasm is of the muscles controlling eye movements. The eyes may roll upward so that only the whites are seen. This is called an *"oculogyric crisis"* and if not understood can be very frightening to parents or patients. Fortunately, it doesn't happen very often. The dystonias may be accompanied by an inability to talk, sweating and pallor. All of these symptoms respond quickly to Congentin or related medications (see next page).

Tardive dyskinesia This subject comes up so often that I have devoted a separate chapter to it[p557].

Other Side Effects

Allergic reactions Any drug can cause allergic or sensitivity type reactions in some people. An increased sensitivity to the sun may occur with haloperidol. Although effects on the liver with jaundice can occur, this is unusual and responds well to stopping the medication and repeated liver function studies are not necessary for patients on haloperidol[1767]. A significant decrease in the white blood cell count is a rare side effect and should be checked for if fever and sore throat occur.

TS due to medication? A woman with schizophrenia was treated with Thorazine for six years. Several months after it was stopped she developed motor and vocal tics[1055]. These symptoms improved with haloperiodol treatment. Since phenothiazines can cause hypersensitivity of postsynaptic dopamine receptors, it was suggested that this provided evidence that Tourette syndrome was due to this hypersensitivity. Other evidence is also consistent with this[p399]. However, because the *Gts* gene is common, it is more likely that this patient was carrying a *Gts* gene than that phenothiazines can cause permanent tics.

A second case of so-called *tardive Tourette's* was described in a boy with autism treated with Mellaril[1851]. When the medication was stopped, at age 24, he developed many motor and vocal tics. However, as a child he made many bizarre noises and as discussed before, many higher functioning autistics actually have Tourette sydrome[p215]. In this case, the TS symptoms may simply have been suppressed until the Mellaril was stopped.

A third case has been described of a 16 year old retarded male with violent behavior who was treated with up to 1,500 mg of Thorazine and then with haloperidol[1808]. When this was discontinued, he developed motor and vocal tics which eventually disappeared after seven months.

There is a question of whether tardive Tourette's actually exists or is simply due to the precipitation of symptoms in individuals carrying the *Gts* gene. If many people carry the gene the latter is much more likely.

Miscellaneous Other occasional side effects include constipation, blurred vision, dry mouth or excessive salivation, slurred speech, dilated pupils with sensitivity to light, nasal congestion, and facial pallor. Females may have menstrual irregularities and, rarely, secretion of some milk from their breasts. Males may have some slight breast tenderness or swelling, although this may occur during puberty in boys who are not on medication.

Treatment of Side Effects

Since many of the side effects of haloperidol are related to their ability to produce Parkinson's disease-like symptoms, the same medications used to treat Parkinson's disease are used to treat the side effects. The following box shows some of these medications.

Anti-Parkinson's Medications

Generic Name	Brand Name	Dose
benztropine	Cogentin	0.5 - 3.0 mg
trihexyphenidyl	Artane	2.0 - 8.0 mg
diphenhydramine	Benadryl	25 - 50 mg
amantadine	Symmetrel	100 - 200 mg

The medicaton I use the most is Cogentin. When the haloperidol is started at a low dose and slowly increased, side effects are sufficiently unusual that I do not routinely give a prescription for Cogentin when haloperidol is started. It is usually sufficient to warn the patient to report if any side effects are occurring and a prescription can be phoned in. The usual dose is 0.5 or 1.0 mg and the side effects usually disappear within a half-hour of taking the Cogentin. The dose can be repeated as necessary, including an hour later if the first dose does not bring relief. In some cases a maintenance dose of 0.5 to 1.0 mg three times a day is necessay until the side effects begin to disappear.

Sometimes Cogentin can be used to determine if certain symptoms are due to the medication or due to TS itself. For example, headaches due to muscle spasms of the temporal muscles sometimes occur with haloperidol. However, headaches themselves, especially migraine headaches, often occur in TS. In those situations where it is unclear which is occurring, a test with Cogentin can be informative.

If an acute oculogyric crisis occurs, it is best to go to the nearest emergency room and get a shot of 0.5 to 1.0 mg of Cogentin intramuscularly. Don't be bashful about telling the emergency room doctor that these are side effects of haloperidol and Cogentin is recommended. It can often save a lot of time. If the patient cannot immediately reach a doctor, Benadryl, which is an over-the-counter medication, is often effective in a dose of 25 or 50 mg.

Cogentin itself may have side effects. These are predominately dry mouth, blurred vision and constipation. Sometimes there is some difficulty urinating or confusion and trouble with short-term memory.

Most of the above medications work by blocking the activity of acetylcholine in the brain and thus re-establishing the balance of dopamine to acetylcholine. By contrast, amantadine works by enhancing the release of dopamine. In a small series of seven patients, amantadine was reported to be more effective than Cogentin in treating the sedation produced by haloperidol[195].

Effects of Haloperidol on Serotonin

In additon to its effect on dopamine receptors, haloperidol has indirect effects on other neurotransmitters, especially serotonin. It both enhances the synthesis and utilization of brain serotonin[1584]. This, through an inhibitory mechanism, decreases the turnover of dopamine and norepinephrine[412,458,914,1684]. Haloperidol also increases the spinal fluid levels of 5-HIAA[968,1599].

Does Haloperiodol Treat ADHD?

While ADHD has been treated with Thorazine and haloperidol[726,2068], in my experience, haloperidol rarely improves the symptoms of ADHD. If some of the "hyperactive" movements are actually tics, they may improve considerably. However, it is unusual for the problem of short attention span to improve significantly. The treatment of ADHD itself is discussed later.

Does Haloperidol Treat the Conduct Disorder?

In contrast to the minimal improvement of the ADHD, when irritability, short temper and poor conduct are a significant problem, they may improve dramatically.

The following are some examples of the effect of haloperidol on behavior:

> A nine year old boy with severe ADHD required 30 mg of Dexedrine 4 times a day. After haloperidol 0.5 mg was added, he was more relaxed, happier, stopped bedwetting and the sounds he was making disappeared. He was less destructive and even showed an increased interest in reading. As his mother said, "His flame is easily lit, but now less so."
>
> A 28 year old woman with TS was intensely jealous of her husband. She

would constantly check his pockets each time he came home and if he just said the word "she" in ordinary conversation she would turn red with anger and often hit him. After taking haloperidol 1.0 mg per day, her tics decreased and the compulsive jealousy disappeared.

Benefits versus Side Effects

While the above list of side effects seems long and forbidding, haloperidol remains one of the major drugs of choice in the treatment of Tourette syndrome. When started at low doses and increased slowly, the side effects are minimized and the beneficial effects of markedly decreasing moderate to severe motor tics and aggressive behavior are oftentimes dramatic.

Summary: Haloperidol is one of the most effective medications for the treatment of Tourette syndrome. When started at low doses, increased slowly, and kept at the minimum dose necessary to bring about a 70 to 80 percent improvement in symptoms, the side effects are minimized.

Chapter 75
Pimozide

Pimozide (Orap) is very closely related to haloperidol and also blocks the postsynaptic dopamine$_2$ receptors[p370]. Although the side effects are the same as those listed for haloperidol, for many patients they are less prominent. Some individuals who cannot tolerate haloperidol can take pimozode without difficulty. The major features of pimozide are shown in the box.

pimozide (Orap)

Action:	blocks dopamine$_2$ receptors
Starting dose:	1 - 2 mg a day
Usual dose:	2 - 10 mg a day
Major side effects:	tiredness headache stomachache weight gain Parkinsonian symptoms
Effectiveness:	70 - 80% of patients

Pimozide is not quite as potent as haloperidol and thus has to be given in higher amounts. A 2 mg dose of pimozide is equivalent to 0.75 to 1.0 mg of haloperidol. Like haloperidol it should be started at a low dose (1 to 2 mg per day) and slowly increased over weekly intervals. Most patients begin to respond with 2 to 10 mg.

Comparison of Pimozide to Haloperidol

All of the studies of the use of pimozide in TS have reported it to be comparable to haloperidol with equal or fewer side effects[1594,1650,1763,1771a]. In one study, 31 patients were treated with pimozidewho had previously been treated with haloperidol and had either inadequate relief of the tics or too many side effects to continue. Bearing in mind that this tends to prejudice the results against haloperidol, since those with a satisfactory response to haldoperidol would not be included, the frequency of side effects for the two were as **shown below**[1764].

This summarizes the essence of pimozide. While the frequency of mild to moderate side effects is less than with haloperidol, for both, the side effects can be severe enough to interfer

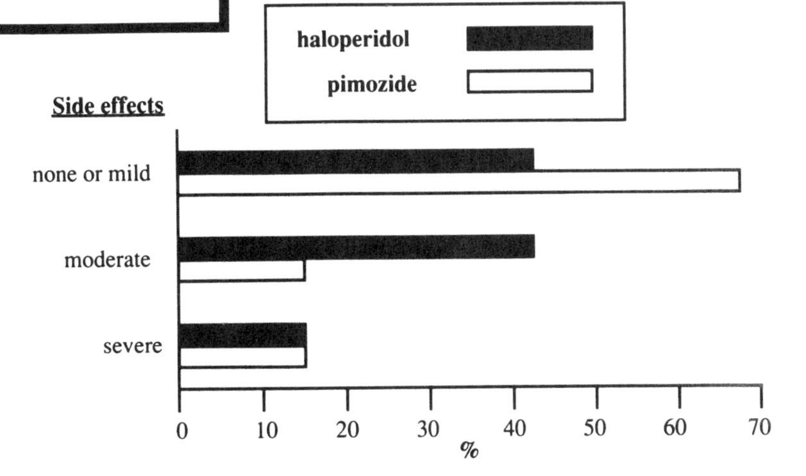

with their use. As with haloperidol, school phobia can also occur with pimozide[1190]. I have had some patients who did not respond to pimozide and had to be switched back to haloperidol.

The Long Half-life of Pimozide

The half-life of a drug refers to the time it takes for the blood level of a single dose to drop to half of the peak dose. The half-life of pimozide averaged 66 hours in children with TS and 111 hours in adults with TS[1682]. It is important to know this since this unusually long half-life means that the total blood level can continue to accumulate over a period of days, as follows:

Effects on the Heart

The only side effect that might be different for pimozide compared to Haldol is a minor change in the electrocardiogram (an increase in the distance between the Q and T waves) [651,974,1762]. This has not been associated with any adverse effects and is believed to be due to the fact that pimozide has some calcium channel blocking effects[p589]. I have seen no heart problems in several hundred patients treated with pimozide. It is generally felt that repeated electrocardiograms are not necessary unless the dose exceeds 20 mg per day, or there is a history of heart problems, or a pre-existing increase in the Q-T interval[651].

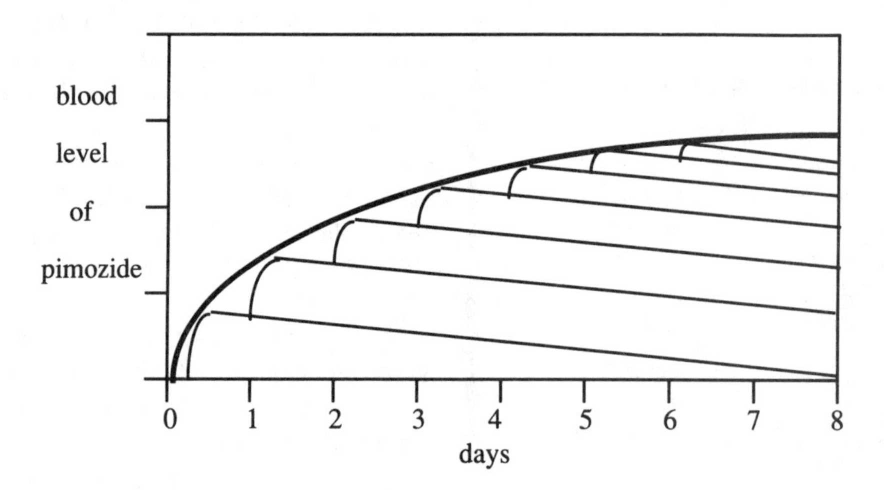

As can be seen, when the medication is very slowly excreted, the next day's dose will keep raising the blood level above the peak of the previous day and it may take a week or more to reach equilibrium. This explains the occasional observation that side effects may develop after an initial period of good response. It also indicates why starting at a low dose and slowly increasing it helps to avoid side effects. Finally, it shows why it may take several days for the symptoms to come back after the medication is stopped.

Summary: Pimozide is similar to haloperidol. Its major advantage is that it causes fewer side effects in some patients.

Chapter 76
Other Phenothiazines

The group of antipsychotic medications that block postsynaptic dopamine receptors are called phenothiazines. The following box shows the ones that are most often used for behavioral disorders in general.

Phenothiazines

Generic Name	Brand Name
chlorpromazine	Thorazine
fluphenazine	Prolixin
haloperidol	Haldol
pimozide	Orap
thioridazine	Mellaril
thiothixene	Navane
trifluoperazine	Stelazine

Prolixin

Next to haloperidol and pimozide, Prolixin (fluphenazine) is the phenothiazine we and others[195,745,974] use most often. It comes in 1, 2.5, 5 and 10 mg tablets and the usual starting dose is 0.5 to 1 mg per day with weekly increases, as for haloperidol. It requires a slightly higher dose than for haloperidol (an average of 4 mg compared to 3 mg for haloperidol). The major advantage of Prolixin is that it is as effective as haloperidol but often has fewer side effects and causes less sedation[195,196,974].

I have had several patients who have responded only to Prolixin. One teenage girl with severe tics and coprolalia had not responded to a wide range of medications over a period of many years. She had an immediate 80% decrease in symptoms as soon as she began taking Prolixin. Another seven year old girl had severe facial grimacing, tongue thrusting and "swinging from the chandeliers" hyperactivity that did not respond to many different medications including haloperidol, pimozide, Ritalin, Dexedrine, imipramine, Prozac, clonidine, Tegretal, Clonopin, and others. After taking 1 mg of Prolixin 90% of her symptoms disappeared. Another patient was a 40 year old male with severe self-abusive behavior consisting of hitting himself on the nose and yelling obscenities every few seconds. This behavior left him depressed and suicidal. Prolixin, 5 mg twice a day, resulted in an 80% improvement in his symptoms.

In summary, Prolixin, which has been overshadowed by its two cousins, haloperidol and pimozide, can be more effective than either with fewer side effects.

If One Medication Doesn't Work, Try Another

In one study, 45 patients were initially treated with haloperidol. Of these 75 percent improved and 25 percent did not. As shown in the following diagram, if one medication did not work a second was tried and if that failed a third was tried[1319]. Clonidine and all of the phenothiazines listed above, except Thorazine and Mellaril, were used.

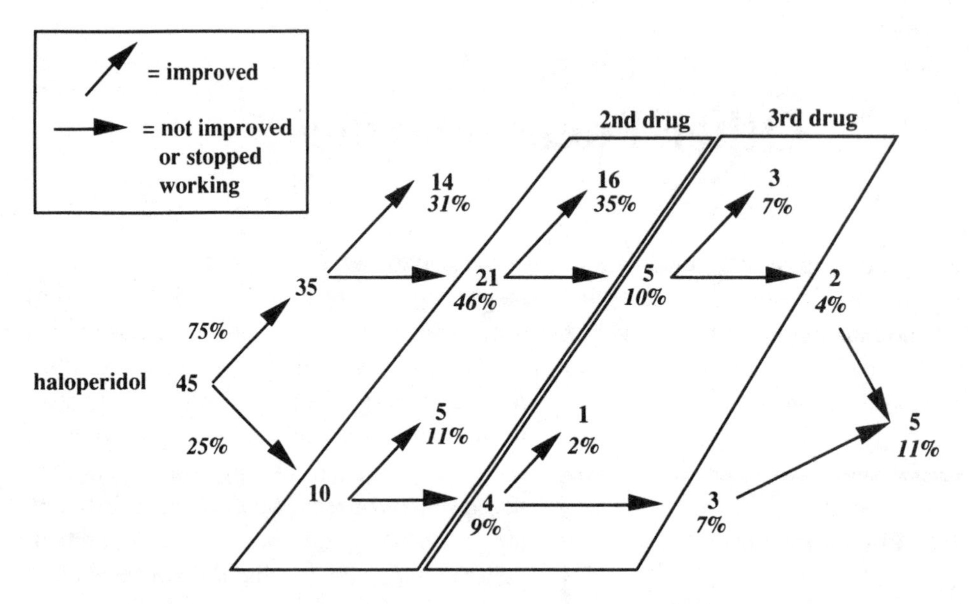

This shows what we have also found — with persistence, the treatment of TS can be quite successful. After a trial of one to three medications, only 11 percent did not have a good response without disturbing side effects.

Other Related Medications

Tetrabenazine (Nitoman) blocks both pre and postsynaptic dopamine receptors. In one study of 15 TS patients, 80 percent had an excellent to moderately good response[974,975]. Others have also had good results[1767,1911]. Piquidone (RO22-1319) also blocks post synaptic D_2 receptors, especially in the mesolimbic[p367] system, and has been benefical in treating TS[1974,1975]. The problem with these two medications is that they are not generally available.

Phenothiazines Not to Use

Children entering the clinic with undiagnosed TS and a conduct disorder are sometimes being treated with Mellaril. This and Thorazine are relatively ineffective in the treatment of tics. Such patients usually do better when switched to Hadol, pimozide or Prolixin. This statement should be modified by the caveat that when behavior is the primary problem, Mellaril may be quite useful.

Summary: Prolixin is similar to haloperidol and pimozide but may have fewer side effects than either of them. In some cases, it has been the only medication that was effective.

If one medication stops working, often a switch to another related drug will once again bring the symptoms under control.

Chapter 77
Tardive Dyskinesia

Tardive dyskinesia (TD) is an involuntary movement disorder that may occur after long periods of treatment with neuroleptics such as Thorazine, haloperidol and pimozide. Because it is the one side effect that parents and patients ask about most often, I have devoted a whole chapter to it. Many patients have come to our clinic with very significant TS symptoms but are petrified of taking haloperidol because they have been told it "can cause permanent brain damage" or something equally frightening. What is tardive dyskinesia and how dangerous is it?

The term *dyskinesia* is used because the major symptoms are muscle movements around the mouth consisting of lip smacking, puckering, pursing, chewing and sticking the tongue out. Rapid eyeblinking or spasms of the eyelids can sometimes occur. It may also involve slow, snake-like (choreic) movements of the arms or legs. The term *tardive* is used because the symptoms come on after long-term use of the medication. They are usually most noticeable after the medication is discontinued.

The syndrome was first described in 1957 and the term tardive dsykinesia was coined in 1964[1579]. Prior to this, because of the prominence of symptoms around the mouth, it was given the difficult name "bucco-linguo-masticatory syndrome," meaning "cheek-tongue-chewing" syndrome. It is mostly described in patients with schizophrenia.

Tardive dystonia Dystonia refers to muscle spasm-like symptoms like those that occur in individuals with a wry neck. When dystonic symptoms of the neck, arms or legs occur in patients who have been treated with neuroleptics, and other neurological diseases are not the cause, it is called *tardive dystonia* and is a variant of tardive dyskinesia[258]. It is somewhat more likely to occur in younger patients. Two TS patients who developed tardive dystonia after treatment with neuroleptics have been described[1808a]. Although the symptoms persisted after the medication was stopped, they improved with tetrabenazine.

The frequency of TD The fact that patients with schizophrenia who have not taken any medications also have these dyskinesia movements[p208] complicates the situation. In determining the frequency with which tardive dyskinesia occurs after treatment of schizophrenia, this background of dyskinesia must be considered. It is estimated that 20% of schizophrenic patients taking neuroleptic drugs for long periods show some dyskinesia, and that 5% of untreated patients have dyskinesia, leaving 15% due to the medication[94,305,981]. The frequency is highly dependent on the definition and whether patients with mild symptoms are included. Most TD is mild to moderate in severity. Severe and disabling dyskinesias are relatively rare[981].

Is TD permanent? While most cases of TD slowly get better after the medication is stopped, the symptoms can be permanent. Younger patients are more likely to improve than older ones[305, 1815].

Risk factors Age, sex and dose are the primary risk factors. The frequency of TD in schizophrenics is lowest in patients less that 40 years of age (10%) and highest in those over 60[1815]. Women are more susceptible than men at a ratio of 1.7 woman to 1 man[1011]. The total accumulated dose of medication plays an important role but there is a lot of variation in this[305].

The cause of TD Since the primary effect of neuroleptics is to block postsynaptic dopamine receptors, it has often been assumed that hypersensitivity of these receptors is somehow involved in TD. However, there is little direct evidence to prove this[305]. A model of TD has been developed in the cebus monkey [304,823,1087] which suggests that a decrease in dopamine turnover and a deficiency of GABA [595,1035] play a role. Treatment with agents that enhance GABA improves symptoms in about half of the cases[981,1935]. The symptoms usually decrease or disappear when the neuroleptics are given again. Tardive dystonia may be treated with tetrabenazine or Cogentin[258].

Does tardive dyskinesia occur in Tourette sydrome? The answer to this important question is — rarely. I have seen it in only one of my TS patients, who had all the risk factors. She was female, 72 years old, and had been treated with 2 to 4 mg of haloperidol daily for 5 years when she began to develop some typical dyskinesia movements around her mouth. Shapiro and Shapiro[1760] had not seen TD in any of their patients. The combined experience of these two clinics includes well over 2,500 patients. The following are some reported cases.

Caine and co-workers[276] described a 15 year old boy with ADHD, motor and vocal tics and "discipline problems." At age 12 he was diagnosed as having TS and treated with haloperiadol at up to 50 mg per day. Eight days after the haloperidol was stopped his tics got much worse and he began to develop oral chewing movements, tooth grinding and choreic movements of the extremities. Four weeks after the medication was discontinued he developed psychotic symptoms. Because of the severe psychosis and tics, the haloperidol was started again and had to be returned to the previous level before all the symptoms improved. Caine and Polinsky[275a] later described three additional TS patients ages 8, 16 and 61 with varying degrees of tardive dyskinesia, after treatment with 4 to 18 mg of haloperidol. In all three the dyskinesias stopped after the medication was decreased or stopped.

Mizrahi and co-workers[1342] reported a seven year old boy with motor and vocal tics, coprolalia, learning disorders and emotional lability. He improved on 0.5 mg of haloperidol twice a day. Five months later he developed protrusions of the tongue and chronic turning of his head to one side. The haloperidol was discontinued and the symptoms of TD completely disappeared within two months. Fortunately, although the tics returned, they were adequately controlled by other medications.

Golden[757,758] reported three cases of TD in Tourette syndrome. In all three the symptoms disappeared one week to three months after the medications were discontinued.

Riddle and co-workers[1610] reported a 32 year old man with severe TS. The symptoms started at age six. He was diagnosed at age 21 and treated with haloperidol. Over a ten year period the dose was gradually increased to seven mg a day. Because of side effects the haloperidol was decreased and then stopped. Over the next ten days his tics became so severe he had to be hospitalized. During this time symptoms of tardive dyskinesia were noted. A trial of clonidine was unsuccessful and haloperidol was given again at a lower dose of 2.5 mg per day. This resulted in a marked decrease in both the tics and the tardive dyskinesia.

Thus, of nine cases of tardive dyskinesia reported among thousands of TS patients treated with phenothiazines, in seven the symptoms disappeared after the medication was stopped, and in two the symptoms of TS became so severe the medication had to be started again and this resulted in a marked improvement of both the tics and the tardive dyskinesis. Thus, statements such as "treating TS with haloperidol can result in permanent brain damage" are, a) untrue, b) unnecessarily frightening, c) primarily reveal that the person making the statement is not really experienced in the treatment of TS, and d) may do more harm than good by preventing a significantly symptomatic TS patient from receiving adequate treatment.

Relevance to treating TS Treatment of TS is not to be taken lightly. A careful assessment of the benefits versus the risks should be undertaken. However, if a TS patient has not responded to alternative medications such as clonidine, and has symptoms that are causing significant problems in their life, haloperidol or pimozide or Prolixin are excellent medications that bring about a significant improvement in 70 to 80% of cases. In regard to tardive dyskinesia all the risk factors are in the TS patient's favor: a) they tend to be young, b) they tend to be males, c) they have a different disorder than schizophrenia, and d) they usually respond to much lower dosages of medication than are used in schizophrenia[547]. All of these factors contribute to why tardive dyskinesia is a rare complication of TS. If care is directed toward keeping the dose as low as possible, and decreasing or discontinuing the medication when the symptoms improve, the phenothiazines can be quite effective medications with a good risk-benefit ratio.

Summary: Tardive dyskinesia is a disorder of involuntary muscle movements, especially around the mouth, that can occur in patients who have taken haloperidol or other phenothiazines for months or years. Despite thousands of TS patients treated with these medications it has been reported less than a dozen times and in most cases the symptoms either disappeared when the medication was stopped or the TS became so severe the medication had be started again and this resulted in disappearance of both the tics and the tardive dyskinesia.When the guidelines for treating TS patients are followed with the dose kept at the lowest effective levels, treatment with phenothiazines presents a minor risk compared to the potential benefits.

Chapter 78
Clonidine

When it works, clonidine is an outstanding drug for the treatment of Tourette syndrome. What makes it so good is the presence of few side effects and in many cases it may improve not only the motor and vocal tics but also the ADHD, oppositional confrontive and obsessive-compulsive behaviors. The major disadvantage is that it only works well in somewhat more than half of TS patients. Clonidine's major features can be summarized as follows:

clonidine (Catapres)

Action:	α_2-noradrenergic receptor agonist decreases brain norepinephrine activity
Starting dose:	0.025 mg (1/4 tablet)
Usual dose:	0.025 - 0.1mg 1 to 5 times/day
Side effects:	tiredness
Effectiveness:	60% of patients

The use of clonidine in Tourette syndrome was first described by Cohen and colleagues in 1979[339,343]. They had noticed that the breakdown products of norepinephrine were elevated in a child with TS[341,342]. This and the observation that stress enhances the activity of norepinephrine neurons in the locus ceruleus[p411] led them to try clonidine. This was known to stimulate the presynaptic a_2-norepinephrine receptors and decrease the release of norepinephrine[p410]. Clonidine is also used for inhibiting the rush of norepinephrine released in the locus ceruleus when addicts are undergoing withdrawal. It has also been used to treat the symptoms of withdrawal from cigarette smoking[731]. The major use of clonidine is to treat high blood pressure in adults.

Initial doses Clonidine comes in 0.1 mg tablets. We have patients cut this into four parts with a razor blade. The starting dose is 1/4 of a tablet (0.025 mg) each morning for the first week. In about 20% of patients this is all that is required. However, if this does not produce a satisfactory improvement, an additional 1/4 tablet is given in the afternoon during the second week. If necessary, a third 1/4 tablet can be given at bedtime for subsequent weeks. I then have the patient report back before there are any further increases in the dose. Usually, if there is no response at a dose of one tablet three times a day, clonidine is unlikely to be successful.

Maintenance dose Most patients require between 1/4 a tablet once a day to 1/2 a tablet several times a day. Unlike haloperidol or pimozide, clonidine is a relatively shortacting medication. We have sometimes found it necessary to give it up to five times a day.

The Clonidine Patch

To avoid these numerous doses and prevent marked changes in blood level we and others[944] have found that *clonidine patches* work well. These are placed on the skin and allow the drug to be adsorbed through the skin to give an even dose throughout the day. We have now used the clonidine patch in over 300 TS patients. The following box summarizes this experience.

Clonidine by Patch

Starting dose:	1/4 -1/2 TTS-1 per week
Effectiveness:	60% effective 25-30% ineffective 10-15% symptoms worse
Side effects:	tiredness (reduce patch size) skin irritation (25%) - change position - use Sween or Aleovera - change to oral form
Problems:	patch falls off or is pulled off

I have often found that *clonidine given by patch may work when clonidine by mouth is ineffective*. The following report from a teacher illustrates this:

> "For about two weeks his behavior was very calm, compliant and appropriate. During this time, medication was administered through a skin patch. It irritated his skin, however, so he was switched back to medication given orally. Since this time his behavior deteriorated. He has great difficulty sitting still, attending to his work, keeping his hands to himself and interacting appropriately. His peers have continually complained about his frequent references to private body parts and their various slang labels."

A mother described what happened when her son's patch fell off:

> "M. is so much better on the patch. He feels better, thinks more clearly and all that horrible arguing and confrontive behavior is gone. One afternoon after he came home the patch fell off. Within 40 minutes he was back to all of his old behaviors which disappeared again within a hour of putting the patch back on."

Because of these and many similar stories I now *start treatment of virtually all new TS patients with the clonidine patch. and find it to be the drug of choice.* It comes in three strengths: Transdermal Treatment System, or TTS-1, -2 and -3, equivalent to a daily dose of 0.1, 0.2 and 0.3 mg. It usually stays on during showers but comes off with a lot of swimming or sweating.

Cutting the patch. As experience with the clonidine patch grew, I found that many patients, even adults, were too tired using even half a TTS-1 patch. Because of this I usually start with 1/4 of a patch in young children and 1/2 of a patch in adolescents or adults. The size of the patch is then adjusted up or down according to their response, i.e., *the dose is adjusted with the scissors*. One 18 year old college student with TS became much more centered and able to concentrate and do her work using 1/8th of a patch. Any higher dose made her too tired.

Pharmacists have sometimes told patients they cannot cut the patch or the medicine will leak out. The drug company assures me this is not true. Also the TTS-2 and TTS-3 patches do not have a higher concentration of medication. They are simply two and three times larger than the TTS-1. Thus, *the dose of clonidine given by patch is adjusted by adjusting the size of the patch*. This may include using, for example, a whole patch plus 1/4th of a patch.

Skin irritation from the patch We have found that significant problems with skin irritation occur in 20 to 30 percent of the patients using the patch. Sometimes this can be severe with blistering and weeping. If the irritation is moderate it can be alleviated by changing the position of the patch every two to three days. In some cases both the irritation or a problem with the patch falling off can be aleviated by first spraying the skin with steroid prepartion such as Vancenase Nasal Spray.

Clonidine Can Improve a Wide Range of TS Symptoms

As described in the next chapter on the treatment of ADHD, one disadvantage of using Ritalin and related medications is that they can sometimes make the tics worse. One of the major advantages of clonidine is that it has the potential of improving both the tics and the ADHD[943,944] with one medication. If distractability persists, small doses of Ritalin may be added to the clonidine[943]. In addition to tics and ADHD, oppositional and confrontive behavior, short temper, obsessive-compulsive behaviors, low frustration tolerance, speech difficulties, poor handwriting, self-mutilation, and other TS symptoms may also improve[339]. Clonidine has been found useful in the treatment of obsessive-compulsive behavior[1063], anxiety[905,1905,1972,1973], panic attacks[905,1183], phobias[772], mania[850,1245], memory defects[597a,1280], schizophrenia[1991a] and narcolepsy[772,1558]. The following statements from parents illustrate some of the positive effects of clonidine.

> "His school work improved. It was clear that the clonidine was responsible because it deteriorated again when I decreased the dose from five to three times a day."
>
> "He is more reasonable now. His teachers said 'He seems to be very happy to be here for the first time all year.'"
>
> "Jim is much calmer. When he doesn't take his medication it is like day and night. Off the medication he again becomes confrontive and thinks he has the right to go through all our stuff."
>
> "Her handwriting is better, she participates more in class and is doing better academically. Her teacher is amazed. The tics are gone, she gets in fewer fights, and the emotional roller coaster has stopped."

The occasional superiority of the brand name Catapres over generic clonidine is indicated by the following example:

> R.H. is an eight year old boy with ADHD and behavior problems. Since he developed motor and vocal tics after treatment with Ritalin he was switched to Catapres. He did very well on a dose of 0.05 mg (1/2 tablet) four times a day. His mother then reported that he again developed trouble concentrating, became more aggressive in school, was suspended several times, once impulsively tried to jump through the window in the classroom, was asking inappropriate questions, and was choking other children. She then noticed that when the prescription was refilled the pharmacist had switched to a generic form of clonidine. Within one day of returning to the non-generic form (Catapres) he was concentrating better, thoughtful, asking appropriate questions and stopped being aggressive.

The effect on compulsive behaviors is illustrated by the following two cases:

> A 5 year old girl with vocal tics whose brother had TS was compulsively masturbating in class. After treatment with 1/4 of a tablet each morning this behavior stopped.

> "He was unable to read because he had a compulsion to count the words. After 1/4 tablet of clonidine twice a day this stopped and he now enjoys reading."

The fact that clonidine is useful for the treatment of ADHD in children, mania in adults, and TS is consistent with a link among these three conditions.

Side Effects

The major side effect of clonidine is *tiredness* or a feeling of being sedated. Fortunately, this is often short-lived and usually disappears by the end of the first week or two, or when the dose is decreased. Occasionally it persists and a different medication has to be used.

Since clonidine is used to treat hypertension, one side effect, especially with higher doses, is the presence of low blood pressure. This is noticed by the onset of a feeling of dizziness when standing up. This has the fancy name, *orthostatic hypotension.* In our experience, less than 5 percent of patients have this problem, and it responds well to decreasing the dose. We have not found it necessary to take blood pressure at home. Other occasional side effects include dry mouth, racing thoughts, headaches, and insomnia.

I have found that *in about 10 percent of patients the tics, hyperactivity, irritability, and conduct problems may actually get worse* after beginning treatment with clonidine. With all the attention paid to the fact that Ritalin can sometimes make the tics worse, this fact tends to be ignored. I'm not sure whether the symptoms get worse more often with Ritalin or with clonidine. If the symptoms get worse, this is an indication to switch to another medication.

Tardive dyskinesia A further advantage of clonidine is that there is no risk of tardive dyskinesia. In fact, clonidine has been reported to be useful in the treatment of tardive dyskinesia[634].

About half of TS patients respond to clonidine In the initial studies 70 percent of patients were reported to respond to clonidine [239,339,502]. Later reports have shown improvement in 20 to 60 percent[1142,1143,1758,1764]. In my experience with hundreds of patients treated wtih clonidine, somewhat over half have responded well.One double blind study showed that 68 percent of TS patients responsed to clonidine[196] while another concluded there was little objective evidence for improvement, although half of the patients preferred to remain on the medication[743]. *Except in the most severe cases, we use clonidine as the initial drug of choice and about 60 percent have had a good response.*

Delayed improvement In their initial report of the use of clonidine in TS, Cohen and co-workers[339] spoke of an immediate effect occurring within hours or days of a decrease in anger and irritability, and feeling more in control. This was followed by a second phase occurring after three to four weeks of decreased compulsive behavior and a decrease in motor and vocal tics. In our experience we find that *for some patients everything, including behavior, attention span, compulsions and tics may improve immediately*, and that there may be some additional further improvement with time. In others the immediate effect may be less striking and the improvement is slower. However, if we do not get some hint of a good response after six weeks of trying with a slow increase in dose, we switch to haloperidol, pimozide or Prolixin. In some patients gradual tolerance develops and the tics[339] or anxiety[1973] return.

Don't stop clonidine suddenly Because of a potential rebound effect of a significant increase in blood pressure and pulse, if clonidine is going to be stopped it should be tapered off slowly over a period of a week or more. Abrupt withdrawal, especially from fairly high doses, may also be associated with a marked increase in tics, restlessness, decreased attention, increased anxiety and talkativeness and difficulty sleeping[1143]. Some of these problems can also occur when the medication is gradually stopped[1144,1145]. Some have reported that if clonidine is stopped, after restarting, it may take several weeks to get the tics back to the same level of control[1143]. For this reason patients should take care that they do not unexpectedly run out of clonidine.

Mode of action While it is natural to assume that the major effect of clonidine is on brain norepinephrine, long-term treatment has little effect on norepinephrine activity[1145,1798]. This suggests that indirect effects on dopamine or serotonin may be more important[1798]. These effects have been discussed elsewhere[P412].

Effect of clonidine on growth Growth of children is dependent upon the presence of growth hormone produced by the pituitary. Clonidine stimulates the release of growth hormone by increasing the release of a growth hormone-releasing peptide[1518]. Although 25 of 35 children with delayed growth showed an increase in their rate of growth when given clonidine[1518,1519], it has no effect on the growth of normal children.

Summary: When it works,clonidine is an excellent medication for the treatment of Tourette syndrome. It can be effective in the treatment of the tics, ADHD, phobias, anxiety, obsessive-compulsive symptoms, depression and aggressive and oppositional behavior. The major side effect is tiredness, which is often temporary or disappears with a decrease in the dose. The disadvantage of clonidine is that it is effective in only about 60% of the cases and about 10% of the time it can actually make the symptoms worse.

Clonidine is maximally effective when given by skin patch and in some cases this has been effective when clonidine by mouth was ineffective. The major side effect of the patch is skin irritation.

Chapter 79

Treatment of ADHD

There are few issues in medicine that evoke a wider diversity of opinion than the treatment of attention deficit hyperactivity disorder (ADHD). The same parents who might pressure their doctor to prescribe penicillin for their child's cold may be adamantly opposed to the treatment of their child's ADHD. Colds are due to viruses which are unaffected by penicillin. Penicillin can cause death by anaphylactic shock and is potentially a far more dangerous drug than Ritalin and related medications used to treat ADHD. Like penicillin, Ritalin and its cousins, can be, and sometimes have been, over-prescribed and overused. The diagnosis of ADHD should be based on DSM-IIIR[490,491,p73] criteria using information from as many sources as possible. Medication should be considered an adjunct to educational and other approaches[p609]. In some cases, tutoring and resource or special education classes are all that is required. In other cases, some children literally cannot function in any environment without the help of medication.

Before discussing some of the details about these medications I would like to cover some general misconceptions about their use.

Misconceptions About the Use of Medications for ADHD

Stimulants cause the child to become addicted to drugs Since stimulants used on the street are called "uppers" or "speed" it is understandable that this would be of some concern. The use of moderate doses of stimulants in children with ADHD does not cause a "high" or euphoria[384,1578]. As discussed in detail elsewhere[p395], the stimulant medications enhance the function of dopamine and norepinephrine neurons. This corrects a chemical imbalance in the brain and allows ADHD children to be less distractible, less hyperactive, and to concentrate better. The medications are not a "chemical straightjacket." If anything, *they cut the strings of the genetically induced chemical straight jacket that ADHD children are locked into.*

Children with ADHD and TS are at a higher risk for alcohol and drug abuse[p84]. This, however, is due to the disorder itself and not the medication. In my experience, ADHD and TS children who have significant symptoms and are properly treated, including medication when indicated, do better in school, have higher self-esteem, are less impulsive, think through the consequences of their actions better, and are less likely to slide down the school social ladder to the lower rungs of the dropouts, truants, "loadies" and "druggies" where peer pressure to use and abuse drugs is the highest. In other words, properly medicated ADHD and TS children are less likely, not more likely, to become drug addicts. Others agree[883,131,1132,1203].

ADHD can be ruled out by a brief inspection in the doctor's office Parents of ADHD and TS-ADHD children not infrequently report, "I took him to my pediatrician. He took one look at him and said 'He is definitely not hyperactive.'" The examination of some ADHD children is like having a cyclone in the room. They are into everything. They take the ophthalmoscope off the rack and bounce it on the floor, open all the drawers, turn all the faucets on and off and rip the paper off the examining table. However, other ADHD children are only having problems in a classroom of 20 other children and can sit like quiet angels during a one-on-one examination in the doctor's office. The diagnosis of ADHD depends on a careful history of how the child is behaving in class and at home, and his school performance over a period of months and years. It cannot be made by a quick glance in the clinic.

An EEG and CAT scan is necessary to diagnose ADHD Again, a careful history is the best tool for diagnosing ADHD. Unless there are other disturbing symptoms such as seizures or excessively explosive behavior, an EEG and CAT scan is usually not necessary.

The medication will just turn my child into a zombie If a child becomes overly quiet or "like a zombie" on medication, that is actually a sign the medication is working. The only problem is he or she is taking too much. Children vary tremendously in the doses of medication required. For some 2.5 mg of Ritalin in the morning is adequate, while others may require 20 mg several times a day. If a child becomes too quiet or seems to lose normal spontaneity, the dose should be decreased or another medication substituted.

The medication will stunt my child's growth Shafer and Allen[1678,1677a-e] reported that the administration of Dexedrine resulted in an average gain of 1.0 pounds over a period when they should have gained 1.5 pounds. This was most marked during the first year and disappeared by the third year of treatment. Slowing of growth was less marked. The effects were also less marked with Ritalin and occurred mainly when the dose was high (greater than 2 mg per kilogram per day). There was a rebound acceleration of growth when the medications were stopped over the summer[1753]. Others have found little or no problems with growth except at high doses of medication[572,809,1056a,b].

Stimulant drugs are overprescribed Stimulant medications should only be given when there is a clear diagnosis of ADHD and the child is clearly disabled by significant symptoms. It is generally accepted that ADHD occurs in about 5 percent of boys. Although the frequency in girls is less, in our experience girls are more likely to have ADHD without hyperactivity and are thus less obvious. While it is not unusual for many children in an educationally handicapped or learning handicapped class to be taking medication, the frequency of its use for the school as a whole should be less than the frequency of ADHD. Thus, on average, only zero to three children per 100 students would be expected to require stimulant medication. Examples of overuse and abuse have been documented. Medication should not be used to help non-handicapped children do better in school. Alternatives to prescription medications are discussed later [p591,p609].

Bradley's Discovery

In the late 1930's Charles Bradley[206] reported on the effects of Benzedrine on 30 children with educational and behavioral disorders severe enough to require hospitalization.

> "Possibly the most striking change in behavior during the week of Benzedrine therapy occurred in the school activities of many of these patients. Fourteen children responded in a spectacular fashion. Different teachers, reporting on these patients, who varied in age and school accomplishment, agreed that a great increase of interest

in school material was noted immediately. There appeared a definite 'drive' to accomplish as much as possible during the school period, and often to spend extra time completing additional work. Speed of comprehension and accuracy of performance were increased in most cases.

"This is the more striking when we note that these patients were of good intelligence and they were receiving adequate attention for any personality disorders which might affect their school progress. Moreover they were already in a school where specially trained, sympathetic teachers dealt with their pupils either on an individual basis or in very small groups. To see a single daily dose of Benzedrine produce a greater improvement in school performance than the combined efforts of a capable staff working in a most favorable setting, would have been all but demoralizing to the teachers, had not the improvement been so gratifying from a practical viewpoint."

Emotionally they were also improved. Children who had expressed their irritability in group activities by noisy, aggressive, domineering behavior became more easy-going and seemed more interested in their surroundings.

He commented on the seeming paradoxical effect of this drug and suggested an explanation that is valid today.

"It appears paradoxical that a drug known to be a stimulant should produce subdued behavior in half of the children. It should be borne in mind, however, that portions of the higher levels of the central nervous system have inhibition as their function, and that stimulation of these portions might indeed produce the clinical picture of reduced activity through increased voluntary control."

Later Studies

Twelve years later, Bradley[207] published a follow-up report on treating children with Benzedrine or Dexedrine and noted that 70% showed some improvement. What usually happens in medicine after such a glowing report of a new form of treatment is that subsequent doctors are far less impressed and the treatment is eventually abandoned or sharply modified. However, as others began to try this group of medications on similar children, the effects were almost universally the same[118].

The following box summarizes some of the reported effects of stimulant medication on ADHD[116,119,125a,283,383-386,391,525,527,724,725,1235,1491a,1838a,1839,1869a,1909a,1913].

Effects of Stimulant Medication on ADHD

Increased	Decreased
attention span	aggressiveness
accuracy	childish behavior
concentration	daydreaming
learning	defiance
memory	demanding behavior
mood	destructiveness
motor coordination	disobedience
on-task behavior	distractibility
self-correcting behavior	fiddling with hands
	hyperactivity
	laziness
	lying
	name calling
	noise making
	oppositional behavior

On average, 75 percent of children respond favorably while the remaining 25 percent are unchanged or made worse[118,207,2048]. The best predictor of whether the medication will be effective are those reports or tests that relate to poor attention span or concentration[118].

Improvement of handwriting One of the more obvious signs that stimulants can improve a wide range of problems in children with ADHD is the effect on tasks requiring fine motor coordination, such as handwriting. The following figure comes from a paper the teacher sent home. She noticed one morning that Scott's handwriting was terrible [left] and asked if he had taken his medication that morning. He had forgotten it. An hour after taking his Ritalin she tested him again and his handwriting was markedly improved [right].

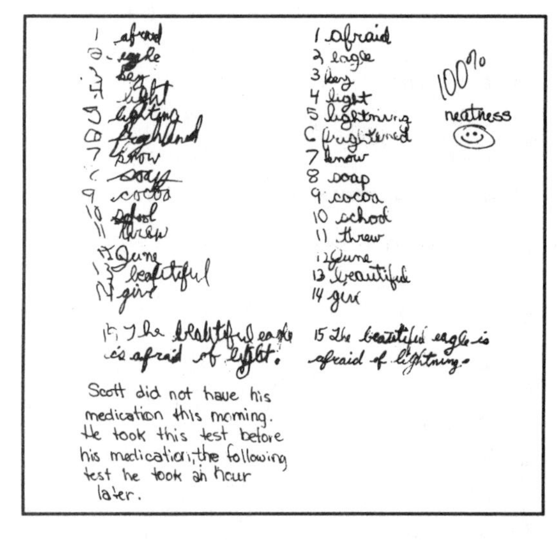

Improvement of frontal lobe functioning The role of a chemical frontal lobe syndrome in ADHD has been discussed [p389]. It is thus not surprising that one of the tests that shows the most striking improvement is the Porteus Maze Test[383,384] which is sensitive to frontal lobe function[p356]. This is true even of children with learning disorders without behavior problems[725].

Improvement of mother-child, teacher-child interactions When mothers care for hyperactive children they show significantly more controlling and directive interactions and attempts at impulse control than when they care for normal children[281,282]. In studies where the observers were blind to who was on medication, the hyperactive children treated with Ritalin made more positive interactions with their mothers and the mothers in turn behaved in a less controlling fashion[117,121,942,2075]. These findings "make it unlikely that the controlling behavior of mothers of hyperactive children is anything other than a response to the way their children act."

The same was observed with teachers. Without knowledge of which children were treated, the teacher was more intense and controlling toward unmedicated hyperactive boys than normal boys or those on medication[2074].

Stimulants do not increase IQ Stimulants are not "smart pills." They simply allow a child's native abilities to be utilized without the interference of distractibility and inability to pay attention. Many ADHD and TS patients have excellent native intelligence and numerous studies have shown that stimulants do not increase the IQ[384].

Long-term effect of stimulants on academic performance It should come as no surprise that stimulant medications that increase attention span and decrease distractability should also result in improved academic performance in the classroom. Grades sometimes improve dramatically. Despite this practical experience, some early studies suggested there was little effect of stimulants on academic performance[116]. Later studies that avoided some of the weaknesses of these earlier reports[506a] and by examining actual classroom performance found increased output, accuracy, efficiency, learning, reading comprehension, spelling word acquisition and on-task and self-correcting behavior for children on Ritalin compared to those on placebo [391,506a,1491a,1869a]. The **top diagram on the next page**, illustrates the results of the study by Pelham and co-workers[1491a].

There was a significant dose-related improvement in math and reading performance. On-task and negative behaviors were also examined (**bottom figure**).

Again there was a significant increase in on-task behavior and a marked decrease in negative behaviors such as destruction of property, disturbing others, inappropriate

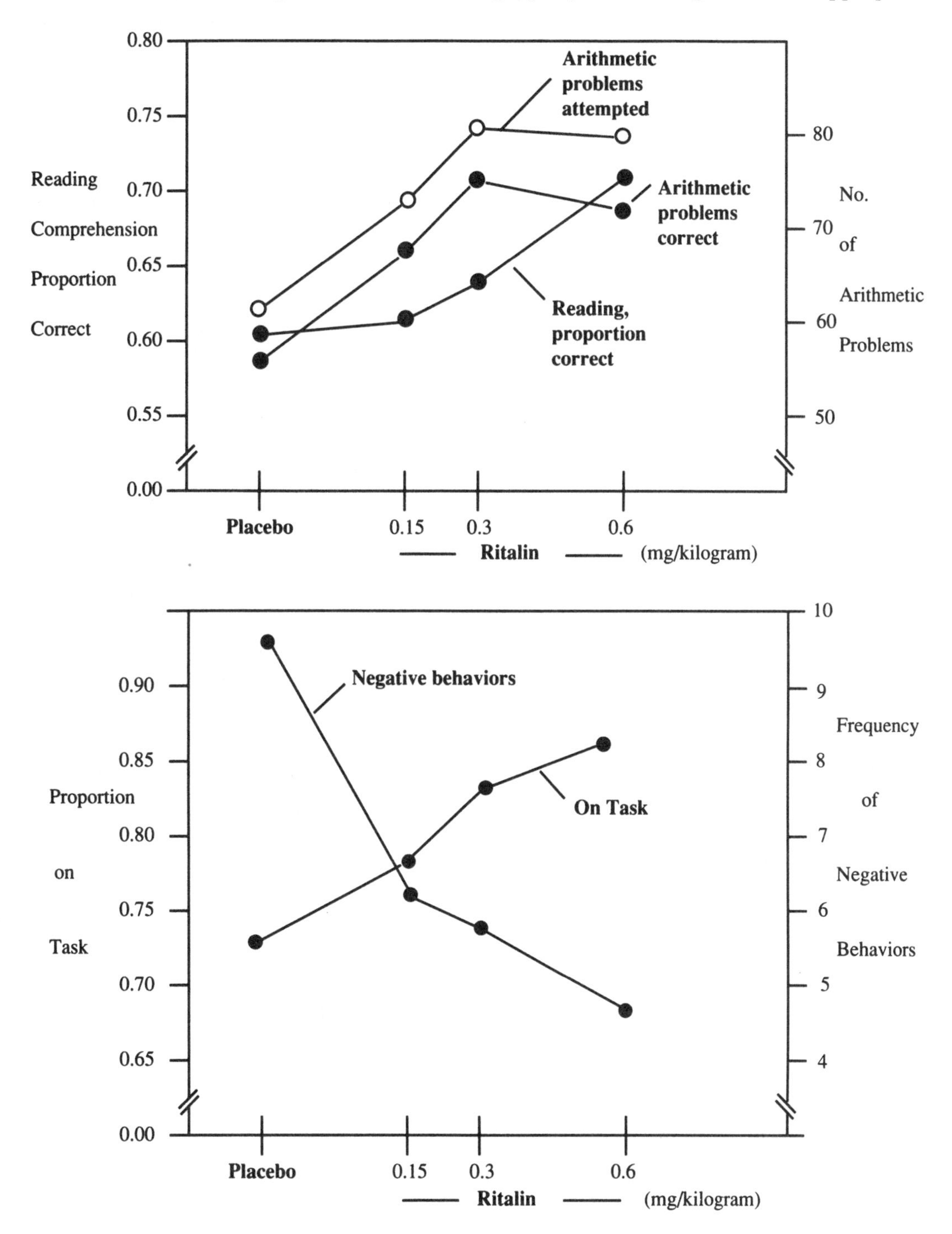

talking, name calling, swearing and teasing.

Is there a paradoxical effect of stimulants? The statement is often made that stimulants have a paradoxical effect of calming down children with hyperactivity while they stimulate normal children[54]. To study this, Judith Rapoport and colleagues at the National Institutes of Health[1578] studied the effects of treating ADHD boys, normal boys, and normal men with Dexedrine. The effect on motor activity was as follows:

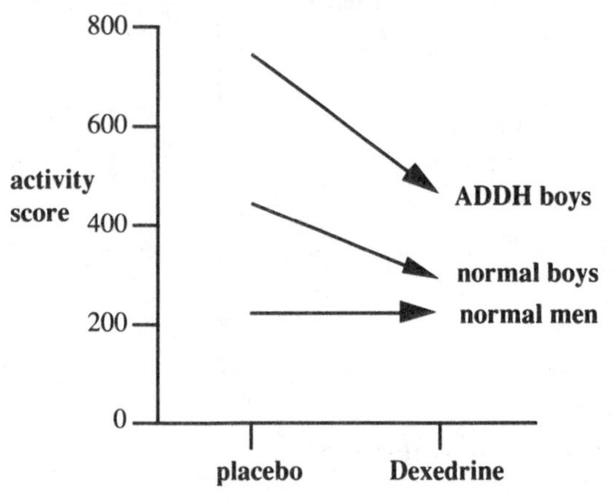

This shows that ADHD boys did indeed show the most striking improvement, but they were also the most hyperactive to begin with and *the medication brought them to about the same level as normal boys, before medication.* Activity was also decreased in normal boys but they were less hyperactive to begin with. Normal men had low activity scores on the placebo and this did not change on the Dexedrine. These results indicate that it is more accurate to state that stimulants have a *"normalizing effect"* rather than a "paradoxical effect" on hyperactive boys. There was an age-related paradoxical effect in regard to feeling "high" or euphoric. Neither the ADHD nor the normal boys had a feeling of euphoria, while the men felt euphoric and more fidgety. The medication improved attention and learning in all three groups.

Medication and psychotherapy The use of medication in the treatment of ADHD is discussed in subsequent chapters. As effective as medication can be, in most cases a multimodal approach of medication, appropriate education, special classes, tutoring, and psychotherapy is required to produce the maximum and most long-lasting effect [1056,1703].

Diet and ADHD

The effectiveness of diet in controlling the symptoms of ADHD has been widely debated. Feingold[584] suggested that many of the symptoms may be due to a reaction to food coloring, salicylates or preservatives in food. Although carefully controlled studies have not verified that the Finegold diet is effective[1395,2005], many parents think it has helped their children. My recommendation to parents who want to try it is to do so, but be objective about its effect. If they feel it is helping, continue it, if not stop.

Oftentimes restriction of a specific type of food rather than the whole diet is effective. The type of foods most mentioned are those rich in sugar. Mothers frequently say, "Sugar really sets him off." Some studies have supported these parents observations[1552], others have not[1840,2103]. A diet designed to improve food allergies may bring about a decrease in some symptoms[521].

Summary: Many of the misconceptions about the use of stimulant medications in the treatment of ADHD are discussed. When properly used they are safe and effective in the treatment of about 75 percent of children with ADHD.

Chapter 80
Ritalin and Dexedrine

Ritalin

Although the first two stimulants used for the treatment of hyperactive children were Benzedrine and Dexedrine, these were soon largely replaced by Ritalin or methyphenidate. The major aspects of Ritalin are shown in the box.

methylphenidate (Ritalin)

Action:	enhances the function of dopamine neurons
Starting dose:	2.5 to 10 mg in the morning
Usual dose:	5 to 30 mg per day
Side effects:	decreased appetite insomnia upset stomach

Dose Depending upon the age and size of the child, the starting dose is usually from 2.5 to 10 mg each morning. The dosage may be slowly increased according to the response. If hyperactivity and attention span improve for the morning but deteriorate by afternoon, then either the academic classes can be moved to the morning time or a second dose may be given at noon. A fairly common regime is 10 mg in the morning, and 5 or 10 mg at noon, or early afternoon, a half an hour before doing homework.

The response to medication depends on the individual, with some children doing well all day on 5 or 10 mg in the morning and others requiring higher doses spread throughout the day. Occasionally there is a roller-coaster effect with significant symptoms returning four to five hours after a dose. This can sometimes be prevented by using a sustained release form of Ritalin (Ritalin 20 mg, S.R.) in the morning. In other patients this is does not work and only multiple doses are effective.

Drug-free periods If Ritain is given primarily for attention span in school, it may be skipped over the weekend, holidays or summertime. If no medication is given during the summertime, it is helpful to hold off returning to medication for the first few weeks of the next school year to determine if it is really needed again. Some children are so out of control off the medication that it must be given continuously with no or only moderate decreases during the summer.

As a further example, the following is a teacher's report on Jack, a nine year old boy with chronic motor tics and ADHD. He was just started on Ritalin, 20 mg sustained release, at the beginning of the school year.

> "Jack is having his best year to date. He is calmer and better behaved in class, accepts responsibility and completes assigned tasks. He does not frequently display the frustration, crying, and refusal to work that we often saw last

year. I have just completed Jack's IEP. He tested well and showed real progress in all areas, especially math, and for the first time this year is having some success in reading. He seems to be grasping the concept of a sound/symbol relationship so is beginning to sound out words instead of guessing wildly and becoming frustrated. He is also retaining a fairly large number of sight words. I am pleased with Jack's progress this year. He seem happier and has a more even temperament. This is having a beneficial effect on his academic progress and school relationships."

[The teacher did not know that Jack had forgotten to take his Ritalin on the day she wrote the following report:]

"I must report that today, he reverted to all his former and worst behaviors. He cried, argued, refused to work, acted silly, fell out of his seat and dropped his pencil repeatedly, defied and disobeyed teachers. In other words, his behavior was intolerable."

This rather dramatically illustrates how successful stimulants can be in *normalizing* behavior.

Other side effects There may be an initial period of irritability, tearfulness and sensitivity to criticism which can resemble a minor depressive episode. This usually responds to a decrease in dosage. Other possible side effects include allergic reactions, visual hallucinations, periods of fearfulness, and outbursts of aggressiveness. In regard to the latter, it must be kept in mind that the usual effect of Ritalin is to decrease the level of frustration and aggressiveness, not increase it.

Dexedrine

The actions, effectiveness and side effects of Dexedrine (dextroamphetamine sulfate) are similar to those of Ritalin. It is available as 5 mg tablets or or as 5, 10 or 15 mg long-acting capsules. Like Ritalin, Dexedrine can also cause an increase in tics. However, in a given patient, if Ritalin has caused an increase in tics, changing to Dexedrine can sometimes avoid this problem.

Cylert

The third stimulant most used for ADHD is Cylert or pemoline. The usual doses are 18.75, 37.5 or 75 mg per day. It is longer acting than Ritalin or Dexedrine and it may take longer to be effective[386]. Unlike Ritalin and Dexedrine, which stimulate the release of both dopamine and norepinephrine, Cylert stimulates the release of only dopamine.

In my experience, *Cylert is much more likely than Ritalin or Dexedrine to cause an increase in tics*. Because of this, I only rarely use Cylert for ADHD in TS.

Summary: In the treatment of ADHD, the usual starting dose of Ritalin is 2.5 to 10 mg each morning. Noon and late afternoon doses are sometimes needed. Dexedrine is comparable to Ritalin in effectiveness and in some cases may be less likely to increase the tics. Cylert is the most likely to adversely affect the tics. The major side effects of stimulant medications are decreased appetite or insomnia.

Chapter 81

The Use of Stimulants in Tourette Syndrome

There are few issues in Tourette syndrome that generate as much divergent opinion as the question of the use of stimulant medications such as Ritalin for the treatment of the ADHD.

One of the questions about the treatment of ADHD is whether stimulants such as Ritalin can "cause" tics or Tourette syndrome. There have been many reports about the onset of tics after children with ADHD have been treated with stimulants[73,194,217,628,752,753,1216,1341,1384,1524,1812]. One study reported the development of TS in 15 ADHD patients treated with stimulants and suggested that stimulants should not be used to treat ADHD in Tourette syndrome or in patients who had a family history of TS[1216]. In a response to that report, Gerald Erenberg[551] pointed out that four of these patients had tics before the stimulants were given and stated, as I have[357], that for many TS patients the ADHD "is an even greater handicap than the tic disorder." We[357,358,360,368] feel that the combined use of haloperidol and Ritalin in TS can be very beneficial. Shapiro and Shapiro[1759] stated that a combination of Ritalin and haloperidol

> "did not increase tics in 42 patients treated by us for concomitant hyperactivity and Tourette syndrome, nor in 62 patients with Tourette syndrome treated with stimulants for various adverse effects from haloperidol."

Thus, the lines were drawn with opinions varying all the way from an absolute contraindication to the use of stimulants in TS to the recommendation that their judicial use can be very helpful. This is an important issue for several reasons.

1. If a significant proportion of ADHD is due to a *Gts* gene in either single or double dose then the statement that stimulants should never be used if there is a personal or family history of tics might prevent the appropriate treatment of many children with ADHD.

2. ADHD is often the overriding problem in children with TS and appropriate treatment with stimulants can lead to significant improvement.

3. Parents need to know if the use of stimulants is going to help or hurt their child. Do stimulants cause permanent tics or can stimulants actually prevent TS?

I will cover the different questions one at a time.

Can Stimulants Make the Tics Worse?

As noted above there are many reports of stimulants making the tics of TS patients worse. These, however, have the potential of being biased. For example, if 100 children with TS were treated with Ritalin but the tics got worse in only five, and these five were reported while the 95 were not, this would give an incorrect impression that stimulants always made the tics worse.

The following table summarizes studies where TS patients, who already had tics at the time the stimulants were given, were asked about the effect on the tics.

Effect of Stimulants on Tics in TS Patients Who Already Had Tics before Receiving Stimulant Medication

Study	N	Percent		
		Increase	No change No opinion	Decrease
Comings and Comings[360]	60	38.3	48.3	13.4
Erenberg[551]	39	28.3	66.6	5.1
Others[551]	110	25.4		
Total	209	29.6		

This indicates that about 30 percent of the time stimulants resulted in some increase in the tics. This ranged from a very minimal to a marked increase. In our experience, the increase in tics is usually controlled by a slight increase in dose of haloperidol or pimozide. We have had only nine patients in whom the increase in tics could not be controlled and the stimulants had to be discontinued. In 13 percent of our cases the tics actually got better.

Thus, it is clear that the stimulant medications can make the tics worse, and if they are to be used, patients and parents should be warned of this possibility. This is usually either not a significant problem or can be controlled with medication.

Can Stimulants Cause Tics or Tourette Syndrome?

This is a more complex question and requires the review of a number of studies to answer. In the above table we asked about the effect of stimulants on individuals who already had tics. Can stimulants cause tics in individuals who never had them? In a study of 1,475 children treated with stimulants, Denckla and co-workers[476] reported the development of new tics in 14 or only 1 percent. In a series of 10 other reports, 28 other cases have been reported of the onset of tics within days to six months after starting treatment with stimulants[549]. Unlike the Denckla report, the denominator is unknown, that is, how many cases of ADHD were treated without the onset of tics? The real question is: Did those children who developed tics have a *Gts* gene to start with, and would the tics have developed anyway, or did the stimulants cause them?

ADHD is often the first symptom of TS It is important to understand that in the majority of children with TS the first symptoms are not tics but ADHD. In our controlled study, the average age of onset of ADHD was 4.3 years while the average age of onset of TS (motor or vocal tics) was 6.6 years[360]. This left an average of 2.4 years between the onset of the ADHD and the development of tics. This natural history of the early years of TS was illustrated earlier [p103]. If the ADHD is severe it is quite likely that stimulant medications will be prescribed during this interval. The development of tics would be expected to occur whether the child was placed on medication or not. Thus, many and possibly all of the reports of the development of tics in children after treatment with stimulants may simply represent the natural development of tics in a child with ADHD due to *Gts* genes. The occasional reports of the onset of tics within days or weeks of giving stimulants unquestionably represent the precipitation of tics in children carrying *Gts* genes.

Do stimulants prevent the onset of tics? Given the above discussion this might seem to be a radical question. However, as I have discussed in some detail in Part IV, ADHD can be thought of as a dopamine deficiency disorder in the frontal lobes and the tics may develop later as a result of a compensatory hypersensitivity of the dopamine neurons in the striatum[P400]. If this is correct then the early administration of stimulant medications may prevent or delay this compensatory hypersensitivity and result in a delay in the onset of tics[360]. To examine this possibility we looked at two types of TS patients — those who were given stimulants for ADHD before they ever had tics (ADHD->stimulant->tics group) and those who were given stimulants after they had already developed tics (ADHD ->tics->stimulant group). We reasoned that if stimulants were precipitating the onset of tics the duration between the onset of ADHD and the onset of tics should be shortest in the ADHD->stimulants->tics group. In fact, we found the opposite[357]. This was such an unexpected result that we restudied it even more carefully in our controlled study[360]. The following diagram illustrates the results of both studies (a total of 117 patients):

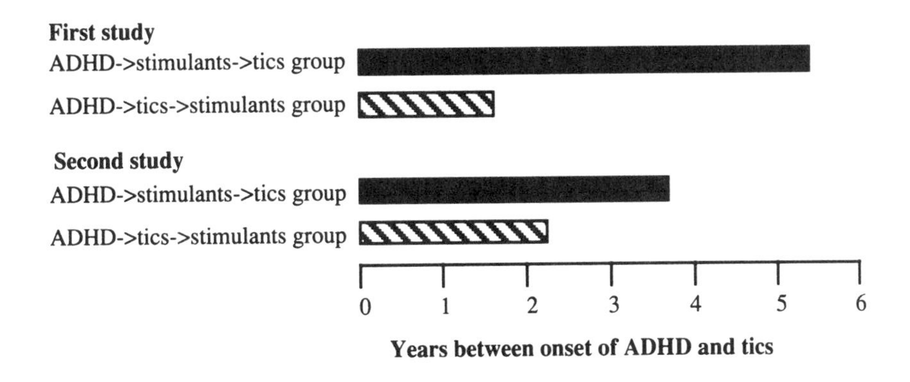

As can be seen, in both studies when stimulants were given before the tics developed, the onset of the tics was delayed. There is a bias here in that if the ADHD and tics started at the same time there would be no time to take stimulants before the tics started. These cases would all be in the ADHD->tics-stimulants group. However, when all cases where the ADHD and tics started within one year of each other were removed, the results were the same, but less striking. There was a family history of TS in 3/4ths of the patients in both groups. The average time between starting the medication and the onset of tics was six months.

Golden[759,760] has objected to our initial study claiming that a) the onset of ADHD is difficult to determine, and b) asking patients or parents to remember back to these dates is difficult. While it is correct that an accurate determination of the onset of ADHD is difficult, it is equally difficult for both groups and the errors in accuracy tend to balance out. The results in our second study verified those in the first. The objection that patients or parents had to look far back into their memories is not valid since most of the subjects on whom these two studies were based were less than 10 years of age and the parents' memories of these events were still fresh in their minds and often validated by medical records. In a study where the patients didn't know whether they were taking Ritalin or a placebo, Sverd and co-workers[1909] examined four boys with ADHD and TS.

> "In all four children, the highest dose resulted in improved classroom ratings of tics compared with initial pla-

cebo treatment. In three cases, mild tic exacerbation was reported for a lower dose. The findings. . .were consistent with the possibility that methylphenidate, a dopamine agonist, might exert its effects on tic status by altering dopmaine receptor sensitivity."

Some of the parents of our TS patients relate striking stories of the *positive effects of stimulants on the tics*. For example, a mother of a nine year old boy with severe TS and ADHD reported that each day as the previous dose of Ritalin began to wear off after about four hours,

> "Bryan began to swear, spit, jerk his head and have temper tantrums. These disappeared within a half an hour of taking his next dose of Ritalin."

These results simply verify what we tell parents — *each child is different and while stimulants can make tics worse in some children, thay may actually decrease the tics in others.*

Twin studies Another way to approach this question is to study identical twins with TS where one received stimulants and the other did not. In such a study of six identical twins[1549] the other twin, who had not received stimulants, developed tics at about the same time as the treated twin. In this study, all six of the treated twins had developed tics before they received stimulant medications. Non-twin TS patients were also studied. In 17 who were treated with stimulants before the onset of tics, the average duration between starting the medication and the onset of tics was 1.2 years.

Do stimulants help ADHD in TS? All of this would be academic if stimulants did not improve the symptoms of ADHD in TS patients. We examined this in our controlled study and found that the effect of stimulants on hyperactivity, concentration and school performance was about evenly divided into 1/4ths. Thus, 1/4 showed marked improvement, 1/4 some improvement, 1/4 no improvement and in 1/4 these problems were worse[360]. Thus, about half showed moderate to significant improvement. Erenberg and coworkers[549] reported a similar figure of improvement of the ADHD in 22 of 42 TS patients (52 percent).

Does giving dopaminergic agonists and antagonists to the same patient make sense? The stimulant medications act as dopamine agonists by enhancing the release of dopamine from neurons[p395]. The tics are probably caused by the hypersensitivity of striatal dopamine neurons and thus, it is not surprising that stimulants might cause the tics to get worse. One of the major medications used in the treatment of TS is haloperidol, which acts as a dopamine antagonist by blocking the action of dopamine at the postsynaptic receptors[p396]. How can we justify the apparent paradox of giving two different medications with seemingly opposite effects to the same patient?

One of the easiest ways to understand this combination is to view TS as due, in part, to a chemical imbalance of dopamine. Some regions of the brain have too much dopamine activity, and some have too little. The result is like a seesaw.

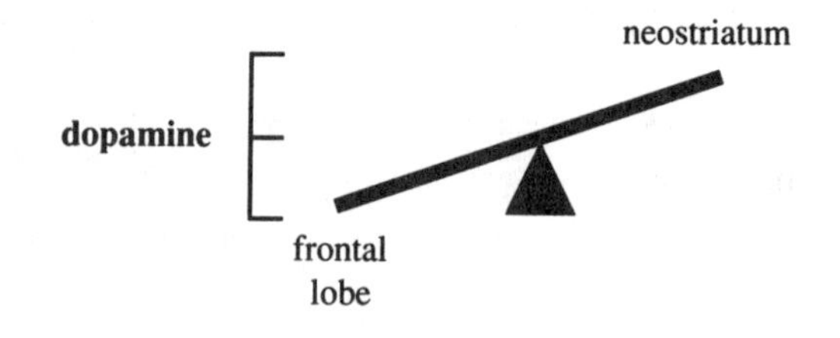

Thus, the frontal lobe syndrome appears to be the result of too little dopamine in the frontal lobes[p389], and the tics are the result of a hypersensitivity to dopamine in the neostriatum. Giving haloperidol can improve the tics but can make the frontal lobe problems worse.

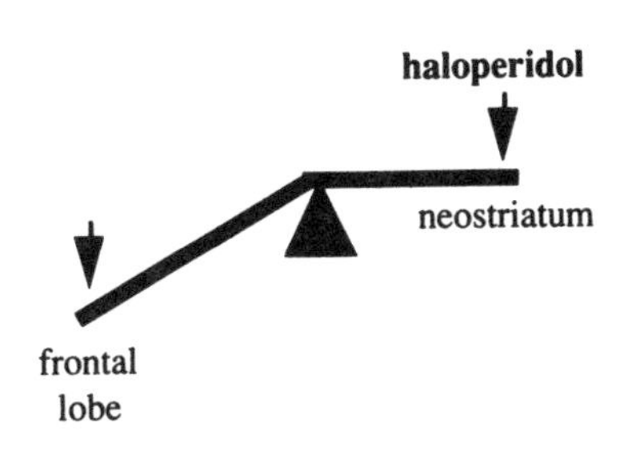

In the same fashion, giving stimulants can improve the frontal lobe problems but make the tics worse.

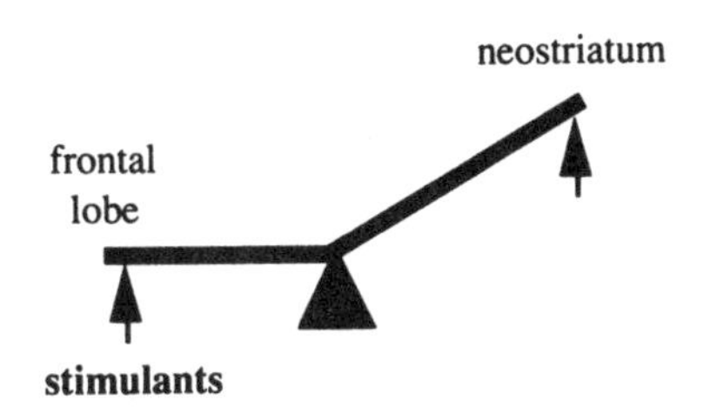

In some cases the balance may be best attained by giving both.

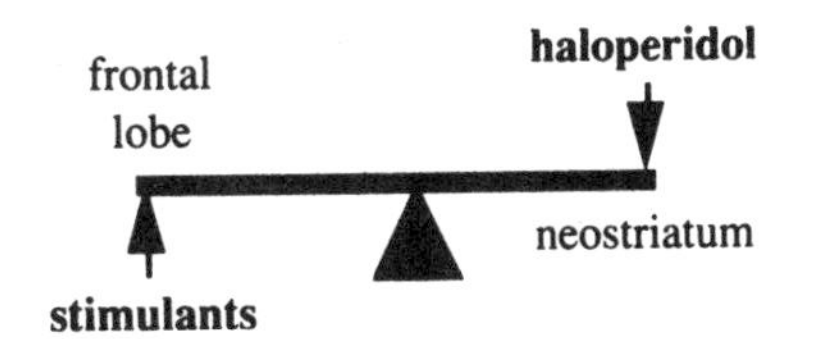

This is the concept; in practice it may not be that simple.

Summary: The evidence indicates that stimulants do not cause Tourette syndrome. If permanent tics develop in a child whose ADHD has been treated with stimulants, it is highly likely that child had ADHD due to a Gts gene. While stimulants can make the tics worse, this is usually not a significant problem. If the ADHD is serious enough to justify treatment with stimulants, a moderate increase in the dose of haloperidol or pimozide is usually enough to control the tics. The administration of both stimulants, which enhance dopamine activity, and haloperidol, which suppresses dopamine hypersensitivity, can best be understood by realizing that in TS children with ADHD there is an imbalance of dopamine with the frontal lobe having too little dopamine and the neostriatum being too sensitive to dopamine. The combination of stimulants and haloperidol, or related medications, are sometimes necessary for optimal treatment of this inbalance.

Chapter 82
Imipramine

The tricyclic antidepressants have frequently been reported to be almost as effective as stimulants in the treatment of ADHD[420,677,801,808,809,1092,1576,1577,1581,1826,2068] . The two that are most frequently used are imipramine (Tofranil) and desipramine (Norpramin). The following is an example of the usefulness of imipramine:

> Tracy was a heavy set, teenage girl whose sister was being seen for TS. She never had any motor or vocal tics, but was having trouble in school, getting D's and F's, and had severe mood swings. She was fine until at 10 years of age the teachers began to complain that she was "starry eyed and spacy" in school. She had trouble understanding simple concepts and her mother said she couldn't retain anything in her head. Later in that year her parents were divorced and all her problems were attributed to the emotional turmoil of that. However, a year later she was still "always crying" and sometimes was so depressed her mother couldn't get her out of the car in the morning to go to school. After treatment with only 10 mg of imipramine each morning her mother stated, "she was a new person." She was able to concentrate, her grades zoomed up to A's and B's. Instead of her mother having to constantly nag her to do her homework she would do it on her own and complain if she was interrupted. She lost 20 pounds, felt calmer, had more self-esteem, and "was much easier to live with."
>
> Another mother said,
>
> "He has stopped lying all the time."

Dose Imipramine and desipramine come in 10, 25, 50 and 100 mg tablets. The usual starting dose, depending on age and size, is 10 to 25 mg each morning. We prefer to error on the side of too little and frequently start with only 10 mg. The dose is then slowly increased over weekly intervals according to the response. Ten to 75 mg per day is usually adequate.

Effect When tricyclic antidepressants are given to adults for depression, it often takes three to four weeks to see an effect. By contrast, when they are used to treat ADHD and behavior problems the effect is usually immediate and may even wear off after four to six hours, just as with Ritalin. Because of this it is unlikely that these medications improve symptoms by relieving depression.

Comparison to stimulants There are both advantages and disadvantages to the use of tricyclic antidepressants compared to Ritalin. The major disadvantage is that they are generally not quite as effective as Ritalin or other stimulants. However, this is often compensated for by a number of advantages. In treating ADHD associated with TS, the most striking advantage is that the *tricyclic antidepressants are less likely to cause an increase in tics*. While some have reported that imipramine can cause an increase in tics, in our experience this is unusual. Thus, in TS patients with only moderately severe ADHD, I prefer to first use clonidine or imipramine. If these are not effective, the combination of haloperidol (or pimozide) and Ritalin may then be used. Other advantages of imipramine (or desipramine) are that they have a longer duration of action, and thus may give greater control of after-school behavior, have a greater tendency to improve mood and self-esteem, eliminate separation anxiety[470], show less disturbance of sleep, and may eliminate co-existing bed wetting[673]. Imipramine has been widely used by pediatricians to treat bed wetting.

Side effects As with most antidepressants the major side effects include dry mouth, constipation, trouble urinating, or insomnia.

A potential side effect of tricyclic antidepressants is a change in the heart rhythm[2067,2068], especially at high doses. Theoretically, similar problems could occur with stimulants[1222]. Fortunately, as mentioned above, we find that in TS the tricyclics are often effective at relatively low doses. They should not be given to patients with heart problems, and if high doses are contemplated, monitoring with an electrocardiogram is recommended. Withdrawal symptoms may occur when imipramine is stopped, after high doses have been given[1136].

Another potential danger of tricyclic antidepressants is that when taken in significant overdose they can be lethal. For this reason it is important that suicidally depressed patients have access to only limited amounts of medication at one time.

Despite these considerations, tricyclics are safely used for a variety of childhood disorders[1577].

Treatment of depression As discussed previously, depression is common in patients with TS. Individuals who carry the *Gts* gene may have problems with depression. In the treatment of a child with TS, it is important not to forget the rest of the family. Other family members who carry the *Gts* gene, may have problems with chronic depression that they are not even aware of. This is because the depression has been with them since childhood. They believe everyone feels as they do and this is how the rest of the world lives. If, in the process of taking a pedigree, it is clear that a member of the family has significant problems with chronic depression, treatment with an antidepressant should be considered. It will often result in a considerable decrease in the overall level of emotional chaos in the family and an improvement in the quality of life.

Summary: Tricyclic antidepressant medications such as imipramine (Tofranil) or desipramine (Norpramin) can be almost as effective as Ritalin and Dexedrine in the treatment of ADHD. They have the advantage that they are less likely to cause an increase in motor or vocal tics. Like the stimulant medications, they can also lead to a significant decrease in oppositional and confrontive behavior.

Chapter 83

Other Medications

As shown throughout this book, there is a wide range of different symptoms in different patients with Tourette syndrome. Oftentimes these become severe enough to be disabling and require treatment. The following is a brief review of a range of medications, some of which are quite valuable and others that have been recommended but in practice are not particularly useful. I will list them in relation to the symptoms they treat.

Obsessive-compulsive Behavior

Some of the milder obsessive-compulsive behaviors may improve following treatment with clonidine and sometimes with haloperidol or pimozide. If they are severe and persist, fluoxetine (Prozac) or clomipramine (Anafronil) are the most effective medications for obsessive-compulsive disorder. These are antidepressants that specifically inhibit serotonin uptake in the synapse[p430,1500a].

Anafronil (clomipramine) Several studies of the use of Anafronil in TS have been published. In the first report[2124,2125], 20 TS patients were treated. Of these, 17 had obsessive-compulsive behavior and there was improvement of both the tics and the obsessive-compulsive behavior. In a later study of six patients, the tics were not improved in four and actually increased in two[278]. There were no statements about its effect on obsessive-compulsive behaviors.

Many of our patients have used Anafronil and found it to be quite beneficial in the control of obsessive-compulsive behaviors, depression and sometimes oppositional behavior, without significantly increasing or decreasing the tics. We recommend starting with 25 mg a day and to slowly increase as necessary according to the response. The following are some examples of its usefulness.

> A 19 year old girl with TS and many compulsive behaviors was constantly hearing voices, a good voice and a bad voice. She often spent hours talking to them. Many different medications had no effect until she was treated with clomipramine, 25 mg twice a day. Over a period of two months the voices slowly went away until she came in one day elated saying, "They're gone for the first time in years!" This effect continued as long as she took the clomipramine.

> A 26 year old mother of a girl with TS had TS herself. She was treated with clomipramine because of anorexia nervosa. On 25 mg twice a day her anorexia disappeared and she stated "I can now read a book for the first time in years since I don't have to go over the same words again and again."

> Mary is 27 years old. Multiple medications were used to try to control her bizarre behaviors with minimal success. Finally, clomipramine at 25 mg per day resulted in a dramatic decrease in swear-

ing, depression, and paranoia.

As discussed earlier, obsessive-compulsive behaviors may be due to hypersensitivity of serotonin receptors[p441]. The occasional exacerbation of the tics may be due to the increased availability of serotonin before the serotonin receptors are "down-regulated."

As discussed elsewhere, clomipramine is also useful in the treatment of depression[p430], panic attacks[p442], trichotillomania[1909d] and agoraphobia[985].

Prozac (fluoxetine) While Prozac has been used primarily for depression[267], it has often been of significant benefit in the treatment of both moderate and severe depression[267], obsessive-compulsive behavior[p442], panic attacks[p442], alcohol abuse[p436] and obesity [p586]. The usual dose is 20 to 40 mg/day. The major side effects have been insomnia and some feelings of nervousness in 14% of patients on the medication compared to 8% taking a placebo. There is occasionally drowsiness, tremor, or dry mouth.

Because of the effect of Prozac on serotonin and the involvement of serotonin in TS, we have used this medication in many TS patients. It has proven to be *very effective*[370]. The depression or mood swings often disappeared and irritability or oppositional behavior have frequently diminished in two to seven days. Obsessive and compulsive behaviors may decrease significantly. While there is only moderate or no effect on the tics, or ADHD, many of the other associated behaviors have shown moderate to marked improvement. The following are some examples:

> "Before Prozac I was constantly scowling at people and never smiled. After one week on the medication I felt good about people, loving and smiled much more. Before my wife was on the verge of leaving me and now we are back together. It has changed my life."
>
> "There was a significant decrease in his tics, he stopped patting his mouth and grinding his teeth."
>
> A 16 year old boy with TS also had severe compulsive behavior consisting of writing hundreds of numbers on sheets of paper. If he was asked to stop and participate in family outings or other things before he "was through" he got very angry and aggressive. This virtually disappeared after treatment with Prozac.
>
> "His depression is gone, his sense of humor is back, he is less paranoid and easier to live with. I can talk with him again."
>
> "The first day I felt more alert and by the third day my chronic depression disappeared. I became more open and talkative and less afraid to mix with people. My memory also improved. I am ecstatic about the results."
>
> A 16 year old girl with TS and autism had such severe symptoms she required a highly structured school. Her behavior often necessitated several hours of timeouts per day. Haloperidol, Tegretal, Ritalin, Cylert, imipramine and other medications had been relatively ineffective. Treatment with Prozac, 20 mg/day, resulted in a decrease in timeouts to less than 15 minutes per day.

In summary, *Prozac is a valuable medication in the treatment of the spectrum of disorders seen in TS. This effectiveness is consistent with a role of low serotonin the TS.*

Aggressive Behavior

Excessively aggressive behavior can be one of the most disturbing aspects of TS. It may be significantly improved by the same medications used to treat the tics and ADHD, that is, haloperidol, pimozide, clonidine, Ritalin, Dexedrine, or imipramine. The following are alternative medications that may be helpful when these medications do not significantly alleviate troublesome aggressive behaviors.

Lithium carbonate Lithium was first described for the treatment of manic-depressive disorder by Cade in Australia in 1945. Thousands of reports have verified its value in significantly modulating the peaks and valleys of manic-depressive individuals, many of whom may carry a *Gts* gene[p191]. In addition to treating mood swings, lithium can also modulate aggressive behavior[280,1781]. We have found that for the treatment of depression, mood swings and aggression in TS, the required dose is often less than that generally used. Thus, while the usual recommended dose for adults is 1,200 to 1,800 mg a day, with blood levels of 0.9 to 1.5 meq/liter, we find that 600 to 900 mg a day (300 mg 2 to 3 times a day), with blood levels of 0.4 to 0.9 meq/liter, is often adequate. With higher doses, frequent tests to monitor the blood level are necessary. With these low doses, after a few initial tests, the requirement for blood testing is much less.

It has been suggested that impulsiveness is on the same dimension as hypomania. Highly impulsive subjects had much in common with hypomanic and manic patients. Lithium changes many of these features in the direction of normality[1371,1372,1959]. This would be consistent with the low serotonin-impulsiveness hypothesis[p447].

The exact way lithium works is not known. Like clonidine, it has been reported to decrease norepinephrine activity[1272] and modulate brain serotonin. In a study of 10 TS boys who also had manic-depressive disorder, five responded to lithium[1033b].

Carbamazepine (Tegretal) The temporal lobes and limbic system have often been implicated in periodic rage attacks. Carbamazepine is an anticonvulsant that is particularly effective in suppressing temporal lobe and limbic system activity[1535]. It is the medication of choice for epilepsy associated with psychological symptoms. There have been many reports of a high frequency of brain wave abnormalities in individuals with periodic, violent, rage outbursts[1272]. These behaviors have improved following treatment with carbamazepine[1351]. Importantly, carbamazepine can diminish violent behavior in individuals without brain wave evidence of epilepsy[1102,1891].

I have often found Tegretal to be useful for the control of severely aggressive behavior. Sometimes it also results in a significant decrease in motor and vocal tics. There is a report of carbamazepine "causing" tics but the clinical history in these two patients indicated they both had a *Gts* gene to begin with[816]. In our experience carbamazepine can often be effective when nothing else seems to work. The starting dose is usually 400 to 600 mg per day in divided doses. If necessary, up to 1,200 mg per day may be given.

The side effect of Tegretal that is of most concern is the very small possibility of affecting the function of the bone marrow. Very rare cases of fatal anemia (aplastic anemia) have been described. Because of this potential danger, blood counts (red cells, white cells and platelets) should be done at periodic intervals, especially for the first several months that the medication is being taken. If significant decreases occur it should be stopped. This potential side effect should not prevent the use of Tegretal when indicated. I have used Tegretal often and have never yet had to discontinue it for this reason. It is considered just as safe as dilantin[856,1603]. Other side effects include tiredness and occasionally skin rashes. In a small percentage of cases, a paradoxical worsening of behavioral problems with irritability, agitation and aggressive outbursts may occur[1799].

Tegretal has come into increasing use as

a substitute for lithium in the treatment of mania[96,1531,1534] and depression[1535]. It may also be useful in the treatment of ADHD[1272].

The combined effectiveness of carbamazepine in aggressive behavior, rage attacks, mania, depression, ADHD and tics, accounts for its usefullness in the treatment of TS.

Propranolol (Inderal) Propranolol is one of a number of medications that block β-adrenergic receptors. In plain English, this means the effects of the adrenalin released during stress and tantrums are blocked. Thus, it is not surprising that a number of reports have indicated that propranolol may decrease the frequency and intensity of temper tantrums and rage attacks in a wide range of disorders including ADHD[1272,2079]. It has not been particularly helpful in the treatment of TS[1905a,1907a]. I have found propanolol to be most useful in alleviating severe *test anxiety* for some TS patients. A 20 or 40 mg dose a half an hour before the test may be quite effective without affecting intellectual performance.

Obesity

Obesity becomes a problem in TS in two ways. The most obvious is that some patients taking haloperidol, pimozide or Prolixin show a marked stimulation of their appetite and gain weight. The second is the presence of obesity as a result of compulsive eating especially in women carrying a *Gts* gene[p236].

The first line of defense in controlling the increased appetite due to haloperidol or pimozide is diet. Since one of the best ways to avoid constant snacking is to have no food in the house, we recommend that parents clean out all carbohydrate-rich foods from the cupboards and refrigerator. What is not there can't be eaten. This can be replaced with "rabbit food" (celery, lettuce, jicama, radishes, etc.) to fill their stomachs. This approach may fail if the children sneak candy bars and junk food when they are out of the house. If alternative medications are not successful in controlling the tics, we have found that Pondimin (fenfluramine) 20 mg one to three times a day can be quite effective. This medication affects the serotonin control of the appetite centers[p495]. I had a woman patient with TS, alcoholism and obesity (375 lbs.) who was able to stop drinking and slim down to 140 lbs. with diet, encouragement and Pondimin.

Prozac also suppresses appetite and has been used in the treatment of obesity[1162,2150] and bulimia[637].

Panic Attacks and Anxiety

For some adults who carry the *Gts* gene, panic attacks are the most crippling symptom. They usually begin between 18 and 35 years of age, often years after the tics have stopped. They may also occur or in *Gts* gene carriers who have never had tics[p175].

Some of the best medications for the treatment of panic attacks include BuSpar (buspirone), MAO inhibitors, imipramine, Prozac, and Anafranil. The latter two, with especial influence on serotonin, are particularly effective[p442]. Most of these medications have already been described for use in treatment of other aspects of TS.

Benzodiazepines As discussed in detail in the chapter on GABA[p481], a group of medications called the benzodiazepines interact with GABA receptors to help alleviate anxiety. As such they are particularly effective in the treatment of anxiety and the acute symptoms of panic attacks. They may also be quite helpful in the treatment of the tics themselves. The benzodiazepines most used in TS are Valium, Xanax, and Clonopin.

Xanax (alprazolam) Xanax, (0.25 to 1.0 mg 1 to 4 times a day) is often helpful in relieving anxiety and the acute symptoms of panic attacks. In some cases I have found it to be the only medication that relieved the tics.

> This 18 year old male had motor and vocal tics, ADHD, dyslexia, depression and conduct disorder. Haloperidol, pimozide, clonidine, imipramine, Tegretal, Valium, Dexedrine, and methylphenidate had all been tried without success. Treatment with Xanax 1 mg three times a day resulted in a significant decrease in his tics, improved ability to concentrate, better grades and he changed from wanting to drop out of high school to wanting to finish.

Extensive studies have demonstrated the effectiveness of Xanax in the treatment of panic attacks[97]. Within one week there was a significant decrease in panic attacks, phobias, and anxiety. Sedation was the major side effect and decreased with continued use or decreasing the dose[1413]. Although there was a tendency for a rebound of symptoms or mild to moderate withdrawal effects, these usually subsided after being off the medication for two weeks[1491].

Clonopin (clonazepam) Clonopin can sometimes be effective in the treatment of tics[776,784] or panic attacks. In my experience it has been most valuable when clonidine or haloperidol were ineffective or produced significant side effects. Clonopin is also effective in the treatment of myoclonus, a disorder with severe tic-like muscle movements[P288]. It has been suggested that brain serotonin is low in this disorder since treatment with 5-hydroxytryptophan, the immediate precursor to serotonin[P418], is very effective and since Clonopin also increases brain serotonin[784]. These observations are all consistent with the serotonin theory of TS[P461].

Valium (diazepam) Finally, the most commonly used benzodiazepine, Valium, can also be useful. I have a few patients who have found it to be the only medication that has given them some relief of symptoms.

All of the benzodiazepines have the potential of being abused. If used at significant doses they may produce withdrawal symptoms if stopped suddenly, and this should be avoided.

Baclofen (lioresal)

Baclofen is effective in decreasing the muscle spasms of individuals with spinal cord injuries. It does this by a direct effect in the spinal cord by inhibiting the release of the excitatory neurotransmitters — glutamic and aspartic acid[626,1536]. Baclofen has a chemical structure that is very similar to GABA and it interacts with $GABA_B$ receptors[203,896]. Unlike GABA, Baclofen can pass across the blood brain barrier. Thus, in addition to its effect on the spinal cord it also acts as a GABA agonist in the brain. In this role it is able to supress the firing rate of dopamine neurons in the substantia nigra and ventral tegmental area[1435]. It has been useful in the treatment of schizophrenia[631] and eye spasms (blepharospasm)[1690] in humans, and myoclonic muscle jerking in mice[1314]. Because of these features, especially its anti-dopaminergic action, I tried Baclofen in a number of patients who tics were not responding well to other medications. It improved the tics in about one-third of the patients, all of whom were still on either haloperidol or pimozide. The following are two examples.

> A 12 year old girl had severe vocal tics that had not responded to 6 mg a day of pimozide. Because she could not tolerate any higher doses, Baclofen was added. At a dose of 20 mg three times a day, the vocal tics disappeared.

> A 24 year old male had been treated elsewhere with pimozide, 6 mg per day. Although this resulted in about 80% supression of his motor tics, it left him feeling tired and he had gained over 80 pounds. Haloperidol was ineffective. He was depressed about the weight gain but everytime he decreased the dose of the pimozide his tics became so severe they

> interfered with his work. After I added Baclofen, 10 mg three times a day, the tics virtually disappeared and the pimozide could be decreased to 2 mg day. At this dose he felt less tired and less hungry and lost the 80 pounds. As a result of his improved self-image he began dating, and soon got married.

The one side effect that seems to have been a problem in some patients is an increased problem with memory.

L-5-hydroxytryptophan

If TS is due to a genetic defect in tryptophan hydroxylasethat causes a relative deficiency of brain serotonin, then treatment with L-5-hydroxytryptophan, the immediate precursor of serotonin, should be effective. This medication has been effectively used in the treatment of myoclonus[2000,2003]. It must be used in conjunction with carbidopa, which inhibits the formation of serotonin outside the brain and thus prevents serotonin-induced diarrhea and other gastrointestinal side effects. Since carbidopa does not enter the brain, formation of serotonin in the brain is not inhibited. L-5-hydroxytryptophan was first used in the treatment of TS by Van Woert and colleagues [2002,2003]. One patient with severe self-mutilation was significantly improved. However, when tried on an additional nine TS patients who had not responded to haloperidol, there was no consistent effect on the tics and in some cases the ADHD got worse.

After obtaining permission from the FDA, I used L-5-hydroxytryptophan (75 to 300 mg) with carbidopa (50 to 100 mg) to treat 20 TS patients who had responded poorly to other medications. In one patient the tics got much worse and the medication was discontinued. About half of the patients had to stop the medication because of diarrhea or abdominal cramps, or because it was ineffective. One patient was markedly improved and has remained on 200 to 300 mg of L-5-hydroxytryptophan for over eight months. After two months the symptoms got somewhat worse but they never returned to the original degree of severity. She was able to stop pimozide altogether.

A second patient, a 13 year old girl, was having severe tics, oppositional behavior and learning problems. The tics were unresponsive to phenothiazines and clonidine. On 150 mg of 5-HTP the tics and the oppositional behavior disappeared and her attention span improved. After seven months the dose was deceased to 100 mg per day and within a week the tics returned. They disappeared again when the dose was returned to 150 mg per day.

In the remaining patients, who were treated for one to six months, there was a mild to moderate improvement in the tics, a significant improvement in mood, decreased irritability and increased attention span. Because of the effectiveness of other medications, the problem with gastrointestinal side effects, and the fact that it is not generally available, L-5-hydroxytryptophan does not play a major role in the treatment of TS. However, it was useful in about 30 percent of patients who had not responded to other medications.

Nicotine Gum

Nicotine gum (Nicorette) has been described as an adjunct to haloperidol in the treatment of TS[1686].

> A six year old boy was described with severe TS consisting of many tics, self-abusive acts and obsessive-compulsive behaviors. Haldol at a dose of 1.5 mg made him tired but didn't help the tics.
>
> "Nicorette chewing gum (2 mg) twice a day was added and a striking improvement in the tics occurred for 40 minutes when the boy was chewing the gum; he was calmer, not hyperactive,

> and able to read, watch television, or sit quietly without fighting."

The Nicorette gum with Haldol controlled the frequent outbursts of jumping from his seat or bothering other children in school.

> A second eight year old boy also with severe TS was taking Haldol 1.0 mg twice a day with minimal improvement and problems with weight gain. His tics were much better for an hour after chewing the nicotine gum and he could concentrate. He used the gum during times when he needed to concentrate at school or do his homework.

Nicotine apparently works by acting on dopamine nerves in the brain. It seems to potentiate the effects of Haldol without causing more side effects.

While the authors note that the side effects of the gum are unusual and that dependence is rare, the one possible problem with this approach could be dependence on the Nicotine gum. It is felt that because the rapid increase in blood nicotine that occurs with smoking does not occur with the gum, the risk of dependence is less than with cigarettes[940]. In my experience nicotine gum has not been helpful.

Calcium Channel Blockers

A group of medications called calcium channel blockers (verapamil and nifedipine) have been proposed as being useful in the treatment of Tourette syndrome[2024]. In my experience, when used alone, these medications have only occasionally been effective in the treatment of TS. However, they may be useful in combination with haloperidol where the two together may be more effective than either one alone[17a]. These medications were also not successful in schizophrenia[791], but have been reported to be effective in the treatment of mania[511] and they have a direct effect on serotonin uptake[1594a]. Pimozide has some calcium channel blocking effects[783]. Verapamil has been reported to be effective in the treatment of stuttering in adults[211a].

Tetrahydrobiopterin

This mouthful is a chemical that acts as a cofactor for two important enzymes, tryosine hydroxyalse[p363] and tryptophan hydoxylase[p417], rate-limiting steps in the formation of dopamine and serotonin respectively. On the assumption that autism was due to a defect in the formation of these two compounds, Japanese workers[1394] gave tetrahydrobiopterin in a blind study to 41 autistic children and noted a marked or definite improvement in 54 percent compared to 31 percent of those given the placebo. They also reported its effectiveness in the treatment of a single patient with depression who had not responded to tricyclic antidepressants[955a]. This medication, which is not presently available, needs much furthur study. To my knowledge it has never been tried in TS but might be helpful.

Adverse Effects of Other Medications

In addition to the question of what helps TS, an equally important question is what medications make TS worse. The role of stimulants in exacerbating the tics in some cases has been discussed[p567]. Over-the-counter medications for colds or medications for asthma, which simulate the effects of adrenalin, may cause problems[1756]. In reading the labels of over-the-counter medications, look for compounds such as phenylephrine hydrochioride, ephedrine, pseudephedrine hydrochloride, phenylpropanolamine hydrochloride, and related substances. While antihistamines (Benadryl, Polaramine and others) may occasionally make the tics worse[1756], they may sometimes actually improve the symptoms and they have often been used for the treatment of ADHD. Like many things in TS there are no blanket rules. The best advice is to simply be aware of the possibility that some medications for colds, allergies and asthma may make TS symptoms worse and if this appears to be happening they should be discontinued.

Certain medications for epilepsy, such as phenobarbital and mephobarbital can increase the hyperactivity of children with ADHD and increase the tics in patients with TS[254].

Clozapine

Clozapine has been found to be very effective in the treatment of schizophrenia [321a,1303a]. It is particularly useful in treating both the positive and negative[p206] symptoms. This may be due to the fact that while it blocks D_2 dopamine receptors in the striatum it also preserves or even enhances dopamine activity in parts of the brain that are is blocked by standard phenothiazines[1303a]. Since the negative symptoms appear to be due to a dopamine deficiency in the frontal lobe[p206], the ability of clozapine to enhance dopamine activity may account for its effect on negative symptoms. Since both hyper- and hypo-functioning of dopamine nerves appear to play a role in TS as well, it would be of interest to evaluate the effect of clozapine on the negative symptoms in some TS patients, especially the lack of motivation.

Summary: A wide range of medications are useful in the treatment of the associated behaviors in Tourete syndrome. One of the most effective of these is Prozac (fluoxetine) ,which specifically increases the efficiency by which the brain uses serotonin. Tegretal, a limbic system anti-convulsant, and lithium carbonate, can be very useful in the treatment of aggressive behavior and severe mood swings. Some of the other effective medications are described.

Chapter 84
Non-prescription Medications

It is very natural for people to want to take control of their own lives and not have to be dependent upon others for their health. This feeling has resulted in many striking improvements in public health, especially through programs to stop smoking, exercise, keep from becoming overweight, and maintain a proper diet. This desire to find effective non-prescription treatments extends to patients with Tourette syndrome and their parents. Over the years I have always kept an open ear and mind to what parents say they have tried in their own efforts to decrease their child's symptoms. This chapter summarizes what I have been told, what I have tried, and some ideas that in theory might work, but have not been rigorously tested.

Diet

The idea that what you eat can affect your behavior has been a popular idea, especially in the area of the treatment of hyperactivity. The box at the upper right lists some of the aspects of the dietary approaches that have been proposed, or that parents have reported to be effective in decreasing their child's hyperactivity.

Many of these either have been incorporated into the Feingold diet[584,p572] or are part of a standard food allergy elimination diet. As discussed before, the effectiveness of these diets has been hotly debated. My feeling is that if they seem to help, use them. If they don't help, don't compulsively keep trying them.

In Some Children Hyperactivity Has Been Decreased by Avoiding:

- sugar
- simple carbohydrates
- food preservatives
- food coloring
- dairy products
- chocolate
- tomatoes

Amino Acids

Given the apparent role of low brain serotonin in Tourette syndrome, the amino acid most likely to be helpful would be **tryptophan**. However, as discussed before, this is a double-edged sword[p470] in that tryptophan also induces tryptophan oxygenase to become more active, thus breaking down tryptophan more rapidly and counteracting the effect of tryptophan itself. While tryptophan has been reported to be effective in the treatment of tics[308a], in general I have only occasionally found it to be that helpful in the treatment of TS. It is more likely to be useful in helping some patients sleep at night. The usual dose is one to four 500 mg tablets at bedtime.

Tyrosine is a precursor of dopamine and part of the defect in TS is the poor functioning of dopamine neurons in the frontal lobe with a subsequent hypersensitivity of dopamine neurons in the striatum[p389]. One of my patients claimed his tics were significantly decreased by taking a combination of tyrosine and tryptophan. Others have not been helped by this combination, or by tyrosine alone.

Phenylalanine increases the levels of endorphins in the brain by inhibiting the rate that they are broken down. Some of our patients have felt this helped their tics, others did not. It has been of no value in the treatment of ADHD[2106,2145].

Vitamins

Vitamins act as cofactors in that they are necessary for certain enzymatic reactions to work properly. Thus, the question of whether vitamins could help in treating TS depends on which chemical reactions might need to be helped. As shown in the diagram of pathways of serotonin and tryptophan metabolism (see Appendix), **vitamin B_6** (pyridoxine) is required for many of these enzymes to work. B_6 is a cofactor in helping tryptophan hydroxylase to produce serotonin and inhibits the breakdown of tryptophan by inhibiting tryptophan oxygenase in rats[148]. High doses of vitamin B_6 have been reported to be useful in the treatment of autistic children[p507] and to be both helpful[351,657a] and not helpful[861] in treating ADHD. Some of our TS patients feel that taking vitiman B_6 has helped them, others have not been convinced. We did a low dose double-blind study several years ago and the results were negative.

Perhaps the most important aspect of the use of vitamins in the treatment of hyperactivity, TS, or any other neurological disorder is that huge doses are often used — so-called mega-vitamin therapy. This has the real potential of causing more harm than good since high doses of many of the vitamins, including vitamin B_6[22,1712], can cause severe disease.

Amino Acids and Vitamins

One amino acid-vitamin combination that theoretically might be helpful for TS is **nicotinamide and tryptophan**. Nicotinamide is one of the B vitamins similar to niacin. In rats it decreases the rate at which tryptophan is degraded by tryptophan oxygenase[2118]. It was unable to prevent the increase in activity of tryptophan oxygenase produced by steroids, but strongly inhibited the increase in activity produced by tryptophan[81a,87,317]. As pointed out before, taking tryptophan by mouth to attempt to increase brain serotonin can be self-defeating because the tryptophan pushes tryptophan oxygenase to higher levels of activity. Because of this, the combination of nicotinamide and tryptophan has been tried in the treatment of depression. In some reports this was successful[320], while for others it was not. When human subjects were given 1,500 mg of nicotinamide a day for a week and then tested with a tryptophan load, the nicotinamide resulted in a decrease in the breakdown of tryptophan as indicated by a 28 percent decrease in blood kynurenine and a 39 percent reduction in urinary kynurenine[320]. This indicated that in humans, as in the rats, nicotinamide was capable of inhibiting the activity of tryptophan oxygenase. I have had a number of patients try this combination with decidedly mixed results. Some felt it helped, some felt it did nothing, some felt it made the symptoms worse.

A second potentially useful combination is **vitamin B_6 and tryptophan**. In rats, vitamin B_6 alone causes no change in brain serotonin, but did increase brain serotonin when it was given in combination with tryptophan[1149]. The results of the four combinations of normal and high amounts of dietary vitamin B_6 and tryptophan are shown on the **next page**.

The usefulness of any of these tryptophan-vitamin combinations in the treatment of TS has yet to be tested by double blind studies.

Over-the-Counter Medications

Tylenol I have now heard from two different sets of parents that Tylenol

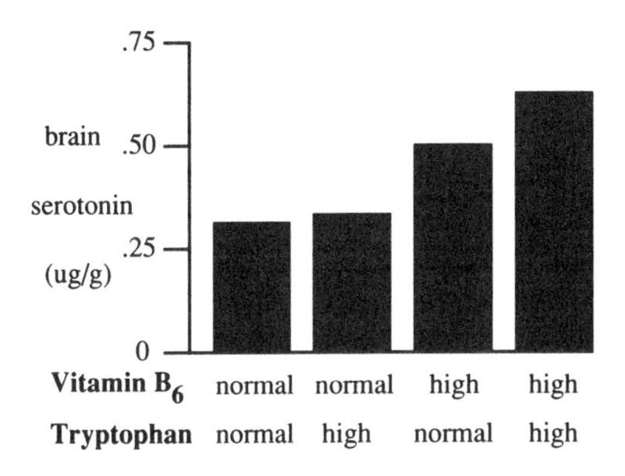

Over-the-Counter Medications

Tylenol I have heard from two different sets of parents that Tylenol (acetaminophen) one to two tablets several times a day, significantly decreased their child's tics. Others have not found this to be helpful.

Aspirin Like Tylenol, an occasional parent has reported a calming effect of aspirin on tics and behavior. One possible mechanism may be that salicylates, produced by taking aspirin, have been reported to decrease tryptophan oxygenase in rats[82a]. Because of this I have recommended patients try 1 to 2 tablets of aspirin three times a days. To date, there have been no dramatic results.

Benadryl is an antihistaminic. It was once available only by prescription but is now given over the counter. It has been occasionally used to treat ADHD. The infant with TS who I described earlier[p272] responded very well to treatment with only 12.5 mg of Benadryl twice a day. On other occasions I have found 12.5 to 50 mg of Benadryl one to three times a day to be useful in the treatment of ADHD and TS. Given at bedtime it can also serve as a safe sleeping pill. If a child needs an antihistaminic, Benadryl alone might be less likely to increase tics than other antihistaminics in combination with decongestants[p589].

Efamol — Evening primrose oil. This oil contains gamma-linolenic acid, a precursor to prostaglandins. This is a complex group of compounds with many different functions. Double blind studies have suggested mildly to significantly beneficial effects in the treatment of premenstral tension and attention deficit disorder[54a]. In our experience it has been more effective for PMS than ADHD.

Symptom Substitution

Occassionally it is possible to eliminate a particularly unpleasant symptom by encouraging the substitution of another — so called symptom substitution[251a]. In one case, wrist banging that was causing a skin lesion was eliminated by having the patient snap a soft wrist band instead, and in another it was possible to eliminate a spitting tic by having the patient snap his fingers each time he had the urge to spit [251a]. In our experience, sometimes this technique works and sometimes it does not.

Summary: Despite many years of keeping an open mind to stories that parents tell me about different non-prescription treatments of TS or ADHD, the best that can be said is that some of the above approaches seemed to help certain children but not others. In my experience, none of them begins to approach the effectiveness of more specific prescription medications.

Chapter 85
Psychotherapy

The symptoms of Tourette syndrome can range from barely sufficient to make a diagnosis to the full spectrum of tics and associated behaviors described in this book. Since the treatment has to be molded to the symptoms in a given individual, it can range from no treatment to a complex combination of medication and psychotherapy. The role and use of medications for the tics and behavior problems in TS were presented in the previous chapters. When these fail or, in the more severe cases, give only partial improvement of the behavior, psychotherapy is necessary. Psychotherapy does nothing for the control of the tics but can be effective in helping to control many of the behavioral problems. Much of this chapter comes from the experience of Brenda Comings, Ph.D., who has been doing therapy with hundreds of TS families over the past 10 years. She sees those children who have not significantly improved on medication alone. I will start with some important basic principles.

Basic Principles of Psychotherapy in TS

First make the proper diagnosis The single greatest necessity in the adequate psychotherapy of Tourette syndrome is to first make a correct diagnosis. This may seem obvious, but of the patients we see, this is one of the most common causes for their failure to improve when seen by other therapists. The most brilliant therapist in the world is likely to fail if the diagnosis of Tourette syndrome has been missed and inappropriate treatments are being used.

Initiate medical treatment when indicated While not all TS patients need to be medically treated, many of those who have problems that are severe enough to bring them to the doctor will benefit from some type of medication. Without medication, attempts at therapy are often impossible since the patient is unable to concentrate on what is being said or is simply too out of control. Since the medication can often result in either a dramatic elimination or a marked decrease in the behavioral and discipline problems, we usually start medication either before or at the same time as starting psychotherapy. If severe behavioral problems persist despite an adequate trial of one or more medications, therapy may then be the only way to bring about any improvement.

What not to treat Therapists who are not familiar with the spectrum of disorders in TS often produce psychological explanations to explain the behaviors they are seeing or hearing about. Tics may be labeled "acting out behaviors," or "the result of psychological problems" or due to "stress in the home." Vocal noises may be labeled "attention-getting behaviors," especially by teachers. Obsessive-compulsive behaviors may be labeled "efforts to protect oneself from some inner conflict." Sexual touching or compulsive masturbation are often assumed to be "the result of sexual abuse in the home." The anger, which is often such an integral part of TS, may be thought of as due to "a reaction to parental conflicts or divorce," to "compensate for feelings of inadequacy," or to a long list of "unresolved conflicts." Phobias are the "result of some trauma in their childhood." Depression is "due to a problem in the family or some external situation in the environment." We have often seen children or adults treated for years under one or more of these assumptions with no relief of symptoms and watched the symptoms promptly disappear after proper medical or medical and psychotherapeutic treatment has been initiated.

What to treat It is important to define what we are treating. Parents often ask what behaviors they should and should not discipline. The rules are fairly clear cut — DO NOT discipline for motor or vocal tics, true coprolalia, short attention span, learning problems or obsessive-compulsive behaviors. DO discipline for antisocial behaviors such as lying; stealing; setting fires; hurting pets; vandalism; hitting parents, siblings or peers; disrespect for authority; insensitivity to rights and feelings of others; refusal to do simple chores; temper tantrums; inappropriate controllable swearing; talking back; oppositional, confrontive behavior; and generally obnoxious antisocial behavior. The following is just one small example of typical oppositional behavior that parents have related to us.

> "Mom, I want a cookie"
>
> "No, you can't have a cookie until you brush your teeth."
>
> "Oh, please Mom, give me a cookie now, I promise I'll brush my teeth right after, I promise, I promise."
>
> "OK, you can have a cookie."

After munching down the cookie his mother naturally said, "Now go brush your teeth."

> "No! I won't brush my teeth and you can't make me," at which point he ran out of the house. His furious mother ran after him, chasing him around the car seven times till on the verge of tears she caught him and dragged him to the bathroom. At which point he said through tightly clamped jaws, "You can't make me brush my teeth, I won't do it."

To the parent who does not have an oppositional child this may seem like a minor skirmish in the life of raising children. However, it is pathological when, instead of being an occasional isolated incident, this is the flavor of every interaction — a confrontive war of manipulation, tears, fighting, defiance and opposition. The TS - ADHD child with oppositional behavior is constantly testing the limits of how far he can go in controlling those around him. Parents rapidly come to their wits end, question what they are doing wrong in child rearing to be raising such a "monster" and at this point seek some type of psychotherapeutic help.

Don't always blame the parents One of the most common complaints of parents of TS children is that friends, relatives, in-laws, teachers, and mental health professionals blame them for the aggressive and other behaviors their children are manifesting. These behaviors cause them to be referred for professional help. The TS or ADHD child, who can be out of control at home or a behavior problem at school, can easily suppress these symptoms while they are in the therapist's office, and appear totally calm and appropriate. After a brief one-to-one evaluation, the parents are often blamed for all of the child's behavioral problems and told that their poor parenting skills are the basis of the trouble. The **next page** shows an example taken from a pediatrician's notes.

Since this child was quiet and acted appropriately in the doctor's office, the blame was shifted to the mother, euphemistically stated as — "Intolerance on the maternal side."

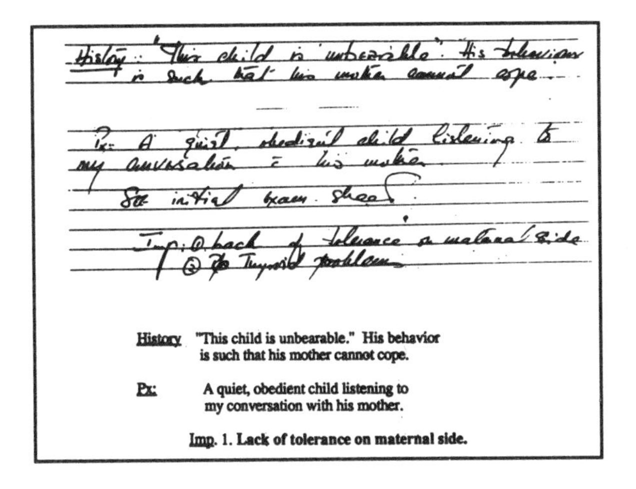

History "This child is unbearable." His behavior is such that his mother cannot cope.

Px: A quiet, obedient child listening to my conversation with his mother.

Imp. 1. Lack of tolerance on maternal side.

Not only did this child have typical TS but a careful family history showed the TS and chronic motor and vocal tics were rampant in the family, shown as follows:

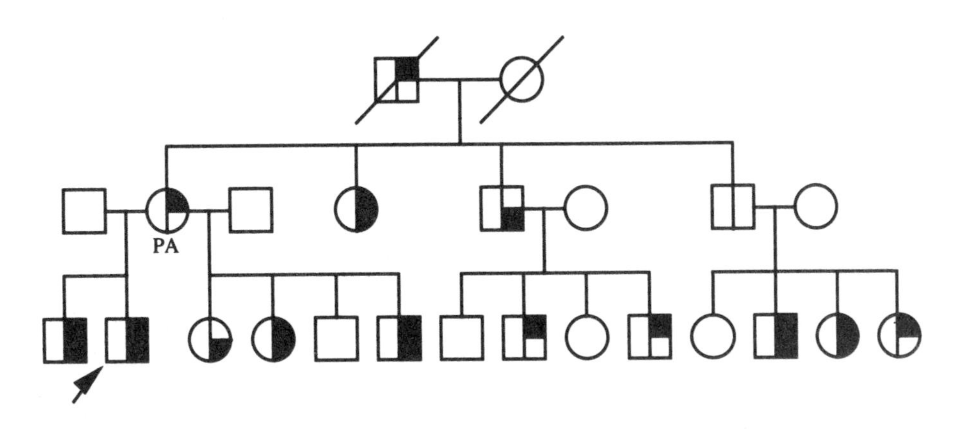

This is not to say the parenting styles may not need changing — they often do. However, many times the therapist has ignored that fact that these same parents have raised other children with outstanding success, before the TS child came on the scene. Parents who inherently "know" that something is wrong or unusual about this particular child are often alienated or even enraged when they are pompously told by a professional the problems are "all their fault." When I see this happen I often think of the phrase, "If you haven't walked a thousand miles in my moccasins, don't judge me."

"My child has Tourette syndrome — he can't help his behavior," or "I have Tourette syndrome — I can't help my behavior." The opposite extreme of blaming the parents for the child's behavior, is either the parents or the patients themselves using the diagnosis of TS as an excuse for inappropriate behavior. Our rule to children and adults is, *we don't care what disorder you have, you are still responsible and accountable for your behavior.* The only exceptions are the direct TS symptoms listed above.

From the easy to the impossible Just as there is a wide range of behaviors in TS, there is a wide range of different types of therapy that are required. We have not approached the psychotherapy of TS with any rigid, preconceived psychotherapeutic dogmas, of which there are many. Our fundamental concept has been — if it works use it; if it doesn't work try something else. As discussed elsewhere, the conduct disorders in the more severe TS patients often have the characteristics of driven, psychopathic behavior. Psychiatrists have for years commented on how enormously difficult these patients can be to treat[1311,490,491,525,1244].

Mild cases Behavioral problems in milder cases of TS usually respond to the same good sense that parents have been using for centuries — consistency, fairness, frequent praise, rewards for good behavior, timeouts, consequences or restrictions for inappropriate behavior, talking out problems, and clear statements of the parents' love and affection for the child. If any psychotherapy were to be recommended to this group it would be of a supportive nature. Learning about what TS is, dealing with other peoples reactions to it, learning they are not the only ones affected with TS, and sharing experiences with others, are all helpful.

More severe cases If the above techniques are working, as they do for most normal children, you really don't have a problem. Parents primarily ask for help when these obvious approaches are not working. The statements we hear over and over from many parents of TS children are — "If I spank him it only makes things worse. Positive rewards don't work. Negative rewards don't work. Nothing works! What shall I do?" Please keep in mind that the following is a discussion of *these* children, where all forms of discipline seem to fail. Recall the sections on frontal lobe dysfunction[p341] and learning disorders[p105]. Children showing a combination of

a) short attention span,

b) impulsivity,

c) learning disorders,

d) failure to appreciate the connection between action and consequence,

e) ease of frustration,

g) short temper and quick to anger,

h) inability to take "no" for an answer,

i) a compulsive resistance to change,

j) disrespect for authority, and

j) a need to be in control,

— can try the patience of Job, parents, teachers and therapist.

What Works and Doesn't Work

The following is a summary of some of our experience with over 1,400 TS patients — techniques that work and don't work.

Spanking A mild swat on the rear end is an age-old disciplinary approach that often serves to bring to a halt some minor degrees of inappropriate behavior in a young child. One of the problems with this approach for the severely oppositional TS child is that it often does not have the desired effect and is then applied harder and more frequently in an ever-increasing escalation of attempts to change behavior. This can quickly pass over the line from a gentle disciplinary maneuver to unintended child abuse, as both the child and the frustrated parent get angrier and angrier. An older child will often fight back, sometimes in an out-of-control rage. A second problem with this approach is that instead of the child reflecting on his own behavior he will turn all the blame onto the parent, thinking or saying what terrible parents they are. They then do not reflect on what behavior of theirs provoked such an episode to occur. They remain angered at the parents, claiming how horrible *they* were for taking such disciplinary measures. Their focus may even turn to how to get back at their parents for such unwarranted abuse instead of contemplating on how to change their inappropriate behavior so that such a scene will not occur again.

"No" is a singular word Many TS children have difficulty accepting the word "no." It is as though this word was stricken from their vocabularies. "No" will often set them off into a temper tantrum or rage attack or set the parent and child off into an escalating swirl of, "No, you can't." "Why can't I?" "You just can't." "Why?" To avoid being drawn into this, make it clear that you will use the word "no" only once and if it is not obeyed, a consequence of some type will follow.

Do not let yourself be drawn into their anger Related to this is the problem of oppositional or confrontive behavior often taking the form of "talking back" or a "having a smart mouth." Again, state your position once and only once. Don't allow yourself to be drawn into constant back and forth haranguing. It is very difficult to carry on a one-person argument if the parent refuses to argue. Remain silent no matter what the child says. Do not say "Lets talk it over," this only makes things worse. The child wants to have things their way regardless of reason and usually no amount of talking will change that. As you continue to hold firm, the child continues to yell, scream and swear all the louder and things simply escalate until everyone is hysterical — parents, siblings, pets and sometimes even the neighbors. Thus, simply say "I refuse to discuss it further." This avoids being drawn into prolonged confrontations that go nowhere.

Rage attacks All children can have an occasional temper tantrum or get very angry when they can't have something they want. However, in many ADHD and TS children this becomes the standard way of reacting and occurs on an hourly, daily or weekly basis. As described previously[p131] these episodes can rapidly escalate into truly fearsome events with the child being totally out of control. One of the most frequent requests we get is how to handle these situations.

The single most basic rule is to *think of your child as temporarily beyond reasoning with*. They are in a blind rage and their brain chemical juices are pumping so out of control that they have lost their ability to reason or to have any insight into their behavior. You are the "bad guy," everything is your fault, you are ruining their social lives, keeping them from having friends, unfair, terrible parents. During these tantrums they hate you and want you dead. They may get very physical and throw or break things, slam doors, kick holes in the wall, and even pull knives or threaten to do bodily harm. *It is important to avoid being drawn into this spiral of mounting confrontation. When a volcano goes off you do not try to put your hands on top of it. Do not take what they are saying or doing personally*; it will only make you more angry, or hurt, and increase the tendency on your part to overreact. *You should make every attempt to disengage.* It is difficult for children to argue with themselves. *Give the volcano time to quiet down.* The following are some typical scenarios.

If the interaction is entirely verbal say things such as:

"That is my decision, I am not going to discuss it further."

"I am not going to dignify this by further discussion."

If the interaction begins to get physical, the following are some alternatives:

One mother of a small boy found that simply holding him tightly until the rage subsided was effective and he then went to sleep. This approach becomes untenable as the child grows older and stronger. For older children, other approaches are necessary such as:

"No TV for the rest of the evening."

"You can't go out Friday night."

"Go to your room until you calm down."

"Go take a hot shower and relax."

If they do go to their room, but then begin to destroy everything in it there are several alternatives. One is to leave them alone, it is their room. If they destroy their own things they have to live with it and explain it when friends visit. Just make sure you don't reward this behavior by replacing what they have destroyed.

If they climb out of the window and run away again, there are several alternatives. One is to let them go. They usually come back within minutes to hours after they have calmed down. Some parents have put bars on the windows. Another effective approach is to have a special timeout room, such as the bathroom, stripped of everything that can be broken. The child is in charge of his or her confinment— "you calm down — you come out."

When you are attempting to handle such out-of-control children, a frequent ploy they use, often with great effectiveness, is to threaten to call the police, claiming "child abuse." They may say, "If you don't let me have my way I'm going to call to call 911" (or what ever applies in you area). These bright children quickly learn to manipulate the system to their advantage. Again there are a number of effective rejoiners. Often the best is to call their bluff and say:

"Go ahead, they will only take you out of the house and put you in Juvenile Hall because if you say you are abused you must be protected from your terrible abusing parents. The police will investigate *after* you are taken away from us."

If this does not work and they do call, we tell our parents to call us and we will explain the situation. What usually happens is that soon the police become familiar with the situation and take the parents' side and the calls from the child then lose their effectiveness. Alternatively, you may wish to stay one jump ahead and call them yourself. Have an officer come. Explain the problem and diagnosis to him or her and even call us for verification. Working with the system in this fashion will get them on your side. Just make sure that you keep your record scrupulously clean and don't get drawn into physically abusing your child where there will be "probable cause" for taking the child seriously.

If all else fails, and you feel your safety or your child's safety is in danger, *you* call the police.

The following is an example of one of many such confrontations, with two different outcomes:

A 16 year old girl with TS was told by her mother that she could not use the car that Saturday evening because she had lied about where she had been the previous weekend. The girl flew into a rage and began hysterically screaming, "You are ruining my social life. I won't have any friends anymore. I am so bored staying home by myself. I hate you. I wish you were dead." The mother became so angry with being told "I hate you. I wish your were dead" that she began shouting back and ended up slapping her daughter in angry retribution, herself yelling, "You don't talk to your mother that way." Very quickly the husband and siblings were drawn into this craziness and it was several weeks before everyone's anger had died down.

After discussing this situation we gave her a few suggestions about what to do next time. The next time was not far off. Two weeks later an almost identical situation occurred and the daughter was again told she could not have the car for a Saturday night. She once again flew into a wild rage, screaming, crying and yelling, "I hate you. You are a goddamn fucking bitch. You fucking whore, I wish you were dead." This time the mother only said "That is my decision. I won't discuss it further" and withdrew from any further contact. This temporarily angered her daughter even more and she tried desperately to drag her mother into further confrontation, even baiting her for not talking by screaming, "Isn't that mature, you aren't even talking to me." However, after five more minutes of yelling and getting no reply she went into her room to sulk. After ten minutes she had calmed down and came back saying, "I am sorry for losing my temper mother." She stayed home that night without any further incidents.

While it takes a great deal of self-control not to react when someone is calling you a "fucking whore," saying "I hate you, I wish you were dead," or similar taunts — silence, after a simple statement that you will not dignify this, can be marvelously effective.

Rewards Rewards of many types, often called positive reinforcement, are wonderful and encouraged, if they work. However, what often happens in the TS child is that initially they are motivated to earn the reward by behaving appropriately, and the reward system works very well. However, as time goes by, parents notice that the poor behavior returns almost immediately after the reward is obtained. Eventually, there is a reversal of its primary purpose and the child begins to use good behavior to elicit bigger and better reward for less and less good behavior. In a contorted sort of way the rewards begin to reinforce the bad behavior. When parents finally realize they have been "had," they react with anger, resentment and further frustration.

We have often observed that the behavior in some TS children actually deteriorates right after they get a reward. In therapy, it often comes out that they feel they do not deserve or are unworthy of any reward and consciously or unconsciously say "I'll show you I don't deserve this." They may react negatively even though they want the reward.

Contracts Contracts are written agreements that if the child will perform or behave in a certain manner certain benefits will accrue. They have the advantage that things are written down and mutually agreed upon ahead of time. This may avoid the complaint, "That's not fair." Contracts may sometimes be outstandingly effective, may work for awhile, or may fail miserably. Reasons for failure are that the child may say he forgot what was on it, never saw that part when he signed his name, accuse the parent of changing the terms, someone forged his name to the contract, or disregard what he agreed to with some illogical justification.

Long-term consequences Placing an ADHD or TS child on restriction for days or weeks for a given act is usually ineffective for the simple reason, what do you now do when they misbehave again an hour later? Never say "you are grounded for a month." As time passes, most parents cannot abide by their own con-sequence, made in the heat of anger, and it clearly looks like "overkill." Such consequences tend to lose their impact as the child forgets why they were grounded in the first place. For the hyperactive, impulsive child the world is played out in terms of minutes or hours and such long-term consequences have little meaning.

Short-term, neutral consequences In our experience, short-term neutral consequences are much more effective. By neutral we mean — nothing physical like a spanking. These include 15 minutes of time out in their room, standing in the corner, doing pushups or sit ups, jogging in place, or hitting a punching bag. The latter may be particularly effective as an outlet for their anger. There are many advantages to this approach. If five minutes later they are misbehaving again, the same consequence can be repeated. If they refuse, the time is doubled. If they refuse again it is doubled again and may have to be put off until both parents are present to enforce it. They can also be instructed to use the time thinking about what they did and why they are receiving a consequence. For the child who has difficulty making the connection between behavior and consequences, this can be as much of a learning experience as a consequence. If the behavior returns soon afterward, the same consequence can be repeated as often as necessary.

If you are in the car or somewhere where the consequence cannot be immediately carried out, statements such as, "for that you owe me 5 minutes," meaning 5 minutes of a consequence when the family gets home.

For older children, the use of consequences must often be used with imagination and finesse. A consequence will be far more effective if it means something to the child. Thus, it is useless to send a child to their room as a consequence, if they like being in their room alone, listening to music. However, if the child does not like being alone in their room, this consequence will be much more effective. Also, because these children are such concrete thinkers and have such difficulty with abstract thoughts, it is often helpful to have the consequences be representative of what might happen to them as adults, if they continue certain behaviors. Thus, time out in a stripped bathroom can represent "jail," and may make a greater impact than time out in their own room. The following are some examples of the use of imaginative consequences.

Food During an evening family session the problem of Jason's swearing at his mother at breakfast was discussed. He would say things to her such as , "Get out of my way you ugly bitch." This was not appreciated as being the way a boy should interact with his mother on a sunny Monday morning. Since he loved corn flakes it was decided that the next time this behavior occurred he would only get bread, butter, and water, symbolic of prison fare. Sure enough, the next morning, despite the promises of the evening before that this would never happen again, he was right back to his old behavior and the consequence was applied.

"Fine, I don't care, I won't eat anything for breakfast," was his predictable reply.

This is often said to maximize the production of guilt in the parent. If the parent does not bite, the consequence is effective.

"Here is your lunch," as she handed him a bread and butter sandwich.

This he promptly threw in the waste basket yelling, "I won't eat that either you bitch. I'll just go hungry all day."

Again no takers as his mother calmly replied, "That's your choice."

He returned from school ready to attack the rich load of goodies always available, and was visibly shaken to find a lock on the cupboard door and anything faintly resembling a delectable morsel gone from the refrigerator. However, he quickly regained an air of bravado and stomped off to his room.

That evening, not entirely by accident, the family chose to dine at his favorite place — the Pizza Hut. A steaming, delicious smelling pepperoni and sausage pizza was dropped in front of him. As he reached out to engulf his share, his mother calmly opened her purse and handed him the bread and butter sandwich he had thrown away that morning. No pizza for him, as promised. Now things were beginning to hurt. This was no longer funny. He wolfed down his sandwich and glowered as the rest of the family enjoyed their meal. The next and subsequent mornings were miraculously calm.

Food again Jim was having bowel movements in his pants and we couldn't get him to stop. He absolutely detested bran. The consequence was that each time he had a bowel movement in his pants, the next day the only food he was to get was bran. The parents only had to do this once and the behavior was completely extinguished and has never returned.

Frequently the consequence must occur more than once before the child realizes the parent is serious. Recall, parents usually learn these methods only after years of putting up with the negative behaviors and feeling helpless about what to do. It most often takes children and parents a number of trials before they begin to switch gears and parental authority is re-established.

Clothes Since Bill, a teenager, loved to

play his stereo, it was mutually decided that taking it away for varying periods of time would be an effective consequence for inappropriate behavior. Wrong! The commonly heard refrain, "I don't care, I can get along without it," was unfortunately true for Bill as his behavior continued despite being out one stereo. The next step was being placed on restriction Saturday night. However, he climbed out his second-story window and stayed out all night. He returned to find himself locked out of the house. After sitting in the backyard for three hours he began to panic because he had to get to work by afternoon. Work was the one thing that really meant something to Bill, because he viewed this as the only way to eventually be free of his "horrible" parents. He was working as a shoe salesman in a local mall and had to be dressed better than his present state of blue jeans and barefeet. He finally climbed onto an upper balcony and broke in through the French doors, figuring he had won again. Much to his dismay, he found that the parents he had always been able to beat down one way or another were one step ahead and had gotten him where it hurt the most — his good clothes were locked up. This finally was a consequence that really meant something and his behavior slowly began to change. Not surprisingly, as his behavior changed, so did his parents attitude toward him, and soon they ceased being "horrible" and "getting away" was no longer a top priority.

Family Therapy

The type of therapy we have found to be most helpful is family therapy. This involves getting all of those who live in the house, or who are involved in caring for the TS child, together. There are multiple reasons why this is the most effective approach.

To get the truth Oftentimes when TS patients are seen in individual therapy they can suppress both the tics and the behavior for an hour and appear to be doing much better than they are. In addition, they tend to project all their problems onto others. Nothing is their fault. It is their mother's fault, their father's fault, the teacher's fault, their friend's fault, the dog's fault, everyone but themselves. They can be marvelous con artists and a therapist unfamiliar with TS can be easily misled. The best way to avoid this is to have all players in the drama present. If the TS child is lying, stealing, confrontive, or oppositional, they rarely admit it and it requires the presence of the whole family to get at the truth.

One imaginative mother found that if her child lied to her about something it was impossible to get him to admit the truth — verbally. However, if she told him to go to his room and write down an explanation of what really happened, he would write the truth. It was as though this bypassed a defense and denial mechanism that existed only when things were expressed verbally.

To see the family dynamics Many therapists do not like to do family therapy because it is so difficult. It is difficult since instead of having only one person to keep track of, there are many. But one of its advantages is the insight it gives into interactions between different members of the family. This can be critical in helping to correct the problems. Many books have been written on this subject[6,201,182,587,611,825,838,1337,1403,1711] and the interested reader may refer to them.

To remove the total focus on the TS child Although the major focus is usually the behavior of the TS child, others in the family may be contributing to the chaos in various ways, and one or more other members of the family usually carry a *Gts* gene. The family approach helps to foster the idea that everyone needs to be involved in the changing process because everyone is affected in some way.

To modify behavior Since much of the behavior of TS children is a chemically driven part of their innate personality, their behavior may be very slow to change. Changing a chaotic family to a coping family often requires giving the parents new techniques for managing the behavior. These skills can often bring about a fairly rapid decompression of family tensions, even though a change in the behavior itself may take much longer.

Cognitive Therapy

An important aspect of therapy, which is useful for TS children, is what has come to be known as cognitive therapy[537,1024]. The essence of cognitive therapy is that *you are what you think* and to change behavior it is necessary to change your thinking processes. Thus, instead of the psychoanalytic approach of attempting to probe the past or the unconscious, this approach is a re-education process, helping the patient to learn more appropriate ways to view situations and react to the world. It is especially effective because it can bring about a much more rapid change than prolonged psychoanalysis. It is important for ADHD and TS children because their impulsive reactions to their environment are often the result of "acting before thinking" and their frontal lobe syndrome[p341] causes them to fail to foresee the consequences of their actions. In essence, *the therapist must initially supply their frontal lobes.* Many years ago Virginia Douglas put it well in her "Stop, Look and Listen" approach to ADHD children[505]. This could also read, "Stop, Look and Think." One must remember that improvement is slow because of the driven impulsiveness, learning problems and short attention span of many TS-ADHD children. For this reason, treatment often progresses much more rapidly if they are also on proper medication that allows them to hear what the therapist is saying and perform in a less impulsive manner.

Goals of Family Therapy

Given the above generalizations, the following is a list of some of the issues that need to be addressed in therapy and some general tips and thoughts.

Getting parents to work together Most children are brought to therapy because of significant behavioral and discipline problems. Since TS and ADHD children are so bright they quickly learn how to exploit any differences in parenting styles. Very often one parent is the disciplinarian and the other is more lenient. These conflicting styles can easily undermine the authority of the parents since children quickly learn how to get the parents arguing between themselves about discipline issues and forget what the issue of disagreement was in the first place. To avoid this, the parents need to sit down alone and hammer out a consistent approach, and then *always support each other*. One of the most destructive elements in parental authority is to have one parent countermand the rules or statements of another, especially in front of the child. If you don't agree with what your spouse has said, support him or her regardless, or suggest that you see your mate for a moment in private, and then discuss the issue with the children, presenting a united front. If there are marital problems, issues over how to discipline the children often become the battleground for parental fighting. Marital therapy for the parents may be necessary before any progress can be made in helping the children change.

Giving parents back their parental role TS children who express a great deal of instant anger and sometimes violence against furniture, walls, and doors, or against either parent (usually the mother), or TS-ADHD children with a great deal of oppositional behavior, often physically or mentally intimidate their parents to such a degree that the children become the parents and the parents become the children. This is called *role reversal*. One of the first goals of therapy is to give parents back their parental role. *The parents may be so worn down that they may have to be given permission to use discipline, rules and structure in the home. They have often forgotten that they, not the children, are the parents and that parents make the rules in the house and children follow them.* We have seen parents leave sessions in abject fear of starting to say "no" to their child. Children must also stop comparing themselves to adults with statements such as, "Dad, you drink why can't I?" The rules for children are different than those for adults and everyone in the home must remember this.

When this authority is taken back, the initial reaction of the child is anger and initially things may seem to deteriorate. As they see their power taken back into their parents hands, the rate and degree of challenging that authority may accelerate. In this initial stage it is often the task of the therapist to simply be there to help the parents weather the storm and to re-emphasize to parents and child together the changes that are going to take place. Not infrequently, the TS child is glad to have some structure return to the family life and to give up the power they didn't really want in the first place.

Resolving feelings of guilt Parents often have many things they feel guilty about in regard to their TS child. They may feel especially guilty if they have punished their child for having tics or making noises, or for coprolalia, before they knew of the diagnosis of TS. They may also have guilt problems about passing this disorder onto their children. Here it helps to point out that TS and especially the *Gts* gene is common and they certainly are not alone. We have found that professionally run support groups are especially helpful in getting the parents and children to realize they are not the only ones struggling with this disorder. Parents need to realize that we all carry several abnormal genes and these just happened to be the ones they drew. There are many genetic disorders that can be more crippling than TS, although not as common.

Initially, when we felt that TS was predominantly a dominant disorder and assumed that only one parent was the carrier, the problem with guilt feelings of the affected or carrier parent could be particularly troublesome. In moments of anger one parent might say to the other one, who had tics or whose relatives had tics, "It all your fault, you gave him the gene!" Now that the evidence indicates that in most of the more symptomatic cases a gene has been contributed by both sides, the guilt is more evenly spread and shared and its use as a weapon has largely disappeared.

Another part of guilt that may cause trouble is the guilt some parents feel for simply applying structure and discipline. We often hear the phrase, "I just can't bring myself to give him any consequences because he has Tourette syndrome." The problem with this is that the child then pushes everything to the limit and is soon out of control, realizing there will be no retribution for even the worst behavior.

Working out rewards for good behavior As mentioned previously, rewards for good behavior have results that can range all the way from highly successful to disastrous. It is important to evaluate where a given child is on this continuum. Sometimes, to prevent the, "I'll show you I don't deserve this" reaction, it may be necessary to give the rewards indirectly. For example, a jar may be kept for rewards, for spontaneous good behaviors, where the reward is time off from consequences recieved later for inappropriate behavior. Thus, a child given 10 minutes of a consequence for inappropriate behavior may reach into the jar and take out a reward saying, "good for 10 minutes off a consequence." This has the added advantage that it rewards spontaneous good behavior rather than good behavior primarily for the purpose of obtaining a reward, and is less likely to provoke an "I'll show you I don't deserve this" reaction.

Work with siblings Brothers or sisters of a TS child may have completely normal behavior or, more often, they may also be carrying one or two *Gts* genes themselves and have various problems. Some of the issues concerning siblings that often need work are their own behavior; their anxiety about developing TS themselves; the stigma they may suffer, or feel they suffer, being the brother or sister of a TS child; problems with bringing friends home if their TS sibling has problems with hitting, lying, stealing, touching, sexual acting out, or other inappropriate behaviors.

Getting divorced parents to work together Children often become the battleground for ex-wives or ex-husbands to vent their anger on each other. If there is joint physical custody or significant visitation time with the other parent, the battle may take on many forms from direct countermanding of the rules of the other parent, denying the child has TS or even denying there is a problem at all, refusing to give medication, refusing to share medical bills, or generally doing anything that will embarrass or make life difficult for the other parent. As painful as it may be, it is often critical to have both divorced parents involved in the family therapy to work out and prevent one parent from sabotaging the efforts of the other. If neither parent has re-married, there have been times when a reconciliation occurred following the realization that most of the marital problems have stemmed from not knowing what was wrong with their child and conflict over how to handle it.

When there has been a re-marriage of one or both parents, all partners must be involved to develop into a cohesive working unit between both homes, working toward what is best for the child.

Getting at the truth Whether it is a problem with simple lying, or something more serious like stealing money from a parent, exhibitionism or sexually abusing a sibling, if the parent or therapist says, "Did you do it?" the automatic answer is "No." If the index of suspicion is high, a more successful approach is to say, in a non-accusing way, "I know you did it, would you like to talk about it?" The usual response is then, "How did you find out," and they proceed to talk about what they did without the shield of angry, self-righteous denial.

Teaching appropriate behavior As mentioned above, under cognitive therapy, one of the major issues with TS-ADHD children is to teach them appropriate ways of behaving and responding. It is not a $5 per "unhuh" type of therapy. It is an active, dynamic, continuously teaching, here and now, type of therapy. TS children are often "picked on" by other children. This is not because they have tics. Many of our patients with very significant tics are some of the most popular students in the school. They usually get picked on because they may have a very short fuse and overreact to insignificant situations. If life gets boring and their peers are looking for a little excitement, they can tease Jimmy. "He always loses his temper and gives a good show." Through therapy, they may learn optimal ways of dealing with peers' cruel taunting. If this can help them to avoid overreacting, the teasing usually stops.

Even talking to a TS-ADHD child is often different from talking to other children. Phrases like "Do you understand?" will get a nod or a "yes" even if nothing went in. Instead the question should be, "What did I say?" If that gets the appropriate answer, and it may take several repetitions to get the answer, the next question is, "What did I mean?" This especially may take repeated statements of the same concept.

It is important to use simple, direct words and short sentences, and to work on one idea at a time. Since TS sufferers are such concrete thinkers, any type of abstract idea will need a lot of explaining and restating.

Work on self-esteem Building self- esteem is always an important goal in treatment. When discussing some of the above aspects of the psychotherapy we sometimes hear the statement, "But doesn't an emphasis on consequences hurt their self-esteem?" First, bear in mind we are discussing children in which all the standard approaches, including the liberal use of rewards, have already been tried and failed. We find that when the appropriate structure and clear, consistent set of consequences are in place, the child's behavior becomes more normal, they are better accepted by their peers, parents and siblings, and then their self-esteem improves. *Low self-esteem may also be a consequence of the intrinsic mood disorder often present and may improve significantly with appropriate combination of medication and psychotherapy.*

Residential treatment There are times when a TS child has such severe behavioral problems that no form of outpatient care is adequate. This is especially true when the TS child or adolescent has significant psychopathic behavior with lying, stealing, and many other antisocial acts done without any sense of remorse. In these cases, full-time residential treatment in a highly structured psychiatric hospital or school may be required.

The Relationship between Behavior and Tics

As I have pointed out before[p136] there is often a poor correlation between the tics and the behavior. Some children with severe tics have no behavior problems; others with minimal tics, have severe behavior problems. Sometimes parents comment that their child's behavior gets worse when the tics get worse; other times they are struck by the reciprocal relation between the two — when the tics are bad the behavior is excellent, when the tics go away the behavior may become unbearable. It sometimes seems like the behavior itself is a tic, waxing and waning just as tics wax and wane.

Like tics, the oppositional behavior can often be quickly suppressed in social circumstances. *Behavior may be wonderful at school and terrible at home.* On more than one occasion we have hospitalized a child because their constant behavioral problems were such a severe strain on the family. As soon as they walked through the door of the hospital any hint of bad behavior disappeared, only to return shortly after coming home. This response to social pressure has led to one trick that has frequently cut off a violent temper tantrum. That is turning on a tape recorder (or video camera) and telling the child you are recording their behavior for others to see. Miraculously the misbehavior suddenly stops.

Summary: If medications alone are not effective in improving behavioral problems in TS, psychotherapy may be very helpful. We have found that family therapy, where all the players are involved, is the most effective. The major goals of therapy are to get the parents to use the same and consistent discipline styles, provide a lot of structure, work out appropriate rewards for good behavior and consequences for bad behavior. There should be no consequences for motor and vocal tics, obsessive-compulsive behaviors, ADHD, learning problems and true coprolalia. However, individuals with Tourette syndrome, whether children or adults, are held responsible for all other inappropriate, antisocial behaviors.

Chapter 86
School

School is the one arena where TS children have some of their greatest problems and greatest assaults to their self-esteem. The following was written by a 10 year old boy with TS.

How It Feels to Have Tourette Syndrome

"The worst thing about Tourette syndrome is that you have to do some things that are embarrassing that you don't want to do — such as having your eyes roll up, your neck twist or making noises when you don't really want to.

"The hardest part of having TS is I can't listen and pay attention and I can't write. Almost all the time I have conflicts with my parents over homework and wanting to do things my way.

"Sometimes my friends do not want to play with me because I act like a baby by jumping on people and sometimes crying over little things that aren't worth crying over. The biggest problem of my life is school. The hardest part is trying to pay attention to the teacher or I get distracted and I daydream and wander off into space. Even when I'm not ticing my writing is very sloppy and kids make comments like, 'Can't you write any better, Chuck?' and I can't. The teachers tell me they can't read my writing.

"It feels awful to sit in class as a bright, intelligent boy and forget things I know like my times tables, and my glasses, and the answers to questions.

"If I could have only one wish granted, I would have my TS cured so that I could get along with my family better."

The multiple areas in which TS impacts on school performance are listed below[360]:

School Problems for TS Children

I. Primary TS Symptoms
   1. Motor tics
   2. Vocal tics
II. ADHD
III. Learning Disorders
   1. Dyslexia
   2. Poor Retention
IV. Obsessive-compulsive Behaviors
   1. Obsessive thoughts
   2. Racing thoughts
   3. Compulsive behaviors
V. Phobias and Panic Attacks
   1. Primary school phobia
   2. Other phobias
   3. Test anxiety
VI. Other Secondary Symptoms
   1. Echolalia and palilalia
   2. Short temper
   3. Coprolalia and copropraxia
   4. Excessive touching and sexual touching
   5. Exhibitionism
VII. Poor Socialization Skills
VIII. Poor Self-esteem

These have all been covered elsewhere and are listed here to show how many different ways TS can impact on school performance. The medical treatment of many of these problems has already been covered. This chapter is devoted to some of the non-medical things that can be done to help TS children in school. It is not meant to be exhaustive or to replace the wisdom of professionals who work with handicapped children every day. It is simply a list of some of the most obvious things we and others have found to be helpful.

Don't Call Me Stupid or Lazy

This may seem so obvious that it shouldn't need emphasis, but it happens all the time in blatant or subtle form. Many children have told us their parents, teachers or "friends" have at various times called them "stupid," "dumb," "retarded or retards," and many other names. For a child who has specific learning disabilities or an ADHD child who can't catch on as rapidly as their peers, such statements are devastating to their self- esteem. Slightly less harsh but sometimes equally wrong is to call them "lazy." TS and ADHD children who passively or actively avoid doing school or homework because they know they can't do it right may quickly be simply categorized as lazy. This doesn't mean that they are totally immune to being lazy. The point is that if your child is constantly characterized as being lazy, consider the possibility that there may be some alternate explanations. The presence of poor performance due to genetic or developmental disorders has variously been referred to as "developmental output failure"[1164] and "failure to strive" (a take-off on the newborn problem of failure to thrive)[1163]. As discussed previously[p341], one of the major symptoms of a chemical frontal lobe syndrome is poor motivation.

Praise Good Performance

This is another "motherhood and apple pie" type of statement but one that is often forgotten in the deluge of poor performance. At some level, every student can do something well and these small goals attained need praise. However, it is equally important not to praise *everything* done well. Intermittent praise is a far better reinforcer than constant praise.

Allow Relief of Hyperactivity

Most schools have a great deal of difficulty dealing with the hyperactivity and motor restlessness of children with ADHD. Teachers often feel that if they let a child stand up and "work off" their restlessness they will abuse the privilege and all the other children will want to do the same thing. However, one approach that has worked is for some children with a professional diagnosis of ADHD or TS is to be given a "pass card" which they can hold up whenever they "need" to get out of their chair and work off their restlessness. The success of this approach depends on teaching the child not to abuse the privilege, teaching their peers to understand and ignore this activity, and most of all helping teachers to be flexible enough to try it.

Work with Strengths

This has sometimes been referred to as a "bypass technique"[125] of bypassing weak developmental areas through stressing stronger areas. TS children are often much better with verbal skills than with handwriting. As a result they may understand the spoken word but have difficulty taking notes, difficulty writing papers and difficulty with written exams. Severe dysgraphia or poor handwriting[p290] is just one component of this. The following are some examples of working with strengths.

Emphasize acquiring knowledge Sometimes teachers have the attitude that they refuse to make exceptions for any child. This ignores the fact that not all children are alike and what one can do with ease another equally bright student may never master. Many of the following recommendations are based on the philosophy that the goal is the acquisition of knowledge rather than the development of specific academic skills and positive attitudes toward learning. Rigid rules such as those requiring "specific timed test goals" and "specific amounts of homework" often interfere with rather than help these goals.

Work with small steps Many TS and ADHD children have told us how they panic or block when presented with a whole page of reading or math problems. However, if the same math problems are given in small groups of 1 to 4, they can be done. If the whole page of text is covered and only the sentence being read is visible, the "whole page panic" can be avoided. This concept of breaking the work into small pieces has many variations.

Avoid complex instructions Complex, multistep instructions will simply be forgotten or avoided. Use simple, short, direct instructions. Better yet, write them down so they will not be forgotten.

Avoid timed tests As mentioned before, many TS patients have special difficulty with timed tests. This may cause them to panic and then block on everything. Simple avoidance of rigid rules like "a specific number of math problems must be done in a specified time," or "a given amount of text must be read" can avoid this type of panic.

Progress at their own speed Another way of saying the above is to allow the student to progress at their own speed.

Reduce work load I have seen some cases where significant tics have gotten better or even disappeared by reducing the child's work- load. TS children are very susceptible to stress and often feel easily overwhelmed. Even a moderate reduction in the workload can sometimes result in a significant decrease in tics.

Reduce rote copying A child with poor handwriting (dysgraphia) is not likely to improve by being required to do what they cannot do over and over. This may simply produce more frustration and make them angry with their handicap. Avoid frustrating tasks requiring a lot of fine motor control, such as "write this sentence 100 times."

Oral examinations Oral examination of a student with severe dyslexia or dysgraphia can bypass a major block to allowing them to show that they really do know the material. This doesn't have to take a lot of time. Usually a few questions of increasing difficulty will give the teacher a good feel for the student's understanding of a subject. This has the distinct advantage that it tests a student's knowledge, not their skills in writing or reading.

Use a tape recorder in class This allows the student to get all the material down and review it at their leisure, or take notes at their own speed. Use of non-word visual aids such as TV and movies can also help.

Use of Braille Institute tapes There are many books on tape for blind people. They are also available for individuals with other types of handicaps, such as dyslexia, and increasingly available for adults stuck in commuter traffic. Their use allows children to utilize their strong verbal or hearing skills and bypass their weak reading skills. Since children have to learn to read, the tapes should not to be used to the exclusion of remedial work on improving their reading.

Read to your child An alternative to the tape-recorded material is simply to read to your child. This has the advantage of the personal parent-child touch.

Use of calculators Some teachers have a computer phobia in math classes. We feel that once the basic addition and multiplication skills are reasonably in hand, the use of computers can often allow a child to appreciate the exciting concepts of math without being bogged down by the drudgery of rote calculations.

Use of computers and word processors Many of our TS children are fascinated with computers and can work with one for hours but have difficulty attending to homework for a few minutes. This ability does not mean they don't have ADHD, it simply re-emphasizes the fact that the attention problems are most severe for tasks that are of borderline interest. Teaching a child with severe handwriting problems how to type and used a word processor may bypass a severe resistance to writing reports. In the absence of a word processor, allowing the child to print rather than obligating them to use cursive writing may help.

Another simple approach is to allow handicapped students to use xerox copies of class notes made by non-handicapped students.

This list could be much longer. However, the general principles should be clear. Work with strengths, bypass weaknesses. Break things into small, non-threatening bites and amounts. Test for knowledge, not skills. Most important of all — be flexible.

Individual Educational Program (IEP)

If a parent, or a teacher, feels a child is doing poorly in school, they may request an evaluation or assessment. This may lead to the development of an individual educational program (IEP) specifically tailored to the student's needs. The recommendations may range all the way from remaining in regular classes with modifications; being placed in special resource classes for 30 minutes or more a day for special help in reading, math or other areas of weakness; or full-time special day classes. Since schools are legally responsible for giving every child an education, in extreme cases they may be required to pay for placement outside of the public school. If you feel your child needs special help, don't be bashful about asking for an evaluation.

Tutor

Sometimes an educational therapist or tutor can do wonders for a child. This approach has the advantage that it does not require any interaction with the school district — an area that can sometimes be enormously frustrating. Paying a tutor to work one-on-one with your child for one or two hours a week, with a focus on specific areas of their weakness, is sometimes all that is required to bring them to grade level. It would serve the child's needs best if the therapist and school can work together.

Summary: Many of the different symptoms of Tourette syndrome cause the TS child to have problems in school. Often these are best handled by non-rigid programs that work with the student's strengths and bypass their weaknesses. Hiring a tutor can provide the TS-ADHD child with enough individual help to allow them to remain in regular classes. If these approaches are not adequate, part-time resource classes or full-time special education classes may be necessary.

Part VI
SOME CONCLUSIONS

This final part serves two purposes: to present new information that has become available while Parts I to V were being written, and to present some general conclusions concerning the impact of Tourette syndrome on the mental health professions and society in general.

Chapter 87

How Common Is Tourette Syndrome?

Tourette syndrome was once thought to be extremely rare and virtually disappeared from the medical journals for the major part of the first half of the 20th century. This was due to that fact that it was either forgotten; or the symptoms were always explained as being due to emotional stress (who needs a syndrome if the cause is obvious?); or that only the most dramatically severe cases were diagnosed. The demonstration that TS responded well to haloperidol, and the evidence that it was a genetic disorder, pushed the psychoanalytical theories back into the closet and slammed the door. With the availability of an effective treatment, the number of cases identified progressively increased. This is well illustrated in the following table, which lists five different studies estimating the frequency of TS. The table is divided into estimates of the *incidence* of TS, or the number of new cases indentified per year, and estimates of the *prevalence*, or the total number of cases in an area of a given population.

Table 1. Frequency of Tourette Syndrome

Area	Case Detection	Frequency	Reference
Incidence			
Mayo Clinic, Rochester, Minnesota	record review	1/100,000 males new cases/yr	Lucas et al., 1982[1220]
Prevalence —Previously Diagnosed Cases			
Children			
North Dakota	questionnaire to all doctors and mental health workers	1/1,075 males 1/10,000 females	Burd et al., 1986[255a]
Adults			
North Dakota	questionnaire to all doctors and mental health workers	1/13,000 males 1/45,000 females	Burd et al., 1986[256]
Prevalence — Previously Diagnosed and Undiagnosed Cases			
Monroe Co., New York	query to all doctors & schools, mass media appeals	1/1,400 males 1/11,000 females	Caine et al., 1988[277]
Prevalence — Diagnosed in One School District			
Los Angeles, California for 2 years	daily monitoring of three K-8 schools	1/95 males 1/760 females	Comings et al., 1989[380]

In a study of the incidence of TS at the Mayo Clinic, all the medical records were examined over a period of 12 years from 1968 to 1979. Only 3 cases, all males, were identified[1220], for an incidence of 1 case per 100,000 males per year. This very low frequency covered the period of time when there was little interest in TS.

In the mid 1980's a study of the frequency of TS in North Dakota showed a 200-fold greater frequency. Based on questionnaires sent to all doctors and mental health workers, the estimated number of previously diagnosed cases was 1 in 1,000 boys and 1 in 10,000 girls[256]. The frequency was considerably lower for adults[256,277].

Very similar results for children were obtained from a study of Monroe County in New York. In this study, in addition to sending queries to all doctors, announcements were made in the mass media to obtain cases that had not seen a physician. In addition to definite TS cases, there were eight probable cases with many TS-like features. There were also eight cases that were later identified and known to have symptoms during the time of the original study. Combined, the prevalence was approximately 1 in 1,400 for males and 1 in 12,000 for females. Since this is the only study that allowed an examination of all cases in the community, whether they had gone to a doctor or not, I will look at it in more detail. Table 2 summarizes the major features.

Table 2. Tourette syndrome in Monroe County, N.Y.

	Total	Males	Females
Number of TS cases studied			
Definite	41	37	4
Probable	8	7*	1*
Later identified	8	7*	1*
Total	57	51	6
Prevalence	1/2,502	1/1,379	1/11,882

* estimate based on sex ratio for examined cases

Other findings

Average age	7.0 years
Average duration of symptoms	5.9 years
Already seen by physicians	59%
Never seen by physicians	41%
Positive family history	56%
Required medical treatment	44%
Treated with haloperidol	32%

In this study about half of the cases had already gone to a doctor while the rest had not. All but one of the cases with behavioral problems had already seen a physician while only one of those who had not seen a physician had behavior problems. This is not unexpected, since parents are more likely to take their TS children to the doctor if there are significant behavioral problems than if there are not.

Grade 1 TS = Community Cases

The objection has sometimes been raised[1475] that studies of TS patients who have sought medical care are not representative of all TS patients. It should be obvious that with any disease, patients whose symptoms are significant enough to push them into seeking medical care will have more severe problems than those who don't need to seek medical care. In our controlled study, we divided the cases into three grades of severity, with grade 1 cases being those that were too mild to treat. We have suggested that the frequency of behavior problems in these cases tends to be representative of all TS cases in the community whether they seek help or not[366]. This is because grade 1 cases share characteristics of both groups reported in the Monroe County study — those that had seen a doctor and those that had not. By definition, all grade 1 cases had seen a doctor, but since the symptoms were too mild to treat they also shared features of the community cases that never needed a doctor. The following table illustrates the validity of this conclusion. When the frequency of various behaviors in all the TS cases from the Monroe County study were compared with the frequency of those same behaviors in our grade 1 cases, the results are virtually identical.

Behaviors	Monroe County Study	Grade 1 Cases (Controlled Study)
ADHD	27%	32%
Sleep problems	27%	21%
Repeated a grade	27%	23%
Learning disabilities	24%	23%
Conduct disorder	17%	16%
Obsessive-compulsive disorder	7%	7%*

* obsessive thoughts occur continuously for greater than 3 weeks.

This supports the hypothesis that in addition to tics the Gts gene causes ADHD, obsessive-compulsive behaviors, conduct, learning, and sleep disorders.

The Frequency of TS in a California School District

The major problem with the approach used in the North Dakota and Monroe County studies is the presumption that all physicians or other individuals surveyed are sufficiently knowledgeable about TS to diagnose all the existing cases. As discussed in a later chapter, the diagnosis of TS is often missed for a wide variety of reasons[p655]. A potentially more accurate method of determining the community frequency of TS would be to have someone thoroughly familiar with all the symptoms of TS intensively monitor a group of school children over a prolonged period of time. This is what we did in Los Angeles[380]. Jim Himes is a school psychologist who is very knowledgeable about TS. Over a period of two years he was responsible for the evaluation of children in three schools of a Southern California school district. This constituted a total of 3,304 students from kindergarden to 8th grade. Half of these were boys. Over the two years, 15 cases of definite TS adhering to rigid research criteria were identified. In addition there were eight cases that fulfilled the same criteria except they moved away before the full year of observation was over. These were termed definite TS < 1 yr. Of these, five cases were discarded to compensate for the somewhat higher frequency of special education classes in the three monitored schools. This gave a frequency of definite TS of 1 in 95 males, 1 in 759 females, and 1 in 169 males and females. There were 12 cases of definite transient tic disorder where tics were observed to be present and, at a later date of less than one year, were observed to be absent. Finally, there were 12 cases of probable or possible TS. These observations are summarized as follows[380]:

Frequency of Tourette Syndrome in a Single School District

Diagnosis	Males	Females	Both
Definite TS	1/152	1/1,517	1/276
Definite TS<1	1/253	1/1,517	1/433
Total Definite	**1/95**	**1/759**	**1/169**
Def. Trans. tics	1/190	1/759	1/303
Prob. + Poss. TS	1/169	------	1/337
Total	**1/45**	**1/379**	**1/82**

If all the diagnostic categories were included, then approximately 1 in 4 children in special education or resource classes had a TS-related disorder. In this epidemiological study, all 10 of the children with definite TS seen in the clinic had ADHD. This further indicates that ADHD is an integral part of TS and is not due to a selective bias.

How Common Are Tics in Children?

When we simply ask whether a child has ever had any tics, even for a brief period of time, the figures are much higher than for TS. In five different studies, the frequency of tics ranged from 5.4 to 18 percent of boys and 1.1

to 11 percent of girls[2142]. In a study comparing teacher and parent reports, the rates were 5.4 to 5.9 percent for boys and 1.1 to 2.9 percent for girls[1667]. Most of these would be diagnosed as *transient tics of childhood*. Our pedigree studies, and those of others[1106b], suggest that *transient tic disorder* is a variant form of expression of the *Gts* gene. If the frequency of *Gts* gene carriers is in the 10 to 18 percent range[p518,p623] this would be consistent with even the higher estimates of the frequency of transient tic disorder. Whether these represent children who carry a *Gts* gene, but only briefly express it, or whether transient tics can occur in individuals without a *Gts* gene, will have to be determined after a genetic marker is available.

Heterozygotes and Homozygotes

If 10 to 18 percent of the population carry the *Gts* gene[p518] and most of these individuals express it in some form, then *this gene is playing a major role in human behavior*. The different effect on behavior of those who do not carry the gene, those who carry a single dose (heterozygotes), and those who carry a double dose (homozygotes) can be **diagrammed as shown below**.

This illustrates the important concept that all individuals, whether they carry a *Gts* gene or not, are at some risk for many of the TS associated behaviors. If they have a single *Gts* gene this risk is increased, while if they have a double dose of the *Gts* gene the risk is markedly increased.

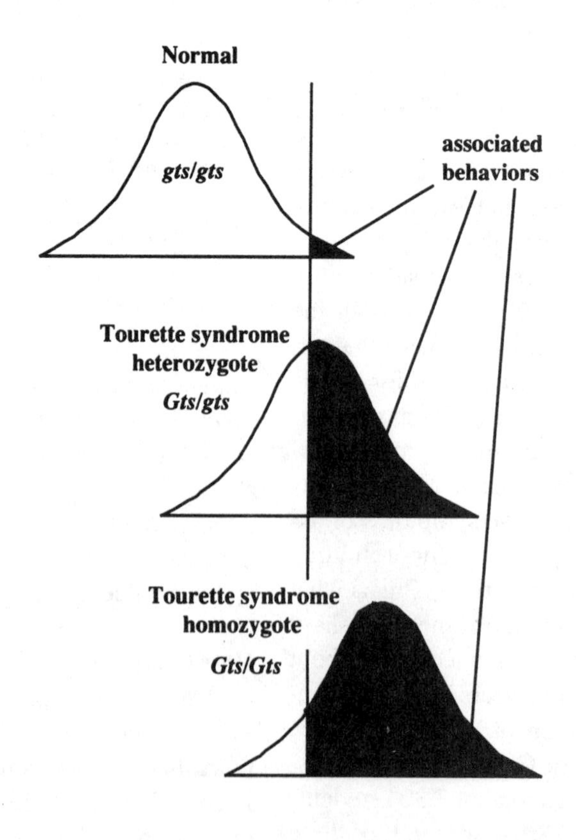

The *Gts* Gene and Psychiatric Diagnoses

The bible of psychiatric diagnoses, the DSM-IIIR[491], lists a large number of different diagnostic categories. The general assumption is that these are distinct entities and most are due to environmental factors under the influence of a large number of different genes. Thus:

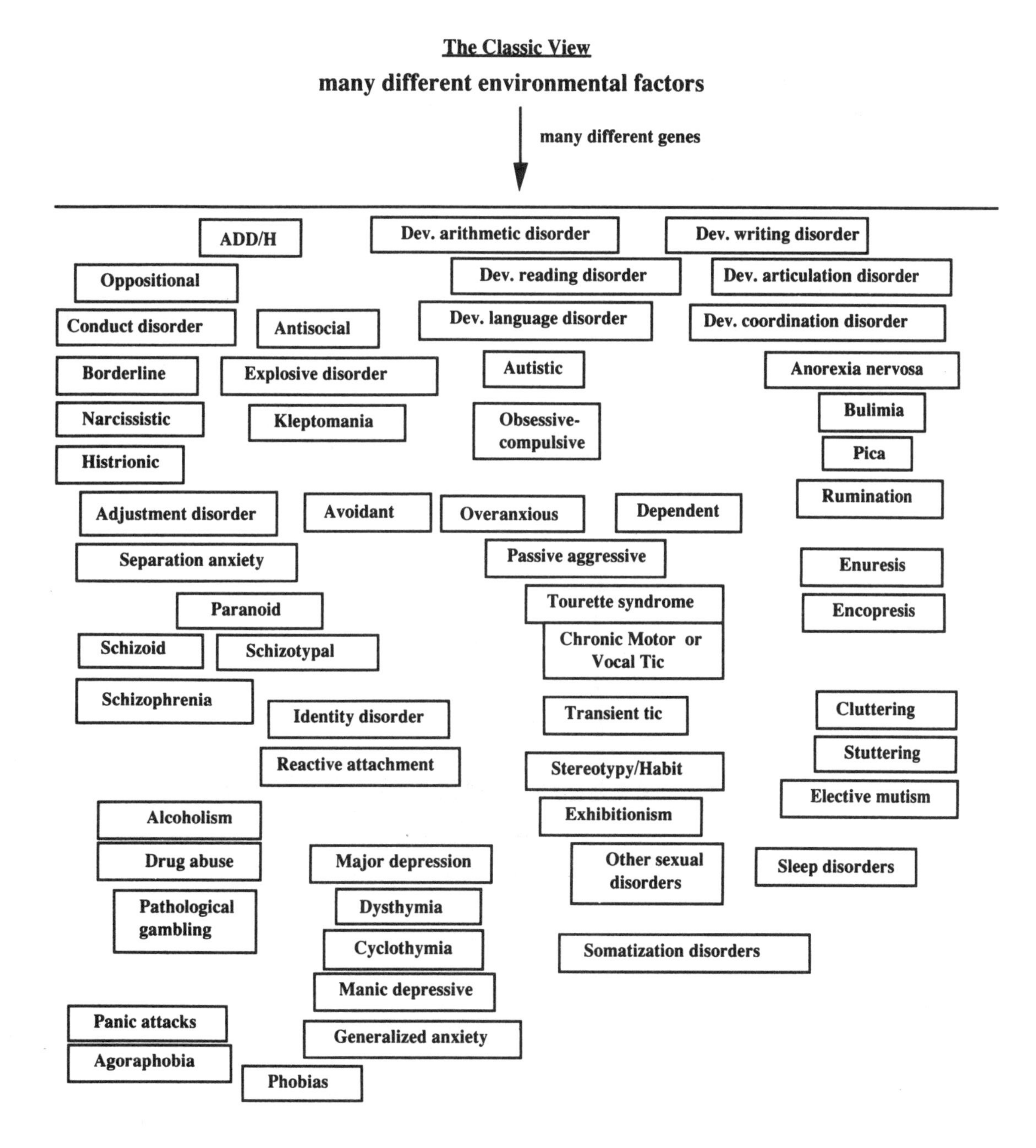

However, as discussed previously[p271], some genes, such as the *Gts* gene, may have been strongly selected. Over a number of generations, this can result in a significant increase in the frequency of the gene[p278]. Thus, instead of a large number of genes influencing the production of the different DSM-IIIR diagnoses, a few may play a much greater role than others. This is **shown as below**.

If some of these common genes, such as the *Gts* gene, have the potential of being expressed in a wide variety of forms, then the diagram on page 619 may have to be redrawn to indicate that most of these diagnoses may not be distinct entities (borders removed), but rather variant forms of expression of the *Gts* and a few other high-frequency genes. Thus:

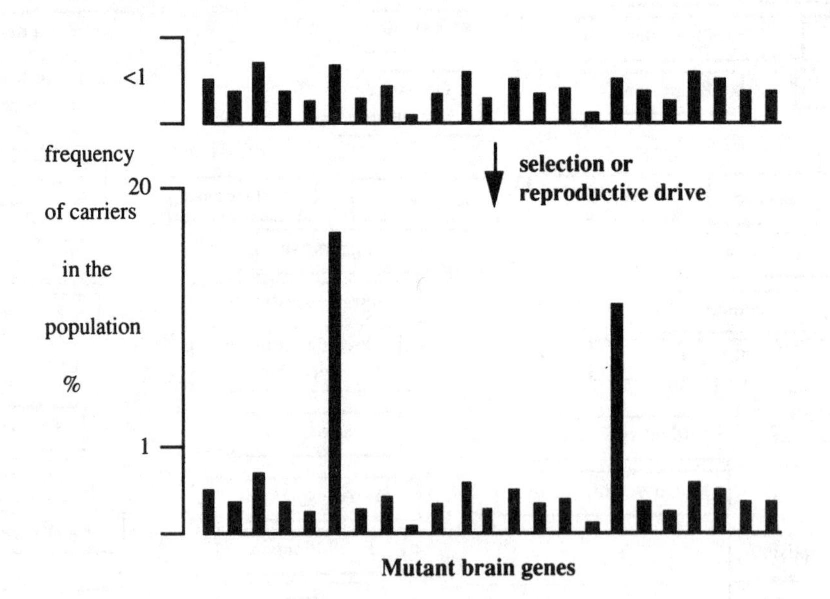

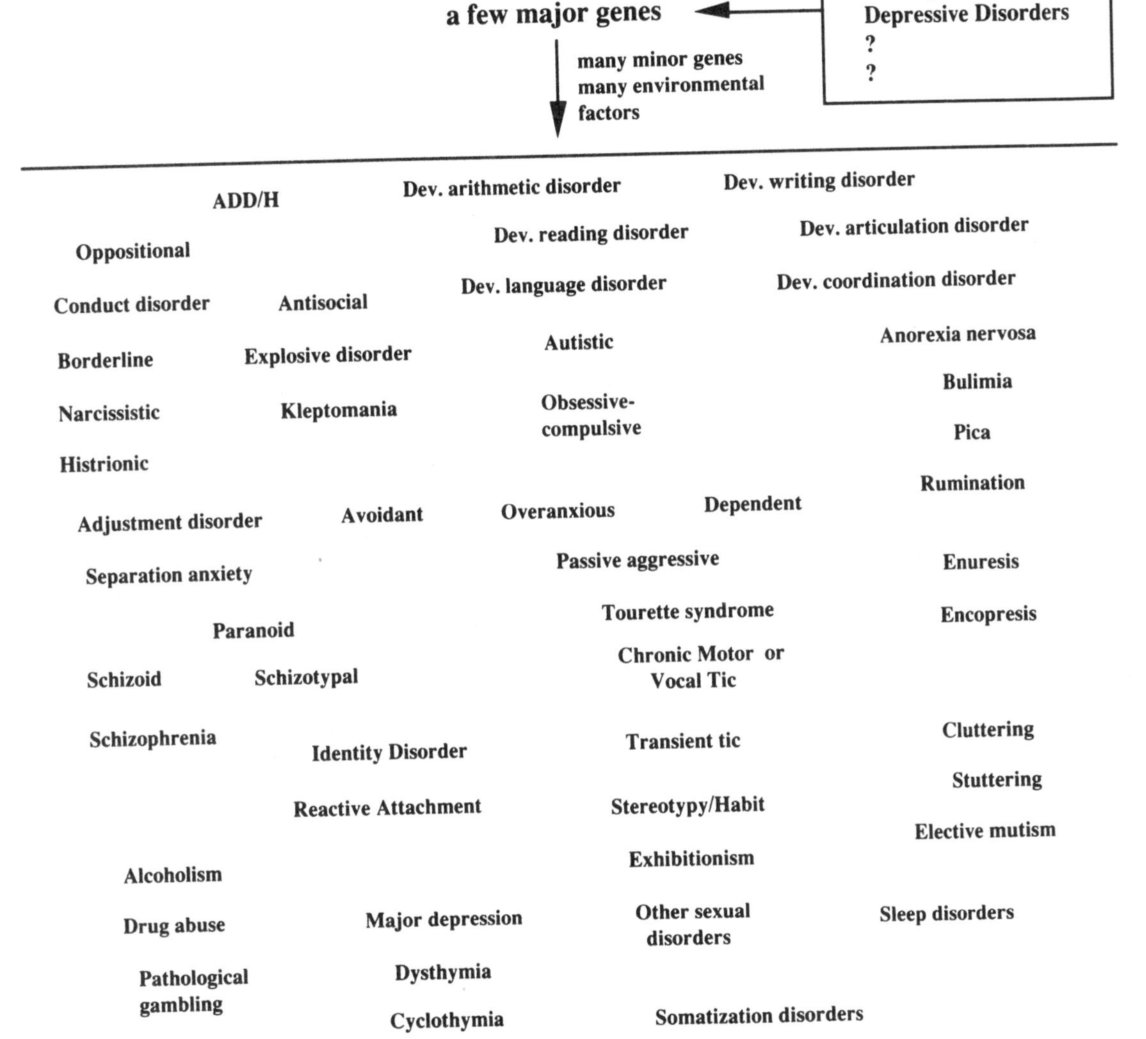

Further evidence for this alternative view comes from the observation that many disorders are frequently associated with a number of other diagnoses. This is called comorbidity or accompanying-illnesses and some examples are shown in the **table on the next page**.

Primary Diagnosis	Comorbid Conditions
ADHD	alcoholism, antisocial personality, conduct disorder, depression, hysteria, oppositional disorder, learning disorders
alcoholism	attention deficit disorder, antisocial personality, conduct disorder, depression, drug abuse, hysteria, panic attacks, pathological gambling, somatization disorder
bipolar II	alcoholism, self-abusive behavior, sexual disorders, panic attacks
borderline personality	antisocial personality, compulsions, depression, impulsive, narcissistic, self-abusive behavior, obsessions, phobias, alcohol and drug abuse
cocaine addiction	attention deficit disorder, depression, manic-depression, alcoholism
conduct disorder	attention deficit disorder, alcoholism and drug abuse, eating disorders, depression, phobias, somatization disorders
obsessive-compulsive	agoraphobia, anxiety, depression, panic attacks, phobias
panic attacks	alcoholism, generalized anxiety, depression, phobias, suicide
schizophrenia	schizoid, schizophreniform, alcoholism, drug abuse
Tourette syndrome	alcoholism , agoraphobia, antisocial personality, conduct disorder, drug abuse, dyslexia, learning disorders, depression, mania, obsessive-compulsive behavior, narcissistic, oppositional disorder, panic attacks, phobias, schizoid behaviors

The *Gts* Gene and the Type II Alcoholism Gene

There are few if any other mutant genes in man that are as common as these estimates for the *Gts* gene. The one exception appears to be the gene for type II alcoholism[p232]. In a genetic study of 109 families with type II alcoholism, using the computer program POINTER[p48], Gilligan, Reich and Cloninger[719] estimated that the frequency of the mutant gene was 0.11. In a similar study of 30 families in which at least two sibs had alcoholism, Hill and colleagues[897a] estimated a gene frequency of 0.15. This translates into 20 to 25% of the population being carriers of a type II alcoholism gene. It is of more than passing interest that estimates of the frequency of *Gts* gene carriers are only slightly lower[p518]. As discussed earlier[p232], I have suggested that the *Gts* gene and the *type II alcoholism* gene could be the same gene. This similarity of gene frequency is consistent with this hypothesis.

[For further evidence suggesting the *Gts* and type II alcoholism gene may be the same, see Chapter 98.]

Summary: Estimates of the frequency of Tourette syndrome have progressively increased as knowledge of the syndrome has increased. When several schools were closely monitored over a period of several years by a school psychologist completely familiar with all aspects of the syndrome, 1 in 95 boys were found to have definite TS.

Selective or reproductive advantages may have produced a few common genes, such as the Gts gene, which have major effects on human behavior. Instead of being distinct entities, many of the DSM-IIIR psychiatric diagnoses may represent the varying ways these common genes can be expressed.

[One of the possible major contributing factors to this marked variation in expression is presented in Chapter 91.]

Chapter 88
The Natural History of Tourette Syndrome

The Tics Usually Decrease with Age

One of the questions parents ask most often is whether their child's symptoms are going to get better as they get older. The answer is — they usually do. However, every TS patient is different and in some the symptoms persist or even get worse. Many of our TS patients tend to be seen for several years and then stop coming to the clinic as the symptoms improve.

Shapiro and co-workers[1771] found that 16% of their patients showed a spontaneous remission and 25% of adolescent patients demonstrated significant improvement with time. Erenberg and colleagues[522] sent a questionnaire to 58 patients between the ages of 15 and 25. This showed that 75% of the patients reported their tics to be either diminished or essentially gone by the time they reached late adolescence or early adulthood. There were twice as many patients under 18 on medication as over 18.

Of 130 patients on medication, followed for 5 to 15 years, Bruun[242] observed that 28% had stopped all medication. All but two of these were over 18 years of age. Of 128 patients treated with haloperidol, 66% had stopped taking this medication. Thus, haloperidol was discontinued at a much higher rate than medication in general. The figure on the right shows the improvement over a period of 5 to 15 years in 136 TS patients followed by Bruun[242]->:

All these results combined suggest that more than half of all TS patients will experience some improvement in their tics as they reach adulthood.

Other Behaviors

Although the tics may get better as TS patients get older, they may be replaced by other symptoms. This aspect of the natural history of TS becomes clear from interviewing family members. Many times a parent will relate that they had tics for several years during their childhood which either disappeared completely during adolescence, or now occur only under stress. However, these individuals may show a wide range of the other symptoms that have been discussed throughout this book; such as obsessive-compulsive behavior, antisocial personality, depression, panic attacks, agoraphobia, alcoholism or compulsive eating with obesity.

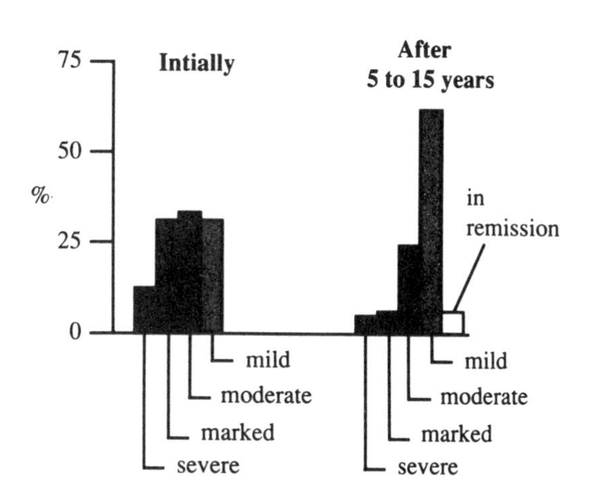

These tend to come on in later years even though the tics may have disappeared. This is shown as follows:

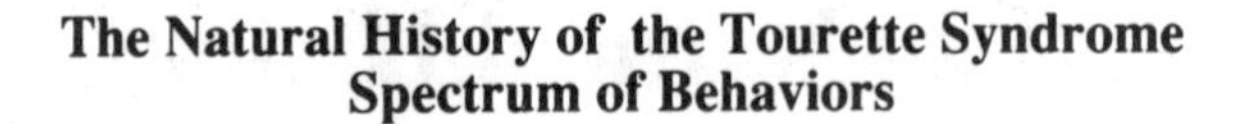

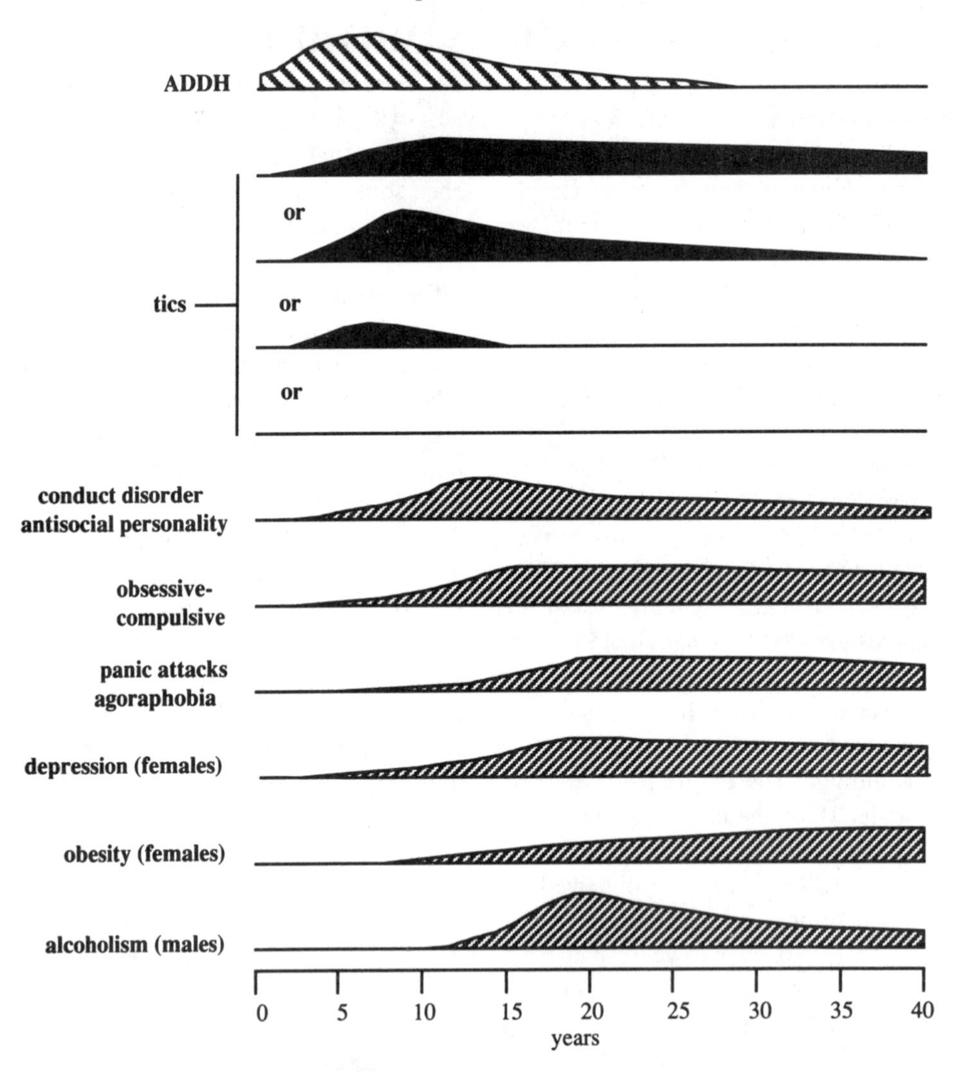

The earliest symptom of the presence of the *Gts* gene is ADHD. On average, the tics start abut 2.5 years later[p103]. The tics are shown in black to indicate that they have always been the most emphasized part of this disorder. After onset in childhood the tics can follow several courses— persistence, slow improvement or compete disappearance. In some cases they may never appear at all. The age trends in the associated disorders are shown. When other behaviors are present, women tend to have more problems with depression and panic attacks, while men tend to have problems with antisocial behavior. Obsessive-compulsive behaviors are common in both sexes. In women they tend to manifest as compulsive eating or shopping; and in men as compulsive drinking, working or gambling.

Occupation versus Unemployment

The effect of Tourette syndrome on life adjustment arises not only from the tics but from the other behavioral problems as well. In the Ohio study of 431 TS patients[1859] only 36 percent were employed full time, 15 percent part time and 34 percent were unemployed. This was based on a statewide survey of all TS patients and the employment figures were strikingly worse than for Ohio residents in general. In a private practice setting in New York, the results were only moderately better with 56 percent employed full time and 5 percent part time and 25 percent unemployed. Although we have not yet studied this in detail, in our clinic, where all patients are seen regardless of their ability to pay, the figures are similar to those in Ohio. Thus, for multiple reasons, TS and its associated behaviors can have a significant impact on life adjustment[p302] and on society at large that must support those who are unable to maintain employment.

Summary: The natural history of TS has some good aspects and some bad aspects. The good is that the tics themselves tend to improve with age and many patients may eventually discontinue their medication. The bad is that even though the tics go away, other problems may arise, including depression, panic attacks, agoraphobia, mood swings, antisocial behaviors, obsessive-compulsive symptoms, alcoholism in men and excessive eating with obesity in women.

Chapter 89
Some Sample Pedigrees

In previous chapters we have seen the evolution of thinking of TS as a dominant disorder to the realization that most cases may be recessive-like in the sense that they may have a double dose of the *Gts* gene. With that philosophical orientation in mind, in this chapter I will present a series of pedigrees illustrating both the recessive tendency of TS and the concept of TS as a *spectrum disorder* in which those who carry a double dose of the gene usually have full-blown TS, while those who carry a single dose of the gene may have a wide range of all the disorders that have been associated with low brain serotonin.

The following pedigrees will differ from those shown previously in that I have added the probable genotype. Thus, individuals in the pedigree who presumably carry a double dose of the *Gts* gene are shown as:

those who presumably carry a single dose of the *Gts* gene are:

and those who presumably don't carry the *Gts* gene are:

| |

The presumed carriers will have a single line through them as follows:

Since genetic markers for the *Gts* gene are not yet available, these are *presumed* genotypes. If there is some question about even the presumed genotype, it will have a question mark.

The other aspects of the legends are as before: **A** - alcoholism, **ADHD** - attention deficit hyperactivity disorder, **LD** - learning disorder, **obc** - **ob**sessive **c**ompulsive behavior, and **PA** - panic attacks. Marked obesity is:

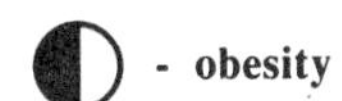

Unless indicated, the probands were all typical TS patients with both motor and vocal tics, usually ADHD, and often other behavior problems. Since the pedigrees are self-explanatory, each will be accompanied with only a brief description. They are presented in the order in which they came to the clinic. Each focus on a different point and they all illustrate the concept that most TS patients probably carry two *Gts* genes.

The pedigree on the **next page (#365)** illustrates the principle of intermarriage or consanguinity in recessive disorders. The rarer a recessive disease the more often parents are related. This rule, however, does not apply if the gene is very common, as is the case in Tourette syndrome. Nevertheless, as illustrated in this pedigree, it does occur. It was not known which common great-grandparent carried the *Gts* gene. I have randomly assigned it to the great-grandfather. This gene was then passed to both of his daughters, and they in turn passed

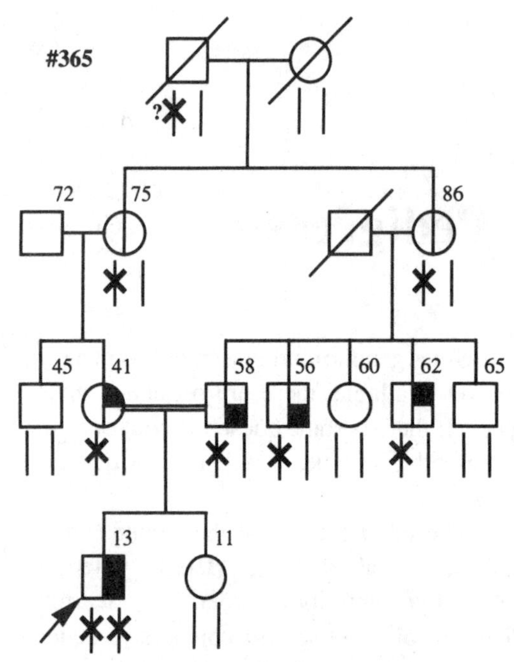

it to a son and a daughter who married each other and had a son with Tourette syndrome. Both parents carrying a single *Gts* gene had mild symptoms (motor or vocal tics), while the son receiving both genes had full blown TS.

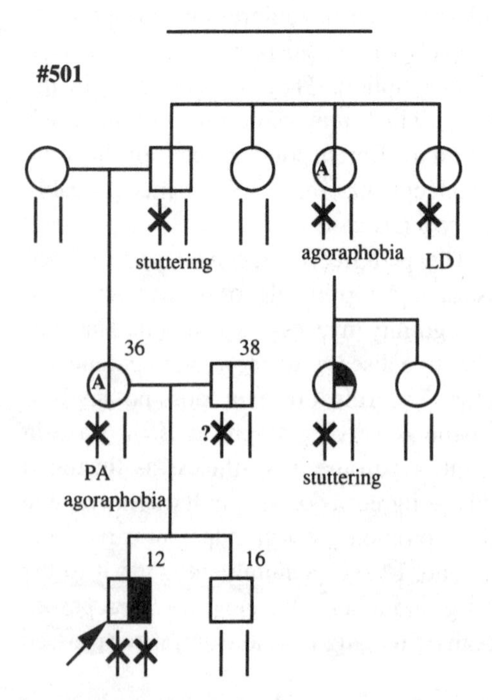

This pedigree **(#501)** illustrates the presence of panic attacks /or agoraphobia in the mother and grandaunt of a TS patient. Both were also alcoholics. The presence of motor tics with stuttering in a second cousin helps to establish the mother, grandfather, and grandaunt as carriers of the *Gts* gene.

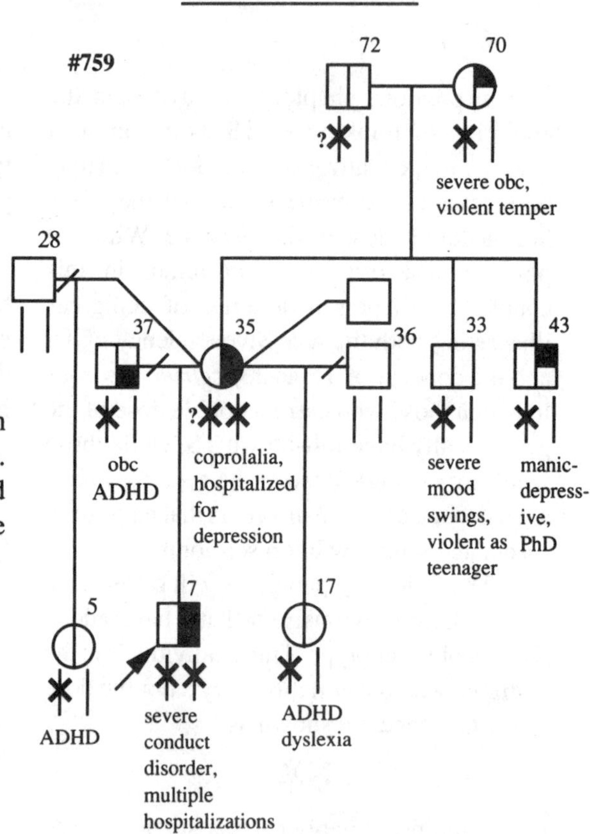

This is a particularly informative pedigree **(#759)** in regard to the recessive nature of TS. In addition to motor and vocal tics, the seven year old proband had such severe conduct disorder that he required almost constant hospitalization[p269]. His mother had a long history of depression requiring hospitalization. She led a chaotic life, and as the pedigree shows, she was on her way to becoming a "spider person"[p275]. Since she is probably a homozygote, she had to pass a *Gts* gene to each of her children. The proband's father had vocal tics and obsessive-compulsive behaviors and the severely affected proband received a double dose. However, his two half-sisters each received only a single *Gts* gene and they were more mildly affected with only ADHD and dyslexia. The mother's 43 year old brother was manic-depressive. The presence of motor tics indicates he carried the gene. He very intelligent and somewhat compulsive and obtained a PhD degree. He probably benefited from his *Gts* gene[p298].

#928

57
56
A
A
died of cirrhosis
mental break-down 3X
manic-depressive
30
33
25
A
A
adult ADHD temper tantrums wife abuse
PA 360 lbs
severe mood swings
ADHD
12

The usual finding is that the male *Gts* gene carriers tend to abuse alcohol while the females tend to abuse food. In pedigree **#928**, the mother of this TS child did both. In addition to her problems with alcohol she weighed 360 pounds and describes herself as a compulsive eater and drinker. She has twin siblings. The twin brother had ADHD as a child and the twin sister had severe mood swings. Her father was diagnosed elsewhere as manic-depressive and her mother had mental problems all her life with three breakdowns. The proband's father was also an alcoholic with vocal tics, temper tantrums and a history of physically abusing his wife. His alcoholic father died of cirrhosis of the liver.

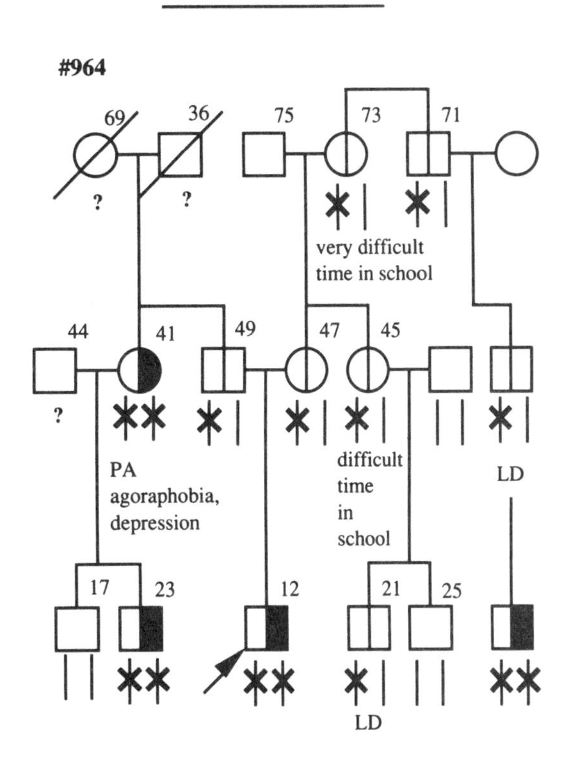

This is an example of a pedigree in which neither parent had any tics or other problems yet both had a suspicious family history. This suggests both parents carried the gene but either did not express it or expressed it to their advantage. For example, the father was a very successful real estate agent and said of himself, "I'm a go-getter, I like to get things done." When the pedigree was initially taken, all he knew of his sister was that she was "very nervous" and had a child with TS. She came to the clinic on the next visit and told of having motor and vocal tics that started at age four and went away at age 12. She said, "I have been depressed since I was a child." At age 35 she developed panic attacks and has subsequently had hundreds of them, causing severe agoraphobia. She had to be dragged into the clinic. This individual presents a classical example of what was illustrated in the chapter on the natural history of TS[p625]. She had motor and vocal tics as a child which went away and were then replaced by depression, panic attacks, and agoraphobia.

The mother's sister had "a difficult time in school" and has a son with learning disorders. The grandmother's brother had a son with learning disorders and he had a son with TS.

---

insightful, without knowing what she said. The proband, Ruth, was a five year old girl with motor and vocal tics, ADHD, and severe oppositional behavior. In asking about the siblings, the mother said of her nine year old daughter, "She is oppositional too, but only half as bad as her sister." I thought, "Half as bad because she carries half the genetic load." All the other siblings, in fact, had some problems. When the mother told me that her 11 year old was gifted and worked very diligently on her studies, I thought "she probably doesn't carry any *Gts* genes." But, in the next breath the mother said, "but she does have a lot of annoying vocal noises." She is probably benefiting from a *Gts* gene causing no ADHD or learning

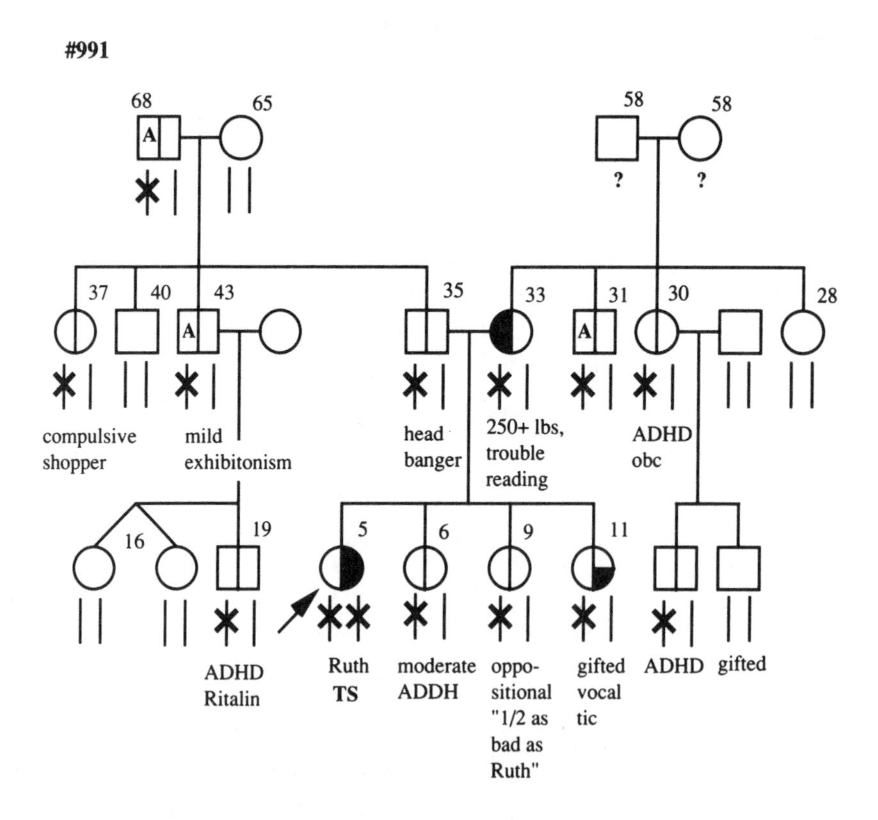

Pedigree **#991** was taken after we began to realize that TS could be inherited as a recessive disorder. I have included it because the mother made a statement that was genetically disorder but mild compulsive behavior.

The mother was markedly overweight and her brother was an alcoholic. The father had no symptoms except a history of being a head banger as a child. However, his father and brother were alcoholics and his brother had a child with ADHD treated with Ritalin.

While these are typical samples of our TS families, the skeptic might complain that out of 1,400 families it would be easy to selectively pick out such a sampling. It is clear that the ultimate proof of the role of the *Gts* gene in these disorders will have to await the development of a specific genetic test. However, as I was writing this chapter, I was seeing family #1007 in the clinic and decided to show the pedigrees of the next five families, *regardless of what showed up*. The following are those five families.

#1008

70
72
67
65
several siblings with sleep apnea
a "worrier" sleep apnea
depression
31
46
41
37
accountant "I'm compulsively good at it."
drinks 2 liters a day
talks very fast
years in psycho-therapy, depression, drug abuse
10

In pedigree **#1008** the proband had severe vocal tics and was thus diagnosed as having a chronic vocal tic disorder. Her father had mild motor and vocal tics all his life. He was a hard-working corporate executive and said of himself, "Everything I do, I do very well." His sister was functioning in her work but "drinks two liters of wine a day." His other sister was an accountant and had said, "I'm compulsively good at it." His mother was a "worrier" and had sleep apnea, as did several of her siblings. (In a sleep study of 34 TS patients, 23 percent were found to have sleep apnea[974]). The proband's mother had no symptoms but "talked a blue streak." She had produced, discussed, and commented on reports from three other doctors before I had a chance to sit down and pick up a pen. Her twin sister had spent years in psychotherapy for depression and drug abuse and her mother had chronic depression.

#1009

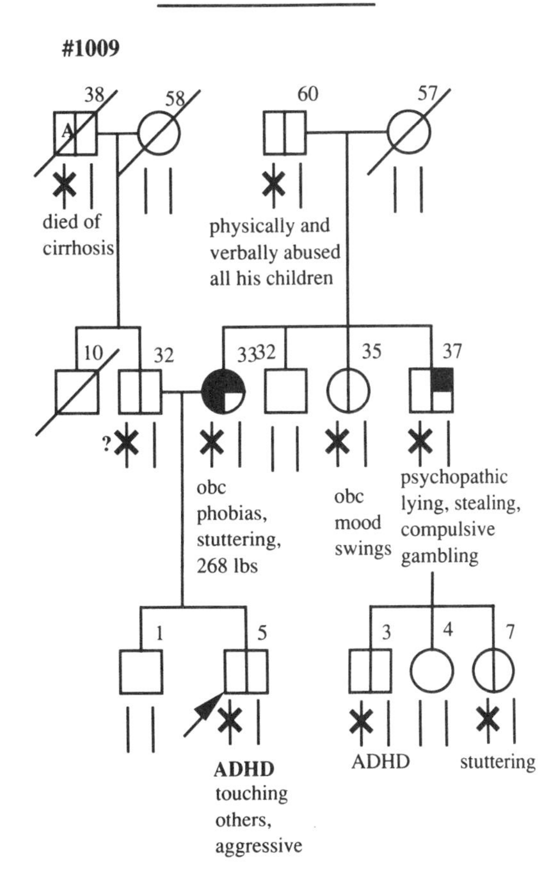

In pdeigree **#1009** the proband did not

have TS [yet]. He had ADHD, was constantly touching other children and was brought in because of his aggressive behavior. The presence of obesity in some members of a TS family occurs so often that as soon as I saw this very overweight mother I suspected that a *Gts* gene was lurking around somewhere. In addition to compulsive eating, she had phobias, eyeblinking tics and stuttering as a child. Sure enough, she described her brother as always having eyeblinking and facial grimacing, and as having a psychopathic personality. "He doesn't care about anyone but himself, compulsively lies, steals and gambles." Of his four children, one had ADHD and the other stuttered. Her sister had mood swings and obsessive-compulsive behaviors. Her father was "a mean SOB" who physically and mentally abused all of his children.

The father of the proband was asymptomatic, but his father was an alcoholic and died of cirrhosis.

#1010

46 48

LD short temper

depression

16 18 21 24 25

short temper

LD short temper

LD

PA

LD homosexual

This 24 year old woman **(#1010)** came to the clinic because of panic attacks. She had no motor or vocal tics, but her brother and father had TS and a learning disorder. Her 18 year old sister had a learning disorder but no tics, and her 16 year old sister had motor tics and a short temper but no learning disorder. Her older brother had motor tics, a learning disorder and was a homosexual.

#1011

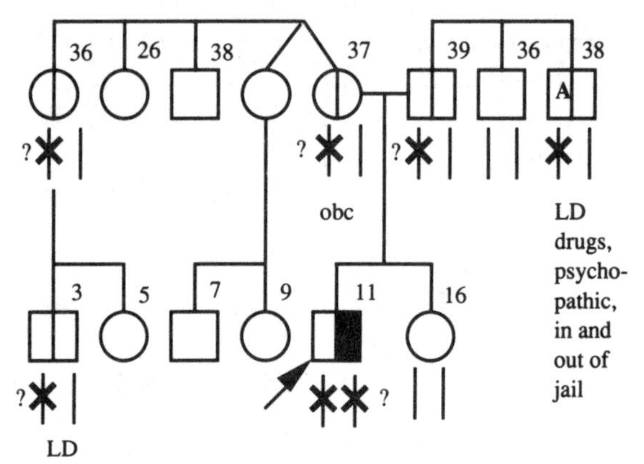

In pedigree **#1011**, other than some mild compulsive behaviors, neither parent of this TS boy had any symptoms. However, the father had a 38 year old psychopathic brother who had a learning disorder as a child, long-term problems with drug abuse and was constantly in and out of jail. The mother's sister's three year old son was diagnosed elsewhere as having a learning disorder.

In pedigree **#1012** **(next page)**, in addition to tics, the proband had such severe behavioral problems he was placed in an SED class and had just come home from spending a year in a residential treatment facility. His mother was divorced, a nurse, and asymptomatic. She had divorced her husband because of his alcoholism. He also had a history of "always being in trouble in school." His brother had problems with both alcohol and drug abuse and his sister, father and grandfather were alcoholics.

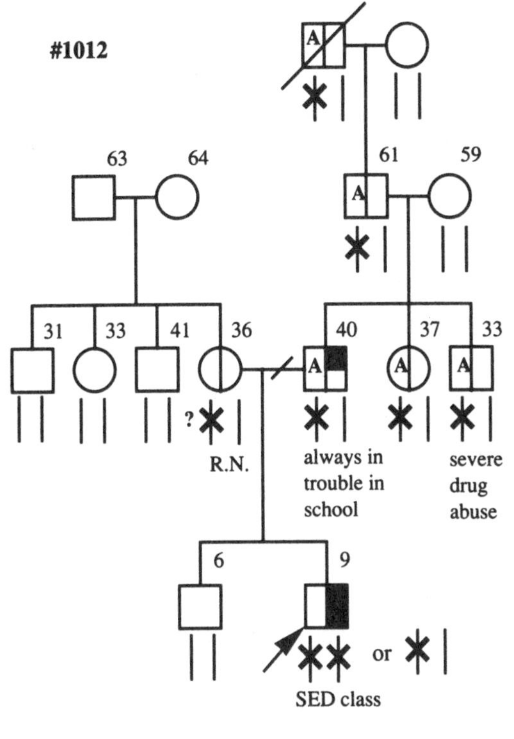

A double dose in two generations In many pedigrees, a parent has TS and appears to be carrying two *Gts* genes and has married a *Gts* gene carrier spouse. Here the risk of having a child carrying two *Gts* genes is high[p657], thus giving rise to two generations of homozygotes. The **pedigree on the right** is an example of this->.

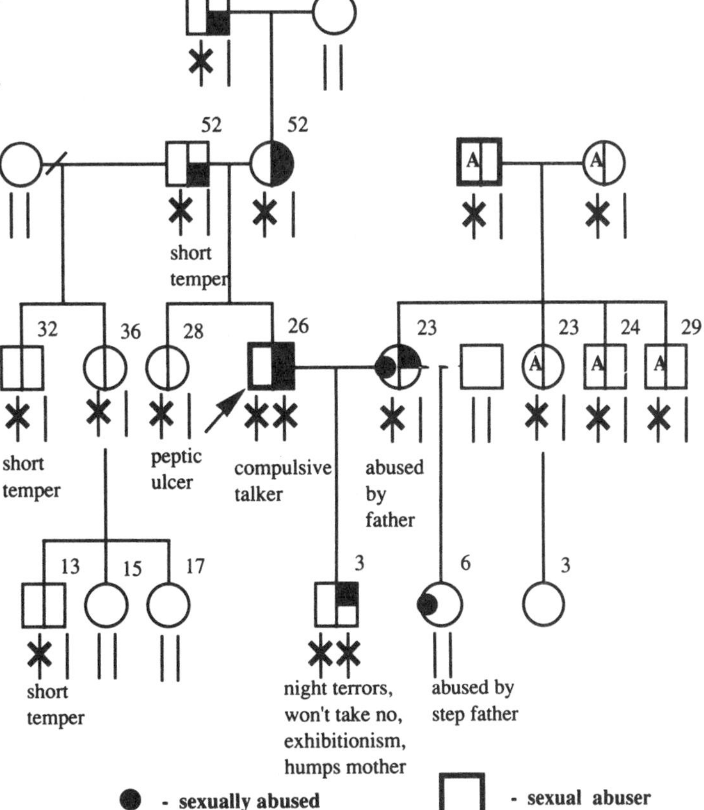

The proband (arrow) had severe TS with many tics. He was also a compulsive talker. His obsessive nature had turned into religious fanaticism. He was presently serving weekends in prision after being convicted of sexually abusing his stepdaughter. Both of his parents had tics. His wife had motor tics and was sexually abused by her alcoholic father. Her mother, sister and two brothers were also alcoholic.Their son started having motor tics at two years of age, was very oppositional, and hypersexual. His mother had frequently woken up to find him humping her leg while she was sleeping and he frequently exposed himself to other children.

These are a representative sampling of TS pedigrees and show the wide range of various behavior problems that can occur in presumptive *Gts* gene carriers. Among other things they illustrate to any mental health worker the value of taking a *careful and complete* family history. Good pedigrees take time and a touch of compulsiveness. It is not adequate to just ask the general question, "Are there any problems in the family?" It works best to initially draw an outline of the pedigree, first asking just the age and sex of all the relatives. This gets the family into thinking about the aunts and uncles, nephews and nieces. Once this is done, go back and ask specific questions about each member, concerning a wide range of behaviors[P30]. Some people are very open about their families, others will respond only to direct, specific questions, or not even then.

The Mennonite Pedigree

Is there evidence from other published pedigrees for the proposal that many cases of TS are homozygous for the *Gts* gene? In an early analysis of the genetics of TS, Eldridge noticed that in a small number of pedigrees where both parents had TS, all the children had TS[531]. If TS was recessive, then both parents would be homozygous *Gts/Gts,* and all the children would have to be homozygous. However, with a dominant trait and full penetrance, 75 percent of the children would be expected to have TS and the numbers were too small to distinguish between the two[371a].

The largest TS pedigree ever published was from a Mennonite kindred in Alberta, Canada[1106,1106a]. It was proposed that the inheritance was autosomal dominant. However, one of the characteristics of religious isolates like the Mennonites is an increase in occurrence of certain recessive disorders[977a] due to the inbreeding. A careful examination of this large pedigree showed that 7 of the 10 individuals with definite TS all came from one set of parents where both had symptoms of TS and the entire pedigree was consistent with a semidominant semirecessive mode of inheritance[377].

Review of Past Genetic Studies

Given the new evidence that many individuals with TS may carry a double dose of the *Gts* gene, it is of some interest to look back at the three genetic studies (**see table below**) that examined the inheritance of TS using the POINTER[P48] program. The first two of these actually predicted semidominant inheritance since *the penetrance was significantly higher for those carrying two Gts genes instead of only one*. Or, to put it differently, individuals who carry two *Gts* genes have more severe symptoms than those who carry a single dose.

The study of Comings and co-workers[376] predicted that all male homozygotes and 89 percent of females homozygotes would have tics, and 68 percent of male and 30 percent of female heterozygotes would have tics. The

Study	**Estimated penetrance % for:**		**% of population estimated to be *Gts* gene carriers**
	Gts/Gts	***Gts/gts***	
Comings et al., 1984[376]			1.2
Males	99	68	
Females	89	30	
Devor, 1984[485]			7.7
Males	99	13	
Females	99	1.5	
Pauls and Leckman, 1986[1473]			1.2
Males	100	100	
Females	71	71	

results of Devor[485] suggested that only 13 percent of male and 1.5 percent of female heterozygotes had tics. While characterized by some[1473] as indicating only recessive inheritance, in true recessives the heterozygotes have 0 percent expression. In a completely dominant form of inheritance the penetrance is the same for those carrying a single or a double dose of the gene. In recessive disorders, carriers have no symptoms. Since TS has characteristics of both (carriers have symptoms but doubling the dose makes the symptoms worse), this was termed semidominant[376,485]. Devor's results also predicted a high percentage of the population would be *Gts* gene carriers, i.e., 7.7 percent . The study of Pauls and Leckman[1473] predicted purely dominant inheritance with the same penetrance in homozygotes as heterozygotes.

How Many Genes?

Throughout this book I have suggested that there is a single major *Gts* gene that is responsible for all cases of TS and this gene causes a wide range of associated behaviors. I have also suggested that this gene is quite common and some individuals with the associated behaviors, such as type II alcoholism, depression, panic attacks, ADHD, learning disorders, and other problems, but no tics, may also be carrying a *Gts* gene. The ultimate proof of this proposal will require the development of a DNA test for the *Gts* gene (or genes). In the mean time, some may find such a sweeping hypothesis that a single gene does so much difficult to accept. For this reason I will examine the evidence in favor of a single major gene and then examine the alternative explanation that multiple genes are involved.

Evidence in Favor of a Single Common *Gts* Gene

1. Selective advantage When a genetic disorder is common, it is natural to assume that perhaps it is caused by several different genes. Cystic fibrosis is a common recessive disease and approximately 1 in 27 people carry a cystic fibrosis gene. It was often assumed that there were actually several different cystic fibrosis genes. However, once the cystic fibrosis gene was found to be on chromosome 7, all studies indicated there was a single major cystic fibrosis gene. By contrast, manic-depressive disorder is even more common but apparently inherited as a dominant disorder. Here, the linkage studies suggest there are at least three different genes[p193]. The intriguing aspect of the *Gts* gene is that it involves disinhibition of the limbic system[p321] which regulates both emotional and sexual behavior. As such, it is almost unique among almost all the estimated 100,000 human genes in its ability to cause a strong selective reproductive advantage[p277]. As pointed out before, even a modest degree of reproductive advantage can result in a striking increase in gene frequency[p278]. This provides the potential of changing a rare *Gts* mutation into a common one.

2. The tics and associated behaviors occur together In genetic terminology this is stated by saying that the tics and associated behaviors *segregate together* rather than separately. For example, if the presence of alcoholism in an individual with TS was due to the fact that he or she inherited one gene for TS and one gene for alcoholism, when the family was examined the two genes should segregate independently of each other. Thus, the families would tend to look like the **pedigree shown on the next page**.

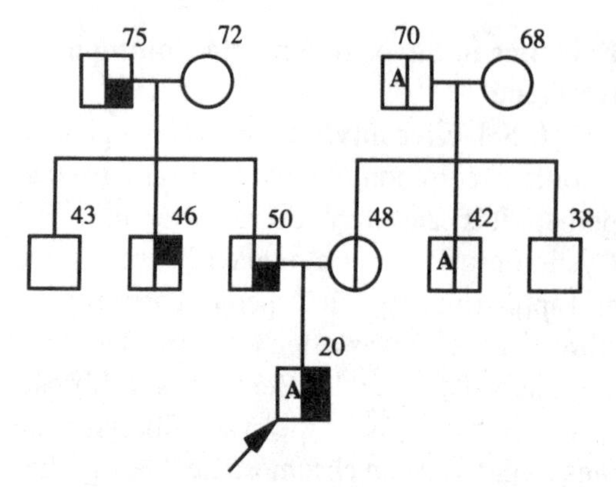

Here the TS gene came from the father's side, the alcoholism gene from the mother's side, and the reason the 20 year old son had TS and alcoholism was that he inherited a gene from both parents. This would represent the conventional wisdom that two problems are caused by two different genes. However, what we actually see is as follows:

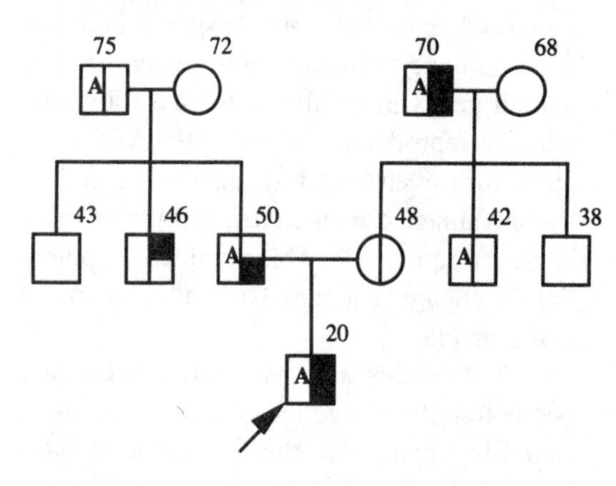

Motor and/or vocal tics and the associated behaviors tend to occur together on both sides of the family and many of the individuals who have the associated behaviors also have chronic tics. Again, in genetic terms, the tics and associated behaviors *segregate together,* not separately. This does not mean, however, that they are always expressed together. As shown in the above pedigree, sometimes the gene may be expressed as alcoholism alone without tics, while other members of the same family express the gene as tics alone or as tics and alcoholism together.

3. Many TS patients are homozygotes The evidence for this has been presented elsewhere[p513]. The reason this favors a single common gene is that *in recessive diseases a double dose of the gene only produces a disease if the two genes are the same.* For example, when only one gene is involved, the enzyme the gene controls is still produced at 50 percent of normal levels and this is enough for normal function.

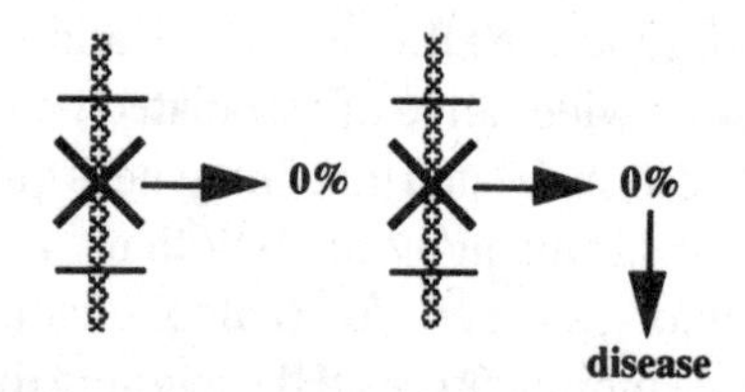

If both the gene from the mother and the father are affected (a double dose) the gene product is completely missing and the disease occurs.

0%
0%
disease

If two different genes are involved the normal counterpart of each can still produce the respective gene products and no disease occurs.

gene 1
0%
50%
no disease
gene 2
50%
0%

As discussed before, Tourette syndrome has many of the characteristics of a recessive disease in that the proband is almost always more severely affected than either parent. If two different genes in single dose were involved and the child got both genes, the result would be equivalent to adding the two disorders together. If neither parent had symptoms, as often occurs, this sum would not amount to much and the child would be unaffected. If a single gene was involved and the child got an abnormal gene from both parents, then the disorder in the child could be many times more severe than in either parent. This is what usually occurs in TS.

Although I have suggested that the *Gts* gene causes an increase in the activity of an enzyme (tryptophan oxygenase) rather than a decrease[p480], the principle is similar.

These are the reasons I favor the idea of a common major gene causing the TS spectrum of behaviors rather than multiple independent genes on different chromosomes. However, there is a second possibility that is often suggested, namely, that there are multiple genes but they are all linked together, next to each other on the same chromosome. Thus, the fourth reason for believing there is a single common gene is:

4. Different but linked genes is an outmoded concept. This concept is illustrated as follows:

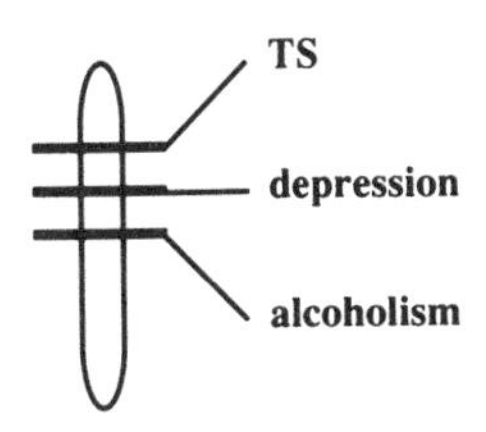

Here an individual with just TS would have a mutation of just the TS gene. Someone with TS and depression would have a mutation of the TS gene and depression gene, and an individual with TS, depression and alcoholism would have a mutation of all three. This is an old idea which has been suggested for many disorders and has long since been disproven in favor of a single gene with variable expression. The reason it is not valid is that it requires that if the proband with TS has depression and alcoholism then all the relatives with TS would also be expected to have depression and alcoholism (three mutant genes), and if the proband only had tics (one mutant gene), the relatives would not be expected to have problems with alcoholism or depression . The family history studies indicate this is not the case. The relatives of TS patients with only tics have just as many problems with alcoholism and depression as the TS patients who also have alcoholism and depression.

Minor *Gts* Genes

None of these considerations is inconsistent with the possibility that in addition to a major *Gts* gene, there may be one or more minor ones. Thus, 70 to 95 percent of TS may be due to a single major gene while the remainder may be due to other mutations on different chromosomes.

Multiple Alleles

Even though a single major gene appears to be involved, there may be many different alleles or mutations of this gene all causing TS. This is illustrated as follows:

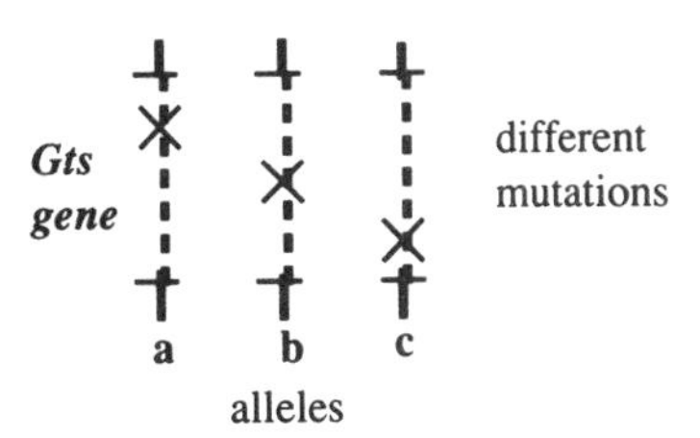

If the severity of TS is greater with the ***a*** allele than the ***b*** allele, and both are worse than the ***c*** allele, then the presence of various combinations of these alleles could account for some of the variability in expression.

Selective Bias

There is general agreement that TS patients who are the probands in the family, that is, the first to seek medical care, have the wide range of associated behaviors discussed in Part III. There is less agreement about why. The view I have presented is that these are all various ways the *Gts* gene can be expressed. Others have suggested that this is all due to selective bias[1472,1480] in that a patient who has TS and just happens to also have ADHD will seek medical care while other members of the family with TS but without ADHD will not. Thus, the argument goes, the only reason ADHD and TS appear together is this bias in selection. While there is no question that the most severely affected family member is usually the first to seek medical care, there are several reasons why I feel this theory is inadequate to account for what is seen in TS patients.

1. There are too many associated behaviors in TS for all to be due to ascertainment bias. For example, in the general population, 1 in 100 boys have TS, 1 in 20 have ADHD, 1 in 50 have dyslexia, 1 in 50 have conduct disorder, 1 in 100 have depression, and 1 in 100 have panic attacks. If these were independent entities, as the selective bias hypothesis states, there would be only a 1 in 50 billion chance a child would have all of the above (100 x 20 x 50 x 50 x 100 x 100). There are not this many children in the entire world. However, the fact that we see this combination of behaviors in many TS children suggests these problems are due to different expressions of the *Gts* gene, not selective bias.

2. TS patients detected through epidemiological studies, where there is no selective bias, show the same frequency of associated behaviors as grade 1 TS patients[p616].

3. All of the associated behaviors in TS have been suspected by other research to be due to defects in serotonin metabolism[p429].

4. If these behaviors were due to selective bias, we would expect them to be absent in the relatives with TS who have not sought medical care. Innumerable pedigrees in this book, and our studies of TS relatives compared to relatives of controls, demonstrate these behaviors occur in both the patients *and* their relatives (see next chapter).

5. An approach that totally avoids selective bias is to evaluate patients coming to a psychiatric clinic for non-TS related disorders. In this regard, Sverd[1905b,1907b] reported a series of patients seeking help for alcohol and drug abuse, major depression, bipolar disorder, obesity, somatization, pervasive developmental disorder and other psychiatric problems in whom TS was diagnosed for the first time in the patients or family members.

In summary, the present evidence favors the proposal of a major, common *Gts* gene which can be expressed as a range of spectrum disorders. However, the best logic in the world is sometimes in error and absolute proof of this hypothesis will require a specific genetic test for the *Gts* gene or genes.

Summary: Twelve different TS pedigrees are presented. They illustrate that relatives of TS patients have many of many of the behavior problems associated with TS including obsessive-compulsive behaviors, attention deficit disorder, learning disorders, dyslexia, stuttering, panic attacks, agoraphobia, short temper, spouse abuse, conduct disorder, depression, mania, alcohol abuse, drug abuse, obesity and others.

While the genetic evidence favors the presence of a major, common gene for the TS spectrum of disorders, specific genetic markers of the Gts gene or genes will be required to prove this hypothesis.

Chapter 90
Family Studies of Tourette Syndrome

On October 13, 1988, in the Presidential Address to the American Society of Human Genetics, I discussed some of the implications that studies of Tourette syndrome seem to hold for the genetics of human behavior. Based on pedigrees like those shown in the previous chapter, I suggested that alcoholism and/or drug abuse in males may be genetically related to compulsive eating and obesity in females[p238], and that a *Gts* gene may predispose individuals to either of these problems. As I mentioned in the previous chapter, Pauls and colleagues at Yale have accepted obsessive-compulsive behavior as a fundamental part of Tourette syndrome but have claimed that other aspects, such as ADHD, are only present in probands because of selective bias. At this meeting they presented a paper[1480a] suggesting that many of the other associated behaviors that we have proposed as integral parts of the expression of the *Gts* gene were also just due to selective bias. Although they observed all the same behavioral disorders in TS probands that we have described[355,360-364] in a study of 86 TS probands and 338 first-degree relatives, they reported that many disorders such as ADHD, depression, and manic-depressive disorder were not significantly increased in first-degree relatives of TS probands compared to relatives in control families. If these behaviors were just due to selective bias they would not be present at increased frequency in the relatives of TS patients. If they are part of the expression of the *Gts* gene, they should be present in other members of the family who also carry the gene but were are not ill enough to seek medical care. This is called a **family study** technique since the relatives were individually interviewed. Although they found that the frequency of each disorder was greater in the TS relatives than in the control relatives, it was not significantly increased for any disorder except obsessive-compulsive behavior.

I felt there were two problems with this study. First, though they found each individual disorder was more common in TS relatives, they did not add them together into an "any behavior disorder" category. If a *Gts* gene predisposes to a spectrum of disorders, each occurring at a relatively low frequency, this may be proven only by adding together the frequency of the individual disorders. If these disorders occur in TS by selective bias, they will not be additive in relatives since some will be more common in the control relatives and some more common in the TS relatives. However, if they all have something in common, such as neurotransmitter defect due to a *Gts* gene, they would be additive.

Second, the number of relatives studied might have been too small to identify significant differences. In statistics there are two types of potential error. One is to claim there is a difference between two groups when in fact the differences are due to chance. The second error, less often discussed, is to assume there are no differences when in fact significant differences are present. Based on our 1,400+

pedigrees, similar to those shown in the previous chapter, I suspected that some of the behaviors Pauls et al. described would show significant differences if their study was larger.

Dr. Arno Motulsky, my professor when I took a genetic fellowship at the University of Washington, was at the meeting and was very vocal in his skepticism that either Tourette syndrome or the *Gts* gene could be associated with such a wide spectrum of behavioral disorders. Since this skepticism was expressed in front of reporters, this controversy was widely reported in the newspapers. It was apparent that I needed to formally analyze the relatives in our TS families to see if our clinical impressions were statistically and scientifically valid.

For several years, a graduate student in my department, Ellen Knell, had been using the family study technique and individually interviewing controls and relatives of 161 TS probands so we would have data on the relatives of our TS patients. However, this work was still in progress at the time of this meeting and she still had at least a year to go before it would be finished.

Because of the controversy I decided to analyze the last 130 TS pedigrees seen in the clinic using the **family history** technique. This uses information on the whole family that is provided by the nuclear family, that is, the parents and other close relatives of the proband. This has proven to be a valuable technique in the study of the genetics of psychiatric disorders[38a]. It has some advantages and disadvantages over the family study technique.

We have found that the accuracy of individuals identified as having problems by the family history approach is quite high. We have rarely found that an individual identified by other members of the family as having a given disorder did not have that disorder when personally interviewed. This has been shown by others as well[38a,1943a]. An additional advantage is completeness. Family members tend to be widely scattered, and as Satterfield and colleagues pointed out[p91], those who are the most difficult to trace tend to have the most severe problems. These problems are usually well known to other family members even though the individuals cannot be found for individual interviews. A related advantage of the family history technique is that noncooperative relatives are included. In our experience, individuals who refuse to be interviewed often have more severe problems than those who are willing to cooperate. The most obvious advantage of the family history approach is time and expense. To track down and individually interview thousands of relatives is extremely expensive and time consuming.

A disadvantage of the family history technique is that it is less likely to pick up certain information than a personal interview or questionnaire. When relatives are individually interviewed, we have often found problems to be present that were not known or obvious to other members of the family. For some disorders, this sensitivity was much greater than for others. The sensitivity was highest for those symptoms that could be seen or heard such as motor or vocal tics, hyperactivity, and marked obesity, and for disorders which clearly interrupted the lives of those concerned such as alcoholism, drug addiction, temper tantrums, severe mood swings, multiple panic attacks, agoraphobia, violent behavior, conduct disorder, truancy, imprisonment, a need to be on medication, hospitalization and related problems[370a]. The sensitivity was lowest for attention deficit disorder without hyperactivity, mild obsessive-compulsive behaviors, depression that did not require medication or hospitalization, occasional panic attacks, and related less severe and less obvious problems. The sensitivity is also highest for first-degree relatives, many of whom were present during the clinic visit, less for second degree relatives, and least for third-degree relatives.

As with our controlled study of the TS probands, in this controlled family history study we used controls from the prenatal diagnosis clinic. The many advantages in using this group as controls were described previously[360]. The lower sensitivity for less severe problems and more distant relatives is compensated for by the use of control families who were asked the identical, detailed questions as the TS families.

Number of subjects in the study There were 130 TS families, 1,851 TS relatives, 25 control families and 541 control relatives in the study. This gave a total of 2,392 relatives. With this number there was less possibility of making the second type of statistical error described above.

As with the study by Pauls and colleagues [1480a], we found a number of specific disorders that were not significantly increased in frequency in the TS first-degree relatives (parents and siblings) compared to the control first-degree relatives. However, each of these occurred more often in TS relatives than in controls, and when *all relatives* were included, providing adequate numbers for statistical analysis, many behavioral disorders (see below) were significantly increased in TS relatives.

ADHD-learning disorders are more common in TS relatives There is a significant overlap in symptoms among ADHD, learning disabilities and dyslexia. Thus, dividing cases into too many categories can eliminate significant differences that really exist. The following shows the frequency in TS relatives versus control relatives of ADHD, learning disorders and/or dyslexia (ADHD-LD). The p values for the comparison of TS versus control relatives are shown.

There was a significant increase in the frequency of ADHD, learning disorders and/or dyslexia in first-degree relatives, cousins and the total of all relatives[370a]. The frequency of these disorders was relatively low in the second-degree relatives because they were much older than the cousins and thus there was less awareness of their childhood problems.

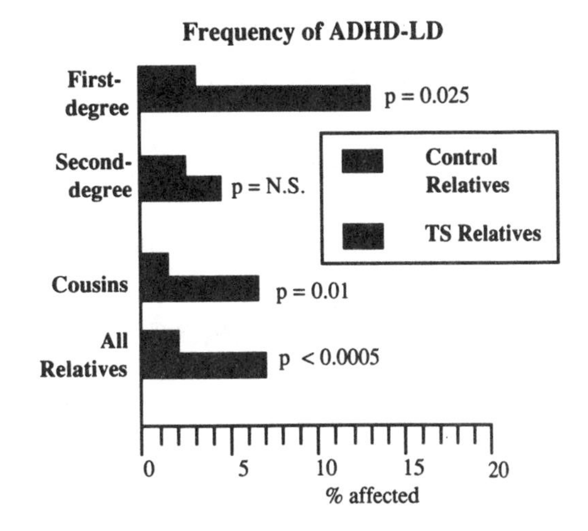

Alcoholism and/or drug abuse and obesity are more common in TS relatives In the chapter on alcoholism I presented the results of a small preliminary study of the frequency of alcoholism and/or drug abuse among the relatives of 20 TS families[374]. This trend persisted after examining 130 TS families. The results were as shown on the **next page**.

Of the fathers of TS patients, 22.8 percent had significant problems with alcoholism and/or drug abuse ($p = 0.05$). These problems were only half as common in the mothers at 11.6 percent. When all relatives were included, 14.4 percent, the difference from controls at 4.4 percent, was highly significant ($p < 0.0005$). Alcoholism or drug abuse was 1.8 times more common in the men than in the women. Alcoholism alone was also significantly more common in all TS relatives than in control relatives.

Obesity was counted only if it was severe and equivalent to an average adult woman being 100 or more pounds overweight. Although present in 10.8 percent of TS moth-

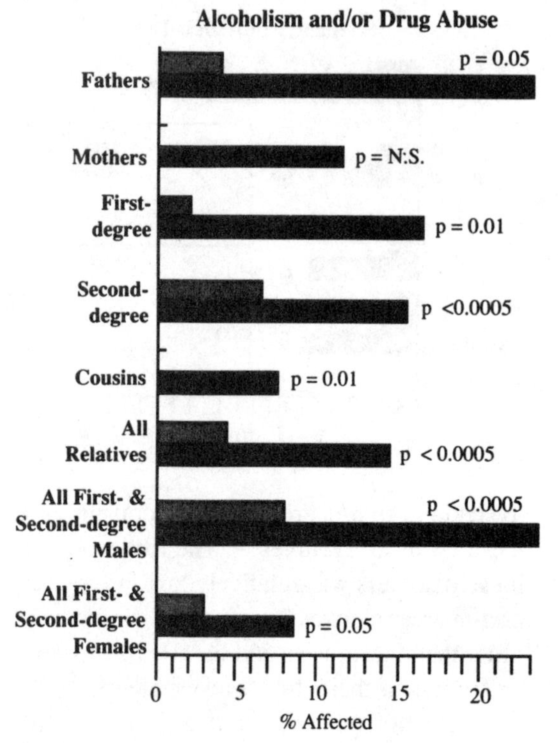

ers compared to none of the control mothers, this was not quite significant due to the relatively small numbers. However, when all relatives were included, 3.2 percent had severe obesity compared to 0.8 percent of control relatives. This was a significant difference ($p = 0.01$). Obesity was 2.9 times more common in the women than the men[370b].

When the category of alcoholism and/or drug abuse was combined with obesity, the differences between TS relatives and control relatives were even more significant. Even more striking was that *now the differences between the sexes disappeared and the number of affected male relatives was comparable to the female relatives*. This supports the suggestion[p238] that both are genetically related appetitive compulsions, and both can be caused by the *Gts* gene which tends to be expressed as alcoholism and/or drug abuse in males and compulsive overeating in females[374].

Depression or manic-depressive disorder are more common in TS relatives Due to the relatively small numbers of first-degree relatives, mood disorders were not significantly more common in the first-degree relatives of TS probands even though they were present in 19.4 percent of mothers compared to 4 percent of controls. As shown below, they were significantly more frequent for all relatives combined ($p < 0.0005$).

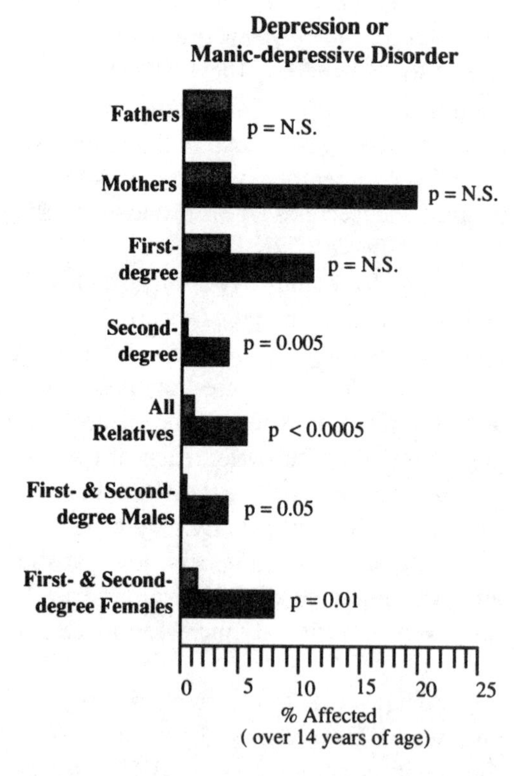

Other behavior problems are more common in TS relatives This category consisted of a number of different behaviors that are frequently seen in TS patients. These include severe phobias, extremely short temper associated with physical abuse of individuals or destruction of objects, paranoia, chronic inability to hold a job, stuttering, multiple migraine headaches, anorexia nervosa, bulimia, and compulsive gambling[370c]. The frequency of these disorders in the TS and control relatives was as follows:

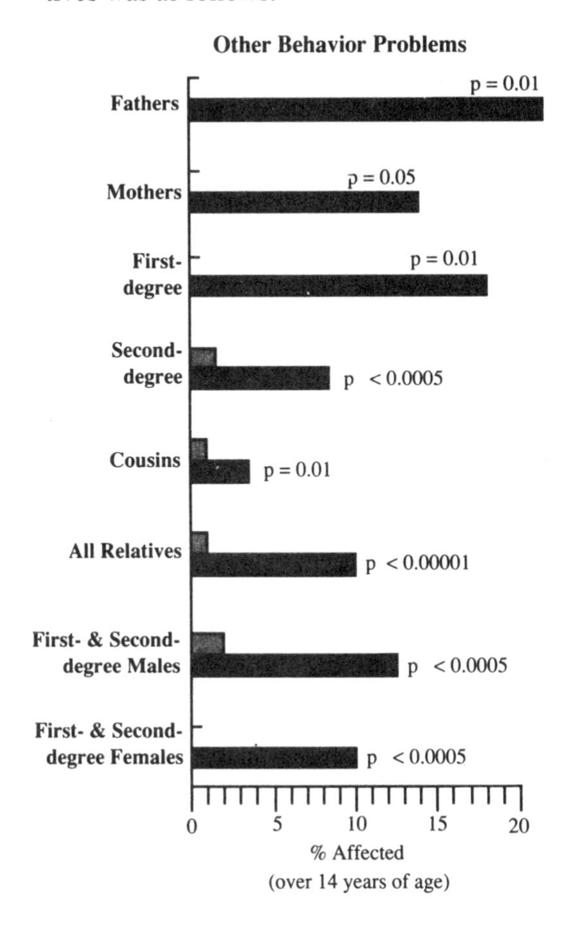

There were striking differences between the occurrence of **any behavior problem** in the relatives of TS patients compared to the relatives of controls. For all relatives, the p values were highly significant ($p < 0.0000001$). The category of mothers and fathers is particularly important since if all TS patients carry a double dose of the *Gts* gene, the mothers and fathers must carry at least a single dose. Because of this, the frequency of occurrence of any behavioral problem in the mothers and fathers gives us an estimate of the penetrance of the *Gts* gene, that is, how often individuals carrying the gene have symptoms. Here 48.4 percent of the fathers and 59.2 percent of the mothers had some type of behavioral problem. This suggests that the penetrance for any behavioral problem is 48 percent in males and 59 percent in females.

Frequency of multiple diagnoses Individual TS patients often have more than one behavior problem. The frequency of multiple diagnoses in a single individual for the TS and control relatives are **shown below**. The statistical comparison showed that these were very different.

The increase in frequency for TS relatives was apparent in all major groups of relatives.

As mentioned above, if a gene is capable of being expressed as a spectrum of behavioral disorders which are all related to the same neurochemical imbalance, then the behaviors should be additive. To examine this, the category of "any behavior problem," except tics, was used. These results are shown on the **next page**.

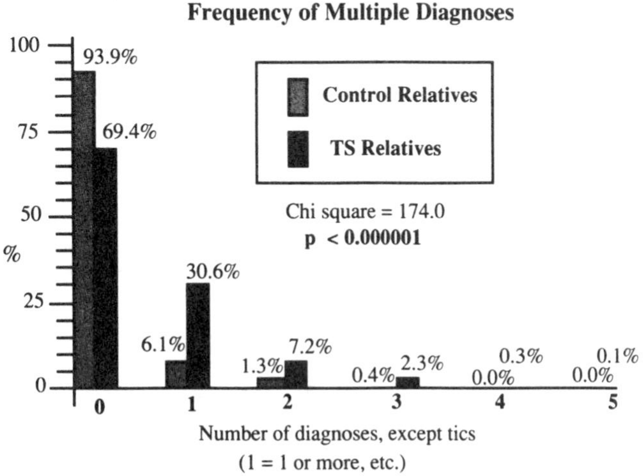

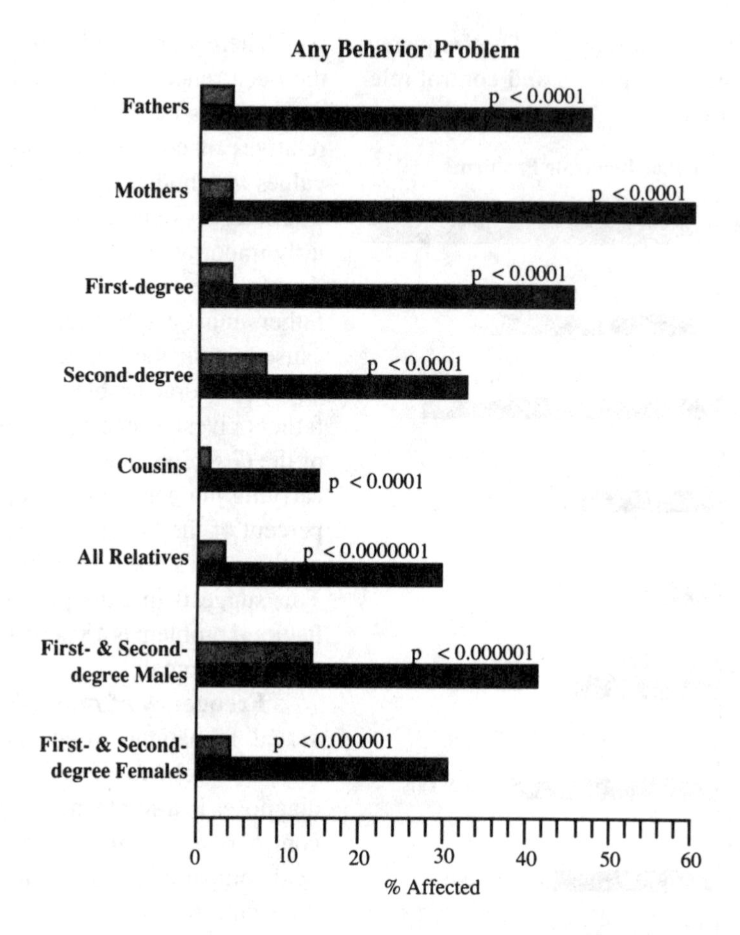

Tics and no tics These family studies indicate that various behavior problems occur significantly more often in the relatives of TS patients than in the relatives of controls. However, this does not address the question of whether tics are always present in those who have these symptoms. To examine this we looked at these behaviors in those with and without tics. The results for ADHD, learning disorders and/or dyslexia are **shown on the right->**.

The p values represent the comparison against control relatives. This indicates that the frequency of ADHD, learning disorders and/or dyslexia was significantly increased in first-degree relatives,

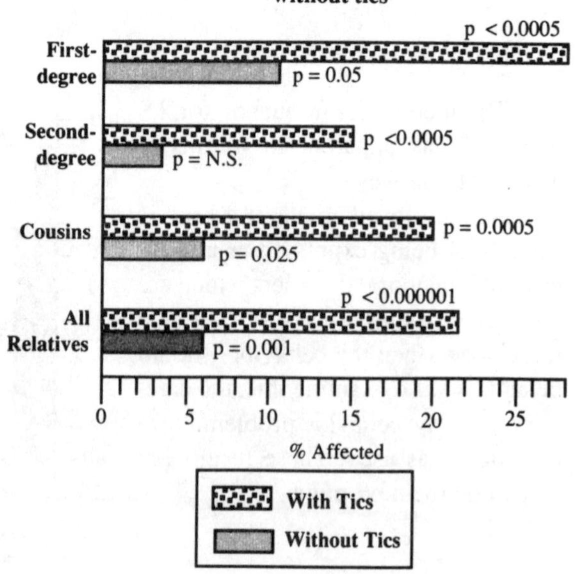

cousins and all relatives whether they had tics or not. The data on the first-degree relatives suggest that when the *Gts* gene is expressed as one of these disorders, in three-fourths of the cases tics are also present and in one-fourth they are not[370a].

The results for alcoholism, drug abuse and/or obesity were as follows:

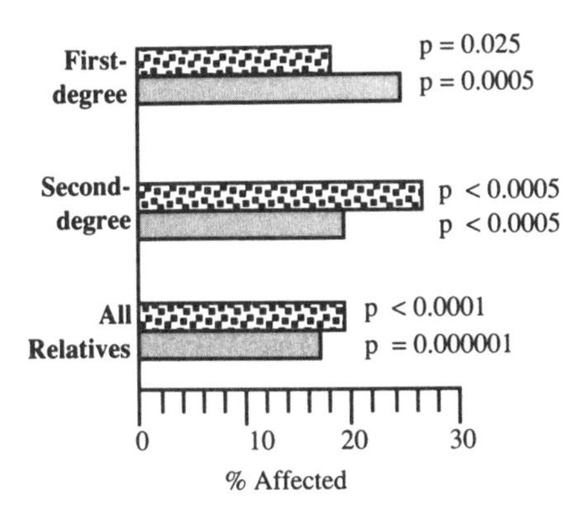

These disorders occurred about as often in the absence of tics as with tics. The following are the results for the category of any behavioral disorder.

Based on the first-degree relatives, *for any behavioral disorder, the Gts gene is just as likely to be expressed with as without tics.*

Blood Serotonin

It is now over a year since the chapters on serotonin and tryptophan in TS were written. During that interval I continued to determine the blood serotonin/platelet ratio and plasma tryptophan on every new TS patient entering the clinic, and where possible, every first-degree relative. Now that many more relatives have been examined it is possible to ask more sophisticated questions than was possible with the more limited data. For example, if all TS patients carry a double dose of the *Gts* gene then every parent should be a *Gts* gene carrier. If the blood serotonin/platelet ratio or plasma tryptophan is a marker for the *Gts* gene, then these values should be low in the parents whether they had symptoms or not. To test this the parents were divided into three types: 1) those who had no symptoms and none of their relatives had symptoms, 2) those who had some symptoms of behavior problems or tics, and, 3) those who had no symptoms themselves but had relatives with symptoms.

We could also ask questions about the sisters and brothers. If both parents are *Gts* gene carriers, then on average 1/4 of the siblings will get a double dose *(Gts/Gts)*, 1/2 will get a single dose *(Gts/gts)*, and 1/4 will be completely normal *(gts/gts)*. If the sibling has symptoms, they are likely to be either *Gts/Gts*

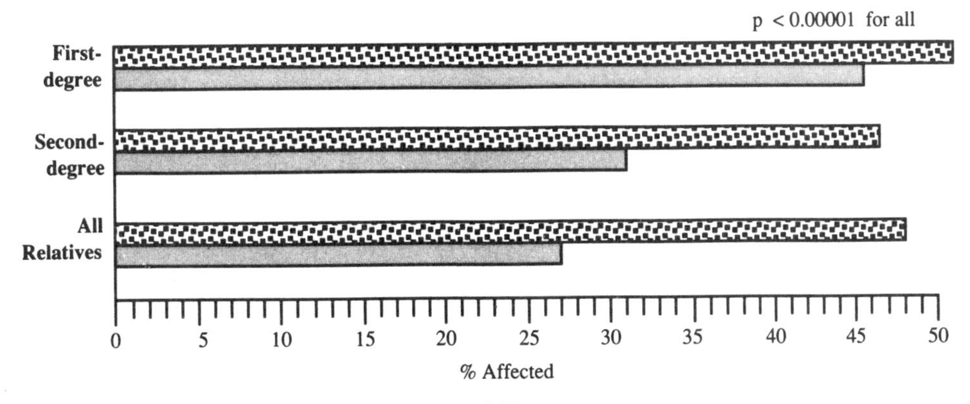

or *Gts/gts*. If they have no symptoms they could still represent *Gts/Gts* or *Gts/gts* states that have not been expressed, or they may not have inherited a *Gts* gene and would be *gts/gts*. Thus, on average, the values should be somewhat higher for the asymptomatic versus the symptomatic sibs but not for the asymptomatic versus the symptomatic parents.

The following figure illustrates the results with these newer serotonin/platelet data. The number of individuals, the average s/p value, and the p value for difference from the controls is given for each type of relative.

Plasma Tryptophan

The results with tryptophan are shown on the **next page**. Here again, the tryptophan (Try) values in the parents were significantly lower than for the controls, the same as for the TS patients, and it made no difference whether the parents had symptoms or not. The siblings' values were also as low as the TS patients. This time, if anything, the symptomatic siblings had slightly higher levels than the asymptomatic siblings. However, more siblings need to be studied since the numbers were lower than for the parents.

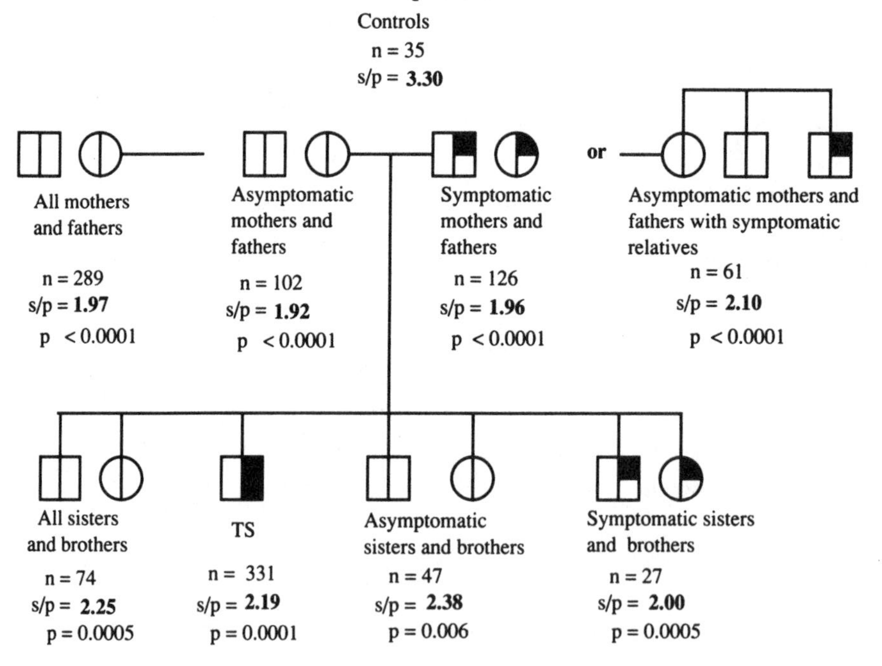

As can be seen, the serotonin/platelet ratio was as low (or lower) for parents as for the TS patients themselves and *the level was the same whether they had symptoms or not*. By contrast, the asymptomatic siblings tended to have somewhat higher levels than either the TS patients or the symptomatic siblings.

These results are consistent with the earlier proposal that the serotonin/platelet ratio or the plasma tryptophan can be a trait marker for the *Gts* gene. They provide especially strong evidence for the suggestion that almost all TS patients are homozygous for the *Gts* gene and that both parents are carriers of a *Gts* gene.

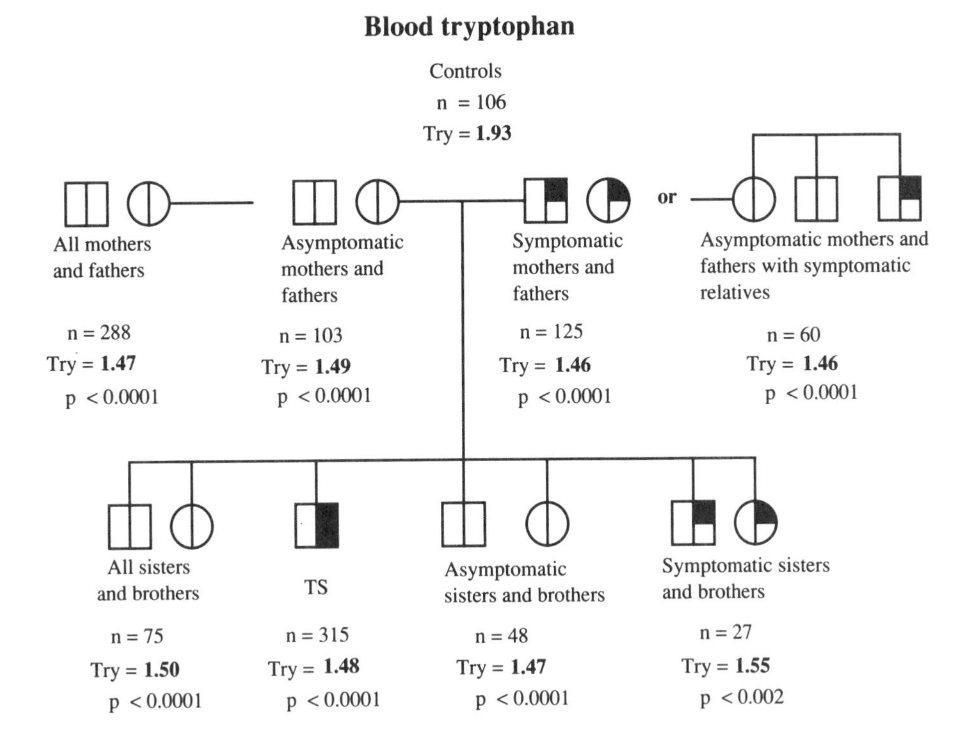

Serotonin and Tryptophan in Parents of ADHD Children

In addition to examining the parents of TS children, in this expanded study we also had the opportunity to examine the parents of ADHD children, both with and without a family history of TS. The remarkable observation was that while the serotonin/platelet ratio was not significantly decreased in the "pure" ADHD children compared to controls, when 30 of the parents were examined, the level was just as low (2.05) as it was for the 289 parents of TS patients (1.97). For those children with ADHD 2° TS, the serotonin/platelet level was significantly lower than controls and the level in the parents was also low (1.91). The levels of tryptophan were significantly decreased for both the children with ADHD 2° TS and "pure" ADHD and the parents of both. This suggests a close biochemical relationship between TS and ADHD, whether there is a family history of TS or not, and implicates a role of abnormal serotonin in many children with ADHD.

Percent of the General Population with Behavior Problems due to a Gts Gene

If the parents are always *Gts* gene carriers then it is not necessary to use complicated computerized analyses to determine the penetrance of the *Gts* gene for various behavioral disorders. All that is necessary is to examine the parents of many TS patients and add up the frequency of various problems. For example, as stated above, if 48 percent of all fathers have some type of behavioral disorder, then the penetrance of the *Gts* gene in males for some type of behavioral disorder is 48 percent. Using this approach, it is possible to estimate the penetrance of the *Gts* gene in males and females for a variety of different disorders. The following table shows the results:

	Penetrance in %	
Behavior	**Males**	**Females**
ADHD	7	2
Learning disorder	4	6
Dyslexia	2	1

Behavior	Penetrance in % Males	Females
ADHD, LD, and/or dyslexia	12	9
Affective (mood) disorder	4	19
Alcoholism and/or drug abuse	23	12
Conduct disorder	5	1
Panic attacks	1	8
Molestation	1	0
Obesity	2	11
Obsessive-compulsive	10	11
Other	21	15
Tics (motor or vocal only)	16	6
Tourette syndrome	7	2
Tics or TS	23	8
Any behavior disorder (except tics)	48	59

Thus, for example, a man carrying a *Gts* gene would be at a 23 percent risk of developing problems with alcoholism or drug abuse sometime during his life, while the comparable risk would be 12 percent for a woman carrying a *Gts* gene.

If we knew exactly how many people in the general population carried a *Gts* gene we could use that figure to determine how many people had one of these problems because they carried this gene. While we don't know this for certain, as discussed before[p518,p623] it may be in the 10 to 20 percent range. Using a guess of 15 percent, the following table was produced simply by multiplying the penetrance by the carrier frequency.

Estimated Percent of the General Population Having a Given Disorder Because They Carry a *Gts* Gene

Behavior	Males	Females
ADHD	1.0	0.3
Learning disorder	0.6	0.9
Dyslexia	0.3	0.2
ADHD, LD, and/or dyslexia	1.8	1.3
Mood disorder	0.6	2.8
Alcoholism and/or drug abuse	3.4	1.8
Conduct disorder	0.7	0.2
Panic attacks	0.2	1.2
Molestation	0.2	0.0
Obsessive-compulsive	1.5	1.6
Other	3.1	2.2
Tics (motor or vocal only)	2.4	0.9
Tourette syndrome	1.0	0.3
Tics or TS	3.4	1.1
Any behavior disorder (except tics)	7.2	8.85

This, for example, would suggest that 7.2 percent of men and 8.8 percent of women have some type of behavior problem because they carry a *Gts* gene. Bear in mind that at the present time these are only estimates.

Summary. If the wide range of behaviors present in probands with Tourette syndrome are due to the Gts gene itself, these behaviors should also be present in the relatives who carry the gene. If a Gts gene can cause these behaviors in the absence of tics, then these behaviors should also be present in relatives without tics. To examine this, 1,851 relatives of 130 TS probands and 541 relatives of 25 control probands were examined. There were significant increases in ADHD, learning disorders and or dyslexia; alcoholism and/or drug abuse; and obesity, mood and other disorders both in those with and without tics. This provides preliminary evidence that the Gts gene can be expressed as a wide variety of disorders with or without tics.

Blood serotonin and tryptophan levels were significantly decreased in the parents of TS patients, whether they had symptoms or not. This supports the proposal that almost all parents of TS children are Gts gene carriers and most TS patients carry a double dose of the Gts gene.

Blood serotonin and tryptophan levels were also decreased in the parents of children with either ADHD 2° TS or "pure" ADHD. This supports the proposed intimate relationship between TS and ADHD.

Chapter 91

Chaos and Human Behavior

One of the fundamental assumptions of science is that if we know the rules or equations that describe how a system works, and have reasonably accurate estimates of all the variables, then we can predict future events. One of the most spectacular affirmations of this view of the world is the extreme accuracy with which astronomers can predict the motions of the planets years in advance and plan the most intricate journeys of space exploring probes. However, in recent years there has been an increasing appreciation of the fact that some systems are so complex that only short-term predictions are possible. Despite knowledge of dozens of variables the systems soon behave in a random manner. The development of this new science has been chronicled in the book *Chaos*, by James Gleick[734].

Attempts at long-range weather prediction provided the first complex system that gave birth to this new science. In the early 1960's, Edward Lorentz was attempting to develop simplifying formulas for weather predictions. When he had the computer printout of the results for a specific variable, such as barometric pressure, it gave a reproducible series of waves. What happened next was described by Gleick[734,p16].

> "One day in the winter of 1961, wanting to examine one sequence at greater length, Lorenz took a shortcut. Instead of starting the whole run over, he started midway through. To give the machine its initial conditions, he typed the numbers straight from the earlier printout. Then he walked down the hall to get away from the noise and drink a cup of coffee. When he returned an hour later, he saw something unexpected, something that planted a seed for a new science.
>
> "The new run should have exactly duplicated the old. Lorenz had copied the numbers into the machine himself. The program had not changed. Yet as he stared at the new printout, Lorenz saw his weather diverging so rapidly from the pattern on the last run that, within just a few [predicted] months, all resemblance had disappeared. He looked at one set of numbers then back at the other. He might as well have chosen two random weathers out of a hat."

What had happened was that in the first run the variable 0.506127 had been entered. In starting again he had rounded this off to 0.506, assuming that the difference of one part in a ten thousand was inconsequential. The differences in predicting a specific result are shown on the **next page**.

In this complex system, after a time, tiny differences in input could result in major differences in output. Unlike the planets, weather was a nonperiodic or nonlinear system. The lesson was that *"any system that behaved nonperiodically would be unpredictable" regardless of how much was known about the variables affecting it.*

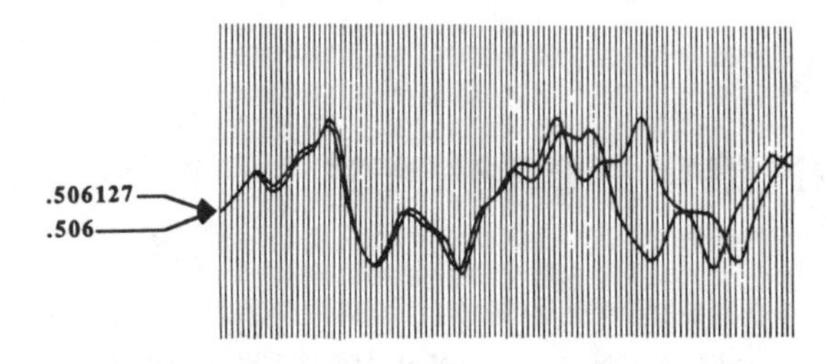

Modified from *Chaos*, by James Gleick, Viking Penguin, Inc. by permission.

These studies led to a series of landmark papers[1213-1215] initiating a new science of chaos. In the world after Newton, many felt that because nature observed precise physical laws everything that happened in the world was predetermined. The science of chaos eliminated these fantasies.

One additional feature of chaos is that despite the apparent randomness, when the patterns are plotted after many cycles of observation, a degree of non-randomness begins to show up. This is illustrated in the **following diagram** produced by a 3-dimensional plot of many cycles of a chaotic system.

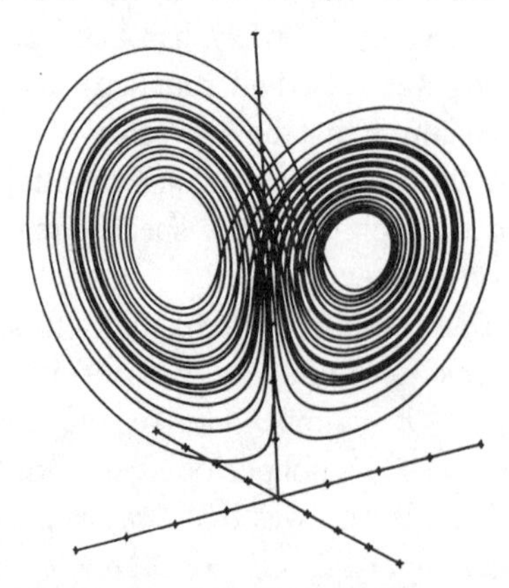

From *Chaos* by James Gleick, Viking Penguin Press, Inc. By permission.

In most biological and physical systems, the assumption has been that small variations in the beginning variables will have only a small effect on the later outcome. For complex systems, the science of chaos produces exactly the opposite conclusion — that tiny changes in the beginning variables can produce giant differences in later output.

This can be put in non-mathematical terms by such statements as — a man who misses a bus that leaves every 10 minutes may miss a plane leaving once a week. Or, two pinball players, each pulling the handle a seemingly identical distance, may obtain drastically different scores.

The Role of Chaos in Human Behavior

In addition to physics, the science of chaos also has wide implications for biology in general and for the genetics of human behavior in particular. As usual, I will use Tourette syndrome as an example. In a previous chapter I showed how a single dose of the *Gts* gene could produce the TS spectrum of behaviors in about half of the gene carriers as follows:

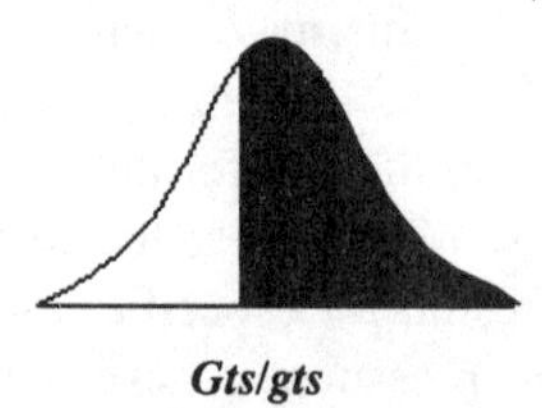

Gts/gts

However, it is unrealistic to present the cutoff of normal versus TS spectrum behavior so sharply. In fact, there is a gradation from normal to spectrum behaviors.

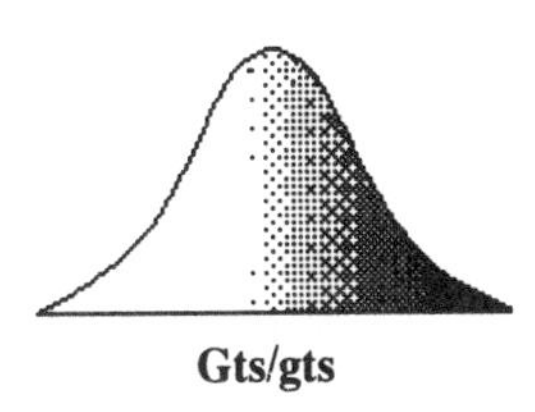

Thus, *Gts* gene carriers can range from having no symptoms to full-blown TS and many of its associated behaviors. It is usually assumed that other genes, and environmental factors, are responsible for this variation in expression of the *Gts* gene. However, we have just seen that in complex nonlinear systems, of which the human brain is a prime example, small differences in variables (such as the level of serotonin in the brain) can result in striking differences in output. Thus, in *Gts* gene carriers, three important factors are regulating behavior:

1. Genetic factors (the *Gts* gene and modifying genes)
2. The environment
3. Chaos

These can be diagramed as follows:

Normal *(gts)* or mutant *(Gts)* genes are put into the black box which is shaken around by the environment and a number of behaviors come out indvidually or in various almost random combinations.

The added factor of chaos has some very important implications. For example, when a genetic test for the *Gts* gene is available, it might be widely used for people with various behavioral disorders. Thus, if we had a patient with alcoholism and the test was positive, we could reasonably say that the presence of this gene probably played an important role in causing the alcoholism. However, the converse does not hold. If we were to test children at birth and found that they carried the *Gts* gene, we could not say that they would become alcoholics. All we could safely say was that their risk of becoming an alcoholic was greater than if they did not have the *Gts* gene. The point is that even if we knew about all the genes known to influence alcoholism, and controlled as much of the environment as possible, a major factor in determining the outcome might still be the uncontrollable one of chaos.

The relative role of chaos would be differ-

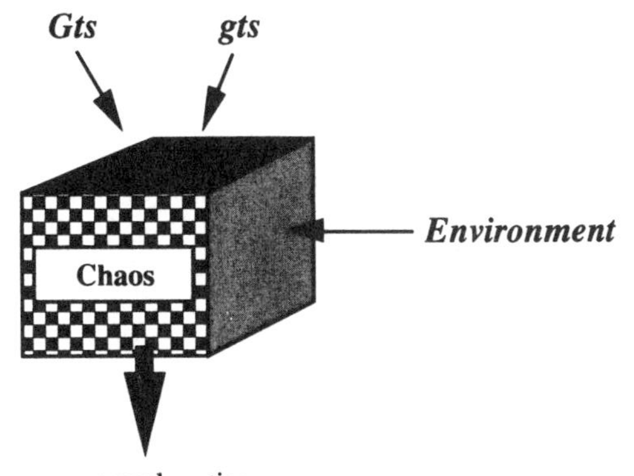

normal tics
ADHD dyslexia depression
antisocial sleep disorders learning disorders
alcoholism obsessive-compulsive panic attacks phobias
obesity short temper other behaviors migraine hypomania

ent for individuals who carried none, one, or two *Gts* genes. In those who carried no mutant *Gts* genes or other behavior-controlling genes, environmental factors would play the greatest role in determining behavior.

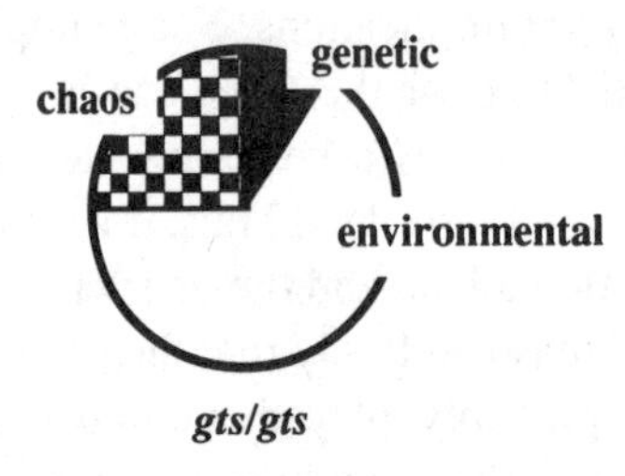

gts/gts

This would be analogous to the studies relating conduct disorder with socioeconomic class, where, in the lower classes, the presence of ADHD genes played less of a role than environment in determing whether there were problems with the law[P91].

In *Gts* gene carriers, a portion of the behavior would be determined by the genotype, a portion by the environment, and a portion by chaos.

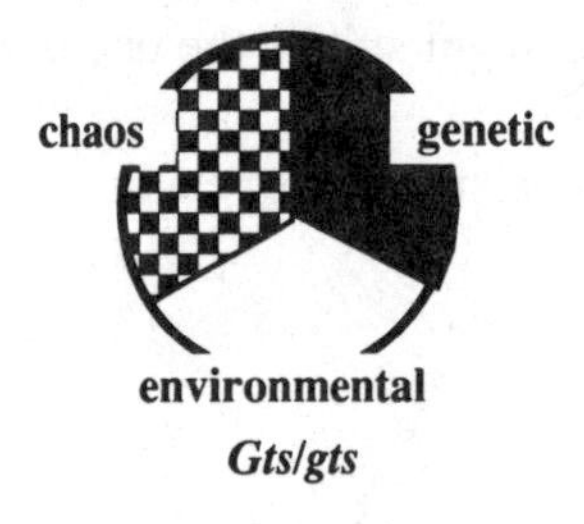

Gts/gts

In the presence of a double dose of the *Gts* gene, the majority of the behavior would be determined by the genotype.

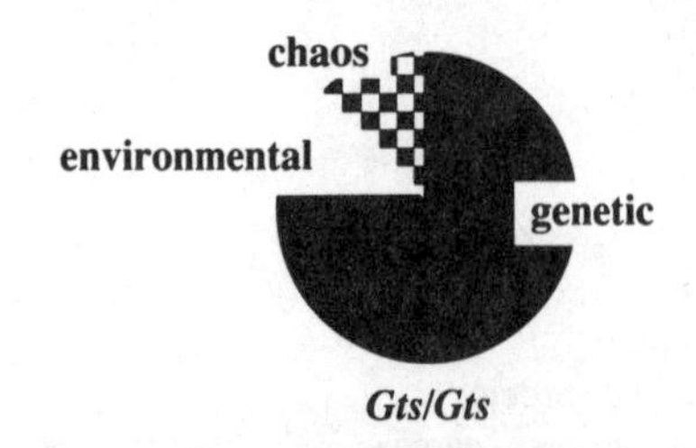

Gts/Gts

The treatment of these three different individuals would be different. In those whose behavior was predominately due to environmental factors, traditional psychotherapeutic methods and community action programs might suffice. For those who carried a *Gts* or a related gene causing an alteration in brain chemistry, both medication and traditional psychotherapy would be required. Finally, for those carrying a double dose of the *Gts* gene, medication would play a particularly important role and psychotherapy would be along the lines described in Chapter 85.

The message is that not all behavioral problems are genetic in origin and those that are not may respond equally well to traditional psychotherapies as to appropriate medication. However, since successful treatment depends upon identifying those that are genetic, *making a correct diagnosis is a critically important part of the therapeutic process.*

It is likely that some genes affecting behavior breed true, that is, they cause predominately a single disorder while some genes breed chaotically, that is, they cause a spectrum of disorders. The *Gts* gene would be an example of a gene that breeds chaotically. It is also likely that behavioral genes that breed chaotically are more likely to be selected for than genes that breed true, since by the very nature of their diversity it is more likely that some aspects of the spectrum will be advantagous to the people carrying the gene.

Summary: The science of chaos indicates that in complex systems, such as the human brain, even if all the variables known to influence behavior are known, it will still be impossible to accurately predict behavior. The corollary of this is that in addition to our genes and environment, a third factor, chaos, plays an important role in determining behavior. A total knowledge of all genetic and environmental factors will allow us to predict the behavior of any given individual only in general terms.

Chapter 92

Why Is the Diagnosis of Tourette Syndrome Often Missed?

Although the criteria for the diagnosis of Tourette syndrome are simple and straightforward[P11], the diagnosis is very often missed and patients often report seeing many doctors before a proper diagnosis is made. Parents frequently report that they had suggested the diagnosis to their physicians only to be told, "No, your child clearly does not have Tourette syndrome." The following are some of the many reasons for this.

Belief that TS patients must swear In the earlier descriptions of TS the symptom of coprolalia was so often stressed that many physicians still believe a TS patient must have compulsive swearing. In fact, this symptom is present in less than 30 percent of cases.

Belief that TS patients must tic in the office The second major misconception is that the patients must be ticing while in the office. This reflects the common feeling of many neurologists that, "If I don't see them tic they don't have TS." This ignores one of the major diagnostic criteria of TS, that patients can suppress the symptoms for minutes to hours. It is very common for parents to relate that their child was ticing or making loud vocal noises in the car but as soon as they entered the clinic all the symptoms stopped. The critical criteria for determining if tics are present and for how long is a careful history from the patients, their parents, and spouses.

Mislabeling TS as "habits" A term common in the neurological literature is "habits" or "habit tics." This is somewhat similar to "transient tics of childhood" which are defined as motor tics present for less than one year. However, the term has also been used for single tics that are present for years. What often happens, however, is that even when more than one tic is present, or vocal tics are also present, or there is a strong family history of motor and vocal tics, and the tics have been present almost every day for a year, the term "habit tics" is still used, even though the correct diagnosis should be Tourette syndrome.

Forgot to ask about tics If a patient comes to the doctor complaining about tics, this problem is avoided. But many children with TS are brought to the doctor because of ADHD, or conduct or oppositional behavior, or dyslexia or learning disorders, or other problems. Here a failure to ask about tics in the patient or the family can lead to an incorrect diagnosis. There are two questions that should be asked: Are tics present now? and, Were tics ever present? As shown in the chapter on natural history, many patients who are 20 years of age or older may present with obsessive-compulsive behavior, depression, mood swings, alcoholism, obesity, panic attacks or other problems and have had motor or vocal tics for several years in childhood only to have them disappear and be replaced by other problems. Here again, the primary diagnosis can be missed is the proper specific questions are not asked.

Don't want to make the diagnosis of

TS Parents have also reported that a doctor has said to them, "It could be Tourette syndrome, but I am reluctant to make that diagnosis." This reflects the view that TS is a rare, severe disorder, and ignores the fact that many cases of TS are very mild, often too mild to treat. This approach leaves the parents with a "no diagnosis" child and often sends them off into a continued search to find out what is wrong. It is far better to make a clear diagnosis of Tourette syndrome if the patient meets all the criteria, and point out that this does not necessarily mean severe problems are ahead, or that treatment is necessary.

Didn't listen Mothers have often related that when they have asked their doctor if their child has Tourette syndrome, they have gotten an immediate reply of "No" even before the doctor has had a chance to listen to the medical history. This also reflects the misconception that a TS patient must be vigorously ticing (and possibly swearing) in the office to make a diagnosis.

Didn't take a family history As I have mentioned repeatedly, taking a careful family pedigree is critical. True TS, chronic motor tics, or chronic vocal tics are less likely to be mislabeled "habits," "habit tics," "allergies," or "just due to stress" if other members of the family are shown to have the same problem.

Incorrect conclusion the symptoms are due to copying This error is best illustrated by quoting directly from an actual psychologist's report:

> "It is interesting to note that the admission history indicated that his father has a history of motor and vocal tics. In light of this fact it is speculated that if these symptoms have been observed by John, they may be the result of exposure to and modeling of his father's behavior."

This is a remarkable testimony to the degree that the psychological community is imbued with the idea that all behavior is learned or environmentally caused. Dozens of times parents have made statements to us like:

> "Billy's younger brother is starting to have tics but I'm sure he is just copying his older brother."

In our experience it has never turned out to be copying. Months to years later these "copy cats" have developed into full-blown Tourette syndrome.

Didn't take a longitudinal history Once an individual meets the DSM-III criteria for TS, even though the tics may disappear as the individual grows older, the *Gts* genes don't go away. Oftentimes the connection between conduct disorder, alcoholism, anti-social personality, depression, panic attacks or other behaviors and Tourette syndrome is missed because of a failure to take a longitudinal history.

Didn't realize the tics were gone because of medication Many times a TS patient on medication is evaluated and the examiner concludes the patient does not have TS because there are no tics. This is a double error. It fails to consider the fact that patients often suppress the tics during the examination and even more seriously fails to take into consideration the fact that the tics are absent because the patient is taking medication.

Haven't heard of TS While this used to be the major reason for missing a diagnosis of TS, it has largely been replaced by knowledge of the disorder but a misconception of what is required to make a diagnosis.

Summary: The major reasons that a diagnosis of TS is missed are lack of awareness of the fact that patients can often completely suppress the symptoms when in public, the belief that patients must compulsively swear, mislabeling true TS as "habit tics," not asking about tics when one or more of the associated behaviors are the prominent symptoms, or simply not listening.

Chapter 93
Will My Child Have Tourette Syndrome?

This is one of questions most commonly asked by individuals with TS. The answer depends upon whether the individual has one or two *Gts* genes, and whether their spouse has none, one, or two *Gts* genes. Let us look at each of these possibilities. The following table gives the risk in percent that a child will carry one or two Gts genes given the status of the parents.

	Number of Gts genes in: Mother	Number of Gts genes in: Father	% Risk for the child to have: 1 *Gts* gene	% Risk for the child to have: 2 *Gts* genes
Neither parent carrier				
	0	0	0	0
One parent carrier				
	1	0	50	0
	0	1	50	0
Both parents carrier				
	1	1	50	25
One parent with 2 *Gts* genes				
	2	0	100	0
	0	2	100	0
One parent with 2 *Gts* genes, other a carrier				
	2	1	100	50
	1	2	100	50
Both parents with 2 *Gts* genes				
	2	2	100	100

It is clear that in order to determine which of these situations applies, a careful family history is necessary and even then it is not always easy to determine whether a parent is carrying none, one, or two *Gts* genes. Even if this is fairly clear, the second question is, what is the penetrance of the expected set of genes? If the child inherits no *Gts* genes the risk of developing TS is negligible. If the child inherits one *Gts* gene, the risk of developing TS or TS-associated behaviors depends on the penetrance[p42] of the single *Gts* gene. While the precise figures for this will not be available until extensive studies are completed using a DNA marker for the *Gts* gene, the family history studies[p650] indicate the risk for some type of symptoms is about 50 percent. If the child inherits two *Gts* genes then the risk for some type of symptoms is probably very close to 100 percent.

When the **right half of the above table** is combined with these penetrance figures they produce the following estimates for the risk of a child having some TS symptoms.

From the previous table:

% Risk for the child to have:		Risk of the child having some
1 Gts gene	**2 Gts genes**	**TS symptoms:**
0	-	low
-	25	25%
50	-	25%
-	50	50%
100	-	50%
-	100	100%

If the semidominant semirecessive theory of transmission is correct, the most common situation would be where a couple has one TS child and are asking what their risk would be to have a second. If both parents are relatively asymptomatic they would probably both be carriers (heterozygotes). From the first table, the risk is 50 percent that the next child would inherit one *Gts* gene and 25 percent they would inherit two *Gts* genes. From the second table, the risk of their next child having some TS symptoms is 25 percent. If the child inherited two *Gts* genes, the symptoms would probably be significant. If the child inherited only one *Gts* gene the symptoms would be milder.

If one of the parents had significant symptoms then then one parent would probably carry two genes and the other would be a carrier. From the first table the risk of the next child inheriting one *Gts* gene would be 100 percent and the risk of inheriting two would be 50 percent. From the second table the risk of the child having some symptoms would be 50 percent, and the symptoms would be severe or mild depending upon whether they had one or two *Gts* genes.

If the inheritance of TS were autosomal dominant, instead of semidominant semirecessive, then the risk of inheriting the *Gts* gene would be 50 percent. If the penetrance was 90 percent then the risk of a TS parent having a TS child would be 0.5 X 0.9 or 45 percent. This is similar to the risk figure given by the semidominant semirecessive mode of inheritance, when one parent is symptomatic.

All of these estimates will become more accurate when specific DNA tests for the *Gts* gene become widely available.

Summary: Determining the risk of passing on TS or its associated symptoms requires two pieces of information:

1) whether the parents carry none, one, or two Gts genes, and

2) the penetrance of the single or double dose of the Gts gene(s) for TS and related problems.

The first requires a careful family history, the second can be estimated from family studies. Both of these will become much more accurate when a DNA test for the Gts gene becomes generally available.

Chapter 94

Genetic versus Psychological Causes

I know that most men, including those at ease with problems of the greatest complexity, can seldom accept even the simplest and most obvious truth if it be such as would oblige them to admit the falsity of conclusions which they have delighted in explaining to colleagues, which they have proudly taught to others, and which they have woven, thread by thread, into the fabric of their lives.

Tolstoy[734]

In medical malpractice suits there is a phrase "res ipsa loquiter," meaning "the result speaks for itself." For example, in a damage suit, a retarded child may be shown to the jury with the implication, "the child is defective; it must be the obstetrician's fault." This may ignore the real possibility that the retardation was due to a genetic defect that would have occurred regardless of who delivered the baby. We often see a similar line of thinking in psychiatry. There are many psychiatric theories to explain why we behave the way we do. Sometimes the presence of a given behavior is used in a "res ipsa loquiter" fashion to prove the theory. Thus, since it is known that emotional conflicts can produce disturbed behavior, the conclusion is often made that if there is disturbed behavior, there must have been emotional conflicts.

The potential trap that this may lead to is well illustrated in David Viscott's *The Making of a Psychiatrist* [2013]. A patient, Mrs. Tuoy, the wife of an elderly physician, complained of hip pain for several weeks. Because of a vague description of the pain it was assumed to be psychosomatic and she ended up in a psychiatry ward. In the resident's presentation to the attending psychiatrist, Dr. Gavin, it was assumed that her complaints of pain were an attempt to capture her husband's attention and sympathy.

> "'I think you'd like to get another doctor interested in the way you feel. Maybe you'd like your husband to show *he* cares more?' Gavin, sharp as ever, was really hammering away.
>
> "'Yes,' said Mrs. Tuoey, and began to cry.
>
> "Gavin could get to a patient's feelings instantly. I really envied that guy. What technique!
>
> "Unfortunately, Mrs. Tuoey limped out of the conference room uncured, wincing with each step."

After reading a nurse's notes indicating that the pain was well localized to the hip, Dr. Viscott pulled back Mrs. Tuoey's nightgown, revealing a large red swelling over the site of a fractured hip. He mused,

> "I learned something very important from Mrs. Tuoey. I learned that no matter how well the logic of my discipline explained the emotional meaning of a patient's symptoms, I could not be sure that what I understood so well was

everything there was to know. I could still be missing something — perhaps the most important thing."

The problem with the assumption that where there is disturbed behavior there must have been emotional conflicts ignores an alternative explanation that the *disturbed behavior may be due to genetically disturbed brain chemistry, in the absence of any deep emotional conflicts.* The following examples illustrate how thinking about behavior in genetic terms may result in strikingly different conclusions than those reached by purely psychological approaches[374]. Many of these examples can also be viewed as *serial versus parallel thinking*. Serial thinking states that the presence of A results in the occurrence of B. Parallel thinking suggests that in some cases C causes both A and B, which may be independent events. This is diagramed as follows:

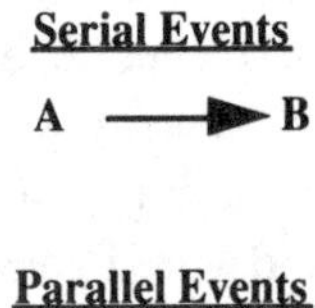

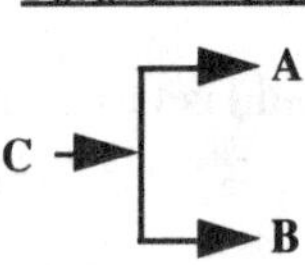

The following are some examples.

Tics For many years the presence of tics in Tourette syndrome was explained in psychoanalytic terms as being the result of psychological conflicts or behavioral problems[1767,1771].

The psychoanalytic explanation of tics was:

behavioral disorder (emotional conflicts) → tics

The behavioral problems and tics are visualized as serial events. By contrast, the the genetic explanation is:

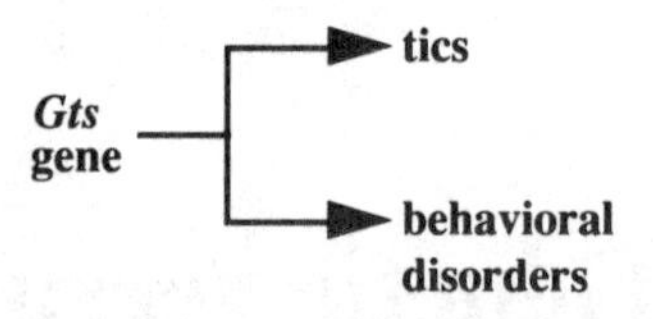

Here an independent event, the *Gts* gene causes both the tics and the behavioral problems.

I once had a psychiatrist call and ask me to see his nine year old daughter who he finally recognized had Tourette syndrome. His reasoning provided insight into these two different views of the world. He said:

"Several years ago she began to develop very confrontive and oppositional behavior and later developed rapid eyeblinking and facial grimacing. Assuming the tics were due to emotional conflicts I sent her to a child analyst for play therapy and psychoanalysis. At first he saw her once a week, but as she slowly got worse the visits were increased to two, three and now four times a week. She has been in therapy for three years and recently began to develop throat clearing noises. I finally had to face the fact that this might all be due to Tourette syndrome rather than to any psychiatric problem."

Depression in TS If patients with TS have depression, the usual explanation is that the depression is simply a result of having the physical problem of tics.

However, as discussed earlier[p183], depression, mood swings and manic-depressive disorder can be an integral part of the general emotional disinhibition caused by the *Gts* gene. Thus, parallel thinking is more appropriate:

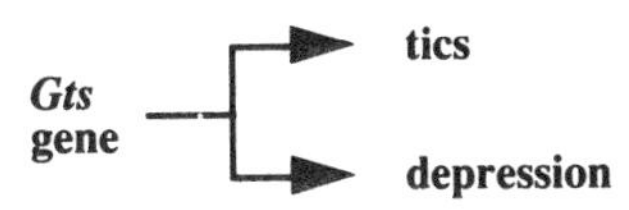

Since other genes, in addition to the *Gts* gene, may be involved, in some of the following examples I have substituted the more general term *disinhibition genes,* meant to include the *Gts* gene and other related genes.

Learned behavior versus driven behavior When a child with behavioral problems responds in a predictable way to a given circumstance, whether it is with manipulation, anger, or tantrums, it is often referred to as *learned behavior,* with the implication that it can be easily unlearned by behavioral modification. While behavioral modification may reverse the behavior, oftentimes it tends to persist despite all types of effort to bring about a change[p595]. This is understandable when the behavior is seen as chemically driven, rather than learned. Correction of the chemical imbalance will be necessary before the behavior can change.

The psychopathic personality In their book *High Risk Children without a Conscience,* Magid and McKelvey[1244] suggest that psychopathic or antisocial personality disorder is due to loss of attachment in early childhood.

> "...the chances for increasing numbers of psychopaths are escalating. We must search for answers to the pressing social problems that are helping to create unattached children. We must learn how to prevent unattached children. The solutions will not be easy or cheap, but they *must* be found" (p 43).

These loss of attachments were proposed to occur through a breakup of the parents' marriage, or by a delay in adopting a child, or by too early use of child care facilities. Thus:

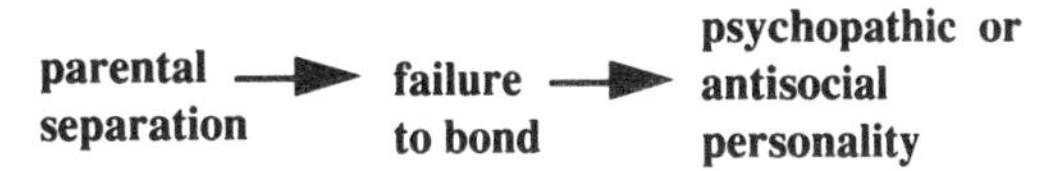

There are a number of problems with this serial approach. It does not explain the large number of children with conduct disorder and adults with antisocial personality who had very loving and caring parents who were never separated or divorced and who raised a number of other perfectly normal children. In a similar fashion it does not explain the significant number of children whose parents were separated or divorced, or who were adopted late, or who had chaotic childhoods, but turned out to be caring, bonding, normal adults. While I am not questioning the tremendous importance of good parent-child bonding in early childhood[1664], the following parallel thinking makes more sense to me:

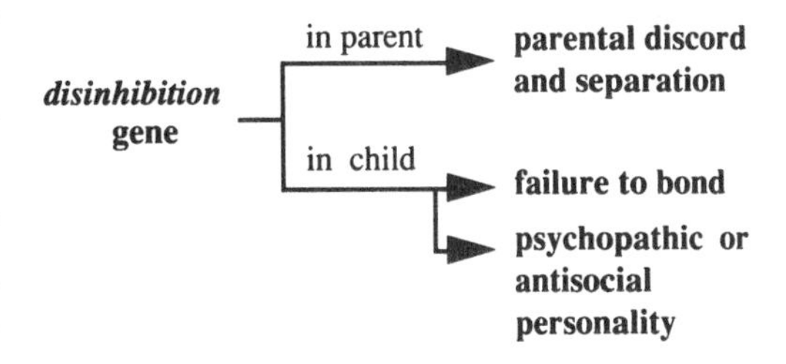

An adult carrying a disinhibition gene is at higher risk of having multiple chaotic relationships with separation and divorce[p263]. The same gene may then be passed to their children who themselves have more problems with bonding and an increased risk of conduct disorder and antisocial personality.

The distinction is more than trivial. The serial approach would lead to the suggestion that billions of dollars spent in avoiding unattached children would rid the world of criminals and psychopaths, while in fact it may have little or no effect. By contrast, the use of appropriate genetic tests to allow the early detection of individuals carrying various disinhibition genes, especially in a homozygous state or in combination with other genes such as for schizophrenia, would allow early and long-term treatment for a small number of very high risk individuals. This might be far more effective[374].

This is not to undervalue the role of the environment. Cloninger's cross-fostering studies[p158] clearly indicated a role of environment in antisocial behavior. Just as appropriate psychotherapy can help the individual with a chemical frontal lobe syndrome to stop and think, and learn to evaluate the consequences of their behavior, an environment rich in other psychopaths could act like "reverse therapy" and simply reinforce rather than improve the problem in an individual carrying a disinhibition gene.

Adoption A similar example concerns adoption. I have noted previously that a higher than expected percentage of our TS, ADHD, and conduct disorder patients were adopted[p279]. The serial explanation[1244] is:

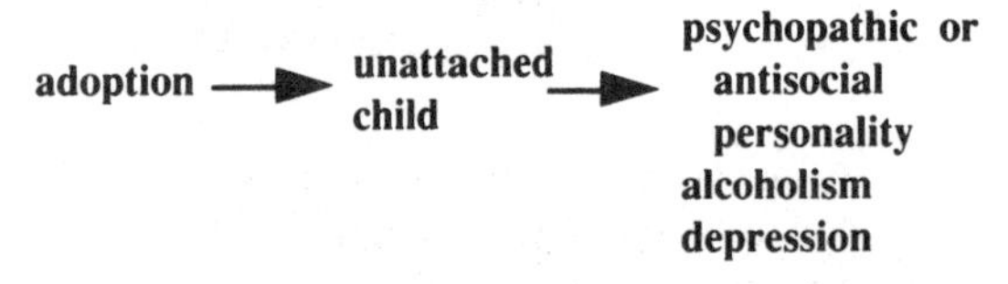

The alternative genetic or parallel approach is:

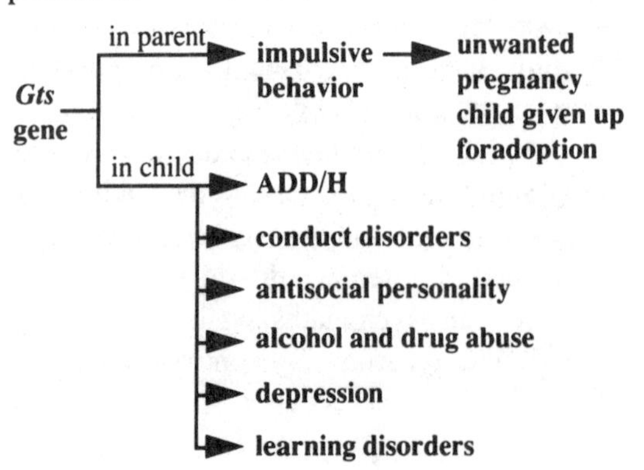

The same gene that led to impulsive behavior by one or both parents, resulting in an unwanted pregnancy and needing to place the child up for adoption, is passed on, resulting in a wide range of impulsive behavioral problems in the child.

It would be a disservice to children awaiting adoption to suggest this is a common problem in all adopted children. Nothing could be further from the truth. Here I am talking about slight increases in the risk. It will require genetic markers to precisely define the degree of the problem. In our TS patients, 4 percent are adopted compared to about 1 percent of children in general being adopted. This suggests the risk of TS in adopted children is approximately four times what it is for non-adopted children.

Child abuse: learned or genetic? Physical and sexual abuse of children is common. Psychologists often speak of the *transgenerational* effect of child abuse in that children who are abused learn this way of behaving from their abusing parents and then abuse their own children.

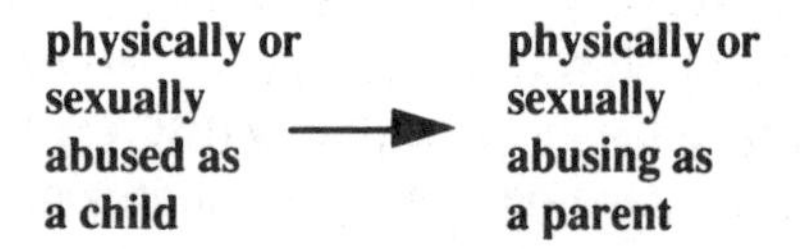

It has been stated that "The cycle of child abuse. . .is passed down, like a genetic defect, from parent to child"[1244]. I would suggest that for many cases the wording could be changed to say, ". . .is passed down *as* a genetic defect, from parent to child" shown as follows:

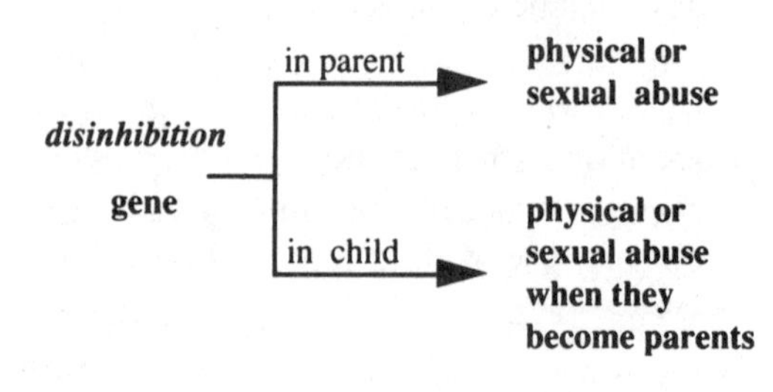

Thus, if one or both parents carry a disinhibition gene they are more likely to be impatient, impulsive, short-tempered, and quick to strike out at a crying or disruptive child. As the child grows older, such a parent, perhaps already possessed with a greater than average libido, is less likely to think through the consequences of his or her behavior and may become embroiled in sexually abusing their children. The offspring who inherit such a gene may themselves contribute to the problem by being more disruptive as children or by having a precocious interest in sex themselves. The presence of one or both parents and one child, all with the same gene, can be especially explosive. I am not claiming that all child or wife abuse is genetically caused. This is clearly not the case. I am suggesting, however, that individuals carrying the various disinhibiton genes are at much greater risk for this behavior and its "transgenerational" passage may not always be due to "learned behavior." In a study of this "cycle of violence," Widom[2076a] concluded that

> "These findings do not show, however, that every abused or neglected child will become delinquent, criminal or a violent crimnal. The linkage between childhood victimization and later antisocial and violent behavior is far from certain, and the intergenerational transmission of violence is not inevitable. Although early child abuse and neglect place one at increased risk for official recorded delinquency, adult criminality, and violent criminal behavior, a large proportion of abused or neglected children do not succumb. Twenty-six percent of child abuse or neglect victims had juvenile offenses; 74 percent did not. Eleven percent had an arrest for a violent criminal act, whereas almost 90 percent did not."

She then wondered what protective factors intervene in the child's development to account for those who succumb and those who are "resistant" and do not. The genetic hypothesis would suggest that the resistance factor is simply the absence of "disinhibition" genes while those who succumb have inherited from their abusing parent the genes that made them abusers.

Inappropriate sexual behavior doesn't always mean sex abuse It is almost a truism that sexual molestation causes a child to undertake inappropriate sexual behavior and thus, res ipsa loquiter, whenever a child displays inappropriate sexual behavior it is the result of being sexually abused. This is shown as follows:

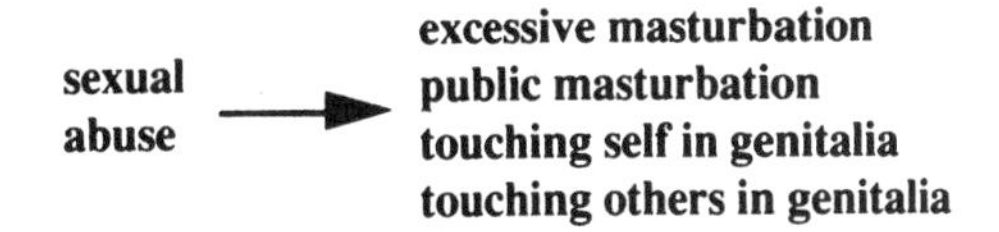

However, as discussed previously[p263], an alternative explanation is:

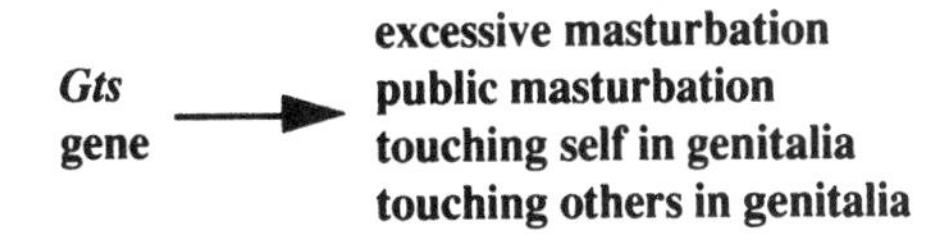

Significant injustice can be done if this possibility is not kept in mind.

MAO and sensation-seeking behavior Going back to the chapter on personality disorders[p530], there are many parallels between the personality scales of sensation-seeking, novelty-seeking, risk-taking behavior and ADHD or antisocial personality. The explanation for the correlation between low platelet MAO and sensation-seeking behavior was based on the following serial events:

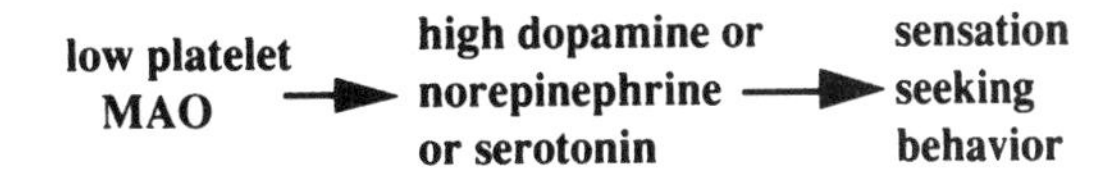

However, an alternate parallel explanation is the following:

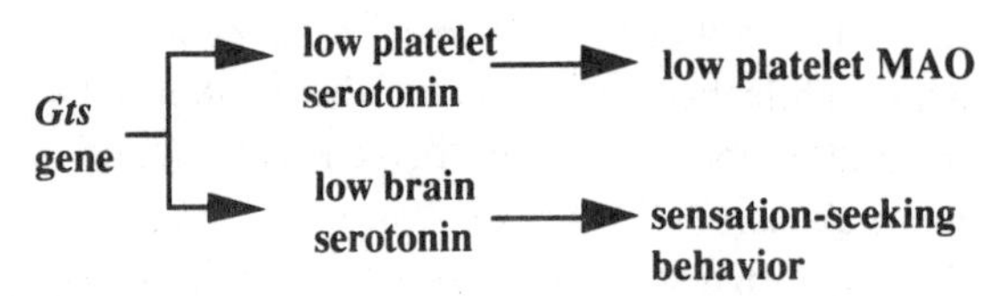

Here the low platelet MAO levels have no cause and effect role in the sensation-seeking behavior. This is an alternative hypothesis that needs to be investigated.

Bad childhood or bad genes? There is a strong propensity to assume that any significant, persistent behavioral problem in a child or adolescent is due to some type of bad experience or bad environment in their childhood. We have often heard the phrase, "Of course he is behaving this way, look at what he had to go through as a child." The "look what he had to go through" usually refers to the divorce of his parents, a "mean" stepparent, or a wide range of other presumably emotionally traumatizing events. However, as mentioned before, this fails to take into account the fact many other children have experienced the same problems in childhood and turned out fine, or the fact that if you dig enough you can find some traumatic events in every persons' childhood. Thus, instead of jumping to the conclusion that:

bad childhood → bad behavior

the possibility that a genetic disorder is in part responsible for the behavior should be considered as well as environmental factors:

***disinhibition* genes → bad behavior**

bad childhood → bad behavior

Anger "Has a lot of anger," "easy to anger," "short temper," "temper tantrums," "set off by trivial things," and other similar statements are commonly used phrases when it comes to describing some TS children. Mental health professionals who have missed the diagnosis of TS, or may know their patient has TS but do not appreciate this aspect of it, often search for "the cause of the anger" by assuming there is some unseen emotional conflict lurking somewhere. This approach is diagramed as follows:

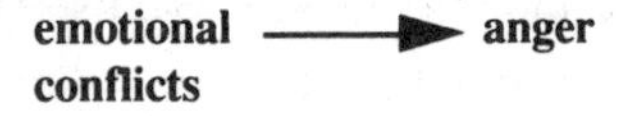

Most of the time the reasons for that anger are usually assumed to be something the parents have done to the child, such as being too strict, not being strict enough, liking a sibling more, or divorcing a spouse. The alternative explanation is rarely considered, that *the anger is purely a symptom of Tourette syndrome, in of itself, and not due to anything other than than the chemical imbalance produced by a Gts gene.*

Alcoholism

A wide range of disorders have been associated with alcoholism often with the implication that they cause the alcohol abuse.

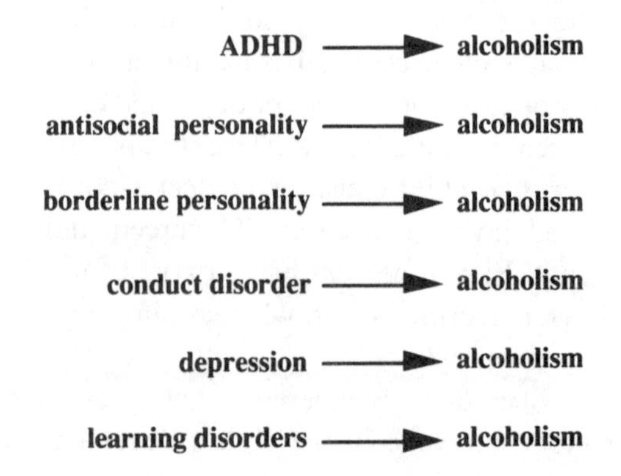

However, an alternative genetic approach is that none of these things "cause" alcoholism. Rather a disinhibition gene such as a *Gts* gene causes all of these things. This is illustrated as follows:

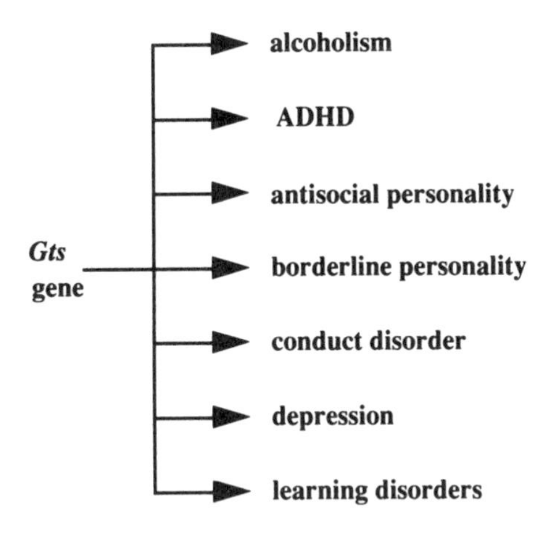

Each entity can occur independent of the others or in any combination.

Summary: There is a long history of the "nature versus nurture" question. Which is more important in regulating our behavior, our genes or our environment? The answer to this is clear — both play a role. Despite the increasing evidence for the important role of genes in our behavior, the major tilt for most mental health professionals is toward environmental factors. However, many things that on the surface seem to have a direct environmental cause and effect may be due to a third, genetic factor that causes both events. Several examples are given. One is the proposal that the loss of attachments in early childhood lead to psychopathic behavior. An alternative is that the child inherited the same gene that caused one or both parents to have a chaotic life style and break up, and it was this gene, not the breakup and disrupted attachment, that led to the behavior problems in the child. A second example is the dictum that physical or sexual abuse by a parent may cause the child to be a physical or sexual abuser when they become parents. An alternative factor that may also play a role is that the parent was physically or sexually abusing because of the presence of a disinhibition gene, and that same gene, passed on to the child, caused them to be physically or sexually abusing when they became parents.

Chapter 95

The Cortex and the Sub-cortex The Superego and the Id

The reader may have noticed a recurrent theme in the past chapters concerning the interaction between cortical and sub-cortical events. By cortical I mean those functions of the nervous system controlled by the neocortex, but most specifically, the prefrontal lobes. By subcortical I mean almost everything below the cortex, but most specifically, those functions controlled by the limbic system and caudate and putamen.

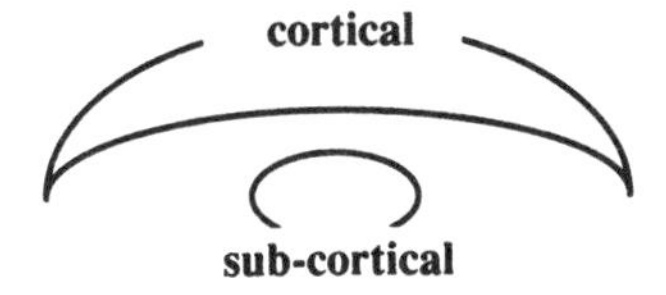

This cortical/sub-cortical theme means different things to people in different professions. For example, to neuroanatomists and evolutionary neurologists like Paul McLean it means frontal or prefrontal lobes versus limbic system and neostriatum.

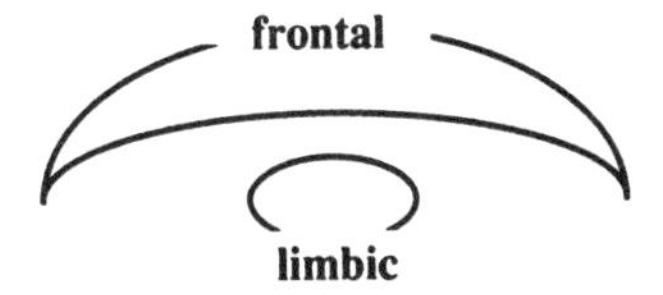

To neurophysiologists, such as Pycock and his colleagues, where destruction of the dopamine neurons of the frontal lobe leads to hypersensitivity of the sub-cortical dopamine neurons, it means decreased cortical dopamine versus to increased sub-cortical dopamine.

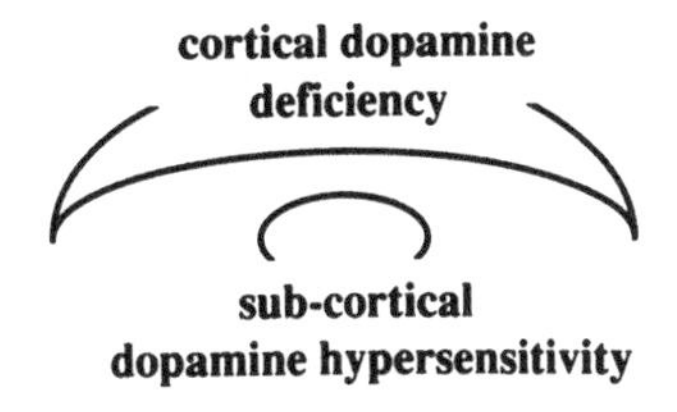

To psychologists, it means cognitive versus affective or emotional or instinctive.

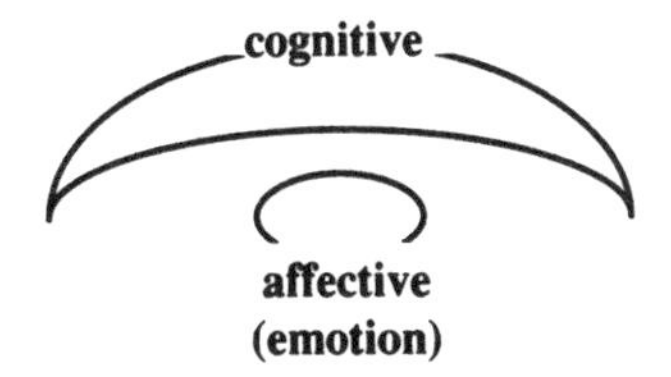

To the psychoanalyist like Freud it means the superego and ego versus the id.

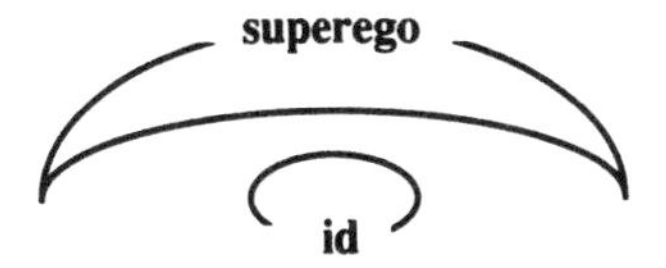

Finally, to students of Tourette syndrome it means frontal lobe syndrome versus motor and vocal tic syndrome.

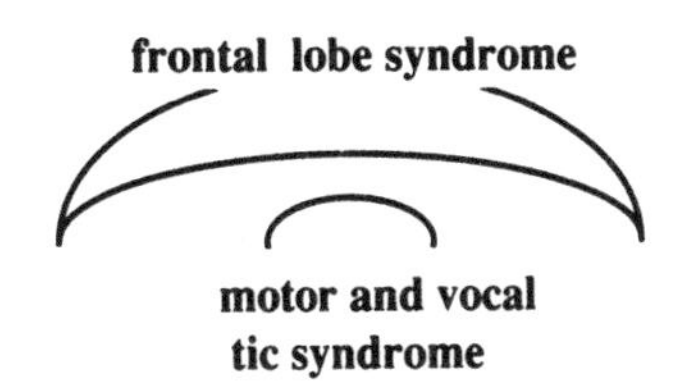

Throughout many chapters, especially those on dopamine and treatment, we have seen that Tourette syndrome is really two distinct syndromes — a *frontal lobe syndrome* with ADHD, impulsivity, conduct disorders, and learning disorders; and a *tic syndrome* with motor and vocal tics and other stereotyped behaviors. Just as it was almost impossible to separate some of the functions of the frontal lobe from the functions of the caudate and putamen[p354], it is equally difficult to make a clear distinction between which of these syndromes is primarily responsible for some of the associated behaviors in Tourette syndrome. Just as the answer for the frontal lobe-caudate problem was that they perform in series, each making a contribution to a given behavior, the cortical and the sub-cortical aspects of Tourette syndrome also work in series, each playing an important role.

However, that still leaves us with the question—How can a single genetic defect cause hypofunctioning of one part of the brain and hyperfunctioning of the other? The best answer to this comes from the results of the studies of Pycock and colleagues[p393] who showed that destruction of frontal lobe dopamine neurons (hypofunctioning) leads to a hypersensitivity of the sub-cortical dopamine neurons (hyperfunctioning). In TS, the former requires dopaminergic agonists to correct and the latter requires dopamine antagonists to correct. A defect in serotonin production, such as a hyperfunctioning tryptophan oxygenase, would lead to a relative deficiency of central serotonin stores causing a hypofunctioning of the frontal lobes (cortical defect) and a compensatory hyperfunctioning of the sub-cortical structures (tic syndrome).

Freud, The Id and Superego

Although throughout this book the emphasis has been on the role of genetics in human behavior, I have also emphasized the interplay with the environment. In the early part of the 20th century many types of human behavior were given a psychoanalytic explanation. Freud postulated the existence of three major parts of the psyche: the id, ego and superego. The *id* is the part of the brain wrapped up in the pleasure principle, finding pleasure and avoiding pain. It wants immediate gratification. It is the impulsive, irrational, pleasure-seeking self. The spoiled child of the personality. The *superego* represents the thinking input to behavior and puts the moral brakes on the hedonistic strivings of the id. The *ego* is the sense of self, the executive, or conscious, or working psyche balanced between the id and the superego. Freud proposed that many forms of abnormal behavior were due to an improper balance between the id and the superego. Too strong of an id would lead to a self-centered, narcissistic, sociopathic individual. Too much superego would lead to an overly moralistic, compulsive, rigid personality. This, of course, is a considerable simplification of Freud's massive contributions to psychology. The point I wish to make is that in conceptual terms the psychoanalytic thoughts of Freud and the conclusions of a genetically-chemically oriented view of human behavior are not that dissimilar. In many ways the id is equivalent to the limbic system, and the superego is equivalent to the prefrontal cortex. The concepts differ mainly in causation. The psychoanalytic view suggests that the inbalance is due to emotional conflicts stemming from some trauma or incompletely traversed stage of early development ("oral," "anal," or "genital phase"). The genetic view is that in many cases the imbalance is in fact chemical in nature and due to the presence of mutant genes.

Jung

Carl Jung positioned himself into a place somewhat intermediate between Freud and a more genetic approach to behavior. While

Freud believed that psychological complexes originated in traumatic experiences in early childhood, Jung believed they must originate from something deeper in human nature. At the time, genetic principles were just being formulated and clarified, and instead of suggesting that these deeper elements might be mutant genes, Jung opted instead for the now disproven view that experiences of earlier generations can be inherited (Lamarckian inheritance). This concept led to the idea of the *collective unconscious* in which experiences of previous generations somehow became incorporated into the consciousness to form primordial images or *archetypes*. An imbalance in the interaction between these archetypes was believed to lead to abnormal behavior.

Genetics and Behavior

Major differences in psychoanalytic versus genetic views come in the area of treatment. Instead of searching for and attempting to work through a hypothetical early childhood trauma, or an archetype imbalance, the emphasis is on the here and now — determine if a specific syndrome is present by a careful longitudinal history, a detailed pedigree, and when available, by biochemical or genetic testing. Then correct the chemical inbalance, and where necessary, consolidate this with supportive or cognitive, individual or family psychotherapy.

Summary: A major concept in this book is that a chemically induced frontal lobe lesion, due to the presence of a common Gts gene, leads to an imbalance in the function of the cortex and sub-cortex and is responsible for a wide range of disturbed human behavior. It is pointed out that this view is not that distant from more classical psychoanalytic theories of Freud who also postulated an imbalance in the cortical superego and the sub-cortical id. The major difference is that Freud proposed that the imbalance was due to early childhood psychological trauma while the genetic view pushes the onset back to the time of conception—i.e., the imbalance is due to genetic rather than psychological trauma.

Chapter 96
Ethical Issues

The genetics of behavior has always been a sensitive subject. The early part of this century saw the development of the eugenics movement. One of its strongest proponents, Charles Davenport, defined eugenics as "the science of improvement of the human race by better breeding"[457]. Various elements of this movement were used[1814] and severely abused by the Nazi's in the 1930's and 1940's. This provided an extraordinarily poor start for the study of the genetics of behavior. The development of modern techniques, including linkage studies with DNA markers[P53], has resulted in far more sophisticated results, none the least of which is the demonstration of the chromosome location of genes for manic-depressive disorder, schizophrenia, dyslexia, Alzheimer's disease and Huntington's disease. As reviewed in this book, modern studies of behavior are replete with references to neurotransmitters, receptors, PET scans, and cloned genes. However, despite the precision with which we can now talk about the biological basis of human behavior, the number of ethical issues, if anything, has increased. The following are just a few.

Don't Stigmatize Them

There have been many times in the course of our studies of Tourette syndrome when we have been told, "You can't talk about that." "You can't publish that." "You can't say that." "You are stigmatizing TS patients." Some of these statements may even be voiced by people reading this book. There are many answers to this complaint.

They are us The concept of Tourette syndrome has been one of constant change. For many years it was considered to be an extremely rare disorder characterized by violent tics, loud noises and constant swearing. This, of course, was the worst possible stigmatization — the concept that every TS patient had to swear. It is clear now that coprolalia is present in less than a third of the patients. As the medical community slowly began to realize there were many TS patients who never swore, and who could suppress their tics for many hours, the number of cases began to increase. Our experience with a Los Angeles school district suggests that 1 in 100 school boys may meet the diagnostic criteria for TS[P617]. If only half of these are homozygotes, then about 15 percent of everyone in the country carries a *Gts* gene and many people who carry the gene express it in some form. Thus, discussing the wide spectrum of behaviors associated with this gene is not stigmatizing some tiny minority — they are us.

The fallacy of generalizing If we hear on the TV news a story about people who steal cars, even though the rest of us are also people, we don't jump to the conclusion that all people steal cars. In a similar vein, if we discuss that fact that a specific behavior is more common in TS patients, most of us are able to understand that this does not mean that *all* TS patients, or *all Gts* gene carriers, show this behavior. As discussed before[p618] the *Gts* gene simply pushes the frequency curve to one side.

It is important to make correct diagnoses One of the most important reasons for openly discussing the spectrum of behaviors in individuals who carry one or two *Gts* genes is to allow clinicians to make a proper diagnosis. Many of our TS patients, and their affected relatives, had endured their symptoms for years without understanding why they were behaving the way they were. They had, in fact, been far more severely stigmatized by themselves or others before a proper diagnosis was made than after. These stigmatizations include their own feelings that, "I'm different," "I'm crazy," "I'm going crazy," "No one could love me," "I have trouble reading therefore I'm stupid," or even worse stigmatization by others that, "It's all in your head," "Since there is nothing physically wrong with you you must see a psychiatrist," "You are just trying to get attention," or, in the worst case senario of exhibitionism associated with TS, "You are a mentally disordered sex offender and we are going to put you in jail and throw away the key."

One need only observe, over and over, the feelings of intense relief individuals express when they finally realize that there is a physical cause of their problems, to understand that it is critical to learn all we can about the effects of the *Gts* and related genes — and openly discuss them. Stigmatization is removed, not added, by making a proper diagnosis.

Legal Issues

The other side of the coin is whether to excuse behaviors because of the presence of a *Gts* gene. TS patients occasionally get into trouble with the law. The most common legal problems in younger patients arise from being incorrigible, stealing, or starting fires. In older patients exhibitionism and physical or sexual abuse and alcohol or drug-related problems lead the list. It is important to point out that the majority of TS patients do not have these problems. But when they do get into trouble is the presence of TS a valid legal excuse? *Our fundamental view is that TS patients are responsible for their behavior regardless of their diagnosis.* None the less, we have often argued that at least for a first offense, it makes more sense to attempt to improve the behavior with medication and psychotherapy than to be incarcerated in jail. However, as with anyone, the threat of jail is often an effective tool in behavior modification, and if the offenses continue, incarceration may be an important aspect of indicating that partial forgiveness of antisocial behavior is a time-limited gift. From a legal defense point of view, the possibility of diminished capacity may arise. The doctrine of *mens rea* means that for a crime to have been committed the individual must be capable of formulating a specific intent to commit the crime. The defense might argue that in some cases a TS patient's past history might have demonstrated an inability to formulate such a specific intent. The presence of obsessive-compulsive behaviors or significant lack of impulse control in a TS patient might be used to suggest that the person was not able to act volitionally. TS individuals in trouble with the law would do well to obtain a booklet entitled *Tourette Syndrome and the Law* from the Tourette Syndrome Association.

Testing versus Screening

The wide range of conditions associated with the *Gts* gene indicate that when a specific genetic test is available it might be widely used in the diagnosis of ADHD, learning disorders, dyslexia, depression, mood swings, conduct disorder, antisocial personality, obesity, alcoholism, panic attacks, phobias, migraine headaches, and sleep disorders, to mention just the most obvious. The voluntary use of such a test by a patient's physician, with the results remaining confidential information, would be no different than any other test. Such a test would have the potential of saving enormous sums of money that might otherwise be spent on expensive diagnostic evaluations or on inappropriate treatments. However, ethical problems arise when such a test is proposed as a screening tool for insurance companies or prior to issuing health or life insurance, or for employers prior to employment. This non-voluntary screening of individuals for the purpose of potentially denying them certain privileges, such as employment and insurance, would clearly fall into the realm of invasion of privacy and would constitute the unethical use of such a test.

By contrast, the screening of groups of individual with specific disorders, such as those listed above, for the purpose of understanding what percentage carry the gene, would be ethical as long as participation in the study was voluntary, the study was clearly explained, written consent to enter such a program was obtained, and the research project was reviewed and approved by the Research Review Board of the institution where the study was being done. Such studies would provide valuable information into the fundamental cause and appropriate treatment of these problems.

Is having a *Gts* gene good for you or bad for you? Although the major portion of this book is devoted to discussing the associated behaviors in individuals carrying a single or double dose of the *Gts* gene, as indicated in Chapter 46, the presence of this gene can carry some unique advantages. In many individuals it results in just enough obsessive-compulsive or hypomanic or aggressive behavior to act as an internal flame, driving that person to do, to be, and to achieve beyond the normal standards. Thus, if one had the opportunity to know ahead of time whether a given employee carried or did not carry the *Gts* gene, it is not clear whether such an individual should be preferentially hired or not hired. The intense competition in the work place may, in fact, have selected for individuals with such characteristics and such genes.

Prenatal Diagnosis

When a genetic marker for the *Gts* gene becomes available it will be possible to do prenatal diagnosis for Tourette syndrome, that is, determine whether a fetus has 0, 1 or 2 *Gts* genes. The question of whether it is a predominately an advantage or a disadvantage to have a single *Gts* gene is still an open question, and there would be no justification in terminating a pregnancy for this reason. By contrast, some parents might prefer not to have a child homozygous for the *Gts* gene. This, however, is purely a decision for the parents themselves to make, based upon their prior experience, religion, and a wide range of other factors.

Summary: The Gts gene is so common that discussions of its effect on human behavior is not "picking on" a small group of individuals—those people are us. The positive aspects of making a correct diagnosis concerning the cause and thus potential treatment of a wide range of disorders far outweigh any perceived negative aspects. Confidential genetic testing for these genes is no different than for any other type of medical test. Obligatory testing for the purpose of any type of discrimination would be unethical. The legal aspects of antisocial behaviors performed under the influence of a Gts gene are complex and need to be addressed on an individual basis. Prenatal diagnosis for Tourette syndrome should be directed only toward Gts/Gts homozygotes.

Chapter 97
The Human Tryptophan Oxygenase Gene Is at 4q31

Once it was apparent that blood serotonin and tryptophan were decreased in patients with Tourette syndrome, it was clear that a mutation of the tryptophan oxygenase gene could be causing Tourette syndrome and all of its associated behavioral problems[p480]. Since the ultimate proof of this required isolating the human tryptophan oxygenase gene, we immediately set about doing that. The job was made easier by the fact that the rat tryptophan oxygenase gene had already been isolated or cloned by Schmid and co-workers[1722a] in Germany. We obtained a copy of that clone. It contained 562 base pairs of DNA that matched part of the sequence of the larger rat tryptophan oxygenase gene. We used it in hybridization experiments[p54] to identify a comparablehuman clone in a "library" of human liver messenger RNA. We found a clone that matched all 562 base pairs of the rat clone indicating we had successfully cloned the human tryptophan oxygenase gene[379b,380a,497a].

Location of the Human Gene

The next objective was to find what chromosome and what part of the chromosome the human tryptophan oxygenase gene was on. To do this we obtained from Dr. Mohandas at UCLA a set of mouse or Chinese hamster cells that contained different human chromosomes. By hybridizing our cloned DNA to the DNA from different sets of these cells we found that the human tryptophan oxygenase gene was on chromosome 4.

The next question was whether the gene was on the short or the long arm of chromosome 4. To examine this we obtained another set of hybrid cells from Dr. John Wasmuth at the University of Irvine. Different sets contained different parts of chromosome 4. We also obtained from Dr. Moyra Smith cells from a baby (B.O.) with a deletion of the tip of chromosome 4 . The results are **shown on the next page**. If the hybridization was positive (+), the gene could be on any part of chromosome 4 that was present. If the hybridization was negative (0), the gene could not be present on any part of the chromosome that was present. The region between the arrows shows where the gene would have to be. Progressively combining the results from all the cells showed that the human tryptophan oxygenase gene was on the long arm of chromosome 4, somewhere between band 4q25 and 31.3[379b].

To localize the gene even more accurately, we gave a copy of the clone to Dr. Tim Donalon at Stanford. He used a technique called *in situ* hybridization which consisted of making the cloned DNA radioactive and adding it directly to chromosome spreads and covering them with photographic film (autoradiograph[p56]). Several weeks later grains appeared over the part of the chromosome containing the gene. The majority of the grains were over bands 4q31-32. Combining the results from the cells and the *in situ* hybridization technique indicated that tryptophan oxygenase was at 4q31.

The next task was to find DNA markers or RFLP's[p.56] to identify the tryptophan oxygenase gene for linkage analysis, and to determine the DNA sequence of the entire gene. This information would allow us to determine if the tryptophan oxygenase gene was the same as the *Gts* gene. Indirect evidence that they may be the same, is given in the next chapter.

Summary: The human tryptophan oxygenase gene is located on the long arm of chromosome 4 at band q31.

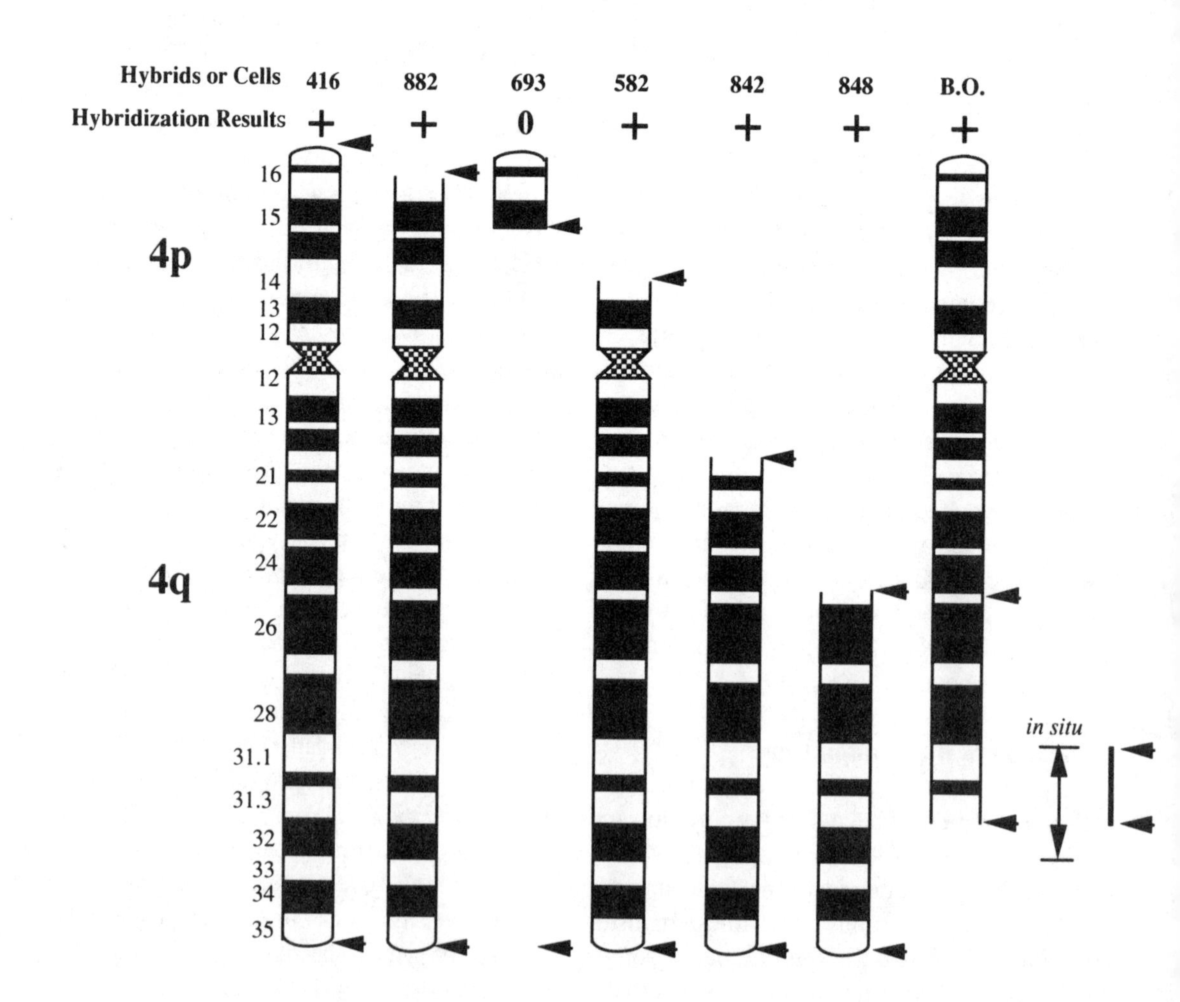

Chapter 98
Closing the Logic Loop

As mentioned previously, new developments are being added to this book as it is being written. It is now February 1989 and I have just become aware of a report by Dr. Shirley Hill and colleagues of the Western Psychiatric Clinic in Pittsburgh providing preliminary evidence for the linkage of type II alcoholism[p233] to the MNS blood group. The following explains why this is potentially important.

The Logic Loop

When we began studying Tourette syndrome in 1980 it was considered to be a rare neurological disorder consisting entirely of motor and vocal tics[1767] and some[2080] even doubted it was a genetic disorder. By 1984, after the first 250 families had been analyzed, it was clear TS was a genetic disorder inherited as a dominant or semidominant trait with reduced penetrance[376] and clearly associated with ADHD, obsessive-compulsive behaviors and conduct disorder[357,358]. It was generally considered to be due to a defect in dopamine balance in the brain[p399]. Thus began the rudiments of the logic loop **see right->**:

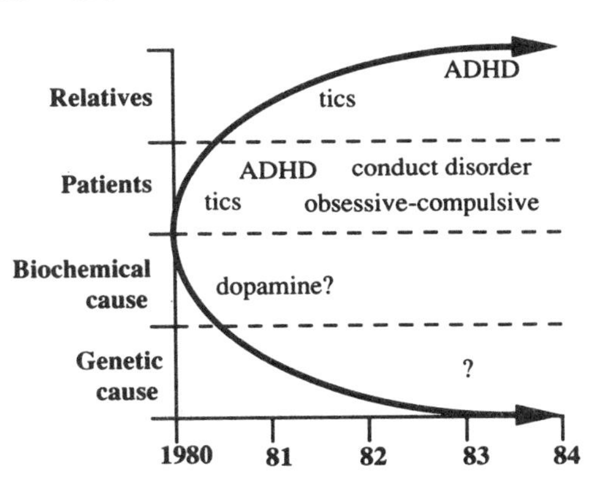

For studies of the next 250 families we realized it was necessary to have a detailed structured questionnaire to assess the prevelance of a wide range of behavioral problems, and a control group[p69]. By the time this study was completed, in 1986[360-364], it was clear that TS was much more common than previously suspected, and was a general behavioral disorder including phobias, panic attacks, depression, mania, alcoholism, sleep disorders and others.

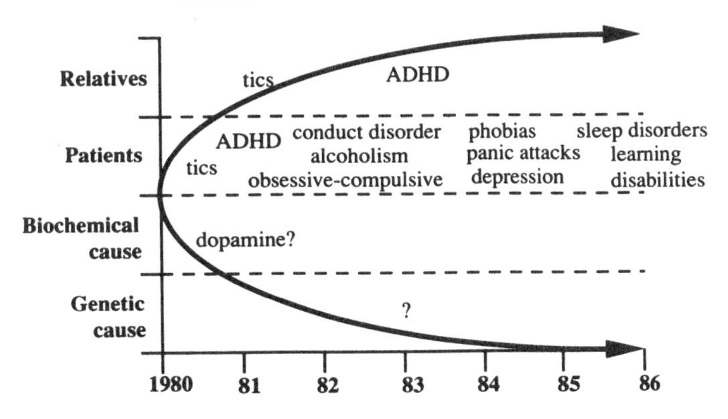

Because of the similarity between these behaviors and those reported to be associated with defects in brain serotonin[p429] I began to examine the blood serotonin and plasma tryptophan in all our TS patients. By August 1987 it was apparent that both these values were decreased and that a mutation causing an increase in tryptophan oxygenase[p467] might be involved.

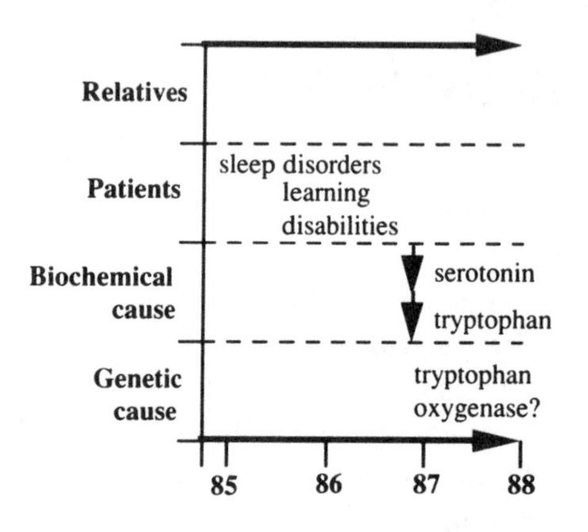

As described in the previous chapter, we obtained a clone of the rat tryptophan oxygenase and began searching for the human tryptophan oxygenase gene. After a clone to the human tryptophan oxygenase gene was isolated we determined that it was located on the long arm of chromosome 4 at 4q31[p676]. By the end of that year we had also completed the families studies[p643] indicating a highly significant increase in the frequency of type II alcoholism in the relatives of TS patients as well as a wide range of additional disorders such as ADHD, learning problems, dyslexia, obsessive-compulsive behaviors, drug abuse, obesity, depression, and others[p641]. Studies were also completed indicating that most TS patients probably carried a double dose of the gene[p513], and that 1 in 100 school boys had TS[p617]. The latter two observations indicated the estimated frequency of the *Gts* gene was comparable to that estimated for the type II alcoholism gene[p623] **->:**

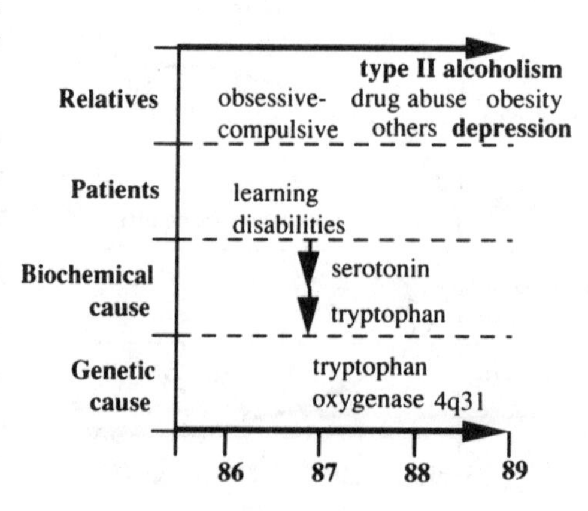

Type II alcoholism and the MNS blood group This brings us back to the observation about type II alcoholism and the MNS blood group. Dr. Hill and colleagues studied 30 families of familial type II alcoholism, each containing at least two affected siblings. Genetic analysis using POINTER[p48] suggested the frequency of the alcohol susceptibility gene was 0.15. They used two methods of examining linkage, the **sib pair method** and the **lod score method**[p53]. Of five blood groups studied only MNS showed evidence for linkage. The lod score method gave a score of 2.02 at a recombination fraction of 0.007 map units. A lod score of 2 means there is a 1 in 100 chance the linkage was by chance. A score of 3, meaning there is only a 1 in 1,000 possibility the results are due to chance is generally accepted as clearly indicating linkage. A recombination fraction of 0.007 means the alcoholism locus was 0.7 map units[p54] from MNS. Both methods showed evidence suggesting linkage, but since a lod score of 3 is desired to rule out chance, this is considered preliminary.

MNS blood group is at 4q31 The exciting aspect of the above study is that the *MNS blood group has been localized to the same region of the long arm of chromosome 4 as the tryptophan oxygenase gene*. The gene responsible for the MNS blood group is the *glycophorin* gene. Studies of the MNS blood group in individuals with chromosome abnormalities have indicated that the MNS gene is located somewhere between bands 4q28 and 4q31[386b]. The gene for fibrinogen, a blood clotting protein, has been localized to band 4q31[1259a]. Linkage studies by Jeffrey Murray and colleagues in Iowa have shown that the order of genes on the long arm of 4 is: centromere - fibrinogen - MNS — telomere with 7.4 map units between the fibrinogen and MNS[1381a]. This suggests the MNS blood group is in band 4q31.

Depression and the MNS blood group Other behavioral disorders have also been examined for linkage to blood groups. It is of interest that in a study by Lynn Goldin and her colleagues of a large number of blood markers in patients with depression only the MNS blood group gave any indication of linkage (lod score of 1.39 at a recombination fraction of 0.20)[762].

While the items in parentheses are still considered tentative and additional studies need to be done, the above findings are consistent with the proposal that the genes for tryptophan oxygenase, type II alcoholism, one form of depression and Tourette syndrome may all be the same gene. This is progress, but not proof, toward *closing the logic loop* for the proposal that TS and its many associated behaviors may due to a defect in serotonin caused by a mutation of tryptophan oxygenase:

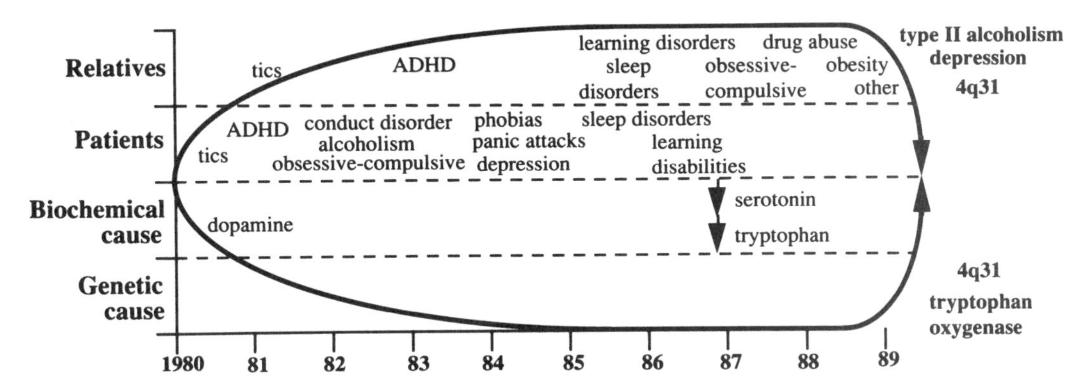

This has been verified by an independent study of Elizabeth Hill and her colleagues. Again using the sib pair method they examined five families with familial pure depressive disease and found evidence for linkage to MNS with less than a 1 in 100 possibility it was by chance[896a]. The long arm of chromosome 4 is thus begining to look like this:

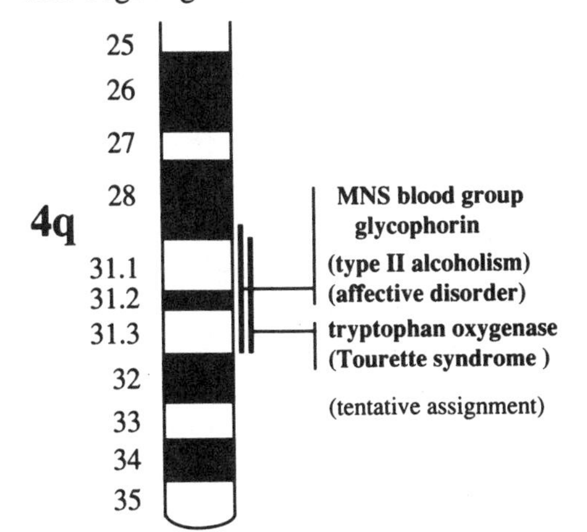

Linkage in Tourette Syndrome

The next important question is whether TS is also linked to the MNS blood group. We had examined this question back in 1986, using lod score studies of MNS blood groups in 27 TS families[379] and found no linkage. However, this was based on the assumption that TS was inherited as a dominant trait with reduced penetrance. Thus, a typical family for linkage might look like this:

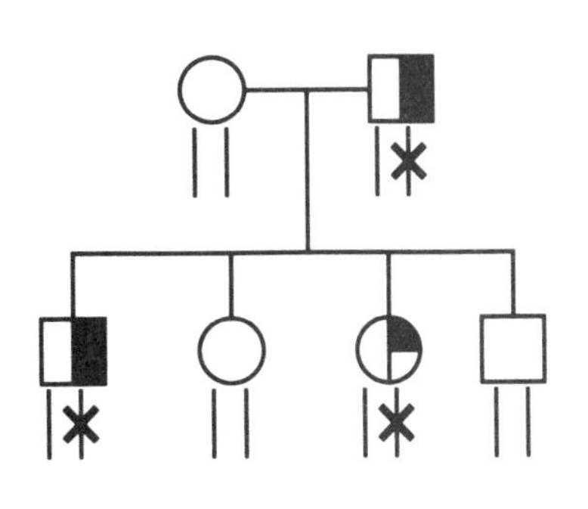

Linkage in a such a family would be easy to detect or rule out and we thought we had ruled it out. However, subsequent studies have suggested that most TS patients carry a double dose of the *Gts* gene. Thus, the gene may actually be distributed as follows:

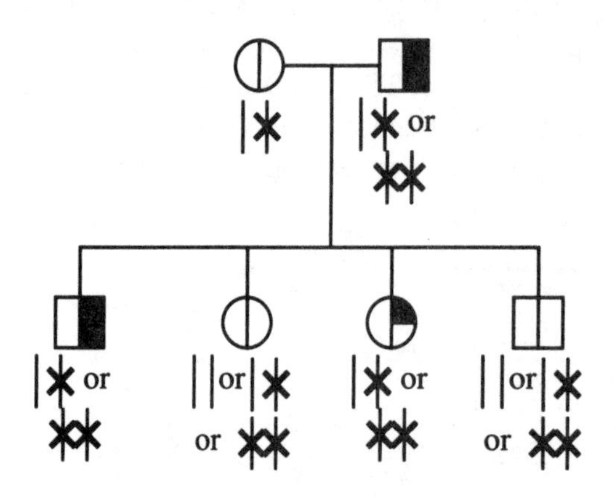

Now, using the lod score technique, linkage is extremely difficult to rule in or out since almost everyone carries at least one copy of the gene and in the brothers or sisters without symptoms it is not clear whether they carry 0, 1 or 2 *Gts* genes. Identification of the *Gts* mutation itself, or alternative linkage methods will be required to definitively prove if the tryptophan oxygenase and the *Gts* gene are the same and if Tourette syndrome is linked to the MNS blood group.

How Common Is TS? — Working Backward

If individuals with either a single or a double dose of the *Gts* gene are at risk for a wide range of behavior problems, as discussed throughout this book, it is important to know how common the gene is. Until this chapter, the best estimate we had was the observation that in our epidemiological study 1 in 100 school boys had TS. It was then conservatively stated that if half of these boys were homozygotes, then the gene frequency would be 0.07 and 13 percent of the population would be *Gts* gene carriers[p518]. However, if they were all homozygotes then the gene frequency would be 0.10 and 18 percent of the population would be gene carriers[p518]. Finally, if not only the definite but also the probable and possible TS students were homozygotes (1/47), then the gene frequency would be 0.15 and 25 percent of the population would be gene carriers. Thus, it is clear that our estimates of the *Gts* gene frequency are still emeshed in considerable uncertainty. It would help if we had an independent estimate of the frequency of the *Gts* gene. *If* the *Gts* gene and the *type II alcohol susceptibility* gene were in fact one and the same, we could use the estimates of the frequency of the alcoholism gene to help refine estimates of the frequency of the *Gts* gene. Two such studies using POINTER have been done. The study by Gillian and colleagues[719] gave a gene frequency of 0.11 and the above study of S. Hill and colleagues gave a frequency of 0.15. If we average these, the estimate from both studies would be 0.13. If this is in fact the frequency of the *Gts* gene then 22.6 percent of the population carry the *Gts* gene and 5% would be homozygotes.

If this is correct, then there are many individuals homozygous for the *Gts* gene who do not have TS but manifest other symptoms and the penetrance for tics in homozygotes is much less than 100 percent. This figure could also explain the serotonin results on the parents of children with "pure" ADHD. Unexpectedly, the values in both parents was just as low as for the parents of the TS children, suggesting that many children with "pure" ADHD may also be homozygous for the *Gts* gene. It would also suggest that many of the parents of TS children who have symptoms are themselves homozygotes. This, in turn, would also explain why the siblings of TS patients had average serotonin and tryptophan levels almost as low as the TS patients themselves. If many of the parents of TS children are homozygous for the *Gts* gene, then all their children would have to carry the gene and thus there would be relatively few homozygous normal *gts/gts* siblings. *I warn the reader that until a Gts marker is available all of this is still speculation.* However one of the most important aspects of this speculation is that it potentially explains another mystery in the genetics of human behavior. Even though there are tentative findings suggesting the linkage of a gene for manic-depression to the X-chromosome[p195], and a gene for schizophrenia to chromosome 5[p210], massive efforts to find other families with linkage to these or other sites have failed. However, most of these efforts have used either the lod score or the sib pair linkage technique, both of which only work well if the gene frequency is relatively low. If the carrier frequency is in the range of 15 to 25 percent they may fail. The proposal presented in this book, of a few very common genes playing a major role in the genetics of behavior, still requires confirmation through the use of genetic markers. However, if correct it could explain the difficulties encountered and would suggest that some alternative approaches may be needed to find major linkages in behavior genetics.

Summary: Many chapters in this book have shown that Tourette syndrome is a complex neurobehavioral disorder presenting as motor and vocal tics plus a wide spectrum of impulsive, compulsive, addictive, mood and anxiety disorders including early onset (type II) alcoholism and depression. A similar spectrum of behavior problems have been found to occur in the relatives of TS patients. The fact that these are the same behaviors previously shown to be associated with a low brain serotonin prompted us to examine serotonin in TS patients. The finding that blood serotonin and tryptophan were significantly decreased in patients and both their parents implicated a genetic defect in the enzyme tryptophan oxygenase. This gene was isolated in humans and shown to be located on the long arm of chromosome 4 at band q31. The independent evidence suggesting that type II alcoholism and familial pure depressive disease are linked to the MNS blood group is significant since the gene for this blood group is located in the same chromosome region as the tryptophan oxygenase gene. Thus, while studies of the chemistry of TS went off in one direction and studies of behavior problems in TS relatives went off in another, this "hard science" linkage evidence tends to close the logic loop and lends some preliminary support to the proposal that the Tourette syndrome gene, the tryptophan oxygenase gene, the type II alcoholism gene, and one of the genes for depression might be the same gene.

Chapter 99

The End of the Beginning, and the Future

This book represents the author's view of Tourette syndrome based on 10 years of involvement with over 1,400 TS and 600 ADHD patients and their families. The ideas and concepts presented occupy a continuum from straight forward clinical observations, case and pedigree reports, to theories and speculations. For the benefit of the reader, the following presents a guide as to which is which.

There is broad agreement among almost all clinicians who care for TS patients that in addition to the motor and vocal tics there is a high frequency of obsessive-compulsive behaviors, ADHD, learning disabilities, conduct disorder, self-abusive behaviors, and anxiety, mood and sleep disorders. These associated behaviors are covered and discussed in Part III.

As mentioned previously, there is less agreement as to why these occur in TS patients. Are they coincidental, due to ascertainment bias, or do they represent varying ways in which the *Gts* gene or genes may be expressed? I have presented a range of reasons based on clinical observations, pedigree and genetic data, and neurotransmitter function, why this clinical geneticist believes the most parsimonious explanation is that these behaviors represent the variable expression of the *Gts* gene.

There is also agreement that TS is a much more common disorder than originally believed. Since it is more common in males and since the tics tend to ameloriate with age, it is most common in school boys. Few would argue that the frequency in this group is at least 1 in 1,500. Our experience with constantly monitoring three schools for two years suggests a frequency in school boys of 1 in 100.

Although we and others have reported that TS is inherited as an autosomal dominant trait, I have presented reasons for believing that a semidominant semirecessive form of inheritance is more likely. This is not a trivial, academic difference. If the inheritance is purely dominant and only one gene is needed to produce TS, the percentage of individuals carrying the *Gts* gene would be in the 0.1 to 2 percent range. However, if it takes two *Gts* genes to produce TS, the percentage of individuals carrying the *Gts* gene would be in the 10 to 20 percent range. If individuals carrying the *Gts* gene express some of the associated behaviors then a significant number of human behavioral disorders could be due to this gene.

Whether there is a single gene or a number of major *Gts* genes is not known. I have presented reasons why the presence of a single major gene and possibly several minor genes seems most likely. There may, however, be many different mutations of this major gene (alleles), each causing different degrees of penetrance and expression.

The question of whether there is a significant increase in the frequency of the associated behaviors in relatives of TS patients is just beginning to receive attention. There is agreement that tics and obsessive-compulsive behaviors are much more frequent in relatives.

Three different studies of ours have shown an increase in ADHD and learning disorders in TS relatives. The presence of the additional disorders is beginning to be reported by others[1905b,1907b], but, in general, these studies are just beginning.

The cause of TS is still unknown. Defects in dopamine, norepinephrine, serotonin, and endorphins have all been proposed. I have presented reasons for believing that a mutation affecting serotonin could be the primary defect and that the other changes are secondary. The observations that tryptophan is often low in TS patients suggests the *Gts* mutation may cause increased levels of tryptophan oxygenase. This is yet to be proven and mutations of other genes may be the cause of TS.

Until a specific genetic marker for the *Gts* gene is found, the proposals about the identity, frequency and mode of inheritance of the *Gts* gene and its relationship to a variety of behavioral disorders will remain this author's interpretations based on the presently available data. Others may have similar or different interpretations and opinions. However, enough clinical, genetic and chemical data are now available to begin to piece together some definite conclusions, and some less definite but logical conclusions. There are enough data to present in this book both the beginning of the story and the end of the beginning.

The next steps will be the application of powerful genetic tools to attempt to identify the specific genes and specific mutations involved in this complex disorder. That will allow the unambiguous examination of the role of Tourette syndrome genes in human behavior.

Summary

To summarize this book in a few paragraphs would be to say that Tourette syndrome is a common, hereditary disorder whose cardinal feature is the presence of supressible motor and vocal tics that occur almost daily for a period of at least one year, and that a wide range of other impulsive, compulsive, learning and mood disorders occur in some, but not all patients. These include short attention span (attention deficit hyperactivity disorder), learning disabilities, dyslexia, obsessive-compulsive behaviors, conduct disorders, inability to take no for an answer, short temper, mood swings, depression, acute anxiety or panic attacks, alcoholism, drug abuse, overeating and others. TS can be considered a disinhibition disorder in which the motor and vocal tics are the most obvious part since they can actually be seen or heard. This concept can be summarized as follows:

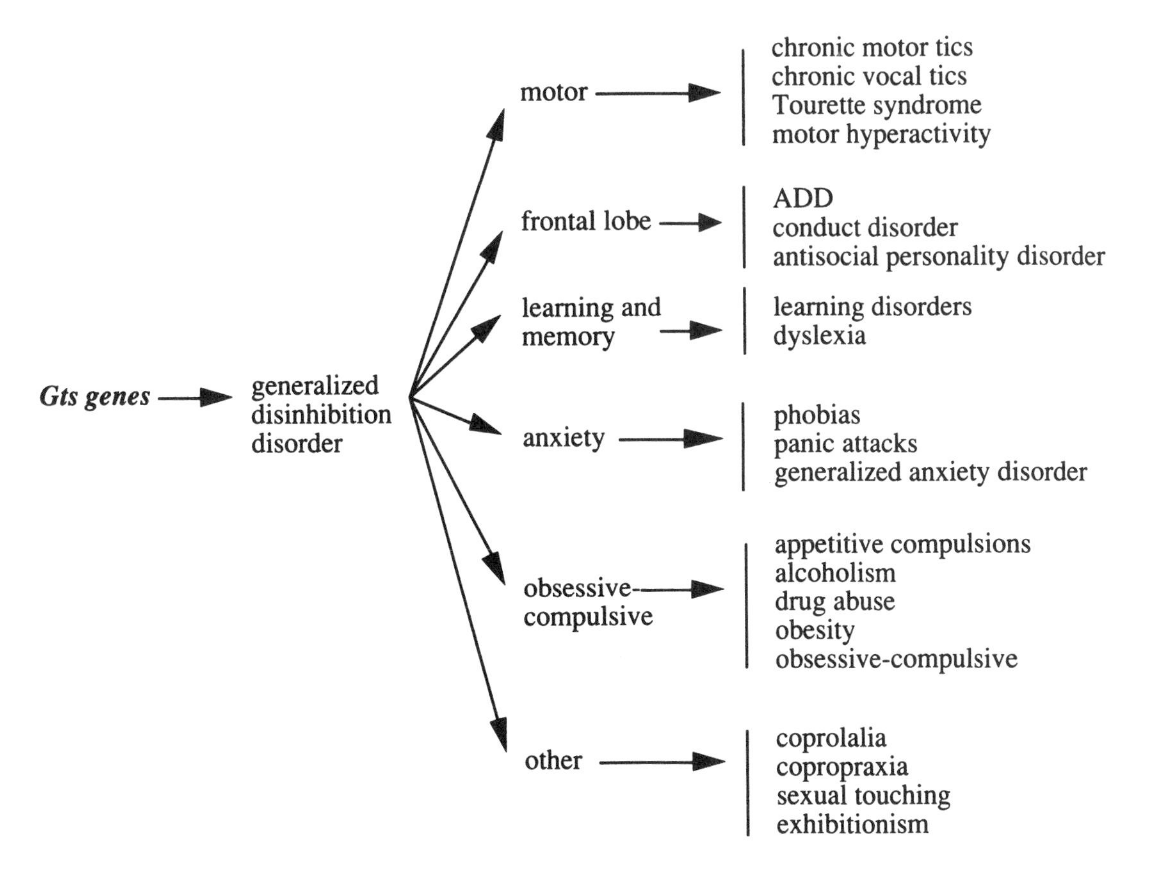

A number of different studies suggest that many individuals with TS have inherited two Gts genes, one from each parent. The parents or other individuals who carry only one Gts gene may either have no symptoms or one or more of behaviors listed above. This concept is ***diagramed as follows:***

Everyone having these disorders does not have a Gts gene, but some may.

Many types of evidence suggest the basic defect in Tourette syndrome may be a genetically caused decrease in brain serotonin. This chemical imbalance results in a combined frontal lobe-limbic system syndrome consist-

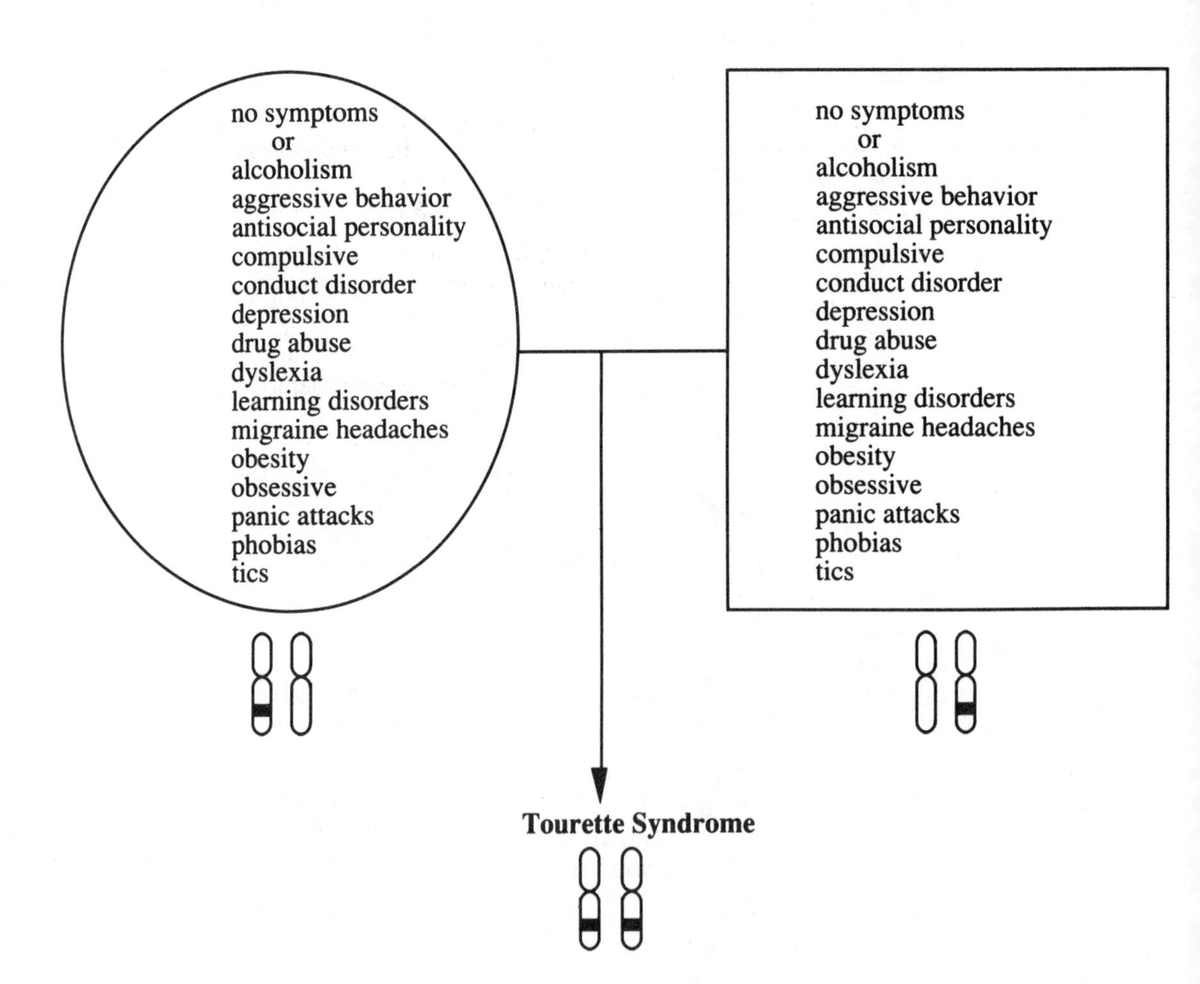

ing of a spectrum of obsessive, compulsive, learning, mood and tic disorders.

One possible explanation for why the Gts gene or genes are so common is that the limbic system disinhibition results in a slight reproductive advantage for those carrying the gene. Over many generations this would result in an increase in frequency of the gene.

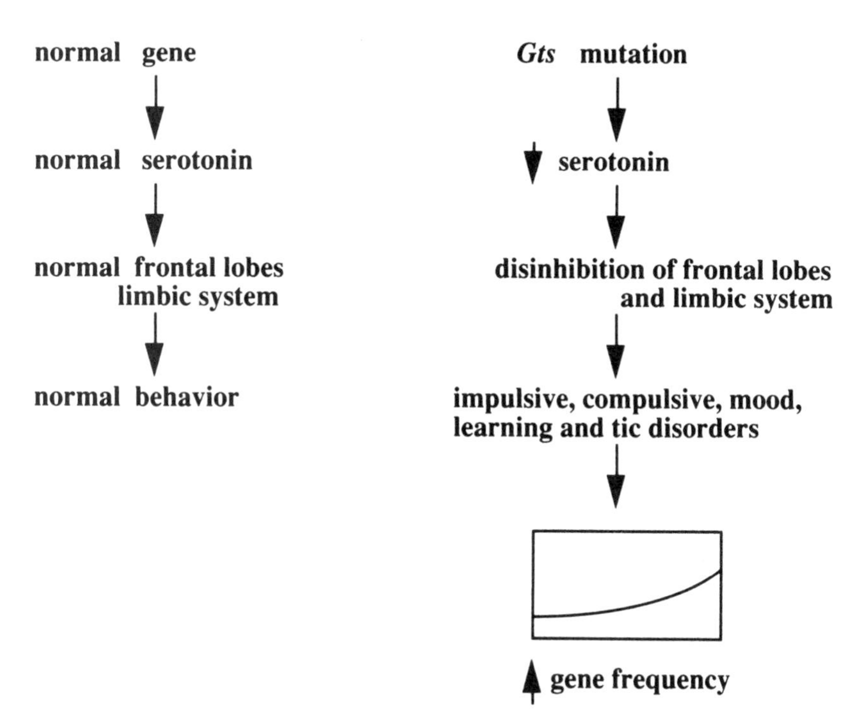

The concept that one or a few common genes can cause a wide spectrum of impulsive, compulsive, learning and mood disorders represents a departure from the prevailing assumption that most behavioral disorders are either due to early childhood psychological trauma or are learned and can thus be unlearned. This genetic concept proposes that, in fact, the human psyche is quite resilient to many environmental stresses unless it is destabilized by the effects of these genes.

Where to Go for Help

If you think you have a child or relative with Tourette syndrome and need help finding a physician familiar with the disorder, the easiest solution is to ask your doctor, look up your local Tourette Syndrome Association chapter, or write to the national office of the Tourette Syndrome Association.

Tourette Syndrome Association
42-40 Bell Boulevard
Bayside, NY 11361-2861
718-224-2999

Mr. Roy Hillard
169 Wickham Way
Welling, Kent, England

Tourette Syndrome Fundation of Canada
173 Owen Boulevard
Willowdale, Ontario,
Canada M2P 1GA

Manitoba Society for Tourette Syndrome
P.O.Box 25064
1650 Main Street
Winnipeg, Canada R2V 4C7

Tourette Syndrome Association of Australia
Victorian Branch
Agnes Zillner
P.O. Box 86 Bundoora, 3083
Australia
(03) 434-3991
(03) 459-2748

Norsk Tourette Forening
Christian Melbye
Minkerudasen 33
1165 Oslo 11
Norway
472-285-043

Those in Southern California can make an appointment at the City of Hope Tourette Syndrome Clnic.

Tourette Syndrome Clinic
City of Hope National Medical Center
1500 E. Duarte Rd.
Duarte, CA 91010
818-359-8111

Glossary

If the word is not in the Glossary, look in the index for the part of the text where it is described.

ADD Attention deficit disorder.

ADD/H Attention deficit disorder with or without hyperactivity.

ADDH Attention deficit disorder with hyperactivity.

ADHD Attention deficit hyperactivity disorder.

adversive stimulus Unpleasant stimulus that a subject will avoid if possible.

adversive Unpleasant.

affect Mood, feelings or emotion.

affective disorder Disorder of mood. May be depression only, mania only, or manic-depressive disorder.

agonist A chemical compound that stimulates a given receptor.

amygdala An almond-shaped part of the limbic system that compares messages from the different sense organs.

anoxia A reduced supply of oxygen. Often refers specifically to the brain.

antagonist A chemical compound that inhibits or blocks a given receptor.

ascertainment The finding of cases or individuals with a given disorder.

association areas Areas of the brain involved in comparing and matching sensations from the different sense organs.

auditory Refers to hearing.

autosomal On a non-sex chromosome, i.e., on chromosomes 1 to 22.

basal ganglia Consists of the caudate, putanem and globus pallidus. These regions contain dopamine-rich neurons which are involved in the control of muscle movement.

base pairs Pairs of guanine and cytosine or adenine and thymine in DNA.

bipolar depression Manic-depressive disorder.

borderline A psychiatric disorder intermediate in severity between neurosis and psychosis.

C Stands for carbon in a chemical formula.

CAT or CT scan Computerized axial tomography or computerized tomography. A computer analyzed series of X-rays of the brain that allow one to see inside the bony structure of the skull.

caudate nucleus A group of neurons in the basal ganglia involved in the control of muscle movements.

central nervous system (CNS) The spinal cord and brain.

cerebellum A part of the brain located at the back of the neck that provides the neuronal feedback loops that allow smooth movements of the extremities.

chi square A statistical test especially useful in determing if the percentage of persons affected in a test group is significantly different from those in the control group.

chromosome Unit of a giant piece of DNA containing many genes. Chromosomes and their associated proteins are most easily seen during cell division (mitosis) when they are stained (*chromo–* colored, *some* –body).

concordance The situation where both of a pair of twins have the same condition or disorder.

consanguineous When the parents are related to each other.

coprolalia Involuntary, compulsive swearing.

correlation coefficient A number that indicates whether two different sets of numbers or variables tend to increase or decrease together. For example, if there is a perfect correlation in that each increase in variable A is associated with a comparable increase in variable B, the correlation coefficient would equal 1.0. If each increase in A was associated with a comparable decrease in B the correlation coefficient would be –1.0. A value of 0 would indicate the two were totally independent of each other.

cyclic AMP A chemical, circular in form, which acts as a messenger inside the cell. Many receptors, when stimulated, result in the release of cyclic AMP.

disinhibition Poorly inhibited. Removing influences that inhibit nerve cell activity. Inability to refrain from inappropriate behaviors.

DNA The genetic material deoxyribonucleic acid.

dominant A trait caused by a gene in which only a single abnormal gene is required to produce the disorder. In strictly dominant conditions, the disorder produced by one abnormal gene is the same as when two abnormal genes are present.

dopamine A chemical that acts as a neurotransmitter in the brain. Dopamine neurons are especially involved in the control of muscle movements and have been implicated in schizophrenia, Tourette syndrome and Parkinson's disease.

dopaminergic Neurons which use dopamine.

DSM-III The Diagnostic and Statistical Manual of the American Psychiatric Association, 1980 edition.

DSM-IIIR The Diagnostic and Statistical Manual of the American Psychiatric Association, Revised, 1987 edition.

dysarthria Difficulty in articulating words.

dysgraphia Difficulty in producing the motor movements required for handwriting.

dyskinesia Abnormal muscle movements, especially refers to movements around the mouth in tardive dyskinesia.

dyslexia A disorder in which reading skills are significantly below grade level in an individual who is not otherwise retarded.

echolalia To repeat words someone else has said.

echopraxia To repeat or copy the actions of other people.

electroencephalogram (EEG) Brain wave test.

electrophoresis A technique by which chemicals are separated in an electrical field.

encopresis Repeated loss of bowel control past the usual age of control (4 years).

endogenous From internal causes, such as endogenous depression due to a brain chemical imbalance.

enuresis Repeated, involuntary bed wetting past the usual age of bladder control (5 years).

enzyme A protein that serves as a catalyst for a specific chemical reaction.

exogenous From external causes such as exogenous depression due to a divorce or death.

exon A segment of DNA within the gene that is read into the nuclear RNA, remains in the messenger RNA and is translated into protein.

expression Refers to manner in which different symptoms of a genetic disorder may be present in an individual carrying the gene(s). In mild expression there are few or mild symptoms; in severe expression there are many or severe symptoms.

expressive language disorder Problems with the ability to express oneself verbally.

extinction Reduction or elimination of a behavior by removing all reinforcement.

familial Occurring in two or more members of a family. May or may not be genetic.

fodrin A protein in the neurons involved in controlling the shape of the synapses. Changes in fodrin are involved in long-term memory.

frontal lobes Major lobes of the brain located behind the forehead.

GABA Gamma-amino-butyric acid. A neurotransmitter that inhibits neurons.

gene The portion of DNA that carries the code for a specific protein or RNA.

genetics The study of hereditary disorders or mechanisms.

globus pallidus A group of neurons in the basal ganglia involved in the control of muscle movements.

Gts When italicized refers to the Tourette syndrome gene (or genes).

H Refers to hydrogen in chemical formulas.

hereditary Inherited by genetic mechanisms.

heterozygote An individual in whom the genes from their mother and their father, at a given site, are different. For example, an individual who has both a normal *gts* and and mutant *Gts* gene is a heterozygote for the *Gts* gene.

heterozygous The state of having both genes at a given site different.

hippocampus The part of the limbic system involved in memory.

5-HIAA 5-hydroxyindole acetic acid. A breakdown product of serotonin. Measurement of 5-HIAA in the spinal fluid gives an estimate of the serotonin activity of the brain.

homozygote An individual in whom both the gene from their mother and their father, at a given site, are the same. For example, an individual with two *Gts* genes is a homozygote for the mutant *Gts* gene.

homozygous The state of having both genes at a given site the same.

HVA Homovanillic acid. A breakdown product of dopamine. Measurement of HVA in the spinal fluid gives an estimate of dopamine activity in the brain.

impulsivity A tendency to respond quickly without thinking of the consequences.

intron A segment of DNA inside the gene that is read into the nuclear RNA but is excised from the messenger RNA.

Korsakoff syndrome A syndrome of poor memory and loss of initiative in chronic alcoholics.

limbic Pertains to the **limbic system**, the part of the brain controlling emotions. It was called limbic because it bends around the central part of the brain like a limb.

linkage When two traits or genetic markers travel together in families they are said to be linked.

locus ceruleus The region of the brain containing the norepinehprine neurons.

mainstreaming Placing a child with handicaps in regular classes.

marker A genetic trait which tends to distinguish between different individuals. For example, ABO blood groups are genetic markers.

mesocortical Refers to pathways between the mesencephalon (especially ventral tegmental area) and the limbic system.

mesolimbic Refers to pathways between the mesencephalon (especially ventral tegmental area) and the limbic system.

microcephaly Small head.

minimal brain dysfunction An old term

for attention deficit hyperactivity disorder. It went out of use because it incorrectly implied there was brain damage.

motor Pertaining to muscle. A motor movement is a muscle movement.

mRNA Messenger RNA.

multifactorial A form of inheritance in which more than one type of gene is involved.

mutation A change in the structure of DNA that can be inherited.

myoclonic Refers to sudden muscle jerks generally much more severe than tics.

N Refers to nitrogen in chemical formulas.

narcissistic To be overly involved in one's own welfare.

narcolepsy A neurological disorder associated with a tendency to fall asleep at inappropriate times and places. It is usually associated with episodes of paralysis around the time of sleep and with cataplexy or sudden falling down due to loss of muscle tone.

neostriatum The basal ganglia containing the caudate, putamen and globus pallidus. It is involved in controlling muscle movements.

neurology The medical specialty dealing with the study and treatment of nervous system disorders.

neurons Nerve cells.

neurotransmitters Chemicals that are involved in the communication between one nerve and another.

norepinephrine A chemical that acts as a neurotransmitter in the brain. It is formed from dopamin, and norepinephrine neurons modulate dopamine neurons.

O Refers to oxygen in chemical formulas.

occipital lobes Major lobes of the brain located at the back of the head.

organic Involves a chemical or structural defect.

palilalia To repeat one's own words, often many times.

parietal lobes Major lobes of the brain located at the middle sides of the brain.

pedigree A recorded family history often diagramed for easy interpretation.

penetrance Refers to whether an abnormal gene has produced any symptoms. If the gene is non-penetrant, there are no symptoms. If it is penetrant, there are some symptoms and the number is described by the degree of expression.

perception The ability to process and interpret stumuli such as sight, hearing, and touch.

perseveration To repeat words or behaviors over and over.

PET scan Positron emission tomography.It is like a CAT scan except that radioactive compounds are injected and the function of different parts of the brain can be examined.

phenylalanine One of the amino acids.

plastic changes Permanent changes in the structure of the nervous system usually in response to a repeated set of stimuli.

polygenic Due to more than one gene.

positive reinforcement Use of rewards to increase the probability that a given behavior will be repeated.

postnatal Occurring after birth.

prefrontal cortex Part of the frontal lobe that has a distinctive agranular appearance under the microscope. It is involved in comparing information from many different areas of the brain.

prenatal Occurring or exisiting before birth.

psychopathology Psychological abnormalities or disorders.

putamen A group of neurons in the basal ganglia involved in the control of muscle movements.

R complex A primitive part of the brain consisting of the basal ganglia and brain stem the contains the nerves controlling instinctive, automatic behaviors.

raphe nuclei Regions of the brain containing the serotonin neurons

receptors Structures on the surface of nerve cells that tightly bind neurotransmitters. They define which nerve cells will respond to which neurotransmitters and whether the effect will be to stimulate or inhibit the nerve.

recessive A trait caused by a gene in which two abnormal genes are required to produce the disorder. Individuals who carry a single abnormal gene have no symptoms.

reinforcement A procedure to strengthen a response by giving immediate rewards. This is positive reinforcement. Negative reinforcement occurs when punishment is given instead of rewards.

replication Process of duplicating DNA.

restriction endonuclease Enzymes that cut DNA at specific sites.

RFLP Restriction fragment length polymorphism (see Chapter 13).

RNA Ribonucleic acid. Consists of several types such as messenger RNA, transfer RNA, ribosomal RNA, and heterogeneous nuclear RNA.

S.D. Standard deviation. A statistical term that provides an estimate of the scatter of a range of values.

segregation analysis A mathematical analysis of how disorders are inherited based on studies of many families.

serotonin A chemical that acts as a neurotransmitter in the brain. It functions as an inhibiting neurotransmitter. The limbic system and frontal lobes are particularly rich in serotonin.

stereotyped Done in a constant, repetitious manner.

stimulus Any event (touch, noise, chemical, emotional) which activates nerves or the nervous system.

substantia nigra A region of the brain rich in dopamine neurons that pass to the basal ganglion (caudate and putamen).

substrate A chemical that is used by an enzyme to form a different chemical.

striatum Caudate and putamem.

syndrome A group of symptoms or signs that identifies a specific disorder, as Tourette syndrome.

tardive Delayed or late in onset.

temporal lobes Major lobes of the brain, near the area of the ears.

tic An involuntary, sudden, rapid muscle movement or vocal noise.

Tourette syndrome A common, hereditary syndrome consisting of chronic motor (muscle) and vocal tics and associated with a wide range of other behaviors.

tryptophan An amino acid required to make serotonin.

TS Tourette syndrome

TSA Tourette Syndrome Association

unipolar depression Depression only, without mania

vocal Pertaining to the voice. Vocal tics are sometimes called phonic tics.

VTA Ventral tegmental area. A region of the brain rich in dopamine nerve fibers that pass to the frontal cortex and the limbic system.

Appendix
Human Behavior Questionnaire

In Chapter 15 and throughout this book I have spoken of the questionnaire we used for our controlled study of Tourette syndrome. After that study was completed, the questionnaire was expanded still further to cover additional areas we felt were more common in individuals with TS. In addition, an adult and a child version were made. In the child version, all the questions that were inappropriate for children were removed but no new ones were added. Thus, it simply represents a shortened version of the adult form.

We now routinely send these questionnaires to patients for them to fill out before they come to the clinic. If the patient is a young child, the parents fill them out. For intermediate-aged children, both the parents and the child participate. Older children and adults fill them out themselves.

Using the questionnaire has many advantages:

1. It gives patients time to think about the answers, and where appropriate, consult parents or spouses for input.

2. It gives the health professional answers to questions about a wide range of symptoms — questions that otherwise might never be asked either because they were not thought of, or time is too limited for such a detailed history.

3. It provides important data concerning the wide range of behavioral problems associated with Tourette syndrome. This is especially valuable for professionals who may have thought that Tourette syndrome was simply a tic disorder. The consistent use of the questionnaire helps to illustrate the frequency with which other disorders are also present.

The questionnaire draws heavily from three sources: The National Institutes of Mental Health Diagnostic Interview Schedule[1630], the DSM-IIIR[491], and our clinical experience with Tourette syndrome. A few questions were taken from other sources[627,1291a]. The questionnaire is not designed to be scored and there are no right or wrong answers. It is used as a guide to make sure all relevant behavioral areas are covered. Specific diagnoses should be made from direct interviews and use of the DSM-IIIR criteria.

The reader is welcome to make free use of the questionnaire and has the author's and publisher's permission to copy it without written permission. For your convenience, printed adult and child questionnaires are available in bulk from the publisher (see backpiece).

____
Name

HUMAN BEHAVIOR QUESTIONNAIRE
(Adult Version)

Department of Medical Genetics
City of Hope Medical Center
Duarte, CA 91010

This questionnaire is designed to elicit detailed information about many aspects of human behavior. It is critical that you take your time and answer the questions as frankly as possible. Because it is quite long, many people find it easier to divide the questions into different sets for different days. You are an important person and your care in filling this out will be of great value in helping us to diagnose problems and to understand the role of genetic factors in controlling how people behave. *If you feel that you cannot carefully and accurately complete the questionnaire, we would prefer that you not start.*

The questionnaire is designed for a wide range of purposes. It is used to help us evaluate people coming to the clinic for possible Tourette syndrome, attention deficit disorder, learning disorders, and a wide range of behavior problems. It is also designed for the parents, siblings and other family members. Finally, it is for controls — that is, people who may have none of the above.

There are two versions of this questionnaire — An adult version and a child version.

The **adult version** is for people 14 years of age or older. If you have the adult version you should answer the questions yourself. Feel free, however, to consult your parents or others if necessary to answer the questions about your childhood.

The **child version** is for children less than 14 years of age. Parents or some other adult very familiar with the child should fill it out. You should consult with the child or read him or her the questions, as necessary.

Again, please take your time and answer all the questions carefully. Whenever you are in doubt, please write out a narrative description or explanation of what you think the question asks. In this questionnaire there is no such thing as wrong answers or too much information. It is easy to condense answers; it is impossible to work with answers that are incorrect or not answered.

Some of the questions may seem a little unusual. Please bear with us they all have an important meaning. All the information in this questionnaire will be treated as confidential material. If it is used in any studies it will be identified by number only.

Personal Data

1. Last name ______________________ First name ____________ Initial ___
2. Address ____
3. City ______________________________ State _____ Zip ______
4. Telephone Home ___ ___ ___ - ___ ___ ___ - ___ ___ ___ ___
5. Office ___ ___ ___ - ___ ___ ___ - ___ ___ ___ ___

6. Present date: Month______Day______Year_____
7. Your birth date: Month______Day______Year_____

8. Sex: Male ☐ Female ☐
9. Your age: Years ______
10. Your weight Pounds ______
11. Your height Feet ______ Inches ______

12. Marital Status: Single ☐ 1 Married ☐ 2 Separated ☐ 3
    Divorced ☐ 4 Widowed ☐ 5 Other ☐ 6

13. Race: White ☐ 1 Black ☐ 2 Oriental ☐ 3
    Hispanic ☐ 4 Other, specify______________ 5

14. Are you adopted? No ☐ Yes ☐

15. I live:
    - At home with my parents ☐ 1
    - With my wife/husband/children ☐ 2
    - With someone of the opposite sex ☐ 3
    - With relatives not my parents ☐ 4
    - Alone ☐ 5
    - Other, specify____________________ ☐ 6
    - In a treatment facility. Specify__________ ☐ 7

____

16. The principal wage earner in my home is:
    - Myself ☐ 1
    - My husband/wife ☐ 2
    - Myself and my spouse ☐ 3
    - My father ☐ 4
    - My mother ☐ 5
    - Other, specify____________________ ☐ 6

17. How many times have you been legally married? _____
18. How many times have you been divorced? _____
19. How many times have you lived with someone for more than 1 month, whether your were married or not? _____
20. How many children do you have? _____
21. How many times have you been pregnant, or if a male, caused someone to become pregnant? _____

22. Are you right or left handed?
    Right handed ☐ Left handed ☐ Use both hands equally ☐

Employment

1. Are you presently employed? No ☐ Yes ☐
2. If you are employed is it:
   Full time ☐ 1 Part time ☐ 2 Off and on ☐ 3
3. What is your occupation? Describe____
____
4. If you have been employed in multiple jobs, describe the last ones:____
____
____
5. Since you finished school, how many years have you been unemployed (add together the unemployment between jobs):
   Haven't finished school ☐ 1
   Less than one year ☐ 2
   More than 1 year (enter number of years) ____
6. Have you ever been fired from a job? No ☐ Yes ☐
   7. If yes, how many times? ________
8. Have you ever quit a job? No ☐ Yes ☐
   9. If yes, how many times? ________
10. Have you ever been on public assistance, welfare, or social security insurance? No ☐ Yes ☐
    11. If yes, how many years were you on such assistance?
    (If more than once add them all together) ________ years.
12. What is your income or, or if you are a student, your parents' present income?

$0 - $10,000	☐ 1.	$30,100 - $40,000	☐ 4
$10,100 - $20,000	☐ 2	$40,100 - $50,000	☐ 5
$20,100 - $30,000	☐ 3	more than $50,000	☐ 6

Early History

1. How old were you when you first talked?
   a. First words? Years ______ Months ______
   b. First sentences? Years ______ Months ______
2. How old were you when you first walked? Years ______ Months ______
3. How old were you when you were first toilet trained?
   a. For daytime? Years ______ Months ______
   b. For nighttime Years ______ Months ______
4. What is the oldest age that you frequently wet the bed at night (more than 2 times a month)?
   Years______ Months______ Still do ☐
5. What is the oldest age that you frequently had a bowel movement in your pants?
   Years______ Months______ Still do ☐
6. After you were two years of age did you ever play with or handle your bowel movement more than was necessary for regular hygiene? No ☐ Yes ☐
   If yes, give latest age______
   If yes, give details____
____
7. When you first went to school (nursery school, kindergarten) was there a problem separating from your mother (or father)?
   No ☐ Moderate ☐ Severe ☐
8. What were the teachers general comments about you in nursery school and kindergarten?
____
____

Sleep

1. Do or did you ever you have any problems getting to sleep at night?
   Rarely ☐ 1-2 times a month ☐ 1-2 times a week ☐ Daily or almost daily ☐
2. Do or did you ever you walk in your sleep?
   Rarely ☐ 1-2 times a month ☐ 1-2 times a week ☐ Daily or almost daily ☐
3. Do or did you ever have night terrors, that is, wake up at night screaming and terrified?
   Rarely ☐ 1-2 times a month ☐ 1-2 times a week ☐ Daily or almost daily ☐
4. Do or did you have problems waking up early and not being able to get back to sleep?
   Rarely ☐ 1-2 times a month ☐ 1-2 times a week ☐ Daily or almost daily ☐
5. Do or did you ever talk in your sleep?
   Rarely ☐ 1-2 times a month ☐ 1-2 times a week ☐ Daily or almost daily ☐
6. Do or did you ever have nightmares?
   Rarely ☐ 1-2 times a month ☐ 1-2 times a week ☐ Daily or almost daily ☐
7. If yes, did they have a particular theme played over and over?
   No ☐ Occasionally ☐ Often ☐
8. When you were a child, between the ages of first born and 5 years of age, did you have problems with being unable to take an afternoon nap and unable to sleep in the evening?
   Rarely ☐ 1-2 times a month ☐ 1-2 times a week ☐ Daily or almost daily ☐
9. As a child, did you ever have a period of time when you were afraid to sleep alone and wanted to sleep with a parent or other person?
   Rarely ☐ 1-2 times a month ☐ 1-2 times a week ☐ Daily or almost daily ☐
10. Are you or were you ever afraid of the dark?
    No ☐ Occasionally ☐ Often ☐

11. If you have or had significant sleep problems how old were you? From ______ to ______

12. If you feel your sleep problems have not been covered by the above, describe them: ________

____

____

Activity

THE FOLLOWING QUESTIONS APPLY TO YOU **BETWEEN THE TIME YOU WERE BORN AND 16 YEARS OF AGE.** IF YOU ARE AN ADULT, ASSUME THE QUESTIONS SAY "DID YOU ....." CONSULT WITH A PARENT IF YOU DO NOT RECALL.

IF YOU ARE A PARENT FILLING OUT THE FORM FOR A CHILD, ASSUME THE QUESTIONS SAY, "DOES HE/SHE..." OR "DID HE/SHE..."

Inattention	No or Don't Know	Occas-ionally	Often
1. Do you fail to finish things you started?............................	☐	☐	☐
2. Do you seem not to listen to your parents or teachers?............	☐	☐	☐
3. Are you easily distracted?...	☐	☐	☐
4. Do you have difficulty concentrating in school or elsewhere?....	☐	☐	☐
5. Do you have difficulty sticking to play activities?.................	☐	☐	☐
6. Do you have difficulty following instructions, or fail to finish chores?..	☐	☐	☐
7. Do you often lose thing necessary for school, home or work activities?..	☐	☐	☐

Impulsivity

1. Do you often act before thinkng?	☐	☐	☐
2. Do you shift excessively from one activity to another?	☐	☐	☐
3. Do you have trouble organizing your work?	☐	☐	☐
4. Do you need a lot of supervision?	☐	☐	☐
5. Do you frequently call out in class or blurt out answers to questions?	☐	☐	☐
6. Do you have difficulty waiting your turn in games or other situations?	☐	☐	☐
7. Do you often do dangerous things without considering the consequences?	☐	☐	☐

Hyperactivity

1. Do you run about or climb on things excessively?	☐	☐	☐
2. Do you have difficulty sitting still, often fidget with your hands or feet or squirm in the seat?	☐	☐	☐
3. Do you have difficulty staying seated?	☐	☐	☐
4. Do you move about excessively in your sleep?	☐	☐	☐
5. Do you often interrupt others or butt into their activities?	☐	☐	☐
6. Are you always on the go?	☐	☐	☐
7. Do you often talk excessively?	☐	☐	☐
8. Do you have difficulty playing quietly?	☐	☐	☐

Other

1. If many of the above are answered "often," at what age did these things first begin?______
2. Has a physician, psychologist or any other professional ever made any of the following diagnoses?
   - a. Minimal brain damage (MBD)........................... No ☐ Yes ☐
   - b. Hyperactive... No ☐ Yes ☐
   - c. Attention deficit disorder (ADD).......................... No ☐ Yes ☐
   - d. Severely emotionally disturbed (SED).................. No ☐ Yes ☐

   If any of the above are answered yes, what is the physician's or professional's name and address?

   ____

   ____
3. Were any of the following medications ever prescribed and taken? If yes, give the dose and ages given.

a. Ritalin	No ☐	Yes ☐	Dose_______	Ages______
b. Cylert	No ☐	Yes ☐	Dose_______	Ages______
c. Dexedrine (amphetamine)	No ☐	Yes ☐	Dose_______	Ages______
d. Mellaril	No ☐	Yes ☐	Dose_______	Ages______
e. Tofranil (imipramine)	No ☐	Yes ☐	Dose_______	Ages______
f. Other (name)______________________	No ☐	Yes ☐	Dose_______	Ages______

IF YOU **NEVER** TOOK RITALIN, CYLERT OR AMPHETAMINES
SKIP TO THE SECTION ON **Activity as an Adult**

4. If Ritalin, Cylert, or amphetamines were taken, what affect did they have on the following?

	No Effect	Some Better	Much Better	Worse
a. Hyperactivity	☐	☐	☐	☐
b. Ability to concentrate	☐	☐	☐	☐
c. School performance	☐	☐	☐	☐
d. Behavior or mood	☐	☐	☐	☐

5. If Ritalin, Cylert, Dexedrine (amphetamines), or Tofranil (imipramine) were taken and then stopped, why were they stopped?
   - Didn't help ☐ 1
   - Just didn't want to take them anymore ☐ 2
   - They made the symptoms worse ☐ 3
   - Symptoms got better and didn't need the medications anymore ☐ 4
   - Stopped during the summer and never started them again ☐ 5
   - Other, describe ______________________ ☐ 6

ANSWER THE FOLLOWING IF YOU EVER HAD MUSCLE TICS OR VOCAL NOISES.
(For a definition of muscle tics and vocal noises see page 29)
IF YOU NEVER HAD TICS SKIP TO SECTION ON **Activity as an Adult**

6. Did the tics and/or vocal noises start before, at the same time, or after you took Ritalin, Cylert or amphetamines?
   - I had tics before I took these medications........... ☐ 1
   - The tics came on at the same time that I took these medications.......................... ☐ 2
   - The tics came on after I took these medications......... ☐ 3

7. If you had tics or vocal noises before taking these medications, how did these medications affect the tics?
   No effect ☐ 1 Better ☐ 2 Slightly worse ☐ 3 Much worse ☐ 4

8. If you had tics or vocal noises after taking these medications, what was the duration of time between starting the medication and the onset of the tics or vocal noises?
   Years ______ Months ______

Activity as an Adult

ANSWER THE FOLLOWING QUESTIONS AS THEY RELATE TO YOU NOW OR MOST OF THE TIME YOU WERE AN ADULT

	No or Don't Know	Occas-ionally	Often
Inattention			
1. Do you have trouble finishing things you start?.....................	☐	☐	☐
2. Do you seem not to listen when people talk to you?................	☐	☐	☐
3. Are you easily distracted?..	☐	☐	☐
4. Do you have difficulty concentrating?...............................	☐	☐	☐
5. Do you have difficulty sticking to one activity?.....................	☐	☐	☐
6. Do you have difficulty following instructions?.....................	☐	☐	☐
7. Do you often lose things necessary for home or work?...........	☐	☐	☐
Impulsivity			
1. Do you often act before thinking?...................................	☐	☐	☐
2. Do you shift excessively from one activity to another?............	☐	☐	☐
3. Do you have trouble organizing your work?........................	☐	☐	☐
4. Do you have difficulty waiting in lines?............................	☐	☐	☐
5. Do you often do dangerous things without considering the consequences?................................	☐	☐	☐

Tests

1. Have you ever had a brain wave test (EEG)? No ☐ Yes ☐
   2. If yes, what was the result? Normal ☐ 1 Borderline ☐ 2 Abnormal ☐ 3

2. Have you ever had a CT or MRI scan? No ☐ 1 Yes ☐ 2
   3. If yes, what was the result? Normal ☐ 1 Abnormal ☐ 2

4. Have you ever had an I.Q. test? No ☐ Yes ☐
   5. If yes, what was the score? ______

School

1. What is the highest grade you reached in school? (grade school 1-6, junior high 7-9, senior high 10-12, college 14-18, post college 19 -)______

2. Are you now a full time student? No ☐ Yes ☐

3. Was there ever a period of time, even a day or two, when you refused to go to school? No ☐ Yes ☐

4. Was there ever a period of time when you didn't want to go to school because you had a headache, stomachache, or other problems, even though you know you were not really ill? No ☐ Yes ☐
   5. If either question #3 or #4 were answered yes, how many days all together did you stay out of school for these reasons? ______

6. Was there a period of time when you really disliked school? No ☐ Yes ☐
   7. If yes, what age did it start? ______

8. Have you ever been placed in any of the following special classes?
   a. Educationally handicapped (EH), learning handicapped (LH), learning disorder (LD)? No ☐ Yes ☐ Ages____________
   b. Resource classes? No ☐ Yes ☐ Ages____________
   c. Severely emotionally disturbed classes (SED)? No ☐ Yes ☐ Ages____________
   d. Aphasia or speech classes? No ☐ Yes ☐ Ages____________
   e. Gifted class? No ☐ Yes ☐ Ages____________
   f. Other - describe________________________________ No ☐ Yes ☐ Ages____________

9. Were you ever assigned a special teacher? No ☐ Yes ☐
10. Were you ever assigned a teacher to come to your home because of behavior problems? No ☐ Yes ☐
    10. If yes, state why ____

____

11. Were you ever held back a grade? No ☐ Yes ☐
12. Have you ever skipped a grade? No ☐ Yes ☐
13. For grades 1 to 6 was your school performance on the whole below average, average, or above average in the following?

	Below Average	Average	Above Average
a. Math	☐ 1	☐ 2	☐ 3
b. Reading	☐ 1	☐ 2	☐ 3
c. Writing	☐ 1	☐ 2	☐ 3
c. Physical education	☐ 1	☐ 2	☐ 3

14. For junior and senior high school, was your school performance on the whole average, below average, or above average in the following?

	Below Average	Average	Above Average
a. Math	☐ 1	☐ 2	☐ 3
b. Reading	☐ 1	☐ 2	☐ 3
c. Writing	☐ 1	☐ 2	☐ 3
d. Physical education	☐ 1	☐ 2	☐ 3

15. Were you ever told that you had a learning disorder? No ☐ Yes ☐

16. Do you feel like your school performance was up to your potential? No ☐ Yes ☐

17. Do you or did you have test anxiety?
Never ☐ 0 Occasionally ☐ 1 Frequently ☐ 2 Always ☐ 3

18. Were you ever suspended or expelled from school? No ☐ Yes ☐
19. If yes, explain why and how often ____________________

19. Is your memory for things:
Very poor ☐ Poor ☐ About average ☐ Excellent ☐

Reading

1. Did you ever have frequent problems with any of the following?

a. Letter reversal (p or q, b for d, etc.)	No ☐	Yes ☐
b. Number reversal (Ɛ for 3, ᴦ for 7, etc.)	No ☐	Yes ☐
c. Word reversal (saw for was, etc.)	No ☐	Yes ☐
d. Drop out or insert words while reading	No ☐	Yes ☐
e. Read very slow, word by word, when your peers were reading normal speed?	No ☐	Yes ☐
f. Unable to retain the meaning of what you just read?	No ☐	Yes ☐

2. What is the greatest number of years you were felt to be behind your peers in reading, if any? For example, if when you were in the 6th grade you were only reading at 4th grade level, you would have been 2 years behind.
Not behind ☐ 1 1 year ☐ 2 2 years ☐ 3 3 or more ☐ 4

3. What is the greatest number of years you were felt to be behind your peers in math, if any?
Not behind ☐ 1 1 year ☐ 2 2 years ☐ 3 3 or more ☐ 4

4. My handwriting is:
Beautiful ☐ 1 Average ☐ 2 Terrible ☐ 3

Speech

1. Have you ever had problems with stuttering? No ☐ Yes ☐
2. If yes, what age did it start?______
What age did it stop?______ Still present ☐

3. Have you ever had problems with speaking so fast or so erratically that people had difficulty understanding you? No ☐ Yes ☐
4. If yes, what age did it start?______
what age did it stop?______ Still present ☐

5. Was there ever a time, after you knew how to talk, when you refused to speak in school or other social situations, for a period of several weeks or months? No ☐ Yes ☐
4. If yes, what age did it start?______
What age did it stop?______ Still present ☐

Social and Other History

1. Have you ever stolen anything (from family, peers, or stores, without confronting the victim)?
Never ☐ Once ☐ 2-5 times ☐ Often ☐

2. Have you ever forged a check?
Never ☐ Once ☐ 2-5 times ☐ Often ☐

3. Have you ever run away from home overnight?
Never ☐ Once ☐ 2-5 times ☐ Often ☐

4. Have you ever lied (other than to avoid unreasonable physical abuse)?
Never ☐ Once ☐ 2-5 times ☐ Often ☐

5. Have you ever persistently lied about not doing something even when it was clear to others that you did it? Never ☐ Once ☐ 2-5 times ☐ Often ☐

6. Have you ever set fires (other than camp or cooking fires)?
Never ☐ Once ☐ 2-5 times ☐ Often ☐

7. Have you ever played hooky from school or missed work without a good reason?
Never ☐ Once ☐ 2-5 times ☐ Often ☐

8. Have your ever broken into someone else's house, building or car?
Never ☐ Once ☐ 2-5 times ☐ Often ☐

9. Have you ever deliberately destroyed property?
Never ☐ Once ☐ 2-5 times ☐ Often ☐

10. Have you ever been cruel to your pets or other animals?
Never ☐ Once ☐ 2-5 times ☐ Often ☐

11. Have you ever forced someone into sexual activity against their will?
Never ☐ Once ☐ 2-5 times ☐ Often ☐

12. Have you ever used a weapon in a fight?
Never ☐ Once ☐ 2-5 times ☐ Often ☐

13. Have you ever initiated physical fights with others?
Never ☐ Once ☐ 2-5 times ☐ Often ☐

14. Have you ever physically attacked your mother?
Never ☐ Once ☐ 2-5 times ☐ Often ☐

15. Have you ever physically attacked your father?
Never ☐ Once ☐ 2-5 times ☐ Often ☐

16. Have you ever stolen anything by confronting the victim (purse-snatching, pick-pocketing, mugging, extortion, armed robbery)?
Never ☐ Once ☐ 2-5 times ☐ Often ☐

17. Have you ever been in trouble with the law?
Never ☐ Once ☐ 2-5 times ☐ Often ☐

18. If yes, describe circumstances:____

____

____

19. Have you ever been arrested?
Never ☐ Once ☐ 2-5 times ☐ Often ☐
If yes, describe circumstances and ages: ____________________________________

____

____

____

20. Have you shouted at your parents?
Never ☐ Once ☐ 2-5 times ☐ Often ☐

21. Do you lose your temper easily?
Never ☐ Occasionally ☐ Often ☐ Very often ☐

22. As a child did you have respect for adults?
A lot ☐ Some respect ☐ No respect ☐

23. As a child did you argue with adults?
Never ☐ Occasionally ☐ Often ☐ Very often ☐

24. As a child did you actively defy or refuse adult requests or rules (such as chores)?
No ☐ Occasionally ☐ Often ☐

25. As a child did you like to be held?
No ☐ Occasionally ☐ Often ☐

26. As a child, did you often not look at people when they were talking to you?
Never ☐ Occasionally ☐ Often ☐ Very often ☐

27. As a child did you like to spin things like jar lids, coins, or other objects?
No ☐ Occasionally ☐ Often ☐

28. As a child did you show an unusual degree of skill for certain things or memory for certain things?
No ☐ Yes ☐

29. If yes, please describe ______________________________

____

____

30. Do you deliberately do things things that annoy other people?
Never ☐ Occasionally ☐ Often ☐ Very often ☐

31. Do you often blame others for your mistakes?
Never ☐ Occasionally ☐ Often ☐ Very often ☐

32. Are you touchy or easily annoyed by others?
Never ☐ Occasionally ☐ Often ☐ Very often ☐

33. Are you often angry or resentful?
Never ☐ Occasionally ☐ Often ☐ Very often ☐

34. Are you spiteful or often say "I'll get even," or "I'll get back at them"?
Never ☐ Occasionally ☐ Often ☐ Very often ☐

35. Do you swear more than most people?
Never ☐ Occasionally ☐ Often ☐ Very often ☐

36. Do you swear compulsively, sometimes saying swear words over and over even when you don't want to and are not angry (coprolalia)?
Never ☐ Occasionally ☐ Often ☐ Very often ☐

37. Do you ever say swear words over and over in your mind even when you don't want to and are not angry (mental coprolalia)?
Never ☐ Occasionally ☐ Often ☐ Very often ☐

38. Do you "give people the finger" a lot?
Never ☐ Occasionally ☐ Often ☐ Very often ☐

39. Would you say you are a competitive person?
Never ☐ Occasionally ☐ Often ☐ Very often ☐

40. Are you a confrontive person?
Never ☐ Occasionally ☐ Often ☐ Very often ☐

41. When involved in fighting with others, did you ever get to a point where you couldn't seem to stop? Never ☐ Occasionally ☐ Often ☐ Very often ☐

42. Have there been periods when you felt full of hate for others?
No ☐ Occasionally ☐ Often ☐

43. Has there ever been a time when you suddenly got so angry that you hit someone?
Never ☐ Occasionally ☐ Often ☐ Very often ☐

44. Can you entertain yourself or are you easily bored?
- ☐ 1 I can entertain myself.
- ☐ 2 I can entertain myself only if I have to.
- ☐ 3 I am very easily bored.

45. Are you
- ☐ 1 Very well coordinated
- ☐ 2 Average coordination
- ☐ 3 Clumsy and poorly coordinated

46. Have you ever been in an automobile accident that was mostly your fault? No ☐ Yes ☐
46. If yes, how many? ______

47. Have you ever felt alone and abandoned when separated from someone close to you, such as a parent or spouse, for several hours?
Never ☐ Occasionally ☐ Often ☐ Very often ☐

48. Everyone likes to receive reassurance, approval or praise. Do you feel that your need for these is: Less than average ☐ Average ☐ More than average ☐ Much more than average ☐

49. Everyone has some concern about their own physical attractiveness. Do you feel your concern is: Less than average ☐ Average ☐ More than average ☐ Much more than average ☐

50. If you suddenly decide that you want something, do you find it easy, moderately difficult, or very difficult to wait to get it?
Easy to wait ☐ Moderately difficult to wait ☐ Very difficult to wait ☐

51. If people criticize you how would you best characterize your reaction.
It doesn't bother me ☐ It bothers me for awhile ☐
I get quite upset ☐ I really get very angry ☐

52. If someone does something to you that you don't like, what is your reaction?
- I immediately shrug it off ☐
- I usually get over it in a day ☐
- It bothers me for several days ☐
- It bothers me for a long time and I often think a lot about how I can get even ☐

53. In relation to making new friends, which most characterizes you?
- I have difficulty making new friends ☐
- I make new friends as well as the next person ☐
- I find it very easy to make new friends ☐

54. In relation to keeping friends, which most characterizes you?
- I usually keep friends for quite a long time ☐
- Oftentimes my friendships are broken up by something I or they said or did ☐

55. In regard to religion, would your consider yourself:
Not religious ☐ Moderately religious ☐ Religious ☐ Very religious ☐

56. What percentage of your income do you contribute to your religion?
Less than 1% ☐ 1 to 2% ☐ 2 to 9% ☐ 10% or more ☐

57. My religion is:
Protestant ☐ Catholic ☐ Jewish ☐ Other , specify ____________________

Discipline

1. Do your parents feel you are (or were) a discipline problem? No ☐ Yes ☐
   2. If yes, describe the nature of the problem:____________________________

3. Do you have a problem accepting "No" for an answer?
   Never ☐ Occasionally ☐ Often ☐ Very often ☐
4. Have you ever poked holes in walls, broken furniture, etc., when you were angry?
   Never ☐ Occasionally ☐ Often ☐ Very often ☐
5. During any period of your life were you were usually aggressive, destructive, rebellious or what would generally be termed a difficult child? No ☐ Yes ☐
   If yes, Age started ______ Age stopped _____ Still present ☐
6. Either as a child or adult have there been times when you got unusually angry or upset over some minor matter to the point that you hit someone or broke things. No ☐ Yes ☐
   If yes, please elaborate____________________________

7. Do you have any other comments on discipline or behavior?____________________

Phobias

1. Some people have phobias, that is, such a strong fear of something or some situation that they try to avoid it, even though they know there is no real danger. Have you ever had such an unreasonable fear of any of the following situations that you tried to avoid it/them.

a. Of heights? .... No ☐ Yes ☐
b. Of tunnels or bridges? .... No ☐ Yes ☐
c. Of being in a crowd? .... No ☐ Yes ☐
d. Of being on any kind of public transportation like airplanes, buses, or elevators? .... No ☐ Yes ☐
e. Of going out of the house alone? .... No ☐ Yes ☐
f. Of being in a closed place? .... No ☐ Yes ☐
g. Of being alone? .... No ☐ Yes ☐
h. Of eating in front of other people (either people you know or in public)? .... No ☐ Yes ☐
i. Of speaking in front of a small group of people you know? .... No ☐ Yes ☐
j. Of speaking to strangers or of meeting new people? .... No ☐ Yes ☐
k. Of storms? .... No ☐ Yes ☐
l. Of being in water, for instance in a swimming pool or lake? .... No ☐ Yes ☐
m. Of spiders, bugs, mice, snakes or bats? .... No ☐ Yes ☐
n. Of being near any other harmless animal that couldn't get near you? .... No ☐ Yes ☐
o. Of people? .... No ☐ Yes ☐
p. Is there anything else you are or were unreasonably terrified to do or be near? .... No ☐ Yes ☐

If yes, specify. ____

____

2. If any of the above were answered yes, how old were you when it/they first started and stopped? Age started ______ Age stopped ______ Still present ☐

3. If any of the above are yes, does attempting to avoid these situations interfere with your life, for example, keep you from going places or doing things you would otherwise do?

No ☐ Yes, but only very minimal interference ☐
Yes, sometimes or many times interfere a lot ☐

4. If any of the above were answered yes, do you feel these are unreasonable fears?

All of my fears are reasonable ☐
I recognize that most of my fears are unreasonable ☐

5. If there is anything else about phobias you wish to describe, please elaborate:_____

____
____
____
____

Panic Attacks

1. Have you ever had a spell or attack (not due to a physical illness) when all of a sudden you felt frightened, anxious, or very uneasy in situation where most people would not be afraid?
   No ☐ Yes ☐

2. If yes, how many such attacks have you had in your life? ______

3. If yes, during one of your worst spells, which of the following problems were present?

a.	Trouble catching your breath?	No ☐	Yes ☐
b.	Pounding of your heart?	No ☐	Yes ☐
c.	Dizziness or light-headed?	No ☐	Yes ☐
d.	Tingling of your hands or feet?	No ☐	Yes ☐
e.	Tightness in your chest?	No ☐	Yes ☐
f.	Feeling of choking or smothering?	No ☐	Yes ☐
g.	Feeling faint or choking	No ☐	Yes ☐
h.	Excessive sweating?	No ☐	Yes ☐
i.	Trembling or shaking?	No ☐	Yes ☐
j.	Hot or cold flashes?	No ☐	Yes ☐
k.	Things seeming unreal?	No ☐	Yes ☐
l.	Afraid you might die or act in a crazy way?	No ☐	Yes ☐

4. If yes, at what age did these attacks start and what age did they stop?
   Age started ______ Age stopped ______ Still present ☐

5. If yes, have you ever had 3 spells like these close together, that is, within a 3-week period?
   No ☐ Yes ☐

6. If yes, and if you also had some of the phobias listed in above, have the phobias and the panic attacks sometimes occurred together (i.e., at the same time)?
   No ☐ Yes ☐

7. Did you ever have periods of time when you were afraid to go out of the house (agoraphobia)?
   No ☐ Yes ☐

8. If yes, Age started ______ Age stopped ______ Still present ☐

9. If there is anything else you want to say about your panic attacks, please elaborate.

____
____
____
____
____
____
____
____
____
____
____

General Anxiety

THE ABOVE QUESTIONS REFERRED TO BRIEF, SELF-CONTAINED EPISODES OF ANXIETY OR PANIC. THE FOLLOWING REFER TO MORE LONG-TERM FEELINGS OF ANXIETY

1. Have you ever had periods of excessive anxiety or worry about various things in your life?
No ☐ Yes ☐

2. If yes, have these feelings persisted for a period of 6 months or more when they were present more than they were absent? No ☐ Yes ☐

3. If yes, Age started ______ Age stopped ______ Still present ☐

4. If yes, during these episodes which of the following problems were present?

a.	Trembling, twitching, or feeling shaky?	No ☐	Yes ☐
b.	Muscle tension, aches, or soreness?	No ☐	Yes ☐
c.	Feeling restless?	No ☐	Yes ☐
d.	Becoming tired easily?	No ☐	Yes ☐
e.	Shortness of breath or smothering sensation?	No ☐	Yes ☐
f.	Palpitations or heart beating rapidly?	No ☐	Yes ☐
g.	Sweating or cold clammy hands?	No ☐	Yes ☐
h.	Dry mouth?	No ☐	Yes ☐
i.	Dizziness or light headed?	No ☐	Yes ☐
j.	Nausea, diarrhea, or other abdominal distress?	No ☐	Yes ☐
k.	Flushes, or hot flashes, chills?	No ☐	Yes ☐
l.	Frequent urination?	No ☐	Yes ☐
m.	Trouble swallowing or a lump in your throat?	No ☐	Yes ☐
n.	Feeling keyed up or on edge?	No ☐	Yes ☐
o.	Overreacting to noises?	No ☐	Yes ☐
p.	Difficulty concentrating or mind going blank?	No ☐	Yes ☐
q.	Trouble falling or staying asleep?	No ☐	Yes ☐
r.	Irritable?	No ☐	Yes ☐

5. If there is anything else you want to say about your feelings of anxiety, please elaborate.____

____
____
____
____
____
____
____
____
____
____

Obsessions

1. Have you ever been bothered by having certain unpleasant thoughts all the time. An example would be the persistent idea that you might harm or kill someone you loved, even though you really didn't want to. Have you ever been bothered by that or by any other unpleasant and persistent thought?
   No ☐ Yes ☐
   2. If yes, please describe.____
   ____
   ____
   3. If yes, was this only for a short time or was it over a period of several weeks?
   Less than 3 weeks ☐ Three weeks or more ☐
   4. If yes, did these thoughts keep coming into your mind no matter how hard you tried to get rid of them?
   No ☐ Yes ☐
5. Other thoughts that keep bothering some people, even when they know they are silly, are that their hands are dirty or have germs on them, no matter how much they wash them, or that relatives who are away have been hurt or killed. Have you ever had any kind of unreasonable thought like that?
   No ☐ Yes ☐
   6. If yes, please describe.____
   ____
   ____
   7. If yes, was this only for a short time or did these thoughts keep coming into your mind over a period of several weeks?
   Less than 3 weeks ☐ Three weeks or more ☐
   8. If yes, did these thoughts keep coming into your mind no matter how hard you tried to get rid of them?
   No ☐ Yes ☐
9. Do you have a tendency to get "hooked" or fixated on one topic?
   No ☐ Occasionally ☐ Often ☐
10. Is it hard to relax because of unwanted thoughts that come into your mind and won't go away?
    No ☐ Occasionally ☐ Often ☐
11. Do you worry about little things?
    No ☐ Occasionally ☐ Often ☐
12. Do you have strong impulses toward doing forbidden or dangerous things?
    No ☐ Occasionally ☐ Often ☐
13. Do you have impulses to hurt yourself or other people?
    No ☐ Occasionally ☐ Often ☐
14. Do dirty words or thoughts come into your head when you are thinking about other things?
    No ☐ Occasionally ☐ Often ☐
15. Do bloody or violent scenes pop into your head when you are thinking about other things?
    No ☐ Occasionally ☐ Often ☐
16. If there is anything else you want to say about obsessive thoughts, please elaborate.
    ____
    ____
    ____
    ____
    ____
    ____

Compulsions and Other Activities

Do you, or did you do, any the following things in a compulsive manner, that is, as the result of an uncontrollable need to do them? If yes, enter the ages when they started and stopped, or mark if still present.

1. Echolalia (repeating over and over words that others have said)? No ☐ Yes ☐
   If yes, Age started ______ Age stopped _____Still present ☐
2. Parilalia (repeating over and over words that you have said)? No ☐ Yes ☐
   If yes, Age started ______ Age stopped _____Still present ☐
3. Perseveration (asking the same question or repeating the same thought over and over) No ☐ Yes ☐
   If yes, Age started ______ Age stopped ______Still present ☐
4. Shouting inappropriately? No ☐ Yes ☐
   If yes, Age started ______ Age stopped ______Still present ☐
5. Touching objects excessively? No ☐ Yes ☐
   If yes, Age started ______ Age stopped _____Still present ☐
6. Touching things a specific number of times, like 2 tor 4 times but not 3 times? No ☐ Yes ☐
   If yes, Age started ______ Age stopped _____Still present ☐
7. Needing to "even up," that is if you touch something with one hand do you have to also touch it with the other hand? No ☐ Yes ☐
   If yes, Age started ______ Age stopped _____Still present ☐
8. Touching other people excessively (without sexual intent)? No ☐ Yes ☐
   If yes, Age started ______ Age stopped _____Still present ☐
9. Touching your crotch excessively? No ☐ Yes ☐
   If yes, Age started ______ Age stopped ______Still present ☐
10. Touching others sexually (breasts, buttock or genitalia)? No ☐ Yes ☐
    If yes, Age started ______ Age stopped _____Still present ☐
11. Biting, picking, scratching or hurting yourself in some way? No ☐ Yes ☐
    If yes, Age started ______ Age stopped _____Still present ☐
12. Head banging? No ☐ Yes ☐
    If yes, Age started ______ Age stopped _____Still present ☐
13. Constant rocking in crib or elsewhere? No ☐ Yes ☐
    If yes, Age started ______ Age stopped ______Still present ☐
14. Mimicking physical actions of others? No ☐ Yes ☐
    If yes, Age started ______ Age stopped _____Still present ☐
15. Count things in your environment like tiles on the floor? No ☐ Yes ☐
    If yes, Age started ______ Age stopped _____Still present ☐
16. Have to step on cracks or avoid stepping on cracks? No ☐ Yes ☐
    If yes, Age started ______ Age stopped ______Still present ☐
17. Check and recheck things like the stove or if the door is locked? No ☐ Yes ☐
    If yes, Age started ______ Age stopped ______Still present ☐
18. Did you ever bite your nails?
    No ☐ Occasionally ☐ Often ☐
    If yes, Age started ______ Age stopped ____Still present ☐
19. Did you ever crack your knuckles?
    No ☐ Occasionally ☐ Often ☐
    If yes, Age started ______ Age stopped ____Still present ☐

20. Before you go to bed at night do you have to do certain things in a certain order, such as brushing your teeth, or washing your face
No ☐ Occasionally ☐ Often ☐
If yes, Age started ______ Age stopped ____ Still present ☐

21. Do you have to have personal belongings arranged in a certain specific way?
No ☐ Occasionally ☐ Often ☐
If yes, Age started ______ Age stopped ____ Still present ☐

22. Have you, or have you ever been told you have a problem with lying, almost in a compulsive way and about things that were not really that important?
No ☐ Occasionally ☐ Often ☐
If yes, Age started ______ Age stopped ____ Still present ☐

Eating

1. Have you ever had a period of your life when you could not maintain your weight at the minimum of what was normal for your age — that is, had or were close to having anorexia nervosa?
No ☐ Yes ☐ Ages________________
2. If yes, what was the minimum weight you obtained? ______

3. Did you ever have an intense fear of getting fat even though you were underweight at the time?
No ☐ Yes ☐ Ages________________

4. Was there ever a time when you had episodes of binge eating (rapid consumption of large amounts of food in a short period of time)?
No ☐ Yes ☐ Ages________________
5. If yes, did you ever have a period of two binge eating episodes a week for at least three months? No ☐ Yes ☐ Ages________________

6. Was there ever a time when you regularly engaged in self-induced vomiting, the use of laxatives, or water pills, or vigorous exercise in order to prevent weight gain?
No ☐ Yes ☐ Ages________________

7. Was there ever a period of at least one month when you repeatedly ate things of no nutritive value (paper, plaster, cloth, pebbles, dirt, etc.)?
No ☐ Yes ☐ Ages________________

8. Was there ever a time when you were a very picky eater and hated most foods?
No ☐ Yes ☐ Ages________________

9. Do you have a craving for any of the following

	No Craving	Moderate Craving	Strong Craving
a. Sugar	☐	☐	☐
b. Chocolate	☐	☐	☐
c. Sweets or carbohydrates	☐	☐	☐
d. Other, Specify________________	☐	☐	☐

10. What is the maximum weight you ever attained? ______ pounds
11. How long were you at or close to this weight? ______ years ______ months

12. Did you ever consider yourself to be overweight?........................No ☐ Yes ☐

13. Did you ever consider yourself a compulsive eater?......................No ☐ Yes ☐

14. Do you drink excessive amounts of water, or other liquids or eat ice?
No ☐ Somewhat ☐ Definitely ☐
15. If yes, please elaborate ____
____

16. How many glasses of water or other liquids do you drink each day? ______

Sexual Behavior

1. Did you ever sexually exhibit yourself by removing part or all of your clothing in public?
   Never ☐ 1 Once ☐ 2 2-5 times ☐ 3 Often ☐ 4
   2. If yes, Age started ____ Age stopped ____ Still present ☐
3. Did you ever have the urge to exhibit yourself, even if you did not do so?
   Never ☐ 1 Once ☐ 2 2-5 times ☐ 3 Often ☐ 4
4. Do you or your parents think that you had a precocious or very early interest in sexual things?
   No ☐ Yes ☐
5. As a child did you draw dirty pictures or write dirty words on things much more than other children your age? No ☐ Yes ☐
   6. If yes, Age started ____ Age stopped ____ Still present ☐
7. Do you think that your sex drive is:
   Less than average ☐ Average ☐ Greater than average ☐ Much greater than average ☐
8. Do you think you have more recurrent thoughts about sex than others your age? No ☐ Yes ☐
9. Do you sexually prefer: The opposite sex ☐ 1 The same sex ☐ 2 Both ☐ 3
10. Did you ever persistently feel, for a period of two years or more, that you were born the wrong sex? No ☐ Yes ☐
11. Did you ever dress up as someone of the opposite sex, other than for Halloween or for a costume party? No ☐ Yes ☐
12. Have you ever had a period of over 6 months when you found yourself sexually aroused by certain objects, such as clothing of the opposite sex? No ☐ Yes ☐
13. Have you ever had a period of over 6 months when you found yourself sexually aroused by a child less than 13 years of age? No ☐ Yes ☐
14. Have you ever had a period, lasting for at least 6 months, when you had recurrent intense sexual urges or arousing fantasies involving being humiliated, beaten, bound, or otherwise made to suffer? No ☐ Yes ☐
15. Have you ever had a period, lasting for at least 6 months, when you had recurrent intense sexual urges or arousing fantasies involving causing others to be sexually or psychologically hurt or humiliated? No ☐ Yes ☐
16. Have you ever had a period of 6 months or more when you had an aversion to being touched, whether sexually or just affectionately? No ☐ Yes ☐
17. Have you ever had a period of 6 months or more when you had an aversion to having any sexual contact with a sexual partner? No ☐ Yes ☐
18. Were either you or anyone in your family ever sexually abused or molested? No ☐ Yes ☐
    If yes, explain or describe.____

____

____

____

Smoking

1. Have you ever smoked cigarettes, cigars or a pipe daily for more than a month or more? No ☐ Yes ☐

2. How old were you when you first smoked daily? _____

3. If you smoke cigarettes now, how many packs a day do you smoke?_____

Gambling

1. Have you ever gambled? No ☐ Yes ☐

   If yes please answer the following:

   a. Have you gambled more than 5 times? No ☐ Yes ☐

   b. Have you ever lost more money than you could afford or intended to? No ☐ Yes ☐

   c. Have you ever needed to bet more and more to achieve the desired excitement? No ☐ Yes ☐

   d. Have you ever felt restless or irritable because you were unable to gamble? No ☐ Yes ☐

   e. After repeatedly losing money, have you ever returned to gambling to win back your losses? No ☐ Yes ☐

   f. Have you repeatedly tried to stop gambling? No ☐ Yes ☐

   g. Have you been gambling when you should be working? No ☐ Yes ☐

   h. Have you ever sacrificed some important social or other obligation in order to gamble? No ☐ Yes ☐

   i. Have you ever continued to gamble despite an inability to pay mounting debts or legal problems? No ☐ Yes ☐

Shopping

1. Have you ever bought more items than you really needed to buy to meet your needs?
   No ☐ Occasionally ☐ Often ☐

2. Have you ever gotten into financial trouble because of buying more things than you could afford? No ☐ Occasionally ☐ Often ☐

3. Have you ever run up a total balance on all your credit cards that was greater than your net monthly income? No ☐ Yes ☐

4. Have you ever shopped to fill a feeling of emptiness ?
   No ☐ Occasionally ☐ Often ☐

5. Do you ever shop to get a high or feeling of happiness?
   No ☐ Occasionally ☐ Often ☐

6. Have you ever taken things without paying for them?
   No ☐ Occasionally ☐ Often ☐

7. If yes, were they things you needed, or things you didn't really need?
   Really needed ☐ Didn't really need ☐

8. If yes, were you ever caught shoplifting? No ☐ Yes ☐

9. If yes, how many times? ________

Work

1. Some people get so involved in their work they are called "workaholics." Would you describe your relation to work as: Slightly involved ☐ I do the work but don't get overly involved ☐
   I would qualify as a full fledged workaholic ☐

Alcohol

1. Do you feel that you are a normal drinker (that is, drink no more than average?) No ☐ Yes ☐
2. Do close relatives ever worry or complain about your drinking? No ☐ Yes ☐
3. Have you ever felt guilty about drinking? No ☐ Yes ☐
4. Are you always able to stop drinking when you want to? No ☐ Yes ☐
5. Have you ever gone on binges or benders where you kept drinking for a couple of days or more without sobering up? No ☐ Yes ☐
   6. If yes, did you neglect some of your responsibilities on two or more of these occasions? No ☐ Yes ☐
7. Have you ever tried to limit or control your drinking by adopting rules, such as avoiding whiskey, or not drinking alone or in bars, or other similar rules? No ☐ Yes ☐
8. Has your drinking ever created problems between you and your wife, husband, parent. or near relative? No ☐ Yes ☐
9. Do you ever drink in the morning? No ☐ Yes ☐
10. Have you ever felt the need to cut down on your drinking? No ☐ Yes ☐
11. Have you ever been told by a doctor to stop drinking? No ☐ Yes ☐
12. Have you ever gotten into physical fights when drinking? No ☐ Yes ☐
13. Was drinking ever part of the problem that resulted in your being hospitalized in a psychiatric or general hospital? No ☐ Yes ☐
14. Have you ever had liver disease, cirrhosis, or yellow jaundice that a doctor said was probably due to your drinking? No ☐ Yes ☐
15. Have you ever gotten into trouble driving an automobile after drinking, like having an accident or being arrested for driving under the influence? No ☐ Yes ☐
    16. If yes, how many times________
17. Have you ever had trouble abstaining entirely from drinking alcohol? For instance, have you ever planned to stop completely or said that you were going to stop, then failed to stick to your original plan? No ☐ Yes ☐
18. Have you ever had any treatment for drinking including joining Alcoholics Anonymous? No ☐ Yes ☐
19. If you have had problems with alcohol abuse, at what age did this start?
    Age started ____ Age stopped ____ Still present ☐
20. Do you feel that any members of your family are alcoholic or have problems with alcohol abuse? No ☐ Yes ☐
    21. If yes, please identify who they are, how they are related to you, and describe the problems.____

____
____
____
____
____

Drugs

1. Have you ever used any of the following drugs to get high, or without a prescription, or more than was prescribed, that is on your own?

a. Marijuana, hashish, pot or grass :
No ☐ Yes ☐ Age started _____ Age stopped _____ Still use ☐

b. Amphetamines, stimulants, uppers, speed:
No ☐ Yes ☐ Age started _____ Age stopped _____ Still use ☐

c. Barbiturates, sedatives, downers, Seconal, Quaaludes:
No ☐ Yes ☐ Age started _____ Age stopped _____ Still use ☐

d. Tranquilizers, Valium, Librium:
No ☐ Yes ☐ Age started _____ Age stopped _____ Still use ☐

e. Cocaine, coke, crack:
No ☐ Yes ☐ Age started _____ Age stopped _____ Still use ☐

f. Heroin:
No ☐ Yes ☐ Age started _____ Age stopped _____ Still use ☐

g. Opiates (other than heroin) codeine, Demerol, morphine, Methadone, Darvon, opium:
No ☐ Yes ☐ Age started _____ Age stopped _____ Still use ☐

h. Psychedelics, LSD, mescaline, peyote, psilocybin, DMT, PCP:
No ☐ Yes ☐ Age started _____ Age stopped _____ Still use ☐

2. Have you ever used any of these drugs or any other illicit drug every day for two weeks or more? No ☐ Yes ☐

3. Have you ever used any of these drugs or other illicit drug enough so that you felt like you needed it or were dependent on it? No ☐ Yes ☐

4. Have you ever tried to cut down on any of these drugs but found you couldn't do it? No ☐ Yes ☐

5. Did you ever find you needed larger amounts of these drugs to get an effect, or that you could no longer get high on the amount you used to use? No ☐ Yes ☐

6. Have you ever had withdrawal symptoms, that is, have you you felt sick because you stopped or cut down on any of these drugs? No ☐ Yes ☐

7. Did you ever have any health problems like fits, an accidental overdose, a persistent cough or an infection as a result of using any of these drugs? No ☐ Yes ☐

8. Did any of these drugs cause you considerable problems with your family, friends, on the job, at school or with the police? No ☐ Yes ☐

9. Did you have any emotional or psychological problems from using drugs, such as feeling crazy, paranoid, or depressed or uninterested in things? No ☐ Yes ☐

10. If any of the questions from #2 to #9 were answered yes, during what period of your life did you have these troubles? Age started _____ Age stopped _____ Still present ☐

11. Were you ever arrested because of drug use or selling drugs? No ☐ Yes ☐

12. If yes, how many times?______

13. If yes, describe the circumstances.____

____

Coffee

1. How many cups of coffee do you drink per day? ________

Mood

1. In your lifetime, have you ever had two weeks or more during which you felt sad, blue, depressed, or when you lost all interest and pleasure in things that you usually cared about or enjoyed?
   No ☐ Yes ☐
2. Have you ever had two years or more in your lifetime when you felt depressed or sad almost all the time, even if you felt OK sometimes? No ☐ Yes ☐
3. Has there ever been a period of two weeks or longer when you lost your appetite?
   No ☐ Yes ☐ Yes, but it was due to a physical illness ☐
4. Have you ever lost weight without trying to — as much as two pounds a week for several weeks?
   No ☐ Yes ☐ Yes, but it was due to a physical illness ☐
5. Have you ever had a period when your eating increased so much that you gained as much as two pounds a week for several weeks? No ☐ Yes ☐
6. Have you ever had a period of two weeks or more when you had trouble falling asleep, staying asleep, or waking up too early? No ☐ Yes ☐
7. Have you ever had a period of two weeks or longer when you were sleeping too much?
   No ☐ Yes ☐ Yes, but it was due to a physical illness ☐
8. Has there ever been a period lasting two weeks or more when you felt tired out all the time?
   No ☐ Yes ☐ Yes, but it was due to a physical illness ☐
9. Has there ever been a period of two weeks or more when you talked or moved more slowly than is normal for you? No ☐ Yes ☐
10. Has there ever been a period of two weeks or more when you had to be moving all the time — that is, you couldn't sit still and paced up and down? No ☐ Yes ☐
11. Has there ever been a period of several weeks when your interest in sex was a lot less than usual? No ☐ Yes ☐
12. Has there ever been a period of two weeks or more when you felt worthless, sinful, or guilty?
    No ☐ Yes ☐
13. Has there ever been a period of two weeks or more when you had a lot more trouble concentrating than is normal for you? No ☐ Yes ☐
14. Has there ever been a period of time when your thoughts came much slower than usual or seemed mixed up? No ☐ Yes ☐ Yes, but it was due to a physical illness ☐
15. Has there ever been a period of two weeks or more when you thought a lot about death — either your own, someone else's, or death in general? No ☐ Yes ☐
16. Has there ever been a period of two weeks or more when you felt like you wanted to die? No ☐ Yes ☐
17. Have you ever felt so low you thought of committing suicide? No ☐ Yes ☐
18. Have you ever attempted suicide? No ☐ Yes ☐
    If yes, explain the circumstances, how you tried, and the number of times. ________

____
____
____
____

19. If you answered yes to several of the above items, how old were you when these feelings or problems first started and stopped? Age started ____ Age stopped ____ Still present ☐

20. If you answered yes to several of the above items, were you only feeling this way after the death of someone close to you?
    I felt this way only during a period of 2 months or less after the death of someone close to me. ☐ 1
    I felt that way at times other than 2 months after someone close to me died ☐ 2

21. If you answered yes to several of the above items, what is the longest period of time that you felt depressed more days than not.
    Less than 1 year ☐ 1-2 years ☐ more than two years ☐

22. If your answered yes to several of the above items, are there times of the year when your depression is much worse No ☐ Yes ☐
    23. If yes, what seasons of the year are the worst. Mark all that apply:
    Summer ☐ Fall ☐ Winter ☐ Spring ☐

24. Has there ever been a period of one week or more when you were so happy or excited or high that you got into trouble, or your family or friends worried about it, or a doctor said you were manic? No ☐ Yes ☐

25. Has there ever been a period of a week or more when you were so much more active than usual that you or your family or friends were concerned about it? No ☐ Yes ☐

26. Has there ever been a period of a week or more when you went on spending sprees — spending so much money that it caused you or your family some financial trouble? No ☐ Yes ☐

27. Has there ever been a period of a week or more when your interest in sex was so much stronger than is typical for you that you wanted to have sex a lot more frequently than is normal for you or with people you normally wouldn't be interested in? No ☐ Yes ☐

28. Has there ever been a period of a week or more when you talked so fast that people said they couldn't understand you? No ☐ Yes ☐

29. Have you ever had a period of a week or more when thoughts raced through your head so fast that you couldn't keep track of them? No ☐ Yes ☐

30. Have you ever had a period of a week or more when you felt that you had a special gift or special powers to do things others couldn't do or that you were an especially important person? No ☐ Yes ☐

31. Has there ever been a period of a week or more when you hardly slept at all but still didn't feel tired or sleepy? No ☐ Yes ☐

32. Was there ever a period of a week or more when you were easily distracted so that any little interruption could get you off the track? No ☐ Yes ☐

33. If several of the above questions from #21 to #29 were answered yes, how old were you when they started and stopped? Age started ____ Age stopped ____ Still present ☐

34. If several of the above questions from #21 to #29 were answered yes, how may total episodes have occurred in your life that lasted more than one week?
    Less than two ☐ 1 Five to ten ☐ 3
    Two to five ☐ 2 More than ten ☐ 4

35. Have you had some periods in your life when you felt very down and blue and depressed, but other times when you felt much more up, excited and active than is normal for most people? No ☐ Yes ☐
    36. If yes, how old were you when these mood swings first began?
    Age started ____ Age stopped ____ Still present ☐

37. Have you ever believed people were watching you or spying on you?
No ☐ Yes ☐
38. If yes, please elaborate ______________________________

39. Was there ever a time when you believed people were following you?
No ☐ Yes ☐
40. If yes, please elaborate.______________________________

41. Have you ever believed that someone was plotting against you or trying to hurt you?
No ☐ Yes ☐
42. If yes, please elaborate.______________________________

43.Have you ever believed someone was reading your mind? No ☐ Yes ☐
44. If yes, please elaborate.______________________________

45. Have you ever believed you could actually hear what another person was thinking, even though he was not speaking, or believed that others could hear your thoughts?
No ☐ Yes ☐
46. If yes, please elaborate.______________________________

47. Have you ever believed that others were controlling how you moved or what you thought against your will? No ☐ Yes ☐
48. If yes, please elaborate.______________________________

49. Have you ever felt that someone or something could put strange thoughts directly into your head or could take or steal your thoughts out of your mind? No ☐ Yes ☐
50. If yes, please elaborate.______________________________

51. Have you ever believed that you were being sent special messages through television or the radio? No ☐ Yes ☐
52. If yes, please elaborate.______________________________

53. Have you ever had the experience of seeing something or someone that others who were present could not see — that is, a vision when you were completely awake?
No ☐ Yes ☐
54. If yes, please elaborate.______________________________

55. Have you, more than once, had the experience of hearing things other people couldn't hear, such as a voice? No ☐ Yes ☐
56. If yes, please elaborate.______________________________

57. If yes, did these voices ever tell you to do bad things or things you wouldn't normally have done? No ☐ Yes ☐
58. If yes, did you ever carry on conversations with the voice?
No ☐ Yes ☐
59. If yes, were you able to identify the voice?
Couldn't identify ☐ 1 Was the devil ☐ 2
Was a male voice ☐ 3 Was a female voice ☐ 4
There was a good voice and a bad voice ☐ 5
Other ☐ 6 Describe ______________________________

60. If any of the question above from #32 to #53 were answered yes, how old were you when they first started and stopped? Age started ____ Age stopped ____ Still present ☐

61. Do you often have dreams or uncomfortable thoughts about violence, that is, someone harming you or you harming someone or something else?

Never or rarely ☐ Moderately often ☐ Often ☐

62. If yes, please elaborate.______________________________

____

63. Have you ever had the experience of feeling detached from your own body as if you were an observer of your own body or thoughts, or like you were a robot in a dream but awake?

Never or rarely ☐ Moderately often ☐ Often ☐

64. Have you ever had the feeling that you were doing something that you had already experienced doing before, that is, *deja vu.*

Never or rarely ☐ Moderately often ☐ Often ☐

Present Feelings About Yourself

IN THE FOLLOWING PART THERE ARE GROUPS OF STATEMENTS. PLEASE READ THE ENTIRE GROUP FOR EACH QUESTION. THEN PICK THE ONE STATEMENT IN THAT GROUP THAT BEST DESCRIBES THE WAY YOU FEEL MOST OF THE TIME. IF SEVERAL STATEMENTS IN A GROUP SEEM TO APPLY EQUALLY WELL, MARK EACH ONE.

1. ☐4 I am so sad or unhappy that I can't stand it.
   ☐3 I am blue or sad all the time and I can't snap out of it.
   ☐2 I feel sad or blue.
   ☐1 I do not feel sad.

2. ☐4 I feel that the future is hopeless and that things cannot improve.
   ☐3 I feel I have nothing to look forward to.
   ☐2 I feel discouraged about the future.
   ☐1 I am not particularly pessimistic or discouraged about the future.

3. ☐4 I am a complete failure as a person.
   ☐3 As I look back on my life, all I can see is a lot of failures.
   ☐2 I feel I have failed more than the average person.
   ☐1 I do not feel like a failure.

4. ☐4 I am dissatisfied with everything.
   ☐3 I don't get satisfaction out of anything anymore.
   ☐2 I don't enjoy things the way I used to.
   ☐1 I am not particularly dissatisfied.

5. ☐4 I feel as though I am very bad or worthless.
   ☐3 I feel quite guilty.
   ☐2 I feel bad or unworthy a good part of the time.
   ☐1 I don't feel particularly guilty.

6. ☐4 I hate myself.
   ☐3 I am disgusted with myself.
   ☐2 I am disappointed in myself.
   ☐1 I don't feel disappointed in myself.

7. ☐4 I would kill myself if I had a chance.
   ☐3 I have definite plans about committing suicide.
   ☐2 I feel I would be better off dead.
   ☐1 I don't have any thought of harming myself.

8. ☐4 I have lost all my interest in other people and don't care about them at all.
☐3 I have lost most of my interest in other people and have little feeling for them.
☐2 I am less interested in other people than I used to be.
☐1 I have not lost interest in other people.

9. ☐4 I can't make any decisions at all anymore.
☐3 I have great difficulty in making decisions.
☐2 I try to put off making decisions.
☐1 I make decisions about as well as ever.

10. ☐4 I feel that I am ugly or repulsive-looking.
☐3 I feel that there are permanent changes in my life.
☐2 I am worried that I look old or unattractive.
☐1 I don't feel that I look any worse than I used to.

11. ☐4 I can't do any work at all.
☐3 I have to push myself very hard to do anything.
☐2 It takes extra effort to get started at doing something.
☐1 I can work about as well as before.

12. ☐4 I get too tired to do anything.
☐3 I get tired from doing anything.
☐2 I get tired more easily than I used to.
☐1 I don't get any more tired than usual.

13. ☐4 I have no appetite at all anymore.
☐3 My appetite is much worse now.
☐2 My appetite is not a good as it used to be.
☐1 My appetite is no worse than usual.

Past Diagnoses

1. Have you ever seen a mental health professional for emotional or psychological problems?
No ☐ Yes ☐
2. If yes, please elaborate and give your age when you had the problems.________

____

____

3. Have you ever been hospitalized for psychiatric reasons? No ☐ Yes ☐
4. If yes, please elaborate and give your age when you were hospitalized. ________

____

____

4. Have you even been told you had any of the following problems?

☐ 1 Agoraphobia
☐ 2 Alcoholism
☐ 3 Anxiety
☐ 4 Aphasia (delayed speech)
☐ 5 Autism
☐ 6 Bulimia or anorexia nervosa
☐ 7 Depression
☐ 8 Drug abuse or addiction
☐ 9 Manic-depression
☐ 10 Obsessive-compulsive
☐ 11 Panic or hyperventilation
☐ 12 Schizophrenia
☐ 13 Other

5. If any of the above were marked yes, describe any details you wish such as date, doctor, and circumstances.______________________________________

____

General Health

PLEASE READ OVER THE FOLLOWING QUESTONS ABOUT GENERAL HEALTH. IF YOU MARK ANY YES, ENTER THE AGES WHEN YOU HAD THESE SYMTPOMS.

1. Migraine headaches? No ☐ Yes ☐
   If yes, how often? Once a month or less ☐ 1-3 times a month ☐ 1-3 times a week ☐ Almost daily ☐
2. Chronic back pain? No ☐ Yes ☐
3. Pain when you urinate? No ☐ Yes ☐
4. Unable to urinate for 24 hours or longer (not due to surgery)? No ☐ Yes ☐
5. Rapid heartbeat or palpitations? No ☐ Yes ☐
6. Chest pains? No ☐ Yes ☐
7. Short of breath when not exerting yourself? No ☐ Yes ☐
8. Trouble swallowing? No ☐ Yes ☐
9. Had a feeling there was a lump in your throat? No ☐ Yes ☐
10. Trouble with excessive gas or bloating? No ☐ Yes ☐
11. Trouble with loose stools or diarrhea other than acute illness? No ☐ Yes ☐
12. Trouble with constipation? No ☐ Yes ☐
13. Trouble with peptic or duodenal ulcer? No ☐ Yes ☐
14. Trouble with nausea (unrelated to pregnancy or car sickness)? No ☐ Yes ☐
15. Unable to tolerate certain foods? No ☐ Yes ☐
16. Periods of vomiting other than pregnancy? No ☐ Yes ☐
17. Have you ever been diagnosed as having a spastic colon? No ☐ Yes ☐
18. Trouble with burning pains of your arms or legs? No ☐ Yes ☐
19. Do you have chronic pains in your joints? No ☐ Yes ☐
20. Have you ever been paralyzed, unable to move some part of your body? No ☐ Yes ☐
21. Have a burning sensation in your sexual organs or rectum (other than during intercourse)? No ☐ Yes ☐
22. Pain during intercourse? No ☐ Yes ☐
23. Have you ever lost your voice for 30 minutes or more? No ☐ Yes ☐
24. Periods of dizziness? No ☐ Yes ☐
25. Problems with seeing double? No ☐ Yes ☐
26. Has your vision ever become blurred for some period, when it wasn't just due to needing glasses? No ☐ Yes ☐
27. Have you ever been blind in one or both eyes where you couldn't see anything at all for a few minutes or more? No ☐ Yes ☐
28. Periods of being unable to hear or deafness (not permanent)? No ☐ Yes ☐
29. Periods of fainting or loss of consciousness? No ☐ Yes ☐
30. Have you ever had a seizure or convulsion of any kind since you were 12, where you were unconscious and your body jerked? No ☐ Yes ☐
31. Have you ever had a period of amnesia, lasting for several hours or days when you couldn't remember anything afterwards? No ☐ Yes ☐
32. Have you ever had a period of a strange feeling or spell when objects seemed much larger or smaller than they usually are? No ☐ Yes ☐
    If yes, Age started ____ Age stopped ____ Still present ☐
    If yes, please elaborate.____________________
33. Have you ever had to give up work, going to school, or other regular activities for at least several weeks because you did not feel well enough to carry on (other than when you were in the hospital)?
    No ☐ Yes ☐ Age started ____ Age stopped ____ Still present ☐

34. Has your physical health been pretty good or have you been sickly for the majority of your life?
Pretty good most of my life .... ☐ 1
Single long-term physical illness explained being sickly... ☐ 2
Sickly most of my life.. ☐ 3

36. Do you often feel hotter than others around you?
No☐ Somewhat ☐ Very much so ☐

37. Do you have any unusual body odors. No☐ Somewhat ☐ Very much so ☐
38. If yes, please elaborate.____

____

39. If you feel any of the above answers need an explanation, please elaborate.______________

____

____

____

THE FOLLOWING QUESTIONS APPLY ONLY IF YOU ARE A WOMAN.

40. Do you have an unusual amount of depression, irritability, or tearfulness during the week or so before your menstrual period? No ☐ Yes ☐
41. Do you have a lot of trouble with painful periods? No ☐ Yes ☐
42. Other than your first year of menstruation, have you ever missed two periods in a row and it was not due to pregnancy or nursing? No ☐ Yes ☐
43. Periods of excessive menstrual bleeding? No ☐ Yes ☐
44. Problems with vomiting throughout pregnancy? No ☐ Yes ☐

THE FOLLOWING QUESTION APPLIES ONY IF YOU ARE A MAN.

45. Have you had periods of time when you were unable to have an erection?
No ☐ Yes ☐

Tics

Muscle tics are involuntary, jerky, muscle movements such as excessive eyeblinking, facial grimaces, jerking the head to one side or up and down, eyes turning of out or upward, shoulder shrugging, arm, or leg jerking, or others.

Vocal noises are involuntary noises including excessive throat clearing, grunting, snorting, barking, spitting, sniffing, squeaking, or other non-word noises.

1. Have you ever had muscle tics? No ☐ Yes ☐
2. Have you ever had vocal noises? No ☐ Yes ☐

IF **BOTH** OF THE ABOVE TWO QUESTIONS ARE ANSWERED NO, THEN SKIP TO THE LAST PAGE.
IF **EITHER** OF THE ABOVE QUESTIONS WERE ANSWERED YES, PLEASE CONTINUE.

3. If you had muscle tics at what age did they first occur? Age started ____
4. Describe the first tics you had.____

____

5. If you had vocal noises at what age did they first occur? Age started ____
6. Describe the first tics you had.____

____

7. Carefully read the following list and for any tics that you ever experienced, mark in the age at which they were first noticed by you or others and if you no longer have those tics, the age at which they went away.

Motor tics:	Age Started	Age Stopped	Still Present
Eyeblinking	______	______	_____
Eyes looking up or sideways	______	______	_____
Facial grimacing	______	______	_____
Head tic (hair out of eyes tic)	______	______	_____
Arm tic	______	______	_____
Diaphragm tic	______	______	_____
Leg or foot tic	______	______	_____
Other tics	______	______	_____

Describe other tics ____

____

____

Vocal tics:			
Repeated throat clearing	______	______	_____
Grunting	______	______	_____
Barking	______	______	_____
Spitting.	______	______	_____
Squeaking	______	______	_____
Sniffing	______	______	_____
Yelling-screaming	______	______	_____
Stopping in mid-sentence	______	______	_____
Other vocal tics	______	______	_____

Describe other tics.____

____

____

8. If you marked yes to any of the above motor or vocal tics, have they been present almost every day for more than a year?..No ☐ Yes ☐

9. Since your tics began list how old you were when they were only mild, moderate, or severe? (If they were never severe, list only for mild or moderate).

My tics were mild	From age ____	to age ____
My tics were moderate	From age ____	to age ____
My tics were severe	From age ____	to age ____

10. List the most severe tics you have ever had: ______________________________

____

11. List the most severe vocal tics you have ever had: __________________________

____

12. List things that make your tics worse, including medications: __________________

____

____

13. List things that make your tics better, including medications: __________________

____

Narrative

1. If there is anything else about yourself you would like to mention, please elaborate.________

____
____
____
____
____
____
____
____
____

2. Sometimes people will deny certain personality traits, behaviors, habits, or excesses (such as drinking). Is there anyone in your family who you feel you would like to add some notes about their behaviors in this regard? If yes, please note who it is, how they are related to you, and the behaviors that bother you or others.

____
____
____
____
____
____
____
____
____
____
____
____
____
____
____
____
____

For further information on the diagnosis, associated behaviors, cause, genetics and treatment of Tourette Syndrome and the meaning and interpretation of this questionnaire, see *Tourette Syndrome and Human Behavior* by David E. Comings, M.D. Hope Press, Box 188, Duarte, CA 91009-0188. To order use attached form.

Human Behavior Questionnaire-Adult ISBN 1-878267-00-0

lescent boys with attention deficit disorder and 88 normal adolescent boys. Am. J. Psychiatry 139:795-798.
1073a. Satterfield, J.H., Satterfield, B.T. and Cantwell, D.P.(1981). Three-year multimodality treatment study of 100 hyperactive boys. Behav. Pediatr. 98:650-655.
1704. Saul, R.C. and Ashby, C.D.(1986). Measurement of whole blood serotonin as a guide in prescribing psychostimulant medication for children with attentional deficits. Clin. Neuropharmacol. 9:189-195.
1705. Scatton, B., Boireau, A., Garret, C., Glowinski, J. and Julow, L.(1977). Action of palmitic ester of pipotiazine on dopamine metabolism in the nigro-striatal, meso-limbic and meso-cortical sytems. Arch. of Pharmacol. 296:169-175.
1706. Scatton, B., Worms, P., Lloyd, K.G. and Bartholini, G.(1982). Cortical modulation of striatial function. Brain Res. 232:331-343.
1707. Schachar, R., Rutter, M. and Smith, A.(1981). The characteristics of situationally and pervasively hyperactive children: Implication for syndrome definition. J. Child. Psychol. Psychiat. 22:375-392.
1708. Schacter, D.L.(1987). Memory, amnesia, and frontal lobe dysfunction. Psychobiology 15:21-36.
1709. Schain, R.J. and Freedman, D.X.(1961). Studies on 5-hydroxyindole metabolism in autistic and other mentally retarded children. J. Pediatr. 58:315-320.
1710. Schalling, D., Edman, G. and Asberg, M.(1983). Impulsive cognitive style and inability to tolerate boredom. In M. Zuckerman (ed) Biological Bases of Sensation Seeking, Impulsivity and Anxiety. Erlbaum Assoc., New York.
1711. Scharff, D.E. and Scharff, J.S.(1987). Object Relations Family Therapy. Jason Aronson, Inc., New York.
1712. Schaumburg, H., Kaplan, J., Windebank, A., Vick, N., Rasmus, S., Pleasure D. and Brown, J.J.(1983). Sensory neuropathy from pyridoxine abuse. A new megavitamin syndrome. N. Engl. J. Med. 309:445-448.
1715. Schickit, M.A.(1973). Alcoholism and sociopathy —Diagnostic confusion. Q. J. Stud. Alcohol. 34:157-164.
1716. Schickit, M.A., Goodwin, D.W. and Winokur, G.(1972). The half-sibling approach in a genetic study of alcoholism. In M. Roff, L.N. Rolins and M. Pollack (eds) Life History Research in Psychopathology (Vol. 2). University of Minneapolis Press, Minneapolis.
1717. Schildkraut, J.J. and Kety, S.S.(1967). Biogenic amines and emotion. Science 156:21-30.
1718. Schildkraut, J.J.(1965). The catecholamine hypothesis of affective disorders: A review of supporting evidence. Am. J. Psychiatry 122:509-522.
1719. Schildkraut, J.J., Dodge, G.A. and Logue, M.A.(1969). Effects of tricyclic antidepressants on the uptake and metabolism of intracisternally administered norepinephrine-^{3}H in rat brain. J. Psychiatr. Res. 7:29-34.
1720. Schlusinger, F.(1972). Psychopathy: Heredity and environment. Int. J. Ment. Health 1:190-206.
1721. Schmale, H. and Richter, D.(1984). Single base deletion in the vasopressin gene is the cause of diabetes insipidis in Brattleboro rats. Nature 308:705-709.
1722. Schmauss, C. and Emrich, H.M.(1985). Dopamine and the action of opiates: A reevaluation of the dopamine hypothesis of schizophrenia. Biol. Psychiatry 20:1211-1231.
1722a. Schmid, W., Scherer, G., Danesch, U., Zentgraf, H., Matthias, P., Strange, C.M., Rowekamp, W. and Schutz, G. (1982). Isolation and characterization of the rat tryptophan oxygenase gene. EMBO Journal 1:1287-1293.
1723. Schmideberg, M.(1947). The treatment of psychopaths and borderline patients. Am. J. Psychotherapy. 1:45-55.
1723a. Schmidt,A.W. and Peroutka,S.J.(1989). 5-Hydroxytryptamine receptor "families." FASEB J. 3:2242-2249.
1724. Schmitt, B.D.(1975). The minimal brain dysfunction myth. Am. J. Dis. Child. 129:1313-1318.
1725. Schmitt, F.O.(1984). Molecular regulation of brain function: A new view. Neuroscience 13:991-1001.
1726. Schmitt, H., Schmitt, H. and Fenard, S.(1971). Evidence for an a-sympathomimetic component in the effects of catapresan on vasomotor centres: Antagonism by piperoxane. Eur. J. Pharmacol. 14:98-100.
1727. Schneider, K.(1957). Primärie und sekundärie Symptom bei der Schizophrenie. Fortschr. Neurol. Psychiatr. 25:487-490.
1728. Schneider, K.(1959). Clinical Psychopathology (translation by Hamilton). Grune & Straton, Inc, New York.
1728a. Schenider, K.(1974). Primary and secondary symptoms in schizophrenia. In S.R. Hirsch and M. Shepherd (eds) Themes and Variations in Psychiatry. J. Wright, Bristol.
1729. Schoch, P., Richards, J.G., Haring, P., Takacs, B., Stahli, C., Staehelin, T., Haefey, W. and Moher, H.(1985). Co-localization of GABAA receptors and benzodiazepine receptors in the brain shown by monoclonal antibodies. Nature 314:168-171.
1730. Schofield, P.R., Darlism, M.G., Fujita, N., Purt, D.R. [8] (1987). Sequence and functional expression of the GABAA receptor shows a ligand-gated receptor super-family. Nature 328:221-227.
1731. Schooler, C., Zahn, T.P., Murphy, D.L. and Buchsbaum, M.S.(1978). Psychological correlates of monamine oxidase activity in normals. J. Nerv. Ment. Dis. 166:177-186.
1732. Schreiner, L and Kling, A.(1953). Behavioral changes following rheincephalic injury in cat. J. Neurophysiol. 16: 643-659.
1733. Schubert, J.(1973). Effect of chronic lithium on monamine metabolism in rat brain. Psychopharmacology 32:301-311.
1734. Schuster, C.R. and Thompson, T.(1969). Self administration and behavioral dependence on drugs. Ann. Rev. Pharmacol. 9:483-502.
1735. Schwarcz, R., Okuno, E., White, R.J., Bird, E.D. and Whetsell, W.O., Jr.(1988). 3-hydroxyanthranilate oxygenase activity is increased in the brains of Huntington disease victims. Proc. Natl. Acad. Sci. U.S.A. 85:4079-4081.
1736. Schwarcz, R., Whetsell, W.O. Jr., and Mangano, R.M.(1983). Quinolinic acid: An endogenous metabolite that produces axon-sparing lesions in rat brain. Science 219:316-318.
1737. Schwartz, G.E., Davidson, R.J. and Maer, F.(1975). Right hemisphere lateralization for emotion in the human brain: Interactions with cognition. Science 190:286-288.
1738. Schwartz, J.-C.(1979). Opiate receptors on catecholaminergic neurons in brain. Trends Neurosci. 2:137-139.
1738a. Scott, D.H.(1959). Delinquency. Adv. Sci. No 61:497-505.
1739. Scoville, W.B. and Milner, B.(1957). Loss of recent memory after bilateral hippocampal lesoins. J. Neurol. Neurosurg. Psychiatry 20:11-21.
1740. Scoville, W.B.(1954). The limbic lobe in man. J. Neurosurg. 11:64-66.
1741. Sedvall, G.C. and Wode-Helgodt, B.(1980). Aberrant monamine metabolite levels in CSH and family history of schizophrenia. Arch. Gen. Psychiatry 37:1113-1116.
1742. Sedvall, G.C., Fyrö, B., Gullberg, B. et al.(1980). Relationships in healthy volunteers between concentrations of monamine metabolites in cerebrospinal fluid and family his

Bibliography

1. Aaron, M. and Thorne, B.M.(1975). Omission training and extinction in rats with septal damage. Physiol. Behav. 15:149-154.
2. Abbott, D., Holman, S., Berman, M.D., Neff, D. and Goy, R.(1984). Effects of opiate antagonists on hormones and behavior of males and female rhesus monkeys. Arch. Sex. Behav. 13:1-25.
3. Abraham, G.E.(1980). Premenstral tension. Curr. Prob. Obst. Gynecol. Fertil. 3:1-8.
4. Achenbach, T.M.(1978). The child behavioral profile. I. Boys aged 6-11. J. Consult. Clin. Psychol. 46:478-488.
5. Achenbach, T.M.(1979). The child behavioral profile. II. Boys aged 12-16 and girls aged 6-11 and 12-16. J. Consult. Clin. Psychol. 47:223-233.
6. Ackerman, N.W.(1958). The Psychodynamics of Family Life. Basic Books, Inc, New York.
7. Ackerman, P.T., Dykman, R.A. and Peters, J.E.(1977). Teenage status of hyperactive and non-hyperactive learning disabled boys. Am. J. Orthopsychiatry 47:577-596.
8. Agarwal, D.P. and Goedde, H.W.(1979). Blood platelet monamine oxidase activity in schizophrenia, affective disorders and alcoholism. In: Monamine Oxidase: Structure, Function and Altered Functions. Academic Press, New York, pp397-402.
9. Aggleton, J.P. and Mishkin, M.(1986). The amygdala: Sensory gateway to the emotions. In R. Plutchik and H. Kellerman (eds) Emotion Theory, Research, and Experience. Academic Press, Inc., New York, pp281-299.
10. Aghajanian, G.K. and Bunney, B.S.(1974). Pre-and postsynaptic feedback mechanisms in central dopaminergic neurons. In P. Seeman and B.M. Brown (eds) Frontiers in Neurology and Neuroscience Research: First International Symposium of the Neuroscience Institute. University of Toronto, Toronto, pp4-11
11. Aghajanian, G.K. and Wang, R.Y.(1978). Physiology and pharmacology of central serotonergic neurons. In M.A. Lipton, A. DiMascio and K.F.Killam. (eds). Psychopharmacology: A Generation of Progress. Raven Press, New York, pp171-183.
12. Aghajanian, G.K.(1972). Influence of drugs on the firing of sertonin-containing neurons in brain. Fed. Proc. 31:91-96.
13. Aghajanian, G.K.(1981). The modulatory role of serotonin at multiple receptors in brain. In B.L. Jacobs and A. Gelperin (eds) Serotonin Neurotransmission and Behavior. MIT Press, Boston, pp156-185.
14. Aghajanian, G.K., Foote, W.E. and Sheard, M.H.(1970). Action of psychotogenic drugs on midbrain raphe neurons. J. Pharm. Exp. Ther. 171:178-187.
15. Aghajanian, G.K. and Wang, R.Y.(1978). Physiology and pharmacology of central serotonergic neurons. In M.A. Lipton, A. DiMascio and K.F. Killam (eds) Psychopharmacology: A Generation of Progress. Raven Press, New York, pp171-183.
16. Agren, H.(1980). Symptom patterns in unipolar and bipolar depression correlating with nonamine metabolites in the cerebrospinal fluid: II. Suicide. Psychiatr. Res 3:225-236.
17. Agren, H.(1983). Life at risk markers of suicidality in depression. Psychiatr. Dev. 1:87-103.
17a. Alessi, N.E., Walden, M.E. and Hsieh, P.S.(1988). Nifedipine augments haloperidol in the treatment of Tourette syndrome. Pediatr. Neurol. 4:191.
18. Akiskal, H.S.(1981). Subaffective disorders: Dysthymic, cyclothymic, and bipolar II disorders in the 'borderline' realm. Psychiatr. Clin. North Am. 1:25-46.
19. Alajouanine, T., Castaigne, P., Sabouraud, O. et al. (1959). Palalalie paroxystique et vocalisations interatives au cours de crises epileptiques par lesion interessant l'aire motrice supplementaire. Rev. Neurol. 101:685-697.
20. Albert, D.J. and Walsh, M.L.(1982). The inhibitory modulation of agonistic behavior in the rat brain: A review. Neurosci. & Behav. Rev. 6:125-143.
21. Alexander, G.E. and Goldman, P.S.(1978). Functional development of the dorsolateral prefrontal cortex: An analysis utilizing reversible cryogenic depression. Brain Res. 143:233-249.
22. Alhadeff, L., Gaultieri, C.T. and Lipton, M.A.(1984). Toxic effects of water soluble vitamins. Nutr. Rev. 42:33-40.
23. Allegri, G., Costa, C. and DeAntoni, A.(1978). A further contribution to the choice of the dose for L-tryptophan load test. Acta Vitamin Enzymol. (Milano) 32:163-166.
24. Almay, B.G.L., vonKnorring, L. and Oreland, L.(1987). Platelet MAO in patients with idiopathic pain disorders. J. Neural Transmission 69:243-253.
25. Alpert, M. and Friedhoff, A.J.(1982). An un-dopamine hypothesis of schizophrenia. Schizophr. Bull. 6:380-387.
26. Altman, K. and Greengard, O.(1966). Correlation of kynurenine excretion with liver tryptophan pyrrolase levels in disease and after hydrocortisone induction. J. Clin. Invest. 45:1527-1534.
27. Aminoff, M.J., Trenchad, O., Turner, P. , Eood, W.G. and Hills, M.(1974). Plasma uptake of dopamine and 5-hydroxytryptamine and plasma catecholamine levels in patients with Huntington's chorea. Lancet 2:1115-1116.
28. Anden, N.E., Atack, C.V., Svensen, T.H.(1973). Release of dopamine from central noradrenaline and dopam-

ine nerves induced by a dopamine b-hydroxylase inhibitor. J. Neural Transmission 34:93-100.

29. Anden, N.E., Corrodi, H., Fuxe, K., Hökfelt, B., Hökfelt, T., Tydin, C. and Svensson, T.(1970). Evidence for a central noradrenalin receptor stimulation by clonidine. Life Sci. 9:513-523.

30. Anden, N.E., Dahlström,A., Fuxe, K., Larsson, K., Olson, L. and Ungersteadt, U.(1966). Ascending monamine neurons to the telencephalon and diencephalon. Acta Physiol. Scand. 67:313-326.

31. Anders, T.F. and Weinstein, P.(1972). Sleep and its disorders in infants and children: A review. Pediatrics 50:312-324.

32. Andersen, P., Sundberg, S.H., Sveen, O., Swann, J.W. and Wigström, H.(1980). Possible mechanisms for long-lasting potentiation of synaptic transmission in hippocampal slices from guinea-pigs. J. Physiol. 302:463-482.

33. Anderson, G.M., Minderaa, R.B., van Benthem, P.-P.G., Volkman, F.R. and Cohen, D.J.(1984). Platelet imipramine binding in autistic subjects. Psychiatry Res. 11:133-141.

34. Andreasen, N.C. and Olsen, S.(1982). Negative vs. positive schizophrenia. Arch. Gen. Psychiatry 39:789-794.

35. Andreasen, N.C. and Winokur, G.(1979). Newer experimental methods for classifying depression. Arch. Gen. Psychiatry 36:447-452.

36. Andreasen, N.C.(1985). Positive vs. negative schizophrenia: A critical evaluation. Schizophr. Bull. 11:380-389.

37. Andreasen, N.C.(1987). Creativity and mental illness: Prevalence rates in writers and their first-degree relatives. Am. J. Psychiatry 144:1288-1292.

38. Andreasen, N.C., Rice, J., Endicott, J., Coryell, W., Grove, W.M. and Reich, T.(1987). Familial rates of affective disorder. Arch. Gen. Psychiatry 44:461-469.

38a. Andreasen, N.C., Rice, J., Endicott, J., Reich, T. and Coryell, W.(1986). The family history approach to diagnosis. Arch. Gen. Psychiatry 43:421-429.

39. Andrulonis, P.A., Glueck, B.C., Stroebel, C.F. and Vogel, N.G.(1982). Borderline personality subcategories. J. Nerv. Ment. Dis. 170:670-679.

40. Andrulonis, P.A., Glueck, B.C., Stroebel, C.F. Vogel, N.G., Shapiro, A.L. and Aldridge, D.M.(1980). Organic brain dysfunction and the borderline syndrome. Psychiatr. Clin. North Am. 4:47-66.

41. Andy, O.J. and Jurko, M.F.(1972). Focal thalamic discharges with visceral disturbance and pain treated by thalamotomy. Clin. Electroenceph. 3:215-223.

42. Ang, L., Borison, R., Dysken, M. and Davis, J.M.(1982). Reduced excretion of MHPG in TS. In A.J. Friedhoff and T.N. Chase (eds) Gilles de la Tourette Syndrome. Raven Press, New York, pp171-175.

43. Angrist, B., Thompson, H., Shopsin, B. and Gershon, S.(1975). Clinical studies with dopamine-receptor stimulants. Psychopharmacologia 44:273-280.

44. Angst, J.(1979). The reliability of morbidity risk figures in affective disorders. In J. Mendlewicz and B. Shopsin (eds) Genetic Aspects of Affective Illness. Spectrum Publications, Inc., New York, pp 21-26.

45. Angst, J., Frey, R., Lohmeyer, B. and Zerbin-Rüdin, E.(1980). Bipolar manic-depressive psychoses: Results of a genetic investigation. Hum. Genet. 55:237-254.

46. Annett, M.(1979). Genetic and non-genetic influences on handedness. Behav. Genet. 8:227-249.

47. Annibali, J.A., Kales, J.D. and Tan, T-L.(1986). Anorexia nerovosa in a young man with Tourette's syndrome. J. Clin. Psychiatry 47:324-326.

48. Antelman, S.M. and Caggiula, A.R.(1977). Norepinephrine-dopamine interactions and behavior. Science 195:646-653.

49. Antelman, S.M., Eichler, A.J., Black, C.A. and Kocan, D.(1980). Interchangeability of stress and amphetamine in sensitization. Science 207:329-331.

50. Anthony, F.J. (1957). An experimental approach to the psychopathology of childhood encopresis. Br. J. Med. Psych. 30:146-175.

51. Anthony, E.J.(1974). The syndrome of the psychologically invulnerable child. In E. Anthony and C. Koupernik (eds).The Child in His Family: Children at Psychiatric Risk. John Wiley & Sons, New York.

51a. Anthony, M. and Lance, J.(1989). Plasma serotonin in patients with chronic tension headaches. J. Neurol. Neurosurg. Psychiatry 52:182-184.

52. Arato, M., Tekes, K., Palkovits, M., Demeter, E. and Falus, A.(1987). Serotonergic split brain and suicide. Psychiatry Res. 21:355-356.

53. Arkonac, O. and Guze, S.B.(1963). A family study of hysteria. N. Eng. J. Med. 266:239-242.

54. Arnold, L.E., Strobl, D. and Weisenberg, A.(1972). Hyperkinetic adult: Study of the "paradoxical" amphetamine response. JAMA 222:693-694.

54a. Arnold, L.E., Kleykamp, D., Votolato, N.A., Taylor, W.A., Kontras, S.B. and Tobin, K.(1989). Gamma-linolenic acid for attention-deficit hyperactivity disorder: Placebo-controlled comparison to d-amphetamine. Biol. Psychiatry 25:222-228.

55. Asarnow, R.F. and MacCrimmon, D.J.(1978). Residual performance deficit in clinically remitted schizophrenics: A marker of schizophrenia? J. Abnorm. Psychol. 87:597-608.

56. Asarnow, R.F., Steffy, R.A., MacCrimmon, D.J. and Cleghorn, J.M.(1977). An attentional assessment of foster children at risk for schizophrenia. J. Abnorm. Psychol. 86:267-275.

57. Asberg, M., Bertilsson, L., Mårtensson, B., Scalia-Tombr, G-P., Thoren, P. Tråskman-Bendz, L.(1984). CSF monamine metabolites in melancholia. Acta Psychiatr. Scand. 69:201-219.

58. Asberg, M., Thorén, P.,and Bertilsson, L.(1982). Clomipramine treatment of obsessive-compulsive disorder: Biochemical and clinical aspects. Psychopharmacol. Bull. 18:13-21.

59. Asberg, M., Thorén, P., Träskman, L, Bertilsson, L. and Ringberger, V.(1976). "Serotonin depression" — A biochemical subgroup within the affective disorders? Science 191:478-450.

60. Asberg, M., Traskman, L. and Thoren, P.(1976). 5-HIAA in the cerebrospinal fluid. A biochemical suicide predictor? Arch. of Gen. Psychiatry 33:1193-1197.

61. Asberg-Wistedt, B., Wistedt, B. and Bertilsson, L.(1985). Higher CSF levels of HVA and 5-HIAA in delusional compared to nondelusional depression. Arch. Gen. Psychiatry 42:925-926.

62. Aschroft, G.W., Crawford, T.B.B., Eccleston, D., Sharman, D.F., MacDougall, E.J., Stanton, J.B. and Binns, J.K.(1966). 5-hydroxyindole compounds in the cerebrospinal flluid of patients with psychiatric or neurological diseases. Lancet 2:1049-1050.

63. Aserinsky, E. and Kleitman, N.(1955). A motility cycle in sleeping infants as manifested by ocular and gross bodily activity. J. Appl. Physiol 8:1-11.
64. Asperger, H.(1944). Die 'autistischen Psychopathen' im Kindesalter. Arch. Psychiatrie NervenKrankheiten 117:76-136.
65. Aston-Jones, G (1985). The locus coeruleus. Behavioral functions of locus coeruleus derived from cellular attributes. Physiol. Psychol. 13:118-126.
66. Aston-Jones, G. and Bloom, F.E.(1981). Activity of norepinephrine-containing locus coeruleus neurons in behaving rats anticipates fluctuations in the sleep-waking cycle. J. Neurosci. 1:876-886.
67. Aston-Jones, G. and Bloom, F.E.(1981). Norepinephrine containing locus coeruleus neurons in behaving rats exhibit pronounced responses to non-noxious environmental stimuli. J. Neurosci. 1:887-900.
68. Aston-Jones, G., Ennis, M., Pieribone, V.A., Nickell, W.T. and Shipley, M.T.(1986). The brain nucleus locus coeruleus: Restricted afferent control of a broad efferent network. Science 234:734-737.
69. August, G.J. and Stewart, M.A.(1983). Familial subtypes of childhood hyperactivity. J. Nerv. Ment. Dis. 171:362-368.
70. Awapara, J., Landua, A.J., Fuerst, R. and Steale, B.(1950). Free γ-aminobutyric acid in brain. J. Biol. Chem. 187:35-39.
71. Axelrod, J. and Weinshilboum, R.(1972). Catecholamines. N. Engl. J. Med. 287:237-242.
72. Azmitia, E.C. and Segal, M.(1978). An autoradiographic analysis of the differential ascending projections of the dorsal and median raphe nuclei in the rat. J. Comp. Neurol. 179:641-668.
73. Bachman, D.S.(1981). Pemoline-induced Tourette's disorder: A case report. Am. J. Psychiatry 138:1116-1117.
74. Bacopoulos, N.C., Spokes, E.G., Bird, E.D. and Roth, R.H.(1979). Antipsychotic drug action in schizophrenic patients: Effect on cortical dopamine metabolism after long term treatment. Science 205: 1405-1407.
75. Badawy, A.B. and Evans, M.(1987). The significance of kynurenine determination in human serum. In D.A. Bender, M.H. Joseph, W. Kochen and H. Steinhart (eds) Progress in Tryptophan and Serotonin Research 1986. Walter de Gruyter, New York, pp27-30.
76. Badawy, A.B. and Evans, M.(1972). Alcohol, addiction, porphyria and mental disorders. Lancet 2:374-375.
77. Badawy, A.B. and Evans, M.(1973). The effects of chemical prophyrogens and drugs on the activity of rat liver tryptophan pyrrolase. Biochem. J. 136:885-892.
78. Badawy, A.B. and Evans, M.(1973). The effects of chronic phenobarbitone administration and subsequent withdrawal on the activity of rat liver tryptophan pyrrolase and their resemblance to those of ethanol. Biochem. J. 135:555-557.
79. Badawy, A.B. and Evans, M.(1973). Tryptophan pyrrolase in ethanol administration and withdrawal. Adv. Exp. Biol. Med. 35:105-123.
80. Badawy, A.B. and Evans, M.(1975). The effects of acute and chronic nicotine hydrogen (+)- tartrate administration and subsequent withdrawal in rat liver tryptophan pyrrolase activity and their comparison with those of morphine, phenobarbitone and ethanol. Biochem. J. 148:425-432.
81. Badawy, A.B. and Evans, M.(1975). The effects of ethanol on tryptophan pyrrolase activity and their comparison with those of phenobarbitone and morphine. Adv. Exp. Biol. Med. 59:229-251.
81a. Badawy, A.B. and Evans, M.(1976). The regulation of rat liver tryptophan pyrrolase activity by reduced nicotinamide-adenine dinucleotide (phosphate). Biochem. J. 156:381-390.
81b. Badawy, A.B. and Evans, M.(1976). The role of free serum tryptophan in the bi-phasic effect of acute ethanol administration on the concentration of rat brain tryptophan, 5-hydroxytryptamine and 5-hydroxyindol-3-acetic acid. Biochem. J. 1690:315-324.
82. Badawy, A.B. and Evans, M.(1981). Inhibition of rat liver tryptophan pyrrolase activity and elevation of brain tryptophan concentration by administration of antidepressants. Biochem. Pharmacol. 30:1211-1216.
82a. Badawy, A.B.(1982). Mechanisms of elevation of rat brain tryptophan concentration by various doses of salicylate. Br. J. Pharmacol. 76:211-213.
83. Badawy, A.B., Punjani, N.F. and Evans, M.(1978). Liver tryptophan pyrrolase and brain-5- hydroxytryptamine: Effects of chronic ethanol administration. Biochem. Soc. Trans. 6:1002-1004.
84. Badawy, A.B., Punjani, N.F. and Evans, M.(1979). Enhancement of rat brain tryptophan metabolism by chronic ethanol administration and possible involvement of decreased liver tryptophan pyrrolase activity. Biochem. J. 178:575-580.
85. Badawy, A.B., Punjani, N.F. and Evans, M.(1981). The role of liver tryptophan pyrrolase in the opposite effects of chronic administration and subsequent withdrawal of drugs of dependence on rat brain tryptophan metabolism. Biochem. J. 196: 161-170.
86. Badawy, A.B., Punjani, N.F., Evans, C.M. and Evans, M.(1980). Inhibition of rat brain tryptophan metabolism by ethanol withdrawal and possible involvement of the enhanced tryptophan pyrrolase activity. Biochem. J. 192:449-455.
87. Badawy, D.L., Warren, W.H. and White, P.J.(1974). Involvement of the liver in the regulation of tryptophan availability. Possible role in the responses of liver and bran to starvation. Life Sci. 15:1443-1445.
88. Baker, E.F.W.(1962). Gilles de la Tourette syndrome treated by bimedial frontal leucotomy. Can. Med. Assoc. J. 86:746-747.
89. Bakwin, H.(1970). Sleep walking in twins. Lancet 2:446-447.
90. Bakwin, H.(1973).The genetics of enuresis. In I. Kolvin, R.C. MacKeith, and S.R. Meadow(eds). Bladder Control and Enuresis. J.B. Lippincott Co., Philadelphia pp73-77.
91. Balagot, R.C., Ehrenpreis, S., Greenberg, J., and Hyodo, M.(1983). D-phenylalanine in human chronic pain. In S. Ehrenpreis and F. Sicuteri (eds) Degradation of Endogenous Opioids: Its Relevance in Human Pathology and Therapy. Raven Press, New York, pp207-216.
92. Balcar, V.J. and Johnston, G.A.R.(1972). The structural specificity of the high affinity uptake of L-glutamate and L-aspartate by rat brain slices. J. Neurochem. 19:2657-2666.
93. Baldessarini, R.J.(1977). Schizophrenia. N. Engl. J. Med. 297:988-995.
94. Baldessarini, R.J., Cole, J.O., Davis, J.M. Gardos, G., Preskorn, S.H., Simpson, G.M. and Tarsy, D.(1980). Tardive Dyskinesia: A Task Force Report. American Psychiatric Press, Washington, DC.

95. Ball, M.J., Hachinski, V., Fox, A., Kirshen, A.J., Fisman, M., Blume, W., Kral, V.A., Fox, H. and Merskey, H.(1985). A new definition of Alzheimer's disease: A hippocampal dementia. Lancet 1:14-16.
96. Ballenger, J.C. and Post, R.M.(1980). Carbamazepine in manic-depressive illness: A new treatment. Am. J. Psychiatry 137:782-790.
97. Ballenger, J.C., Burrows, G.D., DuPont, R.L., Lesser, I.R., Noyes, R., Pecknold, J.C., Rifkin, A. and Swinson, R.P.(1988). Alprazolam in panic disorder and agoraphobia: Results from a multicenter trial. I. Efficacy in short-term treatment. Arch. Gen. Psychiatry 45:413-422.
98. Ballenger, J.C., Goodwin, F.K., Major, L.F. and Brown, G.L.(1979). Alcohol and central serotonin metabolism in man. Arch. Gen. Psychiatry 36:224-227.
99. Ballenger, J.C., Post, R.M. and Goodwin, F.K.(1983). Neurochemistry of cerebrospinal fluid in normal individuals: Relationship between biological and psychological variables. In J.H. Wood (ed) The Neurobiology of Cerebrospinal Fluid Vol II. Plenum, New York, pp24-37.
100. Balsara, J.J., Jadhav, J.H. and Chadorkar, A.G.(1979). Effect of drugs influencing central serotonergic mechanisms on haloperidol-induced catalepsy. Psychopharmacology 62:67-69.
101. Banki, C.(1981). Factors influencing mono-amine metabolites and tryptophan in patients with alcohol dependence. J. Neural Transmission. 50:98-101.
102. Banki, C.M. and Arato, M.(1983). Amine metabolites and neuroendocrine responses related to depression and suicide. J. Affective Disord. 5:223-232.
103. Banki, C.M., Arató, M., Papp, Z. and Kurz, M.(1984). Biochemical markers in suicidal patients. Investigations with cerebrospinal fluid amine metabolites and neuroendocrine tests. J. Affective Disord. 6:341-350.
104. Banki, C.M., Bissette, G., Arato, M., O'Connor, L. and Nemeroff, C.B.(1987). CSF corticotropin-releasing factor-like immunoreactivity in depression and schizophrenia. Am. J. Psychiatry 144:873-877.
105. Banki, C.M., Vojnik, M. Papp, Z., Balla, K.Z. and Arató, M.(1985). Cerebrospinal fluid magnesium and calcium related to amine metabolites, diagnosis, and suicide attempts. Biol. Psychiatry 20:163-171.
106. Bannon, M.J. and Roth, R.H.(1983). Pharmacology of mesocortical dopamine neurons. Pharmacol. Rev. 35:53-68.
107. Bannon, M.J., Michaud, R.L. and Roth, R.H.(1981). Mesocortical dopamine neurons. Lack of autoreceptors modulating dopamine synthesis. Mol. Pharmacol. 19:270-275.
108. Baraban, J.M. and Aghajanian, G.K.(1981). Noradrenergic innervation of serotonergic neurons in the dorsal raphe: Demonstration by electron microscopic autoradiography. Brain Res. 204:1-11.
109. Barabas, G. and Matthews, W.S.(1985). Homogenous clinical subgroups in children with Tourette sydnrome. Pediatrics 75:73-75.
110. Barabas, G., Matthews, W.S. and Ferrari, M.(1984). Tourette's syndrome and migraine. Arch. Neurol. 41:871-872.
111. Barabas, G., Matthews, W.S. and Ferrari, M..(1984). Disorders of arousal in Gilles de la Tourette's syndrome. Neurology 34:815-817.
112. Barabas, G., W.S. Matthews, and M. Ferrari.(1984). Somnambulism in children with Tourette syndrome. Dev. Med. Child Neurol. 26:457-460.
113. Baradan, J. and Aghajanian, G.K.(1980). Suppression of firing activity of 5-HT neurons in the dorsal raphe by alpha-adrenoreceptor antagonists. Neuropharmacology 19:355-363.
114. Barbeau, A.(1970). Dopamine and disease. Can. Med. Assoc. J. 103:824-832.
115. Baribeau-Braun, J. Picton, T.W. and Gosselin, J.-Y.(1983). Schizophrenia: A neurophysiological evaluation of abnormal information processing. Science 219:874-876.
116. Barkley, R.A. and Cunningham, C.E.(1978). Do stimulant drugs improve the academic performance of hyperkinetic children? Clin. Pediatr. 17:85-92.
117. Barkley, R.A. and Cunningham, C.E.(1979). The effects of methyphenidate on the mother-child interactions of hyperactive children. Arch. Gen. Psychiatry 36:201-208.
118. Barkley, R.A.(1976). Predicting the response of hyperkinetic children to stimulant drugs: A review. J. Abnorm. Child Psychol. 4:327-348.
119. Barkley, R.A.(1977). A review of stimulant drug research with hyperactive children. J. Child Psychol. Psychiatry 18:137-165.
120. Barkley, R.A.(1981). Hyperactive Children. A Handbook for Diagnosis and Treatment. The Guilford Press, New York, 458pp.
121. Barkley, R.A.(1985). The social behavior of hyperactive children: Developmental changes, drug effects, and situational variation. In R. McMahon and R. Peters (eds) Childhood Disorders. Brunner/Mazel, New York.
122. Barnes, D.M.(1988). New data intensify the agony over ecstasy. Science 239:864-866.
123. Barofski, A.L., Grier, H.C., and Pradkan, T.K.(1980). Evidence for regulation of water intake by median raphe 5HT neurons. Physiol. Behav. 24:951-955.
124. Baron, M.(1980). Genetic models of affective disorder: Application to twin data. Acta Genet. Med. Gemellol. 29:289-294.
125. Baren, M.(1986). By-pass techniques.(personal communication).
125a. Baren, M.(1989). The case for Ritalin: A fresh look at the controversy. Contemporary Pediatrics 6:16-28.
126. Baron, M., Gruen, R., Asnis, L., Kane, J.(1982). Schizoaffective illness, schizophrenia and affective disorders: Morbidity risk and genetic transmission. Acta Psychiatr. Scand. 65:263-282.
127. Baron, M., Risch, N., Hamburger, R., Mandel, B., Kushner, S., Newman, M., Drumer, D. and Belmaker, R.H.(1987). Genetic linkage between X-chromosome markers and bipolar affective illness. Nature 326:289-292.
128. Baron, M., Shapiro, E., Shapiro, A. and Rainer, J.D.(1981). Genetic analysis of Tourette syndrome suggesting major gene effect. Am. J. Hum. Genet. 33:767-775.
129. Bartak, L., Rutter, M. and Cox, A.(1975). A comparative study of infantile autism and specific developmental receptive language disorders. I. The children. Br. J. Psychiatry 126:127-145.
130. Bartholini, G.(1976). Differential effect of neuroleptic drugs on dopamine turnover in the extraparamidal and limbic system. J. Pharm. Pharmacol. 28:429-433.
131. Basbaum, H.I., and Fields, H.(1978).Endogenous pain control mechanisms: Review and hypothesis. Ann. Neurol. 4:451-462.
132. Bassett, A.S., Jones, B.D., McGullivray, B.C. and Pantzar, J.T.(1988).Partial trisomy chromosome 5 cosegre-

gating with schizophrenia. Lancet 1: 799-800.
133. Baylies, M.K., Bargiello, T.A., Jackson, F.R. and Young, M.W.(1987). Changes in abundance or structure of the per gene product can alter periodicity of the Drosophilia clock. Nature 326:390-392.
134. Bazemore, A.W., Elliott, K.A.C., and Florey, E.(1957). Isolation of factor I. J. Neurochem. 1:334-339.
134a. Bear, D.M.(1983). Hemispheric specialization and the neurology of emotion. Arch. Neurol. 40:195-202.
135. Beard, G.M.(1886). Experiments with the "jumpers" or "jumping" Frenchman of Maine. J. Nerv. Ment. Dis. 7:487-490.
136. Beck, A.T., Ward, C.H., Mendelson, M., Mock, J. and Erbaugh, J.(1961). An inventory for measuring depression. Arch. Gen. Psychiatry 4:53-63.
137. Beckstead, E.M., Domesick, V.B. and Nauta, W.J.H.(1979). Efferent connections of substantia nigra and ventral tegmental area in the rat. Brain Res. 175: 191-199.
138. Beckwith, B.E., Petros, T., Kanaan-Beckwith, S., Couk, D.I. and Haug, R.J.(1982). Vasopressin analog (DDAVP) facilitates concept learning in human males. Peptides 3:627-630.
139. Behar, D. and Stewart, M.A.(1982). Aggressive conduct disorder of children. The clinical history and direct observations. Acta Psychiatr. Scand. 65:210-220.
140. Behar, D. and Stewart, M.A.(1984). Aggressive conduct disorder: The influence of social class, sex and age on the clinical picture. Child Psychol. Psychiatry 25:119-124.
141. Bekhtereva, N.P.(1905-1907). Fundamentals of the Study of Brain Functions, Volumes 1-7. St Petersburg. (see 1225).
142. Belendiuk, K., Belendiuk, G.W. and Freedman, D.X.(1980). Blood monamine metabolism in Huntington's disease. Arch. Gen. Psychiatry 37:325-332.
143. Bell, D.S.(1965). Comparison of amphetamine psychosis and schizophrenia. Br. J. Psychiatry 111:701-707.
144. Bellak, L. (ed) (1979). Psychiatric Aspects of Minimal Brain Dysfunction in Adults. Grune & Stratton, NewYork, p208.
145. Bellman, M.(1966). Studies on encopresis. Acta Paediatr. Scand. Suppl 170:1-137.
146. Bemporad, J.R., Smith, H.F., Hanson, G. and Cicchetti, D.(1982). Borderline syndrome in childhood: Criteria for diagnosis. Am J. Psychiatry 139:596-602.
147. Benassi, C.A., Benassi, P., Allegri, G. and Ballarin, P.(1961). Tryptophan metabolism in schizophrenic patients. J. Neurochem. 7:264-270.
147a. Bender, D.A.(1989). The kynurenine pathway of tryptophan metabolism. In T. W. Stone (ed) Quinolinic Acid and the Kynurenines. CRC Press, Inc, Boca Raton, FL 3-38.
148. Bender, D.A. and Totoe, L.(1984). High doses of vitamin B6 in the rat are associated with inhibition of hepatic tryptophan metabolism and increased uptake of tryptophan into the brain. J. Neuochem. 43:733-736.
149. Bender, L.(1947). Childhood schizophrenia. Am. J. Psychiatry 17:40-56.
150. Bender, L.(1949). Psychological problems of children with organic brain disease. Am. J. Orthopsychiatry. 19:404-415.
151. Beninger, R.J. and Phillps, A.G.(1980). The effect of pimozide on the establishment of conditioned reinforcement. Psychopharmacology 68:147-153.
152. Beninger, R.J.(1983). The role of dopamine in locomotor activity and learning. Brain Res. Rev. 6:173-196.
153. Bennett, J.P. and Snyder, S.H.(1976). Serotonin and lysergic acid diethylamide binding in rat brain membranes: Relationship to postsynaptic serotonin receptors. Mol. Pharmacol. 12:373-389.
154. Bennett, J.P., Enna, S.J., Bylund, D.B., Gillin, J.C., Wyatt, R.J. and Snyder, S.H.(1979). Neurotransmitter receptors in frontal corex of schizophrenics. Arch. Gen. Psychiatry 36:927-934.
155. Benson, D.F. and Geschwind, N.(1975). Psychiatric conditions associated with focal lesions of the central nervous system. In M.F. Reiser (ed) Organic Disorders and Psychosomatic Medicine. Basic Books, New York, pp208-243.
156. Benton, A.L. and Pearl, D. (eds) (1978). Dyslexia An Appraisal of Current Knowledge. Oxford Univ. Press, New York.
157. Berg, I.(1976). School phobia in children of agoraphobic women. Br. J. Psychiatry 128:86-89.
158. Berger, M., Yule, W. and Rutter, M.(1975). Attainment and adjustment in two geographical areas. I: The prevelence of specific reading retardation . Br. J. Psychiatry 126:510-519.
159. Bergler, E.(1936). The psychology of the gambler. Imago 22:409.
160. Berlet, H.H., Bull, C., Himwich, H.E., Kohl, H., Matsumato, K., Pscheidt, G.R., Spaide, J., Tourlentes, T.T. and Valverde, J.M.(1964). Endogenous metabolic factors in schizophrenic behavior. Science 144:311-313.
161. Berlet, H.H., Spaide, J., Kohn, H., Bull, C. and Himwich, H.E.(1965). Effects of reduction of tryptophan and methionine intake on urinary indole compounds and schizophrenic behavior. J. Nerv. Ment. Dis. 140:297-304.
162. Bertelsen, A., Harvald, B. and Hauge, M. (1977). A Danish twin study of manic-depressive disorder. Br. J. Psychiatry 130:330-351.
163. Bertler, A. and Rosengren, E.(1959). Occurrence and distribution of dopamine in brain and other tissues. Experientia 15:10-11.
164. Bhagavan, H.N., Coleman, M., and Cursin, D.B.(1975). The effect of pyridoxine hydrochloride on blood serotonin and pyridoxal phosphate contents in hyperactive children. Pediatrics 55:437-441.
164a. Biederman, J. (1988). Evidence of familial association between attention deficit disorder (ADD) and major depression (MDD). Scientific Proceedings of the Annual Meeting of the American Academy of Child and Adolescent Psychiatry 4:53.
165. Biederman, J., Munir, K., Knee, D., Habelow, W., Anmentano, M., Autor, S., Hoge, S.K. and Waternaux, C.(1986). A family study of patients with attention deficit disorder and normal controls. J. Psychiat. Res. 20:263-274.
166. Biederman, J., Munir, K., Knee, D., Armentano, M., Autor, S., Waternaux, C. and Tsuang, M. (1987). High rate of affective disorders in probands with attention deficit disorder and in their relatives: A controlled family study. Am. J. Psychiatry 144:330-333.
167. Bille, B.S.(1962). Migraine in school children. Acta Paediatr. Scand. 51 (Suppl 136): 1-151.
168. Bioulac, B., Benezich, M., Renaud, B., Noel B. and Roche, D.(1980). Serotonergic functions in the 47 XYY syndrome. Biol. Psychiatry 15:917-923.
169. Bird, E.D.(1980). Chemical pathology of Huntington's disease. Annu. Rev. Pharmacol. Toxicol. 20:533-551.

170. Bisgaard, M.L., Eiberg, H., Niebuhr, M.E. and Mohr, J.(1987). Dyslexia and chromosome 15 heteromorphism: Negative lod score in a Danish material. Clin. Genet. 32:118-119.
171. Bjorklund, A., Divac, I. and Lindvall, O.(1978). Regional distribution of catecholamines in monkey cerebral cortex. Evidence for a dopaminergic innervation of the primate prefrontal cortex. Neurosci. Lett. 7:115-120.
172. Black, F.W.(1976). Cognitive deficits in patients with unilateral war-related frontal lobe lesions. J. Clin. Psychol. 32:366-372.
173. Black, I.B., Adler, J.E., Dryefus, C.F., Friedman, W.F., LaGamma, E.F. and Roach, A.H.(1987). Biochemistry of information storage in the nervous system. Science 236:1263-1268.
174. Blanck G., Hervé, D., Simon, H., Lisoprawski, A., Glowinski, J., and Tassin, J.P.(1980). Response to stress of mesocortical-frontal dopaminergic neurons in rats after long-term isolation. Nature 284:265-267.
175. Bland, R.C., Newman, S.C. and Orn, H.(1986). Recurrent and nonrecurrent depression. A family study. Arch. Gen. Psychiatry 43:1085-1089.
176. Blass, E.M. and Hanson, D.G.(1970). Primary hyperdipsia in the rat following septal lesions. J. Comp. Physiol. Psychol. 70:87-93.
176a. Bleich, A., Bernoit, E. Apter, A. and Tyano, S.(1985). Gilles de la Tourette syndrome and mania in an adolescent. Br. J. Psychiatry 146:664-665.
176b. Bleich, A., Brown, S-L., Kahn, R. and van Praag, H.M.(1988). The role of serotonin in schizophrenia. Schizophr. Bull. 14:297-315.
177. Bleuler, E.(1911). Dementia Praecox oder Gruppe der Schizophrenien. F. Deuticke, Leipzig.
178. Bleuler, E.(1950). Dementia praecox or the group of schizophrenias. J. Zenkin (translator). International University Press, New York.
179. Blier, P., De Montigny, C., and Chaput, Y.(1987). Modifications of the serotonin system by antidepressant treatments: Implications for the therapeutic response in major depression. J. Clin. Psychopharmacol. 7:24S-35S.
179a. Blier, P., De. Montigny, C. and Chaput, Y.(1988). Electrophysiological assessment of the effects of antidepressant treatments on the efficacy of 5-HT neurotransmission. Clin. Neuropharmacol. 11:(Suppl 2) S1-S10.
180. Bliss, J., Cohen, D.J. and Freedman, D.X.(1980). Sensory experiences of Gilles de la Tourette syndrome. Arch. Gen. Psychiatry 37:1343-1347.
181. Bliss, T.V.P. and Lømo, T.(1973). Long-lasting potentiation of synaptic transmission in the dentate area of the anaesthetized rabbit following stimulation of the perforant path. J. Physiol. 232:331-356.
182. Bloch, D. and Simon, R.(1982). The Strength of Family Therapy. Selected Papers of Nathan W. Ackerman. Brunner/Mazel, Publishers, New York.
183. Blomquist, H.K., Bohman, M., Edvinsson, S.O., Gillberg, C., Gustavson, K-H., Holmgren, G. and Wahlström, J.(1985). Frequency of fragile X syndrome in infantile autism. Clin. Genet. 26:513-517.
184. Blondaux, C., Juge, A., Sordet, F., Chouvet, G., Jouvet, M., and Pujol, J-F.(1973). Modification du métabolisme de la sérotonine (5-HT) cérébrale induite chex le rat par administration de 6-hydroxydopamine. Brain Res. 50:101-114.
185. Bloom, F. Segal, D., Ling, N. and Guillemin, R.(1976). Endorphins: Profound behavioral effects in rats suggest new etiological factors in mental illness. Science 215:630-632.
186. Bloom, F.E., Hoffer, B.J., Siggins, G.R., Barker, J.L., and Nicoll, R.A.(1972). Effects of serotonin on central neurons: Micro-intophoretic administration. Fed. Proc. 31:97-106.
187. Blouin, A.G.A., Bornstein, R.A. and Trites, R.L.(1978). Teenage alcohol use among hyperactive children: A five year follow-up study. J. Pediatr. Psychol. 3:188-194.
188. Blundell, J.E.(1984). Serotonin and appetite. Neurpharmacology 23:537-551.
189. Bockner, S.(1959). Gilles de la Tourette's disease. J. Nerv. Ment. Dis. 105:1078-1081.
190. Bohman, M.(1978). Some genetic aspects of alcoholism and criminality: A population of adoptees. Arch. Gen. Psychiatry 35:269-276.
191. Bohman, M., Cloninger, C.R., Sigvardsson, S., von Knorring, A.-L(1982). Predispostion to petty criminality in Swedish adoptees: I. Genetic and environmental heterogenity. Arch. Gen. Psychiatry 39:1233-1241.
192. Bohman, M., Sigvardsson, S., and Cloninger, C.R.(1981). Maternal inheritance of alcohol abuse. Arch.Gen. Psychiatry 38:965-969.
193. Bonnet, K.A.(1982). Neurobiological dissection of Tourette syndrome: A neurochemical focus on a human neuroanatomical model. In A.J. Friedhoff and T.N. Chase (eds) Gilles de la Tourette Syndrome. Raven Press, New York, p77-81.
194. Bonthala, C.M. and West, A.(1983). Pemoline induced chorea and Gilles de la Tourette's syndrome. Br. J. Psychiatry 143:300-302.
195. Borison, R.L., Ang, L., Chang, S., Dysken, M., Comaty, J.E. and Davis, J.M.(1982). New pharmacological approaches in the treatment of Tourette syndrome. In A.J. Friedhoff and T.N. Chase(eds). Gilles de la Tourette Syndrome Raven Press, New York, pp377-382.
196. Borison, R.L., Ang. L., Hamilton, W.J., Diamond, B.I. and David, J.M.(1983). Treatment approaches in Gilles de la Tourette syndrome. Brain Res. Bull. 11:205-208.
197. Borland, B.L., Heckman, H.K.(1976). Hyperactive boys and their brothers. A 25 year follow-up study. Arch. Gen. Psychiatry 33:669-675.
198. Bornstein, R.A., King, G., and Carroll, A.(1983). Neuropsychological abnormalities in Gilles de la Tourette syndrome. J. Nerv. Ment. Disord. 171:497-502.
198a. Botstein, D., White, R.L., Skolnick, M. and Davis, R.W.(1980) Construction of a genetic linkage map in man using restriction fragment length polymorphisms. Am. J. Hum. Genet. 32:314-331.
199. Boullin, D.J. and O'Brien, R.A.(1971). Abnormalities of 5-hydroxytryptamine uptake and binding by blood platelets from children with Down's syndrome. J. Physiol. 212:287-297.
200. Boulin, D.J, Freeman, B.J., Geller, E., Ritvo, E., Rutter, M. and Yuwiler, A.(1982). Toward the resolution of conflicting findings. J. Autism Dev. Disord. 12:97-98.
201. Bowen, M.(1978). Family Therapy in Clinical Practice. Jason Aronson, New York.
202. Bower, E.M., Shellhammer, T.A. and Daily, J.M.(1960). School characteristics of male adolescents who later became schizophrenic. Am. J. Orthopsychiatry 30:712-

729.
203. Bowery, N.G., Hill, D.R., Hudson, A.L., Doble, A., Middlemiss, D.N., Shaw, J. and Turnbull, M.(1980). (-)Baclofen decreases neurotransmitter release in the mammalian CNS by action at a novel GABA receptor. Nature 283:92-94.
204. Bozarth, M.A. and Wise, R.A.(1981). Heroin reward is dependent on a dopaminergic substrate. Life Sci. 29:1881-1886.
205. Braak, H.(1976). A primitive gigantopyramidal field buried in the depth of the cingulate sulcus of the human brain. Brain Res. 109:219-233.
206. Bradley, C.(1937-1939). The behavior of children receiving Benzedrine. Am. J. Psychiatry 99:577-585.
207. Bradley, C.(1950). Benzedrine and Dexedrine in the treatment of children's behavior disorders. Pediatrics 5:24-37.
208. Bradshaw, C.M., Stoker, M.J. and Szabadi, E.(1983). Comparison of the neuronal responses to 5-hydroxytryptamine, noradrenaline and phenylephirine in the cerebral cortex: Effects of haloperidol and methylsergide. Neuropharmacology 22:677-585.
209. Brady, E.S. and Nauta, W.J.H.(1953). Subcortical mechanisms in emotional behavior: Affective changes following septal forebrain lesions in the albino rat. J. Comp. and Physiol. Psychol. 46:339-346.
210. Brady, E.S. and Nauta, W.J.H.(1955). Subcortical mechanisms in control of behavior. J. Comp. Physiol. Psychol. 48:412-420.
211. Brady, K.T., Balster, R.L., and May, E.L.(1982). Stereoisomers of N-allynormetazocine: Phencyclidine-like behavioral effects in squirrel monkeys and rats. Science 215:178-180.
211a. Brady, J.P., Price, T.R., McAllister, T.W. and Dietrich, K.(1989). A trial of verapamil in the treatment of stuttering in adults. Biol. Psychiatry 25:626-630.
212. Braestrup, C. and Nielsen, M.(1982). Anxiety. Lancet 2:1030-1034.
213. Brain, R. and Benton, D.(1979). The integration of physiological correlates of differential housing in laboratory rats. Life Sci. 24:99-116.
214. Branchey, L. and Lieber, C.S.(1982). Activation of tryptophan pyrrolase after chronic alcohol administration. Substance Alcohol Actions/Misuse 2:225-229.
215. Branchey, L., Branchey, M., Shaw, S. and Lieber, C.S.(1984). Depression, suicide, and aggression in alcoholics and their relationship to plasma amino acids. Psychiatry Res. 12:219-226.
215a. Brase, D.A. and Loh, H.H.(1975). Possible role of 5-hydroxytryptamine in minimal brain dysfunction. Life Sci. 16:1005-1015.
216. Breier, A., Charney, D.S. and Heninger, G.R.(1984). Major depression in patients with agoraphobia and panic disorder. Arch. Gen. Psychiatry 41:1129-1135.
217. Bremness, A.B. and Sverd, J.(1979). Methylphenidate-induced Tourette syndrome: Case report. Am. J. Psychiatry 136:1334-1335.
218. Brickner, R.M.(1936). The Intellectual Functions of the Frontal Lobes. Macmillan Co., New York.
219. Broca, P.(1878). Anatomie comparée des circonvolutions cérébrales. Le grande lobe limbique et la scissure limbique dans la série des mammiféres. Rev. Anthrop. Series 2, 1:285-498.
220. Broderick, P. and Lynch, V.(1982). Behavioral and biochemical changes induced by lithium and L-tryptophan in muricidal rats. Neuropharmacology 21:671-679.
221. Broderick, P.(1985). In vivo electrochemical studies of rat striatal dopamine and serotonin release after morphine. Life Sci. 36:2269-2275.
222. Brodie, B.B. and Shore, P.A.(1957). Concept for a role of serotonin and norepinephrine as chemical mediators in the brain. Ann. N.Y. Acad. Sci. 66:631-642.
223. Brodie, J.D., Christman, D.R., Corona, J.F., Fowler, J.S., et. al.(1984). Patterns of metabolic activity in the treatment of schizophrenia. Ann. Neurol. 15 (Suppl): S166-S169.
224. Brody, J.F., Jr.(1970). Behavioral effects of serotonin depletion and of p-chlorophenylalanine (a serotonin depletor) in rats. Psychopharmacologia 17:14-33.
225. Broekkamp, C.L.E., Phillips, A.G. and Cools, A.R.(1979). Stimulant effects of enkephalin microinjection into the dopaminergic A10 area. Nature 278:560-561.
226. Broughton, R.J.(1968). Sleep disorders: Disorders of arousal? Science 159:1071-1078.
227. Brown, F.W.(1942). Heredity in the psychoneurosis. Proc. R. Soc. Med. 35:785-790.
228. Brown, G.A. and Goldman, P.S.(1977). Catecholamines in neocortex of rhesus monkeys: Regional distribution and ontogenetic development. Brain Res. 124:576-580.
229. Brown, G.L., Ebert, M.F., Goyer, P.H., Jimerson, D.C., Klein, W.J., Bunney, W.E., and Goodwin, F.K.(1982). Aggression, suicide and serotonin relationships to CSF amine metabolism. Am. J. Psychiatry 139:741-746.
230. Brown, G.L., Goodwin, F.K., Ballenger, J.C., Goyer, P.F. and Major, L.I.(1979). Aggression in humans correlates with cerebrospinal fluid metabolites. Psychiatry Res. 1:131-139.
231. Brown, W.T., Jenkins, E.C., Cohen, I.L., Fisch, G.S., et al. (1986). Fragile X and autism: A multi center study. Am. J. Med. Genet. 23:341-352.
232. Brown, W.T., Jenkins, E.C., Friedman, E., Brooks, J., Wisniewski, K., Raguthu, S. and French, J.(1982). Autism is associated with the fragile-X syndrome. J. Autism Dev. Disord. 12:303-308.
233. Brozoski, T.J., Brown, R.M., Rosvold, H.E. and Goldman, P.S.(1979). Cognitive deficit caused by regional depletion of dopamine in prefrontal cortex of rhesus monkey. Science 205:929-932.
234. Bruce, C.J. and Goldberg, M.E.(1984). Physiology of the frontal eye fields. Trends Neurosci. 7:436-441.
235. Brune, G.G. and Pscheidt, G.R.(1961). Correlations between behavior and urinary excretion of indole amines and catecholamines in schizophrenic patients as affected by drugs. Fed. Proc. 20:889-893.
236. Brunn, R.D, Dussault, W., Erenberg, G. and Kapp, M.(1987). Tourette Syndrome and the Law. Tourette Syndrome Association Booklet.
237. Brutkowski, S.(1964). Prefrontal cortex and drive inhibition. In J.M. Warren and K. Akert (eds) Frontal Granular Cortex and Behavior. McGraw-Hill, New York.
238. Brutkowski, S., Konorski, J., Lawicka, W., Stephen, I. and Stepien, L.(1957). The effect of the removal of the frontal poles of the cerebral cortex on motor conditioned reflexes. Acta Biol. Exp. (Lodz) 17.
239. Bruun, R.D.(1982). Clonidine treatment of Tourette syndrome. In A.J. Friedhoff and T. N. Chase(eds). Gilles de la Tourette Syndrome. Raven Press, New York. pp 403-405.
240. Bruun, R.D.(1988). Subtle and underrecognized

side effects of neuroleptic treatment in children with Tourette's disorder. Am. J. Psychiatry 145:621-624.
241. Bruun, R.D.(1988). The natural history of Tourette's syndrome. In Cohen, D.J., Bruun, R.D. and Leckman, J.F. (eds) Tourette's Syndrome and Tic Disorders: Clinical Understanding and Treatment. John Wiley & Sons, New York, pp21-39.
242. Bucher, K.D., Elston, R.C., Clayton, P., Green, R., Whybrow, P. Helzer, J., Reich, T. and Winokur, G.(1981). The transmission of manic depressive illness. II. Segregation analysis of three sets of family data. J. Psychiatr. Res. 16:65-78.
243. Buchsbaum, M.S., Cappelletti, J., Ball, R., Hazlett, E., King, A.C., Johnson, J., Wu, J. and De Lisi, L.E.(1984). Positron emission tomographic image measurement in schizophrenia and affective disorders. Ann. Neurol. 15 (Suppl): S157-S165.
244. Bucht, G., Gottfries, C.G., Roos, B.-E. and Winblad, B.(1979). Reduced concentration of 5-hydroxytryptamine in brains from demented schizophrenic patients. In J. Obiois, C. Ballús, E. Gonzalez and J. Pukol (eds) Biological Psychiatry Today. Elsevier/North Holland Biomedical Press, Amsterdam. pp381-388.
245. Buijs, R.M., Geffard, M., Pool, C.W. and Hoorneman, M.D.(1984). The dopaminergic innervation of the supraoptic and paraventricular nucleus. A light and electron microscopical study. Brain Res. 323: 65-72.
246. Bullock, T.H.(1977). Introduction to Nervous Systems. W.H. Freeman and Co., San Francisco, 559pp.
247. Bunney, B.S. and Aghajanian, G.K.(1976). Dopamine and norepinephrine innervated cells in the rat prefrontal cortex: Pharmacological differentiation using microiontophoretic techniques. Life Sci. 19:1783-1792.
248. Bunney, B.S. and DeRiemer, S.(1982). Effect of clonidine on dopaminergic neurons activity in the substantia nigra: Possible indirect mediation by noradrenergic regulation of the serotonergic raphé system. In A.J. Friedhoff and T.N. Chase (eds). Gilles de la Tourette Syndrome. Raven Press, New York, pp99-104.
249. Bunney, W.E.Jr., Pert, C.B., Klee, W., Costa, E., Pert, A., and Davis, G.C.(1979). Basic and clinical studies of endorphins. Ann. Intern. Med. 91:239-250.
250. Bunney, W.E.Jr. and Davis, J.M.(1965). Norepinephrine in depressive reactions. Arch. Gen. Psychiatry 13:483-494.
250a. Burd, L. and Kerbeshian, J. (1984). Tourette syndrome and bipolar disorder. Arch. Neurol. 41:1236.
251. Burd, L. and Kerbeshian, J.(1985). Tourette syndrome, atypical pervasive developmental disorder and Ganser syndrome in a 15-year-old visually impaired, mentally retarded boy. Can. J. Psychiatry 30:74-76.
251a. Burd, L. and Kerbeshian, J.(1988). Symptom substitution in Tourette disorder. Lancet 2:1072, 1988.
252. Burd, L. and Kerbeshian, J.(1988). Familial pervasive development disorder, Tourette disorder and hyperlexia. Neurosci. Biobehav. Rev. 12:233-234.
253. Burd, L., Fisher,W.W., Kerbeshian,J. and Arnold, M.E.(1987). Is development of Tourette disorder a marker for improvement in patients with autism and other pervasive developmental disorders? J. Am. Acad. Child. Adolesc. Psychiatry 26:162-165.
253a. Burd, L., Kerbeshian, J., Cook, J., Bornhoeft, D.M., and Fisher, W.(1988). Tourette disorder in North Dakota. Neurosci. Biobehav. Rev. 12:223-228.
254. Burd, L., Kerbeshian, J., Fisher, W. and Gascon, G.(1986). Anticonvulsant medications: An iatrogenic cause of tic disorders. Can. J. Psychiatry 31:419-423.
255. Burd, L., Kerbeshian, J., Fisher, W., Barcome, D.F. and Lipp, L.(1986). Pseudohemiparesis and Tourette syndrome. J. Child Neurol. 1:369-371.
255a. Burd, L., Kerbeshian, J., Wikenheiser. M. and Fisher, W.(1986). A prevalence study of Gilles de la Tourette syndrome in North Dakota school-age children. J. Am Acad. Child Psychiatry 25:552-553.
256. Burd, L., Kerbeshian, J., Wikenheiser, M., and Fisher, W.(1986). Prevalence of Gilles de la Tourette's syndrome in North Dakota adults. Am. J. Psychiatry 143:787-788.
257. Burg, I.(1976). School phobia in the children of agoraphobic women. Br. J. Psychiatry 128:86-89.
258. Burke, R.E., Fahn, S., Jankovic, J., Mardsen, C.D., Lang, A.E., Gollomp, S. and Ilson, J.(1982). Tardive dystonia: Late-onset and persistent dystonia caused by antipsychotic drugs. Neurology 32:1335-1346.
259. Burns, D.D. and Mendels, J.(1979). Serotonin and affective disorders. Cur. Dev. Psychopharmacol. 5:293-359.
260. Burns, R., Chiueh, C., Markey, S.P., Ebert, M.H., Jacobowitz, D.M. and Kopin, I.J.(1983). A primate model of Parkinsonism: Selective destruction of dopaminergic neruons in the pars compacts of the substantia nigra by N-methyl-4-phenyl-1,2,3,6,-tetrahydropyridine. Proc. Natl. Acad. Sci. U.S.A. 80:4546- 4550.
260a. Buydens-Branchey, L., Branchey, M.H. and Noumair, D.(1989). Age of alcoholism onset. I. Relationship to psychopathology. Arch. Gen. Psychiatry 46:225-230.
260b. Buydens-Branchey, L., Branchey, M.H., Noumair, D. and Lieber, C.S.(1989). Age of alcoholism onset. II. Relationship to susceptibiity to serotonin precursor availability. Arch. Gen. Psychiatry 46:231-236.
261. Buss, A.H., Durkee, A., Baer, M.(1957). The measurement of hostility in clinical situations. J. Abnorm. Psychiatry 21:343-348.
262. Butler, I.J., Koslow, S.H., Seifert, W.E. Jr., Caprioli, R.M. and Singer, H.S.(1979). Biogenic amine metabolism in Tourette syndrome. Ann. Neurol. 6:37-39.
263. Butter, C.M., Mishkin, M. and Rosvold, H.E.(1963). Conditioning and extinction of a food-rewarded response after selective ablations of frontal cortex in rhesus monkeys. Exp. Neurol. 7:65-75.
264. Butters, N. and Cermak, L.S.(1980). Alcoholic Korsakoff's syndrome: An Information Processing Approach to Amnesia. Academic Press, New York, 188pp.
265. Buydens-Branchey, L., Branchey, M., Worner, T.M., Zucker, D., Aramsom-Batdee, E. and Lieber, C.S.(1988). Increase in tryptophan oxygenase activity in alcoholic patients. Alcohol.: Clin. Exp. Res. 12:163-167.
266. Byerley, W.F., McConnell, E.J., McCabe, R.T., Dawson, T.M., Grosser, B.I. and Wamsley, J.K.(1987). Chronic administration of sertraline, a selective serotonin uptake inhibitor, decreased the density of b-adrenergic receptors in rat frontoparietal cortex. Brain Res. 421:377-381.
267. Byerley, W.F., Reimherr, F.W., Wood, D.R. and Grosser, B.I.(1988). Fluoxetine, a selective serotonin uptake inhibitor, for the treatment of outpatients with major depression. J. Clin. Psychopharmacol. 8:112-115.
268. Cadoret, R.J. and Gath, A.(1978). Inheritance of alcoholism in adoptees. Br. J. Psychiatry 132:252-258.
269. Cadoret, R.J.(1978). Evidence for genetic inheri-

tance of primary affective disorder in adoptees. Am. J. Psychiatry 135:463-466.
270. Cadoret, R.J.(1978). Psychopathology in adopted-away offspring of biologic parents with antisocial behavior. Arch. Gen. Psychiatry 35:176-184.
271. Cadoret, R.J., Cain, C.A. and Grove, W.M.(1980). Development of alcoholism in adoptees raised apart from alcoholic biologic relatives. Arch. Gen. Psychiatry 37:561-563.
272. Cadoret, R.J., Cunningham, L., Loftus, R., Edwards, J.(1975). Studies of adoptees from psychiatrically disturbed biological parents. II. Temperament, hyperactive, antisocial, and developmental variables. J. Pediatr. 87:301-306.
273. Cadoret, R.J., O'Gorman, T.W., Troughton, E. and Heywood, E.(1985). Alcoholism and antisocial personality. Arch. Gen. Psychiatry 42:161-167.
274. Cadoret, R.J., Troughton, E. and Widmer, R.(1984). Clinical differences between antisocial and primary alcoholics. Comp. Psychiatry 25:1-8.
275. Caine, E.D. and Polinksy, R.J.(1979). Haloperidol-induced dysphoria in patients with Tourette syndrome. Am. J. Psychiatry 136:1216-1217.
275a. Caine, E.D. and Polinksy, R.J.(1981). Tardive dyskinesia in persons with Gilles de la Tourette's disease. Arch. Neurol. 38:471-472.
276. Caine, E.D., Margolin, P.I., Brown, G.L., and Ebert, M.H.(1978). Gilles de la Tourette's syndrome, tardive dyskinesia and psychosis in an adolescent. Am. J. Psychiatry 135:241-243.
277. Caine, E.D., McBride, M.C., Chiverton, P., Bainford, K.A., Rediess, S., and Shiao, J.(1988). Tourette syndrome in Monroe County school children. Neurology 38:472-475.
278. Caine, E.D., Polinsky, R.J., Ebert, M.H., Papoport, J.L., and Mikkelsen, E.J.(1979). Trial of chlorimipramine and desipramine for Gilles de la Tourette syndrome. Ann. Neurol. 5:305-306.
279. Campbell, M., Friedman, E., DeVito, E., Greenspan, L., and Collins, P.(1974). Blood serotonin in psychotic and brain damaged children. J. Autism Child. Schizophr. 4:33-41.
280. Campbell, M., Small, A.M., Green, W.H., Jennings, S.J., Perry, R., Bennett, W.G. and Anderson, L.(1984). Behavioral efficacy of haloperidol and lithium carbonate. Arch. Gen. Psychiatry 41:650-656.
281. Campbell, S.(1973). Mother-child interactions in reflective, impulsive, and hyperactive children. Dev. Psychol. 8:341-349.
282. Campbell, S.(1975). Mother-child interaction: A comparison of hyperactive, learning disabled, and normal boys. Am. J. Orthopsychiatry 40:51-57.
283. Campbell, S.B., Douglas, V.I. and Morgenstern, G.(1971). Cognitive styles in hyperactive children and the effect of methylphenidate. J. Child. Psychol. Psychiatry 12:55-67.
284. Cancela, L.M., Fulginiti, S. and Ramirez, O.A.(1985). Involvement of a serotonergic control in the regulation of plasma levels of immunoreactive beta-endorphin. Acta Physiol. Pharmacol. Latinoam. 35:409-413.
285. Cantwell, D.P.(1972). Psychiatric illness in the families of hyperactive children. Arch. Gen. Psychiatry 27:414-417.
286. Cantwell, D.P.(1974). Genetic studies of hyperactive children: Psychiatric illness in biologic and adopting parents. In Fieve, R., Rosenthal, D. and Brill, H.(eds) Genetic Research in Psychiatry, Johns Hopkins Press, Baltimore, pp273-280.
287. Cantwell D.P.(1976). Genetic factors in hyperkinetic syndrome. J. Am. Acad. Child Psychiatry 15:214-223.
288. Cantwell, D.P.(1979). Minimal brain dysfunction in adults: Evidence from studies of psychiatric illness in the families of hyperactive. children. In Bellak, L. (ed) Psychiatric Aspects of Minimal Brain Dysfunction in Adults. Grune & Stratton, New York.
289. Caprini, G. and Melotti, V. (1961). Un grave sindrome ticcosa guarita con haloperidol. Riv. Sper. Freniat. 85:191-196.
290. Carlson, G.A. and Cantwell, D.P.(1980). Unmasking masked depression in children and adolescents. Am. J. Psychiatry 137:445-449.
291. Carlsson, A. and Lindqvist, M.(1963). Effect of chlorpromazine or haloperidol on formation of 3-methoxytyramine and normetanephrine in mouse brain. Acta Pharmacol. Toxicol. 20:140-144.
292. Carlsson, A.(1978). Antipsychotic drugs, neurotransmitters, and schizophrenia. Am. J. Psychiatry 135:164-173.
293. Carlsson, M., Svensson, K., Ericksson, E. and Carlsson, A.(1985). Rat brain serotonin: Biochemical and functional evidence for a sex difference. J. Neural Transmission 63:297-313.
294. Carlton, P.L. and Goldstein, L.(1987). Physiological determinants of pathological gambling. In T. Galski (ed) A Handbook of Pathological Gambling. Charles C. Thomas, Springfield, IL.
295. Carlton, P.L. and Manowitz, P.(1987). Physiological factors as determinants of pathological gambling. J. Gambling Behav. 3:274-285.
296. Carlton, P.L., Manowitz, P., McBride, H., Nora, R., Swartzburg, M. and Goldstein, L.(1987). Attention deficit disorder and pathological gambling. J. Clin. Psychiatry 48:487-488.
297. Carrol, E.N. and Zukerman, M.(1977). Psychopathology and sensation seeking in "downers." "speeders," and "trippers": A study of the relationship between personality and drug choice. Int. J. Addict. 12:591-601.
298. Carroll, B.J.(1982). The dexamethasone suppression test for melancholia. Br. J. Psychiatry 140:292-304.
299. Carroll, B.J.(1985). Dexamethasone suppression test: A review of contemporary confusion. J. Clin. Psychiatry 46:13-24.
300. Carter, C.J. and Pycock, C.J.(1978). A study of the sites of interaction between dopamine and 5-hydroxytryptamine for the production of fluphenazine-induced catalepsy. Naunyn-Schmiedebergs Arch. Pharmacol. 304:135-139.
301. Carter, C.J. and Pycock, C.J.(1979). The effects of 5,7-dihydroxytryptamine lesions of extrapyramidal and mesolimbic sites on spontaneous motor behavior and amphetamine-induced stereotypy. Naunyn-Schmiedebergs Arch. Pharmacol. 308:51-54.
302. Carter, C.J. and Pycock, C.J.(1980). Behavioral and biochemical effects of dopamine and noradrenaline depletion within the prefrontal cortex of the rat brain. Brain Res. 192:163-176.
303. Cary, G. and Gottesman, I.I.(1981). Twin and family studies of anxiety, phobic and obsessive disorders. In D.F. Klein and J. Rabkin (eds) Anxiety: New Research and Changing Concepts. Raven Press, New York, pp 117-136.

304. Casey, D.E.(1984). Tardive dyskinesia in animal models. Psychopharmacol. Bull. 20:376-379.
305. Casey, D.E.(1987). Tardive dyskinesia. In H.Y. Meltzer (ed). Psychopharmacology: The Third Generation of Progress. Raven Press, New York, pp1411-1419.
306. Casper, R.C., Eckert, E.D., Halmi, K.A., Goldberg, S.C. and Davis, J.M.(1980). Bulimia. Its incidence and clinical importance in patients with anorexia nervosa. Arch. Gen. Psychiatry 37:1030-1035.
307. Cermak, L.S., and Reale, L.(1978). Depth of processing and retention of words by alcoholic Korsakoff patients. J. Exp. Psychol. 4:165-174.
307a. Chaillet, P., Coulaud, A., Zajac, J.-M., Fournie-Zaluski, M.-C., Costentin, J. and Roques, B.P.(1984). The m rather than the d subtype of opioid receptors appears to be involved in enkephalin-induced analgesia. Eur. J. Pharmacol. 101:83-90.
308. Chamberlain, B., Ervin, F.R., Pihl, R.O. and Young, S.H.(1987). The effect of raising or lowering tryptophan levels on aggression in Vervet monkeys. Pharmacol. Biochem. Behav. 28:503-510.
308a. Chandler, M.L., Barnhill, J.L., and Gualtieri, C.T. (1989) Tryptophan antagonism of stimulant-induced tics. J. Clin. Psychopharmacol. 9:69-70.
309. Chase, T.N. and Murphy, D.L.(1973). Serotonin and central nervous system function. Annu. Rev. Pharmacol. 18:181-197.
310. Chase, T.N., Foster, N.L., Fedio, P., Brooks, R., Mansi, L., Kessler, R. and Chiro, G.D.(1984). Gilles de la Tourette syndrome: Studies with the fluorine-18-labeled fluorodeoxyglucose positron emission tomographic method. Ann. Neurol. (Suppl) 15:S175.
311. Checkley, H.(1982). The Mask of Sanity. C.V.Mosby, New York.
312. Chess, S.(1960). Diagnosis and treatment of the hyperactive child. N. Y. State J. Med. 60:2379-2385.
313. Chess, S.(1971). Autism in children with congenital rubella. J. Autism Child. Schizophr. 1:33-47.
314. Chethik, M. and Fast, I.(1970). A function of 'fantasy' in the borderline child. Am. J. Orthopsychiatry 40:756-765.
314a. Chiara, G.D. and Imperato, A.(1988). Drugs abused by humans preferentially increase synaptic dopamine concentrations in the mesolimbic system of freely moving rats. Proc. Nat. Acad. Sci. U.S.A. 85:5275-5278.
315. Chiara, G.D., Camba, R. and Spano, P.F.(1971). Evidence for inhibition by brain serotonin of mouse killing behavior in rats. Nature 233:272-273.
316. Chiara, G.D., Imperato, A. and Mulas, A.(1987). Preferential stimulation of dopamine release in the mesolimbic system: A common feature of drugs of abuse. In M. Sandler et al.(eds). Neurotransmitter Interactions in the Basal Ganglia. Raven Press, New York. pp171-182.
317. Cho-Chung, Y.S. and Pitot, H.C.(1968). Regulatory effects of nicotinamide on tryptophan pyrrolase synthesis in rat liver in vivo. Eur. J. Biochem. 3:401-406.
318. Cho-Chung, Y.S., and Pitot, H.C.(1967). Feedback control of rat liver tryptophan pyrrolase. J. Biol. Chem. 242:1192-1198.
319. Chouinard, G. and Jones, B.D.(1978). Schizophrenia as dopamine-deficiency disease. Lancet 2:99-100.
320. Chouinard, G., Young, S.N., Annable, L., Sourkes, T.L.(1979). Tryptophan-nicotinamide, imipramine and their combination in depression. Acta Psychiatr. Scand. 59:395-414.
320a. Chouvet, G., Akaoka, H. and Aston-Jones, G. (1988). Diminution sélective par la sérotonine de l'excitation des neurones du locus coeruleus évoquée par le glutamate. C.R. Acad. Sci. Paris 306: Série III, 339-344.
321. Churchland, P.S.(1986). Neurophilosophy. MIT Press, Cambridge, MA.
321a. Claghorn, J., Honigfeld, G., Abuzzahab, F.S., Wang, R. Steinbook, R., Tuason, V., and Klerman, G.(1987). The risks and benefits of clozapine versus chlorpromazine. J. Clin. Psychopharmacol 7:377-384.
322. Clarke, D.J. and Ford, R.(1988). Treatment of refractory Tourette syndrome with haloperidol decanoate. Acta Psychiatr. Scand. 77:495-496.
322a. Clifford, C.A., Murray, R.M. and Fulker, D.W.(1984). Genetic and environmental influences on obsessional traits and symptoms. Psychol. Med. 14:791-800, 1984.
323. Clements, S.D. and Peters, E.(1962). Minimal brain dysfunction in the school-age child. Arch. Gen. Psychiatry 6:185-197.
324. Clements, S.D.(1962). Minimal brain dysfunctions in the school-age child. Arch. Gen. Psychiatry 6:17-29.
325. Clements, S.D.(1966). Minimal Brain Dysfunction. DHEW Publication No. (NIH) 76-349. U.S. Dept. Health, Education and Welfare, Washington, DC.
326. Cloninger, C.R. and Gottesman, I.I.(1987). Genetic and environmental factors in antisocial behavior disorders. In S.A. Mednick, T.E. Moffitt and S.A. Stack (eds) The Causes of Crime. New Biological Approaches. Cambridge Univ. Press, New York, pp6-109.
327. Cloninger, C.R. and Guze, S.B.(1970). Psychiatric illness in female criminality: The role of sociopathy and hysteria in the antisocial woman. Am. J. Psychiatry 127:303-311.
328. Cloninger, C.R.(1986). A unified biosocial theory of personality and its role in the development of anxiety states. Psychiatr. Dev. 3:167-226.
329. Cloninger, C.R.(1987). Neurogenetic adaptive mechanisms in alcoholism. Science 236:410-416.
330. Cloninger, C.R., Bohman, M., Sigvardsson, S. (1981). Inheritance of alcohol abuse: Cross-fostering analysis of adopted men. Arch. Gen. Psychiatry 38:861-868.
331. Cloninger, C.R., Christiansen, K.O., Reich, T. and Gottesman, I.I.(1978). Implications of sex differences in the prevalence of antisocial personality, alcoholism and criminality for familial transmission. Arch. Gen. Psychiatry 35:941-951.
332. Cloninger, C.R., Martin, R.L., Clayton, P., and Guze, S.B.(1981). A blind follow-up and family study of anxiety neuroses: Preliminary analysis of the St. Louis 500. In D.F. Klein and J. Rabkin (eds) Anxiety: New Research and Changing Concepts. Raven Press, New York, pp137-154.
333. Cloninger, C.R., Sigvardsson, S. and Bohman, M.(1988). Childhood personality predicts alcohol abuse in young adults. Alcoholism 12:494-505.
334. Cloninger, C.R., Sigvardsson, S., Bohman, M. and von Knorring, A.L.(1982). Predispositon to petty criminality in Swedish adoptees. II. Cross-fostering analysis of gene-environment interaction. Arch. Gen. Psychiatry 39:1242-1247.
335. Cloninger, C.R.. Bohman, M., Sigvardsson, S. and von Knorring, A.-L.(1985). Psychopathology in adopted-out children of alcoholics. The Stockholm adoption study. Recent

Dev. Alcohol. 3:37-51.

335a. Coccaro, E.F., Siever, L.J., Klar, H.M., Maurer, G., Cochrane, K., Cooper, T.B., Mohs, R.C. and Davis, K.L.(1989). Serotonergic studies in patients with affective and personality disorders. Arch. Gen. Psychiatry 46:587-599.

336. Cohen, B.C., Rosenbaum, G., Luby, E.D., Gottlieb, J.S. and Yelen, D.(1962). Comparison of phenyclidine hydrochloride (sernyl) with other drugs. Arch. Gen. Psychiatry 6:395-401.

337. Cohen, B.M. and Lipinski, J.F.(1986). In vivo potencies of antipsychotic drugs in blocking alpha 1 noradrenergic and dopamine D_2 receptors: Implications for drug mechanisms of action. Life Sci. 39:2571-2580.

338. Cohen, D.J., Caparolo, B.K., Shaywitz, B.A. and Bowers, M.B.(1977). Dopamine and serotonin metabolism in neuropsychiatrically disturbed children: CSF homovanilic acid and 5-hydroxyindolacetic acid. Arch. Gen. Psychiatry 34:545-550.

339. Cohen, D.J., Detlor, J., Young, G. and Shaywitz, B.A.(1980). Clonidine ameliorates Gilles de la Tourette syndrome. Arch. Gen. Psychiatry 37:1350-1357.

340. Cohen, D.J., Leckman, J.F. and Shaywitz, B.A.(1983). Tourette's syndrome and other tics. In D. Shaffer, A.A. Ehrhardt and L. Greenhill (eds) Diagnosis and Treatment in Pediatric Psychiatry. McMillan Free Press, Inc., New York pp3-28.

341. Cohen, D.J., Shaywitz, B.A., Caparulo, B., Young, J.G and Bowers, M.B.(1978). Chronic multiple tics of Gilles de la Tourette's disease: CSF acid monamine metabolites after probenecid administration. Arch. Gen. Psychiatry 35:245-250.

342. Cohen, D.J., Shaywitz, B.A., Young, J.G., Carbonari, C.M., Nathanson, J.A., Lieberman, D., Bowers, M.B. Jr., and Maaš, J.W.(1979). Central biogenic amine metabolism in children with the syndrome of chronic multiple tics of Gilles de la Tourette. J. Am. Acad. Child Psychiatry 18:320-341.

343. Cohen, D.J., Young, J.G., Nathanson, J.A. and Shaywitz, B.A.(1979). Clonidine in Tourette's syndrome. Lancet 2:551-553.

344. Cohen, M.E., Badal, D.W., Kilpatrick, A., Reed, E.W. and White, P.D.(1951). The high familial prevalence of neurocirculatory asthenia (anxiety neurosis, effort syndrome). Am. J. Hum. Genet. 3:126-158.

345. Cohen, N.J., and Douglas, V.I.(1972). Characteristics of the orienting response in hyperactive and normal children. Psychophysiology 9:238-245.

346. Coid, J., Allolio, B. and Rees, L.H.(1983). Raised plasma metenkephalin in patients who habitually mutilate themselves. Lancet 1:545-546.

347. Cole, S.O.(1978). Brain mechansims of ampthemine-induced anorexia, locomotion, and stereotypy: A review. Neurosci. Biobehav. Rev. 2:89-100.

348. Coleman, M. and Gillberg, C.(1985). The Biology of the Autistic Syndromes. Praeger Scientific, New York, p262.

349. Coleman, M.(1971). Serotonin concentrations in whole blood of hyperactive children. J. Pediatr. 78:985-990.

350. Coleman, M., Boulin, D.J. and Davis, M.(1971). Serotonin abnormalities in the infantile spasms syndrome. Neurology 21:421-426.

351. Coleman, M., Steinberg, G., Tippett, J., et al.(1979). A preliminary study of the effect of pyridoxine administration in a subgroup of hyperkinetic children: A double blind crossover comparison with methylphenidate. Biol. Psychiatry 14:741-751.

352. Collins, D.M. and Myers, R.D.(1987). Buspirone attenuates volitional alcohol intake in the chronically drinking monkey. Alcohol 4:49-56.

353. Comb, M., Seeburg, P.H., Adelman, J., Eiden, L., Herbert, E.(1982). Primary structure of the human Met- and Leu-enkephalin precursor and its mRNA. Nature 295:663-667.

354. Comfort, A.(1977). Morphine as antipsychotic drug. Lancet 1:95.

355. Comings, B.G. and Comings, D.E.(1987). A controlled study of Tourette syndrome.V. Depression and mania. Am. J. Hum. Genet. 41:804-821.

356. Comings, D.E. and Comings, B.G.(1982). A case of familial exhibitionism in Tourettte's syndrome successfully treated with haloperidol. Am. J. Psychiatry 139:913-915.

357. Comings, D.E. and Comings, B.G.(1984). Tourette's syndrome and attention deficit disorder with hyperactivity: Are they genetically related? J. Am. Acad. Child Psychiatry 23:138-146.

358. Comings, D.E. and Comings, B.G.(1985). Tourette syndrome: Clinical and psychological aspects of 250 cases. Am. J. Hum. Genet. 37:435-450.

359. Comings, D.E. and Comings, B.G.(1986). Evidence for an X-linked modifier gene affecting the expression of Tourette syndrome and its relevance to the increased frequency of speech, cognitive, and behavioral disorders in males. Proc. Natl. Acad. Sci. U.S.A. 83:2551-2555.

360. Comings, D.E. and Comings, B.G.(1987). A controlled study of Tourette syndrome. I. Attention deficit disorder, learning disorders, and school problems. Am. J. Hum. Genet. 41:701-741.

361. Comings, D.E. and Comings, B.G.(1987). A controlled study of Tourette syndrome. II. Conduct. Am. J. Hum. Genet. 41:742-760.

362. Comings, D.E. and Comings, B.G.(1987). A controlled study of Tourette syndrome. III. Phobias and panic attacks. Am. J. Hum. Genet. 41:761-781.

363. Comings, D.E. and Comings, B.G.(1987). A controlled study of Tourette syndrome. IV. Obsessive-compulsive and schizoid behavior. Am. J. Hum. Genet. 41:782-803.

364. Comings, D.E. and Comings, B.G.(1987). A controlled study of Tourette syndrome. VI. Early development, sleep problems, allergies and handedness. Am. J. Hum. Genet. 41: 822-838.

365. Comings, D.E. and Comings, B.G.(1987). Hereditary agoraphobia with panic attacks and hereditary obsessive-compulsive behavior in relatives of patients with Tourette syndrome. Br. J. Psychiatry 151:195-199.

366. Comings, D.E. and Comings, B.G.(1987). Tourette syndrome and attention deficit disorder with hyperactivity. Letter to the editor. Arch. Gen. Psychiatry 44:1023-1025.

367. Comings, D.E. and Comings, B.G.(1988). A controlled study of Tourette syndrome— Revisited. A reply to the letter of Pauls et al. Am. J. Hum. Genet. 43: 209-217.

368. Comings, D.E. and Comings, B.G.(1988). Tourette's syndrome and attention deficit disorder. In D.J. Cohen, R.D. Bruun, and J.F. Leckman (eds) Tourette's Syndrome and Tic Disorders: Clinical Understanding and Treatment. John Wiley & Sons, New York, pp119-135.

369. Comings, D.E. and Comings, B.G.(1989).The

genetics of Tourette syndrome and its relationship to other psychiatric disorders. In L. Wetterberg (ed) The Genetics of Neuropsychiatric Disorders. MacMillan Press, LTS. London. pp 179-189.
370. Comings, D.E. and Comings, B.G.(1989). Use of fluoxetine in the treatment of behavior disorders in Tourette syndrome. (unpublished manuscript).
370a. Comings, D.E. and Comings, B.G.(1990). A controlled family history study of Tourette syndrome. I. Attention deficit hyperactivity disorder, learning disorders and dyslexia. J.Clin.Psych. 51:275-280.
370b. Comings, D.E. and Comings, B.G.(1990). A controlled family history study of Tourette syndrome. II. Alcoholism, drug abuse and obesity. J.Clin.Psych.51:281-287
370c. Comings, D.E. and Comings, B.G.(1990). A controlled family history study of Tourette syndrome. III. Other psychiatric disorders. J.Clin.Psych.51:288-291.
371. Comings, D.E. and Okada, T.A.(1970). Whole-mount electron microscopy of the centromere region of metacentric and telocentric mammalian chromosomes. Cytogenetics 9:436-449.
371a. Comings, D.E.(1982). The genetics of Tourette syndrome. Tourette Syndrome Association Newsletter.
372. Comings, D.E.(1986). The genetics of Rett syndrome: The consequences of a disorder where every case is a new mutation. Am. J. Med Genet. 24:383-388.
373. Comings, D.E.(1987). A controlled study of Tourette syndrome. VII. Summary: A common genetic disorder due to disinhibition of the limbic system. Am. J. Hum. Genet. 41: 839-866.
374. Comings, D.E.(1989). The genetics of human Behavior: Lessions for two societies. Presidential Address, American Society of Human Genetics. Am. J. Hum. Genet. 44:452-460.
375. Comings, D.E.(1990). Blood serotonin and tryptophan in Tourette syndrome. Amer. J. Med. Genet. 36:418-430.
376. Comings, D.E., Comings, B.G., Cloninger, C.R. and Devor, R.(1984). Detection of major gene for Gilles de la Tourette syndrome. Am. J. Hum. Genet. 36:586-600.
377. Comings, D.E., Comings, B.G. and Knell, E.(1989). Hypothesis: Homozygosity in Tourette syndrome. Am. J. Med. Genet. 35:413-421.
378. Comings, D.E., Comings, B.G., Dietz, G., Muhleman, D., Okada, T.A., Sarinana, F., Simmer, R. and Stock, D.(1986). Evidence the Tourette syndrome gene is at 18q22.1. International Congress of Human Genetics, West Berlin.
379. Comings, D.E., Comings, B.G., Dietz, G., Muhleman, D., Okada, T.A., Sarinana, F., Simmer, R., Sparkes, R., Crist, M. and Stock, D.(1986). Linkage studies in Tourette syndrome. Am. J. Hum. Genet. 39:A151.
379a. Comings, D.E., Comings, B.G., Muhleman, D. and Dietz, G.(1989).Tourette syndrome, serotonin, tryptophan oxygenase, type II alcoholism and depression. First World Congress on Psychiatric Genetics, August 3-5, Churchill College, Cambridge, England, p85.
379b. Comings, D.E., Dietz, G. and Muhlemann, D.(1989). Localization of human tryptophan oxygenase to 4q25-31. Cytogenet. Cell Genet. HGM 10. 51:979.
380. Comings, D.E., Himes, J. and Comings, B.G.(1989). A epidemiological study of Tourette syndrome in a school district. J.Clin.Psych.51:463-469.
380a. Comings, D.E., Muhlemann, D. and Dietz, G.(1991). Localization of human tryptophan oxygenase to 4q31. Possible implications for human behavioral disorders. Genomics 9:301-308.
381. Connell, P.H.(1958). Amphetamine Psychosis. Chapman & Hall, London.
382. Conners, C.(1970). Symptom patterns in hyperactive, neurotic, and normal children. Child. Dev. 41:667-682.
383. Conners, C.K. and Eisenberg, L.(1963). The effects of methylpenidate on symptomatology and learning in disturbed children. Am. J. Psychiatry 120:458-464.
384. Conners, C.K.(1972). Pharmacotherapy of psychopathology in children. In H.C. Quay and H. Wery (eds) Psychopathological Disorders of Childhood. John Wiley & Sons, Inc, New York, pp316-347.
385. Conners, C.K.(1975). Learning disabilities and stimulant drugs in children: Theoretical implications. Presented at the NATO Conference on the Neuropsychology of Learning Disorders, Denmark.
386. Conners, C.K., and Taylor, E.(1980). Pemoline, methylphenidate, and placebo in children with minimal brain dysfunction. Arch. Gen. Psychiatry 37:922-930.
386a. Connick, J.H. and Stone, T.W.(1989). Kynurenines in the CNS: Miscellaneous neurochemical considerations. In T.W. Stone (ed) Quinolinic Acid and the Kynurenines. CRC Press, Inc, Boca Raton, FL, pp77-90.
386b. Cook, P.J., Lindenbaum, R., Salonen, R., de la Chapelle, A., Daker, M.G., Buckton, K.E., Noade, J. Tippett, P.(1981). The MNSs blood groups of families with chromosome 4 rearrangements. Ann. Hum. Genet. 45:39-47.
387. Cools, A.R. and van Rossum J.M.(1976). Excitation-mediating and inhibition-mediating dopamine-receptors: A new concept towards a better understanding of electrophysiological, biochemical, pharmacological, functions and clinical data. Psychopharmacologia 45:243-254.
388. Cools, A.R. and van Rossum, J.M.(1980). Multiple receptors for brain dopamine in behavior regulation: Concept of dopamine-e and dopamine-i receptors. Life Sci. 27:1237-1253.
389. Cools, A.R.(1974). The transsynaptic relationship between dopamine and serotonin in the caudate nuclei of cats. Psychopharmacologia 36: 229-237.
390. Cools, A.R.(1977). Two functionally and pharmacologically distinct dopamine receptors in the rat brain. In E. Costa and G.L. Gessa (eds) Adv. Biochem. Psychopharmacol. 16:215-225.
391. Coons, H.W., Klorman, R. and Bordstedt, A.D.(1987). Effects of methylphenidate on adolescents with a childhood history of attention deficit disorder: II. Information processing. J. Am. Acad. Child Psychiatry 26:368-374.
392. Coppen, A.J. and Wood, K.(1978). Tryptophan and depressive illness. Psychol. Med. 8:49-57.
393. Coppen, A.J.(1967). The biochemistry of affective disorders. Br. J. Psychiatry 113:1237-1264.
394. Coppen, A.J., Brooksbank, B.W.L. and Peet, M.(1972). Tryptophan concentration in the cerebrospinal fluid of depressive patients. Lancet 1:1393.
395. Coppen, A.J., Eccleston, E.G. and Peet, M.(1973). Total and free tryptophan concentration in the plasma of depressive patients. Lancet 2:60-63.
396. Coppen, A.J., Prange, A.J., and Whybrow, P.C.(1972). Abnormalities of indoleamines in affective disorders. Arch. Gen. Psychiatry 26:474-478.
397. Coppen, A.J., Shaw, D.M. and Farrell, J.P.(1963).

Potentiation of the antidepressive effect of nonamine oxidase inhibitor by tryptophan. Lancet 1:79.
398. Coppen, A.J., Swade, C. and Wood, K.(1978). Platelet 5-hydroxytryptamine accumulation in depressive illness. Clin. Chim. Acta. 87:165-168.
399. Corbett, J.A., Mathews, A.M., Connell, P.H. and Shapiro, D.A.(1969). Tics and Gilles de la Tourette's syndrome: A follow-up study and critical review. Br. J. Psychiatry 115:1229-1241.
400. Cornblatt, B.A., Lenzenweger, M.F., Dworkin, R.H. and Erlenmeyer-Kimling, L.(1985). Positive and negative schizophrenic symptoms, attention, and information processing. Schizophr. Bull. 11:397-408.
401. Corrodi, H., Fuxe, K., Ljungdahl, A., and Ögren, S.O.(1970). Studies on the action of some psychoactive drugs on central noradrenaline neurons after inhibition of dopamine b-hydroxylase. Brain Res. 24:451-470.
402. Coryell, W. and Zimmerman, M.(1988). The heritability of schizophrenia and schizoaffective disorder. Arch. Gen. Psychiatry 45:323-327.
402a. Coryell, W., Endicott, J., Keller, M., Andreasen, N., Grove, W., Hirschfeld, R.M.A. and Scheftner, W.(1989) Bipolar affective disorder and high achievement: A familial association. Am. J. Psychiatry 8:983-988.
403. Coscina, D.V., Grant, L.D., Balagura, S. and Grossman, S.P.(1972). Hyperdipsia after serotonin depleting midbrain lesions. Nature New Biol. 235:63-64.
404. Costa, E. and Garattini, S. (eds) (1970). Amphetamine and Related Compounds. Raven Press, New York.
405. Costa, E. and Guidotti, A.(1979). Molecular mechanisms in the receptor action of benzodiazepines. Annu. Rev. Pharmacol. 19:531-545.
406. Costa, E.(1977). Morphine, amphetamine, and noncataleptogenic neuroleptics. In E. Costa and G.L. Gessa (eds) Nonstriatal Dopaminergic Neurons. Advances in Biochemical Psychopharmacology. Raven Press, New York, pp557-563.
407. Costa, E., Fratta, W., Hong, J.S., Moroni, F. and Yong, H.-Y.T (1978). Interactions between enkephalinergic and other neuronal systems. In E. Costa and M. Trabucchi (eds) The Endorphins: Advances in Biochemical Psychopharmacology. Vol 18. Raven Press, New York, pp111-123.
408. Costall, B. and Naylor, R.J.(1974). The involvement of dopaminergic systems with the stereotyped behavior patterns induced by methylphenidate. J. Pharm. Pharmacol. 26:30-33.
409. Costall, B. and Naylor, R.J.(1975). The behavioral effects of dopamine applied intracerebrally to areas of the mesolimbic system. Eur. J. Pharmacol. 32:87-92.
410. Costall, B. and Naylor, R.J.(1976). A comparison of the abilities of typical neuroleptic agents and of thioridazine, clozapine, sulphride and metoclopramide to antagonize the hyperactivity induced by dopamine applied intracerebrally to areas of the extrapyramidal and mesolimbic systems. Eur. J. Pharmacol. 40:9-19.
411. Costall, B., Marsden, C.D., Naylor, R.J. and Pycock, C.J.(1977). Stereotyped behavior patterns and hyperactivity induced by amphetamine and apomorphine after discrete 6-hydroxydopamine lesions of extra-pyramidal and mesolimbic nuclei. Brain Res. 123:89-111.
412. Costall, B., Naylor, J.R., Marsden, C.D. and Pycock, C.J.(1976). Serotoninergic modulation of the dopamine response from the nucleus accumbens. J. Pharmacol. 28:523-526.
413. Costall, B., Naylor, R.J., Cannon, J.G. and Lee, T.(1977). Differentiation of the dopamine mechanism mediating stereotyped behavior and hyperactivity in the nucleus accumbens and caudatoputamen. J. Pharmacol. 29:337-342.
414. Cotton, N.S.(1979). The familial incidence of alcoholism: A review. J. Stud. Alcohol 40:89-116.
415. Cotzias, G.C., Papavasiliou, P.S. and Gellene, R.(1969). Modification of Parkinsonism – chronic treatment with L-dopa. N. Engl. J. Med. 280:337-345.
416. Courchesne, E., Yeung-Courchesne, R., Press, G.A., Hesselink, J.R. and Jernigan, T.L.(1988). Hypoplasia of cerebellar vermal lobules VI and VII in autism. N. Engl. J. Med. 318:1349-1354.
417. Coursey, R.D., Buchsbaum, M.S. and Murphy, D.L.(1979). MAO activity and evoked potentials in the identification of subjects biologically at risk for psychiatric disorders. Br. J. Psychiatry 134:372-381.
418. Coursin, D.B.(1964). Recommendations for standardization of the tryptophan load test. Am. J. Clin. Nutr. 14:56-61.
419. Cox, A., Rutter, M., Newman, S. and Bartak, L.(1975). A comparative study of infantile autism and specific developmental receptive language disorder: II. Parental characteristics. Br. J. Psychiatry 126:146-159.
420. Cox, W.H.(1982). An indication for use of imipramine in attention deficit disorder. Am. J. Psychiatry 139:1059-1060.
421. Crawley, J.C.W., Crow, T.J., Johnstone, E.C., Owen, F., Odland, S.R.D., et al. (1986). Dopamine D2 receptors in schizophrenia studies in vivo. Lancet 2:224.
422. Creese, I., Burt, D.R. and Snyder, S.H.(1976). Dopamine receptor binding predicts clinical and pharmacological potencies of antipsychotic drugs. Science 192:481-483.
423. Creese, I., Burt, D.R. and Snyder, S.H.(1977). Dopamine receptor binding enhancement accompanies lesion induced behavioral sensitivity. Science 197: 596-598.
424. Crescimanno, G., Salerno, M.T., Cortimiglia, R., Amato, G. and Infantellina, F.(1984). Functional relationship between claustrum and pyramidal tract neurons, in the cat. Neurosci. Lett. 44:125-129.
425. Critchley, M.(1970). Developmental Dyslexia. Heinemann, London.
426. Crosley, C.J.(1979). Decreased serotoninergic activity in Tourette syndrome. Ann. Neurol. 5:596.
427. Crow, T.J.(1971).The relation between electrical self-stimulation sites and catecholamine-containing neurones in the rat mesencephalon. Experientia 27:662.
428. Crow, T.J.(1977). Neurotransmitter-related pathways. The structure and function of central monamine neurons. In A.N. Davidson (ed) Biochemical Correlates of Brain Structure and Function. Academic Press, New York.
429. Crow, T.J.(1980). Molecular pathology of schizophrenia: More than one disease process? Br. Med. J. 280:66-68.
430. Crow, T.J., Baker, H.F., Cross, A.J., Joseph, M.H., Lofthouse, R., Longden, A., Owen, F., Riley, G.J., Glover, V. and Killpack, W.S.(1979). Monamine mechanisms in chronic schizophrenia: Postmortem neurochemical findings. Br. J. Psychiatry 134:249-256.
431. Crow, T.J., Johnstone, E.C., Deakin, J.F.W. and

Longden, A.(1976). Dopamine and schizophrenia. Lancet 2:563-566.
432. Crowe, R.R.(1974). An adoption study of antisocial personality. Arch. Gen. Psychiatry 31:785-791.
433. Crowe, R.R.(1975). An adoptive study of psychopathy: Preliminary results from arrest records and psychiatric hospital records. In R.R. Fieve, D. Rosenthal and H. Brill (eds) Genetic Research in Psychiatry. Johns Hopkins Univ. Press, Baltimore.
434. Crowe, R.R., Gaffney, G. and Kerger, R.(1982). Panic attacks in families of patients with mitral valve prolapse. J. Affect. Disord. 4:121-125.
435. Crowe, R.R., Namboodiri, K.K., Ashby, H.B. and Elston, R.C. (1981).Segregation and linkage analysis of a large kindred of unipolar depression. Neuropsychobiology 7:20-25.
436. Crowe, R.R., Noyes, R., Jr., Pauls, D.L. and Slymen, D.(1983). A family study of anxiety disorder. Arch. Gen. Psychiatry 40:1065-1069.
437. Crowe, R.R., Noyes, R., Jr., Wilson, A.F., Elston, R.C. and Ward, L.J.(1987). A linkage study of panic disorder. Arch. Gen. Psychiatry 44:933-937.
438. Crowe, R.R., Pauls, D., Slymen, D. and Noyes, R., Jr.(1980). A family study of anxiety neuroses. Arch. Gen. Psychiatry 37:77-79.
440. Cummings, J.L. and Frankel, M.(1985). Gilles de la Tourette syndrome and the neurological basis of obsessions and compulsions. Biol. Psychiatry 20:1117-1126.
441. Cunningham, L., Cadoret, R.J., Loftus, R. and Edwards, J.(1975). Studies of adoptees from psychiatrically disturbed biological parents. I. Psychiatric conditions in childhood and adolescence. Br. J. Psychiatry 126:534-549.
442. Curran, D.X., Hinterberger, H. and Lance, J.W.(1965). Total plasma serotonin 5-hydroxyindole acetic acid and p-hydroxy-m-methoxymandelic acid concentration in normal and migrainous subjects. Brain 88: 997-1010.
442a. Curzon, G. and Bridges, P.K.(1970). Tryptophan metabolism in depression. J. Neurol. Neurosurg. Psychiatry 33:698-704.
443. Curzon, G. and Green, A.R.(1968). Effect of hydrocortisone on rat brain 5-hydroxytryptamine. Life Sci. 7 (Part I):657-663.
444. Curzon, G. and Green, A.R.(1969). Liver tryptophan pyrrolase activity and brain 5-hydroxy-tryptamine. Biochem. J. 1111:15P.
445. Curzon, G.(1969). Tryptophan pyrrolase — A biochemical factor in depressive illness? Br. J. Psychiatry 115:1367-1374.
445a. Cruzon, G.(1979). Relationship between plasma, CSF and brain tryptophan. J. Neural Transmission Suppl. 15:81-92.
446. Curzon, G.(1980). Tryptophan, brain 5HT and behavior: How are they related? In O. Hayaishi, Y. Ishimura and R. Kido (eds) Biochemical and Medical Aspects of Tryptophan Metabolism. Elsevier/North Holland, New York, pp267-279.
447. Curzon, G., Friedel, J. and Knott, P.J.(1973). The effect of fatty acids on the binding of tryptophan to plasma protein. Nature 242:198-200.
448. Curzon, G., Kantamaneni, B.D., Van Boxel, P., Gillman, P.K., Bartlett, J.F. and Bridges, P.K.(1980). Substances related to 5-hydroxytryptopamine in plasma and in lumbar and ventricular fluids of psychiatric patients. In T.H. Svensson and A. Carlsson (eds) Biogenic Amines and Affective Disorders. Acta Psychiatry. Scand. (Suppl 280) 61:13-17.
449. Custer, R.L. and Custer, L.F.(1981). Soft signs of pathological gambling. Fifth National Conference on Gambling. Univ. of Nevada, Reno.
450. Custer, R.L.(1982). Gambling and addiction. In R.J. Craig and S.L. Baker (eds) Drug Dependent Patients: Treatment and Research. Charles C. Thomas, Springfield, IL.
451. Dahlström, A. and Fuxe, K.(1964). Evidence for the existence of monamine-containing neurons in the central nervous system. I. Demonstration of monamines in the cell bodies of brain stem neurons. Acta Physiol. Scand. 62 (Suppl 232):1-51.
452. Dalland, T.(1970). Response and stimulus perseveration in rats with septal and dorsal hhippocampal lesions. J. Comp. Physiol. Psychol. 71:114-118.
453. Dallard, T.(1974). Stimulus perseveration of rats with septal lesions. Physiol. Behav. 12:1057-1061.
454. Damasio, A.R. and Geschwind, N.(1984). The neural basis of language. Annu. Rev. Neurosci. 7:127-147.
455. Damasio, A.R. and Maurer, R.G.(1978). A neurological model for childhood autism. Arch. Neurol. 35: 777-786.
456. Damasio, A.R. and Van Hoesen, G.W.(1983). Emotional disturbances associated with focal lesions of the limbic frontal lobe. In K. Heilman, and P. Satz (eds), Neuropsychology of Human Emotion. Guilford Press, New York.
457. Davenport, C.B.(1911). Heredity in Relation to Eugenics. Henry Holt and Company, New York.
458. Davies, J. and Tongroach, P.(1978). Neuropharmacological studies on the nigrostriatial and raphe-striatal system in the rat. Eur. J. Pharmacol. 51:91-100.
459. Davis, A.(1971). An objective instrument for assessing hyperkinesis in chldren. J. Learn. Disabil. 4:491-498.
460. Davis, G.C., Wiliams, A.C., Markey, S.P., Ebert, M.H., Caine, E.D., Reichert, C.M. and Kopin, I.J.(1979). Parkinson secondary to intravenouos injection of meperidine analogues. Psychiatry Res. 1:249-254.
461. Davis, M. and Sheard, M.H.(1974). Habituation and sensitization of the rat startle-response. Effects of raphe lesions. Physiol. Behav. 12:425-431.
462. Davis, W.M. and Smith, S.G.(1977). Catecholaminergic mechanisms of reinforcement: Direct assessment by drug self-administration. Life Sci. 20:483-492.
463. Deakin, J.F.W., File, S.E., Hyde, J.R.G. and MacLeod, N.K.(1979). Ascending 5-HT pathways and behavioral habituation. Pharm. Biochem. Behav. 10:687-694.
464. DeFries, J.,C., Fulker, D.W. and LaBuda, M.C.(1987). Evidence for a genetic aetiology in reading disability of twins. Nature 329:537-539.
465. DeFries, J.C, Singer, S., Foch, T. and Lewitter, F.(1978). Familial nature of reading disability. Br. J. Psychiaty 132:361-367.
466. Delay, J., Deniker, P. and Harl, J.-M.(1952). Traitement des états d'excitation et d'agitation par une méthode médicamenteuse dérivéé de l'hiberno thérapie. Ann. Méd. Psychol. 110 (pt 2):267-273.
467. DeLisi, L.E., Neckers, L. M., Weinberg, D.R. and Wyatt, R.J.(1981). Increased whole blood serotonin concentrations in chronic schizophrenic patients. Arch. Gen. Psychiatry 38:647-650.
468. DeLong, G.R., Bean, C. and Brown, F.R.(1981). Acquired reversible autistic syndrome in acute encephalopa-

thic illness in children. Arch. of Neurol. 38:191-194.
469. DeLong, M.R., Georgopoulos, A.P. and Crutcher, M.D.(1983). In J. Massion, J. Paillard, W. Schultz, and M. Wiesendanger, (eds), Neural Coding of Motor Performance. Brain Res. Suppl 7:29-40.
470. Demet, W. and Kleitman, N.(1957). Cyclic variation in EEG during sleep and their relationship to eye movement, body motility and dreaming. Electroenceph. Clin. Neurophysiol. 9:673-678.
471. DeMyer, M., Barton, K., DeMyer, W.E., Norton, J.A., Allen, J. and Steele, R.(1973). Prognosis in autism: A follow-up study. J. Autism Child. Schizophr. 3:199-246.
472. DeMyer, M.K., Hington, J.N. and Jackson, R.K.(1981). Infantile autism reviewed: A decade of research. Schizophr. Bull. 7:388-451.
473. DeMyer, M.K., Shea, P.A., Hendrie, H.C. and Yoshimura, N.N.(1981). Plasma tryptophan and five other amino acids in depressed and normal subjects. Arch. Gen. Psychiatry 38:642-646.
474. Denckla, M.B., Bemporad, J.R., and MacKay, M.C.(1976). Tics following methylphenidate administration: A report of 20 cases. J. Am. Med. Assoc. 235:1349-1351.
475. Denhoff, E.(1973). The natural history of children with minimal brain dysfunction. Ann. N.Y. Acad. Sci. 205:188-206.
476. Denny-Brown, D.(1951). The frontal lobes and their functions. In A. Feiling (ed) Modern Trends in Neurology, Butterworth, London, pp13-89.
477. deObaldia, R and Parsons, O.A.(1984). Relationship of neuropsychological performance to primary alcoholism and self-reported symptoms of childhood minimal brain dysfunction. J. Stud. Alcohol 45:386-392.
478. DePauls, J.R. Jr., and Simpson, S.G.(1987). Therapeutic and genetic prospects of an atypical affective disorder. J. Clin. Psychopharmacol. 7:50S-54S.
478a. Depue, R.A., Arbisi,P., Spoont, M.R., Krauss, S., Leon, A. and Ainsworth, B.(1989). Seasonal and mood independence of low basal prolactin secretion in premenopausal women with seasonal affective disorder. Am. J. Psychiatry 146:989-995.
479. Descarries, L. and Beaudet, A.(1978). The serotonin innervation of adult rat hypothalamus. In J. D. Vincent and C. Kordon (eds) Cell Biology of Hypothalamic Neurosecretion. Colloq. Int. CNRS. 80:135-153.
480. Descarries, L., Beaudet, A., and Watkins, K.C.(1975). Serotonin nerve terminal in adult rat neocortex. Brain Res. 100:563-588.
481. Descarries, L., Watkins, K.C. and Lapierre, Y.(1977). Noradrenergic axon terminal in the cerebral cortex of rat. III. Topographic ultrastructural analysis. Brain Res. 133:197-222.
482. Detera-Wadleigh, S.D., Berrettini, W.H., Goldin, L.R., Boorman, D., Anderson, S. and Gershon, E.S.(1987). Close linkage of c-Harvey-ras-1 and the insulin gene to affective disorder is ruled out in three North American pedigrees. Nature 325:806-808.
483. Deutsch, H.(1942). Some forms of emotional disturbance and their relationship to schizophrenia. Psychoanal. Q. 11:301-321.
484. Devinsky, O.(1983). Neuroanatomy of Gilles de la Tourette's syndrome. Arch. Neurol. 40:508-514.
485. Devor, E.J.(1984). Complex segregation analysis of Gilles de la Tourette syndrome: Further evidence for a major locus mode of transmission. Am. J. Hum. Genet. 36:704-709.
486. deWeid, D.(1984). The importance of vasopressin in memory. Trends Neurosci. 7:62-63.
487. deWeid, D., Bohus, B., Van Wimersma Greidanus, T.B.(1975). Memory defect in rats with hereditary diabetes insipidus. Brain Res. 85:152-156.
488. deWied, D.(1979). Schizophrenia as an inborn error in the degradation of b-endorphin — A hypothesis. Trends Neurol. Sci. 2:79-84.
489. Deyo, R.A., Conner, R.L. and Panksepp, J.(1987). Perinatal leupeptin retards subsequent acquisition of a visual discrimination task in chicks. Behavioral. Neural Biol. 47:219-224.
490. Diagnostic and Statistical Manual of Mental Disorders, Third Edition.(1980). Am. Psychiatric Assoc., Washington, DC.
491. Diagnostic and Statistical Manual of Mental Disorders, Third Edition, Revised.(1987). Am. Psychiatric Ass., Washington, DC.
492. Divac, I.(1971). Frontal lobe system and spatial reversal in the rat. Neurpsychologia 9:175-183.
493. Divac, I.(1972). Neostriatum and functions of the prefrontal cortex. Acta Neurobiol. Exp. 32:461-477.
494. Divac, I., Bjorklund, A., Lindvall, O., and Passingham, R.E.(1978). Converging projections from the mediodorsal thalamic nucleus and mesencephalic dopaminergic neurons to the neocortex in three species. J. Comp. Neurol. 180:59-72.
495. Doenicke, A., Brand, J. and Perrin, V.L.(1988). Possible benefit of GR43175, a novel 5-HT1-like receptor agonist, for the acute treatment of severe migraine. Lancet 1:1309-1311.
496. Domino, E.F. and Krause, R.R.(1974). Free and bound tryptophan in drug free normal controls and chronic schizophrenic patients. Biol. Psychiatry 8: 265-279.
497. Domino, E.F. and Krause, R.R.(1974). Plasma tryptophan tolerance curves in drug free normal controls, schizophrenic patients and prisoner volunteers. J. Psychiatr. Res. 10:247-261.
497a. Donalan, K., Dietz, G., Muhlemann, D., and Comings, D.E.(1989). Localization of human tryptophan oxygenase to 4q31.3 by *in situ* hybridization. Cytogenet. Cell Genet. HMG 10. 51:992.
498. Donaldson, I.M., Dolphin, A., Jenner, P., Mardsen, C.D. and Pycock, C.(1976). The roles of noradrenaline and dopamine in contraversive circling behavior seen after unilateral electrolytic lesions of the locus coeruleus. Eur. J. Pharmacol. 39:179-191.
499. Donnai, D.(1987). Gene localization in Tourette syndrome. Lancet 1:627.
500. Donovick, P.J. and Burright, R.G.(1968). Water consumption in rats with septal lesions following two days of water deprivation. Physiol. Behav. 3:285-288.
501. Donovick, P.J., Burright, R.G., Sikorszky, R.D., Stamato, N.J. and MacLaughlin, W.W.(1978). Cue elimination effects on discrimination behavior of rats with septal lesions. Physiol. Behav. 20:71-78.
502. Dorsey, R.(1981). Clonidine and Gilles de la Tourette syndrome. Arch. Gen. Psychiatry 38:1185.
503. Douglas, V.A. and Peters, K.G.(1979).Toward a clearer definition of the attentional deficit of hyperactive children. In G.A. Hale and M. Lewis (eds) Attention and Cognitive Development, Plenum Press, New York pp173-247.

504. Douglas, V.I.(1963). Children's response to frustration: A developmental study. Can. J. Psychol. 19:161-171.
505. Douglas, V.I.(1972). Stop, look and listen: The problem of sustained attention and impulse control in hyperactive and normal children. Can. J. Behav. Sci. 4:259-282.
506. Douglas, V.I.(1974). Sustained attention and impulse control: Implications for the handicapped child. In Psychology and the Handicapped Child. U.S. Dept. Health, Education and Welfare, Office of Education, Washington,DC, pp149-168.
507. Dowling, J.E. and Ehinger, E.(1978). Synaptic organization of the dopaminergic neurons in the rabbit retina. J. Comp. Neurol. 180:203-220.
508. Dowling, J.E.(1987). The Retina. An Approachable Part of the Brain. Harvard Univ. Press, Boston, pp282.
509. Drewe, E.A.(1974). The effect of type and area of brain lesion on Wisconsin Card Sorting Test performance. Cortex 10:159-170.
510. Drewe, E.A.(1975). Go-no go learning after frontal lobe lesions in humans. Cortex. 11:8-16.
511. Dubovsky, S.L., Franks, R.D., Allen, S. and Murphy, J.(1986). Calcium antagonists in mania: A double-blind study of verapamil. Psychiatry Res. 18:309-320.
512. Dunlap, J.R.(1960). A case of Gilles de la Tourette's syndrome (maladie des tics): A study of intrafamily dynamics. J. Nerv. Ment. Dis. 130:340-344.
513. Dunner, D.L., Gershorn, E.S., and Goodwin, F.K.(1970). Heritable factors in the severity of affective illness. Sci. Proc. Am. Psychiatr. Assoc. 123:187-188.
514. Dunner. D.L.(1983). Sub-types of bipolar affective disorder with particular regard to bipolar II. Psychiatry Dev. 1:75-86.
515. Dykman, R.A and Ackerman, P.T(1980). Long term follow-up studies of hyperactive children: A 5 year follow-up study. In B.W. Camp (ed) Advances in Behavioral Pediatrics, Vol 1. Jai Press, Greenwich, CT.
516. Dykman, R.A., Ackerman, P.T., Clements, S.D., and Peters, J.E.(1971). Specific learning disabilities: An attentional deficit syndrome. In H.R. Myklebust (ed) Progress in Learning Disabilities, Vol 2. Grune & Stratton, New York.
517. Early, T.S., Reiman, E.M., Raichle, M.E. and Spitznagel, E.L.(1987). Left globus pallidus abnormality in never-medicated patients with schizophrenia. Proc. Natl. Acad. Sci. U.S.A. 84:561-563.
518. Ebaugh, F.G.(1923). Neuropsychiatric sequalae of epidemic encephalitis in children. Am. J. Dis. Child. 25:89-97.
519. Editorial.(1986). Akathesia and antipsychotic drugs. Lancet 2:1131-1132.
520. Egeland, J.A., Gerhard, D.S., Pauls, D.L., Sussex, J.N., Kidd, K.K., Allen, C.R., Hostetter, A.M. and Housman, D.E.(1987). Bipolar affective disorders linked to DNA markers on chromosome 11. Nature 325:783-787.
521. Egger, J., Carter, C.M., Graham, P.J., Gumley, D. and Soothill, J.F.(1985). Controlled trial of oligoantigenic treatment in the hyperkinetic syndrome. Lancet 1:540-545.
522. Ehringer, H. and Hornykiewicz, O.(1960). Verteilung von Noradrenalin und Dopamin (3-hydroxytyramin) im Gehirn des Menschen und ihr Verhalten bei Erkrankungen des extrapyramidalen Systems. Klin. Wochenschr. 38:1236-1239.
523. Eichler, A.J., Antelman, S.M. and Black, C.A.(1980). Amphetamine stereotypy is not a homogeneious phenomenon: Sniffing and licking show distinct profiles of sensitization and tolerance. Psychopharmacology 68:289-290.
524. Eisenberg, L.(1957). Psychiatric implication of brain damage in children. Psychiatr. Q. 31:72-92.
525. Eisenberg, L.(1966). The management of the hyperactive child. Dev. Med. Child. Neurol. 8:593-598.
526. Eisenberg, L.(1972). The hyperkinetic child and stimulant drugs. New Engl. J. Med. 287:249-250.
527. Eisenberg, L.(1973). The overactive child. Hosp. Pract. 8:151-160.
528. Eisenberg, L., Ascher, E., Kanner, L.(1959). A clinical study of Gilles de la Tourette's syndrome (maladie des tics) in children. Am. J. Psychiatry 115:715-723.
529. Ekstein, R. and Wallerstein, J.(1954). Observations on the psychology of borderline and psychotic children. Psychoanal. Study Child 9:344-369.
530. Eldridge, R., Sweet, R., Lake, R., Ziegler, M. and Shapiro, A.K.(1977). Gilles de la Tourette's syndrome: Clinical, genetic, psychologic, and biochemical aspects in 21 selected families. Neurology 27:115-124.
531. Eldridge, R., Wassman, E.R., Lee, L. and Koerber, T.(1979). Gilles de la Tourette Syndrome. In R.M. Goodman and A.G. Motulsky (eds) Genetic Diseases Among Ashkenazi Jews. Raven Press, New York.
532. Ellinwood, E.H.,Jr.(1967). Amphetamine psychosis: I. Description of the individuals and process. J. Nerv. Ment. Dis. 144:273-283.
533. Elliott, F.A.(1986). Historical perspective on neurobehavior. Psychiatr. Clin. N. A. 9:225-239.
534. Ellison, G.D.(1977). Animal models of psychopathology: The low norepinephrine and low-serotonin rat. Am. Psychol. 32:1036-1045.
534a. El-Sewedy, S.M. (1989).Pharmacology of the kynurenine pathway. In T.W. Stone (ed) Quinolinic Acid and the Kynurenines. CRC Press, Inc, Boca Raton, FL, pp101-112.
535. Elston, R.C. and Stewart, J.(1971). A general model for the genetic analysis of pedigree data. Hum. Hered. 21:523-542.
536. Elston, R.C.(1973). Ascertainment and age of onset in pedigree analysis. Hum. Hered. 23:105-112.
537. Emery, G., Hollon, S.D. and Bedrosian, R.C.(1981). New Directions in Cognitive Therapy. Guilford Press, New York.
538. Emson, P.C. and Koob, G.F.(1978). The origin and distribution of dopamine-containing afferents to the rat frontal cortex. Brain Res. 142:249-267.
539. Emson, P.C. and Lindvall, O.(1979). Distribution of punitive neurotransmitters in the neocortex. Neuroscience 4:1-30.
539a. Endicott, N.A.(1989). Psychophysiological correlates of 'bipolarlity.' J. Affect. Disord. 17:47-56.
540. Engel, M.(1963). Psychological testing of borderline psychotic children. Arch. Gen. Psychiatry 8:426-434.
541. Enna, S.J.(1981). Brain serotonin receptors and neuropsychiatric disorders. In B. Haber and S. Gabay (eds) Serotonin: Current Apsects of Neurochemistry and Function. Plenum Press, New York, pp347-358.
542. Enoch, J.M., Itzhaki, A., Lakshminarayanan, V., Comerford, J.P., Lieberman, M. and Lowe, T.(1988). Gilles de la Tourette syndrome: Visual effects. Neuro-opthalmology 5:251-257.
543. Enoch, J.M., Itzhaki, A., Lakshmnarayanan. V., Comerford, J., Lieberman, M. and Lowe, T. (1988). Gilles de la Tourette syndrome: Genetic marker? Neuro-opthalmology.

8:259-265.
544. Enoch, J.M., Lakshinarayanan, V., and Itzhaki, A.(1987). Psychophysical studies of neuropsychiatric patients on and off haloperidol. In I. Bodis-Woilner (eds). Dopaminergic Mechanisms in Vision, Roundtable Conference pp1-9.
545. Enoch, J.M., Lakshminarayanan, V., Itzhaki, A., Schechter, G. and Marmor, M.(1987). Layer-by-layer perimetry and haloperidol: Implication for schizophrenia and other diseases. Proc. 6th International Catecholamine Symposium, Jerusalem, June.
546. Erenberg, G. and Rothner, A.D.(1978). Tourette syndrome: A childhood disorder. Cleveland Clin. Q. 45:207-212.
547. Erenberg, G.(1982). Stimulant medications in Tourette's syndrome. J. Am. Med. Assoc. 258:1062.
548. Erenberg, G.(1988). Pharmacologic therapy of tics in childhood. Pediatr. Ann. 17:395-404.
549. Erenberg, G., Cruse, R.P. and Rothner, A.D.(1985). The natural history of Tourette's syndrome. Ann. Neurol. 18:386.
550. Erenberg, G., Cruse, R.P. and Rothner, A.D.(1986). Tourette syndrome: An analysis of 200 pediatric and adolescent cases. Cleveland Clin. Q. 53:127-131.
551. Erenberg, G., Cruse, R.P., and Rothner, A.D.(1985). Gilles de la Tourette's syndrome: Effects of stimulant drugs. Neurology 35:1346-1348.
552. Erenberg, G., Cruse, R.P., Rothner, D.O. and Rothner, A.D.(1987). The natural history of Tourette syndrome: A follow up study. Ann. Neurol. 22:383-385.
553. Ericksson, B. and Persson, T.(1969). Gilles de la Tourette's syndrome. Br. J. Psychiatry 115:351-353.
554. Erinoff, L., MacPhail, R.C., Heller, A. and Seiden, L.S.(1979). Age-dependent effects of 6-hydroxydopamine on locomotion activity in the rat. Brain Res. 164:195-205.
555. Erspamer, V. and Asero, B.(1952). Identification of enteramine, the specific hormone of the enterochromataffin cell system, is 5-hydroxytryptamine. Nature 169:800-801.
556. Erspamer, V.(1940). Pharmakologische Studien über Enteramin. I. Mitteilung: Uber die Wirkung von Acetonextrakten der Kaninchenmagenshcleimhaut auf den Blutdruck und auf isolierte überlebende Organe. Arch. Exp. Pathol. Pharmakol. 196:343-365.
557. Erspamer, V., Testini, A.J.(1959). Observations on the release and turnover rate of 5-hydroxytryptamine in the gastrointestinal tract. J. Pharm. Pharmacol. 11:618-623.
558. Ervin, C.S., Little, R.E., Streissguth, P.A. and Beck, D.E.(1984). Alcoholic fathering and its relation to child's intellectual development: A pilot investigation. Alcohol. Clin. Exp. Res. 8:362-365.
559. Eslinger, P.J. and Damasio, A.R.(1985). Severe disturbance of higher cognition after bilateral frontal lobe ablation. Neurology 35:1731-1741.
560. Evans, L., Best, J., Moore, G. and Cox, J.(1980). Zimelidine — A serotonin uptake blocker in the treatment of phobic anxiety. Prog. Neuropsychopharmacol. 4:75-79.
561. Evans, R.W., Gualtieri, C.T. and Hicks, R.E.(1986). A neuropathic substrate for stimulant drug effects in hyperactive children. Clin. Neuropharmacol. 9:264-281.
562. Evarts, E.V., Kimura, M., Wurtz, R.H. and Hikosaka, O.(1984). Behavioral correlates of activity in basal ganglia neurons. Trends Neurosci. 7:447-453.
563. Evarts, E.V., Shinoda, Y. and Wise, S.P.(1984). Neurophysiological Approaches to Higher Brain Function. John Wiley and Sons, New York.
564. Everett, G. M. and Borcherding, J.W.(1970). L-DOPA: Effect on concentration of dopamine, norepinephrine and serotonin in brains of mice. Science 168:849-850.
565. Everett, G.M.(1973). Genetic factors in the control of biogenic amines and brain excitability. In E. Usdin and S.H. Snyder (eds). Frontiers in Catecholamine Research. Pergamon Press, New York, pp 657-659.
566. Everett, G.M.(1975). Role of biogenic amines in the modulation of aggressive and sexual behavior in animals and man. In M. Sandler and G.L. Gessa Sexual Behavior: Pharmacology and Biochemistry. Raven Press, NewYork, pp81-84.
567. Eysenck, H.J. and Eysenck, M.W.(1984). Personality and Individual Differences. Plenum Press, New York.
568. Eysenck, H.J.(1967). The Biological Basis of Personality. Charles C. Thomas, Springfield, IL.
569. Eysenck, H.J.(1977). Crime and Personality, 3rd Edition. Granada, London..
570. Faillace, L.A., Allen, R.P., McQueen, J.D. and Northrup, B.(1971). Cognitive defects from bilateral cingulotomy for intractable pain in man. Dis. Nerv. Syst. 32:171-175.
571. Falconer, D.S.(1981). Introduction to Quantitative Genetics, 2nd Edition. Longman, New York.
572. Fallon, J.H. and Moore, R.Y.(1978). Catecholamine innervation of the basal forebrain. IV. Topography of the dopamine projection onto the basal fore brain and neostriatum. J. Comp. Neurol. 180:545-580.
573. Fanciullacci, M., Michelacci, S., Baldi, E., Spillantini, M., Pietrini, U., Spolveri, S, Salmon, S. and Sicuteri, F.(1983). Analgesizing activity of captopril in migraine: A clinical and pharmacological approach. In S. Ehrenpreis and F. Sicuteri (eds) Degradation of Endogenous Opioids: Its Relevance in Human Pathology and Therapy. Raven Press, New York, pp217-231.
574. Farkas, T., Wolf, A.P., Jaeger, J., Brodie, J.D., Christman, D.R. and Fowler, J.S.(1984). Regional brain glucose metabolism in chronic schizophrenia. Arch. Gen. Psychiatry 41:293-300.
575. Farley, I.J., Price, K.S., McCullough, E., Deck, J.H.N., Hordynski, W. and Hornykiewicz, O.(1978). Norepinephrine in chronic paranoid schizophrenia: Above-normal levels in limbic forebrain. Science 200:456-458.
576. Fast, I. and Chethik, M.(1972). Some aspects of object relationships in borderline children. Int. J. Psychoanal. 53:479-485.
577. Faull, K.F., King, R.J., Berger, P.A. and Barchas, J.D.(1984). Systems theory as a tool for integrating functional interactions among biogenic amines. In Catecholamines: Neuropharmacology and Central Nervous System — Therapeutic Aspects. Alan R. Liss, Inc., New York, pp143-152.
578. Faurbye, A and Pind, K.(1964). Investigations on the tryptophan metabolism (via kynurenine) in schizophrenic patients. Acta. Psychiatr. Scand. 40:244-248.
579. Faurbye, A., Rasch, P.J., Peterson, P.B., Brandborg, G. and Pakkenberg, H.(1964). Neurological symptoms in pharmacotherapy of psychoses. Acta Psychiatr. Scand. 40:10-27.
580. Feild, J.R., Corbin, K.B., Goldstein, N.P., Klass, D.W.(1966). Gilles de la Tourette's syndrome. Neurology 16:453-462.
581. Feinberg, I. and Carlson, V.(1968). Sleep variables as a function of age in man. Arch. Gen. Psychiat. 18:239-250.

582. Feinberg, M. and Carroll, B.J.(1979). Effects of dopamine agonists and antagonists in Tourette's disease. Arch. Gen. Psychiatry 36:979-985.
583. Feinberg, T.E., Shapiro, A.K. and Shapiro, E.(1986). Paroxysmal myoclonic dystonia with vocalizations: New entity or variant of preexisting syndromes? J. Neurol. Neurosurg. Psychiatry 49:52-57.
584. Feingold, B.F.(1975). Why Is Your Child Hyperactive? Random House, New York.
585. Feldberg, W. and Myers, R.D.(1963). A new concept of temperature regulation by amines in the hyopthalamus. Nature 200:1325.
586. Fenton, W.S. and McGlashan, T.H.(1986). The prognostic significance of obsessive-compulsive symptoms in schizophrenia. Am. J. Psychiatry 143:437-441.
587. Ferber, A., Mendelsohn, M. and Napier, A.(1983). The Book of Family Therapy. Jason Aronson,Inc., New York.
588. Ferguson, H.B., Pappas, B.A., Trites, R.L., Peters, D.A.V., and Taub, H.(1981). Plasma free and total tryptophan, blood serotonin, and the hyperactivity syndrome: No evidence for the serotonin deficiency hypothesis. Biol. Psychiatry 16:213-238.
589. Fernando, S.J.M.(1967). Gilles de la Tourette's syndrome. Br. J. Psychiatry 113:607-617.
590. Fernando, S.J.M.(1976). Six cases of Gilles de la Tourette's syndrome. Br. J. Psychiatry 128:436-441.
591. Fernstrom, J.D. and Wurtman, R.J.(1971). Brain serotonin content: Increase following ingestion of carbohydrate diet. Science 174:1023-1025.
592. Fernstrom, J.D.(1983). Role of precusor availability in control of monamine biosynthesis in brain. Physiol. Rev. 63:484-546.
592a. Fernstrom, J.D.(1988). Carbohydrate ingestion and brain serotonin synthesis: Relevance to punative control loop for regulating carbohydrate ingestion, and effects of aspartame consumption. Appetite 11(Supplement):35-41.
592b. Ferrari, M., Matthews, W.S. and Barabas, G.(1984). Children with Tourette syndrome: Results of psychological tests given prior to drug treatment. Dev. Behav. Pediatr. 5:116-119.
593. Ferrero, P., Guidotti, A., Conti-Tronconi, B. and Costa, E.(1984). A brain octadecaneuropeptide generated by tryptic digestion of DBI (diazepam binding inhibitor) functions as a proconflict ligand of benzodiazepine recognition sites. Neuropharmacology 23:1359-1362 .
594 . Ferron, A., Descartes, L. and Reader, T.A.(1982). Altered neuronal responsiveness to biogenic amines in rat cerebral cortex after serotonin denervation or depletion. Brain Res. 231:93-108.
594a. Ferron, A.(1988). Modified coeruleo-cortical noradrenergic neurotransmission after serotonin depletion by PCPA: Electrophysiological studies in the rat. Synapse 2:532-536.
595. Fibiger, H.C. and Lloyd, K.G.(1984). Neurological substrates of tardive dyskinesis. Trends Neurosci. 7:462-464.
596. Fibiger, H.C.(1978). Drugs and reinforcement mechanisms: A critical review of the catecholamine theory. Ann Rev. Pharmacol Toxicol. 18:37-56.
597. Fibiger, H.C., Carter, D.A. and Phillips, A.G.(1976). Decreased intracranial self-stimulation after neuroleptics of 6-hydroxydopamine. Evidence for mediation by motor deficits rather than reduced reward. Psychopharmacology 47:21-27.
597a. Fields, R.B., van Kammen, D.P., Peters, J.L., Rosen, J., van Kammen, W.B., Nugent, A., Stipetic, M. and Linnoila, M.(1988). Clonidine improves memory function in schizophrenia independently from change in psychosis. Schizophr. Res. 1:417-423.
598. Fieve, R.R.(1975). Moodswing. Bantum Books, New York.
599. File, S.E.(1975). Effects of parachlorophenylalanine and amphetamine on habituation of orienting. Pharmacol. Biochem. Behav. 3:979-983.
600. Filho, N.G.S and Graeff, F.G.(1977). Effect of tryptamine antagonists on self-stimulation. Psychopharmacology 52:87-92.
602. Fink, M. and Taylor, M.A.(1970). Anxiety precipitated by lactate. N. Engl. J. Med. 281:1129.
603. Finucci, J.A. and Childs B.(1983). Dyslexia: Family studies. In C.L. Ludlow and J.A. Cooper (eds) Genetics Aspects of Speech and Language Disorders. Academic Press, New York, pp157-167.
604. Finucci, J.M.(1978). Genetic considerations in dyslexia. In H.R. Myklebust (ed) Progress in Learning disabilities. Grune and Stratton, New York.
605. Finucci, J.M., Guthrie J.T., Childs A.L., Abbey H., Childs, B.(1976). The genetics of specific reading disability. Ann. Hum. Genet. (Lond.) 40:1-23.
606. Fish, B. and Hagin, R.(1973). Visual-motor disorders in infants at risk for schizophrenia. Arch. Gen. Psychiatry 28:900-924.
607. Fish, B.(1971). The "one child, one drug" myth of stimulants in hyperkinesis. Arch. Gen. Psychaitry 25:193-203.
608. Fish, B.(1977). Neurobiologic antecedents of schizophrenia in children. Evidence for an inherited, congenital neurointegrative defect. Arch. Gen. Psychiatry 34:1297-1313.
609. Fish, G.S., Cohen, I.L., Wolf, E.G., Brown, W.T., Jenkins, E.C. and Gross, A.(1986). Autism and the fragile X syndrome. Am. J. Psychiatry 143:71-73.
609a. Fishbein, D.H., Lozovsky, D. and Jaffe, J.H.(1989). Impulsivity, aggression, and neuroendrocrine responses to serotonergic stimulation in substance abusers. Biol. Psychiatry 25:1049-1066.
610. Fisher, W., Kerbeshian, J. and Burd, L.(1986). A treatable language disorder: Pharmacological treatment of pervasive developmental disorder. J. Dev. Behav. Pediatr. 7:73-76.
611. Fishman, H.C. and Rosman, B.L.(1986). Evolving Models for Family Change. Guilford Press, New York.
612. Fishman. E.B., Siek, G.C., MacCallum, R.D., Bird, E.D., Volicer, L., and Marquis, J.K.(1986). Distribution of the molecular forms of acetylcholinesterase in human brain: Alterations in dementia of the Alzheimer type. Ann. Neurol. 19:246-252.
613. Flament, M.F., Rapoport, J.L., Murphy, D.L., Berg, C.J. and Lake, C.R.(1987). Biochemical changes during clomipramine treatment of childhood obsessive-compulsive disorder. Arch. Gen. Psychiatry 44:219-225.
614. Florey, E.(1953). Uber einen nervosen Hemmungsfaktor in Gehirn and Ruckenmark. Naturwisenschaften 40:295-296.
615. Fobes, J.L. and Olds, M.E.(1981). Effects of neonatal 6-hydroxydopamine treatment on catecholamine levels and behavior during development and adulthood. Psychopharmacology 73:27-30.
616. Fog, R. and Pakkenberg, H.(1981). Behavioral effects of dopamine and p-hydroxyamphetamine injected into

corpus striatum of rats. Exp. Neurol. 31:75-86.
617. Fog, R.(1972). On stereotypy and catalepsy: Studies on the effect of dopamine and neuroleptics in rats. Acta Neurol. Scand. 48:1-64.
618. Fog, R.L., Randrup, A. and Pakkenberg, H.(1968). Neuroleptic action of quaternary chlorpromazine and related drugs injects into various brain areas in rats. Psychopharmacologia 12:428-432.
619. Folstein, S. and Rutter, M.(1977). Genetic influences and infantile autism. Nature 265:726-728.
620. Fonnum, F., Storm-Mathisen, J. and Divac, I.(1981). Biochemical evidence for glutamate as neurotransmitter in corticostriatal and corticothalamic fibers in rat brain. Neurosc. 6:863-873.
621. Fonnum, F., Storm-Mathisen, J. and Walberg, F.(1970). Glutamate decarboxylase in inhibitory neurons. A study of the enzyme in Purkinje cell axons and boutons in the cat. Brain Res. 20:259-275.
622. Foote, S.L., Bloom, F.E. and Ashton-Jones, G.(1983). Nucleus locus ceruleus: New evidence of anatomical and physiological specificity. Physiol. Rev. 63:844-914.
623. Fowler, C.J., Tipton, K.F., MacKay, A.V.P., and Youdim, M.B.H.(1982). Human platelet monamine oxidase — A useful enzyme in the study of psychiatric disorders? Neuroscience 7:1577-1594.
624. Fowler, C.J., von Knorring, L., and Oreland, L.(1980). Platelet monamine oxidase activity in sensation seekers. Psychiatry Res. 3:273-279.
625. Fowler, M. Graham, P., Chadwick, O., et al.(1975). Spouses of schizophrenics: A blind comparative study.Compr. Psychiatry 16:339-342.
626. Fox, S., Krnjevic, K., Morris, M.E., Puil, E. and Werman, R.(1978). Action of Baclofen on mammalian synaptic transmission. Neuroscience 3:495-515.
626a. Frankel, M. and Cummings, J.L.(1984). Neuro-ophthalmic abnormalities in Tourette's syndrome: Functional and anatomic implications. Neurology 34:359-361.
627. Frankel, M., Cummings, J.L., Robertson, M.M., Trinble, M.R., Hill, M.A. and Benson, D.F.(1986). Obsessions and compulsions in Gilles de la Tourette's syndrome. Neurology 36:378-383.
627a. Franzel, L.I.S.(1980). Gilles de la Tourette's syndrome. A case report. N. Engl. J. Med. 92:234-235.
628. Fras, I. and Karlavage, J.(1977). The use of methylphenidate and imipramine in Gilles de la Tourette's disease in children. Am. J. Psychiatry 134:195-197.
629. Fras, I.(1978). Gilles de la Tourette's syndrome. Effects of tricyclic antidepressants. N. Y. State J. Med. July:1230-1232.
630. Frazer, A., Pandey, G.N. and Mendels, J.(1973). Metabolism of tryptophan in depressive disease. Arch. Gen. Psychiatry 29:528-535.
631. Frederikson, P.K.(1975). Baclofen in the treatment of schizophrenia. Lancet 1:702.
632. Freedman, D.X., Belendiuk, K., Belendiuk, G.W. and Crayton, J.W.(1981). Blood tryptophan metabolism in chronic schizophrenics. Arch. Gen. Psychiatry 38:655-659.
633. Freedman, M. and Oscar-Berman, M.(1986). Bilateral frontal lobe disease and selective delayed response deficits in humans. Behav. Neurosci. 100:337-342.
634. Freedman, R., Bell, J.,and Kirch, D.(1980). Clonidine therapy for coexisting psychosis and tardive dyskinesia. Am. J. Psychiatry 237:629-630.
635. Freedman, R., Hoffer, B.J., Puro, D. and Woodward, D.J.(1976). Noradrenaline modulation of the responses of the cerebellar Purkinje cell to afferent synaptic activity. Br.J. Pharmacol. 57:603-605.
636. Freedman, R., Kirch, D., Bell, J., Adler, L.E., Pecebich, M., Pachtman, E. and Denver, P.(1982). Clonidine treatment in schizophrenia. Acta Psychiatry Scand. 65:35-45.
637. Freeman, C.P.L. and Hampson, M.(1987). Fluoxetine as a treatment for bulimia nervosa. Int. J. Obesity 11 (Suppl 3):171-177.
638. Freeman, W. and Watts, J.W.(1942). Psychosurgery. Charles C. Thomas, Springfield, IL.
639. Freidman, P.A., Kappelman, A.H. and Kaufman, S.(1972). Partial purification and characterization of tryptophan hydroxylase from rabbit hind brain. J. Biol. Chem. 247:4165-4173.
640. French, G.M. and Harlow, H.F.(1962). Variability of delayed reaction performance in normal and brain damaged rhesus monkeys. J. Neurophysiol. 25:585-599.
641. Fricker, L.D. and Snyder, S.H.(1982). Enkephalin convertase: Purification and characterization of a specific enkephalin-synthesizing carboxypeptidase localized to adrenal chromaffin granules. Proc. Natl. Acad. Sci. U.S.A. 79:3886-3890.
642. Fricker, L.D. and Snyder, S.H.(1983). Purification and characterization of enkephalin convertase, an enkephalin-synthesizing carboxypeptidase. J. Biol. Chem. 258:10950-10955.
643. Friedman, A.P.(1963). Studies in the pharmacology of headache. Neurology 13 (special issue):27-33.
644. Friedman, H.M. and Allen, N.(1969). Chronic effects of complete limbic lobe destruction in man. Neurology 19:679-690.
645. Friedman, M.J., Krstulovic, A.M., Severinghaus, J.M. and Brown, S.J.(1988). Altered conversion of tryptophan to kynurenine in newly abstinent alcoholics. Biol. Psychiatry 23:89-93.
646. Friel, P.B.(1973). Familial incidence of Gilles de la Tourette's disease, with observations on aetiology and treatment. Br. J. Psychiatry 122:655-658.
647. Frijling-Schreuder, E.C.M.(1970). Borderline states in children. Psychoanal. Study Child 24:307-327.
648. Frosch, J. and Wortis, S.B.(1954). A contribution to the nosology of the impulse disorders. Am. J. Psychiatry 111:132-138.
649. Frost, N., Feighner, J. and Schuckit, M.(1976). A family study of Gilles de la Tourette syndrome. Dis. Nerv. Syst. 37:537-538.
650. Fulker, D., Eysenck, S.B. and Zuckerman, M.(1980). A genetic and environmental analysis of sensation seeking. J. Res. Personality 1:261-281.
651. Fulop, G., Philips, R.A., Shapiro, A.K., Gomes, J.A., Shapiro, E., and Nordlie, J.W.(1987). ECG changes during haloperidol and pimozide treatment of Tourette's disorder. Am. J. Psychiatry 144:673-675.
652. Furger, R.(1961). Psychiatric investigation in Cushing's syndrome. Schweiz. Arch. Neurol. Psychiatry 88:9.
653. Fuster, J.M.(1980). The Prefrontal Cortex. Raven Press, New York.
654. Fuxe, K. and Johnston, G.(1974). Further mapping of central 5-hydroxytryptamine neurons: Studies with the neurotoxic dihydrotryptamines. In E. Costa, G.L. Gessa and M. Sandler (eds) Advances in Biochemical Pharmacology ,

Raven Press, New York 10:1-12.
655. Fuxe, K., Agnati, L.F., Kalla, M., Goldstein, M., Anderson, K., and Härfstrand, A.(1985). Dopaminergic systems in the brain and pituitary. In B. Halász et al. (eds)The Dopaminergic System. Springer-Verlag, Berlin, pp11-25.
656. Fuxe,K.(1965). Evidence for the existence of monamine neurons in the central nervous system. IV. Distribution of monamine nerve terminals in the central nervous system. Acta. Physiol. Scand. 64 (suppl 247): 37-85.
657. Gabay, S.(1981). Serotonergic-dopamine interactions: Implications for hyperkinetic disorders. In B. Haber, M.R. Issidorides, and G.A. Alivistos (eds) Serotonin: Current Aspects of Neurochemistry and Function. Plenum Press, NewYork.
657a. Gaby, A.(1985). The Doctor's Guide to Vitamin B6. Rodale Press, Emmaus, PA.
658. Gabrielli, W. and Mednick, S.A.(1983). Intellectual performance in children of alcholics. J. Nerv. Ment. Dis. 171:444-447.
659. Gaddum, J.H.(1954). Drugs antagonistic to 5-hydroxytryptamine. In G.W. Wolstenholme (ed) Ciba Foundation Symposium on Hypertension. Little Brown, Boston, pp75-77.
660. Gailani, S., Murphy, G., Kenny, G., Nussbaum, A. and Silvernail, P.(1973). Studies on tryptophan metabolism in patients with bladder cancer. Cancer Res. 33:1071-1077.
661. Gainotti, G.(1972). Emotional behavior and hemispheric side of the lesion. Cortex 8:41.
662. Gal, E.M. and Sherman, A.D.(1980). L-kynurenine: Its synthesis and possible regulatory function in brain. Neurochem. Res. 5:223-239.
663. Gal, E.M., Armstrong, J.C. and Ginsberg, B.(1966). The nature of in vitro hydroxylation of L-tryptophan by brain tissue. J. Neurochem. 13:643-654.
664. Gal, E.M., Young, R.B. and Sherman, A.D.(1978). Tryptophan loading: Consequent effects on the synthesis of kynurenine and 5-hydroxyindoles in rat brain. J. Neurochem. 31:237-244.
665. Gal, E.M.(1974). Cerebral tryptophan-2,3-dioxygenase (pyrrolase) and its induction in rat brain. J. Neurochem. 22:861-863.
666. Galey, D., Jafford, R, and LeMoal, M.(1976). Spontaneous alternation disturbance after lesions of the ventral mesencephalic tegmentum in the rat. Neuosci. Lett. 3:65-69.
667. Galey, D., R, Jafford, R. and LeMoal, M.(1979). Alteration behavior, spatial discrimination, and reversal after electrocoagulation of the ventral mesencephalic tegmentum in the rat. Behav. Neurol. Biol. 26:81-88.
668. Galey, D., Simon, H. and LeMoal, M(1977). Behavioral effects of lesions in the A10 dopaminergic area of the rat. Brain Res. 124:83-97.
669. Garattini, S., Giacolone, E. and Valzelli, E.L.(1969). Biochemical changes during isolation-induced aggressiveness in mice. In S. Garattini and E.B. Sigg (eds) Aggressive Behavior. Amsterdam, Excerpta Medica, pp179-187.
670. Garbutt, G. and Goldstein, A.(1972). Blind comparison of three methadone maintenance dosages in 180 patients. Proc. 4th Nat. Conf. on Methadone Treatment. U.S. Govt Printing Office, Washington, DC.
671. Garbutt, J.C. and van Kammen, D.P.(1983). The interaction between GABA and dopamine: Implications for schizophrenia. Schizophr. Bull. 9:336-353.
672. Garelis, E., Gillin, J., Wyatt, R.J. and Neff, N.(1975). Elevated blood serotonin concentrations in unmedicated chronic schizophrenic patients: A preliminary study. Am. J. Psychiatry 132:184-186.
673. Garfinkel, B.D., Wender, P.H., Sloman, L., and O'Neill, I.(1983). Tricyclic antidepressant and methylphenidate treatment of attention deficit disorder. J. Am. Acad. Psychiatry 22:343-348.
674. Garfinkel, P.E., Moldofsky, H. and Garner, D.M.(1980). The heterogeneity of anorexia nervosa. Arch. Gen. Psychiatry 37:1036-1040.
675. Garmezy, N.(1974). The study of competence in children at risk for severe psychopathology. In E. Anthony and C. Koupernik (eds) The Child in His Family: Children at Psychiatric Risk. John Wiley & Sons, New York.
676. Garmezy, N.(1978). Observations on research with children at risk for child and adult psychopathology. In M. McMillan and S. Henao (eds) Child Psychiatry: Treatment and Research. New York: Brunner/Mazel.
677. Gastfriend, D.R., Biederman, J. and Jellinek, M.S.(1984). Desipramine in the treatment of adolescents with attention deficit disorder. Am. J. Psychiatry 141:906-908.
678. Gavin, F.H. and Kleber, H.D.(1986). Abstinence symptomatology and psychiatric diagnosis in cocaine abusers. Arch. Gen. Psychiatry 43:107-113.
679. Gawin, F.H. and Ellinwood, E.H.(1988). Cocaine and other stimulants. Actions, abuse, and treatment. N. Engl. J. Med. 318:1173-1182.
680. Gawin, F.H. and Kleber, H.D.(1985). Cocaine use in a treatment population: Patterns and diagnostic distinctions. Natl. Inst. Drug Abuse Res. Monogr. Ser. 61:182-192.
681. Gazzaniga, M.S. and LeDoux, J.E.(1978). The Integrated Mind. Plenum Press, New York.
682. Geleerd, E.R.(1958). Borderline states in childhood and adolescence. Psychoanal. Study Child. 13:279-295.
683. Geller, I. and Blum, K.(1970). The effects of 5-HTP on para-chlorophenylalanine (p-CPA) attenuation of "conflict" behavior. Eur. J. Pharmacol. 9:319-324.
684. Geller, I. and Hartman, R.J.(1982). Effects of buspirone on operant behavior of laboratory rats and cynomologus monkeys. J. Clin. Psychiatry 43:25-32.
685. Geller, I.(1973). Effects of para-chorophenylalanine and 5-hydroxytryptophan on alcohol intake in the rat. Pharmacol. Biochem. Behav. 1:361-365.
686. Geller, I., Hartmann, R.J. and Croy, D.J.(1974). Attenuation of conflict behavior with cinanserin, a serotonin antagonist: Reversal of the effect with 5- hydroxytryptophan and a-methyltryptamine. Res. Commun. Chem. Pathol. Pharmacol. 7:165-174.
687. Geller, I., Purdy, R. and Merrit, J.H.(1973). Alterations in ethanol preference in the rat: The role of brain biogenic amines. Ann. N.Y.Acad. Sci. 215:54-59.
688. Gerbner, M.(1973). Study on the functional mechanism of the dorsolateral frontal lobe cortex. In K.H. Pribram and A.R. Luria (eds) Psychophysiology of the Frontal Lobes. Academic Press, New York, pp237-252.
689. German, D.C. and Bowden, D.M.(1974). Catecholamine systems as the neural substrate for intracranial: A hypothesis. Brain Res. 73:381-419.
690. Gerner, R.H. and Fairbanks, L.(1985). Discriminate function of CSF neurochemistry among normals, depressed, schizophrenia, mania and anorexia subjects. In C. Shagass, R.C. Josiassen, W.H. Bridger, K.J. Weiss, D.S. Staff, and G.M. Simpson (eds) Biological Psychiatry. Elsevier, Amster-

dam. p853.
691. Gerner, R.H., Fairbanks, L., Anderson, G.M., Youjng, J.G., Scheinin, M., Linnoila, M., Hare, T.A., Shaywitz, B.A. and Cohen, D.J.(1984). CSF neurochemistry in depressed, manic, and schizophrenic patients compared with that of normal controls. Am. J. Psychiatry 141:1533-1540.
692. Gershon, E.S., DeLisi, L.E., Hamovit, J., Nurnberger, J.I., Maxwell, M.E., Schreiber, J., Dauphinais, D., Dingman, C.W. and Curoff, J.J.(1988). A controlled family study of chronic psychosis. Schizophrenia and schizoaffective disorder. Arch. Gen. Psychiatry 45:328-336.
693. Gershon, E.S., Hamovit, J., Guroff, J., Dibble, E., Lechman, J.F., Sceery, W., Targum, S.D., Nurberger, J.I., Golden, L.R. and Bunney, W.L.S., Jr.(1982). A family study of schizoaffective, bipolar I, bipolar II, unipolar, and normal controls. Arch. Gen. Psychiatry 39:1157-1167.
694. Gershon, E.S., Targum, S.D., Kessler, L.R., Mazure, C.M. and Bunney, W.E. Jr.(1977). Genetic studies and biologic strategies in the affective disorders. Prog. Med. Genet., New Series 2:101-163.
695. Geschwind, N. and Behan, P. (1982). Left-handedness: Association with autoimmune disease, migraine, and developmental learning disorder. Proc. Natl. Acad. Sci.U.S.A. 79:5097-5100.
696. Geschwind, N.(1964) The development of the brain and the evolution of language. In C.I.J.M. Stuart (ed) Monograph Series on Linguistics and Language Studies. No. 17. Georgetown Univ. Press, Washington, DC, pp155-169.
697. Geschwind, N.(1965). Disconnection syndromes in animal and man. Brain 88:237-294; 585- 644.
698. Gessa, G.L. and Tagiamonte, A.(1974). Role of brain monamines in male sexual behavior. Mini review. Life Sci. 14:425-436.
699. Gessa, G.L. and Tagiamonte, A.(1975). Role of brain serotonin and dopamine in male sexual behavior. In M. Sandler and G.L. Gessa(eds) Sexual Behavior: Pharmacology and Biochemistry. Raven Press, New York, pp117-128.
700. Gessa, G.L., Tagliamonte, A., Tagliamonte, P. and Brodie, B.B.(1970). Essential role of testosterone in the sexual stimulation induced by p-chlorophenylalanine in male animals. Nature 227:616-617.
701. Geyer, M.A. and Segal, D.S.(1974). Shock-induced aggression: Opposite effects of intraventricularly infused dopamine and norepinephrine. Behav. Biol. 10:99-104.
702. Geyer, M.A., Puerto, A., Menkes, D.B., Segal, D.S. and Mandell, A.J.(1976). Behavioral studies following lesions of the mesolimbic and mesostriatal serotonergic pathways. Brain Res. 106:257-270.
703. Giambalvo, C.T. and Snodgrass, S.R.(1978). Biochemical and behavioral effects of serotonin neurotoxins on the nigrostriatal dopamine system: Comparison of injection sites. Brain Res. 152:555-566.
704. Gianturco, D.(1972). Effect of psychiatric and antonomic symptoms on the incidence of fourteen-and-six per second positive spikes among adolescents. Clin. Electroenceph. 3:55-59.
705. Giarman, N.J., Tanaka, C. and Atkins, E.(1968). Serotonin, norepinephrine and fever. Adv. Pharmacol. 6A:307-321.
706. Gibbons, J.L. and Glusman, M.(1979). Effects of quipazine, fluoxetine and fenfluramine on muricide in rats. Fed. Proc. Fedn. Am. Socs. Exp. Biol. 38:257.
707. Gibbons, J.L., Barr, G.A., Bridger, W.H. and Leibowitz, S.F.(1979). Manipulation of dietary tryptophan: Effects on mouse killing and brain serotonin in the rat. Brain Res. 169:139-153.
708. Gibbs, F.A. and Gibbs, E.L.(1963). Borderland of epilepsy. J. Neuropsychiatry 4:287-295.
709. Gibbs, F.A. and Gibbs, E.L.(1963). Fourteen and six per second positive spikes. Electroenceph. Clin. Neurophys. 15:553-558.
710. Gilka, L.(1975). Schizophrenia - A Disorder of Tryptophan Metabolism. Acta Psychiatrica Scandinavica (ed). Minksgaard, Copenhagen, pp1-83.
711. Gillberg, C. and Svennerholm, L.(1987). CSF monamines in autistic syndromes and other pervasive developmental disorders of early childhood. Br. J. Psychiatry 151:89-94.
712. Gillberg, C. and Wahlström J.(1985). Chromosome abnormalities in infantile autism and other childhood psychoses: Population study of 66 cases. Dev. Med. Child Neurol. 27:293-304.
713. Gillberg, C., Terenius, L. and Lonnerholm, G.(1985). Endorphin activity in childhood psychosis. Spinal fluid levels in 24 cases. Arch. Gen. Psychiatry 42:780-783.
714. Gilles de la Tourette, G.(1884). Jumping, latah, myriachit. Arch. Neurol. 8:68-84.
715. Gilles de la Tourette, G.(1885). Étude sur une affection nerveuse caractérisée par de l'incordination motrics accompagnée d'echolalie et de copralalie. Arch. Neurol. 9:19-42, 158-200.
716. Gilles de la Tourette, G.(1899). La maladie des tics convulsifs. La Semaine Médicale 19:153-156.
717. Gillespie, D.D., Manier, D.H., Sanders-Bush, E. and Sulser, F.(1988). The serotonin/norepinephrine-link in brain. II. Role of serotonin in the regulation of beta-adrenoreceptors in the low agonist affinity conformation. J. Pharmacol. Exp. Ther. 244:154-159.
718. Gilliam, T.C., Bucan, M., MacDonald, M.E., Zimmer, M., et al.(1987). A DNA segment encoding two genes very tightly linked to Huntington's disease. Science 238:950-952.
719. Gilligan, S.B., Reich, T. and Cloninger, C.R.(1987). Etiologic heterogenity in alcoholism. Gen. Epidemiol. 4:395-414.
720. Gillman, M.A. and Sandyk, R.(1984). Tourette syndrome: Effect of analgesic concentrations of nitrous oxide and naloxone. Br. Med. J. 288:114.
721. Gillman, M.A., Kok, L., Lichtigfeld, F.J.(1980). The paradoxical effects of naloxone on nitrous oxide analgesia in man. Eur. J. Pharmacol. 61:175-177.
722. Gilmour, D.G., Manowitz, P., Frosch, W.A. and Shopsin, B.(1973). Association of plasma tryptophan levels with clinical change in female schizophrenic patients. Biol. Psychiatry 6:119-128.
723. Gittelman, R., Mannuzza, S., Shenker, R. and Bonagura, N.(1985). Hyperactive boys almost grown up. I. Psychiatric status. Arch. Gen. Psychiatry 42:937-947.
724. Gittelman-Klein, R., Klein, D.F., Katz, S., Saraf, K. and Pollack, E.(1976). Comparative effects of methyphenidate and thioridazine in hyperkinetic children. Arch. Gen. Psychiatry 33:1217-1231.
725. Gittelman-Klein, R. and Klein, D.F.(1976). Methylphenidate effects in learning disabilities. Arch. Gen. Psychiatry 33:655-664.
726. Gittelman-Klein, R., Klein, D.F., Katz, S., et

al.(1976). Comparative effects of methyphenidate and thioridazine in hyperkinetic children. Arch. Gen. Psychiatry 33:1217-1231.
727. Gittleson, N.(1966). The effect of obsessions on depressive psychosis. Br. J. Psychiatry 112:253-259.
728. Gittleson, N.(1966). The fate of obsessions in depressive psychosis. Br. J. Psychiatry 112:705-708.
729. Gittleson, N.(1966). The depressive psychosis in the obsessional neurotic. Br. J. Psychiatry 112:883-887.
730. Gjerris, A., Sørensen, A.S., Rafaelsen, O.J., Werdelin, L., Alling, C. and Linnoila, M.(1988). 5-HT and 5-HIAA in cerebrospinal fluid in depression. J. Affect. Disord. 12:13-22.
731. Glassman, A.H., Jackson, W.K., Walsh, B.T., and Roose, S.P.(1984). Cigarette craving, smoking withdrawal, and clonidine. Science 226:864-866.
732. Glaze, D.G., Jankovic, J. and Frost, J.D., Jr.(1982). Sleep in Gilles de la Tourette syndrome: Disorder of arousal. Neurology 32:153A.
733. Glazer, W.M., Charney, D.S. and Heninger, G.R.(1987). Noradrenergic function in schizophrenia. Arch. Gen. Psychiatry 44:898-904.
734. Gleick, J.(1988). Chaos Making a New Science. Viking, New York, 352pp.
735. Glick, S.D.(1972). Changes in amphetamine sensitivity following frontal cortical damage in rats and mice. Eur. J. Pharmacol. 20:351-356.
736. Gloger, S., Grunhaus, L., Birmacher, B. and Troudart, T.(1981). Treatment of spontaneous panic attacks with clomipramine. Am. J. Psychiatry 138: 1215-1217.
737. Glowinksi, J.(1981). Present knowledge on the properties of the mesocortico-frontal dopaminergic neurons. In Matthysee (ed) Psychiatry and the Biology of the Human Brain: A Symposium Dedicated to Seymour S. Kety. Elsevier/North Holland, Inc., New York, pp15-28.
738. Glowinski, J. and Axelrod, J.(1964). Inhibition of uptake of tritiated noradrenalin in the intact rat brain by imipramine and structurally related compounds. Nature 204:1318-1319.
739. Glowinsky, J., Tassin, J.P. and Thierry, A.M.(1984). The mesocortical-prefrontal dopaminergic neurons. Trends Neurosci. 7:415-418.
740. Glueck, S. and Glueck, E.(1940). Juvenile Delinquents Grown Up. Commonwealth Fund, New York.
741. Goeders, N.E. and Smith, J.E.(1983). Cortical dopaminergic involvement in cocaine reinforcement. Science 221:773-775.
742. Goelet, P., Castellucci, V.F., Schacher, S. and Kandel, E.R.(1986). The long and the short of long-term memory — A molecular framework. Nature 322:419-422.
743. Goetz, C.C., Tanner, C.M., Wilson, R.S., Carroll, C.V.S., Como, P.G. and Shannon, K.M.(1987). Clonidine and Gilles de la Tourette syndrome: Double-blind study using objective rating methods. Ann. Neurol. 21:307-310.
744. Goetz, C.G. and Klawans, H.L.(1982). Gilles de la Tourette on Tourette Syndrome.(A complete English Translation of the original paper). In A.J. Friedhoff and T.N. Chase (eds). Gilles de la Tourette Syndrome. Raven Press, New York, pp1-16.
745. Goetz, C.G., Tanner, C.M., and Klawans, H.L.(1984). Fluphenazine and multifocal tic disorders. Arch. Neurol. 41:271-272.
746. Goforth, E.G.(1974). A single case study. Gilles de la Tourette's syndrome. A 25 year follow-up study. J. Nerv. Ment. Dis. 158:306-309.
747. Gold, M.S., and Pottash, A.C.(1981). The neurobiological implications of clonidine HCl. Ann. N.Y. Acad. Sci. 362:191-202.
748. Gold, M.S., Pottash, A.C., Sweeney, D.R., and Kleber, H.D.(1980). Opiate withdrawal using clonidine: A safe, effective, and rapid non-opiate treatment. J.A.M.A. 243:343-346.
749. Gold, M.S., Redmond, D.E., Jr., and Kleber, H.D.(1978). Clonidine blocks acute opiate-withdrawal symptoms. Lancet 2:599-602.
750. Goldberg, T.E., Weinberger, D.R., Berman, K.F., Pliskin, N.H. and Podd, M.H.(1987). Further evidence of dementia of the prefrontal type in schizophrenia? Arch. Gen. Psychiatry 44:1008-1014.
751. Golden, G.S. and Greenhill, L.(1981). Tourette syndrome in mentally retarded children. Ment. Retar. 1:17-19.
752. Golden, G.S.(1974). Gilles de la Tourette's syndrome following mehtylphenidate administration. Dev. Med. Child Neurol. 16:76-78.
753. Golden, G.S.(1977). The effect of central nervous system stimulants on Tourette syndrome. Ann. Neurol. 2:69-70.
754. Golden, G.S.(1977). Tourette syndrome. The pediatric perspective. Am. J. Dis. of Child. 131:531-534.
755. Golden, G.S.(1978). Tics and Tourette's: A continuum of symptoms? Ann. Neurol. 4:145-148.
756. Golden, G.S.(1979). Tics and Tourette syndrome. Hosp. Pract. 14:91-100.
757. Golden, G.S.(1984). Tardive dyskinesia in Tourette's sydrome. Ann. Neurol. 16:390.
758. Golden, G.S.(1985). Tardive dyskinesia in Tourette syndrome. Pediatr. Neurol. 1:192-194.
759. Golden, G.S.(1988). The relationship between stimulant medication and tics. Pediatr. Ann. 17: 405-408.
760. Golden, G.S.(1988). The use of stimulants in the treatment of Tourette's syndrome. In Cohen, D.J., Bruun, R.D. and Leckman, J.F. (eds) Tourette's Syndrome and Tic Disorders: Clinical Understanding and Treatment. John Wiley & Sons. New York, pp318-325.
761. Goldgaber, D., Lerman, M.I., McBride, O.W., Saffiotti, U., and Gajdusek, D.C.(1987). Characterization and chromosomal localization of a cDNA encoding brain amyloid of Alzheimer's disease. Science 235:877-880.
762. Goldin, L.R., Gershon, E.S., Targum, S.D., Sparkes, R.S. and McGinniss, M.(1983). Segregation and linkage analysis in families of patients with bipolar, unipolar, and schizoaffective mood disorders. Am. J. Hum. Genet. 35:274-287.
763. Goldman-Rakic, P.S. and Schwartz, M.L.(1982). Interdigitation of contralateral and ipsilateral columnar projections to frontal association cortex in primates. Science 216:755-757.
764. Goldman-Rakic, P.S.(1984). Modular organization of prefrontal cortex. Trends Neurosci. 7:419-424.
765. Goldman-Rakic, P.S.(1984). The frontal lobes: Uncharted provinces of the brain. Trends Neurosci. 7:425-429.
766. Goldstein, A. and Goldstein, D.B.(1968). Enzyme expansion theory of drug tolerance and physical dependence. Proc. Assoc. Res. Nerv. Ment. Dis. 46:265-267.
767. Goldstein, A., Tachibana, S., Lowney, L.I., Hunkapiller, M. and Hood, L.(1979). Dynorphin-(1-13), an extraor-

dinarily potent opioid peptide. Proc. Natl. Acad. Sci.U.S.A. 76:6666-6670.
767a. Goldstein, J.A. (1984). Nifedipine treatment of Tourette's syndrome. J. Clin. Psychiatry 45:360.
768. Goldstein, K.(1954). The brain-injured child. In H. Michal-Smith (ed) Pediatric Problems in Clinical Practice. Grune & Stratton, New York, pp97-120.
769. Goldstein, L., Manowitz, P., Nora, R., Swartzburg, M. and Carlton, P.L.(1985). Differential EEG activation and pathological gambling. Biol. Psychiatry 20:1232-1234.
770. Goldstein, M., Battista, A.F., Ohgmotto, T. et al. (1973). Tremor and involuntary movements in monkeys. Effect of L-dopa and a dopamine receptor stimulating agents. Science 179: 816-817.
771. Goldstein, M., Mahanand, D., Lee, J., and Coleman, M.(1976). Dopamine-beta- hydroxylase and endogenous total 5-hydroxyindole levels in autistic patients and controls, In M. Coleman (ed) The Autistic Syndromes. American Elsevier Publishing Co., New York, pp57-63.
772. Goldstein, S.(1987). Treatment of social phobia with clonidine. Biol. Psychiatry 22:369-372.
773. Goldstone, S. and Lhamon, W.T.(1976). The effects of haloperidol upon temporal information processing by patients with Tourette's syndrome. Psychopharmacology 50:7-10.
774. Gomer, F.E. and Goldstein, R. (1974). Attentional rigidity during exploratory and simultaneous discrimination behavior in septal lesioned rats. Physiol. Behav. 12:19-28.
775. Gomes, U.C.R., Shanely, B.C., Potgieter, L. and Roux, J.T.(1980). Noradrenergic overactivity in chronic schizophrenia: Evidence based on cerebrospinal fluid noradrenaline and cyclic nucleotide concentration. Br. J. Psychiatry 137:346-351.
776. Gonce, M. and Barbeau, A.(1977). Seven cases of Gilles de la Tourette's syndrome: Partial relief with clonazepam: A pilot study. Can. J. Neurol. Sci. 4:279-283.
777. Goodwin, D.W.(1979). Alcoholism and heredity. Arch. Gen. Psychiatry 36:57-61.
778. Goodwin, D.W., Schlushinger, F., Hermansen, L., Guze, S.B. and Winokur, G.(1975). Alcoholism and the hyperactive child syndrome. J. Nerv. Ment. Dis. 160:349-353.
779. Goodwin, D.W., Schlusinger, F., Hermansen, L., Guze, S.B. and Winokur, G. (1973). Alcohol problems in adoptees raised apart from alcoholic biological parents. Arch. Gen. Psychiatry 28: 238-243.
780. Goodwin, D.W., Schlusinger, F., Moller, N., Hermansen, L., Winokur, G. and Guze, S.B.(1974). Drinking problems in adopted and non-adopted sons of alcoholics. Arch. Gen. Psychiatry, 31: 164-169.
780a. Gorman, J.M., Liebowitz, M.R., Fyer, A.J. and Sten, J.(1989). A neuroanatomical hypothesis for panic disorder. Am. J. Psychiatry 146:148-161.
781. Gottesman, I.I. and Shields, J.(1972) Schizophrenia and Genetics: A Twin Study Vantage Point. Academic Press, New York.
782. Gottfries, C.G.(1985). Alzheimer's disease and senile dementia: Biochemical characteristics and aspects of treatment. Psychopharmacology 86:245-252.
783. Gould, R.J., Murphy, K.M.M., Reynolds, I.G. and Snyder, S.H.(1983). Antischizophrenic drugs of the diphenylbutylpiperidine type act as calcium channel antagonists. Proc. Natl. Acad. Sci. U.S.A. 80:5122-5125.
784. Gounce, M. and Barbeau, A.(1977). Seven cases of Gilles de la Tourette's syndrome: partial relief with clonazepam, a pilot study. Can. J. Neurol. Sci. 4: 279-283.
784a. Grad, L.R., Pelcovitz, D., Olson, M., Matthews, M. and Grad, G.J.(1987). Obsessive-compulsive symptomatology in children with Tourette's syndrome. J. Am. Acad. Child Adolesc. Psychiatry 26:69-73.
785. Graeff, F.G. and Filho, N.G.S.(1978). Behavioral inhibition induced by electrical stimulation of the median raphe nucleus of the rat. Physiol. Behav. 21:477-484.
786. Graeff, F.G. and Schoenfeld, R.I.(1970). Tryptaminergic mechanisms in punished and nonpunished behavior. J. Pharmacol. Exp. Ther. 173:277-283.
787. Grahame-Smith, D.G.(1971). Studies in vivo on the relationship between brain tryptophan, brain 5-HT synthesis and hyperactivity in rats treated with a monamine oxidase inhibitor and L-tryptophan. J. Neurochem. 18:1053-1066.
788. Grande, T.P., Wolf, A.W., Schubert, D.S.P., Patterson, M.B. and Brocco, K.(1984). Associations among alcoholism, drug abuse, and antisocial personality: A review of literature. Psychol. Rep. 55:455-474.
789. Grant, D.A. and Berg, E.A.(1954). A behavioral analysis of degree of reinforcement and ease of shifting to new responses in a Weigl-type card sorting problem. J. Exp. Psychol. 50:237-244.
790. Grant, L.D., Coscina, D.V., Grossman, S.P. and Freedman, D.X.(1973). Muricide after serotonin depleting lesions of the midbrain raphe nuclei. Pharmacol. Biochem. Behav. 1:77-80.
790a. Gray, A.J.(1982). The Neuropsychology of Anxiety: An Enquiry into the Function of the Septo-hippocampal System. Oxford Univ. Press, Oxford.
791. Grebb, J.A., Shelton, R.C., Taylor, E.H. and Bigelow, L.B.(1986). A negative, double blind, placebo-controlled, clinical trial of verapamil in chronic schizophrenia. Biol. Psychiatry 21:691-694.
792. Green, A.R. and Curzon, G.(1970). The effect of tryptophan metabolites in brain 5-hydroxytryptophan metabolism. Biochem. Pharmacol. 19:2061-2068.
793. Green, A.R. and Heal, D.J.(1985). The effects of drugs on serotonin-mediated behavioral models. In A.R. Green (ed) Neuropharmacology of Serotonin. Oxford University Press, Oxford, pp326-365.
794. Green, A.R., Guy, A.P. and Gardner, C.R.(1984). The behavioral effects of RU 24969, a suggested 5-HT1 receptor agonist in rodents and the effect on the behavior of treatment with antidepressants. Neuropharmacology 23:655-661.
795. Green, A.R., Woods, H.F. and Joseph, M.H.(1976). Tryptophan metabolism in the isolated perfused liver of the rat: Effect of tryptophan concentration, hydrocortisone and allopurinol on tryptophan pyrrolase activity and kynurenine formation. Br. J. Phamacol. 57:103-114.
796. Green, M. and Walker, E.(1986). Attentional performance in positive- and negative-symptom schizophrenia. J. Nerv. Ment. Dis. 174:208-213.
797. Green, R.C and Pitman, R.K.(1986). Tourette syndrome and obsessive compulsive disorder. In M.A. Jenike, L. Baer and W.E. Minichiello (eds) Obsessive-Compulsive Disorders: Theory and Management. PSG Publishing Co., Inc, Littleton, MA, pp147-164.
798. Greenberg, A.(1980). Effects of opiates on male orgasm. Med. Aspects Hum. Sexuality 18:207-210.
799. Greenberg, A.S. and Coleman, M.(1976). Depressed

5-hydroxyindole levels associated with hyperactive and aggressive behavior. Arch. Gen. Psychiatry 33:331-336.
800. Greenberg, H.R.(1980). Psychology of gambling. In H.I. Kaplan, A.M. Freedman and B.J. Sadock (eds) Comprehensive Textbook of Psychiatry Vol. 3. Williams and Wilkins, Baltimore.
801. Greenberg, L., Yellin, A., Spring, C. and Metcalf, M.(1975). Clinical effects of imipramine and methylphenidate in hyperactive children. Int. J. Ment. Health 4:144-156.
802. Greenblatt, D.J., Shader, R.I. and Abernathy, D.R.(1983). Current status of benzodiazepines. N. Engl. J. Med. 309:354-358.
803. Greenwald, B., Marin, D.H., Silverman, S.M.(1986). Serotoninergic treatment of screaming and banging in dementia. Lancet 2:1464-1465.
805. Grim, P.F., Kohlberg, L. and White, S.H.(1968). Some relationship between conscious and attentional processes. J. Pers. Soc. Psychol. 8:239-252.
806. Grinker, R.R., Werble, B., and Dryre, R.C.(1968). The Borderline Syndrome. Basic Books, New York.
807. Gross, M.D. and Wilson, W.C.(1974). Minimal Brain Dysfunction. Brunner/Mazel, New York.
808. Gross, M.D.(1973). Imipramine in the treatment of mnimial brain dysfunction in children. Psychosomatics 14:283-285.
809. Gross, M.D.(1976). Growth of hyperkinetic children taking methyphenidate, dextroamphetamine, or imipramine/desipramine. Pediatrics 58:423-431.
810. Grossman, A., Moult, P., McIntyre, H., Evans, J., Silverstone, T., Rees, L. and Besser, G.(1984). Opiate modulation of amenorrhea in hyperprolactinemia and in weight-loss amenorrhea. Clin. Endocrinol. 17:379-388.
811. Grossman, H.Y., Mostofsky, D.I. and Harrison, R.H.(1986). Psychological aspects of Gilles de la Tourette syndrome. J. Clin. Psychol. 42:228-235.
812. Grossman, S.P., Dacey, D., Halaris, A.E., Collier, T. and Routtenberg, A.(1978). Aphagia and adipsia after preferential destruction of nerve cell bodies in hypothalamus. Science 202:537-539.
813. Growdon, J.H.(1978). Changes in motor behavior following the administration of serotonin neurotoxins. Ann. N.Y. Acad. Sci. 305:510-523.
814. Grueninger, W.E., Kimble, D.P., Grueninger, J. and Levine, S.E.(1965). GSR and corticosteroid response in monkeys with frontal ablations. Neuropsychologia 3:205-216.
815. Grunhaus, L., King, D., Greden, J.F. and Flegel, P.(1985). Depression and panic in patients with borderline personality disorder. Biol. Psychiatry 20:688-692.
816. Gualtieri, C.T. and Evans, R.W.(1984). Carbamazepine-induced tics. Dev. Med. Child Neurol. 26:546-548.
817. Gualtieri, C.T., Ondrusek, M.G. and Finky, C.(1985). Attention deficit disorders in adults. Clin. Neuropharmacol. 8:343-356.
818. Gualtieri, C.T.,Hicks, R.E. and Mayo, J.R.(1983). Hyperactivity and homeostasis. J. Am. Acad. Child. Psychol. 22: 382-384.
819. Guidotti, A., Forchetti, C.M., Corda, M.G., Konkel, D. Bennett, C.D. and Costa, E.(1983). Isolation, characterization, and purification of an endogenous polypeptide with agonistic action on benzodiazepine receptors. Proc. Natl. Acad. Sci. U.S.A. 80:3531-3533.
820. Gunderson, J.G., Kolb, J.E., and Austin, V.(1981). The diagnostic interview for borderline patients. Am. J. Psychiatry 138:896-903.
821. Gunderson, R.R. and Kolb, J.E.(1978). Discriminating features of borderline patients. Am. J. Psychiatry 135:792-796.
822. Gunderson, R.R. and Singer, M.T.(1975). Defining borderline patients: An overview. Am. J. Psychiatry 132:1-10.
823. Gunne, L.M. and Barany, S.(1976).Haloperidol-induced tardive dyskinesia in monkeys. Psychopharmacology 50:237-240.
824. Gur, R.E., Gur, R.C., Skolnick B.E., Caroff, S., Obrist, W.D., Resnick, S. and Reivich, M.(1985). Brain function in psychiatric disorders. III. Regional cerebral blood flow in unmedicated schizophrenics. Arch. Gen. Psychiary. 42:329-334.
825. Gurman, A.S. and Kniskern, D.P.(1981). Handbook of Family Therapy. Brunner/Mazel Publishers, New York.
826. Gusella, J.F., Wexler, N.S., Conneally, P.M., Naylor, S.L.et. al. (1983). A polymorphic DNA marker genetically linked to Huntington's disease. Nature 306:234-238.
827. Guze, S.B., Goodwin, D.W. and Crane, J.B.(1970). A psychiatric study of the wives of convicted felons: An example of assortative mating. Am. J. Psychiatry. 126:1773-1776.
828. Guze, S.B., Wolfgram, E.D., McKinney, J.K.and Cantwell, D.P.(1967). Psychiatric illness in the families of convicted crininals: A study of 519 first degree relatives. Dis. Nerv. Syst. 28:651-659.
829. Gysling, K. and Wang, R.Y.(1983). Morphine-induced activation of A10 dopamine neurons in the rat. Brain Res. 277:119-127.
830. Haber, S.N.(1986). Neurotransmitters in the human and nonhuman primate basal ganglia. Hum. Neurobiol. 5:159-168.
831. Haber, S.N., Kowall, N.W., Vonsattel, J.P., Bird, E.D. and Richardson, E.P.(1986). Gilles de la Tourette's syndrome. A postmortem neuropathological and immunohistochemical study. J. Neurol. Sci. 75:225-241.
832. Hagberg, B., Aicardi, J., Dias, K.,and Ramos, O.(1983). A progressive syndrome of autism, dementia, ataxia, and loss of purposeful hand use in girls: Rett syndrome: Report of 35 cases. Ann. Neurol. 14:471-479.
833. Hagerman, R.J., Jackson, A.W. III, Levitas, A., Rimland, B. and Braden, M.(1986). An analysis of autism in fifty males with the fragile X syndrome. Am. J. Med. Gen. 23:359-374.
834. Hagin, R.A., Beecher, R., Pagano, G. and Kreeger, H.(1982). Effects of Tourette syndrome on learning. In A.J. Friedhoff and T. N. Chase (eds) Gilles de la Tourette Syndrome. Raven Press, New York, pp323-328.
835. Haider, M., Spong, P. and Lindsley, D.B.(1964). Attention, vigilance, and cortical evoked potentials in humans. Science 145:180-182.
836. Haigler, H.J. and Aghaganian, G.K.(1974). Lysergic acid diethylamide and serotonin: A comparison of effects on serotonergic neurons and neurons receiving a serotonergic input. J. Pharmacol. Exp. Ther. 188:688-699.
837. Haley, H.G., Stahl, S.M. and Freedman, D.X.(1977). Hyperserotonemia and amine metabolites in autistic and retarded children. Arch. Gen. Psychiatry 34:521-531.
838. Haley, J.(1981). Problem-Solving Therapy. Jossey-Bass Publishers, San Francisco.
839. Hall, R.L., Hesselbrock, V.M. and Stabenau,

J.R.(1983). Familial distribution of alcohol use: I. Assortative mating in the parents of alcoholics. Behav. Genet. 13:361-372.
840. Hallgren, B.(1950). Specific dyslexia: A clinical and genetic study. Acta Psychiatr. Neurol. Scand. Suppl. 65:1-287.
841. Hallgren, B.(1957). Enuresis — A clinical and genetic study. Acta Psychiatr. Neurol. Scand. Suppl. 114.
842. Hallgren, B.(1960). Nocturnal enuresis in twins. Acta Psychiatr. Neurol. Scand. 35:73-90.
843. Halliday, R.A., Rosenthal, J.H., Naylos, H. et al.(1976). Average evoked potential predictors of clinical improvement in hyperactive children treated with methyphenidate: An initial study and replication. Psychophysiology 13:429-440.
844. Halstead, W.C.(1949). Brain and Intelligence. Chicago Univ. Press, Chicago.
845. Hammond, W.A.(1884). Myriachit, a newly described disease of the nervous system and its analogues. N. Y. Med. J. 39:191-192.
846. Hamon, M., Nelson, D.L., Herbert, A., and Glowinski, J.(1980). Multiple receptors for serotonin in the rat brain. In G. Pepeu, M.J. Kuhar, and S.J. Enna (eds) Receptors for Neurotransmitters and Peptide Hormones. Raven Press, New York, pp223-233.
847. Hamon, Paraire, and Velluz. (1952). Remarques sur l'action du 4560 R.P. sur l'agitation maniaque. Ann. Med. Psychol. 110:331-335.
848. Hanley, H.G., Stahl, S.M., Freedman, D.X.(1977). Hypersertonemia and amine metabolites in autistic and retarded children. Arch. Gen. Psychiatry 34:521-531.
849. Haracz, J.L.(1982). The dopamine hypothesis: An overview of studies with schizophrenic patients. Schizophr. Bull. 8:438-468.
850. Hardy, M.-C., Lecrubier, Y., and Widlöcher, D. (1986). Efficacy of clonidine in 24 patients with acute mania. Am. J. Psychiatry 143:1450-1453.
851. Harik, S.I.(1984). Locus ceruleus lesion by local 6-hydroxydopamine infusion cases marked and specific destruction of noradrenergic neurons, long-term depletion of norepinephrine and enzymes that synthesize it, and enhances dopaminergic mechanisms in the ipsilateral cerebral cortex. J. Neurosci. 4: 699- 707.
852. Harlow, J.M.(1868). Recovery from the passage of an iron bar to the head. Mass. Med. Soc. Publ. 2:328-347.
853. Haroutunian, V., Kanof, P.D., and Davis, K.L.(1987). Effects of mesocortical dopaminergic lesions upon subcortical dopaminergic function. American College of Neuro-spychopharmacology Abstracts 26th Annual Meeting, Dec. 7-11, San Juan, Puerto Rico, p28.
854. Harris, E.L., Noyes, R., Crowe, R.R. and Chaudhry, D.R.(1983). Family study of agoraphobia. Arch. Gen. Psychiatry 40:1061-1064.
855. Harris, R.A. and Allan, A.M.(1985). Functional coupling of g-aminobutyric acid receptors to chloride channels in brain membranes. Science 228:1108-1110.
856. Hart, R.G. and Easton, J.D.(1982). Carbamazepine and hematological monitoring. Ann. Neurol. 11:309-312.
857. Hartman, R.J. and Geller, I.(1971). p-Chlorophenylalanine effects on conditioned emotional response in rats. Life Sci. 10:927-933.
858. Hartocollis, P.(1968). The syndrome of mimimal brain dysfunction in young adult patients. Bull. Menninger Clin. 32:102-114.
859. Harvey, J. and Hunt, H.F.(1965). Effect of septal lesions on thirst in the rat as indicated by water consumption and operant responding for water reward. J. Comp. Physiol. Psychol. 59:49-56.
860. Hashimoto, T., Endo, S., Fukuda, K. and Hiura, K. et al. (1981). Increased body movement during sleep in Gilles de la Tourette syndrome. Brain Dev. 3:31-35.
861. Haslam, R.H.A. and Dalby, J.T.(1983). Blood serotonin levels in attention-deifict disorder. N. Eng. J. Med. 309:1328-1329.
862. Hasselager, E., Rolinksi, A. and Randrup, A.(1972). Specific antagonism by dopamine inhibitors of items of amphetamine induced aggressive behavior. Psychopharmacologia 24:485-495.
863. Hayaishi, O.(1976). Properties and function of indoleamine 2,3-dixoygenase. J. Biochem. 79:13p-21p.
864. Hayaishi, O.(1980). Newer aspects of tryptophan metabolism. In O. Hayaishi, Y. Ishimura and R. Kido (eds) Biochemical and Medical Aspects of Tryptophan Metabolism. Elsevier/North Holland, New York pp15-30.
864a. Hays, P. and Field, L.L.(1989). Postulated genetic linkage between manic-depression and stuttering. J. Affect. Disord. 16:37-40.
865. Healy, D., O'Halloran, A., Carney, P.A. and Leonard, B.E.(1986). Variations in platelet 5-hydroxytryptamine in control and depressed populations. J. Psychiatr. Res. 20:345-353.
866. Healy, J.M., Aram, D.M., Horwitz, S.J. and Kessler, J.W.(1982). A study of hyperlexia. Brain Langu. 17:1-23.
867. Heath, R.G.(1964). The Role of Pleasure in Behavior. Harper and Row, New York.
868. Heatherington, A.W. and Ranson, S.W.(1942). Hypothalamic lesion and adiposity in the rat. Anat. Rec. 78:149-172.
869. Hebb, D.O.(1945). Man's frontal lobes. Arch. Neurol. Psychiatry 54:10-24.
870. Hebb, D.O.(1949). Organization of Behavior. John Wiley and Sons, New York.
871. Hebb, D.O.(1955). Drives and the CNS (conceptual nervous system). Psychol. Rev. 62:243-254.
872. Hechtman, L. and Weiss, G.(1986). Controlled prospective 15 year follow-up of hyperactives as adults: Non-medical drug and alcohol use and antisocial behavior. Can. J. Psychol. 31:557-567.
873. Hechtman, L., Weiss, G. and Perlman, T.(1984). Hyperactives as young adults: Past and current substance abuse and antisocial behavior. Am. J. Orthopsychiatry. 54:415-425.
874. Hechtman, L., Weiss, G., Perlman, T. and Amsel, R.(1984). Hyperactives as young adults: Initial predictors of adult outcomes. J. Am. Acad. Child Psychiatry 23:250-260.
875. Hechtman, L., Weiss, G., Perlman, T., Hopkins, J. Wener, A.(1981). Hyperactive children in young adulthood: A controlled, prospective, ten-year follow-up. In M. Gittelman (ed). Strategic Interventions for Hyperactive Children. Sharpe, Armonk, NY pp186-200.
876. Heeley, A.F. and Roberts, G.E.(1965). Tryptophan metabolism in psychotic children. Dev. Med. Child Neurol. 7:46-49.
877. Hegedus, A., Alterman, A. and Tarter, R.(1984). Learning achievement in sons of alcoholics. Alcohol: Clin. Exp. Res. 8:330-333.
878. Heikkila, R.E., Hess, A., and Duvoisin, R.C.(1984).

Dopaminergic neurotoxicity of 1-methyl-4-phenyl-1,2,5,6-tetrahydropyridine in mice. Science 224:1451-1453.
879. Heilman, K.M. and Valenstein, E.(1972). Frontal lobe neglect in man. Neurology 22:660-664.
880. Heimer, L.(1983). The Human Brain and Spinal Cord. Springer-Verlag, New York.
881. Helzer, J.E. and Pryzbeck, T.R.(1988). The co-occurrence of alcoholism with other psychiatric disorders in the general popualtion and its impact on treatment. J. Stud. Alcohol 49:219-224.
882. Hemsley, D.R. and Zawada, S.L.(1976). "Filtering" and the cognitive deficit in schizophrenia. Br. J. Psychiatry 128:456-461.
883. Henker, B., Whalen, C.K., Bugental, D.B. and Barker, C.(1981). Licit and illicit drug use patterns in stimulant-treated children and their peers. In K.D. Gadow and J. Loney(eds) Psychosocial Aspects of Drug Treatment for Hyperactivity. pp 443-462.
883a. Hermann, K. and Norrie, E.(1958). Is congenital word blindness an hereditary type of Gerstman's syndrome? Psychiatr. Neurol.. 136:59-73.
884. Hermelin, B. and O'Conner, N.(1970). Psychological Experiments with Autistic Children. Pergamon, London.
885. Herschel, M.(1978). Dyslexia revisited. Hum. Genet. 40:115-134.
886. Hertting, G., Axelrod, J. and Whitby, L.G.(1961). Effect of drugs on the uptake and metabolism of H^3-norepinephrine. J. Pharmacol. Exp. Ther. 134:146-153.
887. Herve D., Blanc, G., Glowinski, J. and Tassin, J.P.(1982). Reduction of dopamine utilization in the prefrontal cortex but not in the nucleus accumbens after selective destruction of noradrenergic fibers innervating the ventral tegmental area in the rat. Brain Res. 237:510-516.
888. Hesselbrock, M., Hesselbrock, V., Babor, T., Stabenau, J. Meyer, R. and Weidenman, M.(1984). Antisocial behavior, psychopathology and problem drinking in the natural history of alcoholism. In D. Goodwin, K. Van Dusen, and S. Mednick (eds) Longitudinal Research in Alcoholism. Kluver-Nijhoff, Boston. pp197-214.
889. Hesselbrock, M.N., Meyer, R.E., and Keener, J.J.(1985). Psychopathology in hospitalized alcoholics. Arch. Gen. Psychiatry 42:1050-1055.
890. Heston, L.L. and Denney, D.(1968). Interactions between early life experience and biological factors in schizophrenia. In D. Rosenthal and S.S. Kety (eds)The Transmission of Schizophrenia. Pergamon, Oxford, pp 363-376.
891. Heston, L.L.(1966). Psychiatric disorders in foster home reared children of schizophrenic mothers. Br. J. Psychiat. 112:819-825.
892. Heston, L.L.(1970). The genetics of schizophrenic and schizoid disease. Science 167:249-256.
893. Hewitt, L. and Jenkins, R.L.(1946). Fundamental Patterns of Maladjustment. State of Illinois, Springfield.
894. Higuchi, K. and Hayaishi, O.(1967). Enzymatic formation of D-kynurenine from D-tryptophan. Arch. Biochem. Biophys. 120:397-403.
895. Hill, D.(1944). Amphetamine in psychopathic states. Br. J. Addict. 44:50-54.
896. Hill, D.R. and Bowery, N.G.(1981). ^{3}H-Baclofen and ^{3}H-GABA bind to bicculline-insensitive GABAB sites in rat brain. Nature 290:149-152.
896a. Hill, E.M., Wilson, A.F., Elston, R.C. and Winokur, G.(1988). Evidence for possible linkage between genetic markers and affective disorders. Biol. Psychiatry 24:903-917.
897. Hill, S.Y., Goodwin, D.W., Cadoret, R., Osterland, C.K. and Doner, S.M.(1975). Association and linkage between alcoholism and eleven serological markers. J. Stud. Alcohol 37:981-992.
897a. Hill, S.Y., Aston, C. and Rabin, B.(1988). Suggestive evidence of genetic linkage between alcoholism and the MNS blood group. Alcohol.: Clin. Exp. Res. 12:811-814.
898. Hiller, J.M., Pearson, J. and Simon, E.J.(1973). Distribution of stereospecific binding of the potent narcotic analgesic etorphine in the human brain: Predominance in the limbic system. Res. Commun. Chem. Pathol. Pharmacol. 6:1052-1062.
899. Hirsch, R.(1980). The hippocampus, conditional operations, and cognition. Physiol. Psychol. 8:175-182.
900. Ho, A.K.S., Tsai, C.S., Chen, R.C.A., Begleiter, H. and Kissin, B.(1974). Experimental studies on alcoholism. I. Increase in alcohol preference by 5,6- dihydroxytryptamine and brain acetylcholine. Psychopharmacologia 40:101-107.
901. Hoch, P.H. and Cattell, J.P.(1955). The diagnosis of pseudoneurotic schizophrenia. Psychiatr. Q. 33:17-43.
902. Hock, P.H. and Polatin, P.(1949). Pseudoneurotic forms of schizophrenia. Psychiatr. Q. 23:248-276.
903. Hodgkinson, S.RZ., Sherringtron, R., Gurling, H., Marchbanks, R., Reeders, S., Mallet, J., McInnis, Petursson, H. and Brynjolfsson, J.(1987). Molecular genetic evidence for heterogeneity in manic depression. Nature 325:805-806.
904. Hoehn-Saric, R.(1982). Neurotransmitters in anxiety. Arch. Gen. Psychiatry 39:735-742.
905. Hoehn-Saric, R., Merchant, A.F., Keyser, M.L. and Smith, V.K.(1981). Effects of clonidine on anxiety disorders. Arch. Gen. Psychiatry 38:1278-1282.
906. Hoenig, J.(1983). The concept of schizophrenia Kraepelin-Bleuler-Schneider. Br. J. Psychiatry 142:547-556.
907. Hokfelt, T., Elde, R., Johansson, O., Ljungdahl, A. et al. (1978). Distribution of peptide-containing neurons. In M.A. Lipton, A. DiMascio, and K.F. Killam (eds) Psychopharmacology: A Generation of Progress. Raven Press, New York, pp39-66.
908. Hokfelt, T., Fuxe, K., Golstein, M. and Johansson, O.(1974). Pharmaco-histochemical evidence for the existence of dopamine nerve terminals in the limbic cortex. Eur. J. Pharmacol. 25:108-112.
909. Hokfelt, T., Johansson, O., Ljungdahl, Å., Lundberg, J.M. and Schultzberg, M.(1980). Peptidergic neurones. Nature 284:515-521.
910. Hokfelt, T., Ljungdahl, A., Fuxe, K. and Johansson, O.(1974). Dopamine nerve terminals in the rat limbic cortex: Aspects of the dopamine hypothesis of schizophrenia. Science 184:177-179.
911. Holden, C.(1980). Identical twins reared apart. Science 207:1323-1328.
912. Holland, A.J., Hall, A., Murray, R., Russell, G.F.M. and Crisp, A.H.(1984). Anorexia nervosa: A study of 34 twins and one set of triplets. Br. J. Psychiatry 145:414-419.
913. Hollister, A.S., Breese, G.R. and Cooper, B.R.(1974). Comparison of tyrosine hydroxylase and dopamine beta hydroxylase inhibition with the effects of various 6-hydroxydopamine treatments on d-amphetamine induced motor activity. Psychopharmacology 36:1-16.
914. Hollister, A.S., Breese, G.R., Moreton, C., Cooper, B.R. and Bunney, W.E.(1976). An inhibitory role for brain serotonin-containing systems in the locomotor effects of d-

amphetamine. J. Pharmacol. Exp. Ther. 198:12-22.
915. Hollister, L.(1975). The mystique of social drugs and sex. In M. Sandler and G. Gessa (eds). Sexual Behavior: Pharmacology and Biochemistry. Raven Press, New York, pp85-92.
916. Holmberg, G. and Gershon, S.(1961). Autonomic and psychic effects of yohimbine hydrochloride. Psychopharmacol. 2:93-106.
917. Holmes, G.(1931). Discussion on the mental symptoms associated with cerebral tumors. Proc. R. Soc. Med. 24:997-1008.
918. Holtzman, P.S., Kringlen, E., Levy, D.L. and Haberman, S.L.(1980). Deviant eye tracking in twins discordant for psychosis: A replication study. Arch. Gen. Psychiatry 37:627-631.
919. Holtzman, P.S., Kringlen, E., Levy, D.L. and Proctor, L.R.(1978). Smooth pursuit eye movements in twins discordant for schizophrenia. J. Psychiatr. Res. 14:111-126.
920. Holtzman, P.S., Levy, D.L. and Proctor, L.R.(1976). Smooth pursuit eye movements, attention and schizophrenia. Arch. Gen. Psychiatry 33:1415-1420.
921. Holtzman, P.S., Proctor, L.R., Levy, D.L., Yasillo, N.J., Meltzer, H.J. and Hurt, S.W.(1974). Eye-tracking dysfunctions in schizophrenic patients and their relatives. Arch. Gen. Psychiatry 31:143-151.
922. Holtzman, P.S., Solomon, C.M, Levin, S. and Watemaux, C.S.(1984). Pursuit eye movement dysfunctions in schizophrenia. Arch. Gen. Psychiatry 41:136-140.
923. Hong, J.S., Yang, H-Y.T., and Costa, E.(1977). Determination of methionine-enkephalin in discrete regions of the rat brain. Brain Res. 134:383-386.
925. Hong, J.S., Yang, H-Y.T., Fratta, W. and Costa, E.(1978). Rat striatal methionine-enkephalin content after chronic treatment with cataleptic and noncataleptic antischizophrenic drugs. J. Pharmacol. Exp. Ther. 205:141-147.
926. Horn, J.M., Green, M., Carney, R. and Erickson, M.T.(1975). Bias against genetic hypotheses in adoption studies. Arch. Gen. Psychiatry 32:1365-1367.
927. Hornykiewicz, O.(1982). Brain catecholamines in schizophrenia—A good case for noradrenaline. Nature 299: 484-486.
928. Hornykiewicz, O.(1986). Brain noradrenaline and schizophrenia. In J.M. van Ree and S. Matthysse (eds) Prog. Brain Res. 65:29-39.
929. Horton, A.(1985). Prevelence of attention deficit disorder, residual type in alcoholics: A cross-validation. Int. J. Clin. Neuropsychol. 8:52.
930. Hoshino, Y., Yamamoto, T., Kaneko, M., Tachibana, R., Watanabe, M., Ohno, Y., and Kumashiro, H.(1985). Blood serotonin and free tryptophan concentration in autistic children. Neuropsychobiol. 39:531-536.
931. Howell, D.C. and Huessy, H.R.(1981). Hyperkinetic behavior followed from 7 to 21 years of age. In M. Gittelman (ed) Strategic Interventions for Hyperactive Children. Sharpe, Armonk, NY, pp201-215.
932. Hubel, D.H. and Wiesel, T.N.(1968). Receptive field and functional architecture of monkey striate cortex. J. Physiol. 195:215-243.
933. Huessy, H.R. and Cohen, A.H.(1976). Hyperkinetic behaviors and learning disabilities followed over seven years. Pediatrics 57:4-10.
934. Huessy, H.R.(1974). The adult hyperkinetic. Am. J. Psychiatry 131:724-725.
935. Huessy, H.R., Marshall, C.D. and Gendron, R.A.(1973). Five-hundred children followed from grade 2 through grade 5 for the prevalence of behavioral disorders. Acta Paedopsychiatry 39:301-309.
936. Huessy, H.R., Metoyer, M. and Townsend, M.(1973). Eight-to-ten-year follow-up of children treated in rural Vermont for behavioral disorder. Am. J. Orthopsychiatry 43:236-238.
937. Hughes, J.(1965). A review of the positive spike phenomena, In W. Wilson (ed) Applications of Electroencephalography in Psychiatry. Duke Univ. Press, Durham, NC.
938. Hughes, J.(1975). Isolation of an endogenous compound from the brain with pharmacological properties similar to morphine. Brain Res. 88:295-308.
939. Hughes, J., Smith, T.W., Kosterlitz, H.W., Fothergill, L.A., Morgan, B.A., and Morris, H.R.(1975). Identification of two related pentapeptides from the brain with potent opiate agonist activity. Nature 258:577-579.
940. Hughes, J.R. and Miller, S.A.(1984). Nicotine gum to help stop smoking. J.A.M.A. 252:2855-2858.
941. Hughes, J.R., Gianturco, D. and Stein, W.(1961). Electroclinical correlations in the positive spike phenomenon. Electroenceph. Clin. Neurophys. 13:599-605.
942. Humphries, T., Kinsbourne, M. and Swanson, J.(1978). Stimulant effects on cooperation and social interaction between hyperactive children and their mothers. J. Child Psychol. Psychiatry 19:13-22.
943. Hunt, R.D.(1987). Treatment effects of oral and transdermal clonidine in relation to methyphenidate: An open pilot study in ADD-H. Psychopharmacol. Bull. 23:111-114.
944. Hunt, R.D., Minderra, R. and Cohen, D.J.(1985). Clonidine benefits children with attention deficit disorder and hyperactivity: Report of a double-blind placebo-crossover therapeutic trial. J. Am. Acad. Child Psychiatry 24:617-629.
945. Hutchings, B. and Mednick, S.A.(1977). In S.A. Mednick and K.O. Christiansen (eds) Biosocial Basis of Criminal Behavior. Gardner, New York.
946. Ingram, T.T.S.(1956). A characteristic form of overactive behavior in brain damaged children. J. Ment. Sci. 102:550-558.
947. Inouye, E.(1965). Similar and dissimilar manifestations of obsessive-compulsive neuroses in mono-zygotic twins. Am. J. Psychiatry 121:1171-1175.
948. Insel, T.R. and Murphy, D.L.(1981). The psychopharmacologic treatment of obsessive compulsive disorder. A review. J. Clin. Psychopharmacol. 1:304-311.
949. Insel, T.R. and Pickar, D.(1983). Naloxone administration in obsessive compulsive disorder: Report of two cases. Am. J. Psychiatry 140:1219-1220.
950. Insel, T.R., Mueller, E.A., Alterman, I., Linnoila, M. and Murphy, D.L.(1985). Obsessive-compulsive disorder and serotonin: Is there a connection? Biol. Psychiatry 20:1174-1188.
951. Insel, T.R., Roy, B.F., Cohen, R.M. and Murphy, D.L.(1982). Possible development of the serotonin syndrome in man. Am. J. Psychiatry 139:954-955.
952. Irwin, M., Belenbiuk, K., McCloskey, K. and Freedman, D.X.(1981). Tryptophan metabolism in children with attentional deficit disorder. Am J. Psychiatry 138:1082-1085.
953. Isaacson, R.L.(1982). The Limbic System, Second Edition. Plenum Press, New York. pp327.
954. Israngkun, P.P., Newman, H.A.I. and Patel, S.T.(1986). Potential biochemical markers for infantile au-

tism. Neurochem. Pathol. 5:51-70.
955. Itard, J.M.G.(1825). Memorie sur quelques fonctions involontaires des appareils de la locomotion de la prehension et de la voix. Arch. Gen. Med. 8:385-407.
955a. Ito, T., Fujita, K., Matsuura, S. and Nagatsu, T.(1988). Treatment of depression with (6R)-tetrahydrobiopterin, the natural cofactor of tyrosine hydroxylase and tryptohan hydroxylase. Biogenic Amines 5:489-493.
956. Iversen, S.D. and Koob, G.F.(1977). Behavioral implications of dopaminergic neurons in the mesolimbic system. Adv. Biochem. Psychopharmacol. 16:209-214.
957. Iversen, S.D.(1971). The effect of surgical lesions to frontal cortex and substantial nigra on amphetamine responses in rats. Brain Res. 31:295-311.
958. Iversen, S.D.(1984). 5-HT and anxiety. Neuropharmacology 23:1553-1560.
959. Iversen, S.D., Wilkinson, S., and Simpson, B.(1971). Enhanced amphetamine responses after frontal cortex lesions in the rat. Eur. J. Pharmacol. 13:387-390.
960. Iverson, S.D.(1977). Striatal function and stereotyped behavior. In A.R. Cook, A.H.M. Loman and J.H.L. van den Berken (eds) Physiology of the Striatum. North-Holland, Amsterdam.
961. Jackman, H., Luchins, D. and Meltzer, H.Y.(1983). Platelet serotonin levels in schizophrenia: Relationship to race and psychopathology. Biol. Psychiatry 18:887-902.
962. Jacob, J., Girault, J.M. and Peindaries, R.(1972). Actions of 5-hydroxytryptamine and 5-hydroxytryptophan injected by various routes on the rectal temperature of the rabbit. Neuropharmacology 11:1-16.
963. Jacobs, B.L. and Trulson, M.E.(1979). Dreams, hallucinations, and psychoses — The serotonin connection. Trends Neurosci. 2:276-280.
964. Jacobs, B.L. and Trulson, M.E.(1979). Mechanism of action of LSD. Am. Sci. 67:396-404.
965. Jacobs, B.L.(1974). Evidence for the functional interaction of two central neurotransmitters. Psychopharmacol. 39:81-86.
966. Jacobs, B.L.(1976). An animal behavior model for studying central serotonergic synapses. Life Sci. 19: 777-786.
967. Jacobsen, C.F.(1935). Function of frontal association area in primates. Arch. Neurol. Psychiatry 33:558-569.
968. Jacoby, J.H., and Poulakos, J.J.(1977). The actions of neuroleptic drugs and punative serotonin receptor antagonists on LSD and quipazine-induced reductions of brain 5-HIAA concentration. J. Pharm. Pharmacol. 29:771-773.
969. Jacquet, Y.T., and Marks, N.(1976). The C-fragment of b-lipotropins: A endogenous neuroleptic or antipsychotogen. Science 194:632-634.
970. Jaeken, J. and Van den Berghe, G.(1984). An infantile autistic syndrome characterized by the presence of succinylpurines in body fluids. Lancet 2:1058-1061.
971. Jaffe, J.H., Babor, T.F. and Fishbein, D.H.(1988). Alcoholics, aggression and antisocial personality. J. Stud. Alcohol 49:211-218.
972. Jagger, J., Prusoff, B.A., Cohen, D.J., Kidd, K.K., Carbonari, C.M. and John, K.(1982). The epidemiology of Tourette's syndrome: A pilot study. Schizophr. Bull. 8:267-277.
973. James, W.(1890). The Principles of Psychology. (2 vols.) Holt, New York.
974. Jankovic, J. and Rohaidy, H.(1987). Motor, behavioral and pharmacologic findings in Tourette's syndrome. Can. J. Neurol. Sci. 14:541-546.
975. Jankovic, J., Glaze, D.G. and Frost, J.D.(1984). Effect of tetrabenazine on tics and sleep of Gilles de la Tourette's syndrome. Neurol. 34:688-692.
976. Janowksy, D.S., El-Yousef, M.K., Davis, J.M. and Sekerke, H.J.(1973). Provocation of schizophrenic symptoms by intravenous administration of methylphenidate. Arch. Gen. Psychiatry 28:185-191.
977. Jarvik, L. and Kulbach, M.(1983). Atypical miner: Prototype of the pathological gambler? Compr. Psychiatry 24:213-217.
977a. Jaworski, M.A., Slater, J.D., Severini, A., Henning, K.R., Mansour, G., Mehta, J.G., Jeske, R., Schlaut, J., Pak, C.Y. and Yoon, J.-W.(1988). Unusual clustering of diseases in a Canadian Old Colony (Chortiza) Mennonite kindred and community. Can. Med. Assoc. J. 138:1017-1025.
978. Jenike, M.A., Baer, L., Minichiello, E., Schwartz, C.E. and Carey, R.J.(1986). Concomitant obsessive-compulsive disorder and schizotypal personality disorder. Am. J. Psychiatry 143:530-532.
979. Jenkins, R.L. and Ashby, H.B.(1983). Gilles de la Tourette's syndrome in identical twins. Arch. Neurol. 40:249-251.
980. Jeste, D.V. and Wyatt, R.J.(1982). Therapeutic strategies against tardive dyskinesia. Arch. Gen. Psychiatry 803-816.
981. Jeste, D.V. and Wyatt, R.J.(1982). Understanding and Treating Tardive Dyskinesia Guilford Press, New York.
982. John, E.R., Tang, Y., Brill, A.B., Young, R. and Ono, K.(1986). Double-labeled metabolic maps of memory. Science 233:1167-1175.
983. Johnson, G.F.S. and Leeman, M.M.(1977). Analysis of familial factors in bipolar affective illness. Arch. Gen. Psychiatry 34:1074-1083.
984. Johnson, J.L., Bennett, L.A., Wolin, S.J. and Eckardt, M.J.(1987). Cognitive performance patterns of children from alcoholic and nonalcoholic families. (personal communication).
985. Johnston, D.G., Troyer, I.E. and Whitsett, S.F.(1988). Clomipramine treatment of agoraphobic women. Arch. Gen. Psychiatry 45:453-459.
986. Jolles, J.(1986). Neuropeptides and cognitive disorders. Prog. Brain Res. 65:177-192.
987. Jones, H. and Jones, H.(1977). Sensual Drugs: Deprivation and Rehabilitation of the Mind. Cambridge University Press, Cambridge.
988. Jones, H.E., MacFarlane, J.W., and Eichorn, D.H.(1960). A progress report on growth studies at the Unviersity of California. Vita Humana 3:17-31.
989. Jones, J. and Gold, M.S.(1986). Naltrexone reverses bulemia symptoms. Lancet 1:807.
990. Jones, M.C.(1968). Personality correlates and antecedents of drinking patterns in adult males. J. Consult. Clin. Psychol. 32:2-12.
991. Jones, R.S.G. and Boulton, A.A.(1980). Tryptamine and 5-hydroxytryptamine: Action and interaction on cortical neurons in the rat. Life Sci. 27:1849-1856.
992. Jones, R.S.G.(1982). Responses of cortical neurons to stimulation of the nucleus raphe medianus: A pharmacological analysis of the role of indolemines. Neuropharmacology 21:511-520.
993. Joseph, M.H., Baker, H.F., Crow, T.J., Riley, G.J. and Risby, D.(1979). Brain tryptophan metabolism in schizo-

phrenia: A post mortem study of metabolites on the serotonin and kynurenine pathways in schizophrenic and control subjects. Psychopharmacology 62:279-285.
994. Joseph, M.H., Young, S.N. and Curzon, G.(1976). The metabolism of a tryptophan load in rat brain and liver. The influence of hydrocortisone and allopurinol. Biochem. Pharmacol. 25:2599-2604.
995. Joseph, M.S., Brewerton, T.D., Reus, V.I. and Stebbins, G.T.(1984). Plasma L-tryptophan/neutral amino acid ratio and dexamethasone suppression in depression. Psychiatry Res. 11:185-192.
996. Jouvet, M.(1969). Biogenic amines and the states of sleep. Science 163:32-41.
997. Jouvet, M.(1972). The role of monamines and acetylcholine-containing neurons in the regulation of the sleep-waking cycle. Ergeb. Physiol. Biol. Chem. 64:166-307.
998. Kadikani, H., Furutanin, Y., Takahashi, H., Noda, M., Morimoto, Y., Hirose, T., Asai, M., Inayama, S., Nakaishi, S., Numa, S.(1982). Cloning and sequence analysis of cDNA for procine-neo-endorphin/dynorphin precursor. Nature 298:245-249.
999. Kagan, J.(1966). Reflection-impulsivity: The generality and dynamics of conceptual tempo. J. Abnorm. Psychol. 71:17-24.
1000. Kahn, E. and Cohen, L.H.(1934). Organic drivenness: A brain stem syndrome and an experience. N. Engl. J. Med. 210:748-756.
1001. Kahn, R.J., McNair, D.M., Pipman, R.S., et al.(1986). Imipramine and chlordiazepoxide in depressive and anxiety disorders. II. Efficacy in anxious outpatients. Arch. Gen. Psychiatry 43:79-85.
1002. Kahn, R.S. and Westenberg, H.G.M.(1985). L-5-hydroxytryptophan in the treatment of anxiety disorders. J. Affective Disord. 8:197-200.
1003. Kahn, R.S., van Praag, H.M., Weltzler, S., Asnis, G.M. and Barr, G.(1988). Serotonin and anxiety revisited. Biol. Psychiatry 23:189-208.
1004. Kahn, R.S., Westenberg, H.G.M. and Jolles, J.(1984). Zimeldine treatment of obsessive compulsive disorder. Acta Psychiatr. Scand. 69:259-261.
1005. Kahn, R.S., Westenberg, H.G.M., Verhoeven, W.M.A, Gispen-deWied, C.C., Kamerbeek, W.D.J. (1987). Effect of serotonin precursor and uptake inhibitor in anxiety disorders: A double blind comparison of 5-hydroxytryptophan, clomipramine and placebo. Int. Clin. Psychopharmacol. 2:33-45.
1006. Kaila, K. and Voipio, J.(1987). Postsynaptic fall in intracellular pH induced by GABA-activated bicarbonate conductance. Nature 330:163-165.
1007. Kaim, B.(1983). A case of Gilles de la Tourette's syndrome treated with clonazepam. Brain Res. Bull. 11:213-214.
1008. Kalat, J.W.(1975). Minimal brain dysfunction: Dopamine depletion? Science 194:450.
1009. Kallmann, F.J.(1938). The Genetics of Schizophrenia. J. J. Augustin, New York.
1010. Kandel, E.R. and Schwartz, J.H.(1982). Molecular biology of learning: Modulation of transmitter release. Science 218:433-442.
1011. Kane, J.M. and Smith, J.M.(1982). Tardive dyskinesie. Prevalence and risk factors 1959 to 1979. Arch. Gen. Psychiatry 39:473-481.
1012. Kanner, L.(1943). Autistic disturbances of affective contact. Nerv. Child. 2:217-250.
1013. Kanner, L.(1949). Problems with nosology and psychodynamics in early childhood autism. Am. J. Orthopsychiatry 19:416.
1014. Kantor, J.S., Zitrin, C.M. and Zeldis, S.M.(1980). Mitral valve prolapse syndrome in agoraphobic patients. Am.J. Psychiatry 137:467-469.
1015. Karlsson, J.L.(1968). Genealogic studies of schizophrenia. In D. Rosenthal and S.S. Kety (eds) The Transmission of Schizophrenia Pergamon, Oxford, pp85-94.
1015a. Karno, M., Golding, J.M., Sorenson, S.B. and Burnam, M.A.(1988). The epidemiology of obsessive-compulsive disorder in five US communities. Arch. Gen. Psychiatry 45:1094-1099.
1016. Kauffman, C., Grunebaum, H., Cohler, B. and Gamer, E.(1979). Superkids: Competent children of psychotic mothers. Am. J. Psychiatry 136:1398-1402.
1017. Kaye, W.D., Pickar, D., Naber, D. and Ebert, M.(1982). Cerebrospinal fluid opioid activity in anorexia nervosa. Am. J. Psychiatry 139:643-645.
1018. Kaye, W.H., Ebert, M.H., Gwirtsman, H.E. and Weiss, S.R.(1984). Differences in brain serotonergic metabolism between nonbulemic and bulimic patients with anorexia nervosa. Am. J. Psychiatry 141:1598-1601.
1019. Kebabian, J.W. and Calne, D.B.(1979). Multiple receptors for dopamine. Nature 277:93-96.
1020. Kellam, S.G., Ensminger, M.E. and Simon, M.B.(1980). Mental health in first grade and teenage drug, alcohol and cigarette use. Drug Alcohol Depend. 5:273-304.
1021. Kelly, D., Mitchell-Heggs, N. and Sherman, D.(1971). Anxiety and the effects of sodium lactate assessed clinically and physiologically. Br. J. Psychiatry 119:129-141.
1022. Kelly, P.H. and Iversen, S.D.(1976). Selective 6OHDA-induced destruction of mesolimbic dopamine neurons: Abolution of psychostimulant-induced locomotor activity in rats. Europ. J. Pharmacol. 40:45-56.
1023. Kelly, P.H., Seviour, P.W. and Iversen, S.D.(1975). Amphetamine and apomorphine responses in the rat following 6-OHDA lesions of the nucleus accumbens septi and corpus striatum. Brain Res 94:507-522.
1023a. Kelsoe, J.R., Gings, E.I., Egeland, J.A., Goldstein, A.M., Bale, S.J., Pauls, D.L., Long, R.T., Conte, G., Gerhard, D.S., Housman, D.E. a dn Paul, S.M.(1989). Reevaluation of the linkage between chromosome 11p and the gene for bipolar disorder in the old order amish. First World Congress on Psychiatric Genetics, August 3-5, Churchill College, Cambridge, England, p12.
1024. Kendall, P.C. and Braswell, L.(1985). Cognitive Behavioral Therapy for Impulsive Children. Guilford Press, New York.
1025. Kendell, R. and DiScipio, W.(1970). Obsessional symptoms and obsessional personality traits in patients with depressive illnesses. Psychol. Med. 1:65-72.
1026. Kendell, R.(1968). The classification of depressive illnesses. Oxford Univ. Press, London.
1027. Kendler, K.S.(1985). Diagnostic approaches to schizotypal personality disorder: A historical perspective. Schizophr. Bull. 11:538-553.
1028. Kendler, K.S., Gruenberg, A.M. and Strauss, J.S.(1981). An independent analysis of the Copenhagen sample of the Danish adoption study of schizophrenia. II. Relationship between schizotypal personality disorder and schizophrenia. Arch. Gen. Psychiatry 38: 982-984.

1029. Kendler, K.S., Gruenberg, A.M. and Strauss, J.S.(1982). An independent analysis of the Copenhagen sample of the Danish adoption study of schizophrenia: V. The relationship between childhood social withdrawal and adult schizophrenia. Arch. Gen. Psychiatry 39:1257-1261.
1030. Kennard, M.A., Spencer, S. and Fountain, G.(1941). Hyperactivity in monkeys following lesions of the frontal lobes. J. Neurophysiol. 4:512-524.
1030a. Kennedy, J.L., Giuffra, L.A., Moises, H.W., Cavalli-Sforza, L.L., Pakstis, A.J., Kidd, J.R., Castiglione, C.M., Sjogren,B., Wetterberg, L. and Kidd, K.K. (1988). Evidence against linkage of schizophrenia to markers on chromosome 5 in a northern Swedish pedigree. Nature 336:167-170.
1031. Keogh, B.K.(1971). Hyperactivity and learning disorders: Review and speculation. Except. Child. 38:101-109.
1032. Kerbeshian, J. and Burd, L.(1985). Auditory hallucinosis and atypical tic disorder: Case reports. J. Clin. Psychiatry 46:398-399.
1032a. Kerbeshian, J. and Burd, L.(1985). Asperger's syndrome and Tourette syndrome: The case of the pinball wizard. Br. J. Psychiatry 148:731-736.
1033. Kerbeshian, J. and Burd, L.(1987). Are schizophreniform symptoms present in attenuated form in children with Tourette disorder and other developmental disorders. Can. J. Psychiatry 32:123-135.
1033a. Kerbeshian, J. and Burd, L.(1989). Tourette disorder and bipolar symptomatology in children and adolescence. Can. J. Psychiatry (in press).
1033b. Kerbeshian, J. and Burd, L.(1988). Differential responsiveness to lithium in patients with Tourette syndrome. Neurosci. Biobehav. Rev. 12:247-250.
1033c. Kerbeshian, J. and Burd, L.(1988). A clinical pharmacological approach to treating Tourette syndrome in children and adolescents. Neurosci. Biobehav. Revs. 12:241-245.
1033d. Kerbeshian, J. and Burd, L.(1988). Tourette disorder and schizophrenia in children. Neurosci. Biobehav. Rev. 12:267-270.
1033e. Kerbeshian, J. and Burd, L.(1988). Tourette disorder and mutational falsetto. Neurosci. Biobehav. Rev. 12:271-273.
1034. Kerbeshian, J., Burd, L. and Martsolf, J.T.(1984). Fragile X syndrome associated with Tourette symptomatology in a male with moderate mental retardation and autism. Dev. Behav. Pediatr. 5:201-203.
1035. Kernberg, O.F.(1967). Borderline personality organization. J. Am. Psychoanal. Assoc. 15:641-685.
1036. Kernberg, O.F.(1975). Borderline Conditions and Pathological Narcissism. Jason Aronson, New York.
1037. Kety, S.S., Rosenthal, R., Wender, P.H., Schulsinger, F. and Jacobson, B.(1975). Mental illness in the biological and adoptive families of adopted individuals who have become schizophrenic: A preliminary report based upon psychiatric interviews. In R. Fieve, D. Rosentahl and H. Brill (eds) Genetic Research in Psychiatry. Johns Hopkins Univ. Press, Baltimore. pp147-165.
1038. Kety, S.S., Rosenthal, R., Wender, P.H., Schulsinger, F. and Jacobson, B.(1978). The biologic and adoptive families of adopted individuals who became schizophrenic: Prevalence of mental illness and other characteristics. In L.C. Wynne, R.L. Cromwell and S. Matthysse (eds) The Nature of Schizophrenia. John Wiley & Sons, New York, pp25-37.
1039. Kety, S.S., Rosenthal, R., Wender, P.H., Schulsinger, F.(1968). The types and prevelence of mental illness in the biological and adoptive families of adopted schizophrenics. In D. Rosenthal and S.S. Kety (eds) The Transmission of Schizophrenia Pergamon, Oxford, pp 345-362.
1040. Khantzian, E.J.(1985). The self-medication hypothesis of addictive disorder: Focus on heroin and cocaine dependence. Am. J. Psychiatry 142:1259-1264.
1040a. Khanna, S.(1988). Obsessive-compulsive disorder: Is there a frontal lobe dysfunction? Biol. Psychiatry 24:602-613.
1041. Kidd K.K. and Records, M.A.(1982). Genetic methodologies for the study of speech. In X.O. Breakfield (ed) Neurogenetics: Genetic Approaches to the Nervous System. Elsevier, New York, pp311-343.
1042. Kidd, K.K.(1983). Recent progress in the genetics of stuttering. In C.L. Ludlow and J.A. Cooper (eds) Genetic Aspects of Speech and Language Disorders. Academic Press, New York, pp197-213.
1043. Kidd, K.K., Prussoff, B.A., and Cohen, D.J.(1980). Familial pattern of Gilles de la Tourette syndrome. Arch. Gen. Psychiatry 37:1336-1339.
1044. Kido, H., Fukusen, N. and Katunuma, N. (1987). Epidermal growth factor as a new regulator of induction of tyrosine aminotransferase and tryptophan oxygenase by glucocorticoids. FEBS Lett. 223:223-226.
1045. Kiely, M. and Sourkes, T.L.(1972). Transport of L-tryptophan into slices of rat cerebral cortex. J. Neurochem. 19:2863-2872.
1046. Kimball, R.W., Friedman, A.P. and Vallejo, E.(1960). Effect of serotonin in migraine patients. Neurology 10:107-111.
1047. Kimble, D.P.(1968). Hippocampus and internal inhibition. Psychol. Bull. 70:285-295.
1048. Kimble, D.P., Bagshaw, M.H. and Pribram, K.H.(1965). The GSR of monkeys during orienting and habituation after selective partial ablations of the cingulate and frontal cortex. Neuropsychology 3:121-128.
1049. Kimmel, H.D., Olst, E.H. and Orlebeke, J.F.(1979). The Orienting Reflex in Humans. John Wiley & Sons, New York..
1050. King, R., Faull, K.F., Stahl, S.M., Mefford, I.N., Thiemann, S., Barchas, J.D. and Berger, P.A.(1985). Serotonin and schizophrenia: Correlations between serotonergic activity and schizophrenic motor behavior. Psychiatry Res. 14:235-240.
1051. Kirkegaard-Sorensen, L., Mednick, S.A.(1975). Registered criminality in families with chldren at high-risk for schizophrenia. J. Abnor. Psychol. 84:197-204.
1052. Kishimoto, H. and Hama, Y.(1976). The level and diurnal rhythm of plasma tryptophan and tyrosine in manic-depressive patients. Yokohama Med. Bull. 27:89-97.
1053. Kishner, H.S.(1986). Behavioral Neurology. A Practical Approach. Churchill Livingstone, New York.
1054. Klawans, H.L. and Margolin, D.I.(1975). Amphetamine-induced dopaminergic hypersensitivity in guinea pigs. Arch. Gen. Psychiatry 32:725-732.
1055. Klawans, H.L., Falk, D.K., Nausieda, P.A., and Weiner, W.J.(1978). Gilles de la Tourette syndrome after long-term chlorpromazine therapy. Neurology 28:1064-1068.
1056. Klein, R.G.(1974). Pharmacotherapy of childhood hyperactivity: An update. In H. Meltzer(ed) Psychopharmacology: The Third Generation of Progress. Raven Press, New York. pp1215-1224.

1056a. Klein, R.G. (1988). Methylphenidate and growth in hyperactive children. A controlled withdrawal study. Arch. Gen. Psychiatry 45:1127-1130.
1056b. Klein, R.G. (1988). III. Hyperactivie boys almost grown up. Methylphenidate effects on ultimate height. Arch. Gen. Psychiatry 45:1131-1134.
1057. Kline, N.S. and Sacks, W.(1963). Relief of depression within one day using an M.A.O. inhibitor and intravenous 5-HTP. Am. J. Psychiatry 120:274-265.
1058 Klob, B., Sutherland, R.J. and Whishaw, I.Q.(1983). A comparison of the contributions of the frontal and parietal association cortex to spatial local-ization in rats. Behav. Neurosci. 97:13-27.
1059. Klopf, H.(1987). In Barnes, D.M. Neural models yield data in learning. Science 236:1628-1629.
1060. Klüver, H. and Bucy, P.C.(1939). Preliminary analysis of functions of the temporal lobes in monkeys. Arch. Neurol. Psychiatry 42:979-1000.
1061. Knapp, S. and Mandell, A.J.(1973). Short- and long-term lithium administration effects on the brain's serotonergic biosynthesis systems. Science 180:645-647.
1062. Knapp, S. and Mandell, A.J.(1975). Effects of lithium chloride on parameters of biosynthetic capacity for 5-hydroxytryptamine in rat brain. J. Pharmacol. Exp. Ther. 193:812-823.
1063. Knesevich, J.W.(1982). Successful treatment of obsessive-compulsive disorder with clonidine hydrochloride. Am. J. Psychiatry 139:364-365.
1064. Knight, R.P.(1954).Management and psychotherapy of the borderline schizophrenic patient. In R.P. Knight and C.R. Freidman(eds) Psychoanalytic Psychiatry and Psychology. International Universities Press, New York pp110-122.
1065. Knobel, M., Wolman, M.B. and Mason, E.(1959). Hyperkinesis and organicity in children. A.M.A. Archives of General Psychiatry 1:310-321.
1066. Knop, J., Teasdale, T.W., Schulsinger, F. and Goodwin, D.W.(1985). A prospective study of young men at high risk for alcoholism: School behavior and achievement. J. Stud. Alcohol 46:273-278.
1067. Knopp, J.(1980). Selection of variables in a prospective study of young men at high risk for alcoholsim. Acta Psychiatr. Scand. 285 (Suppl) 347-352.
1068. Knott, P.J. and Curzon, G.(1972). Free tryptophan in plasma and brain tryptophan metabolism. Nature 239:452-453.
1069. Knox, W.E. and Auerbach, V.H.(1955). Hormonal control of tryptophan peroxidase in rat. J. Biol. Chem. 214:307-313.
1070. Koe, B.K., Koch, S.W., Lebel, L.A., Minor, K.W. and Page, M.G.(1987). Sertraline, a selective inhibitor of serotonin uptake, induces subsensitivity of b-adrenoceptor system of rat brain. Eur. J. Pharmacol. 141:187-194.
1071. Koenig, J.I., Gudelsky, G.A. and Meltzer, H.Y.(1987). Stimulation of corticosterone and beta-endorphin secretin in the rat by selective 5-HT receptor subtype activation. Eur. J. Pharmacol. 137:1-8.
1072. Koester, G.(1889).Uber die maladie des tics impulsifs (mimische Kramphneurose). Deutsch Z. Nervenheik 15:147.
1073. Kohut, H.(1971). The Analysis of the Self. International Universities Press, New York.
1074. Kolb, L.C.(1968). Noyes' Modern Clinical Psychiatry. W.B. Saunders, Philadelphia, pp638.
1075. Kondo, K. and Nomura, Y.(1982). Tourette syndrome in Japan: Etiologic considerations based on associated factors and familial clustering. In A.J. Friedhoff and T.N. Chase (eds) Gilles de la Tourette syndrome. Raven Press, New York, pp 271-276.
1076. Konorski, J. and Lawicka, W.(1964). Analysis of errors by prefrontal animals in the delayed response test. In J.W. Warren and K. Akert (eds) The Frontal Granular Cortex and Behavior. McGraw-Hill, New York.
1077. Konorski, J.(1961). The physiological approach to the probles of recent memory. In J.F. Delafresnaye (ed) Brain Mechanisms and Learning. Blackwell, New York.
1078. Koob, G.F., Stinus, L., and LeMoal, M.L.(1981). Hyperactivity and hypoactivity produced by lesions to the mesolimbic dopamine system. Behav. Brain Res. 3:341-359.
1079. Kornetsky, C. and Eliasson, M.(1969). Reticular stimulation and chlorpromazine: An animal model for schizophrenic overarousal. Science 165:1273-1274.
1080. Kornetsky, C.(1972). The use of a simple test of attention as a measure of drug effects in schizophrenic patients. Psychopharmacology 24:99-106.
1081. Korsakoff, S.S.(1887). Disturbance of psychic function in alcoholic paralysis and its relation to the disturbance of the psychic sphere in multiple neuritis of nonalcoholic origin. Vestkik Psichiatrii 4:2.
1082. Korsgaard, S., Gerlach, J. and Christensson, E.(1985). Behavioral aspects of serotonin-dopamine interaction in the monkey. Eur. J. Pharmacol. 118:245-252.
1083. Kosofsky, B.E. and Molliver, M.E.(1987). The serotonergic innervation of cerebral cortex: Different classes of axon terminals arise from dorsal and median raphe nuclei. Synapse 1:153-168.
1084. Kostowski, W.(1975). Interactions between serotonergic and catecholaminergic systems in the brain. Pol. J. Pharmacol. Pharm. 27:15-24.
1085. Kostowski, W., Giacalone, E., Garattini, S. and Valzelli, L.(1968). Studies on behavioral and biochemical changes in rats after lesion of midbrain raphé. Eur. J. Pharmacol. 4:371-376.
1086. Kostowski, W., Samanin, R., Bareggi, S.R., Marc, V., Garattini, S. and Valzelli, L.(1974). Biochemical aspects of the interaction between mid-brain raphe and locus coeruleus in the rat. Brain Res. 82:178-182.
1087. Kovacic, B. and Domino, E.F.(1982). A monkey model of tardive dyskinesia (TD): Evidence that reversible TS may turn into irreversible. J. Clin. Psychopharmacol. 2:305-307.
1088. Kovács, G.L., Veldhuis, H.D., Versteed, D.H. G. and De Weid, D.(1986).Facilitation of avoidance behavior by vasopressin fragments microinjected into limbic-midbrain structures. Brain Res 371:17-24.
1089. Kraepelin, E.(1893). Psychiatrie, 4th Edition. J.A. Barth, Leipzig.
1090. Kraepelin, E.(1909-1913). Psychiatrie, 8th Edition. J.A. Barth, Leipzig.
1091. Kraepelin, E.(1971). Dementia Praecon and Paraphrenia. R.M. Barclay and G.M. Roberston (translator) R.E. Krieger, New York.
1092. Krakowski, A.J.(1965). Amitriptyline in treatment of hyperkinetic children: A double blind study. Psychosomatics 6:355-360.
1093. Kratz, K.E. and Mitchell, J.C.(1977). Internal and external cue use following septal ablation in the rat. Physiol.

Psychol. 5:177-180.
1094. Kretschmer, E.(1970). Physique and Character: An Investigation of the Nature of Constitution and of the Theory of Temperament. E. Miller (translator) Cooper Square Publishers, Inc., New York.
1095. Krieg, W.J.S.(1953). Functional Neuroanatomy. The Blakiston Co, Inc. New York.
1096. Krnjevic, R.J. and Schwartz, S.(1966). Is g-aminobutyric acid an inhibitory transmitter? Nature 211:1372-1374.
1097. Krueger, H., Eddt, N. and Sumwalt, M.(1943). The pharmacology of the opium alkaloids. U.S. Public Health Reports Suppl 165. Part II. U.S. Govt. Printing Office, Washington, DC.
1097a. Kruesi, M.J.P.(1989). Cruelty to animals and CSF-5HIAA. Psychiatry Res. 28:115-116.
1098. Kuczenski, R.(1979). Effects of para-chlorophenylalanine on amphetamine and haloperiodol-induced chages in striatial dopamine turnover. Brain Res. 164:217-225.
1099. Kuffler, S.W. and Edwards, C.(1958). Mechanism of gamma-aminobutyric acid (GABA) and its relation to synaptic inhibition. J. Neurophysiol. 21:589-610.
1100. Kuhar, M.J., Pert, C.B. and Snyder, E.J.(1973). Regional distribution of opiate receptor binding in monkey and human brain. Nature 245:447-451.
1101. Kuhn, R.(1958). The treatment of depressive states with G22355 (imipramine hydrochloride). Am. J. Psychiatry 115:459-464.
1102. Kuhn, R.(1976). The psychotropic effect of carbamazepine in non-epileptic adults, with particular reference to the drug's possible mechanism of action. In W. Birkmayer (ed) Epileptic Seizures — Behavior— Pain University Park Press, London.
1103. Kulonen, E.(1983). Ethanol and GABA. Medical Biology 61:147-167.
1104. Kupfer, D.J., Detre, T.P. and Koral, J.(1975). Relationship of certain childhood "traits" to adult psychiatric disorders. Am. J. Orthopsychiatry 45:74-80.
1105. Kurachi, M., Kobayashi, K., Matsubara, R., Hiramatsu, H.et al.(1985). Regional cerebral blood flow in schizophrenic disorders. Eur. Neurol. 24: 176-181.
1106. Kurlan, R., Behr, J., Medved, L., Shoulson, I., Pauls, D., Kidd, J.R. and Kidd, K.K.(1986). Familial Tourette's syndrome: Report of a large pedigree and potential for linkage studies. Neurology 36:772-776.
1106a. Kurlan, R., Behr, J., Medved, L., Shoulson, I., Pauls, D. and Kidd, K.K. 1987. Severity of Tourette's syndrome in one large kindred. Arch. Neurol. 44:268-269.
1106b. Kurlan, R., Behr, J., Medved, L. and Como, P.(1988). Transient tic disorder and the spectrum of Tourette's syndrome. Arch. Neurol. 45:1200-1201.
1107. Laczi, F., Valkusz, Z., Laszio, F.A., Wagner, A., Jardanhazy, T., Szasz, A., Szilard, J. and Telegdy, G.(1982). Effects of lysine-vasopressin and 1-deamino-8-arginine-vasopressin on memory in healthy individuals and diabetes insipidus patients. Psychoneuroendocrinology 7:185-193.
1108. Laduron, P., De Bie, K., and Leysen, J.(1977). Specific effect of haloperidol on dopamine turnover in the frontal cortex. Arch. Pharmacol. 296:183-185.
1109. Lahey, B.B., Green, K.D. and Forehand, R.(1980). On the independence of ratings of hyperactivity, conduct problems, and attention deficits in children: A multiple regression analysis. J. Consult. Clin. Psychol. 48:566-574.
1110. Lake, C.R., Sternberg, D.E., van Kammen, D.P., Ballenger, J.C., Ziegler, M.G., Post, R.M., Kopin, I.J. and Bunney, W.E.(1980). Schizophrenia: Elevated cerebrospinal fluid norepinephrine. Science 207:331-33.
1111. Lal, S. and Martin, J.B.(1980). Neuroanatomy and neuropharmacological regulation of neuroendocrine function. In H.M. van Praag, M.H. Lader, O.J. Rafaelsen, and E.J. Schar (eds) Handbook of Biological Psychiatry: Part III. Brain Mechamism and Abnormal Behavior — Genetics and Neuroendrocrinology. Marcel Dekker, New York pp101-166.
1112. Lalouel, J-M. and Morton, N.E.(1981). Complete segregation analysis with pointers. Hum. Hered. 31:312-321.
1113. Lalouel, J-M.(1977). Linkage mapping for pairwise recombination data. Heredity 38:61-77.
1114. Lalouel, J.-M., Rao, D.C., Morton, N.E. and Elston, R.C.(1983). A unified model for complex segregation analysis. Am. J. Hum. Genet. 35: 816-826.
1115. Landis, C.L., Zubin, J. and Mettler, F.A.(1950). The functions of the human frontal lobe. J. Psychol. 50:123-138.
1116. Langer, S.Z.(1977). Presynaptic receptors and their role in the regulation of transmitter release. Br. J. Pharmacol. 60:481-497.
1117. Langston, J.W., Ballard, P.A., Tetrud, J.W., Irwin, I.(1983). Chronic Parkinsonism in humans due to a product of meperidine-analog synthesis. Science 219:979-980.
1118. Langston, J.W., Irwin, I., Langston, E.B. and Forno, L.S.(1984). Pargyline prevents MPTP-induced Parkinsonism in primates. Science 225:1480-1482.
1119. Lantz, I. and Terenius, L.(1985). High enkephalyl peptide degradation, due to angiotensin-converting enzyme-like activity in human CSF. FEBS Lett. 193:31-34.
1120. Lapin, I.P. and Oxenkrug, G.F.(1969). Intensification of the central serotoninergic process as a possible determination of the thymoleptic effect. Lancet 1:132-136.
1121. Lapin, I.P.(1973). Kynurenines as probable participants of depression. Pharmakopsychiat. 6:273-279.
1122. Lapin, I.P.(1980). Experimental studies on kynurenines as neuroactive tryptophan metabolites: past, present and future. Trends Neurosci. 3:410-412.
1123. Lapin, I.P.(1981). Antagonism of l-glycine to seizures induced by l-kynurenine, quinolinic acid and strychnine in mice. Eur. J. Pharmacol. 71:495-498.
1124. Lapin, I.P.(1981). Kynurenine and seizures. Eplepsia 22:257-265.
1125. Lapin, I.P.(1982). Convulsant action of intracerebroventricularly administered l-kynurenine sulphate, quinolinic acid and other derivatives of succinic acid, and effect of amino acids: Structure-activity relationships. Neuropharmacol. 21:1227-1233.
1126. Lapin, I.P.(1985). Endogenous antagonists of quinolinic acid and kynurenine as links of the defensive mechanism in epilepsy. Acta Neurol. New Series 40:203-206.
1127. Lapouse, R. and Monk, M.(1958). An epidemiologic study of behavior characteristics in children. Am. J. Public Health 48:1134-1144.
1128. Lapouse, R. and Monk, M.(1964). Behavior deviations in a representative sample of children: Variation by sex, age, race, social class, and family size. Am. J. Orthopsychiatry 34:436-446.
1129. Lassen, N.A., Ingvar, D.H. and Skinhøj, E.(1978). Brain function and blood flow. Sci. Am. 239: (October) 62-71.
1130. Laufer, M.W. and Denhoff, E.(1957). Hyperkinetic

behavior syndrome in children. J. Pediatr. 50: 463-474.
1131. Laufer, M.W.(1971). Long term management and some follow-up findings on the use of drugs with minimal cerebral syndromes. J. Learn. Disabilities 4:519-522.
1133. Laursen, A.M.(1963). Corpus striatum. Acta Physiol. Scand. 59(Suppl 211): 1-109.
1134. Lavaggi, M.V., Politi, V., del Luca, G. and Gorini, A.(1987). Effects of indole-3-pyruvic acid on tryptophan pyrrolase activity "in vitro" and "in vitro." In D.A. Bender, M.H. Joseph, W. Kochen and H. Steinhart (eds) Progress in Tryptophan and Serotonin Research 1986. Walter de Gruyter, New York, pp 51-54.
1135. Lavielle, S., Tassin, J.P., Thierry, A.M., Blanc, G., Hervé, D., Barthelemy, C. and Glowinski, J.(1978). Blockade by benzodiazepines of the selective high increase in dopamine turnover induced by stress in mesocortical dopaminergic neurons of the rat. Brain Res. 168:585-594.
1136. Law, W., Petti, T.A., Kazdin, A.E.(1981). Withdrawal symptoms after graduated cessation of imipramine in children. Am. J. Psychiatry 138:647-650.
1137. Lawson, J., McGhie, A. and Chapman, J.(1967). Distractibility in schizophrenia and organic cerebral disease. Brit. J. Psychiatry113:527-535.
1138. Lazarus, L.H., Ling, N. and Guillemin, R.(1976). b-Lipotropin as a prohormone for the morphinomimetic peptides, endorphins and enkephalins. Proc. Natl. Acad. Sci. U.S.A. 73:2156-2159.
1139. Lechin, F., van der Dijs, B., Gomez, F., Lechin, E., Oramas, O. and Villa, S.(1983). Res. Comm. Psychol. Psychiatry Behav. 8:23-54.
1140. Leckman, J.F., Anderson, G.M., Cohen, D.J., Ort, S., Harcherik, D.F., Hoder, E.L. and Shaywitz, B.A.(1984). Whole blood serotonin and tryptophan levels in Tourette's disorder: Effects of acute and chronic clonidine treatment. Life Sci. 35:2497-2503.
1141. Leckman, J.F., Bowers, M.B. Jr., and Sturges, J.S.(1981). Relationship between estimated premorbid adjustment and CSF homovanillic acid and 5-hydro-xyindoleacetic acid levels. Am. J. Psychiatry 138: 472-477.
1142. Leckman, J.F., Cohen, D.J., Detlor, J., Young, J.G., Harcherik, D. and Shaywitz, B.A.(1982). Clonidine in treatment of Tourette's syndrome: A review of data. In A.J. Friedhoff and T. N. Chase(eds) Gilles de la Tourette Syndrome. Raven Press, New York, pp391-401.
1143. Leckman, J.F., Detlor, J., Harcherik, D.F., Ort, S., Shaywitz, B.A. and Cohen, D.J.(1985). Short- and long-term treatment of Tourette's disorder with clonidine: A clinical perspective. Neurology 35:343-351.
1144. Leckman, J.F., Detlor, J., Harcherik, D.F., Young, J.G., Anderson, G.M., Shaywitz, B.A. and Cohen, D.J.(1983). Acute and chronic clonidine treatment in Tourette's syndrome: A preliminary report on clinical response and effect on plasma and urinary catecholamine metabolites, growth hormone, and blood pressure. J. Am. Acad. Child Psychiatry 22:433-440.
1145. Leckman, J.F., Ort, S., Caruso, K.A., Anderson, G.M., Riddle, M.A., and Cohen, D.J.(1986). Rebound phenomena in Tourette's syndrome after abrupt withdrawal of clonidine. Arch. Gen. Psychiatry 43:1168-1176.
1146. Leckman, J.F., Price, R.A., Ort, S., Pauls, D.L. and Cohen, D.J.(1987). Nongenetic factors in Gilles de la Tourette syndrome. Arch. Gen. Psychiatry 44:100.
1147. Leckman, J.F., Weisman, M.M., Merikangas, K.R., Pauls, D.L. and Prusoff, B.A.(1983). Panic disorder and major depression. Arch. Gen. Psychiatry 40: 1055-1060.
1148. Lecrubier, Y., Puech, A.J., Simon, P. and Widlöcher, D.(1980). Schizophrénie: Hyper ou hypofonctionnement du systéme dopaminergique? Une hypothése bipolaire. Psychol. Medicale 12:2431-2441.
1149. Lee, N.S., Muhs, G., Wagner, G.C., Reynolds, R.D. and Fisher, H.(1988). Dietary pyridoxine interaction with tryptophan or histidine on brain serotonin and histamine metabolism. Pharmacol. Biochem. Behav. 29:559-564.
1150. Lees, A.J., Robertson, M., Trimble, M.R. and Murray, N.M.F.(1984). A clinical study of Gilles de la Tourette syndrome in the United Kingdom. J. Neurol. Neurosurg. Psychiatry 47:1-8.
1151. Leger, L. and Descarries, L.(1978). Serotonin nerve terminals in the locus coeruleus of adult rat: A radioautographic study. Brain Res. 145:1-13.
1152. Leibowitz, S.F.(1980) Neurochemical systems of the hypothalamus. Control of feeding and drinking behavior and water-electrolyte excretion.In P.J. Morgane and J. Ransapp (eds) Handbook of the Hypothalamus, Vol 3. Marcel/ Dekker, New York, pp299-437.
1152a. Leibowitz, S.F., Weiss, G.F. and Shor-Posner, G.(1988). Hypothalamic serotonin: Pharmacological, biochemical, and behavioral analyses of its feeding-suppressive action. Clin. Neuropharmacol. 11 Suppl 1:S51-S71.
1153. Leimkuhler, M.E. and Mesulam, M.-M.(1985). Reversible go-no go deficits in a case of frontal lobe tumor. Ann. Neurol. 18:617-619.
1154. Lelord, G., Callaway, E., Muh, J.P., Arlot, J.C., Sauvage, D., Garreau, B., and Domenech, J.(1978). L'acide homovanilique urinaire et ses modifications par ingestion de vitamine B6: Dans lautisme de l'en-fant. Rev. Neurol. (Paris) 134:797-801.
1155. LeMoal, M., Stinus, L., Simon, H., Tassin, J.P., Thierry, A.M., Blanc, G., Glowinski, J. and Cardo, B.(1977). Behavioral effects of a lesion in the ventral mesencephalic tegmentum: Evidence for involvement of A10 dopaminergic neurons. In E. Costa, and G.L. Gessa, (eds) Nonstriated Dopaminergic Neurons. Adv. Biochem. Psychopharmacology Vol.16. Raven Press, New York, pp237-245.
1156. LeMoal, M.L., Cardo, B. and Stinus, L.(1969). Influence of ventral mesencephalic lesions on various spontaneous and conditioned behaviors in the rat. Physiol. Behav. 4:567-573.
1157. LeMoal, M.L., Galey, D. and Cardo, B.(1975). Behavioral effects of local injection of 6-hydroxydopamine in the medial ventral tegmentum in the rat. Possible role of the mesolimbic dopaminergic system. Brain Res. 88:190-194.
1158. LeMoal, M.L., Stinus, L. and Galey, D.(1976). Radiofrequency lesion of the ventral mesencephalic tegmentum: Neurological and behavioral considerations. Exp. Neurol. 50:521-535.
1159. LeMoal, M.L., Stinus, L. and Simon, H.(1979). Increases sensitivity to (+) amphetamine self-administered by rats following meso-cortico-limbic dopamine neuron destruction. Nature 280:156-158.
1159a. Lerer, B., Ran, A., Blacker, M., Silver, H., Weller, M.P.I., Drummer, D., Ebstein, B. and Calev, A.(1988). Neuroendocrine responses in chronic schizophrenia. Evidence for serotonergic dysfunction. Schizophr. Res. 1:405-410.
1160. Levin, S.(1984). Frontal lobe dysfunctions in schizophrenia - I. Eye movement impairments. J. Psychiatr. Res. 18:27-55.

1161. Levin, S.(1984). Frontal lobe dysfunctions in schizophrenia — II. Impairments of psychological and brain functions. J. Psychiatr. Res. 18:57-72.
1162. Levine, L.R., Rosenblatt, S. and Bosomworth, J.(1987). Use of a serotonin re-uptake inhibitor, fluoxetine, in the treatment of obesity. Int. J. Obesity 11 (Suppl 3): 185-190.
1163. Levine, M.D. and Zallen, B.G.(1984). The learning disorders of adolescence: Organic and nonorganic failure to strive. Pediatr. Clin. N. Am. 31:345-369.
1164. Levine, M.D., Oberklaid, F., and Meltzer, L.(1981). Developmental output failure: A study of low productivity in school-aged children. Pediatrics 67:18-25.
1165. Lewin, R.(1984). Trail of ironies to Parkinson's disease. Science 224:1083-1085.
1166. Lewis, C.C.(1981). The effects of parental firm control: A reinterpretation of findings. Psychol. Bull. 90:547-563.
1167. Lewis, C.E., Rice, J. and Helzer, J.E.(1983). Diagnostic interactions: Alcoholism and antisocial personality. J. Nerv. Ment. Dis. 171:105-113.
1168. Lewitter, F., DeFries, J.C., and Elston, R.C.(1980). Genetic models of reading disability. Behav. Genet. 10:9-30.
1169. Leysen, J.E., De Chaffoy De Courcelles, D., De Clerk, F., Niemegeers, C.J.E., and Van Nueten, J.M.(1984). Serotonin-S2 receptor binding sites and functional correlates. Neuropharmacology 23:1493-1501.
1170. Leysen, J.E., Niemeggers, C.J.E., Tollenaere, J.P. and Laduron, P.M.(1978). Serotonergic component of neuroleptic receptors. Nature 272:168-171.
1171. Lhermitte, F.(1983). 'Utilization behavior' and its relation to lesions of the frontal lobes. Brain 106:237-255.
1172. Lhermitte, F.(1986). Human autonomy and the frontal lobes. Part II. Patient behavior in complex and social situations: The "Environmental dependency syndrome." Ann. Neurol. 19:335-343.
1173. Lhermitte, F., Pillon, B. and Serdaru, M.(1986). Human autonomy and the frontal lobes. Part I: Imitation and utilization behavior: A neuropsychological study of 75 patients. Ann. Neurol. 19:326-334.
1174. Li, C. and Chung, D.(1976). Isolation and structure of an untriakontapeptide with opiate activity from camel pituitary glands. Proc. Natl. Acad. Sci.U.S.A. 73:1145-1148.
1175. Li, C.(1964). Lipotropin, a new active peptide from pituitary glands. Nature 201:924.
1176. Li, T.K., Lumeng, L., McBride, W.J. and Murphy, J.M.(1987). Alcoholism: Is it a model for the study of disorders of mood and consummatory behavior? Ann. N.Y. Acad. Sci. 499:239-249.
1177. Li, T.K., Lumeng, L., McBride, W.J. and Murphy, J.M.(1987). Rodent lines selected for factors affecting alcohol consumption. Alcohol Alcohol. Suppl 1:91-96.
1178. Li, T.K., Lumeng, L., McBride, W.J., Waller, M.B. and Murphy, J.M.(1986). Studies on an animal model of alcoholism. Natl. Inst. Drug Abuse Res. Monogr. Ser. 66:41-49.
1179. Lidberg, L., Tuck, J.R., Asberg, M., Scalia-Tomba, G.P. and L. Bertilsson, L.(1985). Homicide, suicide and CSF 5-HIAA. Acta Psychiatr. Scand. 71: 230-236.
1180. Lieberman, J.(1984). Evidence for a biological hypothesis of obsessive-compulsive disorder. Neuropsychobiology 11:14-21.
1181. Lieblich, I., Gross, R. and Cohen, E.(1977). Effects of testosterone replacement on the recovery from increased emotionality, produced by septal lesions in prepubertal castrated male rats. Physiol. Behav. 18:1159-1164.
1182. Liebowitz, M.R., Fyer, A.J., Gorman, J.M., Dillon, D., Appleby, I.L., Levy, G., Anderson, S., Levitt, M., Palij, M., Davies, S.O. and Klein, D.F.(1984). Lactate provocation of panic attacks. Arch. Gen. Psychiatry 41:764-770.
1183. Liebowtiz, M.R., Fyer, A.J., McGrath, P., and Klein, D.F.(1981). Clonidine treatment of panic disorder. Psychopharmacol. Bull. 17:122-123.
1184. Lightman, S.L.and Forsling, M.(1980). Evidence of dopamine as an inhibtor of vasoprotein release in man. Clin. Endocrinol. 12:39-46.
1185. Lindberg, L., Åsberg, M. and Sundqvist-Stensman, U.B.(1984). 5-hydroxyindoleacetic acid levels in attempted suicides who have killed their children. Lancet 2:928.
1186. Lindgren, T. and Kantak, K.M.(1987). Effects of serotonin receptor agonists and antagonists on offensive aggression in mice. Aggress. Behav. 13:87-96.
1187. Lindström, L.H.(1985). Low HVA and normal 5-HIAA CSF levels in drug-free schizophrenic patients compared to healthy volunteers: Correlations to symptomatology and family history. Psychiatry Res. 14:265-273.
1188. Lindvall, O., Björklund, A., and Divac, I.(1978). Organization of catecholamines projecting to the frontal cortex in the rat. Brain Res. 142:1-24.
1189. Lindvall, O., Björklund, A., Moore, R.Y. and Stenevi, U.(1974). Mesencephalic dopamine neurons projecting to the neocortex. Brain Res. 81:325-331.
1190. Linet, L.S.(1985). Tourette syndrome, pimozide, and school phobia: The neuroleptic separation anxiety syndrome. Am. J. Psychiatry 142:613-615.
1191. Ling, N., Burgus, R. and Guillemin, R.(1976). Isolation, primary structure, and synthesis of a-endorphin and g-endorphin, two peptides of hypothalamic-hypophyseal origin with morphinomimetic activity. Proc. Natl. Acad.Sci.U.S.A. 73:3942-3946.
1191a. Linnoila, M., Dejong, J. and Virkkunen, M.(1989). Family history of alcoholism in violent offenders and impulsive fire stetters. Arch. Gen. Psychiatry 46:613-616.
1192. Linnoila, M., Virkkunen, M., Scheinin, M., Nuutila, A., Rimon, R. and Goodwin, F.K.(1983). Low cerebrospinal fluid 5-hydroxyindolacetic acid concentration differentiates impulsive from nonimpulsive violent behavior. Life Sci. 33:2609-2614.
1193. Lipsett, D., Madras, B.F., Wurtman, R.J. and Munro, H.N.(1973). Serum tryptophan level after carbohydrate ingestion: Selective decline in non-albumin-bound tryptophan coincident with reduction in serum free fatty acids. Life Sci. 12 (Part II): 57-64.
1194. Lipton, M.A.(1974). In Neuropsychopharmacology of Monamines and Their Regulatory Enzymes. E. Usdin (ed) Raven Press, New York, p451.
1195. Litman, D.A. and Correia, M.A.(1985). Elevated brain tryptophan and enhanced 5-hydroxytryptamine turnover in acute hepatic heme deficiency: Clinical implications. J. Pharmacol. Exp. Therap. 232:337-345.
1196. Lloyd, K.G., Morselli, P.L. and Bartholini, G.(1987).GABA and affective disorders. Med. Biol. 65:159-165.
1196a. Lodge, D., Aran, J.A., Chruch, J. Davies, S.N. Martin, D., O'Slaughnessy, C.T. and Zeman, S.(1987) Excitatory amion acids and phencyclidine drugs. In T.P. Hick, D. Lodge and H. McLennan (eds) Excitatory Amiono Acid Trans-

mission. Alan R. Liss. Inc., New York, pp83-90.
1197. Loeber, R.(1982). The stability of antisocial and delinquent child behavior: A review. Child Dev. 53:1431-1446.
1198. Loizou, L.A.(1969). Projections of the nucleus locus coeruleus in the albino rat. Brain Res. 15:563-566.
1199. Loman and J. and van den Berken, L. (eds) (1977) Physiology of the Striatum. North-Holland, Amsterdam.
1200. Loney, J. and Milich, R.(1982). Hyperactivity, inattention, and aggression in clinical practice. In M. Wolraich and D.K. Roth (eds) Adv. Dev. Behav. Pediatrics Vol. 3. Academic Press, New York, pp113-147.
1201. Loney, J., Kramer, J. and Milich, R.(1981). The hyperkinetic child grows up: Predictors of symptoms, delinquency, and achievement at follow-up. In K. Gadow and J. Loney (eds) Psychosocial Apsects of Drug Treatment for Hyperactivity. Westview Press, Boulder, CO, pp381-415.
1202. Loney, J., Langhorne, J.E. and Paternite, C.E.(1978). An emperical basis for subgrouping the hyperkinetic/minimal brain dysfunction syndrome. J. Abnor. Psychol. 87:431-441.
1203. Loney, J., Whaley-Klahn, M.A., Koiser, T. and Conboy, J.(1983). Hyperactive boys and their brothers at 21: Predictors of aggressive and antisocial out-comes. In Prospective Studies of Crime and Delinquency. K.T. Van Dusen and S.A. Mednick (eds) Kluwer Academic Publishers Group, Boston, pp181-206.
1204. Loomer, H.P., Saunders, J.C. and Kline, N.S.(1957). A clinical and pharmacological evaluation of iproniazid as a psychic energizer. Psychiat. Res. Am. Psychiat. Assoc. 8: 129-141.
1205. Lopez, R.(1965). Hyperactivity in twins. Can. Psychol. Assoc. 10:421.
1206. Lopez-Ibor, J.J., Saiz-Ruiz, J. and Perez de los Cobos, J.C.(1985). Biological correlations of suicide and aggressivity in major depressions (with melancholia): 5-hydroxyindoleacetic acid and cortisol in cerebral spinal fluid, dexamethasone suppression test and therapeutic response to 5-hydroxytryptophan. Neuropsychobiol. 14:67-74.
1207. Loranger, A., Oldham, J.M. and Tulis, E.H.(1982). Familial transmission of DSM-III borderline personality disorder. Arch. Gen. Psychiatry 39:795-799.
1208. Loranger, A.W. and Tulis, E.H.(1985). Family history of alcoholism in borderline personality disorder. Arch. Gen. Psychiatry 42:153-157.
1209. Lord, J., Waterfield, A., Hughes, J. and Kosterlitz, H.(1979). Endogenous opioid peptides: Multiple agonists and receptors. Nature 267:495-499.
1210. Lorens, S.A. and Unger, L.M.(1974). Morphine analgesia, two-way avoidance, and consummatory behavior following lesions in the midbrain raphe nuclei of the rat. Pharmacol. Biochem. Behav. 2:215-221.
1211. Lorens, S.A.(1971). Raphe lesions in cats: Forebrain serotonin and avoidance behavior. Physiol. Behav. 7:815-818.
1212. Lorens, S.A., Sorensen, J.P. and Yunger, L.M.(1971). Behavioral and neurochemical effects of lesions in the raphe system of the rat. J. Comp. Physiol. Psychol. 77:48-52.
1213. Lorenz, E.(1963). Deterministic nonperiodic flow. J. Atmospheric Sci. 20:130-141.
1214. Lorenz, E.(1963). The mechanics of vacillation. J. Atmospheric Sci. 20:448-464.
1215. Lorenz, E.(1964). The problems of deducing the climate from governing equations. Tellus 16:1-11.
1216. Lowe, T.L., Cohen, D.J., Detlor, J., Kremenitzer, M.W. and Shaywitz, B.A.(1982). Stimulant medication precipitate Tourette's syndrome. J. Am. Med. Ass. 247:1729-1731.
1217. Lubs, H.A.(1969). A marker X syndrome. Hum. Genet. 21:231-244.
1218. Lucas, A.R. and Robin, E.A.(1973). Electroencephalogram in Gilles de la Tourette's disease. Dis. Nerv. Syst. 34:85-89.
1219. Lucas, A.R.(1973). Report of Gilles de la Tourette's syndrome in two succeeding generations. Child Psychiatry Hum. Dev. 3:231-233.
1220. Lucas, A.R., Beard, C.M., Rajput, A.H. and Kurland, L.T.(1982). Tourette syndrome in Rochester, Minnesota. 1968-1979. In A.J. Friedhoff and T.N. Chase (eds) Gilles de la Tourette Syndrome. Raven Press, New York, pp267-269.
1221. Lucas, A.R., Kaufman, P.E. and Morris, E.M.(1967). Gilles de la Tourette's disease in two succeeding generations. Am. J. Acad. Child. Psychiatry 6:700-722.
1222. Lucas, P.B., Gardner, D.L., Wolkowitz, O.M., Tucker, E.E. and Cowdry, R.W.(1986). Methyphenidate-induced cardiac arrhythmias. N. Eng. J. Med. 315: 1485.
1223. Lucki, I., Nobler, M.S. and Frazer, A.(1984). Differential actions of serotonin antagonists on two behavioral models of serotonin receptor activation in the rat. J. Pharmacol. Exp. Ther. 228:133-139.
1223a. Ludlow, C.L., Polinsky, R.J., Caine, E.D., Bassich, C.J. and Evert, M.H.(1982). In A.J. Friedhoff and T.N. Chase(eds) Gilles de la Tourette Syndrome. Raven Press, New York, pp351-361.
1224. Luria, A.R.(1973). The Working Brain. Basic Books, Inc., New York.
1225. Luria, A.R.(1980). Higher Cortical Functions in Man, Second Edition Basic Books, New York, 634pp.
1226. Luxenburger, H.(1928). Vorläufiger Bericht über psychiatrische Serienuntersuchungen an Zwillingen. Zeitschrift fur die gesamte Neurologie und Psychiatrie 116:297- 326.
1227. Lynch, D.R., Strittmatter, S.M. and Snyder, S.H.(1984). Enkephalin convertase localization by [^{3}H]guanidinoethylmercaptosuccinic acid autoradio-graphy: Selective association with enkephalin-containing neurons. Proc. Natl. Acad. Sci. U.S.A.81:6543-6547.
1228. Lynch, G. and Baudry, M.(1984). The biochemistry of memory: A new and specific hypothesis. Science 224:1057-1063.
1229. Lynch, G.S., Ballantine, P. and Campbell, B.A.(1969). Potentiation of behavioral arousal after cortical damage and subsequent recovery. Exp. Neurol. 23:195-205.
1230. Maas, J.W., Kocsis, J.H., Bowden, C.L., Davis, J.M., Redmond, D.E., Hanin, I. and Robins, E.(1982). Pretreatment neurotransmitter metabolites and response to imipramine or amitritryptyline treatment. Psychol. Med. 12:37-43.
1231. Mabry, P.D. and Campbell, B.A.(1974). Ontogeny of serotonergic inhibition of behavioral arousal in the rat. J. Comp. Physiol. Psychol. 86:193-201.
1232. MacDonald, M.E., Anderson, M.A., Gilliam, T.C., Tranebjaerg, L., Carbenter, N.J., Magenis, E., Hayden, M.R., Healey, S.T., Bonner, T.I. and Gusella, J.F.(1987). A somatic cell hybrid panel for localizing DNA sediments near the Huntington's disease gene. Genomics 1:29-34.
1233. MacKay, , A.V.P., Bird, E.D., Spokes, E.G., Rossor, M., Iversen, L.L., Creese, I. and Snyder, S.H.(1980). Dopam-

ine receptors and schizophrenia: Drug effect or illness? Lancet 2:915-916.
1234. Mackay, A.V.P., Ivesen, L.L., Rossor, M., Spokes, E., Bird, E., Arregui, A., Creese, I. and Snyder, S.H.(1982). Increased brain dopamine and dopamine receptors in schizophrenia. Arch. Gen. Psychiatry 39:991-997.
1235. MacKay, M.C., Beck, I. and Talor, R.(1973). Methylphenidate for adolescents with MBD. N.Y. J. Med. 173:550-554.
1236. MacLean, P.D.(1949). Psychosomatic disease and the "visceral brain." Psychosom. Med. 11:338-353.
1237. MacLean, P.D.(1952). Some psychiatric implications of physiological studies on frontotemporal portion of limbic system (visceral brain). Electroenceph. Clin. Neuophysiol. 4:407-418.
1238. MacLean, P.D.(1954). The limbic system and its hippocampal formation. J. Neurosurg. 11:29-43.
1239. MacLean, P.D.(1958). Contrasting functions of limbic and neocortical systems of the brain and their relevance to psychophysiological aspects of medicine. Am. J. Med. 25: 611-626.
1240. MacLean, P.D.(1973). A triune concept of the brain and behavior. Univ. of Toronto Press, Toronto, 165pp.
1241. MacLean, P.D.(1975). Brain mechanisms of primal sexual functions and related behavior. In M. Sandler and G.L. Gessa (eds) Sexual Behavior: Pharmacology and Biochemistry. Raven Press, New York, pp1-11.
1242. MacLenna, A.J. and Maier, S.F.(1983). Coping and the stress-induced potentiation of stimulant stereotypy in the rat. Science 219:1091-1094.
1243. Maes, M., De Ruyter, M., Hobin, P. and Suy, E.(1987). Relationship between the dexamethasone suppression test and the L-tryptophan/competing amino acids ratio in depression. Psychiatry Res. 21:323-335.
1244. Magid, K. and McKelvey, C.A.(1987). High Risk Children without a Conscience. Bantam Books, New York, 361pp.
1245. Maguire, J.(1987). Clonidine. An effective antimanic agent? Br. J. Psychiatry 150:863-864.
1246. Mahler, H.R. and Cordes, E.H.(1969). Biological Chemistry. Harper & Row, New York, 693pp.
1247. Mahler, M.S.(1972). On the first three phases of the separation-individuation process. Int. J. Psychoanal. 53:333-338.
1248. Malfroy, B., Swerts, J.P., Guyon, A., Roques, B.P. and Schwartz, J.C.(1978). High-affinity enkephalin-degrading peptidase in brain is increased after morphine. Nature 276:523-566.
1249. Malmo, R.B.(1942). Interference factors in delayed response in monkeys after removal of the frontal lobe. J. Neurophysiol. 5:295-308.
1250. Mangoni, A.(1974). The "kynurenine shunt" and depression. Adv. Biochem. Pharmacol. 11:293-298.
1251. Mann, H.B. and Greenspan, S.I.(1976). The identification and treatment of brain dysfunction. Am. J. Psychiatry 133:1013-1017.
1252. Mann, J.J., Stanley, M., McBride, P.A. and McEwen, B.S.(1986). Increased serotonin2 and β-adrenergic receptor binding in the frontal cortices of suicide victims. Arch. Gen. Psychiatry 43:954-959.
1253. Mann, J.J.,Arango, V., Tierney, H. and Stanley, M.(1987). Recent neurochemical findings in suicide victims. Abstracts 26 Ann. Meeting Amer. College Neuropsychopharmacology, San Juan, Puerto Rico. p92.
1254. Mannuzza, S. and Gittelman, R.(1984). The adolescent outcome of hyperactive girls. Psychiat. Res. 13:19-29.
1255. Manowitz, P., Gilmour, D.G. and Racevskis, J.(1973). Low plama tryptophan levels in recently hospitalized schizophrenics. Biol. Psychiatry 6:109-118.
1256. Manschreck, T.C.(1986). Motor abnormalities in schizophrenia. In H.A. Nasrallah and D.R. Weinberger (eds) The Neurology of Schizophrenia. Elsevier, New York, 65-96.
1257. Maranto, G.(1987). Are autistic children high on themselves? Discover Feb:6-7.
1258. Marcus, J., Hans, S.L., Mednick, S.A., Schulsinger, F. and Michelsen, N.(1985). Neurological dysfunctioning in offspring of schizophrenics in Israel and Denmark. Arch. Gen. Psychiatry 42:753-761.
1259. Margules, D.L., Moisset, B., Lewis, M.J., Shibuya, H. and Pert, C.B.(1978). b-Endorphin is associated with overeating in genetically obese mice (ob/ob) and rats (fa/fa). Science 202:988-991.
1259a. Marino, M.W., Fuller, G.M. and Elder, F.F.B.(1986). Chromosomal location of human and rat A alpha, B beta, and gamma fibrinogen genes by in situ hybridization. Cytogenet. Cell Genet. 42:36-41.
1260. Mark, V.H. and Ervin, F.R.(1970). Violence and the brain. Harper & Row, New York.
1261. Marks, I.M., Stern, R.S. and Mawson, D.(1980). Clomipramine and exposure for obsessive compulsive rituals. I. Br. J. Psychiatry, 136:1-25.
1262. Marks, P.C., O'Brien, M. and Paxinos, G.(1978). Chlorimipramine inhibition of muricide: The role of the ascending 5-HT projection. Brain Res. 149:270-273.
1263. Marshall, W. and Ferguson, J.(1939). Hereditary word blindness as a defect of selective attention. J. Nerv. Ment. Ability 89:164-173.
1264. Martin, C.(1966). A New Approach to Noctural Enuresis. H.K. Lewis & Con, Ltd., London.
1265. Martin, J.B. and Barchas, J.D.(1986). Neuropeptides in Neurologic and Psychiatric Disease. Raven Press, New York.
1266. Martin, J.B. and Gusella, J.F.(1986). Huntington's disease. Pathogenesis and management. N. Engl. J. Med. 315:1267-1276.
1267. Mason, S.T. and Fibger, H.C.(1979). NE and selective attention. Life Sci. 25:1949-1956.
1268. Mason, S.T. and Fibiger, H.C.(1979). Anxiety: The locus coeruleus disconnection. Life Sci. 25:2141-2147.
1269. Mason, S.T. and Iversen, S.D.(1975). Learning in the absence of forebrain noradrenaline. Nature 258: 422-424.
1270. Masterson, J.F. and Rinsley, D.B.(1975). The borderline syndrome: The role of the mother in the genesis and psychic structure of the borderline personality. Int. J. Psychoanal. 56:163-177.
1271. Mattes, J.A.(1980). The role of frontal lobe dysfunction in childhood hyperkinesis. Compr. Psychiatry 21:358-369.
1272. Mattes, J.A.(1986). Psychopharmacology of temper outbursts. J. Nerv. Ment. Dis. 174:464-470.
1273. Matthysse, S.(1973). Antipsychotic drug actions: A clue to the neuropathology of schizophrenia? Fed. Proc. 32:200-205.
1273a. Matthysse, S.(1975). Neuronal models of transmitter balance. In E.F. Domino and J.M. Davis (eds) Neurotransmitter Balance Regulating Behavior. NPP Books, Ann Arbor.

1274. Matussek, N.(1966). Neurobiologie und Depression. Med. Monatsschr. 20:109-112.
1275. Maurer, R.G. and Damasio, A.R.(1982). Childhood autism from the point of view of behavioral biology. J. Autism Dev. Disord. 12:195-205.
1276. Maurer, R.G.(1986). Neuropsychology of autism. Psychiatr. Clin. N. Am. 9:367-380.
1276a. McBride, O.W. and Battey, J.(1987). Human gastrin-releasing peptide gene maps to chromosome band 18q21. Somatic Cell Molec. Genet. 13:81-86.
1277. McCord, W. and McCord, J.(1960). Origins of Alcoholism. Stanford Univ. Press, Stanford, CA.
1278. McCord, W., McCord, J. and Howard, A.(1961). Familial correlates of aggression in non-delinguent male children. J. Abnor. Soc. Psychol. 62:79-93.
1279. McCormick, R.A., Russo, A.M., Ramirez, L.F., Taber, J.I.(1984). Affective disorders among pathological gamblers seeking treatment. Am. J. Psychiatry 141:215-218.
1280. McEntee, W.J. and Mair, R.G.(1980). Memory enhancement in Korsakoff's psychosis by clonidine: Further evidence for a nonadrenergic deficit. Ann. Neurol. 7:466-470.
1281. McFie, J., Thompson, J.A.(1982). Picture arrangement: A measure of frontal lobe and temporal lobe lesions in man. Neuropsychology 20:249-262.
1282. McGee, R., Williams, S. and Silva, P.A.(1984). Behavioral and developmental characteristics of aggressive, hyperactive and aggressive-hyperactive boys. J. Am. Acad. Child. Psychol. 23:270-279.
1283. McGeer, E.G. and Singh, E.(1984). Neurotoxic effects of endogenous materials: Quinolinic acid, L-pyroglutamic acid and TRH. Exp. Neurol. 86:410-413.
1284. McGeer, P.L., McGeer, E.G., Scherer, U. and Singh, K.(1977). A glutaminergic cortico-striatal path? Brain Res. 128:369-373.
1285. McGhie, A. and Chapman, J.(1961). Disorders of attention and perception in early schizophrenia. Br. J. Med. Psychol. 34:103-117.
1286. McGuffin, P. and Mawson, D.(1980). Obsessive-compulsive neurosis: Two indentical twin pairs. Br. J. Psychiatry 137:285-287.
1287. McGuinness, D. and Pribram, K.(1980). The neuropsychology of attention: Emotional and motivational controls. In The Brain and Psychology. Academic Press, New York, pp95-139.
1288. McHenry, L.C.(1967). Samuel Johnson's tics and gesticulations. J. Hist. Med. 22:152-168.
1289. McInnes, R.G.(1937). Obsevations on heredity in neuroses. Proc. R. Soc. Med. 30:895-904.
1290. McKay, A.V.P.(1980). Positive and negative schizophrenic symptoms and the role of dopamine: Discussion, 1. Br. J. Psychiatry 137:379-383.
1291. McKeith, I.G., Marshall, E.F., Ferrier, I.N., Armstrong, M.M., Kennedy, W.N., Perry, R.H., Perry, E.K., and Eccleston, D.(1987). 5-HT receptor binding in post-mortem brain from patients with affective disorder. J. Affective Disord. 13:67-76.
1291a. McLellan, A.T., Luborsky, L., Woody, G.E. and O'Brien, C.P.: An improved diagnostic evaluation instrument for substance abuse patients. The Addiction Severity Index. J. Nerv. Ment. Dis. 168:26-33, 1980.
1292. McLeod, W.E. and McLeod, M.F.(1972). Indoleamines and the cerebrospinal fluid. In B.M. Davies, B.J. Carroll and R.M. Mowbrag (eds) Depressive Illness: Some Research Studies. Charles C. Thomas, Springfield, IL. pp209-225.
1293. McMahon, R.C.(1980). Genetic etiology in the hyperactive child syndrome: A critical review. Am. J. Orthopsychiatry 50:145-150.
1294. McTaggart, A. and Scott, M.(1959). A review of twelve cases of encopresis. J. Pediatr. 54:762-768.
1295. Mednick, S.A. and Schulsinger, F.(1968). Some premorbid characteristics related to breakdown in children with schizophrenic mothers. J. Psychiatry Res. 6 (Suppl 1): 267-291.
1296. Mednick, S. A. and Volavka, J.(1980). Biology and crime. In N. Morris and M. Tonry (Eds) Crime and Justice: An Annual Review of Research Vol. 11. Univ. of Chicago Press, Chicago.
1297. Mednick, S.A. (eds). (1983). Prospective Studies of Crime and Delinquency, Martinus Nyhoff, Hingham, MA.
1298. Mednick, S.A., Gabrielli, W.F. and Hutchings, B.(1983). Genetic influences in criminal behavior:Some evidence from an adoption cohort. In K.T. VanDusen and S.A.Mednick (eds) Prospective Studies of Crime and Delinquency. Kluuer-Nijhoff Pub., Boston.
1299. Mednick, S.A., Pollock, V., Volavka, J. and Gabrielli, W.F.(1982). Biology and violence. In M.E. Wolfgang and N.A. Weiner (eds) Criminal Violence. Sage Publications, Beverly Hills, CA.
1300. Mednick, S.A., Schulsinger, H. and Schlusinger, F.(1975). Schizophrenia in children of schizophrenic mothers. In A. Davids (ed) Child Personality and Psychopathology—Current Topics, Vol 2. John Wiley & Sons, New York, pp:217-253.
1301. Meibach, R.C.(1984). Serotonergic receptors. In A. Björklund, T. Hökfelt and M.J. Kuhar (eds) Handbook of Chemical Neuroanatomy, Vol 3. Classical Transmitters and Transmitter Receptors in the CNA. Part II. Elsevier Science Publishers, Vancouver, BC. pp304-324.
1302. Melamed, E., Hefti, F., and Bird, E.D.(1982). Huntington chorea is not associated with hyperactivity of nigrostriatal dopaminergic neurons: Studies in postmortem tissues and in rats with kainic acid lesions. Neurology 32:640-644.
1303. Mellsop, G.(1972). Psychiatric patients seen as children and adults: Childhood predictors of adult illness. J. Child Psychol. Psychiatry 13:91-101.
1303a. Meltzer, H.Y.(1988). New insights into schizophrenia through atypical antipsychotic drugs. Neuro-psychopharmacol. 1:193-196.
1304. Meltzer, H.Y., Umberkoman-Wiita, B., Robertson, A., Tricou, B.J., Lowry, M. and Perline, R.(1984). Effect of 5-hydroxytryptophan on serum cortisol levels in major affective disorders. I. Enhanced response in depression and mania. Arch. Gen. Psychiatry 41:366-374.
1305. Meltzer, H.Y., Wita, B., Tricou, B.J., Simonovic, M., Fang, V. and Manov, G.(1982). Effect of serotonin precursors and serotoin agonists on plasma hormone levels. In B.T. Ho, J.C. Schoolar and E. Usdin (eds) Serotonin in Biological Psychiatry. Advances in Biochemical Psychopharmacology, Vol. 34. Raven Press, New York, pp117-139, .
1306. Mendels, J., Frazer, A., Fitzgerals, R.G., Ramsey, T.A. and Stokes, J.W.(1972). Biogenic amine metabolites in cerebrospinal fluid of depressed and manic patients. Science 175:1380-1382.
1307. Mendelson, W., Johnson, N. and Stewart, M.(1971).

Hyperactive children as teenagers: A follow-up study. J. Nerv. Ment. Dis. 153:273-279.
1308. Mendelson, W.B., Caine, E.D., Goyer P., Ebert, M., and Gillin, J.C.(1980). Sleep in Gilles de la Tourette syndrome. Biol. Psychiatry 15:339-343.
1309. Mendlewicz, J. and Baron, M.(1981). Morbidity risks in subtypes of unipolar depressive illness: Differences between early and late onset forms. Br. J. Psychiatry 139:463-466.
1310. Mendlewicz, J. and Rainer, J.D.(1977). Adoption study supporting genetic transmission in manic-depressive illness. Nature 268:327-329.
1311. Mendlewicz, J., Fleiss, J.L., and Fieve, R.R.(1972). Evidence for X-linkage in the transmission of manic-depressive illness. J. Am. Med. Assoc. 222:1624-1627.
1312. Mendlewicz, J., Linkowski, P., and Wilmotte, J.(1980). Relationship between schizoaffective illness and affective disorders or schizophrenia. Morbidity risk and genetic transmission. J. Affect. Disord. 2:289-302.
1313. Menkes, M.M., Rowe, J.S. and Menkes, J.H.(1967). A twenty-five year follow-up study on the hyperkinetic child with minimal brain dysfunction. Pediatrics 39:393-399.
1314. Menon, M.K. and Vivonia, C.A.(1981). Serotonergic drugs, benzodiazepines and baclofen block muscimol-induced myoclonic jerks in a strain of mice. Eur. J. Pharmacol. 73:155-161.
1315. Merikangas, K.R., Leckman, J.F., Prusoff, B.A., Pauls, D.L. and Weissman, M.M.(1985). Familial transmission of depression and alcoholism. Arch. Gen. Psychiatry 42: 367-372.
1316. Meryash, D.L., Szymanski, L.S. and Gerald, P.S.(1982). Infantile autism is associated with the fragile-X syndrome. J. Autism Dev. Disord. 12:295-301.
1317. Messing, R.B. and Lytle, L.D.(1977). Serotonin-containing neurons: Their possible role in pain and analgesia. Pain 4:1-21.
1318. Mesulam, M.-M. and Geschwind, N.(1978). On the possible role of neocortex and its limbic connections in the process of attention and schizophrenia: Clinical cases of inattention in man and experimental anatomy in monkey. J. Psychiatr. Res. 14:249-259.
1319. Mesulam, M.-M. and Petersen, R.C.(1987). Treatment of Gilles de la Tourette's syndrome: Eight-year, practice-based experience in a predominately adult population. Neurology 37:1828-1833.
1320. Mesulam, M.-M.(1986). Cocaine and Tourette's syndrome. N. Engl. J. Med. 315:398.
1321. Mesulam, M.-M.(1986). Frontal cortex and behavior. Ann. Neurol. 19:320-325.
1322. Mesulam, M.-M., Waxman, S.G., Geschwind, N. and Sabin, T.D.(1976). Acute confusional states with right middle cerebral artery infarctions. J. Neurol. Neurosurg. Psychiatry 39:84-89.
1323. Mettler, F.A.(1944). Physiologic effects of bilateral simultaneous frontal lesions in the frontal cortex. J. Comp. Neurol. 81:105-136.
1324. Michaels, J.J. and Goodman, S.E.(1934). Incidence and intercorrelations of enuresis and other neuropathic traits in so-called normal chldren. Am. J. Orthopsychaitry 4:79.
1325. Middlemiss, D.N.(1984). Stereoselective blockade at [3H] bnding sites and at the 5-HT autoreceptors by propanolol. Eur. J. Pharmacol. 101:289-293.
1326. Mikkelsen, E.J., Detlor, J. and Cohen, D.J.(1981). School avoidance and social phobia triggered by haloperidol in patients with Tourette's disorder. Am. J. Psychiatry 138:1572-1576.
1327. Miller, F.E., Heffner, T.G., Kotake, C. and Seiden, L.S.(1981). Magnitude and duration of hyperactivity following neonatal 6-hydroxydopamine is related to the extent of brain dopamine depletion. Brain Res. 229:123-132.
1328. Miller, J.S.(1978). Hyperactive children: A ten-year study. Pediatrics 61:217-222.
1329. Millington, W.R., McCall, A.L. and Wurtman, R.J.(1979). Chronic lithium administration increases tryptophan's blood brain barrier transport. Fed. Proc. Fed. Am. Soc. Exp. Biol. 38:757.
1330. Milner, B. and Petrides, M.(1984). Behavioral effects of frontal-lobe lesions in man. Trends Neurosci. 7:403-407.
1331. Milner, B.(1963). Effects of different brain lesions on card sorting. The role of the frontal lobes. Arch. Neurol. 9:90-100.
1332. Milner, B.(1964). Some effects of frontal lobectomy in man. In J.M. Warren and K. Akert (eds) The Frontal Granular Cortex and Behavior. McGraw-Hill Book Co., New York, pp313-334.
1333. Minde, K., Lewin, D., Weiss, G., Laviqueur, H., Douglas, V. and Sykes, E.(1971). The hyperactive child in elementary school: A five year controlled follow-up. Except. Child 38:215-221.
1334. Minde, K., Webb, G. and Sykes, D.(1968). Studies on the hyperactive child: VI. Prenatal and paranatal factors associated with hyperactivity. Dev. Med. Child Neurol. 10:355-363.
1335. Minde, K., Weiss, G., Mendelson, N.(1972). A five-year follow-up study of 91 hyperactive school children. J. Am. Acad. Child. Psychiatry 11:595-619.
1336. Minderaa, R.B., Anderson, G.M., Volkmar, F.R., Akkerhuis, G.W. and Cohen, D.J.(1987). Urinary 5-hydroxyindoleacetic acid and whole blood serotonin and tryptophan in autistic and normal subjects. Biol Psychiatry 22:933-940.
1337. Minuchin, S.(1974). Families & Family Therapy. Harvard University Press, Cambridge, MA.
1338. Mirin, S.R., Meyer, R., Mendelson, J. and Ellingboe, J.(1980). Opiate use and sexual function. Am. J. Psychiatry 137:909-915.
1339. Mishkin, M. and Appenzeller, T.(1987). The anatomy of memory. Sci. Am. 256 (June): 80-89.
1340. Mishkin, M., Malamut, B. and Bachevalier, J.(1984). Memories and habits: Two neural systems. In G. Lynch, J.L. McGaugh and N.M. Weinberger (eds) Neurobiology of Learning and Memory. The Guilford Press, New York..
1341. Mitchell, E. and Matthews, K.L.(1980). Gilles de la Tourette's disorder associated with pemoline. Am. J. Psychiatry 137:1618-1619.
1342. Mizrahi, E.M., Holtzman, D., Tharp, B.(1980). Haloperidol-induced tardive dyskinesia in a child with Gilles de la Tourette's syndrome. Arch. Neurol. 37:780.
1343. Mizuno, T. and Yugari, Y.(1975). Prophylactic effect of L-5-hydroytryptophane on self-mutilation in the Lesch-Nyhan syndrome. Neuropaed. 6:13-23.
1343a. Model, J.G., Mountz, J.M., Curtis, G.C. and Greden, J.F.(1989). Neurophysiologic dysfunction in basal ganglia/limbic striatal and thalamocortical circuits as a pathogenetic mechanism of obsessive-compulsive disorder. J. Neuropsychiatry 1:27-36.

1344. Mogenson, G.J., Jones, D.J. and Yim, C.Y.(1980). From motivation to action: Functional interface between the limbic system and the motor system. Prog. Neurobiol. 14:69-97.
1345. Mohler, H. and Okada, T.(1977). Benzodiazepine receptor: Demonstration in the central nervous system. Science 198:849-851.
1346. Moldofsky, H.(1971). A psychophysiology study of multiple tics. Arch. Gen. Psychiatry 25:79-85.
1347. Moldofsky, H., Tullis, C. and Lamon, R.(1974). Multiple tic syndrome (Gilles de la Tourtette's syndrome). Clinical, biological and psychological variables and their influence with haloperidol. J. Nerv. Ment. Dis. 159:282-292.
1348. Moller, S.E.(1981). Pharmacokinetics of tryptophan, renal handling of kynurenine and the effect of nicotinamide on its appearance in plasma and urine following L-tryptophan loading of healthy subjects. Eur. J. Clin. Pharmacol. 21:137-142.
1350. Molliver, M.E.(1987). Serotonergic neuronal systems: What their anatomic organization tells us about function. J. Clin.l Psychopharmacol. 7:3S-23S.
1351. Monroe, R.R.(1970). Episodic Behavioral Disorders: A Psychodynamic and Neurophysiologic Analysis. Harvard Univ. Press, Cambridge, MA .
1352. Monroe, R.R.(1975). Anticonvulsants in the treatment of aggression. J. Nerv. Ment. Dis. 160:119-126.
1353. Montgomery, M.A., Clayton, P.J. and Friedhoff, A.J.(1982). Psychiatric illness in Tourette syndrome patients and first-degree relatives. In A.J. Friedhoff and T.N. Chase (eds) Gilles de la Tourette Syndrome. Raven Press, New York, pp335-339.
1354. Montgomery, S.A. and Montgomery, D.(1982). Pharmacological prevention of sucidal behavior. J. Affective Disord. 4:291-298.
1354a. Moore, R.Y. and Bloom, F.E. (1979). Central catecholamine neuron systems: Anatomy and physiology of the norepinephrine and epinephrine systems. Annu. Rev. Neurosci. 2:113-168.
1355. Moran, E.(1970). Pathological gambling. Br. J. Hosp. Med. 4:59-70.
1356. Morand, C., Young, S.N. and Ervin, F.R.(1983). Clinical response of aggressive schizophrenics to oral tryptophan. Biol. Psychiatry 18:575-577.
1357. Morland, J.(1974). Effect of chronic ethanol treatment on tryptophan oxygenase, tyrosine aminotransferase and general protein metabolism in the intact and perfused rat liver. Biochem. Pharmacol. 23:21-35.
1358. Morland, J.(1974). Hepatic tryptophan oxygenase activity as a marker of changes in protein metabolism during chronic ethanol treatment. Acta Pharmacol. Toxicol. 35:155-168.
1359. Morley, J.E.(1980). The neuroendocrine control of appetite: The role of endogenous opiates, cholecystokinin, TRH, gamma-amino-butyric-acid and the diazepam receptor. Life Sci. 27:355-368.
1360. Morphew, J.A. and Sim, M.(1969). Gilles de la Tourette's syndrome: A clinical and psychopathological study. Br. J. Med. Psychol. 42:292-301.
1361. Morris, H.H., Escoll, P.J. and Wexler, R.(1956). Aggressive behavior disorders of childhood: A follow-up study. Am. J. Psychiatry 112:991-997.
1362. Morris, R.G.M.(1981). Spatial localization does not require the presence of local cues. Learn. Motivat. 12:239-260.
1363. Morrison, J.R. and Stewart, M.A.(1971). A family study of the hyperactive child syndrome. Biol Psychiatry 3:189-195.
1364. Morrison, J.R. and Stewart, M.A.(1973). The psychiatric status of the legal families of adopted hyperactive children. Arch. Gen. Psychiatry 28:888-891.
1365. Morrison, J.R. and Stewart, M.A.(1974). Bilateral inheritance as evidence for polygenicity in the hyperactive child syndrome. J. Nerv. Ment. Dis. 158:226-228.
1366. Morrison, J.R.(1980). Childhood hyperactivity in an adult psychiatric population: Social factors. J. Clin. Psych. 41:40-43.
1367. Morton, N.E.and MacLean, C.J.(1974). Analysis of family resemblance. III. Complex segregation of quantitative traits. Am. J. Hum. Genet. 26:489-503.
1368. Moskowitz, J.A.(1980). Lithium and lady luck. N.Y. State J. Med. 80:785-788.
1369. Mountcastle, V.B.(1957). Modality and topographic properties of single neurons of cats somatic sensory cortex. J. Neurophysiol. 20:408-434.
1370. Mucha, R.F., Van der Kooy, D., O'Shaughnessy, M. and Bucenieks, P.(1982). Drug reinforcement studied by the use of place conditioning in rat. Brain Res. 243:91-105.
1371. Muhlbauer, H.D. and Müller-Oerlinghausen, B.(1985). Fenfluramine stimulation of serum cortisol in patients with major affective disorders and healthy controls: Further evidence for central serotonergic action of lithium in man. J. Neural Transmission 61:81-94.
1372. Muhlbauer, H.D.(1985). Human aggression and the role of central serotonin. Pharmacopsychiatry 18:218-221.
1373. Mulder, A.H., Wardeh, G., Hogenboom, F., Frankhuyzen, A.L.(1984). k- and d-opioid receptor agonists differentially inhibit striatal dopamine and acetylcholine release. Nature 308:278-280.
1374. Muller, J.C., Pryer, W.W., Gibbons, J.E. and Orgain, E.S.(1955). Depression and anxiety occurring during rauwolfia therapy. J.A.M.A. 159:836-839.
1375. Muller-Oerlinghausen, B.(1985). Lithium long-term treatment — Does it act via serotonin? Pharmacopsychiatry 18:214-217.
1375a. Mullis , K., Faloona, F., Scharf, S., Saiki, R., Horn, G., and Erlich, H.(1986). Specific enzymatic amplification of DNA in vitro: The polymerase chain reaction. Cold Spring Harbor Symp. Quant. Biol. 51: 263-272.
1376. Munjack, D.J. and Moss, H.B.(1981). Affective disorder and alcoholism in families of agoraphobics. Arch. Gen. Psychiatry 38:869-871.
1377. Murphy, D.L., Belmaker, R.H., Buchsbaum, M.S., Martin, N.F., Ciaranello, R. and Wyatt, R.J.(1977). Biogenic amine related enzymes and personality variations in normals. Psychol. Med. 7:149-157.
1378. Murphy, D.L., Wright, C., Buchsbaum, M.S., Nicols, A., Costa, J.L. and Wyatt, R.J.(1976). Platelet and plasma amine oxidase activity in 680 normals: Sex and age differences and stability over time. Biochem. Med. 16:254-265.
1379. Murphy, J.M., McBride, W.J., Lumeng, L. and Li, T.K.(1987). Contents of monamines in forebrain regions of alcohol-preferring (P) and nonpreferring (NP) lines of rats. Pharmacol. Biochem. Behav. 26:389-392.
1380. Murray, E.A. and Mishkin, M.(1985). Amygdalectomy impairs crossmodal association in monkeys. Science 228:604-606.

1381. Murray, T.J.(1978). Tourette's syndrome: A treatable tic. Can. Med. Assoc. J. 10:1407-1410.
1381a. Murray, J.C., Buetow, K.H., Smith, M., Carlock, L., Chakravarti, A., Ferrell, R.F., Gedamu, L., Gilliam, C., Shiang, R. and DeHaven, C.R. (1988). Pairwise linkage analysis of 11 loci on human chromosome 4. Am. J. Hum. Genet. 42:490-497.
1382. Murray, T.J.(1982). Doctor Samuel Johnson's abnormal movements. In A.J. Friedhoff and T.N. Chase (eds) Gilles de la Tourette Syndrome. Raven Press, New York, pp25-30.
1383. Murrin, L.C and Kuhar, M.J.(1979). Dopamine receptors in rat frontal cortex: An autoradiographic study. Brain Res. 177:279-285.
1384. Myerhoff, J.H. and Snyder, S.H.(1973). Gilles de la Tourette's disease and minimal brain dysfunction: Amphetamine isomers reveal catecholamine correlates in an affected patient. Psychopharmacology 29:211-220.
1385. Myers, R.D. and Melchior, C.L.(1975). Alcohol drinking in the rat after destruction of serotonergic and catecholaminergic neurons in the brain. Res. Commun. Chem. Pathol. Pharmacol. 10:363-378.
1386. Myers, R.D. and Melchior, C.L.(1977). Alchohol and alcoholism: Role of serotonin. In W.B. Essman (ed) Serotonin in Health and Disease, Volume II: Physiological Regulation and Pharmacological Action. Spectrum Publications, Inc., New York, pp373-430.
1387. Myers, R.D. and Veale, W.L.(1968). Alcohol preference in the rat: Reduction following depletion of brain serotonin. Science 160:1469-1471.
1388. Myers, R.D., Evans, J.E. and Yaksh, T.L.(1972). Ethanol preference in the rat: Interactions between brain serotonin and ethanol, acetaldehyde, paraldehyde, 5-HTP and 5-HTOL. Neuropharmacology 11:539-549.
1389. Naganishi, , S., Ternishi, Y., Noda, M. Notake, M., Watanabe, Y., Kadikani, H., Jingami, H., Numa, S.(1980). The protein-coding sequence of the bovine ACTH-b-LHP precursor gene is split near the signal peptide region. Nature 287:752-755.
1390. Naganishi, N.S., Inove, A., Kita, T., Nakamua, M., Chang, A.C.Y., Cohen, S.N., and Numa, S.(1979). Nucleotide sequence of cloned cDNA for bovine corticotropin-b-lipotropin precursor. Nature 278:423-427.
1391. Nakaya, K.(1976). Serum free tryptophan concentration - The effects on the brain serotonin metabolism and its relationship to the mental diseases. Psychiatr. Neurol. Jpn. 78:119-132.
1392. Nameche, G., Waring, M, and Ricks, D.(1964). Early indicators of outcome in schizophrenia. J. Nerv. Men. Dis. 139:232-240.
1393. Naranjo, C.A., Sellers, E.M., Sullivan, J.I., Woodley, D.V., Kadlec, K.and Sykora.(1987). The serotonin uptake inhibitor citalopram attenuates ethanol intake. Clin. Pharmacol. Ther. 41:266-274.
1394. Naruse, H., Hayashi. T., Takesada, M., Nakane, A., Yamazaki, K., Noguchi, T., Watanabe, Y. and Hayaishi, O.(1987). Therapeutic effect of tetrahydrobiopterin in infantile autism. Proc. Jpn. Acad. 63 Ser. B: 231-233.
1395. National Advisory Committee on Hyperkinesis and Food Additives.(1980). Report to the Nutrition Foundation, New York.
1395a. Naylor, S.L., Sakaguchi, A.Y., Spindel, E. and Chin, W.W.(1987). Human gastrin-releasing peptide gene is located on chromosome 18. Somatic Cell Molec. Genet. 13: 87-91.
1396. Nauta, W.J.H.(1971). The problem of the frontal lobe: A reinterpretation. I. Psychiatr. Res. 8:167-187.
1397. Nee, L.E., Caine, E.D., Polinsky, R.J., Eldridge, R., and Ebert, M.H.(1980). Gilles de la Tourette syndrome: Clinical and family study in 50 cases. Ann. Neurol. 7:41-49.
1398. Nee, L.E., Polinsky, R.J. and Ebert, M.H.(1982). Tourette syndrome: Clinical and family sudies. In A.J. Friedhoff and T.N. Chase (eds) Gilles de la Tourette Syndrome, Raven Press, New York, pp335-339.
1399. Neill, D.B., Grant, L.D. and Grossman, S.P.(1972). Selective potentiation of locomotor effects of amphetamine by midbran raphe lesions. Physiol. Behav. 9:655-657.
1400. Nelson, H.E.(1976). A modified card sorting test sensive to frontal lobe defects. Cortex 12:313-324.
1401. Nemeroff, C.B., Owens, M.J., Bissette, G., Andorn, A.C., and Stanley, M.(1989). Reduced corticotropin-releasing factor (CRF) binding sites in the frontal cortex of suicides. Arch. Gen. Psychiatry (in press).
1402. Nemeroff, C.B., Widerlöv, Bissette, G., Walléus, Karlsson, I., Eklund, K., Kilts, C.D., Loosen, P.T., and Vale, W.(1984). Elevated concentrations of CSF corticotropin-releasing factor-like immunoreactivity in depressed patients. Science 226:1342-1344.
1403. Nichols, M.P.(1984). Family Therapy Concepts and Methods. Gardner Press, Inc., New York.
1404. Nies, A., Robinson, D.S., Harris, L.S. and Lamborn, K.R.(1974). Consequences of monoamine oxidase substrate activities in twins, schizophrenics, depressives and controls. In E. Usdin (ed) Neuropsychopharmacology of Monamines and Their Regulatory Enzymes. Raven Press, New York.
1405. Nikulina, E.M. and Popova, H.K.(1986). Serotonin's influence on predatory behavior of highly aggressive CBA and weakly aggressive DD strains of mice. Aggressive Behav. 12:277-283.
1406. Ninan, P.T., van Kammen, D.P., Scheinin, M., Linnoila, M., Bunney, W.E. and Goodwin, F.K.(1984). CSF 5-hydroxyindoleacetic acid levels in suicidal schizophrenic patients. Am. J. Psychiatry 141:566-569.
1407. Nishizawa, Y., Olsen, T.S., Larsen, B. and Lassen, N.A.(1982). Left-right cortical asymmetries of regional cerebral blood flow during listening to words. J. Neurophysiol. 48:458-466.
1408. Niskanen, P., Huttunen, M., Tamminen, T. and Jääkeläinen, J.(1976). The daily rhythm of plasma tryptophan and tyrosine in depression. Br. J. Psychiatry 128:67-73.
1409. Noda, M., Furutani, Y., Takahashi, H., Toyosato, M., Hirose, T., Inayama, S., Nakanishi, S., and Numa, S.(1982). Cloning and sequence analysis of cDNA for bovine adrenal preproenkephalin. Nature 295:202-206.
1410. Noda, M., Teranishi, Y., Takahashi, H., Toyosato, M., Notake, M., Nakanishi, S., and Numa, S.(1982). Isolation and structural organization of the human preproenkephalin gene. Nature 297:431-434.
1411. Nordin, L., Siwers, B., Bertilsson, L.(1984). Bromocriptine treatment of depressive disorders: Clinical and biochemical effects. Acta Psychiatr. Scand. 64:25-33.
1412. Noyes, R., Clancy, J., Crowe, R., Hoenk, P.R. and Slymen, D.J.(1978). The familial prevalence of anxiety neurosis. Arch. Gen. Psychiatry 35:1057-1059.
1413. Noyes, R., DuPont, R.L., Pecknold, J.C., Rifkin, A., Rubin, R.T., Swinson, R.P., Ballenger, J.C. and Burrows, G.D.(1988). Alprazolam in panic disorder and agoraphobia:

Results from a multicenter trial. Arch. Gen. Psychiatry 45:423-428.
1414. Nuechterlein, K. and Dawson, M.E.(1984). Information processing and attentional functioning in the developmental course of schizophrenic disorders. Schizophr. Bull. 10:160-203.
1415. Nuller, J.L. and Ostroumova, M.N.(1980). Resistance of inhibiting effect of dexamethasone in patients with endogenous depression. Acta Psychiatr. Scand. 61:169-177.
1415a. Nuwer, M.R.(1982).Coprolalia as an organic symptom. In A.J. Friedhoff and T.N. Chase (eds) Gilles de la Tourette Syndrome. Raven Press, New York, pp363-368.
1416. Nyback, H., Walters, J.R., Aghajanian, G.K., and Roth, R.H.(1975). Tricyclic antidepressants: Effects on the firing rate of brain noradrenergic neurons. Eur. J. Pharmacol. 32:302-312.
1417. Nyberg, F., Nordström, K., and Terenius, L.(1985). Endopeptidase in human cerebrospinal fluid which cleaves proenkephalin B opioid peptides at consecutive basic amino acids. Biochem. Biophys. Res. Commun. 131:1069-1074.
1418. Nyhan, W.L.(1972). Clinical features of the Lesh-Nyhan syndrome. Arch. Intern Med. 130:186-192.
1419. Nyhan, W.L., James, J.A., Terberg, A.J., Sweetman, L. and Nelson, L.G.(1969). A new disorder of purine metabolism with behavioral manifestations. J. Pediatr. 74:20-27.
1420. Nylander, I.(1979). A 20-year prospective follow-up study of 2164 cases at the child guidance clinics in Stockholm. Acta Paediatr. Scand. 276:1-45.
1421. O'Brien,J.A.(1883). Latah. J. Br. Asiat. Soc. 1:381-429.
1422. O'Neal, P. and Robins, L.(1957). The relation of childhood behavior problems to adult psychiatric status: A 30 year follow-up study of 150 subjects. Am. J. Psychiatry 114:961-969.
1423. O'Neil, M., Page, N., Adkins, W.N. and Eichelman, B.(1986). Tryptophan-trazedone treatment of aggressive behavior. Lancet 2:859-860.
1424. O'Rourke, D.H., McGuffin, P., and Reich, T.(1983). Genetic analysis of manic-depressive illness. Am. J. Physical Anthropol. 62:51-59.
1425. Obata, K. and Takeda, K.(1969). Release of GABA into the fourth ventricle induced by stimulation of the cat cerebellum. J. Neurochem. 16:1043- 1047.
1426. Obeso, J.A., Rothwell, J.C. and Marsden, C.D.(1981). Simple tics in Gilles de la Tourette's syndrome are not prefaced by a normal premovement EEG potential. J. Neurol. Neurosurg. Psychiat. 44:735-738.
1427. Obler, L.K. and Fein, D.(1988). The Exceptional Brain. The Guilford Press, New York.
1428. Offermeier, J. and Van Rooyen, J.M.(1982). Is it possible to integrate dopamine receptor terminology? Trends Pharmacol. Sci. 3:326-328.
1429. Ogren, S.O.(1982). Central serotonin neurones and learning in the rat. Biology of serotonergic transmission. N.N. Osborne (ed) Biology of Serotonergic Transmission. John Wiley & Sons, New York, pp317-334.
1430. Ogren, S.O.(1985). Evidence for a role of brain serotonergic neurotransmission in avoidance learning. Acta Physiol. Scand. 125(Suppl 544):1-71.
1431. Oikawa, K., Deonauth, J., Breidbart, S.(1978). Mental retardation and elevated serotonin levels in adults. Life Sci. 23:45-48.
1432. Olds, J. and Milner, P.(1954). Positive reinforcement produced by electrical stimulation of septal area and other regions of the rat brain. J. Compar. Physiol. Psychol. 47:419-427.
1433. Olds, J.(1977). Drives and Reinforcements: Behavioral Studies of Hypothalamic Function. Raven Press, New York.
1434. Olds, M.E. and Olds, J.(1963). Approach-avoidance analysis of rat diencephalon. J. Compar. Neurol. 120:259-295.
1435. Olpe, H.-R., Koella, W.P., Wolf, P., and Haas, H.L.(1977). The action of Baclofen on neurons of the substantia nigra and the ventral tegmental area. Brain Res. 134:577-580.
1436. Olsen, R.W.(1981). GABA-benzodiazepine-barbiturate receptor interactions. J. Neurochem. 37:1-13.
1437. Olson, L. and Fuxe, K.(1971). On the projection from the locus coeruleus noradrenalin neurones: The cerebellar innervation. Brain Res. 26:165-171.
1438. Olweus, D.(1979). Stability of aggressive reaction patterns in males: A review. Psychol. Bull. 86:852-875.
1439. Omenn, G.S. and Weber, B.A. (1978). Dyslexia Search for phenotypic and genetic heterogeneity. Am. J. Med. Genet. 1:333-342.
1440. Onari, K. and Spatz, H.(1926). Picks circumscribed atrophy of cerebral cortex (Pick's disease) from anatomic standpoint. Z. ges. Neurol. Psychiatry 101:470-511.
1441. Oppel, W., Harper, P. and Rider, R.(1968). Social, psychological, and neurological factors associated with nocturnal enuresis. J. Pediatr. 42: 627-641.
1442. Oppenheim, G.(1983). Propanolol-induced depression: mechanism and management. Aust. N. Z. J. Psychiatry 17:400-402.
1443. Optiz, J.M. (1986) The Rett Syndrome. Am. J. Med. Genet. (Suppl 1), pp404.
1444. Oreland, L., Wiberg, A., Asberg, M., Träskman, L., Sjöstrand, L., Thorén, P., Bertilsson, L., and Tybring, G.(1981). Platelet MAO activity and monamine metabolites in cerebrospinal fluid in depressed and suicidal patients and healthy controls. Psychiat. Res 4:21-29.
1445. Ornitz, E.M.(1983). The functional neuroanatomy of infantile autism. Intern. J. Neurosc. 19:85-124.
1446. Orzack, M. and Kornetsky, C.(1966). Attention dysfunction in chronic schizophrenia. Arch. Gen. Psychiatry 14:323-326.
1447. Ounsted, C.(1955). The hyperkinetic syndrome in epliptic children. Lancet 2:303-311.
1448. Owen, F., Cross, A.J., Crow, T.J., Deakin, J.F.W., Ferrier, I.N., Lofthouse, R. and Poulter, M.(1983). Brain 5-HT2 receptors and suicide. Lancet 2:1256.
1449. Owen, F.J., Chambers, D.R., Cooper, S.J., Crow, T.J., Johnson, J.A., Lofthouse, R. and Poulter, M.(1986). Serotonergic mechanisms in brains of suicide victims. Brain Res. 362:185-188.
1450. Owens, D.G.C., Johnstone, E.C. and Frith, C.D.(1982). Spontaneous involuntary disorders of movement. Their prevalence, severity, and distribution in chronic schizophrenics with and without treatment with neuroleptics. Arch. Gen. Psychiatry 39:452-461.
1451. Paasonen, M.K., MacLean, P.D. and Giarman, N.J.(1957). 5-hydroxytryptamine (serotonin, enteramine) content of structure of the limbic system. J. Neurochem. 1:326-333.

1452. Paine, R.S.(1962). Minimal chronic brain syndromes in children. Dev. Med. Child. Neurol. 4:21-27.
1453. Pandey, G.N., Heinze, W.J., Brown, B.D. and Davis, J.M.(1979). Electroconvulsive shock treatment decreases β-adrenergic receptor sensitivity in rat brain. Nature 280:234-235.
1454. Pandya, D.N., Van Hoesen, G.W. and Domesick, V.B.(1973). A cingulo-amygdaloid projection in the rhesus monkey. Brain Res. 61:369-373.
1455. Panksepp, J.(1979). A neurochemical theory of autism. Trends Neurosci. 2:174-177.
1456. Panksepp, J.(1982). Toward a general psychological theory of emotions. Behav. Brain Sci. 5:407-422.
1457. Panksepp, J., Bean, N.J., Bishop, P., Vilberg, T. and Sahley, T.L.(1980). Opoid blockage and social comfort in chicks. Pharmacol. Biochem. Behav. 13:673-683.
1458. Papez, J.W.(1937). A proposed mechanism of emotion. Arch. Neurol. Psychiatry 38: 728-743.
1459. Pardes, H., Kaufmann, C.A. and West, A.(1987). Update on research in schizophrenia: Report on the Tarrytown Conference, April 13-15, 1986. Schizophr. Bull. 13:185-197.
1460. Pare, C.M.B. and Sandler, M.A.(1959). A clinical and biochemical study of a trial of iproniazid in the treatment of depression. J. Neurol. Neurosurg. Psychiatry 22:247-251.
1461. Pare, C.M.B.(1965). Some clinical aspects of anti-depressant drugs. In J. Marks and C.M.B. Pare (eds) The Scientific Basis of Drug Therapy in Psychiatry. Oxford Univ. Press, NY.
1462. Pare, C.M.B., Sandler, M. and Stacey, R.S.(1960). 5-hydroxyindoles in mental deficiency. J. Neurol. Neurosurg. Psychiatry 23:341-346.
1463. Paris Conference. Standardization in Human Cytogenetics.(1972) The National Foundation.
1464. Park, C.C.(1967). The Siege: The First Eight Years of an Autistic Child. Atlantic/Little Brown, Boston.
1465. Parker, K.(1985). Helping school-age children cope with Tourette syndrome. J. School Health 55:30-32.
1466. Parkinson, J.(1817). An Essay on the Shaking Palsy. Sherwood, Neely and Jones, London.
1467. Parr, D.(1975). Sexual aspects of drug abuse in narcotic addicts. Br. J. Addict. 71:261-268.
1468. Partington, M.W., Tu, J.B. and Wong, C.Y.(1973). Blood serotonin levels in severe mental retardation. Dev. Med. Child Neurol. 15:616-627.
1469. Pasamanick, B. and Kawi, A.(1956). A study of the association of prenatal and paranatal factors in the development of tics in children. Pediatrics 48:596-601.
1470. Paterson, S., Robson, L., and Kosterlitz, H.(1983). Classification of opioid receptors. Br. Med. Bull. 39:31-36.
1471. Paul, S.M., Rehavi, M., Skolnick, P., Ballenger, J.C., and Goodwin, F.K.(1981). Depressed patients have decreased binding in tritiated imipramine to platelet serotonin 'transporter'. Arch. Gen. Psychiatry 38:1315-1317.
1472. Pauls, D.L., Hurst C.R., Kidd K.K., Kruger S.D., Leckman J.F., Cohen, D.J.(1986). Evidence against a genetic relationship between Tourette syndrome and attention deficit disorder. Arch. Gen. Psychiatry 43: 1177-1179.
1473. Pauls, D.L. and Leckman, J.F.(1986). The inheritance of Gilles de la Tourette syndrome and associated behaviors: Evidence for autosomal dominant transmission. N. Engl. J. Med.315:993-997.
1474. Pauls, D.L. and Leckman, J.F.(1987). Letter to the editor: The inheritance of Gilles de la Tourette syndrome. N. Engl. J. Med. 316:1347-1348.
1475. Pauls, D.L. and Leckman, J.F.(1988). The genetics of Tourette's syndrome. In D.J. Cohen, R.D. Bruun and J.F. Leckman (eds). Tourette's Syndrome and Tic Disorders: Clinical Understanding andTreatment. John Wiley & Sons, New York. pp92-101.
1476. Pauls, D.L., Bucher, K.D., Crowe, R.R. and Noyes, R.Jr.(1980). A genetic study of panic disorder pedigrees. Am. J. Hum. Genet. 32:639-644.
1477. Pauls, D.L., Cohen, D.J., Heimbuch, R., Detlor, J. and Kidd, K.K.(1981). Familial pattern and transmission of Gilles de la Tourette syndrome and multiple tics. Arch. Gen. Psychiatry 38:1091-1093.
1478. Pauls, D.L., Crowe, R.R., and Noyes, R.Jr.(1979). Distribution of ancestral secondary cases in anxiety neurosis (panic disorder). J. Affect. Disord. 1:287-290.
1479. Pauls, D.L., Kruger, S.D., Leckman, J.F., Cohen, D.J. and Kidd, K.K.(1984). The risk of Tourette's syndrome and chronic multiple tics among relatives of Tourette's syndrome patients obtained by direct interview. J. Am. Acad. Child Psychiatry 23:134-137.
1480. Pauls, D.L., Leckman, J.F., Kidd, K.K. and Cohen, D.J. (1988).Tourette syndrome and neuropsychiatric disorders: Is there a genetic relationship. [Letter] Am. J. Hum Genet 43:206-209..
1480a. Pauls, D.L., Leckman, J.F., Raymond, C.LR., Hurst, C.R. and Stevenson,J.M.(1988). A family study of Tourette's syndrome: Evidence against the hypothesis of association with a wide range of psychiatric phenotypes. Am. J. Hum. Genetics 43:A64.
1481. Pauls, D.L., Noyes, R. Jr. and Crowe, R.R.(1979). The familial prevalence in second degree relatives of patients with anxiety neurosis (panic disorder). J. Affect. Disord. 1:279-285.
1482. Pauls, D.L., Shaywitz, S.E., Kramer, P.L., Shaywitz, B.A. and Cohen, D.J.(1983). Demonstration of vertical transmission of attention deficit disorder. Ann. Neurol. 14:363.
1483. Pauls, D.L., Towbin, K.E., Leckman, J.F., Zahner, G.E.P., and Cohen, D.J.(1986). Gilles de la Tourette syndrome and obsessive compulsive disorder. Arch. Gen. Psychiatry 43:1180-1182.
1484. Pavlov, I.P.(1949). Complete Collected Works. Vol. 1-6.Izd. Akad. Nauk. SSSR. Moscow and Leningrad (in Russian).
1485. Payne, I.R., Walsh, E.M. and Whittenberg, E.J.R.(1974). Relationship of dietary tryptophan and niacin to tryptophan metabolism in schizophrenics and non-schizophrenics. Am. J. Clin. Nutr. 27:565-571.
1486. Pazos, A. and Palacios, J.M.(1985). Quantitative autoradiographic mapping of serotonin receptors in the rat brain. I. Serotonin-1 receptors. Brain Res. 346:205-230.
1487. Pazos, A., Cortés, R. and Palacios, J.M.(1985). Quantitative autoradiographic mapping of serotonin receptors in the rat brain. II. Serotonin-2 receptors. Brain Res. 346:231-249.
1488. Pearse, A.G.E.(1969). The cytochemistry and ultrastructure of polypeptide hormone-producing cells of the APUD series and the embryologic, physiologic and pathologic implications of this concept. J. Histochem. Cytochem. 17:303-313.
1489. Pearse, A.G.E.(1976). Peptides in brain and intestine. Nature 262:92-94.
1490. Pecknold, J.C. and Suranyi-Cadotte, B.E.(1986).

Panic disorder and depression: 5HT and imipramine binding studies. Clin. Neurpharmacol. 9(Suppl 4): 46-48.
1491. Pecknold, J.C., Swinson, R.P., Kuch, K. and Lewis, C.P.(1988). Alprazolam in panic disorder and agoraphobia: Results from a multicenter trial. III. Discontinuation effects. Arch. Gen. Psychiatry. 45:429-436.
1491a. Pelham, W.E., Bender, M.E., Caddell, J. Booth, S. and Moorer, S.H.1985. Methylphenidate and children with attention deficit disorder. Arch. Gen. Psychiatry 42:948-952.
1492. Penfield, W. and Milner, B.(1958). Memory deficit produced by bilateral lesions in the hippocampal zone. AMA Arch. Neurol. Psychiatry 79:475-497.
1493. Penna, M.W. and Lion, J.R.(1975). Gilles de la Tourette's syndrome and depression: A case report. Dis. Nerv. Syst. 36:41-43.
1494. Perez-Cruet, J., Chase, T.N. and Murphy, D.L.(1974). Dietary regulation of brain tryptophan metabolism by plasma ratio of free tryptophan and neutral amino acids in humans. Nature 248:693-695.
1495. Perez-Cruet, J., Tagliamonte, A., Tagliamonte, P. and Gessa, G.(1978). Stimulation of serotonin synthesis by lithium. J. Pharmac. Exp. Ther. 178:325-330.
1496. Perkins, M.N. and Stone, T.W.(1982). An iontophoretic investigation of the actions of convulsant kynurenines and their interaction with endogenous excitant quinolinic acid. Brain Res. 247: 184-187.
1497. Perley, M.J. and Guze, S.B.(1962). Hysteria—The stability and usefulness of clinical criteria. N. Engl. J. Med. 266:421-426.
1498. Peroutka, S.J. and Snyder, S.H.(1980). Relationship of neuroleptic drug effects at brain dopamine, serotonin, a-adrenergic, and histamine receptors to clinical potency. Am. J. Psychiatry 137:1518-1522.
1499. Peroutka, S.J. and Snyder, S.H.(1981). Two distinct central serotonin receptors with different physiological functions. Science 212:827-829.
1500. Peroutka, S.J. and Snyder, S.H.(1983). Multiple serotonin receptors and their physiological significance. Fed. Proc. 42:213-217.
1500a. Peroutka, S.J., Sleight, A.J., McCarthy, B.G., Pierce, P.A., Schmidt, A.W. and Hekmatpanah, C.R.(1989) The clinical utility of pharmacological agents that act at serotonin receptors. J. Neuropsychiatry 1:253-262.
1501. Perry, E.K., Marshall, E.F., Blessed, G., Tomlinson, B.E. and Perry, R.H.(1983). Decreased imipramine binding in the brains of patients with depressive illness. Br. J. Psychiatry 142:188-192.
1502. Persson, T. and Roos, B.-E.(1967) 5-hydroxytryptophan for depression. Lancet 2:987-988.
1503. Pert, C.B. and Snyder, S.H.(1973). Opiate receptor: Demonstration in nervous tissue. Science 179:1011-1014.
1504. Pert, C.B., Aposhian, D., and Snyder, S.H.(1974). Phylogenetic distribution of opiate receptor binding. Brain Res. 75:356-361.
1505. Pert, C.B., Kuhar, M.J. and Snyder, S.H.(1976). Opiate receptor: Autoradiographic localization in rat brain. Proc. Natl. Acad. Sci. U.S.A. 73:3729-3733.
1506. Pert, C.B., Pasternak, G. and Snyder, S.H.(1973). Opiate agonists and antagonists discriminated by receptor binding in brain. Science 182:1359-1361.
1507. Petit, T.L. and Markus, E.J.(1987). The cellular basis of learning and memory: The anatomical sequel to neuronal use. In Neuroplasticity, Learning and Memory. Alan R. Liss, Inc. , New York, pp87-124.
1508. Pfaus, J.G. and Gorzalka, B.B.(1987). Opioids and sexual behavior. Neurosci. Biobehav. Rev. 11:1-34.
1509. Pfefferkorn, E.R.(1984). Interferon g blocks the growth of Toxoplasma gondii in human fibroblasts by inducing the host cells to degrade tryptophan. Proc. Natl. Acad. Sci. U.S.A. 81:908-912.
1510. Phillipson, O.T.(1979). A Golgi study of the ventral tegmental area of Tsai and infrafascicular nucleus of the rat. J. Compar. Neurol. 187:99-116.
1511. Pickel, V., Joh, T.H. and Reis, D.J.(1977). A serotonergic innervation of noradrenergic neurons in nucleus locus coeruleus: Demonstration by immunocytochemical localization of transmitter specific enzymes and tryptophan hydroxylase. Brain Res. 131:197-214.
1512. Pierce, C. Lipcon, H., McClary, J. and Noble, H.(1956). Enuresis: Clinical, laboratory and electroencephalographic studies. U.S. Armed Forces Med. J. 7:208-213.
1513. Pierce, C., Whitman, R., Maas, J. and Gay, M.(1961). Enuresis and dreaming. Arch. Gen. Psychiatry 4:166-170.
1514. Pijenburg, A.J.J., Honig, W.M.M., and van Rossum, J.M.(1975). Inhibition of d-amphetamine-induced locomotor activity by injection of haloperidol into the nucleus accumbens of the rat. Psychopharmacology 41:87-95.
1515. Pijenburg, A.J.J., Honig, W.M.M., van den Heyden, J.A.M., and Rossum, J.M.(1976). Effects of chemical stimulation of the mesolimbic dopamine system upon locomotor activity. Eur. J. Pharmacol. 35:45-58.
1516. Pilleri, G.(1966). The Klüver-Bucy syndrome in man. Psychiatr. Neurol. 152:65-103.
1517. Pintar, J.E. and Breakfield, X.O.(1982). Monamine oxidase (MAO) activity as a determinant in human neurophysiology. Behav. Genet. 12:53-68.
1518. Pintor, C., Cella, S.G., Corda, R., Locatelli, V., Puggioni, R., Loche, S. and Müller, E.E.(1985). Clonidine accelerates growth in children with impaired growth hormone secretion. Lancet 1:1482-1484.
1519. Pintor, C., Cella, S.G., Loche, S., Puggioni, R., Corda, E., Locatelli, V. and Müller, E.E.(1987). Clonidine treatment for short stature. Lancet 1:1226-1229.
1520. Pitman, R.K.(1987). Pierre Janet on obsessive-compulsive disorder (1903). Review and commentary. Arch. Gen. Psychiatry 44:226-232.
1520a. Pitman, R.K., Green, R.C., Jenike, M.A. and Mesulam, M.M.(1987). Clinical comparison of Tourette's disorder and obsessive-compulsive disorder. Am. J. Psychiatry 144:1166-1171.
1521. Pitts, F.N. and McClure, J.N.Jr.(1967). Lactate metabolism in anxiety neurosis. N. Engl. J. Med. 277:1328-1336.
1522. Pletscher, A., Shore, P.A. and Brodie, B.B.(1956). Serotonin as a mediator of reserpine action in brain. J. Pharmacol. 116:84-89.
1523. Polites, D.J., Kruger, D. and Stevenson, I.(1965). Sequential treatment in a case of Gilles de la Tourette syndrome. Br. J. Med. Psychol. 38:43-52.
1524. Pollack, M.A., Cohen, D.L. and Friedhoff, A.J.(1977). Gilles de la Tourette's syndrome. Familial occurrence and precipitation by methylphenidate therapy. Arch. Neurol. 34:630-632.
1525. Pontius, A.A.(1972). Neurological aspects in some type of delinquency especially among juveniles: Toward a neurological model of ethical action. Adolescence 7:289-308.

1526. Pontius, A.A.(1973). Dysfunction patterns analogous to frontal lobe system and caudate nucleus syndrome in some groups of minimal brain dysfunction. J. Am. Med. Women's Ass. 26: 285-292.
1527. Porteus, S.D.(1965). Porteus Maze Tests: Fifty Years Application. Pacific Books, Palo Alto, CA.
1528. Porteus, S.D., DeMonbrun, R., and Kepner, M.D.(1944). Mental changes after bilateral prefrontal lobotomy. Genet. Psychol. Monographs 29:3-115.
1529. Poschel, B.P.H. and Ninteman, F.W.(1971). Intracranial reward and the forebrain's serotonergic mechanism: Studies employing para-chlorophenylalanine and para-chloroamphetamine. Psychol. Behav. 7:39-46.
1530. Poschel, B.P.H., Ninteman, F.W., McLean, J.R. and Potoczak, D.(1975). Intracranial reward after 5,6-dihydroxytryptamine: Further evidence for serotonin's inhibitory role. Life Sci. 15:1515-1522.
1531. Post, R.M. and Uhde, T.W.(1985). Are the psychotrophic effects of carbamazepine in manic-depressive illness mediated through the limbic system? Psychiatr. J. Univ. Ottawa 10:205-219.
1532. Post, R.M.(1975). Cocaine psychoses: A continuum model. Am. J. Psychiatry 132:225-231.
1533. Post, R.M., Ballanger, J.C., and Goodwin, F.K.(1974). Effects of amitriptyline and imipramine on amine metabolites in the cerebrospinal fluid of depressed patients. Arch. Gen. Psychiatry 30:234-239.
1534. Post, R.M., Ballenger, J.C., Uhde, T.W. and Bunney, W.E.(1984). Efficacy of carbamazepine in manic-depressive illness: Implications of underlying mechanisms. In R.M. Post and J.C. Ballenger (eds) Neurobiology of Mood Disorder. Williams and Wilkins, Baltimore. pp777-816.
1535. Post, R.M., Uhde, T.W., Roy-Byrne, P.P. and Joffe, R.T.(1986). Antidepressant effects of carbamazepine. Am. J. Psychiatry 143:29-34.
1536. Potashner, S.J.(1979). Baclofen: Effect on amino acid release and metabolism in slices of guinea pig cortex. J. Neurochemistry 32:103-109.
1537. Precht, W. and Yoshida, M.(1971).Blockage of caudate-evoked inhibition of neurons in the substantia nigra by picrotoxin. Brain Res. 32:229-233.
1538. Preziosi, P., Cerrito, F. and Vacca, M.(1983). Effects of naloxone on the secretion of prolactin and corticosterone induced by 5-hydroxytryptophan and a serotonin agonist, mCPP. Life Sci. 32:2423-2430.
1539. Pribram, K.H.(1960). A review of the theory in physiological psychology. Annu. Rev. Psychol. 11:1-40.
1540. Pribram, K.H.(1959). On the Neurology of Thinking. Behavioral Sciences Series, Harper & Row, New York.
1541. Pribram, K.H.(1961). A further analysis of the behavior deficit that follows injury to the primate frontal cortex. J. Exp. Neurol. 3:432-466.
1542. Pribram, K.H.(1971). Languages of the Brain: Experimental Paradoxes and Principles of Neuropsychology. Prentice-Hall, New York.
1543. Pribram, K.H., Mishkin, M., Rosvold, H.E., and Kaplan, S.J.(1952). Effects on delayed response performance of lesions of dorsolateral and ventro-medial frontal cortex of baboons. J. Comp. Physiol. Psychol. 45:565-575.
1544. Price, J.M., Brown, R.R. and Peters, H.A.(1959). Tryptophan metabolism in porphyria, schizophrenia, and a variety of neurologic and psychiatric diseases. Neurology 9:456-468.
1545. Price, J.M., Brown, R.R. and Yess, N.(1965). Testing the functional capacity of the tryptophan-niacin pathway in man by analysis of urinary meta-boiites. Adv. Metab. Disord. 2:159-225.
1546. Price, R.A.(1987). Genetics of human obesity. Ann. Behav. Med. 9:9-14.
1547. Price, R.A., Cadoret, R.J., Stunkard, A.J. and Troughton, E.(1987). Genetic contributions to human fatness: An adoption study. Am. J. Psychiatry 144:1003-1008.
1548. Price, R.A., Kidd, K.K., Cohen, D.J., Pauls, D.L. and Leckman, J.F.(1985). A twin study of Tourette syndrome. Arch. Gen. Psychiatry 42:815-820.
1549. Price, R.A., Leckman, J.F., Pauls, D.L., Cohen, D.J. and Kidd, K.K.(1986). Gilles de la Tourette's syndrome: Tics and central nervous system stimulants in twins and nontwins. Neurology 36:232-237.
1550. Prichep, L.S., Sutton, S., Hakerem, G.(1976). Evoked potentials in hyperkinetic and normal children under certainty and uncertainty: A placebo and methylphenidate study. Psychopharmacology 13:419-428.
1551. Prinz, R.J., Conner, P.A. and Wilson, C.C.(1981). Hyperactive and aggressive behaviors in childhood: Intertwined dimensions. J. Abnorm. Child Psychol. 9:191-202.
1552. Prinz, R.J., Roberts, W.A. and Hantman, E.(1980). Dietary correlates of hyperactive behavior in children. J. Consult. Clin. Psychol. 48:760-769.
1553. Prior, M.(1979). Cognitive abilities and disabilities in infantile autism: A review. J. Abnorm. Child Psychol. 7:357-380.
1554. Proac, C. and Coren, S.(1981). Lateral Preferences and Human Behavior. Springer-Verlag, New York, pp147-156.
1555. Pujol, J.-F., Degueurce, A., Natali, J.-P., Tapaz, M., Wiklund, L. and Lger, L.(1981). The serotonin connection: Some evidence for a specific metabolic connection. Adv. Exp. Med. Biol. 133:417-429.
1556. Pujol, J.-F.(1979). Interactions between serotonin and noradrenalin containing neurons: Link between the raphe system and the locus coeruleus. In P. Simon (ed) Advances in Pharmacological Therapy 2: Neurotransmitter. Pergamon Press, Oxford, pp121-129.
1557. Pulkkinen, L.(1983).Youthful smoking and drinking in a longitudinal perspective. J. Youth Adolesc. 12:253-260.
1558. Pulkkinen, L.(1983). Finland: The search for alternatives to aggression, In A.P. Goldstein and M.H. Segal (eds) Aggression in Global Perspective. Pergamon, New York, pp104-144.
1559. Putkonen, P.T.S., and Bergström, L.(1981). Clonidine alleviates cataleptic symptoms in narcolepsy. In W.P. Koella (ed) Sleep 1980. S. Karger, Basel, pp414-416.
1560. Pycock, C.J., Carter, C.J. and Kerwin, R.W.(1980). Effect of 6-hydroxydopamine lesions of the medial prefrontal cortex on neurotransmitter system in subcortical sites in the rat. J. Neurochem. 34 :91-99.
1561. Pycock, C.J., Donaldson, I.M. and Marsden, C.D.(1975). Circling behavior produced by unilateral lesions in the region of the locus coeruleus in rats. Brain Res. 97:317-329.
1562. Pycock, C.J., Kerwin, R.W. and Carter, C.J.(1980). Effect of lesion of cortical dopamine terminal on subcortical dopamine receptors in rats. Nature 286:74-77.
1563. Quay, H.C. and Peterson, D.(1967). Manual for the

Behavior Problem Checklist. University of Illinois Press, Champaign, IL.
1564. Quay, H.C.(1979).Classification. In H.C. Quay and J.S Werry (eds) Psychopathological Disorders of Childhood, Second Edition. John Wiley & Sons, New York, pp1-42.
1565. Quitkin, F. and Kelin, D.F.(1969). Two behavioural syndromes in young adults related to possible minimal brain dysfunction. J. Psychiatr. Res. 7:131-142.
1566. Rado, S.(1953). Dynamics and classification of disordered behavior. Am. J. Psychiatry 110: 406-416.
1567. Ragusa, N., Sfogliano, L., Calabrese, V. and Rizza, V.(1981). Effects of multivitamin treatment on the activity of rat liver tryptophan pyrrolase during ethanol administration. Acta Vitamin. Enzymol. 3:199-204.
1568. Ragusa, N., Zito, D., Bondi, C., Vanella, A., and Rizzo, V.(1980). Effects of pyridoxine on hepatic tryptophan pyrrolase activity in rat during chronic ethanol administration. Biochem. Exp. Biol. 16:391-396.
1569. Randrup, A. and Munkvad, I.(1965). Special antagonism of amphetamine-induced abnormal behavior. Psychopharmacology 7:416-422.
1570. Randrup, A. and Munkvad, I.(1966). DOPA and other naturally occurring substances as causes of stereotypy and rage in rats. Acta Psychiat. Scand.(Suppl 191):42:193-199.
1571. Randrup, A. and Munkvad, I.(1966). Role of catecholamines in the amphetamine excitatory response. Nature 211:540.
1572. Randrup, A. and Munkvad, I.(1967). Stereotyped activities produced by amphetamine in several animal species and man. Psychopharmacology 11:300-310.
1573. Randrup, A. and Scheel-Kruger, J.(1966). Diethyldithiocarbamate and stereotyped behavior. J. Pharm. Pharmacol. 18:752.
1574. Randrup, A., Munkvad, I., and Udsen, P.(1963). Adrenergic mechanisms and amphetamine induced abnormal behavior. Acta Pharmacol. 20:145-157.
1575. Rapkin, A.J., Edelmuth, E., Chang, L.C., Reading, A.E., McGuire, M.T. and Su, T.-P.S.(1987). Whole-blood serotonin in premenstral syndrome. Obstet. Gynecol. 70:533-537.
1576. Rapoport, J.L.(1965). Childhood behavior and learning problems treated with imipramine. Int. J. Neuropsychiatry 1:635-642.
1576a. Rapoport, J.L.(1986). Childhood obsessive compulsive disorder. J. Child. Psychol. Psychiatry 27: 289-295.
1577. Rapoport, J.L.(1987). Pediatric psychopharmacology: The last decade. In H.Y. Meltzer (ed) Psychopharmacology: The Third Generation of Progress. Raven Press, New York, pp1211-1214.
1578. Rapoport, J.L., Buchsbaum, M.S., Weingartner, H., Zhan, T.P., Ludlow, C. and Mikkelsen, E.J.(1980). Dextroamphetamine. Its cognitive and behavioral effects in normal and hyperactive boys and normal men. Arch. Gen. Psychiatry 37:933-943.
1579. Rapoport, J.L., Mikkelsen, E.J., Zavadil, A., Nee, L., Greunau, C., Mendelson, W. and Gillin, C.(1980). Childhood enuresis: II. Psychopathology tricyclic concentration in plasma and antienuretic effect. Arch. Gen. Psychiatry 37:1146-1152.
1580. Rapoport, J.L., Quinn, P., Schibanu, Q.N. and Murphy, D.L.(1974). Platelet serotonin of hyperactive school age boys. Br. J. Psychiatry 125:138-140.
1581. Rapoport, J.L., Quinn, P.O., Bradbard, G., Riddle, D. and Brooks, E.(1974). Imipramine and methylphenidate treatment of hyperactive boys: A double blind comparison. Arch. Gen. Psychiatry 30:789-793.
1582. Rapp, D. (1979). Allergies and the Hyperactive Child. Cornerstone Library, New York.
1583. Rapport, M.M.(1949). Serum vasoconstrictor (serotonin). V. The presence of creatinine in the complex. A proposed structure of the vasoconstrictor principle. J. Biol. Chem. 180:961-969.
1584. Rastogi, R.B., Singhal, R.L., Lapierre, Y.D.(1981). Effects of short- and long-term neuroleptic treatment on brain sertonin synthesis and turnover: Focus on the serotonin hypothesis of schizophrenia. Life Sci. 29:735-741.
1585. Ravaris, C.L., Nies, A., Robinson, D.S. Ives, J.O., Lamborn, K.R. and Korson, L.(1976). Multiple-dose controlled study of phenelzine in depression-anxiety states. Arch Gen. Psychiatry 33:347-350.
1586. Reader, T.A.(1981). Distribution of catecholamines and serotonin in the rat cerebral cortex: Absolute levels and relative proportions. J. Neural Transmission 50:13-27.
1587. Realmuto, G.M. and Main, B.(1982). Coincidence of Tourette's disorder and infantile autism. J. Autism Dev. Disord. 12:367-372.
1588. Rebec, G.V., Alloway, K.D. and Curtis, S.D.(1981). Apparent serotonergic modulation of the dose-dependent biphasic response of neostriatal neurons produced by d-amphetamine. Brain Res. 210:277-289.
1589. Redmond, D.E.Jr., and Huang, Y.H.(1979). New evidence for a locus coeruleus-norepinephrine connection with anxiety. Life Sci. 25:2149-2162.
1590. Reece, S.A. and Levin, B.(1968). Psychiatric disturbances in adopted children: A descriptive study. Social Work 13:101-111.
1591. Rees, L.(1953). Psychosomatic aspects of the premenstral tension syndrome. J. Mental Sci. 99:62-73.
1592. Reese, S. and Levin, B.(1968). Psychiatric disturbances in adopted children: A descriptive study. Social Work 13:101.
1593. Reeves, J.C., Werry, J.S., Elkind, G.S. and Zametkin, A.(1987). Attention deficit, conduct, oppositional, and anxiety disorders in chldren: II. Clinical characteristics. Acad. Child Adolesc. Psychiatry 26:144-155.
1594. Regeur, L., Pakkenberg, B., Fog, G. and Pakkenberg, H.(1986). Clinical features and long-term treatment with pimozide in 65 patients with Gilles de la Tourette's syndrome. J. Neurol. Neurosurg. Psychiatry 49:791-795.
1594a. Rehaavi, M., Carmi, R. and Weizman, A.(1988). Tricyclic antidepressants and calcium channel blockers: Interaction at the (-)-desmethoxyverapamil binding site and the serotonin transporter. Eur. J. Pharmacol. 155:1-9.
1595. Reich, T., Cloninger, C.R., Eerdewegh, P.V., Rics, J.P. and Mullaney, J.(1989). Secular trends in the familial transmission of alcoholism. Alcoholism: Clin. Exp. Res. (in press).
1596. Reiman, E.M., Raichle, M.E., Butler, K.F., Herscovitch, P., Robins, E.(1984). A focal brain abnormality in panic disorder. Nature 310:683-685.
1597. Reiman, E.M., Raichle, M.E., Robins, E., Butler, K., Herscovitch, P., Fox, P. and Perlmutter, J.(1986). The application of positron emission tomography to the study of panic disorder. Am. J. Psychiatry 143:469-477.
1598. Reinhard, J.F. and Roth, R.H.(1982). Noradrenergic

modulation of serotonin synthesis and metabolism. I. Inhibition by clonidine in vivo. J. Pharmacol. Exp. Ther. 22:541-546.

1599. Reinhard, J.F. and Wurtman, R.J.(1978). Relation between brain 5-HIAA levels and the release of serotonin into brain synapses. Life. Sci. 21:1741-1746.

1600. Reiss, A.L., Feinstein, C. and Rosenbaum, K.N.(1986). Autism and genetic disorders. Schizophr. Bull. 12:724-738.

1601. Renkawitz, R.(1987). Molecular genetics of L-tryptophan 2,3, dioxygenase in rat liver. In D.A. Bender, M.H. Joseph, W. Kochen and H. Steinhart (eds) Progress in Tryptophan and Serotonin Research 1986. Walter de Gruyter, New York, pp1-6.

1602. Rett, A.(1966). Über ein eingartiges hirnatrophisches Syndrome bei Hyperammonämie im Kindsalter. [On an unusual brain atrophy syndrome with hyperamonemia in childhood]. Wien. Med. Wochenschr. 116:723-726.

1603. Reynolds, E.H.(1982). The pharmacological management of epilepsy associated with psychological disorders. Br. J. Psychiatry 141:549-557.

1603a. Reynolds, G.P., Czudek, C., Bzowej, N., and Seeman, P.(1987). Dopamine receptor asymmetry in schizophrenia Lancet 1:979.

1604. Riblet, L.A., Taylor, D.P., Eison, M.S. and Stanton, H.C.(1982). Pharmacology and neurochemistry of buspirone. J. Clin. Psychiatry. 43:11-16.

1605. Ricaurte, G., Bryan, G., Strauss, L., Seiden, L. and Schuster, C.(1985). Hallucinogenic amphetamine selectively destroys brain serotonin nerve terminals. Science 229:986-988.

1606. Richardson, J.S. and Zaleski, W.A.(1983). Naloxone and self-mutilation. Biol. Psychiatry 18:99-101.

1607. Riche, D. and Lanoid, J.(1978). Some claustro-cortical connections in the cat and baboon as studied by retrograde horseradish peroxidase transport. J. Comp. Neurol. 177:435-444.

1608. Richfield, E.K., Twyman, R. and Berent, S.(1987). Neurological syndrome following bilateral damage of the head of the caudate nuclei. Ann. Neurol. 22:768-771.

1609. Riddel, D.C., R. Mallonee, R., Phillips, J.A., Parks, J.S., Sexton, L.A. and Hamerton, J.L.(1985). Chromosomal assignment of human sequences encoding arginine vasopressin-neurophysic II and growth hormone releasing factor. Somatic Cell Molec. Genet. 11:189-195.

1610. Riddle, M.A., Hardin, M.T., Towbin, K.E., Leckman, J.F. and Cohen, D.J.(1987). Tardive dyskinesia following haloperidol treatment in Tourette's syndrome. Arch. Gen. Psychiatry 44:98-99.

1610a. Rie, H., Rie, E., Stewart, S. and Ambuel, J. (1976). Effects of methylphenidate on underachieving children. J. Consult. Clin. Psychol. 44:250-260.

1611. Rieder, R.O. and Nichols, P.L.(1979). Offspring of schizophrenics III. Hyperactivity and neurological soft signs. Arch. Gen. Psychiatry 36:665-674.

1612. Rimland, B. and Fein, D.(1988). Special talents of autistic savants. In L.K. Obler and D. Fein (eds) The Exceptional Brain. The Guilford Press, New York.

1613. Rimland, B., Callaway, E. and Dreyfus, P.(1978). The effect of high doses of vitamin B6 on autistic children: A double-blind crossover study. Am. J. Psychiatry 135:472-475.

1614. Rinsley, D.B. Bordeline psychopathology (1978). A review of etiology, dynamics and treatment. Int. Rev. Psychoanal. 5:45-54.

1614a. Ritvo, E.R. and Garber, H.J.(1988). Cerebellar aplasia and autism. N. Engl. J. Med. 319:1152.

1615. Ritvo, E.R., Freeman, B.J., Yuweiler, A., Geller, E., Schroth, P., Yokota, A., Mason-Brothers, A., August, G.J., Klykylo, W., Leventhal, B., Lewis, K., Piggott, L., Realmuto, G., Stuggs, G. and Umansky, R.(1986). Fenfluramine treatment of autism: UCLA — Collaborative study of 81 patients at nine medical centers. Psychopharm. Bull. 25:133-141.

1616. Ritvo, E.R., Jorde, L.B., Freeman, B.J., McMahon, W.M., Jenson, W.R., Petersen, P.B., Brothers, A.M., Mo, A. and Pingree, C.B.(1987). Population prevalence and recurrence of risk of autism. Am. J. Hum. Genet. 411:A81.

1617. Ritvo, E.R., Spence, M.A., Freeman, B.J., Mason-Brothers, A., Mo, A. and Marazita, M.L.(1985). Evidence for autosomal recessive inheritance in 46 families with multiple incidences of autism. Am. J. Psychiatry 142:187-192.

1618. Ritvo, E.R., Yuwiler, A., Geller, E., Ornitz, E.M., Seager, K. and Plotkin, S.(1970). Increased blood serotonin and platelets in early infantile autism. Arch. Gen. Psychiatry 23:566-572.

1619. Ritvo, E.R., Yuwiler, A., Geller, E., Plotkin, S., Mason, A. and Saeger, K.(1971). Maturational changes in blood serotonin levels and platelet levels. Biochem. Med. 5:90-96.

1620. Roberge, A.G., Missala, K. and Sourkes, T.L.(1972). Alpha-methyltryptophan: Effects on synthesis and degradation of serotonin in the brain. Neuropharmacology 11:197-209.

1621. Roberts, E. and Frankel, S.(1950). g-Aminobutyric acid in brain: Its formation from glutamic acid. J. Biol. Chem. 187:55-63.

1622. Roberts, E.(1972). An hypothesis suggesting that there is a defect in the GABA system in schizophrenia. Neurosci. Res. Program Bull. 10:468-481.

1623. Roberts, E.(1976). Disinhibition as an organizing principle in the nervous system—The role of the GABA system. Application to neurologic and psychiatric disorder. In T. Chase and D. Tower (eds) GABA in Nervous System Function. Raven Press, New York, pp515-539.

1624. Roberts, M.H.T. and Straughan, D.W.(1967). Excitation and depression of cortical neurons by 5-hydroxytryptamine. J. Physiol. 193:269-294.

1625. Robertson, M.M., Trimble, M.R. and Lees, A.J.(1988). The psychopathology of the Gilles de la Tourette syndrome. Br. J. Psychiatry 152:383-390.

1625a. Robertson, M.M.(1989). The Gilles de la Tourette syndrome: The current status. Br. J.Psychiatry 154:147-169.

1626. Robichard, R.C. and Sledge, K.L.(1970). The effects of p-chlorophenylalanine on experimentally induced conflict in the rat. Life Sci. 8:965-969.

1627. Robins, L.N.(1966). Deviant Children Grown Up. Williams & Wilkins, Baltimore.

1628. Robins, L.N.(1978). Sturdy childhood predictors of adult antisocial behavior: Replications from longitudinal studies. Psychol. Med. 8:611-622.

1629. Robins, L.N., Bates, W.M. and O'Neal, P.(1962). Adult drinking patterns of former problem children. In D.J. Pittman and C.R. Snyder (eds) Society, Culture, and Drinking Patterns. John Wiley & Sons, Inc., New York, pp395-412.

1630. Robins, L.N., Helzer, J.E., Croughan, J. and Ratclif, K.S.(1981). National Institute of Mental Health Diagnostic Interview Schedule. Arch. Gen. Psychiatry 38:381-389.

1631. Robins, L.N., Helzer, J.E., Ratcliff, K.S., Seyfried, W.(1982). Validity of the Diagnostic Interview Schedule, Version II. DSMIII Diagnoses. Psychol. Med. 12:855-870.
1632. Robins, L.N., Helzer, J.E., Weissman, M., Orvaschel, H., Gruenberg, E., Burke, J.D., Reigier, D.A.(1984). Lifetime prevalence of specific psychiatric disorders in three sites. Arch. Gen. Psychiatry 41:949-958.
1633. Robinson, A.G.(1985). Disorders of antidiuretic hormone secretion. Clinics Endocrinol. Metab. 14:55-88.
1634. Robinson, A.L., Heaton, R.K., Lehman, R.E.W. and Stilson, D.W.(1980). The ability of the Wisconsin Card Sorting Test in detecting and localizing frontal lobe lesions. J. Consult. Clin. Psychol. 48:605-614.
1635. Robinson, D.S., Davis, J.M., Nies, A., Ravaris, C.L. and Slyvester, D.(1971). Relation of sex and aging to monamine oxidase activity of human brain, plasma, and platelets. Arch. Gen. Psychiatry 24:536-539.
1636. Rochette, L. and Bralet, J.(1975). Effect of the norepinephrine receptor stimulating agent 'clonidine' on the turnover of 5-hydroxytryptamine in some areas of the rat brain. J. Neural Transmission 37:259-267.
1636a. Rogers, K.M. and Zoccolillo, M.S. (1989).Conduct disorder in aldolescent girls. DIS Newsletter, Washington Univ. School of Med., St. Louis, 6:9-10.
1637. Roland, P.E.(1982). Cortical regulation of selective attention in man. A regional cerebral blood flow study. J. Neurophysiol. 48:1059-1078.
1638. Roland, P.E.(1984). Metabolic measurements of the working frontal cortex in man. Trends Neurol. Sci. 7:430-435.
1639. Roland, P.E., Shinhøj, E., Lassen, N.A. and Larsen, B.(1980). Different cortical areas in man in organization of vountary movements in extrapersonal space. J. Neurophysiol. 43:137-150.
1640. Roland, P.E., Skinhøj, E. and Lassen, N.A.(1981). Focal activations of human cerebral cortex during auditory discrimination. J. Neurophysiol. 45:1139-1151.
1641. Root-Bernstein, R.S and Westall, F.C.(1984). Fenfluramine binds 5-hydroxytryptophan. Brain Res. Bull. 12:17-22.
1642. Rose, R. and Ditto, W.B.(1983). A developmental-genetic analysis of common fears from early adolescence to early childhood. Child Dev. 54:361-368.
1643. Rosenbaum, J.F., Biederman, J., Gersten, M., Hirshfeld, D.R., Meminger, S.R., Herman, J.B., Kagan, J., Reznick, J.S. and Snidman, N.(1988). Behavioral inhibition in children of parents with panic disorder and agoraphobia. Arch. Gen. Psychiatry 45:463-470.
1644. Rosenblatt, S., Chanley, J.D., Sobotka, H. and Kaufman, M.R.(1960). Interrelationship between electroshock, the blood-brain barrier, and catecholamines. J. Neurochem. 5:172-176.
1645. Rosenthal,D.(1971). Genetics of Psychopathology. McGraw Hill, New York, 182pp.
1646. Rosenthal, D.(1980). Genetic aspects of schizophrenia. In Van Praag, H.M. (ed) Handbook of Biological Psychiatry, Part III. Marcel Dekker, Inc., New York, pp3-34.
1647. Rosenthal, N., Davenport, Y., Cowdry, R., Webster, M. and Goodwin, F.(1980). Monamine metabolites in cerebrospinal fluid of depressive subgroups. Psychiatry Res. 2:113-119.
1648. Rosenthal, N.E., Sack, D.A., Gillin, J.C., Lewy, A.J., Goodwin, F.K., Davenport, Y., Mueller, P.S., Newsome, D.A. and Wehr, T.A.(1984). Seasonal affective disorder. Arch. Gen. Psychiatry 41:72-80.
1649. Rosenthal, R.H and Allen, T.W.(1978). An examination of attention, arousal, and learning dysfunctions of hyperkinetic children. Psychol. Bull. 85:689-715.
1650. Ross, M.S. and Moldofsky, H.(1978). A comparison of pimozide and haloperidol in the treatment of Gilles de la Tourette's syndrome. Am. J. Psychiatry 135:585-587.
1650a. Ross, H.E., Glaser, F.B. and Germanson, T.(1988). The prevalence of psychiatric disorders in patients with alcohol and other drug problems. Arch. Gen. Psychiatry 45:1023-1031.
1651. Rossier, J.(1982). Opioid peptides have found their roots. Nature 298:221-222.
1652. Rosvold, H.E.(1959). Physiological psychology. Annu. Rev. Psychol. 10:415-453.
1652a. Rosvold, H.E.:The prefrontal cortex and caudate nucleus: A system for effecting correction in response mechanism. In C. Rupp (ed) Mind as a Tissue. Hoeber, New York, 1968, pp21-38.
1653. Roth, M., Gurney, C., Garside, R.F. and Kerr, T.A.(1972). The relationship between anxiety states and depressive illness, part I. Br. J. Psychiatry 121:147-161.
1654. Rothman, S. and Lichter, S.R.(1982). Roots of Radicalism. Jews, Christian, and the New Left. Oxford University Press, New York, 466pp.
1655. Routtenberg, A. and Santos-Anderson, R.(1977). The role of prefrontal cortex in intracranial self-stimulation. In L. L. Iversen, S.D. Iversen and S.H. Snyder (eds) Handbook of Psychopharmacology, Vol 8. Plenum Press, New York.
1656. Routtenberg, A.(1978). The reward system of the brain. Sci. Am. 239:154-165.
1657. Roy, A. and Linnoila, M.(1986). Alcoholism and suicide. Suicide Life-Threatening Behav. 16:244-273.
1658. Roy, A., Adinoff, B., Roehrich, L., Lamparski, D., Custer, R., Lorenz, V., Barbaccia, M., Guidotti, A., Costa, E., and Minnoila, M.(1988). Pathological gambling. A psychobiological study. Arch. Gen. Psychiatry 45:369-373.
1659. Roy, A., Virkkunen, M., Guthrie, S., Poland, R. and Linnoila, M.(1986). Monamines, glucose metabolism, suicidal and aggressive behaviors. Psychopharmacol. Bull. 22:661-665.
1660. Roy, M., Boyer, L. and Barbeau, A.(1983). A prospective study of 50 cases of familial Parkinson's disease. Can. J. Neurol. Sci. 10:37-42.
1660a. Roy, M., DeJong, J., Linnoila, M.(1989). Cerebrospinal fluid monamine metabolites and suicidal behavior in depressed patients. Arch. Gen. Psychiatry 46:609-612.
1661. Rubin, R.T., Mandell, A.J. and Crandall, P.H.(1966). Corticosteroid responses to limbic stimulation in man: Localization of stimulus sites. Science 153:767-768.
1662. Rusak, B. and Zucker, I.(1979). Neural regulation of circadian rhythms. Physiol. Rev. 59: 449-526.
1663. Rutter, M.(1972). Maternal Deprivation Reassessed. Penguin Books, New York.
1664. Rutter, M.(1978). Early sources of security and competence, In J. Brunner and A Garton (eds) Human Growth and Development. Oxford University Press, New York.
1665. Rutter, M.(1982). Syndromes attributed to "minimal brain dysfunction" in childhood. Am. J. Psychiatry 139:21-33.
1666. Rutter, M.(1983). Cognitive deficits in the pathogenesis of autism. J. Child Psychol. Psychiatry 24:513-531.
1667. Rutter, M., Tizard, J. and Whitmore, K.(1970).

(Eds) Education, Health and Behavior, Longman, London.
1668. Rutter, M., Yule, W. and Graham, P.(1973). Enuresis and behavioral deviance: Some epidemiological considerations. In I. Kolvin, R.C. MacKeith and S.R. Meadow (eds) Bladder Control and Enuresis. J.B. Lippincott Co., Philadelphia, pp137-147.
1669. Ryan, N.D., Puig-Antich, J., Ambrosini, P., Rabinovich, H., Robinson, D., Nelson, B., Iyengar, S. and Twomey, J.(1987). The clinical picture of major depression in children and adolescents. Arch. Gen. Psychiatry 44:854-861.
1670. Rydin, E., Schalling, D. and Åsberg, M.(1982). Rorshach rating in depressed and suicidal patients with low levels of 5-hydroxyindoleacetic acid in cerebrospinal fluid. Psychiatry Res. 7:229-243.
1671. Rylander, G.(1939). Personality changes after operations on the frontal lobes. Acta Psychiat. Neurol. Suppl 20:3-327.
1672. Saavedra, J.M., Grobecker, H., and Zivin, J.(1976). Catecholamines in the raphe nuclei of the rat. Brain Res. 114:339-345.
1673. Saaverdra, J.M., Brownstein, M. and Palkovits, M.(1974). Serotonin distribution in the limbic system of the rat. Brain Res. 79:437-441.
1674. Saccuzzo, D.P., Hirt, M. and Spencer, T.J.(1974). Backward masking as a measure of attention in schizophrenia. J. Abnorm. Psychol. 83:512-522.
1675. Sacerdote, P., Brini, A., Mantegazza, P. and Panerai, A.E.(1987). A role for serotonin and beta-endorphin in the analgesia induced by some tricyclic antidepressant drugs. Pharmacol. Biochem. Behav. 26:153-158.
1676. Sacks, O.(1983). Awakenings. E.P. Dutton, New York, 338pp.
1677. Sacks, O.(1985). The Man Who Mistook His Wife for a Hat. Summit Books, New York, 233pp.
1677a. Safer, D.J. and Allen, R.P.(1973). Factors influencing the suppressant effects of two stimulant drugs on the growth of hyperactive children. Pediatrics 51: 660-667.
1677b. Safer, D.J.(1969). Control of enuresis with imipramine. Arch. Dis. Child. 43:665- 671.
1677c. Safer, D.J.(1973). A familial factor in minimal brain dysfunction. Behav. Genet. 3:175-186.
1677d. Safer, D.J.(1973). The association between enuresis and emotional disorder: A review of the literature. In I. Kolvin, R.C. MacKeith and S.R. Meadow (eds) Bladder Control and Enuresis. J.B. Lippincott Co., Philadelphia, pp118-136.
1677e. Safer, D.J., Allen, R. and Barr, E.(1972). Depression of growth in hyperactive children on stimulant drugs. N. Engl. J. Med. 287:217-220.
1678. Safer, D.J., Allen, R.P. and Barr, E.(1975). Growth rebound after termination of stimulant drugs. J. Pediatr. 86:113-119.
1679. Sagan, C.(1977). The Dragons of Eden. Ballantine Books, New York, 271pp.
1680. Sahley, T.L. and Panksepp, J.(1987). Brain opioids and autism: An updated analysis of possible linkages. J. Autism Dev. Dis. 17:201-216.
1681. Saito, K., Barber, R., Wu, J., Matsuda, T., Roberts, E. and Vaughn, J.E.(1974). Immunohistochemical localization of glutamate decarboxylase in rat cerebellum. Proc. Natl. Acad. Sci. U.S.A.71:269-277.
1682. Sallee, F.R., Pollock, B.G., Stiller, R.L., Stull, S., Everett, G. and Perel, J.M.(1987). Pharmacokinetics of pimozide in adults and children with Tourette's syndrome. J. Clin. Pharmacol. 27:776-781.
1683. Salom, F.(1983). A psychological assay of modern day gambling. J. Am. Soc. Psychosom. Dent. Med. 30:66-67.
1684. Samanin, R. and Garattini, S.(1975). The serotonergic system in the brain and its possible functional connections with other aminergic systems. Life Sci. 17:1201-1210.
1685. Samanin, R., Quattrone, A., Consolo, S., Ladinsky, H., and Algeri, S.(1978). Biochemical and pharmacological evidence of the interaction of serotonin with other aminergic systems in the brain. In S. Garattini, J.F. Pujol, and R. Samanin (eds) Interactions between Putative Neurotransmitters in the Brain. Raven Press, New York. pp383-399.
1686. Sanberg, P.R., Fogelson, H.M., Manderscheid, P.Z., Parker, K.W., Norman, A.B. and McConville, B.J.(1988). Nicotine gum and haloperidol in Tourette's syndrome. Lancet 1:592.
1687. Sandberg, S.T., Wieselberg, M. and Shaffer, D.(1980). Hyperkinetic and conduct problems of children in a primary school population: Some epidemiological considerations. J. Child Psychol. Psychiatry 21:293-331.
1688. Sanders, D.G.(1973). Familial occurrence of Gilles de la Tourette syndrome. Report of the syndrome occurring in a father and son. Arch. Gen. Psychiatry 28:326-328.
1689. Sandler, M. and Gessa, L. (eds) (1975). Sexual Behavior Pharmacology and Biochemistry. Raven Press.
1690. Sandyk, R.(1983). Blepharospasm — Successful treatment with baclofen and sodium valproate. S. Africa Med. J. 64:955-956.
1691. Sandyk, R.(1985). Naloxone abolishes self-injuring in a mentally retarded child. Ann. Neurol. 17:520.
1692. Sandyk, R.(1985). The effects of naloxone in Tourette's syndrome. Ann. Neurol. 18:367-368.
1693. Sandyk, R.(1985). The opioid system in Gilles de la Tourette's syndrome. Neurology 35:449-450.
1694. Sandyk, R.(1986). Naloxone ameliorates autoaggressive behavior in Tourette's syndrome. Neurology 36(Suppl 1): 275.
1695. Sandyk, R.(1986). Naloxone withdrawal exacerbates Tourette syndrome. J. Clin. Psychopharmacol. 6:58-59.
1696. Sandyk, R.(1987). Naloxone abolished obsessive-compulsive behavior in Tourette's syndrome. Intern. J. Neurosci. 35:93-94.
1697. Sandyk, R.(1987). Opioid neuronal denervation in Gilles de la Tourette syndrome. Int. J. Neurosci. 35:95-98.
1698. Sangdee, C. and Franz, D.N.(1980). Lithium enhancement of central 5-HT transmission induced by 5-HT precursors. Biol. Psychiatry 15:59-75.
1699. Saper, C.B.(1987). Function of the locus coeruleus. Trends Neurosci. 10:343-344.
1700. Sapun-Malcolm, D, Farah, J.M. and Mueller, G.P.(1986). Serotonin and dopamine independently regulate pituitary β-endorphin release in vivo. Neuro-endocrinol. 42:191-196.
1701. Sarrias, M.J., Artigas, F., Martinez, E., Gelpi, E., Alverez, E., Udina, C. and Casas, M.(1987). Decreased plamsa serotonin in melancholic patients: A study with clomipramine. Biol. Psychiatry 22:1429-1438.
1702. Satterfield, J.H. and Dawson, M.E.(1971). Electrodermal correlates of hyperactivity in children. Psychophysiology 8:191-197.
1703. Satterfield, J.H., Hoppe, C.M. and Schell, A.M.(1982). A prospective study of delinquency in 110 adolescent boys with attention deficit disorder and 88 normal

lescent boys with attention deficit disorder and 88 normal adolescent boys. Am. J. Psychiatry 139:795-798.
1073a. Satterfield, J.H., Satterfield, B.T. and Cantwell, D.P.(1981). Three-year multimodality treatment study of 100 hyperactive boys. Behav. Pediatr. 98:650-655.
1704. Saul, R.C. and Ashby, C.D.(1986). Measurement of whole blood serotonin as a guide in prescribing psychostimulant medication for children with attentional deficits. Clin. Neuropharmacol. 9:189-195.
1705. Scatton, B., Boireau, A., Garret, C., Glowinski, J. and Julow, L.(1977). Action of palmitic ester of pipotiazine on dopamine metabolism in the nigro-striatal, meso-limbic and meso-cortical sytems. Arch. of Pharmacol. 296:169-175.
1706. Scatton, B., Worms, P., Lloyd, K.G. and Bartholini, G.(1982). Cortical modulation of striatial function. Brain Res. 232:331-343.
1707. Schachar, R., Rutter, M. and Smith, A.(1981). The characteristics of situationally and pervasively hyperactive children: Implication for syndrome definition. J. Child. Psychol. Psychiat. 22:375-392.
1708. Schacter, D.L.(1987). Memory, amnesia, and frontal lobe dysfunction. Psychobiology 15:21-36.
1709. Schain, R.J. and Freedman, D.X.(1961). Studies on 5-hydroxyindole metabolism in autistic and other mentally retarded children. J. Pediatr. 58:315-320.
1710. Schalling, D., Edman, G. and Asberg, M.(1983). Impulsive cognitive style and inability to tolerate boredom. In M. Zuckerman (ed) Biological Bases of Sensation Seeking, Impulsivity and Anxiety. Erlbaum Assoc., New York.
1711. Scharff, D.E. and Scharff, J.S.(1987). Object Relations Family Therapy. Jason Aronson, Inc., New York.
1712. Schaumburg, H., Kaplan, J., Windebank, A., Vick, N., Rasmus, S., Pleasure D. and Brown, J.J.(1983). Sensory neuropathy from pyridoxine abuse. A new megavitamin syndrome. N. Engl. J. Med. 309:445-448.
1715. Schickit, M.A.(1973). Alcoholism and sociopathy —Diagnostic confusion. Q. J. Stud. Alcohol. 34:157-164.
1716. Schickit, M.A., Goodwin, D.W. and Winokur, G.(1972). The half-sibling approach in a genetic study of alcoholism. In M. Roff, L.N. Rolins and M. Pollack (eds) Life History Research in Psychopathology (Vol. 2). University of Minneapolis Press, Minneapolis.
1717. Schildkraut, J.J. and Kety, S.S.(1967). Biogenic amines and emotion. Science 156:21-30.
1718. Schildkraut, J.J.(1965). The catecholamine hypothesis of affective disorders: A review of supporting evidence. Am. J. Psychiatry 122:509-522.
1719. Schildkraut, J.J., Dodge, G.A. and Logue, M.A.(1969). Effects of tricyclic antidepressants on the uptake and metabolism of intracisternally administered norepinephrine-^{3}H in rat brain. J. Psychiatr. Res. 7:29-34.
1720. Schlusinger, F.(1972). Psychopathy: Heredity and environment. Int. J. Ment. Health 1:190-206.
1721. Schmale, H. and Richter, D.(1984). Single base deletion in the vasopressin gene is the cause of diabetes insipidis in Brattleboro rats. Nature 308:705-709.
1722. Schmauss, C. and Emrich, H.M.(1985). Dopamine and the action of opiates: A reevaluation of the dopamine hypothesis of schizophrenia. Biol. Psychiatry 20:1211-1231.
1722a. Schmid, W., Scherer, G., Danesch, U., Zentgraf, H., Matthias, P., Strange, C.M., Rowekamp, W. and Schutz, G. (1982). Isolation and characterization of the rat tryptophan oxygenase gene. EMBO Journal 1:1287-1293.
1723. Schmideberg, M.(1947). The treatment of psychopaths and borderline patients. Am. J. Psychotherapy. 1:45-55.
1723a. Schmidt, A.W. and Peroutka, S.J.(1989). 5-Hydroxytryptamine receptor "families." FASEB J. 3:2242-2249.
1724. Schmitt, B.D.(1975). The minimal brain dysfunction myth. Am. J. Dis. Child. 129:1313-1318.
1725. Schmitt, F.O.(1984). Molecular regulation of brain function: A new view. Neuroscience 13:991-1001.
1726. Schmitt, H., Schmitt, H. and Fenard, S.(1971). Evidence for an a-sympathomimetic component in the effects of catapresan on vasomotor centres: Antagonism by piperoxane. Eur. J. Pharmacol. 14:98-100.
1727. Schneider, K.(1957). Primärie und sekundärie Symptom bei der Schizophrenie. Fortschr. Neurol. Psychiatr. 25:487-490.
1728. Schneider, K.(1959). Clinical Psychopathology (translation by Hamilton). Grune & Straton, Inc, New York.
1728a. Schenider, K.(1974). Primary and secondary symptoms in schizophrenia. In S.R. Hirsch and M. Shepherd (eds) Themes and Variations in Psychiatry. J. Wright, Bristol.
1729. Schoch, P., Richards, J.G., Haring, P., Takacs, B., Stahli, C., Staehelin, T., Haefey, W. and Moher, H.(1985). Colocalization of GABAA receptors and benzodiazepine receptors in the brain shown by monoclonal antibodies. Nature 314:168-171.
1730. Schofield, P.R., Darlism, M.G., Fujita, N., Purt, D.R. [8] (1987). Sequence and functional expression of the GABAA receptor shows a ligand-gated receptor super-family. Nature 328:221-227.
1731. Schooler, C., Zahn, T.P., Murphy, D.L. and Buchsbaum, M.S.(1978). Psychological correlates of monamine oxidase activity in normals. J. Nerv. Ment. Dis. 166:177-186.
1732. Schreiner, L and Kling, A.(1953). Behavioral changes following rheincephalic injury in cat. J. Neurophysiol. 16: 643-659.
1733. Schubert, J.(1973). Effect of chronic lithium on monamine metabolism in rat brain. Psychopharmacology 32:301-311.
1734. Schuster, C.R. and Thompson, T.(1969). Self administration and behavioral dependence on drugs. Ann. Rev. Pharmacol. 9:483-502.
1735. Schwarcz, R., Okuno, E., White, R.J., Bird, E.D. and Whetsell, W.O., Jr.(1988). 3-hydroxyanthranilate oxygenase activity is increased in the brains of Huntington disease victims. Proc. Natl. Acad. Sci. U.S.A. 85:4079-4081.
1736. Schwarcz, R., Whetsell, W.O. Jr., and Mangano, R.M.(1983). Quinolinic acid: An endogenous metabolite that produces axon-sparing lesions in rat brain. Science 219:316-318.
1737. Schwartz, G.E., Davidson, R.J. and Maer, F.(1975). Right hemisphere lateralization for emotion in the human brain: Interactions with cognition. Science 190:286-288.
1738. Schwartz, J.-C.(1979). Opiate receptors on catecholaminergic neurons in brain. Trends Neurosci. 2:137-139.
1738a. Scott, D.H.(1959). Delinquency. Adv. Sci. No 61:497-505.
1739. Scoville, W.B. and Milner, B.(1957). Loss of recent memory after bilateral hippocampal lesoins. J. Neurol. Neurosurg. Psychiatry 20:11-21.
1740. Scoville, W.B.(1954). The limbic lobe in man. J. Neurosurg. 11:64-66.
1741. Sedvall, G.C. and Wode-Helgodt, B.(1980). Aberrant monamine metabolite levels in CSH and family history of schizophrenia. Arch. Gen. Psychiatry 37:1113-1116.
1742. Sedvall, G.C., Fyrö, B., Gullberg, B. et al.(1980). Relationships in healthy volunteers between concentrations of monamine metabolites in cerebrospinal fluid and family his

tory of psychiatric morbidity. Br. J. Psychiatry 136:366-374.
1743. Seeman, P. Lee, T., Chau-Wong, M. and Wong, K.(1976). Antipsychotic drug doses and neuroleptic dopamine receptors. Nature 261:717-719.
1744. Seeman, P.(1987). Dopamine receptors and the dopamine hypothesis of schizophrenia. Synapse 1:133-152.
1745. Seeman, P., Chau-Wong, M. and Lee, T.(1974). Dopamine receptor-block and nigral fiber impulse blockade by major tranquilizers. Fed. Proc. 33:246.
1746. Seeman, P., Chau-Wong, M., and Tedesco, J.(1975). Brain receptors for antipsychotic drugs and dopamine: Direct binding assays. Proc. Natl. Acad. Sci. U.S.A. 72:4376-4380.
1747. Segal, M. and Bloom, F.E.(1976). The action of norepinephrine in the rat hippocampus. IV. The effects of locus coeruleus stimulation on evoked hippocampal unit activity. Brain Res. 107:513-525.
1748. Segal, M.(1975). Physiological and pharmacological evidence for a serotonergic projection to the hippocampus. Brain Res. 94:115-131.
1749. Seignot, M.J.N.(1961). Un cas de maladie des tics de Gilles de la Tourette gueri par le R-1163. Ann. Med. Psychol. 119:578-579.
1756. Shafii, M.(1986). The effects of sympathomimetic and antihistaminic agents on chronic motor tics and Tourette's syndrome. N. Engl. J. Med. 315:1228-1229.
1757. Shapiro, A.K. and Shapiro, E.S.(1977). Subcategorizing Gilles de la Tourette's syndrome. Am. J. Psychiatry 134:818-819.
1758. Shapiro, A.K. and Shapiro, E.S.(1981). Clonidine and haloperidol in Gilles de la Tourette syndrome. Arch. Gen. Psychiatry 38:1183-1184.
1759. Shapiro, A.K. and Shapiro, E.S.(1981). Do stimulants provoke, cause, or exacerbate tics and Tourette syndrome? Compr. Psychiatry 22:265-273.
1760. Shapiro, A.K. and Shapiro, E.S.(1981). The treatment and etiology of tics and Tourette's syndrome. Compr. Psychiatry 22:193-205.
1761. Shapiro, A.K. and Shapiro, E.S.(1982). Tourette syndrome: History and present status. In A.J. Friedhoff and T. N. Chase (eds) Gilles de la Tourette Syndrome. Raven Press, New York, pp17-23.
1762. Shapiro, A.K. and Shapiro, E.S.(1984). Controlled study of pimozide vs placebo in Tourette's syndrome. J. Am. Acad. Child Psychiatry 23:161-173.
1763. Shapiro, A.K., Shapiro, E.S. and Eisenkraft, G.J.(1983). Treatment of Gilles de la Tourette syndrome with pimozide. Am. J. Psychiatry 140:1183-1186.
1764. Shapiro, A.K., Shapiro, E.S. and Eisenkraft, G.J.(1983). Treatment of Gilles de la Tourette's syndrome with clonidine and neuroleptics. Arch. Gen. Psychiatry 40:1235-1240.
1765. Shapiro, A.K., Shapiro, E.S. and Wayne, H.(1973). Treatment of Tourette's syndrome with haloperidol, review of 34 cases. Arch. Gen. Psychiatry 28:92-96.
1766. Shapiro, A.K., Shapiro, E.S. and Wayne, H.L.(1973). The symptomology and diagnosis of Gilles de la Tourette's syndrome. J. Am. Acad. Child. Psychiatry 12:702-723.
1767. Shapiro, A.K., Shapiro, E.S., Bruun, R.D. and Sweet, R.D.(1978). Gilles de la Tourette Syndrome. Raven Press, New York, 435pp.
1768. Shapiro, A.K., Shapiro, E.S., Bruun, R.D., Sweet, R., Wayne, H. and Solomon, G.(1976). Gilles de la Tourette's syndrome: Summary of clinical experience with 250 patients and suggested nomenclature for tic syndromes. Adv. Neurol. 14:277-283.
1769. Shapiro, A.K., Shapiro, E.S., Wayne, H. and Clarkin, J.(1972). The psychopathology of Gilles de la Tourette's syndrome. Am. J. Psychiatry 129:427-434.
1770. Shapiro, A.K., Shapiro, E.S., Wayne, H. and Clarkin, J.(1973). Organic factors in Gilles de la Tourette's syndrome. Br. J. Psychiatry 122:659-664.
1771. Shapiro, A.K., Shapiro, E.S., Young, J.G. and Feinberg, T.E.(1988). Gilles de la Tourette Syndrome. Raven Press, New York, 558pp.
1771a. Shapiro, E., Shapiro, A.K., Fulop, G., Hubbard, M., Mandeli, J., Nordie, J. and Phillips, R.A.(1989). Controlled study of haloperidol, pimozide, and placebo for the treatment of Gilles de la Tourette's syndrome. Arch. Gen. Psychiatry 46:722-730.
1772. Sharma, J.N., Sandrew, B.B., and Wang, S.C.(1978). CNS sites of clonidine induced hypotension: A microiontophoretic study of bulbar cardiovascular neurons. Brain Res. 151:127-133.
1773. Sharma, S.K., Klee, W.A. and Nirenberg, M.(1975). Dual regulation of adenylate cyclase accounts for narcotic dependence and tolerance. Proc. Natl. Acad. Sci. U.S.A. 72:3092-3096.
1774. Sharpless, S. and Jasper, H.(1956). Habituation of the arousal reaction. Brain 79:655-680.
1775. Shaw, D.M., Tidmarsh, S.F. and Karajgi, B.M.(1980). Tryptophan, affective disorder and stress. J. Affect. Disord. 2:321-325.
1776. Shaywitz, B.A., Cohen, D.J. and Bowers, M.B.(1977). CSF monamine metabolites in children with minimal brain dysfunction: Evidence for alteration of brain dopamine. J. Pediatr. 90:67-71.
1777. Shaywitz, B.A., Klopper, J.H. and Gordon, J.W.(1978). Methyphenidate in 6-hydroxydopamne-treated developing rat pups. Arch. Neurol. 35:463-469.
1778. Shaywitz, B.A., Klopper, J.H., Yager, R.D. and Gordon, J.W.(1976). Paradoxical response to amphetamine in developing rats treated with 6-hydroxydopamine. Nature 261:153-155.
1779. Shaywitz, B.A., Yager, R.D. and Klopper, J.H.(1976). Selective brain dopamine depletion in developing rats: An experimental model of minimal brain dysfunction. Science 191:305-307.
1780. Sheard, M.(1973). Aggressive behavior: Modification by amphetamine, p-chlorophenylalanine and lithium in rats. Aggressologie 14:323-326.
1781. Sheard, M.H., Marini, J.L., Bridges, C.I. and Wagner, E.(1976). The effects of lithium on impulsive aggressive behavior in man. Am. J. Psychiatry 133: 1409-1413.
1782. Sheinkin, D., Schachter, M. and Hutton, R.(1979). Food, Mind and Mood. Warner Books, New York.
1782a. Shekim, W.O.(1988). Diagnosis & treatment of adults with attention-deficit hyperactivity disorder (AADHD). Scientific Proceedings of the Annual Meeting of the American Academy of Child and Adolescent Psychiatry 4:9.
1783. Shekim, W.O., Davis, L.G., Bylund, D.B., Brunngraber, E., Fikes, L. and Lanham, J.(1982). Platelet MAO in children with attention deficit disorder and hyperactivity: A pilot study. Am. J. Psychiatry 139:936-938.
1784. Shenkenberg, G.E., Dustman, R.E. and Beck, E.C.(1971). Changes in evoked responses related to age, hemisphere and sex. Electroenceph. Clin. Neurophysiol.

30:163-164.
1784a. Sherrington, R., Brynjolfsson, Petursson, H., Potter, M., Dudleston,K., Barraclough, B., Wasmuth, J., Dobbs, M. and Gurling, H.(1988). Localization of susceptibility locus for schizophrenia on chromosome 5. Nature 336:164-167.
1785. Shetty, T. and Chase, T.N.(1976). Central monamines and hyperkinesis of childhood. Neurology 26:1000-1002.
1786. Shields, J.(1975). Some recent developments in psychiatric genetics. Arch. Psychiatr. Nevenkranken 220:347-360.
1787. Shillito, E.E.(1970). The effect of para chlorophenylalanine on social interaction of male rats. Br. J. Pharmacol. 38:305-315.
1788. Shopsin, B., Friedman, E. and Gershon, S.(1976). Parachlorophenylalanine reversal of tranylcypromine effects in depressed patients. Arch. Gen. Psychiatry 33:811-819.
1789. Shopsin, B., Wilk, S., Sathananthan, C., Gershon, S.,and Davis, K.(1974). Catecholamines and affective disorders. A critical assessment. J. Nerv. Ment. Disord. 158:369-383.
1790. Shore, P.A.(1962). Release of serotonin and catecholamines by drugs. Pharmacol. Rev. 14:531-550.
1791. Shrag, P., Divoky, D.(1975). The Myth of the Hyperactive Child. Dell Publishing Co., New York.
1793. Sicuteri, F., Del Bianco, P., Franchi, G., Anselmi, B., Panconesi, A. and Curradi, C.(1983). Loss of opioid homeostasis in idiopathic pain sufferers: The rationale to antienkaphalinase treatment. In S. Ehrenpreis and F. Sicuteri (eds) Degradation of Endogenous Opioids: Its Relevance in Human Pathology and Therapy. Raven Press, New York, pp189-198.
1794. Siever, L. and Gunderson, J.G.(1979). Genetic determinants of borderline conditions. Schizophr. Bull. 5:59-86.
1795. Siever, L.J., Murphy, D.L., Slater, S., de la Vwerga, E. and Lipper, S.(1984). Plasma prolactin changes following fenfluramine in depressed patients compared to controls: An evaluation of central serotonergic responsivity in depression. Life Sci. 34:1929-1939.
1795a. Sigg, E.B.(1959). Pharmacological studies with Tofranil. Can. Psychiat. Assoc. J. 4(Suppl 1):75-85.
1796. Sills, M.A., Wolfe, B.B. and Frazer, A.(1984). Determination of selective and nonselective compounds for the 5-HT1A and 5-HT1B receptor subtypes in rat frontal cortex. J. Pharmacol. Exp. Ther. 231:480-487.
1797. Silver, L.B.(1981). The relationship between learning disabilities, hyperactivity, distractibility, and behavioral problems. J. Am. Acad. Child Psychiatry 20:385-397.
1798. Silverstein, F., Smith, C.B. and Johnston, M.V.(1985). Effect of clonidine on platelet alpha$_2$-adrenoreceptors and plasma norepinephrine of children with Tourette syndrome. Dev. Med. Child Neurol. 27:793-799.
1799. Silverstein, F.S., Parrish, M.A. and Johnston, M.V.(1982).Adverse behavioral reactions in children treated with carbamazepine (Tegretal). J. Pediatr. 5:785-787.
1800. Simantov, R., Kuhar, M.J., Uhl, G. and Syndre, S.H.(1977). Opioid peptide enkephalin: Immunohistochemical mapping in the rat central nervous system. Proc. Natl. Acad. Sci. U.S.A. 74: 2167-2171.
1801. Simon, E.J., Hiller, J.M. and Edelman, I.(1973). Stereospecific binding of the potent narcotic analgesic [^{3}H] etorpine to rat-brain homogenate. Proc. Natl. Acad. Sci. U.S.A. 70:1947-1949.
1802. Simon, H. and LeMoal, M.(1984). Mesencephalic dopaminergic neurons: Functional role. In Catecholamines: Neuropharmacology and Central Nervous System — Theoretical Aspects. Alan R. Liss, Inc., New York, pp293-307.
1803. Simon, H., Scatton, B., and Le Moal, M.L.(1980). Dopaminergic A10 neurons are involved in cognitive functions. Nature 286:150-151.
1804. Simon, H., Stinus, L., Tassin, J.P., Lavielle, S., Blanc, G., Thierry, A.M., Glowinski, J. and Le Moal, M.(1979). Is the dopaminergic mesocorticolimbic system necessary for intracranial self-stimulation? Biochemical and behavioral studies from A10 cell bodies and terminals. Behav. Neural Biol. 27:125-145.
1805. Singer, H.S., Gammon, K. and Quaskey, S.(1985-86). Haloperidol, fluphenazine and clonidine in Tourette syndrome: Controversies in treatment. Pediatr. Neurosci. 12:71-74.
1806. Singer, H.S., Tune, L.E., Butler, I.J., Zaczek, R. and Coyle, J.T.(1982). Clinical symptomatology, CSF neurotransmitter metabolites, and serum haloperidol levels in Tourette syndrome. In A.J. Friedhoff and T.N. Chase (eds) Gilles de la Tourette Syndrome. Raven Press, New York, pp177-183.
1807. Singer, H.S., Wong, D.F., Tiemeyer, M., Whitehouse, P. and Wagner, H.N.(1985). Pathophysiology of Tourette syndrome: A positron emission tomographic and postmortem analysis. Ann. Neurol. 18:416.
1808. Singer, W.D.(1981). Transient Gilles de la Tourette syndrome after chronic neuroleptic withdrawal. Dev. Med. Child. Neurol. 23:518-530.
1808a. Singh, S.K. and Jankovic, J. (1988). Tardive dystonia in patients with Tourette's Syndrome. Movement Disorders 3:274-280.
1809. Skinner, B.F.(1965). Science and Human Behavior. Free Press, New York.
1810. Slater, E. and Shields, J.(1969). Genetical aspects of anxiety In M.H. Lader (ed) Studies of Anxiety. Headly Brothers, Ashford, England.
1811. Slater, E.(1953). Psychotic and Neurotic Illnesses in Twins. Her Majesty's Stationery Office, London.
1812. Sleator, E.K.(1980). Deleterious effects of drugs used for hyperactivity on patients with Gilles de la Tourette syndrome. Clin. Pediatr. 19:452-454.
1813. Smeraldi, E., Negri, F., Heimbuch, R.C. and Kidd, K.K.(1981). Familial patterns and possible modes of inheritance of primary affective disorders. J. Affective Disord. 3:173-182.
1814. Smith, J.D.(1985). Minds Made Feeble: The Myth and Legacy of the Kallikaks. Aspen Systems Corp. Rockville, MD.
1815. Smith, J.M. and Baldessarini, R.J.(1980). Changes in prevalence, severity and recovery in Tardive dyskinesia with age. Arch. Gen. Psychiatry 37:1368-1373.
1816. Smith, S.D., Kinberling, W.J., Penningtion, B.F. and Lubs, H.A.(1983). Specific reading disability: Identification of an inherited from through linkage analysis. Science 219:1345-1347.
1817. Snedden, W., Mellor, C.S. and Martin, J.R.(1983). Familial hypertryptophanemia, tryptophanuria and indoleketonuria. Clin. Chim. Acta 131:247-256.
1818. Snyder, S.H.(1976). The dopamine hypothesis of schizophrenia: Focus on the dopamine receptor. Am. J. Psychiatry 130:61-67.
1819. Snyder, S.H.(1977). Opiate receptors and internal opiates. Sci. Am. 236:44-56.

1820. Snyder, S.H., Banerjee, S.P., Yamamura, H.I. and Greenberg, D.(1974). Drugs, neurotransmitters and schizophrenia. Science 184:1243-1253.
1821. Snyder, S.H., Taylor, K.M., Coyle, J.T., et al.(1970). The role of brain dopamine in behavioral regulation and the actions of psychotropic drugs. Am J. Psychiatry 127:117-125.
1822. Sokolov, E.N.(1960). Neuronal model of the orienting reflex. In M.A.B. Brazier (ed) The Central Nervous System and Behavior. J. Macy Jr. Foundation, New York.
1823. Soloff, P.H. and Millward, J.W.(1983). Psychiatric disorders in the families of borderline patients. Arch. Gen. Psychiatry 40:37-44.
1824. Soloff, P.H., George, A., Nathan, S., Schulz, P.M., Ulrich, R.F. and Perel, J.M.(1986). Progress in pharmchotherapy of borderline disorders. Arch. Gen. Psychiatry 43:691-697.
1825. Soubrié, P.(1986). Reconciling the role of central serotonin neurons in human and animal behavior. Behav. Brain Sci. 9:319-364.
1826. Soubrié, P., Reisine, T.D. and Glowinski, J.(1984). Functional aspects of serotonin transmission in the basal ganglia: A review and an in vivo approach using the push-pull canula technique. Neuroscience 13:605-625.
1827. Soulairac, M.-L. and Soulairac, A.(1975). Monoaminergic and cholinergic controls of sexual behavior in the male rat. In M. Sandler and G.L. Gessa (eds) Sexual Behavior: Pharmacology and Biochemistry. Raven Press, New York, pp99-116.
1828. Sourkes, T.L., Missala, K. and Oravec, M.(1970). Decrease of cerebral serotonin and 5-hydroxyindolacetic acid caused by (-)-a-methyltryptophan. J. Neurochem. 17:111-115.
1829. Sovner, R.D.(1981). The clinical characteristics and treatment of atypical depression. J. Clin. Psychiatry 42:285-289.
1830. Spampinato, U., Esposito, E., Romandini, S. and Sumanin, R.(1985). Changes of serotonin and dopamine metabolism in various brain areas of rats injected with morphine systemically or in the raphe nuclei dorsalis and medianus. Brain Res. 328:88-95.
1831. Spealman, R.D. and Goldberg, S.R.(1978). Drug self-administration by laboratory animals: Control by schedule of reinforcement. Annu. Rev. Pharmacol. Toxicol. 18:313-339.
1832. Spector, S., Shore, P.A. and Brodie, B.B.(1960). Biochemical and pharmacological effects of the monamine oxidase inhibitors, iproniazid, 1-phenyl-2- hydrazinopropane (JB516) and 1-phenyl-3-hydrazinobutane (JB835). J. Pharmacol. Exp. Ther. 128:15-21.
1833. Spencer, H.J.(1976). Antagonism of cortical excitation of striatal neurons by glutamic acid diethyl ester: Evidence of glutamic acid as an excitatory transmitter in the rat striatum. Brain Res. 102:91-101.
1834. Spiegel, E.A., Miller, H.R. and Oppenheimer, M.J.(1940). Forebrain and rage reactions. J. Neurophysiol. 3:538-548.
1834a. Spindel, E.R., Chin, W.W., Price J., Rees, LH, Besser, G.M. and Habener, J.F. (1984). Cloning and characterization of cDNAs encoding human gastrin-releasing peptide. Proc. Natl. Acad. Sci. U.S.A. 81:5699-5703.
1835. Spitzer, R.L., Endicott, J. and Gibbon, M.(1979). Crossing the border into borderline personality and borderline schizophrenia. Arch. Gen. Psychiatry 36:17-24.
1836. Splitter, S.R. and Kaufman, M.(1966). A new treatment for underachieving adolescents: Psychotherapy combined with nortriptyline medication. Psychosomatics 7:171-174.
1837. Spooner, C.E. and Winters, W.D.(1965). Evidence for a direct action of nonamines on the chick central nervous system. Experientia 21:256-258.
1838. Spooner, C.E. and Winters, W.D.(1967). Evoked responses during spontaneous and nonamine-induced states of wakefulness and sleep. Brain. Res. 4: 189-205.
1838a. Sprague, R. and Sleator, E.(1977). Methylphenidate in hyperkinetic children: Differences in dose effects on learning and social behavior. Science 198:1274-1276.
1839. Sprague, R., Barnes, K. and Werry, J.(1970). Methylphenidate and thioridazine: Learning, activity, and behavior in emotionally disturbed boys. Am. J. Orthopsychiatry 40:615-628.
1840. Spring, B., Chiodo, J. and Bowen, D.J.(1987). Carbohydrates, tryptophan, and behavior: A methodological review. Psychol. Bull. 102:234-256.
1841. Spring, C., Greenberg, L., Scott, J. et al.(1974). Electrodermal activity in hyperactive boys who are methylphenidate responders. Psychophysiology 11:436-442.
1842. Squire, L.R.(1980). The anatomy of amnesia. Trends Neurosci. 3:52-54.
1843. Squires, R.F. and Braestrup, C.(1977). Benzodiazepine receptors in rat brain. Nature 266:732-734.
1844. Srebro, B. and Lorens, S.A.(1975). Behavioral effects of selective midbrain raphe lesons in the rat. Brain Res. 89:303-325.
1845. Sroufe, L.A. and Stewart, M.A.(1973). Treating problem children with stimulant drugs. N. Engl. J. Med. 289:407-413.
1846. Stabenau, J.R. and Hesselbrock, V.(1980). Assorative mating, family pedigree and alcoholism. Substance and Alcohol Actions/Abuse. 1:375-382.
1847. Stabenau, J.R.(1977). Genetic and other factors in schizophrenic, manic-depressive, and schizoaffective psychoses. J. Nerv. Ment. Dis. 164:149-167.
1848. Stabenau, J.R.(1984). Implications of family history of alcoholism, antisocial personality, and sex differences in alcohol dependence. Am. J. Psychiatry. 141:1178-1182.
1849. Stabenau, J.R.(1987). Independent additive risk factors in alcoholism. (personal communication).
1850. Stacey, W.A. and Shupe, A.(1983). The Family Secret. Domestic Violence in America. Beacon Press, Boston, 237pp.
1851. Stahl, S.M.(1980). Tardive Tourette syndrome in an autistic patient after long-term neuroleptic administration. Am. J. Psychiatry 137:1267-1269.
1852. Stahl, S.M., Woo, D.J., Mefford, I.N., Berger, P.A., and Ciaranello, R.D.(1983). Hyperserotonemia and platelet serotonin uptake and release in schizophrenia and affective disorders. Am. J. Psychiatry 140:26-30.
1853. Stamm, J.S. and Kreder, S.V.(1979). Minimal brain dysfunction: Psychological and neurophysiological disorders in hyperkinetic children. In Gazzaniga, M.S.(ed)Handbook of Behavioral Neurobiology, Vol 2. Plenum Press, New York. pp119-150.
1854. Stamm, J.S.(1955). The function of the median cerebral cortex in maternal behavior of rats. J. Compar. Physiol. Psychol. 48:347-356.
1855. Stanley, M. and Mann, J.J.(1983). Increased sero-

tonin-2 binding in frontal cortex of suicide victims. Lancet 1:214-216.
1856. Starke, K.(1987). Presynaptic a-autoreceptors. Rev. Physiol. Biochem. 107:74-145.
1857. Start, P., Fuller, R.W. and Wong, D.T.(1985). The pharmacologic profile of fluoxetine. J. Clin. Psychiatry 46:7-13.
1858. Stefl, M.E. and Rubin, M.(1985). Tourette syndrome in the classroom: Special problems, special needs. J. School Health 55:72-75.
1859. Stefl, M.E.(1984). Mental health needs associated with Tourette syndrome. J. Public Health 74:1310-1313.
1860. Stein, G., Milton, F., Bebbington, P., Wood, K. and Coppen, A.(1976). Relationship between mood disturbances and free and total plasma tryptophan in pospartum women. Br. Med. J. 2:457.
1861. Stein, L. and Wise, D.C.(1974). Serotonin and behavioral inhibition. In E. Costa, G.L. Gessa and M. Sandler (eds) Serotonin — New Vistas. Adv. Biochem. Pharmacol. 11:281-292.
1862. Stein, L.(1983). Benzodiazepines and behavioral disinhibition. In E. Usdin, P. Skolnick, J.F. Tallman, Jr., D. Greenblatt and S.M. Paul (eds) Pharmacology of Benzodiazepines. Verlag Chemie, Basel, pp383-390.
1863. Stein, L., Wise, C.D. and Belluzzi, J.D.(1975). Effects of benzodiazepines on central serotonic mechanisms. In E. Costa and P. Greengard (eds) Mechanisms of Action of Benzodiazepines. Adv. Biochem. Psychopharmacol. 14: 29-44.
1864. Stein, L., Wise, C.D. and Berger, B.D.(1973). Antianxiety action of benzodiazepines. Decrease in activity of serotonin neurons in punishment system. In P.S. Garattini, G. Mussini and L.D. Randall (eds) The Benzodiazepines. Raven Press, New York, pp299-326.
1865. Stein, Z. and Susser, M.(1966). Nocturnal enuresis as a phenomenon of institutions. Dev. Med. Child Neurol. 8:677-685.
1866. Stein, Z. and Susser, M.(1967). Social factors in the development of sphincter control. Dev. Med. Child Neurol. 9:692-706.
1867. Steinbusch, H.W.M.(1981). Distribution of serotonin-immunoreactivity in the central nervous system of rat-cell bodies and terminals. Neuroscience 6:557-618.
1868. Steinbusch, H.W.M., Nieuwenhuys, R., Verhofstad, A.A.J. and Van Der Kody, D.(1981). The nucleus raphe dorsalis of the rat and its projection upon the caudatoputamen. A combined cytoarchitectonic, immunohistochemical and retrograde transport study. J. Physiol. (Paris) 77:157-174.
1869. Stengel, E.(1945). A study on some clinical aspects of the relationship between obsessional neurosis and psychotic reaction types. J. Ment. Sci. 91:166-187.
1869a. Stephens, R.S., Pelham, W.E. and Skinner, R. (1984). State-dependent and main effect of methylphenidate and pemoline on paired-associate learning and spelling in hyperactive children. J. Counsult. Clin. Psychol. 52:104-113.
1870. Stephenson, S.(1907). Six cases of congenital word blindness affecting three generations of one family. The Ophthalmoscope 5:482-484.
1871. Stern, A.(1938). Psychoanalytic investigation and therapy in the borderline group of neuroses. Psychoanal. Q. 7:467-489.
1872. Stern, R.S., Marks, I.N., and Mawson, D.(1980). Clomipramine and exposure for compulsive rituals. Plasma levels, sided effects and otucome. Br. J. Psychiatry 135:161-166.
1873. Stevens, C.(1979). The neuron. Sci. Am. 241: (September): 54-65.
1874. Stevens, J. R., Wilson, K. and Foote, W.(1974). GABA blockade, dopamine and schizophrenia: Experimental studies in the cat. Psychopharmacology 39:105-119.
1875. Stevens, J.R. and Livermore, A.(1978). Kindling of the mesolimbic dopamine system: Animal model of psychosis. Neurology 28:36-46.
1876. Stevens, J.R.(1973). An anatomy of schizophrenia? Arch. Gen. Psychiatry 29:177-189.
1877. Stevens, J.R., Sachder, K. and Milstein, V.(1968). Behavior disorders of childhood and the electroencephalogram. Arch. Neurol. 18:160-177.
1878. Stewart, M.A.(1970). Hyperactive children. Sci. Am. 222:94-98.
1879. Stewart, M.A.(1985). Aggressive conduct disorder. A Brief Review. Aggres. Behav. 11:323-331.
1880. Stewart, M.A., Cummings, C., Singer, S. and deBlois, C.S.(1981). The overlap between hyperactive and unsocialized aggressive children. J. Child. Psychol. Psychiatry 22:35-45.
1881. Stewart, M.A., DeBlois, C.S. and Cummings, C.(1980). Psychiatric disorder in the parents of hyper-active boys and those with conduct disorder. J. Child Psychol. Psychiatry 21:283-292.
1882. Stewart, M.A., DeBlois, S., Meardon, J., and Cummings, C.(1980). Aggressive conduct disorder of children. The clinical picture. J. Nerv. Ment. Dis. 168:604-610.
1883. Stewart, M.A., Mendelson, W.B. and Johnson, N.E.(1982). Hyperactive children as adolescents: How they describe themselves. Child Psychiatry Hum. Dev. 4:3-11.
1884. Stewart, M.A., Pitts, F., Graig, A. and Dieruf, W.(1966). The hyperkinetic child syndrome. Am. J. Orthopsychiatry 36:861-867.
1885. StGeorge, P.H., Tanzi, R.E., Polinski, R.J., Neve, R.L.et al.(1987). Absence of duplication of chromosome 21 genes in familial and sporadic Alzheimer's disease. Science 238:664-669.
1886. Still, G.F.(1902). Some abnormal psychical conditions in children. Lancet 1:1077-1082.
1887. Stinus, L., Gaffori, O., Simon, H. and LeMoal, M.L.(1977). Small doses of apomorphine and chronic administration of d-amphetamine reduce locomotor hyperactivity produced by radiofrequency lesions of dopaminergic A10 neurons area. Biol. Psychiatry 12:719-732.
1888. Stinus, L., Koob, G.F., Ling, N., Bloom, F.E. and Le Moal, M.(1980). Locomotor activation induced by infusion of endorphins into the ventral tegmental area: Evidence for opiate-dopamine interactions. Proc. Natl. Acad. Sci. U.S.A. 77:2323-2327.
1889. Stinus, L., Simon, L. and Le Moal, M.(1978). Disappearance of hoarding and disorganization of eating behavior after ventral mesencephalic tegmentum lesion in rats. J. Comp. Physiol. Psychol. 92:289-296.
1890. Stockmeier, C.A., Martino, A.M. and Kellar, K.J.(1985). A strong influence of serotonin axons on beta-adrenergic receptors in rat brain. Science 230:323-325.
1890a. Stokes, M.D., Bawden, H., Camfield, P., Backman, J. and Dooley, J.(1988). Factors associated with the adjustment of children with Tourette's disorder. Scientific Proceedings of the Annual Meeting of the American Academy of Child and Adolescent Psychiatry 4:47.

1891. Stone, J.L., McDaniel, K.D., Hughes, J.R. and Hermann, B.P.(1986). Episodic dyscontrol disorder and paroxysmal EEG abnormalities: Successful treatment with carbamazepine. Biol. Psychiatry 21:208-212.
1892. Stone, M.H.(1980). The Borderline Syndromes. McGraw-Hill Book Co., New York, 553pp.
1893. Stone, T.W. and Connick, J.H.(1985). Quinolinic acid and other kynurenines in the central nervous system. Neuroscience 15:597-617.
1894. Stoof, J.C., Dijkstra, H. and Hillegers, J.P.M.(1978). Changes in the behavioral response to a novel environment following lesions of the central dopaminergic system in rat pups. Psychopharmacology 57:163-166.
1896. Strauss, A.A. and Werner, H.(1942). Disorders of conceptual thinking in the brain-injured child. J. Nerv. Ment. Dis. 96:153-172.
1897. Strauss, A.A.(1944). Ways of thinking in brain-crippled deficient children. Am J. Psychiat. 100:639-647.
1898. Stubbs, E.G.(1978). Autistic symptoms in a child with congenital cytomegalovirus infection. J. Autism Child. Schizophr. 8:37-43.
1899. Stunkard, A.J., Sorensen, T.I.A., Hanis, C., Teasdale, T.W., Chakraborty, R., Schull, W.J. and Schulsinger, F.(1986). An adoption study of obesity. N. Engl. J. Med. 314:193-198.
1900. Stuss, D.T. and Benson, D.F.(1984). Neuropsychological studies of the frontal lobes. Psychol. Bull. 95:3-28.
1901. Sudzak, P.D., Schwartz, R.D., Skolnick, P. and Paul, S.M.(1986). Ethanol stimulates γ-aminobutyric acid receptor-mediated chloride transport in rat brain synaptosomes. Proc. Natl. Acad. Sci. U.S.A. 83:4071-4075.
1901a. Sulser, F., Gillespie, D.D., Mishra, R. and Manier, D.H.: Desensitization by antidepressant of central norepinephrine receptor systems coupled to adenylate cyclase. Ann. N.Y.Acad. Sci. 430:91-101, 1984.
1902. Summers, W.K., Majovski, L.V., Marsh, G.M., Tachiki, K. and Kling, A.(1986). Oral tetrahydroaminoacridine in long-term treatment of senile dementia, Alzheimer type. New Engl. J. Med. 315:1241-1245.
1903. Sutherland, R.J., Kolb, B., Schoel, W.M., I.Q. Whishaw, I.Q. and Davies, D.(1982). Neuropsychological assessment of children and adults with Tourette sydnrome: A comparison with learning disabilities and schiizophrenia. In A.J. Friedhoff and T. N. Chase (eds) Gilles de la Tourette Syndrome. Raven Press, New York, pp311-322.
1904. Svensson, T.H., Bunney, B.S., and Aghajanian, G.K.(1975). Inhibition of both noradrenergic and serotonergic neurons in brain by the a-agonist clonidine. Brain Res. 92:291-306.
1905. Svensson, T.H., Persson, R., Wallin, L. and Walinger, J.(1978). Anxiolytic action of clonidine. Nord Psychiatry Tidsskr 32:439-441.
1905a. Sverd, J. and Kupietz, S. (1984). Effects of high dose propanolol in Tourette syndrome. J. Clin. Psychopharmacol. 4:359-361.
1905b. Sverd, J. and Nolan, E.(1989). Gilles de la Tourette syndrome, autistic disorder, neuropsychiatric disturbance: Is there an etiological relationship? First World Congress on Psychiatric Genetics, August 3-5, Churchill College, Cambridge, England, p85.
1907. Sverd, J.(1988). Imipramine treatment of panic disorder in a prepubertal boy with Tourette syndrome. J. Clin. Psychiatry 49:31-32.
1907a. Sverd, J.(1989). Tourette syndrome associated with pervasive developmental disorder: Is there an etiological relationship? J. Multihandicapped Person (in press).
1907b. Sverd, J.(1989). Clinical presentations of the Tourette's syndrome diathesis. (submitted).
1907b. Sverd, J., Cohen, S. and Camp. J.A. (1983).Brief report. Efects of propanolol in Tourette syndrome. J. Autism Dev. Dis. 13:207-213.
1908. Sverd, J., Curley, A.D., Jandorf, L. and Volker, L.(1988). Behavior disorder and attention deficits in boys with Tourette syndrome. J. Am. Acad. Child Adolesc. Psychiatry 27:413-417.
1909. Sverd, J., Gadow, K.D. and Paolicelli, L.M.(1989). Methylphenidate treatment of attention- deficit hyperactivity disorder in boys with Tourettte syndrome. J. Am. Acad. Child Psychiatry 28:574-579.
1909a. Swanson, J. and Kinsbourne, M. (1979). The cognitive effects of stimulant drugs on hyperactive (inattentive) children. In G. Hale and M. Lewis (eds) Attention and the Development of Cognitive Skills. Plenum, New York, pp249-274.
1909b. Swedo, S.E., Rapoport, J.L., Leonard, H., Lenane, M. and Cheslow, D.(1989). Obsessive-compulsive disorder in children and adolescents. Arch. Gen. Psychiatry 46:335-341.
1909c. Swedo, S.E., Schapiro, M.B., Grady, C.L., Cheslow, D.L., Leonard, H.L., Kumar, A., Friedland, R., Rapoport, S.I. and Rapoport, J.L.(1989). Cerebral glucose metabolism in childhood-onset obsessive-compulsive disorder. Arch. Gen. Psychiatry 46:518-523.
1909d. Swedo, S.E, Leonard,H.L., Rapoport, J.L., Lenane, M.C., Goldberger, E.L. and Cheslow, D.L.(1989). A double-blind comparison of clomipramine and desipramine in the treatment of trichotillomaina (hair pulling). N. Eng. J. Med. 321:497-501.
1910. Sweeney, D., Pickar, D., Redmond, D.E., Jr., and Maas, J.(1978). Noradrenergic and dopaminergic mechanisms in Gilles de la Tourette syndrome. Lancet 1:872.
1911. Sweet, R.D., Bruun, R., Shapiro, E. and Shapiro, A.K.(1974). Presynaptic catecholamine antagonist as treatment for Tourette syndrome: Effects alpha-methyl-para-tyrosine and tetrabenzine. Arch. Gen. Psychiatry 31:857-861.
1912. Sweet, R.D., Solomon, G.E., Wayne, H., Shapior, E. and Shapiro, A.K.(1973). Neurological features of Gilles de la Tourette's syndrome. J. Neurol. Neurosurg. Psychiatry 36:1-9.
1913. Sykes, D.H., Douglas, V. and Morgenstern, G.(1972). The effect of methylphenidate (Ritalin) on sustained attention in hyperactive children. Psychopharmacology 25:262-274.
1914. Tabakoff, B. and Hoffman, P.L.(1980). Alcohol and neurotransmitters. In H. Rigter and J.C. Crabbe (eds) Alcohol Tolerance and Dependence. Elsevier Biomedical Press, Amsterdam, pp201-226.
1915. Tagliamonte, A., Tagliamonte, P., Gessa, G.L. and Brodie, B.B.(1969). Compulsive sexual activity induced by p-chlorophenylalanine in normal and pinealectomized male rats. Science 166:1433-1435.
1916. Takahashi, S., Kanai, H., and Miyamoto, Y.(1976). Reassessment of elevated serotonin levels in blood platelets in early infantile autism. J. Autism Child Schizophr. 6:317-326.
1917. Takahashi, S., Yamane, H., Kondi, H. and Tani, N.(1974). CSF monamine metabolites in alco-holism: A comparative study with depression. Folia Psychiatr. Neurol. Jpn. 28:347-354.
1918. Takeuchi, K., Yamashita, M., Morikiyo, M. et al.(1986). Gilles de la Tourette's syndrome and schizophrenia.

J. Nerv. Ment. Dis. 174:247-248.
1919. Tallman, J.F., Paul, S.M., Skolnick, P. and Gallager, D.W.(1980). Receptors for the age of anxiety: Pharmacology of the benzodiazepines. Science 207:274-281.
1920. Tanna, V.L., Wilson, A.F., Winokur, G. and Elston, R.C.(1988). Possible linkage between alcoholism and esterase-D. (personal communication).
1921. Tanzi, R.E., Gusella, J.F., Watkins, P.C. et al(1987). Amyloid b-protein gene: cDNA, mRNA distribution, and genetic linkage near the Alzheimer locus. Science 235:880-884.
1922. Tarter, R.E., Alterman, A.I. and Edwards, K.I.(1985). Vulnerability to alcoholism in men: A behavioral-genetic perspective. J. Stud. Alcohol. 46:329-356.
1923. Tarter, R.E. and Edwards, K.L.(1986). Multifactorial etiology of neurspsychological impairment in alcoholics. Alcoholism: Clin. Exp. Res. 10:128-135.
1924. Tarter, R.E., McBride, H., Buonpane, N. and Schneider, D.U.(1977). Differentiation of alcoholics. Childhood history of mimimal brain dysfunction, family history, and drinking pattern. Arch. Gen. Psychiatry 34:761-768.
1925. Tassin, J.P., Lavielle, S., Hereve, D., Blanc, G., Thierry, A.M., Alvarez, C., Berger, B. and Glowinski, J.(1979). Collateral sprouting and reduced activity of the rat mesocortical dopaminergic neurons after selective destruction of the ascending bundles. Neuroscience 4:1969-1982.
1926. Tassin, J.P., Stinus, L., Simon, H., Blanc, G., Thierry, A.M., Le Moal, M., Cardo, B. and Glowinski, J.(1978). Relationship between the locomotor hyeractivity induced by A10 lesions and the destruction of the frontocortical dopaminergic innervation in the rat brain. Brain Res. 141:267-281.
1927. Taylor, M.A. and Abrams, R.(1981). Early and late-onset bipolar illness. Arch. Gen. Psychiatry 38:58-61.
1928. Teitelbaum, P. and Epstein, A.N.(1962). The lateral hypothalamic syndrome. Recovery of feeding and drinking after lateral hypothalamic lesions. Psychol. Rev. 69:74-90.
1929. Teitelbaum, P.(1955). Sensory control of hypothalamic hyperphagia. J. Compar. Physiol. Psychol. 48:158-163.
1930. Tenen, S.S.(1967). The effects of p-chlorophenylalanine, a serotonin depletor, on avoidance acquisition, pain sensitivity and related behavior in the rat. Psychopharmacology 10:204-206.
1931. Terenius, L.(1973). Characteristics of the 'receptor' for narcotic analgesics and a synaptic plasma membrane fraction from rat brain. Acta Pharmacol Toxicol. 33:377-384.
1932. Terenius, L., Waldstrom, A., Lindstrom, L., Widerlor, E.(1976). Increased CSF levels of endorphins in chronic psychosis. Neurosci. Lett 3:157-162.
1933. Terzian, H. and Ore, G.O.(1955). Syndrome of Klüver and Bucy. Reproduced in man by bilateral removal of the temproal lobes. Neurology 5:373-380.
1934. Teuber. H.L.(1964). The riddle of frontal lobe function in man. In J.M. Warren and K. Akert, (eds) The Frontal Granular Cortex and Behavior. McGraw-Hill, New York.
1935. Thaker, G.K., Tamminga, C.A., Alphs, L.D., Lafferman, J., Ferraro, T.N. and Hare, T.A.: Brain γ-aminobutyric acid abnormality in tardive dyskinesia. Arch. Gen. Psychiatry 44:522-529.
1936. Thiebot, M-E., Hamon, M. and Soubrié, P.(1984). Serotonerigic neurones and anxiety-related behavior in rats. In M.R. Trimble and E. Zarifian (eds) Psychopharmacology of the Limbic System. Oxford Univ. Press, New York, pp164-173.
1937. Thieme, R.E., Dijkstra, H. and Stoff, J.C.(1980). An evaluation of the young dopamine-lesioned rat as an animal model for minimal brain dysfunction (MBD). Psychopharmacol. 67:165-169.
1938. Thierry, A.M., Stinus, L. and Glowinski, J.(1973). Some evidence for the existence of dopaminergic neruons in the rat cortex. Brain Res. 50:230-234.
1939. Thierry, A.M., Stinus, L., Blanc, G., Sobel, A. Stinus, L. and Glowinski, J.(1973). Dopaminergic terminal in the rat cortex. Science 182:499-501.
1939a. Thierry, A.M., Tassin, J.P., Blanc, G. and Glowinski, J.(1976). Effects of stimulation of the mesocortical dopaminergic system by stress. Nature 263:242-244.
1940. Thomas, R.(1987). Inhibition by acidification? Nature 330:110-111.
1941. Thompson, C.I., Gergland, R.M. and Towfight, J.T.(1977). Social and nonsocial behaviors of adult rhesus monkey after amygdalectomy in infancy or adulthood. J. Comp. Physiol. Psychol. 91:533-548.
1942. Thompson, J., Rankin, H., Aschroft, G.W., Yaes, C.M., McQueen, J.K. and Cummings, S.W.(1982). The treatment of depression in general practice. A comparison of L-tryptophan and amitriptyline, and a combination of L-tryptophan and amitryptyline with placebo. Psychol. Med. 12:741-751.
1943. Thompson, R.F., Berger, T.W. and Madden, J.,IV.(1983). Cellular processes of learning and memory in the mammalian CNS. Am. Rev. Neurosci. 6:447-491.
1944. Thoren, P., Åsberg, M., Bertilsson, L., Mellstrom, B., Sjoquist, F., Traskman, L.(1980). Clomipramine treatment of obsessive-compulsive disorder II: Biochemical aspects. Arch. Gen. Psychiatry 37:1289-1294.
1945. Thoren, P., Åsberg, M., Cronholm, B., Jornesledt, L. and Traskman, L.(1980). Clomipramine treatment of obsessive compulsive disorder. A controlled clinical trial. Arch. Gen. Psychiatry 37:1281-1285.
1946. Thornton, E.W. and Goudie, A.J.(1978). Evidence for the role of serotonin in the inhibiton of specific motor responses. Psychopharmacology 60:73-79.
1947. Titeler, M., List, S. and Seeman, P.(1979). High affinity dopamine receptors (D^3) in rat brain. Commun. Pharmacol. 3:411-420.
1949. Todrick, A., Tait, A.C., and Marshall, E.F.(1960). Blood platelet 5-HT levels in psychiatric patients. Br. J. Psychiatry 106:884-890.
1950. Toh, C.C.(1954). Release of 5-hydroxytryptamine (serotonin) from the dog's gastrointestinal tract. J. Physiol. 126:248-254.
1952. Torgersen, S.(1980). Hereditary-environmental differentiation of general neurotic, obsessive, and impulsive hysterical personality traits. Acta. Genet. Med. Gemell. 29:193-207.
1953. Torgersen, S.(1983). Genetic factors in anxiety disorders. Arch. Gen. Psychiatry 40:1085-1089.
1954. Torgersen, S.(1986). Genetic factors in moderately severe and mild affective disorders. Arch. Gen. Psychiatry 43:222-226.
1955. Torrey, E.F.(1974). Schizophrenia and the limbic system. Lancet 2:942-946.
1956. Torup, E.(1962). A followup study of children with tics. Acta Paediatr. Scand. 51:261-268.
1957. Toussieng, P.W.(1962). Thoughts regarding the etiology of psychological difficulties in adopted children.

Child Welfare 41:59-65.
1958. Traskman, L., Asberg, M. Bertilsson, L. and Sjöstrand, L.(1981). Monamine metabolites in CSF and suicidal behavior. Arch. Gen. Psychiatry 38:631-636.
1959. Treiser, S.L., Casio, C., O'Donohue, S., Jacobowitz, D. and Kellar, K.(1981). Lithium increases serotonin release and decreases serotonin receptors in the hippocampus. Science 231:1529-1531.
1960. Trimble, M.R. and Cummings, J.L.(1981). Neuropsychiatric disturbances following brainstem lesions. Br. J. Psychiatry 138:56-59.
1961. Trousseau, A.(1873). Clinique medical dé l'hotel Dieu de Paris. 2:267.
1962. Trulson, M.E. and Jacobs, B.L.(1979). Long-term amphetamine treatment decreases brain serotonin metabolism: Implicatons for theories of schizophrenia. Science 205:1295-1297.
1963. Trulson, M.E. and Jacobs, B.L.(1979). Raphe unit activity in freely moving cats: Correlation with level of behavioral arousal. Brain Res. 163: 135-150.
1964. Tsuang, M.T.(1979). Schizoaffective disorder: Dead or alive? Arch. Gen. Psychiatry 36:633-634.
1965. Tsuda, H., Noguchi, T. and Kido, R.(1972). 5-hydroxytryptophan pyrrolase in brain. J. Neurochem. 19:887-889.
1966. Tsuji, M.(1985). Measurement of platelet monamine oxidase using three different substrates in patients with alcoholism and schizophrenia. Folia Neurol. Jpn. 39:521-530.
1967. Tu, J. and Partington, M.W.(1972). 5-hydroxyindole levels in the blood and CSF in Down's syndrome, phenylketonuria and severe mental retardation. Dev. Med. Child. Neurol. 14:457-466.
1968. Turner, S.M., Beidel, D.C. and Nathan, R.S.(1985). Biological factors in obsessive-compulsive disorders. Psychol. Bull. 97:430-450.
1969. Twito, T.J.and Stewart, M.A.(1982). A half-sibling study of aggressive conduct disorder. Neuropsychobiology 8:144-150.
1970. Tye, N.C., Everitt, B.J. and Iversen, S.D.(1977). 5-hydroxytryptamine and punishment. Nature 268: 741-742.
1971. U'Prichard, D.C. and Snyder, S.H.(1978). Distinct alpha-noradrenergic receptors differentiated by binding and physiological relationships. Life Sci. 24:79.
1971a. Ugedo, L., Greenhoff, J. and Svensson, T.H.(1989). Ritanserin, a 5-HT_2 receptor antagonist, activates midbrain dopamine neurons by blocking serotonergic inhibiton. Psychopharmacology 98:45-50.
1972. Uhde, T.W., Post, R.M., Siever, L., Bushbaum, M.S., Jimerson, D.C., Silberman, E.K., Murphy, D.L. and Bunney, W.E. Jr.(1981). Clonidine: Effects on mood, anxiety, and pain. Psychopharmacol. Bull. 17:125-126.
1973. Uhde, T.W., Siever, L.J. and Post, R.M.(1984). Clonidine: Acute challenge and clinical trial paradigms for the investigation and treatment of anxiety disorder, affective illness, and pain syndromes. In R. M. Post and J.C. Ballenger (eds) Neurobiology of Mood Disorder. Williams and Wilkins, Baltimore, p554-571.
1974. Uhr, S.B., Berger, P.A., Pruitt, B. and Stahl, S.M.(1985). Treatment of Tourette's syndrome with RO22-1319, a D-2 receptor antagonist. N. Engl. J. Med. 311:989.
1975. Uhr, S.B., Pruitt, B., Berger, P.A. and Stahl, S.M.(1986). Case report of four patients with Tourette syndrome treated with Piquindone, a D_2 receptor antagonist. J. Clin. Psychopharmacol. 6:128-130.
1976. Umberkoman-Wiita, B., Vogel, W.H. and Wiita, P.J.(1981). Some biochemical and behavioral (sensation seeking) correlates in healthy adults. Res. Comm. Psychol. Psychiatry Behav. 6:303-316.
1977. Ungerstedt, U.(1971). Adipsia and aphagia after 6-hydroxydopamine-induced degeneration of the nigro-striatial dopamine system. Acta Physiol. Scand. Suppl 36:795-122.
1978. Vahlquist, B.(1955). Migraine in children. Int. Arch. Allergy 7:348-355.
1979. Vaillant, G.E.(1983). The Natural History of Alcoholism. Harvard Univ. Press, Cambridge, MA.
1980. Valzelli, L.(1967). Drugs and aggressiveness. Adv. Pharmacol. 5:79-108.
1981. Valzelli, L.(1973). The "isolation syndrome" in mice. Psychopharmacologia 31:305-320.
1982. Valzelli, L.(1974). 5-hydroxytryptamine in aggressiveness. Adv. Biochem. Psychopharm. 11:255-263.
1983. Valzelli, L.(1981). Psychobiology of Aggression and Violence. Raven Press, New York. 248pp.
1984. Valzelli, L.(1981). Psychopharmacology of aggression: An overview. Int. Pharmacopsychiatry 16:39-48.
1985. Valzelli, L.(1984). Reflections on experimental and human pathology of aggression. Prog. Neuro-Psychopharm. Biol. Psychiatry 8:311-325.
1986. Valzelli, L.(1985). Animal models of behavioral pathology and violent aggression. Meth. Find. Exp. Clin. Pharmacol. 7:189-193.
1987. VanBuren, J.M. and Fedio, P.(1976). Functional representation of the medial aspect of the frontal lobes in man. J. Neurosurg. 44:275-289.
1988. VanBuren, J.M.(1966). Evidence regarding a more precise localization of the posterior frontal-caudate arrest response in man. J. Neurosurg. 24:416-417.
1989. VanBuren, J.M., and Ojemarin, G.A.(1960). The frontostriatal arrest response in man. Electroencephalogr. Clin. Neurophysiol. 21:117-130.
1990. Vandenberg, S.G., Singer, S.M. and Pauls, D.L.(1986). The Heredity of Behavior Disorders in Adults and Children. Plenum Medical Book Company, New York.
1991. van de Wetering, B.J.M., Van Woerkom, T.C.A.M., Minderaa, R.B., Roos, R.A.C. and Cohen, A.P.(1987). Gilles de la Tourette syndrome in the Netherlands — Behavioral problems. Tourette Syndrome Assoc. Conference on Behavior in Tourette Syndrome, Washington, DC, Sept 12.
1991a. van Kammen, D.P., Peters, J.L., van Kammen, W.B., Rosen, J., Yao, J.K., McAdam, D. and Linnoila, M.(1989). Clonidine treatment of schizophrenia: Can we predict treatment response? Psychiatr. Res. 27:297-311.
1992. vanPraag, H.M.(1978). Psychotropic Drugs. A Guide for the Practitioner. Brunner/Mazel. New York.
1993. vanPraag, H.M.(1981). Management of depression with serotonin precursors. Biol. Psychiatry 16:291-310.
1994. vanPraag, H.M.(1982). Neurotransmitters and CNS disease. Lancet 2:1259-1264.
1995. vanPraag, H.M.(1983). In search of the action mechanism of antidepressants. 5-HTP/tyrosine mixtures in depression. Neuropharmacology 22:433-440.
1996. vanPraag, H.M.(1984). Studies in the mechanism of action of serotonin precursors in depression. Psychopharm. Bull . 20:599-602.
1997. vanPraag, H.M.(1986). Indolemaines in depression and suicide. In J.M. vanRee and S. Matthysee (eds) Prog.

Brain Res. 65:59- 71.
1998. vanPraag, H.M., Kahn, R.S., Asnis, G.M., Wetzler, S., Brown, S.L., Bleich, A. and Korn, M.L. (1987). Denosologication of biological psychiatry or the specificity of 5-HT disturbances in psychiatric disorders. J. Affect. Disord. 13:1-8.
1999. vanWimersma Greidanus, T.B., van Ree, J.M. and de Wied, D.(1983). Vasopressin and memory. Pharmacol. Ther. 20:437-458.
2000. VanWoert, M.H. and Sethy, V.H.(1975). Therapy of intention myoclonus with L-5-hydroxytryptophan and a peripheral decarboxylase inhibitor, MK 486. Neurology 25:135-140.
2001. VanWoert, M.H., Jutkowitz, R., Rosenbaum, D. and Bowers, M.B. Jr.(1976). Gilles de la Tourette syndrome: Biochemical approaches. In M.D. Yahr (ed) The Basal Ganglia Raven Press, New York, pp 459-465.
2002. VanWoert, M.H., Rosenbaum, D., and Enna, S.J.(1982). Overview of pharmacological approaches to therapy for Tourette syndrome. In A.J. Friedhoff and T.N. Chase (eds) Gilles de la Tourette Syndrome. Raven Press, New York, pp 369-375.
2003. VanWoert, M.H., Rosenbaum, D., Howieson, J. and Bowers, M.B.(1977). Long-term therapy of myoclonus and other neurologic disorders with L-5-hydroxytryptophan and carbidopa. N. Engl. J. Med. 296:70-75.
2004. VanWoert, M.H., Yip, L.C., Blais, M.E.(1977). Purine phosphoribosyl transferase in Gilles de la Tourette syndrome. N. Engl. J. Med. 296:210-212.
2005. Varley, C.K.(1984). Diet and the behavior of children with attention deficit disorder. J. Am. Acad. Child Psychiat. 23:182-185.
2006. Vasar, E.E., Maimets, M.O., and Allikmets, L.K.(1986)). Role of the serotonin$_2$ - receptors in regulation of aggressive behavior. Translated from Zhurnal Vysshei Nervcnoi Deyatel 'nosti imeni I.P. Pavlova 34:283-289, 1984. Plenum Press, New York.
2007. Vellutino, F.R.(1987). Dyslexia. Sci. Am. 256:34-41.
2007a Vereby, K., Volavka, J. and Clouet, D.(1978). Endorphins in psychiatry. Arch. Gen. Psychiatry 35:877-888.
2008. Vessie, P.R.(1932). On the transmission of Huntington's chorea for 300 years — The Bures family group. J. Nerv. Ment. Dis. 76:553-573.
2009. Villablanca, J.R., and Marus, R.J.(1975). Effects of caudate nuclei removal in cats: Comparison with effect of frontal cortex ablation. UCLA Form. Med. Sci. 18:273-311.
2010. Vinogradova, O.S.(1975). Functional organization of the limbic system in the process of registration of information: Facts and hypotheses. In R.L. Isaacson and K.H. Pribram (eds) The Hippocampus, Vol 2. Neurophysiology and Behavior. Plenum Press, New York.
2011. Virkkunen, M.(1979). Alcoholism and antisocial personality. Acta Psychiatr. Scand. 59:493-501.
2011a. Virkkunen, M., DeJong, J., Bartko, J., Goodwin, F.K. and Linnoila, M.(1989). Relationship of psychobiological variables to recidivism in violent offenders and impulsive fire starters. Arch. Gen. Psychiatry 46:600-603.
211b. Virkkunen, M., DeJong, J., Bartko, J. and Linnolia, M.(1989). Psychobiological concomitants of history of suicide attempts among violent offenders and impulsive fire setters. Arch. Gen. Psychiatry 46:604-606.
2012. Virkkunen, M., Nuutila, A., Goodwin, F.K. and Linnoila, M.(1987). Cerebral spinal fluid monamine metabolite levels in male arsonists. Arch. Gen. Psychiatry 44:241-247.
2013. Viscott, D.(1972). The Making of a Psychiatrist. Pocket Books, Simon & Schuster, Inc., New York.
2014. Vogel, F. and Kruger, J.(1967). Multifactorial determination of genetic affectations. In Proceedings of the Third International Congress of Human Genetics. Johns Hopkins University Press, Baltimore.
2015. VonEconomo, C.(1931). Encephalitis Lethargica. K.O. Newman (translator). Oxford Univ. Press, Oxford, England, 122pp.
2016. vonHarnack, G.A.(1953). Wesen und soziale Bedingtheit frühkindlicher Verhaltenstörungen. Biblio. Paediatr. Fasc 55.
2017. vonHungen, K. and Roberts, S.(1973). Adenylate-cyclase receptors for adrenergic neurotransmitters in rat cerebral cortex. Eur. J. Biochem. 36:391-401.
2018. vonKnorring, L., Oreland, L. and Winblad, B.(1984). Personality traits related to monamine oxidase (MAO) activity in platelets. Psychiatry Res. 12:11-26.
2019. Vorhees, C.V., Schaeffer, G.J. and Barrett, R.J.(1975). p-Chloramphetamine: Behavioral effects of reduced cerebral serotonin in rats. Pharmacol. Biochem. Behav. 3:279-284.
2020. Waldinger, R.J. and Gunderson, J.G.(1984). Completed psychotherapies with borderline patients. Am. J. Psychotherapy 38:190-202.
2021. Waldmeier, P.C. and Delini-Stula, A.A.(1979). Serotonin-dopamine interaction in the nigrostriatal system. Eur. J. Pharmacol. 55:363-373.
2022. Waldmeier, P.C.(1981). Noradrenergic transmission in depression: Under- or overfunction? Pharmacopsychiatry 14:3-9.
2022a. Walkup, J.T., Leckman, J.F., Price, R.A., Hardin, M., Ort, S.I., and Cohen, D.J. The relationship between obsessive-compulsive disorder and Tourette's syndrome: A twin study. Psychopharmacol. Bull. 24: 375-379.
2023. Wallach, M.B.(1974). Drug-induced stereotyped behavior: Similarities and differences. In E. Usdin (ed) Neuropsychopharmacology of Monamines and Their Regulatory Enzymes. Raven Press, New York, pp241-260.
2024. Walsh, T.L., Lavenstein, B., Licamele, W.L., Bronheim, S. and O'Leary, J.(1986). Calcium antagonists in the treatment of Tourette's disorder. Am. J. Psychiatry 143:1467-1468.
2025. Walter, W.G.(1973). Human frontal lobe function in sensory-motor association. In K.H. Pribram and A.R. Luria (eds) Psychophysiology of the Frontal Lobes. Academic Press, New York, pp109-122.
2026. Wang, H.-Y. and Friedman, E.(1988). Chronic lithium: desensitization of autoreceptors mediating serotonin release. Psychopharmacology 94:312-314.
2027. Wang, R.Y. and Aghajanian, G.K.(1972). Inhibiton of neurons in the amygdala by dorsal raphé stimulation: Mediation through a direct serotonergic pathway. Brain Res. 120:85-102.
2028. Warson, S., Cadwell, M., Warinner, A., Kirk, A. and Jensen, R.(1954). The dynamics of encopresis. Am. J. Orthopsychiatry 24:402-415.
2029. Wassman, E.R., Eldridge, R., Abuzzahab, S. and Nee, L.(1978). Gilles de la Tourette syndrome: Clinical and genetic studies in a midwestern city. Neurology 28:304-307.
2030. Waszczak, B. and Walters, J.(1979). Effects of

GABA mimetics on substantia nigra neurons. Adv. Neurol. 23:727-740.
2031. Watson, J.B.(1924). Behavioralism. Chicago University Press, Chicago.
2032. Watson, J.D. and Crick, F.H.C.(1953). Molecular structure of nucleic acids: A structure for deoxyribose nucleic acid. Nature 171:737-738.
2033. Watson, R.T., Heilman, K.M., Cauthen, J.C. and King, F.A.(1973). Neglect after cingulectomy. Neurology 23:1003-1007.
2034. Watt, N.F.(1978). Patterns of childhood social development in adult schizophrenics. Arch. Gen. Psychiatry 35:160-165.
2035. Weiden, P. and Brunn, R.(1987). Worsening of Tourette's disorder due to neuroleptic-induced akathisia. Am. J. Psychiatry 144:504-505.
2036. Weil, A.P.(1953). Certain severe disturbances of ego development in childhood. Psychoanal. Study Child. 8:271-287.
2037. Weil, A.P.(1953). Clinical data and dynamic considerations in certain cases of chldhood schizophrenia. Am. J. Orthopsychiatry 23:518-529.
2037a. Weilburg, J.B., Mesulam, M-M., Weintraub, S., Buonanno,F., Jenike, M. and Stakes, J.W.(1989). Focal striatal abnormalities in a patient with obsessive-compulsive disorder. Arch. Neurol. 46:233-235.
2038. Weinberg, W. and Rehmet, A.(1983). Childhood affective disorder and school problems. In Cantwekk, D.P. and Carlson, G.A. (eds) Affective Disorders in Childhood and Adolescence, An Update. Pergamon Press, NY, pp90-98.
2039. Weinberger, D.R., Berman, K.F. and Zec, R.F.(1986). Physiologic dysfunctin of dorsolateral prefrontal cortex in schizophrenia: I. Regional cerebral blood flow evidence. Arch. Gen. Psychiatry 43:114-125.
2040. Weingartner, H., Gold, P., Ballenger, J.C., Smallberg, S.A., Summers, R., Rubinow, D.R., Post, R.M. and Goodwin, F.K.(1981). Effects of vasopressin on human memory functions. Science 211:601-603.
2041. Weingartner, H., Rudorfer, M.V., Buchsbaum, M.S. and Linnoila, M.(1983). Effects of serotonin on memory impairments produced by ethanol. Science 221:472-474.
2042. Weinstock, M., Speiser, Z. and Ashkenazi, R.(1978). Psychopharmacology 56:205-209.
2043. Weintraub, S. and Mesulam, M.-M.(1985). Mental state assessment of young and elderly adults in behavioral neurology. In Mesulam, M.-M.(ed) Principles of Behavioral Neurology. Davis, Philadelphia. pp 71-123.
2044. Weiskrantz, L.(1964). Neurological studies and animal behavior. Br. Med. Bull. Vol. 20.
2045. Weiss, G., Hechtman, L. and Perlman, T.(1978). Hyperactives as young adults: School, employers and self-rating scales obtained during ten-year follow-up evaluations. Am. J. Orthopsychiatry 48:438-445.
2046. Weiss, G., Hechtman, L., Milroy, T. and Perlman, T.(1985). Psychiatric status of hyperactives as adults: A controlled prospective 15-year follow-up of 63 hyperactive children. J. Am. Acad. Child. Psychiatry 24:211-220.
2047. Weiss, G., Hechtman, L., Perlman, T., Hopkins, J. and Wener, A.(1979). Hyperactives as young adults: A controlled prospective 10 year follow-up of 75 children. Arch. Gen. Psychiatry 36:675-681.
2048. Weiss, G., Minde, K., Douglas, V., Werry, J. and Sykes, D.(1971). Comparison of the effects of chlorpromazine, dextroamphetamine, and methylphenidate on the behavior and intellectual functioning of hyperactive children. Can. Med. Assoc. J. 104: 20-25.
2049. Weiss, G., Minde, K., Werry, J.S., Douglas, V. and Nemeth, E.(1971). Studies on the hyperactive child. VIII. Five-year follow-up. Arch. Gen. Psychiatry 24:409-414.
2050. Weiss, R.D., Mirin, S.M., Michael, J.L. and Sollogub, A.C.(1986). Psychopathology in chronic cocaine abusers. Am. J. Drug Alcohol Abuse 12:17-29.
2051. Weissman, M., Fox, K. and Klerman, J.L.(1973). Hostility and depression associated with suicide attempts. Am J. Psychiatry 130:450-455.
2052. Weissman, M.M., Gershorn, E.S., Kidd, K.K., Prusoff, B.A., Leckman, J.F., Dibble, E. Hamovit, J., Thompson, D.,Pauls, D.L. and Guroff, J.J.(1984). Psychiatric disorders in the relatives of probands with affective disorders. Arch. Gen. Psychiatry 41:13-21.
2053. Welch, K. and Stuteville, P.(1958). Experimental production of unilateral neglect in monkeys. Brain 81:341-347.
2054. Welner Z., Welner A., Stewart M., Palkes H. and Wish E.(1977). A controlled study of siblings of hyperactive children. J. Nerv. Ment. Dis. 165:110-117.
2055. Wender, P.H. and Kalm, M.(1983). Prevalence of attention deficit disorder, residual type, and other psychiatric disorders in patients with irritable colon syndrome. Am. J. Psychiatry 140:1579-1582.
2056. Wender, P.H. and Klein, D.F.(1981). Mind, Mood, and Medicine. Farrar, Straus, Giroux, New York, 372pp.
2057. Wender, P.H.(1969). Platelet-serotonin level in children with "minimal brain dysfunction." Lancet 2:1012.
2058. Wender, P.H.(1971). Minimal Brain Dysfunction in Children. Wiley-Interscience, New York.
2059. Wender, P.H.(1974). Some speculations concerning a possible biochemical basis of minimal brain dysfunction. Life Sci. 14:1605-1621.
2060. Wender, P.H.(1981). Recent research on attention deficit disorder. In. S. Matthysee (ed) Psychiatry and the Biology of the Human Brain: A symposium Dedicated to Seymour S. Kety. Elsevier North Holland, Inc., New York, pp239-244.
2061. Wender, P.H., Rosenthal, D. and Kety, S.S.(1968). A psychiatric assessment of the adoptive parents of schizophrenics. In D. Rosenthal and S.S. Kety (eds) The Transmission of Schizophrenia. Pergamon Press, London.
2062. Werblin, F.S. and Dowling, J.E.(1969). Organization of the retina of the mudpuppy. Necturus maculosus. II. Intraretinal recordings. J. Neurophysiol. 32:339-355.
2063. Werblin, F.S.(1972). Lateral interaction at the inner plexiform layer of the retina: Antagonistic response to change. Science 175:1008-1009.
2064. Werner, H. and Strauss, A.(1941). Pathology of figure-background relation in the child. J. Abnorm. Psychol. 36:236.
2065. Wernicke, C.(1977). Der Aphasische Symptom Komplex. Breslau: Cohen & Weigart, 1874. Reprinted in Eggert, G. Werkick's Works on Aphasia. A Source Book and Review. Vol 1. Mouton, The Hague, pp92-145,
2066. Werry, J.S. and Aman, G.M.(1975). Methylphenidate and haloperidol in children. Arch. Gen. Psychiatry 32:790-796.
2067. Werry, J.S.(1982). An overview of pediatric psychopharmacology. J. Am. Acad. Child. Psychiatry 21:3-9.

2068. Werry, J.S., Aman, M.G. and Diamond, E.(1980). Imipramine and methylphenidate in hyperactive children. J. Child Psychol. Psychiatry 21:27-35.
2069. Werry,J.S. and Sprague, R.(1970). Hyperactivity, In Costelli (ed) Symptoms of Psychopathology John Wiley and Sons, New York.
2070. West, D.J. and Farrington, D.P.(1977). The Delinquent Way of Life. Heinemann, London.
2071. Westergaard, P., Sørensen, T., Hoppe, E., Rafelsen, O.J., Yates, C.M. and Nicolaou, N.(1978). Biogenic amine metabolism in cerebrospinal fluid of patients with affective disorders. Acta Psychiatr. Scand. 58:88-96.
2072. Westfall, T.C.(1977). Local regulation of adrenergic neurotransmission. Physiol. Rev. 57:659-728.
2073. Westlund, K.N., Denney, R.M., Kochersperger, L.M., Rose, R.M. and Abell, C.W.(1985). Distinct monamine oxidase A and B populations in primate brain. Science 230:181-183.
2074. Whalen, C.K., Henker, B., and Dotemoto, S.(1980). Methylphenidate and hyperactivity: Effects on teacher behaviors. Science 208:1280-1282.
2075. Whaley-Klahn, M.A. and Loney, J.(1977). A multivariate study of the relationship of parental management of self-esteem and initial drug response in hyperkinetic/MBD boys. Psychol. Schools 14: 485-492.
2075a. White, P.J., Cybulski, K.A., Primus, R., Johnson, D.F., Collier, G.H. and Wagner, G.C.(1988). Changes in macronutrient selection as a function of dietary tryptophan. Physiol. Behav. 43:73-77.
2076. Whitehouse, P.J., Price, D.L., Clark, A.W., Coyle, J.T. and DeLong, M.R.(1981). Alzheimer disease: Evidence for selective loss of cholinergic neurons in the nucleus basalis. Ann. Neurol. 10:122-126.
2076a. Widom, C.S.(1989). The cycle of violence. Science 244:160-166.
2077. Wilcock, G.K., Stevens, J. and Perkins, A.(1987). Trazedone/tryptophan for aggressive behavior. Lancet 1:929-930.
2078. Willerman, L.(1973). Activity level and hyperactivity in twins. Child Dev. 44:288-293.
2079. Williams, D.T., Mehl, R., Yudofsky, S., Adams, D. and Roseman, B.(1982). The effects of propanolol on uncontrolled rage outbursts in children and adolescents with organic brain dysfunction. J. Am. Acad. Child Psychiatry 21:129-135.
2080. Wilson, R.S., Garron, D.C. and Klawans, H.L.(1978). Significance of genetic factors in Gilles de la Tourette syndrome: A review. Behav. Genet. 8:503-510.
2081. Wilson, R.S., Garron, D.C., Tanner, C.M. and Klawans, H.L.(1982). Behavior disturbance in children with Tourette syndrome. In A.J. Friedhoff and T. N. Chase (eds) Gilles de la Tourette syndrome. Raven Press, New York, pp 329-333.
2082. Wing, L(1979). The current status of childhood autism. Psychol. Med. 9:9-12.
2083. Wing, L. and Gould, J.(1979). Severe impairments of social interaction and associated abnormal-ities in children: Epidemiology and classification. J. Autism Dev. Dis. 9:11-29.
2084. Wing, L.(1981). Asperger's syndrome: A clinical account. Psychol. Med. 11:115-129.
2085. Winokur, G., Behar, D., Vanvalkenburg, C. and Lowry, M.(1978). Is a familial definition of depression both feasible and valid? J. Nerv. Ment. Dis. 166:764-768.
2086. Winokur, G., Cadoret, R., Dorzab, J. and Baker, M.(1971). Depressive disease. A genetic study. Arch. Gen. Psychiatry 24:135-144.
2087. Winokur, G., Tsuang, M.T. and Crowe, R.R.(1982). The Iowa 500: Affective disorders in relatives of manic and depressed patients. Am. J. Psychiatry 139:209-212.
2088. Winsberg, B.G., Bialer, I., Kupietz, S. and Tobias, J.(1972). Effects of imipramine and dextroamphetamine on behavior of neuropsychiatrically impaired children. Am. J. Psychiatry 128:1425-1431.
2089. Wirz-Justice, A. and Puhringer, W.(1978). Increased platelet serotonin in bipolar depression and hypomania. J. Neurol. Transm. 42:55-62.
2090. Wise, C.D., Berger, B.D. and Stein, L.(1972). Benzodiazepines: Anxiety-reducing activity by reduction of serotonin turnover in the brain. Science 177:180-183.
2091. Wise, C.D., Berger, B.D. and Stein, L.(1973). Evidence of noradrenergic reward receptors and serotonergic punishment receptors in the rat brain. Biol. Psychiatry 6:3-21.
2092. Wise, R.A.(1980). The dopamine synapse and the notion of 'pleasure centers' in the brain. Trends Neurosci. 3:91-94.
2093. Wise, R.A., Spindler, J., DeWitt, H. and Gerber, G.J.(1978). Neuroleptic induced "anhedonia" in rats: Pimozide blocks reward quality of food. Science 201:262-264.
2094. Wise, R.D., Mirin, S.M., Michael, J.L. and Sollogub, A.C.(1986). Psychopathology in chronic cocaine abusers. Am. J. Drug. Alcohol Abuse 12:17-29.
2095. Witkin, H.A., Dyk, R.B., Faterson, H.F., Goodenough, D.R., and Karp, S.A.(1962). Psychological differentiation. John Wiley and Sons, New York.
2096. Woerner, P.I. and Guze, S.B.(1968). A family and marital study of hysteria. Br. J. Psychiatry 114:161-168.
2097. Wohlberg, G. and Kornetsky, C.(1973). Sustained attention in remitted schizophrenics. Arch. Gen. Psychiatry. 28:533-537.
2098. Wolf, P., Olpe, D., Avith, D. and Haas, H.L.(1978). GABAergic inhibition of neurons in the ventral tegmental area. Experientia 34:73-74.
2099. Wolf, W.A., Youdim, M.B.H. and Kuhn, D.M.(1985). Does brain 5-HIAA indicate serotonin release or monoamine oxidase activity? Eur. J. Pharmacol. 109:381-387.
2100. Wolkin, A., Jaeger, J., Brodie, J.D. and Wolf, A.P. [4](1985). Persistence of cerebral metabolic abnormalities in chronic schizophrenia as determined by positron emission tomography. Am. J. Psychiatry 142:564-571.
2101. Wolkowitz, O.M., Roy, A., and Doran, A.R.(1985). Pathologic gambling and other risk-taking pursuits. Psychiatr. Clin. North Am. 8:311-322.
2102. Wolley, D.W. and Shaw, E.(1954). A biochemical and pharmacological suggestion about certain mental disorders. Proc. Natl. Acad. Sci.U.S.A. 40:228-231.
2103. Wolraich, M.L., Stumbo, P., Milich, R., Chenard, C. and Schultz, F.(1986). Dietary characteristics of hyperactive and control boys and their behavioral correlates. J. Am. Dietetic Assoc. 86:500-504.
2104. Wong, D.F., Wagner, H.N., Tune, L.E., Dannals, R.F., et al.(1986). Positron emission tomography reveals elevated D_2 dopamine receptors in drug-naive schizophrenics. Science 234:1558-1563.
2105. Wood, C.D.(1958). Behavioral changes following discrete lesions of temporal lobe structures. Neurology 8:215-220.
2106. Wood, D.R., Reimherr, F.W. and Wender,

P.H.(1985). Treatment of attention deficit disorder with DL-phenylalanine. Psychiatry Res. 16:21-26.
2107. Wood, D.R., Reimherr, F.W., Wender, P.H. and Johnson, G.E.(1976). Diagnosis and treatment of minimal brain dysfunction in adults. Arch. Gen. Psychiatry 33:1453-1460.
2108. Wood, D.R., Wender, P.H. and Reimherr, F.E.(1983). The prevelence of attention deficit disorder, residual type, or minimal brain dysfunction, in a population of male alcoholic patients. Am J. Psychiatry 140: 95-98.
2109. Wood, P.(1983). Opioid regulation of CNS dopaminergic pathways: A review of methodology, receptor types, regional variations and species difference. Peptides 4:595-601.
2110. Woodrow, K.M.(1974). Gilles de la Tourette's disease— A review. Am. J. Psychiat. 131:1000-1003.
2111. Woodruff. R.A., Guze, S.B., Clayton, P.J. and Carr. D.(1973). Alcoholism and depression. Arch. Gen. Psychiatry 28: 97-100.
2112. Wooley, C.F.(1976). Where are the diseases of yesteryear? DaCosta's syndrome, soldier's heart, the effort syndrome, neurocirculatory asthenia, and the mitral valve prolapse syndrome. Circulation 53:749-751.
2113. Wray, I. and Dickerson, MG.(1981). Cessation of high frequency gambling and "withdrawal" symptoms. Br. J. Addict. 76:401-405.
2113a. Wurtman, J.J. and Wurtman, R.J.(1977). Fenfluramine and fluoxetine spare protein consumption while suppressing caloric intake by rats. Science 198:1178-1180.
2113b. Wurtman, J.J. and Wurtman, R.J.(1979). Drugs that enhance central serotonergic transmission diminish elective carbohydrate consumption by rats. Life Sci. 24:895-904.
2113c. Wurtman, R.J. and Wurtman, J.J.(1986). Carbohydrate craving, obesity and brain serotonin. Appetite 7 (Suppl):99-103.
2113d. Wurtman, R.J. and Wurtman, J.J.(1988). Charbohydrates and depression. Sci. Am. 260:68-75.
2114. Wurtman, R.J.(1980). Nutritional control of brain tryptophan and serotonin. In O. Hayaishi, Y. Ishimura and R. Kido (eds) Biochemical and Medical Aspects of Tryptophan Metabolism. Elsevier/North Holland, New York, pp31-46.
2115. Wurtman, R.J.(1982). Nutrients that modify brain function. Sci. Am. 246:42-52.
2116. Wurtman, R.J.(1983). Behavioral effects of nutrients. Lancet 1:1145-1147.
2117. Wyatt, R.J., Potkin, S.G., Bridge, T.P., Phelps, B.H. and Wise, C.D.(1980). Monamine oxidase in schizophrenia: An overview. Schizophr. Bull. 6:199-207.
2118. Yamaguchi, K., Shimoyama, M., and Gholson, R.K.(1967). Measurements of tryptophan pyrrolase in vivo— Induction and feedback inhibition. Biochem. Biophys. Acta. 146:102-110.
2119. Yamamoto, T. and Ueki, S.(1977). Characteristics in aggressive behavior induced by midbrain raphe lesions in rats. Physiol. Behav. 19:105-110.
2120. Yamamoto, T. and Hirano, A.(1985). Nucleus raphe dorsalis in Alzheimer's disease. Neurofibrillary tangles and loss of large neurons. Ann. Neurol. 17:573-577.
2121. Yamazaki, F., Takikawa, O., Kuroiwa, T. and Kido, R.(1987). Purification and characterization of indoleamine 2,3-dioxygenase from human placenta. In D.A. Bender, M.H. Joseph, W. Kochen and H. Steinhart (eds) Progress in Tryptophan and Serotinin Research 1986. Walter de Gruyter, New York, pp37-42.
2122. Yaryura-Tobias, J.A. and Bhagavan, H.N.(1977). L-tryptophan in obsessive-compulsive disorders. Am. J. Psychiatry 134:1298-1299.
2123. Yaryura-Tobias, J.A. and Neziroglu, F.(1975). The action of chlorimipramine in obsessive-compulsive neurosis: A plot study. Curr Ther. Res. 17:111-116.
2124. Yaryura-Tobias, J.A. and Neziroglu, F.A.(1977). Gilles de la Tourette syndrome: A new clinico-therapeutic approach. Prog. Neuro-Psychopharmacol. 1:335-338.
2125. Yaryura-Tobias, J.A.(1975). Chlorimipramine in Gilles de la Tourette syndrome. Am. J. Psychiatry 132:1221.
2126. Yaryura-Tobias, J.A.(1979). Gilles de la Tourette syndrome. Interactions with other neuropsychiatric disorders. Acta Psychiatr. Scand. 59:9-16.
2127. Yaryura-Tobias, J.A., Bebirian, R.J., Neziroglu, F.A., and Bhagaven, H.N.(1977). Obsessive-compulsive disorders as a serotonin defect. Res. Comm. Psychol. Psychiatr. Behav. 2:279-286.
2128. Yaryura-Tobias, J.A., Chang, A. and Neziroglu, F.(1978). A study of the relationship of serum glucose, insulin, free fatty acids, and free and total tryptophan to mental illness. Biol. Psychiatry 13:243-254.
2129. Yaryura-Tobias, J.A., Diamond, B. and Merlis, S.(1970). The action of L-DOPA on schizophrenic patients (a preliminary report). Therap. Res. 12:528-531.
2130. Yeragani, V.K., Blackman, M. and Baker, G.B.(1983). Biological and psychological aspects of a case of Gilles de la Tourette's syndrome. J. Clin. Psychiatry 44:27-29.
2131. Yonehara, N. and Clouet, D.(1984). Effects of delta and mu opiopeptides on the turnover and release of dopamine in rat striatum. J. Pharmacol. Exp. Ther. 231:38-42.
2132. York, P., York, D. and Wachtel, T.(1982). Tough Love. Bantum Books, New York.
2133. Yorkston, N.J., Zaki, S.A., Weller, M.P., Gruzelier, J.H. and Hirsch, S.R.(1981). DL-propanolol and chlorpromazine following admission for schizophrenia. A controlled comparison. Acta Psychiatr. Scand. 63:13-27.
2134. Yoshida, M. and Precht, W.(1971). Monosynaptic inhibition of neurons of the substantia nigra by caudato-nigral fibers. Brain Res. 32:225-228.
2135. Young, D. and Scoville, W.B.(1938). Paranoid psychosis in narcolepsy and the possible danger of Benzedrine treatment. Med. Clin. North Am. 22:637-646.
2136. Young, J.G., Kavanaugh, M.E., Anderson, G.M., Shaywitz, B.A. and Cohen, D.J.(1982). Clinical neurochemistry of autism and associated disorders. J. Autism Dev. Disord. 12:147-165.
2137. Young, S.H.(1981). Mechanism of decline in rat brain 5-hydroxytryptamine after induction of liver tryptophan pyrrolase by hydrocortisone: Roles of tryptophan catabolism and kynurenine synthesis. Br. J. Pharmacology 74:695-700.
2138. Young, S.N. and Sourkes, T.L.(1977). Tryptophan in the central nervous system: Regulation and significance. Adv. Neurochem. 2:133-191.
2139. Young, S.N., Smith, S.E., Pihl, R.O. and Ervin, F.R.(1985). Tryptophan depletion causes a rapid lowering of mood in normal males. Psychopharmacol. 87:173-177.
2140. Yuwiler, A., Ritvo, E., Geller, E., Glousman, R., Schneiderman, G. and Matsuno, D.(1975). Uptake and efflux of serotonin from platelets of autistic and nonautistic children. J. Autism Child. Schizophr. 5:83-98.
2141. Zahn, T.P., Abate, F., Little, B.C. et al.(1975).

Minimal brain dysfunction, stimulant drugs, and autonomic nervous system activity. Arch. Gen. Psychiatry 32:381-387.
2142. Zahner, G.E.P., Clubb, M.M., Leckman, J.F. and Pauls, D.L. (1988). The epidemiology of Tourette syndrome. In D.J. Cohen, R.D. Bruun, and J.F. Leckman,(eds) Tourette's syndrome and Tic Disorders: Clinical understanding and treatment. John Wiley & Sons, New York, pp80-89.
2143. Zambelli, A.J., Stamm, J.S., Maitinsky, S. and Loiselle, D.L. (1977). Auditory evoked potentials and selective attention in formerly hyperactive adolescent boys. Am. J. Psychiatry 134:742-747.
2144. Zametkin, A.J. and Rapoport, J.L.(1987). Neurobiology of attention deficit disorder with hyperactivity: Where have we come in 50 years? J. Am. Acad. Child Adolesc. Psychiatry 26:676-686.
2145. Zametkin, A.J., Karoum, F. and Rapoport, J.L.(1987). Treatment of hyperactive children with D-phenylalanine. Am. J. Psychiatry 144:792-794.
2146. Zarcone, V., Thorpe, B. and Dement, W. (1972). Sleep parameters in two patients with Gilles de la Tourette syndrome. Sleep Res. 1:155.
2147. Zausmer, D.M.(1954). The treatment of tics in childhood: A review and follow-up study. Arch. Dis. Child. 29:537-542.
2148. Zec, R.F. and Weinberger, D.R.(1986). Brain areas implicated in schizophrenia: A selective review. In H.A. Nasrallah and D.R. Weinberger (eds) Handbook of Schizophrenia, Vol 1: The Neurology of Schizophrenia. Elsevier Science Publishers, B.V. pp175-205.
2149. Zemlan, F.P., Trulson, M.E., Howell, R. and Hobel, B.G.(1977). Influences of p-chloramphetamine on female sexual reflexes and brain monamine levels. Brain Res. 123:324-356.
2150. Zerbe, R.L.(1987). Safety of fluoxetine in the treatment of obesity. Int. J. Obesity 11 (Suppl 3): 191-199.
2150a. Zerbin-Rüdin, E.(1967). Endogene psychosen. In P.E. Becker (ed) Human genetik: Ein kurzes Hanbuch in funf Banden. Vol. 2. Theime, Stuttgart.
2151. Zerbin-Rüdin, E.(1972). Genetic research and the theory of schizophrenia. Int. J. Ment. Health 1:42-58.
2152. Zerbin-Rüdin, E.(1980). Genetics of affective psychoses. In H.M. VanPraag (ed) Handbook of Biological Psychiatry Part III. Marcel Dekker, Inc., New York. pp35-58.
2153. Zilboorg, G.(1941). Ambulatory schizophrenias. Psychiatry 4:149-155.
2154. Zilboorg, G.(1957). Further observations on ambulatory schizophrenia. Am. J. Orthopsychiatry 27:677-682.
2155. Zohar, J. and Insel, T.R.(1987). Obsessive-compulsive disorder: Psychobiological approaches to diagnosis, treatment, and pathophysiology. Biol. Psychiatry 22:667-687.
2156. Zohar, J., Mueller, E.A., Insel, T.R., Zohar-Kadouch, R.C. and Murphy, M.D.(1987). Serotonergic responsivity in obsessive-compulsive disorder. Arch. Gen. Psychiatry 44:946-951.
2157. Zrull, J. et al(1966). An evaluation of methodology used in the study of psychoactive drugs. J. Am. Acad. Child Psychiatry 5:284-291.
2158. Zuckerman, M.(1979). Sensation Seeking: Beyond the Optimal Level of Arousal. Lawrence Erlbaum Associates, Hillsdale, New York.
2159. Zuckerman, M.(1984). Sensation seeking: A comparative approach to a human trait. Behav. Brain Sci. 7:413-471.

Index

A = in appendix
f = in a figure
Authors are indexed only if their name appears in the text and not if they are cited by reference number only.
Pedigree refers to illustrated pedigrees containing individuals showing the cited behavior.

Order Form

1. Books: Quantity — Amount

Tourette Syndrome and Human Behavior
_______ 1S Softback $39.95 _______

Search for the Tourette Syndrome and Human Behavior Genes
_______ 8H Hardback $34.00 _______
_______ 8S Softback $29.95 _______

The Gene Bomb Does Higher Education and Advanced Technology Accelerate the Selection of Genes for Learning Disorders, ADHD, Addictive and Disruptive Behaviors?
_______ 9H Hardback $29.95 _______
_______ 9S Softback $25.00 _______

RYAN— A Mother's Story of Her Hyperactive-Tourette Syndrome Child
_______ 2S Softback $9.95 _______

What Makes Ryan Tick? A Family's Triumph over TS and ADHD
_______ 10S Softback $14.95 _______

Hi, I'm Adam - A Child's Book about Tourette Syndrome
_______ 4A Softback $4.95 _______

Adam and the Magic Marble
_______ 4B Softback $6.95 _______

Hi, I'm Adam + Adam and the Magic Marble
_______ 4C Both together $11.50 _______

Echolalia - An Adult's Story of Tourette Syndrome
_______ 5A Softback $11.95 _______

Don't Think About Monkeys - Extraordinary Stories by People with Tourette Syndrome
_______ 6A Softback $12.95 _______

Teaching the Tiger - A Handbook for Individuals Involved in the Education of Students with Attention Deficit Disorder, Tourette Syndrome or Obsessive-Compulsive Disorder
_______ 7A Softback $35.00 _______

A.D.D. Kaleidoscope - The Many Facets of Adult Attention Deficit Disorder
_______ 8A Softback $24.95 _______

Understanding and Treating the Tourette Syndrome/ADHD Spectrum Disorders by Dr. Comings 8 **tapes** 10 hrs
_______ 11A $75.00 _______

Subtotal for Books _______

2. Tax: **California residents please add 8.25% sales tax** _______

3. Mailing and Handling:
- ☐ Fourth Class: $4.00 lst item $1.00 each additional item
- ☐ U.P.S. Ground: $6.00 lst item $1.00 each additional item
- ☐ U.P.S. Air: $10.00 lst item $2.00 each additional item _______

Name: _______________ **Total** _______

Address: _______________

City: _______________ State: _______ Zip: _______

Country (if other than U.S.A.): _______________

Check Enclosed _____ **or** Visa ___ Mastercard ___

CC# _______________ Expiration Date _______

send to: **Hope Press P.O.Box 188, Duarte, CA 91009-0188**

or Fill out this form with credit card # and FAX it to 626-358-3520

or Order by phone **1-800-321-4039** — 24 hr service

[Foreign buyers outside North America please: a) send bank check in U.S. dollars, or b) order by credit card with charge in U.S. dollars, or c) FAX in the form. For surface mail add $6.00 shipping for first book and $1.00 for each additional and allow 4-6 weeks. For air mail add $25.00 shipping and $2.00 for each additonal book and allow 1 week.]

for more details on each book visit our web site: **http://www.hopepress.com**

Order Form

1. Books: Quantity | Amount

Quantity	Item	Price	Amount
	Tourette Syndrome and Human Behavior		
_______	1S Softback	$39.95	_______
	Search for the Tourette Syndrome and Human Behavior Genes		
_______	8H Hardback	$34.00	_______
_______	8S Softback	$29.95	_______
	***The Gene Bomb** Does Higher Education and Advanced Technology Accelerate the Selection of Genes for Learning Disorders, ADHD, Addictive and Disruptive Behaviors?*		
_______	9H Hardback	$29.95	_______
_______	9S Softback	$25.00	_______
	RYAN — A Mother's Story of Her Hyperactive-Tourette Syndrome Child		
_______	2S Softback	$9.95	_______
	What Makes Ryan Tick? A Family's Triumph over TS and ADHD		
_______	10S Softback	$14.95	_______
	Hi, I'm Adam - A Child's Book about Tourette Syndrome		
_______	4A Softback	$4.95	_______
	Adam and the Magic Marble		
_______	4B Softback	$6.95	_______
	Hi, I'm Adam + Adam and the Magic Marble		
_______	4C Both together	$11.50	_______
	Echolalia - An Adult's Story of Tourette Syndrome		
_______	5A Softback	$11.95	_______
	Don't Think About Monkeys - Extraordinary Stories by People with Tourette Syndrome		
_______	6A Softback	$12.95	_______
	Teaching the Tiger - A Handbook for Individuals Involved in the Education of Students with Attention Deficit Disorder, Tourette Syndrome or Obsessive-Compulsive Disorder		
_______	7A Softback	$35.00	_______
	A.D.D. Kaleidoscope - The Many Facets of Adult Attention Deficit Disorder		
_______	8A Softback	$24.95	_______
	Understanding and Treating the Tourette Syndrome/ADHD Spectrum Disorders by Dr. Comings 8 **tapes** 10 hrs		
_______	11A	$75.00	_______

Subtotal for Books _______

2. Tax: **California residents please add 8.25% sales tax** _______

3. Mailing and Handling:

- ☐ Fourth Class: $4.00 lst item $1.00 each additional item
- ☐ U.P.S. Ground: $6.00 lst item $1.00 each additional item
- ☐ U.P.S. Air: $10.00 lst item $2.00 each additional item _______

Name:______________________________ **Total** _______

Address: ______________________________

City: ______________________ State:________ Zip: ________

Country (if other than U.S.A.): ______________________

Check Enclosed _____ **or** Visa ___ Mastercard ___

CC# ______________________ Expiration Date ________

send to: **Hope Press P.O.Box 188, Duarte, CA 91009-0188**

or Fill out this form with credit card # and FAX it to 626-358-3520

or Order by phone **1-800-321-4039** — 24 hr service

[Foreign buyers outside North America please: a) send bank check in U.S. dollars, or b) order by credit card with charge in U.S. dollars, or c) FAX in the form. For surface mail add $6.00 shipping for first book and $1.00 for each additional and allow 4-6 weeks. For air mail add $25.00 shipping and $2.00 for each additonal book and allow 1 week.]

for more details on each book visit our web site: **http://www.hopepress.com**

Order Form

1. Books: Quantity — Amount

Tourette Syndrome and Human Behavior
_______ 1S Softback $39.95 _______

Search for the Tourette Syndrome and Human Behavior Genes
_______ 8H Hardback $34.00 _______
_______ 8S Softback $29.95 _______

The Gene Bomb Does Higher Education and Advanced Technology Accelerate the Selection of Genes for Learning Disorders, ADHD, Addictive and Disruptive Behaviors?
_______ 9H Hardback $29.95 _______
_______ 9S Softback $25.00 _______

RYAN — A Mother's Story of Her Hyperactive-Tourette Syndrome Child
_______ 2S Softback $9.95 _______

What Makes Ryan Tick? A Family's Triumph over TS and ADHD
_______ 10S Softback $14.95 _______

Hi, I'm Adam - A Child's Book about Tourette Syndrome
_______ 4A Softback $4.95 _______

Adam and the Magic Marble
_______ 4B Softback $6.95 _______

Hi, I'm Adam + Adam and the Magic Marble
_______ 4C Both together $11.50 _______

Echolalia - An Adult's Story of Tourette Syndrome
_______ 5A Softback $11.95 _______

Don't Think About Monkeys - Extraordinary Stories by People with Tourette Syndrome
_______ 6A Softback $12.95 _______

Teaching the Tiger - A Handbook for Individuals Involved in the Education of Students with Attention Deficit Disorder, Tourette Syndrome or Obsessive-Compulsive Disorder
_______ 7A Softback $35.00 _______

A.D.D. Kaleidoscope - The Many Facets of Adult Attention Deficit Disorder
_______ 8A Softback $24.95 _______

Understanding and Treating the Tourette Syndrome/ADHD Spectrum Disorders by Dr. Comings 8 **tapes** 10 hrs
_______ 11A $75.00 _______

Subtotal for Books _______

2. Tax: **California residents please add 8.25% sales tax** _______

3. Mailing and Handling:
- ☐ Fourth Class: $4.00 lst item $1.00 each additional item
- ☐ U.P.S. Ground: $6.00 lst item $1.00 each additional item
- ☐ U.P.S. Air: $10.00 lst item $2.00 each additional item _______

Name: _______________ **Total** _______

Address: _______________

City: _______________ State: _______ Zip: _______

Country (if other than U.S.A.): _______________

Check Enclosed _____ or Visa ___ Mastercard ___

CC# _______________ Expiration Date _______

send to: **Hope Press P.O.Box 188, Duarte, CA 91009-0188**

or Fill out this form with credit card # and FAX it to 626-358-3520

or Order by phone **1-800-321-4039** — 24 hr service

[Foreign buyers outside North America please: a) send bank check in U.S. dollars, or b) order by credit card with charge in U.S. dollars, or c) FAX in the form. For surface mail add $6.00 shipping for first book and $1.00 for each additional and allow 4-6 weeks. For air mail add $25.00 shipping and $2.00 for each additonal book and allow 1 week.]

for more details on each book visit our web site: **http://www.hopepress.com**

Order Form

1. Books: Quantity | Amount

Tourette Syndrome and Human Behavior
_______ 1S Softback $39.95 _______

Search for the Tourette Syndrome and Human Behavior Genes
_______ 8H Hardback $34.00 _______
_______ 8S Softback $29.95 _______

The Gene Bomb Does Higher Education and Advanced Technology Accelerate the Selection of Genes for Learning Disorders, ADHD, Addictive and Disruptive Behaviors?
_______ 9H Hardback $29.95 _______
_______ 9S Softback $25.00 _______

RYAN — A Mother's Story of Her Hyperactive-Tourette Syndrome Child
_______ 2S Softback $9.95 _______

What Makes Ryan Tick? A Family's Triumph over TS and ADHD
_______ 10S Softback $14.95 _______

Hi, I'm Adam - A Child's Book about Tourette Syndrome
_______ 4A Softback $4.95 _______

Adam and the Magic Marble
_______ 4B Softback $6.95 _______

Hi, I'm Adam + Adam and the Magic Marble
_______ 4C Both together $11.50 _______

Echolalia - An Adult's Story of Tourette Syndrome
_______ 5A Softback $11.95 _______

Don't Think About Monkeys - Extraordinary Stories by People with Tourette Syndrome
_______ 6A Softback $12.95 _______

Teaching the Tiger - A Handbook for Individuals Involved in the Education of Students with Attention Deficit Disorder, Tourette Syndrome or Obsessive-Compulsive Disorder
_______ 7A Softback $35.00 _______

A.D.D. Kaleidoscope - The Many Facets of Adult Attention Deficit Disorder
_______ 8A Softback $24.95 _______

Understanding and Treating the Tourette Syndrome/ADHD Spectrum Disorders by Dr. Comings 8 **tapes** 10 hrs
_______ 11A $75.00 _______

Subtotal for Books _______

2. Tax: **California residents please add 8.25% sales tax** _______

3. Mailing and Handling:
☐ Fourth Class: $4.00 lst item $1.00 each additional item
☐ U.P.S. Ground: $6.00 lst item $1.00 each additional item
☐ U.P.S. Air: $10.00 lst item $2.00 each additional item _______

Total _______

Name:_______________________________
Address: _______________________________
City: _______________________ State:_________ Zip: _________
Country (if other than U.S.A.): _______________________
Check Enclosed _____ **or** Visa ___ Mastercard ___
CC# _______________________ Expiration Date _________

send to: **Hope Press P.O.Box 188, Duarte, CA 91009-0188**

or Fill out this form with credit card # and FAX it to 626-358-3520

or Order by phone **1-800-321-4039** — 24 hr service

[Foreign buyers outside North America please: a) send bank check in U.S. dollars, or b) order by credit card with charge in U.S. dollars, or c) FAX in the form. For surface mail add $6.00 shipping for first book and $1.00 for each additional and allow 4-6 weeks. For air mail add $25.00 shipping and $2.00 for each additonal book and allow 1 week.]

for more details on each book visit our web site: **http://www.hopepress.com**